T0211244

MODERN CANONICAL QUANTUM GENERAL RELATIVITY

Modern physics rests on two fundamental building blocks: general relativity and quantum theory. General relativity is a geometric interpretation of gravity, while quantum theory governs the microscopic behaviour of matter. According to Einstein's equations, geometry is curved when and where matter is localized. Therefore, in general relativity, geometry is a dynamical quantity that cannot be prescribed • •••••• but is in interaction with matter. The equations of nature are background independent in this sense; there is no space-time geometry on which matter propagates without backreaction of matter on geometry. Since matter is described by quantum theory, which in turn couples to geometry, we need a quantum theory of gravity. The absence of a viable quantum gravity theory to date is due to the fact that quantum (field) theory as currently formulated assumes that a background geometry is available, thus being inconsistent with the principles of general relativity. In order to construct quantum gravity, one must reformulate quantum theory in a background-independent way. • ••••• • •••••••• • •••••• • •••••• • •• ••• ••• is about one such candidate for a background-independent quantum gravity theory: loop quantum gravity.

This book provides a complete treatise of the canonical quantization of general relativity. The focus is on detailing the conceptual and mathematical framework, describing the physical applications, and summarizing the status of this programme in its most popular incarnation: loop quantum gravity. Mathematical concepts and their relevance to physics are provided within this book, so it is suitable for graduate students and researchers with a basic knowledge of quantum field theory and general relativity.

THOMAS THIEMANN is Staff Scientist at the Max Planck Institut für Gravitationsphysik (Albert Einstein Institut), Potsdam, Germany. He is also a long-term researcher at the Perimeter Institute for Theoretical Physics and Associate Professor at the University of Waterloo, Canada. Thomas Thiemann obtained his Ph.D. in theoretical physics from the Rheinisch-Westfälisch Technische Hochschule, Aachen, Germany. He held two-year postdoctoral positions at The Pennsylvania State University and Harvard University. As of 2005 he holds a guest professor position at Beijing Normal University, China.

CAMBRIDGE MONOGRAPHS ON MATHEMATICAL PHYSICS

General editors: P. V. Landshoff, D. R. Nelson, S. Weinberg

W. C. Saslaw *Gravitational Physics of Stellar and Galactic Systems*[†]
H. Stephani, D. Kramer, M. A. H. MacCallum, C. Hoenselaers and E. Herlt *Exact Solutions of Einstein's Field Equations, 2nd edition*
J. M. Stewart *Advanced General Relativity*[†]
T. Thiemann *Modern Canonical Quantum General Relativity*
A. Vilenkin and E. P. S. Shellard *Cosmic Strings and Other Topological Defects*[†]
R. S. Ward and R. O. Wells Jr *Twistor Geometry and Field Theory*[†]
J. R. Wilson and G. J. Mathews *Relativistic Numerical Hydrodynamics*

[†]Issued as a paperback

Modern Canonical Quantum
General Relativity

THOMAS THIEMANN

Max Planck Institut für Gravitationsphysik, Germany

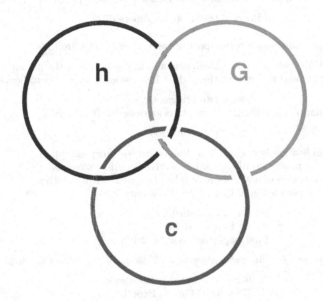

CAMBRIDGE
UNIVERSITY PRESS

CAMBRIDGE
UNIVERSITY PRESS

University Printing House, Cambridge CB2 8BS, United Kingdom

One Liberty Plaza, 20th Floor, New York, NY 10006, USA

477 Williamstown Road, Port Melbourne, VIC 3207, Australia

314-321, 3rd Floor, Plot 3, Splendor Forum, Jasola District Centre, New Delhi - 110025, India

79 Anson Road, #06-04/06, Singapore 079906

Cambridge University Press is part of the University of Cambridge.

It furthers the University's mission by disseminating knowledge in the pursuit of education, learning and research at the highest international levels of excellence.

www.cambridge.org
Information on this title: www.cambridge.org/9780521741873

First published 2007
Reprinted 2008
First paperback edition 2008

A catalogue record for this publication is available from the British Library

ISBN 978-0-521-84263-1 Hardback
ISBN 978-0-521-74187-3 Paperback

Figure 1 Copyright: Max Planck Institute for Gravitational Physics (Albert Einstein Institute), MildeMarketing Science Communication, Exozet. To see the animation, please visit the URL http://www.einstein-online.info/de/vertiefung/Spinnetzwerke/ index.html.

Quantum spin dynamics

This is a still from an animation which illustrates the dynamical evolution of quantum geometry in Loop Quantum Gravity (LQG), which is a particular incarnation of canonical Quantum General Relativity.

The faces of the tetrahedra are elementary excitations (atoms) of geometry. Each face is coloured, where red and violet respectively means that the face carries low or high area respectively. The colours or areas are quantised in units of the Planck area $\ell_P^2 \approx 10^{-66} \, \text{cm}^2$. Thus the faces do not have area as they appear to have in the figure, rather one would have to shrink red and stretch violet faces accordingly in order to obtain the correct picture.

The faces are dual to a four-valent graph, that is, each face is punctured by an edge which connects the centres of the tetrahedra with a common face. These edges are 'charged' with half-integral spin-quantum numbers and these numbers are proportional to the quantum area of the faces. The collection of spins and edges defines a spin-network state. The spin quantum numbers are created and annihilated at each Planck time step of $\tau_P \approx 10^{-43} \, \text{s}$ in a specific way as dictated by the quantum Einstein equations. Hence the name Quantum Spin Dynamics (QSD) in analogy to Quantum Chromodynamics (QCD).

Spin zero corresponds to no edge or face at all, hence whole tetrahedra are created and annihilated all the time. Therefore, the free space not occupied by tetrahedra does not correspond to empty (matter-free) space but rather to space without geometry, it has zero volume and therefore is a hole in the quantum spacetime. *The tetrahedra are not embedded in space, they are the space.* Matter can only exist where geometry is excited, that is, on the edges (bosons) and vertices (fermions) of the graph. Thus geometry is completely discrete and chaotic at the Planck scale, only on large scales does it appear smooth.

In this book, this fascinating physics is explained in mathematical detail.

Contents

II   FOUNDATIONS OF MODERN CANONICAL QUANTUM GENERAL RELATIVITY

Foreword

Over half a century of collective study has not diminished the fascination of searching for a consistent theory of quantum gravity. I first encountered the subject in 1969 when, as a young researcher, I spent a year in Trieste working with Abdus Salam who, for a while, was very interested in the subject. In those days, the technical approaches adopted for quantum gravity depended very much on the background of the researcher: those, like myself, from a theoretical particle-physics background used perturbative quantum field theory; those whose background was in general relativity tended to use relatively elementary quantum theory, but taking full account of the background general relativity (which the other scheme did not).

The perturbative quantum field theory schemes foundered on intractable ultraviolet divergences and gave way to super-gravity – the super-symmetric extension of standard general relativity In spite of initial optimism, this approach succumbed to the same disease and was eventually replaced by the far more ambitious superstring theories. Superstring theory is now the dominant quantum gravity programe in terms of the number of personnel involved and the number of published papers, per year, per unit researcher.

However, notwithstanding my early training as a quantum field theorist, I quickly became fascinated by the "canonical quantization", or "quantum geometry," schemes favored by those coming from general relativity. The early attempts for quantizing the metric variables were rather nave, and took on various forms according to how the intrinsic constraints of classical general relativity are handled. In the most popular approach, the constraints are imposed on the state vectors and give rise to the famous Wheeler–DeWitt equation arguably one of the most elegant equations in theoretical physics, and certainly one of the most mathematically ill-defined. Indeed, it was the very intractability of this equation that first intrigued me and prompted me to see what could be done with more sophisticated quantization methods. After much effort it became clear that the answer was "not much."

The enormous difficulty of the canonical quantum gravity scheme eventually caused it to go into something of a decline, until new life was imparted with Ashtekar's discovery of a set of variables in which the constraint equations simplify significantly. This scheme slowly morphed into "loop quantum gravity:" an approach which has, for the first time, allowed real insight into what a nonperturbative quantisation of general relativity might look like. A number of genuine results were obtained, but it became slowly apparent that the old problems with the Wheeler–DeWitt equation were still there in transmuted form, and the critical Hamiltonian constraint was still ill-defined.

It was at this point that Thomas Thiemann – the author of this book – entered the scene. I can still remember the shock I felt when I first read the papers he put onto the web dealing with the Hamiltonian constraint. Suddenly, someone with a top-rate mathematical knowledge had addressed this critical question anew, and with considerable success. Indeed, Thiemann succeeded with loop quantum gravity where I had failed with the old Wheeler–DeWitt equation, and he has gone on since that time to become one of the internationally acknowledged experts in loop quantum gravity.

Thiemann's deep knowledge of mathematics applied to quantum gravity is evident from the first page of this magnificent book. The subject is explored in considerable generality and with real mathematical depth. The author starts from first principles with a general introduction to quantum gravity, and then proceeds to give, what is by far, the most comprehensive, and mathematically precise, exposition of loop quantum gravity that is available in the literature. The reader should be warned though that, when it comes to mathematics, the author takes no hostages, and a good knowledge of functional analysis and differential geometry is assumed from the outset. Still, that is how the subject is these days, and anyone who seriously aspires to work in loop quantum gravity would be advised to gain a good knowledge of this type of mathematics. In that sense, this is a text that is written for advanced graduate students, or professionals who work in the area.

My graduate students not infrequently ask me what I think of the current status of canonical quantum gravity and, in particular, what I think the chances are of ever making proper mathematical sense of the constraints that define the theory. For some years now I have replied to the effect that, if anybody can do it, it will be Thomas Thiemann and, if he cannot do it, then probably nobody will. Anyone who reads right through this major new work will understand why I place so much trust in the author's ability to crack this central problem of quantum gravity.

Chris Isham,
Professor of Theoretical Physics at
The Blankett Laboratory, Imperial College, London

Preface

Quantum General Relativity (QGR) or Quantum Gravity for short is, by definition, a Quantum (Field) Theory of Einstein's geometrical interpretation of gravity which he himself called General Relativity (GR). It is a theory which synthesises the two fundamental building blocks of modern physics, that is, (1) the generally relativistic principle of background independence, sometimes called general covariance and (2) the uncertainty principle of quantum mechanics.

The search for a viable QGR theory is almost as old as Quantum Mechanics and GR themselves, however, despite an enormous effort of work by a vast amount of physicists over the past 70 years, we still do not have a credible QGR theory. Since the problem is so hard, QGR is sometimes called the 'holy grail of physics'. Indeed, it is to be expected that the discovery of a QGR theory revolutionises our current understanding of nature in a way as radical as both General Relativity and Quantum Mechanics did.

What we do have today are candidate theories which display some promising features that one intuitively expects from a quantum theory of gravity. They are so far candidates only because for each of them one still has to show, at the end of the construction of the theory, that it reduces to the presently known standard model of matter and classical General Relativity at low energies, which is the minimal test that any QGR theory must pass.

One of these candidates is Loop Quantum Gravity (LQG). LQG is a modern version of the canonical or Hamiltonian approach to Quantum Gravity, originally introduced by Dirac, Bergmann, Komar, Wheeler, DeWitt, Arnowitt, Deser and Misner. It is modern in the sense that the theory is formulated in terms of connections ('gauge potentials') rather than metrics. It is due to this fact that the theory was called Loop Quantum Gravity since theories of connections are naturally described in terms of Wilson loops. This also brings GR much closer to the formulation of the other three forces of nature, each of which is described in terms of connections of a particular Yang–Mills theory for which viable quantum theories exist. Consequently, the connection reformulation has resulted in rapid progress over the past 20 years.

The purpose of this book is to provide a self-contained treatise on canonical – and in particular Loop Quantum Gravity. Although the theory is still under rapid development and the present book therefore is at best a snapshot, the field has now matured enough in order to justify the publication of a new textbook. The literature on LQG now comprises more than a thousand articles scattered over a vast number of journals, reviews, proceedings and conference reports. Structures which were believed to be essential initially turned out to be negligible later on and vice versa, thus making it very hard for the beginner to get an overview of the subject. We hope that this book serves as a 'geodesic' through the literature enabling the reader to move quickly from the basics to the frontiers of current research. By definition, a geodesic cannot touch on all the subjects of the theory and we apologise herewith to our colleagues if we were unable to cover their work in this single volume manuscript. However, guides to further reading and a detailed bibliography try to compensate for this incompleteness. A complete listing of all LQG-related papers, which is periodically being updated, can be found in [1, 2].[1]

Loop Quantum Gravity is an attempt to construct a mathematically rigorous, background-independent, non-perturbative Quantum Field Theory of Lorentzian General Relativity and all known matter in four spacetime dimensions, not more and not less. In particular, no claim is made that LQG is a unified theory of everything predicting, among other things, matter content and dimensionality of the world. Hence, currently there is no restriction on the allowed matter couplings although these might still come in at a later stage when deriving the low energy limit. While the connection formulation works only in four spacetime dimensions and in that sense is a prediction, higher p-form formulations in higher dimensions are conceivable. Matter and geometry are not unified in the sense that they are components of one and the same geometrical object, however, they are unified under the four-dimensional diffeomorphism group which in perturbative approaches is broken. LQG provides a universal framework for how to combine quantum theory and General Relativity for all possible matter and in that sense is robust against the very likely discovery of further substructure of matter between the energy scales of the LHC and the Planck scale which differ by 16 orders of magnitude. This is almost the same number of orders of magnitude as between 1 mm and the length scales that the LHC can resolve, and we found a huge amount of substructure there.

The stress on mathematical rigour is here no luxurious extra baggage but a necessity: in a field where, to date, no experimental input is available, mathematical consistency is the only guiding principle to construct the theory. The strategy is to combine the presently known physical principles and to drive them to their logical frontiers without assuming any extra, unobserved structure such as extra dimensions and extra particles. This deliberately conservative approach has the advantage of either producing a viable theory or of deriving which extra structures are needed in order to produce a successful theory. Indeed, it is conceivable that at some point in the development of the theory a 'quantum leap' is necessary, similar to Heisenberg's discovery that the

---

[1] See also the URLs http://www.nucleares.unam.mx/corichi/lqgbib.pdf and http://www.matmor.unam.mx/corichi/lqgbib.pdf.

Bohr–Sommerfeld quantisation rules can be interpreted in terms of operators. The requirement to preserve background independence has already led to new, fascinating mathematical structures. For instance, a fundamental discreteness of spacetime at the Planck scale of 10^{-33} cm seems to be a prediction of the theory which is a first substantial evidence for a theory in which the gravitational field acts as a natural cutoff of the usual ultraviolet divergences of QFT.

Accordingly, the present text tries to be mathematically precise. We will develop in depth the conceptual and mathematical framework underlying LQG, stating exact definitions and theorems including complete proofs. Many of the calculations or arguments used during the proofs cannot be found anywhere in the literature detailed as they are displayed here. We have supplied a vast amount of mathematical background information so that the book can be read by readers with only basic prior knowledge of GR and QFT without having to consult too much additional literature. We have made an effort to stress the basic principles of canonical QGR, of which LQG is just one possible incarnation based on a specific choice of variables.

For readers who want to get acquainted first with the physical ideas and conceptual aspects of LQG before going into mathematical details, we strongly recommend the book by Carlo Rovelli [3]. The two books are complementary in the sense that they can be regarded almost as Volume I ('Introduction and Conceptual Framework') and Volume II ('Mathematical Framework and Applications') of a general presentation of QGR in general and LQG in particular. While this book also develops a tight conceptual framework, the book by Carlo Rovelli is much broader in that aspect. Recent review articles can be found in [4–14]. The status of the theory a decade ago is summarised in the books [15–17].

The present text is aimed at all readers who want to find out in detail how LQG works, conceptually and technically, enabling them to quickly develop their own research on the subject. For instance, the author taught most of the material of this book in a two-semester course to German students in physics and mathematics who were in their sixth semester of diploma studies or higher. After that they could complete diploma theses or PhD theses on the subject without much further guidance. Unfortunately, due to reasons of space, exercises and their solutions had to be abandoned from the book, see [12] for a selection. We hope to incorporate them in an extended future edition. As we have pointed out, LQG is far from being a completed theory and aspects of LQG which are at the frontier of current research and whose details are still under construction will be critically discussed. This will help readers to get an impression of what important open problems there are and hopefully encourage them to address these in their own research.

The numerous suggestions for improvements to the previous online version of this book (http://www.arxiv.org/list/gr-qc/0110034) by countless colleagues is gratefully acknowledged, in particular those by Jürgen Ehlers, Christian Fleischhack, Stefan Hofmann, Chris Isham, Jurek Lewandowski, Robert Oeckl, Hendryk Pfeiffer, Carlo Rovelli, Hanno Sahlmann and Oliver Winkler. Special thanks go to my students Johannes Brunnemann, Bianca Dittrich and Kristina Giesel for a careful reading of the manuscript and especially to Kristina Giesel for her help with the figures.

Посвящаю своей жене Татьяне.
Ebenso gewidmet meinen Söhnen Andreas und Maximilian.

Thomas Thiemann
Berlin, Toronto 2001–2007

Notation and conventions

Symbol	Meaning
$G = 6.67 \times 10^{-11}\,\mathrm{m^3\,kg^{-1}\,s^{-2}}$	Newton's constant
$\kappa = 16\pi G/c^3$	gravitational coupling constant
$\ell_p = \sqrt{\hbar\kappa} \approx 10^{-33}\,\mathrm{cm}$	Planck length
$m_p = \sqrt{\hbar/\kappa}/c \approx 10^{19}\,\mathrm{GeV}/c^2$	Planck mass
Q	Yang–Mills coupling constant
$M,\ \dim(M) = D+1$	spacetime manifold
$\sigma,\ \dim(\sigma) = D$	abstract spatial manifold
Σ	spatial manifold embedded into M
G	compact gauge group
Lie(G)	Lie algebra
$N-1$	rank of gauge group
$\mu, \nu, \rho, .. = 0, 1, \ldots, D$	tensorial spacetime indices
$a, b, c, .. = 1, \ldots, D$	tensorial spatial indices
$\epsilon_{a_1..a_D}$	Levi–Civita totally skew tensor pseudo density of weight -1
$g_{\mu\nu}$	spacetime metric tensor
q_{ab}	spatial (intrinsic) metric tensor of σ
K_{ab}	extrinsic curvature of σ
R	curvature tensor
$\underline{h}$	group elements for general G
$\underline{h}_{mn},\ m, n, o, .. = 1, \ldots, N$	matrix elements for general G
$I, J, K, .. = 1, 2, \ldots, \dim(\mathrm{G})$	Lie algebra indices for general G
$\underline{\tau}_I$	Lie algebra generators for general G
k_{IJ}	$= -\mathrm{tr}(\underline{\tau}_I \underline{\tau}_J)/N := \delta_{IJ}$: Cartan–Killing metric for G
$[\underline{\tau}_I, \underline{\tau}_J] = 2 f_{IJ}{}^K \underline{\tau}_K$	structure constants for G
$\underline{\pi}(\underline{h})$	(irreducible) representations for general G or algebra
h	group elements for SU(2)
$h_{AB},\ A, B, C, .. = 1, 2$	matrix elements for SU(2)
$i, j, k, .. = 1, 2, 3$	Lie algebra indices for SU(2)
τ_i	Lie algebra generators for SU(2)

$k_{ij} = \delta_{ij}$	Cartan–Killing metric for SU(2)
$f_{ij}{}^k = \epsilon_{ijk}$	structure constants for SU(2)
$\pi_j(h)$	(irreducible) representations for SU(2) with spin j
$\underline{A}$	connection on G-bundle over σ
$\underline{A}_a^I$	pull-back of $\underline{A}$ to σ by local section
$\iota_A, o_A; \ A = 1, 2$	$\emptyset_A \iota^A := \epsilon^{AB} \ A \iota_B = 1$: spinor dyad
$\bar{\iota}_{A'}, \bar{o}_{A'}; \ A' = 1, 2$	primed (complex coinjugate) spinor dyad
g	gauge transformation or element of complexification of G
P	principal G-bundle
A	connection on SU(2)-bundle over σ
A_a^i	pull-back of A to σ by local section
$*\underline{E}$	pseudo-$(D-1)$-form in vector bundle associated to G-bundle under adjoint representation
$*\underline{E}_{a_1.., a_{D-1}}^I$	$:= k^{IJ} \epsilon_{a_1.., a_D} \underline{E}_J^{a_D}$: pull-back of $*\underline{E}$ to σ by local section
$*E$	pseudo-$(D-1)$-form in vector bundle associated to SU(2)-bundle under adjoint representation
$*E_{a_1.., a_{D-1}}^i$	$:= k^{ij} \epsilon_{a_1.., a_D} E_j^{a_D}$: pull-back of $*E$ to σ by local section
E_j^a	$:= \epsilon^{a_1 .. a_{D-1}} (*E)_{a_1.. a_{D-1}}^k k_{jk} / ((D-1)!)$: 'electric fields'
e	one-form co-vector bundle associated to the SU(2)-bundle under the defining representation (D-bein)
e_a^i	pull-back of e to σ by local section
Γ_a^i	pull-back by local section of SU(2) spin connection over σ
R, X	right-invariant vector field on G
L	Left-invariant vector field on G
$Y = iX$	momentum vector field
$\mathcal{M}$	phase space
$\mathcal{E}$	Banach manifold or space of smooth electric fields
$T_{(a_1.. a_n)}$	$:= \frac{1}{n!} \sum_{\iota \in S_n} T_{a_{\iota(1)} .. a_{\iota(n)}}$: symmetrisation of indices
$T_{[a_1.. a_n]}$	$:= \frac{1}{n!} \sum_{\iota \in S_n} \mathrm{sgn}(\iota) \ T_{a_{\iota(1)} .. a_{\iota(n)}}$: antisymmetrisation of indices
$\mathcal{A}$	space of smooth connections
$\mathcal{G}$	space of smooth gauge transformations

$\overline{\mathcal{A}}$	space of distributional connections
$\mathcal{A}^{\mathbb{C}}$	space of smooth complex connections
$\mathcal{G}^{\mathbb{C}}$	space of smooth complex gauge transformations
$\overline{\mathcal{G}}$	space of distributional gauge transformations
$\mathcal{A}/\mathcal{G}$	space of smooth connections modulo smooth gauge transformations
$\overline{\mathcal{A}/\mathcal{G}}$	space of distributional connections modulo distributional gauge transformations
$\overline{\mathcal{A}}/\overline{\mathcal{G}}$	space of distributional gauge equivalence classes of connections
$\overline{\mathcal{A}}^{\mathbb{C}}$	space of distributional complex connections
$\overline{\mathcal{A}/\mathcal{G}}^{\mathbb{C}}$	space of distributional complex gauge equivalence classes of connections
$\mathcal{C}$	set of semianalytic curves or classical configuration space
$\overline{\mathcal{C}}$	quantum configuration space
$\mathcal{P}$	set (groupoid) of semianalytic paths or set of punctures
$\mathcal{Q}$	set (group) of semianalytic closed and basepointed paths
$\mathcal{L}$	set of tame subgroupoids of $\mathcal{P}$ or general label set
$\mathcal{S}$	set of tame subgroups of $\mathcal{Q}$ (hoop group) or set of spin-network labels
l	subgroupoid
s	spin-net— spin-network label
$[s]$	(singular) knot-net= diffeomorphism equivalence class of s
Γ_0^ω	set of semianalytic, compactly supported graphs
Γ_σ^ω	set of semianalytic, countably infinite graphs
$\mathrm{Diff}(\sigma)$	group of smooth diffeomorphisms of σ
$\mathrm{Diff}_{\mathrm{sa}}^\omega(\sigma)$	group of semianalytic diffeomorphisms of σ
$\mathrm{Diff}_{\mathrm{sa},0}^\omega(\sigma)$	group of semianalytic diffeomorphisms of σ connected to the identity
$\mathrm{Diff}_0^\omega(\sigma)$	group of analytic diffeomorphisms of σ connected to the identity
$\mathrm{Diff}^\omega(\sigma)$	group of analytic diffeomorphisms of σ
φ	(semi-)analytic diffeomorphism
c	semianalytic curve
p	semianalytic path
e	entire semianalytic path (edge)
α	entire semianalytic closed path (hoop) or algebra automorphism

γ	semianalytic graph
v	vertex of a graph
$E(\gamma)$	set of edges of γ
$V(\gamma)$	set of vertexes of γ
$h_p(A) = A(p)$	holonomy of A along p
$\prec$	abstract partial order
Ω	vector state or symplectic structure or curvature two-form
F	pull-back to σ of 2Ω by a local section
ω	general state on *algebra
$\mathfrak{X}$, X, Y	measure space or topological space
$L(X,Y)$, $L(X)$	linear (un)bounded operators between X, Y or on X
$B(X,Y)$, $B(X)$	bounded operators between X, Y or on X
$K(X)$	compact operators on X
$B_1(X)$	trace class operators on X
$B_2(X)$	Hilbert–Schmidt operators on X
$\mathcal{B}$	σ-algebra
μ, ν, ρ	measure
$\mathcal{H}$	general Hilbert space
Cyl	space of cylindrical functions
$\mathcal{D}$	dense subspace of $\mathcal{H}$ equipped with a stronger topology
$\mathcal{D}'$	topological dual of $\mathcal{D}$
$\mathcal{D}^*$	algebraic dual of $\mathcal{D}$
$\mathcal{H}^0 = L_2(\overline{\mathcal{A}}, d\mu_0)$	uniform measure L_2 space
$\mathcal{H}^\otimes$	infinite tensor product extension of $\mathcal{H}^0$
Cyl_l	restriction of Cyl to functions cylindrical over l
$[.], (.)$	equivalence classes
$\mathfrak{A}$, $\mathfrak{B}$	abstract (*-)algebra or C^*-algebra
$\Delta(\mathfrak{A})$	spectrum on Abelian C^*-algebra
χ	character (maximal ideal) of unital Banach algebra or group or characteristic function of a set
$\mathfrak{I}$, $\mathfrak{J}$	ideal in abstract algebra
$\mathfrak{P}$	classical Poisson*-algebra
$\mathfrak{G}$	automorphism group (of principal fibre bundle)
$\mathfrak{D}$	Dirac or hypersurface deformation algebra
$\mathfrak{M}$	Master Constraint algebra
M	Master Constraint

Introduction: Defining quantum gravity

In the first section of this chapter we explain why the problem of quantum gravity cannot be ignored in present-day physics, even though the available accelerator energies lie way beyond the Planck scale. Then we define what a quantum theory of gravity and all interactions is widely expected to achieve and point out the two main directions of research divided into the perturbative and non-perturbative approaches. In the third section we describe these approaches in more detail and finally in the fourth motivate our choice of canonical quantum general relativity as opposed to other approaches.

Why quantum gravity in the twenty-first century?

It is often argued that quantum gravity is not relevant for the physics of this century because in our most powerful accelerator, the LHC to be working in 2007, we obtain energies of the order of a few 10^3 GeV while the energy scale at which quantum gravity is believed to become important is the Planck energy of 10^{19} GeV. While that is true, it is false that nature does not equip us with particles of energies much beyond the TeV scale; we have already observed astrophysical particles with energy of up to 10^{13} GeV, only six orders of magnitude away from the Planck scale. It thus makes sense to erect future particle microscopes not on the surface of the Earth any more, but in its orbit. As we will sketch in this book, even with TeV energy scales one might speculate about quantum gravity effects in the close future with γ-ray burst physics and the GLAST detector. Next, quantum gravity effects in the early universe might have left their fingerprint in the cosmological microwave background radiation (CMBR) and new satellites such as WMAP and PLANCK which have considerably increased the precision of experimental cosmology might reveal those. Notice that these data have already given us new cosmological puzzles recently, namely they have, for the first time, enabled us to reliably measure the energy budget of the universe: about 70% is a so-called dark energy component which could be a positive[1] cosmological constant, about 25% is a dark matter component which is commonly believed to be due to a weakly interacting massive particle (WIMP) (possibly supersymmetric) and only about 5% is made out of baryonic matter. Here 'dark' means

---

[1] Recent independent observations all indicate that the expansion of the universe is currently accelerating.

that these unknown forms of matter do not radiate, they are invisible. Hence we see that •• ••• •••••••••• •• •• ••• • ••••• •• ••• ••••••• and at least as far as dark energy is concerned, quantum gravity could have a lot to do with it. What we want to argue here is that quantum gravity is not at all of academic interest but possibly touches on brand new observational data which point at ••• •••••••• ••••••• ••• •••••••• • •••• and are of extreme current interest. See, for example, [18–20] for recent accounts of modern cosmology.

But even apart from these purely experimental considerations, there are good theoretical reasons for studying quantum gravity. To see why, let us summarise our current understanding of the fundamental interactions:

Embarassingly, the only quantum fields that we fully understand to date in four dimensions are •••• ••••• ••• ••••• •• ••••••• •• •••••••• • •• ••• ••• •••••. Formulated more provocatively:

In four dimensions we only understand an (infinite) collection of uncoupled harmonic oscillators on Minkowski space!

In order to leave the domain of these rather trivial and unphysical (since non-interacting) quantum field theories, physicists have developed two techniques: perturbation theory and quantum field theory on curved backgrounds. This means the following: with respect to accelerator experiments, the most important processes are scattering amplitudes between particles. One can •••• •••• write down a unitary operator that accounts for the scattering interaction between particles and which maps between the well-understood free quantum field Hilbert spaces in the far past and future. Famously, by Haag's theorem [21] whenever that operator is really unitary, there is no interaction and if it is not unitary, then it is ill-defined giving rise to the ultraviolet divergences of ordinary QFT. In fact, one can only define the operator perturbatively by writing down the formal power expansion in terms of the generator of the would-be unitary transformation between the free quantum field theory Hilbert spaces. The resulting series is divergent order by order but if the theory is 'renormalisable' then one can make these orders artificially finite by a regularisation and renormalisation procedure with, however, no control on convergence of the resulting series. Despite these drawbacks, this recipe has worked very well so far, at least for the electroweak interaction.

Until now, all we have said applies only to free (or perturbatively interacting) quantum fields on Minkowski spacetime for which the so-called Wightman axioms [21] can be verified. Let us summarise them for the case of a scalar field in $(D+1)$-dimensional Minkowski space:

W1 • •••••••• •• •••• •

There exists a unitary and continuous representation $U : \mathcal{P} \to \mathcal{B}(\mathcal{H})$ of the **Poincaré group** $\mathcal{P}$ on a Hilbert space $\mathcal{H}$.

W2 ....

The momentum operators P^μ have spectrum in the forward lightcone:
$\eta_{\mu\nu}P^\mu P^\nu \leq 0; \; P^0 \geq 0$.

W3 ......

There is a unique **Poincaré** invariant vacuum state $U(p)\Omega = \Omega$ for all
$p \in \mathcal{P}$.

W4 ....

Consider the smeared field operator-valued tempered distributions $\phi(f) = \int_{\mathbf{R}^{D+1}} d^{D+1}x\phi(x)f(x)$ where $f \in \mathcal{S}(\mathbf{R}^{D+1})$ is a test function of rapid decrease. Then finite linear combinations of the form $\phi(f_1)\ldots\phi(f_N)\Omega$ lie dense in $\mathcal{H}$ (that is, Ω is a cyclic vector) and $U(p)\phi(f)U(p)^{-1} = \phi(f \circ p)$ for any $p \in \mathcal{P}$.

W5 ....

Suppose that the supports (the set of points where a function is different from zero) of f, f' are **spacelike separated** (that is, the points of their supports cannot be connected by a non-spacelike curve) then $[\phi(f), \phi(f')] = 0$.

The most important objects in this list are those that are highlighted in boldface letters: the fixed, non-dynamical Minkowski background metric η with its well-defined causal structure, its Poincaré symmetry group $\mathcal{P}$, the associated representation $U(p)$ of its elements, the invariant vacuum state Ω and finally the fixed, non-dynamical topological, differentiable manifold $\mathbf{R}^{D+1}$. Thus the Wightman axioms assume the existence of a non-dynamical, Minkowski background metric which implies that we have a preferred notion of causality (or locality) and its symmetry group, the Poincaré group from which one builds the usual Fock Hilbert spaces of the free fields. We see that the whole structure of the theory is heavily based on the existence of these objects which come with a fixed, non-dynamical background metric on a fixed, non-dynamical topological and differentiable manifold.

For a general background spacetime, things are already under much less control: we still have a notion of causality (locality) but generically no symmetry group any longer and thus there is no obvious generalisation of the Wightman axioms and no natural perturbative Fock Hilbert space any longer. These obstacles can partly be overcome by the methods of algebraic quantum field theory [22] and the so-called microlocal analysis [23–26] (in which the locality axiom is taken care of pointwise rather than globally), which recently have also been employed to develop perturbation theory on arbitrary background spacetimes [27–33] by invoking the mathematically more rigorous implementation of the renormalisation programme developed by Epstein and Glaser in which no divergent expressions ever appear at least order by order (see, e.g., [34]). This way one manages to construct the interacting fields, at least perturbatively, on arbitrary backgrounds.

In order to go beyond a fixed background one can consider 'all backgrounds simultaneously' [35, 36]. Namely, the notion of a local quantum field theory $\mathfrak{A}(M,g)$ (thought of as a unital C^*-algebra for convenience) on a given curved background spacetime (M,g) can be generalised in the following way:[2] given an isometric embedding $\varphi : (M,g) \to (M',g')$ of one spacetime into another, one relates $\mathfrak{A}(M,g)$, $\mathfrak{A}(M',g')$ by asking that there is a $*$-algebraic homomorphism $\alpha_\varphi : \mathfrak{A}(M,g) \to \mathfrak{A}(M',g')$. The homomorphisms α_ψ could for instance just act geometrically by pulling back the fields. More abstractly, what one has then is the category Man whose objects are globally hyperbolic spacetimes (M,g) and whose morphisms are isometric embeddings with unit $1_{(M,g)} := \mathrm{id}_M$, the identity diffeomorphism. On the other hand, we have the category Alg whose objects are unital C^*-algebras $\mathfrak{A}$ and whose morphisms are injective $*$-homomorphisms with unit $1_{\mathfrak{A}} = \mathrm{id}_{\mathfrak{A}}$, the identity element in the algebra. A local quantum field is then a covariant functor $\mathsf{A} : \mathsf{Man} \to \mathsf{Alg}; (M,g) \mapsto \mathfrak{A}(M,g)$, $\varphi \mapsto \alpha_\varphi$ which relates objects and morphisms of Man with those of Alg. The functor is called causal if those quantum field theories $\mathfrak{A}(M_j,g_j)$ for which there exist isometric embeddings $\varphi_j : (M_j,g_j) \to (M,g)$; $j = 1,2$ so that $\varphi_1(M_1)$, $\varphi_2(M_2)$ are spacelike separated with respect to g satisfy the causality axiom $[\alpha_{\varphi_1}(\mathfrak{A}(M_1,g_1)), \alpha_{\varphi_2}(\mathfrak{A}(M_2,g_2))] = \{0\}$. The functor is said to obey the time slice axiom when $\alpha_\varphi(\mathfrak{A}(M,g)) = \mathfrak{A}(M',g'))$ for all isometries $\varphi : (M,g) \to (M',g')$ such that $\varphi(M)$ contains a Cauchy surface for (M',g'). This framework is background-independent because the functor A considers all backgrounds (M,g) simultaneously.

Unfortunately, QFT on curved spacetimes, even stated in this background-independent way, is only an approximation to the real world because it completely neglects the backreaction between matter and geometry which classically is expressed in Einstein's equations. Moreover, it neglects the fact that the gravitational field must be quantised as well, as we will argue below. One can try to rescue the framework of ordinary QFT by studying the quantum excitations around a given classical background metric, possibly generalised in the above background-independent way. However, not only does this result in a non-renormalisable theory without predictive power when treating the gravitational field in the same fashion, it is also unclear whether the procedure leads to (unitarily) equivalent results when using backgrounds which are physically different, such as two Schwarzschild spacetimes with different mass (the corresponding spacetimes are ••• •••• •••••). More seriously, it is expected that especially in extreme astrophysical or cosmological situations (black holes, big bang) the notion of a classical, smooth spacetime **breaks down altogether!** In other words, the fluctuations of the metric operator become deeply quantum and there is no semiclassical notion of a spacetime any more, similarly to the

---

energy spectrum of the hydrogen atom far away from the continuum limit. It is precisely here where a full-fledged quantum theory of gravity is needed: we must be able to treat all backgrounds on a common footing, otherwise we will never understand what really happens in a Hawking process when a black hole loses mass due to radiation. Moreover, we need a background-independent theory of GR where the lightcones themselves start fluctuating and hence locality becomes a fuzzy notion. Let us phrase this again, provocatively, as:

The whole framework of ordinary quantum field theory breaks down once we make the gravitational field (and the differentiable manifold) dynamical, once there is no background metric any longer!

Combining these issues, one can say that we have a working understanding of scattering processes between elementary particles in arbitrary spacetimes as long as the backreaction of matter on geometry can be neglected and that the coupling constant between non-gravitational interactions is small enough (with QCD being an important exception) since then the classical Einstein equation, which says that curvature of geometry is proportional to the stress energy of matter, can be approximately solved by neglecting matter altogether. Thus, in this limit, it seems fully sufficient to have only a classical theory of general relativity and perturbative quantum field theory on curved spacetimes.

From a fundamental point of view, however, this state of affairs is unsatisfactory for many reasons among which we have the following:

(i) • •• •••••• • •••• • • ••••••• • • ••••• •• ••• •• •••••• ••

There are two kinds of problem with the idea of keeping geometry classical while matter is quantum:

(i1) • ••••••• •••• •

At a fundamental level, the backreaction of matter on geometry cannot be neglected. Namely, geometry couples to matter through • •• •• •• •• •• ••• ••• •• •

$$R_{\mu\nu} - \tfrac{1}{2} R \cdot g_{\mu\nu} = \kappa\, T_{\mu\nu}[g]$$

and since matter underlies the rules of quantum mechanics, the right-hand side of this equation, the stress–energy tensor $T_{\mu\nu}[g]$, becomes an operator. One has tried to keep geometry classical while matter is quantum mechanical by replacing $T_{\mu\nu}[g]$ by the Minkowski vacuum Ω_η expectation value $< \Omega_\eta, \hat{T}_{\mu\nu}[\eta]\Omega_\eta >$, but the solution of this equation will give $g \neq \eta$ which one then has to feed back into the definition of the vacuum expectation value, and so on. Notice that the notion of vacuum itself depends on the background metric, so that this is a highly non-trivial iteration process. The resulting iteration does not converge in general [37]. Thus, such a procedure is also inconsistent, whence we • ••• ••• • •••• ••• ••• •••• ••• ••• • ••• •• • •••. This leads to the ••• ••• • •• ••• •• ••• • ••• • •

$$\hat{R}_{\mu\nu} - \tfrac{1}{2}\hat{R} \cdot \hat{g}_{\mu\nu} = \kappa \, \hat{T}_{\mu\nu}[\hat{g}]$$

Of course, this equation is only formal at this point and must be embedded into an appropriate Hilbert space context.

(i2) • • •• ••• •

There is another piece of evidence for the need to quantise geometry: recall that in perturbative QFT one integrates over virtual particles in higher loop diagrams with arbitrarily large energy. Suppose that such a particle has energy E and momentum $P \approx E/c$ in some rest frame. According to quantum mechanics, such a particle has a lifetime $\tau \approx \hbar/E$ and a spatial extension given by the Compton radius $\lambda \approx \hbar c/E$. According to classical GR, such a lump of energy collapses to a black hole if the Compton radius drops below the Schwarzschild radius $r \approx GE/c^4$, in other words, when the energy exceeds the Planck energy $E_p = \sqrt{\hbar c/G c^2}$. The problem is now not only that in ordinary QFT this general relativistic effect is neglected, but moreover that this effect leads to new processes: according to the Hawking effect, after the lifetime τ the black hole evaporates. However, it evaporates into particles of all possible species. Suppose for instance that the original particle was a neutrino. All that the resulting black hole remembers is its mass and spin. Now while the neutrino only interacts electroweakly according to the standard model, the black hole can produce gluons and quarks, which is impossible within the standard model.

Of course, all of these arguments are only heuristic, however, they reveal that it is problematic to combine classical geometry with quantum matter. They suggest that it is problematic or even inconsistent to resolve spacetime distances below the Planck scale $\ell_p = \sqrt{\hbar c G/c^2}$. It is due to considerations of this kind that one expects that gravity provides a natural UV cutoff for QFT. If that is the case, then it is natural to expect that the quantum spacetime structure reveals a discrete structure at Planck scale. We will see a particular incarnation of this idea in LQG.

(ii) •• • ••••• • •• ••••••• •••• • •••• •• ••• ••••••• ••

Even without quantum theory at all Einstein's field equations predict space-time singularities (black holes, big bang singularities, etc.) at which the equations become meaningless. In a truly fundamental theory, there is no room for such breakdowns and it is suspected by many that the theory cures itself upon quantisation in analogy to the hydrogen atom whose stability is classically a miracle (the electron should fall into the nucleus after a finite time lapse due to emission of Bremsstrahlung) but is easily explained by quantum theory which bounds the electron's energy from below.

(iii) •••••••• •••••••• • •••••• •• ••••• •••••• ••

As outlined above, perturbative quantum field theory on curved spacetimes is itself also ill-defined due to its UV (short distance) singularities which can be cured only with an ad hoc recipe order by order which lacks a fundamental explanation; moreover, the perturbation series is usually divergent. Besides that, the corresponding infinite vacuum energies being usually neglected in such a procedure contribute to the cosmological constant and should have a large gravitational backreaction effect. That such energy subtractions are quite significant is maybe best demonstrated by the Casimir effect. Now, since general relativity possesses a fundamental length scale, the Planck length $\ell_p \approx 10^{-33}$ cm, it has been argued ever since that gravitation plus matter should give a finite quantum theory since gravitation provides the necessary, built-in, short distance cutoff.

(iv) • ••• ••••••••• ••• ••••• • •••••••

However, that cutoff cannot work naively: consider for simplicity a free massless scalar field on Minkowski space. The difference between the Hamiltonian and its normal ordered version is given by the divergent expression

$$\hat{H} - :\hat{H}: = \hbar \int d^3x [\sqrt{-\Delta}\delta(x,y)]_{y=x} = \hbar \int d^3x \int d^3k \, |k|$$

where Δ is the flat space Laplacian. If we assume a naive momentum cutoff due to quantum gravity at $|k| \le 1/\ell_P$ the divergent momentum integral becomes proportional to ℓ_P^{-4}. Comparing this with the cosmological constant Hamiltonian $\frac{\Lambda}{G} \int d^3x \sqrt{\det(q)}$ where Λ is the cosmological constant, G is Newton's constant and q is the spatial metric (which is flat on Minkowski space) then we conclude that $\Lambda \ell_P^2 \approx 1$ where $\hbar G = \ell_P^2$ was used. However, experimentally we find $\Lambda \ell_P^2 \approx 10^{-120}$. Thus the cosmological constant is unnaturally small and presents the worst fine-tuning problem ever encountered in physics. Notice that the cosmological constant is a possible candidate for dark energy.

(v) • ••••• ••• •••• ••• •••• ••• ••••••• •• ••• •••••••• ••

Given the fact that perturbation theory works reasonably well if the coupling constant is small for the non-gravitational interactions on a background metric it is natural to try whether the methods of quantum field theory on curved spacetime work as well for the gravitational field. Roughly, the procedure is to write the dynamical metric tensor as $g = \eta + h$ where η is the Minkowski metric and h is the deviation of g from it (the graviton) and then to expand the Lagrangian as an infinite power series in h. One arrives at a formal, infinite series with finite radius of convergence which becomes meaningless if the fluctuations are large. Although the naive power counting argument implies that general relativity so defined is a non-renormalisable theory, it was hoped that due to cancellations of divergences the perturbation theory could actually be finite. However, that this hope was unjustified was shown in [38, 39] where calculations demonstrated the appearance of divergences at the two-loop level, which suggests that at every order of perturbation theory one must introduce new coupling constants which the classical theory did not know about and one loses predictability.

It is well known that the (locally) supersymmetric extension of a given non-supersymmetric field theory usually improves the ultraviolet convergence of the resulting theory as compared with the original one due to fermionic cancellations [40]. It was therefore natural to hope that quantised supergravity might be finite. However, in [41] a serious argument against the expected cancellation of perturbative divergences was raised and recently even the again popular (due to its M-theory context) most supersymmetric 11D 'last hope' supergravity theory was shown not to have the magical cancellation property [42–44].

Summarising, although a definite proof is still missing up to date (mainly due to the highly complicated algebraic structure of the Feynman rules for quantised supergravity) it is today widely believed that perturbative quantum field theory approaches to quantum gravity are meaningless.

The upshot of these considerations is that our understanding of quantum field theory and therefore fundamental physics is quite limited unless one quantises the gravitational field as well. Being very sharply critical one could say:

The current situation in fundamental physics can be compared with the one at the end of the nineteenth century: while one had a successful theory of electromagnetism, one could not explain the stability of atoms. One did not need to worry about this from a practical point of view since atomic length scales could not be resolved at that time but from a fundamental point of view, Maxwell's theory was incomplete. The discovery of the mechanism for this stability, quantum mechanics, revolutionised not only physics. Similarly, today we still have no thorough understanding for the stability of nature in the sense discussed above and it is similarly expected that the more complete theory of quantum gravity will radically change our view of the world. That is, considering the metric as a quantum operator will bring us beyond standard model physics even without the discovery of new forces, particles or extra dimensions.

The role of background independence

The twentieth century has dramatically changed our understanding of nature: it revealed that physics is based on two profound principles, quantum mechanics and general relativity. Both principles revolutionise two pivotal structures of

Newtonian physics. First, the determinism of Newton's equations of motion evaporates at a fundamental level, rather dynamics is reigned by probabilities underlying the Heisenberg uncertainty obstruction. Second, the notion of absolute time and space has to be corrected; space and time and distances between points of the spacetime manifold, that is, the metric, become themselves dynamical, geometry is no longer just an observer. The usual Minkowski metric ceases to be a distinguished, externally prescribed, background structure. Rather, the laws of physics are •••••••••••••••••••••••••, mathematically expressed by the classical Einstein equations which are •••••••••• ••• •••••••• ••• ••••••• •••••••••. As we have argued, it is this new element of •••••••••• •••••••••••• brought in with Einstein's theory of gravity which completely changes our present understanding of quantum field theory.

A satisfactory physical theory must combine both of these fundamental principles, quantum mechanics and general relativity, in a consistent way and will be called 'Quantum Gravity'. However, the quantisation of the gravitational field has turned out to be one of the most challenging unsolved problems in theoretical and mathematical physics. Although numerous proposals towards a quantisation have been made since the birth of general relativity and quantum theory, none of them can be called successful so far. This is in sharp contrast to what we see with respect to the other three interactions whose description has culminated in the so-called standard model of matter, in particular, the spectacular success of perturbative quantum electrodynamics whose theoretical predictions could be verified •• ••• •••••• • ••• •• ••• ••••••• ••••••••• •••• until today.

Today we do not have a theory of quantum gravity, what we have is:

1. The Standard Model, a quantum theory of the non-gravitational interactions (electromagnetic, weak and strong) or • ••••• which, however, completely ignores General Relativity.
2. Classical General Relativity or •••• ••••, which is a background-independent theory of all interactions but completely ignores quantum mechanics.

What is so special about the gravitational force that it has persisted in its quantisation for about 70 years already? As outlined in the previous section, •••
••••• •• •• ••• • •• •••• •••••• •• • ••• •• •••• ••• •• •• • •• •• ••••• •••••••••••
• •••••. The whole formalism of ordinary QFT relies heavily on this background structure and collapses to nothing when it is missing. It is already much more difficult to formulate a QFT on a non-Minkowski (curved) background but it seems to become a completely hopeless task when the metric is a dynamical, even fluctuating quantum field itself. This underlines once more the source of our current problem of quantising gravity: •• ••••• •• ••••• ••• •• •• • • • ••
• ••• •••• •••• • • •• ••• •• ••• ••• •••••• • •••• • ••• ••••• •• ••••• ••• •••• •••••• •
•• • ••••• ••••••••••• •• ••• •••• •••• • •• ••• ••• • ••••• •••• • ••••••• •.

In order to proceed, today a high-energy physicist has the choice between the following two, extreme approaches. Either the •••••••• ••••••••••, who prefers to take over the well-established mathematical machinery from QFT

on a background at the price of dropping background independence altogether
to begin with and then tries to find the true background-independent theory
by summing the perturbation series (summing over all possible backgrounds).
Or the •• •••••••• •••• ••••••, who believes that background independence lies at
the heart of the solution to the problem and pays the price to have to invent
mathematical tools that go beyond the framework of ordinary QFT right from
the beginning. Both approaches try to unravel the truly deep features that are
unique to Einstein's theory associated with background independence from different ends.

The particle physicist's language is perturbation theory, that is, one writes
the quantum metric operator as a sum consisting of a background piece and a
perturbation piece around it, the graviton, thus obtaining a graviton QFT on a
Minkowski background. • • ••• •••• ••••••••••••••• •••••••• •• ••• •••• •••••••••••
•••••• ••••••••••• ••••••••••••• ••• •••••• •••••••• ••••••••••• •• •••••• •• •••
••••• •• •••••••••••• •••••••. Thus one can restore background independence
only by summing up the entire perturbation series, which is of course not easy.
Not surprisingly, as already mentioned, since $\hbar\kappa = \ell_p^2$ has negative mass dimension in Planck units, applying this programme to Einstein's theory itself results
in a mathematical disaster, a so-called non-renormalisable theory without any
predictive power. In order to employ perturbation theory, it seems that one has
to go to string theory which, however, requires the introduction of new additional
structures that Einstein's classical theory did not know about: supersymmetry,
extra dimensions and an infinite tower of new and very heavy particles next to
the graviton. This is a fascinating but extremely drastic modification of general
relativity and one must be careful not to be in conflict with phenomenology as
superparticles, Kaluza Klein modes from the dimensional reduction and those
heavy particles have not been observed until today. On the other hand, string
theory has a good chance to be a unified theory of the perturbative aspects of
all interactions in the sense that all interactions follow from a common object,
the string, thereby explaining the particle content of the world.

The quantum geometer's language is a non-perturbative one, keeping background independence as a guiding principle at every stage of the construction of
the theory, resulting in mathematical structures drastically different from the
ones of ordinary QFT on a background metric. One takes Einstein's theory
absolutely seriously, uses only the principles of General Relativity and quantum
mechanics and lets the theory build itself, driven by mathematical consistency.
Any theory meeting these standards will be called • •••••• • ••••••• ••• ••••••
•• • • •. Since QGR does not modify the matter content of the known interactions, QGR is therefore not in conflict with phenomenology but also it does not
obviously explain the particle content of the world. However, it tries to unify all
interactions in a different sense: all interactions must transform under a common gauge group, the four-dimensional diffeomorphism group which on the other
hand is almost completely broken in perturbative approaches.

Let us remark that even without specifying further details, any QGR theory is a promising candidate for a theory that is free from two divergences of the so-called perturbation series of Feynman diagrams common to all perturbative QFTs on a background metric: (1) each term in the series diverges due to the ultraviolet (UV) divergences of the theory which one can cure for renormalisable theories through so-called renormalisation techniques and (2) the series of these renormalised, finite terms diverges, one says the theory is not finite. The first, UV, problem has a chance to be absent in a background-independent theory for a simple but profound reason: in order to say that a momentum becomes large one must refer to a background metric with respect to which it is measured, but there simply is no background metric in the theory. The second, convergence, problem of the series might be void as well since there are simply no Feynman diagrams! Thus, the mere existence of a consistent background-independent quantum gravity theory could imply a finite quantum theory of all interactions. Of course, a successful quantum gravity theory must recover all the results that have been obtained by perturbative techniques ••• that have been verified in experiments.

Approaches to quantum gravity

The aim of the previous section was to convince the reader that background independence is, maybe, the **Key Feature** of quantum gravity to be dealt with. No matter how one deals with this issue, whether one starts from a perturbative (= background-dependent) or from a non-perturbative (= background-independent) platform, one has to invent something drastically new in order to quantise the gravitational field. Roughly speaking, if one wants to keep perturbative renormalisability as a criterion for a meaningful theory, then one has to increase the amount of symmetries, resulting in superstring theory which hopefully has General Relativity and the standard model as an effective low-energy limit. (Compare the historically similar case of the non-renormalisable Fermi model of the weak interaction with massive gauge bosons which was replaced by the more symmetric and renormalisable electroweak Yang–Mills theory.) If one considers General Relativity as a fundamental theory then one cannot introduce extra structure, one has to give up the renormalisability principle and instead has to invent a new mathematical framework which can deal with background independence. (Compare the historically similar case of the bizarre ether model based on the Newtonian notion of absolute spacetime which was abandoned by the special relativity principle.)

We will now explain these approaches in more detail.

1. ••••••••••• ••••••••••• ••••• • ••••••

    The only known consistent perturbative approach to quantum gravity is string theory which has good chances to be a theory that unifies all interactions. String theory [45] is not a field theory in the ordinary sense of the word.

•• ••••• •••••• •• •• ••• •• ••• •••• •• •• •••

Originally, it was a two-dimensional field theory of worldsheets embedded into a fixed, • -dimensional pseudo-Riemannian manifold (M, g) of Lorentzian signature which is to be thought of as the spacetime of the physical world. The Lagrangian of the theory is a kind of non-linear σ-model Lagrangian for the associated embedding variables X (and their supersymmetric partners in case of the superstring). If one perturbs $g(X) = \eta + h(X)$ as above and keeps only the lowest order in X one obtains a free field theory in two dimensions which, however, is consistent (Lorentz covariant) only when $D + 1 = 26$ (bosonic string) or $D + 1 = 10$ (superstring), respectively. Strings propagating in those dimensions are called critical strings, non-critical strings exist but have so far not played a significant role due to phenomenological reasons. Remarkably, the mass spectrum of the particle-like excitations of the closed worldsheet theory contains a massless spin-two particle which one interprets as the graviton. Until recently, the superstring was favoured since only there was it believed to be possible to get rid of an unstable tachyonic vacuum state by the GSO projection. However, one recently also tries to construct stable bosonic string theories [46].

Moreover, if one incorporates the higher-order terms $h(X)$ of the string action, sufficient for one-loop corrections, into the associated path integral one finds a consistent quantum theory up to one loop only if the background metric satisfies the Einstein equations. These are the most powerful outcomes of the theory: although one started out with a fixed background metric, the background is not arbitrary but has to satisfy the Einstein equations up to higher loop corrections, indicating that the one-loop effective action for the low-energy quantum field theory in those D dimensions is Einstein's theory plus corrections. Finally, only recently has it been shown [47] that at least the type II superstring theories are one- and two-loop and, possibly, to all orders, •• •••. String theorists therefore argue to have found candidates for a consistent theory of quantum gravity with the additional advantage that they do not contain any free parameters (like those of the standard model) except for the string tension.

These facts are very impressive, however, some cautionary remarks are appropriate, see also the beautiful review [48]:
− •• ••••• ••••• •••••

  Dimension $D + 1 = 10, 26$ is not the dimension of everyday physics so that one has to argue that the extra $D − 3$ dimensions are 'tiny' in the Kaluza–Klein sense although nobody knows the mechanism responsible for this 'spontaneous compactification'. According to [49] there exist at least 10^4 consistent, distinct Calabi–Yau compactifications (other compactifications such as toroidal ones seem to be inconsistent with phenomenology), each of which has an order of 10^2 free, continuous parameters (moduli) like the vacuum expectation value of the Higgs field in the standard model. For each compactification of each of the five string theories in $D = 10$ dimensions and for each choice of the moduli one obtains a distinct low-energy effective theory. This is clearly not what one expects from a theory that aims to unify all the interactions, the 18 (or more for massive neutrinos) free, continuous parameters of the standard model have been replaced by 10^2 continuous plus at least 10^4 discrete ones.

This vacuum degeneracy problem is not cured by the M-theory interpretation of string theory but it is conceptually simplified if certain conjectures are indeed correct: string theorists believe (bearing on an impressively huge number of successful checks) that so-called T (or target space) and S (or strong–weak coupling) duality transformations between all these string theories exist, which suggests that we do not have 10^4 unrelated 10^2-dimensional moduli spaces but that rather these 10^2-dimensional manifolds intersect in singular, lower-dimensional submanifolds corresponding to certain singular moduli configurations. This typically happens when certain masses vanish or certain couplings diverge or vanish (in string theory the coupling is related to the vacuum expectation value of the dilaton field). Crucial in this picture are so-called • -branes, higher-dimensional objects additional to strings which behave like solitons ('magnetic monopoles') in the electric description of a string theory and like fundamental objects ('electric degrees of freedom') in the S-dual description of the same string theory, much like the electric–magnetic duality of Maxwell theory under which strong and weak coupling are exchanged. Further relations between different string theories are obtained by compactifying them in one way and decompactifying them in another way, called a T-duality transformation. The resulting picture is that there exists only one theory which has all these compactification limits just described, called M-theory. Curiously, M-theory is an 11D theory whose low energy limit is 11D supergravity and whose weak coupling limit is type IIA superstring theory (obtained by one of these singular limits since the size of the 11th compactified dimension is related to the string coupling again). Since 11D supergravity is also the low-energy limit of the 11D supermembrane, some string theorists interpret M-theory as the quantised 11D supermembrane (see, e.g., [50, 51] and references therein).

Until today, no conclusive proof exists that for any of the compactifications described above we obtain a low-energy effective theory which is experimentally consistent with the data that we have for the standard model [52], although one seems to get at least rather close. The challenge in string phenomenology is to consistently and spontaneously break supersymmetry in order to get rid of the so far non-observed superpartners. There is also an infinite tower of very massive (of the order of the Planck mass and higher) excitations of the string, but these are too heavy to be observable. More interesting are the Kaluza–Klein modes whose masses are inverse proportional to the compactification radii and which have recently given rise to speculations about 'sub-mm-range' gravitational forces [53], which one must make consistent with observation also.

— •• ••••• •• ••• ••••••••• •••• •

Even before the M-theory revolution, string theory has always been a theory without Lagrangian description, S-matrix element computations have been guided by conformal invariance but there is no 'interaction Hamiltonian', string theory is a first-quantised theory. Second quantisation of string theory, called string field theory [54], has so far not attracted as much attention as it possibly deserves. However, a fascinating possibility is that the 11D supermembrane, and thus M-theory, is an already second-quantised theory [55].

— • ••••••••• ••••••••••

As mentioned above, string theory is best understood as a free 2D field theory propagating on a 10D Minkowski target space plus perturbative corrections for scattering matrix computations. This is a heavily background-dependent description, issues like the action of the 10D diffeomorphism group, the fundamental symmetry of Einstein's action, or the probability amplitude for the quantum evolution of one background into another cannot be questioned. Perturbative string theory, as far as quantum gravity is concerned, can describe graviton scattering in a background space-time, however, the most interesting problems near classical singularities require a non-perturbative description, such as the fundamental description of Hawking radiation. As a first step in that direction, recently stringy black holes have been discussed [56]. Here one uses so-called BPS • -brane configurations which are so special that one can do a perturbative calculation and extend it to the non-perturbative regime since the results are protected against non-perturbative corrections due to supersymmetry. So far this works only for extremely charged, supersymmetric black holes which are astrophysically not very realistic. But still these developments are certainly a move in the right direction since they use for the first time non-perturbative ideas in a crucial way and have been celebrated as one of the triumphs of string theory.

— • •• ••••••••

Coming back to the D-branes mentioned above, these are surfaces on which open strings must end (D stands for Dirichlet boundary conditions). Since these D-branes are completely arbitrary and not constrained by the theory, M-theory contains as many vacua as there are D-brane configurations (sometimes called charges or fluxes), which of course have to be gauge-invariant, in particular supersymmetric. This makes the number of string vacua plain infinite [57] and the number of physically relevant (e.g., consistent with cosmological observations, supersymmetry, topology and/or stable) vacua has been estimated to be of the order of

10^{100}–10^{500} [58,59] or even infinite [60] depending on one's assumptions (all analyses count compactification possibilities as well). Whether this number is infinite or just very large seems to be currently under debate, however, the number seems to be robustly above the 10^{80} particles contained in the observable (causally connected, i.e., of Hubble radius size) universe. This number of vacua, called the landscape, is so vast that some string theorists [61] employ the anthropic principle in order to rescue predictability of string theory, which is not unproblematic [62]. From the point of view of a background-independent theory which in some sense describes all background-dependent quantum field or string theories (i.e., vacua) simultaneously, the landscape could be an artifact of trying to describe quantum gravity by a collection of background-dependent theories which are not connected to each other while they should be. See [63] for more details.

In a celebrated paper [64, 65], Maldacena conjectured that string theory on an anti-de Sitter (AdS) background can be described by a conformal quantum field theory (CFT)[3] on the boundary of the AdS space. For an introduction to CFT, see [66]. This is yet another duality conjecture of string theory whose most studied incarnation is string theory on an $AdS_5 \times S^5$ background and $\mathcal{N} = 4$ Super–Yang–Mills theory (SYM). The latter is finite order by order in perturbation theory. The AdS/CFT correspondence can be considered as a concrete application of the holographic principle (see, e.g., [67]).

Unfortunately, so far this conjecture has mostly been checked at the level of the low-energy limit of string theory, that is, the corresponding supergravity theory, while there has been recent progress [68] as far as the conformal field theory side of the correspondence is concerned, based on the discovery of certain integrability structures. Moreover, in a mathematically precise formulation of the conjecture [69–72] one can show by the methods of algebraic QFT (local quantum physics) that if the theory in the bulk is described by a local Lagrangian then the boundary theory is non-local and vice versa. There is no contradiction because the full low-energy effective action of string theory is non-local (containing an infinite tower of α' corrections), however, it then becomes hard to verify the conjecture just using the tree term.

---

[3] That is, a QFT on D-dimensional Minkowski space whose underlying Lagrangian is not only invariant under the Poincaré group $\text{ISO}(1, D - 1)$ but also under conformal transformations. The resulting enlarged group is called the conformal group and its elements g satisfy $g^*\eta = \Omega^2\eta$ where η is the Minkowski metric and Ω is an arbitrary function. For isometries $\Omega = 1$, for non-trivial conformal transformations $\Omega \neq 1$. The AdS/CFT correspondence or conjecture is based on the fact that the isometry groups on an AdS space in $D + 1$ spacetime dimensions, as well as the conformal group of Minkowski space in D spacetime dimensions, have (locally) the structure of $\text{SO}(2, D)$.

Notice that current observations indicate that our universe is in a de Sitter phase (positive cosmological constant). However, a de Sitter background, in contrast to an anti-de Sitter background, does not have a positive energy supersymmetric extension of the de Sitter algebra (the analogue of the Poincaré algebra). One way to see this is to note that in supersymmetric theories the energy is always positive while de Sitter space does not admit a global timelike Killing field and hence no positive energy. String theories based on de Sitter space, if they exist, thus tend to be unstable since the corresponding low-energy supergravity theories are. In general it is hard to formulate string theory on time-dependent backgrounds which, however, are the most relevant ones for cosmology. Quite generally it is simply not true that every solution of Einstein's equations without Rarita–Schwinger fields has a supersymmetric extension including Rarita–Schwinger fields, that is, not every Einstein space is compatible with supergravity (local supersymmetry).

2. • • •• •• •• ••• •••• • •• •• •• •• •

The non-perturbative approaches to quantum gravity can be grouped into the following five main categories.

2a. • • • • •• •• • •• •• •• • •• •• •• • •• ••• •••

If one wanted to give a definition of this theory then one could say the following:

> **Canonical Quantum General Relativity is an attempt to construct a mathematically rigorous, non-perturbative, background-independent Quantum Field Theory of four-dimensional, Lorentzian general relativity plus all known matter in the continuum.**

This is the oldest approach and goes back to the pioneering work by Dirac [73–76] started in the 1940s and was further developed by Bergmann and Komar [77–80] as well as Arnowittt, Deser and Misner [81] in the 1950s and especially by Wheeler and DeWitt [82–85] in the 1960s. The idea of this approach is to apply the Legendre transform to the Einstein–Hilbert action by splitting spacetime into space and time and to cast it into Hamiltonian form. The resulting 'Hamiltonian' H is actually a so-called Hamiltonian constraint, that is, a Hamiltonian density which is constrained to vanish by the equations of motion. A Hamiltonian constraint must occur in any theory that, like general relativity, is invariant under local reparametrisations of time. According to Dirac's theory of the quantisation of constrained Hamiltonian systems, one is now supposed to impose the vanishing of the quantisation $\hat{H}$ of the Hamiltonian constraint H as a condition on states ψ in a suitable

Hilbert space $\mathcal{H}$, that is, formally

$$\hat{H}\psi = 0$$

This is the famous Wheeler–DeWitt equation or **quantum Einstein equation** of canonical quantum gravity and resembles a Schrödinger equation, only that the familiar $\partial\psi/\partial t$ term is missing, one of several occurrences of the 'absence or problem of time' in this approach (see, e.g., [86] and references therein).

Since the status of this programme, that is, its Loop Quantum Gravity (LQG) incarnation, is the subject of the present book we will not go too much into details here. The successes of LQG are a mathematically rigorous framework, manifest background independence, a manifestly non-perturbative language, an inherent notion of quantum discreteness of spacetime which is •••••• rather than postulated, certain UV finiteness results, a promising path integral formulation (spin foams) and finally a consistent formulation of quantum black hole physics. A conceptually very similar but technically different canonical programme has been launched by Klauder [87–91] to which the following remarks apply as well.

The following issues are at the moment unresolved within this approach:

* • ••• •• • ••• •• ••• •• • ••• ••• • ••• •• ••• ••

The Wheeler–DeWitt operator is, in the so-called ADM formulation, a functional differential operator of second order of the worst kind, namely with non-polynomial, not even analytic (in the basic configuration variables) coefficients. To even define such an operator rigorously has been a major problem for more than 60 years. What should be a suitable Hilbert space that carries such an operator? It is known that a Fock–Hilbert space is not able to support it. Moreover, the structure of the solution space is expectedly very complicated. Thus we see that one meets a great deal of mathematical problems before one can even start addressing physical questions. As we will describe in this book, there has been a huge amount of progress in this direction since the introduction of new canonical variables due to Ashtekar [92, 93] in 1986. However, the physics of the Wheeler–DeWitt operator is still only poorly understood.

* • ••• •• • •• ••••• •• • •••• •• ••• •• • ••• ••• ••• • ••• ••• •• •• •• ••

Due to the split of spacetime into space and time the treatment of spatial and time diffeomorphisms is somewhat different and the original four-dimensional covariance of the theory is no longer manifest. Classically one can prove (and we will in fact do that later on) that four-dimensional diffeomorphism covariance is encoded in a precise sense into the canonical formalism, although it is deeply hidden. In the quantum theory the issue reappears in the form of possible anomalies of the constraint algebra. We will show how to avoid those anomalies but possibly at the price of having a physical Hilbert space which is too small, which affects the classical limit, see below.

Let us clarify an issue that comes up often in debates between quantum geometers and string theorists: what one means by $(D + 1)$-dimensional covariance in string theory on a Minkowski target space is just $(D + 1)$-dimensional • •• •••• covariance but not ••• ••• •••• ••• covariance. Clearly the Poincaré group is not even a subgroup of the diffeomorphism group (for asymptotically flat spacetimes). The Poincaré group is a group of symmetries of asymptotically flat spacetimes while the diffeomorphisms, which are asymptotically trivial by definition, are gauge transformations. The latter group is completely broken in string theory, the former is also present in General Relativity.

* •• •••• ••• ••• ••• •• •••• •••• •• •• ••••• ••

Once one has found the solutions of the quantum Einstein equations one must find a complete set of Dirac observables (operators that leave the space of solutions invariant), which is a hard task to achieve even in classical General Relativity. One must therefore find suitable approximation methods, which is a development that has just recently started. However, even if one had found those (approximate) operators, which would be in some sense even time-independent and therefore extremely non-local, one would need to deparametrise the theory, that is, one must find an explanation for the local dynamics in our world. There are technically precise proposals for dealing with the classical part of this issue, but there is no rigorous quantum framework available at the moment.

* • •• ••••• • ••• ••

As we will see, our Hilbert space is of a new (background-independent) kind, operators are regulated in a non-standard (background-independent) way. It is therefore no longer clear that the theory that has been constructed so far indeed has General Relativity as its classical limit. The issue must be settled by a semiclassical analysis for canonical QGR, a programme that has only been launched recently.

2b. • • • ••• •• • •• •• ••••• •• •• •••••• •••••••

Here one tries to give meaning to the sum over histories of e^{-S_E} where S_E denotes the Euclidean Einstein–Hilbert action [94]. It is extremely hard to do the path integral and apart from semiclassical approximations and steepest descent methods in simplified models with a finite number of degrees of freedom one could not get very far within this framework yet [95–98]. There are at least the two following reasons for this:

1. The action functional S_E is unbounded from below. Therefore the path integral is badly divergent from the outset and although rather

sophisticated proposals have been made on how to improve the convergence properties, none of them has been fully successful to the best knowledge of the author.

2. The Euclidean field theory underlying the functional integral and the quantum theory of fields propagating on a Minkowski background are related by Wick rotating the Schwinger distributions of the former into the Wightman distributions of the latter (see, e.g., [99]). However, in the case of quantum gravity the metric itself becomes dynamical and is being integrated over, therefore the concept of Wick rotation becomes ill-defined. In other words, there is no guarantee that the Euclidean path integral even has any relevance for the quantum field theory underlying the Lorentzian Einstein–Hilbert action.

Nevertheless, one can try to define such a Euclidean path integral non-perturbatively by looking for non-Gaußian fixed points in Wilson's renormalisation analysis corresponding to an interacting microscopic theory (an asymptotically safe theory in Weinberg's terminology [100]). This line of thought has recently again picked up momentum due to non-trivial new results by Reuter and coworkers [101–108] and Niedermaier [109–111].

2c.

This approach can be subdivided into two main streams (see [112] for a review):

(a) Regge calculus [113–115]. Here one introduces a fixed triangulation of spacetime and integrates with a certain measure over the lengths of the links of this triangulation. The continuum limit is reached by refining the triangulation.

(b) Dynamical triangulations [116]. Here one takes the opposite point of view and keeps the lengths of the links fixed but sums over all triangulations. The continuum limit is reached by taking the link length to zero.

In both approaches one has to look for critical points (second-order phase transitions). An issue in both approaches is the choice of the correct measure. Although there is no guideline, it is widely believed that the dependence on the measure is weak due to universality in the statistical mechanical sense. The reason for the possibility that the path integral exists although the Euclidean action is unbounded from below is that the configurations with large negative action have low volume (measure) so that 'entropy wins over energy'. Especially in the field of dynamical triangulations there has been a major breakthrough recently [117–120]: the convergence of the partition function could be established in two dimensions (the action is basically a cosmological constant term) and the relation between the Lorentzian and Euclidean theory becomes transparent. This opens the possibility that similar results hold in higher dimensions, in particular, it seems as if the Lorentzian theory is much better behaved than the Euclidean theory because one has to sum over fewer configurations (those that are compatible with quantum causality). There are also promising new results concerning a non-perturbative Wick rotation [121–125] as well as a dynamical explanation for why the world is four-dimensional [126–128].

What is still missing within this approach (in more than two dimensions), as in any path integral approach for quantum gravity that has been established so far, is a clear physical interpretation of the expectation values of observables as transition amplitudes or expectation values in a physical Hilbert space. A possible way out could be proposed if one were able to establish reflection positivity of the measure (see [99]) from which the existence of a Hilbert space structure follows automatically.

2d. • •••• •• • ••• •••• •••• •••••••••

As already mentioned, the standard canonical formalism as being used in canonical QGR needs, almost by definition, a notion of time in order that one can obtain the momentum phase space underlying the Hamiltonian formulation from the velocity phase space of the Lagrangian formulation through the Legendre transform. While the Lagrangian formulation is • •• ••••••• covariant, the Hamiltonian formulation is not, in order to establish covariance one has to do some extra work, even at the classical level. At the quantum level the issue of the covariance of the measurement process appears [129]. On the other hand, for generic interacting systems only the canonical formulation allows for a straightforward quantisation by well-defined axioms, as we will see later on. The covariant canonical approaches try to combine the virtues of both formulations, manifest covariance on the one hand and a well-defined quantisation procedure on the other. They can roughly be grouped as follows:

2d(i) • •••• ••• • ••••• ••••• • ••••••

If the time evolution is well-defined, then there exists a bijection between the initial data (instantaneous or canonical phase space) and the space of solutions (covariant phase space) which can be turned into an isometry of the associated symplectic structures by simply pulling back the canonical one. One can imagine basing a quantisation on this procedure [130]. However, it is very likely that such an approach is in a sense too classical because by construction the singularities of the classical theory (e.g., big bang) are imported into the quantum theory. More generally, the path integral approach suggests that one has to deal with all possible histories in quantum theory, not only with the classical ones. See [131, 132] for the most advanced results within this approach based on the so-called 'Peierls bracket' which uses the classical solutions in the definition of propagators.

2d(ii) • • •••••• • ••• ••• • • •• •••

One way to get rid of a preferred time direction is to use as many canonical momenta as there are spacetime coordinates. In other words, there are as many momenta as there are velocities, which is why such an approach has been coined multisymplectic [133–139]. While the classical theory is well under control and equivalent to the standard canonical formalism, the quantisation of the multisymplectic Poisson bracket turns out to be rather difficult. To the best of our knowledge, major advances have only been obtained by Kanatchikov, see [140–142] for the state of the art in this subject.

2d(iii) • •••••• •••••••• ••••• • •• ••• •

The history bracket formulation grew out of the consistent histories formulation of quantum mechanics due to Gell-Mann, Griffiths, Hartle, Omnés and others [143–155] which is in many senses superior over the Copenhagen interpretation of quantum mechanics, especially when it comes to closed systems (cosmologies) for which there is no 'outside observer' any more. This theory is closely related to the path integrals in that it is based on chains of propositions, within the standard canonical Hilbert space, that is, projection operators onto states at certain points of time, sandwiched between the corresponding unitary time evolution operators. An obstacle for a long time had been that these propositions are no longer projections and therefore lack probabilistic features because projection operators do not necessarily commute. The final form was reached by Isham, Linden, Savvidou and Schreckenberg, now called the history projection operator approach [156–158], by blowing up the instantaneous Hilbert space into a ••• ••• •••• •• ••• • ••• ••• ••• ••••••• Hilbert space for which now projections at different points of time are uncorrelated (they live in different copies of the standard Hilbert space) and thus define projections again. Savvidou then realised that this structure suggests a new classical canonical formulation, namely a history bracket [159–163] phase space, which allows us to compute Poisson brackets between functions at different points of time • •••••• •••• • ••• ••••• •••, it is a purely kinematical construction as it should be. This observation allows us to clearly distinguish between the kinematical four-dimensional diffeomorphism invariance of General Relativity, which is always there (even in the standard canonical formalism) and the invariance group generated by the instantaneous constraints [221–226] which is not obviously a subgroup thereof, as we will see. These findings have been further developed by Kuchař

and Koutlesis in [164]. The classical time evolution is generated
by the action (four-dimensional integral over the Lagrangian density) rather than by the Hamiltonian (three-dimensional integral
over the Hamiltonian density) and thus manifestly covariant. One
should now proceed and quantise the history bracket formulation,
see [165, 166] for first promising steps in that direction.

2e. •

Approaches belonging to this category start by questioning standard quantum field theory at an even more elementary level. Namely, if the ideas
about spacetime foam (discrete structure of spacetime) are indeed true
then •
• • • • • • • • • • • • • • • • • but rather something intrinsically discrete. Maybe
we even have to question the foundations of quantum mechanics and to
depart from a purely binary logic. Among theories of this kind we find
• by Alain Connes [167, 168] also considered
recently by string theorists [169], • • • • • • • • • • • • by Chris Isham [170–174],
• • • • • • • • • • • • • • by Roger Penrose [175–180], the •
by Rafael Sorkin [181–186] and finally the •
• • • • • • • • • • by Gerard 't Hooft [187, 188]. These approaches are, maybe,
the most radical reformulations of fundamental physics but they are also
the most difficult ones because the contact with standard quantum field
theory is, a priori, very small. These programmes are in some sense 'far-
thest' from observation and are consequently least developed so far.[4]
However, the ideas spelt out in these programmes could well reappear
in the former approaches as well once these have reached a sufficiently
high degree of maturity in order to take the 'quantum leap' to a more
fundamental formulation.

All of these five non-perturbative programmes are mutually loosely connected: roughly, the operator formulation of the standard canonical approach
is equivalent to the continuous path integral formulation through some kind of
Feynman–Kac formula, a concrete implementation of which are the so-called
spin foam models of LQG to be mentioned later, path integrals are even closer
to the covariant canonical approaches, lattice quantum gravity is a discretisation of the path integral formulation and finally both the canonical and the
lattice approach seem to hint at discrete structures on which the non-orthodox
programmes are based.

Finally, every non-perturbative programme better contains a sector which
is well described by perturbation theory and therefore string theory, which
then provides an interface between the two big research streams. A more
immediate connection could be provided through the so-called **Pohlmeyer**

---

[4] It would take us too far apart to even describe the basics of these rather abstract theories,
however, the references listed provide excellent introductions to the subject.

string [189–204], which is based on a reduced phase space quantisation of the algebra of Dirac observables for the string which can be explicitly constructed in this case.

This ends our survey of the existent quantum gravity programmes.

Motivation for canonical quantum general relativity

In the previous section we have tried to give a very rough overview of the available approaches to quantum gravity, their main successes and their major unresolved problems. We will now motivate our choice to follow the non-perturbative, canonical approach. Of course, our discussion cannot be entirely objective.

I. • •••••••••• ••• •••••• • •••••• ••• ••••

Our preference for a non-perturbative approach is twofold:

The first reason is certainly a matter of taste, a preference for a certain methodology. Try to combine the two fundamental principles, General Relativity and quantum mechanics • ••• •• •••••••• •• ••••• •• ••, explore the logical consequences and push the framework until success or until there is a contradiction (inconsistency) either within the theory or with the experiment. In the latter case, examine the reason for failure and try to modify the theory appropriately. The reason for not allowing additional structure (principle of minimality) is that unless we only use structures which have been confirmed to be a property of nature, then we are standing in front of an ocean of possible new theories which a priori could be equally relevant. In a sense we are saying that if gravity cannot be quantised perturbatively without extra structures such as necessary in string theory, then one should try a non-perturbative approach. If that still fails then maybe we find out why and exactly which extra structures are necessary rather than guessing them. Such a methodology has proved to be very successful in the history of science.

The second reason, however, is maybe more serious: •• •• ••• •• ••• •••• • ••• • •••••••• •• ••••• ••• ••• ••••• •• • •• ••• •• • •• ••• ••• • •••• • •• •••• ••••••••• •••••. To quote an example from [10], consider the harmonic oscillator Hamiltonian $H = p^2 + \omega^2 q^2$ and let us treat the potential $V = \omega^2 q^2$ as an interaction Hamiltonian perturbing the free Hamiltonian $H_0 = p^2$ at least for low frequencies ω. The exact spectrum of H is discrete while that of H_0 is continuous. The point is now that ••• •• ••••• ••••• •• •••• ••• ••• •• ••• •• •• ••• ••••••••• •••••••• •• $\omega > 0$• ••• •••••••••• •••• •• ••• •••••••• •• ••• ••••••••••• • ••• •••• •••• •• • •••••• ••• •• ••••• ••••••••••• ••• • ••••••• ••• •••• ••• ••• •••••• • ••••• ••• •••••••• •••••••!

Finally, borrowing from [15], let us exhibit a calculation which demonstrates the •••• •• •••••• • •••••• ••• ••• • ••••••••••••• •••• ••••••• ••• •• •••• •••• •••••• •••••• • ••• •••• •••••• ••• •••••• •••• ••••••••. Consider the self-energy of a homogeneously charged and massive ball of radius r with bare charge e_0 and bare rest mass m_0 due to static electromagnetic and gravitational interaction. From the point of view of Newtonian physics, this energy is of the form ($\hbar = c = 1$, the bare Newton's constant is denoted by G_0 and we have absorbed numerical multiples of π into e_0, m_0)

$$m(r) = m_0 + e_0^2/r - G_0 m_0^2/r$$

and diverges as $r \to 0$ unless e_0, m_0, G_0 are fine-tuned. However, General Relativity tells us that all of the mass of the charge, that is rest mass plus field energy within a shell of radius r couples to the gravitational field, which is why the above equation should be replaced by

$$m(r) = m_0 + e_0^2/r - G_0 m(r)^2/r$$

which can be solved for

$$m(r) = \frac{r}{2G_0}\left[-1 + \sqrt{1 + \frac{4G_0}{r}\left(m_0 + \frac{e_0^2}{r}\right)}\right]$$

Notice that now the bare mass $m(r = 0) = e_0/\sqrt{G_0}$ is •• ••• • ••• ••• • •••• •• •• •• • • •••••••• ••• ••• •• •• • ••••••••••••• •• • •• ••••• ••• ••••• G_0 ••• •• • •• ••••••••• •• •• •••••••••••• ••••••• •• •••••••••••••• ••• •••• • ••• •• •• •••

•••••••• •• ••• •••• • •••! Of course, this calculation should not be taken too seriously since, for example, no quantum effects have been brought in, it merely serves to illustrate our point that General Relativity could serve as a natural regulator of field theory divergences. (However, a proper general relativistic treatment can be performed, see also [15] for more details.)

These arguments can be summarised by saying that there is a good chance that perturbative quantum gravity ••• • ••• ••• • ••••• ••• ••••• although, of course, there is no proof!

II. • • •• ••• •••••• ••• •• ••• •••••••••••• ••••• ••••••••••

Here our motivation is definitely just a matter of taste, that is, we take a practical viewpoint:

Path integrals have the advantage that they are manifestly four-dimensionally diffeomorphism-invariant but their huge disadvantage is that they are hard to compute analytically, even in quantum mechanics. While numerical methods will certainly enter the canonical approach as well in the close future, one gets further with analytical methods. However, it should be stressed that path integrals and canonical methods are very closely related and usually one can derive one from the other through some kind of Feynman–Kac formula.

The non-orthodox approaches have the advantage of starting from a discrete/non-commutative spacetime structure from scratch, while in canonical quantum gravity one begins with a smooth spacetime manifold and obtains discrete structures as a derived concept only, which is logically less clean: the true theory is the quantum theory and if the world is discrete one should not begin with smooth structures at all. Our viewpoint is here that, besides the fact that again the canonical approach is more minimalistic, at some stage in the development of the theory there must be a quantum leap and in the final reformulation of the theory everything is just combinatorial. This can actually be done in $2 + 1$ gravity, as we will describe later on!

Outline of the book

In this section we briefly describe what is covered in this book. We will drop all references here, they will be properly supplied as we move on. The road map is as follows:

(A) • •• ••••••• •• ••• • •• ••• •

It is mandatory to start with the classical theory. That is, we explain in detail what exactly is meant by the canonical formulation of General Relativity. Roughly speaking, we take the Einstein–Hilbert action S for a differentiable four-manifold M and foliate M into a one-parameter family of hypersurfaces $t \mapsto \Sigma_t$ which is always possible classically. The Einstein–Hilbert action is an integral over M of a Lagrangian which involves the metric tensor g and its first and second derivative. The parameter t is one of the four coordinates of M and serves to identify the velocities $v = \partial q / \partial t$ of the components q of the metric tensor so that we can perform the Legendre transformation $p = \delta S / \delta \dot{q}$ from the velocity phase space to the momentum phase space. The functions q, p then have canonical equal time brackets, that is, if we denote coordinates on Σ_t by $\vec{x}$ then roughly $\{q(t, \vec{x}), p(t, \vec{x}')\} = \delta(\vec{x}, \vec{x}')$. The parameter t, however, is not to be identified with a distinguished time variable. Indeed, coordinates have no a priori physical meaning since the action is invariant under diffeomorphisms, that is, arbitrary smooth bijections of M so that the inverse is also smooth.

The split of the manifold M into space and time is not unique and in fact there are as many foliations as there are diffeomorphisms of M. Since the Einstein–Hilbert action is diffeomorphism-invariant, all the foliations are physically equivalent. Now for each foliation we can perform the Legendre transform and obtain a phase space and a Hamiltonian. It turns out that the phase space $\mathcal{M}$ together with its Poisson bracket does not depend on the choice of the foliation, they are all mutually isomorphic. What does depend on the choice of the foliation is the form of the Hamiltonian. Since the action does not depend on the choice of the foliation, the variation of the action with respect to the foliation must vanish. As a result, one gets an infinite number of local constraints, one for each point of Σ_t for all $t \in \mathbb{R}$, which together are equivalent to the condition that the Hamiltonian is constrained to vanish for every foliation. The vanishing of the Hamiltonian is in fact a consequence of the Einstein equations, that is, the Euler–Lagrange equations derived from the Einstein–Hilbert action.

The vanishing of the Hamiltonian may seem strange at first sight but is in fact a logical consequence of diffeomorphism invariance. Namely, normally the Hamiltonian flow of the Hamiltonian generates time translations. In GR these time translations are diffeomorphisms, that is, gauge transformations and physical observables must not depend on the choice of the coordinate system. Thus we arrive at the conclusion that physical observables must have vanishing Poisson brackets with the Hamiltonian. One can also understand this from the fact that the Hamiltonian and hence its flow is foliation-dependent, which must not be the case for physical observables. Moreover, we are only interested in the constraint submanifold of the phase space where the Hamiltonian vanishes. We see that the physical phase space is parametrised by those functions on the constraint submanifold which have vanishing Poisson brackets with all the Hamiltonians.

Naively one expects that the Hamiltonian flow of the Hamiltonians simply corresponds to diffeomorphisms of M. It turns out that this is indeed the case, however, only in the solutions to the equations of motion. The reason for this is that the Hamiltonian is a specific functional of the canonical variables which depends on the action that one started from. There are an infinite number of algebraically independent action functionals of the metric tensor field, all of which are diffeomorphism-invariant. Their canonical formulation gives rise to the same phase space (if the number of higher derivatives is the same). Yet their dynamics, encoded in the Euler–Lagrange equations, is different. Hence, while the motions are gauge motions in all cases, they are generated by different functionals and if they are to be interpreted as diffeomorphisms then this can hold only on the corresponding trajectories. We will also give a simple, more technical explanation for this phenomenon later on. For the same reason, the Poisson algebra of the various Hamiltonians for different foliations reduces only on shell to the algebra of infinitesimal diffeomorphisms of M.

The appearance of an infinite number of Hamiltonian constraints rather than a single Hamiltonian is a particular feature of diffeomorphism-invariant field theories and gives rise to much confusion, summarised under the abbreviation 'problem of time'. Namely, since there is no Hamiltonian there is no physical time, which one would interpret as the parameter that enters the definition of the Hamiltonian flow of the Hamiltonian. Instead, physical observables are completely 'frozen' because they are supposed to have trivial flow under all the Hamiltonian constraints. There seems to be no time notion at all, in sharp contrast to what we observe in everyday life. The resolution of the puzzle is the relational point of view: consider two non-observables T, O, sometimes called partial observables, which are •••

invariant under the gauge flow of all the Hamiltonian constraints. Consider their flow $\alpha_t(T)$ and $\alpha_t(O)$ with respect to one of the Hamiltonians. Then fix some parameter τ and invert, if possible, the condition $\alpha_t(T) = \tau$ for t. Insert the solution $t_T(\tau)$ into the function $t \mapsto \alpha_t(O)$ and obtain $O_T(\tau)$. It is easy to check that $O_T(\tau)$ is invariant under the flow, $\alpha_t(O_T(\tau)) = O_T(\tau)$, and it has a simple interpretation: it is the value of O when T has the value τ. Thus, although both T, O are not invariant, we can construct a simple invariant $O_T(\tau)$ which is frozen but still has a dynamical interpretation. A moment of thought reveals that this is precisely how we perceive time in physics: time is not itself an object that we can grasp, rather we observe relative motions such as the distance that one has travelled after the pointer of a clock has changed by a certain angle. Moreover, one can show that the evolution with respect to τ has a canonical generator, that is, a • • • • • • • • • • • •• • ••• • •• • which is completely independent of the Hamiltonian constraint: it is itself gauge-invariant and does not need to vanish on the constraint surface.

This is a beautiful idea and yet there is a flaw in this argument: we have an infinite number of Hamiltonian constraints rather than a single one. Thus we must consider the flow of all of them and we seem to obtain an infinite number of times and physical Hamiltonians. Moreover, the different flows do not commute and hence the above idea to construct an observable does not work. We will show how to overcome this problem, for instance, by combining all the constraints into a single one which is called the • • •••••
• • •• ••• •• •. Also other conceptual problems which are particular to quantum gravity, such as how to interpret quantum mechanics in cosmological circumstances when the observer is part of the system, will be addressed and a consistent picture will be proposed.

(B) • • • • • •••• • •• •• • •• ••• •

Classical GR is a dynamical theory of metrics on a differential manifold M. However, it has been known for a long time that one can recast the Einstein–Hilbert action, which involves the metric and its first and second derivative, into the Palatini action, which involves a connection for an $SO(1,3)$ gauge theory and its first derivative as well as a Vierbein field. The two actions are equivalent when the spacetime is orientable and this is precisely the case when one can consistently couple spinor fields, which is necessary anyway. What was not known is that one can choose the connection and the Vierbein field (or rather their pull-backs to the three-dimensional leaves of the foliation) as configuration and momentum variable of a canonical pair with canonical Poisson brackets, just like in Yang–Mills theory. This key observation really enabled the huge amount of progress that occurred over the past almost 20 years in LQG. Hence we will derive in detail how the traditional Hamiltonian formulation in terms of metrics is related to the connection formulation.

(C)

Having cast GR into canonical form as a dynamical system with constraints we will explain how to quantise such a system. Roughly speaking, one selects a *-Poisson subalgebra of the Poisson algebra on the unconstrained phase space $\mathcal{M}$ which separates the points of $\mathcal{M}$ and which is closed under complex conjugation. Every function on the phase space can then be expressed in terms of (limits of) elements of this algebra.

One then defines an abstract *-algebra $\mathfrak{A}$ by promoting bounded functions of the generators of the *-subalgebra formally to abstract operators which satisfy the canonical commutation relations and the adjointness relations. That is, commutators are given by the Poisson bracket of the corresponding functions times $i\hbar$ and adjoint algebra elements are given by the algebra elements corresponding to the complex conjugate functions. Notice that at this point these operators have not been implemented on any particular Hilbert space, we have just defined an abstract algebra from the Poisson bracket and the complex conjugation on the phase space.

In the next step one must study the representation theory of $\mathfrak{A}$. That is, one looks for all irreducible representations of $\mathfrak{A}$ as bounded operators on some Hilbert space. The additional requirement is that this Hilbert space also allows us to represent the constraints as operators and that their algebra is not anomalous. By this we mean the following: classically the constraints form a first-class system, that is, the Poisson bracket of the constraints among themselves is a linear combination of constraints. Since the constraints simultaneously define the constraint surface in the phase space and the gauge motions, geometrically this means that the constraint surface is gauge-invariant. This is important as otherwise it would be inconsistent to impose the constraints, physical quantities must be gauge-invariant. Now in quantum theory we must have a similar consistency condition: the commutator of two constraint operators must be again a linear combination of constraint operators. If that were not the case then the following would happen: suppose that $\{C_1, C_2\} = C_3$ but that $[\hat{C}_1, \hat{C}_2] = i\hbar\hat{C}_3 + \hbar^2\hat{A}$. Here C_j are some first-class constraint functions and the term proportional to $\hat{A}$ is a quantum correction. It is a quantum correction because the classical limit $\hbar \to 0$ of the commutator divided by $i\hbar$ gives the correct classical result. Now suppose that we have found a simultaneous solution ψ to all constraints $\hat{C}_j\psi = 0, j = 1, 2, 3$. Then it is easy to see that also $\hat{A}\psi = 0$. Now unless $\hat{A}$ is a linear combination of constraint operators, this is an extra condition on ψ which has no classical counterpart. It follows that the quantum theory has less physical states than the classical theory has observables, whence the quantum theory does not have the correct limit. The operator $\hat{A}$ is then called an anomaly, and such anomalies must be avoided by all means.

The origin of such anomalies are ordering and regularisation ambiguities in the definition of the operators $\hat{C}_j$. Namely, unless the C_j are linear in q, p there is an issue with the ordering of products such as qp. If one wants to have a symmetric operator one will choose a symmetric ordering $(\hat{q}\hat{p} + \hat{p}\hat{q})/2$. Next, in field theories q, p are functions of $t, \vec{x}$ and in quantum theory become operator-valued distributions rather than operators. Since the product of distributions is generically singular, one must regularise products such as $(q \cdot p)(t, \vec{x})$, for instance by point splitting $(\hat{q} \cdot \hat{p})(t, \vec{x}) = \lim_{\vec{\epsilon} \to 0}(\hat{q}(t, \vec{x}) \cdot \hat{p})(t, \vec{x} + \vec{\epsilon})$. Then one writes this in the form $\sum_I c_I(\vec{x}, \vec{x} + \vec{\epsilon})\hat{O}_I(t, \vec{x})$ called an operator product expansion (OPE) where the operators $\hat{O}_I$ are well-defined and the coefficients c_I are singular. Finally one applies some procedure, called renormalisation, to remove the singular terms (in $\vec{\epsilon}$) as we remove the UV regulator. This usually cannot be done in a unique fashion. The most familiar OPEs are normal orderings of products of normal-ordered operators on Fock space. It is clear that ordering prescriptions and regularisation ambiguities strongly affect the issue of anomalies.

Next one must solve the quantum constraints by asking that physical states be annihilated by them. One can show that this amounts to selecting gauge-invariant states. If zero lies in the continuous part of the spectrum of the constraint operators then the solutions to the constraints do not lie in the Hilbert space that one started with. In this case one must construct a new Hilbert space with a new scalar product with respect to which the physical states are normalisable. There are several constructive procedures available to do this, which we will discuss and apply in great detail.

Finally, one must represent the gauge-invariant observables as self-adjoint operators on this physical Hilbert space. As mentioned, in field theories all of these steps are to be supplemented by sophisticated regularisation procedures and it is therefore a non-trivial question whether the theory that one ends up with does in fact have the original classical theory as its classical limit. Hence one must prove that there exist semiclassical physical states with respect to which gauge-invariant operators have the correct expectation values.

(D) • • •••• •••• •• • •• ••••• • •• ••• •••

We then apply this quantisation programme to General Relativity. As we indicated above, it turns out to be important to reformulate the theory in terms of connections and frame fields. This can be achieved, for instance, starting from Palatini's reformulation of classical GR. The canonical formulation of this theory is based on electric and magnetic fluxes familiar from an SU(2) Yang–Mills theory with the crucial difference that there is no Hamiltonian but instead four Hamiltonian constraints per spacetime point. The canonical flow of three of them generates diffeomorphisms of the spatial hypersurfaces while the fourth one generates time reparametrisations (on shell, i.e., when the equations of motion hold). We will refer to them as the spatial diffeomorphism and Hamiltonian constraint respectively.

As the algebra $\mathfrak{A}$ we choose exponentials of the electric and magnetic fluxes times the imaginary unit with canonical commutation relations and adjointness relations imposed. We then study the representation theory of this algebra. Since we want to solve in particular the spatial diffeomorphism constraint, we impose as a restriction on the class of representations to be considered that the corresponding 'ground state' can be chosen spatially diffeomorphism-invariant. Under some mild technical assumptions one can show that there is a unique representation of $\mathfrak{A}$. In this representation, states are labelled by loops, or more generally by graphs. These arise from the holonomy (path-ordered exponential) of the connection along paths which is related to the magnetic flux via (the non-Abelian version of the) Stokes' theorem and which become operators on the Hilbert space. They generate a dense subset of the Hilbert space by acting on the 'vacuum' just like the usual creation operators on Fock space generate a dense set of states by acting on the vacuum.

One can then explicitly solve the spatial diffeomorphism constraint and construct the Hilbert space of spatially diffeomorphism-invariant states. These turn out to be labelled by knot classes. This is not surprising because a knot class is the spatial diffeomorphism equivalence class of a loop.

Next one implements the Hamiltonian constraint on the Hilbert space of spatially diffeomorphism-invariant states. It turns out that this is not directly possible when one works with the infinite number of constraints. However, one can use the above-mentioned reformulation in terms of a single constraint in order to achieve this. Surprisingly one does not encounter UV divergences. One can trace this back to background independence: UV divergences are a short distance phenomenon of background-dependent theories, where 'short' is meant as a qualifier with respect to the background metric in question. In background-independent theories 'short' has no meaning, hence the theory protects itself against UV divergences. One can then show that the total physical Hilbert space exists and is non-trivial.

What is still missing is a representation of the gauge-invariant observables on the physical Hilbert space and the demonstration that the theory has classical GR as one of its semiclassical sectors. This brings us to the frontiers of current research.

(E) • • • ••• ••• ••

Before one has analysed the physical Hilbert space in sufficient detail so that one can tell whether LQG has classical GR as its classical limit, it is not possible to make reliable physical predictions for physics beyond the standard model. A lot of effort within LQG is currently devoted to completing this final missing step. Nevertheless, one can probe the theory in various aspects before one solves it completely. These provide either consistency checks or development of tools for later use. In what follows we list the most important of these applications.

1. • •• •• •••••• •••• ••••••• •••••••• •• ••

It turns out that operators corresponding to length, area and volume of curves, surfaces and regions in the spatial hypersurfaces can be constructed on the unconstrained Hilbert space. This is very surprising because none of them exists on the usual background-dependent Fock spaces. Even more beautifully, their spectra are entirely discrete, given as multipla of Planck length, area and volume respectively. This provides a first concrete hint that the theory is fundamentally discrete or combinatorial. It means that in order to build the area of a piece of A4 paper we must use a state which is a complicated weave of at most 10^{68} loops. This number is huge but finite. Moreover, the area eigenvalue jumps by a discrete amount of the order of ℓ_P^2 if we add or remove a loop from the state, just like the energy eigenvalues of the hydrogen atom do.

It would be wrong to say that this proves that the physical area of a surface has discrete spectrum because the area operator constructed does not commute with the constraints. However, it is conceivable that the Dirac observable constructed from it as a partial observable does.

2. • • • • ••• • •• • • ••••

It turns out that what we have said above in item D can also be applied to the matter of the standard model (including possible supersymmetric extensions). One gets as far with the quantisation programme as without matter and there are no UV singularities. Up to this point there is no hint that there is any restriction on the possible matter couplings. However, these could still show up in the final step of the programme because switching on and off matter degrees of freedom affects the spectrum of the single Master Constraint, which is always a subset of the non-negative real numbers but does not automatically contain zero.

3. • • • • ••• • ••• •• •• •• •••••••

If one arranges that the manifold M has an inner boundary and that the classical phase space is supplemented with boundary conditions there that are suitable for black hole formation then one can treat quantum black holes within LQG. Doing this it was possible to isolate and count the microcosmic states that account for the celebrated Bekenstein-Hawking entropy $S = A/(4\ell_P^2)$ where A is the area of the corresponding black hole. Due to the boundary conditions the area of the black hole is a truly gauge-invariant observable. It turns out that the entropy is due to loops which intersect the horizon. It is lack of knowledge of all possible ways to intersect the horizon while producing a given area that accounts for the entropy. This works in particular for all astrophysically relevant rotating, charged, dilatonic black holes and those with Yang–Mills hair.

4. •••• •••• • •• ••

For systems with a Hamiltonian one can construct a corresponding path integral which has the physical meaning of a transition amplitude between initial and final states. For systems without a Hamiltonian but a single Hamiltonian constraint the corresponding quantity is a (generalised) projector on physical states which solve the constraints. The two objects are related by an integral over the unphysical time t mentioned above that produces from the time evolution operator $\exp(itH)$ the generalised projector $\delta(H)$, which can be given rigorous meaning by the spectral theorem. For GR the Hamiltonian constraint acts by creating and annihilating loops to or from a given graph. The time evolution of a loop is a surface and hence the dynamics looks like that of a foam where surfaces are created and annihilated all the time. This is how one intuitively expects to derive a covariant path integral over spacetime fields from the canonical formulation of fields at a given time. To date this connection is not very well understood, mainly because so far spin foam models have not really been derived from the Hamiltonian formulation but rather start with the postulate that the final path integral should be $\exp(iS)$ times some measure on the space of spacetime fields (a spacetime connection and a Vierbein in Palatini's first-order formulation) where S is the Palatini action for $M \cong [0, t] \times \sigma$ and σ is any three-manifold. A lot of activity in LQG is currently devoted towards making this connection more concrete, one possibility given by the Master Constraint.

5. ••• •• •• •••••• ••• •• •••

It turns out that it is possible to find a precise, background-independent analogue of the usual coherent states for free quantum field theories on Minkowski spacetime for non-Abelian gauge theories. These can then be applied, in particular, to GR as far as the kinematical, that is, unconstrained Hilbert space is concerned. For linearised gravity, which is a free field theory of gravitons on Minkowski space, one can even do this, using the Minkowski background, at the level of the physical Hilbert space which is not a Fock space but rather LQG inspired. What we need for the full theory are coherent states on the spatially diffeomorphism-invariant Hilbert space and on the physical Hilbert space. The former are needed in order to test whether the Hamiltonian constraint, which generates the physical Hilbert space from the spatially diffeomorphism-invariant one, has the correct classical limit (there is no way to test the Hamiltonian constraint on the physical Hilbert space where it is the zero operator by definition). The physical coherent states are needed in order to test the classical limit of gauge-invariant observables. Work is in progress at both fronts.

These are technically hard problems and it is worthwhile to think about approximation schemes which are technically simpler. One such scheme consists of •• •••• • •• ••• ••••: kinematical coherent states are labelled by a point on the constraint surface of the phase space and that point will lie on some gauge orbit. Doing this for all gauge orbits is equivalent to saying that kinematical coherent states are labelled by points on the constraint surface on a gauge cut of the orbit space. We call this quantum gauge fixing because all the degrees of freedom, also the unphysical ones, are fluctuating, however, their fluctuations are suppressed by the semiclassical state. It turns out that one can define an alternative Master Constraint on the kinematical Hilbert space and its expectation value with respect to those states, as well as its fluctuation, is close to zero.

6. •• •••• •• • ••• •••• •••••• •

LQG is a background-independent, non-perturbative QFT of GR and due to the non-linearities of GR it is the most complicated, most non-linearly interacting QFT that was ever studied. It is therefore, in the absence of a perturbative scheme like the quantum gauge fixing just mentioned, very hard to do any practical computations. A huge simplification occurs if we artificially reduce the number of degrees of freedom. In classical GR this is done either by dimensional reduction or by Killing symmetry reduction. In both cases one imposes a symmetry on the metric tensor and thus reduces the number of degrees of freedom. The most extensively studied symmetry reduction is $2 + 1$ gravity, which is a topological field theory (a background-independent theory with only a finite number of physical degrees of freedom; $1 + 1$ gravity is trivial since the Einstein–Hilbert action is a topological invariant in this case). The most familiar forms of Killing reduction are (a) spherical symmetry leading to Schwarzschild spacetimes, (b) homogeneity leading to Bianchi cosmologies, (c) additional isotropy leading to Friedman–Robertson–Walker universes, (d) cylindrical symmetry leading to Gowdy universes (two modes) or cylindrically symmetric waves (one mode). All these models, with a few exceptions among the Bianchi-type models and the two-mode Gowdy model, are classically completely integrable and hence quantisation is straightforward.

These models thus provide an ideal testing ground for LQG. Of course, a model never has all aspects in common with the real theory and, in particular, switching off modes, which is no problem in the classical theory, is not necessarily quantum mechanically stable due to quantum fluctuations so that a model can never prove or predict anything about the full theory. In other words, the quantum model is not embedded in the full quantum theory. However, a model can probe technical methods and conceptual ideas used in the full theory in a much simpler context, which provides important insights and intuition for the full theory.

We will show that one can successfully quantise homogeneous models by LQG methods, which leads to a spectacular new picture of the early universe if confirmed by later calculations in full LQG, some of which have already started.

7. • •• • ••   ••• •• •••   •• •• • •   •• •• ••

Due to the weakness of the gravitational interaction it is widely believed that we cannot see any quantum gravity effects in the close future. Even if there was an effect linear in E/E_p where E is the energy of a probe and E_P is the Planck energy, today we are at least 15 orders of magnitude away with collider energies. However, recently physicists have started to speculate how to get around this problem. The idea is quite similar to the one that was used to probe the lifetime of the proton: a proton is expected to decay after some 10^{30} years. Nobody can wait that long to see whether a given proton decays. But we can observe 10^{30} protons over a year to see whether there is at least one decay. Similarly, for quantum gravity effects the general idea is to use the accumulation of a large number of tiny effects to something that lies within reach of present-day detector sensitivity. More specifically, these Gedankenexperimente start from the general idea of a discrete structure at Planck scale which should have some effect on matter propagation, just like a crystal does compared with a vacuum. We will report on some of these ideas, all of which point to an effective modification of the Poincaré group at high energies when the gravitational state is concentrated around Minkowski space. More generally, in this branch of the theory one should make contact with the physics of the standard model and beyond.

(F) • ••• •• •••• • ••• ••

In an effort to make the book self-contained we have supplied a large amount of mathematical background material. The experience from teaching the subject to (under)graduate students is that this material is most welcome. This material is not specific to our LQG applications but rather is helpful for any (quantum) field theory. Also physical applications of this mathematical theory, relevant to the main topics of this book, are contained in this last part. For instance, geometric quantisation, which is needed for black hole physics, uses in an important way differential, Riemannian, symplectic and complex geometry as well as fibre bundle theory. Another example is the Bohr compactification of the real line, which is a baby version of the distributional space of connections that underlies LQG and illustrates Gel'fand's abstract theory of Abelian C^*-algebras, another application of which is an elegant proof of the spectral theorem. Operator algebraic techniques, especially the GNS construction, are reviewed and used heavily in the representation theoretic part of the book. The direct integral decomposition of a Hilbert space adapted to a given self-adjoint operator is explicitly derived as it is needed for the construction of the physical Hilbert space of LQG (Master Constraint Programme). Finally, harmonic analysis on compact Lie groups is developed and in particular the Peter and Weyl theorem is proved and then applied to the explicit construction of spin-network functions for SU(2).

I

Classical foundations, interpretation and the canonical quantisation programme

1

Classical Hamiltonian formulation
of General Relativity

In this chapter we provide a self-contained exposition of the classical Hamiltonian formulation of General Relativity. It is mandatory to know all the details of this classical work as it lays the ground for the interpretation of the theory, the understanding of the problem of time and its implication for the interpretation of quantum mechanics, the meaning of observables, the relation between spacetime diffeomorphisms and gauge transformations and finally (Poincaré) symmetries versus gauge transformations. It also defines the platform on which the quantum theory is based. Only a solid knowledge of topology and differential geometry is necessary for this chapter, of which we give an account in Chapters 18 and 19.

1.1 The ADM action

The contents of this section were developed by Arnowitt •• •• [206]. Modern treatments can be found in the beautiful textbooks by Wald [207] (especially appendix E and chapter 10) and by Hawking and Ellis [208]. We will treat only the vacuum case. Matter and cosmological terms can be treated the same way.

What we are going to do in what follows seems to be a dangerous enterprise in a generally covariant theory: we will split the spacetime manifold into space and time. While this is necessary in a canonical approach, as otherwise we cannot define velocities and hence momenta conjugate to the configuration variables, this seems to break diffeomorphism invariance. However, this is not the case because we do not fix the split into space and time, rather we keep it arbitrary, we •• ••• •• • •••••• ••• •••••• . The arbitrariness in fact exhausts the full diffeomorphism group. Since the action is diffeomorphism-invariant it does not depend on this auxiliary split and varying with respect to it leads, not surprisingly, to the generators of this invariance group subject to an important reservation which we will derive.

The object of interest is the Einstein–Hilbert action for metric tensor fields $g_{\mu\nu}$ of Lorentzian ($s = -1$) or Euclidean ($s = +1$) signature which propagate on a $(D + 1)$-dimensional manifold M

$$S = \frac{1}{\kappa} \int_M d^{D+1}X \sqrt{|\det(g)|} R^{(D+1)} \tag{1.1.1}$$

In this book we will mostly be concerned with $s = -1, D = 3$ but since the subsequent derivations can be done without extra effort we will be more general here.

Our signature convention is 'mostly plus', that is, $(-, +, \ldots, +)$ or $(+, +, \ldots, +)$ in the Lorentzian or Euclidean case respectively so that timelike vectors have negative norm in the Lorentzian case. Here $\mu, \nu, \rho, \ldots = 0, 1, \ldots, D$ are indices for the components of spacetime tensors and X^μ are the coordinates of M in local trivialisations. $R^{(D+1)}$ is the curvature scalar associated with $g_{\mu\nu}$ and $\kappa = 16\pi G$ where G is Newton's constant (in units where $c = 1$). The definition of the Riemann curvature tensor is in terms of one-forms given by

$$[\nabla_\mu, \nabla_\nu] u_\rho = ^{(D+1)} R_{\mu\nu\rho}{}^\sigma u_\sigma \qquad (1.1.2)$$

where ∇ denotes the unique, torsion-free, metric-compatible, covariant differential associated with $g_{\mu\nu}$. To make the action principle corresponding to (1.1.1) well-defined one has, in general, to add boundary terms (unless one assumes that M is spatially compact without boundary) which we will explicitly derive below.

In order to cast (1.1.1) into canonical form one makes the assumption that M has the special topology $M \cong \mathbb{R} \times \sigma$ where σ is a fixed three-dimensional manifold of arbitrary topology. By a theorem due to Geroch [209], if the spacetime is globally hyperbolic (existence of Cauchy surfaces[1] in M; loosely speaking, everywhere spacelike surfaces which are connected to any point in M by a causal curve – in accordance with the determinism of classical physics) then it is necessarily of this kind of topology. Therefore, for classical physics our assumptions about the topology of M seem to be no restriction at all, at least in the Lorentzian signature case. In quantum gravity, however, different kinds of topologies and, in particular, •••••••• •••••• are conceivable. Our philosophy will be first to construct the quantum theory of the gravitational field based on the classical assumption that $M \cong \mathbb{R} \times \sigma$ and then to •••• •••• •••••••••••• in the quantum theory. It will turn out that in LQG topology change is ••• •••• ••• ••••• in the sense that typical states, but not semiclassical states, correspond to a completely degenerate spatial geometry. For more information on topology change in quantum gravity see, for example, [210–215] and references therein.

Having made this assumption, one knows that M foliates into hypersurfaces $\Sigma_t := X_t(\sigma)$, that is, for each fixed $t \in \mathbb{R}$ we have an embedding (a globally injective immersion) $X_t : \sigma \to M$ defined by $X_t(x) := X(t, x)$ where x^a, $a, b, c, \ldots = 1, 2, \ldots, D$ are local coordinates of σ. Likewise we have a diffeomorphism $X : \mathbb{R} \times \sigma \mapsto M$; $(t, x) \mapsto X(t, x) := X_t(x)$. Any diffeomorphism $\varphi \in \mathrm{Diff}(M)$ of M is of the form $\varphi = X' \circ X^{-1}$ where X, X' are two different foliations and any two foliations are related by a diffeomorphism via $X' = \varphi \circ X$. It follows that up to this point the freedom in the choice of the foliation is equivalent to $\mathrm{Diff}(M)$. In fact, since the action (1.1.1) is invariant under all diffeomorphisms of M the foliations X are not specified by it and we must allow them to be completely arbitrary.

---

[1] Any inextendible causal (= nowhere spacelike) curve intersects the Cauchy surface in precisely one point.

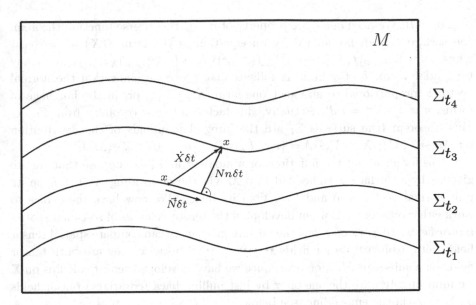

Figure 1.1 Foliation of spacetime into spacelike hypersurfaces and the meaning of lapse and shift.

We now use these foliations in order to give a $D+1$ (space and time) decomposition of the action (1.1.1). A useful parametrisation of the embedding and its arbitrariness can be given through its deformation vector field

$$T^\mu(X) := \left(\frac{\partial X^\mu(t,x)}{\partial t}\right)_{|X=X(x,t)} =: N(X)n^\mu(X) + N^\mu(X) \qquad (1.1.3)$$

Here n^μ is a unit normal vector to Σ_t, that is, $g_{\mu\nu}n^\mu n^\nu = s$ and N^μ is tangential, $g_{\mu\nu}n^\mu X^\nu_{,a} = 0$. Clearly, the vector field n^μ is completely determined as a function of g, X by these two requirements. The coefficients of proportionality N and N^μ respectively are called •• ••• function and •• ••• vector field respectively. See Figure 1.1 for an illustration of the geometrical situation. Notice that implicitly information about the metric $g_{\mu\nu}$ has been invoked into (1.1.3), namely we are only dealing with spacelike embeddings and metrics of the above-specified signature. Hence the foliation T is required to be timelike everywhere, which leads to the constraint $-N^2 + g_{\mu\nu}N^\mu N^\nu < 0$ and in particular implies that the lapse is nowhere vanishing. Moreover, we take N to be •• •• ••• everywhere as we want a future directed foliation (negative sign would give a past directed one and mixed sign would not give a foliation at all since then necessarily the leaves of the foliation would intersect). Hence at this point we are dealing with a proper subset of all embeddings and this subset is dynamically constrained as it depends on the metric tensor $g_{\mu\nu}$. This will have important consequences for what follows. A more precise characterisation of these 'dynamical foliations' as compared to Diff(M) can be found in [216, 217]. We need one more property of n : by the inverse function theorem, the surface Σ_t can be defined by an equation of the form $f(X) = t = \text{const}$. Thus, $0 = \lim_{\epsilon \to 0} [f(X_t(x + \epsilon b)) - f(X_t(x))]/\epsilon = b^a X^\mu_{,a}(f_{,\mu})_{X = X_t(x)}$ for any tangential vector b of σ in x. It follows that up to normalisation the normal vector is proportional to an exact one-form, $n_\mu = F f_{,\mu}$ or, in the language of forms, $n = n_\mu dX^\mu = F df$. Actually, this fact is an easy corollary from Frobenius' theorem (the surfaces Σ_t are the integral manifolds of the distribution $v : M \to T(M); X \mapsto V_X(n) = \{v \in T_X(M); i_v(n) = 0\} \subset T_X(M))$.

Let us forget about the foliation for a moment and just suppose that we are given a hypersurface σ embedded into M via the embedding X. Let n be its unit normal vector field and $\Sigma = X(\sigma)$ its image. We now have the choice to work either on σ or on Σ when developing the tensor calculus of so-called spatial tensor fields. To work on Σ has the advantage that we can compare spatial tensor fields with arbitrary tensor fields $\bullet \bullet \bullet \bullet \bullet \bullet \bullet \bullet \bullet \ \bullet \bullet$ Σ because they are both tensor fields on a subset of M. Moreover, once we have developed tensor calculus on Σ we immediately have the one on σ by just pulling back (covariant) tensor fields on Σ to σ via the embedding, see below.

Consider then the following tensor fields, called the $\bullet \bullet \bullet \bullet \ \bullet \bullet \bullet \ \bullet \bullet \bullet \bullet \bullet \bullet \ \bullet \bullet \bullet \bullet \bullet \bullet$ $\bullet \ \bullet \bullet \bullet \bullet \ \bullet \bullet \bullet$ of Σ

$$q_{\mu\nu} := g_{\mu\nu} - s n_\mu n_\nu \quad \text{and} \quad K_{\mu\nu} := q^\rho_\mu q^\sigma_\nu \nabla_\rho n_\sigma \qquad (1.1.4)$$

where all indices are moved with respect to $g_{\mu\nu}$. Notice that both tensors in (1.1.4) are 'spatial', that is, they vanish when either of their indices is contracted with n^μ. A crucial property of $K_{\mu\nu}$ is its symmetry: we have $K_{[\mu\nu]} = q^\rho_\mu q^\sigma_\nu ((\nabla_{[\rho} \ln(F)) n_{\sigma]} + F \nabla_{[\rho} \nabla_{\sigma]} f) = 0$ since ∇ is torsion free. The square brackets denote antisymmetrisation defined as an idempotent operation. From this fact one derives another useful differential geometric identity by employing the relation between the covariant differential and the Lie derivative:

$$\begin{aligned} 2K_{\mu\nu} &= q^\rho_\mu q^\sigma_\nu \left(2\nabla_{(\rho} n_{\sigma)} \right) \\ &= q^\rho_\mu q^\sigma_\nu (\mathcal{L}_n g)_{\rho\sigma} = q^\rho_\mu q^\sigma_\nu (\mathcal{L}_n q + s \mathcal{L}_n n \otimes n)_{\rho\sigma} \\ &= q^\rho_\mu q^\sigma_\nu (\mathcal{L}_n q)_{\rho\sigma} = (\mathcal{L}_n q)_{\mu\nu} \end{aligned} \qquad (1.1.5)$$

since $n^\mu \mathcal{L}_n q_{\mu\nu} = -q_{\mu\nu}[n, n]^\mu = 0$. Using $n^\mu = (T^\mu - N^\mu)/N$ we can write (1.1.5) in the form

$$2K_{\mu\nu} = \frac{1}{N}(\mathcal{L}_{T-Nq})_{\mu\nu} - 2n^\rho q_{\rho(\mu} \ln(N)_{,\nu)} = \frac{1}{N}(\mathcal{L}_{T-Nq})_{\mu\nu} \qquad (1.1.6)$$

Next we would like to construct a covariant differential associated with the metric $q_{\mu\nu}$. We would like to stress that this metric is non-degenerate as a bijection between $\bullet \bullet \bullet \bullet \bullet \ \bullet \bullet \bullet \bullet \bullet \ \bullet \bullet \bullet$ and not as a metric between arbitrary tensors defined on Σ. Recall that, by definition, a differential ∇ is said to be covariant with respect to a metric g (of any signature) on a manifold M if it is (1) metric compatible, $\nabla g = 0$ and (2) torsion free, $[\nabla_\mu, \nabla_\nu] f = 0 \ \forall \ f \in C^\infty(M)$. According to a classical theorem reviewed in Section 19.2, these two conditions fix ∇ uniquely in terms of the Christoffel symbols (which defines the so-called Levi–Civita connection), which in turn are defined by the action of ∇ on one-forms through $\nabla_\mu u_\nu := \partial_\mu u_\nu - \Gamma^\rho_{\mu\nu} u_\rho$. Since the tensor q is a metric of Euclidean signature on Σ we can thus apply these two conditions to q and we are looking for a covariant differential D on spatial tensors only such that (1) $D_\mu q_{\nu\rho} = 0$ and (2) $D_{[\mu} D_{\nu]} f = 0$ for scalars f. Of course, the operator D should preserve the set of spatial tensor fields. It is easy to verify that $D_\mu f := q^\nu_\mu \nabla_\nu \tilde{f}$ and $D_\mu u_\nu := q^\rho_\mu q^\sigma_\nu \nabla_\rho \tilde{u}_\sigma$, for $u_\mu n^\mu = 0$ and extended to arbitrary tensors by linearity and Leibniz' rule, does the job and thus, by the above-mentioned theorem, is the unique choice. Here, $\tilde{f}$ and $\tilde{u}$ denote arbitrary smooth extensions of f and u respectively into a neighbourhood of Σ in M, necessary in order to perform the ∇ operation. The covariant differential is independent of that extension as derivatives not tangential to Σ are projected out by the q tensor (go into a local, adapted system of coordinates to see this) and we will drop the tilde again. One can convince oneself that the action of D on arbitrary spatial tensors is then given by acting with ∇ in the usual way followed by spatial projection of all appearing indices including the one with respect to which the derivative was taken.

We now ask what the Riemann curvature $R^{(D)\,\sigma}_{\mu\nu\rho}$ of D is in terms of that of ∇. To answer this question we need the second covariant differential of a spatial co-vector u_ρ which, when carefully using the definition of D, is given by

$$D_\mu D_\nu u_\rho = q^{\mu'}_\mu q^{\nu'}_\nu q^{\rho'}_\rho \nabla_{\mu'} D_{\nu'} u_{\rho'}$$
$$= q^{\mu'}_\mu q^{\nu'}_\nu q^{\rho'}_\rho \nabla_{\mu'} q^{\nu''}_{\nu'} q^{\rho''}_{\rho'} \nabla_{\nu''} u_{\rho''} \tag{1.1.7}$$

The outer derivative hits either a q tensor or ∇u, the latter of which will give rise to a curvature term. Consider then the ∇q terms.

Since ∇ is g compatible we have $\nabla q = s \nabla n \otimes n = s[(\nabla n) \otimes n + n \otimes (\nabla n)]$. Since all of these terms are contracted with q tensors and q annihilates n, the only terms that survive are proportional to terms either of the form

$$(\nabla_{\mu'} n_{\nu'})(n^{\rho''}(\nabla_{\nu''} u_{\rho''})) = -(\nabla_{\mu'} n_{\nu'})(\nabla_{\nu''} n^{\rho''}) u_{\rho''}$$

where $n^\mu u_\mu = 0 \Rightarrow \nabla_\nu(n^\mu u_\mu) = 0$ was exploited, or of the form $(\nabla_{\mu'} n_{\nu'})(\nabla_n u_{\rho'})$. Concluding, the only terms that survive from ∇q terms can be transformed into terms proportional to $\nabla n \otimes \nabla n$ or $\nabla n \otimes \nabla_n u$ where the ∇n factors, since contracted with q tensors, can be traded for extrinsic curvature terms (use $u_\mu = q^\nu_\mu u_\nu$ to do that).

It turns out that the terms proportional to $\nabla_n u$ cancel each other when computing the antisymmetrised second D derivative of u due to the symmetry of K, and we are thus left with the famous • ••• •• ••• ••

$$R^{(D)\,\sigma}_{\mu\nu\rho} u_\sigma := 2D_{[\mu} D_{\nu]} u_\rho$$
$$= \left[2s K_{\rho[\mu} K^\sigma_{\nu]} + q^{\mu'}_\mu q^{\nu'}_\nu q^{\rho'}_\rho q^\sigma_{\sigma'} R^{(D+1)\,\sigma'}_{\mu'\nu'\rho'} \right] u_\sigma$$
$$R^{(D)}_{\mu\nu\rho\sigma} = 2s K_{\rho[\mu} K_{\nu]\sigma} + q^{\mu'}_\mu q^{\nu'}_\nu q^{\rho'}_\rho q^{\sigma'}_\sigma R^{(D+1)}_{\mu'\nu'\rho'\sigma'} \tag{1.1.8}$$

Using this general formula we can specialize to the Riemann curvature scalar which is our ultimate concern in view of the Einstein–Hilbert action. Employing the standard abbreviations $K := K_{\mu\nu}q^{\mu\nu}$ and $K^{\mu\nu} = q^{\mu\rho}q^{\nu\sigma}K_{\nu\sigma}$ (notice that indices for spatial tensors can be moved either with q or with g) we obtain

$$R^{(D)} = R^{(D)}_{\mu\nu\rho\sigma}q^{\mu\rho}q^{\nu\sigma}$$
$$= s[K^2 - K_{\mu\nu}K^{\mu\nu}] + q^{\mu\rho}q^{\nu\sigma}R^{(D+1)}_{\mu\nu\rho\sigma} \qquad (1.1.9)$$

Equation (1.1.9) is not yet quite what we want since it is not yet purely expressed in terms of $R^{(D+1)}$ alone. However, we can eliminate the second term in (1.1.9) by using $g = q + sn \otimes n$ and the definition of curvature $R^{(D+1)}_{\mu\nu\rho\sigma}n^\sigma = 2\nabla_{[\mu}\nabla_{\nu]}n_\rho$ as follows:

$$R^{(D+1)} = R^{(D+1)}_{\mu\nu\rho\sigma}g^{\mu\rho}g^{\nu\sigma}$$
$$= q^{\mu\rho}q^{\nu\sigma}R^{(D+1)}_{\mu\nu\rho\sigma} + 2sq^{\rho\mu}n^\nu[\nabla_\mu, \nabla_\nu]n_\rho$$
$$= q^{\mu\rho}q^{\nu\sigma}R^{(D+1)}_{\mu\nu\rho\sigma} + 2sn^\nu[\nabla_\mu, \nabla_\nu]n^\nu \qquad (1.1.10)$$

where in the first step we used the antisymmetry of the Riemann tensor to eliminate the term quartic in n and in the second step we used again $q = g - sn \otimes n$ and the antisymmetry in the $\mu\nu$ indices. Now

$$n^\nu([\nabla_\mu, \nabla_\nu]n^\mu) = -(\nabla_\mu n^\nu)(\nabla_\nu n^\mu) + (\nabla_\mu n^\mu)(\nabla_\nu n^\nu) + \nabla_\mu(n^\nu\nabla_\nu n^\mu - n^\mu\nabla_\nu n^\nu)$$

and using $\nabla_\mu s = 2n^\nu\nabla_\mu n_\nu = 0$ we have

$$\nabla_\mu n^\mu = g^{\mu\nu}\nabla_\nu n_\mu = q^{\mu\nu}\nabla_\nu n^\mu = K$$
$$(\nabla_\mu n^\nu)(\nabla_\nu n^\mu) = g^{\nu\sigma}g^{\rho\mu}(\nabla_\mu n_\sigma)(\nabla_\nu n_\rho) = q^{\nu\sigma}q^{\rho\mu}(\nabla_\mu n_\sigma)(\nabla_\nu n_\rho) = K_{\mu\nu}K^{\mu\nu}$$
$$(1.1.11)$$

Combining (1.1.9), (1.1.10) and (1.1.11) we obtain the • •••••• •••••••

$$R^{(D+1)} = R^{(D)} - s[K_{\mu\nu}K^{\mu\nu} - K^2] + 2s\nabla_\mu(n^\nu\nabla_\nu n^\mu - n^\mu\nabla_\nu n^\nu) \quad (1.1.12)$$

Inserting this differential geometric identity back into the action, the third term in (1.1.12) is a total differential which we drop for the time being as one can retrieve it later on when making the variational principle well-defined.

At this point it is useful to pull back various quantities to σ. Consider the D spatial vector fields on Σ_t defined by

$$X^\mu_a(X) := X^\mu_{,a}(x,t)_{|X(x,t)=X} \qquad (1.1.13)$$

Then we have due to $n_\mu X^\mu_a = 0$ that

$$q_{ab}(t,x) := (X^\mu_{,a}X^\nu_{,b}q_{\mu\nu})(X(x,t)) = g_{\mu\nu}(X(t,x))X^\mu_{,a}(t,x)X^\nu_{,b}(t,x) \quad (1.1.14)$$

and

$$K_{ab}(t,x) := (X^\mu_{,a}X^\nu_{,b}K_{\mu\nu})(X(x,t)) = (X^\mu_{,a}X^\nu_{,b}\nabla_\mu n_\nu)(t,x) \quad (1.1.15)$$

Using q_{ab} and its inverse $q^{ab} = \epsilon^{aa_1...a_{D-1}}\epsilon^{bb_1...b_{D-1}}q_{a_1b_1}\cdots q_{a_{D-1}b_{D-1}}/[\det((q_{cd}))$
$(D-1)!]$ we can express $q_{\mu\nu}, q^{\mu\nu}, q_\mu^\nu$ as

$$q^{\mu\nu}(X) = \left[q^{ab}(x,t)X_{,a}^\mu X_{,b}^\nu\right](x,t)_{|X(x,t)=X}$$
$$q_\mu^\nu(X) = g_{\mu\rho}(X)q^{\rho\nu}(X)$$
$$q_{\mu\nu}(X) = g_{\nu\rho}(X)q_\mu^\rho(X) \tag{1.1.16}$$

To verify that this coincides with our previous definition $q = g - sn \otimes n$ it is sufficient to check the matrix elements in the basis given by the vector fields n, X_a. Since for both definitions n is annihilated we just need to verify that (1.1.16) when contracted with $X_a \otimes X_b$ reproduces (1.1.14), which is indeed the case.

Next we define $N(x,t) := N(X(x,t)), \vec{N}^a(x,t) := q^{ab}(x,t)(X_b^\mu g_{\mu\nu}N^\nu)$ $(X(x,t))$. Then it is easy to verify that

$$K_{ab}(x,t) = \frac{1}{2N}(\dot{q}_{ab} - (\mathcal{L}_{\vec{N}}q)_{ab})(x,t) \tag{1.1.17}$$

We can now pull back the expressions quadratic in $K_{\mu\nu}$ that appear in (1.1.12) using (1.1.16) and find

$$K(x,t) = (q^{\mu\nu}K_{\mu\nu})(X(x,t)) = (q^{ab}K_{ab})(x,t)$$
$$(K_{\mu\nu}K^{\mu\nu})(x,t) = (K_{\mu\nu}K_{\rho\sigma}q^{\mu\rho}q^{\nu\sigma})(X(x,t)) = (K_{ab}K_{cd}q^{ac}q^{bd})(x,t) \tag{1.1.18}$$

Likewise we can pull back the curvature scalar $R^{(D)}$. We have

$$R^{(D)}(x,t) = (R_{\mu\nu\rho\sigma}^{(D)}q^{\mu\rho}q^{\nu\sigma})(X(x,t))$$
$$= (R_{\mu\nu\rho\sigma}^{(D)}X_a^\mu X_b^\nu X_c^\rho X_d^\sigma)(X(x,t))q^{ac}(x,t)q^{bd}(x,t) \tag{1.1.19}$$

We would like to show that this expression equals the curvature scalar R as defined in terms of the Christoffel symbols for q_{ab}. To see this it is sufficient to compute $(X_a^\mu D_\mu f)(X(x,t)) = \partial_a f(X(x,t)) =: (D_a f)(x,t)$ with $f(x,t) := F(X(x,t))$ and with $u_a(x,t) := (X_a^\mu u_\mu)(X(x,t)),\ u^a(x,t) = q^{ab}(x,t)u_b(x,t)$

$$(D_a u_b)(x,t) := (X_a^\mu X_b^\nu D_\mu u_\nu)(X(x,t))$$
$$= X_{,a}^\mu(x,t)X_{,b}^\nu(x,t)(\nabla_\mu u_\nu)(X(x,t))$$
$$= (\partial_a u_b)(x,t) - X_{,ab}^\mu u_\mu(X(x,t))$$
$$- u^c(x,t)\Gamma_{\rho\mu\nu}^{(D+1)}(X(x,t))X_{,c}^\rho(x,t)X_{,a}^\mu(x,t)X_{,b}^\nu(x,t)$$
$$= (\partial_a u_b)(x,t) - \Gamma_{cab}^{(D)}(x,t)u^c(x,t) \tag{1.1.20}$$

where in the last step we have used the explicit expressions of the Christoffel symbols $\Gamma^{(D+1)}$ and $\Gamma^{(D)}$ in terms of $g_{\mu\nu}$ and q_{ab} respectively. Now since every tensor field W is a linear combination of tensor products of one-forms and since D_μ satisfies the Leibniz rule we easily find $(X_a^\mu X_b^\nu \ldots D_\mu W_\nu...)(X(x,t)) =: (D_a W_{b...})(x,t)$ where now D_a denotes the unique torsion-free covariant differential associated with q_{ab} and $W_{a...}$ is the pull-back of $W_{\mu...}$. In particular, we have $X_a^\mu X_b^\nu X_c^\rho D_\mu D_\nu u_\rho = D_a X_b^\mu X_c^\nu D_\mu u_\rho = D_a D_b u_c$, from which our assertion follows since

$$
\begin{aligned}
\left(R_{abcd} u^d\right)(x,t) &:= ([D_a, D_b] u_c)(x,t) = \left(X_a^\mu X_b^\nu X_c^\rho [D_\mu, D_\nu] u_\rho\right)(X(x,t)) \\
&= \left(X_a^\mu X_b^\nu X_c^\rho X_d^\sigma R_{\mu\nu\rho\sigma}^{(D)}\right)(X(x,t)) u^d(x,t)
\end{aligned}
\tag{1.1.21}
$$

From now on we will move indices with the metric q_{ab} only.

One now expresses the line element in the new system of coordinates x, t using the quantities q_{ab}, N, N^a (we refrain from displaying the arguments of the components of the metric)

$$
\begin{aligned}
ds^2 &= g_{\mu\nu} dX^\mu \otimes dX^\nu \\
&= g_{\mu\nu}(X(t,x)) \left[X_{,t}^\mu dt + X_{,a}^\mu dx^a\right] \otimes \left[X_{,t}^\nu dt + X_{,b}^\nu dx^b\right] \\
&= g_{\mu\nu}(X(t,x)) \left[Nn^\mu dt + X_{,a}^\mu (dx^a + N^a dt)\right] \otimes \left[Nn^\nu dt + X_{,b}^\nu (dx^b + N^b dt)\right] \\
&= [sN^2 + q_{ab} N^a N^b] dt \otimes dt + q_{ab} N^b [dt \otimes dx^a + dx^a \otimes dt] + q_{ab} dx^a \otimes dx^b
\end{aligned}
\tag{1.1.22}
$$

and reads off the components g_{tt}, g_{ta}, g_{ab} of X^*g in this frame. Since the volume form $\Omega(X) := \sqrt{|\det(g)|} d^{D+1} X$ is covariant, that is, $(X^*\Omega)(x,t) = \sqrt{|\det(X*g)|} dt d^D x$, we just need to compute $\det(X^*g) = sN^2 \det(q_{ab})$ in order to finally cast the action (1.1.1) into $D+1$ form. The result is (dropping the total differential in (1.1.12)) the • • • ••••••

$$
S = \frac{1}{\kappa} \int_{\mathbb{R}} dt \int_\sigma d^D x \sqrt{\det(q)} |N| (R - s[K_{ab} K^{ab} - (K_a^a)^2])
\tag{1.1.23}
$$

We could drop the absolute sign for N in (1.1.23) since we took N positive but we will keep it for the moment to see what happens if we allow arbitrary sign. Notice that (1.1.23) ••• •••••• ••• •• •••••• for $D = 1$, indeed in two spacetime dimensions the Einstein action is proportional to a topological charge, the so-called Euler characteristic of M, and in what follows we concentrate on $D > 1$.

1.2 Legendre transform and Dirac analysis of constraints

We now wish to cast this action into canonical form, that is, we would like to perform the Legendre transform from the Lagrangian density appearing in (1.1.23) to the corresponding Hamiltonian density. The action (1.1.23) depends on the velocities $\dot{q}_{ab}$ of q_{ab} but not on those of N and N^a. Therefore we obtain for the conjugate momenta (use (1.1.17) and the fact that R does not contain time derivatives)

$$
P^{ab}(t,x) := \frac{\delta S}{\delta \dot{q}_{ab}(t,x)} = -s \frac{|N|}{N\kappa} \sqrt{\det(q)} [K^{ab} - q^{ab}(K_c^c)]
$$

$$
\Pi(t,x) := \frac{\delta S}{\delta \dot{N}(t,x)} = 0
$$

$$
\Pi_a(t,x) := \frac{\delta S}{\delta \dot{N}^a(t,x)} = 0
\tag{1.2.1}
$$

The Lagrangian in (1.1.23) is therefore a •••••••• Lagrangian, one cannot solve all velocities for momenta [218]. In order to further analyse the system one must therefore apply Dirac's algorithm [219] for constrained Hamiltonian systems, which we summarise in Chapter 24.

We can solve $\dot{q}_{ab}$ in terms of q_{ab}, N, N^a and P^{ab} using (1.1.17) but this is not possible for $\dot{N}, \dot{N}^a$, rather we have the so-called •••• ••• ••• •••• ••

$$C(t,x) := \Pi(t,x) = 0 \text{ and } C^a(t,x) := \Pi^a(t,x) = 0 \qquad (1.2.2)$$

According to [219] we are supposed to introduce Lagrange multiplier fields $\lambda(t,x)$, $\lambda_a(t,x)$ for the primary constraints and to perform the Legendre transform as usual with respect to the remaining velocities which can be solved for. We have

$$\dot{q}_{ab} = 2NK_{ab} + (\mathcal{L}_{\vec{N}}q)_{ab}$$

$$\dot{q}_{ab}P^{ab} = (\mathcal{L}_{\vec{N}}q)_{ab}P^{ab} - 2\frac{s|N|}{\kappa}\sqrt{\det(q)}[K_{ab}K^{ab} - K^2]$$

$$P_{ab}P^{ab} = \frac{\det(q)}{\kappa^2}(K_{ab}K^{ab} + (D-2)K^2)$$

$$P^2 := \left(P_a^a\right)^2 = \frac{(1-D)^2}{\kappa^2}\det(q)K^2 \qquad (1.2.3)$$

and by means of these formulae we obtain the canonical form of the action (1.1.23)

$$S = \int_{\mathbb{R}} dt \int_{\sigma} d^D x \{\dot{q}_{ab}P^{ab} + \dot{N}\Pi + \dot{N}^a\Pi_a - [\dot{q}_{ab}(P,q,N,\vec{N})P^{ab} + \lambda C + \lambda^a C_a$$

$$- \sqrt{\det(q)}\frac{|N|}{\kappa}(R - s[K_{ab}K^{ab} - K^2])(P,q,N,\vec{N})]\}$$

$$= \int_{\mathbb{R}} dt \int_{\sigma} d^D x \{\dot{q}_{ab}P^{ab} + \dot{N}\Pi + \dot{N}^a\Pi_a - [(\mathcal{L}_{\vec{N}}q)_{ab}P^{ab} + \lambda C + \lambda^a C_a$$

$$- \sqrt{\det(q)}\frac{|N|}{\kappa}(R + s[K_{ab}K^{ab} - K^2])(P,q,N,\vec{N})]\}$$

$$= \int_{\mathbb{R}} dt \int_{\sigma} d^D x \left\{ \dot{q}_{ab}P^{ab} + \dot{N}\Pi + \dot{N}^a\Pi_a - \left[(\mathcal{L}_{\vec{N}}q)_{ab}P^{ab} + \lambda C + \lambda^a C_a \right.\right.$$

$$\left.\left. + \frac{|N|}{\kappa}\left(-\frac{s\kappa^2}{\sqrt{\det(q)}}\left[P_{ab}P^{ab} - \frac{1}{D-1}P^2\right] - \sqrt{\det(q)}R\right)\right]\right\} \qquad (1.2.4)$$

Upon performing a spatial integration by parts (whose boundary term we drop for the moment, it will be recovered later on) one can cast it into the following more compact form

$$S = \int_{\mathbb{R}} dt \int_{\sigma} d^D x \{\dot{q}_{ab}P^{ab} + \dot{N}\Pi + \dot{N}^a\Pi_a - [\lambda C + \lambda^a C_a + N^a H_a + |N|H]\}$$

$$(1.2.5)$$

where

$$H_a := -2q_{ac}D_b P^{bc}$$

$$H := -\left(\frac{s\kappa}{\sqrt{\det(q)}} \left[q_{ac}q_{bd} - \frac{1}{D-1} q_{ab}q_{cd} \right] P^{ab}P^{cd} + \sqrt{\det(q)}R/\kappa \right) \qquad (1.2.6)$$

are called the •••• ••• •• • •• ••• ••• •••• •• ••••••• •• and • •• •••• ••• ••• ••••• ••• respectively, for reasons that we will derive below.

The geometrical meaning of these quantities is as follows. At fixed t the fields $(q_{ab}(t,x), N^a(t,x), N(t,x); P^{ab}(x,t), \Pi_a(t,x), \Pi(t,x))$ are points (configuration; canonically conjugate momenta) in an infinite-dimensional phase space $\mathcal{M}$ (or symplectic manifold). Strictly speaking, we should now specify on what Banach space this manifold is modelled [220], however, we will be brief here as we are not primarily interested in the metric formulation of this chapter but rather in the connection formulation of the next chapter for which we will give more details in Chapter 33. For the purpose of this section it is sufficient to say that we can choose the model space to be the direct product of the space $T_2(\sigma) \times T_1(\sigma) \times T_0(\sigma)$ of smooth symmetric covariant tensor fields of rank $2,1,0$ on σ respectively and the space $\tilde{T}^2(\sigma) \times \tilde{T}^1(\sigma) \times \tilde{T}^0(\sigma)$ of smooth symmetric contravariant tensor density fields of weight one and of rank $2,1,0$ on σ respectively, equipped with some Sobolev norm. In particular, one shows that the action (1.2.5) is differentiable in this topology. The precise functional analytic description is somewhat more complicated in case that σ has a boundary; we postpone boundary conditions until Section 1.5.

The phase space carries the strong (see [220] or Chapter 33) symplectic structure Ω or Poisson bracket

$$\{P(f^2), F_2(q)\} = \kappa F_2(f^2), \quad \{\vec{\Pi}(\vec{f}^1), \vec{F}_1(\vec{N})\} = \kappa \vec{F}_1(\vec{f}^1), \quad \{\Pi(f), F(N)\} = \kappa F(f)$$

$$(1.2.7)$$

(all other brackets vanishing) where we have defined the following pairing, invariant under diffeomorphisms of σ, for example

$$\tilde{T}^2(\sigma) \times T_2(\sigma) \to \mathbb{R}; \ (F_2, f^2) \to F^2(f_2) := \int_\sigma d^D x F_2^{ab}(x) f_{ab}^2(x) \qquad (1.2.8)$$

and similar for the other fields. Physicists use the following shorthand notation for (1.2.7)

$$\{P^{ab}(t,x), q_{cd}(t,x')\} = \kappa \delta^a_{(c} \delta^b_{d)} \delta^{(D)}(x,x') \qquad (1.2.9)$$

In the language of symplectic geometry, of which we give an account in Chapter 19, the first term in the action (1.2.5) is a symplectic potential for the symplectic structure (1.2.7).

The Poisson bracket between arbitrary functionals $G, G' : \mathcal{M} \to \mathbb{C}$ follows from the basic ones (1.2.7) by imposing the Leibniz rule, that is, that the Poisson bracket is a derivation. It then follows that

$$\{G, G'\} = \kappa \int_\sigma d^D x \left[\frac{\delta G}{\delta P^{ab}(x)} \frac{\delta G'}{\delta q_{ab}(x)} - \frac{\delta G'}{\delta P^{ab}(x)} \frac{\delta G}{\delta q_{ab}(x)} \right] \qquad (1.2.10)$$

where we observe the appearance of functional derivatives. For a precise definition of the functional derivative we would again need to make use of the theory of Banach manifolds, however, the following definition will be sufficient for our purposes.

Definition 1.2.1. ••• ϕ ••••••• •• ••••• • •• •••• •• Φ •• • •••• •••••••• •••••
•• •• ••••• ••••••• ••••• ••••• • $(D+1)$•• •• •••• •• M • ••• ••••••••••• •• •••••••
•••••••••• ••• ••• ••• $\delta\phi \in T_\phi(\Phi)$ •••••• •• ••• ••••••••• ••••• •• •••• • •• ••• ••
•• ϕ •••• •••••• ••••• •• •••••• •• •••••••••• •••• • ••• •• ••• •••• • ••• ••••• ••
$G: \Phi \to \mathbb{C}$ •• •••••••• • ••• ••• •••• ••••• •• •• $\phi \in \Phi$ •• ••••• •••••• • ••• ••• ••• ••••
DG_ϕ •••• •••••• • ••• •••••••• •• ••• ••••• •••••• ••• •• ϕ •••• •• •••

$$\left(\frac{d}{ds}\right)_{r=0} G[\phi + s\delta\phi] = \int_M d^{D+1}x \, (DG)_\phi(x) \cdot \delta\phi(x)$$

• ••• •• •••• $DG_\phi(x)$ •• •• • •••••••• •••• • •••• $T_\phi(\Phi)$• •••• ••• •• •• ••• ••••••••
•••• •••••• • •• •••• ••• ••• •••• •••••••• •• •• ••••• •••••••••••••• ••• • •• ϕ•

It is easy to check that with this definition, for example, $\delta F_2(q)/\delta q_{ab}(x) = F^{ab}(x)$ and hence we indeed reproduce (1.2.9). The motivation for introducing the Poisson bracket as above in field theory, as in classical mechanics, is that it reproduces the Euler–Lagrange equations of motion $\delta^{D+1}S/\delta g_{\mu\nu}(X) = G^{(D+1)}_{\mu\nu}(X) = 0$, where $G_{\mu\nu}$ is the Einstein tensor, as the Hamiltonian equations of motion $\dot{q} := \{\mathbf{H}, q\}$ and similar for P. Here we have written $\delta^{(D+1)}$ instead of δ in order to stress that with respect to the action we perform a variation over M while in Poisson brackets we perform a variation with respect over σ. We will see this correspondence between the Lagrangian and Hamiltonian formulation explicitly in the next section. However, let us point out that these equations, while often called 'evolution equations', just describe infinitesimal gauge transformations, they do not correspond to the physical evolution with respect to a physical (gauge-invariant) Hamiltonian.

We now turn to the meaning of the term in square brackets in (1.2.5), that is, the 'Hamiltonian'

$$\kappa\mathbf{H} := \int_\sigma d^D x [\lambda C + \lambda^a C_a + N^a H_a + |N|H]$$
$$=: C(\lambda) + \vec{C}(\vec{\lambda}) + \vec{H}(\vec{N}) + H(|N|) \qquad (1.2.11)$$

of the action and the associated equations of motion.

The variation of the action with respect to the Lagrange multiplier fields $\vec{\lambda}, \lambda$ reproduces the primary constraints (1.2.2). If the dynamics of the system is to be consistent, then these constraints must be preserved under the evolution of the

• •• •••••• • •• •••• • ••• •• •••• • •• •••• ••• • •• ••••• • •• •••••••

system, that is, we should have, for example, $\dot{C}(t,x) := \{\mathbf{H}, C(t,x)\} = 0$ for all $x \in \sigma$, or equivalently, $\dot{C}(f) := \{\mathbf{H}, C(f)\} = 0$ for all (t-independent) smearing fields $f \in T_0(\sigma)$. However, we do not get zero but rather

$$\{\vec{C}(\vec{f}), \mathbf{H}\} = \vec{H}(\vec{f}) \text{ and } \{C(f), \mathbf{H}\} = H\left(\frac{N}{|N|}f\right) \qquad (1.2.12)$$

which is supposed to vanish for all $f, \vec{f}$. Thus, consistency of the equations of motion asks us to impose the •••• •••• •• •••••• ••

$$H(x,t) = 0 \text{ and } H_a(x,t) = 0 \qquad (1.2.13)$$

for all $x \in \sigma$. Since these two functions appear next to the C, C_a in (1.2.11), in General Relativity the 'Hamiltonian' is constrained to vanish! General Relativity is an example of a so-called constrained Hamiltonian system with no true Hamiltonian. The reason for this will become evident in a moment.

Now one might worry that imposing consistency of the secondary constraints under evolution results in tertiary constraints and so on, but fortunately, this is not the case. Consider the smeared quantities $H(f), \vec{H}(\vec{f})$ where, for example, $\vec{H}(\vec{N}) := \int_\sigma d^3x N^a H_a$ (notice that indeed H, Π and H_a, Π_a are, respectively, scalar and co-vector densities of weight one on σ). Then we obtain

$$\{\mathbf{H}, \vec{H}(\vec{f})\} = \vec{H}(\mathcal{L}_{\vec{N}}\vec{f}) - H(\mathcal{L}_{\vec{f}}|N|)$$
$$\{\mathbf{H}, H(f)\} = H(\mathcal{L}_{\vec{N}}f) + \vec{H}(\vec{N}(|N|, f, q)) \qquad (1.2.14)$$

where $\vec{N}(f, f', q)^a = q^{ab}(ff'_{,b} - f'f_{,b})$. Equations (1.2.14) are equivalent to the • •••• ••••••• $\mathfrak{D}$ [219]

$$\{\vec{H}(\vec{f}), \vec{H}(\vec{f}')\} = -\kappa\vec{H}(\mathcal{L}_{\vec{f}}\vec{f}')$$
$$\{\vec{H}(\vec{f}), H(f)\} = -\kappa H(\mathcal{L}_{\vec{f}}f)$$
$$\{H(f), H(f')\} = s\kappa\vec{H}(\vec{N}(f, f', q)) \qquad (1.2.15)$$

also called the •••••••• •••• •••••• ••••• •••••• which we will derive in full detail below. The meaning of (1.2.12) and (1.2.15) is that the constraint surface $\overline{\mathcal{M}}$ of $\mathcal{M}$, the submanifold of $\mathcal{M}$ where the constraints hold, is preserved under the motions generated by the constraints. In the terminology of Dirac [219], all constraints are of first class (determine co-isotropic constraint submanifolds [218] of $\mathcal{M}$) rather than of second class (determine symplectic constraint submanifolds [218] of $\mathcal{M}$). See Figure 1.2 for a sketch of these notions and Sections 19.3, 24.2 for the explanation of the terminology.

1.3 Geometrical interpretation of the gauge transformations

We turn now to the study of the equations of motion of the canonical coordinates on the phase space. Since $C = \Pi, C_a = \Pi_a$ it remains to study those of

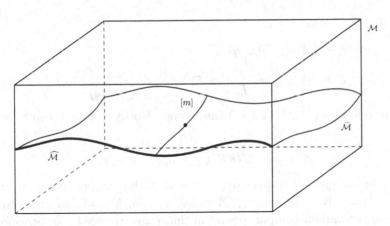

Figure 1.2 Unconstrained phase space, constraint surface, gauge orbit and reduced phase space.

N, N^a, q_{ab}, P^{ab}. For shift and lapse we obtain $\dot{N}^a = \lambda^a, \dot{N} = \lambda$. Since λ^a, λ are •••••••••• ••••••••••• •• ••••••••• we see that also the trajectory of lapse and shift is completely arbitrary. Moreover, the equations of motion of q_{ab}, P^{ab} are completely unaffected by the term $\vec{C}(\vec{\lambda}) + C(\lambda)$ in $\mathbf{H}$. It is therefore completely straightforward to solve the equations of motion as far as N, N^a, Π, Π_a are concerned: simply treat N, N^a as •••••••• • •••••• •••••• and drop all terms proportional to C, C_a from the action (1.2.5). The result is the reduced action

$$S = \frac{1}{\kappa} \int_{\mathbb{R}} dt \int_{\sigma} d^D x \{ \dot{q}_{ab} P^{ab} - [N^a H_a + |N|H] \} \tag{1.3.1}$$

called the ••••••••• • ••• •••• •••••• ••••• •• • • •••••• [206]. It is straightforward to check that as far as q_{ab}, P^{ab} are concerned, the actions (1.2.5) and (1.3.1) are completely equivalent.

The equations of motion of q_{ab}, P^{ab} then finally allow us to interpret the motions that the constraints generate on $\mathcal{M}$ geometrically. Since the reduced Hamiltonian (using the same symbol as before)

$$\mathbf{H} = \frac{1}{\kappa} \int_{\sigma} d^D x [N^a H_a + |N|H] \tag{1.3.2}$$

is a linear combination of constraints, we obtain the equations of motion once we know the Hamiltonian flow of the functions $\vec{H}(\vec{f}), H(f)$ for any $\vec{f}, f$ separately. Denoting, for any function J on $\mathcal{M}$,

$$\delta_{\vec{f}} J := \{\vec{H}(\vec{f}), J\} \text{ and } \delta_f J := \{H(f), J\} \tag{1.3.3}$$

it is easiest to begin with the corresponding equations for $J = F_2(q)$ since upon integration by parts we have $\vec{H}(\vec{f}) = \int d^D x P^{ab} (\mathcal{L}_{\vec{f}} q)_{ab}$ so that both constraint functions are simple polynomials in P^{ab} not involving their derivatives. We then readily find

$$\delta_{\vec{f}} F_2(q) = \kappa F_2(\mathcal{L}_{\vec{f}} q)$$

$$\delta_f F_2(q) = -2s\kappa \int_\sigma d^D x \, F_2^{ab} \, f \frac{P_{ab} - P q_{ab}/(D-1)}{\sqrt{\det(q)}} \tag{1.3.4}$$

Using the relations (1.2.1), (1.1.17) the second identity in (1.3.4) can be written as

$$\delta_{|N|} q_{ab} = 2N\kappa K_{ab} = \kappa(\dot{q}_{ab} - (\mathcal{L}_{\vec{N}} q)_{ab})$$

In order to interpret this quantity, notice that the components of n_μ in the frame t, x^a are given by $n_t = n_\mu X^\mu_{,t} = sN$, $n_a = n_\mu X^\mu_{,a} = 0$. In order to compute the contravariant components n^μ in that frame we need the corresponding contravariant metric components. From (1.1.22) we find the covariant components to be $g_{tt} = sN^2 + q_{ab}N^a N^b$, $g_{ta} = q_{ab}N^b$, $g_{ab} = q_{ab}$ so that the inverse metric has components $g^{tt} = s/N^2$, $g^{ta} = -sN^a/N^2$, $g^{ab} = q^{ab} + sN^a N^b/N^2$. Thus $n^t = 1/N$, $n^a = -N^a/N$ and since $q_{at} = q_{tt} = 0$ we finally obtain

$$\delta_{|N|} F_2(q) = \kappa F_2(\mathcal{L}_{Nn} q) \tag{1.3.5}$$

which of course we guessed immediately from the $(D+1)$ dimensional identity (1.1.6). Concluding, as far as q_{ab} is concerned, H_a generates diffeomorphisms of M that preserve Σ_t while H generates diffeomorphisms of M orthogonal to Σ_t, however, only when the equations of motion $\dot{q}_{ab} = \{\mathbf{H}, q_{ab}\}$ are satisfied which we used in re-expressing P in terms of $\dot{q}$.

The corresponding computation for $P(f^2)$ is harder by an order of magnitude due to the curvature term involved in H and due to the fact that the identity corresponding to (1.3.5) holds •• •• •••••, that is, when the (vacuum) Einstein equations $G^{(D+1)}_{\mu\nu} := R^{(D+1)}_{\mu\nu} - \frac{g_{\mu\nu}}{2} R^{(D+1)} = 0$ hold. The variation with respect to $\vec{H}(\vec{f}) = -\int_\sigma d^D x q_{ab} (\mathcal{L}_{\vec{f}} P)^{ab}$ (notice that P^{ab} carries density weight one to verify this identity) is still easy and yields the expected result

$$\delta_{\vec{f}} P(f^2) = \kappa(\mathcal{L}_{\vec{f}} P)(f^2) \tag{1.3.6}$$

We will now describe the essential steps for the analogue of (1.3.5). The ambitious reader who wants to fill in the missing steps should expect to perform at least one Din A4 page of calculation in between each of the subsequent formulae.

We start from formula (1.2.6). Then

$$\{H(|N|), P^{ab}\} = \frac{\delta H(|N|)}{\delta q_{ab}}$$

$$= \frac{s|N|}{\sqrt{\det(q)}} [2(P^{ac} P^b_c - P^{ab} P/(D-1))$$

$$- \frac{q^{ab}}{2}(P^{cd} P_{cd} - P^2/(D-1))] + \frac{\delta}{\delta q_{ab}} \int d^D x |N| \sqrt{\det(q)} R \tag{1.3.7}$$

where the second term comes from the $\sqrt{\det(q)}^{-1}$ factor and we used the well-known formula $\delta \det(q) = \det(q) q^{ab} \delta q_{ab}$. To perform the remaining variation in (1.3.7) we write

$$\delta \sqrt{\det(q)} R = [\delta \sqrt{\det(q)}] R + \sqrt{\det(q)} [\delta q^{ab}] R_{ab} + \sqrt{\det(q)} q^{ab} [\delta R_{ab}]$$

use $\delta \delta_b^a = \delta[q^{ac} q_{cb}] = 0$ in the second variation and can simplify (1.3.7)

$$\{H(|N|), P^{ab}\} = \frac{s|N|}{\sqrt{\det(q)}} [P^{ac} P_c^b - P^{ab} P/(D-1)]$$

$$+ \frac{q^{ab}|N|H}{2} + |N| \sqrt{\det(q)} (q^{ab} R - R^{ab})$$

$$+ \int d^D x |N| \sqrt{\det(q)} q^{cd} \frac{\delta}{\delta q_{ab}} R_{cd} \qquad (1.3.8)$$

The final variation is the most difficult one since R_{cd} contains second derivatives of q_{ab}. Using the explicit expression of R_{abcd} in terms of the Christoffel symbol Γ_{ab}^c and observing that, while the connection itself is not a tensor, its variation in fact $\bullet\bullet$ a tensor, we find after careful use of the definition of the covariant derivative

$$q^{cd} \delta R_{cd} = q^{cd} \left[- D_c \delta \Gamma_{ed}^e + D_e \delta \Gamma_{cd}^e \right] \qquad (1.3.9)$$

We now use the explicit expression of Γ_{bc}^a in terms of q_{ab} and find

$$\delta \Gamma_{bc}^a = \frac{q^{ad}}{2} [D_c \delta q_{bd} + D_b \delta q_{cd} - D_d \delta q_{bc}] \qquad (1.3.10)$$

Next we insert (1.3.9) and (1.3.10) into the integral appearing in (1.3.8) and integrate by parts twice using the fact that for the divergence of a vector v^a we have $\sqrt{\det(q)} D_a v^a = D_a(\sqrt{\det(q)} v^a) = \partial_a(\sqrt{\det(q)} v^a)$ where we keep all boundary terms for later use and find

$$\int d^D x |N| \sqrt{\det(q)} q^{cd} \delta R_{cd} = \int d^D x \sqrt{\det(q)} q^{cd} [(D_c|N|) \delta \Gamma_{ed}^e - (D_e|N|) \delta \Gamma_{cd}^e]$$

$$+ \int_{\partial \sigma} \sqrt{\det(q)} \, q^{cd} |N| [-dS_c \delta \Gamma_{ed}^e + dS_e \delta \Gamma_{cd}^e]$$

$$= \int d^D x \sqrt{\det(q)} q^{cd} q^{ef} [(D_c|N|)(D_d \delta q_{ef})$$

$$- (D_e|N|)(D_c \delta q_{df})] + \int_{\partial \sigma} \sqrt{\det(q)} \, q^{cd} |N|$$

$$\times [-dS_c \delta \Gamma_{ed}^e + dS_e \delta \Gamma_{cd}^e]$$

$$= \int d^D x \sqrt{\det(q)} [-(D_c D^c|N|) q^{ab} + (D^a D^b|N|)] \delta q_{ab}$$

$$+ \int_{\partial \sigma} \sqrt{\det(q)} q^{cd} q^{ef} [(D_c|N|)(dS_d \delta q_{ef})$$

$$- (D_e|N|)(dS_c \delta q_{df})] + \int_{\partial \sigma} \sqrt{\det(q)} q^{cd} |N|$$
$$\times \left[-dS_c \delta \Gamma^e_{ed} + dS_e \delta \Gamma^e_{cd} \right] \tag{1.3.11}$$

Collecting all contributions and neglecting the boundary term which we will take care of in the next section we obtain the desired result

$$\{H(|N|), P^{ab}\} = \frac{2s|N|}{\sqrt{\det(q)}} \left[P^{ac} P^b_c - P^{ab} P/(D-1) \right] + \frac{q^{ab}|N|H}{2}$$
$$+ |N| \sqrt{\det(q)} (q^{ab} R - R^{ab}) + \sqrt{\det(q)} [-(D_c D^c|N|) q^{ab}$$
$$+ (D^a D^b |N|)] \tag{1.3.12}$$

which does not look at all as $\mathcal{L}_{Nn} P^{ab}$!

In order to compute $\mathcal{L}_{Nn} P^{ab}$ we need an identity for $\mathcal{L}_{Nn} K_{\mu\nu} = N \mathcal{L}_n K_{\mu\nu}$ which we now derive. Using the definition of the Lie derivative in terms of the covariant derivative ∇_μ and using $g = q + sn \otimes n$ one finds first of all

$$\mathcal{L}_n K_{\mu\nu} = -K K_{\mu\nu} + 2K_{\rho\mu} K^\rho_\nu + \left[\nabla_\rho(n^\rho K_{\mu\nu}) + 2s K_{\rho(\mu} n_{\nu)} \nabla_n n^\rho \right] \tag{1.3.13}$$

Using the Gauß equation (1.1.8) we find for the Ricci tensor $R^{(D)}_{\mu\nu}$ the following equation (use again $g = q + sn \otimes n$ and the definition of curvature as $R = [\nabla, \nabla]$)

$$R^{(D+1)}_{\rho\sigma} q^\rho_\mu q^\sigma_\nu - R^{(D)}_{\mu\nu} = s \left[-K_{\mu\nu} K + K_{\mu\rho} K^\rho_\nu + q^\rho_\mu q^\sigma_\nu n^\lambda [\nabla_\rho, \nabla_\lambda] n_\sigma \right] \tag{1.3.14}$$

We claim that the term in square brackets on the right-hand side of (1.3.13) equals $(-s)$ times the sum of the left-hand side of (1.3.14) and the term $-s(D_\mu D_\nu N)/N$. In order to prove this we manipulate the commutator of covariant derivatives appearing in (1.3.14) making use of the definition of the extrinsic curvature. One finds

$$q^\rho_\mu q^\sigma_\nu n^\lambda [\nabla_\rho, \nabla_\lambda] n_\sigma = q^\rho_\mu q^\sigma_\nu n^\lambda (\nabla_\rho \nabla_\lambda n_\sigma) + K K_{\mu\nu} - \nabla_\rho(n^\rho K_{\mu\nu})$$
$$- s(\nabla_n n^\rho) n_\nu K_{\mu\rho} - s(\nabla_n (n_\mu n^\rho))(\nabla_\rho n_\nu) \tag{1.3.15}$$

Using this identity we find for the sum of the term in square brackets on the right-hand side of (1.3.13) and s times the sum of the right-hand side of (1.3.14) the expression (dropping the obvious cancellations)

$$K_{\mu\rho} K^\rho_\nu + q^\rho_\mu q^\sigma_\nu n^\lambda (\nabla_\rho \nabla_\lambda n_\sigma) + s [K_{\rho\nu} n_\mu (\nabla_n n^\rho) - (\nabla_n (n_\mu n^\rho))(\nabla_\rho n_\nu)]$$
$$= K_{\mu\rho} K^\rho_\nu + q^\rho_\mu q^\sigma_\nu n^\lambda (\nabla_\rho \nabla_\lambda n_\sigma) + s [n_\mu (\nabla_n n^\rho) \{ q^\sigma_\rho - \delta^\sigma_\rho \} (\nabla_\sigma n_\nu) - (\nabla_n n_\mu)(\nabla_n n_\nu)]$$
$$= K_{\mu\rho} K^\rho_\nu + q^\rho_\mu q^\sigma_\nu (\nabla_\rho \nabla_n n_\sigma) - q^\rho_\mu q^\sigma_\nu (\nabla_\rho n^\lambda)(\nabla_\lambda n_\sigma) - s(\nabla_n n_\mu)(\nabla_n n_\nu)$$
$$= + q^\rho_\mu q^\sigma_\nu (\nabla_\rho \nabla_n n_\sigma) - s(\nabla_n n_\mu)(\nabla_n n_\nu) \tag{1.3.16}$$

where in the second step use has been made of the fact that the curly bracket vanishes since it is proportional to n_ρ and contracted with the spatial vector $\nabla_n n^\rho$, in the third step we moved n^λ inside a covariant derivative and picked up a correction term and in the fourth step one realises that this correction term is just the negative of the first term using $K_{\mu\nu} = q_\mu^\rho \nabla_\rho n_\nu$. Our claim is equivalent to showing that the last line of (1.3.16) is indeed given by $-s(D_\mu D_\nu N)/N$.

To see this notice that if the surface Σ_t is defined by $t(X) = t = \text{const.}$ then $1 = T^\mu \nabla_\mu t$. Since $\nabla_\mu t$ is orthogonal to Σ_t we have $n_\mu = sN\nabla_\mu t$ as one verifies by contracting with T^μ and thus $N = 1/(\nabla_n t)$. Thus

$$D_\mu N = -N^2 D_\mu(\nabla_n t) = -N^2 q_\mu^\nu n^\rho (\nabla_\rho \nabla_\nu t)$$
$$= -sN(\nabla_n n_\mu) = -sN\nabla_n n_\mu \qquad (1.3.17)$$

where in the first step we interchanged the second derivative due to torsion freeness and could pull n^ρ out of the second derivative because the correction term is proportional to $n_\rho \nabla n^\rho = 0$ and in the second we have pulled in a factor of N, observed that the correction is annihilated by the projection, used once more $sN\nabla t = n$ and finally used that $\nabla_n n_\nu$ is already spatial. The second derivative then gives simply

$$D_\mu D_\nu N = -s(D_\mu N)\nabla_n n_\nu - sNq_\mu^\rho q_\nu^\sigma \nabla_\rho \nabla_n n_\sigma$$
$$= N(\nabla_n n_\mu)(\nabla_n n_\nu) - sNq_\mu^\rho q_\nu^\sigma \nabla_\rho \nabla_n n_\sigma \qquad (1.3.18)$$

which is indeed N times (1.3.16) as claimed. Notice that in (1.3.18) we cannot replace N by $|N|$ if N is not everywhere positive so ••• •• ••••••••• •••• •••• •• ••• •••••• •• •• ••• ••• ••• •• • •• ••• ••• ••• $N = |N|$ •••••• •••. It is at this point that we must take N positive in all that follows.

We have thus established the key result

$$\mathcal{L}_{Nn} K_{\mu\nu} = N\left(-KK_{\mu\nu} + 2K_{\rho\mu}K_\nu^\rho\right) - s\left[D_\mu D_\nu N + N\left(R_{\rho\sigma}^{(D+1)} q_\mu^\rho q_\nu^\sigma - R_{\mu\nu}^{(D)}\right)\right]$$
$$(1.3.19)$$

In order to finish the calculation for $\mathcal{L}_{Nn} P^{\mu\nu}$ we need to know $\mathcal{L}_{Nn}\sqrt{\det(q)}, \mathcal{L}_{Nn} q^{\mu\nu}$. So far we have defined $\det(q)$ in the ADM frame only, its generalisation to an arbitrary frame is given by

$$\det((q_{\mu\nu})(X)) := \frac{1}{D!}[(\nabla_{\mu_0} t)(X)\epsilon^{\mu_0\cdots\mu_D}][(\nabla_{\nu_0} t)(X)\epsilon^{\nu_0\cdots\nu_D}]q_{\mu_1\nu_1}(X)\cdots q_{\mu_D\nu_D}(X)$$
$$(1.3.20)$$

as one can check by specialising to the ADM coordinates $X^\mu = (t, x^a)$. Here $\epsilon^{\mu_0\cdots\mu_D}$ is the metric-independent, totally skew Levi–Civita tensor pseudo-density of weight one. One can verify that with this definition we have $\det(g) = sN^2 \det(q)$ by simply expanding $g = q + sn \otimes n$. It is important to see that $\mathcal{L}_T \nabla_\mu t = \mathcal{L}_N \nabla_\mu t = 0$, from which then it follows immediately that

$$\mathcal{L}_{Nn}\sqrt{\det(q)} = \frac{1}{2}\sqrt{\det(q)}q^{\mu\nu}\mathcal{L}_{Nn} q_{\mu\nu} = N\sqrt{\det(q)}K \qquad (1.3.21)$$

where (1.1.6) has been used. Finally, using once more (1.3.17) we find indeed

$$\mathcal{L}_{Nn} q^{\mu\nu} = -q^{\mu\rho}q^{\nu\sigma}\mathcal{L}_{Nn} q_{\rho\sigma} = -2NK^{\mu\nu} \qquad (1.3.22)$$

• •• •••••• • •• •••• • ••• •• ••• • •• ••• • ••• • •• ••• •••

We are now in a position to compute the Lie derivative of $P^{\mu\nu} = -s\sqrt{\det(q)}[q^{\mu\rho}q^{\nu\sigma} - q^{\mu\nu}q^{\rho\sigma}]K_{\rho\sigma}$. Putting all six contributions carefully together and comparing with (1.3.12) one finds the non-trivial result

$$\{H(N), P^{\mu\nu}\} = \frac{q^{\mu\nu}NH}{2} - N\sqrt{\det(q)}[q^{\mu\rho}q^{\nu\sigma} - q^{\mu\nu}q^{\rho\sigma}]R_{\rho\sigma}^{(D+1)} + \mathcal{L}_{Nn}P^{\mu\nu}$$

$$(1.3.23)$$

that is, only on the constraint surface and only when the (vacuum) equations of motion hold, can the Hamiltonian flow of $P^{\mu\nu}$ with respect to $H(N)$ be interpreted as the action of a diffeomorphism in the direction perpendicular to Σ_t. Now, using again the definition of curvature as the commutator of covariant derivatives it is not difficult to check that

$$G_{\mu\nu}n^{\mu}n^{\nu} = \frac{sH}{2\sqrt{\det(q)}}$$

$$G_{\mu\nu}n^{\mu}q_{\rho}^{\nu} = -\frac{sH_{\rho}}{2\sqrt{\det(q)}} \qquad (1.3.24)$$

so that the constraint equations actually are equivalent to $D+1$ of the Einstein equations. Since (1.3.23) contains, besides H, all the spatial projections of $G_{\mu\nu}$ we see that our interpretation of $\{H(N), P^{\mu\nu}\}$ holds only on shell, $G_{\mu\nu} = 0$. This finishes our geometrical analysis of the Hamiltonian flow of the constraints.

1.4 Relation between the four-dimensional diffeomorphism group and the transformations generated by the constraints

The following issue has caused much confusion in the literature: the Einstein–Hilbert action is invariant under four-dimensional, passive diffeomorphisms, that is, the group $\mathrm{Diff}(M)$. This is the case whether or not the equations of motion hold. On the other hand, we have just seen that the gauge transformations generated by the constraints can be interpreted as infinitesimal diffeomorphisms only when the equations of motion hold. Thus, off-shell the two groups are genuinely different. This is already obvious from the form of the Dirac algebra $\mathfrak{D}$ (1.2.15) which, in contrast to the infinitesimal diffeomorphisms, is not a Lie algebra because, while $\{H(N_1), H(N_2)\}$ is of the form $\vec{H}(\vec{N}_3)$, the field $\vec{N}_3$ depends on the phase space, it is not a structure constant (independent of the phase space) but a so-called structure function. Bergmann and Komar [221–226] have studied in detail the canonical structure of constraints whose Lagrange multipliers depend on the phase space and the group (rather: enveloping algebra) they generate. We will refer to this 'group' as the Bergmann–Komar group $\mathrm{BK}(M)$.

The fact that $\mathrm{Diff}(M) \neq \mathrm{BK}(M)$ is often taken by critics as a manifestation that four-dimensional diffeomorphism invariance is • •••••• • ••• •••••• ••• •••••• in the canonical formulation. We will now explain why this interpretation is •• •••.

1. First of all, on solutions $\mathrm{Diff}(M) = \mathrm{BK}(M)$, therefore classically we would never see any difference. In quantum theory we will of course implement $\mathrm{BK}(M)$ as an off-shell gauge symmetry, however, in the semiclassical regime

when the theory becomes on-shell we will again recover $\mathrm{Diff}(M)$. Hence it is legitimate to disregard $\mathrm{Diff}(M)$ altogether as the fundamental symmetry.

2. Next, it is simply wrong that the canonical formulation does not admit a representation of $\mathrm{Diff}(M)$. After all, given a foliation X and an element $\varphi \in \mathrm{Diff}(M)$ we get a new foliation $\varphi \circ X$. Given a metric $g_{\mu\nu}$ one can work out explicitly how the spacetime quantities $n^\mu, N, N^\mu, q_{\mu\nu}, K_{\mu\nu}$ derived from the structures $X, g_{\mu\nu}$ change as we switch from X to $\varphi \circ X$ and not surprisingly it is precisely the expected action of $\mathrm{Diff}(M)$. It is just that this action of $\mathrm{Diff}(M)$ does not coincide with that of $\mathrm{BK}(M)$ unless we are on-shell. Hence we have an action of two different groups on our tensor fields, the kinematical group $\mathrm{Diff}(M)$ and the dynamical group $\mathrm{BK}(M)$ and GR is invariant under both with the peculiar feature that $\mathrm{Diff}(M) = \mathrm{BK}(M)$ when the equations of motion hold.

3. Let us explain the origin of the difference between $\mathrm{Diff}(M)$ and $\mathrm{BK}(M)$. Notice that $\mathrm{Diff}(M)$ is a •••••• •••••• symmetry of any diffeomorphism-invariant action. Here by kinematical we mean that the invariance group is insensitive to the form of the Lagrangian. For instance, the Einstein–Hilbert action and a higher-derivative action of the form

$$\int_M d^{D+1}X \ \sqrt{|\det(g)|} \ R^{\mu\nu\rho\sigma} \ R_{\mu\nu\rho\sigma}$$

both have the group $\mathrm{Diff}(M)$ as a kinematical symmetry group. But of course the equations of motion (Euler–Lagrange equations) they generate are completely different. The Einstein–Hilbert action leads to second-order partial differential equations while the above action leads to fourth-order partial differential equations. Even the number of degrees of freedom of the two theories is in general different because not only does one have to prescribe the D metric q_{ab} and its first-time derivative (velocity) as initial data, moreover one has to prescribe accelerations and higher-time derivatives. Now the equations of motion are obtained in the canonical formulation by calculating Poisson brackets with the constraints. Since the constraints therefore know about the •••••• •••, even though it is considered as an infinitesimal gauge transformation, the constraints must know about the •••••• • •••• •• ••• •••••••••••. Thus we see that the Bergmann–Komar group that the constraints generate is a •••••• •••• symmetry group. It is therefore not at all surprising that $\mathrm{Diff}(M) \neq \mathrm{BK}(M)$. In fact, from our discussion it is clear that there can be a relation only after taking the equations of motion into account and this is precisely what happens for the Einstein–Hilbert action.

Given these observations it is hard to believe that there is an exact relation between the canonical formulation and a fully $\mathrm{Diff}(M)$-invariant path integral of the form

$$Z = \int [Dg] \exp(iS)$$

where we integrate over four metric histories weighted by the exponential of the Einstein–Hilbert action. The symmetry of the path integral is $\text{Diff}(M)$ while that of the canonical theory would be $\text{BK}(M)$. One would expect a relation only in the semiclassical regime. We will come back to this point in the spin foam models of Chapter 14.

Related to this are the following issues:

I. Can one construct, in the canonical formulation, a phase space functional $V[u]$, where u is a vector field on M, which generates infinitesimal four-diffeomorphisms of M on the full phase space, that is, off-shell? It is easy to see that this is hard to achieve, if it can be done at all: since $V[u]$ would have to implement the Lie algebra of $\text{Diff}(M)$ we would need that $\{V[u], V[v]\} = V[\mathcal{L}_u v]$. Now suppose that $V[u]$ depends on temporal derivatives of u up to some finite order, say n. Then $V[\mathcal{L}_u v]$ depends on temporal derivatives of both u, v up to order $n + 1$ since $\mathcal{L}$ is a spacetime Lie derivative rather than a spatial one. On the other hand, $\{V[u], V[v]\}$ depends only on temporal derivatives of both u, v up to order n because the Poisson bracket does not generate new temporal derivatives. Hence we get a contradiction unless $n = \infty$. On the other hand, if we take the equations of motion into account then the Poisson bracket does create an additional time derivative if V involves the constraints. This explains, at a completely non-technical level, why the equations of motion have to enter the canonical representation of $\text{Diff}(M)$.

There is an apparent exception to this caveat: suppose we have $D + 1$ scalar fields ϕ^μ with conjugate momenta π_μ. Let u^μ be a vector field on M and consider the functional

$$D[u] := \int_\sigma d^D x \; u^\mu(X)_{X^\nu = \phi^\nu(x)} \; \pi_\mu(x)$$

Then it is easy to verify that $\{V(u), V(v)\} = V([u, v])$. The reason why we now have an honest representation is that we have identified spacetime coordinates with canonical fields. The catch is that $D[u]$ is not related at all to the contribution of $D + 1$ scalar fields to the spatial and Hamiltonian constraints derived in Chapter 12. Hence, while it is possible to obtain a canonical representation of $\text{Diff}(M)$ at least for certain types of matter, again this group has nothing to do with $\text{BK}(M)$ off-shell. See also the discussion in [164].

II. One often hears the statement that a spacetime diffeomorphism-invariant functional $F[g_{\mu\nu}]$ evaluated on a history, that is, a solution of the equations of motion, is a full Dirac observable, that is, it commutes with all the constraints. In order to make this precise, what we have to do in order to evaluate $F[g_{\mu\nu}]$ on solutions is to •••••• lapse and shift functions $N(x, t)$, $N^a(x, t)$ as well as a point in phase space $x \mapsto (q_{ab}(x), P^{ab}(x))$ and then construct a solution $g_{ab}(x, t; N, \vec{N}, q, P]$ where the square bracket is to denote functional dependence and g_{ab} is the pull-back to σ of $g_{\mu\nu}$ in some foliation.

The quantities $N, \vec{N}, g_{ab}$ are of course explicitly foliation-dependent. Since M is diffeomorphic to $\mathbb{R} \times \sigma$ by means of a foliation $X(t, x)$ and $F[g_{\mu\nu}]$ is spacetime diffeomorphism-invariant, we can equivalently think of $F[g_{\mu\nu}]$ as a functional $F[N, N^a, g_{ab}]$. Evaluating $F[g_{\mu\nu}]$ on a solution now means setting $F[N, N^a, q_{ab}, P^{ab}] := F[N, N^a, g_{ab}(N, \vec{N}, q, P)]$, which for fixed N, N^a is a concrete functional on phase space. Let us fix $N = N_0, N^a = N_0^a$, then the above statement amounts to the claim that $F[q, P] := F[N_0, \vec{N}_0, q, P]$ has (weakly) vanishing Poisson brackets with all the constraints $H(N), \vec{H}(\vec{N})$ $\bullet\bullet \ \bullet\bullet \ N, \vec{N}$. The argument that is usually brought forward in favour of this is that on-shell the constraints generate infinitesimal spacetime diffeomorphisms and since $F[g_{\mu\nu}]$ is spacetime diffeomorphism-invariant, the claim seems to follow. However, there is a catch: that $H(N), \vec{H}(\vec{N})$ generate spacetime diffeomorphisms on a solution $N_0, N_0^a, g_{ab}(N_0, N_0^a, q, P)$ requires $N = N_0, \ N^a = N_0^a$ as we have seen explicitly in Section 1.3. Therefore, the only way that the claim can hold is to show that $F[N_0, \vec{N}_0, q, P]$ is independent of $N_0, \vec{N}_0$. One might think that this is possible due to the foliation independence of $F[g]$, because changing the foliation (1) is equivalent to a spacetime diffeomorphism and (2) induces a change in $N_0, \vec{N}_0$. However, a change in foliation also induces a change in g_{ab}, which is just the pull-back to σ of $g_{\mu\nu}$. This, purely kinematical change of g_{ab}, is not the same as the full change of the solution $g_{ab}[N_0, \vec{N}_0, q, P]$ at fixed q, P which takes the dynamics into account. To illustrate these subtleties, consider the simplest example, a cosmological term

$$F[g_{\mu\nu}] := \int_M d^{D+1}X \ \sqrt{|\det(g_{\mu\nu}(X))|}$$

$$= \int_{\mathbb{R}} dt \ \int_\sigma d^3x \ N(x,t)\sqrt{\det(g_{ab}(x,t))} =: F[N, \vec{N}, g_{ab}]$$

Let us fix a static choice of lapse and shift $N(x,t) = N_0(x), \vec{N}(x,t) = \vec{N}_0(x)$. Then the Hamiltonian $\mathbf{H}(N_0, \vec{N}_0) = H(N_0) + \vec{H}(\vec{N}_0)$ is not explicitly time-dependent and we can explicitly write the solution as

$$g_{ab}(x, t; N_0, \vec{N}_0, q, P] = \sum_{n=0}^{\infty} \frac{t^n}{n!} \ \{\mathbf{H}(N_0, \vec{N}_0)[q, P], q_{ab}(x)\}_{(n)}$$

Inserting this into $F[N_0, \vec{N}_0, g_{ab}]$ and taking the functional derivative, say, with respect to N_0 one sees that the result is non-vanishing.

More details are contained in [227]. Hence, in order to construct Dirac observables, more sophisticated work is required which we will describe in the next chapter.

III. One of the reasons for why the Hamiltonian constraints can only generate diffeomorphisms along the timelike vector field Nn when the equations of motion hold and why the Dirac algebra involves structure functions is as follows. We have restricted attention to spacelike embeddings X with respect to a given metric g. However, since the kinematical group $\text{Diff}(M)$ only depends on M but not on g, for every spacelike embedding X we find an element $\varphi \in \text{Diff}(M)$ such that $\varphi \circ X$ is no longer spacelike with respect to g. Therefore, the kinematical group $\text{Diff}(M)$ is incompatible with the dynamical dependence of X on g. In order to make it compatible, $\text{Diff}(M)$ must be restricted to those diffeomorphisms[2] which preserve the spacelike nature of the embedding, that is, it must depend on g. This is why the Hamiltonian constraint as the canonical generator of these timelike diffeomorphisms can only close with structure functions depending on the spatial metric and why its algebra mirrors that of $\text{Diff}(M)$ only when the equations of motion hold. See [216, 217] for a deeper elaboration on this point.

Let us close this section by referring to work which clarifies these issues from various points of view. For more information on the relation between $\text{Diff}(M)$ and $\text{BK}(M)$ from the Hamiltonian point of view see [216, 217]. For an explanation from the Lagrangian point of view in terms of Noether currents see [228–232]. The relation between the Dirac algebra and the Lie algebra of spacetime diffeomorphisms and its geometrical interpretation as a hypersurface deformation algebra has been further elaborated on in [233], which exhibits beautifully that the Dirac algebra has a purely geometrical origin (subject to rather mild assumptions) which holds for any possible covariant matter coupling.

1.5 Boundary conditions, gauge transformations and symmetries

So far we have been rather careless about the boundary conditions on the fields for the case that σ is not compact without boundary. Hence strictly speaking all we have said so far is valid only when the Lagrange multipliers have compact support or are of rapid decrease. We will now generalise this to the important case of asymptotic flatness, which will allow us to derive a true Hamiltonian and hence solve the problem of time at least in that case.

1.5.1 Boundary conditions

A mathematically precise treatment would again lead us into the realm of Sobolev spaces, which would really go beyond the scope of this book. See, for example, [234, 235] for an account and [236] for a pedagogical application thereof to the proof of the positive gravitational energy theorem [237–240] which is strongly related to this section. See also Chapter 33 for a sketch of the infinite-dimensional symplectic geometry underlying gauge field theories.

We will content ourselves with the following simpler definition (for $D = 3$).

---

[2] This is not a group because iteration of diffeomorphisms may boost a spacelike surface more and more until it receives null or timelike portions. This is why $\text{BK}(M)$ is not a group.

Definition 1.5.1. • •••••••• • (M, g) •• •••• •• •• •••• ••• •••• ••• ••• •••••• ••

•• • •••• •• • ••• •••• •••••• B ••• ••• •••••• •• • ••• •••• ••••• •• $\mathbb{R}^4$ •••• ••••

$M - B = \cup_{n=1}^{N} E_n$ •• •••• ••• • ••• ••• •••••• •• • •• ••• •• • E_n • •••• •••• ••

••• ••• •••••• •• ••• ••• • ••• •••• ••• •• •••• •• $\mathbb{R}^4$•

•• •• •••• ••• E_n ••• • ••••• g •••••••••• ••• • •• •••• ••• •• •••••• • • •••••

η •• •••••• •• •••• •• ••••• •• •••••• •• • •••• •••• • •••••••• •••••••••• $(t, \vec{x})$

••• η •• • •••• •• •••• ••• •••••• ••• $\eta = $ •••• $(s, 1, 1, 1)$ ••• ••••• ••• ••••••

$r^2 = \vec{x} \cdot \vec{x}$ •• •••••• ••• •••• •• •• ••• •• •••••• •• ••• 3•• •• ••• •• •••••• ••

$r = const. \rightarrow \infty$ •••• •• ••• ••• •••••• •• $\mathbb{R} \times S^2$• •••• •• •••• ••• ••••

$$g_{\mu\nu}(x) = \eta_{\mu\nu} + \frac{f_{\mu\nu}\left(t, \frac{\vec{x}}{r}\right)}{r} + O(r^{-2}) \tag{1.5.1}$$

••• $r \rightarrow \infty$ •• •••• E_n • •••• $f_{\mu\nu}$ •• • •• •••• •••••• •• ••• •••• •• ••• ••••••
S^2•

The fall-off behaviour (1.5.1) is motivated by the Schwarzschild metric which, in the exterior region where the Cartesian coordinates are valid, takes the form (for $D = 3$)

$$ds^2 = -dt^2 \phi(r) + dr^2/\phi(r) + r^2(d\theta^2 + \sin^2(\theta)d\varphi^2)$$

where $\phi(r) = 1 + s\frac{2GM}{r}$. Every gravitating system approaches a Schwarzschild metric if one observes it from sufficient distance, which provides the physical motivation for definition (1.5.1).

Definition (1.5.1) does not make any reference to the fall-off behaviour of the extrinsic curvature, which is what we need in order to formulate the canonical action principle. In ADM coordinates we require

$$q_{ab}(x) = \delta_{ab} + \frac{f_{ab}\left(t, \frac{x}{r}\right)}{r}, \quad P^{ab}(x) = \frac{F^{ab}\left(t, \frac{x}{r}\right)}{r^2} \tag{1.5.2}$$

where f_{ab}, F^{ab} are again smooth tensor fields on the sphere at spatial infinity. We will now derive and refine these requirements.

The first condition in (1.5.2) of course follows directly from (1.5.1). Now further conditions are that the action be finite and that it is functionally differentiable. The integrand of the kinetic term $\int_\sigma d^3x P^{ab} \dot{q}_{ab}$ should therefore decay as $r^{3+\epsilon}$, $\epsilon > 0$ which would lead to $P = O(r^{2+\epsilon})$, however, this would make the ADM momentum vanish as we will see. Moreover, finiteness of the ADM momentum requires precisely the r^{-2} decay. Hence a possible way out is that the functions f_{ab}, F^{ab} be of •••• •••• •• ••••, that is,

$$f_{ab}\left(t, -\frac{x}{r}\right) = f_{ab}\left(t, \frac{x}{r}\right), \quad F^{ab}\left(t, -\frac{x}{r}\right) = -F^{ab}\left(t, \frac{x}{r}\right) \tag{1.5.3}$$

The logarithmically divergent radial integral in the kinematical term is then in fact cancelled because $d^3x = r^2 dr d\Omega_2$ is a measure, that is, even under parity. To see this we take the r integral up to finite $r \leq R$ and only then take the limit $R \to \infty$. As we will see one cannot switch the parity behaviour (1.5.3) because this would again make the ADM energy momentum vanish.

Given these boundary conditions let us consider the constraint functionals $\vec{H}(\vec{N})$, $H(N)$. As we have seen explicitly in the previous section, the Hamiltonian flow of the constraints generates spacetime diffeomorphisms on the phase space when the equations of motions are satisfied and when $N, \vec{N}$ are at least of rapid decrease. Infinitesimally the spacetime diffeomorphism is given by $\varphi_{\vec{N},N}(X) = X^\mu + Nn^\mu + N^\mu$, hence asymptotically $(t, \vec{x}) \mapsto (t + (N(t, \vec{x}), \vec{x} + \vec{N}(t, \vec{x}))$. In the asymptotically flat context it makes sense to try to allow more general decay behaviour of the smearing functions $N, \vec{N}$ corresponding to infinitesimal Poincaré transformations. Thus we would like to allow for the following 10-parameter set of functions

$$N(t, \vec{x}) = b_0 + \beta_b x^b + S(t, \vec{x}), \quad N^a(t, \vec{x}) = b^a + \omega_{ab} x^b + S^a(t, \vec{x}) \quad (1.5.4)$$

where $\omega_{ab} = \epsilon_{acb}\varphi^c$ and all indices are moved with the flat Euclidean spatial metric δ_{ab}. Here the constant parameters b^μ correspond to an infinitesimal translation, $\vec{\beta}$ to an infinitesimal boost and $\vec{\varphi}$ to an infinitesimal rotation of the asymptotic system of coordinates. The functions $S, \vec{S}$ will be specified below.

The constraint functions $H(N)$, $\vec{H}(\vec{N})$ diverge for the decay behaviour (1.5.4). They are also not functionally differentiable. The idea to cure both problems in one stroke is to add a boundary term to both functionals which on the one hand cancels the boundary term picked up in a variation and on the other hand makes the volume integral converge. We will now derive these boundary terms. We have

$$\delta \vec{H}(\vec{N}) = \int_\sigma d^3x[(\delta P^{ab})\,[\mathcal{L}_{\vec{N}}q_{ab}] - [\mathcal{L}_{\vec{N}}P^{ab}]\,(\delta q_{ab}) - 2\int_{\partial\sigma} dS_b N^a \delta P_a^b \quad (1.5.5)$$

where at spatial infinity $dS_a = \epsilon_{abc} dx^b \wedge dx^c = R^2 d\Omega r n_a(\Omega), R \to \infty$. Here $\Omega = (\theta, \varphi)$ denote angular coordinates, $d\Omega = \sin^2(\theta)d\theta d\varphi$ is the standard measure on S^2 and $n_a = x^a/r$ is the unit normal on S^2. The volume term in (1.5.1) is finite: since $\vec{N}$ is an asymptotic Killing field of δ_{ab} we have $\mathcal{L}_{\vec{N}}q_{ab} = O(r^{-2})$ odd or $O(r^{-1})$ even respectively for asymptotic translation or boost and rotation respectively while δP^{ab} is $O(r^{-2})$ odd, hence the first volume term is either $O(r^{-4})$ even or $O(r^{-3})$ odd. Likewise $\mathcal{L}_{\vec{N}}P^{ab} = O(r^{-3})$ even or $O(r^{-2})$ odd respectively while $\delta q_{ab} = O(r^{-1})$ even. Now the boundary term in (1.5.1) is exact, that is,

$$2\int_{\partial\sigma} dS_b N^a \delta P_a^b =: \kappa \delta \vec{P}(\vec{N}), \quad \vec{P}(\vec{N}) := \frac{2}{\kappa}\int_{\partial\sigma} dS_b N^a \delta P_a^b \quad (1.5.6)$$

We thus define an improved generator

$$\vec{J}(\vec{N}) := \kappa^{-1}\vec{H}(\vec{N}) + \vec{P}(\vec{N}) \quad (1.5.7)$$

whose variation reduces to the volume term in (1.5.5) and moreover by applying
Stokes' theorem backwards

$$\kappa \vec{J}(\vec{N}) = \int_\sigma d^3x \, P^{ab} \, \mathcal{L}_{\vec{N}} q_{ab} \qquad (1.5.8)$$

which is finite by an argument similar to the one just outlined. For an asymptotic
spatial translation we get

$$\vec{P}(\vec{N}) = b^a P_a^{ADM}, \quad P_a^{ADM} = \frac{2}{\kappa} \int_{\partial\sigma} dS_b \, P_a^b \qquad (1.5.9)$$

called the •• •   • •• ••••• . Notice that the parity and decay behaviour of P^{ab}
are precisely such that $\vec{P}^{ADM}$ is well-defined.

We will now repeat the procedure for $H(N)$. Schematically its integrand is
of the form $P^2 - \partial\Gamma + \Gamma^2$ where contractions with q_{ab} and powers of $\sqrt{\det(q)}$
were neglected since they are $O(1)$ even. Both P^2, Γ^2 are $O(r^{-4})$ even and hence
convergent for both even translations and odd boosts while $\partial\Gamma$ is only $O(r^{-3})$
even and hence divergent even for a translation. Next, as far as variations are
concerned, $\delta(P^2)$ contains terms of the form $P^2\delta q, P\delta P$ all of which are well-
defined while $\delta(\Gamma^2)$ contains terms of the form $\Gamma^2\delta q, \Gamma\delta\Gamma \propto \Gamma\partial\delta q$. The former is
well-defined since $\Gamma = O(r^{-2})$ odd while the latter gives rise to a boundary term
with integrand $\Gamma\partial q$ which vanishes identically. Finally the term $\delta\partial\Gamma$ gives rise to
terms of the form $(\partial\Gamma)\delta q, \partial\delta\Gamma$ of which only the latter is not well-defined. Thus
the term which makes $H(N)$ divergent is also the one that spoils its variation
and we may therefore concentrate on the boundary term of the variation $\delta C(N)$.
We just need to specialise (1.3.11) to $D = 3$ and get

$$-[\delta H(N)]_{|\text{ Bdry Term}} = \int_\sigma d^3x \sqrt{\det(q)}[-(D_cD^cN)q^{ab} + (D^aD^bN)]\delta q_{ab}$$

$$+ \int_{\partial\sigma} \sqrt{\det(q)} q^{cd} q^{ef} [(D_cN)(dS_d\delta q_{ef}) - (D_eN)(dS_c\delta q_{df})]$$

$$+ \int_{\partial\sigma} \sqrt{\det(q)} \, q^{cd} N [-dS_c\delta\Gamma_{ed}^e + dS_e\delta\Gamma_{cd}^e] \qquad (1.5.10)$$

The volume term is well-defined because DN vanishes for a translation and is
constant for a boost in leading order, hence D^2N is $O(r^{-2})$ odd for a boost and
$O(r^{-3})$ even for a translation while δq is $O(r^{-1})$ even so the integral converges.

We must now write the boundary term as a total differential. In the second
term proportional to $\delta\Gamma$ we can immediately pull the variation out of the surface
integral because the correction terms would be of the form $N\Gamma\delta q$ which is $O(r^{-3})$
odd for a translation and $O(r^{-2})$ even for a boost and thus would vanish in both
cases because dS is $O(r^2)$ odd. In the first boundary term we cannot immediately
pull out the variation from the surface integral because the correction terms
would be of the form $(\partial N)\delta q$ which is non-vanishing only for a boost and in that
case would be ill-defined since $O(r)$ and even which thus would give rise to an
expression of the form $0 \cdot \infty$. However we notice that $\delta q_{ab} = \delta(q_{ab} - \delta_{ab})$ and so

we may replace δq_{ab} by $\delta(q_{ab} - \delta_{ab})$ in the first surface term. After having done this we may pull the variation out of the surface integral because the correction terms are now of the form $(\partial N)(\delta q)(q - \delta)$ which is non-vanishing for a boost and in that case is $O(r^{-2})$ even and thus does not contribute. We thus define the improved generator

$$J(N) = \kappa^{-1}H(N) + E(N)$$

$$\kappa E(N) = \int_{\partial\sigma} \sqrt{\det(q)}q^{cd}q^{ef}[(D_cN)(dS_d[q_{ef} - \delta_{ef}]) - (D_eN)(dS_c[q_{df} - \delta_{df}])]$$

$$+ \int_{\partial\sigma} \sqrt{\det(q)} \; q^{cd}N\left[-dS_c\Gamma^e_{ed} + dS_e\Gamma^e_{cd}\right] \tag{1.5.11}$$

which now is functionally differentiable. Applying Stokes' theorem in reverse order we may write $E(N)$ in (1.5.11) as a volume integral and we should check whether its combination with the divergence-causing second-order derivative term $\partial\Gamma$ in $H(N)$ is finite. We have

$$\kappa^{-1}H(N)_{|2\text{ndord.}} + E(N) = \frac{1}{\kappa}\int_\sigma d^3x \left\{ \sqrt{\det(q)}q^{cd}(\partial_c\Gamma^e_{ed} - \partial_e\Gamma^e_{cd}) \right.$$

$$+ (\partial_d(\sqrt{\det(q)}q^{cd}q^{ef}(D_cN)[q_{ef} - \delta_{ef}])$$

$$- \partial_c(\sqrt{\det(q)}q^{cd}q^{ef}(D_eN)[q_{df} - \delta_{df}]))$$

$$+ \left(-\partial_c[\sqrt{\det(q)}\;q^{cd}N\Gamma^e_{ed}] + \partial_e[\sqrt{\det(q)}\;q^{cd}N\Gamma^e_{cd}]\right) \Big\}$$

$$= \frac{1}{\kappa}\int_\sigma d^3x \left\{ (\partial_d(\sqrt{\det(q)}q^{cd}q^{ef}(D_cN)[q_{ef} - \delta_{ef}]) \right.$$

$$- \partial_c(\sqrt{\det(q)}q^{cd}q^{ef}(D_eN)[q_{df} - \delta_{df}]))$$

$$+ \left(-\partial_c[\sqrt{\det(q)}\;q^{cd}N]\Gamma^e_{ed} + \partial_e[\sqrt{\det(q)}\;q^{cd}N]\Gamma^e_{cd}\right) \Big\}$$

$$= \frac{1}{\kappa}\int_\sigma d^3x \left\{ (\partial_d(\sqrt{\det(q)}q^{cd}q^{ef}(D_cN))[q_{ef} - \delta_{ef}] \right.$$

$$- \partial_c(\sqrt{\det(q)}q^{cd}q^{ef}(D_eN))[q_{df} - \delta_{df}])$$

$$+ (\sqrt{\det(q)}q^{cd}q^{ef}[(D_cN)\partial_d q_{ef} - (D_eN)\partial_c q_{df}])$$

$$+ N(-\partial_c[\sqrt{\det(q)}\;q^{cd}]\Gamma^e_{ed} + \partial_e[\sqrt{\det(q)}\;q^{cd}]\Gamma^e_{cd})$$

$$+ \sqrt{\det(q)}\;q^{cd}(-(D_cN)\Gamma^e_{ed} + (D_eN)\Gamma^e_{cd}) \Big\}$$

$$= \frac{1}{\kappa}\int_\sigma d^3x \left\{ (\partial_d(\sqrt{\det(q)}q^{cd}q^{ef}(D_cN))[q_{ef} - \delta_{ef}] \right.$$

$$- \partial_c(\sqrt{\det(q)}q^{cd}q^{ef}(D_eN))[q_{df} - \delta_{df}])$$

$$+ N(-\partial_c[\sqrt{\det(q)}\;q^{cd}]\Gamma^e_{ed} + \partial_e[\sqrt{\det(q)}\;q^{cd}]\Gamma^e_{cd})$$

$$+ \sqrt{\det(q)}\;q^{cd}((D_cN)[-\Gamma^e_{ed} + q^{ef}\partial_d q_{ef}]$$

$$+ (D_eN)[\Gamma^e_{cd} - q^{ef}\partial_c q_{df}]) \Big\} \tag{1.5.12}$$

where in the first step we cancelled the $\partial\Gamma$ terms, in the second step we separated out some of the derivative contributions of the first and second term of the first step and in the third step we combined the third and fourth term of the second step.

Consider first the third term in the third step of (1.5.12). We have up to terms of order $O(r^{-3})$ odd

$$q^{cd}\left[-\Gamma^e_{ed} + q^{ef}\partial_d q_{ef}\right] = \delta^{cd}\delta^{ef}\left(q_{ef,d} - \frac{1}{2}(q_{fe,d} + q_{fd,e} - q_{ed,f})\right) = \frac{1}{2}\delta^{cd}\delta^{ef}q_{ef,d}$$

$$q^{cd}\left[\Gamma^e_{cd} - q^{ef}\partial_c q_{df}\right] = \delta^{cd}\delta^{ef}\left(\frac{1}{2}(q_{fc,d} + q_{fd,c} - q_{cd,f}) - q_{df,c}\right)$$

$$= \frac{1}{2}\delta^{cd}\delta^{ef}(q_{fc,d} - q_{fd,c} - q_{cd,f}) = -\frac{1}{2}\delta^{cd}\delta^{ef}q_{cd,f} \quad (1.5.13)$$

Hence the whole third term in the third step of (1.5.13) becomes up to $O(r^{-3})$ odd terms

$$\frac{1}{2}\int_\sigma d^3x\ \delta^{cd}\delta^{ef}[q_{ef,d}\,N_{,c} - q_{cd,f}N_{,e}] = 0 \quad (1.5.14)$$

(relabel $(cd) \leftrightarrow (ef)$ in the second term). Therefore the third term in the last step of (1.5.12) is of the form $O(r^{-3})$ odd times DN which is either $O(r^{-3})$ odd for a boost or $O(r^{-4})$ even for a translation. The first two terms in the last step of (1.5.12) are already finite by inspection, hence we have shown that (1.5.11) is indeed finite and functionally differentiable even for a boost.

In case that $N = a^0$ generates a translation the boundary term is finite and becomes a^0 times the $\bullet\ \bullet\ \bullet \quad \bullet\bullet\bullet\bullet\bullet\bullet$

$$E^{ADM} = \frac{1}{\kappa}\int_{\partial\sigma}\sqrt{\det(q)}\ q^{cd}\left[-dS_c\Gamma^e_{ed} + dS_e\Gamma^e_{cd}\right]$$

$$= \frac{1}{2\kappa}\int_{\partial\sigma}\delta^{cd}\delta^{ef}[-dS_c(q_{fe,d} + q_{fd,e} - q_{ed,f}) + dS_e(q_{fc,d} + q_{fd,c} - q_{cd,f})]$$

$$= \frac{1}{2\kappa}\int_{\partial\sigma}dS_c\delta^{cd}\delta^{ef}[-(q_{fe,d} + q_{fd,e} - q_{ed,f}) + (q_{de,f} + q_{df,e} - q_{ef,d})]$$

$$= \frac{1}{\kappa}\int_{\partial\sigma}dS_c\delta^{cd}\delta^{ef}[q_{ed,f} - q_{ef,d}] \quad (1.5.15)$$

It is instructive to evaluate (1.5.15) for the Schwarzschild solution $N^2 = \phi_M(r)$, $N^a = 0$, $q_{ab} = \delta_{ab} + [\phi_M(r)^{-1} - 1]\frac{x^a x^b}{r^2}$, $\phi_M(r) = 1 - 2GM/r$ which will also ensure that $\kappa = 16\pi G$ has the correct normalisation. Noticing that $dS_a = r^2 n_a d\Omega$ with $n_a = x^a/r$, that $n_a\, n_{a,b} = 0$, that $\phi_M(r)^{-1} - 1 = 2GM/r + O(r^{-2})$ and that $\int_{S^2} d\Omega = 4\pi$ one checks that $E^{ADM} = M$, that is, the ADM energy equals the Schwarzschild mass when evaluated on the Schwarzschild solution.

1.5.2 Symmetries and gauge transformations

Let us summarise: we discovered that if the constraints are to generate asymptotic Poincaré transformations then we have to supplement them by boundary terms as otherwise their Hamiltonian vector fields are not well-defined. Hence the full generator is of the form $J(N) = \kappa^{-1}H(N) + B(N)$ where $B(N)$ is the boundary term. Notice that on the constraint surface we have $J(N) = B(N)$, however, $B(N)$ itself is neither finite (for boosts or rotations) nor is it functionally differentiable. Thus, in Poisson brackets one must always use $J(N)$ even when restricting to $H(N) = 0$ later on. The question now arises whether we should attribute to $J(N)$ the role of a gauge transformation generator or not. To answer this question, notice that Dirac's constraint analysis has unambiguously resulted in the functionals $H_a(x), H(x)$ as (secondary) constraints. The Hamiltonian $\mathbf{H} = C(N) + \vec{H}(\vec{N})$ is well-defined for asymptotically vanishing $N, \vec{N}$ and generates unphysical motions on the phase space because the functions N, N^a are unspecified. Hence there is no question that $H(N), \vec{H}(\vec{N})$ generate gauge transformations for asymptotically trivial N, N^a. The crucial point is now that for asymptotically non-trivial N, N^a the functionals $H(N), \vec{H}(\vec{N})$ are $\bullet\bullet\bullet\bullet\bullet\ \bullet\bullet\bullet\bullet$. Hence the well-defined functionals $J(N), \vec{J}(\vec{N})$ must be considered as $\bullet\bullet\ \bullet\bullet\bullet\bullet\bullet$ $\bullet\bullet\bullet\bullet\ \bullet\bullet\ \bullet\bullet\bullet\bullet\bullet\ \bullet\bullet$ on phase space for asymptotically non-trivial N, N^a. Indeed they are different from $H(N), \vec{H}(\vec{N})$ since they do not vanish on the constraint surface for asymptotically non-vanishing N, N^a. In other words, since in functional analysis the smearing fields N, N^a serve as labels for phase space functions we should distinguish $J(N)$ for different N's. For asymptotically trivial N the functional $J(N)$ equals $H(N)$ but not otherwise. Dirac's analysis only forces us to interpret the $J(N)$ for asymptotically trivial N as generators of gauge transformations. Hence the issue is open for asymptotically non-trivial N.

To settle the question we ask whether the $J(N)$ for asymptotically non-trivial N transform between physically distinct solutions of the field equations, that is, those that correspond to distinct physical observations. The key to the answer lies in working out the algebra of the functionals $J(N), \vec{J}(\vec{N})$. So far we have not specified the functions S, S^a in (1.5.4). These correspond to the so-called $\bullet\bullet\bullet\bullet\bullet\bullet\bullet\ \bullet\bullet\bullet\bullet\bullet\bullet$. They are odd $O(1)$ functions on the asymptotic S^2 and it is clear that $\vec{B}(\vec{S}) = B(S) = 0$ because the integrand, modulo the smearing functions, with respect to the measure $d\Omega$ on S^2 is an even function. Thus while $S, \vec{S}$ are not asymptotically trivial, they still generate gauge transformations because $J(S) = H(S), \vec{J}(\vec{S}) = \vec{H}(\vec{S})$ and hence they vanish on the constraint surface. We absorb all higher orders of r^{-1} into $S, \vec{S}$ as well. This supertranslation ambiguity has been analysed in great detail in [241, 242] using Penrose's powerful conformal techniques. Including the supertranslations we arrive at the most general decay behaviour of N, N^a which still allows for well-defined and differentiable $J(N), \vec{J}(\vec{N})$. In addition we require that $\mathcal{L}_{\vec{S}} q_{ab}$ is not only $O(r^{-1})$ but actually $O(r^{-2})$ so that the dominant part of $\mathcal{L}_{\vec{N}} q_{ab}$ comes from the rotation $\vec{N}^0 = \vec{N} - \vec{S}$.

Let us now work out the algebra for the functionals or 'currents' $J(N, \vec{N}) = J(N) + \vec{J}(\vec{N})$. By definition of the Poisson bracket

$$\{J(N_1, \vec{N}_1), J(N_2, \vec{N}_2)\} = \int_\sigma d^3x [\{J(N_1, \vec{N}_1), q_{ab}(x)\} \{P^{ab}(x), J(N_2, \vec{N}_2)\}$$
$$- \{J(N_2, \vec{N}_2), q_{ab}(x)\} \{P^{ab}(x), J(N_1, \vec{N}_1)\}] \quad (1.5.16)$$

Luckily, all the appearing Poisson brackets have been worked out already. Indeed, we computed so far those between $H(N, \vec{N}) = H(N) + \vec{H}(\vec{N})$ and $q_{ab}(x)$, $P^{ab}(x)$ for asymptotically trivial N, N^a. Now the boundary term $B(N, \vec{N}) = B(N) + \vec{B}(\vec{N})$ by construction is such that the functional derivatives of $J(N, \vec{N})$ coincide with those of $H(N, \vec{N})$ even when extending N, $\vec{N}$ non-trivially to spatial infinity. In other words, we may use formulae (1.3.3), (1.3.4), (1.3.6) and (1.3.12) with H replaced by J. We split the task into the three different kinds of brackets that appear in (1.5.16).

1. $\{\vec{J}(\vec{N_1}), \vec{J}(\vec{N_2})\} = \kappa^{-1} \int_\sigma d^3x \left[-(\mathcal{L}_{\vec{N_1}} q_{ab})(\mathcal{L}_{\vec{N_2}} P^{ab})(\mathcal{L}_{\vec{N_2}} q_{ab})(\mathcal{L}_{\vec{N_1}} P^{ab}) \right]$

$= \kappa^{-1} \int_\sigma d^3x \, P^{ab} \left[\mathcal{L}_{\vec{N_2}}(\mathcal{L}_{\vec{N_1}} q_{ab})) - \mathcal{L}_{\vec{N_1}}(\mathcal{L}_{\vec{N_2}} q_{ab})) \right.$

$\left. + D_c \left(P^{ab} \left(-N_2^c(\mathcal{L}_{\vec{N_1}} q_{ab}) + N_1^c(\mathcal{L}_{\vec{N_2}} q_{ab}) \right) \right) \right]$

$= \kappa^{-1} \int_\sigma d^3x \, P^{ab}(\mathcal{L}_{[\vec{N_2}, \vec{N_1}]} q_{ab}) = \vec{J}(\mathcal{L}_{\vec{N_2}} \vec{N_1}). \qquad (1.5.17)$

where we have used $[\mathcal{L}_u, \mathcal{L}_v] = \mathcal{L}_{[u,v]}$ and the boundary term vanishes because both $\vec{N_1}$, $\vec{N_2}$ are at best asymptotic symmetries.

2. $\{\vec{J}(\vec{N_1}), J(N_2)\}$

$= \kappa^{-1} \int_\sigma d^3x \left\{ -(\mathcal{L}_{\vec{N_1}} q_{ab}) \left(\frac{2sN_2}{\sqrt{\det(q)}} [P^{ac} \Gamma_c^b - P^{ab} \Gamma/2] \right. \right.$

$+ \frac{q^{ab} N_2 H}{2} + N_2 \sqrt{\det(q)}(q^{ab} R - R^{ab})$

$\left. + \sqrt{\det(q)}[-(D_c D^c N_2)q^{ab} + (D^a D^b N_2)] \right)$

$\left. + \left(-2sN_2 \frac{P_{ab} - P q_{ab}/2}{\sqrt{\det(q)}} \right) (\mathcal{L}_{\vec{N_1}} P^{ab}) \right\}$

$= \kappa^{-1} \int_\sigma d^3x \left\{ N_2 \left\{ \left(\frac{\partial}{\partial q_{ab}} \left[-\frac{s}{\sqrt{\det(q)}} (P^{cd} P_{cd} - P^2/2) \right] \right) (\mathcal{L}_{\vec{N_1}} q_{ab}) \right. \right.$

$\left. + \left(\frac{\partial}{\partial P^{ab}} \left[-\frac{s}{\sqrt{\det(q)}} (P^{cd} P_{cd} - P^2/2) \right] \right) (\mathcal{L}_{\vec{N_1}} P^{ab}) \right\}$

$\left. - \sqrt{\det(q)} \left[N_2 \left(\frac{1}{2} q^{ab} R - R^{ab} \right) - q^{ab} D_c D^c N_2 + D^a D^b N_2 \right] (\mathcal{L}_{\vec{N_1}} q_{ab}) \right\}$

$= \kappa^{-1} \int_\sigma d^3x \left\{ N_2 \left(\mathcal{L}_{\vec{N_1}} \left[-\frac{s}{\sqrt{\det(q)}} (P^{cd} P_{cd} - P^2/2) \right] \right) \right\}$

$- N_2 R_{ab}(\mathcal{L}_{\vec{N_1}}(\sqrt{\det(q)} q^{ab})) + \sqrt{\det(q)}[q^{ab} D_c D^c N_2 - D^a D^b N_2] (\mathcal{L}_{\vec{N_1}} q_{ab})$

$= \kappa^{-1} \int_\sigma d^3x \{ N_2(\mathcal{L}_{\vec{N_1}} H) + N_2 \sqrt{\det(q)} q^{ab} (\mathcal{L}_{\vec{N_1}} R_{ab}) + \sqrt{\det(q)}[q^{ab} D_c D^c N_2$

$- D^a D^b N_2] (\mathcal{L}_{\vec{N_1}} q_{ab}) \} \qquad (1.5.18)$

Now by the very definition of the Lie derivative we have

$$\mathcal{L}_{\vec{N}_1} R_{ab} = (\delta R_{ab})_{\delta q = \mathcal{L}_{\vec{N}_1} q}$$

and thus we may use (1.3.11) in order to rewrite the last line of (1.5.18) as

$$-\int_\sigma d^3x \, N_2 \sqrt{\det(q)} q^{ab} \left(\mathcal{L}_{\vec{N}_1} R_{ab}\right)$$

$$= \left\{ -\int_\sigma d^3x (-q^{ab} D_c D^c N_2 + D^a D^b N_2) \, \delta q_{ab} - \kappa \delta E(N_2) \right\}_{\delta q = \mathcal{L}_{\vec{N}_1} q} \tag{1.5.19}$$

where we have observed in the second line that the variation in the first line is precisely the one that gives rise to the boundary term picked up in $\delta H(N_2)$ and that boundary term is the negative of (1.5.11) by definition. Inserting (1.5.19) into (1.5.18) now gives

$$\{\vec{J}(\vec{N}_1), J(N_2)\} = \kappa^{-1} \left\{ -H(\mathcal{L}_{\vec{N}_1} N_2) + \int_{\partial\sigma} dS_a \, N_1^a \, N_2 H + [\delta E(N_2)]_{\delta q = \mathcal{L}_{\vec{N}_1} q} \right\}$$

$$\tag{1.5.20}$$

where we have performed an integration by parts in order to write the first term as a smeared Hamiltonian constraint.

It remains to check that the two additional boundary terms combine to $-E(\mathcal{L}_{\vec{N}_1} N_2)$. The easiest way to do this is to realise that upon defining the quantity $Q_{cd} := q_{cd} - \delta_{cd}$ we have

$$\kappa E(N) = \int_{\partial\sigma} dS_c [N_{,c} Q_{dd} - N_{,d} Q_{cd} + N(Q_{cd,d} - Q_{dd,c})]$$

$$\int_{\partial\sigma} dS_a \, N_1^a \, N_2 H = \int_{\partial\sigma} dS_a \, N_1^a \, N_2 (Q_{dd,cc} - Q_{cd,cd}) \tag{1.5.21}$$

where all indices are raised and lowered with δ_{ab}. To simplify the calculation we notice that the supertranslation part in (1.5.4) drops out in both the variation of the first term and the second term. To see this, notice that $\partial^2 Q$ is $O(r^{-3})$ even while $S^a{}_b x^b, S\omega_{ab} x^b$ are both $O(r)$ even, thus their combination drops out of the second integral in (1.5.20). Now when splitting $\vec{N} = \vec{N}^0 + \vec{S}$ we have

$$\delta Q = \delta q = \mathcal{L}_{\vec{N}_1} q = \mathcal{L}_{\vec{N}^0{}_1} q + \mathcal{L}_{\vec{S}} q = \mathcal{L}_{\vec{N}^0{}_1} Q + \mathcal{L}_{\vec{S}} q$$

where in the last step we have used that $\mathcal{L}_{\vec{N}^0{}_1} \delta_{ab} = 0$. Since by definition of the supertranslations $\mathcal{L}_{\vec{S}} q$ is $O(r^{-2})$ even we have that both $(\partial N) \mathcal{L}_{\vec{S}} q, N \partial(\mathcal{L}_{\vec{S}} q)$ are at most $O(r^{-2})$ even, which drops again out of the surface integral. Notice that for similar reasons $E(N) = E(N^0)$, $\vec{P}(\vec{N}) = \vec{P}(\vec{N}^0)$ is independent of $S, \vec{S}$ as we noticed before.

We may therefore replace $N_2, \vec{N}_1$ by $N_2^0 = b^0 + \beta_b x^b$, $(\vec{N}_1^0)^a = b^a + \omega_{ab} x^b$ in the boundary terms of (1.5.20). It follows (dropping the superscript '0')

$$\kappa \delta_{\vec{N}_1} E(N_2) = \int_{\partial \sigma} dS_c \{[\beta_c N^a Q_{dd,a} - \beta_d (N^a Q_{cd,a} + 2Q_{a(c}\omega^a{}_{d)})]$$
$$+ N_2[\omega_{ad}(Q_{cd,a} + Q_{ca,d}) + \omega_{ac}(Q_{da,d} + Q_{dd,a})$$
$$+ N^a(Q_{cd,ad} - Q_{dd,ac})]\} \tag{1.5.22}$$

Since $E(N)$ does not have $\partial^2 Q$ terms we will first manipulate the $\partial^2 Q$ terms of $\int dS_a N_1^a N_2 H + \kappa \delta_{\vec{N}_1} E(N_2)$ which is given by

$$- \int N_1^a N_2 \{[([Q_{dd,c}),_a dS_c - ([Q_{dd,c}),_c dS_a] - [([Q_{cd,d}),_a dS_c - ([Q_{cd,d}),_c dS_a]\}$$
$$= - \int N_1^a N_2 \epsilon_{acb}[([Q_{dd,c}),_e - ([Q_{cd,d}),_e]\epsilon^{bef} dS_f$$
$$= - \int N_1^a N_2 \epsilon_{acb}[Q_{dd,ce} - Q_{cd,de}]dx^b \wedge dx^e$$
$$= \int N_1^a N_2 \epsilon_{acb} d([Q_{dd,c} - Q_{cd,d}]dx^b)$$
$$= - \int \epsilon_{acb} d(N_1^a N_2) \wedge dx^b [Q_{dd,c} - Q_{cd,d}]$$
$$= - \int \epsilon_{acb}(N_1^a N_2)_{,e} dx^e \wedge dx^b [Q_{dd,c} - Q_{cd,d}]$$
$$= - \int dS_f \epsilon_{acb} \epsilon^{ebf}(N_1^a N_2)_{,e}[Q_{dd,c} - Q_{cd,d}]$$
$$= \int [Q_{dd,c} - Q_{cd,d}] [(N_1^a N_2)_{,a} dS_c - (N_1^a N_2)_{,c} dS_a]$$
$$= \int dS_c \{(N_1^a N_2)_{,a}[Q_{dd,c} - Q_{cd,d}] - (N_1^c N_2)_{,a}[Q_{dd,a} - Q_{ad,d}]\} \tag{1.5.23}$$

Here we applied Stokes' theorem in the fourth step (remember that we keep $r = R$ finite and then take the limit $R \to \infty$ so that we may apply Stokes' theorem) exploiting that $\partial^2 \sigma = \emptyset$ and made frequent use of the identities $dS_a = \epsilon_{abc} dx^b \wedge dx^c/2$, $dx^a \wedge dx^b = \epsilon^{abc} dS_c$.

Reinserting (1.5.23) into (1.5.22) we see that all the terms without derivatives of N_2 cancel each other and we are left with

$$\kappa^{-1} \int dS_c \{(\mathcal{L}_{\vec{N}_1} N^2)[Q_{dd,c} - Q_{cd,d}] + \beta_d[Q_{ac}\omega_{ad} + Q_{ad}\omega_{ac}]$$
$$+ Q_{dd,a}[N_a\beta_c - N_c\beta_a] + \beta_a[N_c Q_{da,d} - N_d Q_{ca,d}]\} \tag{1.5.24}$$

This expression contains four square brackets of which the first is already of the required form and contain the terms proportional to ∂Q. Hence we must manipulate the remaining terms such that no derivatives of Q appear any more. This is already the case for the second square bracket. The third square bracket can be written as

$$\int dS_c Q_{dd,a}[N_a\beta_c - N_c\beta_a] = \int dS_c(\omega_{ca}\beta_a)Q_{dd} \qquad (1.5.25)$$

where again Stokes' theorem was applied using manipulations similar to those in (1.5.23). By the same token

$$\int dS_c\beta_a[N_cQ_{da,d} - N_dQ_{ca,d}] = -\int dS_c\beta_a\omega_{cd}Q_{da} \qquad (1.5.26)$$

Inserting (1.5.25) and (1.5.26) back into (1.5.24) and noticing that $(\mathcal{L}_{\vec{N}_1}N_2)_{,a} = \omega_{ba}\beta_b$ all the terms indeed combine to $-E(\mathcal{L}_{\vec{N}_1}N_2)$. We summarise

$$\{\vec{J}(\vec{N}_1), J(N_2)\} = -J(\mathcal{L}_{\vec{N}_1}N_2) \qquad (1.5.27)$$

3.

$$\{J(N_1), J(N_2)\}$$

$$= \kappa^{-1}\int d^3x \left\{ \left[2sN_1\frac{P_{ab} - Pq_{ab}/2}{\sqrt{\det(q)}} \right] \left[N_2 \left\{ \frac{2s}{\sqrt{\det(q)}}(P^{ac}P_c^b - q^{ab}P/2) \right. \right. \right.$$

$$+ q^{ab}H/2 + \sqrt{\det(q)}(q^{ab}R - R^{ab}) \right\}$$

$$\left. + \sqrt{\det(q)}(-q^{ab}D_cD^cN_2 + D^aD^bN_2) \right] - (N_1 \leftrightarrow N_2) \bigg\}$$

$$= 2s\kappa^{-1}\int d^3x \left\{ N_1\frac{P_{ab} - Pq_{ab}}{\sqrt{\det(q)}}\sqrt{\det(q)}[-q^{ab}D_cD^cN_2 + D^aD^bN_2] - (N_1 \leftrightarrow N_2) \right\}$$

$$= 2s\kappa^{-1}\int d^3x \left\{ N_1\frac{P_{ab} - Pq_{ab}}{\sqrt{\det(q)}}\sqrt{\det(q)}[-q^{ab}D_cD^cN_2 + D^aD^bN_2] - (N_1 \leftrightarrow N_2) \right\}$$

$$= 2s\kappa^{-1}\int d^3x \left\{ N_1P^{ab}D_aD_bN_2 - N_2P^{ab}D_aD_bN_1 \right\}$$

$$= 2s\kappa^{-1}\int d^3x \left\{ N_1\left[(D_aP^{ab}D_bN_2) + \frac{1}{2}H_aD^aN_2\right] - (N_1 \leftrightarrow N_2) \right\}$$

$$= 2s\kappa^1\int d^3x \left\{ [\partial_a(N_1P^{ab}D_bN_2) - (D_aN_1)P^{ab}(D_bN_2) \right.$$

$$\left. + \frac{1}{2}N_1H_aD^aN_2] - (N_1 \leftrightarrow N_2) \right\}$$

$$= s\kappa^{-1}\int d^3x\, H_a[N_1D^aN_2 - N_2D^aN_1]$$

$$+ 2s\kappa\int dS_a; P^{ab}[N_1D_bN_2 - N_2D_bN_1] \qquad (1.5.28)$$

Defining

$$N_{12}^a(q) := q^{ab}[N_1\partial_bN_2 - N_2\partial_bN_1] \qquad (1.5.29)$$

we may write (1.5.28) as

$$\{J(N_1), J(N_2)\} = s\kappa^{-1}\left[\int d^3x \left(-2D_a(P_b^a)\right)N_{12}^b(q) + 2\int dS_a; P_b^a N_{12}^b(q)\right]$$

$$= s\kappa^{-1}\int d^3x\, P^{ab}[D_a N_{b12}(q) + D_b N_{a12}(q)]$$

$$= s\kappa^{-1}\int d^3x\, P^{ab}\left(\mathcal{L}_{\vec{N}_{12}(q)}q\right)_{ab}$$

$$= sJ(\vec{N}_{12}(q)) \tag{1.5.30}$$

We may summarise our findings in the compact expression

$$\{J(N_1, \vec{N}_1), J(N_2, \vec{N}_2)\} = J(\mathcal{L}_{\vec{N}_2}N_1 - \mathcal{L}_{\vec{N}_1}N_2, \mathcal{L}_{\vec{N}_2}\vec{N}_1 + s\vec{N}_{12}(q))$$

$$=: J(N_3, \vec{N}_3) \tag{1.5.31}$$

From (1.5.31) we may read off the following properties:

(i) If both $(N_1, \vec{N}_1)$ and $(N_2, \vec{N}_2)$ are supertranslations then $(N_3, \vec{N}_3)$ is again a supertranslation. Since we have identified the supertranslation generators as gauge transformation generators already, this is the statement that the supertranslation gauge algebra closes.

(ii) If one of $(N_1, \vec{N}_1)$ and $(N_2, \vec{N}_2)$ is a supertranslation while the other contains an asymptotically non trivial piece then $(N_3, \vec{N}_3)$ is still a supertranslation. This is the statement that a gauge transformation transforms an asymptotically non-trivial generator into a gauge generator which vanishes on the constraint surface.

(iii) If both $(N_1, \vec{N}_1)$ and $(N_2, \vec{N}_2)$ contain an asymptotically non-trivial piece then so does $(N_3, \vec{N}_3)$. On the constraint surface the dependence of $J(N_3, \vec{N}_3)$ reduces to the surface term where all supertranslation dependence drops out and $\vec{N}_{12}(q)$ may be evaluated at $q_{ab} = \delta_{ab}$. Let us write

$$J(N_I, \vec{N}_I) =: b_I^0 E + \beta_I^a B_a + b_I^a P_a + \varphi_I^a J_a \tag{1.5.32}$$

Then on the one hand we read off

$$b_3^0 = \vec{b}_2 \cdot \vec{\beta}_1 - \vec{b}_1 \cdot \vec{\beta}_2$$
$$\vec{\beta}_3 = \vec{\beta}_1 \times \vec{\beta}_2 - \vec{\beta}_2 \times \vec{\beta}_1$$
$$\vec{b}_3 = \vec{\varphi}_1 \times \vec{b}_2 - \vec{\varphi}_2 \times \vec{b}_1 + s(b_1^0\vec{\beta}_2 - b_2^0\vec{\beta}_1)$$
$$\vec{\varphi}_3 = \vec{\varphi}_1 \times \vec{\varphi}_2 + s\beta_1 \times \beta_2 \tag{1.5.33}$$

and on the other we may simply calculate the Poisson brackets among E, P_a, B_a, J_a. Let us introduce the notation

$$P^0 := E,\ M^{0a} := -M^{0a} := B^a,\ M^{ab} := \epsilon^{acb}J_c,\ \omega_{0a} := \beta_a,\ \omega_{ab} := \epsilon_{acb}\varphi^c \tag{1.5.34}$$

then

$$J(N, \vec{N}) = b_\mu P^\mu + \frac{1}{2}\omega_{\mu\nu} M^{\mu\nu} \tag{1.5.35}$$

and we arrive at the compact expression

$$\{J(N_1, \vec{N}_1), J(N_2, \vec{N}_2)\} = b^1_\mu b^2_\rho \{P^\mu, P^\rho\} + \frac{1}{2} b^1_\mu \omega^2_{\rho\sigma} \{P^\mu, M^{\rho\sigma}\}$$
$$+ \frac{1}{2}\omega^1_{\mu\nu} b^2_\rho \{M^{\mu\nu}, P^\rho\} + \frac{1}{4}\omega^1_{\mu\nu}\omega^2_{\rho\sigma} \{M^{\mu\nu}, M^{\rho\sigma}\} \tag{1.5.36}$$

and by comparing coefficients we conclude that the $P^\mu, M^{\mu\nu}$ satisfy the Euclidean or Lorentzian • • •• •• •• • •• ••••

$$\{P^\mu, P^\rho\} = 0$$
$$\{M^{\mu\nu}, P^\rho\} = 2\eta^{\rho[\mu} P^{\nu]}$$
$$\{M^{\mu\nu}, M^{\rho\sigma}\} = 2s\left(\eta^{\rho[\mu} M^{\nu]\sigma} - \eta^{\sigma[\mu} M^{\nu]\rho}\right) \tag{1.5.37}$$

Interpretation: We have seen that P^0 defines the mass for the Schwarzschild solution and thus measures gravitational energy at spatial infinity. This energy of course depends on the observer at spatial infinity and must transform non-trivially under a boost. This boost is an observable, that is, measurable transformation. This is precisely accommodated in (iii) as we just saw. It follows that, for example, the boost generator B^a **must not be considered as a generator of a gauge (unobservable) transformation.** Similar arguments show that all the ten Poincaré generators P^μ, $M^{\mu\nu}$ must be considered as observable quantities and hence they define ••• •••• which generate ••• • ••••••. They are called symmetries because they transform solutions to the equations of motions into, possibly different, solutions of the equations of motion which are themselves determined up to a gauge transformation. This follows from the fact that, as we have seen in the previous section, on-shell the transformations generated by $J(N, \vec{N})$ are simply diffeomorphisms and the equations of motion are covariant under diffeomorphisms. For example, in vacuum the Ricci flatness condition is unaffected.

This interpretation fits quite nicely with (ii) because on the constraint surface the gauge generators have vanishing Poisson brackets with these charges whence they define ••• ••••••• ••••••• •• • •••• • •••• •••••••••••. In fact, up to now these are the only Dirac observables for General Relativity which are known to be globally defined on the phase space (see below for weak Dirac observables which presumably are only locally defined). They exist only in the asymptotically flat case and not in the compact case without boundary. Generically, boundaries lead to conserved charges because the then necessary boundary conditions impose certain restrictions on the allowed gauge transformations which gives birth to physical degrees of freedom which would otherwise be considered as gauge. We will see a concrete realisation of this effect in the isolated horizon framework for quantum black holes in Chapter 15.

Summarising, we have derived that $J(N, \vec{N})$ is a gauge generator if $(N, \vec{N})$ is a supertranslation but otherwise a symmetry generator of the equations of motion which in particular is a weak Dirac observable. By (i), the gauge generators close among themselves, the constraint system is first class. We remark that the algebra (1.5.31) and its interpretation holds irrespective of which matter we couple to gravity, which means that it has a purely geometric origin. This origin, the geometry of hypersurface deformations, has been beautifully worked out in [233].

In summary, General Relativity can be cast into Hamiltonian form, however, its equations of motion are complicated non-linear partial differential equations of second order and very difficult to solve. Nevertheless, the Cauchy problem is well-posed and the classical theory is consistent up to the point where singularities (e.g., black holes) appear [207,208]. This is one instance where it is expected that the classical theory is unable to describe the system appropriately any longer and that the more exact theory of quantum gravity must take over in order to remove the singularity. This is expected to be quite in analogy to the case of the hydrogen atom whose stability was a miracle to classical electrodynamics but was easily explained by quantum physics. Of course, the quantum theory of gravity is expected to be even harder to handle mathematically than the classical theory, however, as a zeroth step an existence proof would already be a triumph. Notice that up to date a similar existence proof for, say, QCD is lacking as well [99]. Before we dive into the quantum theory we further develop the physical interpretation of the formalism in the next chapter.

2

The problem of time, locality and
the interpretation of quantum mechanics

In this chapter we are going to address the famous 'problem of time' which has become the headline for all the physical interpretational problems of the mathematical formalism. Roughly speaking the problem of time is that there is none in GR: at least in the spatially compact case without boundaries the Hamiltonian vanishes on the physical, constraint surface. This is physically relevant because we seem to live in a universe with precisely that spatial topology. Since the Hamiltonian generates time translations in any canonical theory we arrive at the conclusion that 'nothing moves' in GR, which is in obvious contradiction to experiment. Since there is no time also the usual interpretation of quantum mechanical measurements at given moments of time breaks down. One can fill books about this issue and we will not even try to cover a substantial amount of the existing literature. A superb source of information on these conceptual problems is Carlo Rovelli's book [3]. Rather, what we will do in what follows is to collect various proposals for solutions to the problem of time taken from other authors, especially Rovelli's relational approach to classical and quantum physics and Hartle •• ••.'s consistent history interpretation, and combine them into a consistent picture. We do not want to suggest that the resulting picture is to be accepted, rather we want to draw attention to the problems involved and to develop a working hypothesis. The discussion on the interpretation of quantum mechanics is very alive and some authors such as Penrose [243] not only propose to alter the interpretational aspect of quantum mechanics but also the mathematical framework.

On the other hand, if one accepts this proposal, then we want to stress that •• •• •• •• ••• ••• •• •••••• •• ••• • ••• •• ••••• ••• •••••• • ••• •• ••• •••••• •••• •• •••. There remain technical challenges, which much of this book is about, and many of them have already been addressed as we will see but conceptually one knows precisely what to do. This is one of the strengths of the canonical approach to quantum gravity, namely that there is a precise programme which one has to implement technically for the case of GR and which we will present in the next chapter.

Notice that in the asymptotically flat case discussed in the previous chapter the problem of time does not arise, there the ADM Hamiltonian generates classical and quantum time evolution and even the Copenhagen interpretation of quantum mechanics is applicable where the system is the isolated gravitational system interacting with matter in the bulk while the external measurement apparatus is located at the spatial infinity boundary. Hence in what follows we will discuss mostly the case that there is no Hamiltonian but rather an infinite number of Hamiltonian constraints, which is especially relevant for (quantum) cosmology.

2.1 The classical problem of time: Dirac observables

Let us summarise the structure at which we have arrived so far. The Hamiltonian of GR is not a true Hamiltonian but a linear combination of constraints. Rather than generating time translations it generates spacetime diffeomorphisms at least on-shell and specific canonical transformations otherwise. Since the parameters of these canonical transformations N, N^a are completely arbitrary unspecified functions, the corresponding motions on the phase space have to be interpreted as ••••• •••••••• •••••. This is quite similar to the gauge motions generated by the Gauß constraint in Maxwell theory [219]. The basic variables of the theory q_{ab}, P^{ab} are not observables of the theory because they are not gauge-invariant. Let us count the number of kinematical and dynamical (true) degrees of freedom: the basic variables are both symmetric tensors of rank two and thus have $D(D+1)/2$ independent components per spatial point. There are $D+1$ independent constraints so that $D+1$ of these phase space variables can be eliminated. $D+1$ of the remaining degrees of freedom can be gauged away by a gauge transformation leaving us with $D(D+1) - 2(D+1) = (D-2)(D+1)$ phase space degrees of freedom or $(D-2)(D+1)/2$ configuration space degrees of freedom per spatial point. For $D=3$ we thus recover the two graviton degrees of freedom.

The further classical analysis of this system could now proceed as follows:

(i) One determines a complete set of gauge-invariant observables on the constraint surface $\overline{\mathcal{M}}$ and computes the induced symplectic structure $\overline{\Omega}$ on the so-reduced symplectic manifold $\hat{\mathcal{M}}$. Equivalently, one obtains the full set of solutions to the equations of motion, the set of Cauchy data are then the searched-for observables. This programme of 'symplectic reduction' could never be completed due to the complicated appearance of the Hamiltonian constraint. In fact, until today one does not know any such so-called Dirac observable for full General Relativity rigorously (with exception of the generators of the Poincaré group as derived in the previous chapter in the asymptotically flat case [244, 245]; notice, however, that formal Dirac observables will be constructed in what follows). The results of [246, 247] reveal that such Dirac observables are necessarily highly non-local, involving an infinite number of spatial derivatives of the canonical variables. This will be confirmed in the constructions that follow.

(ii) One fixes a gauge and solves the constraints. Decades of research in the field of solving the Cauchy problem for General Relativity reveal that such a procedure works at most locally, that is, there do not exist, in general, global gauge conditions. This is reminiscent of the Gribov problem in non-Abelian Yang–Mills theories.

However, the problem of time is not concerned with these technical obstacles. Rather, it addresses the following problem: suppose that we have found a complete set of Dirac observables O_α, $\alpha \in \mathcal{J}$ which at least weakly (that is, on the constraint surface) Poisson commute with the constraints, that is, $\{H(N), O_\alpha\} = \{\vec{H}(\vec{N}), O_\alpha\} = 0$ for all $N, \vec{N}$ when $H(N) = \vec{H}(\vec{N}) = 0$ for all $N, \vec{N}$. The problem of time is now that there is no time in this picture. The formalism is completely frozen, nothing moves. This is certainly very strange and in contradiction with experiment.

To resolve this issue let us analyse how one measures movements physically. Usually, that is in the presence of a true Hamiltonian rather than a Hamiltonian constraint, we have a measurable quantity T, called a clock, and another measurable quantity S, called a system. For instance, T could be the position of a pointer on a real clock and S could be the distance covered by a runner. We register the movement of S by recording the values of S in relation to the values of T. Of course, in the presence of a Hamiltonian there is no problem of time because the parameter of the Hamiltonian flow of that Hamiltonian is a natural time parameter. However, that parameter may not be the one that is directly related to the readings of a physical clock that we are interested in. If we translate the time parameter into the readings of a clock under investigation, we discover a mechanism that can be generalised to the case without a true Hamiltonian.

We will now describe this process mathematically, working our way upwards while increasing the complexity of the system.

1. • ••• ••• •••• • •• •••• • •••

If there is a true Hamiltonian $\mathbf{H}$ we therefore may obtain a curve $\tau \mapsto S_T(\tau)$ as follows. The Hamiltonian $\mathbf{H}$ generates a Hamiltonian flow on the phase space generated by its Hamiltonian vector field $\chi_{\mathbf{H}}$. It transforms any function on phase space as (see Section 19.3 for an account on symplectic geometry)

$$F \mapsto \alpha_t^{\mathbf{H}}(F) := e^{t\mathcal{L}_{\chi_{\mathbf{H}}}} \cdot F = \sum_{n=0}^{\infty} \frac{t^n}{n!} \{\mathbf{H}, F\}_{(n)} \tag{2.1.1}$$

where $\{G, F\}_{(0)} := F$, $\{G, F\}_{(n+1)} := \{G, \{G, F\}_{(n)}\}$ is the iterated Poisson bracket. One can check that the map (2.1.1) defines an automorphism on the Poisson algebra $C^\infty(\mathcal{M})$ of functions on the phase space $\mathcal{M}$, hence $\alpha_t^{\mathbf{H}}(F + G) = \alpha_t^{\mathbf{H}}(F) + \alpha_t^{\mathbf{H}}(G)$ and $\alpha_t^{\mathbf{H}}(FG) = \alpha_t^{\mathbf{H}}(F)\,\alpha_t^{\mathbf{H}}(G)$. Moreover, the collection $t \mapsto \alpha_t^{\mathbf{H}}$ forms an Abelian 1-parameter group of automorphisms, that is, $\alpha_t^{\mathbf{H}} \circ \alpha_s^{\mathbf{H}} = \alpha_{s+t}^{\mathbf{H}}$.

The physical process of measuring the movement of S relative to the movement of T may then be described mathematically as follows. We are interested in the value of S when T takes the value τ. Thus we should solve the equation $\alpha_t^{\mathbf{H}}(T) = \tau$ for t, which is always possible locally unless T is a constant of the motion. Denote the solution by $t_T(\tau)$, which is now phase space-dependent. Then

$$S_T(\tau) := \left[\alpha_t^{\mathbf{H}}(S)\right]_{t=t_T(\tau)} = \left[\alpha_t^{\mathbf{H}}(S)\right]_{\alpha_t^{\mathbf{H}}(T)=\tau} \tag{2.1.2}$$

Now not very surprisingly, (2.1.2) is a constant of the motion. The intuitive reason is that $S_T(\tau)$ is the value of S frozen at the point of parameter t time when T takes the value τ, hence $S_T(\tau)$ cannot move in parameter t time. However, it can move in clock τ time. The mathematical reason is as follows. From the identity

$$\left[\alpha_t^{\mathbf{H}}(T)\right]_{t=t_T(\tau)} = \tau \tag{2.1.3}$$

we derive

$$0 = \{\mathbf{H}, \tau\} = \{\mathbf{H}, \alpha_t^{\mathbf{H}}(T)\}_{t=t_T(\tau)} + \left[\frac{d}{dt}\alpha_t^{\mathbf{H}}(T)\right]_{t=t_T(\tau)} \{\mathbf{H}, t_T(\tau)\}$$
$$= \{\mathbf{H}, \alpha_t^{\mathbf{H}}(T)\}_{t=t_T(\tau)} \left[1 + \{\mathbf{H}, t_T(\tau)\}\right] \tag{2.1.4}$$

by definition of $\alpha_t^{\mathbf{H}}$, hence $\{\mathbf{H}, t_T(\tau)\} = -1$ unless T is a constant of the motion. The same calculation then reveals

$$\{\mathbf{H}, S_T(\tau)\} = \{\mathbf{H}, \alpha_t^{\mathbf{H}}(S)\}_{t=t_T(\tau)} \left[1 + \{\mathbf{H}, t_T(\tau)\}\right] = 0 \tag{2.1.5}$$

Of course, the constant of the motion (2.1.2) might be trivial (e.g., if time evolution is ergodic in which case the only constants of the motion are numerical constants) or it might not be well-defined on the whole phase space (e.g., if there simply is no $\alpha_t^{\mathbf{H}}$-invariant function of S, T in which case (2.1.2) cannot be globally defined), however, in principle this recipe gives a constructive algorithm to find constants of the motion:
(i) Take any two non-constants T, S such that $t \mapsto \alpha_t^{\mathbf{H}}(T)$ is locally invertible.
(ii) Construct $S_T(\tau)$.

2. • •••• ••• •• ••• •••••• •

The only difference between a single constraint H and a true Hamiltonian $\mathbf{H}$ is that the only physically interesting quantities are now the constants of the motion. We now call them Dirac observables. In the case of a true Hamiltonian all the functions on phase space were observables. An observable is a quantity which is gauge-invariant. Therefore only the Dirac observables are actually observable. The role of the quantities S, T is now that they can be mathematically determined at any point of unphysical parameter time $t := \lambda$, which is here just a Lagrange multiplier. However, the value $\alpha_\lambda^H(T)$ depends non-trivially on the gauge parameter λ and thus is gauge-dependent. In other words, we must choose a gauge parameter λ to assign a definite value, namely $\alpha_\lambda^H(T)$, to T which is like fixing a coordinate system. Physical quantities are coordinate-independent, see below. The difference with the case of a Hamiltonian $\mathbf{H}$ is that the time t parameter there does not have the status of a gauge parameter but actually the time parameter defined by the Hamiltonian which corresponds to the notion of time of a physical observer. For instance, in the context of GR in the asymptotically flat case the time t parameter corresponding to the ADM Hamiltonian is actually the time parameter used by an observer in an asymptotic inertial system in Minkowski space.

Such gauge-dependent quantities are called, following Rovelli [248–251], •• • ••• • •••••••• ••. According to Rovelli, they can be measured but are not predictable. However, we can now use the same mathematics as before to construct a gauge-invariant quantity

$$S_T(\tau) := \left[\alpha_\lambda^H(S)\right]_{\alpha_\lambda^H(T)=\tau} \tag{2.1.6}$$

which now has the interpretation of the value of S in that gauge λ in which T takes the value τ. Following again Rovelli, we call (2.1.6) an •• • ••• • ••• •••• • •• ••• • ••• •••••••••.

This terminology is perhaps a bit misleading because by definition an observable is something that can be measured in physics. The following interpretation is maybe more appropriate: a measurable quantity is always a complete observable, even pointers of a clock are observables and not partial observables. Now complete observables are defined with respect to non-measurable (since gauge-dependent) quantities T and S which we will simply call •• • • ••• •••••• •••. From these we construct two complete observables, namely $S_T(\tau)$ and $T_T(\tau) = \tau$. These are Dirac observables and they are measurable. In this book we will conform with Rovelli's terminology, however, we stress that the partial observables S, T cannot be measured, only $S_T(\tau)$, τ are measurable. In physics we are not aware of the non-observable T, rather we construct two observables $S_T(\tau)$, $S_T'(\tau)$ and may treat τ for the value of $S_T'(\tau)$ in order to determine the value of $S_T(\tau)$ when $S_T'(\tau)$ has a given value. In this sense the observable τ or the unobservable T is a 'hidden clock' and an open issue is whether and how physics depends on the choice of those hidden clocks.

For readers who find this construction awkward we mention here that the case of a Hamiltonian $\mathbf{H}$ can be phrased in the language of a single Hamiltonian constraint as follows. Assign to the phase space $\mathcal{M}$ an additional canonical pair (q^0, p_0) and extend the Poisson bracket in such a way that q^0, p_0 have vanishing Poisson bracket with any function on $\mathcal{M}$. Now define the constraint

$$H := p_0 + \mathbf{H}(p, q) \tag{2.1.7}$$

where (p, q) collectively denote the phase space coordinates of $\mathcal{M}$. Now define a partial clock observable $T := q^0$ and a partial system observable S which does not depend on q^0, p_0. Then $\alpha_\lambda^H(T) = T + \lambda$, $\alpha_\lambda^H(S) = \alpha_\lambda^{\mathbf{H}}(S)$ and $S_T(\tau) = \alpha_{\tau-T}^{\mathbf{H}}(S)$. Thus we see that in the gauge $T = 0$ the Dirac observables are described precisely by the usual evolution with respect to $\mathbf{H}$.

Hence the formalism described above can be viewed as an extension of the formalism when a true Hamiltonian is available. In general, if the Hamiltonian constraint can be split as in (2.1.7) then we say that we can •••••••• •••••• the system. Unfortunately, for most physically interesting systems it is not known how to deparametrise them nor whether it is possible at all. However, as we will see in the next section, the τ evolution is automatically Hamiltonian, that is, a canonical transformation, and therefore has a generator. That generator is what one could call a •••• ••• •••• •••• for the gauge system. In contrast to unconstrained systems, that Hamiltonian is not defined by a Legendre transformation but rather is selected by a choice of clock variable.

3. • •••• ••••••• •• • ••• ••• • ••••• •••• • • ••• • ••• •••• •• ••

If we have several constraints H_I, $I \in \mathcal{I}$ which however are in involution $\{H_I, H_J\} = 0$ then we may define their respective flows $\alpha_{\lambda_I}^{H_I}$ and introduce several clocks T_I and parameters τ_I. We now define

$$S_{\{T\}}(\{\tau\}) = \left(\left[\circ_{I \in \mathcal{I}} \ \alpha_{\lambda_I}^{H_I} \right](S) \right)_{\alpha_{\lambda_I}^{H_I}(T_I) = \tau_I} \tag{2.1.8}$$

Notice that the sequence in which we apply the respective gauge evolutions is irrelevant due to the commutativity of the constraints assumed. It is only for this reason that (2.1.8) indeed defines an, even strong, Dirac observable. The quantity (2.1.8) has a physical interpretation analogous to (2.1.6) just that one has to use several gauges and clocks.

4. • ••••• •••••• •• • • ••• ••• •••• •• ••••• •••• • • •••• • ••• •••• •• ••

As we have seen, CR does not fall in either of the categories just described. We have an infinite number of constraints $H(N)$, one for each choice of lapse function. The space of lapse functions is now infinite-dimensional, hence we have an infinite number of gauge parameters or Lagrange multipliers. Moreover, $\{H(N), H(N')\} \neq 0$. In order to apply the framework of the third case just mentioned one would like to work on the space of spatially diffeomorphism-invariant functions, that is, functions satisfying $\{\vec{H}(\vec{N}), O\} = 0$ because then we have $\{\{H(N), H(N')\}, O\} = 0$ at least on the surface defined by $H_a(x) = 0$ for all $x \in \sigma$ so that one might be able to show that it does not matter in which sequence we apply the $\alpha_{t_N}^{H(N)}$. However, unfortunately $\alpha_{t_N}^{H(N)}(O)$ is no longer spatially diffeomorphism-invariant because $\{\vec{H}(\vec{N}), H(N)\} = -\kappa H(\mathcal{L}_{\vec{N}} N)$ is not invariant, hence we have

$$\alpha_{t_{N_1}}^{H(N_1)} \circ \alpha_{t_{N_2}}^{H(N_2)} \circ \alpha_{t_{N_3}}^{H(N_3)}(O) \neq \alpha_{t_{N_2}}^{H(N_2)} \circ \alpha_{t_{N_1}}^{H(N_1)} \circ \alpha_{t_{N_3}}^{H(N_3)}(O)$$

even if $\{\vec{H}(\vec{N}), O\} = 0$.

Thus, in this case we need a new idea. One possibility is the • •••••
••••••••• ••••••• •• introduced in [252] and tested in [253–257]. See also [87–91] for related proposals and Chapter 30 for the mathematical implementation. The currently preferred proposal [258–260], closer to Rovelli's original idea, will be the subject of the next section. A third proposal, also based on a perturbative expansion like [258], is given in [261, 262], which we will not outline in this book for reasons of space.

The classical part of the Master Constraint Programme, to which one is naturally led in GR as we will show later, consists of the following: given a collection of constraints H_I, $I \in \mathcal{I}$ which may be first class or not and which may involve structure functions, consider the associated **Master Constraint**

$$\mathbf{M} = \frac{1}{2} \sum_{I,J \in \mathcal{I}} H_I K^{IJ} H_J \tag{2.1.9}$$

where K^{IJ} is a positive operator on the space of square summable sequences over the index set $\mathcal{I}$. It may depend on the phase space. Similar conditions hold when we are dealing with continuous label sets. Then the constraint surface defined by $\mathbf{M} = 0$ coincides with the one defined by $H_I = 0$, $\forall I \in \mathcal{I}$. This is why it is called the Master Constraint. Now we are in the situation of a single Hamiltonian constraint and we can again apply the mathematics from above. However, notice the following subtlety: for any function F on the phase space we have $\{\mathbf{M}, F\}_{\mathbf{M}=0} = 0$. Thus, the Master Constraint is qualitatively different from the usual single Hamiltonian constraints in that it does not generate gauge transformations on the constraint surface. In particular, it seems that it does not detect weak Dirac observables at all because F could be completely arbitrary. However, notice that

$$\{F, \{F, \mathbf{M}\}\}_{\mathbf{M}=0} = \sum_{I,J} \{F, H_I\}_{\mathbf{M}=0} K^{IJ} \{F, H_J\}_{\mathbf{M}=0} \tag{2.1.10}$$

Thus the single Master Equation $\{F, \{F, \mathbf{M}\}\}_{\mathbf{M}=0} = 0$ is equivalent to the infinite number of equations $\{F, H_I\}_{\mathbf{M}=0} = 0 \,\forall I$ and therefore (2.1.10) precisely detects weak Dirac observables. Now obviously (2.1.10) is identically satisfied if

$$\{F, \mathbf{M}\} = 0 \tag{2.1.11}$$

holds on the full phase space. Functions F with this property are called strong Dirac observables (with respect to $\mathbf{M}$). Thus, as far as strong Dirac observables are concerned, we would again construct

$$S_T(\tau) = \left(\alpha_\lambda(S) \right)_{\alpha_\lambda(T)=\tau} \tag{2.1.12}$$

However, we must now be careful with its interpretation: it is the value of S in the gauge, with respect to $\mathbf{M}$, in which T takes the value τ •• •• •••• ••• •• •• •••• •• •••••••. Now, $S_T(\tau)$ formally commutes everywhere with $\mathbf{M}$ by construction, however, it may be discontinuous there, see the next section. If it is continuous, then we can continue it to the surface $\mathbf{M} = 0$ and $S_T(\tau)$ keeps its τ-dependence. In terms of the individual constraints H_I the interpretation of (2.1.12) would then be the value of S in the gauge when T takes the value τ and where gauge now means that we are considering simultaneous gauge transformations generated by the constraint $H(\lambda) = \sum_I \lambda^I H_I$ and where the 'Lagrange multipliers' are now specified phase space functions $\lambda^I = \sum_J K^{IJ} H_J$ which actually vanish on the constraint surface.

We will elaborate more on the Master Constraint Programme in the concrete case of GR in Chapter 10.

This solves the problem of time classically in terms of evolving constants which have a clear physical interpretation in terms of partial observables. We see that we can regain a notion of time as the measurable parameter τ in all cases, at least in principle. On the other hand, the Dirac observables $S_T(\tau)$ are completely non-local in the unphysical time t since by construction, for example

$$S_T(\tau) = \lim_{R \to \infty} \frac{1}{2R} \int_{-R}^{R} dt \, \alpha_t \, (S_T(\tau)) \tag{2.1.13}$$

The expression on the right-hand side of (2.1.13) is called an •••••• • ••• with respect to the unphysical time t.

An interesting question is whether we can extract from $S_T(\tau)$ a physical Hamiltonian $\mathbf{H}_{\text{phys}}$ which itself is a Dirac observable by defining $S_T(\tau) =: \alpha_\tau^{\mathbf{H}_{\text{phys}}}(S_T(0))$. Taking the derivative with respect to τ results in the equation

$$\{\mathbf{H}_{\text{phys}}, S_T(\tau)\} = \left(\frac{\{H, \alpha_t^H(S)\}}{\{T, \alpha_t^H(S)\}} \right)_{t=t_T(\tau)} \tag{2.1.14}$$

which can be solved for $\mathbf{H}_{\text{phys}}$ since this is a first-order, linear partial differential equation for $\mathbf{H}_{\text{phys}}$ (although possibly in infinite dimensions). However, the solution should be independent of S, it may depend on T. It is easy to check that in the deparametrised case $T = q^0$, $H = p_0 + \mathbf{H}$ we find $\mathbf{H}_{\text{phys}} = \bullet$ which indeed is a Dirac observable, $\{\mathbf{H}_{\text{phys}}, H\} = 0$. In general, whether a suitable $\mathbf{H}_{\text{phys}}$ can be found at least locally in τ-time evolution will depend crucially on the choice of the clock variable T. We will have to say more on this point in the next section.

2.2 Partial and complete observables for general constrained systems

As we will see shortly, given partial observables S, T one can formally solve equation (2.1.12) by

$$S_T(\tau) = \sum_{n=0}^{\infty} \frac{(\tau - T)^n}{n!} \left(\frac{1}{\{\ ,T\}} \chi \right)^n \cdot S \tag{2.2.1}$$

provided the series converges in a neighbourhood of the constraint surface. Here χ is the Hamiltonian vector field of   . We see that (2.2.1) is very likely not to converge unless S, T are carefully chosen. The reason is that on the constraint surface both the Hamiltonian vector field of   and the quantity $\{\ , T\}$ vanish. Hence, the vector field $X := \frac{1}{\{\ , T\}} \chi$ becomes ill-defined at $= 0$, unless the two zeros cancel unambiguously in the sense of de l'Hospital's theorem when expanding numerator and denominator in terms of the individual constraints C_I.

This is equivalent to assigning an unambiguous value to the quantity $C_I/\sqrt{M}$ at $M = 0$ for all the indices I appearing in that quotient which is ambiguous, in general, if there are more than two. This means that in most cases one has to resort to the stronger condition $\{O, \{O, \quad \}\}_{=0} = 0$ which in general only selects weak Dirac observables.

This condition, however, cannot be solved by the partial observable Ansatz since the very definition of the partial observable uses the Hamiltonian vector field of the constraint. The advantage of the above Master Equation is that it is a single equation, its disadvantage is that it is a non-linear condition on O. Since it is equivalent to the infinite number of conditions $\{C_I, O\}_{=0}$ the question is whether one cannot make progress with these linear equations, even if the constraints C_I do not mutually commute. One of the achievements of [258, 259] is to notice that, at least locally, the constraints C_I can be replaced by equivalent ones C_I' which have the property that their Hamiltonian vector fields commute weakly. This means that the structure functions of the new constraints vanish on the constraint surface, which is not the case for the C_I. It turns out that this condition is sufficient in order to use the partial observable Ansatz because the Hamiltonian flows of the C_I' weakly commute and we are back to case (3) in the previous section. We will now describe elements of [258, 259], using the notation of [260] which contains additional ideas concerning the quantisation of the resulting complete (Dirac) observables. In particular, we will derive a formal power series in the general case which is as explicit as (2.2.2). Notice, however, that the results will generically be at most valid locally in phase space.

2.2.1 Partial and weak complete observables

We begin with a more geometrical description of the situation: let C_j, $j \in \mathcal{I}$ be a system of first-class constraints on a phase space $\mathcal{M}$ with (strong) symplectic structure given by a Poisson bracket $\{.,.\}$ where the index set has countable cardinality. This includes the case of a field theory for which the constraints are usually given in the local form $C_\mu(x)$, $x \in \sigma$, $\mu = 1, \ldots, n < \infty$ where σ is a spatial, D-dimensional manifold corresponding to the initial value formulation and μ are some tensorial and/or Lie algebra indices. This can be seen by choosing a basis b_I of the Hilbert space $L_2(\sigma, d^D x)$ consisting of smooth functions of compact support and defining $C_j := \int_\sigma d^D x \, b_I(x) \, C_\mu(x)$ with $j := (\mu, I)$. We assume the most general situation, namely that $\{C_j, C_k\} = f_{jk}{}^l C_l$ closes with structure functions, that is, $f_{jk}{}^l$ can be non-trivial functions on $\mathcal{M}$.

The partial observable Ansatz to generate Dirac observables is now as follows. Take as many functions on phase space $T_j, j \in \mathcal{I}$ as there are constraints. These functions have the purpose of providing a local (in phase space) coordinatisation of the gauge orbit $[m]$ of any point m in phase space, at least in a neighbourhood of the constraint surface $\overline{\mathcal{M}} = \{m \in \mathcal{M}; \; C_j(m) = 0 \; \forall j \in \mathcal{I}\}$. The gauge orbit $[m]$

of m is given by $[m] := \{\alpha_{\beta_1} \circ \ldots \circ \alpha_{\beta_N}(m); \ N < \infty, \ \beta_k^j \in \mathbb{R}, \ k = 1, \ldots, N, \ j \in \mathcal{I}\}$. Here α_β is the canonical transformation (automorphism of $(C^\infty(\mathcal{M}), \{.,.\})$ generated by the Hamiltonian vector field χ_β of $C_\beta := \beta^j C_j$, that is $\alpha_\beta(f) := \exp(\chi_\beta) \cdot f$. (Notice that if the system had structure constants, then it would be sufficient to choose $N = 1$.)

In other words, we assume that it is possible to find functions T_j such that each $m \in \mathcal{M}$ is completely specified by $[m]$ and by the $T_j(m)$. This means that if the value τ_j is in the range of T_j then the gauge fixing surface $\mathcal{M}_\tau := \{m \in \overline{\mathcal{M}}; \ T_j(m) = \tau_j\}$ intersects each $[m]$ in precisely one point. In practice this is usually hard to achieve globally on $\overline{\mathcal{M}}$ due to the possibility of Gribov copies, but here we are only interested in local considerations. It follows that the matrix $A_{jk} := \{C_j, T_k\}$ must be locally invertible so that the condition $[\alpha_\beta(T_j)](m) = T_j(\alpha_\beta(m)) = \tau_j$ can be inverted for β (given $m' \in [m]$ we may write it in the form $[\alpha_\beta(m)]_{|\beta = B(m')}$ for some $B(m')$ which may depend on m').

Take now another function f on phase space. Then the weak Dirac observable $F_{f,T}^\tau$ associated with the partial observables $f, T_j, j \in \mathcal{I}$ is defined by

$$(F_{f,T}^\tau)(m) := [f(\alpha_\beta(m))]_{|\beta = B_T^\tau(m)}, \quad [T_j(\alpha_\beta(m))]_{|\beta = B_T^\tau(m)} = \tau_j \qquad (2.2.2)$$

The physical interpretation of $F_{f,T}^\tau$ is that it is the value of f at those 'times' β_j when the 'clocks' T_j take the values τ_j.

We will now derive an explicit expression for (2.2.2) from an Ansatz for a Taylor expansion. Namely, on the gauge cut $\overline{\mathcal{M}}_\tau$ the function $F_{f,T}^\tau$ equals f since then $B_T^\tau(m) = 0$. Away from this section, $F_{f,T}^\tau$ can be expanded into a Taylor series.[1] Thus we make the Ansatz

$$F_{f,T}^\tau = \sum_{\{k_j\}_{j \in \mathcal{I}} = 0}^\infty \prod_{j \subset \mathcal{I}} \frac{(\tau_j - T_j)^{k_j}}{k_j!} f_{\{k_j\}_{j \in \mathcal{I}}} \qquad (2.2.3)$$

with $f_{\{k_j\} - \{0\}} = f$. We assume that (2.2.3) converges absolutely on an open set S and is continuous there, hence is uniformly bounded on any compact set contained in S. We may then interchange summation and differentiation on S and compute

$$\{C_l, F_{f,T}^\tau\} = \sum_{\{k_j\}_{j \in \mathcal{I}} = 0}^\infty \prod_{j \in \mathcal{I}} \frac{(\tau_j - T_j)^{k_j}}{k_j!} \left[\sum_{m \in \mathcal{I}} -A_{l,m} f_{\{k_j'(m)\}_{j \in \mathcal{I}}} + \{C_l, f_{\{k_j\}_{j \in \mathcal{I}}}\} \right]$$

$$(2.2.4)$$

where $k_j'(m) = k_j$ for $j \neq m$ and $k_m'(m) = k_m + 1$. Setting (2.2.4) (weakly) to zero leads to a recursion relation with the formal solution

$$f_{\{k_j\}_{j \in \mathcal{I}}} = \prod_{j \in \mathcal{I}} (X_j')_j^k \cdot f, \ X_j' \cdot f = \sum_{k \in \mathcal{I}} (A^{-1})_{jk} \{C_k, f\} \qquad (2.2.5)$$

---

[1] In other words, $F_{f,T}^\tau$ is the gauge-invariant extension of the restriction of f to $\overline{\mathcal{M}}_\tau$ mentioned in [263] for which however no explicit expression was given there.

Expression (2.2.5) is formal, among other things, because we did not specify the order of application of the vector fields X'_j. We will now show that, as a weak identity, the order in (2.2.5) is irrelevant. To see this, let us introduce the equivalent constraints (at least on S)

$$C'_j := \sum_{k \in \mathcal{I}} (A^{-1})_{jk} C_k \qquad (2.2.6)$$

and notice that with the Hamiltonian vector fields $X_j \cdot f = \{C'_j, f\}$ we have $X'_{j_1} \dots X'_{j_n} \cdot f \approx X_{j_1} \dots X_{j_n} \cdot f$ for any $j_1, \dots, j_n$ due to the first-class property of the constraints. Here and in what follows we write $\approx$ for a relation that becomes an identity on $\overline{\mathcal{M}}$. Then we can make the following surprising observation.

Theorem 2.2.1. $\cdots$ C_j $\cdots$ $\cdots$ $\cdots$ $\cdots$ $\cdots$ $\cdots$ $\cdots$ T_j $\cdots$ $\cdots$
$\cdots$ $\cdots$ $\cdots$ $\cdots$ $\cdots$ $\cdots$ A $\cdots$ $\cdots$ $\cdots$ $A_{jk} := \{C_j, T_k\}$ $\cdots$ $\cdots$
$\cdots$ $\cdots$ $\cdots$ S $\cdots$ $\cdots$ $\cdots$ $\cdots$ $\cdots$ C'_j
$\cdots$ $\cdots$ $\cdots$ $\cdots$ $\cdots$ $X_j := \chi_{C'_j}$ $\cdots$ $\cdots$
$\cdots$ $\cdots$ $\cdots$

$\cdots$ The proof consists of a straightforward computation and exploits the Jacobi identity. Abbreviating $B_{jk} := (A^{-1})_{jk}$ we have

$$\{C'_j, \{C'_k, f\}\} - \{C'_k, \{C'_j, f\}\}$$
$$\approx \sum_{m,n} B_{jm} \{C_m, [B_{kn}\{C_n, f\} + C_n\{B_{kn}, f\}]\} - j \leftrightarrow k$$
$$\approx \sum_{m,n} B_{jm}[\{C_m, B_{kn}\}\{C_n, f\} + B_{kn}\{C_m, \{C_n, f\}\}] - j \leftrightarrow k$$
$$= \sum_{m,n} B_{jm}\left[-\sum_{l,i} B_{kl}\{C_m, A_{li}\}B_{in}\{C_n, f\} + B_{kn}\{C_m, \{C_n, f\}\}\right] - j \leftrightarrow k$$
$$= \sum_{m,n} B_{jm}\left[-\sum_{l,i} B_{kl} B_{in}\{C_n, f\}(\{C_m, \{C_l, T_i\}\} - \{C_l, \{C_m, T_i\}\})\right.$$
$$\left. + B_{kn}(\{C_m, \{C_n, f\}\} - \{C_n, \{C_m, f\}\})\right]$$
$$= \sum_{m,n} B_{jm}\left[\sum_{l,i} B_{kl} B_{in}\{C_n, f\}\{T_i, \{C_m, C_l\}\} - B_{kn}(\{f, \{C_m, C_n\}\}\right]$$
$$\approx \sum_{m,n} B_{jm}\left[-\sum_{l,i,p} B_{kl} B_{in}\{C_n, f\}f_{ml}{}^p A_{pi} + B_{kn}\sum_{l} f_{mn}{}^l\{C_l, f\}\right]$$
$$= \sum_{m,n,l} B_{jm}\left[-B_{kl}\{C_n, f\}f_{ml}{}^n + B_{kn}f_{mn}{}^l\{C_l, f\}\right]$$
$$= 0 \qquad (2.2.7)$$

Due to

$$\{C'_j, \{C'_k, f\}\} - \{C'_k, \{C'_j, f\}\} = \{\{C'_j, C'_k\}, f\} \approx f'^{\ l}_{jk}\{C'_l, f\} \approx 0 \qquad (2.2.8)$$

this means that the structure functions $f'^{\ l}_{jk}$ with respect to the C'_j are weakly vanishing, that is, themselves proportional to the constraints. □

We may therefore write the Dirac observable generated by f, T_j indeed as

$$F^T_{f,T} = \sum_{\{k_j\}_{j\in\mathcal{I}}=0}^{\infty} \prod_{j\in\mathcal{I}} \frac{(\tau_j - T_j)^{k_j}}{k_j!} \prod_{j\in\mathcal{I}} (X_j)^{k_j} \cdot f \qquad (2.2.9)$$

Expression (2.2.9) is, despite the obvious convergence issues to be checked in the concrete application, remarkably simple. Of course, especially in field theory it will not be possible to calculate it exactly and already the computation of the inverse A^{-1} may be hard, depending on the choice of the T_j. However, for points close to the gauge cut, expression (2.2.9) is rapidly converging and one may be able to do approximate calculations.

••• •••: Let $\alpha'_\beta(f) := \exp(\sum_j \beta_j X_j) \cdot f$ be the gauge flow generated by the new constraints C'_j for real-valued gauge parameters β_j. We easily calculate $\alpha'_\beta(T_j) \approx T_j + \beta_j$. The condition $\alpha'_\beta(T_j) = \tau_j$ can therefore be easily inverted to $\beta_j \approx \tau_j - T_j$. Hence the complete observable prescription with respect to the new constraints C'_j

$$F^T_{f,T} := [\alpha'_\beta(f)]_{|\alpha'_\beta(T)=\tau} \qquad (2.2.10)$$

weakly coincides with (2.2.9).

2.2.2 Poisson algebra of Dirac observables

In [263] we find the statement that the Poisson brackets among the Dirac observables obtained as the gauge-invariant extension of $\overline{\mathcal{M}}_\tau$ of the respective restrictions to the gauge cut of functions f, g is weakly given by the gauge-invariant extension of their Dirac bracket with respect to the associated gauge fixing functions. Expression (2.2.9) now enables us to give an explicit, local proof (modulo convergence issues). See [258] for an alternative one.

Theorem 2.2.2. ••• $F^T_{f,T}$ •• •• ••• •• •• •••••• • ••• •••••••• •• •••••• ••••••••
••••• T_j. •• ••••••• ••• ••••• •••••••• $G_j := T_j - \tau_j$ ••• ••• ••••• ••• •••••• ••
••••• ••• •• ••• •••••• •• $C_{1j} := C_j, C_{2j} := G_j$ ••• ••••••••• $\mu = (I, j), I = 1, 2$•
•• ••••• ••• ••• • •••• •••••••

$$\{f, f'\}^* := \{f, f'\} - \{f, C_\mu\}K^{\mu\nu}\{C_\nu, f'\} \qquad (2.2.11)$$

• ••••• $K_{\mu\nu} = \{C_\mu, C_\nu\}, \ K^{\mu\rho}K_{\rho\nu} = \delta^\mu_\nu$• • ••••

$$\{F^T_{f,T}, F^T_{f',T}\} \approx F^T_{\{f,f'\}^*,T} \qquad (2.2.12)$$

•••••• Let us introduce the abbreviations

$$Y_{\{k\}} = \prod_j \frac{(\tau_j - T_j)^{k_j}}{k_j!}, \quad f_{\{k\}} = \prod_j (X_j)^{k_j} \cdot f, \quad \sum_{\{k\}} = \sum_{k_1, k_2, \ldots = 0}^{\infty} \tag{2.2.13}$$

We have

$$
\begin{aligned}
\{F_{f,T}^\tau, & F_{f',T}^\tau\} \\
&= \sum_{\{k\}, \{l\}} \{Y_{\{k\}} f_{\{k\}}, Y_{\{l\}} f'_{\{l\}}\} \\
&\approx \sum_{\{k\}, \{l\}} Y_{\{k\}} Y_{\{l\}} \left[\{f_{\{k\}}, f'_{\{l\}}\} - \sum_j (X_j \cdot f)_{\{k\}} \{T_j, f'_{\{l\}}\} \right. \\
&\qquad\qquad \left. + \sum_j (X_j \cdot f')_{\{l\}} \{T_j, f_{\{k\}}\} + \sum_{j,m} (X_j \cdot f)_{\{k\}} (X_m \cdot f')_{\{l\}} \{T_j, T_m\} \right] \\
&= \sum_{\{n\}} Y_{\{n\}} \sum_{\{k\}; k_l \le n_l} \prod_l \binom{n_l}{k_l} \left[\{f_{\{k\}}, f'_{\{n-k\}}\} - \sum_j (X_j \cdot f)_{\{k\}} \{T_j, f'_{\{n-k\}}\} \right. \\
&\qquad\qquad \left. + \sum_j (X_j \cdot f')_{\{n-k\}} \{T_j, f_{\{k\}}\} + \sum_{j,m} (X_j \cdot f)_{\{k\}} (X_m \cdot f')_{\{n-k\}} \{T_j, T_m\} \right]
\end{aligned}
\tag{2.2.14}
$$

By definition of a Hamiltonian vector field we have $X_j\{f, f'\} = \{X_j f, f'\} + \{f, X_j f'\}$. Thus, by the (multi) Leibniz rule

$$\prod_l (X_l)_l^n \{f, f'\} = \sum_{\{k\}; k_l \le n_l} \prod_l \binom{n_l}{k_l} \{f_{\{k\}}, f'_{\{n-k\}}\} \tag{2.2.15}$$

is already the first term we need. It therefore remains to show that

$$
\begin{aligned}
\prod_l (X_l)_l^n & [\{f, f'\}^* - \{f, f'\}] \\
&\approx \sum_{\{k\}; k_l \le n_l} \prod_l \binom{n_l}{k_l} \left[-\sum_j (X_j \cdot f)_{\{k\}} \{T_j, f'_{\{n-k\}}\} \right. \\
&\qquad\qquad \left. + \sum_j (X_j \cdot f')_{\{n-k\}} \{T_j, f_{\{k\}}\} + \sum_{j,m} (X_j \cdot f)_{\{k\}} (X_m \cdot f')_{\{n-k\}} \{T_j, T_m\} \right]
\end{aligned}
\tag{2.2.16}
$$

We will do this by multi-induction over $N := \sum_l n_l$.

The case $N = 0$ reduces to the claim

$$\{f, f'\}^* - \{f, f'\} \approx -\sum_j (X_j \cdot f)\{T_j, f'\} + \sum_j (X_j \cdot f')\{T_j, f\}$$

$$+ \sum_{j,m} (X_j \cdot f)(X_m \cdot f')\{T_j, T_m\} \qquad (2.2.17)$$

To compute the Dirac bracket explicitly we must invert the matrix $K_{Jj,Kk}$ with entries $K_{1j,1k} = \{C_j, C_k\} = f_{jk}{}^l C_l \approx 0$, $K_{1j,2k} = \{C_j, T_k\} = A_{jk} = -K_{2k,1j}$ and $K_{2j,2k} = \{T_j, T_k\}$. By definition $\sum_{L,l} K^{Jj,Ll} K_{Ll,Kk} = \delta_K^J \delta_k^j$ therefore $K^{1j,1k} \approx \sum_{m,n} (A^{-1})_{mj} \{T_m, T_n\}(A^{-1})_{nk}$, $K^{1j,2k} \approx -(A^{-1})_{kj} \approx -K^{2k,1j}$ and $K^{2j,2k} \approx 0$. It follows

$$-\{f, f'\}^* + \{f, f'\} = \{f, C_j\} K^{1j,1k} \{C_k, f'\} + \{f, C_j\} K^{1j,2k} \{T_k, f'\}$$

$$+ \{f, T_j\} K^{2j,1k} \{C_k, f'\} + \{f, T_j\} K^{2j,2k} \{T_k, f'\}$$

$$\approx \sum_{m,n} \{f, C_j\}(A^{-1})_{mj} \{T_m, T_n\}(A^{-1})_{nk} \{C_k, f'\}$$

$$- \{f, C_j\}(A^{-1})_{kj} \{T_k, f'\} + \{f, T_j\}(A^{-1})_{jk} \{C_k, f'\}$$

$$\approx -\sum_{m,n} (X_m \cdot f)\{T_m, T_n\}(X_n \cdot f') + (X_k \cdot f)\{T_k, f'\}$$

$$- (X_k \cdot f')\{T_k, f\} \qquad (2.2.18)$$

which is precisely the negative of (2.2.17).

Suppose then that we have proved the claim for every configuration $\{n_l\}$ such that $\sum_l n_l \leq N$. Any configuration with $N + 1$ arises from a configuration with N by raising one of the n_l by one unit, say $n_j \to n_j + 1$. Then, by assumption

$$X_j \prod_l (X_l)_l^n [\{f, f'\}^* - \{f, f'\}]$$

$$\approx X_j \sum_{\{k\};\, k_l \leq n_l} \prod_l \binom{n_l}{k_l} \left[-\sum_l (X_l \cdot f)_{\{k\}} \{T_l, f'_{\{n-k\}}\} \right.$$

$$\left. + \sum_l (X_l \cdot f')_{\{n-k\}} \{T_l, f_{\{k\}}\} + \sum_{l,m} (X_l \cdot f)_{\{k\}} (X_m \cdot f')_{\{n-k\}} \{T_l, T_m\} \right]$$

$$\approx \sum_{\{k\};\, k_l \leq n_l} \prod_l \binom{n_l}{k_l} \left[-\sum_l [(X_l \cdot f)_{\{k^j\}} \{T_l, f'_{\{n-k\}}\} + (X_l \cdot f)_{\{k\}} \{T_l, f'_{\{n^j-k\}}\} \right.$$

$$+ (X_l \cdot f)_{\{k\}} \{X_j \cdot T_l, f'_{\{n-k\}}\}] + \sum_l [(X_l \cdot f')_{\{k^j\}} \{T_l, f_{\{n-k\}}\}$$

$$\left. + (X_l \cdot f')_{\{k\}} \{T_l, f_{\{n^j-k\}}\} + (X_l \cdot f')_{\{k\}} \{X_j \cdot T_l, f_{\{n-k\}}\}] \right.$$

$$+ \sum_{l,m} \left[(X_l \cdot f)_{\{k^j\}} (X_m \cdot f')_{\{n-k\}} \{T_l, T_m\} \right.$$

$$+ (X_l \cdot f)_{\{k\}} (X_m \cdot f')_{\{n^j - k\}} \{T_l, T_m\}$$

$$\left. + (X_l \cdot f)_{\{k\}} (X_m \cdot f')_{\{n-k\}} (\{X_j T_l, T_m\} + \{T_l, X_j T_m\}) \right] \Bigg] \tag{2.2.19}$$

where $\{k^j\}$ coincides with $\{k\}$ except that $k_j \to k_j + 1$ and similarly for $\{n^j\}$. By the multi-binomial theorem the first two terms in each of the three sums in the last equality combine precisely to what we need. Hence it remains to show that

$$0 \approx \sum_{\{k\}; \, k_l \leq n_l} \prod_l \binom{n_l}{k_l} \left[-\sum_l (X_l \cdot f)_{\{k\}} \{X_j \cdot T_l, f'_{\{n-k\}}\} \right.$$

$$+ \sum_l (X_l \cdot f')_{\{k\}} \{X_j \cdot T_l, f_{\{n-k\}}\} + \sum_{l,m} (X_l \cdot f)_{\{k\}} (X_m \cdot f')_{\{n-k\}} (\{X_j T_l, T_m\}$$

$$\left. + \{T_l, X_j T_m\}) \right] \tag{2.2.20}$$

We have

$$X_j \cdot T_l = \delta_{jl} + \sum_m C_m \{(A^{-1})_{jm}, T_l\} =: \delta_{jl} + \sum_m C_m B_{jlm} \tag{2.2.21}$$

Hence

$$\{X_j \cdot T_l, g\} \approx \sum_{m,n} B_{jlm} A_{mn} (X_n \cdot g) =: \sum_n D_{jln} (X_n \cdot g) \tag{2.2.22}$$

Next, using (2.2.21) and (2.2.22)

$$\{X_j T_l, T_m\} + \{T_l, X_j T_m\} \approx \sum_n (B_{jln} A_{nm} - B_{jmn} A_{nl}) = D_{jlm} - D_{jml} \tag{2.2.23}$$

We can now simplify the right-hand side of (1.1.17)

$$\sum_{\{k\}; \, k_l \leq n_l} \prod_l \binom{n_l}{k_l} \sum_{l,m} D_{jlm} \left[-(X_l \cdot f)_{\{k\}} (X_m \cdot f'_{\{n-k\}}) \right.$$

$$\left. + (X_l \cdot f')_{\{k\}} (X_m \cdot f_{\{n-k\}}) + [D_{jlm} - D_{jml}](X_l \cdot f)_{\{k\}} (X_m \cdot f')_{\{n-k\}} \right]$$

$$\sum_{l,m} D_{jlm} \prod_i (X_i)^{n_i} [-(X_l \cdot f)(X_m \cdot f') + (X_l \cdot f')(X_m \cdot f)$$

$$+ (X_l \cdot f)(X_m \cdot f') - (X_m \cdot f)(X_l \cdot f')] = 0 \tag{2.2.24}$$

as claimed. Notice that by using the Jacobi identity we also have $D_{jkl} = D_{jlk}$ so the two terms in the second and third line of (2.2.24) even vanish separately (important for the case that $\{T_j, T_k\} = 0$). $\qquad \square$

We can now rephrase Theorem 2.2.2 as follows: consider the map

$$F_T^\tau : (C^\infty(\mathcal{M}), \{.,.\}_T^*) \to (D^\infty(\mathcal{M}), \{.,.\}_T^*); \; f \mapsto F_{f,T}^\tau \qquad (2.2.25)$$

where $D^\infty(\mathcal{M})$ denotes the set of smooth, weak Dirac observables and $\{.,.\}_T^*$ is the Dirac bracket with respect to the gauge fixing functions T_j. Then Theorem 2.2.2 says that F_T^τ is a weak Poisson homomorphism (i.e., a homomorphism on the constraint surface). To see this, notice that for (weak) Dirac observables the Dirac bracket coincides weakly with the ordinary Poisson bracket. Moreover, the map F_T^τ is linear and trivially

$$F_{f,T}^\tau \, F_{f',T}^\tau = \sum_{\{k\},\{l\}} Y_{\{k\}} Y_{\{l\}} f_{\{k\}} f'_{\{l\}}$$

$$= \sum_{\{n\}} Y_{\{n\}} \sum_{\{k\}; \, k_l \leq n_l} \prod_l \binom{n_l}{k_l} f_{\{k\}} f'_{\{n-k\}}$$

$$\approx \sum_{\{n\}} Y_{\{n\}} \prod_l (X_l)^{n_l} (f \, f') = F_{ff',T}^\tau \qquad (2.2.26)$$

[We can make the homomorphism exact by dividing both $C^\infty(\mathcal{M})$ and $D^\infty(\mathcal{M})$ by the ideal (under pointwise addition and multiplication) of smooth functions vanishing on the constraint surface.] Notice that F_T^τ is onto because $F_{f,T}^\tau \approx f$ if f is already a weak Dirac observable.

2.2.3 Evolving constants

The complete or Dirac observable $F_{f,T}^\tau$ has the physical interpretation of giving the value of f when the T_j assume the values τ_j. In constrained field theories we thus arrive at the multi-fingered time picture, there is no preferred time but there are infinitely many. Accordingly, we define a multi-fingered time evolution on the image of the maps F_T^τ by

$$\alpha^\tau : F_T^{\tau^0}(C^\infty(\mathcal{M})) \to F_T^{\tau+\tau^0}(C^\infty(\mathcal{M})); \; F_{f,T}^{\tau^0} \mapsto F_{f,T}^{\tau+\tau^0} \qquad (2.2.27)$$

As defined, α^τ forms a weakly Abelian group. However, it has even more interesting properties:

$$F_{f,T}^{\tau+\tau^0} = \sum_{\{n\}} \prod_j \frac{(\tau_j + \tau_j^0 - T_j)^{n_j}}{n_j!} \prod_j X_j^{n_j} \cdot f$$

$$\approx \sum_{\{n\}} \sum_{\{k\}; \, k_l \leq n_l} \prod_l \frac{1}{n_l!} \binom{n_l}{k_l} \prod_j (\tau_j^0 - T_j)^{k_j} \tau_j^{n_j - k_j} \prod_j X_j^{k_j} X_j^{n_j - k_j} \cdot f$$

$$\approx \sum_{\{k\}} \prod_j \frac{(\tau_j^0 - T_j)^{k_j}}{k_j!} \prod_j X_j^{k_j} \cdot \left[\sum_{\{l\}} \frac{\tau_j^{l_j}}{l_j!} \prod_j X_j^{l_j} \right] \cdot f$$

$$= F_{\alpha_\tau'(f),T}^{\tau^0} \qquad (2.2.28)$$

where $\alpha'_\tau(f)$ is the automorphism on $C^\infty(\mathcal{M})$ generated by the Hamiltonian vector field of $\sum_j \tau_j C'_j$ with the equivalent constraints $C'_j = \sum_k (A^{-1})_{jk} C_k$. This is due to the multinomial theorem

$$
\begin{aligned}
\alpha'_\tau(f) &= \sum_{n=0}^\infty \frac{1}{n!} \left(\sum_j \tau_j X_j \right)^n \cdot f \\
&= \sum_{n=0}^\infty \frac{1}{n!} \sum_{j_1,\dots,j_n} \prod_{k=1}^n \tau_{j_k} X_{j_k} \cdot f \\
&= \sum_{n=0}^\infty \frac{1}{n!} \sum_{\{k\}; \sum_j k_j = n} \frac{n!}{\prod_j (k_j)!} \prod_j \tau_j^{k_j} \prod_j X_j^{k_j} \cdot f \\
&= \sum_{\{k\}} \prod_j \frac{\tau_j^{k_j}}{k_j!} \prod_j X_j^{k_j} \cdot f
\end{aligned}
\tag{2.2.29}
$$

Thus, our time evolution on the observables is induced by a gauge transformation on the partial observables. From this observation it follows, together with the weak homomorphism property, that

$$
\begin{aligned}
\left\{ \alpha^\tau (F_{f,T}^{T_0}), \alpha^\tau (F_{f',T}^{T_0}) \right\} &= \left\{ F_{f,T}^{T_0+\tau}, F_{f',T}^{T_0+\tau} \right\} \\
&\approx F_{\{f,f'\}^*,T}^{T_0+\tau} = \alpha^\tau \left(F_{\{f,f'\}^*,T}^{T_0} \right) \\
&\approx \alpha^\tau \left(\left\{ F_{f,T}^{T_0}, F_{f',T}^{T_0} \right\} \right)
\end{aligned}
\tag{2.2.30}
$$

In other words, $\tau \mapsto \alpha^\tau$ is a weakly Abelian, multi-parameter group of automorphisms on the image of each map $F_{f,T}^{T_0}$. This is in strong analogy to the properties of the one-parameter group of automorphisms on phase space generated by a true Hamiltonian.

2.2.4 Reduced phase space quantisation of the algebra of Dirac observables and unitary implementation of the multi-fingered time evolution

In this section we present an idea for how to combine the observations of the previous section with quantisation. Moreover, we will derive equations for the physical Hamiltonians that drive the physical time τ evolution.

We assume that it is possible to choose the functions T_j as canonical coordinates. In other words, we choose a canonical coordinate system consisting of canonical pairs (q^a, p_a) and (T_j, P^j) where the first system of coordinates has vanishing Poisson brackets with the second so that the only non-vanishing brackets are $\{p_a, q^b\} = \delta_a^b$, $\{P^j, T_k\} = \delta_k^j$. (In field theory the label set of the $a, b, \dots$ will be of countably infinite cardinality corresponding to certain smeared quantities of the canonical fields.) The virtue of this assumption is that the Dirac bracket reduces to the ordinary Poisson bracket on functions which depend only on q^a, p_a. We will shortly see why this is important. We define with $F_T := F_T^0$ the weak Dirac observables at multi-fingered time $\tau = 0$ (or any other fixed allowed value of τ)

$$Q^a := F_T(q^a), \; P_a := F_T(p_a) \qquad (2.2.31)$$

Notice that $F^\tau_{T_j,T} \approx \tau_j$, so the Dirac observable corresponding to T_j is just a constant and thus not very interesting (but evolves precisely as a clock). Likewise $F^\tau_{C_j,T} \approx 0$ is not very interesting. Since at least locally we can solve the constraints C_j for the momenta P^j, that is $P^j \approx E_j(q^a, p_a, T_k)$ and F_T is a homomorphism with respect to pointwise operations we have

$$F_T(P_j) \approx E_j(F_T(q^a), F_T(p_a), F_T(T_k)) \approx E_j(Q^a, P_a, \tau_k) \qquad (2.2.32)$$

and thus also does not give rise to a Dirac observable which we could not already construct from Q^a, P_a. The importance of our assumption is now that due to the homomorphism property

$$\{P_a, Q^b\} \approx F^0_{\{p_a,q^b\}^*,T} = F^0_{\delta^b_a,T} = \delta^b_a, \; \{Q^a, Q^b\} \approx \{P_a, P_b\} \approx 0 \; (2.2.33)$$

In other words, even though the functions P_a, Q^a are very complicated expressions in terms of q^a, p_a, T_j they have nevertheless canonical brackets at least on the constraint surface. If we had to use the Dirac bracket then this would not be the case and the algebra among the Q^a, P_a would be too complicated and no hope would exist towards its quantisation. However, under our assumption there is now a chance.

Now reduced phase space quantisation consists in quantising the subalgebra of $\mathcal{D}$, spanned by our preferred Dirac observables Q^a, P_a evaluated on the constraint surface. As we have just seen, the algebra $\mathcal{D}$ itself is given by the Poisson algebra of the functions of the Q^a, P_a evaluated on the constraint surface. Hence all the weak equalities that we have derived now become exact. We are therefore looking for a representation $\pi : \mathcal{D} \to \mathcal{L}(\mathcal{H})$ of that subalgebra of $\mathcal{D}$ as self-adjoint, linear operators on a Hilbert space such that $[\pi(P_a), \pi(Q^b)] = i\hbar \delta^b_a$.

At this point it looks as if we have completely trivialised the reduced phase space quantisation problem of our constrained Hamiltonian system because there is no Hamiltonian to be considered and so it seems that we can just choose any of the standard kinematical representations for quantising the phase space coordinatised by the q^a, p_a and simply use it for Q^a, P_a because the respective Poisson algebras are (weakly) isomorphic. However, this is not the case. In addition to satisfying the canonical commutation relations we want that the multi-parameter group of automorphisms α^τ on $\mathcal{D}$ be represented unitarily on $\mathcal{H}$ (or at least a suitable, preferred one-parameter group thereof). In other words, we want that there exists a multi-parameter group of unitary operators $U(\tau)$ on $\mathcal{H}$ such that $\pi(\alpha^\tau(Q^a)) = U(\tau)\pi(Q^a)U(\tau)^{-1}$ and similarly for P_a.

Notice that due to the relation (which is exact on the constraint surface)

$$\alpha^\tau(Q^a) = F_{\alpha'_\tau(q^a),T} = \sum_{\{k\}} \prod_j \frac{\tau_j^{k_j}}{k_j!} F_{\Pi_j X_j^{k_j} \cdot q^a, T} \qquad (2.2.34)$$

and where on the right-hand side we may replace any occurrence of P_j, T_j by functions of Q^a, P_a according to the above rules. Hence the automorphism α^τ preserves the algebra of functions of the Q^a, P_a, although it is a very complicated map in general and in quantum theory will suffer from ordering ambiguities. On the other hand, for short time periods (2.2.34) gives rise to a quickly converging perturbative expansion. Hence we see that the representation problem of $\mathcal{D}$ will be severely constrained by our additional requirement to implement the multi-time evolution unitarily, if at all possible. Whether or not this is feasible will strongly depend on the choice of the T_j.

A possible way to implement the multi-fingered time evolution unitarily is by quantising the Hamiltonians H_j that generate the Hamiltonian flows $\tau_j \mapsto \alpha^\tau$ where $\tau_k = \delta_{jk}\tau_j$. This can be done as follows: the original constraints C_j can be solved for the momenta P^j conjugate to T_j and we get equivalent constraints $\tilde{C}_j = P^j + E_j(q^a, p_a, T_k)$. These constraints have a strongly Abelian constraint algebra.[2] We may write $C'_j = K_{jk}\tilde{C}_k$ for some regular matrix K. Since $\{C'_j, T_k\} \approx \delta_{jk} = \{\tilde{C}_j, T_k\}$ it follows that $K_{jk} \approx \delta_{jk}$. In other words $C'_j = \tilde{C}_j + O(C^2)$ where the notation $O(C^2)$ means that the two constraint sets differ by terms quadratic in the constraints. It follows that the Hamiltonian vector fields $X_j, \tilde{X}_j$ of $C'_j, \tilde{C}_j$ are weakly commuting. We now set $H_j(Q^a, P_a) := F^0_{E_j,T} \approx E_j(F^0_{q^a,T}, F^0_{p_a,T}, F^0_{T_k,T}) \approx E_j(Q^a, P_a, 0)$. Now let f be any function which depends only on q^a, p_a. Then we have

$$\{H_j, F^0_{f,T}\} \approx F^0_{\{E_j,f\}^*,T} = F^0_{\{E_j,f\},T} = F^0_{\{\tilde{C}_j,f\},T}$$

$$= \sum_{\{k\}} \prod_l \frac{(\tau_l - T_l)^{k_l}}{k_l!} \prod_l X_l^{k_l} \cdot \tilde{X}_j \cdot f$$

$$\approx \sum_{\{k\}} \prod_l \frac{(\tau_l - T_l)^{k_l}}{k_l!} \tilde{X}_j \cdot \prod_l X_l^{k_l} \cdot f$$

$$\approx \tilde{X}_j \cdot F^0_{f,T} - \sum_{\{k\}} \left(\tilde{X}_j \cdot \prod_l \frac{(\tau_l - T_l)^{k_l}}{k_l!} \right) \prod_l X_l^{k_l} \cdot f$$

$$\approx + \sum_{\{k\}} \prod_l \frac{(\tau_l - T_l)^{k_l}}{k_l!} X_j \cdot \prod_l X_l^{k_l} \cdot f$$

$$= \left(\frac{\partial}{\partial \tau_j} \right)_{\tau=0} \alpha^\tau(F_T(f)) \tag{2.2.35}$$

where we have used in the second step that $\{T_j, E_k\} = \{T_j, f\} = 0$, in the third we have used that $\{P_j, f\} = 0$, in the fifth we have used that the $X_j, \tilde{X}_k$ are

---

[2] *Proof:* We must have $\{\tilde{C}_j, \tilde{C}_k\} = \tilde{f}_{jk}{}^l\tilde{C}_l$ for some new structure functions $\tilde{f}$ by the first-class property. The left-hand side is independent of the functions P^j, thus must be the right-hand side, which may therefore be evaluated at any value of P^j. Set $P^j = -E_j$. □

weakly commuting, in the seventh we have used that $F_{f,T}^0$ is a weak observable, and in the last the definition of the flow. We conclude that the Dirac observables H_j generate the multi-fingered flow on the space of functions of the Q^a, P_a when restricted to the constraint surface. The algebra of the H_j is weakly Abelian because the flow α^τ is a weakly Abelian group of automorphisms.

Thus, the problem of implementing the flow unitarily can be reduced to finding a self-adjoint quantisation of the functions H_j. Preferred one-parameter subgroups will be those for which the corresponding Hamiltonian generator is bounded from below. Notice, however, that in (2.2.35) we have computed the infinitesimal flow at $\tau = 0$ only. For an arbitrary value of τ the infinitesimal generator $H_j(Q^a, P_a, \tau)$ defined by[3]

$$\{H_j(\tau), F_{f,T}^\tau\} := \frac{\partial}{\partial \tau_j} \alpha^\tau (F_T(f)) \tag{2.2.36}$$

may not coincide with $F_{E_j,T}^0$ since the Hamiltonian could be explicitly time τ-dependent. In particular, the calculation (2.2.35) does not obviously hold any more even by setting $H_j(\tau) := F_{E_j,T}^\tau$ because even if f depends on q^a, p_a only, $\alpha'_\tau(f), \alpha'_\tau(E_j)$ may depend on P_j as well.

Finally let us remark that the physical Hamiltonians defined in this section are not only required for reduced phase space quantisation but also for the Dirac quantisation (quantisation before reducing) that will be chosen for the remainder of the book. For instance, in [264] a one-parameter family $s \mapsto r_j(s)$ is obtained, using suitable matter, such that the Hamiltonian $H(s) = \sum_j \dot{r}_j(s) H_j(\tau(s))$ is actually time s-independent and bounded from below. This physical Hamiltonian turns out to be close to the Hamiltonian of the standard model when the metric is close to being flat. When properly quantised its ground state could be a candidate for a physical vacuum state for General Relativity. This would be an improvement of the situation for QFT on curved, non-stationary spacetimes (such as our universe) where no natural candidate for a vacuum state exists.

2.3 Recovery of locality in General Relativity

The relational point of view also resolves another puzzle about the mathematical formalism: the apparent lack of locality. The observables (2.1.14) are completely smeared out over the unphysical coordinate time t which contradicts our physical intuition that we can make local measurements in spacetime. In GR the Dirac observables will also be smeared out over all of space as we will see, therefore Dirac observables are not local with respect to the unphysical coordinates t, x^a at all. However, the resolution lies precisely in what one means by 'local'. By local

---

[3] Notice that, due to the Jacobi identity and the fact that the τ derivative of $F_{f,T}^\tau$ is a Dirac observable, the Hamiltonian (2.2.36) must be a complete observable, too.

we mean that some property of a system S is measured over a finite time interval in a finite region of space. This spacetime region in which the measurement takes place however is not specified in terms of some coordinates but rather in terms of other measurements. Usually one does this in terms of matter degrees of freedom. For instance, we could use lightrays and mirrors to measure the spatial extension of the laboratory and the decay time of some radioactive element to measure time durations. Abstractly we are using $D+1$ partial observables, one for each spacetime direction, and their values specify a spacetime region. Mathematically these correspond to $D+1$ scalar fields $T(t,\vec{x})$, $Y^a(t,\vec{x})$ on our spacetime manifold M which we can describe in any unphysical coordinates $X = (t, x^a)$ and we assume that we have constructed them in such a way that the spacetime region of interest is defined precisely as the set of coordinates R where these fields are simultaneously non-vanishing. Suppose now that we have one more field $S(t,\vec{x})$ which mathematically is a scalar density on M. An example would be $S = \sqrt{|\det(g)|}$ for the spacetime metric g. Then

$$S_{T,Y} := \int_R d^{D+1}X \, S(X) = \int_M d^{D+1}X \, \chi_R(x) \, S(x)$$

$$= \int_M d^{D+1}x \, S(x) \tilde{\theta} \left(\left| T(x) \prod_{a=1}^{D} Y^a(x) \right| \right) \tag{2.3.1}$$

is Diff(M)-invariant and would measure, in the example just given, the spacetime volume of the spacetime region specified by the support of the matter fields T, Y^a. Here χ_R denotes the characteristic function of a set and $\tilde{\theta}(x) = 1 - \theta(-x)$ where θ is the step function. Hence the integrand is a scalar density and the integral is invariant under passive diffeomorphisms. Notice that the integral (2.3.1) is over all of M, it is therefore completely non-local in the unphysical coordinates $t, \vec{x}$. However, its actual support is possibly compact and is determined by the dynamical fields T, Y^a. It is therefore local in the physical, relational sense.

Notice, however, that (2.3.1) does not define a Dirac observable because the symmetry group of our theory is not Diff(M) but rather BK(M), which coincide only on-shell. Hence what one should do is something along the following lines: first construct a spatially diffeomorphism-invariant, local observable of the form

$$S_Y := \int_\sigma d^x \, S(x) \, \tilde{\theta} \left(\left| \prod_{a=1}^{D} Y^a(x) \right| \right) \tag{2.3.2}$$

which Poisson commutes with the $\vec{H}(\vec{N})$. Then define the gravitational Master Constraint as

$$:= \int_\sigma d^3x \frac{H(x)^2}{\sqrt{\det(q(x))}} \tag{2.3.3}$$

which also Poisson commutes with the $\vec{H}(\vec{N})$. Next we take any other spatially diffeomorphism-invariant clock variable such as the total volume of σ (which is finite in the spatially compact case)

$$T := \int_\sigma d^3x \sqrt{\det(q(x))} \qquad (2.3.4)$$

and finally define the corresponding $S_{T,Y}(\tau)$ (or $F_{S,(T,Y)}^\tau$ with the method of Section 2.2 using more than one clock T) which is ultralocal (since defined at the physical moment of time τ) or the smeared out version

$$S_{T,Y}(I) = \int_I d\tau\, S_T(\tau) \qquad (2.3.5)$$

where I is a bounded interval. In summary, physical locality can easily be accommodated in quantum gravity while coordinate locality is completely lost. Notice the importance of matter in (2.3.5). In fact, while mathematically in loop quantum gravity we seem to be able to quantise geometry without matter, when it comes to physical observables it seems that matter becomes mandatory.

• •• •••: Canonical QFT is often criticised on the basis that by Haag's theorem, about which more will be said later, the fields on a spatial slice (unsmeared in time) do not exist in the interacting case. This is often stated as the fact that the interaction picture does not exist. In GR this 'no-go-theorem' is evaded due to two reasons. First, Haag's theorem only applies to Wightman fields on Minkowski space. The fields in a background-independent QFT are not Wightman fields by definition. Second, the Dirac observables $F_{f,T}^\tau$ are completely smeared out in the unphysical time anyway. Hence on the physical Hilbert space the argument does not apply.

2.4 Quantum problem of time: physical inner product and interpretation of quantum mechanics

There are two sides to the quantum problem of time. The first deals again with the problem of the frozen picture that one obtains in quantum gravity, the absence of time. The second is more a problem with the interpretation of quantum mechanics itself which, however, becomes especially acute in the context of quantum gravity or, more specifically, quantum cosmology. We will discuss them separately.

2.4.1 Physical inner product

In the classical theory we are supposed to find the gauge-invariant Dirac observables. In the quantum theory, in addition we are supposed to find the states annihilated by all the (Hamiltonian) constraint operators $\hat{H}_I$ or equivalently by the single Master Constraint operator $\hat{}$. We will make this mathematically precise in the next chapter, for the purposes of this section we focus on the conceptual issues which arise from this quantum constraint equation

$$\hat{H}_I \Psi = 0 \,\forall\, I \quad \Leftrightarrow \quad \hat{} \Psi = 0 \tag{2.4.1}$$

Here Ψ belongs to some Hilbert space on which we constructed the operators $\hat{H}_I$ or $\hat{}$. We will call this the kinematical Hilbert space $\mathcal{H}_{\text{kin}}$ because its states, called kinematical states, typically do not solve (2.4.1). Those that do are called physical states.

If there was a Hamiltonian then rather we would solve a Schrödinger equation

$$i\hbar \frac{\partial \Psi}{\partial t} + \hat{H}\Psi = 0 \tag{2.4.2}$$

The difference between (2.4.1) and (2.4.2) is again striking, there is no quantum evolution at all!

To see how this comes about we will consider the case of a single constraint or more generally the Master Constraint. Heuristically, given a state in the $\psi \in \mathcal{H}_{\text{kin}}$ we can produce a physical state by the so-called •••••• • • ••

$$\eta : \psi \mapsto \Psi := \delta(\hat{})\psi := \int_{\mathbb{R}} \frac{dt}{2\pi}\, e^{-it\,\hat{}\,/\hbar}\, \psi \tag{2.4.3}$$

That this solves the constraint formally is due to the identity $x\delta(x) = 0$. More precisely, since (2.4.3) is t-time translation-invariant, $\exp(\text{••} \,\hat{}\, /\text{\textasteriskcentered})\text{•} = \text{•}$ due to the translation invariance of the measure dt. The solutions (2.4.3) to (2.4.1) usually are not normalisable, that is, they do not belong to $\mathcal{H}_{\text{kin}}$ due to the fact that $\delta(\hat{})^2$ is ill-defined. What one does, heuristically, is to observe that formally $\delta(\hat{})^2 = \delta(\hat{})\delta(0)$ and to 'renormalise' by dividing by $\delta(0)$. The result is the physical inner product

$$< \Psi, \Psi' >_{\text{phys}} := < \psi, \delta(\hat{})\psi' >_{\text{kin}} \tag{2.4.4}$$

which is well-defined.

Therefore we see what happens: a physical state is produced by actually integrating the quantum evolution generated by the constraint over the associated unphysical time. Its physical interpretation is thus to be a coherent superposition of all unphysical time evolutions of ψ, that is, of solutions to the would-be Schrödinger equation (2.4.2). Ψ is completely non-local in the unphysical time parameter t which is to be expected in a time reparametrisation-invariant theory where time evolution is to be considered as a gauge transformation. One often calls (2.4.4) a 'transition amplitude'. This terminology is misleading because transition amplitudes are matrix elements of $\hat{U}(t)$. But (2.4.4) is actually the time average of all transition amplitudes between the unphysical states ψ, ψ'. One should really abandon this notion and call (2.4.4) the physical inner product between the physical states Ψ, Ψ'.

What can we do with (2.4.4) and how do we make contact with everyday life where we actually do compute transition amplitudes? How can it be that in quantum gravity there is no time while it makes perfect sense to compute transition amplitudes, say in atomic physics? First of all, it makes sense to define the operators $\hat{S}_T(\tau)$ corresponding to the classical Dirac observables $S_T(\tau)$ on $\mathcal{H}_{\text{phys}}$. In particular, we may compute

$$< \Psi, \hat{S}_T(\tau)\Psi' >_{\text{phys}} \qquad (2.4.5)$$

It is tempting to attribute to (2.4.5) the following interpretation: if we had a true Hamiltonian $\mathbf{H}$ then in the Schrödinger picture the states are evolved unitarily by the operator $\hat{U}(t) = \exp(-i\tau\hat{\mathbf{H}}/\hbar)$, that is, $\Psi_H \mapsto \hat{U}(\tau)\Psi_H =: \Psi_S(\tau)$, while the observables are time-independent. Conversely, in the Heisenberg picture the observables evolve unitarily, that is $\hat{O}_S \mapsto \hat{U}(\tau)^{-1}\hat{O}_S\hat{U}(\tau) =: \hat{O}_H(\tau)$ while the states are time-independent. The two pictures are equivalent in the sense that the expectation value

$$< \psi_S(\tau), \hat{O}_S\psi_S(\tau) >=< \psi_H, \hat{O}_H(\tau)\psi_H > \qquad (2.4.6)$$

is interpreted as the mean value of repeated measurements of the classical quantity $\alpha_\tau^H(O)$ in the state ψ_H.

Equations (2.4.5) and (2.4.6) suggest interpreting the Dirac observables $S_T(\tau)$ as Heisenberg operators with respect to some physical Hamiltonian, if it exists, see above. The physical states are then simply states in the Heisenberg picture. They are not annihilated by the physical Hamiltonian, just by the Hamiltonian constraints. We could then define •••••••• ••• •••••• •• •••••••• with respect to the corresponding physical Hamiltonian. This works perfectly in the case that we can exactly deparametrise the system and presumably in more general cases, at least locally. Therefore, upon finding a suitable clock variable T one can recast the frozen picture totally in terms of the usual picture of quantum mechanics. Thus, the reason for why we can do effective computations in everyday life using the usual notion of time and usual Hamiltonians is that these are indeed physical Hamiltonians generated by a proper choice of partial observable T which we actually do not know in detail. In cosmological models it seems to be related to the total three-volume of space.

Notice that the discussion reveals that both the classical and the quantum, physical time evolution is not absolute but a relative notion: for the same system variable S it depends on the choice of clock variable T.

How should we interpret physical states? As one sees from simple examples of time reparametrisation-invariant systems such as the relativistic particle, the interpretation is simply that they are gauge (time reparametrisation)-invariant states. They form an honest, infinite-dimensional Hilbert space with respect to whose inner product physical states must have non-vanishing and finite norm. They can be labelled by the simultaneous (generalised) eigenvalues of a maximal ideal (i.e., set of mutually commuting operators) of Dirac observables and thus acquire a definite physical interpretation as (generalised) eigenstates of those Dirac observables. We are just mentioning this here because, especially when it comes to cosmology, one might think that the Hilbert space should be one-dimensional, given by the state Ω that was somehow born at the big bang and which evolves unitarily. This is mathematically wrong, because unless the physical time evolution leaves all physical states invariant (up to a phase) the physical states do evolve under time evolution so the time evolution will produce an infinite number of distinct physical states from a given initial one and in order to do expectation value computations with those we need the full infinite-dimensional Hilbert space inner product. For instance, among the Dirac observables we are especially interested in the relational ones $S_T(\tau)$ which, as we just said, should be interpreted as physical time evolutions with respect to a physical Hamiltonian $\mathbf{H}(\tau)$ (which could be explicitly physical time-dependent) in the Heisenberg picture. In the simplest case of a time-independent physical Hamiltonian we just have $S_T(\tau) = \exp(i\tau\mathbf{H}/\hbar) \, S_T(0) \exp(-i\tau\mathbf{H}/\hbar)$ and thus the initial state of the big bang Ω is a Heisenberg picture state which monitors the time evolution of the system S as measured by the clock T by $\tau \mapsto < \Omega, S_T(\tau)\Omega >$ as the universe expands. Notice that we are abusing the notation here as in GR we need an infinite number of clocks and we must select a suitable one-parameter time evolution among all the possible bubble time evolutions.

In any case, the interpretation of Ω is clear once we know its decomposition in terms of the generalised eigenstates of a maximal set of mutually commuting Dirac observables. This problem is not specific to GR but arises also in usual quantum mechanics: given an L_2 function for the hydrogen atom, what is its interpretation? One way of answering this question is certainly by decomposing it with respect to energy and angular momentum eigenstates $e_{n,l,m}$ (and generalised unbounded energy eigenstates).

The big question in cosmology really is what that initial physical state Ω is and how to do quantum mechanics in the case of closed systems. This brings us to the next section.

2.4.2 Interpretation of quantum mechanics

The fact that there is no natural Hamiltonian which drives the quantum time evolution in quantum gravity poses several problems with the usual Copenhagen interpretation of quantum mechanics. Recall that in the Copenhagen interpretation of quantum mechanics or quantum field theory one artificially subdivides the available observables into quantum observables S of the system that one wants to get information about and classical observables T which are associated with the measurement apparatus. A state ψ in the system Hilbert space undergoes unitary evolution $\psi \mapsto \hat{U}(t)\psi$, $\hat{U}(t) = \exp(-it\hat{\mathbf{H}}/\hbar)$ with respect to the dynamics generated by the Hamiltonian $\mathbf{H}$ until it is measured. When this happens, the state collapses to an eigenfunction ψ_λ of the system operator $\hat{S}$ and the corresponding eigenvalues λ are the possible measurement outcomes. (This can be generalised to the case that we are dealing with mixed rather than pure states and that the spectrum of $\hat{S}$ has a continuous part, see below.) The probability for measuring λ is $|<\psi_\lambda, \hat{U}(t)\psi>|^2$. In a strict Copenhagen interpretation one would make a stronger statement, namely that this is the probability for the outcome λ in a repeated number of experiments. If λ was measured then ψ has collapsed to ψ_λ and hereafter evolves again unitarily until the next measurement.

1. • • ••• • • •• • ••

The unitary evolution with respect to a Hamiltonian plays a crucial role in the Copenhagen interpretation. Again, in the a priori absence of any Hamiltonian as in the case of a universe without spatial boundary no Hamiltonian is available a priori.

2. • •• ••• •• •••• ••• • ••

When we are dealing with quantum gravity in a cosmological context then we are talking about measurements taking place on observables of the whole universe. Thus the whole universe is the system and all measurement devices and potential human observers are part of the system. Since there is no outside of the universe by definition, the zeroth step in the Copenhagen interpretation, to talk about an outside measurement apparatus, is invalidated.

Moreover, in cosmology we might ask about probabilities for or expectation values of properties of the universe as a whole such as its lifetime, etc. which depend on the initial conditions (initial in the relational sense, that is at the unphysical time when the spatial extension of the universe was close to zero). It is clear that such answers cannot be addressed in the strict Copenhagen context because we can hardly repeat the big bang 'experiment'.

3. • • ••• • •• ••• •••

Even if there is a Hamiltonian, people discuss the collapse problem or, in other words, Schrödinger's cat problem. Is the system really in a coherent superposition of eigenstates of the system operator $\hat{S}$ until it is measured? Or does the system have a definite reality even without any measurement? What causes the wave function to collapse?

4. • •• • ••• •• •••• ••• • ••

The usual Einstein–Podolsky–Rosen Gedankenexperiment shows that there is a fundamental non-locality in quantum mechanics. This seems to contradict the axioms of usual (algebraic) quantum field theory which requires algebras of local operators supported in spacelike separated, that is, causally disconnected, regions to (anti)commute.

5. • • •••• •• ••• • •• ••• ••

If there is unitary evolution by a Hamiltonian, how can it be that there is obviously a time asymmetry in nature towards increasing entropy, sometimes called the (thermodynamic) arrow of time.

6. •• •• •• • ••• • •• ••• •• ••• •

In connection with black holes the following problem arises. Complex systems containing a lot of information such as a star can collapse to a black hole but the black hole is completely described by a few parameters such as mass, charge and spin. There is possibly Hawking radiation coming out of the black hole but it is completely thermal and does not carry any information. In fact, as long as the black hole has not completely evaporated the total system is described by a tensor product of Hilbert spaces for the inside and the outside of the black hole and by taking the partial trace[4] of a given pure state with respect to the inside Hilbert space one artificially produces a density matrix for the outside Hilbert space. The full state is still pure, until the black hole evaporates and the inside Hilbert space is gone, now there is only the density matrix left and we do have information loss. What happened with that information? Obviously, unitary evolution cannot create a mixed state from a pure state.

Let us now attempt a resolution of these problems. Many of these ideas again follow Rovelli [3].

1. •• • ••• • •• •••••

As we have shown in Section 2.4.1 one ••• regain a notion of time. However, this is with respect to a clever choice of partial clock observable giving rise to a •••••••• Hamiltonian. This description should be valid at least locally in physical τ-time. Thus, the 'absence of time problem' can be solved, at least in principle, although it may not be easy to find suitable clock variables. One is then back in the conceptually easier realm of a quantum mechanical system with a Hamiltonian.

2. • •• ••• •• •••• ••• ••

The whole idea of separating the world into a classical and a quantum part is anyway fundamentally wrong. The world is uniformly described by quantum mechanics. Therefore one should discard the classicality of the measurement device altogether and just speak about the compound aggregate consisting of the system under study interacting with the measurement device, be it macroscopic or microscopic.

The reason for why the Copenhagen interpretation works so well is that macroscopic objects display an interesting feature: they decohere, that is, quantum mechanical interference is negligible. Thus, for those the decoherence condition derived below is satisfied with an extremely high precision. That

---

[4] Let $\mathcal{H}_I, I = 1, 2$ be separable Hilbert spaces with orthonormal bases $(b_\alpha^{(1)}), (b_j^{(2)})$ respectively. A general vector state in the tensor product Hilbert space $\mathcal{H} = \mathcal{H}_1 \otimes \mathcal{H}_2$ is given by $\Psi = \sum_{\alpha,j} \Psi_{\alpha,j} b_\alpha^{(1)} \otimes b_j^{(2)}$ where $\Psi_{\alpha,j} \in \mathbb{C}, \sum_{\alpha,j} |\Psi_{\alpha,j}|^2 = 1$. The partial trace of Ψ with respect to $\mathcal{H}_1$ is given by the operator $\rho := \sum_{i,j} \rho_{ij} < b_i^{(2)}, . >_{\mathcal{H}_2} b_j^{(2)}$ on $\mathcal{H}_2$ where $\rho_{ij} := \sum_\alpha \overline{\Psi_{\alpha i}} \Psi_{\alpha j}$. The operator ρ is bounded and of unit trace, hence trace class. It appears naturally in the form $\text{Tr}_{\mathcal{H}_2}(\rho A) = < \Psi, [1_{\mathcal{H}_1} \otimes A] \Psi >_{\mathcal{H}}$ where A is an operator on $\mathcal{H}_2$. These are the kind of operators that one considers when one does not have information about part of the system in question, here encoded by the degrees of freedom described by $\mathcal{H}_1$.

is why their classical description is valid. Technically this happens because macroscopic objects are described by a large number of degrees of freedom and while every single degree of freedom displays the fundamental quantum mechanical coherence, the interference terms of the ensemble become small. More specifically, the Hilbert space of the macroscopic system is a large tensor product of Hilbert spaces for the individual microscopic systems and even if we have a coherent superposition of those large tensor product states, the interference terms (inner products between different states of that linear combination) drop out because they are nearly orthogonal, being the large product of numbers of modulus less than one (by the Schwarz inequality). To illustrate this, consider a crude example, namely the interaction of a single spin system S_0 with a measurement device S_N which we assume to be a system of N spins where N is large. As an, of course physical, interaction Hamiltonian we take $\hat{H} = \hbar \sum_{n=1}^{N} g_n \hat{s}_3^0 \otimes \hat{s}_3^n$ where $\hat{s}_j^n$ is the jth component of the spin operator for the individual spin degree of freedom and g_n is a coupling constant. The Hilbert space of the compound system is the large tensor product $\mathcal{H} = \otimes_{n=0}^{N} \mathcal{H}_n$ where $\mathcal{H}_n = \mathbb{C}^2$. Let us assume that the ground state is such that all spins are down. One can easily solve the associated Schrödinger equation for any given initial (pure) state $\Psi = \otimes_{n=0}^{N} \psi_n$. When computing the decoherence functional for alternative values of projections for the spin operator $n_j \hat{s}_j^0$ for the system spin S_0 one finds that the off-diagonal, interference, entries of the decoherence matrix to be defined below are proportional to

$$\prod_{n=1}^{N} [\cos(2g_n t) + i\sin(2g_n t)[2|b_n|^2 - 1]]$$

where $t = t_2 - t_1$ is the time interval for only two branchings and b_n is determined by $\psi_n = a_n|+>_n + b_n|->_n$, $|a_n|^2 + |b_n|^2 = 1$. The modulus of this term decays to zero exponentially fast with t for sufficiently random distribution of the b_n.

This is the mathematical explanation for why in the famous double slit experiment the quantum mechanical interference is destroyed once there is an interaction between some macroscopic measurement device which detects through which slit the electron has passed as compared with the situation when no such detector exists. Surprisingly, decoherence is rather hard to model for realistic physical situations although it is apparently such a widespread phenomenon and is the underlying reason why the macroscopic world is so well described by classical physics. See, for example, [265] for more information.

It is worth pointing out that the relational approach to observables is especially well suited to the closed system problem because when we talk about the complete observable $S_T(\tau)$ built from the partial ones S, T, we just •• •• S the system and T the clock. Nevertheless, both are treated as operators in the quantum theory and in that sense there is no classical measurement apparatus any longer, both partial observables are subject to quantum fluctuations, the quantum system and classical measurement device separation is absent.

3. • ••• ••• ••• •••

An interesting reinterpretation of quantum mechanics is the consistent history approach [143–155] sometimes called post-Everett interpretation. It arose from the desire to get rid of the artificial separation of the world into classical and quantum that we just discussed. It resolves the collapse problem in the following way.

(i) There is no separation between a quantum system and a classical measurement apparatus, everything is fundamentally quantum. There is only one system.

(ii) Consequently, there is no such thing as a classical measurement of a quantum property. Rather, there is interaction between all components of the system all the time.

(iii) Given a self-adjoint operator $\hat{A}$ on the full Hilbert space $\mathcal{H}$ describing some property of the full system, we know its spectral projections $\lambda \mapsto E(\lambda)$ where $\lambda \in \mathbb{R}$, satisfying $\lim_{\lambda \to -\infty} E(\lambda) = 0$, $\lim_{\lambda \to +\infty} E(\lambda) = 1_{\mathcal{H}}$, $\lim_{\lambda \to \lambda_0+} E(\lambda) = E(\lambda_0)$, $E(\lambda)E(\lambda') = E(\min(\lambda, \lambda'))$. At any given time we can decompose the spectrum of $\hat{A}$ into mutually disjoint intervals $I = (a_I, b_I]$ and define $\hat{P}_I = E(b_I) - E(a_I)$ so that $\sum_I \hat{P}_I = 1_{\mathcal{H}}$ and $\hat{P}_I \hat{P}_J = \delta_{IJ} \hat{P}_I$.

(iv) In the Copenhagen interpretation, given an initial density matrix $\hat{\rho}$, that is, a trace class operator of unit trace, of the full system at $t = 0$ it evolves unitarily according to $t \mapsto \hat{U}(t)\hat{\rho}\hat{U}(t)^{-1}$ until some measurement of some property, corresponding to some operator $\hat{A}$, of the system takes place. If one measures the system property to be in the range of the interval I (taking care of the fact that in physics we can never make absolute precision measurements, measurement values always take some finite range) then the probability for that measurement is

$$\text{Tr}(\hat{P}_I \hat{U}(t)\hat{\rho}\hat{U}(t)^{-1}\hat{P}_I) \tag{2.4.7}$$

and the density matrix gets reduced to

$$\frac{\hat{P}_I \hat{U}(t)\hat{\rho}\hat{U}(t)^{-1}\hat{P}_I}{\text{Tr}(\hat{P}_I \hat{U}(t)\hat{\rho}\hat{U}(t)^{-1}\hat{P}_I)} \tag{2.4.8}$$

and hereafter the reduced density matrix again evolves unitarily.

In the consistent history interpretation of quantum mechanics one takes a radical further step: there are no measurements, simply the system branches out, that is, at each moment of time it has mutually exclusive possibilities of alternatives to decide for each of its properties $\hat{A}$. Suppose that at time t_n, $n = 1, \ldots, N$ we consider a decomposition of the Hilbert space into alternative projections $\hat{P}_{I_n}^n$, $I_n \in \mathcal{I}_n$ corresponding to self-adjoint operators $\hat{A}_n$.

The probability for taking the branch I_n at time t_n • •••• having gone through the branches $I_1, \ldots, I_{n-1}$ at $t_1, \ldots, t_{n-1}$ is given by (2.4.7), that is

$$\mathrm{Tr}\big(\hat{P}^n_{I_n}\hat{U}(t_n - t_{n-1})\hat{\rho}_{(I_1,t_1)\ldots(I_{n-1},t_{n-1})}\hat{U}(t_n - t_{n-1})^{-1}\hat{P}^n_{I_n}\big) \qquad (2.4.9)$$

and when this branch is chosen, the density matrix for that branch is given by (2.4.10), that is

$$\hat{\rho}_{(I_1,t_1)\ldots(I_n,t_n)} = \frac{\hat{P}^n_{I_n}\hat{U}(t_n - t_{n-1})\hat{\rho}_{(I_1,t_1)\ldots(I_{n-1},t_{n-1})}\hat{U}(t_n - t_{n-1})^{-1}\hat{P}^n_{I_n}}{\mathrm{Tr}\big(\hat{P}^n_{I_n}\hat{U}(t_n - t_{n-1})\hat{\rho}_{(I_1,t_1)\ldots(I_{n-1},t_{n-1})}\hat{U}(t_n - t_{n-1})^{-1}\hat{P}^n_{I_n}\big)}$$
$$(2.4.10)$$

We conclude that the joint probability for having followed the • •••••• , that the range I_n of property $\hat{A}_n$ was realised at t_n, is given by the product of the conditional probabilities (2.4.9) which is

$$D(\{I\}, \{I\}; \{t\}; \hat{\rho}) := \mathrm{Tr}\big(\hat{P}^n_{I_n}\hat{U}(t_n - t_{n-1}) \ldots \hat{P}^1_{I_1}\hat{U}(t_2 - t_1)\hat{\rho}$$
$$\times \, [\hat{P}^n_{I_n}\hat{U}(t_n - t_{n-1}) \ldots \hat{P}^1_{I_1}\hat{U}(t_2 - t_1)]^\dagger\big) \quad (2.4.11)$$

There is, however, a potential problem with the interpretation of (2.4.11) as the probability of a history $(\{I\}, \{t\})$: if it was a probability then the sum of the probabilities for all branches must add up to unity. However, this does not follow automatically from (2.4.11). Thus, one must impose an additional condition on the choice of branching that one considers: define the • ••••••••• •• •• • •••• • •

$$D(\{I\}, \{J\}; \{t\}; \hat{\rho}) := \mathrm{Tr}\big(\hat{P}^n_{I_n}\hat{U}(t_n - t_{n-1}) \ldots \hat{P}^1_{I_1}\hat{U}(t_2 - t_1)\hat{\rho}$$
$$\times \, [\hat{P}^n_{J_n}\hat{U}(t_n - t_{n-1}) \ldots \hat{P}^1_{J_1}\hat{U}(t_2 - t_1)]^\dagger\big) \quad (2.4.12)$$

We say that the set of histories is consistent whenever the decoherence functional is close to $\delta_{\{I\},\{J\}}$. This means that all interference terms are subdominant and implies that

$$\sum_{\{I\}} D(\{I\}, \{I\}; \{t\}; \hat{\rho}) \approx \sum_{\{I\},\{J\}} D(\{I\}, \{J\}; \{t\}; \hat{\rho}) = 1 \qquad (2.4.13)$$

where $\sum_{I_n \in \mathcal{I}_n} \hat{P}^n_{I_n} = 1\mathcal{H}$ was used.

How well the decoherence condition is satisfied depends on both the initial density matrix, the set of chosen alternatives and the amount of •••••• •••••• • ••• (i.e., the number of alternatives per chosen point of time and the number of branching points of time t_n per physical unit time interval). As shown in [154, 155] it is not at all straightforward or granted to satisfy the decoherence condition. However, decoherence is a phenomenon obviously satisfied for macroscopic systems in nature as illustrated above, hence it a reasonable condition to assume.

Given this set-up one may then attribute the following interpretation to quantum mechanics: the evolution of the density matrix is not unitary but rather follows a particular path or history in the set of all possible histories. The probability for a particular history is computable. If we know the density matrix at a given point of time we can compute what is most likely to happen next but there is no certainty. The collapse of the density matrix is no longer a mysterium of the measurement process but rather an integral part of the quantum mechanical evolution of the density matrix of the full system, consisting of both the would-be Copenhagen system and the would-be Copenhagen measurement apparatus, as it proceeds along its history of alternatives. It is interaction between the components of the complete system which makes complex sets of alternatives possible, as otherwise the density matrix of the system would not change with probability one.

At this point one may debate whether (a) all the alternative histories are realised but no history knows about the other (many-world or Everett interpretation) or (b) there is always only one history realised, the one that we experience, just that we can never determine with certainty what happens next. The present author prefers the second possibility as the first could never be distinguished experimentally from the second anyway. Notice that the special role of a human observer has completely disappeared from this picture. Schrödinger's cat, as a macroscopic and hence decoherent system, is either alive or dead at any given moment of time and not in a coherent superposition, whether or not a human verifies it and we can compute the probability for the time evolution of either alternative. This also resolves the other interpretational problem with quantum cosmology: there is no repeated experiment interpretation of expectation values necessary.

The real challenge for quantum cosmology lies in understanding the physics of the initial conditions, that is, whether some initial states are preferred over others or whether all are equally probable. In other words, do we happen to live in a universe which somehow is generic or is it one of zillions of possibilities some of which produce life? Is there a big bang in quantum theory at all and if not what replaces it? Is there a time before the big bang? In either case, in our interpretation this would be a prediction about one universe rather than many branches of universes, that is, a multiverse.

Notice that this resolution of the collapse problem is beautifully compatible with the relational resolution of the closed system problem as in both cases the fundamental distinction between system and measurement device is absent. Moreover, in GR of course the projections must be those of Dirac observables, the time is a physical time selected by partial clock observables and the Hamiltonian must be the corresponding physical one. All three can be derived using the partial observable approach. This way the decoherence functional is also a relational object.

4. •••• •••• •••••••

In algebraic QFT [22] one talks about algebras of local observables $\mathfrak{A}(\mathcal{O})$ where $\mathcal{O}$ are certain spacetime regions. Since in algebraic QFT one works on a background spacetime (M, η) this formulation can only be valid in the semiclassical limit of quantum gravity in which the fluctuations of the gravitational field are small and concentrated around η. It then makes sense to talk about spacelike separated regions. Next, from the point of view of the full quantum gravity theory as we discussed before, locality should really be understood as dynamical locality and not as coordinate locality. In QFT one implicitly assumes that this has been done so that the coordinates actually do have physical meaning, that is, a name.

Assuming that all of this has been achieved, one can make the EPR objection, namely that there are quantum mechanical correlations between causally disconnected regions, which seems to contradict the axioms of algebraic QFT, namely that the algebras $\mathfrak{A}(\mathcal{O}), \mathfrak{A}(\mathcal{O}')$ (anti)commute. However, there is no contradiction for two reasons. First, the axioms of algebraic QFT are purely algebraic, they do not even refer to any particular representation of the algebras. On the other hand, the EPR type of paradoxes refer to a particular state that has been prepared in such a way that when a measurement has been made in $\mathcal{O}$ the measurement in $\mathcal{O}'$ has a definite outcome. This has something to do with the non-locality of the state and not with that of the algebras. Second, even if there is a correlation between the outcomes in $\mathcal{O}$, $\mathcal{O}'$ the corresponding observers actually have no way to find out about it before they have communicated. But this requires extending $\mathcal{O}$ and $\mathcal{O}'$ such that they are no longer causally disconnected and now the associated algebras are no longer required to (anti)commute.

5. • •••• •• ••• • •••••••

If we accept the consistent history point of view of quantum mechanics then it is clear from where the direction of time comes: from the fact that there is an initial density matrix but no final one. This could be generalised to a decoherence functional with a final density matrix as well and the question of time symmetry becomes a question of choice of final and initial density matrix. Thus, the time asymmetry is related to the boundary conditions while the mathematical framework is completely time symmetric.

6. •• ••••• ••••• •• •• •••••••

In the mind of the author there is no convincing argument which speaks for or against information loss in quantum gravity. After all, the reasoning which leads to the black hole information loss problem is a semiclassical one in which matter is treated quantum mechanically, geometry classically and the backreaction of geometry on matter is completely neglected. Moreover, any Hawking radiation mode which reaches us at future null infinity is infinitely blue shifted at the black hole horizon and thus reaches the Planck energy close to the horizon at which new physics should happen and a fundamental quantum gravity theory must take over. In other words, one should not trust these arguments since they are made outside the domain of their validity and it is not clear whether there is any problem at all in the full quantum gravity theory. In fact, in LQG it seems that spacetime singularities can indeed be resolved as we will see, hence there just may not be any information paradox if the physical Hamiltonian stays well-defined all the time so that an initially pure state remains pure and no information loss (entropy) emerges.

3
The programme of canonical quantisation

In this chapter we give a systematic description of which steps the method of canonical quantisation consists of. The basic idea, due to Dirac, is that one quantises the unconstrained phase space, resulting in a kinematical Hilbert space and then imposes the vanishing of the constraints as operator equations on physical states. The motivation behind this 'quantisation before constraining' is that in the opposite procedure one would need to know the full set of Dirac observables. This may not only be practically hard even classically as in the case of interacting field theories such as GR but, even if the full set of Dirac observables could be found, it could be very hard to find representations of their corresponding Poisson algebra, see, for example, [189, 205, 260] and the previous chapter.

Thus, Dirac quantisation is a way to enter the quantum regime even if the underlying classical system is too complicated in order to find all its gauge invariants. While it will be even harder to find all the quantum Dirac observables, the real advantage is that (1) in a concrete physical situation we only need a few of these invariants rather than all of them and (2) starting from the kinematical representation of non-observable quantities we automatically arrive at an induced representation of the invariants which can be expressed in terms of the non-observables.

The canonical approach is ideally suited to constructing background metric independent representations of the canonical commutation relations as is needed, for example, in quantum gravity. Dirac's original work was subsequently refined by many authors, see, for example, [16, 17, 266–278]. In what follows we present a modern account. As we will see, in its modern form Dirac's programme uses some elements of the theory of operator algebras and algebraic QFT (AQFT) [22] but what is different from AQFT is that the canonical approach is, by definition, a quantum theory of the initial data, that is, operator-valued distributions are smeared with test functions supported in $(D-1)$-dimensional slices rather than D-dimensional regions. While needed in order to define a background-independent quantum theory, this is usually believed to be a bad starting point in AQFT because of the singular behaviour of the n-point Wightman distributions of interacting scalar fields in perturbation theory when smeared with 'test functions' supported in lower-dimensional submanifolds. The way out of this 'no-go theorem' is twofold. (1) In usual perturbation theory one uses very specific (Weyl) algebras and corresponding representations (of Fock type in perturbation theory) to formulate the canonical commutation relations, but the singular

• • • •••••••• • • •• ••••••• •••• •••• ••••• ••••• •

behaviour might be different for different algebras and their associated representations. (2) In a reparametrisation-invariant theory such as General Relativity
the •••••••••• are by definition time-independent, hence already smeared out
in unphysical time, see our discussion in Chapter 2. Thus the AQFT criticism is
certainly removed at the physical level.

3.1 The programme

Assume we are given an (infinite-dimensional) constrained symplectic manifold
$(\mathcal{M}, \Omega)$ modelled on a Banach space E with strong symplectic structure Ω and
first-class constraint functionals $H_I(N^I)$ (in case of second-class constraints one
should replace Ω by the corresponding Dirac bracket [219]). Here I takes values
in some finite index set and $H_I(N^I)$ is an appropriate pairing as in the previous
chapter between the constraint density $H_I(x)$, x a point in the D-dimensional
manifold of the Hamiltonian framework, and its corresponding Lagrange multiplier N^I. Unless otherwise specified no summation over repeated indices I is
assumed. We may or may not have a physical Hamiltonian $\mathbf{H}$ which Poisson
commutes with the constraints.

The quantisation algorithm for this system consists of the following.

I. • •••••••• • •••••• *••••• •••••• $\mathfrak{P}$

The phase space can be coordinatised in many ways by a set $\mathfrak{S}$ of 'elementary
variables', that is, global coordinates such that all functions on $\mathcal{M}$ can be
expressed in terms of them. Since we want to quantise the system by asking
that commutators be represented as $i\hbar$ times the quantisation of the Poisson
bracket, we must ask that the elementary variables form a closed Poisson
subalgebra of the full Poisson algebra $C^\infty(\mathcal{M})$. It may be convenient to use
complex coordinates and in that case we require that the Poisson subalgebra
is closed under complex conjugation because we want that operator adjoints
are represented by the quantisations of the complex conjugates. We can
guarantee all that by starting with any set $\mathfrak{S}$ of functions which separates the
points of $\mathcal{M}$ and construct from it and their complex conjugates the smallest
Poisson algebra they generate. Mathematically speaking, the resulting object
is a separating Poisson *-subalgebra $\mathfrak{P}$ on $\mathcal{M}$. In the field theory context it
is certainly necessary to smear the fields in order that the Poisson brackets
be non-distributional.

The choice of $\mathfrak{P}$ is guided by physical considerations: $\mathfrak{P}$ should be minimal, in the sense that removing members would violate the definition of
a separating *-algebra, because the quantisation of a redundant function is
already determined by that of a smaller set of functions. One set of elementary variables may be more convenient than another in the sense that
the equations of motion or the constraint functions C_I look more or less
complicated in terms of them. Moreover, it is convenient if the members of

$\mathfrak{P}$ have a simple transformation behaviour under the gauge transformations generated by the constraints because otherwise invariant functions, that is, Dirac observables will be complicated functions of the elementary variables and hence difficult to quantise. The most important condition is that the symplectic structure among the members of $\mathfrak{P}$ should be as simple as possible, ideally the Poisson brackets should be independent of $\mathcal{M}$, so that one has a chance to find representations of $\mathfrak{P}$ as operators on a Hilbert space.

Further complications may arise in case that the phase space does not admit an independent set of global coordinates. In this case it may be necessary to work with an overcomplete set of variables and to impose their relations among each other as conditions on states on the Hilbert space. For example, suppose we want to coordinatise the cotangent bundle over the sphere S^2. The sphere cannot be covered by a single coordinate patch, but we can introduce Cartesian coordinates on $\mathbb{R}^3$ and impose the condition $(\hat{x}^1)^2 + (\hat{x}^2)^2 + (\hat{x}^3)^2 - 1 = 0$ on states depending on $\mathbb{R}^3$.

If $\mathcal{M}$ has the structure of a cotangent bundle $T^*\mathcal{Q}$ over some configuration space $\mathcal{Q}$ then a natural candidate for $\mathfrak{P}$ is as follows. Select a suitable algebra $\mathrm{Fun}(\mathcal{Q})$ of (smeared) functions on $\mathcal{Q}$, say suitable functions of the $F(q)$ for GR. The Hamiltonian vector fields on $\mathcal{M}$ of the (smeared) momentum functions, say the $P(f)$ for GR, preserve the space $\mathrm{Fun}(\mathcal{Q})$ if chosen sufficiently smooth, hence they define elements of the space $V(\mathcal{Q})$ of vector fields on $\mathcal{Q}$. The product space $\mathrm{Fun}(\mathcal{Q}) \times V(\mathcal{Q})$ carries a Lie algebra structure according to $\{(f, u), (f', u')\} := (u[f'] - u'[f], [u, u'])$ where $u[f]$ is the action of vector fields on functions and $[u, u']$ is the Lie bracket of vector fields. In slight abuse of notation we still refer to this Lie algebra bracket on $\mathrm{Fun}(\mathcal{Q}) \times V(\mathcal{Q})$ as a 'Poisson bracket'. The Poisson algebra $\mathfrak{P}$ can then be identified as the closed Lie subalgebra of $\mathrm{Fun}(\mathcal{Q}) \times V(\mathcal{Q})$ generated by the chosen functions of the configuration variables and the Hamiltonian vector fields of the momentum functions. Notice that the Lie bracket between Hamiltonian vector fields of functions is the Hamiltonian vector field of the Poisson bracket between the corresponding functions. It may be easier to compute the Lie bracket rather than the Poisson bracket and the former determines the latter up to a constant. See Section 19.3.

II. ••••••• *•••••••• $\mathfrak{A}$

Given the classical Poisson *-algebra $\mathfrak{P}$ of elementary kinematical variables (kinematical in the sense that they do not Poisson commute with the constraints) we want to define an abstract *-algebra $\mathfrak{A}$ based on $\mathfrak{P}$ which in a precise sense implements $i\hbar$ times the Poisson bracket structure on $\mathfrak{P}$ as commutation relations in $\mathfrak{A}$ and the reality structure on $\mathfrak{P}$ as involution relations in $\mathfrak{A}$. Recall that an involution in an algebra is defined as an antilinear automorphism which reverses order and squares to the identity, that is, $(z_1 a + z_2 b)^* := \bar{z}_1 a^* + \bar{z}_2 b^*$, $(ab)^* := b^* a^*$ and $(a^*)^* = a$ for $a, b \in \mathfrak{A}$, $z_1, z_2 \in \mathbb{C}$. The involution should not be confused with the adjoint operation in a Hilbert space, in fact, we have not talked about a Hilbert space yet which arises only when we consider representations of the algebra $\mathfrak{A}$. For the same reason we will not denote elements of $\mathfrak{A}$ with 'operator hats', we will use operator hats only when considering specific representations of $\mathfrak{A}$ on a Hilbert space.

In order to construct $\mathfrak{A}$ from $\mathfrak{P}$ one proceeds as follows. We consider the (free) tensor algebra $T(\mathfrak{P})$ over $\mathfrak{P}$ defined as

$$T(\mathfrak{P}) := \mathbb{C} \oplus \oplus_{n=1}^{\infty} \otimes_{k=1}^{n} \mathfrak{P} \tag{3.1.1}$$

with elements $a = (a_0, a_1, \ldots, a_n, \ldots)$ where $a_0 \in \mathbb{C}$ and a_n is a finite linear combination of monomials $a_n = a_{1n} \otimes \ldots \otimes a_{nn}$ of elements $a_{kn} \in \mathfrak{P}$ of which all but finitely many vanish. The associative product, addition, multiplication by scalars and involution are defined in the obvious way

$$(a \otimes b)_n = \sum_{k+l=n} a_k \otimes b_l; \quad a_k \otimes b_l = a_{1k} \otimes \ldots \otimes a_{kk} \otimes b_{1l} \otimes \ldots \otimes b_{ll}$$

$$(a+b)_n = a_n + b_n$$

$$(za)_n = za_n; \quad za_n = (za_{1n}) \otimes \ldots \otimes a_{nn} = a_{1n} \otimes \ldots \otimes (za_{nn})$$

$$a^* = \bar{a}_0 \oplus \oplus_{n=1}^{\infty} a_n^*; \quad a_n^* = \bar{a}_{nn} \otimes \ldots \otimes \bar{a}_{1n} \tag{3.1.2}$$

We now divide $T(\mathfrak{P})$ by the two-sided ideal generated from elements of the form

$$a_1 \otimes b_1 - b_1 \otimes a_1 - i\hbar \{a_1, b_1\} \tag{3.1.3}$$

with $a_1, b_1 \in \mathfrak{P}$. This results in the enveloping algebra $\mathfrak{A}$ of the Lie algebra $\mathfrak{P}$ which, in contrast to $T(\mathfrak{P})$, does not carry a (tensor degree) grading.

In view of the representation theory of $\mathfrak{A}$ to which we turn in the subsequent item, when defining $\mathfrak{A}$ from $\mathfrak{P}$ one will usually consider not directly the elements of $\mathfrak{P}$ but rather bounded functions of them, usually called • ••• •••• •••••, which, considered as functions, would still separate the points of $\mathcal{M}$. Thus, for real-valued but unbounded $a \in \mathfrak{P}$ one will consider the one-parameter family of unitary operators $\mathbb{R} \ni t \mapsto W_t(a) := \exp(ita)$ with $a \in \mathfrak{P}$ which for $t \to 0$ approximates $1 + \bullet\bullet a$ where $\mathbf{1}$ is the unit operator in $\mathfrak{A}$. Now (3.1.3) is replaced for $a, b \in \mathfrak{P}$ by

$$W_s(a) W_t(b) W_{-s}(a) := W_t \left(\sum_{n=0}^{\infty} \frac{(\bullet\bullet \hbar)^n}{n!} \{a, b\}_{(n)} \right)$$

$$(W_s(a))^* := W_{-s}(a) = (W_s(a))^{-1} \tag{3.1.4}$$

where $\{a, b\}_{(0)} = b$, $\{a, b\}_{(n+1)} := \{a, \{a, b\}_{(n)}\}$ is the iterated Poisson bracket.

The reason for dealing with Weyl elements rather than the a in case that a is classically unbounded is that in physically interesting representations the operators corresponding to a will be unbounded, hence one can define them only on a dense domain. But it is not at all clear that different self-adjoint operators can be defined on a common and invariant dense domain in which case (3.1.3) would be ill-defined. Dealing with bounded operators which are everywhere defined avoids these so-called ••• ••• •••••••••.

Thus, it is often mathematically more convenient to construct instead of $T(\mathfrak{P})$ the free tensor algebra generated by the Weyl elements and then to quotient it by the two-sided ideal generated by (3.1.4) in order to obtain $\mathfrak{A}$.

III. •••••••••••••••••• $\mathfrak{A}$

A representation of an abstract *-algebra $\mathfrak{A}$ is a *-morphism $\pi : \mathfrak{A} \to \mathcal{L}(\mathcal{H}_{\text{kin}})$ into a subalgebra of linear operators on a Hilbert space $\mathcal{H}_{\text{kin}}$. That is, we have $\pi(1) = 1_{\mathcal{H}_{\text{kin}}}$, $\pi(z_1 a + z_2 b) = z_1 \pi(a) + z_2 \pi(b)$, $\pi(a\,b) = \pi(a)\pi(b)$ and $\pi(a^*) = (\pi(a))^\dagger$ where the latter is the adjoint operation on $\mathcal{H}_{\text{kin}}$ and $1_{\mathcal{H}_{\text{kin}}}$ is the unit operator on $\mathcal{H}_{\text{kin}}$. Here we have made it clear that the Hilbert space $\mathcal{H}_{\text{kin}}$ is the kinematical representation space of the kinematical algebra $\mathfrak{A}$, it is not the physical Hilbert space. See Section 29.1.

Operator algebra theoretic methods such as the GNS construction are of great importance for constructing representations, see Section 29.1. In general, the theory of representations of a given $\mathfrak{A}$ is very rich and not under much control unless one imposes further physical restrictions. Guiding principles here are again gauge invariance, see Section 29.1 and (weak) continuity with respect to $t \mapsto W_t(a)$. For GR the requirement of background independence turns out to be a very tight condition as we will see. Moreover, the representation should be irreducible on physical grounds (otherwise we have superselection sectors, that is, closed invariant subspaces of $\mathfrak{A}$, implying that the physically relevant information is already captured in any one of the closed subspaces). Sometimes one is even able to invoke uniqueness results if one invokes dynamical information such as that the representation should support (i.e., allow a representation of) a Hamiltonian operator [279] or the constraint operators.

Among all possible representations π we are, of course, only interested in those which support the constraints H_I as operators. Since, by assumption, $\mathfrak{A}$ separates the points of $\mathcal{M}$ it is possible to write every H_I as a function of the $a \in \mathfrak{P}$, however, that function is far from unique due to operator ordering ambiguities and in field theory usually involves a limiting procedure (regularisation and renormalisation). We must make sure that the resulting limiting operators $\pi(H_I)$ are densely defined and closable (i.e., their adjoints are also densely defined) on a suitable domain of $\mathcal{H}_{\text{kin}}$. This step usually severely restricts the abundance of representations. (Alternatively, in rare cases it is possible to quantise the finite gauge transformations generated by the classical constraints provided they exponentiate to a group.) In more detail: by assumption we can write the classical constraint functions $H_I(N^I)$ as certain functions $H_I(N^I) = h_I(N^I, \{a\})$ of the elementary variables where the curly brackets denote dependence on an, in general, infinite collection of variables. A naive quantisation procedure would be to define its quantisation as $\hat{H}_I(N^I) = h_I(N^I, \{\hat{a}\})$ where from now on we abbreviate $\hat{a} := \pi(a)$ for all $a \in \mathfrak{A}$. This will in general not work, at least not straightforwardly, for several reasons.

(a) As is well known, the quantisation of a phase space function is not unique, to a given candidate we can add arbitrary $\hbar$ corrections and still the classical limit of the corrected operator will be the original function. This is called the •• •••• ••••••• • •• ••• •••.

(b) While such corrections in quantum mechanics are relatively harmless, in quantum field theory they tend to be disastrous, a simple example is quantum Maxwell theory where the straightforward quantisation of the Hamiltonian gives a divergent nowhere-defined operator. It is only after normal ordering that one obtains a densely defined operator. This is what is called a •• •••• ••••••• • ••• •• •• •••••.

(c) More seriously, in general the singularities of an operator are of an even worse kind and cannot be simply removed by a judicious choice of factor ordering. One has to introduce a regularisation of the operator and subtract its divergent piece as one removes the regulator again. This is called the ••• ••• • •••• •••• of the operator. The end result must be a densely defined operator on $\mathcal{H}_{\text{kin}}$.

(d) If $H_I(N^I)$ is classically a real-valued function then one would like to implement $H_I(N^I)$ as a self-adjoint operator on $\mathcal{H}_{\text{kin}}$, the reason being that this would guarantee that its spectrum (and therefore its measurement values) is contained in the set of real numbers. While this would certainly be a necessary requirement if $H_I(N^I)$ was a true Hamiltonian (i.e., not a constraint), in the case of a constraint this condition can be relaxed as long as the value 0 is contained in its spectrum because this is the only point of the spectrum that we are interested in. On the other hand, a self-adjoint constraint operator is sometimes of advantage when it comes to actually solving the constraints [267,268,276–278], see below.

IV. •• ••• •• • ••• ••• •••• •• ••• ••••••• •• ••• ••••••• •••• • ••• • •• •••••••• •••

We would now like to solve the constraints in the quantum theory. A first guess of how to do that is by saying that a state $\psi \in \mathcal{H}_{\text{kin}}$ is physical provided that $\hat{H}_I(N^I)\psi = 0$ for all N^I. The study of the simple example of a particle moving in $\mathbb{R}^2$ with the constraint $H = p_2$ reveals that this does not work in general: in the momentum representation $\mathcal{H} = L_2(\overline{\mathcal{C}} := \mathbb{R}^2, d\mu_0 := d^2p)$ the physical state condition becomes $p_2\psi(p_1, p_2) = 0$ with the general solution $\psi_f(p_1, p_2) = \delta(p_2)f(p_1)$ for some function f. The problem is that ψ_f is not an element of $\mathcal{H}_{\text{kin}}$. This is a necessary feature of an operator with continuous spectrum: such an operator does not have eigenfunctions in the ordinary sense. However, it has so-called 'generalised eigenfunctions' of which ψ_f is an example [280].

There are essentially two different strategies for dealing with this problem, the first one is called 'Group Averaging' and the second one is called 'Direct Integral Decomposition'. The first method makes additional assumptions about the structure of the quantum constraint algebra while the second does not and is therefore of wider applicability. We will discuss both methods in more detail in Chapter 30 and can be brief here.

(a) For 'Group Averaging' [280] the first assumption is that the $\pi(H_I(N_I))$ are actually self-adjoint operators on $\mathcal{H}_{\text{kin}}$, defined on a common, dense and invariant domain $\mathcal{D}$ (that is, $\pi(H_I(N^I))\mathcal{D} \subset \mathcal{D}$ so that we may define polynomials of constraint operators such as commutators) and that the structure functions of the constraints are actually constants on $\mathcal{M}$.

As a second assumption, we require that there is no anomaly: recall that by assumption the constraint algebra is first class. To be specific, consider the case of a field theory based on three-dimensional manifold σ in the $3+1$ decomposition of the action. Consider a real-valued basis e_α of $\mathfrak{h} :=$ $L_2(\sigma, d^3x)$ consisting of smooth functions or rapid decrease and let $H_{I\alpha} :=$ $H_I(e_\alpha)$ hence $H_I(x) = \sum_\alpha H_{I\alpha}e_\alpha(x)$ since the constraints are elements of $\mathfrak{h}$. Then the first-class property means that there exist so-called structure functions $f^{K\gamma}_{I\alpha,J\beta}$ on $\mathcal{M}$ such that

$$\{H_{I\alpha}, H_{J\beta}\} = f^{K\gamma}_{I\alpha,J\beta}H_{K\gamma}$$

The quantum version of this condition is, in the case that the structure functions are structure constants (do not depend on $\mathcal{M}$)

$$[\pi(H_{I\alpha}), \pi(H_{J\beta})] = i\hbar f^{K\gamma}_{I\alpha,J\beta}\pi(H_{K\gamma}) \tag{3.1.5}$$

which makes sense because all operators are defined on the common dense and invariant domain $\mathcal{D}$. This condition could be somewhat relaxed, sometimes it may be useful to allow for projective representations of the corresponding Lie group (representations up to a multiplier [281]). More on anomalies will be said below.

Since we are now in the position of a proper (infinite-dimensional) Lie algebra we can define the unitary operators

$$U(t) := \exp\left(i \sum_{I\alpha} t^{I\alpha}\pi(H_{I\alpha})\right) \tag{3.1.6}$$

where the parameters take a range in a subset of $\mathbb{R}$ depending on the $\pi(H_I(N_I))$ in such a way that the $U(t)$ define a unitary representation of the Lie group G determined by the Lie algebra generators $H_{I\alpha}$.

The third assumption is that G has an invariant (not necessarily finite) bi-invariant Haar measure μ_H. This is guaranteed if G is a finite-dimensional, locally compact group with respect to a suitable topology. In this case we may define an anti-linear rigging map

$$\eta : \mathcal{D} \to \mathcal{H}_{\text{phys}}; \; \psi \mapsto \int_G d\mu_H(t) \; < U(t)\psi, . >_{\mathcal{H}_{\text{kin}}} \qquad (3.1.7)$$

with physical inner product

$$< \eta(\psi), \eta(\psi') >_{\mathcal{H}_{\text{phys}}} := [\eta(\psi')](\psi) \qquad (3.1.8)$$

Notice that $\eta(\psi)$ defines a distribution on $\mathcal{D}$ and solves the constraints in the sense that

$$[\eta(\psi)](U(t)\psi') = [\eta(\psi)](\psi') \; \forall \, t \in G, \; \psi' \in \mathcal{D} \qquad (3.1.9)$$

Moreover, given any kinematical algebra element $O \in \mathfrak{A}$ we may define a candidate for a corresponding Dirac observable by

$$[O] := \int_G d\mu_H(t) \, U(t) \, O \, U(t)^{-1} \qquad (3.1.10)$$

which formally commutes with the $U(t)$.

(b) Let us now come to the 'Direct Integral Method' [252–257] which is developed in more detail in Section 30.2. Here we do not need to assume that the $\pi(H_{I\alpha})$ are self-adjoint. Also the structure functions $f^{K\gamma}_{I\alpha,J\beta}$ may have non-trivial dependence on $\mathcal{M}$. This is actually the case in GR and hence only this method is available there, see below. Consider an operator-valued positive definite matrix $\hat{Q}^{I\alpha,J\beta}$ such that the operator

$$\widehat{} := \frac{1}{2} \sum_{I\alpha,J\beta} [\pi(H_{I\alpha})]^\dagger \hat{Q}_{I\alpha,J\beta} [\pi(H_{J\beta})] \qquad (3.1.11)$$

is densely defined. Obvious candidates for $\hat{Q}_{I\alpha,J\beta}$ are quantisations $\pi(Q^{I\alpha,J\beta})$ of positive definite, possibly $\mathcal{M}$-valued, matrices with suitable decay behaviour in the space of labels $I\alpha$. Then, since $\widehat{}$ is positive by construction it has self-adjoint extensions (e.g., its Friedrich extension, see the first volume of [282] and Theorem 26.8.1) and its spectrum is supported on the positive real line. Let $\lambda_0 = \inf \sigma(\widehat{})$ be the minimum of the spectrum of $\widehat{}$ and redefine $\widehat{}$ by $\widehat{} - \lambda_0 \, \text{id}_{\mathcal{H}_{\text{kin}}}$. Notice that $\lambda_0 < \infty$ by assumption and proportional to $\hbar$ by construction. We now use the well-known fact that $\mathcal{H}_{\text{kin}}$, if separable, can be represented as a direct integral of Hilbert spaces

$$\mathcal{H}_{\text{kin}} \cong \int_{\mathbb{R}^+}^\oplus d\mu(\lambda) \mathcal{H}_{\text{kin}}^\oplus(\lambda) \qquad (3.1.12)$$

where $\widehat{}$ acts on $\mathcal{H}_{\text{kin}}^\oplus(\lambda)$ by multiplication by λ. The measure μ and the scalar product on $\mathcal{H}_{\text{kin}}^\oplus(\lambda)$ are induced by the scalar product on $\mathcal{H}_{\text{kin}}$. Moreover, μ is unique up to an equivalent measure (with the same measure zero sets, see Chapter 25) and the $\mathcal{H}_{\text{kin}}^\oplus(\lambda)$ are, in fact, unique up to measure theoretical niceties which are explained in detail in Section 30.2. The physical Hilbert space is then simply $\mathcal{H}_{\text{phys}} := \mathcal{H}_{\text{kin}}^{\oplus}(0)$ and candidates for Dirac observables constructed from bounded self-adjoint operators O on $\mathcal{H}_{\text{kin}}$ can be given by the •••••• • •••

$$[O] = \lim_{T \to \infty} \frac{1}{2T} \int_{-T}^{T} dt \, e^{it\,\widehat{}} \, O \, e^{-it\,\widehat{}} \tag{3.1.13}$$

and they induce bounded self-adjoint operators on $\mathcal{H}_{\text{phys}}$.

Notice that both methods can be combined. Indeed, it may happen that a subset of the constraints can be solved by group averaging methods while the remainder can only be solved by direct integral decomposition methods. In this case, one will construct an intermediate Hilbert space which the first set of constraints annihilates and which carries a representation of the second set of constraints. This is actually the procedure followed in LQG and it will be convenient to adopt this 'solution in two steps'.

V. ••••••• •••• •••• ••• •• •••••••• ••• ••

Especially if, as in GR (see 1.2.15), the $f_{I\alpha,J\beta}^{K\gamma}$ depend on the phase space coordinates, we are not guaranteed that the right-hand side of (3.1.5) can actually be written in the form $\sum_{K\gamma} \pi(H_{K\gamma})\pi(f_{I\alpha,J\beta}^{K\gamma})$ with the $\hat{H}_{K\gamma}$ •••••••
•• ••• ••••. If that is not the case then the following inconsistency might arise. For any solution $\eta(\psi)$ we find

$$0 = [\eta(\psi)]([\pi(H_{I\alpha}), \pi(H_{J\beta})] \, \psi') = \sum_{K\gamma} [\eta(\psi)](\pi(f_{I\alpha,J\beta}^{K\gamma} H_{K\gamma})\psi') \tag{3.1.14}$$

for all $\psi' \in \mathcal{D}, I\alpha, J\beta$ if π is a representation of the classical constraint algebra. Thus, every $\eta(\psi)$ not only satisfies the constraints (3.1.9) but also the additional constraints (3.1.14). Depending on how $\pi(f_{I\alpha,J\beta}^{K\gamma} H_{K\gamma})$ is ordered in terms of the individual operators $\pi(f_{I\alpha,J\beta}^{K\gamma})$ and $\pi(H_{K\gamma})$ one possibly obtains additional conditions which are absent in the classical theory. Since (3.1.14) will in general be new constraints, algebraically independent from the original ones, the number of physical degrees of freedom in the classical and the quantum theory would differ from each other. In other words, the physical Hilbert space would have a too small number of semiclassical states in order to qualify as a viable quantisation of the classical theory.

Hence, for the group averaging proposal one must make sure that (3.1.14) is automatically satisfied once (3.1.9) holds, which puts additional restrictions on the freedom to order the constraint operators, if at all possible. Such a requirement is not necessary for the direct integral method as we have explained. In particular, the requirement that the $\pi(H_{I\alpha})$ be to the left of the structure function operators is in conflict with the requirement that the $\pi(H_{I\alpha})$, $\pi(f_{I\alpha,J\beta}^{K\gamma})$ be symmetric operators if they do not commute [283, 284]. This is because the only way that a classical relation of the form $\{a, b\} = cd$ between real-valued functions a, b, c, d can hold as an operator condition between at least symmetric, mutually non-commuting operators is $[\hat{a}, \hat{b}] = i\hbar(\hat{c}\hat{d} + \hat{d}\hat{c})/2$. Thus, in order to avoid quantum anomalies in GR it seems that we would need non-symmetric operators. We will see that this is precisely the case in LQG. But then the group averaging method to solve the constraints clearly breaks down and we must use the direct integral method.

Now in the direct integral decomposition method we just have a single constraint so that anomalies for the Master Constraint itself cannot arise. Yet, anomalies within the individual constraints could still be present and would express themselves in the fact that the spectrum of the Master Constraint does not contain zero. Hence, if we did not subtract the spectral gap, the physical Hilbert space would be empty. Subtracting the gap makes it non-empty but so far there is no proof that the resulting physical Hilbert space is then large enough. So far this proposal has just been successfully tested in a few examples. The simplest is the following: consider a phase space described by canonical pairs (p_α, q_α) and (y_j, x_j) subject to the constraints $H_{1\alpha} = p_\alpha$, $H_{2\alpha} = q_\alpha$; $\alpha = 1, 2, \ldots$. Then $\{H_{I\alpha}, H_{J\beta}\} = \epsilon_{IJ}\delta_{\alpha\beta}$, hence the constraints are second class. Let $c_\alpha > 0$, $\sum_\alpha c_\alpha = \epsilon < \infty$ and define the Master Constraint by $\quad := \sum_\alpha c_\alpha (q_\alpha^2 + p_\alpha^2)/2$. Consider the kinematical Hilbert space $\mathcal{H}_{\text{kin}} = \mathcal{H}_F \otimes \mathcal{H}'_F$ where $\mathcal{H}_F$, $\mathcal{H}'_F$ are Fock spaces based on the annihilation operators $a_\alpha = (q_\alpha - ip_\alpha)/\sqrt{2\hbar}$ and $a_j = (x_j - iy_j)/\sqrt{2\hbar}$ respectively. Now $\hat{\quad}$ is a weighted sum of harmonic oscillator Hamiltonians and the minimum of its spectrum is the 'zero point energy' $\lambda_0 = \sum_\alpha c_\alpha \hbar/2 = \epsilon\hbar/2$. Now $\hat{\quad}' = \hat{\quad} - \lambda_0 =: \hat{\quad} :$ is the normal ordered constraint with pure point spectrum and its unique zero eigenvectors are of the form $\Omega \otimes \psi'$ where $\Omega \in \mathcal{H}_F$ is the vacuum of the first Fock space and $\psi' \in \mathcal{H}'_F$ is an arbitrary vector in the second. Hence the physical Hilbert space is isomorphic to $\mathcal{H}'_F$, which is obviously the correct answer for this example.

Thus, nothing is swept under the rug. Hence, the real advantage of the direct integral method is that it allows us to construct the physical inner product even in the case of structure functions, however, while there is some physical intuition from selected examples, there is no proof yet that the semiclassical limit of the theory is the correct one.

The issue of the semiclassical limit is a non-trivial one also from another perspective. Notice that our construction is entirely non-perturbative, there are no (at least not necessarily) Fock spaces and there is no perturbative expansion (Feynman diagrams) even if the theory is interacting. While this is attractive, the price to pay is that the representation $\mathcal{H}_{\text{kin}}$ to begin with and also the final physical Hilbert space $\mathcal{H}_{\text{phys}}$ will in general be far removed from any physical intuition. Hence, we must make sure that what we have constructed is not just some mathematical object but has, at the very least, the classical theory as its classical limit. In particular, if classical Dirac observables are known, then the quantum Dirac observables (3.1.10) and (3.1.13) should reduce to them in the classical limit. To address such questions one must develop suitable semiclassical tools, in particular the construction of suitable semiclassical or coherent states.

We see that the construction of the quantum field theory in AQFT as well as in LQG is nicely separated: first one constructs the algebra and then its representations. In fact, the kinematical algebra from which one starts is the only input if one can prove later on uniqueness results concerning the representation theory. What is new in LQG as compared with AQFT is that it also provides a framework for dealing with constraints. However, solving the constraints and hence the physical Hilbert space is more or less tightly prescribed by the kinematical analysis already.

We will see that in LQG this programme could so far be systematically carried out until step IV, except for the construction of Dirac observables. Work is now in progress regarding the quantum Dirac observables; there are already proposals for the classical ones as we have seen in Chapter 2. Furthermore, in Chapter 11 semiclassical tools are developed by means of which the correctness of the infinitesimal dynamics of LQG could already be verified. To show that the theory has classical General Relativity as a classical limit would mean showing in addition that the quantum Dirac observables have the correct classical limit. Once these missing steps have been completed (at least in some approximation) and GR has been confirmed as a semiclassical limit of theory, LQG will be in a position to make falsifiable physical predictions.

4

The new canonical variables of Ashtekar for General Relativity

One would now like to apply the programme outlined in Chapter 3 to canonical GR in the ADM formulation of Chapter 1. Unfortunately, to date it has not been possible to go beyond the first two steps in a mathematically rigorous fashion. In other words, while suitable algebras $\mathfrak{P}$, $\mathfrak{A}$ can be defined, nobody succeeded in finding a rigorously defined, background-independent representation of those which also support the Hamiltonian constraint operator. As a consequence, all the other steps of the programme could be addressed only formally, except in situations with a lot of symmetries (called midi- or minisuperspace models respectively depending on whether there are still an infinite or finite number of physical degrees of freedom). Nevertheless, these early investigations resulted in important physical intuition and culminated in DeWitt's three seminal papers [82–84]. Moreover, Wheeler and DeWitt formally quantised the Hamiltonian constraint, and the associated quantum constraint equation is now known as the • •••••• • •• ••• •••••••. By inspection of (1.2.6) it is not surprising that it is hard to give mathematical meaning to the Hamiltonian constraint operator because it depends not even polynomially on the field variables, which in quantum theory become operator-valued distributions, thus non-polynomial expressions of these are hopelessly divergent, at least in the usual Fock representations of QFT.

It is not clear whether it is impossible to make progress with the ADM variables, maybe there simply is no representation that satisfies all our requirements. In any case, the field was more or less stuck for two decades. The situation changed dramatically when Ashtekar introduced new canonical variables for GR [92, 93], which cast the theory into the language of gauge theories of the Yang–Mills type. This chapter derives the classical Ashtekar variables in detail from the ADM framework.

4.1 Historical overview

The history of the classical aspects of the new variables is approximately 20 years old and we wish to give a brief account of the developments (the history of the quantum aspects will be given in Section 5.1):

• ••••••

The starting point was a series of papers due to Sen [285–287] who generalised the covariant derivative ∇_μ of Chapter 1 for $s = -1$ to $SL(2, \mathbb{C})$ spinors of left (right)-handed helicity resulting in an (anti)self-Hodge-dual connection which is therefore complex-valued. An exhaustive treatment on spinors and spinor calculus can be found in [288, 289]. See also Section 15.1.4 for a brief introduction.

Sen was motivated in part by a spinorial proof of the positivity of energy theorem of General Relativity [237–240]. But it was only Ashtekar [15, 16, 92, 93] who realised that modulo a slight modification of his connection, Sen had stumbled on a new canonical formulation of General Relativity in terms of the (spatial projection of) this connection, which turns out to be a generalisation of D_μ to this class of spinors, and a conjugate electric field kind of variable, such that the initial value constraints of General Relativity (1.2.6) can be written in ••• ••• •• • ••• if one rescales H by $H \mapsto \tilde{H} = \sqrt{\det(q)}H$ (which looks like a harmless modification at first sight). In fact, $\tilde{H}$ is only of fourth order in the canonical coordinates, not worse than non-Abelian Yang–Mills theory and thus a major roadblock on the way towards quantisation seemed to be removed. Ashtekar also noted the usefulness of the connection for $s = +1$ in which case it is actually real-valued [290, 291].

Ashtekar's proofs were in a Hamiltonian context. Samuel as well as Jacobson and Smolin discovered independently that there exists in fact a Lagrangian formulation of the theory by considering only the (anti)self-dual part of the curvature of Palatini gravity [292–294]. Jacobson also considered the coupling of fermionic matter [295] and an extension to supergravity [296]. Coupling to standard model matter was considered by Ashtekar •• • •• [297]. All of these developments still used a spinorial language which, although not mandatory, is of course quite natural if one wants to treat spinorial matter.

A purely tensorial approach to the new variables was given by Goldberg [298] in terms of triads and by Henneaux •• • •• in terms of tetrads [299].

While the Palatini formulation of General Relativity uses a connection and a tetrad field as independent variables, Capovilla, Dell and Jacobson realised that there is a classically equivalent action which depends only on a connection and a scalar field; moreover, they were able to solve both initial value constraints of General Relativity • •• ••• •• • ••• for a huge (but not the complete) class of field configurations. Unfortunately, there is a third constraint besides the diffeomorphism and Hamiltonian constraint in this new formulation of General Relativity, the so-called Gauß constraint, which is not automatically satisfied by this so-called 'CDJ-Ansatz' [300–303].

This line of thought was further developed by Bengtsson and Peldan [304–306] culminating in the discovery that in the presence of a cosmological constant the just-mentioned scalar field can be eliminated by a field equation, resulting in a pure connection Lagrangian for General Relativity (but not a polynomial one). For an overview of these ideas see [307] and for ideas towards gauge group unification see [308, 309].

As mentioned above, for Lorentzian (Euclidean) signature one considered complex (real)-valued connection variables. Meanwhile, it turned out that it is very hard to implement the reality conditions for the complex-valued case as adjointness conditions on the measure in the quantum theory while for the real valued case it is relatively easy. This motivated Barbero [310, 311] to consider real-valued connections also for Lorentzian signature. Barbero discovered that one can give a Hamiltonian formulation even ••• • •• ••• •••• •••••• of a parameter considered earlier by Immirzi [312–314] for either choice of signature. However, in order to keep polynomiality of the Hamiltonian constraint when using real-valued connections one has to multiply it by an even higher power of $\det(q)$. Moreover, the constraint becomes algebraically much more complicated.

This caveat is removed by a so-called 'phase space Wick rotation' introduced in [315, 316] and later considered also in [317] where one can work with real connections • •••• •••••••• ••• ••••••••• •••• •• ••• •••••••••• ••• •••. This line of development was motivated by a seminal paper due to Hall [318–320] who constructed a unitary transform from a Hilbert space of square integrable functions on a compact gauge group to a Hilbert space of square integrable, holomorphic functions on the complexification of that gauge group and this transform was generalised in [321] to gauge theories for compact gauge groups. Mena Marugán clarified the relation between this phase space Wick rotation and the usual one (analytic continuation in the time parameter) [322, 323]. A toy model test was performed in [324].

The last development in this respect is the result of [325], which states that polynomiality of the constraint operator is not only unimportant in order to give a rigorous meaning to it in quantum theory, it is in fact ••••••••••. The important condition is that the constraint be a scalar of ••••••• ••• •••••••••. This forbids rescaling of H, which is already a density of weight one, by any non-trivial power of $\det(q)$. It is only in that case that the quantisation of the operator can be done in a ••••••••• •••• ••••••••••• ••• • •••••• ••••••• ••• •• •••••••••• ••• •• ••• •••• ••••••• • •••••• •••••. For this reason, real connection variables are currently favoured as far as quantum theory is concerned. In retrospect, what is really important is that one bases the quantum theory on connections and canonically conjugate electric fields (which is dual in a metric-independent way to a two-form). The reason is that n-forms can be naturally integrated over n-dimensional submanifolds of σ without requiring a background structure, this is not possible for the metric variables of the ADM formulation and has forbidden progress for such a long time. We will come back to this point in the next chapter.

So far a Lagrangian action principle had been given only for the following values of signature s and Immirzi parameter β, namely Lorentzian General Relativity $s = -1, \beta = \pm i$ and Euclidean General Relativity $s = +1, \beta = \pm 1$. For arbitrary complex β and either signature a Lagrangian formulation was discovered by Holst, Barros e Sá and Capovilla •••• [326–328]. Roughly speaking the action is given by a modification of the Palatini action

$$S = \int_M \text{tr}(F \wedge [* - \beta^{-1}](e \wedge e)) \tag{4.1.1}$$

(it results for $\beta = \infty$) where $*$ denotes the Hodge dual with respect to the internal Minkowski metric, $F = F(\omega)$ is the curvature of some connection ω which is considered as an independent field next to the tetrad e. This action should be considered in analogy with the θ angle modification of (the bosonic contribution of the action to) QCD

$$S = \int_M \text{tr}(F \wedge [* + \theta]F) \tag{4.1.2}$$

where $*$ denotes the Hodge dual with respect to the background spacetime Minkowski metric. In the gravitational case the β term drops out by an equation of motion, in the QCD case the variation of the θ term is exact and also drops out of the equations of motion. This holds for the classical theory, but it is well known that in the quantum theory the actions with different values of θ do not result in unitarily equivalent theories. A similar result holds for general relativity [312].

Samuel [329,330] criticised the use of real connection variables for Lorentzian gravity because of the following reason: the Hamiltonian analysis of the action (4.1.1) leads, unless $\beta = \pm i$ for $s = -1$, to constraints of second class which one has to solve by imposing a gauge condition. It eliminates the boost part of the original SO(1, 3) Gauß constraint and one is left with an SO(3) Gauß constraint (which also appears in the case $\beta = \pm i$). That gauge condition fixes the direction of an internal SO(1, 3) vector which is automatically preserved by the remaining SO(3) subgroup and by the evolution derived from the associated Dirac bracket, so that everything is consistent. Now, while for $\beta = \pm i, s = -1$ the spatial connection is simply the pull-back of the (anti)self-dual part of the four-dimensional spin connection to the spatial slice, for real β its spacetime interpretation is veiled due to the appearance of the second-class constraints and the gauge fixing.

Samuel now asked the following question: for any value of β can it be shown that every SO(3) gauge-invariant function of the spatial connection and the triad can be expressed in terms of the (pull-back to the spatial slice of the) spacetime fields $q_{\mu\nu}, K_{\mu\nu}$? In the previous chapter we have shown that the Hamiltonian evolution of these fields under the Hamiltonian constraint coincides, on the constraint surface, with their infinitesimal transformation under a timelike diffeomorphism. Is it then true that the induced Hamiltonian transformation of SO(3) gauge-invariant functions of the connection (such as traces of its holonomy around a loop in a spatial slice) coincides with that of (the pull-back to the spatial slice of) a spacetime connection? He found that this is the case if and only if $\beta = \pm i$. The simple algebraic reason is that only for an (anti)self-dual connection A^{IJ}, $I, J = 0, 1, 2, 3$ are the components A^{0j} already determined by $A^j = \frac{1}{2}\epsilon_{jkl}A^{kl}$ so that the pull-back to the spatial slice of A^j determines the pull-back of an SO(1, 3) connection with its full spacetime interpretation only then.

It should be stressed, however, that Samuel's criticism is purely aesthetical in nature, for interpretational reasons it is certainly convenient to have a spacetime interpretation of the spatial connection but it is by no means mandatory, one just has to bear in mind that the connection does not have the naive transformation behaviour under Hamiltonian evolution on the constraint surface. In fact, to date a satisfactory quantum theory has been constructed only for β real (which in turn does not mean that it is impossible to do for $\beta = \pm i$). In fact, as we will show in this section, at the classical level • •• ••• • ••• •• •• •• ••

•• • •• • •••• •• ••• •••• ••• •• • •• •••• • ••• •••• ••• • •• ••• • ••• • ••••• •• ••• ••• ••• •• •

•• ••• • • • •••• • •• •••• .

• • • • • • • • • •

In [331–334] Alexandrov and coworkers used the results of [326–328] and tried to set up a simultaneously canonical and covariant formulation of General Relativity in terms of connection variables. This involves working with real-valued SL(2, ℂ) connections rather than SU(2) connections. In order to implement the necessarily arising second-class constraint one must use the Dirac bracket rather than the canonical brackets, as a result of which one ends up with connections that have non-vanishing Poisson brackets among themselves. As a result, in the quantum theory the quantum connection operators must not be mutually commuting, which is why it is impossible to construct a connection representation in this approach (by definition, in such a representation the connection operator acts by multiplication; in fact, so far no non-trivial representation could be constructed this way). Notice that the criticism by Samuel as spelt out is really just aesthetical in nature and not an obstruction to implementing spacetime covariance. Namely, as we will prove in this chapter, the SU(2)-connection formulation of General Relativity is completely equivalent to the ADM formulation, which is as manifestly four-dimensionally covariant as any canonical approach can possibly be. One just must not commit the mistake of thinking that the SU(2) connection is the pull-back of a spacetime connection. The transformation properties of the connection under the flow of the Hamiltonian constraint take this explicitly into account, remembering how the connection is built out of the extrinsic curvature and the triads. Moreover, the real complication with four-dimensional diffeomorphism invariance is that it is not implemented everywhere on the phase space (just on-shell) as a canonical transformation induced by the Hamiltonian constraint, as we emphasised in Section 1.4. In their 'covariant' formulation the authors of [331–334] face this issue of a mixture of dynamics and gauge symmetries as well.

4.2 Derivation of Ashtekar's variables

This concludes our historical digression and we now come to the actual derivation of the new variable formulation. We decided for the extended phase space approach to use triads as this makes the contact and equivalence with the ADM formulation most transparent and quickest and avoids the introduction of additional SL(2, ℂ) spinor calculus which would blow up our exposition unnecessarily. Furthermore, the method displayed here, namely to extend a given phase space and reduce its extension to the original one by imposing constraints, can be generalised and is therefore of independent interest, for example, in trying to construct higher-dimensional analogues of what we will do here. Also we do this for either signature and any complex value of the Immirzi parameter. What is no longer arbitrary is the dimension of σ: we will be forced to work with $D = 3$ as will become clear in the course of the derivation if one uses the type of extension of the phase employed here.

The construction actually consists of two steps: first an extension of the ADM phase space and second a canonical transformation on the extended phase space. We will first assume that the SU(2)-bundle to be introduced is trivial and later explain why this is not a restriction. See Chapter 21 for an introduction to fibre bundle theory. For a derivation from an action principle, see the paragraph around equation (12.1.5) in Chapter 12.

4.2.1 Extension of the ADM phase space

We would like to consider the phase space described in Section 1.1 as the symplectic reduction of a larger symplectic manifold with co-isotropic constraint surface, see [218] or Chapter 23. One defines a so-called co-D-Bein field e_a^i on σ where the indices $i, j, k, \ldots$ take values $1, 2, \ldots, D$. The D-metric is expressed in terms of e_a^i as

$$q_{ab} := \delta_{jk} e_a^j e_b^k \tag{4.2.1}$$

Notice that this relation is invariant under local SO(D) rotations $e_a^i \mapsto O^i_j e_a^j$ and we can therefore view e_a^i, for $D = 3$, as an su(2)-valued one-form (recall that the adjoint representation of SU(2) on its Lie algebra is isomorphic with the defining representation of SO(3) on $\mathbb{R}^3$ under the isomorphism $\mathbb{R}^3 \to \mathrm{su}(2)$; $v^i \to v^i \tau_i$ where τ_i is a basis of su(2) (also called 'soldering forms' [288])). This observation makes it already obvious that we have to get rid of the $D(D-1)/2$ rotational degrees of freedom sitting in e_a^i but not in q_{ab}. Since the Cartan–Killing metric of so(D) is just the Euclidean one we will in the sequel drop the δ_{ij} and also do not need to care about index positions.

Next we introduce yet another, independent one-form K_a^i on σ which for $D = 3$ we also consider as su(2)-valued and from which the extrinsic curvature is derived as

$$-sK_{ab} := K_{(a}^i e_{b)}^i \qquad (4.2.2)$$

We see immediately that K_a^i cannot be an arbitrary $D \times D$ matrix but must satisfy the constraint

$$G_{ab} := K_{[a}^j e_{b]}^j = 0 \qquad (4.2.3)$$

since K_{ab} was a symmetric tensor field. Consider the quantity

$$E_j^a := \text{sgn}\big(\det\big((e_a^i)\big)\big) \frac{1}{(D-1)!} \epsilon^{aa_1\ldots a_{D-1}} \epsilon_{jj_1\ldots j_{D-1}} e_{a_1}^{j_1} \cdots e_{a_{D-1}}^{j_{D-1}} = \sqrt{\det(q)} e_j^a$$

$$(4.2.4)$$

where the D-Bein is defined by the relations $e_j^a e_a^k = \delta_j^k$, $e_j^a e_b^j = \delta_b^a$.

•• •••• At this point an important remark about the sign $\text{sgn}(\det((e_a^i)))$ of $\det((e_a^j))$ is appropriate. In classical GR the three-metric q_{ab} is assumed to be everywhere non-degenerate and of Euclidean signature. Therefore $\det(q) = |\det(e)|^2 > 0$. Since one also assumes that the fields e_a^j, q_{ab} are everywhere smooth, it follows that $\det(e)$ has constant sign. It follows that we are implicitly imposing that σ is orientable because then $\epsilon_{j_1\ldots j_D} e^{j_1} \wedge \ldots \wedge e^{j_D}$ is a globally defined D-form. We will see that this classical condition can be completely relaxed in the quantum theory, thus allowing for topology change!!!

With the help of (4.2.4) one can equivalently write (4.2.3) in the form

$$G_{jk} := K_{a[j} E_{k]}^a = 0 \qquad (4.2.5)$$

Consider now the following functions on the extended phase space

$$q_{ab} := E_a^j E_b^j \big| \det\big((E_l^c)\big) \big|^{2/(D-1)}, \quad P^{ab} := 2 \big| \det\big((E_l^c)\big) \big|^{-2/(D-1)} E_k^a E_k^d K_{[d}^j \delta_{c]}^b E_j^c$$

$$(4.2.6)$$

where E_a^j is the inverse of E_j^a. It is easy to see that when $G_{jk} = 0$, the functions (4.2.6) precisely reduce to the ADM coordinates. Inserting (4.2.6) into (1.2.6) we can also write the diffeomorphism and Hamiltonian constraint as functions on the extended phase space, which one can check to be explicitly given by

$$H_a := 2sD_b\big[K_a^j E_j^b - \delta_a^b K_c^j E_j^c\big]$$

$$H := -\frac{s}{\sqrt{\det(q)}}\big(K_a^l K_b^j - K_a^j K_b^l\big) E_j^a E_l^b - \sqrt{\det(q)} R \qquad (4.2.7)$$

where $\sqrt{\det(q)} := |\det((E_j^a))|^{1/(D-1)}$ and $q^{ab} = E_j^a E_j^b / \det(q)$ by which $R = R(q)$ is considered as a function of E_j^a. Notice that, using (4.2.2), (4.2.4), expressions (4.2.7) indeed reduce to (1.2.6) up to terms proportional to G_{jk}.

Let us equip the extended phase space coordinatised by (K_a^i, E_i^a) with the symplectic structure (formally, that is without smearing) defined by

$$\{E_j^a(x), E_k^b(y)\} = \{K_a^j(x), K_b^k(y)\} = 0, \ \{E_i^a(x), K_b^j(y)\} = \frac{\kappa}{2}\delta_b^a\delta_i^j\delta(x,y) \tag{4.2.8}$$

We claim now that the symplectic reduction with respect to the constraint G_{jk} of the constrained Hamiltonian system subject to the constraints (4.2.5), (4.2.7) results precisely in the ADM phase space of Section 1.1 together with the original diffeomorphism and Hamiltonian constraint.

To prove this statement we first of all define the smeared 'rotation constraints'

$$G(\Lambda) := \int_\sigma d^D x \Lambda^{jk} K_{aj} E_k^a \tag{4.2.9}$$

where $\Lambda^T = -\Lambda$ is an arbitrary antisymmetric matrix, that is, an so(D)-valued scalar on σ. They satisfy the Poisson algebra, using (4.2.8)

$$\{G(\Lambda), G(\Lambda')\} = \frac{\kappa}{2}G([\Lambda, \Lambda']) \tag{4.2.10}$$

in other words, $G(\Lambda)$ generates infinitesimal SO(D) rotations as expected. Since the functions (4.2.6) are manifestly SO(D)-invariant by inspection, they Poisson commute with $G(\Lambda)$, that is, they comprise a complete set of rotational-invariant Dirac observables with respect to $G(\Lambda)$ for any Λ. As the constraints defined in (4.2.7) are in turn functions of these, $G(\Lambda)$ also Poisson commutes with the constraints (4.2.7), whence the total system of constraints consisting of (4.2.9), (4.2.7) is of first class.

Finally we must check that Poisson brackets among the q_{ab}, P^{cd}, considered as the functions (4.2.6) on the extended phase space with symplectic structure (4.2.8), are equal to the Poisson brackets of the ADM phase space (1.2.7), at least when $G_{jk} = 0$. Since q_{ab} is a function of E_j^a only it is clear that $\{q_{ab}(x), q_{cd}(y)\} = 0$. Next we have

$$\{P^{ab}(x), q_{cd}(y)\} = ([q^{a(e}q^{bf)} - q^{ab}q^{ef}]E_f^j)(x)\{K_e^j(x), (|\det(E)|^{2/(D-1)}E_c^k E_d^k)(y)\}$$

$$= ([q^{a(e}q^{bf)} - q^{ab}q^{ef}]E_f^j)(x)\left[\frac{2}{D-1}q_{cd}(y)\frac{\{K_e^j(x), |\det(E)|(y)\}}{|\det(E)|(x)}\right.$$

$$\left. + 2(\det(q)E_{(c}^k(x)\{K_e^j(x), E_{d)}^k(y)\}\right]$$

$$= \kappa\left([q^{a(e}q^{bf)} - q^{ab}q^{ef}]\left[-\frac{1}{D-1}q_{cd}q_{ef} + q_{e(c}q_{d)f}\right]\right)(x)\delta(x,y)$$

$$= \kappa\delta_{(c}^a\delta_{d)}^b\delta(x,y) \tag{4.2.11}$$

where we used $\delta E^{-1} = -E^{-1}\delta E E^{-1}$, $[\delta|\det(E)|]/|\det(E)| = [\delta\det(E)]/\det(E)$
$= E_a^j \delta E_j^a$. The final Poisson bracket is the most difficult one. By carefully
inserting the definitions, making use of the relations $E_j^a = |\det(e)|e_j^a$, $E_a^j = e_a^j/|\det(e)|$, $e_j^a = q^{ab}e_b^j$ at various steps one finds after two pages of simple but
tedious algebraic manipulations that

$$\{P^{ab}(x), P^{cd}(y)\} = -\kappa \left(\frac{\det(e)}{4}[q^{bc}G^{ad} + q^{bd}G^{ac} + q^{ac}G^{bd} + q^{ad}G^{bc}]\right)(x)\delta(x,y)$$

$$(4.2.12)$$

where $G^{ab} = q^{ac}q^{bd}G_{cd}$ and so (4.2.12) vanishes only at $G_{ab} := G_{jk}e_a^j e_b^k = 0$.

Let us summarise: the functions (4.2.6) and (4.2.7) reduce at $G_{jk} = 0$ to the
corresponding functions on the ADM phase space, moreover, their Poisson brackets among each other reduce at $G_{jk} = 0$ to those of the ADM phase space. Thus,
as far as rotationally invariant observables are concerned, the only ones we are
interested in, both the ADM system and the extended one, are completely equivalent and we can as well work with the latter. This can be compactly described
by saying that the symplectic reduction with respect to G_{jk} of the constrained
Hamiltonian system described by the action

$$S := \frac{1}{\kappa} \int_{\mathbb{R}} dt \int_{\sigma} d^D x \left(2\dot{K}_a^j E_j^a - [-\Lambda^{jk}G_{jk} + N^a H_a + NH]\right) \quad (4.2.13)$$

is given by the system described by the ADM action of Section 1.1. Notice that,
in accordance with what we said before, there is no claim that the Hamiltonian
flow of K_a^j, E_j^a with respect to H_a, H is a spacetime diffeomorphism. However,
since the Hamiltonian flow of H, H_a on the constraint surface $G_{jk} = 0$ is the
same as on the ADM phase space for the gauge-invariant observables q_{ab}, P^{ab},
a representation of $\text{Diff}(M)$ (on-shell) is still given on the constraint surface of
$G_{jk} = 0$.

4.2.2 Canonical transformation on the extended phase space

Up to now we could work with arbitrary $D \geq 2$, however, what follows works
only for $D = 3$.[1] First we introduce the notion of the •• •• ••• • •••••• , which is
defined as an extension of the spatial covariant derivative D_a from tensors to
generalised tensors with so(D) indices. One defines

$$D_a u_{b...} v_j := (D_a u_b)...v_j + \cdots + u_{b...}(D_a v_j) \text{ where } D_a v_j := \partial_a v_j + \Gamma_{ajk}v^k$$

$$(4.2.14)$$

---

[1] One can introduce higher p-form fields, see [335] and references therein, however, it is not
clear that such a reformulation preserves the simplicity of the symplectic structure, which is
mandatory in order to have a chance to find kinematical Hilbert space representations.

extends by linearity, Leibniz rule and imposes that D_a commutes with contractions, see Chapters 19, 21. Moreover, we extend the metric compatibility condition $D_a q_{bc} = 0$ to e_a^j, that is

$$D_a e_b^j = 0 \Rightarrow \Gamma_{ajk} = -e_k^b[\partial_a e_b^j - \Gamma_{ab}^c e_c^j] \qquad (4.2.15)$$

Then $D_a \delta_{jk} = D_a e_j^b e_b^k = 0$, which implies that $D_a v^j = \partial_a v^j + \Gamma_{ajk} v^k$ since $\Gamma_{a(jk)} = 0$. Obviously Γ_a takes values in so(D), that is, (4.2.15) defines an antisymmetric matrix.

Our aim is now to write the constraint G_{jk} in such a form that it becomes the Gauß constraint of an SO(D) gauge theory, that is, we would like to write it in the form $G_{jk} = (\partial_a E^a + [A_a, E^a])_{jk}$ for some so(D) connection A. It is here where $D = 3$ is singled out: what we have is an object of the form E_j^a which transforms in the defining representation of so(D) while Γ_{jk}^a transforms in the adjoint representation of so(D). It is only for $D = 3$ that these two are equivalent. Thus from now on we take $D = 3$.

The canonical transformation that we have in mind consists of two parts: (1) a constant Weyl (rescaling) transformation and (2) an affine transformation.

Observe that for any non-vanishing complex number $\beta \neq 0$, called the •• • •••• ••••• ••••, the following rescaling $(K_a^j, E_j^a) \mapsto (^{(\beta)}K_a^j := \beta K_a^j, {}^{(\beta)}E_j^a := E_j^a/\beta)$ is a canonical transformation (the Poisson brackets (4.2.8) are obviously invariant under this map). We will use the notation $K = {}^{(1)}K, E = {}^{(1)}E$. In particular, the rotational constraint, which we write in $D = 3$ in the equivalent form,

$$G_j = \epsilon_{jkl} K_a^k E_l^a = \epsilon_{jkl}\left(^{(\beta)}K_a^k\right)\left(^{(\beta)}E_l^a\right) \qquad (4.2.16)$$

is invariant under this rescaling transformation. We will consider the other two constraints (4.2.7) in a moment.

We notice from (4.2.15) that $D_a E_j^b = 0$. In particular, we have

$$D_a E_j^a = [D_a E^a]_j + \Gamma_{aj}{}^k E_k^a = \partial_a E_j^a + \epsilon_{jkl}\Gamma_a^k E_l^a = 0 \qquad (4.2.17)$$

where the square bracket in the first identity means that D acts only on tensorial indices, which is why we could replace D by ∂ as E_j^a is an SU(2)-valued vector density of weight one. We also used the isomorphism between antisymmetric tensors of second rank and vectors in Euclidean space to define $\Gamma_a =: \Gamma_a^l T_l$ where $(T_l)_{jk} = \epsilon_{jlk}$ are the generators of SO(3) in the defining – or, equivalently, of SU(2) in the adjoint representation if the structure constants are chosen to be ϵ_{ijk}. Next we explicitly solve the spin connection in terms of E_j^a from (4.2.15) by using the explicit formula for Γ^a_{bc} and find

$$\Gamma^i_a = \frac{1}{2}\epsilon^{ijk}e^b_k\left[e^j_{a,b} - e^j_{b,a} + e^c_j e^l_a e^l_{c,b}\right]$$

$$= \frac{1}{2}\epsilon^{ijk}E^b_k\left[E^j_{a,b} - E^j_{b,a} + E^c_j E^l_a E^l_{c,b}\right]$$

$$+ \frac{1}{4}\epsilon^{ijk}E^b_k\left[2E^j_a\frac{(\det(E))_{,b}}{\det(E)} - E^j_b\frac{(\det(E))_{,a}}{\det(E)}\right] \qquad (4.2.18)$$

where in the second line we used that $|\det(E)| = [\det(e)]^2$ in $D = 3$. Notice that the second line in (4.2.18) explicitly shows that Γ^j_a is a homogeneous rational function of degree zero of E^a_j and its derivatives. Therefore we arrive at the important conclusion that

$$\left({}^{(\beta)}\Gamma^j_a\right) := \Gamma^j_a\left({}^{(\beta)}E\right) = \Gamma^j_a = \Gamma^j_a\left({}^{(1)}E\right) \qquad (4.2.19)$$

is itself invariant under the rescaling transformation. This is obviously also true for the Christoffel symbol Γ^a_{bc} since it is a homogeneous rational function of degree zero in q_{ab} and its derivatives and $q_{ab} = |\det(E)|E^j_a E^j_b \mapsto {}^{(\beta)}q_{ab} = (\beta^2/\sqrt{\beta^2}^3)\,({}^{(1)}q_{ab})$. Thus the derivative D_a is, in fact, independent of β and we therefore have in particular $D_a({}^{(\beta)}E^a_j) = 0$. We can then write the rotational constraint in the form

$$G_j = 0 + \epsilon_{jkl}\left({}^{(\beta)}K^k_a\right)\left({}^{(\beta)}E^a_l\right) = \partial_a\left({}^{(\beta)}E^a_j\right) + \epsilon_{jkl}\left[\Gamma^k_a + \left({}^{(\beta)}K^k_a\right)\right]\left({}^{(\beta)}E^a_l\right)$$

$$=: {}^{(\beta)}\mathcal{D}_a\,{}^{(\beta)}E^a_j \qquad (4.2.20)$$

This equation suggests introducing the new connection

$$\left({}^{(\beta)}A^j_a\right) := \Gamma^j_a + \left({}^{(\beta)}K^j_a\right) \qquad (4.2.21)$$

This connection could be called the Sen–Ashtekar–Immirzi–Barbero connection (names in historical order) for the historical reasons mentioned at the beginning of this chapter. More precisely, the Sen connection arises for $\beta = \pm i, G_j = 0$, the Ashtekar connection for $\beta = \pm i$, the Immirzi connection for complex β and the Barbero connection for real β. For simplicity we will refer to it as the $\bullet\bullet\bullet$ connection which now replaces the spin connection Γ^j_a and gives rise to a new derivative ${}^{(\beta)}\mathcal{D}_a$ acting on generalised tensors as the extension by linearity of the basic rules ${}^{(\beta)}\mathcal{D}_a v_j := \partial_a v_j + \epsilon_{jkl}({}^{(\beta)}A^k_a)v_l$ and ${}^{(\beta)}\mathcal{D}_a u_b := D_a u_b$. Notice that (4.2.20) has $\bullet\bullet\bullet\bullet\bullet\bullet\bullet$ the structure of a Gauß law constraint for an SU(2) gauge theory, although ${}^{(\beta)}A$ qualifies as the pull-back to σ by local sections of a connection on an SU(2) fibre bundle over σ only when β is real. Henceforth we will call G_j the $\bullet\,\bullet\bullet\bullet\,\bullet\bullet\bullet\bullet\bullet\bullet\,\bullet\bullet$.

Given the complicated structure of (4.2.18) it is quite surprising that the variables $({}^{(\beta)}A, {}^{(\beta)}E)$ form a canonically conjugate pair, that is

$$\left\{{}^{(\beta)}A^j_a(x), {}^{(\beta)}A^k_b(y)\right\} = \left\{{}^{(\beta)}E^a_j(x), {}^{(\beta)}E^b_k(y)\right\} = 0,\ \left\{{}^{(\beta)}E^a_j(x), {}^{(\beta)}A^k_b(y)\right\}$$

$$= \frac{\kappa}{2}\delta^a_b\delta^k_j\delta(x, y) \qquad (4.2.22)$$

This is the key feature for why these variables are at all useful in quantum theory: if we did not have such a simple bracket structure classically then it would be very hard to find Hilbert space representations that turn these Poisson bracket relations into canonical commutation relations.

To prove (4.2.22) by means of (4.2.8) (which is invariant under replacing K, E by $^{(\beta)}K, ^{(\beta)}E$) we notice that the only non-trivial relation is the first one since $\{E_j^a(x), \Gamma_b^k(y)\} = 0$. That relation is explicitly given as

$$\beta[\{\Gamma_a^j(x), K_b^k(y)\} - \{\Gamma_b^k(y), K_a^j(x)\}] = \beta\frac{\kappa}{2}\left[\frac{\delta\Gamma_a^j(x)}{\delta E_k^b(y)} - \frac{\delta\Gamma_b^k(y)}{\delta E_j^a(x)}\right] = 0 \quad (4.2.23)$$

which is just the integrability condition for Γ_a^j to have a generating potential F. A promising candidate for F is given by the functional

$$F = \int_\sigma d^3x\, E_j^a(x)\Gamma_a^j(x) \quad (4.2.24)$$

since if (4.2.23) holds we have

$$\frac{\delta F}{\delta E_j^a(x)} - \Gamma_a^j(x) = \int d^3y\, E_k^b(y)\frac{\delta\Gamma_b^k(y)}{\delta E_j^a(x)} = \int d^3y\, E_k^b(y)\frac{\delta\Gamma_a^j(x)}{\delta E_k^b(y)}$$

$$= \frac{2}{\kappa}\{\Gamma_a^j(x), \int d^3y\, K_b^k(y)E_k^b(y)\} = 0 \quad (4.2.25)$$

because the function $\int d^3y\, K_b^k(y)E_k^b(y)$ is the canonical generator of •••••• scale transformations under which Γ_a^j is invariant as already remarked above. To show that F is indeed a potential for Γ_a^j we demonstrate (4.2.25) in the form $\int d^3x\, E_j^a(x)\delta\Gamma_a^j(x) = 0$. Starting from (4.2.18) we have (using $\delta e_j^a e_j^b = \delta e_b^a e_k^b = 0$ repeatedly)

$$E_i^a\delta\Gamma_a^i = \frac{1}{2}\epsilon^{ijk}|\det(e)|e_i^a\delta(e_k^b[e_{a,b}^j - e_{b,a}^j - e_j^c e_a^l e_{c,b}^l])$$

$$= \frac{1}{2}\epsilon^{ijk}|\det(e)|[e_i^a\delta(e_k^b(e_{a,b}^j - e_{b,a}^j)) - \delta(e_k^b e_j^c e_{c,b}^i) + (\delta e_i^a)e_j^c e_a^l e_k^b e_{c,b}^l]$$

$$= \frac{1}{2}\epsilon^{ijk}|\det(e)|[e_i^a\delta(e_k^b(e_{a,b}^j - e_{b,a}^j)) - \delta(e_k^b e_j^a e_{a,b}^i) - (\delta e_a^l)e_i^a e_j^c e_k^b e_{c,b}^l]$$

$$= \frac{1}{2}\epsilon^{ijk}|\det(e)|[\delta(e_i^a e_k^b(e_{a,b}^j - e_{b,a}^j) - e_k^b e_j^a e_{a,b}^i) + (\delta e_i^a)e_k^b(e_{a,b}^j - e_{b,a}^j)$$

$$\quad - (\delta e_a^l)e_i^a e_j^c e_k^b e_{c,b}^l]$$

$$= \frac{1}{2}\epsilon^{ijk}|\det(e)|[\delta(e_k^b(e_j^a e_{a,b}^i + e_i^a e_{a,b}^j) - e_i^a e_k^b e_{b,a}^j) + (\delta e_k^b)e_i^a e_{b,a}^j$$

$$\quad + (\delta e_i^a)e_k^b e_{b,a}^j - (\delta e_a^l)e_i^a e_j^c e_k^b e_{c,b}^l]$$

$$= -\frac{1}{2}\epsilon^{abc}[e_c^j\delta e_{b,a}^j - (\delta e_a^j)e_{c,b}^j]\operatorname{sgn}(\det(e))$$

$$= -\frac{1}{2}\epsilon^{abc}\partial_a[(\delta e_b^j)e_c^j]\operatorname{sgn}(\det(e)) = -\frac{1}{2}\epsilon^{abc}\partial_a[(\delta e_b^j)e_c^j\operatorname{sgn}(\det(e))]$$

$$(4.2.26)$$

From the first to the second line we pulled e_i^a into the variation of the third term of $\delta\Gamma_i^a$ resulting in a correction proportional to δe_a^i, in the next line we relabelled the summation index c into a in the third term and traded the variation of e_i^a for that of e_a^l in the fourth term, in the next line we pulled again e_i^a inside a variation resulting in altogether six terms, in the next line we collected the total variation terms and reordered them and in the fourth term we relabelled the summation indices a, b into b, a and i, k into k, i or i, j into j, i resulting in a minus sign from the ϵ^{ijk}, in the next line we realised that the first two terms are symmetric in i, j which thus drop out due to the ϵ^{ijk} and that the e_i^a and e_k^b variation pieces of the third term cancel against the fourth and fifth term, in the next line we made use of the relations $\det(e)\epsilon^{ijk}e_j^b e_k^c = \epsilon^{abc}e_a^i$, $\det(e)\epsilon^{ijk}e_i^a e_j^b e_k^c = \epsilon^{abc}$ and relabelled j for l and in the last line finally we relabelled a for b in the second term resulting in a minus sign which allows us to write the whole expression as a derivative. We also exploited that $\mathrm{sgn}(\det(e)) = \mathrm{const.}$ (classically).

It follows that

$$\int_\sigma d^3x\, E_j^a \delta\Gamma_a^j = -\frac{1}{2}\int_\sigma d^3x\, \partial_a\left(\epsilon^{abc}\left(\delta e_b^j\right)e_c^j\mathrm{sgn}(\det(e))\right)$$

$$= -\frac{1}{2}\int_{\partial\sigma} dS_a\, \epsilon^{abc} e_b^j\left(\delta e_c^j\right)\mathrm{sgn}(\det(e)) \tag{4.2.27}$$

which vanishes if $\partial\sigma = \emptyset$. If σ has a boundary such as spatial infinity then we must improve (4.2.24). To do this we must use the boundary conditions on $({}^{(\beta)}A_a^j, {}^{(\beta)}E_j^a)$ which were derived from the ADM boundary conditions stated in Section 1.1 in [244, 245, 336] by simply carefully reinserting the definitions of the new variables in terms of the ADM variables and the Gauß constraint. These considerations can be summarised as follows.

Recall (1.5.2), that for the ADM variables we had asymptotically $q_{ab} = \delta_{ab} + f_{ab}(n)/r + O(r^{-2})$ and $P^{ab} = F^{ab}(n)/r^2 + O(r^{-3})$ where $n^a = x^a/r$ is an asymptotically flat Cartesian coordinate system and f_{ab}, F^{ab} have even and odd parity respectively on the asymptotic S^2. It follows that $K_{ab} = F_{ab}(n)/r^2 + O(r^{-3})$ and F_{ab} has odd parity. For the co-triad we make the Ansatz $e_a^j = \delta_a^j + f_a^j(n)/r$ which leads to $f_{ab}(n) = f_{(a}^j(n)\delta_{b)j}$, hence f_a^j has even parity (a potentially antisymmetric contribution can be excluded). Thus, $E_j^a = \delta_j^a + f_j^a(n)/r + O(r^{-2})$ and f_j^a has even parity. Next, from $-sK_{ab} = K_{(a}^j e_{b)}^j$ we conclude that $K_a^j = F_a^j(n)/r^2 + O(r^{-3})$ and F_a^j has odd parity. Finally, since by (4.2.18) Γ_a^j is a homogeneous function of e_a^j and its first derivatives which appear in first power only it follows that $\Gamma_a^j = \gamma_a^j(n)/r^2 + O(r^{-3})$ where γ_a^j has odd parity. Thus ${}^{(\beta)}A_a^j = {}^{(\beta)}F_a^j(n)/r^2 + O(r^{-3})$ and ${}^{(\beta)}F_a^j(n)$ has odd parity.

We see that $\int_\sigma d^3x\, E_j^a \Gamma_a^j$ diverges linearly. To cure this we add a boundary term just as in Section 1.1.6. Notice that

$$E_j^a \Gamma_a^j = -\frac{1}{2}\mathrm{sgn}(\det(e))\epsilon^{abc} e_a^j \partial_b e_c^j \tag{4.2.28}$$

Hence with $\delta^j = \delta^j_a dx^a$

$$\int_\sigma d^3x\, E^a_j \Gamma^j_a = -\frac{1}{2}\int_\sigma \text{sgn}(\det(e))e^j \wedge de^j$$

$$= -\frac{1}{2}\int_\sigma \text{sgn}(\det(e))e^j \wedge d(e^j - \delta^j)$$

$$= \frac{1}{2}\int_\sigma \text{sgn}(\det(e))de^j \wedge (e^j - \delta^j) - \frac{1}{2}\int_{\partial\sigma} \text{sgn}(\det(e))e^j \wedge (e^j - \delta^j)$$

$$(4.2.29)$$

The integrand of the bulk term in (4.2.27) is $O(r^{-3})$ odd and hence convergent, so we define the improved generator

$$F := \int_\sigma d^3x E^a_j \Gamma^j_a + \frac{1}{2}\int_{\partial\sigma} \text{sgn}(\det(e))e^j \wedge (e^j - \delta^j) \qquad (4.2.30)$$

Using the boundary conditions it is easy to see that the surface term (4.2.27) arising from the variation of the bulk term of F precisely cancels the variation of the surface term of F. Thus we have shown that (4.2.30) is indeed a potential for Γ^j_a even in the asymptotically flat case. This completes the proof that the map $(E^a_j, K^j_a) \mapsto ({}^{(\beta)}E^a_j, {}^{(\beta)}A^j_a)$ is a canonical transformation.

• •• •• •••• • •• ••• ••• •• ••• •• •• ••• ••• ••• •••• •••

It remains to write the constraints (4.2.7) in terms of the variables ${}^{(\beta)}A, {}^{(\beta)}E$. To that end we introduce the curvatures

$$R^j_{ab} := 2\partial_{[a}\Gamma^j_{b]} + \epsilon_{jkl}\Gamma^k_a\Gamma^l_b$$

$${}^{(\beta)}F^j_{ab} := 2\partial_{[a}\,{}^{(\beta)}A^j_{b]} + \epsilon_{jkl}\,{}^{(\beta)}A^k_a\,{}^{(\beta)}A^l_b \qquad (4.2.31)$$

whose relation with the covariant derivatives is given by $[D_a, D_b]v_j = R_{abjl}v^l = \epsilon_{jkl}R^k_{ab}v^l$ and $[{}^{(\beta)}D_a, {}^{(\beta)}D_b]v_j = {}^{(\beta)}F_{abjl}v^l = \epsilon_{jkl}\,{}^{(\beta)}F^k_{ab}v^l$. Let us expand ${}^{(\beta)}F$ in terms of Γ and ${}^{(\beta)}K$

$${}^{(\beta)}F^j_{ab} = R^j_{ab} + 2\beta D_{[a}K^j_{b]} + \beta^2\epsilon_{jkl}K^k_a K^l_b \qquad (4.2.32)$$

Contracting with ${}^{(\beta)}E$ yields

$${}^{(\beta)}F^j_{ab}\,{}^{(\beta)}E^b_j = \frac{R^j_{ab}E^b_j}{\beta} + 2D_{[a}\left(K^j_{b]}E^b_j\right) + \beta K^j_a G_j \qquad (4.2.33)$$

where we have used the Gauß constraint in the form (4.2.16). We claim that the first term on the right-hand side of (4.2.33) vanishes identically. To see this we first derive from (4.2.15) due to torsion freeness of the Levi–Civita connection in the language of forms the • •• •••• •• • ••• ••• •• ••• ••••

$$dx^a \wedge dx^b D_a e^j_b = de^j + \Gamma^j_k \wedge e^k = 0$$

$$\Rightarrow 0 = -d^2 e^j = d\Gamma^j_k \wedge e^k - \Gamma^j_l \wedge de^l = [d\Gamma^j_k + \Gamma^j_l \wedge \Gamma^l_k] \wedge e^k = \Omega^j_k \wedge e^k$$

$$(4.2.34)$$

Now $\Omega^j_k = \Omega^i(T_i)_{jk} =: (\Omega)_{jk}$ and we see that

$$\Omega = d\Gamma + \Gamma \wedge \Gamma = d\Gamma^i\, T_i + \frac{1}{2}[T_j, T_k]\Gamma^j \wedge \Gamma^k = \frac{1}{2}dx^a \wedge dx^b R^i_{ab}T_i$$

Thus the Bianchi identity can be rewritten in the form

$$\epsilon_{ijk}\epsilon^{efc}R^j_{ef}e^k_c = 0 \;\Rightarrow\; \frac{1}{2}\epsilon_{ijk}\epsilon^{efc}R^j_{ef}e^k_c e^i_a = \frac{1}{2}E^b_j\epsilon_{cab}\epsilon^{efc}R^j_{ae} = R^j_{ab}E^b_j = 0$$

$$(4.2.35)$$

as claimed. Now we compare with the first line of (4.2.7) and thus arrive at the conclusion

$$^{(\beta)}F^j_{ab}\,^{(\beta)}E^b_j = -sH_a + {}^{(\beta)}K^j_a G_j \qquad (4.2.36)$$

Next we contract (4.2.32) with $\epsilon_{jkl}\,^{(\beta)}E^a_k\,^{(\beta)}E^b_l$ and find

$$^{(\beta)}F^j_{ab}\epsilon_{jkl}\,^{(\beta)}E^a_k\,^{(\beta)}E^b_l = -\det(q)\frac{R_{abkl}e^a_k e^b_l}{\beta^2} - 2\frac{E^a_j D_a G_j}{\beta}$$

$$+ \left(K^j_a E^a_j\right)^2 - \left(K^j_b E^a_j\right)\left(K^k_a E^b_k\right) \qquad (4.2.37)$$

Expanding $v_j = e^a_j v_a$, $v_a = e^j_a v_j$, using $D_a e^j_b = 0$ and comparing $[D_a, D_b]v_j$ with $[D_a, D_b]v_c$ for any v_j we find $R_{abij} = R_{abcd}e^c_i e^d_j$ and so (4.2.37) can be rewritten as

$$^{(\beta)}F^j_{ab}\epsilon_{jkl}\,^{(\beta)}E^a_k\,^{(\beta)}E^b_l = -\det(q)\frac{R}{\beta^2} - 2\,^{(\beta)}E^a_j D_a G_j$$

$$+ \left(K^j_a E^a_j\right)^2 - \left(K^j_b E^a_j\right)\left(K^k_a E^b_k\right) \qquad (4.2.38)$$

and comparing with the second line of (4.2.7) we conclude that modulo a polynomial in the Gauß constraint coming from $-sK_{ab} = K^j_a e^j_b + G_{ab}$

$$^{(\beta)}F^j_{ab}\epsilon_{jkl}\,^{(\beta)}E^a_k\,^{(\beta)}E^b_l + 2\,^{(\beta)}E^a_j D_a G_j$$

$$= \sqrt{\det(q)}\left[-\sqrt{\det(q)}\frac{R}{\beta^2} - \frac{\left(K^j_b E^a_j\right)\left(K^k_a E^b_k\right) - \left(K^j_a E^a_j\right)^2}{\sqrt{\det(q)}}\right]$$

$$= \frac{\sqrt{\det(q)}}{\beta^2}\left[-\sqrt{\det(q)}R - \beta^2\frac{\left(K^j_b E^a_j\right)\left(K^k_a E^b_k\right) - \left(K^j_a E^a_j\right)^2}{\sqrt{\det(q)}}\right]$$

$$= \frac{\sqrt{\det(q)}}{\beta^2}\left[H + (s - \beta^2)\frac{\left(K^j_b E^a_j\right)\left(K^k_a E^b_k\right) - \left(K^j_a E^a_j\right)^2}{\sqrt{\det(q)}}\right]$$

$$= s\sqrt{\det(q)}\left[-\frac{s}{\sqrt{\det(q)}}\left[\left(K^j_b E^a_j\right)\left(K^k_a E^b_k\right) - \left(K^j_a E^a_j\right)^2\right] - \frac{s}{\beta^2}\sqrt{\det(q)}R\right]$$

$$= s\sqrt{\det(q)}\left[H + \left(1 - \frac{s}{\beta^2}\right)\sqrt{\det(q)}R\right] \qquad (4.2.39)$$

We see that the left-hand side of (4.2.39) is proportional to H if and only if $\beta = \pm\sqrt{s}$, that is, imaginary (real) for Lorentzian (Euclidean) signature. We prefer, for reasons that become obvious only in a later chapter, to solve (4.2.39) for H as follows

$$H = \frac{\beta^2}{\sqrt{|\det(^{(\beta)}E\beta)|}} \left[^{(\beta)}F_{ab}^j \epsilon_{jkl} \, ^{(\beta)}E_k^a \, ^{(\beta)}E_l^b + 2 \, ^{(\beta)}E_j^a D_a G_j \right]$$

$$+ (\beta^2 - s) \frac{\left(^{(\beta)}K_b^j \, ^{(\beta)}E_j^a \right)\left(^{(\beta)}K_a^j \, ^{(\beta)}E_j^b \right) - \left(^{(\beta)}K_c^k \, ^{(\beta)}E_k^c \right)^2}{\sqrt{|\det(^{(\beta)}E\beta)|}} \qquad (4.2.40)$$

In formula (4.2.40) we wrote everything in terms of $^{(\beta)}A, ^{(\beta)}E$ if we understand $^{(\beta)}K = ^{(\beta)}A - \Gamma$.

We notice that both (4.2.36) and (4.2.40) still involve the Gauß constraint. Since the transformation $K_a^j \mapsto ^{(\beta)}A_a^j, E_j^a \mapsto ^{(\beta)}E_j^a$ is a canonical one, the Poisson brackets among the set of first-class constraints given by G_j, H_a, H are unchanged. Let us write symbolically $H_a = H_a' + f_a^j G_j, H = H' + f^j G_j$ where H_a', H' are the pieces of H_a, H respectively not proportional to the Gauß constraint. Since G_j generates a subalgebra of the constraint algebra it follows that the modified system of constraints given by G_j, H_a', H' not only defines the same constraint surface of the phase space but also gives a first-class system again, of course, with somewhat modified algebra (which, however, coincides with the Dirac algebra on the submanifold $G_j = 0$ of the phase space). In other words, it is completely equivalent to work with the set of constraints G_j, H_a', H' which we write once more, dropping the prime, as

$$G_j = ^{(\beta)}D_a \, ^{(\beta)}E_j^a = \partial_a \, ^{(\beta)}E_j^a + \epsilon_{jkl} \, ^{(\beta)}A_a^j \, ^{(\beta)}E_j^a$$

$$H_a = -s \, ^{(\beta)}F_{ab}^j \, ^{(\beta)}E_j^b$$

$$H = \left[\beta^2 \, ^{(\beta)}F_{ab}^j - (\beta^2 - s)\epsilon_{jmn} \, ^{(\beta)}K_a^m \, ^{(\beta)}K_b^n \right] \frac{\epsilon_{jkl} \, ^{(\beta)}E_k^a \, ^{(\beta)}E_l^b}{\sqrt{|\det(^{(\beta)}E\beta)|}} \qquad (4.2.41)$$

For easier comparison with the literature we also write (4.2.41) in terms of $^{(\beta)}A_a^j, K_a^j, E_j^a$, which gives

$$G_j = \left[^{(\beta)}D_a E_j^a \right]/\beta = \left[\partial_a E_j^a + \epsilon_{jkl} \, ^{(\beta)}A_a^j E_j^a \right]/\beta$$

$$H_a = -s\left(^{(\beta)}F_{ab}^j E_j^b \right)/\beta$$

$$H = \left[^{(\beta)}F_{ab}^n - (\beta^2 - s)\epsilon_{jmn}K_a^m K_b^n \right] \frac{\epsilon_{jkl} E_k^a E_l^b}{\sqrt{\det(q)}} \qquad (4.2.42)$$

Summarising, we have rewritten the Einstein–Hilbert action in the following equivalent form

$$S = \frac{1}{\kappa} \int_{\mathbb{R}} dt \int_\sigma d^3x \left(2 \, ^{(\beta)}\dot{A}_a^i \, ^{(\beta)}E_i^a - [\Lambda^j G_j + N^a H_a + N H] \right) \qquad (4.2.43)$$

where the appearing constraints are the ones given by either (4.2.42) or (4.2.41).

We close this chapter with some remarks.

Let us try to give a four-dimensional meaning to $^{(\beta)}A$. To that end we must complete the Dreibein e^a_i to a Vierbein e^μ_α where μ is a spacetime tensor index and $\alpha = 0, 1, 2, 3$ an index for the defining representation of the Lorentz (Euclidean) group for $s = -1(+1)$. By definition $g_{\mu\nu}e^\mu_\alpha e^\nu_\beta = \eta_{\alpha\beta}$ is the flat Minkowski (Euclidean) metric. Thus e^μ_0, e^μ_i are orthogonal vectors and we thus choose $e^\mu_0 = n^\mu$ and in the ADM frame with $\mu = t, a$ we choose $(e^\mu_i)_{\mu=a} = e^a_i$. Using the defining properties of a tetrad basis and the explicit form of $n^\mu, g_{\mu\nu}$ in the ADM frame derived earlier, the above choices are sufficient to fix the tetrad components completely to be $e^t_0 = 1/N$, $e^a_0 = -N^a/N$, $e^t_i = 0$, e^a_i. Inversion gives (notice that $e^0_\mu = se_{\mu 0} = sg_{\mu\nu}e^\nu_0 = sg_{\mu\nu}n^\mu = sn_\mu$) $e^0_t = N$, $e^0_a = 0$, $e^i_t = N^a e^i_a$, e^i_a. Finally we have for $q^\mu_\nu = \delta^\mu_\nu - sn^\mu n_\nu = \delta^\mu_\nu - e^\mu_0 e^0_\nu$ in the ADM frame $q^t_t = 0$, $q^t_a = 0$, $q^a_t = N^a$, $q^a_b = \delta^a_b$. Thus we obtain, modulo $G_j = 0$

$$K^j_a = -se^b_j K_{ab} = -se^b_j q^\mu_a q^\nu_b \nabla_\mu n_\nu = -e^b_j (\nabla_a e_b)^0 = e^b_j (\omega_a)^0{}_\alpha e^\alpha_b$$
$$= 2e^b_j (\omega_a)^0{}_k e^k_b = 2(\omega_a)^0{}_j \qquad (4.2.44)$$

where in the second identity the bracket denotes that ∇ only acts on the tensorial index and in the third we used the definition of the four-dimensional spin connection $\nabla_\mu e^\alpha_\nu = (\nabla_\mu e_\nu)^\alpha + (\omega_\mu)^\alpha{}_\beta e^\beta_\nu = 0$. On the other hand we have

$$(\Gamma_a)^j_k e^k_b = -(D_a e_b)^j = -q^\mu_a q^\nu_b (\nabla_\mu e_\nu)^j = -(\nabla_a e_b)^j = (\omega_a)^j{}_k e^k_b \qquad (4.2.45)$$

whence $\omega_{ajk} = \Gamma_{ajk}$. It follows that

$$^{(\beta)}A_{ajk} = \omega_{ajk} - \beta s \omega_{a0l}\epsilon_{jkl} \qquad (4.2.46)$$

The Hodge dual of an antisymmetric tensor $T_{\alpha\beta}$ is defined by $*T_{\alpha\beta} = \frac{1}{2}\epsilon_{\alpha\beta\gamma\delta}\eta^{\gamma\gamma'}\eta^{\delta\delta'}T_{\gamma'\delta'}$. Since $\epsilon_{0ijk} = \epsilon_{ijk}$ we can write (4.2.46) in the form

$$^{(\beta)}A_{ajk} = \omega_{ajk} - s\beta * \omega_{ajk} \qquad (4.2.47)$$

Now an antisymmetric tensor is called (anti)self-dual provided that $*T_{\alpha\beta} = \pm\sqrt{s}T$ with $\sqrt{s} := i^{[1-s]/2}$ and the (anti)self-dual piece of any $T_{\alpha\beta}$ is defined by $T^\pm = \frac{1}{2}[T \pm *T/\sqrt{s}]$ since $* \circ * = s$ id. An (anti)self-dual tensor therefore has only three linearly independent components. This case happens for (4.2.47) provided that either $s = 1, \beta = \mp 1$ or $s = -1, \beta = \pm i$ and •• •• •• •••• ••• • •••
••• • •••••• •• •• ••• ••• •••• ••• ••• •••• •••••• ••• ••••• •• ••• •• •••••• •• σ •• •••
•• • •• •• •• ••••• •• •••• ••• •••• •••••••. In all other cases (4.2.47) is only half of the information needed in order to build a four-dimensional connection and therefore we do not know how it transforms under internal boosts. That is, from this perspective, the reason why one has to gauge fix the boost symmetry of the action (4.1.1) (by the time gauge $e^\alpha_\mu n^\mu = \delta^\alpha_0$) in order to remove the then present second-class constraints and to arrive at the present formulation. Obviously, this is no obstacle because there •••• ••••• a four-dimensional

interpretation even in that case, since the new variables capture the same information as the ADM variables on the constraint surface defined by the Gauß constraint and the latter do have a four-dimensional interpretation. From an aesthetic point of view it would be desirable to work with these complex variables (for Lorentzian signature), however, to date we do not know how to quantise a theory based on complex connections in a rigorous way. We will comment on that later.

When β is real-valued $^{(\beta)}A$, $^{(\beta)}E$ are both real-valued and can be interpreted directly as the canonical pair for the phase space of an SU(2) Yang–Mills theory. If β is complex then these variables are complex-valued. However, they cannot be arbitrary complex functions on σ but are subject to the following reality conditions

$$^{(\beta)}E/\beta = \overline{^{(\beta)}E/\beta}, \quad [^{(\beta)}A - \Gamma]/\beta = \overline{[^{(\beta)}A - \Gamma]/\beta} \tag{4.2.48}$$

where $\Gamma = \Gamma(^{(\beta)})$ is a non-polynomial, not even analytic function. These reality conditions guarantee that there is no doubling of the number of degrees of freedom and one can check explicitly that they are preserved by the Hamiltonian flow of the constraints provided that Λ^j, the Lagrange multiplier of the Gauß constraint, is real-valued. Thus, only SU(2) gauge transformations are allowed but not general SL(2, $\mathbb{C}$) transformations. These non-polynomial reality conditions are difficult to implement in the quantum theory, which is one of the reasons why dealing with complex connections is so far out of reach.

The original motivation to introduce the new variables was that for the quantisation of General Relativity it seemed mandatory to simplify the algebraic structure of the Hamiltonian constraint, which for $s = -1$ requires $\beta = \pm i$ since then the constraint becomes polynomial after multiplying by a factor proportional to $\sqrt{\det(q)}$. On the other hand, then the reality conditions become non-polynomial. Finally, if one wants polynomial reality conditions then one must have β real and then the Hamiltonian constraint is still complicated. Thus it becomes questionable what has been gained. The answer is the following: for any choice of β one can actually make both the Hamiltonian constraint ••• the reality conditions polynomial by multiplying by a sufficiently high power of $\det(q)$. But the real question is whether the associated classical functions will become well-defined operator-valued distributions in quantum theory while keeping background independence. As we will see in later chapters, the Hilbert space that we choose does not support any quantum versions of these functions rescaled by powers of $\det(q)$ and there are abstract arguments that suggest this is a representation-independent statement. The requirement seems to be that the Hamiltonian constraint is a scalar density of weight one and thus we must keep the factor of $1/\sqrt{\det(q)}$ in (4.2.42) whatever the choice of β (and therefore the motivation for polynomiality is lost completely). The motivation to have a connection formulation rather than a metric formulation is then that •• • ••• •• • • ••• ••••••• •• ••• •••••••••••• ••• •••••••• •••• •••• • ••••• •••• •••••••• • • •••••• •• •••• β •• ••••. For instance, a connection formulation enables us to employ the powerful arsenal of techniques that have been developed for the canonical quantisation of Yang–Mills theories, specifically Wilson loop techniques.

• • • ••• •• • ••• •••• ••

In the whole exposition so far we have assumed that we have a trivial principal SU(2) bundle over σ (see, e.g., [337] for a good textbook on fibre bundle theory and Chapter 21) so that we can work with a globally defined connection potential and globally defined electric field $^{(\beta)}A$, $^{(\beta)}E$ respectively. What about different bundle choices?

Following the notation of Chapter 21 our situation is that we are dealing with a principal SU(2) bundle over σ with pull-backs $^{(\beta)}A_I$ by local sections of a connection and local sections $^{(\beta)}E_I$ of an associated (under the adjoint representation) vector bundle of two-forms and would like to know whether these bundles are trivial. Since the latter is built out of the Dreibein we can equivalently look also at the frame bundle of orthonormal frames in order to decide for triviality. Triviality of the frame bundle is equivalent to the triviality of its associated principal bundle and in turn to σ being parallelisable. But this is automatically the case for any compact, orientable three-manifold provided that G = SU(2) (see [338], paragraph 12, exercise 12-B). More generally, in order to prove that a principal fibre bundle is trivial one has to show that the cocycle h_{IJ} of transition functions between charts of an atlas of σ is a coboundary, that is, its (non-Abelian) Čech cohomology class is trivial. In [338] one uses a different method, obstruction theory, where triviality can be reduced to the vanishing of the coefficients (taking values in the homotopy groups of G) of certain cohomology groups of σ related to Stiefel–Whitney classes.

So far we did not make the assumption that σ is compact but we used the fact that σ is orientable. If σ is not compact but orientable then one usually requires that there is a compact subset B of σ such that $\sigma - B$ has the topology of the complement of a ball in $\mathbb{R}^3$. Then the result holds in B and trivially in $\sigma - B$ and thus all over σ. Thus, compactness is not essential. If σ is not orientable then a smooth nowhere singular frame cannot exist and the above quoted result does not hold, there are no smooth Dreibein fields in this case. We explicitly exclude such σ as it does not allow us to couple (chiral) Weyl spinor fields which do appear in the standard model and do require orientability as well as time orientability of M and thus orientability of σ.

However, as we will see, the choice of the bundle will become completely irrelevant even before solving the Gauß constraint in the quantum theory because the distributional space of connections contains connections on all bundles and even many, many more than those.

We have shown that the symplectic reduction of the new phase space by the Gauß constraint reproduces the ADM phase space. Moreover, the constraints H, H_a are SU(2)-invariant. Hence, for the boundary terms necessary in order to make these constraints functionally differentiable and finite we just need to take the boundary terms of Section 1.5 and write them in the new variables. The fact that modulo the Gauß constraint the difference between the ADM variables and the new variables is just a canonical transformation guarantees that the Poisson brackets between these functionals on the new variable phase space continue to be well-defined and reproduce all the results of Section 1.5. This has been verified explicitly in [244, 336].

As far as the functional differentiability and finiteness of the Gauß constraint itself is concerned, let Λ^j be any test function. Then the boundary term of the variation of $\int_\sigma d^3x \Lambda^j G_j$ equals $\int_{\partial\sigma} dS_a \Lambda^j \delta E_j^a = \delta \int_{\partial\sigma} dS_a \Lambda^j E_j^a$. Now subtracting the SU(2) charge $Q_j = \int_{\partial\sigma} dS_a \Lambda^j E_j^a$ from the bulk gives

$$- \int_\sigma d^3x \left[\Lambda^j_{,a} + \epsilon_{jkl}\,^{(\beta)}A_k \Lambda^l \right] E_j^a$$

The first term is finite provided $\partial\Lambda$ is $O(r^{-3})$ odd while the second is finite provided Λ is $O(r^{-1})$ even. The only solution is that Λ is $O(r^{-2})$ even, but then $Q_j = 0$. There is no SU(2) charge (Dirac observable) in GR because in the ADM formulation we would never see any SU(2) gauge degrees of freedom anyway.

In view of these considerations we will from now on only consider positive β unless otherwise specified. In order to simplify the notation in what follows we will drop the label β but will mean by the field E really the field $^{(\beta)}E$ for $\beta = 1$ while A is actually $^{(\beta)}A$ for arbitrary β.

II

Foundations of modern canonical Quantum General Relativity

5

Introduction

5.1 Outline and historical overview

In the first part of this book we have derived a canonical connection formulation of classical General Relativity. We have defined precisely what one means by the canonical quantisation of a field theory with constraints and have emphasised the importance of n-form fields for a background-independent quantisation of generally covariant theories. In this part we will

and almost complete it. In more detail we will show that:

1. There exists a mathematically rigorous and, under natural physical assumptions, unique kinematical platform from which constraint quantisation is launched.
2. There exists at least one, consistent, well-defined quantisation of the Wheeler–DeWitt constraint operator whose action is explicitly known.
3. A corresponding physical inner product is known to exist.
4. There is a concrete proposal for constructing Dirac observables and physical Hamiltonians.

What is left to do is to check whether this solution of the quantisation problem has the correct semiclassical limit (semiclassical states at the kinematical level are already under control, however, not yet on the space of solutions to the constraints) and to construct quantisations of the classical formula for complete Dirac observables explicitly. This will involve, besides the improvement of the already available semiclassical techniques, the development of appropriate approximation schemes because the exact theory is too complicated to be solvable explicitly. After these steps have been completed one is ready to make physical predictions from the theory. The task will then be to identify quantum gravity effects, which lie in the realm of today's experimental precision, and to falsify the theory.

In the remainder of this chapter we sketch the history of the subject and the results obtained so far, which serves as a guideline as well and will help the reader to bring earlier publications into the context of the present-day adopted point of view.

We have seen that for $\beta = \pm i, s = -1$ the Hamiltonian constraint greatly simplifies, up to a factor of $1/\sqrt{\det(q)}$ it becomes a fourth-order polynomial in

${}^{\mathbb{C}}A_a^j, E_j^a$. In order to find solutions to the quantum constraint we chose a connection representation, that is, wave functions are functionals of ${}^{\mathbb{C}}A$ but not of $\overline{{}^{\mathbb{C}}A}$, the connection itself becomes a multiplication operator while the electric field becomes a functional differential operator. In formulae for the choice $\beta = -i$,

$$\left({}^{\mathbb{C}}\hat{A}_a^j(x)\psi\right)[{}^{\mathbb{C}}A] = {}^{\mathbb{C}}A_a^j(x)\psi[{}^{\mathbb{C}}A] \text{ and } \left(\hat{E}_j^a(x)\psi\right)[{}^{\mathbb{C}}A] = \frac{\ell_p^2}{2}\frac{\delta\psi[{}^{\mathbb{C}}A]}{\delta {}^{\mathbb{C}}A_a^j(x)} \quad (5.1.1)$$

(notice that $2iE/\kappa$ is conjugate to ${}^{\mathbb{C}}A$, $\ell_p^2 = \hbar\kappa$ is the Planck area). With this definition, which is only formal at this point since one does not know what the functional derivative means without specifying the function space to which the ${}^{\mathbb{C}}A$ belong, the canonical commutation relations

$$\left[{}^{\mathbb{C}}\hat{A}_a^j(x), {}^{\mathbb{C}}\hat{A}_b^k(y)\right] = \left[\hat{E}_j^a(x), \hat{E}_k^b(y)\right] = 0, \ \left[\hat{E}_j^a(x), {}^{\mathbb{C}}\hat{A}_b^k(y)\right] = \frac{\ell_p^2}{2}\delta_b^a\delta_j^k\delta(x,y)$$

$$(5.1.2)$$

are formally satisfied. However, the adjointness relations

$$\left(\hat{E}_j^a(x)\right)^\dagger = \hat{E}_j^a(x), \ {}^{\mathbb{C}}\hat{A}_a^j(x) + \left({}^{\mathbb{C}}\hat{A}_a^j(x)\right)^\dagger = 2\hat{\Gamma}_a^j(x) \quad (5.1.3)$$

could not be checked because no scalar product was defined with respect to which (5.1.3) should hold. Besides simpler mathematical problems such as domains of definitions of the operator-valued distributions (5.1.1), equation (5.1.3) looks disastrous in view of the explicit formula (4.2.18) for the spin connection where operator-valued distributions would appear multiplied not only at the same point but also in the denominator, which would be extremely difficult to define if possible at all and could prevent one from defining a positive definite scalar product with respect to which the adjointness conditions should hold.

The implementation of the adjointness relations (which one can make polynomial by multiplying ${}^{\mathbb{C}}A$ by a sufficiently high power of the operator corresponding to $\det(q)$) continues to be the major obstacle with the complex connection formulation even today, which is why the real connection formulation is favoured at the moment. However, in these pioneering years at the end of the 1990s nobody thought about using real connections since the simplification of the Hamiltonian constraint seemed to be the most important property to preserve, which is why researchers postponed the solution of the adjointness relations and the definition of an inner product to a later stage and focused first on other problems. There was no concrete proposal at that time on how to do that, but the fact that the complex connection ${}^{\mathbb{C}}A = \Gamma - iK$ is reminiscent of the harmonic oscillator variable $z = x - ip$ made it plausible that one could possibly make use of the technology known from geometric quantisation concerning complex Kähler polarisations [218] and the relevant Bargmann–Segal transformation theory. A concrete proposal in terms of the phase space Wick rotation transformation mentioned earlier appeared only later in [315], but until today these ideas have not been mathematically rigorously implemented.

Still there was a multitude of results that one could obtain by formal manipulations even in absence of an inner product. The most important observation at that time, in the opinion of the author, is the discovery of the importance of the use of holonomy variables (also known as Wilson loop functions). We will drop the superscripts $\beta, \mathbb{C}$ in what follows.

Already in the early 1980s Gambini . [339–341] pointed out the usefulness of Wilson loop functions for the canonical quantisation of Yang–Mills theory. Given a directed loop (closed path) α in σ and a G-connection A for some gauge group G one can consider the holonomy $h_\alpha(A)$ of A along α. The holonomy of a connection is abstractly defined via principal fibre bundle theory, but physicists prefer the formula $h_\alpha(A) := \mathcal{P} \exp(\oint_\alpha A)$ where $\mathcal{P}$ stands for path-ordering the power expansion of the exponential in such a way that the connection variables are ordered from left to right with the parameter along the loop on which they depend increasing. We will give a precise definition later on. The connection can be taken in any representation of G but we will mostly be concerned with $G = SU(2)$ and will choose the fundamental representation (in case of $G = SL(2, \mathbb{C})$ one chooses one of its two fundamental representations). The Wilson loop functions are then given by

$$T_\alpha(A) := \mathrm{tr}(h_\alpha(A)) \qquad (5.1.4)$$

where tr denotes the corresponding trace. The importance of such Wilson loop functions is that, at least for compact groups, one knows that they capture the full gauge-invariant information about the connection [342]. For the case at hand, $SL(2, \mathbb{C})$, an independent proof exists [343].

After the introduction of the new variables which display General Relativity as a special kind of Yang–Mills theory, Jacobson, Rovelli and Smolin independently rediscovered and applied Gambini .'s ideas to canonical quantum gravity [344, 345]. Since the connection representation was holomorphic, one needed only one of the fundamental representations of $SL(2, \mathbb{C})$ (and not its complex conjugate).

We do not want to go into very much detail about the rich amount of formal and exact results that were obtained by working with these loop variables before 1992, but just list the most important ones. An excellent review of these issues is contained in the book by Gambini and Pullin [346], which has become the standard introductory reference on the loop representation.

1.

By ordering the operators $\hat{E}$ to the right in the quantisation of the rescaled density weight-two operator corresponding to $\tilde{H}$, one can show [344, 345] that formally $\hat{\tilde{H}} T_\alpha = 0$ for every non-intersecting smooth loop α (see also [347–349]

144

for an extension to more complicated loops). The formal character of this argument is due to the fact that this is a regulated calculation where in the limit as the regulator is removed one multiplies zero by infinity. An important role is played by the notion of a so-called 'area-derivative'.

2.

Since the diffeomorphism constraint maps a Wilson loop function to a Wilson loop function for a diffeomorphic loop one immediately sees that knot invariants should play an important role. Let μ be a diffeomorphism-invariant measure on some space of connections, α a loop and ψ any state. One can then define a loop transformed state by $\psi'(\alpha) := \int d\mu(A)\overline{T_\alpha(A)}\psi(A)$. The state $\Psi = 1$ is annihilated by the diffeomorphism constraint if we define the action of an operator $\hat{O}'$ in this loop representation by $(\hat{O}'\psi')(\alpha) := \int d\mu(A)(\hat{O}T_\alpha)(A)\psi(A)$ where $\hat{O}$ is its action in the connection representation. Likewise one sees, at least formally, that if α is a smooth non-self-intersecting loop then $\psi'(\alpha)$ is annihilated by the Hamiltonian constraint. Of course, again this is rather formal because a suitable diffeomorphism-invariant measure μ was not known to exist.

3.

If one considers, in particular, the loop transform with respect to the formal measure given by Lebesgue measure times the exponential of i/λ times the Chern–Simons action where λ is the cosmological constant then one can argue to obtain particular knot invariants related to the Jones polynomial [346, 350], the coefficients of which seem to be formal solutions to the Hamiltonian constraint in the loop representation with a cosmological term. Since the exponential of the Chern–Simons action is also a formal solution to the Hamiltonian constraint with a cosmological term in the connection representation [351] with momenta ordered to the left, one obtains solutions to the Hamiltonian constraint (provided a certain formal integration by parts formula holds) which correspond to arbitrary, possibly intersecting, loops.

4.

Also, commutators of constraints were studied formally in the loop representation reproducing the Poisson algebra up to quantities which become singular as the regulator is removed (see [346]). These singular coefficients will later be seen to come from the fact that $\tilde{H}$ is a density of weight two rather than one. Such singularities must be removed, but this could be done for $\tilde{H}$ only by breaking diffeomorphism invariance, which is unacceptable in quantum gravity. We will come back to this point later.

5.

One could confirm the validity of the connection representation in exactly solvable model systems such as the familiar mini- and midisuperspace models based on Killing or dimensional reduction for which the reality conditions can be addressed and solved quantum mechanically [352–361].

These developments in the years 1987–92 confirmed that using Wilson loop functions was something extremely powerful and a rigorous quantisation of the theory should be based on them. Unfortunately, all the nice results obtained so far in the full theory, especially concerning the dynamics as, for example, the existence of solutions to the constraints, were only formal because there was no Hilbert space available that would enable one to say in which topology certain limits might exist or not.

The time had come to invoke rigorous functional analysis in the approach. Unfortunately, this was not possible so far for quantum theories of connections for non-compact gauge groups such as $SL(2, \mathbb{C})$ but only for arbitrary compact gauge groups. The motivation behind pushing these developments anyway at that time had been, again, that by using Bargmann–Segal transformation theory one would be able to transfer the results obtained to the physically interesting case. Luckily, due to the results of [325] one could avoid this additional step and make the results of this chapter directly available for Lorentzian quantum gravity, although in the real connection formulation rather than the complex one.

We describe the developments of the time period 1992–2006 in chronological order where we only quote the main papers. Additional papers will be quoted as we move along in the main text.

(i)

The first functional analytic ideas appeared in the seminal paper by Ashtekar and Isham [362] in which they constructed a quantum configuration space of distributional connections $\overline{\mathcal{A}}$ by using abstract Gel'fand–Naimark–Segal (GNS) theory for Abelian C^* algebras, see Chapter 27. In quantum field theory it is generic that the measure underlying the scalar product of the theory is supported on a distributional extension of the classical configuration space and therefore it was natural to look for something similar, although in a background-independent context. Rendall [363] was able to show that the classical configuration space of smooth connections $\mathcal{A}$ is topologically densely embedded into $\overline{\mathcal{A}}$.

(ii)

Ashtekar and Lewandowski [364] then succeeded in providing $\overline{\mathcal{A}}$ with a σ-algebra of measurable subsets of $\overline{\mathcal{A}}$ and giving a cylindrical definition of a measure μ_0 which is invariant under G gauge transformations and invariant under the spatial diffeomorphisms of $\mathrm{Diff}(\sigma)$. In [365, 366] Marolf and Mourão established that this cylindrically defined measure has a unique σ-additive extension to the just mentioned σ-algebra. Moreover, they proved that, expectedly, $\mathcal{A}$ is contained in a measurable subset of $\overline{\mathcal{A}}$ of measure zero and introduced projective techniques into the framework. In [367, 368] Ashtekar and Lewandowski developed the projective techniques further and used them in [369] to set up integral and differential calculus on $\overline{\mathcal{A}}$. Also Baez [370, 371] had constructed different spatially

146

diffeomorphism-invariant measures on $\overline{\mathcal{A}}$, however, they are not faithful (do not induce positive definite scalar products).

(iii)

The Segal–Bargmann representation in ordinary quantum mechanics on the phase space $\mathbb{R}^2$ is a representation in which wave functions are holomorphic, square integrable (with respect to the Liouville measure) functions of the complex variable $z = q - ip \in \mathbb{C}$. One can obtain this representation by heat kernel evolution followed by analytic continuation from the usual position space representation. In [318] Hall generalised this unitary, so-called Segal–Bargmann transformation, to phase spaces which are cotangent bundles over arbitrary compact gauge groups based on the observation that a natural Laplace operator (generator of the heat kernel evolution) exists on such groups. The role of $\mathbb{C}$ is then replaced by the complexification $G^{\mathbb{C}}$ of G. Since it turns out that the Hilbert space of functions on $\overline{\mathcal{A}}$ labelled by a piecewise analytic (semianalytic)[1] loop reduces to $SU(2)^N$ for some finite natural number N, one can just apply Hall's construction to quantum gravity which would seem to map us from the real connection representation to the complex one. This was done in [321]. The question remained whether the so obtained inner product incorporates the correct adjointness – and canonical commutation relations among the complexified holonomies. In [315, 316] this was shown not to be the case but at the same time a proposal was made for how to modify the transform in such a way that the correct adjointness – and canonical commutation relations – are guaranteed to hold. This so-called Wick rotation transformation is a special case of an even more general method, the so-called complexifier method, which consists in replacing the Laplacian by a more general operator (the complexifier) and can be utilised, as in the case of quantum gravity, to keep the algebraic structure of an operator simple while at the same time trivialising the adjointness conditions on the inner product. Unfortunately, the Wick rotation generator for quantum gravity is very complicated, which is why there is no rigorous proof to date for the existence and the unitarity of the proposed transform. The complexifier method, however, plays a central role for the semiclassical analysis as we will see.

(iv)
$$(\quad)$$
One may wonder whether the techniques associated with $\overline{\mathcal{A}}$ can be applied to ordinary Yang–Mills theory on a background metric. The rigorous quantisation of Yang–Mills theory on Minkowski space is still one of the

---

[1] Any smooth manifold admits a real analytic structure. Roughly speaking, a piecewise analytic manifold is composed out of finitely many entire analytic pieces which intersect in lower-dimensional submanifolds where it is only $C^{(m)}$ with $m \geq 0$. A semianalytic manifold is almost the same thing as a piecewise analytic manifold, however, the gluing of the entire analytic pieces must be $C^{(m)}$ with $m > 0$.

major challenges of theoretical and mathematical physics [372]. There is a vast literature on this subject [373–393] and the most advanced results in this respect are undoubtedly due to Balaban   ., which are so difficult to understand '. . . that they lie beyond the limits of human communicational abilities . . . ' [394]. Technically the problem has been formulated in the context of constructive (Euclidean) quantum field theory [99], which is geared to scalar fields propagating on Minkowski space. In [395, 396] a proposal for a generalisation of the key axioms of the framework, the so-called Osterwalder–Schrader axioms [397, 398], has been given. These were then successfully applied in [399] to the completely solvable Yang–Mills theory in two dimensions by making explicit use of $\overline{\mathcal{A}}$, μ_0 and spin-network techniques which so far had not been done before, although the literature on Yang–Mills theory in two dimensions is rather vast [400–414]. These results have been refined by Fleischhack [415, 416]. It became clear that these axioms apply only to background-independent gauge field theories, which is why it works in two dimensions only (in two dimensions Yang–Mills theory is not background-independent but almost: it is invariant under area-preserving diffeomorphisms, which turns out to be sufficient for the constructions to work out). However, it is possible to generalise the Osterwalder–Schrader framework to general diffeomorphism-invariant quantum field theories [417] Surprisingly, the key theorem of the whole approach, the Osterwalder–Schrader reconstruction theorem that allows us to obtain the Hilbert space of the canonical quantum field theory from the Euclidean one, can be straightforwardly adapted to the more general context. Unfortunately, we do not have space to describe these findings in more detail and must refer the reader to the literature cited.

One of the Osterwalder–Schrader axioms is the uniqueness of the vacuum, which is stated in terms of the ergodicity property of the underlying measure with respect to the time translation subgroup of the Euclidean group (see, e.g., [282]) which in turn has consequences for the support properties of the measure. In [418] these issues were analysed for μ_0 and ergodicity with respect to any infinite, discrete subgroup of the diffeomorphism group was found, which implied a refinement of the support properties established in [365, 366].

(v)

In [266] it was shown that the Hilbert space $\mathcal{H}_0 = L_2(\overline{\mathcal{A}}, d\mu_0)$ in fact solves the adjointness – and canonical commutation relations for any canonical quantum field theory of connections that is based on a compact gauge group provided one represents the connection as a multiplication operator and the electric field as a functional derivative operator. The results of [266] demonstrated that the Hilbert space $\mathcal{H}_0$ provides in fact a physically correct, kinematical representation for such theories. Kinematical here means that this Hilbert space carries a representation of the constraint operators but its vectors are not annihilated by them, that is, they are not physical (or dynamical) states. Moreover, the complete set of solutions of the spatial diffeomorphism constraint (labelled by singular (intersecting) knot classes) and a natural class of scalar products thereon using group averaging methods [277, 278] (Gel'fand triple techniques) could be given which showed, as a side result, that the Husain–Kuchař [361] model is a completely integrable, diffeomorphism-invariant quantum field theory.

(vi)

Quite independently, Rovelli and Smolin as well as Gambini and Pullin . had pushed another representation of the canonical commutation relations, the so-called loop representation already mentioned above for which states of the Hilbert space are to be thought of as functionals of loops rather than connections. Since the Wilson loop functionals (polynomials of traces of holonomies) are not linearly independent, they are subject to the so-called Mandelstam identities; it was mandatory to first find a set of linearly independent functions. Using older ideas due to Penrose [419], Rovelli and Smolin [420] were able to write down such loop functionals, later called spin-network functions, that are labelled by a smooth $SU(2)$ connection. They then introduced an inner product between these functions by simply   them to be orthonormal. Baez [421] then proved that, using the fact that spin-network functions (considered as functionals of connections labelled by loops) can in fact be extended to $\overline{\mathcal{A}}$, the spin-network functions are indeed orthonormal with respect to $\mathcal{H}_0$, moreover, they form a basis, the two Hilbert spaces defined by Ashtekar and Lewandowski on the one hand and Rovelli and Smolin are indeed unitarily equivalent. In [422] a Plancherel theorem [282] was proved, saying that, expectedly, the loop representation and the connection representation are like mutual, non-Abelian Fourier transforms (called the loop transform as mentioned above) of each other where the role of the kernel of the transform is played by the spin-network functions as one would intuitively expect because they are labelled by both loops and connections. The same was established by De Pietri [423], using a graphical language to relate the two representations.

(vii)

It is possible to define operators on $\mathcal{H}_0$ that measure length [424], area and volume [425–428] and angles [429, 430] of curves, surfaces and regions and between curves intersecting at vertices in σ respectively. It turns out that their spectrum is pure point (discrete), the eigenvectors are essentially just spin-network functions and the eigenvalues are multipla of the appropriate power of the Planck length. Notice that these operators are not Dirac observables, they are partial observables and while one can make them commute with the spatial diffeomorphism constraint using matter without affecting their spectral properties [227], it will depend crucially on the choice of the time partial observable whether the spectrum continues to be discrete when we construct the corresponding Dirac observable. In fact, in simple systems it is easy to see that all 16 combinations $(\sigma(C), \sigma(T), \sigma(P), \sigma(D_T(P))) \in \{d, c\}^4$ are possible [227] where $C, T, P, D_T(P)$ denote constraint, clock variable, partial observable and corresponding Dirac observable respectively, $\sigma(A)$ denotes the spectrum of A and c, d means continuous or discrete respectively.

(viii)   ( )

In all these developments it was crucial, for reasons that will be explained below, that σ is an analytic manifold and that the loops were piecewise analytic (semianalytic). Baez and Sawin [431, 432] were able to transfer much of the structure to the case that the loops are only piecewise smooth and intersect in a controlled way (a so-called web) and some of their results were strengthened in [433, 434]. In [435, 436] Zapata introduced the concept of piecewise linear loops. The motivation for these modifications was that the analytical category is rather unnatural from a physical viewpoint, although it is a great technical simplification. For instance, in the smooth category there is no spin-network basis any longer. Both in the analytic and smooth category the Hilbert space is non-separable after modding out by analytic or smooth diffeomorphisms respectively, while in the piecewise linear category one ends up with a separable Hilbert space. The motivation for the piecewise linear category is, however, unclear from a classical viewpoint (for instance the classical action is not invariant under piecewise linear diffeomorphisms). In [437] arguments were given to support the fact that the (mutually orthogonal, unitarily equivalent) Hilbert spaces labelled by the continuous moduli that still appear in the diffeomorphism-invariant (semi)analytic and smooth category are superselected. If one fixes the moduli, the Hilbert space becomes separable.

(ix)

In [325] it was observed that it is impossible to provide a well-defined, background-independent quantisation for the rescaled Hamiltonian constraint. The basic reason is, as we will explain in more detail as we proceed, that only integrals of density weight   valued scalars can be quantised without encountering ultraviolet problems. However, the rescaled constraint is of density weight two.

The way out is a new, background-independent regularisation and factor ordering technique which was then applied in [437–442] to define the quantum Hamiltonian constraint $\hat{H}(N)$ on $\mathcal{H}_0$ for arbitrary N. The result is well-defined operators on $\mathcal{H}_0$ whose constraint algebra closes in the sense that the commutator annihilates spatially diffeomorphism-invariant states as it should. It is possible to systematically construct all of its solutions. This works for arbitrary matter coupling whose corresponding background-independent representations were defined in [443]. See also [444–447] for earlier and later related work on matter field representations. The matrix elements of this operator were studied, for the simplest cases in [451] and some of its quantisation ambiguities were analysed in [452].

It is only due to the results [437–443] that all the mathematical work invested before was actually of any relevance for Lorentzian GR. They put the quantum dynamics on a solid mathematical footing and made $\mathcal{H}_0$ a carrier space of the Wheeler–DeWitt constraint operator.

It was previously believed that polynomiality of the constraints is mandatory in order to have any chance to give them mathematical meaning as operators. This in turn required that one used complex-valued connections for Lorentzian signature, which meant that one must (1) find Hilbert space representations for the non-compact group $SL(2, \mathbb{C})$ and (2) solve the non-polynomial reality conditions (5.1.3) so that non-polynomiality re-entered through the back-door. Even today there is only representation theory for compact gauge groups available and hence both problems 1 and 2 remain unsolved. The above-mentioned results removed both obstacles in one stroke by making the work for compact gauge groups relevant and, moreover, provided a rigorous quantisation of the Hamiltonian constraint for Lorentzian signature in spite of its tremendous non-polynomiality.

The application of the results [437–443] in cosmological minisuperspace truncations of LQG, called Loop Quantum Cosmology (LQC) which will be summarised below, is nowadays celebrated as one of the major results of LQG.

(x)

Reisenberger and Rovelli [453] used the Hamiltonian constraint operator just mentioned in order to provide a heuristic path integral representation of the 'projector' onto the space of physical states. This seminal work gave birth to the so-called   . The name comes from the fact that the coordinate time translation of a graph sweeps out a worldsheet of faces that intersect in edges, the resulting picture is that of a foam. Spin foam models are closely related to state sum models well known in topological QFTs (TQFT). Until today it has not been possible to make the ideas of [453] mathematically precise, and therefore one started from an independent definition of the path integral resulting in a whole set of models, the most well-studied of which is the Barrett–Crane model [454, 455] for both the Euclidean and the Lorentzian signature. However, the connection with the Hamiltonian theory is presently not very well understood, see [456–460] for first steps, which is why it is not yet established that these models implement the dynamics of GR.

So far spin foam models are triangulation-dependent, in other words, they are defined only with a cutoff. For these cutoff models finiteness results have been obtained by Perez and Rovelli [461] and Perez [462]. Group field theory methods [463, 464] have been proposed by De Pietri [465] in order to remove the triangulation dependence.

(xi)

Any quantum theory of gravity must explain the microscopic origin of the Bekenstein–Hawking black hole entropy $S_{\mathrm{BH}} = \mathrm{Ar}(H)/(\hbar G)$ where $\mathrm{Ar}(H)$ is the area of the event horizon H. The idea [467, 468], due to Krasnov, is to count the number of eigenstates of the area operator of the horizon whose eigenvalues are in the interval $[A - \ell_p^2, A + \ell_P^2]$ where A is a given area value. Unless carefully done this entropy is infinite. However, when making use of a sufficient amount of classical input through boundary conditions at H and through Einstein's classical equations, the counting gives precisely the correct answer [469]. This has become possible through the development of a very powerful new notion of horizon, so-called isolated horizons, which are locally defined in contrast to event horizons which require global information. Event horizons and also cosmological horizons are special cases of isolated horizons. Moreover, due to the boundary conditions necessary at H, the function $\mathrm{Ar}(H)$ actually becomes a Dirac observable. Hence the framework incorporates all the usual black hole solutions such as those from the Schwarzschild–Kerr–Reissner–Nordstrøm family and black holes with Yang–Mills and dilatonic hair [470–472].

These are very convincing and encouraging results and future work will address the issue of Hawking radiation from first principles where one must take into account the backreaction of geometry on matter.

(xii) $\overline{\mathcal{A}}$

Following an earlier idea due to Baez [473], Velhinho [474] gave a nice categorical and purely algebraic characterisation of $\overline{\mathcal{A}}$ and all the structure that comes with it without using C^* techniques. The technical simplifications that are involved rest on the concept of a groupoid of piecewise analytic (semianalytic) paths in σ rather than (base-pointed) loops.

In [475, 476] Fleischhack, motivated by his results in [415, 416], discussed a new notion of 'loop independence' which has the advantage of being independent of the differentiability category of the graphs under consideration and in particular includes the analytical and smooth category. The new type of collections of loops are called hyphs. A hyph is a finite collection of piecewise C^r paths together with an ordering $\alpha \mapsto p_\alpha$ of its paths p_α where α belongs to some linearly ordered index set such that p_α is independent of all the paths $\{p_\beta; \beta < \alpha\}$. Here a path p is said to be independent of another path p' if there exists a free point x on p (which may be one of its boundary points), that is, there is a segment of p incident at x which does not overlap with a segment of p' (although p, p' may intersect in x). That is, path independence is based on the germ of a path. In contrast to graphs or webs (collections of piecewise analytical (semi-analytical) or smooth paths), a hyph requires an ordering. Nevertheless, one can get as far with hyphs as with webs but not as far as with graphs.

Fleischhack also investigated the issue of Gribov copies in $\overline{\mathcal{A}}$ [477, 478] with respect to SU(2) gauge transformations. It should be noted that fortunately Gribov copies are not a problem in our context: the measure is a probability measure and the gauge group therefore has finite volume. Integrals over gauge-invariant functions are therefore well-defined and gauge fixing is not necessary.

(xiii)

The Hilbert space $\mathcal{H}_0$ is sufficient for semiclassical applications of quantum General Relativity only if σ is compact. In the non-compact case an extension from compactly supported to non-compactly supported, piecewise analytic (semianalytic) paths becomes necessary. In [479] it was discovered that the framework of the infinite tensor product of Hilbert spaces, developed by von Neumann more than 60 years ago, is ideally suited to deal with this problem. In contrast to $\mathcal{H}_0$ the extended Hilbert space $\mathcal{H}^{\otimes}$ is not L_2 space any longer. Considerations along similar lines have been performed by Arnsdorf [480].

As already mentioned, the loop representation is unitarily equivalent to the Hilbert space $\mathcal{H}_0$ but so far the loop representation had not been displayed as an L_2 space over 'the space of loops' (more precisely: hoops). An investigation to what extent that is possible has been started in [481]. Connected with this is a mathematically rigorous analysis of the spectrum of the holonomy algebras for non-trivial fibre bundles [482, 483].

(xiv)

To date we only have kinematical semiclassical states, that is, they are not annihilated by the Hamiltonian constraints. The motivation for considering such states at all is that one would like to test with them the semiclassical properties of the Hamiltonian constraint, which is obviously not possible with states that are in the kernel of the constraint. Ultimately, of course, one should construct physical semiclassical states.

Weave states had been considered in [484] by Ashtekar, Rovelli and Smolin as candidate semiclassical states for probing the properties of the kinematical operators mentioned above, which depend only on the electric field. However, they do not probe very well operators that depend non-trivially on the connection. In [485–487] the complexifier method was introduced in order to define coherent states which behave semiclassically for both kinds of variables. This programme was then carried out in [488–490] for a specific choice of complexifier and applied to QFT on curved spacetime situations in [637, 638]. Varadarajan [491–494] showed how to define the Fock states of Maxwell theory and linearised gravity as distributions over (a dense subspace of) $\mathcal{H}_0$. Ashtekar and Lewandowski [495] extended this map to the case of coherent states for these two theories and gave a measure theoretic interpretation of [491]. The complexifier method can be applied to these theories and reproduces the results of [491–495].

Due to the non-separability of $\mathcal{H}_0$ these semiclassical states turn out not to be normalisable, they are distributions (sometimes even measures [495]). Moreover, they only solve at most the Gauß constraint. In order to work with them one has to use certain graph-dependent versions which are normalisable. These cutoff states [487] or shadow states [495] are, however, only suited for operators which do not change the graph underlying the spin-network state on which they act. The Hamiltonian constraint does not have this property. An idea for overcoming this problem is to construct spatially diffeomorphism-invariant coherent states because the spatially diffeomorphism-invariant Hilbert space decomposes into separable Hilbert spaces each of which is an invariant subspace of the Hamiltonian constraint. The resulting semiclassical states could then be both graph-independent and normalisable.

On the other hand, in [487] it was argued that the state on the holonomy algebra used in LQG is universal in the sense that any other state can be obtained as a weak limit from vector states in the LQG Hilbert space. This was proved in [496] based on the notion of cutoff states.

(xv)

Bojowald and Kastrup [497–499] used the LQG type of Hilbert space representations for minisuperspace models of GR, in particular for cosmological models. The effect of this is a very drastic departure from the properties of the standard approach for these models, at least at very small scales, while at large scales the properties of the standard approach are recovered. For instance, as expected from the finiteness results of the full theory for the Hamiltonian constraint, there is no big bang singularity in the quantum theory. Moreover, one can propagate, with respect to a partial observable time, through the would-be big bang singularity. There are certain implications of the modified short distance behaviour at short scales for inflation, which one might even be able to see in the WMAP data [500]. For a more precise analysis of possible LQG effects on the spectrum of anisotropies derived from an LQG-inspired cosmological model using the ADM variables [501], see [502].

Of course, these calculations have to be backed up by calculations in the full theory in order to qualify as predictions. However, at the very least this work already now positively tests the validity of certain technical aspects of the full theory.

(xvi)

In [503–506] the idea was put forward that quantum gravity effects, although tiny when measured over short time periods, might accumulate to a more realistic size over large time scales. The basic idea is that matter would react to the existence of the discrete Planck scale structure by modified dispersion relations just as if it was propagating through a crystal rather than the vacuum. A nice overview of possible signature experiments which always probe violation of Lorentz invariance (which in standard QFT is an exact symmetry) can be found in [507]. This triggered the field of LQG phenomenology [508–512]. Recently it was found that present experimental accuracy already rules out the existence of a preferred reference frame which would lead to Lorentz invariance breaking. It could, however, be that the Lorentz group is realised in a non-standard way at very short distances [513].

(xvii)

Up to now one had always been working in the representation $\mathcal{H}_0$, which seemed to be a rather natural choice. However, usually in QFT the famous Stone–von Neumann uniqueness theorem for quantum mechanics fails due to the infinite number of degrees of freedom for QFT and therefore one has an abundance of unitarily inequivalent representations of the canonical commutation relations. In special cases one can select a unique representation if one asks in addition that a Hamiltonian operator can be densely defined [279]. In [514, 515] Sahlmann systematically investigated the general representation theory of the holonomy–flux algebra which underlies LQG by using tools from algebraic QFT. While it is not easy to classify all possible representations, if one makes natural additional physical assumptions, namely (1) the same that underlie the Stone–von Neumann theorem of quantum mechanics and (2) that the spatial diffeomorphism group is unitarily represented then one gets that $\mathcal{H}_0$ is the unique representation of the kinematical algebra. The second assumption is precisely the additional dynamical input that is expected to give rise to such a strong result. Sahlmann's original proof worked only for Abelian gauge groups but was later simplified and extended to the non–Abelian case [516–521]. The essential idea, however, is due to Sahlmann.

The importance of this result lies in the fact that once we decided to base quantisation on the holonomy–flux algebra and have the spatial diffeomorphism group unitarily implemented, there is no choice in the kinematical Hilbert space (if it is a cyclic representation; every representation decomposes into cyclic ones). Since one is naturally led to the holonomy–flux algebra from physical considerations, we have very strong reasons to believe that we chose the only reasonable starting point for quantisation. On the other hand, once a kinematical representation has been chosen in which the constraints can be defined as operators, the physical representation follows by a rather tight procedure discussed in Chapter 3 and Section 30.2. Hence, the whole quantisation programme is altogether put on a very robust footing.

(xviii)

As mentioned above, while the Hamiltonian constraint can be defined on $\mathcal{H}_0$, so far there was no idea for defining the physical inner product and Dirac observables. One of the obstacles is that the Hamiltonian constraints do not form a Lie algebra, there are structure functions rather than structure constants, which is why group averaging techniques, discussed in Chapter 3, cannot be applied. In [252] the Master Constraint programme was initiated. What it does, as discussed in Chapter 3, is to replace the constraint algebra by an equivalent one to which group averaging or direct integral techniques can be applied. Luckily the results mentioned in item (ix) can be used to define the new quantum algebra [522] (see also [523]). These ideas have been successfully tested in other constrained systems such as Maxwell theory, linearised gravity and for the Gauß constraint of Yang–Mills theory coupled to gravity [253–257].

Thus, once we have agreed on a quantisation of the Hamiltonian constraints (rather: Master Constraint) we are granted that there exists a (unique, up to unitary equivalence and up to measure zero issues) physical inner product on the space of solutions of the Master Constraint and what is left to do is to construct Dirac observables by using averaging techniques and to check the correctness of the classical limit of the theory.

In the above list the following items provide complete or almost complete, robust results: (i)–(viii), (xii), (xiii), (xvii). On the other hand, items (ix), (x), (xi), (xiv), (xvi) and (xviii) comprise the current active research in the field where the main ideas have probably been spelt out already while the details are still in flow. As we see, all of these research programmes of the second category are related to the quantum dynamics of the theory. This is not surprising as the solution of the quantum dynamics of a highly interacting QFT such as GR is the final and most difficult step in any such theory. It is at least as difficult as rigorously proving the existence of QCD as a mathematical theory and showing that the theory describes confinement.

In the remainder of this part of the book we will summarise the status of the quantisation programme. The third part focuses on the applications. The fourth part injects mathematical physics techniques into the main text that may be useful for some readers in order to follow the proofs.

In fact, we could keep the discussion in the subsequent chapters shorter if we just wanted to prove the statements made in the above summary. However, in order to equip the reader with the technology necessary in order to experiment with possible modifications of the quantisation programme in places where certain mathematical or physical choices have to be made, we keep the discussion very general so that the results proved can be applied in a maximally broad context.

6

Step I: the holonomy–flux algebra $\mathfrak{P}$

For steps I, II, III of the quantisation programme the choice of the compact group G and the dimensionality of σ will be unimportant, hence we keep the discussion quite general.

6.1 Motivation for the choice of

Before we dive into the mathematical details, let us motivate our choice for the classical algebra $\mathfrak{P}$ on which the quantum theory is to be based and which has been defined in mathematically precise terms for the first time in [524]. For the sake of these heuristic considerations we will not pay attention to mathematical details, these will be supplemented as we move along.

Remember that we are interested in a background-independent formulation, therefore we are not allowed to use any background metric in defining $\mathfrak{P}$. Next, since the Poisson brackets among the fields A_a^j, E_j^a are singular, we should smear them with test functions as we did for the fields q_{ab}, P^{ab} in Section 1.2. Hence our first guess would be to use

$$E(f) := \int_\sigma d^3x\, f_a^j E_j^a,\quad F(A) := \int_\sigma d^3x\, F_j^a A_a^j \tag{6.1.1}$$

which gives rise to the non-distributional Poisson brackets (remember our convention at the end of Section 4.2.2)

$$\{F(A), F'(A)\} = \{E(f), E(f')\} = 0,\quad \{E(f), F(A)\} = \beta\frac{\kappa}{2}\int_\sigma d^3x\, F_j^a\, f_a^j \tag{6.1.2}$$

The functionals (6.1.1) certainly satisfy our requirement to separate the points of (A, E) on the phase space as one can see by restricting the support of the smearing fields f_a^j, F_j^a. However, the problem with the choice (6.1.1) is that these variables do not transform nicely under gauge transformations. Denoting the smeared Gauß constraint by $G(\Lambda) = \int d^3x \Lambda^j G_j$ we find

$$\{G(\Lambda), F(A)\} = -\beta\frac{\kappa}{2}\int d^3x F_j^a\left[\Lambda_{,a}^j + \epsilon_{jkl}A_a^k\Lambda^l\right] \text{ and } \{G(\Lambda), E(f)\}$$

$$= \beta\frac{\kappa}{2}\int d^3x f_a^j \epsilon_{jkl}\Lambda^k E_l^a \tag{6.1.3}$$

This is precisely the infinitesimal version of the transformation law of an SU(2) connection one-form and a Lie algebra-valued vector density transforming in the adjoint representation of SU(2), that is, $A \mapsto -dg\, g^{-1} + gAg^{-1}$ and $E \mapsto gEg^{-1}$ as one can check by introducing a basis τ_j of su(2) and $g = \exp(\Lambda^j \tau_j)$. In this book we will be using $\tau_j = -i\sigma_j$ where σ_j are the Pauli matrices. As we are eventually interested in gauge-invariant objects, it is clear that it will be very difficult to use (6.1.1) because of the non-local dependence of say $\{G(\Lambda), F(A)\}$ on Λ.

A second idea would be to use the magnetic field of A given by $B^a_j := \epsilon^{abc} F^j_{bc}$ and to construct from these directly the gauge-invariant combinations $\mathrm{Tr}(B^a B^b)$, $\mathrm{Tr}(B^a E^b)$, $\mathrm{Tr}(E^a E^b)$. However, since the functions are quartic in the basic fields A, E, the Poisson bracket among them does not close, we do not get a subalgebra. Next one can try to use the non-gauge-invariant function $B(f) := \int d^3x f^j_a B^a_j$ but again the algebra does not close unless we consider also $F(A)$ as elements of $\mathfrak{P}$. Thus we are led to look for a more suitable choice of $\mathfrak{P}$.

The problem we just described is not unique to gravity but appears, of course, in any non-Abelian Yang–Mills theory. Hence we can take advantage of the experience gained there. The only known solution to the problem just mentioned is to work with so-called Wilson loops. Given a curve $c : [0,1] \to \sigma$ in σ we denote by the $h_c(A) \in$ SU(2) of the connection A along c the unique solution to the differential equation

$$\frac{d}{ds} h_{c_s}(A) = h_{c_s}(A) A(c(s)), \quad h_{c_0} = 1_2, \quad h_c(A) := h_{c_1}(A) \qquad (6.1.4)$$

where $c_s(t) := c(st)$, $s \in [0,1]$ and $A(c(s)) := A^j_a(c(s))\tau_j/2\dot{c}^a(s)$. One can write this explicitly as

$$h_c(A) = \mathcal{P} \exp\left(\int_c A\right) = 1_2 + \sum_{n=1}^{\infty} \int_0^1 dt_1 \int_{t_1}^1 dt_2 \ldots \int_{t_{n-1}}^1 dt_n\, A(c(t_1)) \ldots A(c(t_n))$$

$$(6.1.5)$$

where $\mathcal{P}$ denotes the path ordering symbol which orders the smallest path parameter to the left. Using the fact that the gauge transformed connection is $A^g = -dgg^{-1} + gAg^{-1}$ it is easy to check that $h_c(A^g) = g(c(0))h_c(A)g(c(1))^{-1}$. Hence the holonomy transforms   under gauge transformations. As is obvious from (6.1.5), the connection gets smeared only along the curve c, that is, in one dimension, which is very natural because a connection is, in particular, a one-form and as known from differential geometry, there is a natural pairing, called Poincaré duality, between p-forms and p-dimensional submanifolds (see, e.g., Chapter 19).

Next we turn to the conjugate electric field E. Since A is smeared in one dimension only, the field E must be smeared in at least $D - 1 = 2$ dimensions in order that the Poisson bracket with the holonomy be non-distributional. Can it be smeared in $D = 3$ dimensions? The answer is   because otherwise we do not get a closed algebra. Indeed one can check that

$$\{E(f), (h_c(A))_{mn}\}$$

$$= \beta\frac{\kappa}{2}\left[\int_0^1 dt\, \dot{c}^a(t)\, f^j_a(c(t))\, (h_{c_t}(A)(\tau_j/2)h_{c_t}(A)^{-1})_{mk}\right] (h_c(A))_{kn} \qquad (6.1.6)$$

where $m, n = \pm\frac{1}{2}$ denotes the matrix indices of the 2×2 matrix $h_c(A)$. The right-hand side of (6.1.6) is not a polynomial in some holonomies any longer but rather a   , that is, an integral. Hence the right-hand side of (6.1.6) depends on an infinite number of holonomies rather than a finite number. Therefore, with a $D = 3$-dimensional smearing the algebra would not close.

Thus we are forced to work with $D - 1 = 2$-dimensional smearing of E. This is again very natural because the vector density E_j^a is dual to the pseudo-$(D-1)$-form

$$(*E)_{a_1...a_{D-1}}^j := \epsilon_{aa_1...a_{D-1}} E_j^a \left(= \text{sgn}(\det(e))\epsilon_{jkl}e_{a_1}^k e_{a_{D-1}}^l\right) \qquad (6.1.7)$$

where $\epsilon_{a_1...a_D}$ is the background metric-independent totally skew symbol and the last identity holds for $D = 3$. The appearance of the sign factor explains the word pseudo-form. Since a two-form is naturally integrated in two dimensions we are naturally led to consider the quantities

$$E_{\{p\}}(S, A) := \int_S \text{Tr}(\text{Ad}_{h_{p_x}(A)}(*E)(x)) \qquad (6.1.8)$$

Here $*E = (*E)_{a_1...a_{D-1}}^j dx^{a_1} \wedge ... \wedge dx^{a_{D-2}}\tau_j$, $\{p\}$ denotes a system of paths $x \mapsto p_x$, $x \in S$ within S from a fixed interior point $x_0 \in S$ to $x \in S$ and $\text{Ad}_g(.) := g(.)g^{-1}$ is the adjoint action of the Lie group on its own Lie algebra.

It is easy to check that with $E^g = gEg^{-1}$ we have a local gauge transformation behaviour $E_{\{p\}}^g(S, A^g) = g(x_0), E_{\{p\}}(S, A)g(x_0)^{-1}$. Moreover, as we will show in detail later, the algebra of (6.1.5) and (6.1.8) closes and separates the points of the phase space. However, (6.1.8) is too complicated to work with because we have to choose, next to S, the path system $\{p\}$ and, moreover, (6.1.8) also depends on A. It is easier to work with the (electric)

$$E_n(S) := \int_S n_j \, (*E)^j \qquad (6.1.9)$$

where $n = n^j$ is a Lie algebra-valued scalar function. While the gauge transformation of (6.1.9) is again non-local, it will turn out later that all gauge-invariant functions that we will be ultimately interested in can be written in terms of limits of the fluxes as the surfaces S shrink to points. Hence, from this perspective and since (6.1.5) and (6.1.9) separate the points, the algebra $\mathfrak{P}$ generated from holonomies and fluxes satisfies all our requirements of Chapter 3 for it to be a classical starting point of quantisation and actually nobody has found a more natural one.

As a bonus, from the Poincaré duality between chains and forms, the holonomies and fluxes also have a simple transformation behaviour under spatial diffeomorphisms. To see this, let $V_a := H_a - A_a^j G_j$ then

$$\{\vec{V}(\vec{N}), A\} = \beta\frac{\kappa}{2}\mathcal{L}_{\vec{N}}A, \quad \{\vec{V}(\vec{N}), E\} = \beta\frac{\kappa}{2}\mathcal{L}_{\vec{N}}E \qquad (6.1.10)$$

which is precisely the transformation law under infinitesimal spatial diffeomorphisms of a one-form and vector density respectively. Now recall from

Chapter 19 the definition of a one-parameter family of diffeomorphisms $t \mapsto \varphi_t^{\vec{N}}$ generated by the integral curves of a vector field $\vec{N}$. With the pull-back action $A^{\varphi} = \varphi^* A$ and $(*E)^{\varphi} = \varphi^*(*E)$ under finite diffeomorphisms it is easy to check that, for example, $[d/dt(\varphi_t^{\vec{N}})^* A]_{t=0} = \mathcal{L}_{\vec{N}} A$. It follows that

$$\exp(t\mathcal{L}_{\chi_{\vec{V}(\vec{N})}})F[A,(*E)] = \sum_{n=0}^{\infty} \frac{t^n}{n!}\{\vec{V}(\vec{N}), F\}_{(n)} = F\left[\left(\varphi_{\kappa\beta t/2}^{\vec{N}}\right)^* A, \left(\varphi_{\kappa\beta t/2}^{\vec{N}}\right)^*(*E)\right]$$

$$\exp(t\mathcal{L}_{\chi_{G(\Lambda)}})F[A,(*E)] = \sum_{n=0}^{\infty} \frac{t^n}{n!}\{G(\Lambda), F\}_{(n)} = F\left[A^{g_{\kappa\beta t/2}^{\Lambda}}, \left(*E^{g_{\kappa\beta t/2}^{\Lambda}}\right)\right]$$

(6.1.11)

where $g_t^{\Lambda} = \exp(t\Lambda^j \tau_j)$. Here $F[A, (*E)]$ is an arbitrary function on phase space $\mathcal{M}$ and $\chi_{(.)}$ denotes the Hamiltonian vector field of $(.)$. To check this it is sufficient to check the equation and its first derivative at $t = 0$ and to rely on uniqueness of the solutions of ordinary differential equations.

We may therefore generalise the transformations (6.1.11) which are connected to the identity of the gauge group and the spatial diffeomorphism group respectively to arbitrary transformations. Since transformations on phase space generated by the flow of Hamiltonian vector fields are canonical transformations (they preserve Poisson brackets; see, e.g., Section 19.3) we see that SU(2) gauge transformations and spatial diffeomorphisms are implemented on $\mathfrak{P}$ as (they preserve the algebraic structure of $\mathfrak{P}$)

$$\alpha_g(h_c(A), E_n(S)) = (h_c(A^g), E_n^g(S)) = (g(b(c))h_c(A)g(f(c))^{-1}, E_{\mathrm{Ad}_{g^{-1}n}}(S))$$
$$\alpha_{\varphi}(h_c(A), E_n(S)) = (h_c(A^{\varphi}), E_n^{\varphi}(S)) = (h_{\varphi(c)}(A), E_{(\varphi^{-1})_*n}(\varphi(S))) \quad (6.1.12)$$

The transformation of the holonomy–flux variables is therefore quite natural, it consists simply in a geometrical change of the label.

A few technical remarks are in order:

1. Since $\mathcal{M} = T^*\mathcal{A}$ has a cotangent bundle structure we can make use of the framework displayed in Chapter 3 in order to define $\mathfrak{P}$ and $\mathfrak{A}$. A natural choice of functions on $\mathcal{A}$, the so-called cylindrical functions Cyl, are complex-valued functions of a finite collection of holonomies along mutually non-intersecting paths. In order that this class of functions forms an algebra it must be closed under pointwise multiplication. Now the following issue arises: consider, for example, the product $h_p(A) h_{p'}(A)$. If p, p' are arbitrary (piecewise) smooth paths then it is possible that they intersect in a Cantor set and hence the result is a function depending on an infinite number of mutually non-intersecting paths. While it is possible to deal with these complications as we mentioned, it is easier to restrict the paths to be (piecewise) analytic (semianalytic). Two analytic paths intersecting an infinite number of times are actually analytic continuations of each other, hence the product of the corresponding cylindrical

functions is a cylindrical function again. In order to define an analytic path we must equip σ with a real analytic structure. Fortunately [525], any paracompact smooth manifold admits an analytic structure which is smooth diffeomorphic to the given differentiable structure and any two such chosen analytic structures are then, of course, smooth diffeomorphic.

2. Likewise we must be careful that the Hamiltonian vector field $Y_n(S) := \chi_{E_n(S)}$ of the fluxes $E_n(S)$ defined by $Y_n(S) \cdot A(p) = \{E_n(S), A(p)\}$ preserves the cylindrical functions. Now a given surface S cuts a path $p \in \mathcal{P}$ into pieces and the Poisson bracket will receive a contribution from every possible intersection point. Hence, in order that the space of cylindrical functions be preserved it is necessary that the number of these intersection points be finite for every $p \in \mathcal{P}$. This is clearly impossible for paths p some of whose segments lie inside a given S, but it will turn out from the detailed calculation that those segments do not contribute. Hence, all that is important is that the number of isolated intersection points is finite. It follows that the set of allowed surfaces includes the piecewise analytic (semianalytic) ones because in case that a given analytic segment of a path intersects S in an infinite number of isolated points, it must actually lie inside S along an analytic curve and hence does not have isolated intersection points at all.

3. Interestingly, the Lie bracket $[Y_n(S), Y_{n'}(S')]$ does not vanish if $S \cap S' \neq \emptyset$ although classically we have $\{E(f), E(f')\} = 0$ by (6.1.2). The reason for this unexpected non-commutativity is the singular smearing in the definition of $\mathfrak{P}$. Indeed, $\{E_j^a(x), E_k^b(y)\}$ may be non-vanishing and still be compatible with $\{E(f), E(f')\} = 0$ if the right-hand side vanishes when integrated against $\int d^D x f_a^j(x) \int d^D(y)(f')_b^j(y)$. This is indeed the case: introduce one-parameter families of surfaces $t \mapsto S_t$ which fill out a D-dimensional region. Then consider $\int dt \int dt' [Y_n(S_t), Y_{n'}(S'_{t'})]$ applied to a cylindrical function. The integrand turns out to be non-vanishing provided that the cylindrical function depends on holonomies along paths p such that the set of isolated points in $S_t \cap S'_{t'} \cap p$ is not empty. The set of values t, t' for which this happens has $dt\, dt'$ measure zero, hence the integral of the commutator vanishes as expected. We are therefore forced not to set $\{E_n(S), E_{n'}(S')\} = 0$, otherwise the Jacobi identity between, say, $E_n(S)$, $E_{n'}(S')$, $A(p)$ would be violated. Rather, we need to　　　　　this bracket via the Hamiltonian vector field $\chi_{\{E_n(S), E_{n'}(S')\}} := [\chi_{E_n(S)}, \chi_{E_{n'}(S')}]$ where $\chi_{E_n(S)} = Y_n(S)$. Of course, only vector fields of the form $Y_n(S)$ have an immediate classical interpretation.

4. Given that both the paths and the surfaces are restricted to be piecewise analytic (semianalytic), (6.1.12) defines an automorphism of the corresponding algebra only if the class of allowed diffeomorphisms preserves the piecewise analytic (semianalytic) paths and surfaces respectively. The set of these diffeomorphisms includes the analytic diffeomorphisms of the chosen analytic structure of σ but there is an extension which is larger: these are the piecewise

analytic (semianalytic) diffeomorphisms which are analytic everywhere except on lower-dimensional submanifolds of σ where they are only differentiable of class $C^{(n_0)}$, $n_0 > 0$ as advocated by Zapata [435, 436]. We will construct them in detail in Chapter 20. These not entire analytic diffeomorphisms are important because the analytic ones are rather global: an entire analytic function is already determined by its values in an arbitrarily small neighbourhood of any point while this is not the case for smooth functions and our piecewise analytic (semianalytic) diffeomorphisms. This global aspect would prevent, for instance, the uniqueness result of the representation theory of $\mathfrak{A}$ that we are going to prove on general manifolds different from $\sigma = \mathbb{R}^D$ by using the piecewise (or more precisely semi-) analytic structure. The largest possible extension of the diffeomorphism group preserving the given structure is the automorphism group $\mathrm{Aut}(\mathcal{P})$ of the groupoid of paths $\mathcal{P}$, to be defined below, as has recently been advocated in [526].

This ends our motivational remarks. We will now proceed to the mathematical details.

6.2 Definition of   : (1) paths, connections, holonomies and cylindrical functions

In what follows, holonomies play a fundamental role. For a fibre bundle theoretic definition see Chapter 21. In this section we will follow closely Velhinho [474]. For simplicity we stick to the piecewise analytic, or more precisely, semianalytic category to which an introduction is given in Chapter 20. For generalisation to the other categories discussed above, please refer to the literature cited there. So in what follows, σ is a semianalytic, connected and orientable D-dimensional manifold which is locally compact (every point has an open neighbourhood with compact closure, automatic if σ is finite-dimensional) and paracompact (for finite-dimensional σ equivalent to the condition of being the countable union of compact sets). The generalisation to non-connected and non-orientable σ is straightforward.

We will actually develop more structure than is strictly necessary in order to define 𝔓. However, we will need this additional technology later on when we study representations anyway and this is a good place to introduce it.

6.2.1 Semianalytic paths and holonomies

In all that follows we work with connection potentials, thus we assume that in each fibre of the principal G-bundle a reference point has been chosen. A change of reference point corresponds to a gauge transformation, thus upon passage to the gauge-invariant sector nothing will depend on that choice any more, as shown in Chapter 22 where holonomies in non-trivial fibre bundles are discussed in detail.

We notice that all the developments that follow use a concrete manifold σ and that the loops or paths are embedded into it. However, in order to describe topology change within quantum gravity it would be desirable to formulate a Hilbert space using non-embedded (algebraic) graphs [527–531]. The state of the abstract Hilbert space itself should tell us into which σ's the algebraic graph on which it is based can be embedded. For some ideas in that direction in connection with semiclassical issues, see [490].

Definition 6.2.1. $\mathcal{C}$

σ

$c \in \mathcal{C}$

$$c : [0,1] \to \sigma; \ t \mapsto c(t) \tag{6.2.1}$$

n $[0,1] =$

$[t_0 = 0, t_1] \cup [t_1, t_2] \cup \ldots \cup [t_{n-1}, t_n = 1]$ $(\ )\ c$

$t_k, \ k = 1, \ldots, n-1 \quad (\ )$ $[t_{k-1}, t_k], \ k = 1, \ldots, n$ $(\ )$

$c((t_{k-1}, t_k)), \ k = 1, \ldots, n-1$ σ

σ c

For a precise definition of semianalyticity, see Chapter 20. However, intuitively, a semianalytic curve c is a finite composition of entire analytic curves c_k where the differentiability class at the boundaries $p_k = c_k \cap c_{k+1}$ is C^{m_k} with $m_k > 0$. More precisely, a semianalytic curve is an oriented semianalytic submanifold of dimension one with a two-point boundary. Roughly speaking, the difference between a piecewise analytic curve and a semianalytic curve is that at points of non-analyticity a piecewise analytic curve just has to be continuous while a semianalytic curve has to be at least $C^{(1)}$. See Figure 6.1 for examples.

Recall that a differentiable map $\phi : M_1 \to M_2$ between finite-dimensional manifolds M_1, M_2 is called an immersion when ϕ has everywhere rank $\dim(M_1)$. An immersion need not be injective but when it is, it is called an embedding. For an embedding, the map $\phi : M_1 \to \phi(M_1)$ is a bijection and the manifold structure induced by ϕ on $\phi(M_1)$ is given by the atlas $\{\phi(U_I), \varphi_I \circ \phi^{-1}\}$ where $\{U_I, \varphi_I\}$ is an atlas of M_1. This differentiable structure need not be equivalent to the submanifold structure of $\phi(M_1)$ which is given by the atlas $\{V_J \cap \phi(M_1), \phi_J\}$ where $\{V_J, \phi_J\}$ is an atlas of M_2. When both differential structures are equivalent (diffeomorphic in the chosen differentiability category, say $C^r, \ r \in \mathbb{N} \cup \{\infty\} \cup \{\omega\}$ where ∞, ω denotes smooth and analytic respectively) the embedding is called regular. The above definition allows a curve to have self-intersections and self-overlappings so that it is only an immersion, but on the open intervals (t_{k-1}, t_k) a curve c is a regular $C^m, \ m > 0$ embedding, in particular, it does not come arbitrarily close to itself.

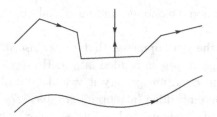

Figure 6.1 Top figure: curve with points of non-differentiability and a retracing. Bottom figure: a (differentiable) edge.

Definition 6.2.2

$()$ $c \in C$

$$b(c) := c(0), \quad f(c) := c(1), \quad r(c) := c([0,1]) \qquad (6.2.2)$$

$()$ $\circ : C \times C \to C$ $c_1, c_2 \in C$ (
$f(c_1) = b(c_2))$ $^{-1} : C \to C$ $c \in C$

$$(c_1 \circ c_2)(t) \begin{cases} := \quad c_1(2t) & t \in [0, \tfrac{1}{2}] \\ c_2(2t-1) & t \in [\tfrac{1}{2}, 1] \end{cases}, \quad c^{-1}(t) := c(1-t) \qquad (6.2.3)$$

Notice that the operations (6.2.3) do not equip C with the structure of a group for several reasons. First of all, not every two curves can be composed. Second, composition is not associative because $(c_1 \circ c_2) \circ c_3$, $c_1 \circ (c_2 \circ c_3)$ differ by a reparametrisation. Finally, the retraced curve $c \circ c^{-1}$ is not really just given by $b(c)$ so that c^{-1} is not the inverse of c and anyway there is no natural 'identity' curve in C.

Definition 6.2.3. $c, c' \in C$ $c \sim c'$

$b(c) = b(c'), f(c) = f(c')$ (   )
c' c

It is easy to see that $\sim$ defines an equivalence relation on C (reflexive: $c \sim c$, symmetric: $c \sim c' \Rightarrow c' \sim c$, transitive: $c \sim c'$, $c' \sim c'' \to c \sim c''$). The equivalence class of $c \in P$ is denoted by p_c and the set of equivalence classes is denoted by P. In order to distinguish the equivalence classes from their representative curves we will refer to them as paths. As always, the dependence of P on σ will not be explicitly displayed. The second condition means that $c' = c'_1 \circ \tilde{c}'_1 \circ (\tilde{c}'_1)^{-1} \circ \ldots \circ c'_{n-1} \circ \tilde{c}'_{n-1} \circ (\tilde{c}'_{n-1})^{-1} \circ c'_n$ for some finite natural number n and curves $c'_k, \tilde{c}'_l$, $k = 1, \ldots, n$, $l = 1, \ldots, n-1$ and that there exists a diffeomorphism $f : [0,1] \to [0,1]$ such that $c \circ f = c'_1 \circ \ldots \circ c'_n$.

Definition 6.2.3 has the following fibre bundle theoretic origin (see, e.g., [337] and Chapter 21): recall that a connection ω on a principal G bundle P may be defined in terms of local connection potentials $A_I(x)$ over the chart U_I of an atlas $\{U_I, \varphi_I\}$ of σ which are the pull-backs to σ by local sections $s_I^\phi(x) := \phi_I(x, 1_G)$ of ω where $\phi_I : U_I \times G \to \pi^{-1}(U_I)$ denotes the system of local trivialisations of P adapted to the U_I and π is the projection of P. The holonomy $h_{cI} := h_{cI}(1)$ of A_I along a curve in the domain of a chart U_I is uniquely defined by the differential equation

$$\dot{h}_{cI}(t) = h_{cI}(t) A_{Ia}(c(t)) \dot{c}^a(t); \quad h_{cI}(0) = 1_G \tag{6.2.4}$$

and one may check that under a change of trivialisation within $U_I \cap U_J$

$$A_I(x) \mapsto A_J(x) = -dh_{JI}(x) h_{JI}(x)^{-1} + \mathrm{Ad}_{h_{JI}(x)}(A_I(x)) \tag{6.2.5}$$

the holonomy transforms as

$$h_{cI} \mapsto h_{cJ} = h_{JI}(b(c)) h_{cI} h_{JI}(f(c))^{-1} \tag{6.2.6}$$

Denote by $\mathcal{A}$ the space of smooth connections (abusing the notation by identifying the collection of potentials with the connection itself) over σ (the dependence on the bundle is not explicitly displayed) and in what follows we will write $h_c(A)$ for the holonomy of A along c, understood as an element of G which is possible once a reference set of points in each fibre is fixed, see Chapter 22. We will denote by $A^g := -dgg^{-1} + \mathrm{Ad}_g(A)$ a gauge transformed connection for a function $g : x \mapsto G$ (which corresponds to a change of reference points) and have

$$h_c^g(A) := h_c(A^g) = g(b(c)) h_c(A) g(f(c))^{-1} \tag{6.2.7}$$

which can be checked directly from (6.2.4) if c is in the domain of a chart but also holds in general bundles. Besides these transformation properties, the holonomy has the following important algebraic properties, even in a non-trivial bundle:

1. $h_{c_1 \circ c_2}(A) = h_{c_1}(A) h_{c_2}(A)$
2. $h_{c^{-1}}(A) = h_c(A)^{-1}$

as may easily be checked by using the differential equation (6.2.4). Furthermore, one can verify that the differential equation (6.2.4) is invariant under reparametrisations of c. These three properties guarantee that $h_c(A)$ does not depend on $c \in \mathcal{C}$ but only on the equivalence class $p_c \in \mathcal{C}$.

One might therefore also have given the following definition of equivalence of curves:

Definition 6.2.4. $c, c' \in \mathcal{C}$ $c \sim c'$

$b(c) = b(c'), \ f(c) = f(c') \ ($ $)$
$h_c(A) = h_{c'}(A) \qquad A \in \mathcal{A}$

In fact, Definitions 6.2.4 and 6.2.3 are equivalent if G is compact and non-Abelian [433] since then every group element can be written as a commutator, that is, in the form $h = h_1 h_2 h_1^{-1} h_2^{-1}$ so that curves of the form $c_1 \circ c_2 \circ c_1^{-1}$ are not equivalent with c_2. In the Abelian case, Definition 6.2.4 is stronger than Definition 6.2.3. In what follows we will work with Definition 6.2.3.

Property (1) of Definition 6.2.3 implies that the functions b, f can be extended to $\mathcal{P}$ by $b(p_c) := b(c)$, $f(p_c) =: f(c)$, the right-hand sides are independent of the representative. However, the function r can be extended only to special elements which we will call edges.

Definition 6.2.5. $\quad\quad e \in \mathcal{P} \quad\quad\quad\quad\quad\quad\quad c_e \in \mathcal{C}$
$$[0,1] \quad\quad\quad\quad\quad r(e) := r(c_e)$$

For a semianalytic curve we may find an equivalent one which is not entire semianalytic but contains a retracing. However, we do not allow such representatives in the definition of $r(e)$. The difference between a generic curve c and c_e is that apart from retracings c may be a composition of entire analytic segments such that at the endpoints of those segments the curve is only continuous but not differentiable. It may be checked that $p_{c_1} \circ p_{c_2} := p_{c_1 \circ c_2}$ and $p_c^{-1} := p_{c^{-1}}$ are well-defined. The advantage of dealing with paths $\mathcal{P}$ rather than curves is that we now have almost a group structure since composition becomes associative and the path $p_c \circ p_c^{-1} = b(p_c)$ is trivial (stays at its beginning point). However, we still do not have a natural identity element in $\mathcal{P}$ and not all of its elements can be composed. The natural structure behind this is that of a ____ . Let us recall the slightly more general definition of a category.

Definition 6.2.6

$()\quad\quad\quad \mathcal{K} \quad\quad\quad\quad (\quad\quad\quad\quad\quad\quad\quad\quad\quad\quad\quad)$
$$x, y, z, \ldots,$$
$$(x,y) \quad\quad\quad\quad\quad\quad\quad\quad\quad (x,y) \quad\quad\quad\quad\quad M(\mathcal{K})$$

$$\circ : \quad\quad (x,y) \times \quad\quad (y,z) \to \quad\quad (x,z); \quad (f,g) \mapsto f \circ g \quad\quad\quad\quad (6.2.8)$$

$()\quad\quad\quad\quad\quad\quad\quad f \circ (g \circ h) = (f \circ g) \circ h \quad\quad\quad\quad f \in \quad\quad (w,x), g \in$
$\quad\quad (x,y), h \in \quad\quad (y,z)$
$()\quad\quad\quad\quad\quad\quad\quad\quad\quad x \in \mathcal{K} \quad\quad\quad\quad\quad\quad\quad\quad\quad\quad x \in \quad\quad (x,x)$
$\quad\quad\quad\quad\quad\quad y \in \mathcal{K} \quad\quad\quad\quad x \circ f = f \quad\quad\quad f \in$
$\quad\quad f \circ \quad_x = f \quad\quad\quad f \in \quad\quad (x,y) \quad\quad\quad\quad (y,x)$
$()\quad\quad\quad\quad\quad \mathcal{K}' \subset \mathcal{K}$
$\quad\quad\quad\quad \mathcal{K} \quad\quad\quad\quad\quad\quad\quad\quad\quad\quad (x,y) \quad\quad \mathcal{K}'$
$\quad\quad\quad\quad\quad '(x,y) \subset \quad\quad (x,y)$

( ) $f \in$ (x, y)
   $g \in$ (y, x) $f \circ g =$ $_{y}, g \circ f =$ $_x$
( ) $\mathcal{K}_1, \mathcal{K}_2$
  $M(\mathcal{K}_1),\ M(\mathcal{K}_2)$ $F : [\mathcal{K}_1, M(\mathcal{K}_1)] \to [\mathcal{K}_2, M(\mathcal{K}_2)]$
                                            $F_* \ [F^*]$

       $f \in$ $(x, y) \Rightarrow F(f) \in$ $(F(x), F(y))\ [$ $(F(y), F(x))]$
       $F(f \circ g) = F(f) \circ F(g)\ [F(g) \circ F(f)]$
       $F(\ _x) =$ $_{F(x)}$
( )

This definition obviously applies to our situation with the following identifications:

Category: σ.
Objects: points $x \in \sigma$.
Morphisms: paths between points $\hom(x, y) := \{p \in \mathcal{P};\ b(p) = x,\ f(p) = y\}$.
   Obviously, every morphism is an isomorphism.
Collection of sets of morphisms: all paths $M(\sigma) = \mathcal{P}$.
Composition: composition of paths $p_{c_1} \circ p_{c_2} = p_{c_1 \circ c_2}$.
Identities: $\mathrm{id}_x = p \circ p^{-1}$ for any $p \in \mathcal{P}$ with $b(p) = x$.

We will call this category σ the category of points and paths and denote it synonymously by $\mathcal{P}$ as well.

Subcategories: $l \subset \mathcal{P}$ consisting of a subset of σ as the set of objects and for each two such objects x, y a subset $\hom'(x, y) \subset \hom(x, y)$.

It is clear that every path is a composition of edges, however, $\mathcal{P}$ is not freely generated by edges (free of algebraic relations among edges) because the composition $e \circ e'$ of two edges e, e' defined as the equivalence class of semianalytic curves $c_e, c_{e'}$ which are semianalytic continuations of each other defines a new edge e'' again. Notice that $\hom(x, y) \neq \emptyset$ for any $x, y \in \sigma$ because we have assumed that σ is connected, one says that $\mathcal{P}$ is connected. Moreover, $\hom(x, x)$ is actually a group with the identity element id_x being given by the trivial path in the equivalence class of the curve $c(t) = x$, $t \in [0, 1]$. The groups $\hom(x, x)$ are all isomorphic: fix an arbitrary path $p_{xy} \in \hom(x, y)$, then $\hom(x, x) = p_{xy} \circ \hom(y, y) \circ p_{xy}^{-1}$.

Definition 6.2.7. $x_0 \in \sigma$ $\mathcal{Q} :=$ (x_0, x_0)

The name 'hoop' is an acronym for 'holonomy equivalence class of a loop based at x_0'. We use the word hoop to distinguish a hoop (a closed path) from its representative loop (a closed curve).

Lemma 6.2.8. $p_x \in$ (x_0, x)

$p_{x_0} =$ x_0 $p \in \mathcal{P}$ $\alpha \in \mathcal{Q}$

$$p = p_{b(p)}^{-1} \circ \alpha \circ p_{f(p)} \tag{6.2.9}$$

The proof consists in solving equation (6.2.9) for α.

Lemma 6.2.9. $l \subset \mathcal{P}$ x_0

$\iota(x_0, x_0)$ $\mathcal{Q}$ l

$\mathcal{Q}'$ $\mathcal{Q}$ $X \subset \sigma$ x_0

$l := \{p_x^{-1} \circ \alpha \circ p_y;\ x, y \in X,\ \alpha \in \mathcal{Q}'\}$ $\mathcal{P}$ $(p_x$

$)$ $\mathcal{Q}' =$ $\iota(x_0, x_0)$

(i) l is a connected subgroupoid: given $p \in l$ there exist $x, y \in X$, $\alpha \in \mathcal{Q}'$ such that $p = p_x^{-1} \circ \alpha \circ p_y$. Thus $p^{-1} = p_y^{-1} \circ \alpha^{-1} \circ p_x \in l$ since $\mathcal{Q}'$ is a subgroup. Also given $p' = p_y^{-1} \circ \beta \circ p_z \in l$ we have $p \circ p' = p_x^{-1} \circ \alpha \circ \beta \circ p_z \in l$ since $\mathcal{Q}'$ is a subgroup. l is trivially connected since by construction every $x \in X$ is connected to $x_0 \in X$ through the path $p_x^{-1} \circ \alpha \circ p_y$ with $y = x_0, \alpha = \mathrm{id}_{x_0}$.

(ii) We have

$$\mathrm{hom}_l(x_0, x_0) = \{p \in \mathcal{Q};\ p \in l\} = \{p_{x_0}^{-1} \circ \alpha \circ p_{x_0};\ \alpha \in \mathcal{Q}'\} = \mathcal{Q}' \tag{6.2.10}$$

since $p_{x_0} = \mathrm{id}_{x_0}$. $\square$

6.2.2 A natural topology on the space of generalised connections

We have noticed above that for an element $A \in \mathcal{A}_P$ its holonomy $h_c(A)$ (understood as taking values in G, subject to a fixed trivialisation of the bundle P) depends only on p_c. We have momentarily explicitly displayed the dependence of the space of smooth connections on the bundle for clarity. To express this we will use the notation

$$A(p_c) := h_c(A) \tag{6.2.11}$$

It follows then that

$$A(p \circ p') = A(p)A(p'),\ A(p^{-1}) = A(p)^{-1} \tag{6.2.12}$$

in other words, every $A \in \mathcal{A}_P$ defines a   .

Definition 6.2.10. $(\mathcal{P}, G)$ $($

$)$ σ

What we have just shown is that $\mathcal{A}_P$ can be understood as a subset of $\mathrm{Hom}(\mathcal{P}, G)$ via the injection $H : \mathcal{A}_P \to \mathrm{Hom}(\mathcal{P}, G);\ A \mapsto H_A$ where $H_A(p) := A(p)$. That H is an injection ($H_A = H_{A'}$ implies $A = A'$) is the content of Giles' theorem [342] and can easily be understood from the fact that for a smooth connection $A \in \mathcal{A}_P$ we have for short curves $c_\epsilon : [0, 1] \to \sigma;\ c_\epsilon(t) = c(\epsilon t)$,

$0 < \epsilon < 1$ an expansion of the form $h_{c_\epsilon}(A) = 1_G + \epsilon \dot{c}^a(0) A_a(c(0)) + o(\epsilon^2)$ so that $(\frac{d}{d\epsilon})_{\epsilon=0} h_{c_\epsilon}(A) = \dot{c}^a(0) A_a(c(0))$, that is, by varying the curve c we can recover A from its holonomy.

We now show that $\mathcal{A}_P$ is certainly not all of $\mathrm{Hom}(\mathcal{P}, G)$, that is, H is not a surjection, suggesting that $\mathrm{Hom}(\mathcal{P}, G)$ is a natural distributional extension of $\mathcal{A}_P$: first of all, as we have said before, unless σ is three-dimensional and $G = SU(2)$ the bundle P is not necessarily trivial and the classical spaces $\mathcal{A}_P$ are all different for different bundles. However, the space $\mathrm{Hom}(\mathcal{P}, G)$ depends only on σ and not on any P, which means that it contains all possible classical spaces $\mathcal{A}_P$ at once and thus is much larger. Beyond this union of all the $\mathcal{A}_P$ it contains distributional elements, for instance the following: let $f : S^2 \to G$ be any map, $x \in \sigma$ any point. Given a path p choose a representative c_p. The curve c_p can pass through x only a finite number of times, say N times, due to piecewise (semi)analyticity (see below). At the kth passage denote by $n_k^\pm$ the direction of $\dot{c}_p(t)$ at x when it enters (leaves) x. Then define $H(p) := [f(-n_1^-)^{-1} f(n_1^+)] \ldots [f(-n_N^-)^{-1} f(n_N^+)]$ (for $N = 0$ defined to be 1_G). Notice that a retracing through x does not affect this formula because in that case $n_k^+ = -n_k^-$ and since we are taking only the direction of a tangent, also reparametrisations do not affect it. It follows that it depends only on paths rather than curves. It is easy to check that this defines an element of $\mathrm{Hom}(\mathcal{P}, G)$. It is not of the form $H_A, A \in \mathcal{A}_P$ because H has support only at x, it is distributional. However, it is   a Schwarz distribution due to its direction dependence. More examples of distributional elements can be found in [364].

Having motivated the space $\mathrm{Hom}(\mathcal{P}, G)$ as a distributional extension of $\mathcal{A}_P$, the challenge is now to equip this so far only algebraically defined space with a topology. The reason is that, being distributional, it is a natural candidate for the support of a quantum field theory measure as we have stressed before, but measure theory becomes most powerful in the context of topology. In order to define such a topology, projective techniques [532] suggest themselves. We begin quite generally.

Definition 6.2.11

( ) $\mathcal{L}$ (   ) $\prec$ $\mathcal{L}$

$\mathcal{L} \times \mathcal{L}$ $(l \prec l)$ $(l \prec l', l' \prec$

$l \Rightarrow l = l')$ $(l \prec l', l' \prec l'' \Rightarrow l \prec l'')$

$\mathcal{L}$ $\mathcal{L}$

( ) $\mathcal{L}$ $l, l' \in \mathcal{L}$

$l'' \in \mathcal{L}$ $l, l' \prec l''$

( ) $\mathcal{L}$

$(X_l, p_{l'l})_{l \prec l' \in \mathcal{L}}$ X_l $\mathcal{L}$

$$p_{l'l} : X_{l'} \to X_l \; \forall \, l \prec l' \tag{6.2.13}$$

$$p_{l'l} \circ p_{l''l'} = p_{l''l} \; \forall \, l \prec l' \prec l'' \tag{6.2.14}$$

$$(\quad) \qquad\qquad \overline{X} \qquad\qquad\qquad (X_l, p_{l'l})$$

$$X_\infty := \prod_{l \in \mathcal{L}} X_l$$

$$\overline{X} := \{(x_l)_{l \in \mathcal{L}}; \; p_{l'l}(x_{l'}) = x_l \; \forall \, l \prec l'\} \tag{6.2.15}$$

The idea of using this definition for our goal to equip $\mathrm{Hom}(\mathcal{P}, \mathrm{G})$ with a topology is the following: we will readily see that $\mathrm{Hom}(\mathcal{P}, \mathrm{G})$ can be displayed as a projective limit. The compactness of the Hausdorff space G will be responsible for the fact that every X_l is compact and Hausdorff. Now on a direct product space (independent of the cardinality of the index set) in which each factor is compact and Hausdorff one can naturally define a topology, the so-called Tychonov topology, such that X_∞ is compact again. If we manage to show that $\overline{X}$ is closed in X_∞ then $\overline{X}$ will be compact and Hausdorff as well in the subspace topology (see, e.g., [533]). However, for compact Hausdorff spaces powerful measure theoretic theorems hold which will enable us to equip $\mathrm{Hom}(\mathcal{P}, \mathrm{G})$ with the structure of a σ-algebra and to develop measure theory thereon.

In order to apply Definition 6.2.11 then to our situation, we must decide on the label set $\mathcal{L}$ and the projective family.

Definition 6.2.12

$$(\quad) \qquad\qquad \{e_1, \ldots, e_n\}$$
$$e_k \qquad\qquad\qquad\qquad\qquad\qquad b(e_k), f(e_k)$$
$$(\quad) \qquad\qquad \{e_1, \ldots, e_n\}$$
$$e_k \qquad\qquad\qquad\qquad\qquad\qquad e_1, \ldots, e_{k-1}, e_{k+1}, \ldots, e_n$$

$$(\quad) \qquad\qquad\qquad \{e_1, \ldots, e_n\} \qquad\qquad\qquad\qquad \gamma$$
$$\gamma := \cup_{k=1}^n r(e_k) \qquad r(e_k) \subset \gamma \qquad\qquad\qquad e_k \;\; (e \cup e' :=$$
$$p_{c_e \cup c_{e'}}) \qquad\qquad \gamma \qquad\qquad\qquad\qquad E(\gamma) = \{e_1, \ldots, e_n\}$$
$$\gamma$$
$$\gamma \quad V(\gamma) = \{b(e), f(e); \; e \in E(\gamma)\} \qquad\qquad \Gamma_0^\omega$$

$$(\quad) \qquad\qquad \gamma \qquad\qquad\qquad l(\gamma) \subset \mathcal{P} \qquad\qquad\qquad \gamma$$
$$V(\gamma) \qquad\qquad\qquad\qquad\qquad e \in E(\gamma)$$

Notice that independence of sets of edges implies algebraic independence but not vice versa (consider independent e_1, e_2 with $f(e_1) = b(e_2)$ and define $e_1' = e_2$, $e_2' = e_1 \circ e_2$; then e_1', e_2' is algebraically independent but not independent) and that $l(\gamma)$ is freely generated by the $e \in E(\gamma)$ due to their algebraic independence. Also, $l(\gamma)$ does not depend on the orientation of the graph since $e_1, \ldots, e_n$ and $e_1^{s_1}, \ldots, e_n^{s_n}$, $s_k = \pm 1$ generate the same subgroupoid. The labels

$\omega, 0$ in Γ_0^ω stand for 'semianalytic' and 'of compact support' respectively for obvious reasons.

The following theorem finally explains why it was important to stick with the analytic, compact category. The proof is elementary, see Chapter 20 for a more abstract proof using semianalyticity.

Theorem 6.2.13. $\qquad\mathcal{L}\qquad\qquad\qquad\qquad\qquad\qquad\qquad l(\gamma)\quad \mathcal{P}$

$\qquad\qquad\qquad\qquad\qquad \gamma \in \Gamma_0^\omega \qquad\qquad\qquad\qquad\qquad l \prec l' \quad l$

$l' \qquad \mathcal{L}$

Since l is a subgroupoid of l' iff all objects of l are objects of l' and all morphisms of l are morphisms of l' it is clear that $\prec$ defines a partial order. To see that $\mathcal{L}$ is directed consider any two graphs $\gamma, \gamma' \in \Gamma_0^\omega$ and consider $\gamma'' := \gamma \cup \gamma'$. We claim that γ'' has a finite number of edges again, that is, it is an element of Γ_0^ω. For this to be the case it is obviously sufficient to show that any two edges $e, e' \in \mathcal{P}$ can only have a finite number of isolated intersections or they are semianalytic extensions of each other. Clearly they are semianalytic extensions of each other if $e \cap e'$ is a common finite segment. Suppose then that $e \cap e'$ is an infinite discrete set of points. We can assume without loss of generality that e, e' are entire analytic, otherwise apply the following argument to each of the finite entire analytic segments out of which semianalytic edges are composed. We may choose parametrisations of their representatives c, c' such that each of its component functions $f(t)^a := e'(t)^a - e(t)^a$ vanishes in at least a countably infinite number of points t_m, $m = 1, 2, \ldots$. We now show that for any function $f(t)$ which is real analytic in $[0, 1]$ this implies $f = 0$. Since $[0, 1]$ is compact there is an accumulation point $t_0 \in [0, 1]$ of the t_m (here the compact support of the $c \in \mathcal{C}$ comes into play) and we may assume without loss of generality that t_m converges to t_0 and is strictly monotonous. Since f is analytic we can write the absolutely convergent Taylor series $f(t) = \sum_{n=0}^\infty f_n(t - t_0)^n$ (here analyticity comes into play). We show $f_n = 0$ by induction over $n = 0, 1, \ldots$. The induction start $f_0 = f(t_0) = \lim_{m \to \infty} f(t_m) = \lim_{m \to \infty} 0 = 0$ is clear. Suppose we have shown already that $f_0 = \ldots = f_n = 0$. Then $f(t) = f_{n+1}(t - t_0)^{n+1} + r_{n+1}(t)(t - t_0)^{n+2}$ where $r_{n+1}(t)$ is uniformly bounded in $[0, 1]$. Thus $0 = f(t_m)/(t_m - t_0)^{n+1} = f_{n+1} + r_{n+1}(t_m)(t_m - t_0)$ for all m, hence $f_{n+1} = \lim_{m \to \infty}[f_{n+1} + r_{n+1}(t_m)(t_m - t_0)] = 0$. $\qquad\square$

Notice that the subgroupoids $l \in \mathcal{L}$ also conversely define a graph up to orientation through its edge generators.

Now that we have a partially ordered and directed index set $\mathcal{L}$ we must specify a projective family.

Definition 6.2.14. $\qquad\qquad\qquad l \in \mathcal{L}\qquad\qquad X_l := \qquad (l, G)$

$\qquad\qquad\qquad\qquad\qquad\qquad\qquad l$

Notice that for $l = l(\gamma)$ any $x_l \in X_l$ is completely determined by the group elements $x_l(e)$, $e \in E(\gamma)$ so that we have a bijection

$$\rho_\gamma : X_l \to G^{|E(\gamma)|}; \quad x_l \mapsto (x_l(e))_{e \in E(\gamma)} \tag{6.2.16}$$

Since G^n for any finite n is a compact Hausdorff space (here compactness of G comes into play) in its natural manifold topology we can equip X_l with a compact Hausdorff topology through the identification (6.2.16). This topology is independent of the choice of edge generators of l since any map $(e_1, \ldots, e_n) \mapsto (e_{\pi(1)}^{s_1}, \ldots, e_{\pi(n)}^{s_n})$ for any element $\pi \in S_n$ of the permutation group of n elements and any $s_1, \ldots, s_n = \pm 1$ induces a homeomorphism (topological isomorphism) $G^n \to G^n$.

Next we must define the projections.

Definition 6.2.15. $l \prec l'$

$$p_{l'l} : X_{l'} \to X_l; \quad x_{l'} \mapsto (x_{l'}) \tag{6.2.17}$$

$$x_{l'} \qquad\qquad\qquad\qquad l'$$

$l \prec l'$

It is clear that the projection (6.2.17) satisfies the consistency condition (6.2.14) since for $l \prec l''$ we have $(x_{l''}) = ((x_{l''})_{l'})_l$ for any intermediate $l \prec l' \prec l''$. Surjectivity is less obvious.

Lemma 6.2.16. $p_{l'l}$, $l \prec l'$

Let $l = l(\gamma) \prec l' = l(\gamma')$ be given. Since l is a subgroupoid of l' we may decompose any generator $e \in E(\gamma)$ in the form

$$e = \circ_{e' \in E(\gamma')} (e')^{s_{ee'}} \tag{6.2.18}$$

where $s_{ee'} \in \{\pm 1, 0\}$. Notice that $|s_{ee'}| > 1$ is not allowed and that any e' appears at most once in (6.2.18) because e is an edge (cannot overlap itself).

Surjectivity: we must show that for any $x_l \in X_l$ there exists an $x_{l'} \in X_{l'}$ such that $p_{l'l}(x_{l'}) = x_l$. Since x_l is completely determined by $h_e := x_l(e) \in G$, $e \in E(\gamma)$ and $x_{l'}$ is completely determined by $h'_{e'} := x_{l'}(e') \in G$, $e' \in E(\gamma')$ and since h_e could be any value in G, what we have to show is that there exist group elements $h'_{e'} \in G$, $e' \in E(\gamma')$ such that for any group elements $h_e \in G$, $e \in E(\gamma)$ we have

$$h_e = \circ_{e' \in E(\gamma')} (h_{e'})^{s_{e,e'}} \tag{6.2.19}$$

However, since the $e \in E(\gamma)$ are disjoint up to their boundaries we have $s_{ee'} s_{\tilde{e}\tilde{e}'} = 0$ for any $e \neq \tilde{e}$ in $E(\gamma')$ so that we may select for each $e \in E(\gamma)$ one of the $e' \in E(\gamma')$ with $s_{e,e'} \neq 0$, say $e'(e) \in E(\gamma')$. These $e'(e)$ are then disjoint up to their boundaries. Since also the $h'_{e'}$ can independently take any value we may choose $h'_{e'(e)} = h_e$, $h'_{e'} = 1_G$ for $e' \notin \{e'(e)\}_{e \in E(\gamma)}$.

Continuity: under the identification (6.2.16) the projections are given as maps

$$p_{l'l} : \ G^{(\gamma')} \to G^{(\gamma)}; \ (h'_{e'})_{e' \in E(\gamma')} \mapsto \left(\prod_{e' \in E(\gamma')} (h'_{e'})^{s_{ee'}} \right)_{e \in E(\gamma)} \qquad (6.2.20)$$

By definition, a net $(h_k^\alpha)_{k=1}^n$ converges in G^n to $(h_k)_{k=1}^n$ if and only if every net $\lim_\alpha (h_k^\alpha) = h_k$, $k = 1, \ldots, n$ individually converges (i.e., $(h_k^\alpha)_{AB} - (h_k)_{AB} \to 0$ for all matrix elements AB). Suppose then that $(h'^\alpha_{e'})_{e' \in E(\gamma')}$ converges to $(h'_{e'})_{e' \in E(\gamma')}$. By definition, in a Lie group inversion and finite multiplication are continuous operations. Therefore $(\prod_{e' \in E(\gamma')} (h'^\alpha_{e'})^{s_{ee'}})_{e \in E(\gamma)}$ converges to $(\prod_{e' \in E(\gamma')} (h'_{e'})^{s_{ee'}})_{e \in E(\gamma)}$ (as one can also check explicitly). $\qquad \square$

We can now form the projective limit $\overline{X}$ of the X_l. In order to equip it with a topology we start by providing the direct product X_∞ with a topology. The natural topology on the direct product is the Tychonov topology.

Definition 6.2.17. $\qquad\qquad\qquad\qquad\qquad\qquad\qquad\qquad X_\infty = \prod_{l \in \mathcal{L}} X_l \qquad\qquad X_l$

$$p_l : \ X_\infty \to X_l; \quad (x_{l'})_{l' \in \mathcal{L}} \mapsto x_l \qquad\qquad (6.2.21)$$

$$l \in \mathcal{L} \qquad (\qquad x^\alpha = (x_l^\alpha)_{l \in \mathcal{L}} \qquad\qquad) \ \mathcal{L} \qquad x = (x_l)_{l \in \mathcal{L}} \qquad x_l^\alpha \to x_l$$

We then have the following non-trivial result.

Theorem 6.2.18 (Tychonov). $\qquad \mathcal{L}$

$$l \in \mathcal{L} \qquad\qquad\qquad\qquad\qquad\qquad\qquad X_l$$

$$X_\infty = \prod_{l \in \mathcal{L}} X_l$$

An elegant proof of this theorem in terms of universal nets is given in Chapter 18, where also other relevant results from general topology including proofs can be found.

Since $\overline{X} \subset X_\infty$ we may equip it with the subspace topology, that is, the open sets of $\overline{X}$ are the sets $U \cap \overline{X}$ where $U \subset X_\infty$ is any open set in X_∞.

Lemma 6.2.19. $\qquad\qquad\qquad\qquad\qquad \overline{X} \qquad\qquad\qquad\qquad\qquad X_\infty$

Let $(x^\alpha) := ((x_l^\alpha)_{l \in \mathcal{L}})$ be a convergent net in X_∞ such that $x^\alpha := (x_l^\alpha)_{l \in \mathcal{L}} \in \overline{X}$ for any α. We must show that the limit point $x = (x_l)_{l \in \mathcal{L}}$ lies in $\overline{X}$. By Lemma 6.2.16, the projections $p_{l'l} : X_{l'} \to X_l$ are continuous, therefore

$$p_{l'l}(x_{l'}) = \lim_\alpha p_{l'l}(x_{l'}^\alpha) = \lim_\alpha x_l^\alpha = x_l \qquad\qquad (6.2.22)$$

where the second equality follows from $x^\alpha \in \overline{X}$. Thus, the point $x \in X_\infty$ qualifies as a point in $\overline{X}$. $\qquad \square$

Since closed subspaces of compact spaces are compact in the subspace topology (see Chapter 18), we conclude that $\overline{X}$ is compact in the subspace topology induced by X_∞.

Lemma 6.2.20. $X_\infty, \overline{X}$

By assumption, G is a Hausdorff topological group. Thus G^n for any finite n is a Hausdorff topological group as well and since X_l is topologically identified with some G^n via (6.2.16) we see that X_l is a topological Hausdorff space for any $l \in \mathcal{L}$. Let now $x \neq x'$ be points in X_∞. Thus, there is at least one $l_0 \in \mathcal{L}$ such that $x_{l_0} \neq x'_{l_0}$. Since X_{l_0} is Hausdorff we find disjoint open neighbourhoods $U_{l_0}, U'_{l_0} \subset X_{l_0}$ of x_{l_0}, x'_{l_0} respectively. Let $U := p_{l_0}^{-1}(U_{l_0}), U' := p_{l_0}^{-1}(U'_{l_0})$. Since the topology of X_∞ is generated by the continuous functions $p_l : X_\infty \to X_l$ from the topology of the X_l, it follows that U, U' are open in X_∞. Moreover, U, U' are obviously neighbourhoods of x, x' respectively since $p_l(U) = X_l = p_l(U')$ for any $l \neq l_0$. Finally, $U \cap U' = \emptyset$ since $p_{l_0}(U \cap U') = U_{l_0} \cap U'_{l_0} = \emptyset$ so that U, U' are disjoint open neighbourhoods of $x \neq x'$ and thus X_∞ is Hausdorff.

Finally, to see that $\overline{X}$ is Hausdorff, let $x \neq x'$ be points in $\overline{X}$, then we find respective disjoint open neighbourhoods U, U' in X_∞ whence $U \cap \overline{X}, U' \cap \overline{X}$ are disjoint open neighbourhoods in $\overline{X}$ by definition of the subspace topology. $\square$

Let us collect these results in the following theorem.

Theorem 6.2.21. $\overline{X}$ $X_l =$ $(l, G), l \in \mathcal{L}$
          $\mathcal{L}$ $\mathcal{P}$

The purpose of our efforts was to equip $\mathrm{Hom}(\mathcal{P}, G)$ with a topology. Theorem 6.2.21 now enables us to do this provided we manage to identify $\mathrm{Hom}(\mathcal{P}, G)$ with the projective limit $\overline{X}$ via a suitable bijection. Now an elementary exercise is that any point of $\mathrm{Hom}(\mathcal{P}, G)$ defines a point in $\overline{X}$ if we define $x_l := H_{|l}$ since the projections $p_{l'l}$ encode the algebraic relations that are induced by asking that H be a homomorphism. That this map is actually a bijection is the content of the following theorem.

Theorem 6.2.22.

$$\Phi : \mathrm{Hom}(\mathcal{P}, G) \to \overline{X}; \ H \mapsto (H_{|l})_{l \in \mathcal{L}} \tag{6.2.23}$$

Injectivity: suppose that $\Phi(H) = \Phi(H')$, in other words, $H_{|l} = H'_{|l}$ for any $l \in \mathcal{L}$. Thus, if $l = l(\gamma)$ we have $H(e) = H'(e)$ for any $e \in E(\gamma)$. Since l is arbitrary we find $H(p) = H'(p)$ for any $p \in \mathcal{P}$, that is, $H = H'$.

Surjectivity: suppose we are given some $x = (x_l)_{l \in \mathcal{L}} \in \overline{X}$. We must find $H_x \in \mathrm{Hom}(\mathcal{P}, G)$ such that $\Phi(H_x) = x$. Let $p \in \mathcal{P}$ be any path, then we can always find a graph γ_p such that $p \in l := l(\gamma_p)$. We may then define

$$H_x(p) := x_{l(\gamma_p)}(p) \tag{6.2.24}$$

Of course, the map $p \mapsto \gamma_p$ is one to many and therefore the definition (6.2.24) seems to be ill-defined. We now show that this is not the case, that is, (6.2.24) does not depend on the choice of γ_p. Thus, let γ'_p be any other graph such that $p \in l' := l(\gamma'_p)$. Since $\mathcal{L}$ is directed we find l'' with $l, l' \prec l''$. But then by the definition of a point x in the projective limit

$$x_l(p) = [p_{l''l}(x_{l''})](p) = (x_{l''})_{|l}(p) \equiv x_{l''}(p) \equiv (x_{l''})_{|l'}(p) = [p_{l''l'}(x_{l''})](p) = x_{l'}(p) \tag{6.2.25}$$

It remains to check that H_x is indeed a homomorphism. We have for any $p, p', p \circ p' \in l$ with $f(p) = b(p')$

$$H_x(p^{-1}) = x_l(p^{-1}) = (x_l(p))^{-1} = H_x(p)^{-1} \text{ and } H_x(p \circ p') = x_l(p \circ p')$$
$$= x_l(p)x_l(p') = H_x(p)H_x(p') \tag{6.2.26}$$

since $x_l \in \mathrm{Hom}(l, G)$. □

Definition 6.2.23. $\quad \overline{\mathcal{A}} := \quad (\mathcal{P}, G)$

$$X_l = \quad (l, G) \quad l \in \mathcal{L} \qquad\qquad\qquad\qquad \overset{X}{}$$
$$\sigma \qquad\qquad \mathcal{P}$$
$$X_\infty$$

Once again it is obvious that the space of distributions $\overline{\mathcal{A}}$ no longer carries any sign of the bundle P, it depends only on the base manifold σ via the set of embedded paths $\mathcal{P}$.

6.2.3 Gauge invariance: distributional gauge transformations

The space $\overline{\mathcal{A}}$ contains connections (from now on considered as morphisms $\mathcal{P} \to G$) which are nowhere continuous as we will see later on, and these turn out to be measure-theoretically much more important than the smooth ones contained in $\mathcal{A}$. Therefore we are motivated to generalise also the space of smooth gauge transformations $\mathcal{G} := C^\infty(\sigma, G)$ to the space of   functions

$$\overline{\mathcal{G}} := \mathrm{Fun}(\sigma, G) \tag{6.2.27}$$

with no restrictions (e.g., continuity). It is clear that $g \in \overline{\mathcal{G}}$ may be thought of as the net $(g(x))_{x \in \sigma}$ and thus $\overline{\mathcal{G}}$ is just the continuous infinite direct product $\overline{\mathcal{G}} = \prod_{x \in \sigma} G$.

The transformation property of $\mathcal{A}$ under $\mathcal{G}$ (6.2.7) can be understood as an action $\lambda: \mathcal{G} \times \mathcal{A} \to \mathcal{A}; (g, A) \mapsto A^g := \lambda_g(A) := \lambda(g, A)$ where $A^g(p) :=$

$g(b(p))A(p)g(f(p))^{-1}$ for any $p \in \mathcal{P}$ which we may simply lift to $\overline{\mathcal{A}}, \overline{\mathcal{G}}$ as

$$\lambda \; : \; \overline{\mathcal{G}} \times \overline{\mathcal{A}} \to \overline{\mathcal{A}}; \; (g, A) \mapsto A^g := \lambda_g(A) := \lambda(g, A) \text{ where}$$
$$A^g(p) := g(b(p))A(p)g(f(p))^{-1} \; \forall \; p \in \mathcal{P} \tag{6.2.28}$$

Notice that this is really an action, that is, A^g is really an element of $\overline{\mathcal{A}} = \text{Hom}(\mathcal{P}, G)$ – it satisfies the homomorphism property

$$A^g(p^{-1}) = g(b(p^{-1}))A(p^{-1})g(f(p^{-1}))^{-1} = g(f(p))A(p)^{-1}g(b(p))^{-1} = (A^g(p))^{-1}$$
$$A^g(p)A^g(p') = [g(b(p))A(p)g(f(p))^{-1}][g(b(p'))A(p')g(f(p'))^{-1}]$$
$$= g(b(p))A(p)A(p')g(f(p'))^{-1}$$
$$= g(b(p \circ p'))A(p \circ p')g(f(p \circ p'))^{-1} = A^g(p \circ p') \tag{6.2.29}$$

because $f(p) = b(p'), b(p) = b(p \circ p'), f(p') = f(p \circ p')$. The action (6.2.28) is also continuous on $\overline{\mathcal{A}}$, that is, for any $g \in \overline{\mathcal{G}}$ the map $\lambda_g : \overline{\mathcal{A}} \to \overline{\mathcal{A}}$ is continuous. To see this, let (A^α) be a net in $\overline{\mathcal{A}}$ converging to $A \in \overline{\mathcal{A}}$. Then $\lim_\alpha(\lambda_g(A_\alpha)) = \lambda_g(A)$ if and only if $\lim_\alpha(p_l(\lambda_g(A_\alpha))) = p_l(\lambda_g(A))$ for any $l \in \mathcal{L}$. Identifying $\overline{\mathcal{A}}_{|l}$ with some G^n via (6.2.16) and using the bijection (6.2.23) we have for any $p \in l$

$$[p_l(\lambda_g(A^\alpha))](p) = [(\lambda_g(A^\alpha))_{|l}](p) = [\lambda_g(A^\alpha)](p) = g(b(p))A^\alpha(p)g(f(p))^{-1}$$
$$= g(b(p))[p_l(A^\alpha)](p)g(f(p))^{-1} \tag{6.2.30}$$

Since group multiplication and inversion are continuous in G^n we easily get $\lim_\alpha[p_l(\lambda_g(A^\alpha))](p) = [p_l(\lambda_g(A))](p)$ for any $p \in l$, that is, $\lim_\alpha p_l(\lambda_g(A^\alpha)) = p_l(\lambda_g(A))$, thus λ_g is continuous for any $g \in \overline{\mathcal{G}}$.

Since $\overline{\mathcal{A}}$ is a compact Hausdorff space and λ is a continuous group action on $\overline{\mathcal{A}}$ it then follows immediately from abstract results (see Chapter 18) that the quotient space

$$\overline{\mathcal{A}}/\overline{\mathcal{G}} := \{[A]; \; A \in \overline{\mathcal{A}}\} \text{ where } [A] := \{A^g; \; g \in \overline{\mathcal{G}}\} \tag{6.2.31}$$

is a compact Hausdorff space in the quotient topology. The quotient topology on the quotient $\overline{\mathcal{A}}/\overline{\mathcal{G}}$ is defined as follows: the open sets in $\overline{\mathcal{A}}/\overline{\mathcal{G}}$ are precisely those whose pre-images under the quotient map

$$[\;] : \overline{\mathcal{A}} \to \overline{\mathcal{A}}/\overline{\mathcal{G}}; \; A \mapsto [A] \tag{6.2.32}$$

are open in $\overline{\mathcal{A}}$, that is, the quotient topology is generated by asking that the quotient map be continuous.

Now as $\overline{\mathcal{G}}$ is a continuous direct product of the compact Hausdorff spaces G it is a compact Hausdorff space in the Tychonov topology by the theorems proved in Section 6.2.2. More explicitly, the projective construction of $\overline{\mathcal{G}}$ proceeds as follows: given $l \in \mathcal{L}$ with $l = l(\gamma)$ we define $\overline{\mathcal{G}}_l := \prod_{v \in V(\gamma)} G$ and extend the surjective projection $p_l : \overline{\mathcal{A}} \to \overline{\mathcal{A}}_l; \; A \mapsto A_{|l}$ to $p_l : \overline{\mathcal{G}} \to \overline{\mathcal{G}}_l; \; g \mapsto g_{|l}$ and for $l \prec l'$ the surjective projection $p_{l'l} : \overline{\mathcal{A}}_{l'} \to \overline{\mathcal{A}}_l; \; A_{l'} \mapsto (A_{l'})_{|l}$ to $p_{l'l} : \overline{\mathcal{G}}_{l'} \to \overline{\mathcal{G}}_l; \; g_{l'} \mapsto$

$(g_{l'})_{|l}$. These projections are obviously surjective again because $\overline{\mathcal{G}}$ is actually a direct product of copies of G, one for every $x \in \sigma$.

Notice that the projective limit $\overline{\mathcal{G}} = \{(g_l)_{l \in \mathcal{L}}; \ p_{l'l}(g_{l'}) = g_l\}$ is a group since $p_{l'l}(g_{l'} g'_{l'}) = g_l g'_l = p_{l'l}(g_{l'}) p_{l'l}(g'_{l'})$ and $p_{l'l}((g^{-1})_{l'}) = (g^{-1})_l = g_l^{-1} = p_{l'l}(g_{l'})^{-1}$ so that actually the $p_{l'l}$ are surjective group homomorphisms. Since the $\overline{\mathcal{G}}_l$ are compact Hausdorff topological groups it follows that $\overline{\mathcal{G}}$ is also a compact Hausdorff topological group.

Summarising: $\overline{\mathcal{A}}/\overline{\mathcal{G}}$ is the quotient of two projective limits both of which are compact Hausdorff spaces.

On the other hand, observe that for $l \prec l'$ we have

$$p_{l'l}(\lambda_{g_{l'}}(A_{l'})) = \lambda_{g_l}(A_l) \qquad (6.2.33)$$

for any $A \in \overline{\mathcal{A}}, g \in \overline{\mathcal{G}}$, one says the group action λ is   . Consider then the quotients

$$[\overline{\mathcal{A}}_l]_l := \overline{\mathcal{A}}_l/\overline{\mathcal{G}}_l := \{[A_l]_l; \ A_l \in \overline{\mathcal{A}}_l\} \text{ where } []_l : \overline{\mathcal{A}}_l \to \overline{\mathcal{A}}_l/\overline{\mathcal{G}}_l; \ A_l \mapsto [A_l]_l$$
$$:= \{\lambda_{g_l}(A_l); \ g_l \in \overline{\mathcal{G}}_l\} \qquad (6.2.34)$$

Due to the equivariance property for $l \prec l'$

$$p_{l'l}([A_{l'}]_{l'}) = \{p_{l'l}(\lambda_{g_{l'}}(A_{l'})); \ g_{l'} \in \overline{\mathcal{G}}_{l'}\} = \{\lambda_{g_l}(A_l); \ g_l \in \overline{\mathcal{G}}_l\} = [A_l]_l \quad (6.2.35)$$

since the projections $p_{l'l} : \overline{\mathcal{G}}_{l'} \to \overline{\mathcal{G}}_l$ are surjective. Now $\overline{\mathcal{A}}_l$ is a compact Hausdorff space and λ a continuous group action on $\overline{\mathcal{G}}_l$ thereon, thus $[\overline{\mathcal{A}}_l]_l$ is a compact Hausdorff space in the quotient topology induced by $[]_l$. By the results proved in Section 6.2.2 we find that the projective limit of these quotients, denoted by $\overline{\mathcal{A}/\mathcal{G}}$, is again a compact Hausdorff space in the induced Tychonov topology.

We therefore have two compact Hausdorff spaces associated with gauge invariance, on the one hand the quotient of projective limits $\overline{\mathcal{A}}/\overline{\mathcal{G}}$ and on the other hand the projective limit of the quotients $\overline{\mathcal{A}/\mathcal{G}}$. The question arises of what the relation between the spaces $\overline{\mathcal{A}}/\overline{\mathcal{G}}, \overline{\mathcal{A}/\mathcal{G}}$ is. In what follows we will show by purely algebraic and topological methods (without using C^* algebra techniques) that they are homeomorphic.

We begin by giving a characterisation of $\overline{\mathcal{A}/\mathcal{G}}$ similar to the characterisation of $\overline{\mathcal{A}}$ as $\mathrm{Hom}(\mathcal{P}, G)$. Here the role of edges will be replaced by the role of hoops as they allow us to take the quotient with respect to the gauge transformations straightforwardly.

Definition 6.2.24. $\qquad\qquad\qquad\qquad\qquad\qquad\qquad\qquad x_0 \in \sigma$
$$\mathcal{Q} := \qquad (x_0, x_0) \qquad\qquad\qquad\qquad\qquad \sigma$$

( ) $\qquad\qquad \{\alpha_1, \ldots, \alpha_n\} \qquad\qquad\qquad\qquad\qquad\qquad\qquad \alpha_k$

$$\alpha_l, \ l \neq k$$

( )
$$\{\alpha_1,\ldots,\alpha_n\}$$
$\check\gamma \qquad \check\gamma := \bigcup_{k=1}^n r(\alpha_k) \ (\alpha \cup \alpha' := p_{c_\alpha \cup c_{\alpha'}}) \qquad x_0$

x_0

x_0 $\qquad\qquad\qquad\qquad\qquad\qquad \gamma \qquad\qquad\qquad H(\gamma) =$
$\{\beta_1,\ldots,\beta_n\}$ $\qquad\qquad\qquad\qquad\qquad\qquad\qquad\qquad \pi_1(\gamma)$
γ ($\qquad\qquad\qquad\qquad\qquad\qquad\qquad \{\alpha_k\} \neq \{\beta_k\}$

n $\qquad\qquad\qquad\qquad)$ $\qquad\qquad\qquad\qquad\qquad \gamma$
$V(\gamma) = \{b(e), f(e); \ e \in E(\gamma)\}$ $\qquad\qquad\qquad\qquad \pi_1(\gamma)$
$\qquad\qquad\qquad\qquad \gamma$

( ) $\qquad\qquad \gamma \qquad\qquad\qquad s(\gamma) \subset \mathcal{Q}$
$\qquad\qquad \pi_1(\gamma) \qquad\qquad\qquad s(\gamma) = \pi_1(\gamma)$

We now have an analogue of Theorem 6.2.13, the proof of which is similar and will be omitted.

Theorem 6.2.25. $\quad \mathcal{S} \qquad\qquad\qquad\qquad\qquad s(\gamma) \quad \mathcal{Q}$
$\qquad\qquad\qquad\qquad \gamma \in \Gamma_0^\omega \qquad\qquad\qquad s \prec s' \quad s$
$s' \qquad \mathcal{Q}$

Let now $Y_s := \mathrm{Hom}(s, G)$. As with $X_l = \mathrm{Hom}(l, G)$ we can identify Y_s with some G^n displaying it as a compact Hausdorff space. Likewise we have surjective projections for $s \prec s'$ given by the restriction map, $p_{s's} : Y_{s'} \to Y_s$; $x_{s'} \mapsto (x_{s'})_{|s}$ which satisfy the consistency condition $p_{s's} \circ p_{s''s'} = p_{s''s}$ for any $s \prec s' \prec s''$. We therefore can form the direct product $Y_\infty = \prod_{s \in \mathcal{S}} Y_s$ and its projective limit subset

$$\overline{Y} = \{y = (y_s)_{s \in \mathcal{S}}; \ p_{s's}(y_{s'}) = y_s \ \forall \ s \prec s'\} \qquad (6.2.36)$$

which in the Tychonov topology induced from Y_∞ is a compact Hausdorff space. Repeating step by step the proof of Theorem 6.2.21 we find that the map

$$\Phi : \mathrm{Hom}(\mathcal{Q}, G) \to \overline{Y}; \ H \mapsto (H_{|s})_{s \in \mathcal{S}} \qquad (6.2.37)$$

is a bijection so that we can identify $\mathrm{Hom}(\mathcal{Q}, G)$ with $\overline{Y}$ and equip it with the topology of $\overline{Y}$ (open sets of $\mathrm{Hom}(\mathcal{Q}, G)$ are the sets $\Phi^{-1}(U)$ where U is open in $\overline{Y}$). This topology is the weakest one so that all the projections $p_s : \overline{Y} \to Y_s$; $y \mapsto y_s$ are continuous.

The action λ of $\overline{G}$ on $\overline{A} = \overline{X}$ reduces on $\overline{Y}$ to

$$\lambda : \overline{G} \times \overline{Y} \to \overline{Y}; \ (g, y) \mapsto \lambda(g, y) = \lambda_g(y) = \mathrm{Ad}_g(y); \ [\mathrm{Ad}_g(y)]_s = \mathrm{Ad}_{g(x_0)}(y_s) \qquad (6.2.38)$$

where for $\alpha \in s$ we have $[\mathrm{Ad}_{g(x_0)}(y_s)](\alpha) = \mathrm{Ad}_{g(x_0)}(y(\alpha))$ and $\mathrm{Ad} : G \times G \to G$; $(g, h) \mapsto ghg^{-1}$ is the adjoint action of G on itself. In other words, $(\lambda_{\overline{G}})_{|\overline{Y}} = \mathrm{Ad}_G$ where G can be identified with the restriction of $\overline{G}$ to x_0. Clearly Ad acts continuously on $\overline{Y}$.

Consider then the quotient space $\mathrm{Hom}(\mathcal{Q}, G)/G$ (notice that we mod out by G and not $\overline{\mathcal{G}}$!) which by the results obtained in the previous section is a compact Hausdorff space in the quotient topology. Now the action Ad on $\overline{Y}$ is completely independent of the label s, that is

$$\mathrm{Ad}_g \circ p_{s's} = p_{s's} \circ \mathrm{Ad}_g \qquad (6.2.39)$$

so that the points in $\overline{Y}/G$ are given by the equivalence classes

$$(y) := \{\mathrm{Ad}_g(y); \ g \in G\} = \{(\mathrm{Ad}_g(y_s))_{s \in \mathcal{S}}; \ g \in G\} = ((y_s)_s)_{s \in \mathcal{S}} \qquad (6.2.40)$$

where $()_s : Y_s \to (Y_s)_s \ y_s \mapsto (y_s)_s = \{\mathrm{Ad}_g(y_s); \ g \in G\}$ denotes the quotient map in Y_s. It follows that $\mathrm{Hom}(\mathcal{Q}, G)/G$ is the projective limit of the $(Y_s)_s$. On the other hand, consider the quotients $[X_l]_l$ discussed above. If $l' = l(\gamma')$ and γ' is not a closed graph then by the action of $\overline{\mathcal{G}}$ on $X_{l'}$ we get $[X_{l'}]_{l'} = [X_l]_l$ where $l = l(\gamma)$ and γ is the closed graph obtained from γ' by deleting its open edges (monovalent vertices). Next, if $x_0 \notin \gamma$ then we add a path to γ connecting any of its points to x_0 without intersecting γ otherwise and obtain a third graph γ'' where again $[X_{l''}]_{l''} = [X_l]_l$ with $l'' = l(\gamma'')$ due to quotienting by the action of the gauge group. But now γ'' is a closed graph up to x_0. Thus we see that the projective limit of the $[X_l]_l$, $l \in \mathcal{L}$ and of the $[Y_s]_s$, $s \in \mathcal{S}$ coincides, in other words we have the identity

$$\overline{\mathcal{A}/\mathcal{G}} = \mathrm{Hom}(\mathcal{Q}, G)/G \qquad (6.2.41)$$

Our proof of the existence of a homeomorphism between $\overline{\mathcal{A}/\mathcal{G}}$ and $\overline{\mathcal{A}}/\overline{\mathcal{G}}$ will be based on the identity (6.2.41) and the fact that $\overline{\mathcal{A}} = \mathrm{Hom}(\mathcal{P}, G)$. We will break this proof into several lemmas.

Fix once and for all a system of edges

$$\mathcal{E} := \{e_x \in \mathrm{Hom}(x_0, x); \ x \in \sigma\} \qquad (6.2.42)$$

where e_{x_0} is the trivial hoop based at x_0. Let $\overline{\mathcal{G}}_{x_0} := \{g \in \overline{\mathcal{G}}; \ g(x_0) = 1_G\}$ be the subset of all gauge transformations that are the identity at x_0 and consider the following map

$$f_{\mathcal{E}} : \mathrm{Hom}(\mathcal{P}, G) \to \mathrm{Hom}(\mathcal{Q}, G) \times \overline{\mathcal{G}}_{x_0}; \ A \mapsto (B, h) \text{ where}$$
$$B(\alpha) := A(\alpha) \ \forall \ \alpha \in \mathcal{Q} \text{ and } h(x) := A(e_x) \ \forall x \in \sigma \qquad (6.2.43)$$

Clearly $h(x_0) = A(e_{x_0}) = 1_G$. From the known action λ of $\overline{\mathcal{G}}$ on $\overline{\mathcal{A}}$ we induce the following action of $\overline{\mathcal{G}}$ on $\mathrm{Hom}(\mathcal{Q}, G) \times \overline{\mathcal{G}}_{x_0}$

$$\lambda' : \overline{\mathcal{G}} \times (\mathrm{Hom}(\mathcal{Q}, G) \times \overline{\mathcal{G}}_{x_0}) \to (\mathrm{Hom}(\mathcal{Q}, G) \times \overline{\mathcal{G}}_{x_0}); \ (g, (B, h)) \mapsto (B^g, h^g) = \lambda'_g(B, h)$$
$$\text{where } B^g(\alpha) = \mathrm{Ad}_{g(x_0)}(B(\alpha)); \ \forall \ \alpha \in \mathcal{Q} \text{ and } h^g(x)$$
$$= g(x_0)h(x)g(x)^{-1} \ \forall \ x \in \sigma \qquad (6.2.44)$$

The action (6.2.44) evidently splits into a G-action by Ad on $\mathrm{Hom}(\mathcal{Q}, G)$ (with $G \equiv \overline{\mathcal{G}}_{|x_0}$) as already observed above and a $\overline{\mathcal{G}}$-action on $\overline{\mathcal{G}}_{x_0}$ (indeed $h^g(x_0) = 1_G$).

Theorem 6.2.26. $\mathcal{E}$ $f_{\mathcal{E}}$ (   )

λ

$$f_{\mathcal{E}} \circ \lambda = \lambda' \circ f_{\mathcal{E}} \tag{6.2.45}$$

Bijection: the idea is to construct explicitly the inverse $f_{\mathcal{E}}^{-1}$. The Ansatz is, of course, that given any $p \in \mathcal{P}$ we can construct a hoop based at x_0 by using $\mathcal{E}$, namely $\alpha_p := e_{b(p)} \circ p \circ e_{f(p)}^{-1}$, which we can use in order to evaluate a given $B \in \mathrm{Hom}(\mathcal{Q}, G)$. Since we want that $A^g(p) = g(b(p))A(p)g(f(p))^{-1}$ we see that given $h \in \overline{\mathcal{G}}_{x_0}$ the only possibility is

$$f_{\mathcal{E}}^{-1} : \mathrm{Hom}(\mathcal{Q}, G) \times \overline{\mathcal{G}}_{x_0} \to \mathrm{Hom}(\mathcal{P}, G); \ (B, h) \mapsto A \text{ where}$$

$$A(p) := h(b(p))^{-1} B\left(e_{b(p)} \circ p \circ e_{f(p)}^{-1}\right) h(f(p)) \tag{6.2.46}$$

One can verify explicitly that this is the inverse of (6.2.43).

Equivariance: trivial by construction.

Continuity: by definition of the topology on the spaces $\mathrm{Hom}(\mathcal{P}, G)$, $\mathrm{Hom}(\mathcal{Q}, G)$, $\overline{\mathcal{G}}$ respectively, a corresponding net $(A^\alpha), (B^\alpha), (g^\alpha)$ converges to A, B, g iff the nets $(A_l^\alpha) = (p_l(A^\alpha)), (B_s^\alpha) = (p_s(B^\alpha)), (g_x^\alpha) = (p_x(g^\alpha))$ converge to $A_l = p_l(A), B_s = p_s(B), g_x = p_x(g)$ where $g_x = g(x)$ for all $l \in \mathcal{L}, s \in \mathcal{S}, x \in \sigma$.

Continuity of $f_{\mathcal{E}}$ then means that $(p_s \times p_x) \circ f_{\mathcal{E}}$ is continuous for all $s \in \mathcal{S}, x \in \sigma$ while continuity of $f_{\mathcal{E}}^{-1}$ means that $p_l \circ f_{\mathcal{E}}^{-1}$ is continuous for all $l \in \mathcal{L}$. Recalling the map (6.2.16) it is easy to see that

$$p_x \circ f_{\mathcal{E}} = \rho_{e_x} \circ p_{l(e_x)} \tag{6.2.47}$$

and since the ρ_γ are by definition continuous we easily get continuity of $p_x \circ f_{\mathcal{E}}$ as the composition of two continuous maps.

To establish the continuity of $p_s \circ f_{\mathcal{E}}, p_l \circ f_{\mathcal{E}}^{-1}$ requires more work.

Lemma 6.2.27

( ) $s \in \mathcal{S}$ $l \in \mathcal{L}$ s

l $s \prec l$ ($s \in \mathcal{L}$)

$$p_{ls} : X_l \to Y_s; \ x_l \mapsto (x_l)|_s \tag{6.2.48}$$

$$p_s \circ f_{\mathcal{E}} = p_{ls} \circ p_l \qquad\qquad \mathcal{E}$$

( ) $l \in \mathcal{L}$ $s \in \mathcal{S}$ $l' \in \mathcal{L}$

$\qquad\qquad l = l(\gamma), l' = l(\gamma') \qquad V(\gamma') = V(\gamma) \cup \{x_0\} \qquad l \prec$

$l' \qquad\qquad l'(x_0, x_0) = s \qquad \overline{\mathcal{G}}_{x_0}(l') := \qquad (V(\gamma'), G) \cap \overline{\mathcal{G}}_{x_0} \qquad \pi_{l'} :$

$\overline{\mathcal{G}}_{x_0} \to \overline{\mathcal{G}}_{x_0}(l') \qquad\qquad\qquad\qquad\qquad\qquad p_{l'l} : X_{l'} \to X_l$

$$p_{sl} : Y_s \times \overline{\mathcal{G}}_{x_0}(l') \to X_l$$

$$p_l \circ f_{\mathcal{E}(l)}^{-1} = p_{sl} \circ (p_s \times \pi_l) \tag{6.2.49}$$

$\mathcal{E}(l) \qquad \mathcal{E}$

$$f_{\mathcal{E}} \circ f_{\mathcal{E}'}^{-1} : \overline{Y} \times \overline{\mathcal{G}}_{x_0} \to \overline{Y} \times \overline{\mathcal{G}}_{x_0} \tag{6.2.50}$$

(i) Let $s \in \mathcal{S}$ be freely generated by the independent hoops $\alpha_1, \ldots, \alpha_m$, let $\check{\gamma}$ be the unoriented graph they determine and choose some orientation for it. Then every α_k is a finite composition of the edges $e_1, \ldots, e_n \in E(\gamma)$ demonstrating that s is a subgroup of $l = l(\gamma)$ consisting of hoops based at $x_0 \in V(\gamma)$. We have bijections $\rho_{\alpha_1, \ldots, \alpha_m} : Y_s \to G^m$ and $\rho_{e_1, \ldots, e_n} : Y_s \to G^n$ as in (6.2.16) which can be used to define the projection $p_{ls} : X_l \to Y_s$. In particular we get $X_s = Y_s$ so that p_{ls} is continuous. It follows that $p_s \circ f_{\mathcal{E}}(A) = A_s = p_{ls}(A_l) = (p_{ls} \circ p_l)(A)$ so that $p_s \circ f_{\mathcal{E}}$ is continuous.

(ii) Let $l \in \mathcal{L}$ be freely generated by independent edges $e_1, \ldots, e_n$ and let γ be the oriented graph they determine. If $x_0 \in V(\gamma)$ invert the orientation of e_k if necessary in order to achieve that $f(e_k) \neq x_0$ for any $k = 1, \ldots, n$. For every vertex $v \in V(\gamma)$ not yet connected to x_0 through one of the edges $e_1, \ldots, e_n$ add another edge e_v connecting x_0 with v to the set $\{e_1, \ldots, e_n\}$ so that the extended set remains independent. The extended set $\{e_1, \ldots, e_{n'}\}$ determines an oriented graph γ' with $x_0 \in V(\gamma')$ and every vertex of γ' is connected to x_0 through at least one edge. Given $v \in V(\gamma')$ choose one edge $e_v^l \in \hom(x_0, v)$ from $e_1, \ldots, e_{n'}$ with the convention that $e_{x_0}^l$ be the trivial hoop. Define $\mathcal{E}'(l) := \{e_v^l; \ v \in V(\gamma) \cup \{x_0\}\}$ and let $\{e_1', \ldots, e_m'\} := \{e_1, \ldots, e_{n'}\} - \mathcal{E}'(l)$. The hoops based at x_0 given by $\alpha_k := e_{h(e_k')}^l \circ e_k' \circ (e_{f(e_k')}^l)^{-1}$, $k = 1, \ldots, m$ are independent due to the segments e_k' traversed precisely once and which are intersected by the other α_l in only a finite number of points (namely the end points). Let s be the subgroup of $\mathcal{Q}$ generated by the α_k and let $l' \in \mathcal{L}$ be the subgroupoid generated by the $(e_x^l)^{-1} \circ \alpha_k \circ e_y^l$, $x, y \in V(\gamma) \cup \{x_0\}, k = 1, \ldots, m$ (we know that it is a connected subgroupoid with $\hom_{l'}(x_0, x_0) = s$ from Lemma 6.2.9). We claim $l \prec l'$. To see this, consider the original set of edges $\{e_1, \ldots, e_n\}$. Each e_k, $k = 1, \ldots, n$ is either one of the e_v^l, $v \in V(\gamma) \cup \{x_0\}$ or one of the e_j', $j = 1, \ldots, m$. In the first case we have $e_k = e_v^l = e_{x_0}^{-1} \circ e_{x_0} \circ e_v^l \in l'$ where e_{x_0} is the trivial hoop. In the latter case by definition $e_k = e_j' = (e_{b(e_j')}^l)^{-1} \circ \alpha_j \circ e^l(f(e_j')) \in l'$.

Consider now the bijection

$$f_{\mathcal{E}'(l)}^{l'} : X_{l'} \to Y_s \times \overline{\mathcal{G}}_{x_0}(l') \tag{6.2.51}$$

defined exactly as in (6.2.43) but restricted to $X_{l'}$ so that only the system of edges $\mathcal{E}'(l)$ is needed in order to define it. We can define now

$$p_{sl} := p_{l'l} \circ \left(f_{\mathcal{E}'(l)}^{l'}\right)^{-1} : Y_s \times \overline{\mathcal{G}}_{x_0}(l') \to X_l \tag{6.2.52}$$

which is trivially continuous again because both X_l and $Y_s \times \overline{\mathcal{G}}_{x_0}(l')$ are identified with powers of G.

Let finally $\mathcal{E}(l)$ be any system of paths $e_x \in \mathrm{hom}(x_0, x)$ that contains $\mathcal{E}'(l)$. Then for any $B \in \mathrm{Hom}(Q, G), g \in \overline{\mathcal{G}}_{x_0}, p \in l$ we have

$$[(p_l \circ f_{\mathcal{E}(l)}^{-1})(B, g)](p) = [f_{\mathcal{E}(l)}^{-1}(B, g)](p) = g(b(p))^{-1} B\big(e_{b(p)}^l \circ p \circ (e_{f(p)}^l)^{-1}\big) g(f(p))$$

$$= (\pi_l \circ g)(b(p))^{-1}(p_s \circ B)\big(e_{b(p)}^l \circ p \circ (e_{f(p)}^l)^{-1}\big)(\pi_l \circ g)(f(p))$$

$$= \big[(f_{\mathcal{E}'(l)}^{l'})^{-1}(p_s \circ B, \pi_l \circ g)\big](p) = \big(p_{l'l} \circ (f_{\mathcal{E}'(l)}^{l'})^{-1}\big)(p_s \circ B, \pi_l \circ g)(p)$$

$$= [p_{sl} \circ (p_s \times \pi_l)(B, g)](p) \tag{6.2.53}$$

where in the second line we exploited that $b(p), f(p) \in V(\gamma)$ and that $e_{b(p)}^l \circ p \circ (e_{f(p)}^l)^{-1} \in s$, in the third we observed that only the subset $\mathcal{E}'(l) \subset \mathcal{E}(l)$ is being used and that $p \in l \prec l'$ and finally we used (6.2.52). Thus, $p_l \circ f_{\mathcal{E}(l)}^{-1} = p_{sl} \circ (p_s \times \pi_l)$ is a composition of continuous maps and therefore continuous.

(iii) Let $\mathcal{E} = \{e_x, \; x \in \sigma\}, \mathcal{E}' = \{e_x', \; x \in \sigma\}$ and $\alpha \in Q, x \in \sigma$, then

$$[f_{\mathcal{E}} \circ f_{\mathcal{E}'}^{-1}(B, g)](\alpha, x) = \big([f_{\mathcal{E}'}^{-1}(B, g)](\alpha), [f_{\mathcal{E}'}^{-1}(B, g)](e_x)\big)$$

$$= \big(g(b(\alpha))^{-1} B\big(e_{b(\alpha)}' \circ \alpha \circ (e_{f(\alpha)}')^{-1}\big) g(f(\alpha)), g(b(e_x))^{-1}$$

$$\times B\big(e_{b(e_x)}' \circ e_x \circ (e_{f(e_x)}')^{-1}\big) g(f(e_x))\big)$$

$$= \big(B(\alpha), B\big(e_x \circ (e_x')^{-1}\big) g(x)\big) \tag{6.2.54}$$

where in the last step we noticed that $f(e_x) = x$, $b(\alpha) = f(\alpha) = b(e_x) = x_0$, $g(x_0) = 1_G$ because $g \in \overline{\mathcal{G}}_{x_0}$ and that e_{x_0}' is the trivial hoop based at x_0. It follows that the map (6.2.50) is given by $(B, g) \mapsto (B', g')$ with $B' = B, g'(.) = B(e_x \circ (e_x')^{-1}) g(.)$. The inverse map is given similarly by $(B, g) \mapsto (B', g')$ with $B' = B, g'(.) = B(e_x' \circ (e_x)^{-1}) g(.)$ so that it will be sufficient to demonstrate continuity of the former.

To show that $f_{\mathcal{E}} \circ f_{\mathcal{E}'}^{-1}$ is continuous requires to show that $(p_s \times p_x) \circ f_{\mathcal{E}} \circ f_{\mathcal{E}'}^{-1}$ is continuous for all $s \in \mathcal{S}, x \in \sigma$. Now obviously $p_s \circ f_{\mathcal{E}} \circ f_{\mathcal{E}'}^{-1} = p_s$ is continuous by definition. Next $[p_x \circ f_{\mathcal{E}} \circ f_{\mathcal{E}'}^{-1}(B, g)](x) = B(e_x \circ (e_x')^{-1}) g(x)$. Define the restriction map

$$f_x^{\mathcal{E}, \mathcal{E}'} := p_{e_x \circ (e_x')^{-1}} \times p_x : \overline{Y} \times \overline{\mathcal{G}}_{x_0} \to Y_{e_x \circ (e_x')^{-1}} \times (\overline{\mathcal{G}}_{x_0})|_x \tag{6.2.55}$$

and denote by $m : G \times G \to G; \; (g_1, g_2) \to g_1 g_2$ multiplication in G. Then

$$p_x \circ f_{\mathcal{E}} \circ f_{\mathcal{E}'}^{-1} = m \circ (p_{e_x \circ (e_x')^{-1}} \times p_x) \tag{6.2.56}$$

is a composition of continuous maps and therefore continuous. Hence, $f_{\mathcal{E}} \circ f_{\mathcal{E}'}^{-1}$ is a homeomorphism. $\square$

We can now complete the proof of continuity of both $f_{\mathcal{E}}$ and $f_{\mathcal{E}}^{-1}$ for a given, fixed $\mathcal{E}$. We showed already that $p_x \circ f_{\mathcal{E}}$ is continuous for all $x \in \sigma$ and by Lemma 6.2.27 (i) we have that $p_s \circ f_{\mathcal{E}}$ is continuous for all $s \in \mathcal{S}$, hence $f_{\mathcal{E}}$ is continuous. Next

$$p_l \circ f_{\mathcal{E}}^{-1} = \left[p_l \circ f_{\mathcal{E}(l)}^{-1}\right] \circ \left[f_{\mathcal{E}(l)} \circ f_{\mathcal{E}}^{-1}\right] \tag{6.2.57}$$

is a composition of two continuous functions since the function in the first bracket is continuous by Lemma 6.2.27(ii) and the second by Lemma 6.2.27(iii), thus $f_{\mathcal{E}}^{-1}$ is continuous.   □

Theorem 6.2.28. $\overline{\mathcal{A}/\mathcal{G}} =$ $(\mathcal{P}, \mathrm{G})/\overline{\mathcal{G}}$ $\overline{\mathcal{A}/\mathcal{G}} =$ $(\mathcal{Q}, \mathrm{G})/\mathrm{G}$

By Theorem 6.2.26 we know that

1. $\mathrm{Hom}(\mathcal{P}, \mathrm{G})$ and $\mathrm{Hom}(\mathcal{Q}, \mathrm{G}) \times \overline{\mathcal{G}}_{x_0}$ are homeomorphic and
2. $\overline{\mathcal{G}}$ acts equivariantly on both spaces via λ, λ' respectively.

We now use the abstract result that if a group acts (not necessarily continuously) equivariantly on two homeomorphic spaces then the corresponding spaces continue to be homeomorphic in their respective quotient topologies (see Chapter 18). We therefore know that $\mathrm{Hom}(\mathcal{P}, \mathrm{G})/\overline{\mathcal{G}}$ and $(\mathrm{Hom}(\mathcal{Q}, \mathrm{G}) \times \overline{\mathcal{G}}_{x_0})/\overline{\mathcal{G}}$ are homeomorphic. But $\overline{\mathcal{G}}$ is a direct product space, that is, $\overline{\mathcal{G}} = \overline{\mathcal{G}}_{x_0} \times \mathrm{G}$ whence $(\mathrm{Hom}(\mathcal{Q}, \mathrm{G}) \times \overline{\mathcal{G}}_{x_0})/\overline{\mathcal{G}} = \mathrm{Hom}(\mathcal{Q}, \mathrm{G})/\mathrm{G}$. More explicitly, recalling the action of λ' in (6.2.44) and writing $g \in \overline{\mathcal{G}}$ as $g = (g_1, g_0) \in \overline{\mathcal{G}}_{x_0} \times \mathrm{G}$ where $g(x) = g_1(x)$ for $x \neq x_0$ and $g(x_0) = g_0$ we see that $B^g(\alpha) = \mathrm{Ad}_{g_0}(B(\alpha))$ and $h^g(x) = g_0 h(x) g(x)^{-1}$ which gives $h^g(x_0) = h(x_0) = 1_\mathrm{G}$ and $h^g(x) = g_0 h(x) g_1(x)^{-1}$ for $x \neq x_0$. It follows that, given $h \in \overline{\mathcal{G}}_{x_0}$, for any choice of g_0 we can gauge $h^g(x) = 1_\mathrm{G}$ for all $x \in \sigma$ by choosing $g_1(x) = g_0 h(x)$. The remaining gauge freedom expressed in g_0 then only acts by Ad on $\mathrm{Hom}(\mathcal{Q}, \mathrm{G})$.   □

6.2.4 The C^* algebraic viewpoint and cylindrical functions

In the previous sections we have defined the quantum configuration spaces of (gauge equivalence classes of) distributional connections $\overline{\mathcal{A}}$ ($\overline{\mathcal{A}/\mathcal{G}}$) as $\mathrm{Hom}(\mathcal{P}, \mathrm{G})$ ($\mathrm{Hom}(\mathcal{P}, \mathrm{G})/\overline{\mathcal{G}}$) and equipped them with the Tychonov topology through projective techniques. We could be satisfied with this because we know that these spaces are compact Hausdorff spaces and this is a sufficiently powerful result in order to develop measure theory on them as we will see later.

However, the result that we want to establish in this section, namely that both spaces can be seen as the Gel'fand spectra of certain C^* algebras, has the advantage of making the connection with so-called cylindrical functions on these spaces explicit, which then helps to construct (a priori only cylindrically defined) measures on them. Moreover, it has a wider range of applicability in the sense that it does not make use of the concrete label sets used in the previous section. It therefore establishes a concrete link with constructive quantum gauge field theories. A brief introduction to Gel'fand–Naimark–Segal theory can be found in Chapter 27. The constructions that follow will be based on that theory. For a simpler, illustrative application to the case of the algebra of almost periodic functions on the real line, which provides good intuition about the mathematical concepts such as the spectrum of an algebra, please refer to Chapter 28. We will follow closely Ashtekar and Lewandowski [369].

We begin again quite generally and suppose that we are given a partially ordered and directed index set $\mathcal{L}$ which labels compact Hausdorff spaces X_l and that we have surjective and continuous projections $p_{l'l} : X_{l'} \to X_l$ for $l \prec l'$ satisfying the consistency condition $p_{l'l} \circ p_{l''l'} = p_{l''l}$ for $l \prec l' \prec l''$. Let $X_\infty, \overline{X}$ be the corresponding direct product and projective limit respectively with Tychonov topology with respect to which we know that they are Hausdorff and compact from the previous sections.

Definition 6.2.29

( ) $C(X_l)$ X_l

$$'(\overline{X}) := \cup_{l \in \mathcal{L}} C(X_l) \tag{6.2.58}$$

$$f, f' \in \quad '(\overline{X}) \quad\quad l, l' \in \mathcal{L} \quad\quad\quad f \in C(X_l), f' \in C(X_{l'})$$
$$f, f' \quad\quad\quad\quad\quad\quad\quad\quad\quad f \sim f'$$

$$p^*_{l''l}f = p^*_{l''l'}f' \; \forall \, l, l' \prec l'' \tag{6.2.59}$$

(   )
( ) $\overline{X}$

$$(\overline{X}) := \quad '(\overline{X})/ \sim \tag{6.2.60}$$

$$f \in \quad '(\overline{X}) \quad\quad [f]_\sim$$

Notice that we are actually abusing the notation here since an element $f \in \mathrm{Cyl}(\overline{X})$ is not a function on $\overline{X}$ but an equivalence class of functions on the X_l. We will justify this later by showing that $\mathrm{Cyl}(\overline{X})$ can be identified with $C(\overline{X})$, the continuous functions on $\overline{X}$.

Condition (6.2.59) seems to be very hard to check but it is sufficient to find just one single l'' such that (6.2.59) holds. For suppose that $f_{l_1} \in C(X_{l_1}), f_{l_2} \in C(X_{l_2})$ are given and that we find some $l_1, l_2 \prec l_3$ such that $p^*_{l_3 l_1} f_{l_1} = p^*_{l_3 l_2} f_{l_2}$. Now let any $l_1, l_2 \prec l_4$ be given. Since $\mathcal{L}$ is directed we find $l_1, l_2, l_3, l_4 \prec l_5$ and due to the consistency condition among the projections we have

(i) $p_{l_4 l_1} \circ p_{l_5 l_4} = p_{l_5 l_1} = p_{l_3 l_1} \circ p_{l_5 l_3}$ and (ii) $p_{l_4 l_2} \circ p_{l_5 l_4} = p_{l_5 l_2} = p_{l_3 l_2} \circ p_{l_5 l_3}$

$$\tag{6.2.61}$$

whence

$$p^*_{l_5 l_4} p^*_{l_4 l_1} f_{l_1} =_{i)} p^*_{l_5 l_3} p^*_{l_3 l_1} f_{l_1} = p^*_{l_5 l_3} p^*_{l_3 l_2} f_{l_2} =_{ii)} p^*_{l_5 l_4} p^*_{l_4 l_2} f_{l_2} \tag{6.2.62}$$

where in the middle equality we have used (6.2.59) for $l'' = l_3$. We conclude that $p_{l_5 l_4}^* [p_{l_4 l_1}^* f_{l_1} - p_{l_4 l_2}^* f_{l_2}] = 0$. Now for any $f_{l_4} \in C(X_{l_4})$ the condition $f_{l_4}(p_{l_5 l_4}(x_{l_5})) = 0$ for all $x_{l_5} \in X_{l_5}$ means that $f_{l_4} = 0$ because $p_{l_5 l_4} : X_{l_5} \to X_{l_4}$ is surjective.

Lemma 6.2.30. $\qquad f, f' \in \quad (\overline{X}) \qquad\qquad\qquad\qquad\qquad l \in \mathcal{L}$
$f_l, f_l' \in C(X_l) \qquad\qquad f = [f_l]_\sim, f' = [f_l']_\sim$

By definition we find $l_1, l_2 \in \mathcal{L}$ and representatives $f_{l_1} \in C(X_{l_1}), f_{l_2} \in C(X_{l_2})$ such that $f = [f_{l_1}]_\sim, f' = [f_{l_2}]_\sim$. Choose any $l_1, l_2 \prec l$ then $f_l := p_{ll_1}^* f_{l_1} \sim f_{l_1}$ (choose $l'' = l$ in (6.2.59) and use $p_{ll} = \mathrm{id}_{X_l}$) and $f_l' := p_{ll_2}^* f_{l_2} \sim f_{l_2}$. Thus $f = [f_l]_\sim, f' = [f_l']_\sim$. $\qquad\qquad\qquad\square$

Lemma 6.2.31

$(\;)\quad f, f' \in \quad (\overline{X}) \qquad\qquad\qquad\qquad\qquad\qquad\qquad ($
$\qquad\qquad\qquad\qquad\qquad)$

$$f + f' := [f_l + f_l']_\sim, \quad f f' := [f_l f_l']_\sim, \quad zf := [z f_l]_\sim, \quad \bar{f} := [\overline{f_l}]_\sim \quad (6.2.63)$$

$$l, f_l, f_l' \qquad\qquad\qquad\qquad\qquad z \in \mathbb{C} \qquad \overline{f_l}$$

$(\;)\quad (\overline{X})$
$(\;)\qquad\qquad\qquad\qquad f = [f_l]_\sim$

$$\|f\| := \sup_{x_l \in X_l} |f_l(x_l)| \qquad\qquad\qquad\qquad (6.2.64)$$

(i) We consider only pointwise multiplication, the other cases are similar. Let l, f_l, f_l' and $l', f_{l'}, f_{l'}'$ be as in Lemma 6.2.30. We find $l, l' \prec l''$ and have $p_{l''l}^* f_l = p_{l''l'}^* f_{l'}$ and $p_{l''l}^* f_l' = p_{l''l'}^* f_{l'}'$. Thus

$$p_{l''l}^*(f_l f_l') = p_{l''l}^*(f_l) p_{l''l}^*(f_l') = p_{l''l'}^*(f_{l'}) p_{l''l'}^*(f_{l'}') = p_{l''l'}^*(f_{l'} f_{l'}') \quad (6.2.65)$$

so $f_l f_l' \sim f_{l'} f_{l'}'$.

(ii) The function $f_l^z : X_l \to \mathbb{C}; x_l \to z$ for any $z \in \mathbb{C}$ certainly is an element of $C(X_l)$ and for any $l, l' \prec l''$ we have $z = (p_{l''l}^* f_l^z)(x_{l''}) = (p_{l''l'}^* f_{l'}^z)(x_{l''})$ for all $x_{l''} \in X_{l''}$ so $f^z := [f_l^z]_\sim$ is well-defined.

(iii) If $f = [f_l]_\sim = [f_{l'}]_\sim$ is given, choose any $l, l' \prec l''$ so that we know that $p_{l''l}^* f_l = p_{l''l'}^* f_{l'}$. Then from the surjectivity of $p_{l''l}, p_{l''l'}$ we have

$$\sup_{x_l \in X_l} |f_l(x_l)| = \sup_{x_{l''} \in X_{l''}} |(p_{l''l}^* f_l)(x_{l''})| = \sup_{x_{l''} \in X_{l''}} |(p_{l''l'}^* f_{l'})(x_{l''})|$$
$$= \sup_{x_{l'} \in X_{l'}} |f_{l'}(x_{l'})| \qquad\qquad\qquad (6.2.66)$$

$\qquad\qquad\qquad\qquad\qquad\qquad\qquad\qquad\qquad\qquad\qquad\qquad\qquad\square$

Lemma 6.2.31(i) tells us that $\mathrm{Cyl}(\overline{X})$ is an Abelian, *-algebra defined by the pointwise operations (6.2.63). Lemma 6.2.31(ii) tells us that $\mathrm{Cyl}(\overline{X})$ is also unital, the unit being given by the constant function $1 = [1_l]_\sim$, $1_l(x_l) = 1$. Finally, Lemma 6.2.31(iii) tells us that $\mathrm{Cyl}(\overline{X})$ is a normed space and that the norm is correctly normalised, that is, $||1|| = 1$. Notice that here the compactness of the X_l comes in since the norm (6.2.64) certainly does not make sense any longer on $C(X_l)$ for non-compact X_l. If X_l is at least locally compact we can replace the $C(X_l)$ by $C_0(X_l)$, the continuous complex-valued functions of compact support and still would get an Abelian *-algebra with norm although no longer a unital one. One can always embed an algebra isometrically into a larger algebra with identity (even preserving the C^* property, see below) but this does not solve all problems in C^*-algebra theory. Fortunately, we do not have to deal with these complications in what follows.

Recall that a norm induces a metric on a linear space via $d(f, f') := ||f - f'||$ and that a metric space is said to be complete whenever all its Cauchy sequences converge. Any incomplete metric space can be uniquely (up to isometry) embedded into a complete metric space by extending it by its non-converging Cauchy sequences (see, e.g., [282] and Chapter 26). We can then complete $\mathrm{Cyl}(\overline{X})$ in the norm $||.||$ in this sense and obtain an Abelian, unital Banach *-algebra $\overline{\mathrm{Cyl}(\overline{X})}$. But we notice that not only the submultiplicativity of the norm $(||ff'|| \leq ||f|| \, ||f'||)$ holds but in fact the C^* property $||f\bar{f}|| = ||f||^2$. Thus $\overline{\mathrm{Cyl}(\overline{X})}$ is in fact a unital, Abelian C^*-algebra. This observation suggests applying Gel'fand–Naimark–Segal theory, to which an elementary introduction can be found in Chapter 27.

Denote by $\Delta(\overline{\mathrm{Cyl}(\overline{X})})$ the spectrum of $\overline{\mathrm{Cyl}(\overline{X})}$, that is, the set of (algebraic, i.e., not necessarily continuous) homomorphisms from $\overline{\mathrm{Cyl}(\overline{X})}$ into the complex numbers and denote the Gel'fand isometric isomorphism by

$$\bigvee : \overline{\mathrm{Cyl}(\overline{X})} \rightarrow C(\Delta(\overline{\mathrm{Cyl}(\overline{X})})); \; f \mapsto \check{f} \text{ where } \check{f}(\chi) := \chi(f) \qquad (6.2.67)$$

where the space of continuous functions on the spectrum is equipped with the sup-norm. The spectrum is automatically a compact Hausdorff space in the Gel'fand topology, the weakest topology in which all the $\check{f}$, $f \in \mathrm{Cyl}(\overline{X})$ are continuous.

Notice the similarity between the spaces $\overline{\mathrm{Cyl}(\overline{X})}$ and $C(\Delta(\overline{\mathrm{Cyl}(\overline{X})}))$: both are spaces of continuous functions over compact Hausdorff spaces and on both spaces the norm is the sup-norm. This suggests that there is a homeomorphism between the projective limit space $\overline{X}$ and the spectrum $\mathrm{Hom}(\overline{\mathrm{Cyl}(\overline{X})}, \mathbb{C})$. This is what we are going to prove in what follows.

Consider the map

$$\mathcal{X} : \overline{X} \rightarrow \Delta(\overline{\mathrm{Cyl}(\overline{X})}); x = (x_l)_{l \in \mathcal{L}} \mapsto \mathcal{X}(x)$$
$$\text{where } [\mathcal{X}(x)](f) := f_l(p_l(x)) \text{ for } f = [f_l]_\sim \qquad (6.2.68)$$

Notice that (6.2.68) is well-defined since $f = p_l^* f_l = p_{l'}^* f_{l'}$ for any $f_l \sim f_{l'}$, which follows from

$$p_l^* f_l(x) = f_l(x_l) = (p_{l''l}^* f_l)(x_{l''}) = (p_{l''l'}^* f_{l'})(x_{l''}) = f_{l'}(x_{l'}) = p_{l'}^* f_{l'}(x) \quad (6.2.69)$$

for any $x \in X$, $l, l' \prec l''$. Notice also that (6.2.68) a priori defines $\mathcal{X}(x)$ only on $\mathrm{Cyl}(\overline{X})$ and not on the completion $\overline{\mathrm{Cyl}(\overline{X})}$. We now show that every $\mathcal{X}(x)$ is actually continuous: let (f^α) be a net converging in $\mathrm{Cyl}(\overline{X})$ to f, that is, $\lim_\alpha \|f_\alpha - f\| = 0$. Then $(f^\alpha = [f_{l_\alpha}^\alpha]_\sim, \ f = [f_l]_\sim, \ l, l_\alpha \prec l_{\alpha,l})$

$$|[\mathcal{X}(x)](f^\alpha) - [\mathcal{X}(x)](f)| = |(p_{l_\alpha}^* f_{l_\alpha}^\alpha - p_l^* f_l)(x)| = |(p_{l_{\alpha,l}}^* {}_{l_\alpha} f_{l_\alpha}^\alpha - p_{l_{\alpha,l}}^* {}_l f_l)(x_{l_{\alpha,l}})|$$

$$= |(f_{l_{\alpha,l}}^\alpha - f_{l_{\alpha,l}})(x_{l_{\alpha,l}})| \leq \sup_{x_{l_{\alpha,l}} \in X_{l_{\alpha,l}}} |(f_{l_{\alpha,l}}^\alpha - f_{l_{\alpha,l}})(x_{l_{\alpha,l}})|$$

$$= \|f^\alpha - f\| \quad (6.2.70)$$

hence $\lim_\alpha [\mathcal{X}(x)](f^\alpha) = [\mathcal{X}(x)](f)$ so $\mathcal{X}(x)$ is continuous. It follows that $\mathcal{X}(x)$ is a continuous linear (and therefore bounded) map from the normed linear space $\mathrm{Cyl}(\overline{X})$ to the complete, normed linear space $\mathbb{C}$. Hence, by the bounded linear transformation theorem [282] (or BLT theorem, see Chapter 26) each $\mathcal{X}(x)$ can be uniquely extended to a bounded linear transformation (with the same bound) from the completion $\overline{\mathrm{Cyl}(\overline{X})}$ of $\mathrm{Cyl}(\overline{X})$ to $\mathbb{C}$ by taking the limit of the evaluation on convergent series in $\overline{\mathrm{Cyl}(\overline{X})}$ which are only Cauchy in $\mathrm{Cyl}(\overline{X})$. We will denote the extension of $\mathcal{X}(x)$ to $\overline{\mathrm{Cyl}(\overline{X})}$ by $\mathcal{X}(x)$ again and it is then easy to check that this extended map $\mathcal{X}$ is an element of $\Delta(\overline{\mathrm{Cyl}(\overline{X})})$ (a homomorphism), for example, if $f_n \to f, f_n' \to f'$ then

$$[\mathcal{X}(x)](f f') := \lim_{n \to \infty} [\mathcal{X}(x)](f_n f_n') = \lim_{n \to \infty} ([\mathcal{X}(x)](f_n)) \, ([\mathcal{X}(x)](f_n'))$$

$$= ([\mathcal{X}(x)](f)) \, ([\mathcal{X}(x)](f')) \quad (6.2.71)$$

The map $\mathcal{X}$ in (6.2.68) is to be understood in this extended sense.

Theorem 6.2.32. $\mathcal{X}$ ( )

Injectivity: suppose $\mathcal{X}(x) = \mathcal{X}(x')$, then in particular $[\mathcal{X}(x)](f) = [\mathcal{X}(x')](f)$ for any $f \in \mathrm{Cyl}(\overline{X})$. Hence $f_l(x_l) = f_l(x_l')$ for any $f_l \in C(X_l)$, $l \in \mathcal{L}$. Since X_l is a compact Hausdorff space, $C(X_l)$ separates the points of X_l by the Stone–Weierstrass theorem [282] (see Chapter 18), hence $x_l = x_l'$ for all $l \in \mathcal{L}$. It follows that $x = x'$.

Surjectivity: let $\chi \in \mathrm{Hom}(\overline{\mathrm{Cyl}(\overline{X})}, \mathbb{C})$ be given. We must construct $x^\chi \in \overline{X}$ such that $\mathcal{X}(x^\chi) = \chi$. In particular for any $f = [f_l]_\sim \in \mathrm{Cyl}(\overline{X})$ we have $f_l(x_l^\chi) = \chi([f_l]_\sim)$. Given $l \in \mathcal{L}$ the character χ defines an element $\chi_l \in \mathrm{Hom}(C(X_l), \mathbb{C})$ via $\chi_l(f_l) := \chi([f_l]_\sim)$ for all $f_l \in C(X_l)$. Since X_l is a compact Hausdorff space, it is the spectrum of the Abelian, unital C^*-algebra $C(X_l)$, hence $X_l = \mathrm{Hom}(C(X_l), \mathbb{C})$ (see Chapter 27). It follows that there exists $x_l^\chi \in X_l$ such that $\chi_l(f_l) = f_l(x_l^\chi)$ for all $f_l \in C(X_l)$. We define $x^\chi := (x_l^\chi)_{l \in \mathcal{L}}$ and must check that it defines an element of the projective limit.

Let $l \prec l'$ and $f = [f_l]_\sim$. Then $f_l \sim f_{l'} := p^*_{l'l} f_l$ (choose $l'' = l'$ and use $p_{l'l'} = \mathrm{id}_{X_{l'}}$) and therefore

$$f_l(x^X_l) = \chi_l(f_l) = \chi([f_l]_\sim) = \chi([f_{l'}]_\sim) = \chi_{l'}(f_{l'}) = f_{l'}(x^X_{l'}) = f_l(p_{l'l}(x^X_{l'}))$$
$$(6.2.72)$$

for any $f_l \in C(X_l)$, $l \in \mathcal{L}$. Since $C(X_l)$ separates the points of X_l we conclude $x^X_l = p_{l'l}(x^X_{l'})$ for any $l \prec l'$, hence $x^X \in \overline{X}$.

: we have established that $\mathcal{X}$ is a bijection. We must show that both $\mathcal{X}, \mathcal{X}^{-1}$ are continuous.

The topology on $\Delta(\overline{\mathrm{Cyl}(\overline{X})})$ is the weakest topology such that the Gel'fand transforms $\check{f}$, $f \in \overline{\mathrm{Cyl}(\overline{X})})$ are continuous while the topology on $\overline{X}$ is the weakest topology such that all the projections p_l are continuous, or equivalently that all the $p^*_l f_l$, $f_l \in C(X_l)$ are continuous.

$\mathcal{X}$: let (x^α) be a net in $\overline{X}$ converging to x, that is, every net (x^α_l) converges to x_l. Let first $f = [f_l]_\sim \in \mathrm{Cyl}(\overline{X})$. Then

$$\lim_\alpha [\mathcal{X}(x^\alpha)](f) = \lim_\alpha (p^*_l f_l)(x^\alpha) = (p^*_l f_l)(x) = [\mathcal{X}(x)](f) \qquad (6.2.73)$$

for any $f \in \mathrm{Cyl}(\overline{X})$. Now given $\epsilon > 0$ for general $f \in \overline{\mathrm{Cyl}(\overline{X})}$ we find $f_\epsilon \in \mathrm{Cyl}(\overline{X})$ such that $\|f - f_\epsilon\| < \epsilon/3$ because $\mathrm{Cyl}(\overline{X})$ is dense in $\overline{\mathrm{Cyl}(\overline{X})}$. Also, by (6.2.73), we find $\alpha(\epsilon)$ such that $|[\mathcal{X}(x^\alpha)](f_\epsilon) - [\mathcal{X}(x)](f_\epsilon)| \le \epsilon/3$ for any $\alpha(\epsilon) \prec \alpha$. Finally, since $\mathcal{X}(x^\alpha), \mathcal{X}(x)$ are characters they are bounded (by one) linear functionals on $\overline{\mathrm{Cyl}(\overline{X})}$ as we have shown above (continuity of the $\mathcal{X}(x)$). It follows that

$$|[\mathcal{X}(x^\alpha)](f) - [\mathcal{X}(x)](f)| \le |[\mathcal{X}(x^\alpha)](f - f_\epsilon)| + |[\mathcal{X}(x)](f - f_\epsilon)|$$
$$+ |[\mathcal{X}(x^\alpha)](f_\epsilon) - [\mathcal{X}(x)](f_\epsilon)|$$
$$\le 2\|f - f_\epsilon\| + \epsilon/3 \le \epsilon \qquad (6.2.74)$$

for all $\alpha(\epsilon) \prec \alpha$. Thus

$$\lim_\alpha \check{f}(\mathcal{X}(x^\alpha)) = \check{f}(\mathcal{X}(f)) \qquad (6.2.75)$$

for all $f \in \overline{\mathrm{Cyl}(\overline{X})}$, hence $\mathcal{X}(x^\alpha) \to \mathcal{X}(x)$ in the Gel'fand topology.

$\mathcal{X}^{-1}$: let (χ^α) be a net in $\Delta(\overline{\mathrm{Cyl}(\overline{X})})$ converging to χ, so $\chi^\alpha(f) \to \chi(f)$ for any $f \in \overline{\mathrm{Cyl}(\overline{X})}$ and so in particular for $f = [f_l]_\sim \in \mathrm{Cyl}(\overline{X})$. Therefore

$$\chi^\alpha(f) = \chi^\alpha(p^*_l f_l) = (p^*_l f_l)(x^{\chi_\alpha}) = (p^*_l f_l)(\mathcal{X}^{-1}(\chi_\alpha)) \to (p^*_l f_l)(\mathcal{X}^{-1}(\chi)) = \chi(f)$$
$$(6.2.76)$$

for all $f_l \in C(X_l), l \in \mathcal{L}$. Hence $\mathcal{X}^{-1}(\chi_\alpha) \to \mathcal{X}^{-1}(\chi)$ in the Tychonov topology. $\qquad \square$

Corollary 6.2.33. $\overline{(X)}$

$C(\overline{X})$

$\overline{X}$

This follows from the fact that via Theorem 6.2.32 we may identify $\overline{X}$ set-theoretically and topologically with the spectrum $\Delta(\overline{\mathrm{Cyl}(\overline{X})})$ and the fact that the Gel'fand transform between $\overline{\mathrm{Cyl}(\overline{X})}$ and $C(\Delta(\overline{\mathrm{Cyl}(\overline{X})}))$ is an (isometric) isomorphism. This justifies in retrospect the notation $\mathrm{Cyl}(\overline{X})$ although cylindrical functions are not functions on $\overline{X}$ but rather equivalence classes of functions on the X_l under $\sim$.

Next we give an abstract and independent C^*-algebraic proof for the fact that the spaces $\overline{X}/G$ and $\overline{X/G}$ are homeomorphic whenever a topological group G acts continuously and equivariantly on the projective limit $\overline{X}$, that is, we reprove Theorem 6.2.28.

Suppose then that for each $l \in \mathcal{L}$ we have a group action

$$\lambda^l : \mathrm{G} \times X_l \to X_l; \; (g, x_l) \mapsto \lambda^l_g(x_l) \tag{6.2.77}$$

where λ^l_g is a continuous map on X_l which is equivariant with respect to the projective structure, that is,

$$p_{l'l} \circ \lambda^{l'} = \lambda^l \circ p_{l'l} \; \forall l \prec l' \tag{6.2.78}$$

Due to continuity of the group action and since X_l is Hausdorff and compact, the quotient space X_l/G is again compact and Hausdorff in the quotient topology (see Chapter 18) and due to equivariance the net of equivalence classes $([x_l]_l)_{l \in \mathcal{L}}$ is a projective net again (with respect to the same projections $p_{l'l}$) so that we can form the projective limit $\overline{X/G}$ of the X_l/G which is then a compact Hausdorff space again. Here $[.]_l : X_l \to X_l/G$ denotes the individual quotient maps with respect to the λ^l.

On the other hand, we may directly define an action of G on $\overline{X}$ itself by

$$\lambda : \overline{X} \times \mathrm{G} \to \overline{X}; \; x = (x_l)_{l \in \mathcal{L}} \mapsto \lambda_g(x) := (\lambda^l_g(x_l))_{l \in \mathcal{L}} \tag{6.2.79}$$

Since $\overline{X}$ is compact and Hausdorff and λ_g is a continuous map on $\overline{X}$ (since it is continuous iff all the λ^l_g are continuous) it follows that the quotient space $\overline{X}/G$ is again a compact Hausdorff space.

We now want to know what the relation between $\overline{X}/G$ and $\overline{X/G}$ is. Let $[.] : \overline{X} \to \overline{X}/G$ be the quotient map with respect to λ. We may then define a map

$$\Phi : \overline{X}/G \to \overline{X/G}; \; [x] = [(x_l)_{l \in \mathcal{L}}] \mapsto ([x_l]_l)_{l \in \mathcal{L}} \tag{6.2.80}$$

as follows: we have

$$[x] = \{\lambda_g(x); \; g \in \mathrm{G}\} := \{(\lambda^l_g(x_l))_{l \in \mathcal{L}} \; g \in \mathrm{G}\} \tag{6.2.81}$$

Now take an arbitrary representative in $[x]$, say $\lambda_{g_0}(x)$ for some $g_0 \in \mathrm{G}$ and compute its class in $\overline{X/G}$, that is,

$$\Phi([x]) := ([p_l(\lambda_{g_0}(x))]_l)_{l \in \mathcal{L}} = (\{\lambda^l_g(\lambda^l_{g_0}(x_l)); \; g \in \mathrm{G}\})_{l \in \mathcal{L}} = (\{\lambda^l_g(x_l); \; g \in \mathrm{G}\})_{l \in \mathcal{L}} \tag{6.2.82}$$

which shows that Φ is well-defined, that is, independent of the choice of g_0.

Theorem 6.2.34. $\qquad \Phi \qquad (\quad)$

The strategy of the proof is to (1) show that the pull-back map

$$\Phi^* : C(\overline{X/G}) \to C(\overline{X}/G) \tag{6.2.83}$$

is a bijection and then (2) show that for any compact Hausdorff spaces A, B such that $\Phi^* : C(B) \to C(A)$ is a bijection it follows that $\Phi : A \to B$ is a homeomorphism.

Step 1: let $f \in C(\overline{X/G})$ be given. Via Corollary 6.2.33 we may think of f as an element of $\overline{\mathrm{Cyl}(\overline{X/G})}$ and elements of $\mathrm{Cyl}(\overline{X/G})$ lie dense in that space. Now any $f \in \mathrm{Cyl}(\overline{X/G})$ is given by $f = [f_l]_\sim$ where f_l is a λ^l invariant function on X_l. Then

$$f_l([x_l]_l) = p_l^* f_l(\Phi([x])) \tag{6.2.84}$$

Thus the functions on $\mathrm{Cyl}(\overline{X/G})$ are obtained as $p_l^* f_l$ for some $l \in \mathcal{L}$ where f_l is λ^l invariant and then $\Phi^* p_l^* f_l$ is a λ-invariant function on $\overline{X}$. But such functions are precisely those that lie dense in $C(\overline{X}/G)$ because a function $f \in C(\overline{X}/G)$ is simply a λ-invariant function in $C(\overline{X})$, that is, via Corollary 6.2.33 a λ-invariant function in $\overline{\mathrm{Cyl}(\overline{X})}$ in which the λ-invariant functions in $\mathrm{Cyl}(\overline{X})$ lie dense and the latter are of the form $p_l^* f_l$ for some $l \in \mathcal{L}$ and λ-invariant.

To see that Φ^* is injective on $\mathrm{Cyl}(\overline{X/G})$ suppose that $\Phi^* p_l^* f_l = \Phi^* p_{l'}^* f_{l'}'$ for some l, l'. Then trivially $p_l^* f_l(x) = p_{l'}^* f_{l'}'(x)$ for all $x \in \overline{X}$. Let $l, l' \prec l''$ then

$$p_l^* f_l(x) = f_l(x_l) = p_{l''l}^* f_l(x_{l''}) = p_{l''}^* f_{l'}'(x) = f_{l'}(x_{l'}) = p_{l''l'}^* f_{l'}(x_{l''}) \ \forall \ x_{l''} \in X_{l''} \tag{6.2.85}$$

which shows that $f_l \sim f_{l'}'$, hence $[f_l]_\sim = [f_{l'}']_\sim$ define the same element of $\mathrm{Cyl}(\overline{X/G})$.

To see that Φ^* is a surjection we notice that it maps the dense set of functions in $\mathrm{Cyl}(\overline{X/G})$ of the form $p_l^* f_l$ (f_l being λ^l-invariant) into the dense set of functions in $\mathrm{Cyl}(\overline{X}/G)$ of the form $\Phi^* p_l^* f_l$ that are λ-invariant. If we can show that $\Phi^* : \mathrm{Cyl}(\overline{X/G}) \to \mathrm{Cyl}(\overline{X}/G)$ is continuous then it can be uniquely extended as a continuous map to the completion $\Phi^* : \overline{\mathrm{Cyl}(\overline{X/G})} \to \overline{\mathrm{Cyl}(\overline{X}/G)}$ by the bounded linear transformation theorem and it will be a surjection since any $f \in \overline{\mathrm{Cyl}(\overline{X}/G)}$ can be approximated arbitrarily well by elements in $\mathrm{Cyl}(\overline{X}/G)$ which we know to lie in the image of Φ^* already. To prove that Φ^* is continuous (bounded), we show that it is actually an isometry and therefore has unity bound.

$$\|\Phi^* p_l^* f_l\|_{\mathrm{Cyl}(\overline{X}/G)} = \sup_{[x] \in \overline{X}/G} |f_l(p_l(\Phi([x])))|$$

$$= \sup_{([x_{l'}]_{l'})_{l' \in \mathcal{L}} \in \overline{X}/G} |f_l(p_l(([x_{l'}]_{l'})_{l' \in \mathcal{L}}))| = \|p_l^* f_l\|_{\overline{\mathrm{Cyl}(\overline{X/G})}}$$

$$\tag{6.2.86}$$

Step 2: let $\Phi : A \to B$ be a map between compact Hausdorff spaces such that $\Phi^* : C(B) \to C(A)$ is a bijection.

Injectivity: suppose $\Phi(a) = \Phi(a')$. Then for any $F \in C(B)$ we have $(\Phi^* F)(a) = (\Phi^* F)(a')$. Since Φ^* is a surjection and $C(A)$ separates the points of A it follows that $a = a'$.

Surjectivity: since A, B are the Gel'fand spectra $\mathrm{Hom}(C(A), \mathbb{C}), \mathrm{Hom}(C(B), \mathbb{C})$ of $C(A), C(B)$ respectively and Φ^* is a bijection we obtain a corresponding bijection between A, B (since the spectrum can be constructed algebraically from the algebras) via

$$\Phi_* : A = \Delta(C(A)) \to B = \Delta(C(B)); a \mapsto a \circ \Phi^* \qquad (6.2.87)$$

where

$$f(a) \equiv a(f) = a(\Phi^* F) = (a \circ \Phi^*)(F) = F(\Phi(a)) = (\Phi(a))(F) \qquad (6.2.88)$$

for any $f = \Phi^* F \in C(A)$, $F \in C(B)$. It follows that any $b \in B$ can be written in the form $b = \Phi(a)$ for some $a \in A$.

Continuity: we know that both $\Phi^{-1}, (\Phi^*)^{-1}$ exist. Then $(\Phi^*)^{-1} = (\Phi^{-1})^*$ since

$$\begin{aligned} f(a) &= [(\Phi^* \circ (\Phi^*)^{-1})f](a) = [(\Phi^*)^{-1} f](\Phi(a)) \\ &= f((\Phi^{-1} \circ \Phi)(a)) = [(\Phi^{-1})^* f](\Phi(a)) \end{aligned} \qquad (6.2.89)$$

for any $f \in C(A), a \in A$. Let now (a^α) be a net in A converging to a. This is equivalent with $\lim_\alpha f(a^\alpha) = f(a)$ for all $f \in C(A)$, which in turn implies $\lim_\alpha F(\Phi(a^\alpha)) = F(\Phi(a))$ for all $F \in C(B)$ since any f can be written as $\Phi^* F$, which is then equivalent with the convergence of the net $\Phi(a^\alpha)$ to $\Phi(a)$ in B. The proof for Φ^{-1} is analogous. $\qquad \square$

This completes the detailed investigation of the quantum configuration space. We now turn to the quantum momentum space.

6.3 Definition of : (2) surfaces, electric fields, fluxes and vector fields

Holonomies of connections were labelled by semianalytic paths. The analogue labelling set for the electric fields are semianalytic surfaces.

Definition 6.3.1. S

$(D - 1)$ S_I $\quad \sigma$ (

)

$(D - 2)$

$(D - 1)$

$C^{(0)}$

analytic surfaces

joining in analytic curves such that
the combined surface is still C^m

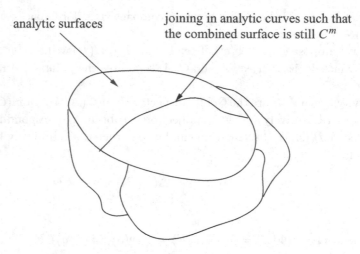

Figure 6.2 Semianalytic surface composed of faces.

$$S \qquad\qquad (D-1) \qquad\qquad C^{(0)}$$

$$S \qquad\qquad\qquad\qquad\qquad\qquad\qquad U$$

$$U - S = U_+ \cup U_-, \qquad U_+, U_-$$

For a generalisation see [521]. Let us explain the ingredients of this definition: the analyticity of the entire analytic patches will ensure the finite intersection property with piecewise analytic edges discussed before. (1) and (2) describe how the analytic patches are glued together, the gluing is continuous but not necessarily differentiable. An entire analytic surface will be transformed into a piecewise (semi)analytic surface with the gluing properties indicated under a piecewise (semi)analytic diffeomorphism, defined below. (3) makes sure that the surface is 'finite', that is, contained in a compact set. Finally, (4) makes sure that S is orientable in the sense that we know which sides of the surface are up or down. Notice that S is a $C^{(0)}$ submanifold without boundary. The openness of S removes the necessity to discuss what happens if an edge intersects the boundary of a surface. A typical example of a surface in $D = 3$ is the boundary of a solid cube with one of the closed square faces removed. See Figure 6.2 for an illustration.

The above definition is intuitive but rather complicated. An equivalent and simpler, however, less intuitive definition is as follows (see Chapter 20 for more details):

Definition 6.3.2. C^0

(   )

We will work with Definition 6.3.2 and it will turn out that it is sufficient to restrict attention to faces in most applications. Notice that the difference between a semianalytic and a piecewise analytic manifold is, roughly speaking, that both are finite unions of entire analytic patches but those are glued in the former case in an at least $C^{(1)}$ fashion while in the latter case the gluing is possibly only $C^{(0)}$. This is in complete analogy to the difference between semianalytic and piecewise analytic edges.

Since E_j^a is a vector density of weight one, the function $(*E)_{a_1...a_{D-1}} := E_j^c \epsilon_{ca_1...a_{D-1}} \tau_j$ is a pseudo-$(D-1)$-form which we may integrate in a background-independent way over S. That is

Definition 6.3.3. n

E_j^a S

$$E_n(S) := -\frac{1}{2} \int_S (n (*E)) = \int_S n^j (*E)_j \qquad (6.3.1)$$

These functions certainly separate the space $\mathcal{E}$ of smooth electric fields on σ: to see this consider a face of the form $S : (-1/2, 1/2)^{D-1} \to \sigma$; $(u_1, \ldots, u_{D-1}) \mapsto S(u_1, \ldots, u_{D-1})$ with semianalytic but at least once differentiable functions $S(u_1, \ldots, u_{D-1})$ and let $S_\epsilon(u_1, \ldots, u_{D-1}) := S(\epsilon u_1, \ldots, \epsilon u_{D-1})$. Choose $n^k = \delta_j^k$. Then (29.1.1) becomes

$$E_n(S_\epsilon) = \int_{(-\epsilon/2,\epsilon/2)^{D-1}} du_1 \ldots du_{D-1} \epsilon_{aa_1...a_{D-1}} (\partial S^{a_1}/\partial u_1)(u_1, \ldots, u_{D-1}) \cdots$$
$$\times (\partial S^{a_{D-1}}/\partial u_{D-1})(u_1, \ldots, u_{D-1}) E_j^a(S(u_1, \ldots, u_{D-1}))$$
$$= \epsilon^{D-1} \epsilon_{aa_1...a_{D-1}} (\partial S^{a_1}/\partial u_1)(0, \ldots, 0) \cdots (\partial S^{a_{D-1}}/\partial u_{D-1})$$
$$\times (0, \ldots, 0) E_j^a(S(0, \ldots, 0)) + O(\epsilon^D) \qquad (6.3.2)$$

where we have written the lowest-order term in the Taylor expansion in the second line. It follows that

$$\lim_{\epsilon \to 0} \frac{E_n(S_\epsilon)}{\epsilon^{D-1}} = \epsilon_{aa_1...a_{D-1}} (\partial S^{a_1}/\partial u_1)(0, \ldots, 0) \cdots (\partial S^{a_{D-1}}/\partial u_{D-1})$$
$$\times (0, \ldots, 0) E_j^a(S(0, \ldots, 0)) \qquad (6.3.3)$$

and by varying S we may recover every component of $E_j^a(x)$ at $x = S(0, \ldots, 0)$.

In Section 6.2.3 we had introduced the distributional gauge transformations on the space of generalised connections. These do not have a natural action on the classical functions (6.3.1), only the smooth ones do. However, the vector fields on the cylindrical functions which we are going to derive do admit an extension to a $\overline{\mathcal{G}}$ action. We are therefore going to postpone the discussion of gauge transformation to a later section.

6.4 Definition of   : (3) regularisation of the holonomy–flux Poisson algebra

The reality conditions are simply that $A(p)$ is G-valued and that $E_n(S)$ is real-valued. The Poisson brackets among $A(p), E_n(S)$ are, however, a priori ill-defined because the Poisson brackets that we derived in Chapter 1 required that the fields A, E be smeared in D directions by smooth functions while the functions $A(p), E_n(S)$ involve one- and $(D-1)$-dimensional smearings only. Therefore it is not possible to simply compute their Poisson brackets: the aim to have a background-independent formulation of the quantum theory forces us, as we heuristically derived above, to consider such singular smearings and prevents us from using the Poisson brackets on $\mathcal{M}$ directly. The strategy will therefore be to regularise the functions $A(p), E(S)$ in order to arrive at a D-dimensional smearing, then to compute the Poisson brackets of the regulated functions and finally we will remove the regulator and hope to arrive at a well-defined symplectic structure for the $A(p), E_n(S)$.

The simplest way to do this is to define a T_p^ϵ with central path p to be a smooth function of the form

$$T_p^{\epsilon t} : \mathbb{R}^{D-1} \times [0,1] \to \sigma; T_p^{\epsilon t}(s_1, \ldots, s_{D-1}, t')$$
$$:= \delta^\epsilon(t' - t)\delta^\epsilon(s_1, \ldots, s_{D-1})p_{s_1, \ldots, s_{D-1}}(t') \qquad (6.4.1)$$

where $p_{s_1, \ldots, s_{D-1}}$ is a smooth assignment of mutually non-intersecting paths diffeomorphic to $p := p_{0, \ldots, 0}$ (a congruence) and δ^ϵ is a smooth regularisation of the δ-distribution in $\mathbb{R}^{D-1}$ and $\mathbb{R}$ respectively. See Figure 6.3. We then define (recall formula (21.2.14) for the holonomy)

$$h_p^\epsilon(A) := \mathcal{P}e^{\int_{\mathbb{R}^{D-1}} d^{D-1}s \, \delta^\epsilon(s_1, \ldots, s_{D-1}) \int_0^1 dt \int_{p_{s_1, \ldots, s_{D-1}}} dt' \, \delta_t^\epsilon A} \qquad (6.4.2)$$

where path ordering is with respect to the t parameter. We obviously have $\lim_{\epsilon \to 0} h_{T_p^\epsilon} = h_p$ pointwise in $\mathcal{A}$ for any choice of δ^ϵ.

Likewise we define a D_S^ϵ with central surface S to be a smooth function of the form

$$D_S^\epsilon : \mathbb{R} \times U \to \sigma; \; D_p^\epsilon(s; u_1, \ldots, u_{D-1}) := \delta^\epsilon(s)S_s(u_1, \ldots, u_D) \qquad (6.4.3)$$

where S_s is a smooth assignment of mutually non-intersecting surfaces diffeomorphic to $S := S_0$ (a congruence). See Figure 6.4. Here U denotes the subset of $\mathbb{R}^{D-1}$ in the pre-image of S. We then define

$$E_n^\epsilon(S) := \int_{\mathbb{R}} ds \, \delta^\epsilon(s)E_n(S_s) \qquad (6.4.4)$$

We obviously have $\lim_{\epsilon \to 0} E(D_S^\epsilon)_n = E_n(S)$ pointwise in $\mathcal{E}$, the space of smooth electric fields over σ. Next recall that the Poisson bracket algebra among the functions $F(A) = \int d^D x A_a^j F_j^a$, $E(f) = \int d^D x E_j^a f_a^j$ of Chapter 1 is isomorphic with a subalgebra of the Lie algebra $C^\infty(\mathcal{A}) \times V^\infty(\mathcal{A})$ of smooth functions and

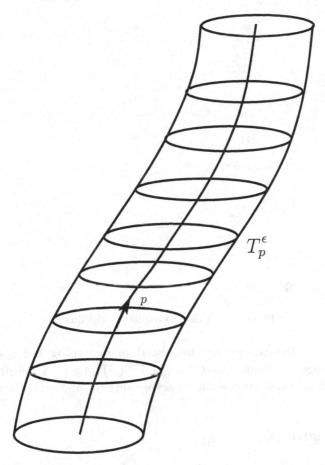

Figure 6.3 A tube to regularise the holonomy.

vector fields (derivatives on functions) on $\mathcal{A}$ respectively. This Lie algebra is defined by

$$[(\phi, \nu), (\phi', \nu')] := (\nu(\phi') - \nu'(\phi), [\nu, \nu']) \qquad (6.4.5)$$

where $\nu(\phi)$ denotes the action of the vector field ν on the function ϕ and $[\nu, \nu']$ denotes the Lie bracket of vector fields. The subalgebra of $C^\infty(\mathcal{A}) \times V^\infty(\mathcal{A})$ which is isomorphic to the Poisson subalgebra generated by the functions $F(A), E(f)$ is given by the elements $(F(A), E(f)) \mapsto (\phi_F, \beta\kappa/2\nu_f)$ with algebra

$$[(\phi_F, \nu_f), (\phi_{F'}, \nu_{f'})] := (F'(f) - F(f'), 0) \qquad (6.4.6)$$

and if one would like to quantise the system based on the real-valued functions and vector fields ϕ_F, ν_f respectively, then one would ask to promote them to self-adjoint operators with commutator algebra isomorphic with (6.4.6).

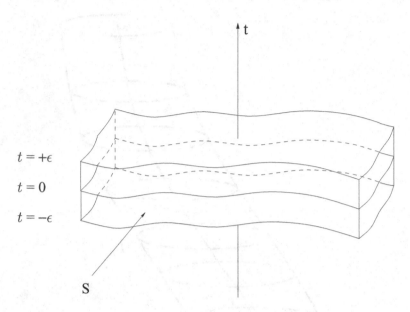

Figure 6.4 A disc to regularise the flux.

As already motivated, we are interested in quantising the system based on another algebra similar to $C^\infty(\mathcal{A}) \times V^\infty(\mathcal{A})$ given by $(A(p), E_n(S)) \mapsto (\phi_p, \beta\kappa/2Y_n(S))$, which we now must derive using the above regularisation. Let

$$F_p^{\epsilon kt}(x)_j^a$$
$$:= \delta_j^k \int_{\mathbb{R}^{D-1}} d^{D-1}s\, \delta^\epsilon(s_1, \ldots, s_{D-1})$$
$$\int_0^1 dt'\delta^\epsilon(t'-t)\dot{p}_{s_1,\ldots,s_{D-1}}^a(t')\delta(x, p_{s_1,\ldots,s_{D-1}}(t'))$$

$$f_S^{\epsilon n}(x)_a^j$$
$$:= n^j(x) \int_{\mathbb{R}} ds\, \delta^\epsilon(s) \int_U d^{D-1}u\epsilon_{aa_1\ldots a_{D-1}}$$
$$\times \frac{\partial S_s^{a_1}(u_1,\ldots,u_{D-1})}{\partial u_1} \cdots \frac{\partial S_s^{a_{D-1}}(u_1,\ldots,u_{D-1})}{\partial u_{D-1}}\delta(x, S_s(u_1,\ldots,u_{D-1})) \quad (6.4.7)$$

then we trivially have

$$h_p^\epsilon(A) = \mathcal{P}e^{\int_0^1 dt\, F_p^{\epsilon jt}(A)\tau_j/2}$$
$$E_n^\epsilon(S) = E(f_S^{\epsilon n}) \quad\quad\quad (6.4.8)$$

Notice that the smearing functions (6.4.7) are not quite smooth due to the sharp cutoff at the boundary of the family of paths and surfaces respectively but this does not cause any trouble, the smeared functions are still functionally differentiable with respect to the phase space coordinates because the functional derivatives (6.4.7) define a bounded linear functional on $\mathcal{M}$ (see Chapter 33).

Formula (6.4.8) thus enables us to map our regulated holonomy and surface variables into the Lie algebra $C^\infty(\mathcal{A}) \times V^\infty(\mathcal{A})$ via

$$h_p^\epsilon(A) \mapsto \phi_p^\epsilon := \mathcal{P}e^{\int_0^1 dt\phi_{F_p^{\epsilon j}} t\, T_j/2} \quad \text{and} \quad E_n^\epsilon(S) \mapsto Y_n^\epsilon(S) := \nu_{f_S^{\epsilon n}} \qquad (6.4.9)$$

compute their algebra and then take the limit $\epsilon \to 0$ where we may use the known action of ν_f on ϕ_F.

Now the following issue arises: by (6.4.6) the vector fields $\nu_{f_S^{\epsilon j}}$ are Abelian at finite ϵ. On the other hand, we will compute a vector field $Y_n(S)$ by $Y_n(S)[\phi_p] := \lim_{\epsilon \to 0} \nu_{Sn}^\epsilon(\phi_p^\epsilon)$. But taking the limit $\epsilon \to 0$ and computing Lie brackets of vector fields does not commute in our case. This is no cause of trouble because, as already mentioned, we will take the resulting limit Lie algebra as a starting point for quantisation.

Let us then actually compute $\phi_p, Y_n(S)$: to simplify the analysis, we notice that, given a piecewise (semi)analytic surface S we can decompose it into faces. A piecewise (semi)analytic path p can be decomposed into a finite number of entire semianalytic edges e, some of which appear with opposite orientation in that decomposition, of the following four types (subdivide edges into two halves at an interior point if necessary; the sets $U_\pm$ are defined in Definition 6.3.1). See Figure 6.5.

$e \cap S = b(e)$ is an isolated intersection point and the beginning segment of e lies in U_+.

$e \cap S = b(e)$ is an isolated intersection point and the beginning segment of e lies in U_-.

$e \cap \overline{S} = e$, that is, e is contained in the closure of an entire analytic patch of S.

$e \cap S = \emptyset$, that is, e does not intersect S at all. This includes the case that e intersects the boundary $\partial S = \overline{S} - S$ of the closure of S because S has no boundary, it is open.

To see that the number of edges of either type is finite, it will be sufficient to show that this is the case for each of the finite number of entire semianalytic pieces of S. Hence, assuming that S is a semianalytic face, notice that if a given analytic segment s of an edge e of p intersects S in an infinite number of isolated points, then we can draw a curve c within S through this chain of points which is analytic (choose an analytic coordinate system for the domain of a chart in which a piece of S, containing an infinite number of intersection points, coincides with a piece of the $x^D = 0$ plane. Then use $c(t) := (e^1(t), \ldots, e^{D-1}(t), 0))$. But then $e = c$ by analyticity, hence e is actually of the inside type. Hence the number of 'up' and 'down' type edges is finite. Next, suppose that there are an infinite

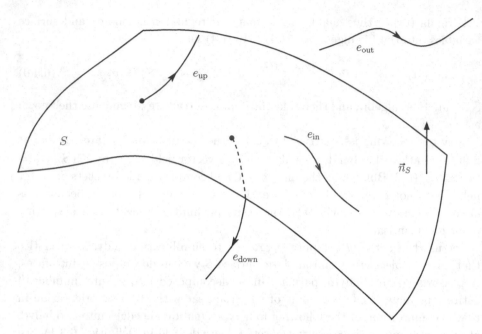

Figure 6.5 Types of edges with respect to a face.

number of 'inside' or 'outside' type edges, then there must be an infinite number of 'outside' or 'inside' type edges as well since the number of 'up' and 'down' types is finite. But then e must infinitely often leave and re-enter S through the boundary of S, which is a piecewise (semi)analytic $(D-1)$-manifold, intersecting it in an infinite number of isolated points. Applying the same argument as above, we conclude that all but a finite number of those 'inside' and 'outside' segments must lie on the boundary of S and hence combine to a finite number of edges of the 'outside' type. The more abstract version of this elementary reasoning is the content of Theorem 20.2.1.

Thus, if $p = e_1^{\sigma_1} \circ \ldots \circ e_n^{\sigma_n}$, $\sigma_k = \pm 1$, $k = 1, \ldots, n$ is a decomposition of p with respect to S into edges of definite type then we   use the identity $h_p(A) = h_{e_1}(A)^{\sigma_1} \cdot \ldots \cdot h_{e_n}(A)^{\sigma_n}$,   regularise $h_p^\epsilon(A) = h_{e_1}^\epsilon(A)^{\sigma_1} \cdot \ldots \cdot h_{e_n}^\epsilon(A)^{\sigma_n}$ use the Leibniz rule

$$\{E_n^{\epsilon'}(S), h_p^\epsilon(A)\} = \sum_{k=1}^{n} \sigma_k h_{e_1}^\epsilon(A)^{\sigma_1} \ldots h_{e_{k-1}}^\epsilon(A)^{\sigma_{k-1}}$$

$$\times \{E_n^{\epsilon'}(S), h_{e_k}^\epsilon(A)^{\sigma_k}\} \, h_{e_{k+1}}^\epsilon(A)^{\sigma_{k+1}} \ldots h_{e_n}^\epsilon(A)^{\sigma_n} \qquad (6.4.10)$$

Finally use

$$\{E_n^{\epsilon'}(S), h_e^\epsilon(A)^{-1}\} = -h_e^\epsilon(A)^{-1}\{E_n^{\epsilon'}(S), h_e^\epsilon(A)\}h_e^\epsilon(A)^{-1} \qquad (6.4.11)$$

in order to reduce all our calculations to expressions of the form $\{E_n^{\epsilon'}(S), h_e^\epsilon(A)\}$ where e is an edge of a definite type. We will also first assume that S is entire analytic and then extend the result to arbitrary piecewise (semi)analytic S later on.

The following calculation is quite lengthy and involves expanding out carefully the path-ordered exponential in (6.4.9) and using the known action $\nu_f(\phi_F) = F(f) = \int d^D x F_j^a(x) f_a^j(x)$. We find

$$
Y_n^{\epsilon'}(S)[\phi_e^\epsilon] = \sum_{n=1}^{\infty} \int_0^1 dt_n \int_0^{t_n} dt_{n-1} \cdots \int_0^{t_2} dt_1
$$

$$
\times \sum_{k=1}^{n} \left(\phi_{F_e^{\epsilon j_1 t_1}} \tau_{j_1}/2\right) \cdots \left(\phi_{F_e^{\epsilon j_{k-1} t_{k-1}}} \tau_{j_{k-1}}/2\right) \left[\nu_{f_S^{\epsilon'} n}\left(\phi_{F_e^{\epsilon j_k t_k}}\right)\tau_{j_k}/2\right]
$$

$$
\times \left(\phi_{F_e^{\epsilon j_{k+1} t_{k+1}}} \tau_{j_{k+1}}/2\right) \cdots \left(\phi_{F_e^{\epsilon j_n t_n}} \tau_{j_n}/2\right) \tag{6.4.12}
$$

Using

$$
\nu_{f_S^{\epsilon'} n}(\phi_{F^{\epsilon k t}}) = \int_{\mathbb{R}^{D-1}} d^{D-1}s\, \delta^\epsilon(s_1,\ldots,s_{D-1}) \int_{\mathbb{R}} ds\, \delta^{\epsilon'}(s)
$$

$$
\times \int_0^1 dt'\, \delta^\epsilon(t'-t) \int_U d^{D-1}u\, \dot{e}^a_{s_1,\ldots,s_{D-1}}(t')\epsilon_{aa_1\ldots a_{D-1}}
$$

$$
\times \frac{\partial S_s^{a_1}(u_1,\ldots,u_{D-1})}{\partial u_1} \cdots \frac{\partial S_s^{a_{D-1}}(u_1,\ldots,u_{D-1})}{\partial u_{D-1}}
$$

$$
\times \delta(S_s(u_1,\ldots,u_{D-1}), e_{s_1,\ldots,s_{D-1}}(t'))n^k(S_s(u_1,\ldots,u_{D-1})) \tag{6.4.13}
$$

we can now take   the limit $\epsilon \to 0$ and $\epsilon' \to 0$ (the reason for doing this will become transparent below). The result is

$$
Y_n^{\epsilon'}(S)[\phi_e] := \sum_{n=1}^{\infty} \int_0^1 dt_n \int_0^{t_n} dt_{n-1} \cdots \int_0^{t_2} dt_1
$$

$$
\times \sum_{k=1}^{n} A(t_1)\cdots A(t_{k-1})\left[\lim_{\epsilon \to 0} \nu_{f_S^{\epsilon'} n}\left(\phi_{F_e^{\epsilon j_k t_k}}\right)\tau_{j_k}/2\right] A(t_{k+1})\cdots A(t_n) \tag{6.4.14}
$$

with $A(t) = A_a^j(e(t))\dot{e}^a(t)\tau_j/2$ and where the limit in the square bracket is given by the distribution

$$
\int_{\mathbb{R}} ds\, \delta^{\epsilon'}(s) \int_U d^{D-1}u\, \dot{e}^a(t_k)n^{jk}(p(t_k))\epsilon_{aa_1\ldots a_{D-1}}
$$

$$
\times \frac{\partial S_s^{a_1}(u_1,\ldots,u_{D-1})}{\partial u_1} \cdots \frac{\partial S_s^{a_{D-1}}(u_1,\ldots,u_{D-1})}{\partial u_{D-1}}\delta(S_s(u_1,\ldots,u_{D-1}),p(t_k)) \tag{6.4.15}
$$

Luckily, there is an additional t_k integral involved in (6.4.14) so that the end result will be non-distributional. Let $t \mapsto F(t)$ be any (integrable) function and consider the integral

$$\int_{\mathbb{R}} ds \delta^{\epsilon'}(s) \int_{U} d^{D-1}u \int_{0}^{t_{k+1}} dt F(t) \, \dot{e}^{a}(t) \epsilon_{aa_1 \ldots a_{D-1}}$$

$$\times \frac{\partial S_s^{a_1}(u_1, \ldots, u_{D-1})}{\partial u_1} \ldots \frac{\partial S_s^{a_{D-1}}(u_1, \ldots, u_{D-1})}{\partial u_{D-1}} \delta(S_s(u_1, \ldots, u_{D-1}), e(t))$$

$$(6.4.16)$$

Notice first of all that the derivative $\dot{e}$ is well-defined since e is entire analytic. We can now discuss the integral (6.4.16) according to the type of edge e.

Case   : this case is trivial, since for sufficiently small ϵ' the δ-distribution vanishes identically.

Case   : since $s \mapsto S_s$ is a congruence it is clear that $\delta(S_s(u_1, \ldots, u_{D-1}), e(t))$ has support at $s = 0$ and the unique solution $u_1(t), \ldots, u_{D-1}(t)$ (which are interior points of U since S is open) of the equation $S(u) = e(t)$. Thus (6.4.16) becomes

$$\delta^{\epsilon'}(0) \int_0^{t_k} dt F(t) \frac{\dot{e}^a(t) \epsilon_{aa_1 \ldots a_{D-1}} \left[\frac{\partial S^{a_1}}{\partial u_1} \cdots \frac{\partial S^{a_{D-1}}}{\partial u_{D-1}} \right]_{u(t)}}{|\det(\partial S_s(u)/\partial(s, u_1, \ldots, u_{D-1}))_{s=0, u=u(t)}|} \qquad (6.4.17)$$

which vanishes at finite ϵ' since the denominator is finite while the numerator vanishes by definition of an   edge which is everywhere tangential to the surface. Since (6.4.17) vanishes at finite ϵ' its limit $\epsilon' \to 0$ vanishes as well. Expression (6.4.17) is the precise reason for why we have not synchronised the limits $\epsilon \to 0, \epsilon' \to 0$ as otherwise we would have obtained an ill-defined result of the form $0 \cdot \infty$.

Case   : in this case, for sufficiently small ϵ' and for every $s > 0$ the edge e cuts the surface S_s transversally in a single interior point $q_s = e(t_s) = S_s(u_s)$. Let $T_{q_s}(S_s)$ be the $(D-1)$-dimensional subspace of the tangent space $T_{q_s}(\sigma)$ at q_s spanned by the vectors $\partial S_s/\partial u_k(u_1, \ldots, u_{D-1})_{S_s(u)=q_s}$ tangential to S_s at q_s carrying the orientation induced from S_s, that is,

$$n_a^s(u) := \epsilon_{aa_1 \ldots a_{D-1}} \frac{\partial S_s^{a_1}(u_1, \ldots, u_{D-1})}{\partial u_1} \ldots \frac{\partial S_s^{a_{D-1}}(u_1, \ldots, u_{D-1})}{\partial u_{D-1}} \qquad (6.4.18)$$

is the outward normal direction. Since $\dot{e}(t_s)$ does not lie in $T_{q_s}(S_s)$ for $s > 0$, the combination $\dot{e}^a(t_s) n_a^s(u_s)$ is positive for $s > 0$. For $s = 0$ it may happen that $\dot{e}(t_0) = \dot{e}(0)$ and all of its higher derivatives lie in $T_{q_0}(S_0)$ without that e is then automatically of the inside type (consider for instance the case that S is a sphere in $\mathbb{R}^3$ and that e is a straight line in the tangent plane of the north pole). Hence this combination could actually vanish, however, the point $s = 0$ is of ds measure zero. Thus we can perform the t, u integral in (6.4.16) by changing to new coordinates $X_s(t, u) = S_s(u) - e(t)$, the Jacobean of which is $|n_a^s(u)\dot{e}^a(t)|$ and then evaluate the δ-distribution $\delta(X_s(t, u))$. The result is

$$\int_{\mathbb{R}} ds \delta^{\epsilon'}(s) \theta(t_{k+1} - t_s) \theta(s) F(t_s) \frac{n_a^s(u_s) \dot{e}^a(t_s)}{|n_a^s(u)\dot{e}^a(t)|} \qquad (6.4.19)$$

where $\theta(x) = 1$ for $x \geq 0$ and zero otherwise denotes the step function. The factor $\theta(s)$ comes from the fact that $\delta(X_s(t, u)) = 0$ for all $s < 0$. The fraction in (6.4.19) equals $+1$ except possibly at $s = 0$. Thus we may replace it by $+1$ at finite ϵ' because the point $s = 0$ is of ds measure zero. We may then perform the limit $\epsilon' \to 0$ in (6.4.19) with the result (notice that $t_0 = 0$)

$$F(0) \int_0^\infty ds\delta(s) = rF(0) \qquad (6.4.20)$$

where $0 < r < 1$ is a number that results from integrating the δ-distribution only over $\mathbb{R}^+$ rather than $\mathbb{R}$.

Case   : this case is completely analogous to the 'up' case, the difference being that now $n_a^s(u_s)\dot{e}^a(t_s)$ equals -1 for $s < 0$, vanishes for $s > 0$ and takes the value -1 or 0 at $s = 0$. The result of the integral is then

$$F(0) \int_{-\infty}^0 ds\delta(s) = (1 - r)F(0) \qquad (6.4.21)$$

It is possible to fix the parameter r to be $r = 1/2$ as follows: under a change of orientation of S the up and down type edges interchange their role. Now the area operator for a surface S, to be derived in a later chapter, should be invariant under change of orientation of S. Since the parameter r enters the formula for the area operator as we will see, we must fix $r = 1 - r$ to achieve orientation independence.

We can summarise the analysis by defining $\epsilon(e, S)$ to be $+1, -1, 0$ whenever e has type   ,   or   (   )   respectively whence the value of (6.4.16) is given by

$$\frac{1}{2}\epsilon(e, S)F(0) \qquad (6.4.22)$$

Inserting (6.4.22) into (6.4.14) we obtain

$$Y_n(S)[h_e] := \lim_{\epsilon' \to 0} \lim_{\epsilon \to 0} \nu_{Sn}^{\epsilon'}(\phi_e^\epsilon)$$

$$= \frac{1}{2}\epsilon(e, S) \sum_{n=1}^\infty \sum_{k=1}^n \int_0^1 dt_n \int_0^{t_n} dt_{n-1} \cdots \int_0^{t_{k+2}} dt_{k+1} \int_0^{t_{k+2}} dt_{k+1} \int_0^0 dt_{k-1}$$

$$\times \int_0^{t_{k-1}} dt_{k-2} \cdots \int_0^{t_2} dt_1 A(t_1) \ldots A(t_{k-1}) \frac{n(b(e))}{2} A(t_{k+1}) \ldots A(t_n)$$

$$= \frac{1}{2}\epsilon(e, S)\frac{n(b(e))}{2}\left(1 + \sum_{n=2}^\infty \int_0^1 dt_n \int_0^{t_n} dt_{n-1} \cdots \int_0^{t_3} dt_2 A(t_2) \ldots A(t_n)\right)$$

$$= \frac{1}{2}\epsilon(e, S)\frac{n(b(e))}{2}h_e \qquad (6.4.23)$$

where $n(x) = n^j(x)\tau_j$. Here in the second step we saw that the sum $\sum_{k=1}^n$ collapses to the term $k = 1$ because $\int_0^0 dt F(t) = 0$ and in the third step we have relabelled terms.

Formula (6.4.23) is our end result. Notice that the details of the regularisation of the delta-distributions did not play any role. It was seemingly important that we smeared via congruences of curves and surfaces as compared with more general smearings, however, any 'reasonable' smearing admits a foliation via curves and surfaces respectively. Thus, the result (6.4.23) is general. Finally, recall that (6.4.23) was derived under the assumption that S is entire analytic. However, the formula is insensitive to this assumption since the type function $\epsilon(e, S)$ can simply be extended to piecewise (semi)analytic surfaces. Hence we may simply   the formula to the general case.

6.5 Definition of   : (4) Lie algebra of cylindrical functions and flux vector fields

The amazing feature of expression (6.4.23) and its generalisation to arbitrary paths is that it is again a product of a finite number of holonomies, the harvest of having started from a manifestly background-independent formulation. If we had started from a function of E which is smeared in all D directions then this would no longer be true, (6.4.23) would be replaced by a more complicated expression in which an additional integral over the extra dimension would appear.

The fact that (6.4.23) is again a product of holonomies enables us to generalise the action of $Y_n(S)$ to arbitrary cylindrical functions, restricted to smooth connections. Let $f \in \mathrm{Cyl}^1(\overline{\mathcal{A}})$, then we find a subgroupoid $l = l(\gamma) \in \mathcal{L}$ and $f_l \in C^1(X_l)$ such that $f = p_l^* f_l = [f_l]_\sim$ and a complex-valued function F_l on $G^{|E(\gamma)|}$ such that $f(A) = f_l(p_l(A)) = F_l(\rho_l(p_l(A)))$ with $\rho_l(A_l) = \{A_l(e)\}_{e \in E(\gamma)} = \{A(e)\}_{e \in E(\gamma)}$. We may choose γ in such a way that it is   to a given surface S, that is, each edge of γ has a definite type with respect to S. This will make the following computation simpler. Notice that every graph can be chosen to be adapted by subdividing edges appropriately. Let us now restrict f to $\mathcal{A}$ then

$$[Y_n(S)(f)](A) = \frac{1}{2} \sum_{e \in E(\gamma)} \epsilon(e, S) \left[\frac{n(b(e))}{2} A(e) \right]_{AB} \frac{\partial F_l}{\partial A(e)_{AB}} (\{A(e')\}_{e' \in E(\gamma)})$$

(6.5.1)

Evidently, (6.5.1) leaves $C^\infty(X_l)$ restricted to $\mathcal{A}$ invariant
$\overline{\mathcal{A}}$!

More precisely: define the so-called right- and left-invariant vector fields on G by

$$(R_j f)(h) := \left(\frac{d}{dt} \right)_{t=0} f(e^{t\tau_j} h) =: \left(\frac{d}{dt} \right)_{t=0} [L^*_{e^{t\tau_j}} f](h)$$

$$(L_j f)(h) := \left(\frac{d}{dt} \right)_{t=0} f(h e^{t\tau_j}) =: \left(\frac{d}{dt} \right)_{t=0} [R^*_{e^{t\tau_j}} f](h) \qquad (6.5.2)$$

where $R_h(h') = h'h$, $L_h(h') = hh'$ denotes the right and left action of G on itself. The right (left) invariance of R_j (L_j), that is, $(R_h)_* R_j = R_j$ $((L_h)_* L_j = L_j)$, follows immediately from the commutativity of left and right translations $L_h R_{h'} = R_{h'} L_h$. Notice, however, that the right-invariant field generates left translations and vice versa. Then we can write (6.5.1) in the compact form

$$Y_l^n(S)[f_l] = \frac{1}{4} \sum_{e \in E(\gamma)} \epsilon(e, S) n_j(b(e)) R_e^j f_l \tag{6.5.3}$$

where R_e^j is R^j on the copy of G labelled by e and where from now on we just identify X_l with $G^{|E(\gamma)|}$ via ρ_l. Expression (6.5.3) obviously does not require us to restrict $f = p_l^* f_l$ to $\mathcal{A}$ any more. Notice that while $Y_l^n(S)$, just as $E_n(S)$ does not have a simple transformation behaviour under gauge transformations, R_e^j, L_e^j in fact do

$$[(\lambda_g^e)^* ([(\lambda_g^e)_* R_e^j](f_e))](h_e) = [R_e^j((\lambda_g^e)^* f_e)](h_e)$$
$$= \left(\frac{d}{dt}\right)_{t=0} f_e\big(g(b(e))e^{t\tau_j} h_e g(f(e))^{-1}\big)$$
$$= \left(\frac{d}{dt}\right)_{t=0} f_e\big(e^{t\mathrm{ad}_{g(b(e))}(\tau_j)} g(b(e)) h_e g(f(e))^{-1}\big)$$
$$= [(\lambda_g^e)^* (R^{\mathrm{ad}_{g(b(e))}(\tau_j)} f_e)](h_e) \tag{6.5.4}$$

so that $(\lambda_g^e)_* R_e^j = [\mathrm{Ad}_{g(b(e))}]_{jk} R_e^k$ where $\mathrm{Ad}_{g(b(e))}(\tau_j) =: [\mathrm{Ad}_{g(b(e))}]_{jk} \tau_k$. Similarly $(\lambda_g^e)_* L_e^j = [\mathrm{Ad}_{g(f(e))}]_{jk} L_e^k$. This shows once more that R_e^j (L_e^j) is right (left)-invariant.

We thus have found a family of vector fields $Y_l^n(S)$ whenever l is adapted to S. If $l = l(\gamma)$ is not adapted then we can produce an adapted one $l_S = l(\gamma')$, for example, by choosing $r(\gamma) = r(\gamma')$ and by subdividing edges of γ into those with definite type with respect to S and where the edges of γ' carry the orientation induced by the edges of γ. Since $p_{l_S l}^* f_l \sim f_l$ we then simply

$$p_{l_S l}^*(Y_l^n(S)(f_l)) := Y_{l_S}^n(p_{l_S l}^* f_l) \tag{6.5.5}$$

We must check that (6.5.5) does not depend on the choice of an adapted subgroupoid. Hence, let l_S' be another adapted subgroupoid then we find $l_S, l_S' \prec l_S''$ which is still adapted (take for instance the union of the corresponding graphs and subdivide edges as necessary). Since (6.5.5) is supposed to be a cylindrical function and $p_{l_S l} \circ p_{l_S'' l_S} = p_{l_S' l} \circ p_{l_S'' l_S'}$ we must show that

$$p_{l_S'' l_S}^* Y_{l_S}^n(S)(p_{l_S l}^* f_l) = p_{l_S'' l_S'}^* Y_{l_S'}^n(S)(p_{l_S' l}^* f_l) \tag{6.5.6}$$

As usual, if (6.5.6) holds for one such adapted l_S'' then it holds for all. To see that (6.5.6) holds, it will be sufficient to show that for any adapted subgroupoids $l_S \prec l_S''$ we have

$$p_{l_S'' l_S}^* Y_{l_S}^n(S)(f_{l_S}) = Y_{l_S''}^n(S)(p_{l_S'' l_S}^* f_{l_S}) \tag{6.5.7}$$

from which then (6.5.6) will follow due to $p_{l_s l} \circ p_{l''_S l_S} = p_{l'_s l} \circ p_{l''_S l'_S}$. We again need to check three cases:

(a) $e \in E(\gamma''_S)$ but $e \notin E(\gamma_S)$, then (6.5.7) holds because $p^*_{l''_S l_S} f_{l_S}$ does not depend on $A(e)$ so that the additional terms proportional to R^j_e, L^j_e in (6.5.3) drop out.

(b) $e \in E(\gamma''_S)$ but $e^{-1} \in E(\gamma_S)$. By definition of an adapted subgroupoid, this case is only allowed for edges of the inside and outside type because all edges with $\epsilon(e, S) \neq 0$ must be outgoing from S. However $\epsilon(e, S) = 0 \Leftrightarrow \epsilon(e^{-1}, S) = 0$.

(c) $e_1, e_2 \in E(\gamma''_S)$ but $e = e_1 \circ e_2 \in E(\gamma_S)$. Then $e_1 \cap S = b(e_1)$ and $\epsilon(e, S) = \epsilon(e_1, S)$ while $e_2 \cap S = \emptyset$ and $\epsilon(e_2, S) = 0$ (recall that $\epsilon(e, S) \neq 0$ implies that e, S intersect in only one point). Let $f_1(h_1) := f_2(h_2) = f(h_1 h_2)$ then due to right invariance

$$(R^j f_1)(h_1) = (R^j f)(h_1 h_2) \tag{6.5.8}$$

hence

$$\sum_{I=1,2} \epsilon(e_I, S) R^j_{e_I} p^*_{(e_1, e_2), e_1 \circ e_2} f_e = \epsilon(e_1, S) R^j_{e_1} p^*_{(e_1, e_2), e_1 \circ e_2} f_e$$

$$= \epsilon(e, S) R^j_e f_e \tag{6.5.9}$$

as claimed.

Hence our family of vector fields $(Y^n_l(S))_{l \in \mathcal{L}}$ is now defined for all possible $l \in \mathcal{L}$, in the language of Section 8.2.2 we have the co-final set $l_0 := l(\emptyset) \prec \mathcal{L}$. Let us check that it is a consistent family, that is

$$p^*_{l'l}\big([Y^n_l(S)](f_l)\big) = [Y^n_{l'}(S)](p^*_{l'l} f_l) \tag{6.5.10}$$

for all $l \prec l'$ which are not necessarily adapted. Given $l \prec l'$ we find always an adapted subgroupoid $l, l' \prec l_S$. Now by the just established independence on the adapted graph we may equivalently show that

$$p^*_{l_S l'} p^*_{l'l}(Y^n_l(S)(f_l)) = p^*_{l_S l'} Y^n_{l'}(p^*_{l'l} f_l) \tag{6.5.11}$$

Now since $p^*_{l_S l'} p^*_{l'l} = p^*_{l_S l}$ the left-hand side equals $p^*_{l_S l}(Y^n_l(S)(f_l)) \equiv Y^n_{l_S}(p^*_{l_S l} f_l)$ by definition of Y^n_l on arbitrary, not necessarily adapted graphs and the right-hand side equals $Y^n_{l_S}(p^*_{l_S l'} p^*_{l'l} f_l) = Y^n_{l_S}(p^*_{l_S l} f_l)$ for the same reason.

We have thus established that the family of vector fields $(Y^n_l(S))_{l \in \mathcal{L}}$ is a consistent family and defines a vector field $Y_n(S)$ on $\overline{\mathcal{A}}$. Notice moreover that $Y_n(S)$ is real-valued: from (6.5.3) this will follow if R_j is real-valued. Now we have embedded G into a unitary group which means that $\bar{h}^T = h^{-1}$, in particular

$\bar{\tau}_j^T = -\tau_j$. Hence

$$\overline{R_h^j} = \overline{(\tau_j h)_{AB} \partial/\partial \bar{h}_{AB}} = -(h^{-1}\tau_j)_{BA} \partial/\partial h_{BA}^{-1}$$
$$= -(h^{-1}\tau_j)_{AB} (\partial h_{CD}/\partial h_{AB}^{-1}) \partial/\partial h_{CD} = (h^{-1}\tau_j)_{AB} h_{CA} h_{BD} \partial/\partial h_{CD}$$
$$= R_h^j \tag{6.5.12}$$

where use was made of $\delta h^{-1} = -h^{-1}hh^{-1}$ and the fact that the symbol $\partial/\partial h_{AB}$ acts as if all components of h_{AB} were independent by definition of $R_j(f) = (\tau_j h)_{AB} \partial f/\partial h_{AB}$.

The definition of $\mathfrak{P}$ is now complete:

Definition 6.5.1. $\mathfrak{P}$ $*$

$\infty \times V(\quad \infty)$

$$\overline{Y_n(S)} \quad \infty \qquad \mathfrak{P}$$
$$\overline{A(e)} = A(e^{-1})^T \qquad \overline{Y_n(S)} = Y_n(S)$$

Here a cylindrical function $f = p_l^* f_l$ is smooth if any of its representatives f_l is smooth on the respective power on G^n, see Section 8.2.2. Notice that, as we have seen, $\mathfrak{P}$ can be thought of as an algebra of the form $C^\infty(X) \times V^\infty(X)$ with either choice of space $X = \mathcal{A}$ or $X = \overline{\mathcal{A}}$ respectively.

7

Step II: quantum *-algebra $\mathfrak{A}$

Since (generalised) holonomies take values in a compact group, cylindrical functions $f = p_l^* f_l$ with f_l a bounded function on a corresponding power of G are bounded functions of (generalised) connections and will be promoted to bounded operators in the quantum theory. However, the flux vector fields will be promoted to unbounded operators and therefore domain questions will arise when we study representations later on. In order to avoid the complications that come with this unboundedness we will pass to an abstract *-algebra of operators which will be promoted to bounded operators in any representation by exponentiating the vector field elements of $\mathfrak{P}$. The result $\mathfrak{A}$ could be called a • • • •• •• •••• • • ••• ••• •••••.
We will follow [518, 521].

7.1 Definition of

Definition 7.1.1. ••• $t \in \mathbb{R}$ ••••• ••• • ••• ••• ••••

$$W_t^n(S) := e^{t\beta \ell_p^2 / 2 Y_n(S)} = e^{-it\left[i\beta \ell_P^2 / 2 Y_n(S)\right]} \tag{7.1.1}$$

• •••• $\ell_P^2 = \hbar\kappa$ ••• ••• ••• •• ••• ••• ••••• •••• •• $\mathfrak{P}$ •••••••• •• ••• •••••
$Y_n(S)$• ••• ••••••• $\mathfrak{A}$ •• •••••••••• •••• ••• •••• •••• $f \in$ •••••• ••• • ••• •••
• •••• •••••• •• ••• •• ••• • •• ••• * •••• ••• •

$$f^* := \overline{f} \quad \bullet\bullet\bullet \quad \left(W_t^n(S)\right)^* := W_{-t}^n(S) = W_t^{-n}(S) = \left(W_t^n(S)\right)^{-1} \tag{7.1.2}$$

••• ••• •• ••• ••• • ••• ••• •••• •• ••• •••• •• $f = p_l^* f_l$, $l = l(\gamma)$ ••••

$$[f, f'] := 0$$

$$W_t^n(S)\, f\, \left(W_t^n(S)\right)^{-1} := \left(W_t^n(S)\right)\cdot f = p_l^*\, f_l(\{e^{t\beta \ell_p^2 n^j (b(e))\tau_j / 8} A(e)\}_{e \in E(\gamma)})$$

$$W_t^n(S)\, W_{t'}^{n'}(S')\, \left(W_t^n(S)\right)^{-1} = \exp\left(t'\beta \ell_p^2 / 2 \sum_{m=0}^{\infty} \frac{(t\beta \ell_P^2 / 2)^m}{m!} [Y_n(S), Y_{n'}(S')]_{(m)}\right)$$

$$\tag{7.1.3}$$

••• ••• ••• •• ••• ••• • ••• •••• •••• •• ••• ••••• •••••• •••• •• $\mathfrak{P}$• • ••• •••
• • •••• •• ••• • •••••• •• •• ••••••• •••••• •• $[Y, Y']_{(0)} := Y'$ ••• $[Y, Y']_{(m+1)} :=$
$[Y, [Y, Y']_{(m)}]$•

That the right-hand side in the second equation of (7.1.3) is really the action of the exponentiated flux on a cylindrical function follows from Section 6.5, where we showed that the fluxes generate infinitesimal left translations on the group. One can verify this by taking the derivative at $t = 0$ for smooth f and rely on the uniqueness theorem for ordinary differential equations. One can then actually lift this relation derived on Cyl^∞ to all of Cyl but we will not need that. The third equation in (7.1.3) is the exponential of a vector field element of $\mathfrak{P}$ again and one can actually write it down explicitly. We will do that below for completeness, although we do not need that expression in the sequel.

Definition 7.1.2

··· · ··· $x \in \sigma$ ·· ······· · ·· ···· $[e]_x$ ···· ·· ····· ··· ·· ··· ···· e · ··· $b(e) =$
$e(0) = x$ ·· ······· ·· ··· ·· · · ··· ··· ··· ·· · ····· ···· ······ ·· $e^{(n)}(0)$ ··
··· · ····· ······· ··· ·

···· · ·· ···· $[e]_x$ ········ ··· ····· ····· ·· e ··· ··· ···· ···· · ···· · ··
·· ···· ···· ··· $e(t)$ ··· · x ·· ·· ······· ······· ···· ··· ·· ·· ········· · ·
··· · ···· ·· · ···· · ··· x ·· ···· ···· ······ ··· ··· · ···· ···· · x·
····· · ·· ··· ·· · ·· ···· · $[e]_x$ ·· ····· $x \in \sigma$ ···· ··· ····· ·· · x ··· · ··· ··
··· ····· ·· $\mathcal{K}$·

···· · ·· $x \in \sigma,\ [e]_x \in \mathcal{K}$· · ····· ······ · ···· $R^j_{x,[e]_x}$ ·· ··· ·· ··· ··· · ·······
····· ·· · · ·∞ ······ · · ····· ···· $l = l(\gamma)$ ·· ········ ·· x ·· ··· ····· ·····
···· ···· ·· ······ ····· ······· ·· ······· ···· x·

$$R^j_{x,[e]_x} p_l^* f_l := p_l^* \sum_{e' \in E(\gamma)} \delta_{x,b(e')} \delta_{[e]_x,[e']_x} R^j_{e'} f_l \tag{7.1.4}$$

··· ··· $x \in \sigma,\ [e]_x \in \mathcal{K}$ ··· S · ········· · · ·····

$$\epsilon(S, [e]_x) := \epsilon(S, e') ··· ··· e' ···· [e]_x = [e']_x \tag{7.1.5}$$

Lemma 7.1.3

··· · ··· ······· · ··· $R^j_{x,[e]_x}$ ··· ····· ··· ·· ··· ··· ··· · · ··· ··· · ··· ··· ···· ·

$$\left[R^j_{x,[e]_x}, R^k_{x',[e']_x} \right] = -f^{jk}{}_l \delta_{[e]_x,[e']_x} \delta_{x,x'} R^l_{[x],[e]_x} \tag{7.1.6}$$

· ···· $[\tau_j, \tau_k] = f_{jk}{}^l \tau_l$ ····· ··· ···· ··· ·· ·····[1]
···· · ·· ··· ······ ··· $Y_n(S)$ ··· ·· ········· ·· ···· · ··· ··· $R^j_{x,[e]_x}$ ·· ····
··· · ··

$$Y_n(S) = \sum_{x \in S} \sum_{[e]_x \in \mathcal{K}} \epsilon(S, [e]_x) n_j(x) R^j_{x,[e]_x} \tag{7.1.7}$$

The proof of the lemma is straightforward by using the formulae and definitions of Section 6.5. Although the definition of $Y_n(S), R^j_{x,[e]_x}$ involves sums over

---

[1] Since G is a compact, connected Lie group, we have $G/D \cong A \times S$ where D is a central discrete subgroup and A, S are Abelian and semisimple Lie groups respectively. Indices are dragged w.r.t. the Cartan–Killing metric $\mathrm{Tr}(T_j T_k) = -\delta_{jk}$ where $(T_j)^k_l = f^k_{lj}, f_{jkl}$ totally skew for the semisimple generators.

an uncountably infinite number of terms, when applying these vector fields to elements of Cyl^∞ these sums reduce to a finite number of terms. The importance of Lemma 7.1.3 is that it shows that the vector fields $R^j_{x,[e]_x}$ are the basic building blocks of the $Y_n(S)$ and, in contrast to them, •••• • •• ••• • •• •••••. In particular, we may use them in order to define ••• ••• •••• ••• •••••• ••••:
classically the electric flux is additive, that is, if $S = \cup^n_{k=1} S_k$ is a disjoint union of $(D-1)$-dimensional subsets of S then $E_n(S) = \sum^n_{k=1} E_n(S_k)$. The problem in transferring this to the $Y_n(S)$ is that if the sets S_k are really disjoint, then some of them must contain part of their common boundaries and others do not, that is, they are partly open and/or closed. However, the definition of $Y_n(S)$ assumes that S has no boundary. For the definition of $E_n(S)$ this is irrelevant because the boundary points are sets of measure zero in the corresponding integral over S. The expression (7.1.7) now shows how to fix this. We have

$$Y_n(S) = \sum^n_{k=1} Y_{n,S}(S_k), \quad Y_{n,S}(S_k) := \sum_{x \in S_k} \sum_{[e]_x \in \mathcal{K}} \sigma(S,[e]_x) n_j(x) R^j_{x,[e]_x} \qquad (7.1.8)$$

that is, we simply restrict the sum over $x \in S$ to the sum over $x \in S_k$. The information about S, however, sits in the type indicator function $\epsilon(S, [e]_x)$. Alternatively we may define the action of $Y_{n,S}(S_k)$ on cylindrical functions to equal the action of $Y_n(S)$ on all edges of the inside or outside type and on those edges of the up and down type which intersect S_k and to disregard those edges of the up and down type which do not intersect S_k. Here the type is defined with respect to S.

The vector fields $R^j_{x,[e]_x}$ have no physical meaning at all; only those vector field elements of $\mathfrak{P}$ which can be written as linear combinations of some $Y_n(S)$ have physical meaning. This does not mean that one cannot get fluxes of more general surfaces than faces: take for instance an open disc D and another, smaller one D' which is contained in it, $D' \subset D$. Then $Y_n(D) - Y_n(D') =: Y_{n,D}(D-D')$ is the flux through the annulus $D-D'$ which has one open and one closed boundary. Hence, $Y_{n,D}(D-D')$ can be expressed directly in terms of fluxes through faces. However, the vector field elements of $\mathfrak{P}$ form a complicated closed subalgebra of the simpler closed algebra of the $R^j_{x,[e]_x}$ which enables us to compute the right-hand side of (7.1.3) explicitly. We find after some algebra

$$W^n_t(S) W^{n'}_{t'}(S') \left(W^n_t(S)\right)^{-1}$$

$$= \exp\left(t'\beta\ell^2_p/8 \left[\sum_{x \in S'-S} n'_k(x) \sum_{[e]_x \in \mathcal{K}} \epsilon(S',[e]_x) + \sum_{x \in S \cap S'} n'_j(x) \right.\right.$$

$$\left.\left. \times \sum_{[e]_x \in \mathcal{K}} \epsilon(S',[e]_x)[e^{-t\beta\ell^2_p/8\epsilon(S,[e]_x)\mathrm{ad}_{n(x)}}]_{jk} R^k_{x,[e]_x}\right]\right) \qquad (7.1.9)$$

where $\mathrm{ad}_\tau(\tau') := [\tau, \tau']$ is the adjoint action of the Lie algebra on itself and $n(x) = n^j(x)\tau_j$. This is again easy to check by differentiation with respect to t, t'.

7.2 (Generalised) bundle automorphisms of

Automorphisms of the principal G-bundle (P, G, Π, σ), where $\Pi : P \to \sigma$ is the bundle projection are invertible pairs of maps (F, f) with $F : P \to P$, $f : \sigma \to \sigma$ such that $\Pi \circ F = f \circ \Pi$, meaning that F maps entire fibres to entire fibres and such that F is G-equivariant, that is, $F \circ \rho = \rho \circ F$ where ρ is the right action on P. The group of such maps will be denoted by $\mathfrak{G} := \mathrm{Aut}(P)$. Here we will restrict f to semianalytic diffeomorphisms φ of σ. Writing $p = \phi(x, h)$ and conversely $(X(p), H(p)) = \phi^{-1}(p)$ in a local trivialisation $\phi : U \times G \to P$, $U \subset \sigma$ and provided that $\Pi(F(p)) \in U$ we find from $\Pi \circ F = \varphi \circ \Pi$ that $F(\phi(x, h)) = \phi(\varphi(x), H(F(\phi(x, h))))$. Using the equivariance condition and the action $\rho_g(\phi(x, h)) = \phi(x, hg)$ we find that $H(F(\phi(x, hg))) = H(F(\phi(x, h)))g$. Setting $h = 1_G$ we find $H(F(\phi(x, g))) = H(F(\phi(x, 1_G)))g =: g(x)^{-1}g$. Hence, in a local trivialisation ϕ, $F(\phi(x, h)) = \phi(\varphi(x), g(x)h)$ and thus F is completely characterised by the diffeomorphism φ and a map $g : \sigma \to G$ which we restrict to be semianalytic in the classical theory. It follows that

$$\mathfrak{g} \cdot \mathfrak{g}' \cdot p = \phi([\varphi \circ \varphi'](x), [g \circ \varphi' \cdot g'](x)h) \tag{7.2.1}$$

Thus, in a local trivialisation $\mathfrak{G}$ is isomorphic to the semidirect product[2] $\mathfrak{G} \cong \mathcal{G} \rtimes \mathrm{Diff}(\sigma)$ with normal subgroup $\mathcal{G}$.

The group $\mathfrak{G}$ has a natural action on our basic variables $A(e), E_n(S)$, if we replace smooth diffeomorphisms by semianalytic ones as already pointed out in Section 6.1 and the action of its connected identity component is in fact generated by the Hamiltonian flow of the Hamiltonian vector fields of the Gauß constraint and diffeomorphism constraint respectively. Since canonical transformations preserve the Poisson brackets (more precisely: Lie brackets) among the basic variables, we actually get an action of $\mathfrak{G}$ by Poisson automorphisms on $\mathfrak{P}$ as displayed in formula (6.1.11). Specifically for $f = p_l^* f_l$, $l = l(\gamma)$

$$\alpha_g((f, Y_n(S))) := (\lambda_g^* f, (\lambda_g)_* Y_n(S))$$
$$= (p_l^* f_l(\{g(b(e))A(e)g(f(e))^{-1}\}_{e \in E(\gamma)}), Y_{\mathrm{Ad}_{g^{-1}}(n)}(S))$$
$$\alpha_\varphi((f, Y_n(S))) := (\delta_\varphi^* f, (\delta_\varphi)_* Y_n(S))$$
$$= (p_l^* f_l(\{A(\varphi(e))\}_{e \in E(\gamma)}), Y_{\varphi^{-1}(n)}(\varphi(S))) \tag{7.2.2}$$

---

[2] A normal subgroup N of a group G is a subgroup invariant under conjugation, that is, $gng^{-1} \in N$ for all $n \in N$, $g \in G$. One writes $N \lhd G$. Let H be a subgroup of G then $G = N \rtimes H$ is said to be the semidirect product of N, H if $G = NH$ and $N \cap H = 1_G$. Conversely given a group action (an automorphism of N) $H \times N \to N$; $(h, n) \mapsto h \cdot n$ we may form the semidirect product $G := N \rtimes H$ by $(n_1, h_1)(n_2, h_2) := (n_1 h_1 \cdot n_2, h_1 h_2)$ or $(n_1, h_1)(n_2, h_2) := (h_2 \cdot n_1 n_2, h_1 h_2)$. We may identify $N \cong (N, 1_H), H \cong (1_N, H), 1_G = (1_N, 1_H)$ and verify that N is a normal subgroup.

Amazingly, this action of $\mathfrak{G}$ by automorphisms on $\mathfrak{P}$, considered as the Lie algebra of smooth cylindrical functions on $\mathcal{A}$ and smooth derivatives thereon, can be generalised in two respects. First of all the algebra can be considered as smooth cylindrical functions on $\overline{\mathcal{A}}$ and smooth derivatives thereon. Hence we have the generalisation $\mathcal{A} \to \overline{\mathcal{A}}$ and the action (7.2.2) simply lifts. Second, we may generalise $\mathfrak{G}$ itself: the semianalytic gauge transformations G can be replaced by arbitrarily discontinuous ones: $\overline{\mathcal{G}} := \mathrm{Fun}(\sigma, G)$. The entire analytic diffeomorphisms $\mathrm{Diff}^\omega(\sigma)$ can be replaced by semianalytic ones $\mathrm{Diff}^\omega_{\mathrm{sa}}(\sigma)$ which are defined in Chapter 20 as maps between semianalytic charts. Equivalently:

Definition 7.2.1. •• •• •••••• •• • ••• ••• • •• ••• ••• ••• ••••••• • • •• $^\omega_{\mathrm{sa}}(\sigma)$ •• •••
•• •••••• •• ••• ••• ••••••• • •• σ • •••• ••••••••••• ••• ••• •• ••• ••• •••••• •••
••••• ••• ••• ••• ••• •• ••••• •••••• • ••• ••• • ••••• ••••• ••••••• •• ••••
•••• ••• •••••••• •• σ •• • ••• ••• ••• • ••••• ••• ••• •••• •••• •

A typical example of an element of $\mathrm{Diff}^\omega_{\mathrm{sa}}(\sigma)$ is as follows: consider two open regions $U_0 \subset U_1 \subset \sigma$ where U_1 has compact closure such that the boundaries $\partial U_0, \partial U_1$ are finite unions of lower-dimensional analytic submanifolds of σ. Next, choose a vector field v on σ which (1) vanishes identically on $\sigma - \overline{U_1}$, (2) is analytic on both $\overline{U_0}$ and $\overline{U_1 - U_0}$ and (3) is at least continuous at the boundaries $\partial U_0, \partial U_1$, say $C^{(n_0)}$ with $0 \le n_0 < \infty$. ($n_0 = \infty$ is not allowed because then v could be analytically continued from $\sigma - \overline{U_0}$ and hence would have to vanish identically.) Finally, construct the one-parameter family $t \mapsto \varphi^v_t$ of homeomorphisms defined by the integral curves of v. These diffeomorphisms exist because v has compact support, hence φ^v_t is the identity map on $\sigma - \overline{U_0}$.[3] These diffeomorphisms are precisely those that arise in applications: we want a diffeomorphism that has a certain property in region U_0 but leaves everything outside a second region U_1 untouched. This requires some locality of that action and this is why entire analytic diffeomorphisms are not general enough: if we specify it in U_0 then it is specified everywhere and we are not sure what it does globally. For instance, in an asymptotically flat context the specification of an entire analytic diffeomorphism in some compact region could be such that the analytically extended map fails to be the identity map at spatial infinity and thus would be no longer an element of the diffeomorphism group.

To be even more specific, suppose that U_1 is contained in the domain of a chart. Choose an analytic coordinate system $x^1, \ldots, x^D$ and define spherical coordinates. Suppose that both U_0, U_1 are solid, open balls in this coordinate system of radius r_0, r_1 respectively. Consider functions $\xi_0, \xi_{12} : \mathbb{R}^D \to \mathbb{R}$ which are some polynomials of $x^1, \ldots, x^D$ and define ξ to equal ξ_0 for $0 \le r \le r_0$, to

---

[3] Notice that given a classical connection $A_0 \in \mathcal{A}$ we may construct the horizontal lifts $\tilde{c}^v_p(t)$, $p \in P$ of the integral curves $c^v_x(t)$ with $\Pi(p) = x$ which then defines a piecewise (semi)analytic bundle automorphism $\tilde{\varphi}^v_t$ which projects to φ^v_t, hence we can make the construction global in the bundle P.

equal ξ_{12} on $r_0 \leq r \leq r_1$ and to vanish for $r \geq r_1$. Given ξ_0 we always find ξ_{12} such that ξ is $C^{(n_0)}$. Now finish the construction by defining the radial vector field $v(x) := r\xi(x)\partial_r$.

This shows by elementary means that semianalytic diffeomorphisms exist and the ones just displayed are precisely those that arise in applications. See Chapter 20 for more details and precise definitions.

••• •••: the maximal possible extension of the diffeomorphism or automorphism group appears to be simply the group of all, not necessarily continuous, bundle automorphisms [534] but they have not yet found an application in the framework.

8

Step III: representation theory of $\mathfrak{A}$

In this chapter we will show that, under reasonable physical assumptions, there is a unique representation of $\mathfrak{A}$. This means that, once the algebra $\mathfrak{A}$ has been chosen, we can be confident to use that unique, kinematical representation as a basis for the constraint quantisation programme.

8.1 General considerations

We are interested in a representation, that is a *-morphism, between $\mathfrak{A}$ and a subalgebra of the set of linear operators on a Hilbert space $\mathcal{H}$. See Section 29.1 for a dictionary on the representation theory of *-algebras. Our *-algebra is generated by unitary elements, that is, those which satisfy $a^* = a^{-1}$ (take exponentials $\exp(itf)$ of cylindrical functions to see this) and hence in any representation these generating elements will become unitary, that is, bounded operators. Moreover, since $\mathfrak{A}$ is unital and contains invertible elements, it follows from the representation property that $\pi(1) = \mathrm{id}_{\mathcal{H}}$. Therefore any representation of $\mathfrak{A}$ is not degenerate. This brings us into the position to apply the following result.

Lemma 8.1.1. • •••• ••••••••• •••• •••••••••••• •• ••• •••••••• •• •
*•• •••••• •• ••••••• •••••••• •• • •••••• ••• •• •••••• ••••••••• •• ••• ••

• ••••• The proof is usually made in the context of C^*-algebras [535–537] but works as well under the assumptions made in the lemma without a Banach or C^*-norm. By Zorn's lemma we are granted that there exists a maximal set of vectors Ω_I in the representation space $\mathcal{H}$ such that $< \Omega_I, \pi(a)\Omega_J >= 0$ for all $a \in \mathfrak{A}$ unless $I = J$. Here $I, J, \ldots$ are taken from some, in general uncountably infinite, index set. Notice that no domain questions arise because the elements of $\mathfrak{A}$ are polynomials of the generators and hence $\pi(a)$ is a bounded operator. Let $\mathcal{H}_I$ be the completion of $\pi(\mathfrak{A})\Omega_I$ and let $P_I : \mathcal{H} \to \mathcal{H}_I$ be the orthogonal projection. Define $\pi_I(a) := P_I\pi(a)P_I$ on $\mathcal{H}_I$. Then $\mathcal{H} = \oplus_I \mathcal{H}_I$ because the set of Ω_I is maximal and the representation is non-degenerate. Hence $\pi = \oplus_I \pi_I$ and $(\pi_I, \mathcal{H}_I, \Omega_I)$ is a cyclic representation of $\mathfrak{A}$. $\qquad\square$

It follows that cyclic representations are the basic building blocks of all representations of our concrete $\mathfrak{A}$ and therefore it is no loss of generality to consider the latter. Now given a cyclic representation $(\pi, \mathcal{H}, \omega)$ we can construct a positive linear functional, that is a state, on $\mathfrak{A}$ by $\omega(a) :=< \Omega, \pi(a)\Omega >_{\mathcal{H}}$. Conversely, a state ω on a *-algebra, even when not generated by unitary elements, determines a unique, up to unitary equivalence, cyclic representation $(\pi_\omega, \mathcal{H}_\omega, \Omega_\omega)$ via the GNS construction, see Section 29.1. Hence, studying cyclic representations on *-algebras is completely equivalent to studying states.

In general there are an infinite number of states on *-algebras and in order to control this abundance of representations one must make additional physical assumptions. For instance, the uniqueness theorem due to Stone and von Neumann [538] for the Weyl algebra of quantum mechanics generated by $(a, b \in \mathbb{R})$

$$U(a) := \exp(iaq/\hbar), \ V(b) := \exp(-ibp/\hbar)$$
$$U(a)U(a') = U(a + a'), \ V(b)V(b') = V(b + b'), \ V(b)U(a) = e^{iab/\hbar}U(a)V(b)$$
$$U(a)^* = U(-a), \ V(b)^* = V(-b) \tag{8.1.1}$$

that the only possible representation is the Schrödinger representation on $\mathcal{H} := L_2(\mathbb{R}, dx)$ defined by $(\pi(U(a))\psi)(x) = \exp(iax)\psi(x)$, $(\pi(V(b))\psi)(x) = \psi(x + b)$ holds only under the assumption that the representation is irreducible and weakly continuous. Irreducible means that every vector is cyclic and weakly continuous means that $\lim_{a \to 0} <\psi, \pi(U(a))\psi'> = <\psi, \psi'>$ for all $\psi, \psi' \in \mathcal{H}$ and similarly for $V(b)$. The Bohr compactification of the real line constructed in Chapter 28 demonstrates that weak continuity cannot be dropped: define a non-separable Hilbert space with orthonormal states $T_x, x \in \mathbb{R}$ and set $\pi(U(a))T_x := T_{x+a}, \pi(V(b))T_x := e^{ibx}T_x$. This defines a representation by unitary operators but $\lim_{a \to 0} <T_x, \pi(U(a))T_x> = \lim_{a \to 0} \delta_{x,x+a} = 0 \neq <T_x, T_x> = 1$.

Even worse than in quantum mechanics (finite number of degrees of freedom) is the situation in quantum field theory (infinite number of degrees of freedom) where the Stone–von Neumann theorem is no longer correct. Rather, there are an uncountably infinite number of unitarily inequivalent representations of the (analogue of the) Weyl algebra. Take for instance a scalar field ϕ on $(D + 1)$-dimensional Minkowski space with canonically conjugate momentum π and consider the following Weyl algebra generated by the Weyl elements $(a, b$ are real-valued test functions of rapid decrease on $\mathbb{R}^D)$

$$U(a) := \exp(i\phi(a)), \ V(b) := \exp(-i\pi(b))$$
$$U(a)U(a') = U(a + a'), \ V(b)V(b') = V(b + b'), \ V(b)U(a) = e^{i<a,b>}U(a)V(b)$$
$$U(a)^* = U(-a), \ V(b)^* = V(-b) \tag{8.1.2}$$

where $\phi(a) = <a, \phi>$, $\pi(b) = <b, \pi>$, $<a, b> = \int d^D x \, a(x) \, b(x)$. Take two cyclic Fock representations $(\pi_m, \mathcal{H}_m, \Omega_m)$ corresponding to the free massive Hamiltonian $H_m = \int d^D x \, [\pi^2 + \phi(-\Delta + m^2)\phi]/2$, where Δ is the Laplacian, with different masses. It is easy to see that any Fock representation is weakly continuous and irreducible. Moreover, one can show that the vacuum state Ω_m is the only spatial translationally and rotationally invariant state in $\mathcal{H}_m$. The Euclidean group E is implemented unitarily on $\mathcal{H}_m$ by $u((\vec{c}, R))$ $\pi_m(U(a))\Omega_m = \pi_m(\alpha_{\vec{c},R}(U(a)))\Omega_m = \pi_m(U(a_{-\vec{c},R^{-1}}))\Omega_m$ and similarly for $V(b)$

where $a_{\vec{c},R}(\vec{x}) = a(R\vec{x} + \vec{c})$. This brings us into the situation of the following result [21, 22].

Theorem 8.1.2 (Haag's theorem). •••••••• •••••• ••• ••• • •••••• •••••••••
••• •••••••••• •••••••••••••••• $(\pi_I, \mathcal{H}_I), I = 1, 2$ •• ••• • ••• ••••••• $\mathfrak{A}$ ••• •
•••• ••[1] •••• •••••• ••• •••••••••• ••• ••• • ••••••••• •••••• E •• •••• ••••• •••• •• ••• •••• •
••• • •• •••••••• •• •• ••• •••• ••••• ••••••••• ••• • ••••• ••• •••• •••••• •• ••••••••••
••••• u_I •• $\mathcal{H}_I$ •••• •••• $u_I(e)\pi_I(a)u_I^{-1}(e) = \pi_I(\alpha_e(a))$ ••• • ••• $e \in E, a \in \mathfrak{A}$ •••
••• •• •••• •• • •• •••• • • •••••• •• ••••••• • •••• $\Omega_I \in \mathcal{H}_I$• ••• • ••• $u_I(e)\Omega_I = \Omega_I$•
•• ••• •• • •••••••••• ••••••• ••• • •• •• ••••• ••• ••••• •• •••••••••• •• •••• •••
•• ••• ••••••• • •• •• ••• ••••••• $W : \mathcal{H}_1 \to \mathcal{H}_2$ •••• •••• $W\pi_1(a)W^{-1} = \pi_2(a)$ •••
••• $a \in \mathfrak{A}$• •••• $Wu_1(e)W^{-1} = u_2(e)$ ••• ••• $e \in E$ ••• $V\Omega_1 = c\Omega_2$ ••••• c •• •
••• •••• ••• ••• ••• •••••• ••••

Notice that the notion of unitary equivalence does not require the representations to be cyclic and even if they are it does not mean that the cyclic states are related by W. Applied to our case we conclude that scalar field theories with different masses correspond to unitarily inequivalent representations of the Weyl algebra because if they were equivalent then by Haag's theorem $W\Omega_m = c\Omega_{m'}$ and $W\pi_m(a)W^{-1} = \pi_{m'}(a)$ for all $a \in \mathfrak{A}$. We would conclude, in particular, $\omega_m(a) = < \Omega_m, \pi_m(a)\Omega_m > = \omega_{m'}(a)$ for all $a \in \mathfrak{A}$. However, for instance $\omega_m(U(a)) = \exp(- < a, \sqrt{-\Delta + m^2}^{-1} a > /2)$ clearly depends on m, hence the representations are unitarily inequivalent. What is going on here is that the following canonical transformation, a special case of a Bogol'ubov transformation, between the annihilation and creation functions

$$\alpha_{mm'}(z_m) = \frac{1}{2} \left(\left[\sqrt{\frac{K_m}{K_{m'}}} + \sqrt{\frac{K_{m'}}{K_m}} \right] z_{m'} + \left[\sqrt{\frac{K_m}{K_{m'}}} - \sqrt{\frac{K_{m'}}{K_m}} \right] \overline{z_{m'}} \right) \quad (8.1.3)$$

with $z_m = (\sqrt{K_m}\phi - i\sqrt{K_m}^{-1}\pi)$, $K_m = \sqrt{-\Delta + m^2}$ cannot be implemented unitarily because its classical generator

$$C = \frac{i}{2} \int d^D k [z_{m'}\chi_{mm'}z_{m'} - \overline{z_{m'}}\chi_{mm'}\overline{z_{m'}}], \cosh(\chi_{mm'})$$

$$= \left[\sqrt{\frac{K_m}{K_{m'}}} + \sqrt{\frac{K_{m'}}{K_m}} \right] / 2 \quad (8.1.4)$$

is too singular, it is not even defined on the vacuum vector as one can easily check.

In particular, Haag's theorem implies that representations of interacting and free field theories are unitarily inequivalent and hence means that the interaction picture underlying perturbative QFT of Wightman fields strictly speaking does not exist, it exists only if there is no interaction.

---

[1] This can be generalised to arbitrary spin.

The discussion shows that there are many unitarily inequivalent representations of the Weyl algebra in QFT and the choice of the physically correct one needs additional dynamical input. In fact, in some cases one can prove [279] that the requirement that the automorphisms on the Weyl algebra, generated by the Hamiltonian flow on the classical phase space of a given Hamiltonian function, be implemented unitarily leads to the selection of a unique representation of the Weyl algebra. Hence we expect that for LQG we can arrive at a reasonably small, physically interesting subset of representations only if we impose (1) irreducibility, (2) weak continuity and (3) the unitary implementability of a physically interesting automorphism group of the Weyl algebra. The natural automorphism group to consider is, of course, the group $\mathfrak{G}$ of bundle automorphisms considered in the previous chapter.

Now we have already shown that there is no loss of generality in considering cyclic representations and that cyclic representations come from states. Cyclic representations are not necessarily irreducible because irreducibility means that •••• vector is cyclic, however, they provide good candidates for irreducible representations. Next, as shown in Section 29.1, if the state ω is $\mathfrak{G}$-invariant, that is, $\omega \circ \alpha_\mathfrak{g} = \omega$ for all $\mathfrak{g} \in \mathfrak{G}$ then $\mathfrak{G}$ is unitarily implementable in the corresponding GNS representation by $U_\omega(\mathfrak{g})\pi_\omega(a)\Omega_\omega = \pi_\omega(\alpha_\mathfrak{g}(a))\Omega_\omega$. Thus, to require $\mathfrak{G}$ invariance of the state is natural and sufficient to guarantee a unitary representation of $\mathfrak{G}$. Finally, as in the Stone–von Neumann theorem one might want to require that the Weyl algebra is represented weakly continuously. States whose GNS representation has this property are called regular.

Actually we will consider representations in which the continuity assumption on the Weyl algebra is slightly relaxed in one direction and slightly tightened in the other. Discontinuous representations were studied in QED, for instance in [539], and in string theory [205]. These representations are physically motivated, among other things, by the fact that they avoid the negative norm states (ghosts) of the more commonly known Gupta–Bleuler construction. These are representations which are unitarily inequivalent to Fock-type representations. Here we are forced to study such representations by background independence.

An important test of validity of such type of representations turns out to be parametrised field theory (PFT) [216, 540, 541] and the just mentioned LQG string [205]. Let us mention the salient features of these investigations.

(A) •••• •••••• ••• •••••• This is just a free (massive) scalar field theory on Minkowski space in any dimension on a manifold of topology $M \cong \mathbb{R} \times T^d$ but such that in the canonical $d+1$ split of the action, arbitrary space-like foliations are allowed. To do this, one turns the one-parameter family of embeddings $t \mapsto X_t^\mu = X^\mu(t,.)$ of the d-torus T^d of the spacetime manifold into a dynamical variable with conjugate momentum P_μ. These extra degrees of freedom are eliminated by imposing $d+1$ first-class constraints $C_\mu := P_\mu + H_\mu = 0$ where $X^\mu_{,a}H_\mu$ would be the scalar field contribution to the spatial diffeomorphism constraint while $n^\mu H_\mu$ would be the contribution to the Hamiltonian constraint if the metric were considered as a dynamical field (see Chapter 12). The Poisson algebra of these constraints is isomorphic to the diffeomorphism algebra diff(M). The interesting question is now whether the Schrödinger picture and the Heisenberg picture are related by a unitary transformation for an arbitrary foliation as we are used to from foliations that are generated by Poincaré transformations.

The surprising answer is that for $d > 1$ this is not the case [542–544]. More in detail, consider some initial slice Σ_0 of a foliation $t \mapsto X_t$. Then the Heisenberg representation is the one obtained by classically evolving the embedding-dependent creation and annihilation operators according to this foliation while leaving the Fock vacuum at $t = 0$ invariant. The Schrödinger picture is obtained by evolving the Fock vacuum at $t = 0$ with the canonical Hamiltonian[2] operator corresponding to that foliation while leaving the creation and annihilation operators at $t = 0$ invariant. Interestingly, these explicitly embedding X-dependent states solve the quantum constraint equations $C_\mu = 0$ ⋯ ⋯. By formally we mean that we represent X^μ as a multiplication operator and P_μ by functional differentiation with respect to X^μ while H_μ is represented on Fock space as usual (notice that H_μ involves X but not P).

However, no rigorous Dirac quantisation of the embedding variables is provided in [542–544]. The beauty is now that a rigorous quantisation which precisely uses this type of discontinuous representations mentioned before and application of the group averaging techniques to solve the constraints provides a physical Hilbert space which is unitarily equivalent to the Fock space representation on flat slices [545], such that on the corresponding kinematical Hilbert space (a certain enlargement of a restriction of) Diff(M) is represented without anomalies. This underlines the power of background-independent quantisation techniques, which naturally lead to discontinuous representations[3] and removes this so-called ⋯⋯⋯ obstruction which otherwise would seem to imply that Dirac quantisation has no chance to deliver representations that are unitarily equivalent to usual background-dependent (Fock) representations.

(B) ⋯⋯ Using discontinuous representations one can quantise the closed bosonic string in any spacetime dimension without encountering

---

[2] That is, the linear combination of the contributions to the Hamiltonian and spatial diffeomorphism constraint of the scalar field with metric fixed to be the Minkowski metric.

[3] In order to achieve this one must exponentiate diff(M) which would result in the connected component of Diff(M). However, this group does not preserve the set of spacelike embeddings [216, 217]. Hence one should restrict to the subgroup, if it exists, of diffeomorphisms which act transitively on the space of spacelike embeddings. Since it is far from obvious that such a group exists it is safer to extend Diff(M) to the symmetric (permutation) group of all spacelike embeddings. This enlargement is similar to the combinatorial extension discussed in [436].

anomalies, ghosts (negative norm states) or a tachyon state (instabilities). The representation-independent and purely algebraic no-go theorem of [546] that the Virasoro anomaly is unavoidable is circumvented by quantising the Witt group $\mathrm{Diff}(S^1)\times \mathrm{Diff}(S^1)$ rather than its algebra $\mathrm{diff}(S^1)\oplus\mathrm{diff}(S^1)$. Since the representation of the Witt group is discontinuous, the infinitesimal generators do not exist and there is no Virasoro algebra in this discontinuous representation, exactly like in LQG. However, as in LQG, a unitary representation of the Witt group is sufficient in order to obtain the Hilbert space of physical states via group averaging techniques and even a representation of the invariant charges [189, 204] of the closed bosonic string.

This representation of the string has been much discussed and criticised in various physics forums. We discuss here two of the most debated questions.

(i) A folklore statement that seems to have entered several physics blogs is that weakly discontinuous representations of the kind used in LQG do not work for the harmonic oscillator so why should they work for more complicated theories? This is also the conclusion reached in [547]. As we will now show, while [547] is technically correct, its physical conclusion is false. In [547] one used a representation discussed first for QED [539] in order to avoid the negative norm states of the Gupta–Bleuler formulation. In this representation neither position q nor momentum p operators are well-defined, only the Weyl operators $U(a) = \exp(iaq), V(b) = \exp(ibp)$ exist. Hence the usual harmonic oscillator Hamiltonian $H = q^2 + p^2$ does not exist in this representation. Consider the substitute $H_\epsilon = [\sin^2(\epsilon q) + \sin^2(\epsilon p)]/\epsilon^2$. What is pointed out in [547] is that this operator is ill-defined as $\epsilon \to 0$. This is no surprise, we knew this without calculation, the representation is not weakly continuous after all. However, what is physically much more interesting is the following. Fix an energy level E_0 above which the harmonic oscillator becomes relativistic and thus becomes inappropriate to model the correct physics. Let[4] $a_\epsilon^\dagger := [\sin(q\epsilon) + i\sin(p\epsilon)]/\epsilon$. Consider the finite number of observables

$$b_{\epsilon,n} := \frac{1}{n!}(a_\epsilon)^n \, (a_\epsilon^\dagger a_\epsilon)(a_\epsilon^\dagger)^n, \ n = 0, \ldots, N = E_0/\hbar \qquad (8.1.5)$$

Let Ω_0 be the Fock vacuum in the Schrödinger representation and ω the state underlying the discontinuous representation. Fix a finite measurement precision δ. Since the Fock representation is weakly continuous we find $\epsilon_0(N, \delta)$ such that $|<\Omega_0, b_{\epsilon,n}\Omega_0> -n\hbar| < \delta/2$ for all $\epsilon \le \epsilon_0$. On the other hand, by Fell's theorem, Theorem 29.1.4, applicable to the unique C^*-algebra [548] generated by the Weyl operators $U(a)$, $V(b)$ and the faithful representation considered in [547], we find a trace class

---

[4] Notice that classically $H_\epsilon = |a_\epsilon|^2$.

operator $\rho_{N,\delta}$ in the GNS representation determined by ω such that $|\text{Tr}(\rho_{N,\delta} b_{\epsilon_0,n}) - <\Omega_0, b_{\epsilon_0,n}\Omega_0>| < \delta/2$ for all $n = 0, 1, \ldots, N$. It follows that with arbitrary, finite precision $\delta > 0$ we find states in the Fock and discontinuous representations respectively whose energy expectation values are given with precision δ by the usual value $n\hbar$. This implies that the two states cannot be physically distinguished.

In [549,550] even more was shown:[5] there the spectrum of the operator H_ϵ was studied and the eigenvalues and eigenvectors were determined explicitly. One could show that by tuning ϵ according to N, δ even the first N eigenvalues do not differ more than δ from $(n+1)\hbar$. Moreover, having fixed such an ϵ, the non-separable Hilbert space is a direct sum of separable H_ϵ-invariant subspaces and if we just consider the algebra generated by a_ϵ each of them is superselected. Hence we may restrict to any one of these irreducible subspaces and conclude that the physics of the discontinuous representation is indistinguishable from the physics of the Schrödinger representation within the error δ. This should be compared with the statement found in [547] that in discontinuous representations the physics of the harmonic oscillator is not correctly reproduced.

(ii) Does this mean that the magical dimension $D = 26$ cannot be seen in this representation? Of course it can: one way to detect it in the usual Fock representation of the string is by considering the Poincaré algebra (in the lightcone gauge) and ask that it closes. For the LQG string [205] again the Poincaré group is represented unitarily but weakly discontinuously. However, we can approximate the generators as above in terms of the corresponding Weyl operators using some tiny but finite parameter ϵ. Since these are a finite number of operators in the corresponding C*-algebra, an appeal to Fell's theorem and using continuity of the Weyl operators in the Fock representation guarantees that we find a state in the folium of the LQG string with respect to which the expectation values of the approximate Poincaré generators coincide with their vacuum (or higher excited state) expectation values in the Fock representation to arbitrary precision δ.

Thus $D = 26$ is also hidden in this discontinuous representation, it is just that for no D there is a quantisation obstruction. Of course, much still has to be studied for the LQG string, for example, a formulation of scattering theory, however, the purpose of [205] was not to propose a phenomenologically interesting model but rather to indicate that $D = 26$ is not necessarily sacred but rather a feature of the specific Fock quantisation used.

---

[5] In a representation which was continuous in one of p or q but discontinuous in the other. But similar results hold in this completely discontinuous representation considered here.

This provides sufficient motivation for allowing discontinuous representations. More precisely, in LQG on the one hand we •• ••• require that the Abelian subalgebra Cyl is represented weakly continuously. By this we mean that matrix elements of cylindrical functions do not need to be continuous under continuous deformations (homotopies) of edges to points. On the other hand, we not only require that the exponentiated fluxes (and the exponentials of more general vector fields in $\mathfrak{P}$) are represented weakly continuously but actually •• •• •• ••. By this we mean that the cyclic GNS vector Ω_ω is a common C^∞-vector for all the $W_t^n(S)$ as S, n vary. Recall that a vector ψ is said to be a C^∞-vector for a weakly continuous, one-parameter group of unitarities $t \mapsto U(t)$ iff it is in the common invariant domain of all powers of the corresponding self-adjoint generator A defined by $U(t) = \exp(itA)$. One can show that the set of C^∞-vectors for •• • generator is dense. What is less trivial to show is that this is still the case when we have an infinite number of generators. We call a representation of this kind a semi-weakly smooth representation. Then the following theorem holds.

Theorem 8.1.3. • •••• •• • •• •• ••• ••• ••• •• •• ••• ••• ••• $\mathfrak{G}$••• •• ••• • ••• •• ••
••• ••• ••• ••• • ••• •••• ••• ••• •• •••• ••• $\mathfrak{A}$• • •• •• •• ••• •• ••• ••• •• ••• •••
• • • •• ••• •• ••• •• •• •••• • •• ••• ••• •

In the existence part of the proof we follow [364–369] and describe the representation in great detail, since it is a fundamental building block of the current formulation of LQG. We develop a general framework which is useful, for instance, for studying states on $\mathfrak{A}$ which are not necessarily $\mathfrak{G}$-invariant. Namely, while $\mathfrak{G}$-invariant states are natural in LQG, other representations might eventually be required if one cannot complete the programme based on the invariant one.[6] After all, since even the invariant state only results in a kinematical representation in which none of the constraints has been implemented one might want to consider other (cyclic) representations which also do not implement the constraints and in which $\mathfrak{G}$ is not implemented unitarily or maybe projectively (no invariant vector Ω).

We also add supplementary material which is not needed on a first reading, marked by $^+$. In the uniqueness and irreducibility part we follow [519].

8.2 Uniqueness proof: (1) existence

The aim of this section is to prove the following theorem.

Theorem 8.2.1. • •• •• ••• ••• • • •• •• ••• •• ••• $\mathcal{H}_0 := L(\overline{\mathcal{A}}, d\mu_0)$ • •• •• $\overline{\mathcal{A}}$ •• •••
••• •• •• ••• ••• ••• ••• •• ••• • ••• •• ••• •• • ••• ••• ••• ••• ••• • ••• •••

---

[6] To avoid confusion, notice that a $\mathfrak{G}$-invariant state does not mean that the corresponding GNS representation contains only $\mathfrak{G}$-invariant vectors, it just means that there is a unitary representation of $\mathfrak{G}$.

•• •••• ••• • •••• σ•••••••• ••• μ_0 •• ••• • • ••• •• • •••••• •••••• •• • ••••••••

• •• ••• • • •••••• ••• •• •••• ••• • •••••••••• •••••• •• ••• ••••• •••••••• • $\mathcal{D}$ ••••

•• ••• • •••• •••••• ••• • •• •• ••• •• •••• •• •• ••• •• •••••• •••••• • • ••• ••••• ••••••

••• ••• ••••• • •• •••• •••• ••• •• •••• •••••• •• • ••••• ••

$$\pi_0(f) \cdot \psi := f\,\psi$$
$$\pi_0(Y_n(S)) \cdot \psi := i\hbar\kappa\beta Y_n(S)[\psi] \tag{8.2.1}$$

• •••• ••• ••••••••• • •••• •• ••• ••••• •• ••• •••••• ••• $Y_n(S)$ •• ••• •• ••••

$\psi \in \mathcal{D}$• • ••• π_0 •• • •••••••••••• •• $\mathfrak{A}$• •• ••••• • ••••• •••• • •••• ••

••• •• • ••••• •••• •••• • •••••• ••••• • •• •••••••••• ••• •••••• •• ••••• •• ••• •••

•••• •••• ••• • ••••••• •••• •••• • •••• ••• * ••••••• ••

That the canonical commutation relations hold is immediately clear because the definition of π_0 is such that the operators $\pi_0(Y_n(S))$ act as the derivations in $\mathfrak{P}$. The non-trivial part of the proof is in the rigorous construction of the measure μ_0 and to show that $\pi_0(Y_n(S))$ is (essentially) self-adjoint.

The proof of this theorem again naturally breaks into several steps which we provide in the following subsections. For the existence proof alone not all of this material is needed, but it is natural to provide it here because we will need properties of the measure μ_0 in later chapters of the book.

8.2.1 Regular Borel measures on the projective limit: the uniform measure

In this subsection we describe a simple mechanism, based on the Riesz representation theorem, of how to construct σ-additive measures on the projective limit $\overline{X}$ starting from a so-called self-consistent family of (so-called cylindrical) measures μ_l on the various X_l. See Chapter 25 for some useful measure-theoretic terminology and the references cited there for further reading. We follow again closely Ashtekar and Lewandowski [364, 367, 368].

Our spaces X_l are compact Hausdorff spaces and in particular topological spaces and are therefore naturally equipped with the σ-algebra $\mathcal{B}_l$ of Borel sets (the smallest σ-algebra containing all open (equivalently closed) subsets of X_l). Let μ_l be a positive, regular, Borel, probability measure on X_l, that is, a positive semidefinite, σ-additive function on $\mathcal{B}_l$ with $\mu_l(X_l) = 1$ and regularity means that the measure of every measurable set can be approximated arbitrarily well by open and compact sets (hence closed since X_l is compact Hausdorff) respectively. Since the measure is Borel, the continuous functions $C(X_l)$ are automatically measurable.

Definition 8.2.2. • ••• ••• •• • •••••••• $(\mu_l)_{l \in \mathcal{L}}$ •• ••• •••••••••••• X_l •• •
••• ••••••• ••• •• ••• $(X_l, p_{ll'})_{l \prec l' \in \mathcal{L}}$ • •••• ••• $p_{l'l} : X_{l'} \to X_l$ ••• ••• ••• •••• ••

$$(p_{l'l})_* \mu_{l'} := \mu_{l'} \circ p_{l'l}^{-1} = \mu_l \qquad (8.2.2)$$

$$\cdots \cdots l \prec l' \cdots \cdots (p_{l'l})_* \mu_{l'} \cdots X_l \cdots \cdots \cdots \cdots \cdots \cdots \cdots$$
$$\cdots \cdots \mu_{l'}.$$

The meaning of condition (8.2.2) is the following: let $\mathcal{B}_l \ni U_l \subset X_l$ be measurable. Since $p_{l'l}$ is continuous the pre-images of open sets in X_l are open in $X_{l'}$ and therefore measurable, hence $p_{l'l}$ is measurable. Since U_l is generated from countable unions and intersections of open sets it follows that $p_{l'l}^{-1}(U_l)$ is measurable. Then we require that

$$\mu_{l'}\left(p_{l'l}^{-1}(U_l)\right) = \mu_l(U_l) \qquad (8.2.3)$$

for any measurable U_l. We can rewrite condition (8.2.3) in the form

$$\int_{X_{l'}} d\mu_{l'}(x_{l'}) \chi_{p_{l'l}^{-1}(U_l)}(x_{l'}) = \int_{X_l} d\mu_l(x_l) \chi_{U_l}(x_l) \qquad (8.2.4)$$

where χ_S denotes the characteristic function of a set S. Here it is strongly motivated to have surjective projections $p_{l'l}$ as otherwise $p_{l'l}^{-1}(X_l)$ is a proper subset of $X_{l'}$ so that $1 = \mu_l(X_l) = \mu_{l'}(p_{l'l}^{-1}(X_l))$ could give a contradiction with the μ_l being probability measures if $X_{l'} - p_{l'l}^{-1}(X_l)$ is not a set of measure zero with respect to $\mu_{l'}$.

Condition (8.2.4) extends linearly to linear combinations of characteristic functions, so-called simple functions (see Chapter 25) and the (Lebesgue) integral of any measurable function is defined in terms of simple functions (see Chapter 25). Therefore we may equivalently write (8.2.2) as

$$\int_{X_{l'}} d\mu_{l'}(x_{l'}) [p_{l'l}^* f_l](x_{l'}) = \int_{X_l} d\mu_l(x_l) f_l(x_l) \qquad (8.2.5)$$

for any $l \prec l'$ and any $f_l \in C(X_l)$ since every measurable function can be approximated by simple functions and measurable simple functions can be approximated by continuous functions by Lusin's theorem, Theorem 25.1.14 (which are automatically measurable). In the form (8.2.5) the consistency condition means that integrating out the degrees of freedom in $X_{l'}$ on which $p_{l'l}^* f_l$ does not depend, we end up with the same integral as if we had integrated over X_l only.

To summarise: let $f = [f_l]_\sim \in \mathrm{Cyl}(\overline{X})$ with $f_l \in C(X_l)$. Then (8.2.5) ensures that the linear functional

$$\Lambda: \mathrm{Cyl}(\overline{X}) \to \mathbb{C}; \ f = [f_l]_\sim \mapsto \Lambda(f) := \int_{X_l} d\mu_l(x_l) f_l(x_l) \qquad (8.2.6)$$

is well-defined, that is, independent of the representative $f_l \sim p_{l'l}^* f_l$ of f. Moreover, it is a positive linear functional (integrals of positive functions are positive) because the μ_l are positive measures. Since $\mathrm{Cyl}(\overline{X}) \subset \overline{\mathrm{Cyl}(\overline{X})}$ is a subset of a unital C^*-algebra, Λ is automatically continuous (see the end of Section 25.1) and therefore extends uniquely and continuously to the completion $\overline{\mathrm{Cyl}(\overline{X})}$ by the bounded linear transformation theorem, Theorem 26.1.8. Now in Sections 6.2.2, 6.2.3 we showed that the Gel'fand isomorphism applied to $\overline{\mathrm{Cyl}(\overline{X})}$ leads to an (isometric) isomorphism of $\overline{\mathrm{Cyl}(\overline{X})}$ with $C(\overline{X})$ given by

$$\bigvee : \ \mathrm{Cyl}(\overline{X}) \to C(\overline{X}); \ f = [f_l]_\sim \mapsto p_l^* f_l \tag{8.2.7}$$

(and extended to $\overline{\mathrm{Cyl}(\overline{X})}$ using that $\mathrm{Cyl}(\overline{X})$ is dense). It follows that we may consider (8.2.5) as a positive linear functional on $C(\overline{X})$. Since $\overline{X}$ is a compact Hausdorff space we are in a position to apply the Riesz–Markov (or representation) theorem.

Theorem 8.2.3. $\bullet \bullet \bullet$ $(X_l, p_{l'l})_{l \prec l' \in \mathcal{L}}$ $\bullet \bullet$ $\bullet$ $\bullet \bullet \bullet$ $\bullet \bullet \bullet \bullet$ $\bullet \bullet \bullet \bullet \bullet \bullet \bullet \bullet$ $\bullet \bullet \bullet \bullet \bullet \bullet \bullet \bullet \bullet$ $\bullet \bullet \bullet$ $\bullet \bullet \bullet$ $\bullet \bullet \bullet$ $\bullet \bullet \bullet \bullet \bullet \bullet \bullet$ $\bullet \bullet \bullet \bullet \bullet \bullet \bullet \bullet \bullet \bullet$ $p_{l'l} : \ X_{l'} \to X_l \bullet$ $\bullet \bullet \bullet \bullet \bullet \bullet \bullet \bullet$ $\bullet \bullet \bullet \bullet$ $\overline{X}$ $\bullet \bullet \bullet$ $\bullet \bullet \bullet \bullet \bullet \bullet \bullet \bullet \bullet \bullet$ $p_l : \ \overline{X} \to X_l \bullet$

$\bullet \bullet \bullet$ $\bullet \bullet$ μ $\bullet \bullet$ $\bullet$ $\bullet \bullet \bullet \bullet \bullet \bullet \bullet$ $\bullet$ $\bullet \bullet \bullet \bullet$ $\bullet \bullet \bullet \bullet \bullet \bullet \bullet \bullet \bullet$ $\bullet$ $\bullet \bullet \bullet \bullet \bullet \bullet \bullet$ $\bullet \bullet$ $\overline{X}$ $\bullet \bullet \bullet$ $(\mu_l := \mu \circ p_l^{-1})_{l \in \mathcal{L}}$
$\bullet \bullet \bullet \bullet \bullet \bullet$ $\bullet$ $\bullet \bullet \bullet \bullet \bullet \bullet \bullet \bullet \bullet$ $\bullet \bullet \bullet$ $\bullet \bullet \bullet$ $\bullet \bullet$ $\bullet \bullet \bullet \bullet \bullet \bullet$ $\bullet$ $\bullet \bullet \bullet \bullet$ $\bullet \bullet \bullet \bullet \bullet \bullet \bullet \bullet \bullet$ $\bullet$ $\bullet \bullet \bullet \bullet \bullet \bullet \bullet$ $\bullet \bullet$ $X_l \bullet$
$\bullet \bullet \bullet \bullet$ $\bullet \bullet$ $(\mu_l)_{l \in \mathcal{L}}$ $\bullet \bullet \bullet \bullet \bullet \bullet$ $\bullet$ $\bullet \bullet \bullet \bullet \bullet \bullet \bullet \bullet$ $\bullet \bullet \bullet$ $\bullet \bullet \bullet$ $\bullet \bullet$ $\bullet \bullet \bullet \bullet \bullet \bullet$ $\bullet$ $\bullet \bullet \bullet \bullet$ $\bullet \bullet \bullet \bullet \bullet \bullet \bullet$ $\bullet$ $\bullet \bullet \bullet \bullet \bullet \bullet \bullet$
$\bullet \bullet$ X_l $\bullet \bullet \bullet \bullet$ $\bullet \bullet \bullet \bullet \bullet$ $\bullet \bullet \bullet \bullet \bullet \bullet$ $\bullet$ $\bullet \bullet \bullet \bullet \bullet \bullet \bullet \bullet$ $\bullet \bullet \bullet \bullet \bullet \bullet$ $\bullet$ $\bullet \bullet \bullet \bullet$ $\bullet \bullet \bullet \bullet \bullet \bullet \bullet \bullet \bullet$ $\bullet$ $\bullet \bullet \bullet \bullet \bullet \bullet \bullet$ $\mu \bullet \bullet$ $\overline{X}$
$\bullet \bullet \bullet \bullet$ $\bullet \bullet \bullet \bullet$ $\mu \circ p_l^{-1} = \mu_l \bullet$
$\bullet \bullet \bullet \bullet \bullet \bullet$ $\bullet \bullet \bullet$ $\bullet$ $\bullet \bullet \bullet \bullet \bullet \bullet$ μ $\bullet \bullet$ $\bullet \bullet \bullet \bullet \bullet \bullet$ $\bullet \bullet$ $\bullet \bullet \bullet$ $\bullet \bullet \bullet$ $\bullet \bullet \bullet \bullet \bullet$ μ_l $\bullet \bullet$ $\bullet \bullet \bullet \bullet \bullet \bullet$ $\bullet \bullet$

$\bullet \bullet \bullet \bullet \bullet$

(i) Define the positive linear functional on $C(X_l)$

$$\Lambda_l : \ C(X_l) \to \mathbb{C}; \ f_l \mapsto \int_{\overline{X}} d\mu(x)(p_l^* f_l)(x) \tag{8.2.8}$$

which satisfies $\Lambda_l(1) = 1$. Since X_l is a compact Hausdorff space, by the Riesz representation theorem there exists a unique, positive, regular Borel probability measure μ_l on X_l that represents Λ_l, that is

$$\Lambda_l(f_l) = \int_{X_l} d\mu_l(x_l) f_l(x_l) \tag{8.2.9}$$

Since $p_{l'l} \circ p_{l'} = p_l$, the consistency condition (8.2.5) is obviously met.
(ii) As was shown above, the positive linear functional on $C(\overline{X})$

$$\Lambda : \ C(\overline{X}) \to \mathbb{C}; \ f = p_l^* f_l \equiv [f_l]_\sim \mapsto \int_{X_l} d\mu_l(x_l) f_l(x_l) \tag{8.2.10}$$

is well-defined due to the consistency condition and satisfies $\Lambda(1) = 1$. Since $\overline{X}$ is a compact Hausdorff space the Riesz representation theorem guarantees the existence of a unique, positive, regular Borel probability measure μ on $\overline{X}$ representing Λ, that is

$$\Lambda(f) = \int_{\overline{X}} d\mu(x) f(x) \tag{8.2.11}$$

(iii) Consider $f \in C(\overline{X})$ of the form $f = p_l^* f_l$ for some $l \in \mathcal{L}$, $f_l \in C(X_l)$. Functions of the form $p_l^* f_l$ lie dense in $C(\overline{X})$. Now $f = p_l^* f_l$ is non-negative iff f_l is non-negative because p_l is a surjection. It follows that we can restrict attention to all non-negative functions of the form $f = p_l^* f_l$ for arbitrary $f_l \in C(X_l)$, $l \in \mathcal{L}$ as far as faithfulness is concerned. Let $\Lambda_\mu, \Lambda_{\mu_l}$ be the positive linear functionals determined by μ, μ_l respectively. Then: μ faithful $\Leftrightarrow \Lambda_\mu(p_l^* f_l) = \Lambda_{\mu_l}(f_l) = 0$ for any non-negative $f_l \in C(X_l)$ and any $l \in \mathcal{L}$ implies $f = p_l^* f_l = 0 \Leftrightarrow$ for any $l \in \mathcal{L}$ and any non-negative $f_l \in C(X_l)$ the condition $\Lambda_{\mu_l}(f_l)$ implies $f_l = 0 \Leftrightarrow$ all μ_l are faithful. $\qquad\square$

We now define a natural measure on the spectrum of interest namely $\overline{A}$, the so-called uniform measure. To do this we must specify the space of cylindrical functions. Given a subgroupoid $l \in \mathcal{L}$ with $l = l(\gamma)$ we think of an element $x_l \in X_l$ as a collection of group elements $\{x_l(e)\}_{e \in E(\gamma)} = \rho_l(x_l)$ and X_l can be identified with $G^{|E(\gamma)|}$ (see (6.2.16)). Thus, an element $f_l \in C(X_l)$ is simply given by

$$f_l(x_l) = F_l(\{x_l(e)\}_{e \in E(\gamma)}) = (\rho_l^* F_l)(x_l) \tag{8.2.12}$$

where F_l is a continuous complex-valued function on $G^{|E(\gamma)|}$. For $l \prec l'$ with $l = l(\gamma), l' = l(\gamma')$ we define $\rho_{l'l} : G^{|E(\gamma')|} \to G^{|E(\gamma)|}$ by $\rho_l \circ p_{l'l} = \rho_{l'l} \circ \rho_{l'}$ (recall that ρ_l is a bijection).

Definition 8.2.4. $\cdots \mathcal{L} \cdots \cdots \cdots \cdots \cdots \cdots \cdots \cdots \cdots \cdots \cdots$
$\cdots \cdots \cdots \cdots \mathcal{P} \cdots \sigma \cdots X_l = \cdots (l, G) \cdots \cdots \cdots \cdots G^{|E(\gamma)|} \cdots l = l(\gamma) \cdots$
$\cdots \cdots \cdots \cdots \cdots \cdots \cdots \cdots \cdots f \in C(X_l)$

$$\mu_{0l}(f_l) = \int_{X_l} d\mu_{0l}(x_l) \rho_l^* F_l(x_l) := \int_{G^{|E(\gamma)|}} \left[\prod_{e \in E(\gamma)} d\mu_H(h_e) \right] F_l(\{h_e\}_{e \in F(\gamma)})$$

$$\tag{8.2.13}$$

$\cdots \cdots \mu_H \cdots \cdots \cdots \cdots \cdots \cdots \cdots \cdots \cdots \cdots \cdots \cdots \cdots \cdots \cdots \cdots \cdots$
$\cdots \cdots \cdots \cdots \cdots \cdots \cdots \cdots \cdots \cdots \cdots \cdots \cdots \cdots \cdots \cdots \cdots \cdots \cdots \cdots$

Lemma 8.2.5. $\cdots \cdots \cdots \cdots \cdots \cdots \cdots \mu_{0l} \cdots \cdots \cdots \cdots \cdots \cdots \cdots \cdots \cdots$
$\cdots \cdots \cdots \cdots \cdots$

$\bullet \cdots \bullet$ That μ_l defines a positive linear functional follows from the explicit formula (8.2.12) in terms of the positive Haar measure on G^n. That $(\mu_{0l})_{l \in \mathcal{L}}$ defines a consistent family follows from the observation that if $l \prec l'$ with $l = l(\gamma), l' = l(\gamma')$ then we can reach l from l' by a finite combination of the following three steps:

(a) $e_0 \in E(\gamma')$ but $e_0 \cap \gamma \subset \{b(e_0), f(e_0)\}$ (deletion of an edge).
(b) $e_0 \in E(\gamma')$ but $e_0^{-1} \in E(\gamma)$ (inversion of an edge).
(c) $e_1, e_2 \in E(\gamma')$ but $e_0 = e_1 \circ e_2 \in E(\gamma)$ (composition of edges).

It therefore suffices to establish consistency with respect to all of these elementary steps.

In general we have

$$p^*_{l'l} f_l = p^*_{l'l} \rho^*_l F_l = \rho^*_{l'} \rho^*_{l'l} F_l \tag{8.2.14}$$

whence

$$\mu_{0l'}(p^*_{l'l} f_l) = \mu_{0l'}(\rho^*_{l'} [\rho^*_{l'l} F_l]) = \int_{G^{|E(\gamma')|}} \left[\prod_{e \in E(\gamma')} d\mu_H(h_e) \right] [\rho^*_{l'l} F_l](\{h_e\}_{e \in E(\gamma')}) \tag{8.2.15}$$

In what follows we will interchange freely orders of integration and break the integral over G^n in integrals over G^m, G^{n-m}. This is allowed by Fubini's theorem (see Theorem 25.1.6) since the integrand, being bounded, is absolutely integrable in any order.

(a) We have $\rho_{l'l}(\{h_e\}_{e \in E(\gamma')}) = \{h_e\}_{e \in E(\gamma)}$ thus

$$\mu_{0l'}(p^*_{l'l} f_l) = \left\{ \int_{G^{|E(\gamma)|}} \left[\prod_{e \in E(\gamma)} d\mu_H(h_e) \right] F_l(\{h_e\}_{e \in E(\gamma)}) \right\}$$

$$\times \left\{ \int_G d\mu_H(h_{e_0}) \, 1 \right\} = \mu_{0l}(f_l) \tag{8.2.16}$$

since μ_H is a probability measure.

(b) We have $\rho_{l'l}(\{h_e\}_{e \in E(\gamma')}) = \{\{h_e\}_{e \in E(\gamma)-\{e_0\}}, h^{-1}_{e_0}\}$ thus

$$\mu_{0l'}(p^*_{l'l} f_l)$$

$$= \int_{G^{|E(\gamma)|-1}} \left[\prod_{e \in E(\gamma)-\{e_0\}} d\mu_H(h_e) \right] \int_G d\mu_H(h_{e_0}) F_l(\{h_e\}_{e \in E(\gamma)-\{e_0\}}, h^{-1}_{e_0})$$

$$= \int_{G^{|E(\gamma)-1|}} \left[\prod_{e \in E(\gamma)-\{e_0\}} d\mu_H(h_e) \right] \int_G d\mu_H(h^{-1}_{e_0}) F_l(\{h_e\}_{e \in E(\gamma)-\{e_0\}}, h^{-1}_{e_0})$$

$$= \int_{G^{|E(\gamma)|}} \left[\prod_{e \in E(\gamma)} d\mu_H(h_e) \right] F_l(\{h_e\}_{e \in E(\gamma)}) = \mu_{0l}(f_l) \tag{8.2.17}$$

since the Jacobian of the Haar measure with respect to the inversion map on G equals unity and where we have defined a new integration variable $h_{e_0^{-1}} := h^{-1}_{e_0}$.

(c) We have $\rho_{l'l}(\{h_e\}_{e\in E(\gamma')}) = \{\{h_e\}_{e\in E(\gamma)-\{e_0\}}, h_{e_1}h_{e_2}\}$ thus

$\mu_{0l'}(p^*_{l'l}f_l)$

$$= \int_{G^{|E(\gamma)|-1}} \left[\prod_{e\in E(\gamma)-\{e_0\}} d\mu_H(h_e)\right] \int_{G^2} d\mu_H(h_{e_1})d\mu_H(h_{e_2})$$
$$\times F_l(\{h_e\}_{e\in E(\gamma)-\{e_0\}}, h_{e_1}h_{e_2})$$

$$= \int_{G^{|E(\gamma)|-1}} \left[\prod_{e\in E(\gamma)-\{e_0\}} d\mu_H(h_e)\right] \int_G d\mu_H(h_{e_1}) \int_G d\mu_H(h_{e_1}^{-1}h_{e_1\circ e_2})$$
$$\times F_l(\{h_e\}_{e\in E(\gamma)-\{e_0\}}, h_{e_1\circ e_2})$$

$$= \int_{G^{|E(\gamma)|-1}} \left[\prod_{e\in E(\gamma)-\{e_0\}} d\mu_H(h_e)\right] \int_G d\mu_H(h_{e_1\circ e_2})F_l(\{h_e\}_{e\in E(\gamma)-\{e_0\}}, h_{e_1\circ e_2})$$
$$\times \left[\int_G d\mu_H(h_{e_1})1\right]$$

$$= \int_{G^{|E(\gamma)|}} \left[\prod_{e\in E(\gamma)} d\mu_H(h_e)\right] F_l(\{h_e\}_{e\in E(\gamma)}) = \mu_{0l}(f_l) \tag{8.2.18}$$

since the Jacobian of the Haar measure with respect to the left or right translation map on G equals unity and where we have defined a new integration variable by $h_{e_1\circ e_2} := h_{e_1}h_{e_2}$. □

It follows from Theorem 8.2.3 that the family (μ_{0l}) defines a regular Borel probability measure on $\overline{X}$.

We can now equip the quantum configuration space $\overline{\mathcal{A}}$ with a Hilbert space structure.

Definition 8.2.6. ••• • •••••• ••••• $\mathcal{H}_0$ •• •••••• •• ••• ••••• •• ••••• •• •••
•••• •• ••• ••••• •• •••• $\overline{\mathcal{A}}$ • ••• •••••••• •• ••• •• •••••• • ••••••• μ_0 ••••• ••

$$\mathcal{H}_0 := L_2(\overline{\mathcal{A}}, d\mu_0) \tag{8.2.19}$$

Notice that since we have identified cylindrical functions over $\overline{\mathcal{A}/\mathcal{G}}$ with gauge-invariant, cylindrical functions over $\overline{\mathcal{A}}$ the measure μ_0 can also be defined as a measure on $\overline{\mathcal{A}/\mathcal{G}}$: simply restrict the μ_{0l} to the invariant elements, which still defines a positive linear functional on $C([X_l]_l)$, and then use the Riesz representation theorem. It is easy to check that the obtained measure coincides with the restriction of μ_0 to $\overline{\mathcal{A}/\mathcal{G}}$ with σ-algebra given by the sets $U\cap\overline{\mathcal{A}/\mathcal{G}}$ where U is measurable in $\overline{\mathcal{A}}$. We will denote the restricted and unrestricted measure by the same symbol μ_0.

In the next section we introduce useful machinery which allows us to define momentum operators from derivations.

8.2.2 Functional calculus on a projective limit

This subsection rests on the simple but powerful observation that in the case of interest the projections $p_{l'l}$ are not only continuous and surjective but also analytic. This can be seen by using the bijection (6.2.16) between X_l and $\mathbf{G}^n$ for some n and using the standard differentiable structure on $\mathbf{G}^n$. We follow closely Ashtekar and Lewandowski [369] once more.

•••••••••

We have seen that we can identify $C(\overline{X})$ with the (completion of the) space of cylindrical functions $f = [f_l]/\sim= p_l^* f_l$, $f_l \in C(X_l)$. This suggests proceeding analogously with the other differentiability categories. Let $n \in \{0, 1, 2, \ldots\} \cup \{\infty\} \cup \{\omega\}$, then we define

$$\mathrm{Cyl}^n(\overline{X}) := \left(\bigcup_{l \in \mathcal{L}} C^n(X_l) \right) \Big/ \sim \tag{8.2.20}$$

That is, a typical element $f = [f_l]_\sim \in \mathrm{Cyl}^n(\overline{X})$ can be thought of as an equivalence class of elements of the form $f_l \in C^n(X_l)$ where $f_l \sim f_{l'}$ iff there exists $l, l' \prec l''$ such that $p_{l''l}^* f_l = p_{l''l'}^* f_{l'}$. As in the previous subsection, the existence of one such l'' implies that this equation holds for all $l, l' \prec l''$. Notice that $f_l \in C^n(X_l)$ implies $p_{l'l}^* f_l \in C^n(X_{l'})$ due to the analyticity of the projections, this is where their analyticity becomes important. Notice that differentiability here means differentiability of the representatives f_l on the respective power of $\mathbf{G}^n$. The space $\overline{\mathcal{A}}$ does not carry a natural manifold structure by itself, hence this notion of differentiability is as close as we can get.

• •• •••• •••• •••• •

In fact, since the Grassman algebra of differential forms on X_l is generated by finite linear combinations of monomials of the form $f_l^{(0)} df_l^{(1)} \wedge \ldots \wedge df_l^{(p)}$ with $0 \leq p \leq \dim(X_l)$, $f_l^{(0)} \in C^n(X_l)$, $f_l^{(k)} \in C^{(n+1)}(X_l)$, $k = 1, \ldots, p$ we can define the space of cylindrical p-forms and the cylindrical Grassman algebra by

$$\overset{p}{\bigwedge}(\overline{X}) = \left(\bigcup_{l \in \mathcal{L}} \overset{p}{\bigwedge}(X_l) \right) \Big/ \sim \tag{8.2.21}$$

because the pull-back commutes with the exterior derivative, that is, $p_{l''l}^* f_l = p_{l''l'}^* f_{l'}$ implies $p_{l''l}^* df_l = d(p_{l''l'}^* f_{l'})$. In other words, the exterior derivative is a well-defined operation on the Grassmann algebra. Notice that if $\omega = [\omega_l]_\sim \in \bigwedge(\overline{X})$ and ω_l has degree p then also $p_{l'l}^* \omega_l$ has degree p, hence the degree of forms on $\overline{X}$ is well-defined.

• ••• • •••• •

The case of volume forms is slightly different because a volume form on an orientable X_l is a nowhere vanishing differential form of degree $\dim(X_l)$ so that the degree varies with the label l. However, volume forms on $\overline{X}$ (even in the non-orientable case) are nothing else than cylindrically defined measures satisfying the consistency condition $\mu_{l'} \circ p_{l'l}^{-1} = \mu_l$ for all $l \prec l'$. If they are probability measures we can extend them to σ-additive measures on $\overline{X}$ using the Riesz–Markow theorem as in the previous section.

Differentiable vector fields $V^n(X_l)$ on X_l are conveniently introduced algebraically on X_l as derivatives, that is, they are linear functionals $Y_l :$ $C^{n+1}(X_l) \to C^n(X_l)$ annihilating constants and satisfying the Leibniz rule. We want to proceed similarly with respect to $\overline{X}$ and the first impulse would be to define

$$V^n(\overline{X}) = \left(\bigcup_{l \in \mathcal{L}} V^n(X_l) \right) \Big/ \sim$$

where the equivalence relation is given through the push-forward map. The push-forward is defined by

$$(p_{l'l})_* : \ V^n(X_{l'}) \to V^n(X_{l'}); p_{l'l}^*([(p_{l'l})_* Y_{l'}](f_l)) := Y_{l'}(p_{l'l}^* f_l) \quad (8.2.22)$$

and we could try to define $Y_l \sim Y_{l'}$ iff for any $l'' \prec l, l'$ we have $(p_{l'l''})_* Y_{l'} = (p_{ll''})_* Y_l$. The problem with this definition is that the push-forward moves us 'down' in the directed label set $\mathcal{L}$ instead of 'up' as is the case with the pull-back, so it is not guaranteed that, given l, l', there exists any l'' at all that satisfies $l'' \prec l, l'$ whence the consistency condition might be empty. This forces us to adopt a different strategy, namely to define $V^n(\overline{X})$ as projective nets $(Y_l)_{l_0 \prec l \in \mathcal{L}}$ with the consistency condition

$$(p_{l'l})_* Y_{l'} = Y_l \Leftrightarrow p_{l'l}^*[Y_l(f_l)] = Y_{l'}(p_{l'l}^* f_l) \ \forall \ f_l \in C^n(X_l), \ l_0 \prec l \prec l' \quad (8.2.23)$$

The necessity of restricting attention to $l_0 \prec l$ is that it may not be possible or necessary to define Y_l for all $l \in \mathcal{L}$ or to have (8.2.23) satisfied. This question never came up of course for the pull-back. Notice that (8.2.23) means that if $f_{l'} = p_{l'l}^* f_l$ then $Y_{l'}(f_{l'}) = p_{l'l}^* Y_l(f_l)$ for $l_0 \prec l \prec l'$, that is consistently defined vector fields map cylindrical functions to cylindrical functions.

It is clear that for $f = [f_l]_\sim = p_l^* f_l$ with $l_0 \prec l$ the formula

$$Y(p_l^* f_l) := p_l^* Y_l(f_l) =: p_l^*[(p_l)_* Y](f_l) \quad (8.2.24)$$

is well-defined, for suppose that $f_l \sim f'_{l'}$ with $l_0 \prec l'$ then we find $l_0 \prec l, l' \prec l''$ such that $p_{l''l}^* f_l = p_{l''l'}^* f'_{l'}$ whence, using $p_{l''l} \circ p_{l''} = p_l, p_{l''l'} \circ p_{l''} = p_{l'}$

$$p_{l'}^* Y_{l'}(f'_{l'}) = p_{l''}^* p_{l''l'}^* Y_{l'}(f'_{l'}) = p_{l''}^* Y_{l''}(p_{l''l'}^* f'_{l'}) = p_{l''}^* Y_{l''}(p_{l''l}^* f_l) = p_l^* Y_l(f_l) \quad (8.2.25)$$

Suppose that $Y = (Y_l)_{l_0 \prec l \in \mathcal{L}}, Y' = (Y'_l)_{l'_0 \prec l \in \mathcal{L}} \in V^n(\overline{X})$ are consistently defined vector fields. We certainly find $l_0, l'_0 \prec l''_0$ and claim that $[Y, Y'] :=$

$([Y_l, Y_l])_{l_0'' \prec l \in \mathcal{L}} \in V^{n-1}(\overline{X})$ is again consistently defined. To see this, consider $l_0'' \prec l \prec l'$ then for any $f_l \in C^n(X_l)$ we have due to $l_0 \prec l$ and $l_0' \prec l$

$$p_{l'l}^*([Y_l, Y_l'](f_l)) = Y_{l'}[p_{l'l}^*(Y_l'(f_l))] - Y_{l'}'[p_{l'l}^*(Y_l(f_l))] = [Y_{l'}, Y_{l'}'](p_{l'l}^* f_l) \quad (8.2.26)$$

•••••• •••• ••••••••• •••

Recall that the Lie derivative of an element $\omega_l \in \bigwedge^n(X_l)$ with respect to a vector field $Y_l \in V^n(X_l)$ is defined by $L_{Y_l}\omega_l = [i_{Y_l}d + d i_{Y_l}]\omega_l$ where

$$i_{Y_l} f_l^{(0)} \, df_l^{(1)} \wedge \ldots \wedge df_l^{(p)}$$

$$= f_l^{(0)} \sum_{k=1}^{p}(-1)^{k+1} Y_l(f_l^{(k)}) \, df_l^{(1)} \wedge \ldots df_l^{(k-1)} \wedge df_l^{(k+1)} \wedge \ldots \wedge df_l^{(p)}$$

denotes contraction of forms with vector fields, annihilating zero forms. Let now μ_l be a volume form on X_l. Since X_l is finite-dimensional, all smooth volume forms are absolutely continuous with respect to each other and there exists a well-defined function, called the divergence of Y_l with respect to μ_l, uniquely defined by

$$L_{Y_l}\mu_l =: [\mathrm{div}_{\mu_l} Y_l]\mu_l \quad (8.2.27)$$

We say that a vector field $Y = (Y_l)_{l_0 \prec l \in \mathcal{L}}$ is compatible with a volume form $\mu = (\mu_l)_{l \in \mathcal{L}}$ provided that the family of divergences defines a cylindrical function, that is

$$p_{l'l}^*[\mathrm{div}_{\mu_l} Y_l] = \mathrm{div}_{\mu_{l'}} Y_{l'} \; \forall l_0 \prec l \prec l' \quad (8.2.28)$$

Hence there exists a well-defined cylindrical function $\mathrm{div}_\mu Y := [\mathrm{div}_{\mu_l} Y_l]_\sim$, called the divergence of Y with respect to μ.

Lemma 8.2.7. ••• μ •• • •• •••• ••••• • ••••• • Y, Y' μ•••• •••••• •••••• •••••
••• $f, f' \in \bullet \bullet^1(\overline{X})$ ••••• •••••• ••••••••• •• $\overline{X}$•

••• •• $\partial X_l = \emptyset$ ••• •• ••••••••• ••••

$$\int_{\overline{X}} \mu \, f \, Y(f') = -\int_{\overline{X}} \mu \, (Y(f) + f \, [\mathrm{div}_\mu Y]) f' \quad (8.2.29)$$

•••• • ••• • •• •••••••• $[Y, Y']$ •• ••••• μ•••• ••••••• •••

$$\mathrm{div}_\mu[Y, Y'] = Y(\mathrm{div}_\mu Y') - Y'(\mathrm{div}_\mu Y) \quad (8.2.30)$$

• ••••

(i) We find $l_0, l_0' \prec l$ such that $f = p_l^* f_l, f' = p_l^* f_l'$. Then

$$\mu(fY(f')) = \mu([p_l^* f_l][p_l^* Y_l(f_l')]) = \mu_l(f_l L_{Y_l}[f_l'])$$

$$= \int_{X_l} \{L_{Y_l}[\mu_l f_l f_l'] - (L_{Y_l}[\mu_l f_l]) f_l'\}$$

$$= \int_{X_l} \{d \, i_{Y_l}[\mu_l f_l f_l'] - \mu_l(Y_l(f_l) + f_l \, [\mathrm{div}_{\mu_l} Y_l]) f_l')\}$$

$$= -\mu((Y(f) + f \, [\mathrm{div}_\mu Y]) f') \quad (8.2.31)$$

where in the third line we have applied Stokes' theorem and that the Lie derivative satisfies the Leibniz rule.

(ii) We find $l_0, l_0' \prec l_0''$ so that $([Y_l, Y_l'])_{l_0'' \prec l \in \mathcal{L}}$ is consistently defined as shown above. From the fact that the Lie derivative is an isomorphism between the Lie algebra of vector fields and the derivatives respectively on $C^n(X_l)$, $L_{[Y_l, Y_l']} = [L_{Y_l}, L_{Y_l'}]$, and the fact that Lie derivation and exterior derivation commute, $[d, L_{Y_l}] = 0$, we have

$$(\operatorname{div}_{\mu_l}[Y_l, Y_l'])\mu_l = [L_{Y_l}, L_{Y_l'}](\mu_l) = L_{Y_l}([\operatorname{div}_{\mu_l} Y_l']\mu_l) - L_{Y_l'}([\operatorname{div}_{\mu_l} Y_l]\mu_l)$$
$$= [Y_l(\operatorname{div}_{\mu_l} Y_l') - Y_l'(\operatorname{div}_{\mu_l} Y_l)]\mu_l \qquad (8.2.32)$$

It follows from the consistency of the Y_l and the compatibility with the μ_l that for $l \prec l'$

$$p_{ll'}^* Y_l(\operatorname{div}_{\mu_l} Y_l') = Y_{l'}(p_{ll'}^*(\operatorname{div}_{\mu_l} Y_l')) = Y_{l'}(\operatorname{div}_{\mu_{l'}} Y_{l'}') \qquad (8.2.33)$$

$\square$

Let Y be a vector field compatible with σ-additive measure (volume form) μ such that it is together with its divergence $\operatorname{div}_\mu Y$ real-valued. We consider the Hilbert space $\mathcal{H}_\mu := L_2(\overline{X}, \mu)$ and define the momentum operator

$$P(Y) := i\left(Y + \frac{1}{2}(\operatorname{div}_\mu Y)1_{\mathcal{H}_\mu}\right) \qquad (8.2.34)$$

with dense domain $D(P(Y)) = \operatorname{Cyl}^1(\overline{X})$. From (8.2.29) we conclude that for $f, f' \in D(P(Y))$

$$< f, P(Y)f' >_\mu = \mu(\overline{f}P(Y)f) = \mu(\overline{P(Y)f}f) =< P(Y)f, f' >_\mu \qquad (8.2.35)$$

from which we see that

$$D(P(Y)) \subset D(P(Y)^\dagger)$$
$$:= \{f \in \mathcal{H}_\mu; \sup_{\|f'\|>0} | < f, P(Y)f' > |/\|f'\| < \infty\} \text{ and } P(Y)^\dagger|_{D(P(Y))}$$
$$= P(Y)$$

whence $P(Y)$ is a symmetric unbounded operator.

Finally we notice that if Y, Y' are both μ-compatible then

$$[P(Y), P(Y')] = iP([Y, Y']) \qquad (8.2.36)$$

by a straightforward computation using Lemma 8.2.7.

That $\operatorname{div}_{\mu_l} Y_l$ is a cylindrical function is a sufficient criterion for $P(Y)$ to be well-defined, but it is too strong a requirement because it means that for given l on any other $l \prec l'$ the function $\operatorname{div}_{\mu_{l'}} Y_{l'} \equiv p_{l'l}^*(\operatorname{div}_{\mu_l} Y_l)$ does not depend on the additional degrees of freedom contained in $X_{l'}$. That is, if some special graphs are not to be distinguished then $\operatorname{div}_\mu Y = \operatorname{const.}$ is the only possibility.

So compatibility between μ and Y is only sufficient but has not been shown to be necessary in order to define interesting momentum operators. It would be important to replace the compatibility criterion by a weaker one. See [514, 515] for first steps in that direction.

• •• •••• • ••• •• ••

More generally we have the following abstract situation: we have a partially ordered and directed index set $\mathcal{L}$, a family of Hilbert spaces $\mathcal{H}_l := \mathcal{H}_{\mu_l} := L_2(X_l, d\mu_l)$ and isometric monomorphisms (linear injections)

$$\hat{U}_{ll'} : \mathcal{H}_l \to \mathcal{H}_{l'} \tag{8.2.37}$$

for every $l \prec l'$ which in our special case is given by $\hat{U}_{ll'} f_l := p^*_{l'l} f_l$. The isometric monomorphisms satisfy the compatibility condition

$$\hat{U}_{l'l''}\hat{U}_{ll'} = \hat{U}_{ll''} \tag{8.2.38}$$

for any $l \prec l' \prec l''$ due to $p_{l'l} \circ p_{l''l'} = p_{l''l}$. A system $(\mathcal{H}_l, \hat{U}_{ll'})_{l \prec l' \in \mathcal{L}}$ of this sort is called a directed system of Hilbert spaces. A Hilbert space $\mathcal{H}$ is called the inductive limit of a directed system of Hilbert spaces provided that there exist isometric monomorphisms

$$\hat{U}_l : \mathcal{H}_l \to \mathcal{H} \tag{8.2.39}$$

for any $l \in \mathcal{L}$ such that the compatibility condition

$$\hat{U}_{l'}\hat{U}_{ll'} = \hat{U}_l \tag{8.2.40}$$

holds. In our case, obviously $\hat{U}_l f_l := p^*_l f_l$ provides these monomorphisms so that we have displayed $\mathcal{H}_\mu$ as the inductive limit of the $\mathcal{H}_{\mu_l}$.

Likewise we have a family of operators $\hat{O}_l = P(Y_l)$ with dense domain $D(\hat{O}_l) = C^1(X_l)$ in $\mathcal{H}_l$ which are defined for a co-final subset $\mathcal{L}(\hat{O}) = \{l \in \mathcal{L}; l_0 \prec l\}$ (that is, for any $l \in \mathcal{L}$ there exists $l \prec l' \in \mathcal{L}(\hat{O})$) of $\mathcal{L}$. These families of domains and operators satisfy the compatibility conditions

$$\hat{U}_{ll'} D(\hat{O}_l) \subset D(\hat{O}_{l'}) \tag{8.2.41}$$

for any $l \prec l' \in \mathcal{L}(\hat{O})$ since $p^*_{l'l} C^1(X_l) \subset C^1(X_{l'})$ (the pull-back of functions is C^1 with respect to the X_l arguments but C^ω with respect to the remaining arguments in $X_{l'}$). Furthermore

$$\hat{U}_{ll'} \hat{O}_l = \hat{O}_{l'} \hat{U}_{ll'} \tag{8.2.42}$$

for any $l \prec l' \in \mathcal{L}(\hat{O})$ since $p^*_{l'l}(Y_l(f_l) + [\text{div}_{\mu_l} Y_l] f_l/2) = (Y_{l'}(p^*_{l'l} f_l) + [\text{div}_{\mu_{l'}} Y_{l'}] p^*_{l'l} f_l/2)$ due to consistency and compatibility. A structure of this kind is called a directed system of operators. An operator $\hat{O}$ with dense domain $D(\hat{O})$ is called the inductive limit of a directed system of operators provided the above-defined isometric isomorphisms interact with domains and operators in the expected way, that is,

$$\hat{U}_l D(\hat{O}_l) \subset D(\hat{O}) \tag{8.2.43}$$

and

$$\hat{U}_l \hat{O}_l = \hat{O} \hat{U}_l \tag{8.2.44}$$

In our case this is by definition satisfied since $p_l^* C^1(X_l) \subset \mathrm{Cyl}^1(\overline{X})$ and $p_l^*(Y_l(f_l) + [\mathrm{div}_{\mu_l} Y_l] f_l/2) \equiv (Y(p_l^* f_l) + [\mathrm{div}_\mu Y] p_l^* f_l/2)$.

It turns out that directed systems of Hilbert spaces and operators always have an inductive limit which is unique up to unitary equivalence.

Lemma 8.2.8

... • .... •....... ...... • .. • .... $(\mathcal{H}_l, \hat{U}_{ll'})_{l \prec l' \in \mathcal{L}}$ ... •.........

$(\hat{O}_l, D(\hat{O}_l), \hat{U}_{ll'})_{l \prec l' \in \mathcal{L}(\hat{O})}$ • ... • .... $\mathcal{L}(\hat{O})$. ....

.. .. $(\mathcal{H}, \hat{U}_l)_{l \in \mathcal{L}}$ ... • ...

... . $(\hat{O}, D(\hat{O}), \hat{U}_l)_{l \in \mathcal{L}(\hat{O})}$ ....

... . • ....

.... $\hat{O}_l$ ... . • ... . $D(\hat{O}_l)$ .... $\hat{O}$ .. ..

.. .. • .. .. $D(\hat{O})$.

.... $\hat{U}_l$ ... . $(\hat{O}'_l, D(\hat{O}'_l), \hat{U}_{ll'})_{l \prec l' \in \mathcal{L}(\hat{O})}$ .. •

.... O'_l .... • ....

$\hat{O}_l$.

•....

(i) In the case of bounded operators, that is $D(\hat{O}_l) = \mathcal{H}_l$, part (i) is standard in operator theory, see, for example, vol. 2 of [535] for more details and an extension of the theorem to directed systems of C^*-algebras and von Neumann algebras which have a unique inductive limit up to algebra isomorphisms.

We consider the vector space V of equivalence classes of nets $f = (f_l)_{l_0 \prec l \in \mathcal{L}(f)}$ for some co-final $\mathcal{L}(f) \subset \mathcal{L}$ with $f_l \in \mathcal{H}_l$ satisfying $\hat{U}_{ll'} f_l = f_{l'}$ for any $l_0 \prec l \prec l'$ and where $f \sim f'$ are equivalent if $f_l = f'_l$ for all $l \in \mathcal{L}(f) \cap \mathcal{L}(f')$. Let us write $[f]_\sim$ for the equivalence class of f. We define

$$\hat{U}_l : \mathcal{H}_l \to V; \ f_l \mapsto [(\hat{U}_{ll'} f_l)_{l \prec l' \in \mathcal{L}}]_\sim \tag{8.2.45}$$

Due to isometry of the $\hat{U}_{ll'}$ the norm on V given by $\|[f]_\sim\| := \|f_l\|_l$ is independent of the choice of $l \in \mathcal{L}(f)$, in particular, $\hat{U}_l$ becomes an isometry. We have for $l \prec l'$

$$\hat{U}_{l'} \hat{U}_{ll'} f_l = [(\hat{U}_{l'l''} \hat{U}_{ll'} f_l)_{l' \prec l''}]_\sim = [(\hat{U}_{ll''} f_l)_{l' \prec l''}]_\sim = [(\hat{U}_{ll''} f_l)_{l \prec l''}]_\sim = \hat{U}_l f_l$$

Finally we consider the subspace of V given by the span of elements of the form $\hat{U}_l f_l$ with $f_l \in \mathcal{H}_l$ and complete it to arrive at the Hilbert space

$$\mathcal{H} := \overline{\bigcup_l \hat{U}_l \mathcal{H}_l} \tag{8.2.46}$$

to which $\hat{U}_l$ can be extended uniquely as an isometric monomorphism by continuity. To see the uniqueness one observes that given another inductive limit $(\mathcal{H}', \hat{V}_l)$ we may define $W_l := \hat{V}_l \hat{U}_l^{-1} : \hat{U}_l \mathcal{H}_l \to \hat{V}_l \mathcal{H}_l$ which one checks to be an isometry. Also for $l \prec l'$ we have $W_{l'} \hat{U}_l = \hat{W}_{l'} \hat{U}_{l'} \hat{U}_{ll'} = \hat{V}_{l'} \hat{U}_{ll'} = V_l = \hat{W}_l \hat{U}_l$, in other words, $W_{l'}$ is an extension of W_l for $l \prec l'$. This means that we have a densely defined isometry $\hat{W} : \bigcup \hat{U}_l \mathcal{H}_l \to \bigcup \hat{V}_l \mathcal{H}_l$ defined by $\hat{W}_{|\hat{U}_l \mathcal{H}_l} = W_l$ which extends by continuity uniquely to an isometry between the two Hilbert spaces.

Next, define an operator on the dense subspace of $\mathcal{H}$ given by $D(\hat{O}) := \bigcup_{l \in \mathcal{L}(\hat{O})} \hat{U}_l D(\hat{O}_l)$

$$\hat{O}\big[(f_l)_{l \in \mathcal{L}(\hat{O})}\big]_\sim := \big[(\hat{O}_l f_l)_{l \in \mathcal{L}(\hat{O})}\big]_\sim \tag{8.2.47}$$

Since $\mathcal{L}(\hat{O}) \cap \{l' \in \mathcal{L}; \, l \prec l'\} = \{l' \in \mathcal{L}(\hat{O}); \, l \prec l'\}$ is co-final we have

$$\hat{O}\hat{U}_l f_l = \hat{O}\big[(\hat{U}_{ll'} f_l)_{l \prec l' \in \mathcal{L}}\big]_\sim = \hat{O}\big[(\hat{U}_{ll'} f_l)_{l \prec l' \in \mathcal{L}(\hat{O})}\big]_\sim = \big[(\hat{O}_{l'} \hat{U}_{ll'} f_l)_{l \prec l' \in \mathcal{L}(\hat{O})}\big]_\sim$$
$$= \big[(\hat{U}_{ll'} \hat{O}_l f_l)_{l \prec l' \in \mathcal{L}(\hat{O})}\big]_\sim = \big[(\hat{U}_{ll'} \hat{O}_l f_l)_{l \prec l' \in \mathcal{L}}\big]_\sim$$
$$= \hat{U}_l \hat{O}_l f_l \tag{8.2.48}$$

(ii) By the basic criterion of essential self-adjointness we know that $(\hat{O}_l \pm i \cdot 1_{\mathcal{H}_l})D(\hat{O}_l)$ is dense in $\mathcal{H}_l$. It follows that

$$(\hat{O} \pm i \cdot 1_{\mathcal{H}})D(\hat{O}) = \bigcup_{l \in \mathcal{L}(\hat{O})} (\hat{O} \pm i \cdot 1_{\mathcal{H}})\hat{U}_l D(\hat{O}_l)$$
$$= \bigcup_{l \in \mathcal{L}(\hat{O})} \hat{U}_l(\hat{O}_l \pm i \cdot 1_{\mathcal{H}_l})D(\hat{O}_l) \tag{8.2.49}$$

hence $(\hat{O} \pm i \cdot 1_{\mathcal{H}})D(\hat{O})$ is dense in $\mathcal{H}$ so that $\hat{O}$ is essentially self-adjoint by the basic criterion of essential self-adjointness.

(iii) Recall that the self-adjoint extension $\hat{O}'_l$ of an essentially self-adjoint operator $\hat{O}_l$ with core $D(\hat{O}_l)$ is unique and given by its closure, that is, the set $D(\hat{O}'_l)$ given by those $f_l \in \mathcal{H}_l$ such that $(f_l, \hat{O}_l f_l) \in \overline{\Gamma}_{\hat{O}_l}$, the closure in $\mathcal{H}_l \times \mathcal{H}_l$ of the graph $\Gamma_{\hat{O}_l} = \{(f_l, \hat{O}_l f_l); \, f_l \in D(\hat{O}_l)\}$ of $\hat{O}_l$ with respect to the norm $\|(f_l, f'_l)\|^2 = \|f_l\|^2 + \|f'_l\|^2$.

To see that $\hat{U}_{ll'} D(\hat{O}'_l) \subset D(\hat{O}'_{l'})$ we notice that $\hat{U}_{ll'} D(\hat{O}_l) \subset D(\hat{O}_{l'})$. Hence, the closure $D(\hat{O}'_{l'})$ of $D(\hat{O}_{l'})$ will contain the closure of $\hat{U}_{ll'} D(\hat{O}_l)$ which coincides with $\hat{U}_{ll'} D(\hat{O}'_l)$ because $\hat{U}_{ll'}$ is bounded.

To see that $\hat{U}_{ll'} \hat{O}'_l = \hat{O}'_{l'} \hat{U}_{ll'}$ holds on $D(\hat{O}'_l)$ we notice that $\hat{U}_{ll'} \hat{O}_l = \hat{O}_{l'} \hat{U}_{ll'}$ holds on $D(\hat{O}_l)$. Since $\hat{O}'_l, \hat{O}'_{l'}$ are just the extensions of $\hat{O}_l, \hat{O}_{l'}$ from

$D(\hat{O}_l), D(\hat{O}_{l'})$ to $D(\hat{O}_l'), D(\hat{O}_{l'}')$ and since $\hat{U}_{ll'} D(\hat{O}_l') \subset D(\hat{O}_{l'}')$ the claim follows. $\qquad\qquad\square$

Finally we consider the case of interest, namely the quotient space $\overline{\mathcal{A}/\mathcal{G}}$ projective limit. The significance of the result $\overline{\mathcal{A}/\mathcal{G}} = \overline{\mathcal{A}}/\overline{\mathcal{G}}$ is that we can identify cylindrical functions on $\overline{\mathcal{A}/\mathcal{G}}$ simply with $\overline{\mathcal{G}}$-invariant functions on $\overline{\mathcal{A}}$. More precisely, if $\lambda : \overline{\mathcal{G}} \times \overline{\mathcal{A}} \to \overline{\mathcal{A}}; \ A \mapsto \lambda_g(A)$ is the $\overline{\mathcal{G}}$-action and $f \in \mathrm{Cyl}^n(\overline{\mathcal{A}})$ is $\overline{\mathcal{G}}$-invariant then we may define $\tilde{f} \in \mathrm{Cyl}^n(\overline{\mathcal{A}/\mathcal{G}})$ by $\tilde{f}([A]) := f(A) = f(\lambda_g(A))$ for all $g \in \overline{\mathcal{G}}$ where $[.] : \overline{\mathcal{A}} \to \overline{\mathcal{A}}/\overline{\mathcal{G}} \equiv \overline{\mathcal{A}/\mathcal{G}}$ denotes the quotient map. Thus we define zero-forms on $\overline{\mathcal{A}/\mathcal{G}}$ as zero-forms on $\overline{\mathcal{A}}$ which satisfy $f = \lambda_g^* f$ for any $g \in \overline{\mathcal{G}}$. Notice that this is possible for any differentiability category because the $\overline{\mathcal{G}}$-action is evidently not only continuous but even analytic!

Since pull-backs commute with exterior derivation we can likewise define the Grassman algebra $\bigwedge(\overline{\mathcal{A}/\mathcal{G}})$ as the subalgebra of $\bigwedge(\overline{\mathcal{A}})$ given by the $\overline{\mathcal{G}}$-invariant differential forms, that is, those that satisfy $\lambda_g^* \omega = \omega$ for all $g \in \overline{\mathcal{G}}$ (if f is $\overline{\mathcal{G}}$-invariant, so is df because $\lambda_g^* df = d\lambda_g^* f = df$).

Next, volume forms on $\overline{\mathcal{A}/\mathcal{G}}$ are just $\overline{\mathcal{G}}$-invariant volume forms on $\overline{\mathcal{A}}$, that is $(\lambda_g)_* \mu = \mu \circ \lambda_g^{-1} = \mu \circ \lambda_{g^{-1}} = \mu$ for all $g \in \overline{\mathcal{G}}$. Given any volume form μ on $\overline{\mathcal{A}}$ we may derive a measure $\overline{\mu}$ on $\overline{\mathcal{A}/\mathcal{G}}$ by $\overline{\mu}(f) := \mu(f)$ for all $\overline{\mathcal{G}}$-invariant functions f on $\overline{\mathcal{A}}$. If we denote the Haar probability measure on $\overline{\mathcal{G}} = \prod_{x \in \sigma} G$ by μ_H then from $\mu(f) = \mu(\lambda_g^* f) = [(\lambda_g)_* \mu](f)$ for all $\overline{\mathcal{G}}$-invariant measurable functions we find

$$\overline{\mu}([A]) = \int_{\overline{\mathcal{G}}} d\mu_H(g) \, [(\lambda_g)_* \mu](A) \qquad (8.2.50)$$

Finally, we define vector fields on $\overline{\mathcal{A}/\mathcal{G}}$ as $\overline{\mathcal{G}}$-invariant vector fields on $\overline{\mathcal{A}}$, that is, those satisfying $(\lambda_g)_* Y = Y$ for all $g \in \overline{\mathcal{G}}$, more precisely, if $Y = (Y_l)_{l_0 \prec l}$ then

$$(\lambda_g^l)^* \left([(\lambda_g^l)_* Y_l](f_l)\right) := Y_l[(\lambda_g^l)^* f_l] = (\lambda_g^l)^* (Y_l(f_l)) \qquad (8.2.51)$$

for any $f_l \in C^n(\overline{\mathcal{A}})$ and $l_0 \prec l$.

8.2.3 $^+$ Density and support properties of $\mathcal{A}, \mathcal{A}/\mathcal{G}$ with respect to $\overline{\mathcal{A}}, \overline{\mathcal{A}/\mathcal{G}}$

In this subsection we will see that $\mathcal{A}$ lies topologically dense, but measure theoretically thin in $\overline{\mathcal{A}}$ (similar results apply to $\mathcal{A}/\mathcal{G}$ with respect to $\overline{\mathcal{A}/\mathcal{G}} = \overline{\mathcal{A}}/\overline{\mathcal{G}}$) with respect to the uniform measure μ_0. More precisely, there is a dense embedding (injective inclusion) $\mathcal{A} \to \overline{\mathcal{A}}$ but $\mathcal{A}$ is embedded into a measurable subset of $\overline{\mathcal{A}}$ of measure zero. The latter result demonstrates that the measure is concentrated on non-smooth (distributional) connections so that $\overline{\mathcal{A}}$ is indeed much larger than $\mathcal{A}$. We follow closely Rendall [363], Marolf and Mourão [365, 366] and [418].

We have seen in Section 6.2.2 that every element $A \in \mathcal{A}$ defines an element of $\mathrm{Hom}(\mathcal{P}, G)$ and that this space can be identified with the projective limit $\overline{X} \equiv \overline{\mathcal{A}}$. Now via the C^*-algebraic framework we know that $\mathrm{Cyl}(\overline{X})$ can be identified with $C(\overline{X})$ and the latter space of functions separates the points of $\overline{X}$ by the Stone–Weierstrass theorem since it is Hausdorff and compact. The question is whether the smaller set of functions $\mathrm{Cyl}(\overline{X})$ separates the smaller set of points $\mathcal{A}$. This is almost obvious and we will do it for $G = \mathrm{SU}(\bullet)$, other compact groups can be treated similarly.

Let $A \neq A'$ be given then there exists a point $x \in \sigma$ such that $A(x) \neq A'(x)$. Take $D = \dim(\sigma)$ edges $e_{x,\alpha} \in \mathcal{P}$ with $b(e_{x,\alpha}) = x$ and linearly independent tangents $\dot{e}_{x,\alpha}(0)$ at x. Consider the cylindrical function

$$F_x^\epsilon : \ \mathcal{A} \to \mathbb{C}; \ A \mapsto \frac{1}{\epsilon^2} \sum_{\alpha, j} \left[\mathrm{tr}\!\left(\tau_j A(e_{x,\alpha}^\epsilon) \right) \right]^2 \qquad (8.2.52)$$

where τ_j is a basis of $\mathrm{Lie}(G)$ with normalisation $\mathrm{tr}(\tau_j \tau_k) = -N\delta_{jk}$ and $e_{x,\alpha}^\epsilon(t) = e_{x,\alpha}(\epsilon t)$. Using smoothness of A it is easy to see that (8.2.11) can be expanded in a convergent Taylor series with respect to ϵ with zeroth-order component $\sum_{j, e_\alpha} |A_a^j(x) \dot{e}_{x,\alpha}^a(0)|^2$ whence $F_x^\epsilon \in \mathrm{Cyl}(\overline{X})$ separates our given $A \neq A'$. The proof for $\mathcal{A}$ replaced by $\mathcal{A}/\mathcal{G}$ is similar and was given by Giles [342] and will not be repeated here. In that proof it is important that G is compact.

We thus have the following abstract situation: a collection $\mathcal{C} = \mathrm{Cyl}(\overline{X})$ of bounded complex-valued functions on a set $X = \mathcal{A}$ including the constants which separate the points of X. The set X may be equipped with its own topology (e.g., the Sobolov topology that we defined in Chapter 33) but this will be irrelevant for the following result which is an abstract property of Abelian unital C^*-algebras.

Theorem 8.2.9. ••• $\mathcal{C}$ •• • •• ••••••• • •• ••• •••• •• ••• •••• ••• •• ••••• •• •
••• X • •••• ••• •• ••• ••• •• •••••• •• ••• •••••••••• ••• •• ••• •• •• X • ••• $\overline{\mathcal{C}}$ •• •••
• •• ••• • ••• •••• C^* •• •• •••• ••• •• •••• •••• • $\mathcal{C}$ •• •• ••• •• ••• ••• •••••• • • •••• •••• ••••
••• •• • • •••• •••• •••• ••• ••• • ••• •• •• •• •••• •• ••• ••••• •• ••••••• • • ••••
••• •• ••• •• X • ••••• ••• • •••••• •• ••••••• • •• •• ••• • • •••••• • •••••••• • $\overline{X}$ •• $\overline{\mathcal{C}}$
•• •• ••• • ••• •••••• •• ••• • •••••• •••• ••••• •• ••• ••••••••• •

• •• •••: Actually the theorem also holds if $\mathcal{C}$ does not separate the points, this is just convenient in order that we may naturally identify X with a subset of $\overline{X}$. Also that $\mathcal{C}$ is unital is inessential because we may always add a unit to a C^*-algebra.

• ••••• Consider the following map

$$J : \ X \to \overline{X}; \ x \mapsto J_x \text{ where } J_x(f) := f(x) \ \forall \ f \in \overline{\mathcal{C}} \qquad (8.2.53)$$

This is an injection since $J_x = J_{x'}$ implies in particular $f(x) = f(x')$ for all $f \in \mathcal{C}$, thus $x = x'$ since $\mathcal{C}$ separates the points of X by assumption, hence J provides an embedding.

Let $\overline{J(X)}$ be the closure of $J(X)$ in the Gel'fand topology on $\overline{X}$ of pointwise convergence on $\overline{\mathcal{C}}$. Suppose that $\overline{X} - \overline{J(X)} \neq \emptyset$ and take any $\chi \in \overline{X} - \overline{J(X)}$. Since $\overline{X}$ is a compact Hausdorff space we find $a \in C(\overline{X})$ such that $1 = a(\chi) \neq a(J_x) = 0$ for any $x \in X$ by Urysohn's lemma. (In Hausdorff spaces one-point sets are closed, hence $\{\chi\}$ and $\overline{J(X)}$ are disjoint closed sets and finally compact Hausdorff spaces are normal spaces, see Chapter 18.)

Since the Gel'fand map $\vee : \overline{\mathcal{C}} \to C(\overline{X})$ is an isometric isomorphism we find $f \in \overline{\mathcal{C}}$ such that $\check{f} = a$. Hence $0 = a(J_x) = \check{f}(J_x) = J_x(f) = f(x)$ for all $x \in X$, hence $f = 0$, thus $a \equiv 0$ contradicting $a(\chi) = 1$. Therefore χ in fact does not exist whence $\overline{X} = \overline{J(X)}$. $\qquad\qquad\square$

Of course in our case $\overline{\mathcal{C}} = \overline{\mathrm{Cyl}(\mathcal{A})}$ and $\overline{X} = \overline{\mathcal{A}}$.

Our next result is actually much stronger than merely showing that $\mathcal{A}$ is contained in a measurable subset of $\overline{\mathcal{A}}$ of μ_0-measure zero. Let e be an edge and if $e(t)$ is a representative curve then consider the family of segments e_s with $e_s(t) := e(st)$, $s \in [0,1]$. Consider the map

$$h^e : \overline{\mathcal{A}} \to \mathrm{Fun}([0,1], G); \quad A \mapsto h_A^e \text{ where } h_A^e(s) := A(e_s) \qquad (8.2.54)$$

The set $\mathrm{Fun}([0,1], G)$ of all functions from the interval $[0,1]$ into G (no continuity assumptions) can be thought of as the uncountable direct product $G^{[0,1]} := \prod_{s \in [0,1]} G$ via the bijection $E : \mathrm{Fun}([0,1], G) \to G^{[0,1]}; \; h \to (h_s := h(s))_{s \in [0,1]}$. The latter space can be equipped with the Tychonov topology generated by the open sets on $G^{[0,1]}$ which are generated from the sets $P_s^{-1}(U_s) = [\prod_{s' \neq s} G] \times U_s$ (where $U_s \subset G$ is open in G) by finite intersections and arbitrary unions. Here $P_s : G^{[0,1]} \to G$ is the natural projection. Now the pre-image of such sets under h^e is given by

$$\begin{aligned}
(h^e)^{-1}\big(P_s^{-1}(U_s)\big) &= \{A \in \overline{\mathcal{A}}; \; h_A^e \in P_s^{-1}(U_s)\} \\
&= \{A \in \overline{\mathcal{A}}; \; h_A^e(s) \in U_s, \; h_A^e(s') \in G \text{ for } s' \neq s\} \\
&= \{A \in \overline{\mathcal{A}}; \; A(e_s) \in U_s\} = p_{e_s}^{-1}(U_s) \qquad (8.2.55)
\end{aligned}$$

where $p_{e_s} : \overline{\mathcal{A}} \to \mathrm{Hom}(e_s, G)$ is the natural projection in $\overline{\mathcal{A}}$. Since $\overline{\mathcal{A}}$ is equipped with the Tychonov topology, the maps p_{e_s} are continuous and since $\overline{\mathcal{A}}$ is equipped with the Borel σ-algebra, continuous functions (pre-images of open sets are open) are automatically measurable (pre-images of open sets are measurable). Hence we have shown that h^e is a measurable map.

Let f be a function on $G^{[0,1]}$, that is, a complex-valued function $h \mapsto f(\{h_s\}_{s \in [0,1]})$. We have an associated map of the form (6.2.16), that is, $\rho_{l^e} : X_{l^e} \to G^{[0,1]}; \; A_{l^e} \mapsto (A_{l^e}(e_s) = h_A^e(s))_{s \in [0,1]}$ where l^e is the subgroupoid generated by the algebraically independent edges e_s. Thus $h^e = \rho_{l^e} \circ p_{l^e}$. The push-forward of the uniform measure $\nu := h_*^e \mu_0 = \mu_0 \circ (h^e)^{-1}$ is then the measure on

$G^{[0,1]}$ given by

$$\int_{G^{[0,1]}} d\nu(h) f(h) = \mu_0((h^e)^* f) = \mu_{0l^e}(\rho_{l^e}^* f)$$

$$= \int_{G^{[0,1]}} \prod_{s \in [0,1]} d\mu_H(h_{e_s}) f(\{h_{e_s}\}_{s \in [0,1]})$$

$$\equiv \int_{G^{[0,1]}} \prod_{s \in [0,1]} d\mu_H(h_s) f(\{h_s\}_{s \in [0,1]}) \qquad (8.2.56)$$

Theorem 8.2.10. •··· • ······ μ_0 ·· ·········· ·· ··· ······ D_e ·· $\overline{A}$ ······
·· ··· ··· ·· ····· $A \in \overline{A}$ ···· ···· h_A^e ·· ··· ···· ··········· ·· $[0,1]$·

• ····· Trivially

$$D_e = \{A \in \overline{A}; h_A^e \text{ nowhere continuous in } [0,1]\}$$
$$= (h^e)^{-1}(\{h \in G^{[0,1]}; s \mapsto h_s \text{ nowhere continuous in } [0,1]\}) =: (h^e)^{-1}(D) \qquad (8.2.57)$$

If we can show that D contains a measurable set of ν-measure one or that $G^{[0,1]} - D$ is contained in a measurable set D' of ν-measure zero then we have shown that D_e contains a measurable set $D'_e = (h^e)^{-1}(G^{[0,1]} - D')$ of measure one because $\mu_0(D_e) = [\mu_0 \circ (h^e)^{-1}](G^{[0,1]} - D') = \nu(G^{[0,1]} - D') = 1$ and because h^e is measurable (since $G^{[0,1]}$ is equipped with the Borel σ-algebra). In other words, D_e will be a support for μ_0.

Let us then show that $G^{[0,1]} - D = \{h \in G^{[0,1]}; \exists s_0 \in [0,1] \ni h$ continuous at $s_0\}$ is contained in a measurable set of ν-measure zero. Let $h_0 \in G^{[0,1]} - D$, then we find $s_0 \in [0,1]$ such that h_0 is continuous at s_0. Fix any $0 < r < 1$ and consider an open cover of G by sets U with Haar measure $\mu_H(U) = r$. Since G is compact, we find a finite subcover, say $U_1, \ldots, U_N$. Now there is $k_0 \in \{1, \ldots, N\}$ such that $h_0(s_0) \in U_{k_0}$. By definition of continuity at a point we find an open interval $I \subset [0,1]$ such that $h(I) \subset U_{k_0}$. This motivates us to consider the subsets $S_k := \{h \in G^{[0,1]}; \exists I \subset [0,1]$ open $\ni h(I) \subset U_k\} \subset G^{[0,1]}$ and obviously $h_0 \in S_{k_0}$. Our aim is to show that these sets are contained in measure-zero sets.

Let $B(q, 1/m) := \{s \in [0,1]; |s - q| < 1/m\}$ with $q \in \mathbb{Q}$, $m \in \mathbb{N}$. It is easy to show that these sets are a countable basis for the topology for $[0,1]$ (every open set can be obtained by arbitrary unions and finite intersections). Hence any open interval is given as a countable union of these open balls, that is, $I = \bigcup_{B(q,m) \subset I} B(q, m)$. Since $h(I \cup J) = h(I) \cup h(J)$ we have

$$S_k = \left\{ h \in G^{[0,1]}; \exists I \subset [0,1] \ni \bigcup_{B(q,m) \subset I} h(B(q,m)) \subset U_k \right\}$$

$$= \bigcup_{(q,m) \in (\mathbb{Q} \times \mathbb{N})_k} S_{k,q,m}$$

$$S_{k,q,m} := \{h \in G^{[0,1]}; h(B(q,m)) \subset U_k\} \qquad (8.2.58)$$

where $(\mathbb{Q} \times \mathbb{N})_k$ are defined to be the subsets of rational and natural numbers (q, m) respectively such that $S_{U_k, q, m} \neq \emptyset$. (We could also remove that restriction.)

We now show that $S_{k,q,m}$ is contained in a measure-zero set. Let (s_n) be a sequence of points in $B(k, q, m)$. Then $S_{k,q,m} \subset \{h \in G^{[0,1]}; h(s_n) \in U_k \; \forall s_n\} = \cap_n \{h \in G^{[0,1]}; h(s_n) \in U_k\}$. Now the sets $\{h \in G^{[0,1]}; h(s_n) \in U_k\} = P_{s_n}^{-1}(U_k)$ are measurable because P_s is continuous and U_k is open, hence so is $\cap_n \{h \in G^{[0,1]}; h(s_n) \in U_k\}$. But

$$\nu\left(\cap_n \{h \in G^{[0,1]}; h(s_n) \in U_k\}\right) = \nu\left(\left[\prod_{s \neq s_n} G\right] \times \left[\prod_n U_k\right]\right)$$

$$= \prod_n \mu_H(U_k) = \prod_n r = 0 \qquad (8.2.59)$$

since $r < 1$. Hence $S_{k,q,m}$ is contained in a measure-zero subset and since ν is σ-additive also S_k is since (8.2.58) is a countable union.

Finally, any $h_0 \in G^{[0,1]} - D$ is contained in one of the S_k, thus $G^{[0,1]} - D \subset \cup_{k=1}^N S_k$ is contained in a measurable subset of measure zero. $\qquad \square$

8.2.4 Spin-network functions and loop representation

In order to study the properties of μ_0 we need to introduce an important concept, the so-called spin-network basis. We will distinguish between gauge-variant and gauge-invariant spin-network states. For representation theory on compact Lie groups, the Peter and Weyl theorem and Haar measures the reader is referred to [551], an extract of which is given in Chapter 31. We will follow closely Baez [421, 422].

Definition 8.2.11. $\bullet$ $\bullet\bullet$ $\bullet\bullet\bullet$ $\bullet\bullet\bullet$ $\bullet\bullet\bullet$ $\bullet\bullet\bullet$ $\bullet\bullet\bullet\bullet\bullet\bullet\bullet\bullet\bullet$ $\bullet\bullet\bullet\bullet$ $\bullet\bullet\bullet\bullet$ $\bullet\bullet\bullet\bullet\bullet\bullet\bullet\bullet\bullet\bullet$
$\bullet\bullet$ $\bullet\bullet$ $\bullet\bullet\bullet\bullet\bullet\bullet\bullet\bullet\bullet\bullet$ $\bullet\bullet\bullet\bullet\bullet\bullet\bullet\bullet\bullet\bullet\bullet$ $\bullet\bullet$ $\bullet\bullet\bullet$ $\bullet\bullet\bullet$ $\bullet\bullet\bullet\bullet$ $\bullet\bullet\bullet$ $\bullet\bullet\bullet\bullet\bullet$ $\bullet$ $\bullet\bullet\bullet$ $\bullet\bullet\bullet\bullet\bullet$ $\bullet\bullet\bullet$
$\bullet\bullet\bullet\bullet\bullet\bullet\bullet\bullet$ $\bullet\bullet$ $\bullet\bullet\bullet\bullet\bullet$ $\bullet\bullet\bullet\bullet\bullet\bullet\bullet\bullet$ $\bullet\bullet\bullet\bullet\bullet\bullet$ $\bullet\bullet$ $\Pi\bullet$ $\bullet\bullet\bullet$ $l = l(\gamma)$ $\bullet\bullet$ $\bullet\bullet\bullet\bullet\bullet\bullet$ $\bullet$ $\bullet\bullet\bullet\bullet\bullet\bullet\bullet\bullet$ $\bullet$ $\bullet\bullet\bullet$
$\bullet\bullet\bullet\bullet\bullet$ $\bullet\bullet\bullet\bullet$ $e \in E(\gamma)$ $\bullet$ $\bullet\bullet\bullet\bullet\bullet\bullet\bullet\bullet\bullet$ $\bullet\bullet\bullet\bullet\bullet\bullet\bullet\bullet\bullet\bullet$ $\bullet\bullet\bullet\bullet\bullet\bullet\bullet\bullet\bullet\bullet\bullet$ $\pi_e \in \Pi$ $\bullet\bullet\bullet\bullet$ $\bullet\bullet$
$\bullet\bullet\bullet\bullet\bullet$ $\bullet\bullet\bullet$ $\bullet\bullet$ $\bullet$ $\bullet\bullet\bullet\bullet\bullet\bullet$ $\vec{\pi} = (\pi_e)_{e \in E(\gamma)} \bullet$
$\bullet$ $\bullet$ $\bullet\bullet\bullet$ $\bullet\bullet\bullet\bullet\bullet$ $\bullet\bullet\bullet$ $\bullet\bullet$ $\bullet\bullet\bullet\bullet\bullet$ $\bullet$

$$T_{\gamma, \vec{\pi}, \vec{m}, \vec{n}} : \overline{\mathcal{A}} \to \mathbb{C}; \quad A \mapsto \prod_{e \in E(\gamma)} \sqrt{d_{\pi_e}} [\pi_e(A(e))]_{m_e n_e} \qquad (8.2.60)$$

$\bullet$ $\bullet\bullet\bullet\bullet$ d_π $\bullet\bullet\bullet\bullet\bullet\bullet\bullet$ $\bullet\bullet\bullet$ $\bullet\bullet\bullet$ $\bullet\bullet\bullet\bullet\bullet$ $\bullet\bullet$ π $\bullet\bullet\bullet$ $\vec{m} = \{m_e\}_{e \in E(\gamma)}, \vec{n} = \{n_e\}_{e \in E(\gamma)}$ $\bullet$ $\bullet\bullet\bullet$
$m_e, n_e = 1, \ldots, d_{\pi_e}$ $\bullet\bullet$ $\bullet\bullet\bullet\bullet$ $\bullet$ $\bullet\bullet\bullet\bullet$ $\bullet\bullet\bullet$ $\bullet\bullet\bullet$ $\bullet\bullet$ $\bullet\bullet\bullet$ $\bullet\bullet\bullet\bullet\bullet\bullet\bullet\bullet\bullet\bullet\bullet$ $\bullet\bullet\bullet\bullet\bullet\bullet$ $\bullet$
$\bullet$ $\bullet\bullet\bullet\bullet$ $\bullet$ $\bullet\bullet\bullet\bullet\bullet\bullet$ $v \in V(\gamma)$ $\bullet\bullet\bullet\bullet\bullet\bullet\bullet\bullet$ $\bullet\bullet\bullet$ $\bullet\bullet\bullet\bullet\bullet\bullet$ $\bullet\bullet$ $\bullet\bullet\bullet\bullet\bullet$ $\bullet\bullet\bullet\bullet\bullet$ $\bullet\bullet$ $E_v^b(\gamma) := \{e \in$
$E(\gamma); b(e) = v\}$ $\bullet\bullet\bullet$ $E_v^f(\gamma) := \{e \in E(\gamma); f(e) = v\}\bullet$ $\bullet\bullet\bullet$ $\bullet\bullet\bullet\bullet$ $v \in V(\gamma)\bullet$ $\bullet\bullet\bullet\bullet$
$\bullet\bullet\bullet$ $\bullet\bullet\bullet$ $\bullet\bullet\bullet\bullet\bullet\bullet$ $\bullet\bullet\bullet\bullet\bullet\bullet\bullet$ $\bullet\bullet\bullet\bullet\bullet\bullet\bullet\bullet$ $\bullet\bullet\bullet\bullet\bullet$

$$\left(\otimes_{e \in E_v^b(\gamma)} \pi_e\right) \otimes \left(\otimes_{e \in E_v^f(\gamma)} \pi_e^c\right) \qquad (8.2.61)$$

• •••• $h \mapsto \pi^c(h) := \pi(h^{-1})^T$ •••••••• ••• ••••••••••••• •••• ••• ••••••••••• ••• •• $\pi \cdot (.)^T$
••• ••••• • ••••• •••••••••• •• ••••••• •• •• ••• ••••• ••••• ••••••••••• •• ••• •
•• ••••• •••••••• ••• •••• ••• •••• •• •••••••••• ••• ••• •• •••••••••• •••••••
••• ••••• • •••• ••••••••• • •• ••••• •• •••• ••• ••• •• $I_v(\vec{\pi}, \pi'_v)$ •• ••• •••• ••
• ••• ••••••••• •••• •••• •• •••• •• ••••• •••••••• •• ••••••••• ••• • ••••• •••
•••••• ••• •• $\pi'_v \in \Pi$ • ••• $\pi_t \in \Pi$ • ••••••••••••• •••• • ••••••••••••••• •
• ••• ••• $I_v \in \mathcal{I}_v(\vec{\pi}, \pi'_v)$ •• ••••• •• ••••••• ••• ••• • • ••••• ••• • ••••• ••••
•• •• ••••• •• ••• •• • •••••• $\vec{I} = (I_v)_{v \in V(\gamma)}$• • • •••••••• ••••• •• • ••• •••••••
••• •••••••• •• •••• • •••••••• •• •• ••• $I_v \in \mathcal{I}_v(\vec{\pi}, \pi'_v)$ •• ••• ••• ••• •
••• •• ••• • ••• • ••••••••••• •• •••••• ••• ••• •• •••••

$$A \mapsto \left(\otimes_{e \in E^b_v(\gamma)} \pi_e(A(e))\right) \otimes \left(\otimes_{e \in E^f_v(\gamma)} \pi_e(A(e))\right) \tag{8.2.62}$$

•••• •••• •• • •• ••• ••••••••••• •••••••• ••••• •••• ••• •••••• ••••• •• v ••
•• •••••• •• •••• •••• •• •• ••• •••••••••••• •••• I_v •• v •••• •••••••• • ••• ••• •• ••••
•• •••• $I_v \in \mathcal{I}_v(\vec{\pi}, \pi'_v)$• • • ••• •••• ••• •••••••

$$A \mapsto \otimes_{e \in E(\gamma)} \pi_e(A(e)) \tag{8.2.63}$$

••• ••• •••• •••••• v •••••••• ••• •••••••• ••••• •••••••• •••• •••••••• • ••• ••
•••••••••• •• ••••• •••• I_v• ••• •••••• •• • ••••••••••••• •• $\overline{\mathcal{A}}$ •••• $l = l(\gamma)$
• •••• •• • •••••• •• $T_{\gamma, \vec{\pi}, \vec{I}}(A)$ ••• • •••• ••••••••• • •• ••• ••••••••••••• I_v •• v•
•• • • •••• ••• π'_v, I_v •••• ••• ••• •• •••••• $T_{\gamma, \vec{\pi}, \vec{I}}$ •••• ••• ••• • ••••• •••••
•• ••• ••••• •• •••••• $T_{\gamma, \vec{\pi}, \vec{m}, \vec{n}}$• •• •••••••• ••• •• •• •••• ••••• ••••• •• ••
•• •••• •••• • ••• ••••••• •• $\mathcal{H}_0$•

••• • •••• •••• ••••• ••••• ••• ••• ••••••••••• ••••••••• ••• ••••• ••••
•• ••• ••••• • •••• •••••• •• ••••• ••• ••••• •••••• •• •• • •••• ••• ••••
• •• ••••••• •••• ••••• •• ••••• •••••••• ••• ••• •••• $C^{(1)}$ •••• ••• •• •••• • ••
•••• • ••••• • •••• •• •••••• •• •••• •••••• •• • ••• •••••••••••• π_t••
• •••• •• ••• ••••••••• $T_{\gamma, \vec{\pi}, \vec{I}}$ •• •• •• • ••••• ••••• •• •••• I_v •• ••• •••••••••
•••• •• •••••• •••••• ••••• ••••••• ••• ••• •• •••• $C^{(1)}$ •••• ••• • •••••
•••••• •••••••••• ••• •••• ••••• •• •••••••• •••••••••••••• π_e ••• • ••• •••••••
••••••• ••• $\pi_e = \pi_t$ •• • •• •••••••• •• $E(\gamma) - \{e\}$•
•••• ••• •••••••••••• •••• •••••• ••• •••••••• ••• •••
••••••••••••• •••• •• ••••• • ••• π'_v •••••• •• •••• ••• •••••• •• π_t • ••• •••
••••••••• •••• $T_{\gamma, \vec{\pi}, \vec{I}}$ •••••••• •• $\mathcal{I}_v(\vec{\pi}, \pi^t) = \emptyset$ ••• ••• $v \in V(\gamma)$• • ••••
••••• •••••• •••••••••••••••• •• • ••••• ••••• •••• •••••

$T_{\gamma, \vec{\pi}, \vec{I}} : \overline{\mathcal{A}/\mathcal{G}} \to \mathbb{C}$•

For the concrete example of G = SU(2), spin-network functions are analysed in more detail in Chapter 32.

The importance of spin-network functions is that they provide a basis for $\mathcal{H}^0$.

Theorem 8.2.12

••• • ••• •••••••••••••• •• ••• •• ••• ••• ••••• •••••• •• •••••••••• •••••••• ••• •••

• •••••• •••••• $L_2(\overline{\mathcal{A}}, d\mu_0)$•

$L_2(\overline{\mathcal{A}/\mathcal{G}}, d\mu_0)$.

(i) The inner product on $L_2(\overline{\mathcal{A}}, d\mu_0)$ is defined by

$$< f, f' >_{L_2(\overline{\mathcal{A}}, d\mu_0)} := \Lambda_{\mu_0}(\overline{f} f') \qquad (8.2.64)$$

where Λ_{μ_0} is the positive linear functional on $C(\overline{\mathcal{A}})$ determined by μ_0 via the Riesz representation theorem. The cylinder functions of the form $p_l^* f_l$, $f_l \in C(X_l)$ are dense in $C(\overline{\mathcal{A}})$ (in the sup-norm) and since $\overline{\mathcal{A}}$ is a (locally) compact Hausdorff space and μ_0 comes from a positive linear functional on the space of continuous functions on $\overline{\mathcal{A}}$ (of compact support), these functions are dense in $L_2(\overline{\mathcal{A}}, d\mu_0)$ (in the L_2 norm $||f||_2 = < f, f >^{1/2}$). This follows again from Lusin's theorem, Theorem 25.1.14 (see, e.g., [552]). It follows that $L_2(\overline{\mathcal{A}}, d\mu_0)$ is the completion of $\mathrm{Cyl}(\overline{\mathcal{A}})$ in the L_2 norm. Now

$$\mathrm{Cyl}(\overline{\mathcal{A}}) = \bigcup_{l \in \mathcal{L}} p_l^* C(X_l) \qquad (8.2.65)$$

and since by the same remark $C(X_l)$ is dense in $L_2(X_l, d\mu_{0l})$ it follows that

$$L_2(\overline{\mathcal{A}}, d\mu_0) = \overline{\bigcup_{l \in \mathcal{L}} p_l^* L_2(X_l, d\mu_{0l})} \qquad (8.2.66)$$

Now by definition $(\rho_l)_* \mu_{0l} = \otimes_{e \in E(\gamma)} \mu_H$ for $l = l(\gamma)$ so that $L_2(X_l, d\mu_{0l})$ is isometric isomorphic with $L_2(G^{|E(\gamma)|}, \otimes^{|E(\gamma)|} d\mu_H)$, which in turn is isometric isomorphic with $\otimes_{e \in E(\gamma)} L_2(G, d\mu_H)$ since $\otimes^{|E(\gamma)|} \mu_H$ is a finite product of measures. By the Peter and Weyl theorem proved in Chapter 31 the matrix element functions

$$\pi_{mn} : G \to \mathbb{C}; h \mapsto \sqrt{d_\pi} \pi_{mn}(h), \pi \in \Pi, m, n = 1, \ldots, d_\pi \qquad (8.2.67)$$

form an orthonormal basis of $L_2(G, d\mu_H)$ for any compact gauge group G, that is,

$$< \pi_{mn}, \pi'_{m'n'} > := \int_G d\mu_H(h) \overline{\pi_{mn}(h)} \pi'_{m'n'}(h) = \frac{\delta_{\pi\pi'} \delta_{mm'} \delta_{nn'}}{d_\pi} \qquad (8.2.68)$$

This shows that functions of the form (8.2.60) span $L'_2(X_l, d\mu_{0l})$, which by definition is isomorphic to $\otimes^{|E(\gamma)|} L'_2(G, d\mu_H)$ restricted to non-trivial intertwiners for two-valent vertices whose adjacent edges are analytical continuations of each other. Here $L'_2(G, d\mu_H)$ is the closed linear span of the functions π_{mn} with $\pi \neq \pi_t$ (only non-trivial representations allowed).

It remains to prove (1) that $p_l^* L'_2(X_l, d\mu_{0l}) \perp p_{l'}^* L'_2(X_{l'}, d\mu_{0l'})$ unless $l = l'$ and (2) that $L_2(X_l, d\mu_{0l}) = \oplus_{l' \prec l} L'_2(X_{l'}, d\mu_{0l'})$ where completion is with respect to $L_2(X_l, d\mu_{0l})$.

(1) To see the former, notice that if $l = l(\gamma) \neq l' = l(\gamma')$ there is $l, l' \prec l'' := l(\gamma \cup \gamma')$. Since $\gamma \neq \gamma'$ are semianalytic, there must be either (A) an edge $e \in E(\gamma)$ which contains a segment $s \subset e$ that is disjoint from γ' and

this segment is certainly contained in $\gamma \cup \gamma'$ or (B) the ranges of γ and γ' actually coincide but there is at least one two-valent vertex v of γ such that the adjacent edges are at least $C^{(1)}$ continuations of each other and such that the corresponding intertwiner is non-trivial (reverse the roles of γ, γ' if necessary) while v is simply an interior point of an edge of γ' and thus carries a trivial intertwiner.

Let $f_l \in L'_2(X_l, d\mu_{0l}), f_{l'} \in L'_2(X_{l'}, d\mu_{0l'})$ then

$$< p_l^* f_l, p_{l'}^* f_{l'} > = \mu_{0l''}(\overline{p_{l''l}^* f_l} p_{l''l'}^* f_{l'}) = 0 \qquad (8.2.69)$$

This follows since $p_{l''l}^* f_l, p_{l''l'}^* f_{l'}$ are (Cauchy sequences of) functions $T_{\gamma, \vec{\pi}, \vec{I}}$ over $\gamma \cup \gamma'$ where either (A) the dependence on s of the former function is through a non-trivial representation and of the latter through a trivial representation or (B) the dependence of the former is through a non-trivial intertwiner at v but through a trivial one for the latter. Hence, in case (A) the claim follows from formula (8.2.68). In case (B) the claim follows from the fact that due to gauge-invariance of $\mu_{0l''}$ under gauge transformations at v (to be demonstrated below) we have

$$\int d\mu_{0l''} \overline{T_{\gamma, \vec{\pi}, \vec{I}}} T_{\gamma', \vec{\pi}', \vec{I}'} = \int d\mu_H(g) \int d\mu_{0l''} \circ \lambda_g^{l''} \overline{T_{\gamma, \vec{\pi}, \vec{I}}} T_{\gamma', \vec{\pi}', \vec{I}'}$$

$$= \int d\mu_H(g) \int d\mu_{0l''} \overline{T_{\gamma, \vec{\pi}, \vec{I}} \circ \lambda_{g-1}^l} T_{\gamma', \vec{\pi}', \vec{I}'} \circ \lambda_{g-1}^{l'}$$

$$= \int d\mu_{0l''} \left(\left[\int d\mu_H(g) \, \pi_{I_v}(g) \right] \cdot \overline{T_{\gamma, \vec{\pi}, \vec{I}}} \right) T_{\gamma', \vec{\pi}', \vec{I}'}$$

$$= 0 \qquad (8.2.70)$$

again due to (8.2.68) where λ^l denotes the gauge group action on X_l as before, which reduces to π_{I_v} by construction when g is non-trivial at v only.

(2) To see the latter, observe that $L_2(G, d\mu_H) = \overline{L'_2(G, d\mu_H) \oplus \text{span}(\{1\})}$ and that a function cylindrical over γ which depends on $e \in E(\gamma)$ through the trivial representation is cylindrical over $\gamma - e$ as well.

Summarising, if we define $\mathcal{H}_l^0 := p_l^* L_2(X_l, d\mu_{0l})$, $\mathcal{H}^{0l} := p_l^* L'_2(X_l, d\mu_{0l})$ then

$$\mathcal{H}^0 = \overline{\bigcup_{l \in \mathcal{L}} \mathcal{H}_l^0} = \overline{\oplus_{l \in \mathcal{L}} \mathcal{H}^{0l}} \qquad (8.2.71)$$

(ii) The assertion follows easily from (i) and the fact that $L_2(\overline{\mathcal{A}/\mathcal{G}}, d\mu_0)$ is simply the restriction of $L_2(\overline{\mathcal{A}}, d\mu_0)$ to the gauge-invariant subspace: that subspace is the closed linear span of gauge-invariant spin-network states by (i) and the specific choice that we have made in Definition 8.2.11 shows that they form an orthonormal system since we have chosen them to be normalised and the intertwiners to be projections onto mutually orthogonal subspaces of a tensor product representation space of G. More specifically, the inner product between two spin-network functions $T_{\gamma, \vec{\pi}, \vec{I}}, T_{\gamma', \vec{\pi}', \vec{I}'}$ is non-vanishing only if

$\gamma = \gamma'$ and $\vec{\pi} = \vec{\pi}'$. In that case, consider $v \in V(\gamma)$ and assume w.l.g. that all edges $e_1, \ldots, e_N$ incident at v are outgoing. An intertwiner $I_v \in \mathcal{I}_v(\vec{\pi}, \pi_t)$ can be thought of as a vector $I_v^{n_1,\ldots,n_N} := (I_v)_{m_1^0,\ldots,m_N^0;n_1,\ldots,n_N}$ in the representation space of the representation $\otimes_{I=1}^N \pi_I$ where m_I^0 are some matrix elements that we fix once and for all. Since I_v is a trivial representation and in particular represents $1_G = (1_G)^T$ we have $(I_v)_{m_1^0,\ldots,m_N^0;n_1,\ldots,n_N} = I_v^{n_1,\ldots,n_N} := (I_v)_{n_1,\ldots,n_N;m_1^0,\ldots,m_N^0}$, moreover the intertwiners are real-valued because the functions $\pi_{mn}(h)$ depend analytically on h and 1_G is real-valued. Now the spin-network state restricted to its dependence on $e_1, \ldots, e_N$ is of the form

$$I_v^{n_1,\ldots,n_N} \left[\otimes_{I=1}^N \pi_I(A(e_I)) \right]_{n_1,\ldots,n_N;k_1,\ldots,k_N} \tag{8.2.72}$$

It follows from (8.2.68) that the inner product between $T_{\gamma,\vec{\pi},\vec{I}}, T_{\gamma,\vec{\pi},\vec{I}'}$ will be proportional to

$$I_v^{n_1,\ldots,n_N} (I')_v^{n_1,\ldots,n_N} = [(I_v)(I_v')]_{m_1^0,\ldots,m_N^0;m_1'^0,\ldots,m_N'^0} \propto \delta_{I_v I_v'} \tag{8.2.73}$$

(if $I_v = I_v'$ then $m_I^0 = m_I^{0'}$ by construction) since the I_v are representations on mutually orthogonal subspaces. $\qquad\square$

We remark that the spin-network basis is not countable because the set of graphs in σ is not countable, whence $\mathcal{H}^0$ is not separable. We will see that this is even the case after modding out by spatial semianalytic diffeomorphisms, although one can show that after modding out by diffeomorphisms the remaining space is an orthogonal, uncountably infinite, almost direct sum of mutually isomorphic, separable Hilbert spaces [437] which might be superselected in terms of the full algebra of observables (i.e., they are separately left-invariant).

Definition 8.2.13. • •• •••••••••• ••• •• ••• ••• ••••••• •• ••• •
••••• $\tilde{\mathcal{H}}^0$ •• •••• •••••••• •• •• •••••• •

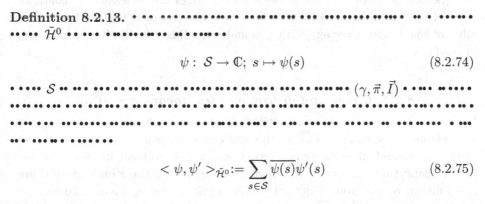

$$\psi: \mathcal{S} \to \mathbb{C}; \ s \mapsto \psi(s) \tag{8.2.74}$$

• ••••• $\mathcal{S}$ •• ••• ••• •••• •• •••• •••••• •••• ••• ••• ••• •• ••••• ••• $(\gamma, \vec{\pi}, \vec{I})$ • •••• •• •••••
••• •• ••• ••• •••••• ••• •••• •••• ••• •••• •••••••••••• •• ••• ••••••••••••• ••••• •
• ••• ••• •••••••••• ••••• •••••• ••• ••••••••• • ••• •••••• ••••• •• ••• •••• • •••
•• • ••• •• • •••••••

$$< \psi, \psi' >_{\tilde{\mathcal{H}}^0} := \sum_{s \in \mathcal{S}} \overline{\psi(s)} \psi'(s) \tag{8.2.75}$$

•••• ••• ••••• ••• ••• • ••• ••••••• ••

Clearly the uncountably infinite sum (8.2.75) converges if and only if $\psi(s) = 0$ except for countably many $s \in \mathcal{S}$. The next corollary shows that the connection representation that we have been dealing with so far and the spin-network representation are in a precise sense Fourier transforms of each other where the role of the kernel of the transform is played by the spin-network functions.

Corollary 8.2.14. • •• ••• ••• •• ••• ••• ••• •••• •••• •• •• ••

$$T : \mathcal{H}^0 \to \tilde{\mathcal{H}}^0; \; f \mapsto \tilde{f}(s) :=< T_s, f >_{\mathcal{H}^0} \qquad (8.2.76)$$

•• • •• ••• •• •••• ••• •• ••••• •••• ••• • ••••• •••••• • ••• •• ••••••

$$(T^{-1}\psi)(A) := \sum_{s \in S} \psi(s) T_s(A) \qquad (8.2.77)$$

• ••••• If $f \in \mathcal{H}^0$ then

$$f = \sum_{s \in S} < T_s, f > T_s \qquad (8.2.78)$$

since the T_s form an orthonormal basis (Bessel's inequality is saturated). Since the T_s form an orthonormal system we conclude that $||f||^2 = \sum_s | < T_s, f > |^2$ converges, meaning in particular that $< T_s, f >= 0$ except for countably many $s \in S$. It follows that $||Tf||^2 := \sum_s |\tilde{f}(s)|^2 = ||f||^2$ which shows that T is a partial isometry. Comparing (8.2.77) and (8.2.78) we see that $T^{-1}\tilde{f} = f$ is indeed the inverse of T. Finally, again by the orthogonality of the T_s, we have $||T^{-1}\psi||^2 = \sum_s |\psi(s)|^2 = ||\psi||^2$ so that T^{-1} is a partial isometry as well. Since T is a bijection, T is actually an isometry. Notice that $\tilde{T}_s(s') = \delta_{s,s'}$. □

Whenever it is convenient we may therefore think of states either in the loop or the connection representation. In this book we will work entirely in the connection representation.

• •• •••: As we have seen, the $T_{\gamma,\vec{\pi},\vec{m},\vec{n}}$ with all π_e non-trivial almost form an orthonormal basis, we just have to be careful to contract with non-trivial intertwiners at two-valent vertices whose adjacent edges are at least $C^{(1)}$ continuations of each other. If we contract with a trival intertwiner, that vertex is actually not counted as a vertex. With this understanding we will often use $T_{\gamma,\vec{\pi},\vec{m},\vec{n}}$ instead of $T_{\gamma,\vec{\pi},\vec{I}}$.

8.2.5 Gauge and diffeomorphism invariance of μ_0

In the previous subsection we investigated the topological and measure theoretical relation between $\mathcal{A}$ and $\overline{\mathcal{A}}$. In this subsection we will investigate the action of the gauge and diffeomorphism group on $\overline{\mathcal{A}}$. The uniform measure has two further important properties: it is invariant under both the gauge group $\overline{\mathcal{G}}$ and the diffeomorphism group $\mathrm{Diff}^\omega_{\mathrm{sa}}(\sigma)$ (semianalytic diffeomorphisms). To see this, recall the action of $\overline{\mathcal{G}}$ on $\overline{\mathcal{A}}$ defined through its action on the subspaces X_l by $x_l \mapsto \lambda_g(x_l)$ with $[\lambda_g(x_l)](p) = g(b(p))x_l(p)g(f(p))^{-1}$ for any $p \in l$. This action has the feature of leaving the X_l invariant for any $l \in \mathcal{L}$ and therefore lifts to $\overline{X}$ as $x \mapsto \lambda_g(x)$ with $[\lambda_g(x)](p) = g(b(p))x(p)g(f(p))^{-1}$ for any $p \in \mathcal{L}$. Likewise we have an action of $\mathrm{Diff}^\omega_{\mathrm{sa}}(\sigma)$ on $\overline{X}$ defined by

$$\delta^l : \; \mathrm{Diff}^\omega_{\mathrm{sa}}(\sigma) \times X_l \to X_{\varphi(l)}; \; (\varphi, x_l) \mapsto \delta^l_\varphi(x_l) = x_{\varphi(l)} \qquad (8.2.79)$$

where $\varphi(l) = l(\varphi(\gamma))$ if $l = l(\gamma)$. This action does not preserve the various X_l. The action on all of $\overline{X}$ is then evidently defined by

$$\delta : \text{Diff}^\omega_{sa}(\sigma) \times \overline{X} \to \overline{X}; \ (\varphi, x = (x_l)_{l \in \mathcal{L}}) \mapsto \delta_\varphi(x) = \left(\delta^l_\varphi(x_l)\right)_{l \in \mathcal{L}} \qquad (8.2.80)$$

Clearly $\delta_\varphi(x)$ is still an element of the projective limit since it just permutes the various x_l among each other. Moreover, $l \prec l'$ iff $\varphi(l) \prec \varphi(l')$ so the diffeomorphisms preserve the partial order on the label set. Therefore

$$p_{\varphi(l')\varphi(l)}\left(\delta^{l'}_\varphi(x_{l'})\right) = x_{\varphi(l)} = \delta^l_\varphi(p_{l'l}(x_{l'})) \qquad (8.2.81)$$

for any $l \prec l'$, so we have equivariance

$$p_{\varphi(l')\varphi(l)} \circ \delta^{l'}_\varphi = \delta^l_\varphi \circ p_{l'l} \qquad (8.2.82)$$

It is now easy to see that for the push-forward measures we have $(\lambda_g)_*\mu_0 = \mu_0$, $(\delta_\varphi)_*\mu_0 = \mu_0$. For any $f = p_l^* f_l \in C(\overline{X})$, $f_l = \rho_l^* F_l \in C(X_l)$, $F_l \in C(G^{|E(\gamma)|})$, $l = l(\gamma) \in \mathcal{L}$ we have

$$\mu_0(\lambda^*_g f) = \mu_0\left(p_l^*\left(\lambda^l_g\right)^* f_l\right) = \mu_{0l}\left(\left(\lambda^l_g\right)^* f_l\right)$$

$$= \int_{G^{|E(\gamma)|}} \left[\prod_{e \in E(\gamma)} d\mu_H(h_e) \right] F_l\left(\{g(b(e))h_e g(f(e))^{-1}\}_{e \in E(\gamma)}\right)$$

$$= \int_{G^{|E(\gamma)|}} \left[\prod_{e \in E(\gamma)} d\mu_H(g(b(e))^{-1}h_e g(f(e))) \right] F_l\left(\{h_e\}_{e \in E(\gamma)}\right)$$

$$= \int_{G^{|E(\gamma)|}} \left[\prod_{e \in E(\gamma)} d\mu_H(h_e) \right] F_l\left(\{h_e\}_{e \in E(\gamma)}\right) = \mu_0(f) \qquad (8.2.83)$$

where we have made a change of integration variables $h_e \to g(b(e))h_e g(f(e))^{-1}$ and used the fact that the associated Jacobian equals unity for the Haar measure (translation invariance). Next

$$\mu_0(\delta^*_\varphi f) = \mu_0\left(p^*_{\varphi(l)}\left(\delta^l_\varphi\right)^* f_l\right) = \mu_{0\varphi^{-1}(l)}\left(\left(\delta^l_\varphi\right)^* f_l\right)$$

$$= \int_{G^{|E(\varphi(\gamma))|}} \left[\prod_{e \in E(\varphi(\gamma))} d\mu_H(h_e) \right] F_l\left(\{h_e\}_{e \in E(\varphi(\gamma))}\right)$$

$$= \int_{G^{|E(\gamma)|}} \left[\prod_{e \in E(\gamma)} d\mu_H(h_e) \right] F_l\left(\{h_e\}_{e \in E(\gamma)}\right) = \mu_0(f) \qquad (8.2.84)$$

where we have written $\{h_e\}_{e \in E(\varphi(\gamma))} = \{h_{\varphi(e)}\}_{e \in E(\gamma)}$ and have performed a simple relabelling $h_{\varphi(e)} \to h_e$. It is important to notice that in contrast to other measures on some space of connections the 'volume of the gauge group is finite':

the space $C(\overline{\mathcal{A}/\mathcal{G}})$ is a subspace of $C(\overline{\mathcal{A}})$ and we may integrate them with the measure μ_0, which is the same as integrating them with the restricted measure. We do not have to fix a gauge and never have to deal with the problem of Gribov copies.

One may ask now why one does not repeat with the diffeomorphism group what has been done with the gauge group: passing from semianalytic diffeomorphisms $\mathrm{Diff}_{\mathrm{sa}}^\omega(\sigma)$ to distributional ones $\overline{\mathrm{Diff}(\sigma)}$ and passing to the quotient space $(\overline{\mathcal{A}/\mathcal{G}})/\overline{\mathrm{Diff}(\sigma)}$. There are two problems:

First, in the case of $\overline{\mathcal{G}}$ there was a natural candidate for the extension $\mathcal{G} \to \overline{\mathcal{G}}$ but this is not the case for diffeomorphisms because distributional diffeomorphisms will not lie in any differentiability category any more, they might be arbitrarily discontinuous bijections (e.g., arbitrary permutations of points) and hence much of the structure of present LQG does not generalise, for example, paths and fluxes would not remain semianalytic. The most general structure-preserving extension would be the set of all maps that preserve piecewise analyticity (rather than semianalyticity) of paths and surfaces, however it is not clear that such maps are invertible and thus form a group. The structure of a group would be desirable in order to be able to solve the spatial diffeomorphism constraint by group averaging.

Second, as we will now show, even the entire analytic diffeomorphisms act ergodically on the measure space, which means that there are no non-trivial invariant functions.

Thus, one either has to proceed differently (e.g., downsizing rather than extending the diffeomorphism group), change the representation or solve the diffeomorphism constraint differently. We will select the third option in Chapter 9. It should be pointed out, however, that the last word of how to deal with diffeomorphism invariance has not been spoken yet. In a sense, it is one of the ••• •••••••• for the following reason: the concept of a smooth spacetime should not have any meaning in a quantum theory of the gravitational field where probing distances beyond the Planck length must result in black hole creation, which then evaporate in Planck time, that is, spacetime should be fundamentally discrete. But clearly smooth diffeomorphisms or even homeomorphisms have no room in such a discrete quantum spacetime. The fundamental symmetry is probably something else, maybe a combinatorial one, that looks like a diffeomorphism group at large scales. See Section 10.6.5 for a proposal.

Also, if one wants to allow for topology change in quantum gravity then talking about the diffeomorphism group for a fixed σ does not make much sense. We see that there is a tension between classical diffeomorphism invariance and the discrete structure of quantum spacetime which in the opinion of the author has not been satisfactorily resolved yet and which we consider as one of the most important conceptual problems left open so far. A first step in overcoming this tension might be the extended Master Constraint proposal sketched in Section 10.6.4.

8.2.6 $^+$ Ergodicity of μ_0 with respect to spatial diffeomorphisms

We show that μ_0 has an important ergodicity property with respect to spatial diffeomorphisms which is the underlying reason for why the solutions to the spatial diffeomorphism constraint, in contrast to the solutions to the Gauß constraint, do not lie in $\mathcal{H}_0$. We follow closely [418].

The above discussion reveals that as far as $\overline{\mathcal{G}}$ and $\mathrm{Diff}(\sigma)$ are concerned we have the following abstract situation (see Chapter 25): we have a measure space with a measure-preserving group action of both groups (so that the pull-back maps $\lambda_g^*, \delta_\varphi^*$ provide unitary actions on the Hilbert space) and the question is whether that action is ergodic. That is certainly not the case with respect to $\overline{\mathcal{G}}$ since the subspace of gauge-invariant functions is by far not the span of the constant functions as we have shown.

Theorem 8.2.15. $\cdots \bullet \bullet \bullet \bullet \bullet \bullet \bullet \bullet_0^\omega(\sigma) \cdots \cdots \cdots \cdots \cdots \cdots \cdots \cdots$
$\cdots \cdots \cdots \sigma \cdots \cdots \cdots \cdots \cdots \cdots \cdots \cdots \cdots \cdots$
$\overline{A} \cdots \cdots \cdots \cdots \cdots \cdots \cdots \cdots \cdots \mu_0.$

$\bullet \bullet \bullet \bullet \bullet$ The diffeomorphism group acts unitarily on $\mathcal{H}^0$ via

$$[\hat{U}(\varphi)f](A) = f(\delta_\varphi(A)) \qquad (8.2.85)$$

which means for spin-network states that $\hat{U}(\varphi)T_s = T_{\varphi(s)}$ where

$$\varphi(s) = \left(\varphi(\gamma),\, \{\pi_{\varphi(e)} = \pi_e\}_{e \in E(\gamma)},\, \{m_{\varphi(e)} = m_e\}_{e \in E(\gamma)},\, \{n_{\varphi(e)} = n_e\}_{e \in E(\gamma)}\right) \qquad (8.2.86)$$

for $s = (\gamma, \vec{\pi}, \vec{m}, \vec{n})$. Let now $f = \sum_{s \in S} c_s\, T_s \in \mathcal{H}^0$ be given with $c_s = 0$ except for countably many. Suppose that $\hat{U}(\varphi)f = f$ μ_0-a.e. for any $\varphi \in \mathrm{Diff}_0^\omega(\sigma)$. Since S is left-invariant by diffeomorphisms, this means that

$$\sum_s c_s T_{\varphi(s)} = \sum_s c_{\varphi^{-1}(s)} T_s = \sum_s c_s T_s \qquad (8.2.87)$$

for all φ. Since the T_s are mutually orthogonal we conclude that $c_s = c_{\varphi(s)}$ for all $\varphi \in \mathrm{Diff}_0^\omega(\sigma)$. Now for any $s \neq s_0 = (\emptyset, \vec{0}, \vec{0}, \vec{0})$ the orbit $[s] = \{\varphi(s);\ \varphi \in \mathrm{Diff}_0^\omega(\sigma)\}$ contains infinitely many different elements (take any vector field that does not vanish in an open set which contains the graph determined by s and consider the one-parameter subgroup of diffeomorphisms determined by its integral curve – this is where we can make the restriction to the identity component). Therefore $c_s = \mathrm{const.}$ for infinitely many s. Since f is normalisable, this is only possible if $\mathrm{const.} = 0$, hence $f = c_{s_0} T_{s_0}$ is constant μ_0-a.e. and therefore δ ergodic. $\qquad \square$

We see that the theorem would still hold if we replaced $\mathrm{Diff}_0^\omega(\sigma)$ by any infinite subgroup D with respect to which each orbit $[s]$, $s \neq s_0$ is infinite. An example would be the case $\sigma = \mathbb{R}^D$ and D a discrete subgroup of the translation group given by integer multiples of translations by a fixed non-zero vector.

The theorem shows that the only vectors in $\mathcal{H}^0$ invariant under diffeomorphisms are the constant functions, hence we cannot just pass to that trivial subspace in order to solve the diffeomorphism constraint. The solution to the problem lies in passing to a larger space of functions, distributions over a dense subspace of $\mathcal{H}^0$ in which one can solve the constraint. The proof of the theorem shows already how that distributional space must look: it must allow for uncountably infinite linear combinations of the form $\sum_s c_s T_s$ where c_s is a generalised knot invariant (i.e., $c_s = c_{\varphi(s)}$ for any φ, generalised because $\gamma(s)$ has in general self-intersections and is not a regular knot). This is already the basic idea for how to solve the diffeomorphism constraint in step IV of the quantisation programme.

8.2.7 Essential self-adjointness of electric flux momentum operators

We had established in Section 6.5 that the flux vector fields $Y_n(S)$ are well-defined derivatives on Cyl^∞. To finally finish the proof that $L_2(\overline{\mathcal{A}}, d\mu_0)$ is a representation of $\mathfrak{A}$ we must show that the corresponding electric flux momentum operators are essentially self-adjoint (or their exponentials unitary). To do this, consider the family of divergences of $Y_n(S)$ with respect to the uniform measure μ_0. Now the projection μ_{0l} is simply the Haar measure on $G^{|E(\gamma)|}$. Since the Haar measure is right- and left-invariant, that is, $(L_h)_* \mu_H = \mu_H = (R_h)_* \mu_H$ we have $\mathrm{div}_{\mu_H} R_j = \mathrm{div}_{\mu_H} L_j = 0$ as the following calculation shows:

$$-\int_{\mathrm{G}} \mu_h [\mathrm{div}_{\mu_H} R_j] f = +\int_{\mathrm{G}} \mu_H R_j(f) = \left(\frac{d}{dt}\right)_{t=0} \int_{\mathrm{G}} \mu_H L^*_{e^{t\tau_j}} f$$

$$= \left(\frac{d}{dt}\right)_{t=0} \int_{\mathrm{G}} (L_{e^{t\tau_j}})_* \mu_H f = 0 \tag{8.2.88}$$

It follows that $\mathrm{div}_{\mu_{0l}} Y_l^n(S) = 0$ so that $Y_n(S)$ is automatically μ_0-compatible (and the divergence is real-valued).

Since $Y_n(S)$ is a consistently defined smooth vector field on $\overline{\mathcal{A}}$ which is μ_0-compatible, all the results from Section 8.2.2 with respect to the definition of corresponding momentum operators apply and the remaining question is whether the family of symmetric operators $P_l^n(S) := i Y_l^j(S)$ with dense domain $D(P_l^j(S)) = C^1(X_l)$ is an essentially self-adjoint family.

Looking at (6.5.3), essential self-adjointness of $P_l^n(S)$ on $L_2(X_l, d\mu_{0l})$ will follow if we can show that iR_j is essentially self-adjoint on $L_2(\mathrm{G}, d\mu_H)$ with core $C^1(\mathrm{G})$. That they are symmetric operators we know already. Now we invoke the Peter and Weyl theorem that tells us

$$L_2(\mathrm{G}, d\mu_H) = \overline{\oplus_{\pi \in \Pi} L_2(\mathrm{G}, d\mu_H)_{|\pi}} \tag{8.2.89}$$

where Π is a collection of representatives of irreducible representations of G, one for each equivalence class, and $L_2(\mathrm{G}, d\mu_H)_{|\pi}$ is the closed subspace of $L_2(\mathrm{G}, d\mu_H)$

spanned by the matrix element functions $h \mapsto \pi_{mn}(h)$. The observation is now that R_j leaves each $L_2(G, d\mu_H)_{|\pi}$ separately invariant. For instance

$$(R_j \pi_{mn})(h) = \left(\frac{d\pi_{mm'}(e^{t\tau_j})}{dt} \right)_{t=0} \pi_{m'n}(h) \tag{8.2.90}$$

It follows that iR_j are symmetric operators on the finite-dimensional Hilbert space $L_2(G, d\mu_H)_{|\pi}$ of dimension $\dim(\pi)^2$ and therefore are self-adjoint. Since the matrix element functions are smooth, by the basic criterion of essential self-adjointness it follows that $(i(R_j)_{|\pi} \pm i \cdot 1_\pi)C^\infty(G)_{|\pi}$ is dense in $L_2(G, d\mu_H)_{|\pi}$, hence so is $(i(R_j)_{|\pi} \pm i \cdot 1_\pi)C^1(G)_{|\pi}$. Correspondingly,

$$(iR_j \pm i \cdot 1)C^\infty(G) = \oplus_{\pi \in \Pi} (i(R_j)_{|\pi} \pm i \cdot 1_\pi)C^\infty(G)_{|\pi} \tag{8.2.91}$$

is dense in $L_2(G, d\mu_H)$ and thus iR_j is essentially self-adjoint.

This completes the existence proof.

8.3 Uniqueness proof: (2) uniqueness

The goal of this section is to show that the representation defined in the previous section is the unique GNS representation which derives from a $\mathfrak{G}$-invariant state.

Notice that by the Weyl relations (7.1.3) we may write any element a of $\mathfrak{A}$ as a finite linear combination of elements of the form $f_{t_1 \ldots t_n} \cdot W_{t_1} \ldots W_{t_n}$ where W_{t_k} is a (generalised) Weyl element and $f_{t_1 \ldots t_n} \in \mathrm{Cyl}^\infty$ depend smoothly on $t_1, \ldots, t_n$. Hence, if Ω_ω is a common C^∞-vector for all the $\pi_\omega(Y)$, $Y \in \mathfrak{P}$ defined as the self-adjoint generators of the corresponding $\pi(W_t)$, then all the elements $\pi_\omega(a)\Omega_\omega \in \mathcal{H}_\omega$ are common C^∞-vectors for all the $\pi_\omega(Y)$. By using the commutation relations $\pi_\omega(Y)\pi_\omega(f) = \pi_\omega(Y \cdot f) + \pi_\omega(f)\pi_\omega(Y)$ we see that by multiple differentiation with respect to the parameters t_k at $t_k = 0$ the most general expressions we get are finite linear combinations of elements of the form $\pi_\omega(f \cdot Y_1 \ldots Y_n)$.

We must show that the representation defined in Section 8.2 is the only one satisfying the assumptions of Theorem 8.1.3. The corresponding positive linear functional is defined by

$$\omega_0(f \, Y_1 \ldots Y_n) = \delta_{n,0} \mu_0(f) \tag{8.3.1}$$

for any $f \in \mathrm{Cyl}^\infty$ in terms of the fluxes or equivalently

$$\omega_0(f \, W_1 \ldots W_n) = \mu_0(f) \tag{8.3.2}$$

in terms of the Weyl elements. Equation (8.3.2) can actually be extended to all of Cyl with respect to f. That ω_0 satisfies all the requirements of Theorem 8.1.3 is obvious: semi-weak smoothness is manifest because $\omega_0(W_t^n(S)) = 1$ is trivially smooth in t for all n, S. Moreover, the $\mathfrak{G}$-invariance reduces to $\mathfrak{G}$-invariance of μ_0, which we established in the previous chapter.

Thus, our aim is to show that $\omega = \omega_0$ once ω satisfies the assumptions of Theorem 8.1.3 . We will break the proof into several steps. We will denote 'semianalytic' by s.a., not to be confused with self-adjoint, for the rest of this section.

Step I

Lemma 8.3.1. ••• ω ••• • $\mathfrak{G}$•••••••••• •••••• •• $\mathfrak{A}$• S • ••••• n • $Lie(G)$••• •• •••
•••• ••• •• •• • ••• •••• •••••••• • ••• $[Y_n(S)] = 0$•

Here $[a] := \{a + b : b \in \mathfrak{A} \text{ s.t. } \omega(b^*b) = 0\}$ denotes the equivalence class of $a \in \mathfrak{A}$ with respect to the Gel'fand ideal of null vectors, see Section 29.1.

• •••• For any $p \in \text{supp}(n)$, by the definition of a face S we find a neighbourhood U_p of p and a chart x_p whose domain contains U_p such that

$$x_p(S \cap U_p) = \{(x^1, \ldots, x^D) \in \mathbb{R}^D : x^D = 0; \, 0 < x^1, \ldots, x^{D-1} < 1\} \quad (8.3.3)$$

The U_p define an open cover of $\text{supp}(n) =: K$ which is compact and thus we find a finite subcover $U_I, I = 1, \ldots, N$ with associated charts x_I. By the results of Chapter 20 we may construct a s.a. partition of unity e_I, that is $\text{supp}(e_I) \subset U_I$ and $\sum_{I=1}^N e_I = 1$ on K. Hence $n = \sum_{I=1}^N n_I$ everywhere on σ where $n_I = n \cdot e_I$. Furthermore, we may decompose $n_I = \sum_j n_I^j \tau_j$ where τ_j is a basis in the Lie algebra of G and set $n_{Ij} = n_I^j \tau_j$ (no summation). It follows that $[Y_n(S)] = \sum_{I=1}^N \sum_{j=1}^{\dim(G)} [Y_{n_{Ij}}(S)]$ and it suffices to show that $[Y_{n_{Ij}}(S)] = 0$.

Consider for fixed I, j the functional

$$(n_{Ij}, n'_{Ij})_S := \, < [Y_{n_{Ij}}(S)], [Y_{n'_{Ij}}(S)] > := \omega(Y_{n_{Ij}}(S)^* \, Y_{n'_{Ij}}(S)) \quad (8.3.4)$$

which for $n = n'$ equals $||[Y_{n_{Ij}}]||^2$. So we must show that (8.3.4) vanishes for $n = n'$. We will show this for each fixed I, j separately.

(8.3.4) is obviously bilinear and, due to the reality of the n, n', also symmetric. Furthermore, it is invariant under s.a. diffeomorphisms φ which preserve S and have support in U_I. This follows immediately from the $\mathfrak{G}$-invariance of ω since, dropping the label I, j

$$(n_{Ij}, n'_{Ij})_S = \omega\left(\alpha_\varphi\left[Y_{n_{Ij}}(S)^* Y_{n'_{Ij}}(S)\right]\right) := \omega\left(Y_{n_{Ij} \circ \varphi^{-1}}(\varphi(S))^* Y_{n'_{Ij} \circ \varphi^{-1}}(\varphi(S))\right)$$

$$= \omega(Y_{n_{Ij} \circ \varphi^{-1}}(S)^* Y_{n'_{Ij} \circ \varphi^{-1}}(S)) = (n_{Ij} \circ \varphi^{-1}, n'_{Ij} \varphi^{-1})_S \quad (8.3.5)$$

Using the coordinate system x_I associated with U_I we set $U'_I = x_I(U_I)$, $S'_I = x_I(S \cap U_I) = \{x \in \mathbb{R}^D : x^D = 0, \, 0 < x^1, \ldots, x^{D-1} < 1\}$ and construct $n'_{Ij} := n_{Ij} \circ x_I^{-1} : S'_I \to \mathbb{R}$. Notice that while n may be defined everywhere on σ, as far as $Y_n(S)$ is concerned we only know its restriction to S. In particular, n'_{Ij} has compact support in S'_I. To extend n'_{Ij} to U'_I, let $f' : \mathbb{R} \to \mathbb{R}$ be an arbitrary s.a. function subject to $f'(0) := 1$ and such that $\tilde{n}'_{Ij}(x^1, \ldots, x^D) := n'_{Ij}(x^1, \ldots, x^{D-1})f'(x^D)$ has compact support in U'_I. Finally, for real t we define

$$\varphi'_t(x^1, \ldots, x^D) := (x^1 + t\tilde{n}'_{Ij}(x^1, \ldots, x^D), x^2, \ldots, x^D) \quad (8.3.6)$$

We now show that there exists $t_0 > 0$ such that for all $0 < t < t_0$ the map φ'_t defines a s.a. diffeomorphism of $\mathbb{R}^D$ which equals the identity outside of U'_I and preserves U'_I. To see this we compute

$$\det\left(\frac{\partial\varphi'_t(x)}{\partial x}\right) = 1 + t\frac{\tilde{n}'_{Ij}(x)}{\partial x^1} = 1 + tf'(x^D)\frac{n'_{Ij}(x^1,\ldots,x^{D-1})}{\partial x^1} \qquad (8.3.7)$$

The function $f'\partial n'_{Ij}/\partial x^1$ has compact support in U'_I and is at least continuous there. Thus, it is uniformly bounded whence there exists $t_0 > 0$ such that $1 + tf'\partial n'_{Ij}/\partial x^1 > 0$ for all $0 < t < t_0$. Hence φ'_t is locally (i.e., within U_I) a s.a. (since f', n'_{Ij}, x^k_I are s.a.) diffeomorphism for $0 < t < t_0$ which is also globally defined because it restricts to the identity outside of U'_I by inspection. That it preserves U'_I follows also from the fact that it is a diffeomorphism, in particular a bijection, which is the identity outside of U'_I, thus it must preserve the complement.

Let now N'_{Ij} be a s.a. function with support in U'_I such that $N'_{Ij}(x) = x^1$ whenever $x \in \text{supp}(\tilde{n}'_{Ij})$. To construct such a function, one may use a s.a. partition of unity. We compute

$$[(\varphi'_t)^* N'_{Ij}](x^1,\ldots,x^D) = N'_{Ij}(x^1 + t\tilde{n}'_{Ij}(x^1,\ldots,x^D), x^2,\ldots,x^D)$$

$$= \begin{cases} N'_{Ij}(x^1 + t\tilde{n}'_{Ij}(x), x^2,\ldots,x^D) & x \in \text{supp}(\tilde{n}'_{Ij}) \\ N'_{Ij}(x^1, x^2,\ldots,x^D) & x \notin \text{supp}(\tilde{n}'_{Ij}) \end{cases}$$

$$= \begin{cases} x^1 + t\tilde{n}'_{Ij}(x) & x \in \text{supp}(\tilde{n}'_{Ij}) \\ N'_{Ij}(x^1, x^2,\ldots,x^D) & x \notin \text{supp}(\tilde{n}'_{Ij}) \end{cases}$$

$$= \begin{cases} N'_{Ij}(x) + t\tilde{n}'_{Ij}(x) & x \in \text{supp}(\tilde{n}'_{Ij}) \\ N'_{Ij}(x) & x \notin \text{supp}(\tilde{n}'_{Ij}) \end{cases}$$

$$= N'_{Ij}(x) + t\tilde{n}'_{Ij}(x) \qquad (8.3.8)$$

Let us denote by $N_{Ij}, n_{Ij}, f, \varphi_t$ the pull-back by x_I of $N'_{Ij}, n'_{Ij}, f', \varphi'_t$. Since x_I is a bijection and $N'_{Ij}, \tilde{n}'_{Ij}$ have compact support in U'_I, it follows that $N_{Ij}, \tilde{n}_{Ij} = fn_{Ij}$ have compact support $U_I = x_I^{-1}(U'_I)$. We may thus extend them to all of σ by setting them equal to zero outside of U_I. Likewise, φ_t equals the identity outside of U_I and preserves U_I for $0 < t < t_0$. Furthermore, (8.3.8) translates into

$$(\varphi_t)^* N_{Ij} = N_{Ij} + tfn_{Ij} \qquad (8.3.9)$$

Notice also that $[\varphi'_t(x)]^D = x^D$ preserves $x^D = 0$, hence it preserves S'_I and therefore φ_t preserves $S_I = U_I \cap S$. Since it is the identity outside of U_I, φ_t and its inverse are diffeomorphisms which preserve S. We may therefore apply (8.3.5) to (8.3.9) and obtain

$$(N_{Ij}, N_{Ij})_S = (\varphi_t^* N_{Ij}, \varphi_t^* N_{Ij})_S = (N_{Ij} + tfn_{Ij}, N_{Ij} + tfn_{Ij})_S$$

$$= (N_{Ij} + tn_{Ij}, N_{Ij} + tn_{Ij})_S$$

$$= (N_{Ij}, N_{Ij})_S + 2t(N_{Ij}, n_{Ij})_S + t^2(n_{Ij}, n_{Ij})_S \qquad (8.3.10)$$

where in the second step we used that $f = 1$ on S_I since $f' = 1$ when $x^D = 0$ and in the third we used symmetry of $(.,.)_S$. Since (8.3.10) holds for all $0 < t < t_0$ we may divide by $t > 0$ and find

$$2(N_{Ij}, n_{Ij})_S + t(n_{Ij}, n_{Ij})_S = 0 \tag{8.3.11}$$

for all $0 < t < t_0$. Subtracting equations (8.3.11) evaluated at $0 < t_1 < t_2 < t_0$ reveals $(n_{Ij}, n_{Ij})_S = 0$ which we intended to show. □

It is instructive to see how the semianalyticity of all the structures involved went crucially into the proof: for instance, had we worked with analytical diffeomorphisms, we would not have been able to establish the existence of an analytical diffeomorphism φ_t with the properties displayed, because we would have no control over what φ_t would do outside the domain of a chart. For s.a. diffeomorphisms we can simply 'switch off' their action outside a compact region, much like for smooth diffeomorphisms.

Step II

We already saw that the GNS Hilbert space is the closed linear span of vectors of the form $[fY_1 \ldots Y_N], N = 0, 1, \ldots$ where Y_k are flux vector fields associated with faces and $f \in \text{Cyl}^\infty$. Now by the GNS representation for $N > 0$ we have $[fY_1 \ldots Y_N] = \pi_\omega(fY_1 \ldots Y_{N-1})[Y_N] = 0$ since $[Y] = 0$ as we just showed. It follows that the state ω is already determined by its restriction to Cyl^∞. Now we make use of the following elementary result:

Lemma 8.3.2. •••••• (X) •• ••• • •• ••• *••• •• •••• •• ••• • ••••• • ••••••• •• ••• • •••••• X • ••• •••• •• ••• •••••••••• • •••• •• •••• ••• •••••• •••••• •• •••••• •••• • •••••••••• •••• •••••• ••• •••• •• ••••••• ••• Λ •• ••• (X) •• ••••• •••••• ••• ••• •••• •••••••• •• ••• •••••••• •• ••••• $||\cdot||_\infty$ ••• Λ •••••• •• •• ••• C^*•• •••••• ••• •••••• $\overline{(X)}$ ••• •• (X)•

Remarkably, there are no topological restrictions on X. The condition on the square root closure is just to ensure that $\Lambda(f) \geq 0$ for $f \geq 0$ is equivalent with $\Lambda(|f|^2) \geq 0$.

• ••••• The proof is usually given for C^*-algebras of continuous functions on compact spaces X [282, pp. 106, 107] but works also in our more general situation. If f is real-valued then obviously $-||f||_\infty \leq f \leq ||f||_\infty$ hence $-||f||_\infty \Lambda(1) \leq \Lambda(f) \leq \Lambda(1)||f||_\infty$ by positivity and linearity, that is, $|\Lambda(f)| \leq \Lambda(1) ||f||_\infty$. If f is complex-valued and $\Lambda(f) = re^{i\phi}$ then $|\Lambda(f)| = r = e^{-i\phi}\Lambda(f) = \Lambda(e^{-i\phi}f)$. We have $0 \leq \Lambda(e^{-i\phi}f) = \Lambda(\Re(e^{-i\phi}f)) + i\Lambda(\Im(e^{-i\phi}f))$, hence $\Lambda(\Im(e^{-i\phi}f)) = 0$. Thus by the result established for real-valued functions $|\Lambda(f)| \leq \Lambda(1) ||\Re(e^{-i\phi}f)||_\infty \leq \Lambda(1) ||f||_\infty$ as claimed. □

We can apply Lemma 8.3.2 to the subalgebra $\text{Cyl}^\infty \subset \text{Cyl}$ and the restriction Λ of ω to Cyl^∞. By the bounded linear function theorem the functional Λ can then be extended, by continuity, to the algebra completion $\overline{\text{Cyl}}$ of Cyl^∞ with respect to the sup-norm. By the results of Section 6.2 $\overline{\text{Cyl}}$ is isometric isomorphic to $C(\overline{\mathcal{A}})$ and[7] by the results of Section 8.2 the functional Λ comes from a regular Borel and probability measure μ on $\overline{\mathcal{A}}$. Now for $f \in \text{Cyl}^\infty$ we have

$$\omega(|f|^2) = ||[f]||_{\mathcal{H}_\omega}^2 = \Lambda(|f|^2) = \mu(|f|^2) = ||f||_{L_2(\overline{\mathcal{A}},d\mu)}^2 \leq ||f||_\infty^2 \quad (8.3.12)$$

since $\Lambda(1) = 1$. Here we have used the GNS notation $[a] = \pi_\omega(a)\Omega_\omega$ of Section 29.1. It follows that the Hilbert space-norm topology of $\mathcal{H}_\omega$ on Cyl^∞ is weaker than the sup-norm topology on Cyl^∞ so that $f \mapsto \pi_\omega(f)$ can be extended to $\overline{\text{Cyl}}$. The Hilbert space $\mathcal{H}_\omega$ thus contains $\overline{[\text{Cyl}]}$. (Notice that the Hilbert space completion $\mathcal{H}_\omega$ of $\overline{[\text{Cyl}]}$ is non-trivial because every $C(\overline{\mathcal{A}})$ function is an L_2 function because continuous functions on compact spaces are uniformly bounded, but this is not necessarily the case for L_2 functions.)

Step III

We compute for $f, f' \in \text{Cyl}^\infty$

$$
\begin{aligned}
< [f], [Y_n(S)f'] >_\omega &= < [f], [[Y_n(S), f']] - [f'Y_n(S)] >_\omega = < [f], [[Y_n(S), f']] >_\omega \\
&= i\hbar < [f], [Y_n(S) \cdot f'] >_\omega \\
&= \omega(\overline{f}Y_n(S)f') = \omega([\overline{f}, Y_n(S)]f') + \omega(Y_n(S)\overline{f}f') \\
&= -i\hbar\omega((Y_n(S) \cdot f)f') + \omega(Y_n(S)\overline{f}f') \\
&= -i\hbar\omega((Y_n(S) \cdot f)^*f') + \omega(Y_n(S)^*\overline{f}f') \\
&= -i\hbar < [Y_n(S) \cdot f], [f'] >_\omega + < [Y_n(S)], [\overline{f}f'] >_\omega \\
&= -i\hbar < [Y_n(S) \cdot f], [f'] >_\omega \quad (8.3.13)
\end{aligned}
$$

where in the first step we used $[f'Y_n(S)] = 0$, in the second we used the commutation relations, in the third we employed the defintion of the scalar product, in the fourth we used the commutation relations again, in the fifth we employed the adjointness relations and in the last we used $[Y_n(S)] = 0$. This shows that $i\pi_\omega(Y_n(S))$ is a symmetric operator and that $\pi_\omega(Y_n(S))[f] = i\hbar[Y_n(S) \cdot f]$. In terms of the measure μ in (8.3.12) this means that μ is invariant under the flow generated by the $Y_n(S)$, that is, the divergence of $Y_n(S)$ with respect to μ vanishes.

Step IV

Given γ, in Lemma 8.4.1 we construct a linear combination $Y_\gamma(t_\gamma)$ out of vector fields of the form $Y_{n_0}(S_{v,e})$ or commutators thereof where $n_0 = \text{const.}$ on $S_{v,e}$. Here $S_{v,e}$ is a face contained in a compact set intersecting γ in $v \in V(\gamma)$ and which has the property to contain a beginning segment of $e \in E(\gamma)$, $b(e) = v$

---

[7] One might suspect that one only gets $C^\infty(\overline{\mathcal{A}})$ this way but this space is not complete, its completion being $C(\overline{\mathcal{A}})$ by the Weierstrass theorem.

and to intersect all other edges adjacent to v transversally in v, except possibly for one other edge which is an analytic continuation of e.

We now claim that any of the vector fields $Y_\gamma(t_\gamma)$ has vanishing divergence with respect to the measure μ. For this it will be sufficient to show that any vector field of the form $[Y_n(S), Y_{n'}(S')]$ has zero μ-divergence. However, this follows from the already established symmetry of the $\pi_\omega(Y_n(S))$

$$< [f], [[Y_n(S), Y_{n'}] \cdot f'] >_\omega = - < [[Y_n(S), Y_{n'}] \cdot f], f' >_\omega \qquad (8.3.14)$$

The corresponding Weyl operator $\pi_\omega(W_\gamma(t_\gamma))$, see (8.4.22), therefore simply acts as $\pi_\omega(W_\gamma(t_\gamma))[f] = [W_\gamma(t_\gamma) \cdot f]$. By construction, it generates arbitrary left translations on cylindrical functions over γ and is unitary. Hence the push-forward measure $\mu_l = \mu \circ p_l^{-1}$, $l = l(\gamma)$ is a left translation-invariant measure on $G^{|E(\gamma)|}$. Since $G^{|E(\gamma)|}$ is compact and μ is a probability measure, this measure must be the product Haar measure by Theorem 31.1.4.

We conclude that μ is actually the uniform measure, hence $\overline{[\text{Cyl}^\infty]} \cong L_2(\overline{\mathcal{A}}, \mu_0) = \mathcal{H}_0$. This concludes the uniqueness proof.

Notice that we did not even use the full group $\mathfrak{G}$, only the subgroup $\text{Diff}^\omega_{\text{sa}}(\sigma)$. However, it should be noted that a more desirable result would be if the representation or state ω_0 was already determined if it was regular with respect to the fluxes, not only smooth. An extension of that sort has been achieved in [520], however, it uses an additional assumption of a different kind whose physical significance is unclear, so we refrain from displaying any of those details here.

8.4 Uniqueness proof: (3) irreducibility

The irreducibility proof must be made in terms of the exponentiated flux operators, that is, the Weyl elements rather than the fluxes themselves. The reason for this can be explained already for the Weyl algebra of ordinary quantum mechanics that we mentioned at the beginning of this chapter. By the Stone–von Neumann theorem every irreducible, weakly continuous representation of the • ••• algebra is unitarily equivalent to the Schrödinger representation. However, the Schrödinger representation contains many invariant subspaces in the common dense domain of the • •••••• •••• algebra generated by the unbounded self-adjoint operators $\hat{q}, \hat{p}$ which exist thanks to weak continuity and Stone's theorem. To see this, consider a closed interval I and the set $C_I^\infty(\mathbb{R})$ of C^∞-functions which vanish on $\mathbb{R} - I$. It is clear that $C_I^\infty(\mathbb{R})$ is left-invariant by any polynomial in the $\hat{q}, \hat{p}$ and it is non-empty because it contains functions of the form (set $I = [-a, a]$ for simplicity) $f(x) := \exp(-1/(x - a)^2 - 1/(x + a)^2)$ for $|x| \leq a$ and $f(x) = 0$ for $|x| \geq a$. Hence the Heisenberg algebra can never change the support of these functions. On the other hand, $(V(b)f)(x) = f(x + b)$ changes the support and hence there is a chance that the Weyl algebra is represented in an irreducible fashion. For the Hilbert space $\mathcal{H}_0$ and the representation π_0 of $\mathfrak{A}$ one can make a similar argument on any cylindrical subspace by considering smooth functions on the group which vanish outside a compact subset. The Weyl elements generate arbitrary left translations which are transitive, hence only for the Weyl elements is there a chance for irreducibility.

The proof that follows is based on [519] and is analogous to the original proof by von Neumann for the Schrödinger representation of the standard Weyl algebra [538]. A different proof is given in [538].

Before we prove the theorem, we first need two preparational results. Let γ be a graph. Split each edge $e \in E(\gamma)$ into two halves $e = e_1' \circ (e_2')^{-1}$ and replace the e's by the e_1', e_2'. This leaves the range of γ invariant but changes the set of edges in such a way that each edge is outgoing from the vertex $b(e') = v \in V(\gamma)$ (notice that by a vertex we mean a point in γ which is not the interior point of a semianalytic curve so that the break points $e_1' \cap e_2'$ do not count as vertices). We call a graph refined in this way a $\bullet\bullet\bullet\bullet\bullet\bullet\ \bullet\bullet\bullet\bullet\bullet$. Every cylindrical function over a graph is also cylindrical over its associated standard graph so there is no loss of generality in sticking with standard graphs in what follows.

With this understanding, the following statement holds.

Lemma 8.4.1. $\bullet\bullet\bullet\ \gamma\ \bullet\bullet\ \bullet\ \bullet\bullet\bullet\bullet\bullet\bullet\bullet\ \bullet\bullet\bullet\bullet\bullet\bullet\ \bullet\ \bullet\bullet\bullet\bullet\ \bullet\bullet\ \bullet\bullet\bullet\bullet\ e \in E(\gamma)\ \bullet\ \bullet\bullet\bullet\bullet\bullet\bullet$
$t_e = (t_e^j)_{j=1}^{\dim(G)}\ \bullet\bullet\bullet\ \bullet\bullet\bullet\bullet\bullet\bullet\ \bullet\bullet\bullet\bullet\ \bullet\bullet\ \bullet\bullet\ \bullet\ \bullet\bullet\bullet\bullet\ t_\gamma = (t_e)_{e \in E(\gamma)}\bullet$
$\bullet\ \bullet\bullet\bullet\ \bullet\bullet\bullet\bullet\bullet\ \bullet\bullet\bullet\bullet\bullet\ \bullet\ \bullet\bullet\bullet\bullet\bullet\ \bullet\bullet\bullet\ Y(t_\gamma, \gamma)\ \bullet\bullet\ \bullet\bullet\bullet\ \bullet\ \bullet\bullet\ \bullet\bullet\bullet\bullet\bullet\bullet\ \bullet\bullet\bullet\bullet\bullet\bullet\bullet\bullet\ \bullet\bullet\ \bullet\bullet\bullet$
$\bullet\bullet\bullet\ \bullet\bullet\bullet\bullet\bullet\ \bullet\bullet\bullet\bullet\ Y_n(S)\ \bullet\bullet\bullet\bullet\ \bullet\bullet\bullet\bullet\ \bullet\bullet\bullet\ \bullet\bullet\bullet\ \bullet\bullet\bullet\bullet\bullet\bullet\bullet\bullet\bullet\ \bullet\bullet\bullet\bullet\bullet\bullet\ f = p_\gamma^* f_\gamma\ \bullet\bullet\bullet\bullet\ \gamma$
$\bullet\bullet\ \bullet\bullet\bullet\bullet$

$$Y_\gamma(t_\gamma) p_\gamma^* f_\gamma = p_\gamma^* \sum_{e \in E(\gamma)} t_e^j R_j^e f_\gamma \tag{8.4.1}$$

$\bullet\ \bullet\bullet\bullet\bullet\bullet$ Any compact connected Lie group G has the structure $G/Z = A \times S$ where Z is a discrete central subgroup, A is an Abelian Lie group and S is a semisimple Lie group.

We will first construct an appropriate vector field Y_e^j for each j and each $e \in E(\gamma)$. The construction is somewhat different for the Abelian and non-Abelian generators respectively so that we distinguish the two cases.

Abelian factor

Let j label only Abelian generators for this paragraph. Consider any $e \in E(\gamma)$ and take any compactly supported face S_e which intersects γ only in an interior point of e and such that the orientation of S_e agrees with that of e_2 where $e = e_1^{-1} \circ e_2$, $e_1 \cap e_2 = S_e \cap \gamma$. Then for any cylindrical function $f = p_\gamma^* f_\gamma$ we have

$$Y_j(S_e) p_\gamma^* f_\gamma = p_\gamma^* [R_{e_2}^j - R_{e_1}^j] f_\gamma \tag{8.4.2}$$

Due to gauge-invariance $[R_{e_1}^j + R_{e_2}^j] f_\gamma = 0$, thus

$$Y_e^j p_\gamma^* f_\gamma = \frac{1}{2} Y_j(S_e) p_\gamma^* f_\gamma \tag{8.4.3}$$

is an appropriate choice.

Non-Abelian factor

Let j label only non-Abelian generators for this paragraph. Given γ select a vertex v and one $e \in E(\gamma)$ with $b(e) = v$. We will have to distinguish two cases: the case where no $e \in E(v)$ is (a segment of) the analytic extension through v of another edge $e' \in E(v)$, and the case where at least one pair $e, \tilde{e} \in E(v)$ of edges exists that are analytic extensions of one another through v. The latter case will require some special consideration and therefore we begin with the:

• •••• ••••

We will prove that in this case, for any $e \in E(v)$, there exists an analytic surface $S_{v,e}$ through v such that

1. $s_e \subset S_{v,e}$ for some beginning segment s_e of e, and the other edges $e' \in E(v), e' \neq e$ intersect $S_{v,e}$ transversally in v.
2. For $e' \in E(v), e' \neq e$ we have $e' \cap S_{v,e} = v$, and for $e' \notin E(v)$, $e' \cap S_{v,e} = 0$.

To start with, we note that if we can find a surface that satisfies (1), we can always make it smaller in such a way that it will also satisfy (2). Therefore we focus on (1): an analytic surface S is completely determined by its germ $[S]_v$, that is, the Taylor coefficients in the expansion of its parametrisation (we consider the case $D = 3$, the case $D \geq 2$ is similar)

$$S(u, v) = \sum_{m,n=0}^{\infty} \frac{u^m \, v^n}{m! \, n!} S^{(m,n)}(0,0) \tag{8.4.4}$$

Likewise, consider the germ $[e]_v$ of e

$$e(t) = \sum_{n=0}^{\infty} \frac{t^n}{n!} e^{(n)}(0) \tag{8.4.5}$$

In order that $s_e \subset S_{v,e}$ we just need to choose a parametrisation of S such that, say, $S(t, 0) = e(t)$ which fixes the Taylor coefficients

$$S^{(m,0)}(0,0) = e^{(m)}(0) \tag{8.4.6}$$

for any m. By choosing the range of t, u, v sufficiently small we can arrange that $s_e \subset S$.

We now choose the freedom in the remaining coefficients to satisfy the additional requirements. We must avoid that for finitely many, say n, edges $e_1', \ldots, e_n'$ there is any beginning segment s_k of e_k' with $s_k \subset S$. If s_k was contained in S then there would exist an analytic function $t \mapsto v_k(t)$, such that $s_k(t) = S(t, v_k(t))$. Notice that v_k must be different from the zero function in a sufficiently small neighbourhood around $t = 0$ as otherwise we would have $s_k = s_e$, which is not the case. For each k let $n_k > 0$ be the first derivative such that $v_k^{(n_k)}(0) \neq 0$. By relabelling the edges we may arrange that $n_1 \leq n_2 \leq \ldots \leq n_N$. Consider $k = 1$ and take the n_1th derivative at $t = 0$. We find

$$s_1^{(n_1)}(0) = S^{(n_1,0)}(0,0) + S^{(0,1)}(0,0)v_1^{(n_1)}(0) \tag{8.4.7}$$

Since $v_1^{(n_1)}(0) \neq 0$ we can use the freedom in $S^{(0,1)}(0,0)$ in order to violate this equation. Now consider $k = 2$ and take the $(n_2 + 1)$th derivative. We find

$$s_2^{(n_2+1)}(0) = S^{(n_2+1,0)}(0,0) + 2S^{(1,1)}(0,0)v_2^{(n_2)}(0) + S^{(0,1)}(0,0)v_2^{(n_2+1)}(0)$$

$$\text{(8.4.8)}$$

Since $v_2^{(n_2)}(0) \neq 0$ we can use the freedom in $S^{(1,1)}$ in order to violate this equation. Proceeding this way we see that we can use the coefficients $S^{(k-1,1)}(0,0)$ in order to violate $s_k(t) = S(t, v_k(t))$ for $k = 1, \ldots, N$.

Having constructed the surfaces $S_{v,e}$ we can compute the associated vector field applied to a cylindrical function over γ

$$Y_j(S_{v,e})p_\gamma^* f_\gamma = p_\gamma^* \sum_{e' \in E(\gamma) - \{e\}, b(e') = v} \sigma(S_{v,e}, e') R_{e'}^j f_\gamma \qquad \text{(8.4.9)}$$

where by construction $|\sigma(S_{v,e}, e')| = 1$ for any $e' \neq e$, $b(e) = v$. Taking the commutator

$$[Y_j(S_{v,e}), Y_k(S_{v,e})]p_\gamma^* f_\gamma = f_{jkl} p_\gamma^* \sum_{e' \in E(\gamma) - \{e\}, b(e') = v} R_{e'}^j f_\gamma \qquad \text{(8.4.10)}$$

Using the Cartan–Killing metric normalisation for the totally skew structure constants $f_{jkl} f_{lmi} = -\delta_{km}$ and writing

$$R_v^j := \sum_{e' \in E(\gamma), \, b(e') = v} R_{e'}^j \qquad \text{(8.4.11)}$$

we get

$$U_e^j p_\gamma^* f_\gamma := f_{jkl}[Y_k(S_{v,e}), Y_l(S_{v,e})]p_\gamma^* f_\gamma = p_\gamma^* [R_v^j - R_e^j] f_\gamma \qquad \text{(8.4.12)}$$

Thus, if $n_v = |\{e \in E(\gamma); \, b(e) = v\}|$ denotes the valence of v

$$Y_e^j p_\gamma^* f_\gamma := \left\{ -f_{jkl}[Y_k(S_{v,e}), Y_l(S_{v,e})] + \frac{1}{n_v - 1} \sum_{e \in E(\gamma)} (f_{jkl}[Y_k(S_{v,e}), Y_l(S_{v,e})]) \right\} p_\gamma^* f_\gamma$$

$$= p_\gamma^* R_e^j f_\gamma \qquad \text{(8.4.13)}$$

Now we return to the case where there is at least one pair of edges $e, \tilde{e} \in E(v)$ that are (segments of) analytic continuations of each other through v. We will denote the set of these special edges as P, and for $e \in P$, $\tilde{e}$ stands for its 'partner'. We start by observing that now, for $e \in P$, we cannot construct a surface $S_{v,e}$ with the property (1) as above, because if a beginning segment of e is contained in an analytic surface (without boundary) then so is at least part of $\tilde{e}$. We can, however, still construct a surface $S_{e,v}$ such that $s_e, s_{\tilde{e}} \subset S_{v,e}$ for beginning segments $s_e, s_{\tilde{e}}$ of $e, \tilde{e}$, such that for $e' \in E(v), e' \neq e, \tilde{e}$; e' intersect $S_{v,e}$ transversally, and such that (2) holds, with exactly the same method as above. For edges $e \in E(v) - P$, we construct the analytic surfaces as above. Then with the definition $U_e^j :=$ $f_{jkl}[Y_k(S_{v,e}), Y_l(S_{v,e})]$, we find in analogy with (8.4.12):

$$U_e^j p_\gamma^* f_\gamma = p_\gamma^* \begin{cases} [R_v^j - R_e^j - R_{\tilde{e}}^j] f_\gamma & \text{if } e \in P \\ [R_v^j - R_e^j] f_\gamma & \text{if } e \in E(v) - P \end{cases} \tag{8.4.14}$$

Consequently we can form the linear combination

$$V^j p_\gamma^* f_\gamma := \left(\frac{1}{2} \sum_{e \in P} U_e^j + \sum_{e \in E(v) - P} U_e^j \right) p_\gamma^* f_\gamma \tag{8.4.15}$$

$$= p_\gamma^* \left(\frac{1}{2} |P| R_v^j - \sum_{e \in P} R_e^j + (|E(v)| - |P|) R_v^j - \sum_{e \in E(v) - P} R_e^j \right) f_\gamma \tag{8.4.16}$$

$$= p_\gamma^* \left(|E(v)| - \frac{1}{2} |P| - 1 \right) R_v^j f_\gamma \tag{8.4.17}$$

Note that since $|E(v)| \geq |P|$, the prefactor of R_v^j can at most be zero if $|E(v)| = 2 = |P|$. But that type of vertex is excluded due to our conventions. So we can define

$$\tilde{Y}_e^j p_\gamma^* f_\gamma := \left\{ -U_e^j + \left(|E(v)| - \frac{1}{2} |P| - 1 \right)^{-1} V^j \right\} p_\gamma^* f_\gamma$$

$$= p_\gamma^* \begin{cases} (R_e^j + R_{\tilde{e}}^j) f_\gamma & \text{if } e \in P \\ R_e^j f_\gamma & \text{if } e \in E(v) - P \end{cases} \tag{8.4.18}$$

We see that for $e \in E(v) - P$, $\tilde{Y}_e^j$ is already what we need, and consequently we set $Y_e^j := \tilde{Y}_e^j$ in these cases.

For $e \in P$, we observe that we can certainly construct an analytic surface S_e' such that $S_e' \cap \gamma = v$ and which is intersected transversally by e. Choosing the orientation of such a surface appropriately, we have

$$\frac{1}{2} (Y_j(S_e') + \tilde{Y}_e^j) = p_\gamma^* (R_e^j + \cdots) f_\gamma \tag{8.4.19}$$

where '...' stands for terms that contain derivatives only with respect to edges other than e and $\tilde{e}$. Therefore

$$Y_e^j := f_{jkl} \left[\frac{1}{2} (Y_k(S_e') + \tilde{Y}_e^k), \tilde{Y}_e^l \right] p_\gamma^* f_\gamma = p_\gamma^* R_e^j f_\gamma \tag{8.4.20}$$

Thus for any configuration of edges beginning at any vertex v of γ we have now constructed vector fields Y_e^j that act as R_e^j on functions cylindrical on γ. Collecting the vector fields Y_e^j for the Abelian and non-Abelian labels j respectively and contracting them with t_e^j and summing over $e \in E(\gamma)$ yields an appropriate vector field

$$Y_\gamma(t_\gamma) = \sum_{e \in E(\gamma)} t_j^e Y_e^j \qquad (8.4.21)$$

$\square$

Lemma 8.4.1 has the following important implication: the algebra $\mathfrak{P}$ also contains the vector field $Y_\gamma(t_\gamma)$ and therefore $\mathfrak{A}$ contains the corresponding Weyl element

$$W_\gamma(t_\gamma) := e^{Y_\gamma(t_\gamma)} \qquad (8.4.22)$$

Also, let us write $I_\gamma = (\{\pi_e\}, \{m_e\}, \{n_e\})_{e \in E(\gamma)}$ for a spin-network $s = (\gamma, I_\gamma)$ over γ. Denoting by $T_s = T_{\gamma, I_\gamma}$ the corresponding spin-network function (where we also allow trivial π_e for any e) we define for any two $\psi, \psi' \in \mathcal{H}_0$ the function

$$(t_\gamma, I_\gamma) \mapsto M_{\psi, \psi'}(t_\gamma, I_\gamma) := <\psi, T_{\gamma, I_\gamma} \, W_\gamma(t_\gamma) \psi' >_{\mathcal{H}_0} \qquad (8.4.23)$$

We now exploit that for a connected Lie group the exponential map is onto. Thus, there exists a region $D_G \subset \mathbb{R}^{\dim(G)}$ such that $\exp: D_G \to G$; $t \mapsto \exp(t^j \tau_j)$ is a bijection. Consider the measure μ on D_G defined by $d\mu(t) - d\mu_H(\exp(t^j \tau_j))$ where μ_H is the Haar measure on G. Finally, let $D_\gamma = \prod_{e \in E(\gamma)} D_G$ and let L_γ be the space of the I_γ. We now define an inner product on the functions of the type (8.4.23) by

$$(M_{\psi_1, \psi_1'}, M_{\psi_2, \psi_2'})_\gamma := \int_{D_\gamma} d\mu(t_\gamma) \sum_{I_\gamma} \overline{M_{\psi_1, \psi_1'}(t_\gamma, I_\gamma)} \, M_{\psi_2, \psi_2'}(t_\gamma, I_\gamma) \qquad (8.4.24)$$

where $d\mu(t_\gamma) = \prod_{e \in E(\gamma)} d\mu(t_e)$.

The inner product of the type (8.4.24) is a crucial ingredient in an elementary irreducibility proof of the Schrödinger representation of ordinary quantum mechanics [267,268] and we can essentially copy the corresponding argument. Of course, we must extend the proof somewhat in order to be able to deal with an infinite number of degrees of freedom. The following result prepares for that.

Lemma 8.4.2

$\bullet\bullet\bullet \;\; \bullet\bullet\bullet \;\; \bullet\bullet\bullet \;\; \psi_1, \psi_1', \psi_2, \psi_2' \in \mathcal{H}_0 \;\; \bullet\bullet \;\; \bullet\bullet\bullet\bullet$

$$|(M_{\psi_1, \psi_1'}, M_{\psi_2, \psi_2'})_\gamma| \leq \|\psi_1\| \; \|\psi_1'\| \; \|\psi_2\| \; \|\psi_2'\| \qquad (8.4.25)$$

$\bullet\bullet\bullet\bullet \;\; \bullet\bullet\bullet \;\; \bullet\bullet\bullet \;\; \psi_1, \psi_1', \psi_2, \psi_2' \in \mathcal{H}_{0,\gamma} \;\; \bullet\bullet \;\; \bullet\bullet\bullet\bullet$

$$(M_{\psi_1, \psi_1'}, M_{\psi_2, \psi_2'})_\gamma = <\psi_2, \psi_1 >_{\mathcal{H}_0} \; <\psi_1', \psi_2' >_{\mathcal{H}_0} \qquad (8.4.26)$$

$\bullet \; \bullet\bullet\bullet\bullet \;\; \mathcal{H}_{0,\gamma} \;\; \bullet\bullet\bullet\bullet\bullet\bullet \;\; \bullet\bullet\bullet \;\; \bullet\bullet\; \bullet\bullet\bullet \;\; \bullet\bullet \;\; \bullet\bullet\bullet \;\; \bullet\bullet\bullet\bullet \bullet\bullet\bullet\bullet\bullet \;\; \bullet\bullet \bullet\bullet\bullet\bullet \bullet \;\; \bullet\bullet\bullet\bullet \;\; \gamma \bullet$

• • •• •• We simply compute

$$(M_{\psi_1,\psi_1'}, M_{\psi_2,\psi_2'})_\gamma = \int_{D_\gamma} d\mu(t_\gamma) \sum_{I_\gamma} \int_{\overline{A}} d\mu_0(A) \int_{\overline{A}} d\mu_0(A') \overline{T_{\gamma,I_\gamma}(A)} T_{\gamma,I_\gamma}(A')$$

$$\times \psi_1(A) \overline{[W_\gamma(t_\gamma)\psi_1'](A)} \psi_2(A') [W_\gamma(t_\gamma)\psi_2'](A')$$

$$= \int_{D_\gamma} d\mu(t_\gamma) \int_{\overline{A}} d\mu_0(A) \int_{\overline{A}} d\mu_0(A') [\sum_{I_\gamma} \overline{T_{\gamma,I_\gamma}(A)} T_{\gamma,I_\gamma}(A')]$$

$$\times \psi_1(A) \overline{[W_\gamma(t_\gamma)\psi_1'](A)} \psi_2(A') [W_\gamma(t_\gamma)\psi_2'](A')$$

$$= \int_{\overline{A}} d\mu_0(A) \int_{\overline{A}} d\mu_0(A') \int_{D_\gamma} d\mu(t_\gamma) \delta_\gamma(A, A')$$

$$\times \psi_1(A) \overline{[W_\gamma(t_\gamma)\psi_1'](A)} \psi_2(A') [W_\gamma(t_\gamma)\psi_2'](A') \qquad (8.4.27)$$

where we have defined the cylindrical δ-distribution

$$\delta_\gamma(A, A') = \prod_{e \in E(\gamma)} \delta_{\mu_H}(A(e), A'(e)) \qquad (8.4.28)$$

which arises due to the Plancherel formula

$$\delta_{\mu_H}(g, g') = \sum_{\pi,m,n} \overline{T_{\pi,m,n}(g)}\, T_{\pi,m,n}(g') \qquad (8.4.29)$$

The interchange of integrals over $\overline{A} \times \overline{A}$ and the sum over L_γ in (8.4.27) is justified by the Plancherel theorem which here is equivalent to the Peter and Weyl theorem proved in Section 31.2.

(i) In order to evaluate the cylindrical δ-distribution in (8.4.28) we subdivide the degrees of freedom $A \in \overline{A}$ into the set $\overline{A}_\gamma = \overline{A}_{|\gamma}$ and the complement $\overline{A}_{\bar\gamma} = \overline{A} - \overline{A}_\gamma$ in the following sense: each of the functions f_1, f_1', f_2, f_2' is a countable linear combination of spin-network functions T_s, each of which is cylindrical over some graph $\gamma(s)$. We may consider those functions as cylindrical over the graph $\gamma \cup \gamma(s)$ and since the edges $e \in E(\gamma)$ are holonomically independent, we can express each edge $\tilde{e} \in E(\gamma(s))$ as a finite composition of the edges of $E(\gamma)$ and some other edges e' of $\gamma(s) \cup \gamma$ such that no segment of any of the e' is a beginning segment of one of the e. Thus, each $T_s(A)$ depends on the $A(e)$, $e \in E(\gamma)$ and some other $A(e')$ which are not finite compositions of the $A(e)$. We can thus write symbolically for any $f \in \mathcal{H}_0$

$$f(A) = F(A_{|\bar\gamma}, A_{|\gamma}) \qquad (8.4.30)$$

where the separation of the degrees of freedom is to be understood in the sense just discussed, that is, $A_{|\gamma} \in \overline{A}_\gamma$, $A_{\bar\gamma} \in \overline{A}_{\bar\gamma}$. It just means that when expanding out inner products of L_2 functions into those of spin-network functions, one can perform the integrals over the degrees of freedom $A(e) \in \overline{A}_\gamma$ and $A(\tilde{e}) \in \overline{A}_{\bar\gamma}$ independently. Given a function of the type (8.4.30) we define the measure on

$\overline{A}_\gamma$ by $\mu_{0\gamma} = \mu_0 \circ p_\gamma^{-1}$ and the (effective) measure on $\overline{A}_{\bar\gamma}$ by

$$\int_{\overline{A}_{\bar\gamma}} d\mu_{0\bar\gamma}(A_{|\bar\gamma}) \left[\int_{\overline{A}_\gamma} d\mu_{0\gamma}(A_{|\gamma}) F(A_{|\bar\gamma}, A_{|\gamma})\right] := \int_{\overline{A}} d\mu_0(A) f(A) \quad (8.4.31)$$

In order to perform concrete integrals of $f \in L_1(\overline{A}, d\mu_0)$ over either $\overline{A}_\gamma$ or $\overline{A}_{\bar\gamma}$ we notice that all our occurring f are countable linear combinations of spin-network functions. Thus either integral can be written as a countable linear combination of integrals over spin-network functions T_s and then the prescription is to integrate only either over the degrees of freedom $A(e), e \in E(\gamma)$ or $A(e'), e' \in E(\gamma(s) \cup \gamma) - E(\gamma)$ for each individual integral with the corresponding product Haar measure. It follows that $\mu_0 = \mu_{0\bar\gamma} \otimes \mu_{0\gamma}$ is a product measure.[8]

We may therefore neatly split (8.4.31) as

$$(M_{\psi_1,\psi_1'}, M_{\psi_2,\psi_2'})_\gamma = \int_{D_\gamma} d\mu(t_\gamma) \int_{\overline{A}_{\bar\gamma}} d\mu_{0\bar\gamma}(A_{|\bar\gamma}) \int_{\overline{A}_{\bar\gamma}} d\mu_{0\bar\gamma}(A_{|\bar\gamma}') \int_{\overline{A}_\gamma} d\mu_{0\gamma}(A_{|\gamma})$$

$$\times \Psi_1(A_{|\bar\gamma}, A_\gamma)\overline{[W_\gamma(t_\gamma)\Psi_1'](A_{|\bar\gamma}, A_\gamma)}\Psi_2(A_{|\bar\gamma}', A_\gamma)$$

$$\times [W_\gamma(t_\gamma)\Psi_2'](A_{|\bar\gamma}', A_\gamma) \quad (8.4.32)$$

In order to evaluate the Weyl operators, consider a spin-network function T_s cylindrical over $\gamma(s)$ which we write in the form

$$T_s(A) = F(\{A(e')\}_{e' \in E(\gamma \cup \gamma(s)) - E(\gamma)}, \{A(e)\}_{e \in E(\gamma)}) \quad (8.4.33)$$

Our concrete vector field $Y_\gamma(t_\gamma)$ involves a finite collection of surfaces to which the edges $e \in E(\gamma)$ are already adapted in the sense that they are all of a definite type ('in', 'out', 'up' or 'down') and we may w.l.g. assume that the same is true for the e'. Then it is easy to see that the action of $Y_\gamma(t_\gamma)$ on T_s is given by

$$Y_\gamma(t_\gamma)T_s = p_{\gamma(s)\cup\gamma}^* \left[\sum_{e' \in E(\gamma \cup \gamma(s)) - E(\gamma)} t_j^{e'}(t_\gamma)R_{e'}^j + \sum_{e \in E(\gamma)} t_j^e R_e^j\right] F \quad (8.4.34)$$

where $t_j^{e'}(t_\gamma)$ is a certain linear combination of the t_j^e depending on e' and the concrete surfaces $S_e, S_{v,e}$ used in the construction of $Y_\gamma(t_\gamma)$. Since the beginning segments of the e', e are mutually independent, the corresponding vector fields commute and it follows that

$$(W_\gamma(t_\gamma)T_s)(A) = F(\{e^{t_j^{e'}(t_\gamma)\tau_j}A(e')\}_{e' \in E(\gamma\cup\gamma(s))-E(\gamma)}, \{e^{t_j^e\tau_j}A(e)\}_{e \in E(\gamma)})$$

$$= F(\{W_\gamma(t_\gamma)A(e')W_\gamma(t_\gamma)^{-1}\}_{e' \in E(\gamma\cup\gamma(s))-E(\gamma)},$$

$$\{W_\gamma(t_\gamma)A(e)W_\gamma(t_\gamma)^{-1}\}_{e \in E(\gamma)}) \quad (8.4.35)$$

---

[8] That $\overline{A} = \overline{A}_{|\gamma} \times \overline{A}_{|\bar\gamma}$ and $\mu_0 = \mu_{0\gamma} \times \mu_{0\bar\gamma}$ can also be described more formally by using projective language [519] but it is equivalent to our reasoning here.

Consider now any L_2 function ψ. Since it is a countable linear combination of spin-network functions we can generalise (8.4.35) to

$$(W_\gamma(t_\gamma)\psi)(A) = \Psi(W_\gamma(t_\gamma)A_{|\bar\gamma}W_\gamma(t_\gamma)^{-1}, W_\gamma(t_\gamma)A_{|\gamma}W_\gamma(t_\gamma)^{-1}) \quad (8.4.36)$$

where the crucial point is that for each $t_\gamma \in D_\gamma$ the map $\alpha_{t_\gamma} : \overline{\mathcal{A}} \to \overline{\mathcal{A}}$; $A \mapsto W_\gamma(t_\gamma)AW_\gamma(t_\gamma)^{-1}$ is just some right or left translation. We can thus estimate (notice that we can interchange the sequence of integration w.r.t. the factors of a product measure)

$$|(M_{\psi_1,\psi_1'}, M_{\psi_2,\psi_2'})_\gamma| \leq \int_{D_\gamma} d\mu(t_\gamma) \int_{\overline{\mathcal{A}}_\gamma} d\mu_{0\gamma}(A_{|\gamma})$$

$$\times \left[\int_{\overline{\mathcal{A}}_{\bar\gamma}} d\mu_{0\bar\gamma}(A_{|\bar\gamma})|\Psi_1(A_{|\bar\gamma}, A_\gamma)| \; |\Psi_1'(\alpha_{t_\gamma}(A_{|\bar\gamma}), \alpha_{t_\gamma}(A_\gamma))| \right]$$

$$\times \left[\int_{\overline{\mathcal{A}}_{\bar\gamma}} d\mu_{0\bar\gamma}(A_{|\bar\gamma}')|\Psi_2(A_{|\bar\gamma}', A_\gamma)| \; |\Psi_2'(\alpha_{t_\gamma}(A_{|\bar\gamma}'), \alpha_{t_\gamma}(A_\gamma))| \right]$$

$$\leq \int_{D_\gamma} d\mu(t_\gamma) \int_{\overline{\mathcal{A}}_\gamma} d\mu_{0\gamma}(A_{|\gamma}) \; ||\Psi_1(A_\gamma)||_{\bar\gamma} \; ||\Psi_1'(\alpha_{t_\gamma}(A_\gamma))||_{\bar\gamma}$$

$$\times ||\Psi_2(A_\gamma)||_{\bar\gamma} \; ||\Psi_2'(\alpha_{t_\gamma}(A_\gamma))||_{\bar\gamma} \quad (8.4.37)$$

where we have used the Cauchy–Schwarz inequality applied to functions such as $\Psi_1(A_\gamma)$ on $L_2(\overline{\mathcal{A}}_{\bar\gamma}, d\mu_{0\bar\gamma})$ defined by $[\Psi_1(A_\gamma)](A_{|\bar\gamma}) = \Psi_1(A_{|\bar\gamma}, A_\gamma)$. Here it was crucial to note that due to the bi-invariance of the measure $\mu_{0\bar\gamma}$ we have, for example,

$$\int_{\overline{\mathcal{A}}_{\bar\gamma}} d\mu_{0\bar\gamma}(A_{|\bar\gamma})|\Psi_1'(\alpha_{t_\gamma}(A_{|\bar\gamma}), \alpha_{t_\gamma}(A_\gamma))|^2 = \int_{\overline{\mathcal{A}}_{\bar\gamma}} d\mu_{0\bar\gamma}(A_{|\bar\gamma})|\Psi_1'(A_{|\bar\gamma}, \alpha_{t_\gamma}(A_\gamma))|^2$$

$$= ||\Psi_1'(\alpha_{t_\gamma}(A_\gamma))||_{\bar\gamma}^2 \quad (8.4.38)$$

To see this, expand ψ_1' into spin-network functions. Then the integral is of the form

$$\sum_{m,n=1}^\infty \bar{z}_m z_n \int_{\overline{\mathcal{A}}_{\bar\gamma}} d\mu_{0\bar\gamma}(A_{|\bar\gamma})\overline{T_{s_m}(\alpha_{t_\gamma}(A))} \; T_{s_n}(\alpha_{t_\gamma}(A))$$

$$= \sum_{m,n=1}^\infty \bar{z}_m z_n \int_{\overline{\mathcal{A}}_{\bar\gamma}} d\mu_{0\bar\gamma}(A_{|\bar\gamma})\overline{T_{s_m}(\alpha_{t_\gamma}(A))} T_{s_n}(\alpha_{t_\gamma}(A))$$

$$= \sum_{m,n=1}^\infty \bar{z}_m z_n \int_{G^{|E(\gamma(s_m)\cup\gamma(s_n)\cup\gamma)-E(\gamma)|}} \left[\prod_{e'\in E(\gamma(s_m)\cup\gamma(s_n)\cup\gamma)-E(\gamma)} d\mu_H(h_{e'}) \right]$$

$$\times \overline{T_{s_m}(\{e^{t_j^{e'}(t_\gamma)\tau_j}A(e')\}, \{e^{t_j^e\tau_j}A(e)\})} \; T_{s_n}(\{e^{t_j^{e'}(t_\gamma)\tau_j}A(e')\}, \{e^{t_j^e\tau_j}A(e)\})$$

$$= \sum_{m,n=1}^{\infty} \bar{z}_m z_n \int_{\mathcal{G}^{|E(\gamma(s_m)\cup\gamma(s_n)\cup\gamma)-E(\gamma)|}} \left[\prod_{e'\in E(\gamma(s_m)\cup\gamma(s_n)\cup\gamma)-E(\gamma)} d\mu_H(h_{e'}) \right]$$

$$\times \overline{T_{s_m}(\{A(e')\},\{e^{t_j^e \tau_j}A(e)\})} \ T_{s_n}(\{A(e')\},\{e^{t_j^e \tau_j}A(e)\})$$

$$= \sum_{m,n=1}^{\infty} \bar{z}_m z_n \int_{\overline{\mathcal{A}}_{\bar{\gamma}}} d\mu_{0\bar{\gamma}}(A_{|\bar{\gamma}}) \overline{T_{s_m}(A_{|\bar{\gamma}}, \alpha_{t_\gamma}(A_{|\gamma}))} \ T_{s_m}(A_{|\bar{\gamma}}, \alpha_{t_\gamma}(A_{|\gamma}))$$

$$= \int_{\overline{\mathcal{A}}_{\bar{\gamma}}} d\mu_{0\bar{\gamma}}(A_{|\bar{\gamma}}) |\Psi_1'(A_{|\bar{\gamma}}, \alpha_{t_\gamma}(A_{|\gamma}))|^2 \qquad (8.4.39)$$

We now exploit that

$$\alpha_{t_\gamma}(A_{|\gamma}) = \{e^{t_j^e \tau_j}A(e)\}_{e\in E(\gamma)} \qquad (8.4.40)$$

and introduce new integration variables $A'(e) := g(t_e)A(e)$ where $g(t_e) = \exp(t_j^e \tau_j)$. Since by definition

$$d\mu(t_\gamma) = \prod_{e\in E(\gamma)} d\mu(t_e) = \prod_{e\in E(\gamma)} d\mu_H(g(t_e)) \qquad (8.4.41)$$

we can estimate further

$$|(M_{\psi_1,\psi_1'}, M_{\psi_2,\psi_2'})_\gamma| \leq \int_{\mathcal{G}^{|E(\gamma)|}} \prod_{e\in E(\gamma)} d\mu_H(g_e) \int_{\overline{\mathcal{A}}_\gamma} d\mu_{0\gamma}(A_{|\gamma})$$

$$\times \|\Psi_1(A_{|\gamma})\|_{|\bar{\gamma}} \ \|\Psi_1'(\{g_e A(e)\}_{e\in E(\gamma)})\|_{\bar{\gamma}}$$

$$\times \|\Psi_2(A_{|\gamma})\|_{\bar{\gamma}} \ \|\Psi_2'(\{g_e A(e)\}_{e\in E(\gamma)})\|_{\bar{\gamma}}$$

$$= \left[\int_{\overline{\mathcal{A}}_\gamma} d\mu_{0\gamma}(A_{|\gamma}) \|\Psi_1(A_{|\gamma})\|_{|\bar{\gamma}} \ \|\Psi_2(A_{|\gamma})\|_{\bar{\gamma}} \right]$$

$$= \left[\int_{\overline{\mathcal{A}}_\gamma} d\mu_{0\gamma}(A_{|\gamma}') \|\Psi_1'(A_{|\gamma}')\|_{\bar{\gamma}} \ \|\Psi_2'(A_{|\gamma}')\|_{\bar{\gamma}} \right]$$

$$\leq \| \ \|\Psi_1\|_{\bar{\gamma}} \ \|_\gamma \ \| \ \|\Psi_1'\|_{\bar{\gamma}} \ \|_\gamma \ \| \ \|\Psi_2\|_{\bar{\gamma}} \ \|_\gamma \ \| \ \|\Psi_2'\|_{\bar{\gamma}} \ \|_\gamma$$

$$(8.4.42)$$

where we have used Fubini's theorem and have again applied the Cauchy–Schwarz inequality to functions in $L_2(\overline{\mathcal{A}}_\gamma, d\mu_{0\gamma})$. But

$$\| \ \|\Psi_1\|_{\bar{\gamma}} \ \|_\gamma^2 = \int_{\overline{\mathcal{A}}_\gamma} d\mu_{0\gamma}(A_{|\gamma}) | \ \|\Psi_1(A_{|\gamma})\|_{\bar{\gamma}} \ |^2$$

$$= \int_{\overline{\mathcal{A}}_\gamma} d\mu_{0\gamma}(A_{|\gamma}) \int_{\overline{\mathcal{A}}_{\bar{\gamma}}} d\mu_{0\bar{\gamma}}(A_{|\bar{\gamma}}) |\Psi_1(A_{|\bar{\gamma}}, A_{|\gamma})|^2$$

$$= \int_{\overline{\mathcal{A}}} d\mu_0(A) |\psi_1(A)|^2 = \|\psi_1\|_{\mathcal{H}_0}^2 \qquad (8.4.43)$$

so we get (8.4.25).

(ii) If all functions in question are cylindrical L_2-functions over γ then the integrals over $\overline{\mathcal{A}}_{|\bar{\gamma}}$ are trivial and (8.4.32) simplifies to

$$
(M_{\psi_1,\psi_1'}, M_{\psi_2,\psi_2'})_\gamma = \int_{D_\gamma} d\mu(t_\gamma) \int_{\overline{\mathcal{A}}_\gamma} d\mu_{0\gamma}(A_{|\gamma})
$$

$$
\times \Psi_1(A_\gamma)\overline{[W_\gamma(t_\gamma)\Psi_1'](A_\gamma)}\Psi_2(A_\gamma)[W_\gamma(t_\gamma)\Psi_2'](A_\gamma)
$$

$$
= \int_{\overline{\mathcal{A}}_\gamma} d\mu_{0\gamma}(A_{|\gamma}) \int_{\overline{\mathcal{A}}_\gamma} d\mu_{0\gamma}(A_{|\gamma}')\Psi_1(A_\gamma)\overline{\Psi_1'(A_\gamma')}\Psi_2(A_\gamma)\Psi_2'(A_\gamma')
$$

$$
= \left[\int_{\overline{\mathcal{A}}} d\mu_0(A)\overline{\psi_2(A)}\psi_1(A)\right] \left[\int_{\overline{\mathcal{A}}} d\mu_0(A')\overline{\psi_1'(A')}\psi_2'(A')\right]
$$

$$
= <\psi_2, \psi_1>_{\mathcal{H}_0} <\psi_1', \psi_2'>_{\mathcal{H}_0} \tag{8.4.44}
$$

that is, (8.4.26). $\qquad\qquad\qquad\qquad\qquad\qquad\qquad\qquad\qquad\qquad\qquad\square$

We may now complete the irreducibility part of the uniqueness Theorem 8.1.3 : suppose that the representation π_0 of $\mathfrak{A}$ is not irreducible, that is, not every vector is cyclic. Thus, we find non-zero vectors $\psi, \psi' \in \mathcal{H}_0$ such that

$$
<\psi, a\psi'> = 0 \ \forall \ a \in \mathfrak{A} \tag{8.4.45}
$$

Since the cylindrical functions lie dense in $\mathcal{H}_0$, for any $\epsilon > 0$ we find a graph γ and functions f, f' cylindrical over γ such that

$$
||\psi - f|| < \epsilon, \quad ||\psi' - f'|| < \epsilon \tag{8.4.46}
$$

Notice that due to the Cauchy–Schwarz inequality (6.2.45) implies $|\ ||\psi|| - ||f||\ | < \epsilon$ hence $||\psi|| - \epsilon \leq ||f||$. Since $\psi, \psi' \neq 0$ we may assume $\epsilon < ||\psi||, ||\psi'||$ so that

$$
|\ ||\psi|| - \epsilon| \leq ||f||, \quad |\ ||\psi'|| - \epsilon| \leq ||f'|| \tag{8.4.47}
$$

From (8.4.45) we have in particular that $M_{\psi,\psi'}(t_\gamma, I_\gamma) = 0$ for all $t_\gamma \in D_\gamma$, $I_\gamma \in L_\gamma$, hence

$$
0 = (M_{\psi,\psi'}, M_{\psi,\psi'})_\gamma
$$

$$
= (M_{\psi-f,\psi'}, M_{\psi,\psi'})_\gamma + (M_{f,\psi'-f'}, M_{\psi,\psi'})_\gamma + (M_{f,f'}, M_{\psi-f,\psi'})_\gamma
$$

$$
+ (M_{f,f'}, M_{f,\psi'-f'})_\gamma + (M_{f,f'}, M_{f,f'})_\gamma
$$

$$
= (M_{\psi-f,\psi'}, M_{\psi,\psi'})_\gamma + (M_{f,\psi'-f'}, M_{\psi,\psi'})_\gamma + (M_{f,f'}, M_{\psi-f,\psi'})_\gamma
$$

$$
+ (M_{f,f'}, M_{f,\psi'-f'})_\gamma + ||f||^2\ ||f'||^2 \tag{8.4.48}
$$

where (8.4.26) has been used. Using (8.4.47) and (8.4.25) we have

$$(||\psi|| - \epsilon)^2 \, (||\psi'|| - \epsilon)^2 \leq ||f||^2 \, ||f'||^2$$
$$\leq ||\psi - f|| \, ||\psi'|| \, ||\psi|| \, ||\psi'|| + ||f|| \, ||\psi' - f'|| \, ||\psi|| \, ||\psi'||$$
$$+ ||f|| \, ||f'|| \, ||\psi - f|| \, ||\psi'|| + ||f|| \, ||f'|| \, ||f|| \, ||\psi' - f'||$$
$$\leq \epsilon\{||\psi'||^2 \, ||\psi|| + (||\psi|| + \epsilon) \, ||\psi|| \, ||\psi'||$$
$$+ (||\psi|| + \epsilon) \, (||\psi'|| + \epsilon) \, ||\psi'|| + (||\psi|| + \epsilon)^2 \, ||f'||\} \qquad (8.4.49)$$

Since this inequality holds for all $\epsilon \leq ||\psi||, \, ||\psi'||$ we can take $\epsilon \to 0$ and find

$$||\psi||^2 \, ||\psi'||^2 = 0 \qquad (8.4.50)$$

that is, either $\psi = 0$ or $\psi' = 0$ in contradiction to our assumption. Hence π_0 is irreducible.

This completes the irreducibility proof and hence proves Theorem 8.1.3.

9

Step IV: (1) implementation and solution of the kinematical constraints

In this chapter we implement the kinematical constraints on the Hilbert space $\mathcal{H}_0$. By kinematical we mean here the Gauß and spatial diffeomorphism constraints which will be the same for any background-independent gauge field theory. The feature that distinguishes such different theories is the Hamiltonian constraint which is the only one that depends on the Lagrangian of the classical theory. The Hamiltonian constraint will be treated in a separate chapter. We will also describe the complete set of solutions to the kinematical constraints and derive an inner product on the combined solution space.

9.1 Implementation of the Gauß constraint

We do not really need to implement the Gauß constraint since we can work directly with gauge-invariant functions (that is, one solves the constraint classically and quantises only the phase space reduced with respect to the Gauß constraint). However, we will nevertheless show how to get to gauge-invariant functions starting from gauge-variant ones by using the technique of refined algebraic quantisation outlined in Chapter 30.

9.1.1 Derivation of the Gauß constraint operator

We proceed similarly as in the case of the electric flux operator and start from the classical expression

$$G(\Lambda) := - \int d^D x [D_a \Lambda^j] E_j^a \equiv -E(D\Lambda) \tag{9.1.1}$$

where $D_a \Lambda^j = \partial_a \Lambda^j + f^j{}_{kl} A_a^k \Lambda^l$ is the covariant derivative of the smearing field Λ^j. Notice that (9.1.1) is almost an electric field smeared in D dimensions except that the smearing field $D\Lambda$ depends on the configuration space. Nevertheless, the vector field on $\mathcal{A}$ corresponding to it is given by $-\kappa\beta/2\nu_{D\Lambda}$. Next we apply it to $\mathrm{Cyl}(\mathcal{A})$ by first computing its action on the special functions ϕ_p (see the notation in Section 6.3) and then use the chain rule for general cylindrical functions. In order to compute its action on ϕ_p we must regulate it similarly to what we did for the flux operators and then define $\nu_{D\Lambda}(\phi_p) := \lim_{\epsilon \to 0} \nu_{D\Lambda}(\phi_p^\epsilon)$. Finally we hope that the end result is again a cylindrical function which we may then extend to $\overline{\mathcal{A}}$ and thus derive a cylindrical family of hopefully consistent vector fields on $\overline{\mathcal{A}}$.

We will not write all the steps, the details are precisely as for the regularisation of $\mathfrak{P}$, just that the additional limit $\epsilon' \to 0$ is missing. For the same reason a split of p into edges of different type is not necessary because E is smeared in D directions. One finds

$$\nu_{D\Lambda}(\phi_p) = \beta\kappa \int_0^1 dt \ddot{p}^a(t)(D_a\Lambda^j)(p(t))h_{p([0,t])}(A)\frac{\tau_j}{2}h_{p([t,1])}(A) \quad (9.1.2)$$

Let us use the notation $\Lambda = \Lambda^j\tau_j$ and $A(p(t)) = \dot{p}^a(t)A_a^j(p(t))\tau_j/2$. Using $[\tau_j, \tau_k] = 2f_{jk}{}^l\tau_l$ we can then recast (9.1.2) into the form

$$\nu_{D\Lambda}(\phi_p) = \frac{\beta\kappa}{4} \int_0^1 dt\, h_{p([0,t])}(A)\left\{\frac{d}{dt}\Lambda(p(t)) + [A(p(t)), \Lambda(p(t))]\right\} h_{p([t,1])}(A)$$

$$(9.1.3)$$

Now we invoke the parallel transport equation for the holonomy

$$\frac{d}{dt}h_{p([0,t])}(A) = h_{p([0,t])}(A)\,\dot{c}(t) \cdot A(p(t)) \quad (9.1.4)$$

and use $h_{p([t,1])}(A) = h_{p([0,t])}(A)^{-1}h_p(A)$. Then it is easy to see that (9.1.3) becomes

$$\nu_{D\Lambda}(\phi_p) = \frac{\beta\kappa}{2} \int_0^1 dt \frac{d}{dt}\{h_{p([0,t])}(A)\Lambda(p(t))h_{(p([t,1])}(A)\}$$

$$= \frac{\beta\kappa}{2}[-\Lambda(b(p))h_p(A) + h_p(A)\Lambda(f(p))] \quad (9.1.5)$$

where we have performed an integration by parts in the last step. So indeed we $\bullet\bullet\bullet$ lucky: (9.1.5) is a cylindrical function again. Let us write $\nu_\Lambda := -\nu_{D\Lambda}$, then for any $f_l \in C^\infty(X_l)$ for any subgroupoid $l = l(\gamma)$ we have

$$[\nu_\Lambda(f_l)](A) = \frac{\beta\kappa}{4} \sum_{e \in E(\gamma)} [\Lambda(b(e))A(e) - A(e)\Lambda(f(e))]_{AB}(\partial f_l/\partial A(e)_{AB})(A)$$

$$= \frac{\beta\kappa}{4} \sum_{e \in E(\gamma)} \left([\Lambda_j(b(e))R_e^j - \Lambda_j(f(e))L_e^j]f_l\right)(A) \quad (9.1.6)$$

Finally we write this as a sum over vertices in the compact form

$$G_l(\Lambda)[f_l] := \nu_\Lambda(f_l) = \frac{\beta\kappa}{4} \sum_{v \in V(\gamma)} \Lambda_j(v)\left[\sum_{e \in E(\gamma);\ v=b(e)} R_e^j - \sum_{e \in E(\gamma);\ v=f(e)} L_e^j\right] f_l$$

$$(9.1.7)$$

Hence we have successfully derived a family of vector fields $G_l(\Lambda) \in V^\infty(X_l)$ for any $l \in \mathcal{L}$. No adaption of the graph was necessary this time. Since Λ_j is real-valued for compact G, it follows from our previous analysis that $G_l(\Lambda)$ is real-valued. Using the steps (a), (b) and (c) of Section 6.5 one quickly verifies that it is a consistent family and that it is trivially μ_0-compatible because it is divergence-free, since it is a linear combination of left- and right-invariant vector fields. For the same reason, the associated momentum operator

$$\hat{G}_l(\Lambda)[f_l] = \frac{i\beta\ell_p^2}{2} \sum_{v \in V(\gamma)} \Lambda_j(v) \left[\sum_{e \in E(\gamma); \, v=b(e)} R_e^j - \sum_{e \in E(\gamma); \, v=f(e)} L_e^j \right] f_l \qquad (9.1.8)$$

is essentially self-adjoint with dense domain $C^1(\overline{\mathcal{A}})$.

9.1.2 Complete solution of the Gauß constraint

Using the Lie algebra of the left- and right-invariant vector fields on X_l given by

$$[R_e^j, R_{e'}^k] = -2\delta_{ee'} f^{jk}{}_l R^l, \quad [L_e^j, L_{e'}^k] = 2\delta_{ee'} f^{jk}{}_l L^l, \quad [R^j, L^k] = 0 \qquad (9.1.9)$$

(e.g., $([R^j, R^k]f)(h) = (\frac{\partial^2}{\partial s \partial s'})_{s=s'=0} f([e^{s'\tau_k}, e^{s\tau_j}]h))$ we find

$$[G_l(\Lambda), G_l(\Lambda')] = \left(\frac{\beta\kappa}{4}\right)^2 \sum_{e \in E(\gamma)} \{\Lambda_j(b(e))\Lambda_k'(b(e)) [R_e^j, R_e^k]$$

$$+ \Lambda_j(f(e))\Lambda_k'(f(e)) [L_e^j, L_e^k]\}$$

$$= -\beta\kappa/2 G([\Lambda, \Lambda']) \qquad (9.1.10)$$

where we have defined $\Lambda(x) := \Lambda_j(x)\tau_j/2$. We see that the Lie algebra of $G_l(\Lambda)$ represents the Lie algebra Lie(G) for each $l \in \mathcal{L}$ separately and also represents the classical Poisson brackets among the Gauß constraints, see Chapter 1. This is already a strong hint that the condition $\hat{G}(\Lambda) = 0$ for all smooth Λ_j really means imposing gauge invariance.

Let us see that this is indeed the case. According to the programme of RAQ we must choose a dense subspace of $\mathcal{H}^0$, which we choose to be $\mathcal{D} := \mathrm{Cyl}^\infty(\overline{\mathcal{A}})$. Let $f = [f_l]_\sim$ be a smooth cylindrical function, that is, $f_l \in C^\infty(X_l)$, then $\hat{G}(\Lambda)f = p_l^*(\hat{G}_l(\Lambda)f_l)$. We are looking for an algebraic distribution $L \in \mathcal{D}^*$ such that

$$L(p_l^* \hat{G}_l(\Lambda)f_l) = 0 \qquad (9.1.11)$$

for all Λ_j, $l \in \mathcal{L}$, $f_l \in C^\infty(X_l)$. Since, given l, the smooth function Λ is still arbitrary, we may restrict its support to one of the vertices of γ with $l = l(\gamma)$ and see that (9.1.11) is completely equivalent to

$$L\left(p_l^* \left[\sum_{e \in E(\gamma); \, v=b(e)} R_e^j - \sum_{e \in E(\gamma); \, v=f(e)} L_e^j \right] f_l \right) = 0 \qquad (9.1.12)$$

for any $v \in V(\gamma)$, $l \in \mathcal{L}$, $f_l \in C^\infty(X_l)$.

We now use the fact that any function in $\mathcal{D} = C^\infty(\overline{\mathcal{A}})$ is a finite linear combination of spin-network functions T_s (or can be approximated by those). Therefore, an element $L \in \mathcal{D}^*$ is completely specified by the complex values $L(T_s)$ with no growth condition on these complex numbers (an algebraic distribution is well-defined if it is definedpointwise in $\mathcal{D}$). We conclude that any element $L \in \mathcal{D}^*$

can be written in the form

$$L = \sum_{s \in S} L_s < T_s, \cdot >$$ (9.1.13)

where $< \cdot, \cdot >$ denotes the inner product on $L_2(\overline{\mathcal{A}}, d\mu_0)$ and $\mathcal{S}$ denotes the set of all spin-network labels. Now, first of all (9.1.12) is therefore completely equivalent to

$$L\left(p^*_{l(\gamma(s))} \left[\sum_{e \in E(\gamma(s)); \, v=b(e)} R^j_e - \sum_{e \in E(\gamma(s)); \, v=f(e)} L^j_e \right] T_s \right) = 0$$ (9.1.14)

for any $v \in V(\gamma(s))$, $s \in \mathcal{S}$ where $\gamma(s)$ is the graph that underlies s. Since the operator involved in (9.1.14) leaves $\gamma(s), \vec{\pi}(s)$ invariant and spin-network functions are mutually orthogonal we find that

$$\sum_{s' \in S, \, \gamma(s')=\gamma(s); \vec{\pi}(s')=\vec{\pi}(s)} L_{s'} < T_{s'}, \left[\sum_{e \in E(\gamma(s)); \, v=b(e)} R^j_e - \sum_{e \in E(\gamma(s)); \, v=f(e)} L^j_e \right] T_s >= 0$$ (9.1.15)

for any $v \in V(\gamma(s))$, $s \in \mathcal{S}$. Effectively the sum over s' is now reduced over all $\vec{m}, \vec{n}$ with $m_e, n_e = 1, \ldots, d_{\pi_e}$ for any $e \in E(\gamma(s))$ and is therefore finite. From this it follows already that the most general solution L is an arbitrary linear combination of solutions of the form $< \psi, \cdot >$ where ψ is actually normalisable.

Consider now an infinitesimal gauge transformation $g_t(x) = e^{t\Lambda_j(x)\tau_j}$ for some function $\Lambda_j(x)$ with $t \to 0$. Since $\overline{\mathcal{G}} \cong G^\sigma$ we may arrange that $g = 1$ at all vertices of $\gamma(s)$ except for v. Our spin-network function is of the form

$$T_s = \left[\prod_{e \in E(\gamma(s)); \, b(e)=v} f_e(h_e) \right] \left[\prod_{e \in E(\gamma(s)); \, f(e)=v} f_e(h_e) \right] F_s$$ (9.1.16)

where F_s is a cylindrical function that does not depend on the edges incident at v. Then under an infinitesimal gauge transformation the spin-network function changes as

$$\left(\frac{d}{dt} \right)_{t=0} \lambda^*_{g_t} T_s$$

$$= \left(\frac{d}{dt} \right)_{t=0} \left[\prod_{e \in E(\gamma(s)); b(e)=v} f_e(g_t(v)h_e) \right] \left[\prod_{e \in E(\gamma(s)); f(e)=v} f_e(h_e g_t(v)^{-1}) \right] F_s$$

$$= \left(\frac{d}{dt} \right)_{t=0} \left[\circ_{e \in E(\gamma(s)); b(e)=v} \left(L^e_{g_t(v)} \right)^* \right] \circ \left[\circ_{e \in E(\gamma(s)); f(e)=v} \left(R^e_{g_t(v)^{-1}} \right)^* \right] T_s$$

$$= \Lambda_j(v) \left[\sum_{e \in E(\gamma(s)); b(e)=v} R^j_e - \sum_{e \in E(\gamma(s)); f(e)=v} L^j_e \right] T_s$$

$$= G_{l(\gamma(s))}(\Lambda)[T_s]$$ (9.1.17)

which proves that $G_l(\Lambda)$ is the infinitesimal generator of $\lambda^l_{e^{t\Lambda}}$. It is therefore clear that the general solution L is a linear combination of solutions of the form $<\psi, . >$ where $\psi \in \mathcal{H}^0$ is gauge-invariant. Strictly speaking, ψ has to be invariant under infinitesimal gauge transformations only but since G is connected there is no difference with requiring it to be invariant under all gauge transformations (the exponential map between Lie algebra and group is surjective since there is only one component, that of the identity).

We could therefore also have equivalently required that

$$L(\lambda_g^* f) = L(f) \qquad (9.1.18)$$

for all $g \in \overline{\mathcal{G}}$ and all $f \in \mathcal{D} := C^\infty(\overline{\mathcal{A}})$. In passing we recall that we have defined in the previous section a unitary representation of $\overline{\mathcal{G}}$ on $\mathcal{H}^0$ defined densely on $C(\overline{\mathcal{A}})$ by $\hat{U}(g)f := \lambda_g^* f$. Let $t \mapsto g_t$ be a continuous one-parameter subgroup of $\overline{\mathcal{G}}$, meaning that $\lim_{t \to 0} g_t(x) = g_0(x) \equiv 1_G$ for any $x \in \sigma$, meaning that $t \mapsto g_{tx} := g_t(x)$ is a continuous one-parameter subgroup of G for any $x \in \sigma$ (if g_t is continuous at $t = 0$ then also at every s since $\lim_{t \to s} g_t = \lim_{t \to 0} g_t g_s = g_s$ since group multiplication is continuous). We claim that the one-parameter subgroup of unitary operators $\hat{U}(t) := \hat{U}(g_t)$ is strongly continuous, that is, $\lim_{t \to 0} ||\hat{U}(t)\psi - \psi|| = 0$ for any $\psi \in \mathcal{H}^0$. Since any $\hat{U}(t)$ is bounded and $C^\infty(\overline{\mathcal{A}})$ is dense in $\mathcal{H}^0$ it will be sufficient to show that strong continuity holds when restricted to $\mathcal{D}$. Also, strong continuity follows already from weak continuity (i.e., $<\psi, \hat{U}(t)\psi' > \to <\psi, \psi'>$ for any $\psi, \psi' \in \mathcal{H}^0$) since $||\hat{U}(t)\psi - \psi||^2 = 2(||\psi||^2 - \Re(<\psi, \hat{U}(t)\psi>)$. Since $\mathcal{D}$ is spanned by finite linear combinations of mutually orthonormal spin-network functions (they are in fact smooth), it will then be sufficient to show that $<T_s, \hat{U}(t)T_{s'}> \to <T_s, T_{s'}> = \delta_{ss'}$. If $s = (\gamma, \vec{\pi}, \vec{m}, \vec{n})$, $s' = (\gamma', \vec{\pi}', \vec{m}', \vec{n}')$ then a short computation, using that λ_g leaves $\gamma(s), \vec{\pi}(s)$ invariant, shows that

$$<T_s, \hat{U}(t)T_{s'}> = \delta_{\gamma,\gamma'} \delta_{\vec{\pi},\vec{\pi}'} \prod_{e \in E(\gamma)} [\pi_e(g_t(b(e)))_{m'_e m_e} \pi_e(g_t(f(e))^{-1})]_{n_e n'_e} \qquad (9.1.19)$$

and since the matrix element functions are smooth, the claim follows. We conclude therefore from Stone's theorem that for $g_t(x) = \exp(t\Lambda(x))$ the operator $\hat{G}(\Lambda)$ is the self-adjoint generator of $\hat{U}(t)$.

Finally we display the corresponding rigging map. Since $\overline{\mathcal{G}}$ is a group, the obvious Ansatz is

$$\eta(f) := < \int_{\overline{\mathcal{G}}} d\mu_H(g) <\lambda_g^* f, . > \qquad (9.1.20)$$

which, since λ_g^* preserves $C(C_l)$, is actually a map $\mathcal{D} \to \mathcal{D}$. Since μ_0 is a probability measure we could therefore immediately take the inner product on $\mathcal{H}^0$ for the solutions $\eta(f)$. But let us see where the rigging map proposal takes us. By definition

$$< \eta(f), \eta(f') >_\eta := \eta(f')[f] = \int_{\overline{\mathcal{G}}} \mu_H(g) < \lambda_g^* f, f' >$$

$$= \int_{\overline{\mathcal{G}}} \mu_H(g) \int_{\overline{\mathcal{G}}} \mu_H(g') < \lambda_g^* f, \lambda_{g'}^* f' >=< \eta(f)^\dagger, \eta(f')^\dagger >$$

$$(9.1.21)$$

where in the second equality we have observed that $< \lambda_g^* f, f' >$ is invariant under gauge transformations of f' and $\eta(f)^\dagger :=< ., \int_{\overline{\mathcal{G}}} \mu_H(g)\lambda_g^* f >$. So, indeed the gauge-invariant inner product is just the restricted gauge-variant inner product. Finally, for any gauge-invariant observable we trivially have $\hat{O}'\eta(f) = \eta(\hat{O}f)$.

9.2 Implementation of the spatial diffeomorphism constraint

Again we could just start from the fact that we have a unitary representation of the diffeomorphism group already defined, but we wish to make the connection to the classical diffeomorphism constraint more clear in order to show that the representation defined really comes from the classical constraint. We will work at the gauge-variant level in this section for convenience, however, we could immediately work at the gauge-invariant level and all formulae in this section go through with obvious modifications. The reason for this is that the Gauß constraint not only forms a subalgebra in the full constraint algebra but actually an ideal, that is, since the diffeomorphism and Hamiltonian constraint are actually gauge-invariant, the corresponding operators leave the space of gauge-invariant cylindrical functions invariant. Hence one can solve the Gauß constraint independently before or after solving the other two constraints.

9.2.1 Derivation of the spatial diffeomorphism constraint operator

The representation $\hat{U}(\varphi)$ of $\mathrm{Diff}(\sigma)$ was densely defined on spin-network functions as

$\hat{U}(\varphi)T_s := T_{\varphi \cdot s}$ where

$$\varphi \cdot s := (\varphi \cdot e := \varphi(e), (\varphi \cdot \vec{\pi}(s))_{\varphi(e)} := \pi_e, (\varphi \cdot \vec{m}(s))_{\varphi(e)} := m_e, (\varphi \cdot \vec{n}(s))_{\varphi(e)}$$

$$:= n_e)_{e \in E(\gamma(s))} \tag{9.2.1}$$

Let u be a semianalytic vector field on σ and consider the one-parameter subgroup $t \mapsto \varphi_t^u$ of $\mathrm{Diff}_{\mathrm{sa}}^\omega(\sigma)$ (semianalytic diffeomorphisms) determined by the integral curves of u, that is, solutions to the differential equation $\dot{c}_{u,x}(t) = u(c(t))$, $c_{u,x}(0) = x$ with $\varphi_t^u(x) := c_{u,x}(t)$. The classical diffeomorphism constraint is given by

$$V_a = H_a - A_a^j G_j = 2(\partial_{[a} A_{b]}^j) E_j^b - A_a^j \partial_b E_j^b \tag{9.2.2}$$

Smearing it with u gives

$$V(u) = \int d^3x (\mathcal{L}_u A^j)_a(x) E_j^a(x) = E(\mathcal{L}_u A) \tag{9.2.3}$$

where $\mathcal{L}$ denotes the Lie derivative. Since the constraint is again linear in momenta we can associate with it a vector field $\beta \kappa \nu_{\mathcal{L}_u A}$ on $\mathcal{A}$ which again depends on A as well. Proceeding similarly as with the Gauß constraint we find for its action on holonomies of smooth connections

$$\nu_{\mathcal{L}_u A} \phi_p = \int_0^1 ds\, h_{p([0,s])}(A)(\mathcal{L}_u A)(p(s)) h_{p([s,1])}(A) \tag{9.2.4}$$

We claim that (9.2.4) equals

$$\left(\frac{d}{dt}\right)_{t=0} h_p((\varphi_t^u)^* A) \tag{9.2.5}$$

To see this, one uses the expansion $(\varphi_t^u)^* A = A + t(\mathcal{L}_u A) + O(t^2)$ and the fact that with $p = p_1 \circ \ldots \circ p_N$ we have $h_p = h_{p_1} \ldots h_{p_N}$ with $p_k = p([t_{k-1}, t_k])$, $0 = t_0 < t_1 < \ldots < t_N = 1$, $t_k - t_{k-1} = 1/N$. Denote $\delta h_{p_k} := h_{p_k}(A + \delta A) - h_{p_k}(A)$. Hence

$$h_p(A + \delta A) - h_p(A)$$

$$= \sum_{n=1}^N \sum_{1 \le k_1 < \ldots < k_n \le N} \left(h_{p_1 \circ \ldots \circ p_{k_1-1}}(A)[\delta h_{p_{k_1}}]\right)\left(h_{p_{k_1+1} \circ \ldots \circ p_{k_2-1}}(A)[\delta h_{p_{k_2}}]\right) \ldots$$

$$\times \left(h_{p_{k_n-1}+1 \circ \ldots \circ p_{k_n-1}}(A)[\delta h_{p_{k_n}}]\right)\left(h_{p_{k_n}+1 \circ \ldots \circ p_N}(A)\right) \tag{9.2.6}$$

which holds at each finite N. Now using the formula $h_{p_k}(A) = \mathcal{P} \exp(A(p_k))$ where $A(p_k) = \int_{p_k} A^j \tau_j / 2$ we obtain

$$\delta h_{p_k} = \mathcal{P}\{e^{[A+\delta A](p_k)} - e^{A(p_k)}\} \tag{9.2.7}$$

so that δh_{p_k} is at least linear in δA and therefore in t for $\delta A = (\varphi_t^u)^* A - A$. Thus, dividing (9.2.6) by t and taking the limit $t \to 0$ we find

$$\left(\frac{d}{dt}\right)_{t=0} h_p((\varphi_t^u)^* A) = \sum_{k=1}^N h_{p_1 \circ \ldots \circ p_{k-1}}(A)\left[\left(\frac{d}{dt}\right)_{t=0} h_{p_k}((\varphi_t^u)^* A)\right] h_{p_{k+1} \circ \ldots \circ p_N} \tag{9.2.8}$$

Finally we have $h_{p_k}(A + \delta A) - h_{p_k}(A) = \delta A(p_k) + O(1/N^2)$ so that in the limit $t \to 0$ indeed (9.2.8) turns into (9.2.4).

Unfortunately, (9.2.4) is no longer a cylindrical function and therefore we cannot construct a consistent family of cylindrically defined vector fields on $\overline{\mathcal{A}}$, in other words, (9.2.4) cannot be extended to $\overline{\mathcal{A}}$. Of course for each s the functions $h_{p([0,s])}(A) = A(p([0,s]))$ can be extended directly to $\overline{\mathcal{A}}$, however, $\mathcal{L}_u A$ only makes sense for smooth A. Moreover, we recall from Section 6.2.4 that the measure μ_0 is supported on connections A such that for any $p \in \mathcal{P}$ the function $s \mapsto A([0, s])$ is nowhere continuous and therefore unlikely to be measurable with respect to ds. Thus, we are not able to define an operator that corresponds to the infinitesimal diffeomorphism constraint.

The way out is the observation that •• •••••• •• •••••• ••• ••• •••••••• • can be extended to $\overline{\mathcal{A}}$. In fact, the identity $\nu_{\mathcal{L}_u} h_p(A) = (\frac{d}{dt})_{t=0} h_p((\varphi_t^u)^* A)$ suggests considering the exponentiation of the vector field $\nu_{\mathcal{L}_u A}$ which then gives the action $h_p(A) \mapsto h_p((\varphi_t^u)^* A)$. Since classically we can always recover the infinitesimal action from the exponentiated one, we do not lose any information. Moreover, we may consider general finite diffeomorphisms φ which unlike the φ_t^u are not necessarily connected to the identity. Now, by the duality between p-*chains* and p-•••• • we have for smooth A

$$h_p(\varphi^* A) = \mathcal{P} e^{\int_p \varphi^* A} = \mathcal{P} e^{\int_{\varphi(p)} A} = h_{\varphi(p)}(A) \qquad (9.2.9)$$

which is exactly as we defined the action of the diffeomorphism group in Section 8.2.5. In the form (9.2.9), it is clear that the finite action of $\text{Diff}_{\text{sa}}^\omega(\sigma)$ on $\mathcal{A}$ can be extended to $\overline{\mathcal{A}}$ when considering it as a map between homomorphisms. We can also generalise it to semianalytic diffeomorphisms $\text{Diff}_{\text{sa}}^\omega(\sigma)$ not connected to the identity. Hence

$$\delta : \text{Diff}_{\text{sa}}^\omega(\sigma) \times \overline{\mathcal{A}} \to \overline{\mathcal{A}}; \quad (\varphi, A) \mapsto \delta_\varphi(A) \text{ where } [\delta_\varphi(A)](p) := A(\varphi(p)) \quad (9.2.10)$$

This furnishes the derivation of the action (9.2.10) already defined in the previous chapter from the classical diffeomorphism constraint. Notice that by construction •• • ••• ••• ••• •• ••• ••• •• ••• •••••• •• •••••• •• •••• •• ••••• • ••••

$$\hat{U}(\varphi)\hat{U}(\varphi')\hat{U}(\varphi^{-1})\hat{U}((\varphi')^{-1}) = \hat{U}(\varphi \circ \varphi' \circ \varphi^{-1} \circ (\varphi')^{-1}) \qquad (9.2.11)$$

9.2.2 General solution of the spatial diffeomorphism constraint

We have seen that we can define a unitary representation of $\text{Diff}_{\text{sa}}^\omega(\sigma)$ on $\mathcal{H}^0$ by (9.2.1) and that it is impossible to construct an action of the Lie algebra of $\text{Diff}_{\text{sa}}^\omega(\sigma)$ on $\overline{\mathcal{A}}$. We will now see that this has a counterpart for the representation $\hat{U}(\varphi)$: if there is a quantum operator $\hat{V}(u)$ which generates infinitesimal diffeomorphisms, then it would be the self-adjoint generator of the one-parameter subgroup $t \mapsto \hat{U}(\varphi_t^u)$, that is, we would have $\hat{U}(\varphi_t^u) = e^{it\hat{V}(u)}$. However, that generator exists only if the one-parameter group is strongly continuous by Stone's theorem, Theorem 26.7.3. We will now show that it is not strongly continuous. To see this, take any non-zero vector field and find an open subset $U \subset \sigma$ in which it is non-vanishing. We find a non-trivial graph γ contained in U and an infinite decreasing sequence (t_n) with limit 0 such that the graphs $\varphi_{t_n}^u(\gamma)$ are mutually different. Take any spin-network state T_s with $\gamma(s) = \gamma$. Since spin-network states over different graphs are orthogonal we have $||\hat{U}(\varphi_{t_n}^u)T_s - T_s||^2 = 2$ for all n, thus proving our claim. This small computation demonstrates once again how distributional $\overline{\mathcal{A}}$ in fact is: once a path just differs infinitesimally from a second one, they are algebraically independent and a distributional homomorphism is able to assign to them completely independent values, there is no continuity at all. This behaviour is drastically different from that of Gaußian measures and is deeply rooted in the background independence of our formalism: the covariance of a Gaußian measure depends on a background metric which is able to tell us $\bullet\bullet\bullet$ $\bullet\bullet$ $\bullet\bullet\bullet\bullet\bullet\bullet$ $\bullet\bullet$ $\bullet$ $\bullet\bullet\bullet\bullet\bullet\bullet$ $\bullet\bullet\bullet$. However, in a diffeomorphism-invariant theory there is no distinguished background metric, in contrast, there are diffeomorphisms which, with respect to $\bullet\bullet\bullet$ background metric, can take the two points as far apart or as close together as we desire, the positions of the two points are not gauge-invariant.

The absence of an infinitesimal generator of diffeomorphisms is not necessarily bad because we can still impose diffeomorphism invariance via finite diffeomorphisms, in fact finite diffeomorphisms are even better suited to constructing a rigging map as we will see. However, it should be kept in mind that the passage from the connected component of $\mathrm{Diff}^{\omega}_{0,sa}(\sigma)$ of $\mathrm{Diff}^{\omega}_{sa}(\sigma)$ to all of $\mathrm{Diff}^{\omega}_{sa}(\sigma)$ is a non-trivial step which is not forced on us by the formalism. Since the so-called mapping class group $\mathrm{Diff}^{\omega}_{sa}(\sigma)/\mathrm{Diff}^{\omega}_{0,sa}(\sigma)$ is huge and not very well understood (see, e.g., [553–556]), to take all of $\mathrm{Diff}^{\omega}_{sa}(\sigma)$ is at least the most practical option then. Furthermore, one should stress once more that while entire analytic diffeomorphisms are not too bad (every smooth paracompact manifold admits a real analytic differentiable structure which is unique up to smooth diffeomorphisms, see, e.g., [525]) they are at least rather unnatural because the classical action has smooth diffeomorphisms as its symmetry group and also because an entire analytic diffeomorphism is determined already by its restriction to an arbitrarily small open subset U of σ. In particular, an entire analytic diffeomorphism cannot be the identity in U and non-trivial elsewhere. This is why the generalisation to $\mathrm{Diff}^{\omega}_{sa}(\sigma)$ is forced on us: they are local in contrast to the entire analytic ones.

On the other hand, there is an important difference between allowing semi-analytic $C^{(n_0)}$-diffeomorphisms with $n_0 > 0$ and $n_0 = 0$ (piecewise analytic homeomorphisms): we will see that in $D = 3$ vertices of valence five or higher contain continuous diffeomorphism-invariant information, so called moduli, provided the diffeomorphisms are at least $C^{(1)}$ while these moduli can be changed if we allow piecewise analytic homeomorphisms. Using homeomorphisms is attractive because then the diffeomorphism-invariant Hilbert space would be separable as advertised in [557, 558], on the other hand, some operators of the theory such as the volume operator [425, 427, 428, 559] depend on some $C^{(1)}$ structure and would therefore forbid us to use homeomorphisms, thus keeping the spatially diffeomorphism-invariant Hilbert space $\mathcal{H}_{\mathrm{diff}}$ which we construct below, non-separable. Fortunately, $\mathcal{H}_{\mathrm{diff}}$ is the uncountable (almost direct) sum over the moduli of mutually isomorphic separable Hilbert spaces which appear to be individually preserved by all diffeomorphism-invariant (also invariant under the action of the Hamiltonian constraint) operators constructed so far and in that sense are superselected. In other words, an irreducible representation of the algebra of diffeomorphism-invariant observables would pick one of these sectors and which one is irrelevant because they are unitarily equivalent.

As we see, some fine points related to spatial diffeomorphisms still need to be fixed within the formalism and hopefully these details will no longer be important in the final picture of the theory in which diffeomorphisms of any differentiability category should have at most a semiclassical meaning anyway. First steps are: for a combinatorial extension of the diffeomorphism group using piecewise linear structures, see [435, 436]. For an extension to the full automorphism group of the groupoid $\mathcal{P}$, see [526, 534]. See also Section 10.6.4 for an algebraic generalisation independent of the manifold structure.

Let us then go ahead and solve the finite diffeomorphism constraint. We do this for the case of semianalytic diffeomorphisms which have the important locality property as compared with the entire analytic ones. Notice that under semianalytic $C^{(m)}$-diffeomorphisms, gauge-invariant two-vertices, at which the adjacent edges meet in some $C^{(n)}$-fashion with $n \geq m$ are to be identified with no vertex at all. Hence, by the methods of RAQ we are looking for algebraic distributions $L \in \mathcal{D}^*$ with $\mathcal{D} = C^\infty(\overline{\mathcal{A}})$ such that

$$L(\hat{U}(\varphi)f) = L(f) \; \forall \varphi \in \text{Diff}^\omega_{\text{sa}}(\sigma), \; f \in \mathcal{D} \tag{9.2.12}$$

Here we have explicitly written out the invariance condition in terms of semianalytic diffeomorphisms. Since the span of spin-network functions is dense in $\mathcal{D}$, (9.2.12) is equivalent to

$$L(\hat{U}(\varphi)T_s) = L(T_s) \; \forall \varphi \in \text{Diff}^\omega_{\text{sa}}(\sigma), \; s \in \mathcal{S} \tag{9.2.13}$$

In order to solve (9.2.12), recall from Section 9.1.2 that every element of $\mathcal{D}^*$ can be written in the form $L = \sum_s L_s < T_s, . >$ where L_s are some complex numbers. Then (9.2.13) becomes a very simple condition on the coefficients L_s given by

$$L_{\varphi \cdot s} = L_s \; \forall \varphi \in \text{Diff}^\omega_{\text{sa}}(\sigma), \; s \in \mathcal{S} \tag{9.2.14}$$

Equation (9.2.14) suggests introducing the orbit $[s]$ of s given by

$$[s] = \{\varphi \cdot s; \; \varphi \in \text{Diff}^\omega_{\text{sa}}(\sigma)\} \tag{9.2.15}$$

and therefore (9.2.14) means that $s \mapsto L_s$ is constant on every orbit. Obviously, $\mathcal{S}$ is the disjoint union of orbits which motivates us to introduce the space of orbits $\mathcal{N}$ whose elements we denote by ν. Introducing the elementary distributions $L_\nu := \sum_{s \in \nu} < T_s, . >$ we may write the general solution of the diffeomorphism constraint as

$$L = \sum_{\nu \in \mathcal{N}} c_\nu L_\nu \tag{9.2.16}$$

for some complex coefficients c_ν which depend only on the orbit but not on the representative. Notice that $L_\nu(T_s) = \chi_\nu(s)$ where χ denotes the characteristic function.

We still do not have a rigging map but the structure of the solution space suggests we define

$$\eta(T_s) := \eta_{[s]} L_{[s]} \qquad (9.2.17)$$

for some complex numbers η_ν for each $\nu \in \mathcal{N}$ and extend (9.2.17) by linearity to all of $\mathcal{D}$, that is, one writes a given $f \in \mathcal{D}$ in the form $f = \sum_s f_s T_s$ with complex numbers $f_s = 0$ except for finitely many s and then defines $\eta(f) = \sum_s f_s \eta(T_s)$. This way the map η is tied to the spin-network basis. The crucial question is now whether the coefficients can be chosen in such a way that η satisfies all requirements to be a rigging map.

First we demand that the coefficients $\eta_{[s]}$ are such that the rigging inner product is well-defined. By definition

$$< \eta(T_s), \eta(T_{s'}) >_\eta := \eta(T_{s'})[T_s] = \eta_{[s']} \chi_{[s']}(s) \qquad (9.2.18)$$

Thus, positivity requires that $\eta_{[s]} > 0$. Imposing hermiticity then requires that

$$\eta_{[s']} \chi_{[s']}(s) = \overline{< \eta(T_{s'}), \eta(T_s) >} = \overline{\eta(T_s)[T_{s'}]} = \eta_{[s]} \chi_{[s]}(s') \qquad (9.2.19)$$

Now both the right- and left-hand side are non-vanishing if and only if $[s] = [s']$ so that (9.2.19) is correct with no extra condition on the $\eta_{[s]}$.

Notice that η is almost an integral over the diffeomorphism group: one could have considered instead of η the following transformation

$$T_s \mapsto \sum_{\varphi \in \mathrm{Diff}^\omega_{\mathrm{sa}}(\sigma)} < \hat{U}(\varphi)T_s, . > \qquad (9.2.20)$$

and the right-hand side is certainly diffeomorphism-invariant. The Haar measure that is being used here is a •••• ••• • • •••• •• which is trivially translation-invariant.

Unfortunately (9.2.20) does not even define an element of $\mathcal{D}^*$ because there are uncountably infinitely many semianalytic diffeomorphisms which leave $\gamma(s)$ invariant, simply because semianalytic diffeomorphisms can be chosen to be non-trivial only in regions which do not contain the range of the graph $\gamma(s)$ due to locality. This is even true for entire analytic diffeomorphisms as one can show [266] with more work.

In a sense then, η is a group averaging map in which these •••••• • •••••• diffeomorphisms have been factored out. These form a subgroup but not an invariant one. Unfortunately, the corresponding coset depends on $[s]$ so there is no universal coset. Put differently, while one can find a subset (coset) $\mathrm{Diff}^\omega_{[s],\mathrm{sa}}(\sigma)$ of $\mathrm{Diff}^\omega_{\mathrm{sa}}(\sigma)$ such that

$$\eta(T_s) = \eta_{[s]} \sum_{\varphi \in \mathrm{Diff}^\omega_{\mathrm{sa},[s]}(\sigma)} < \hat{U}(\varphi)T_s, . > \qquad (9.2.21)$$

(just choose once and for all a representative s_ν in ν and precisely one diffeo-morphism that maps s_ν to a given $s \in \nu$), unfortunately these cosets depend on $[s]$ so one cannot view (9.2.21) as a regularised rigging map.

On the other hand, the formal action of (9.2.20) on s' with $[s] \neq [s']$ van-ishes. This justifies doing the group averaging in each of the Hilbert spaces $\mathcal{H}_{[\gamma]}$ separately where $[\gamma]$ is the orbit of γ and $\mathcal{H}_{[\gamma]}$ is the closed linear span of SNWFs over $\gamma' \in [\gamma]$. In order to do this appropriately we need to con-sider diffeomorphism-invariant observables. Their hermiticity properties then will impose further restrictions on the rigging map.

We call an operator $\hat{O}$ a strong observable if $\hat{U}(\varphi)\hat{O}\hat{U}(\varphi)^{-1} = \hat{O}$. We call it a weak observable if $\hat{O}'$ leaves the solution space invariant, in other words

$$L(\hat{U}(\varphi)f) = L(f) \; \forall \, \varphi \in \mathrm{Diff}_{\mathrm{sa}}^\omega(\sigma) \Rightarrow [\hat{O}'L](\hat{U}(\varphi)f) = L(\hat{O}^\dagger \hat{U}(\varphi)f)$$
$$= L(\hat{U}(\varphi)^{-1}\hat{O}^\dagger \hat{U}(\varphi)f) = \hat{O}'L(f) \tag{9.2.22}$$

We first show that restricting attention to strong observables would lead to superselection sectors. Namely, suppose that $\hat{O}$ is a densely defined, closed, strongly diffeomorphism-invariant operator and consider any two spin-network functions $T_s, T_{s'}$ with $r(\gamma(s)) \neq r(\gamma(s'))$. Then by the above remark we find an at least countably infinite number of semianalytic diffeomorphisms φ_n with $\varphi_n(\gamma(s)) = \gamma(s)$ but such that the ranges of the $\varphi_n(\gamma(s'))$ are mutually different. Hence for any n

$$< T_{s'}, \hat{O}T_s > = < T_{s'}, \hat{U}(\varphi_n)^{-1}\hat{O}\hat{U}(\varphi_n)T_s > = < \hat{U}(\varphi_n)T_{s'}, \hat{O}T_s > \tag{9.2.23}$$

Since the states $\hat{U}(\varphi_n)T_{s'}$ are mutually orthogonal and since

$$||\hat{O}T_s||^2 = \sum_{s'' \in \mathcal{S}} | < T_{s''}, \hat{O}T_s > |^2 \geq \sum_{n=1}^{\infty} | < \hat{U}(\varphi_n)T_{s'}, \hat{O}T_s > |^2$$

$$= | < T_{s'}, \hat{O}T_s > |^2 \sum_{n=1}^{\infty} 1 \tag{9.2.24}$$

we conclude that $< T_{s'}, \hat{O}T_s > = 0$. In other words, strongly diffeomorphism-invariant, closed and densely defined operators cannot have matrix elements between spin-network states defined over graphs with different ranges so that the Hilbert space would split into mutually orthogonal superselection sectors. If σ is compact, the total spatial volume would be an operator of that kind, it actually preserves the graph on which it acts. More generally, operators which are built entirely from electric field operators will have this property. However, classically the theory contains many strongly diffeomorphism-invariant functions which are not built entirely from electric fields but depend on the curvature of the connection (for instance the Hamiltonian constraint) and hence, as operators, do not necessarily leave the graph on which they act invariant (see the next chapter). This means that such operators simply cannot be defined on $\mathcal{H}_0$ but must in fact be constructed directly on the spatially diffeomorphism-invariant Hilbert space where graph (rather knot class) changing operators ••• be defined, an example being the Master Constraint. Since we presumably need those graph-changing, diffeomorphism-invariant operators in order to encode information about the connection, very likely no superselection takes place [560].

As presently graph-changing spatially diffeomorphism-invariant operators have not been constructed, we focus on the strongly diffeomorphism-invariant ones to begin with. We now show that there exists a choice of the $\eta_{[s]}$ such that $\hat{O}'\eta(f) = \eta(\hat{O}f)$ at least for strongly invariant operators which then, by the general theory of Chapter 30, implies that the reality conditions $(\hat{O}')^\star = (\hat{O}^\dagger)'$ are satisfied where * denotes the adjoint on $_{\text{diff}}$.

To see this we must discuss the so-called graph symmetry groups. Let $k \in [\Gamma]$ be a graph orbit. Select a representative $\gamma_k \in k$ and choose for each $\gamma \in k$ a semianalytic diffeomorphism $\varphi_{k,\gamma}$ such that $\varphi_{k,\gamma}(\gamma_k) = \gamma$. Furthermore, consider the subgroup P_k of the permutation group of the edges of γ_k such that for each $p \in P_k$ there exists at least one semianalytic diffeomorphism which preserves γ_k as a set but permutes the edges among each other.[1] For each $p \in P_k$ fix such a diffeomorphism $\varphi_{k,p}$. These permutation diffeomorphisms are important for the following reason: let, for instance, γ_k be the figure-eight loop (with intersection) and let e, e' be its two edges. Then the orbit size of $s = (\gamma_k, \pi_e = \pi_{e'}, m_e = m_{e'}, n_e = n_{e'})$ is half of the orbit size of s' with $\gamma(s') = \gamma_k$ but, for example, $\pi_e \neq \pi_{e'}$. (In the gauge-invariant case choose $m_e = n_e$, $m_{e'} = n_{e'}$ and sum over $m_e, m_{e'}$ to get a gauge-invariant intertwiner.) This demonstrates that the orbit-generating sets $\text{Diff}^\omega_{[s],\text{sa}}(\sigma)$ can have different sizes for $[s] \neq [s']$ even if $\gamma(s), \gamma(s')$ are diffeomorphic. The orbit size of $[s]$ is the larger, the less symmetrically the graph is charged with spin labels.

We now define for $[\gamma(s)] = k$

$$\eta(T_s) := \eta_{[s]}T_{[s]} := \eta_k \sum_{\gamma \in k} \sum_{p \in P_k} < \hat{U}(\varphi_{k,\gamma})\hat{U}(\varphi_{k,p})\hat{U}(\varphi_{k,\gamma(s)})^{-1}T_s, . > \quad (9.2.25)$$

where η_k are positive numbers. It is clear that $\eta_{[s]}$ is just η_k times the ratio of the orbit size of the least symmetric $[s']$ with $[\gamma(s')] = k$ divided by the orbit size of $[s]$, which is always a finite natural number.

---

[1] The correspondence to the terminology used in [266] is as follows: given $\gamma \in k$ let $\text{IDiff}^\omega_{\gamma,\text{sa}}(\sigma)$ be the subgroup of $\text{Diff}^\omega_{\text{sa}}(\sigma)$ which maps γ to itself ('isotropy group') and let $\text{TDiff}^\omega_{\gamma,\text{sa}}(\sigma)$ be the subgroup of $\text{IDiff}^\omega_{\gamma,\text{sa}}(\sigma)$ which maps each $e \in E(\gamma)$ to itself ('trivial action subgroup'). This subgroup is a normal subgroup because for each graph isotropy φ_2, each trivial action isotropy φ_1 and each $e \in E(\gamma)$ we have $\varphi_2(e) \in E(\gamma)$ thus $\varphi_2 \circ \varphi_1 \circ \varphi_2^{-1}(e) = \varphi_2 \circ \varphi_2^{-1}(e) = e$ hence $\varphi_2 \circ \varphi_1 \circ \varphi_2^{-1}$ is a trivial action isotropy. Then P_k and the quotient group (rather than just a coset) $\text{GS}_\gamma := \text{IDiff}^\omega_{\gamma,\text{sa}}(\sigma)/\text{TDiff}^\omega_{\gamma,\text{sa}}(\sigma)$ ('graph symmetries') are naturally isomorphic. In [266] one was working with the analytic category and due to the non-locality of analytic structures had to work with equivalence classes of maximal analytic extensions of graphs and in particular with graphs of type I (there exists an analytic function which vanishes exactly on the maximal analytic extension and nowhere else) or type II (such a function does not exist). In the semianalytic case these subtleties drop out due to the locality properties of semianalytic structures.

Let now $\hat{O}$ be a strong observable, then with $k = [\gamma(s)]$ we have

$$< \eta(f), \hat{O}'\eta(T_s) >_\eta = [\hat{O}'\eta(T_s)](f) = [\eta(T_s)](\hat{O}^\dagger f)$$

$$= \eta_k \sum_{\gamma \in k, p \in P_k} < \hat{U}(\varphi_{k,\gamma})\hat{U}(\varphi_p)\hat{U}(\varphi_{k,\gamma(s)})^{-1} T_s, \hat{O}^\dagger f >$$

$$= \eta_k \sum_{\gamma \in k, p \in P_k} < \hat{U}(\varphi_{k,\gamma})\hat{U}(\varphi_p)\hat{U}(\varphi_{k,\gamma(s)})^{-1} \hat{O}T_s, f >$$

$$= < \eta(f), \eta(\hat{O}T_s) >_\eta \qquad (9.2.26)$$

where in the last step we have used that $\hat{O}T_s$ is a countable linear combination of spin-network states $T_{s'}$ with $\gamma(s) = \gamma(s')$ on each of which the averaging is performed in exactly the same way as on T_s. This was the point of making (9.2.25) depend only on k and not on $[s]$.

There are no additional conditions on η_k as far as non-graph-changing, strong observables are concerned. It follows that the relative normalisations between the $\eta(T_s)$ are only determined for those s with the same $[\gamma(s)]$. The ambiguity is encoded in the freedom to choose the positive numbers η_k. In order to fix those, what we need is to study knot class-changing spatially diffeomorphism-invariant operators and require that they be symmetric (if their classical counterpart is real-valued). One expects that among the infinite number of inner products on the space of solutions to the spatial diffeomorphism constraint a relatively small number survives when implementing self-adjointness of operators corresponding to real-valued, classical, strongly spatially diffeomorphism-invariant observables which are knot class-changing as operators. See, for example, [561] for a systematic investigation in a simplified context.

One can question, however, why we bother about existence or non-existence of a spatially diffeomorphism-invariant inner product at all. The reason is the following: remember that the classical constraint algebra between the Hamiltonian constraint $H(N)$ and diffeomorphism constraint $\vec{H}(\vec{N})$ respectively has the structure

$$\{\vec{H}(\vec{N}), \vec{H}(\vec{N}')\} \propto \vec{H}([\vec{N}, \vec{N}']),$$
$$\{\vec{H}(\vec{N}), H(N)\} \propto H(\vec{N}[N]), \quad \{H(N), H(N')\} \propto \vec{H}(q^{-1}(N dN' - N' dN)) \quad (9.2.27)$$

Thus, the Poisson Lie algebra of diffeomorphism constraints is actually a subalgebra (the first identity) of the full constraint algebra but it is not an ideal (the second identity). It is therefore not possible to solve the full constraint algebra in two steps by first solving the diffeomorphism constraint and then solving the Hamiltonian constraint in a second step: as (9.2.27) shows, the dual Hamiltonian constraint operator must not leave the space of diffeomorphism-invariant distributions invariant and it is therefore meaningless to try to construct an inner product that solves only the diffeomorphism constraint. Rather, one has to construct the space of solutions of all constraints first before one can tackle the issue of the physical inner product. The only way out of this fact and to make use of $\mathcal{H}_{\text{diff}}$ during the quantisation process as a carrier space of the constraint operators is to replace the non-diffeomorphism-invariant Hamiltonian constraints by an equivalent set of constraints which are spatially diffeomorphism-invariant. This is the • ••••• • ••••••••• • ••••••• to which we turn in the next chapter.

10

Step IV: (2) implementation and solution of the Hamiltonian constraint

We come now to the 'Holy Grail' of Canonical Quantum General Relativity, the implementation and solution of the Hamiltonian constraint. It is the benchmark which decides whether all the previous efforts were in vain or not. Without an admissible implementation of the Hamiltonian constraint no progress can be made and no reliable predictions of LQG are possible.

10.1 Outline of the construction

The Hamiltonian constraint is technically and conceptually much more difficult than the kinematical constraints because:

• • •• ••• •

The Hamiltonian constraint is tremendously non-linear.

• • •• ••• •

The Dirac algebra $\mathfrak{D}$ is not a Lie algebra due to the structure functions.

The first issue is bound to create UV problems while the second prohibits solving the constraints by the method of refined algebraic quantisation.

Actually the new complex variables $A^{\mathbb{C}} = \Gamma + iK$, $E^{\mathbb{C}} = -iE$ were originally introduced precisely in order to deal with Problem 1. Namely the •••• •• ••• Hamiltonian constraint $\sqrt{\det(q)}H \propto \mathrm{Tr}(F^{\mathbb{C}}[E^{\mathbb{C}}, E^{\mathbb{C}}])$ is at least polynomial in these variables. Moreover, the degree of this polynomial is only four, no worse than for non-Abelian Yang–Mills theory. However, as already mentioned in Chapter 5 there are two obstacles to using complex variables:

• • •• ••• •• •

All the machinery that we have used in order to arrive at $\mathcal{H}^0$ makes crucial use of the fact that the connection is real-valued so that the corresponding holonomies are valued in a compact gauge group. To date there is no representation theory available for the case of a non-compact gauge group, in this case SL(2, $\mathbb{C}$). By this we mean that, while it is actually possible to define positive linear functionals on the corresponding spaces of cylindrical functions (see, e.g., [562] and [456]) none of them is a representation space for the corresponding *-algebra which must implement the •••• ••• •• ••• ••• relation $A + \bar{A} = 2\Gamma(E)$. Hence, non-polynomiality enters through the backdoor.

•••• •• ••• •• ••• •••• •• •••••••• ••• •• •• •••• • •• ••• • •• •••• •• ••• •• ••••••• •

•• • •••••• •• •

It turns out that it is impossible, on general grounds, to construct a UV-finite, background-independent operator-valued distribution corresponding to $\sqrt{\det(q)}H$. The reason is that the rescaled Hamiltonian constraint is a density of weight two while we will see that •• •• ••• •••••• •• • •••••• ••• •••• • •••••• •• •••• •• •• • •••••••• ••• •••••• ••••. Thus, one is forced to work with the •• ••••• ••• original, density one-valued, Hamiltonian constraint H. However, H is not polynomial and hence the whole virtue of the complex variables is questioned. In fact, all the solutions to the Hamiltonian constraint which were constructed in the late 1980s and early 1990s were only formally solutions, the result of the calculation was of the form $0 \cdot \infty$ and hence vanishes only at finite regularisation which, however, introduces a background.

There are two proposals to deal with Obstacle 1:

• •••••••• •

One works with real rather than complex connections and thus simplifies the representation problem as has been pointed out in [310, 311].
• •••••••• •

One tries to give rigorous meaning to the Wick transform [315] which maps us from spaces of real connections to spaces of complex connections while automatically implementing the correct reality conditions. We will describe this briefly in a later subsection.

However, both proposals still do not cure Obstacle 2. Therefore, currently complex variables are somewhat disfavoured compared with the real variables for which at least we can use the results from steps I, II, III.

Thus we are back to both problems mentioned above, where it is understood that we will be using real-valued variables from now on. The idea to solve the first problem is to exploit spatial diffeomorphism invariance: in a background-independent theory such as LQG it is a priori meaningless to talk about 'short' and 'long' distances because these notions depend on a (spatial) background metric. In other words, short and long distances are fundamentally[1] spatially diffeomorphism equivalent. Therefore, there should not be any ultraviolet divergence if we manage to implement the Hamiltonian constraint on the Hilbert space $\mathcal{H}_{\text{diff}}$ of spatially diffeomorphism-invariant states. That, however, is again prohibited by the structure of the algebra $\mathfrak{D}$ which imposes that the spatial diffeomorphism constraints do not form an ideal, or in other words, that the Hamiltonian constraint operator must not leave the Hilbert space $\mathcal{H}_{\text{diff}}$ invariant. In order to still

---

[1] This does not mean that we cannot talk about short and long distances at all. It just means that this is a background-dependent concept. Thus, in order to make contact with these notions we must construct a physical semiclassical state which approximates a given background 4-metric and then we can talk about physical spatial distances between, say, lumps of matter. However, these physical distances have nothing to do with the kinematical, coordinate distances that are important for the UV behaviour of the operator algebra and which in turn are gauge-dependent.

use spatial diffeomorphism invariance as a UV regulator one therefore has to proceed differently.

• • •• •••• • ••

The first solution to Problem 1 is to implement regulated Hamiltonian operators $\hat{H}_\epsilon(N)$ on the kinematical Hilbert space $\mathcal{H}^0$ and to use an operator topology which uses spatially diffeomorphism-invariant states and in which these nets of regulated operators converge as we remove the regulator ϵ. There is a natural operator topology which suggests itself: recall that a net $\hat{O}_\epsilon$ of (unbounded) operators on a Hilbert space $\mathcal{H}$ with common dense domain $\mathcal{D}$ is said to converge in the weak * operator topology to an operator $\hat{O}$ with dense domain $\mathcal{D}$ provided that $l[(\hat{O}_\alpha f)]$ converges to $l[(\hat{O}f)]$ for all $f \in \mathcal{D}$ and $l \in \mathcal{D}^*$ where $\mathcal{D}^*$ is the algebraic (i.e., not necessarily bounded) linear functional on $\mathcal{D}$. Now we have seen in the previous chapter that the solutions to the diffeomorphism constraint are elements of $\mathcal{D}^*$ where $\mathcal{D}$ is the dense, finite linear span of spin-network functions. Thus, in order to make use of $\mathcal{H}_{\text{diff}}$ we are naturally led to consider the weak *-topology for the $\hat{H}_\epsilon(N)$ where $\mathcal{D}^*$ is restricted to the spatially diffeomorphism-invariant subspace $\mathcal{D}^*_{\text{diff}} \subset \mathcal{D}^*$ and it turns out that this ••• •• ••• • •••••. Moreover, the operators $\hat{H}(N)$ on $\mathcal{H}^0$ are consistent in the sense that their commutator annihilates the elements of $\mathcal{D}^*_{\text{diff}}$, that is, $l([\hat{H}(N), \hat{H}(N')]f) = 0$ for all $f \in \mathcal{D}$ and $l \in \mathcal{D}^*_{\text{diff}}$ as it should according to the algebra $\mathfrak{D}$ since $\{H(N), H(N')\}$ is proportional to a spatial diffeomorphism constraint. The way the calculation works is actually interesting because the generator of spatial diffeomorphisms does not exist as we have seen in the previous chapter. Hence the only way that $l([\hat{H}(N), \hat{H}(N')]f) = 0$ can hold is if $[\hat{H}(N), \hat{H}(N')]f$ is proportional to a finite linear combination of terms of the form $(U(\varphi) - 1)f'$ and this is precisely what happens.

• • •• •••• • ••

The second solution to Problem 1 is to use the • • ••••• • •• •••••• •• • •••••• • • • (MCP), the classical part of which was used in Section 2.1 already. Basically one replaces the infinite number of Hamiltonian constraints by a single Master Constraint which is the weighted sum (actually integral) of the squared Hamiltonian constraints. The weight is carefully chosen in such a way that the Master Constraint is spatially diffeomorphism-invariant. Since, as we show in Chapter 30, the Master Constraint encodes the same reduced phase space as the infinite number of Hamiltonian constraints, no relevant information is lost and we are now able to implement the Master Constraint operator $\hat{}$ on $\mathcal{H}_{\text{diff}}$. It turns out that the same mechanism that makes the $\hat{H}_\epsilon(N)$ converge in the aforementioned topology, that is, background independence, leads to a UV-finite Master Constraint operator on $\mathcal{H}_{\text{diff}}$.

Hence Problem 1 mentioned above can be successfully dealt with and so we managed to •••• ••• •••• • ••••••• • ••• • •••••• • •• • ••• •• ••••••. We will display both Solutions 1A, 1B in this chapter. The second solution, however, is preferred in the sense that it ••••• •••••••• ••••• •• •• ••••••••• •••• •• •••

•••• •• • ••• •• • ••: •• •••••• ••• •• •• •••: •• •• ••• • •• ••••• ••• ••• •••••• •• •

••• ••••••••• • •••••• ••••• which only uses the spectral theory of the Master Constraint operator. This is out of reach with Solution 1A due to Problem 2 mentioned above. For both methods it is possible to systematically construct a huge class of solutions to all constraints, but only Solution 1B provides us with an induced physical inner product on the space of these solutions and a new handle on Dirac observables. It follows that Problem 2 can also be dealt with.

While this is promising, it should be pointed out that this does not yet mean that the mathematical construction of LQG is completed. The reason for this is three open issues.

••••• •

We have seen that $< .,. >_{\mathrm{diff}}$ is ambiguous due to the unspecified normalisation of the $\eta(T_s)$. This ambiguity carries over to $< .,. >_{\mathrm{phys}}$.

••••• •

The limit of the regularisations of both $\hat{H}(N)$ and $\hat{}$ is not unique, they depend, not surprisingly, on certain spatially diffeomorphism-inequivalent characteristics that survive the removal of the regulator, as we will see.

••••• •

Also the inner product $< .,. >_{\mathrm{phys}}$ can be fixed only if one insists on an irreducible representation of the algebra of Dirac observables. Hence, before we have these at our disposal, $< .,. >_{\mathrm{phys}}$ is ambiguous just like $< .,>_{\mathrm{diff}}$.

Hence for any given choice of $< .,. >_{\mathrm{diff}}$, $\hat{}$ we obtain a different induced physical Hilbert space $\mathcal{H}_{\mathrm{phys}}$ with induced inner product $< .,. >_{\mathrm{phys}}$. The correct $< .,. >_{\mathrm{diff}}$ will be selected by implementing a suitable algebra of self-adjoint and spatially diffeomorphism-invariant graph-changing operators which are classically real-valued. The correct $\hat{}$ will be selected by constructing semiclassical states on $\mathcal{H}_{\mathrm{diff}}$ with respect to which $\hat{}$ has admissible expectation values and with respect to which the semiclassical sector of $\mathcal{H}_{\mathrm{phys}}$ captures classical GR. Hence these issues will be solved in step V.

To summarise, while not all the problems with the Hamiltonian constraint have been solved yet, not only is there a large class of consistent proposals but moreover we have explicit control over the freedom involved and for each possible choice we know what the physical Hilbert space is. Hence it is fair to say that step IV of the programme is completed while the restriction of the amount of freedom is reserved for step V. This should be contrasted with the situation before the mid-1980s when one could not even complete step III of the programme.

10.2 Heuristic explanation for UV finiteness due to background independence

Looking at the explicit, complicated expression of the Hamiltonian constraint it is truly astonishing, and even more so for the Master Constraint, that one can make sense out of it at all. Such a result would not hold in a Fock space

representation. The underlying reason is the • •• •••• ••••••••••• •• •••••••••••
of the LQG approach which by definition excludes the background-dependent
Fock space representations. In this section we give a heuristic explanation before
we go into mathematical details: there is a very simple, geometric mechanism at
work which directly relies on background independence.

It is simplest to exhibit this mechanism by the example of Einstein–Klein–
Gordon theory, see Chapter 12 for more details on matter coupling. The matter
phase space is determined by a canonically conjugate pair (ϕ, π) with non-trivial
equal time Poisson brackets $\{\pi(x), \phi(y)\} = \lambda \delta(x, y)$ and the kinetic matter contribution to the Hamiltonian constraint

$$H_{\text{kin}}^{\text{KG}}(N) = \frac{1}{2\lambda} \int_\sigma d^3 x N \frac{\pi^2}{\sqrt{\det(q)}} \qquad (10.2.1)$$

where N is the lapse test function. We take ϕ to be dimensionless and hence
$\ell_s^2 := \hbar \lambda$ has dimension cm^2. For simplicity we disregard the potential term and
the Einstein–Hilbert term to which the subsequent analysis equally applies.

Crucial for what follows is that the function $x \mapsto \pi(x)$, in contrast to $x \mapsto \phi(x)$,
is not a scalar on σ but rather a •••••• ••••••• •• • •••••• ••• which transforms
like $\sqrt{\det(q)}$ under diffeomorphisms of σ. This is reflected, for example, in the
Poisson bracket $\{\pi(x), \phi(y)\}$ because the δ-distribution $\delta(x, y)$ on σ is a scalar
density of weight one in x and a scalar in y. Consequently, in quantum theory
the density weight finds its way into the associated canonical commutation rela-
tions of the corresponding operator-valued distributions $[\hat{\pi}(x), \hat{\phi}(y)] = i\ell_s^2 \delta(x, y)$,
any representation of which must implement the density weight of π, $\sqrt{\det(q)}$.
Notice that the integrand of (10.2.1) comes out automatically with density weight
one as is required by any background-independent theory that derives from a
diffeomorphism-invariant action on M.

We will now compare ordinary QFT and LQG in the way they quantise
(10.2.1).

1. • •••••••• •••••••••• •••• ••• • ••

We choose Minkowski spacetime $(M, g) = (\mathbb{R}^4, g^0)$ with $g^0 = \text{diag}(-1, 1, 1, 1)$
as a background. Then (10.2.1) becomes the kinetic Klein–Gordon energy on
Minkowski space

$$H_{\text{kin},0}^{\text{KG}} = \frac{1}{2\lambda} \int_\sigma d^3 x \pi^2 \qquad (10.2.2)$$

In ordinary QFT we quantise this functional on Fock space $\mathcal{H}_F$ and find the
usual normal ordering correction

$$\hat{H}_{\text{kin},0}^{\text{KG}} - :\hat{H}_{\text{kin},0}: = \frac{\hbar}{4} \int_{\mathbb{R}^3} d^3 x \left(\sqrt{-\Delta_x} \delta(x, y) \right)_{x=y} \qquad (10.2.3)$$

where $\Delta = \delta^{ab} \partial_a \partial_b$ is the Laplacian on flat Euclidean space which enters the
definition of the annihilation operators $\hat{a} = [\sqrt[4]{-\Delta}\hat{\phi} - i(\sqrt[4]{-\Delta})^{-1}\hat{\pi}]/(\sqrt{2}\ell_s^2)$.
Expression (10.2.3) explicitly displays the short distance singularity as $x \to y$.

(The potential term would give the same singularity.) The presence of the singularity is not surprising on geometrical grounds because $\hat{\pi}(x)^2$ is a density of weight two which transforms as the ill-defined expression $\delta(x,0)^2$. Notice that subtractions of the vacuum energy are not allowed in LQG: first of all it contributes to the cosmological constant term and therefore cannot be discarded, second it evidently depends on a background metric and hence is not allowed.

2. •• •••••••• • ••• •••••• ••••• •• •

This time we have to keep the field q_{ab} in (10.2.1) dynamical and we must turn it into an operator. This has two consequences: first, the net density weight of the integrand of (10.2.1) remains unity. Indeed, switching off gravity by locking the dynamical metric field q_{ab} at the fixed value $q_{ab}^0 = \delta_{ab}$ as in (10.2.2) is a crime from a geometrical point of view because one has replaced the scalar density $\sqrt{\det(q)}$ of weight one by a constant of density weight zero, a drastic modification of the geometrical character of (10.2.1) which is responsible for the singularity (10.2.3) as we will show. Second, for the matter sector we cannot use a Fock space representation because $\mathcal{H}_F$ is background-dependent, for example, through the Laplacian which enters the annihilation operators.

Consequently, in LQG entirely new, background-independent representations appear. Skipping the mathematical details, which we supply later, they can be described as follows. The matter Hilbert space $\mathcal{H}^{KG}$ is a space of certain square integrable functionals, on the space of scalar fields ϕ, of the form $\psi_S^{KG}[\phi]$, depending on ϕ only through the field values $\phi(v), v \in S$ where S is an arbitrary finite set of points v of σ. Similarly, the Hilbert space $\mathcal{H}^E$ for the gravitational degrees of freedom consists of certain square integrable functionals, on a space of (SU(2)) connections A, of the form $\psi_\gamma^E[A]$, depending on A only through the holonomies $A(e), e \in \gamma$ where γ is an arbitrary finite set of paths, that is a graph, in σ. The operator-valued distributions corresponding to $\pi(x), \sqrt{\det(q)}(x)$ are respectively represented by

$$\hat{\pi}(x)\psi_S^{KG} = i\ell_s^2 \sum_{v \in S} \delta(x,v)\, \hat{Y}(v)\psi_S^{KG}$$

$$\widehat{\sqrt{\det(q)}}(x)\psi_\gamma^E = \ell_P^3 \sum_{v \in V(\gamma)} \delta(x,v)\hat{V}_v\psi_\gamma^E \tag{10.2.4}$$

Here $\hat{Y}(v) = \partial/\partial\phi(v)$ is a scalar operator on $\mathcal{H}^{KG}$, $V(\gamma)$ denotes the set of vertices (endpoints of paths) of γ and $\hat{V}_v$ is a local, self-adjoint, positive, dimensionless, scalar operator on $\mathcal{H}^E$ which is closely related to the volume operator of LQG. Notice that in (10.2.4) the distributional features are neatly separated from the non-distributional ones and the density weight is explicit on both sides of the equations. Finally, the Hilbert space of the coupled system is the subspace of $\mathcal{H}^{EKG} = \mathcal{H}^E \otimes \mathcal{H}^{KG}$ consisting of states of the form $\psi_\gamma^E \otimes \psi_{V(\gamma)}^{KG}$ where the automatic restriction $S = V(\gamma)$, which can be derived, implements the physical fact that matter can be excited only in regions with non-zero volume.

The existence of a UV singularity as in (10.2.3) is tested by formally inserting (10.2.4) into (10.2.1) resulting in (these heuristics can be justified by a rigorous background-independent regularisation procedure, see Chapter 12).

$$\hat{H}_{\text{kin}}^{\text{KG}}(N)\psi^{\text{E}} \otimes \psi^{\text{KG}}$$

$$= -m_P \left(\frac{\ell_s}{\ell_P}\right)^2 \sum_{v \in V(\gamma)} \int_\sigma d^3x \left[\underbrace{\frac{1}{\hat{V}_v} \underbrace{\frac{1}{\delta(x,v)}}_{\uparrow}} \psi^{\text{E}}_\gamma\right] \otimes [\underbrace{\delta(x,v)}_{\uparrow} \delta(y,v) \hat{Y}(v)\hat{Y}(v)\psi^{\text{KG}}_{V(\gamma)}]_{y \to x}$$

$$\underbrace{}_{\text{Cancellation}}$$

(10.2.5)

where $m_P = \sqrt{\hbar/G}$ is the Planck mass and $1/\hat{V}_v$ is defined by the spectral theorem.

Formula (10.2.5) precisely unveils the regularising mechanism of quantum gravity: the matter part of (10.2.5), as before, displays the short distance singularity stemming from the product of two densities of weight one, hence 'nothing is swept under the rug'. However, one of these δ-distributions in the numerator coming from • •••• gets precisely cancelled by the one in the denominator coming from •••• ••••, leaving us with only one δ-distribution, correctly accounting for the fact that the net density weight of the integrand is +1, • •••• •• •• •••••• •••• ••• •• •• ••• ••••••••••• ••• •••••• ••• •••••.
The integral can then be performed, resulting in the • • ••• expression

$$\hat{H}_{\text{kin}}^{\text{KG}}(N)\psi^{\text{E}} \otimes \psi^{\text{KG}} = -m_P \left(\frac{\ell_s}{\ell_P}\right)^2 \sum_{v \in V(\gamma)} \left[\frac{1}{\hat{V}_v}\psi^{\text{E}}_\gamma\right] \otimes [\hat{Y}(v)^2 \psi^{\text{KG}}_{V(\gamma)}] \quad (10.2.6)$$

the zero modes of $\hat{V}_v$ being taken care of in the rigorous derivation.

The finiteness result (10.2.6) is quite remarkable because, in a formal Fock space quantisation of the gravitational sector using perturbations of the Minkowski metric, the highly interacting operator corresponding to (10.2.1) would have been hopelessly divergent. Indeed, (10.2.6) is a non-perturbative result because the eigenvalues of (10.2.6) scale with ℓ_P^{-3} which is not analytic in Newton's constant and in fact the short distance singularity is recovered in the $G \to 0$, that is, $\ell_P \to 0$ limit.

In summary, in LQG there is a simple, geometrical mechanism, directly relying on background independence, which avoids certain short distance singularities. This does not prove that LQG is ultraviolet finite because the above calculations are not carried out at the level of physical states. However, LQG here succeeds where every other approach has failed so far, which can be taken as a promising hint.

We will now make these heuristics precise, following [252,253,315,325,437–439].

10.3 Derivation of the Hamiltonian constraint operator

The importance of the density weight spelt out in the previous section was noted by many working on formal solutions to the Hamiltonian constraint (see, e.g., [344–349, 563–566] and references therein). In order to solve the associated problem with the density two-valued rescaled constraint $H\sqrt{\det(q)}$ even multiplicative renormalisations were considered, that is, one multiplies the operator by a regulator which vanishes in the limit. While this removes the background dependence one now has a quantum operator whose classical limit is zero.

Another suggestion was to take the square root of the Hamiltonian constraint $\tilde{H}$ since this reduces the density weight to one and to quantise this square root (see [567], in particular in connection with matter coupling [568, 569]). However, since $\tilde{H}$ is famously indefinite it is unclear how to define the square root of an infinite number of non-self-adjoint, non-positive and non-commuting operators, moreover, classically the square root of a constraint has an ill-defined Hamiltonian vector field and therefore does not generate gauge transformations.

A brute force method finally to remove the singularities is to go to a lattice formulation but the problem must undoubtedly reappear when one takes the continuum limit (see, e.g., [570, 571] and references therein).

For those reasons, ••• •• •••• $1/\sqrt{\det(q)}$ •• H ••• •••••• • ••• $\tilde{H} = \sqrt{\det(q)}H$ ••• •• •• ••• •••••• ••• ••• •••••••• • • ••• ••• •••••• ••• ••• •••••• •• •. Now surprisingly, by means of a novel quantisation technique the non-polynomial prefactor can be absorbed into a commutator between well-defined operators. Since a commutator is essentially a derivation one can intuitively understand that this operation will express a denominator in terms of a numerator which has a better chance of being well-defined as an operator. Even more is true: the new technique turns out to be so general that it applies to any kind of field theory for which a Hamiltonian formulation exists [437–443]. The series of these papers is entitled 'Quantum Spin Dynamics (QSD)' for the following reason: the Hamiltonian constraint $\hat{H}$ acting on a spin-network state creates and annihilates the spin quantum numbers with which the edges of the underlying graph are coloured. On the other hand, the ADM energy surface Hamiltonian operator [442] is essentially diagonal on spin-network states where its eigenvalue is also determined by the spin quantum numbers. Thus, we may interpret the spin-network representation as the **non-linear Fock representation of Quantum General Relativity**, the spin quanta playing the role of the occupation numbers of momentum excitations of the usual Fock states of, say, Maxwell theory. The excitations of the gravitational quantum field are string-like, labelled by the edges of a graph, and the degree of freedom corresponding to an edge can be excited only according to half-integral spin quantum numbers.

The rest of this section is devoted to a hopefully pedagogical explanation of the main idea on which [438] is based (see also [325, 572]).

Usually, the Hamiltonian constraint is written in terms of the real connection variables as follows [310, 311, 570, 571] (we set $\beta = 1$ in this section, the generalisation to arbitrary positive values is trivial, and drop the label β from all formulae)

$$H = \frac{1}{\kappa\sqrt{\det(q)}}\mathrm{tr}([F_{ab} - R_{ab}][E^a, E^b]) \qquad (10.3.1)$$

(we have a trace and a commutator for the Lie algebra-valued quantities and kept explicitly a factor of $1/\kappa$ coming from an overall factor of $1/\kappa$ in front of the action). The reason for this is clear: since A, E are the elementary variables one better avoids the appearance of $K_a^i = A_a^i - \Gamma_a^i$. We, however, will work paradoxically with the following identical formula (up to an overall numerical factor)

$$H = \frac{4}{\kappa\sqrt{\det(q)}}\mathrm{tr}([K_a, K_b][E^a, E^b]) - H_{\mathrm{E}} \qquad (10.3.2)$$

where

$$H_{\mathrm{E}} = \frac{2}{\kappa\sqrt{\det(q)}}\mathrm{tr}(F_{ab}[E^a, E^b]) \qquad (10.3.3)$$

is called the • • •••••• • •• ••••• ••• ••• •••••• • because it would be the Hamiltonian constraint of canonical Euclidean gravity. Its natural appearance here is not a coincidence as we will see. The reason for doing this will become clear in a moment. Notice that we have correctly introduced the overall factor $1/\kappa$ in front of the action into H_{E}, H which will get the dimensionalities right and we have used the notation $F_{ab} = F_{ab}^j \tau_j/2, E^a = E_j^a \tau_j/2, K_a = K_a^j \tau_j/2, A_a = A_a^j \tau_j/2, \tau_j = -i\sigma_j$.

Consider the following two quantities:

(i) The volume of an open region R of σ

$$V(R) := \int_R d^3 x \sqrt{|\det(q)|} \qquad (10.3.4)$$

(ii) The integrated densitised trace of the extrinsic curvature

$$K := \int_\sigma d^3 x K_a^i E_i^a \qquad (10.3.5)$$

(the latter of which is nothing else than the generator of the Wick transform up to a factor of $-\pi/(2\kappa)$, see Section 10.7.1). Notice that in (10.3.5) we have taken absolute values under the square root. However, $\det((q_{ab})) = [\det((e_a^i))]^2$ is anyway positive so that we can drop the absolute value at the classical level. However, since $E_j^a = e_j^a\sqrt{\det(q)}$ we have $\det(E) = \mathrm{sgn}(\det(e))\det(q)$ so that we only have $\det(q) = |\det(E)|$ if we allow both signs of $\det(e)$. In the classical theory the sign of $\det(e)$ is constant, however, in the quantum theory, which is an extension of the classical theory, we must allow for both signs although semiclassical states will be peaked on constant sign. If we do not allow for both

•••• •• • ••• •• • ••• •• •••••• • ••• •• •• •••• •• ••• • •• •••• ••• • •• •••• •• •

signs then E_j^a cannot become a derivative operator in the quantum theory. Hence we will be using $\sqrt{\det(q)} := \sqrt{|\det(E)|}$ in order to allow for this possibility.

The following two classical identities are ••• for all that follows:

$$\left(\mathrm{sgn}(\det(e))\frac{E_k^a E_l^b \epsilon_{jkl}}{\sqrt{\det(q)}}\right)(x) = \epsilon^{abc}e_c^j(x) = 2\epsilon^{abc}\frac{\delta V(R)}{\delta E_j^a(x)} = 2\epsilon^{abc}\{V(R), A_a^j(x)\}/(\kappa/2)$$

$$(10.3.6)$$

for any region R such that $x \in R$ and

$$K_a^j(x) = \frac{\delta K}{\delta E_j^a(x)} = \{K, A_a^j(x)\}/(\kappa/2) \tag{10.3.7}$$

where (10.3.7) relies on $\{\Gamma_a^i, K\} = 0$ which follows from the fact that K canonically generates constant rescalings while Γ is a homogeneous, rational function of E and its first spatial derivatives of order zero. In the sequel we will use the notation R_x for any open neighbourhood of $x \in \sigma$.

Using these key identities the reader can quickly convince herself that

$$(\mathrm{sgn}(\det(e))[H - H_E])(x) = -8\epsilon^{abc}\mathrm{tr}(\{A_a, K\}\{A_b, K\}\{A_c, V(R_x)\})/(\kappa/2)^4 \tag{10.3.8}$$

$$(\mathrm{sgn}(\det(e))H_E)(x) = -2\epsilon^{abc}\mathrm{tr}(F_{ab}\{A_c, V(R_x)\})/(\kappa/2)^2 \tag{10.3.9}$$

or, in integrated form, $H(N) = \int_\sigma d^3x N(x)H(x)$, and so on for some lapse function N and any smooth neighbourhood-valued function $R: x \mapsto R_x$

$$(H - H_E)(N) = -8\int_\sigma N'\mathrm{tr}(\{A, K\} \wedge \{A, K\} \wedge \{A, V(R)\})/(\kappa/2)^4 \tag{10.3.10}$$

$$H_E(N) = -2\int_\sigma N'\mathrm{tr}(F \wedge \{A, V(R)\})/(\kappa/2)^2 \tag{10.3.11}$$

Here we have absorbed the classical constant $\mathrm{sgn}(\det(e))$ into N and denoted it by N'. In what follows we will drop the prime again.[2] What we have achieved in (10.3.8), (10.3.9) or (10.3.10), (10.3.11) is to remove the problematic $1/\sqrt{\det(q)}$ from the denominator by means of Poisson brackets.

The reader will now ask what the advantage of all this is. The idea behind these formulae is the following: what we want to quantise is $H(N)$ on $\mathcal{H}^0$ and since $\mathcal{H}_0$ is defined in terms of generalised holonomy variables $A(e)$ we first need to write (10.3.10), (10.3.11) in terms of holonomies. This can be done by introducing a triangulation $T(\epsilon)$ of σ by tetrahedra which fill all of σ and intersect each other only in lower-dimensional submanifolds of σ. The small parameter ϵ is to indicate how fine the triangulation is, the limit $\epsilon \to 0$ corresponding to tetrahedra of vanishing volume (the number of tetrahedra grows in this limit so as to always fill out σ). So let $e_I(\Delta)$ denote three edges of an analytic tetrahedron $\Delta \in T(\epsilon)$ and let $v(\Delta)$ be their common intersection point with outgoing orientation (the quantities $\Delta, e_I(\Delta), v(\Delta)$, of course, also depend on ϵ but we do not display this in

---

[2] Alternatively we may actually quantise $\mathrm{sgn}(\det(e))$ along the lines of the volume operator, see Chapter 13 and [573, 574].

order not to clutter the formulae with too many symbols). The matrix consisting of the tangents of the edges $e_1(\Delta), e_2(\Delta), e_3(\Delta)$ at $v(\Delta)$ (in that sequence) has non-negative determinant, which induces an orientation of Δ. Furthermore, let $a_{IJ}(\Delta)$ be the arc on the boundary of Δ connecting the endpoints of $e_I(\Delta), e_J(\Delta)$ such that the loop $\alpha_{IJ}(\Delta) = e_I(\Delta) \circ a_{IJ}(\Delta) \circ e_J(\Delta)^{-1}$ has positive orientation in the induced orientation of the boundary for $(I, J) = (1, 2), (2, 3), (3, 1)$ and negative in the remaining cases. One can then see that in the limit as $\epsilon \to 0$ the quantities

$$
(H^\epsilon - H_E^\epsilon)(N) = \frac{8}{3(\kappa/2)^4} \sum_{\Delta \in T(\epsilon)} \epsilon^{IJK} N(v(\Delta)) \mathrm{tr}\big(h_{e_I(\Delta)}\{h_{e_I(\Delta)}^{-1}, K\} h_{e_J(\Delta)}
$$
$$
\times \{h_{e_J(\Delta)}^{-1}, K\} h_{e_K(\Delta)} \{h_{e_K(\Delta)}^{-1}, V(R_{v(\Delta)})\}\big) \tag{10.3.12}
$$

$$
H_E^\epsilon(N) = \frac{2}{3(\kappa/2)^2} \sum_{\Delta \in T(\epsilon)} N(v(\Delta)) \epsilon^{IJK} \mathrm{tr}\big(h_{\alpha_{IJ}(\Delta)} h_{e_K(\Delta)}
$$
$$
\times \{h_{e_K(\Delta)}^{-1}, V(R_{v(\Delta)})\}\big) \tag{10.3.13}
$$

converge to (10.3.10), (10.3.11) respectively pointwise on $\mathcal{M}$ ••• • ••• • •••••• •• ••••• • •• •••• •! This independence of the limit, for the classical theory, from the choice of the family of triangulations enables us to choose the triangulations state-dependent just as for the area operator, see below.

In order to verify (10.3.13) one makes use of the following facts: let e, e' be arbitrary paths which are images of the interval $[0, 1]$ under the corresponding embeddings, which we also denote by e, e' such that $v = e(0) = e'(0)$. For any $0 < \epsilon < 1$ set $e_\epsilon(t) := e(\epsilon t)$ for $t \in [0, 1]$ and likewise for e'. Then we expand $h_{e_\epsilon}(A)$ in powers of ϵ. It is not difficult to see that $h_{e_\epsilon}(A) = 1_2 + \epsilon \dot{e}^a(0) A_a^j(v) \tau_j/2 + O(\epsilon^2)$. Next, consider the loop $\alpha_{e_\epsilon, e'_\epsilon}$ where in a coordinate neighbourhood

$$
\alpha_{e_\epsilon, e'_\epsilon}(t) = \begin{cases} e_\epsilon(4t) & 0 \le t \le 1/4 \\ e_\epsilon(1) + e'_\epsilon(4t - 1) - v & 1/4 \le t \le 1/2 \\ e'_\epsilon(1) + e_\epsilon(3 - 4t) - v & 1/2 \le t \le 3/4 \\ e'_\epsilon(4 - 4t) & 3/4 \le t \le 1 \end{cases} \tag{10.3.14}
$$

Now expanding again in powers of ϵ we easily find $h_{\alpha_{e_\epsilon, e'_\epsilon}} = 1_2 + \epsilon^2 F_{ab}^j \dot{e}^a \times (0) \dot{e}'^b(0) \tau_j/2 + O(\epsilon^3)$. Due to the unimodularity of SU(2) and the fact that constants drop out of Poisson brackets we see that the Poisson bracket in (10.3.13) is of order ϵ while the loop contribution is proportional to $\epsilon^{IJK} h_{\alpha_{IJ}(\Delta)} = \epsilon^{IJK}[h_{\alpha_{IJ}(\Delta)} - h_{\alpha_{IJ}(\Delta)}^{-1}]/2$ and thus to ϵ^2. Thus these two terms together are already of order ϵ^3 in lowest order, which is precisely the order that we need in order to recast (10.3.12) into a Riemann sum approximation of the continuum integral.

Suppose now that we can turn $V(R)$ and K into well-defined operators on $\mathcal{H}$, densely defined on cylindrical functions. Then, according to the rule that upon quantisation one should replace Poisson brackets by commutators times $1/(i\hbar)$

(10.3.12), (10.3.13) •••• •• ••••• • ••• ••• •••• ••• •••• •• ••• •••••••• •• $\mathcal{H}^0$ • ••• •
••• ••• ••••••••••• ••• • •••••••• •••••• •• •••••• •••••••• •! We will discuss the
issue of what happens upon removal of the regulator ϵ later.

Is it then true that $\hat{V}(R)$ and $\hat{K}$ exist? We will see in Chapter 13 that the
answer is affirmative for the case of the volume operator. We use the version
of the volume operator that was constructed in [427] compared with the one
in [425] because it turns out that only the operator [427] gives a densely defined
Hamiltonian constraint operator in the regularisation scheme that we advertise
here: it is important that the volume vanishes on planar vertices (that is, the
tangent space at the vertex spanned there by the tangents of the edges incident
at it is at most two-dimensional). We will describe in Chapter 13 that also a
purely kinematical consistency check [573, 574] leads to this conclusion.

We will see in Chapter 13 that the volume operator of [427] acts on a function
cylindrical over a graph γ as follows:

$$\hat{V}(R)f_\gamma := \frac{\ell_p^3}{8} \sum_{v \in V(\gamma) \cap R} \sqrt{\left| \frac{i}{3!} \sum_{e \cap e' \cap \tilde{e} = v} \epsilon(e, e', \tilde{e}) \epsilon_{ijk} R_e^i R_{e'}^j R_{\tilde{e}}^k \right|} f_\gamma \qquad (10.3.15)$$

where the sum is over the set $V(\gamma)$ of all vertices v of the graph γ that lie in R and
over all unordered triples of edges that start at v (we can take the orientation
of each edge incident at v to be outgoing by suitably splitting an edge into
two halves if necessary). The function $\epsilon(e, e', \tilde{e})$ takes the values $+1, -1, 0$ if the
tangents of the three edges at v (in that sequence) form a matrix of positive,
negative or vanishing determinant and the right-invariant vector fields R_e^i were
defined in Chapter 9. The absolute value $|\hat{B}|$ of the operator $\hat{B}$ indicates that
one is supposed to take the square root of the operator $\hat{B}^\dagger \hat{B}$. The dense domain
of this operator consists of the thrice differentiable cylindrical functions. Notice
that planar vertices of arbitrary valence do not contribute. Surprisingly, also
arbitrary tri-valent vertices do not contribute[3] [575] if the corresponding state is
gauge-invariant.

Thus, it seems that one can make sense out of a regulated operator corresponding to (10.3.12) for each N, in particular for $N = 1$. Now recall the classical identity that the integrated densitised trace of the extrinsic curvature is the
'time derivative' of the total volume

$$K = -\{H_E(1), V(\sigma)\} = \{H(1), V(\sigma)\} \qquad (10.3.16)$$

where $N = 1$ is the constant lapse equal to unity and $s = -1$. This formula makes
sense even if σ is not compact (see [438] for the details; basically one takes the
Poisson bracket at finite volume and then takes the limit to infinite volume).

---

[3] *Proof.* We have $-(R_1^j + R_2^j) = R_3^j$ due to gauge invariance where $R_I^j = R_{e_I}^j, I = 1, 2, 3$.
Substituting this into $\epsilon_{jkl} R_1^j R_2^k R_3^l$ and using $[R_I^j, R_J^k] = -2\delta_{IJ} \epsilon_{jkl} R_I^l$ completes the proof.

But if we then replace again Poisson brackets by commutators times $1/(i\hbar)$ and define

$$\hat{K}^\epsilon := \frac{i}{\hbar}[\hat{H}_E^\epsilon(1), \hat{V}(\sigma)] \qquad (10.3.17)$$

using the already defined quantities $\hat{H}^\epsilon(1), \hat{V}(\sigma)$ it seems that we can also define a regulated operator corresponding to (10.3.12)!

This concludes the explanation of the main idea. The next section displays a concrete implementation.

10.4 Mathematical definition of the Hamiltonian constraint operator

Obviously, central questions regarding the concrete implementation of the technique are:

I. What are the allowed, physically relevant choices for a family of triangulations $T(\epsilon)$?

II. How should one treat the limit $\epsilon \to 0$ for the operator $\hat{H}^\epsilon(N)$? That is, should one keep ϵ finite and just refine $\gamma \to \sigma$ for cylindrical functions or is there an operator topology such that this limit can be given a meaning? Secondly, does the refined or limit operator remember something about the choice of the family $T(\epsilon)$ or is there some notion of universality?

III. What is the commutator algebra of these (limits of) operators, is it free of anomalies?

We will address these issues separately.

10.4.1 Concrete implementation

A natural choice for a triangulation turns out to be the following (we simplify the presentation drastically, the details can be found in [438]): given a graph γ one constructs a triangulation $T(\gamma, \epsilon)$ of σ ••• •••• to γ which satisfies the following basic requirements.

(a) The graph γ is embedded in $T(\gamma, \epsilon)$ for all $\epsilon > 0$.

(b) The valence of each vertex v of γ, viewed as a vertex of the infinite graph $T(\epsilon, \gamma)$, remains constant and is equal to the valence of v, viewed as a vertex of γ, for each $\epsilon > 0$.

(c) Choose a system of semianalytic arcs $a_{\gamma,v,e,e'}^\epsilon$, one for each pair of edges e, e' of γ incident at a vertex v of γ, which do not intersect γ except in its endpoints where they intersect transversally. These endpoints are interior points of e, e' and are those vertices of $T(\epsilon, \gamma)$ contained in e, e' closest to v for each $\epsilon > 0$ (i.e., no others are in between). For each $\epsilon, \epsilon' > 0$ the arcs $a_{\gamma,v,ee'}^\epsilon, a_{\gamma,v,e,e'}^{\epsilon'}$ are diffeomorphic with respect to semianalytic diffeomorphisms. The segments of

e, e' incident at v with outgoing orientation that are determined by the end-points of the arc $a^\epsilon_{\gamma,v,e,e'}$ will be denoted by $s^\epsilon_{\gamma,v,e}, s^\epsilon_{\gamma,v,e'}$ respectively. Finally, if φ is a semianalytic diffeomorphism then $s^\epsilon_{\varphi(\gamma),\varphi(v),\varphi(e)}, a^\epsilon_{\varphi(\gamma),\varphi(v),\varphi(e),\varphi(e')}$ and $\varphi(s^\epsilon_{\gamma,v,e}), \varphi(a^\epsilon_{\gamma,v,e,e'})$ are semianalytically diffeomorphic.

(d) Choose a system of mutually disjoint neighbourhoods $U^\epsilon_{\gamma,v}$, one for each vertex v of γ, and require that for each $\epsilon > 0$ the $a^\epsilon_{\gamma,v,e,e'}$ are contained in $U^\epsilon_{\gamma,v}$. These neighbourhoods are nested in the sense that $U^\epsilon_{\gamma,v} \subset U^{\epsilon'}_{\gamma,v}$ if $\epsilon < \epsilon'$ and $\lim_{\epsilon\to 0} U^\epsilon_{\gamma,v} = \{v\}$.

(e) Triangulate $U^\epsilon_{\gamma,v}$ by tetrahedra $\Delta(\gamma, v, e, e', \tilde{e})$, one for each ordered triple of distinct edges $e, e', \tilde{e}$ incident at v, bounded by the segments $s^\epsilon_{\gamma,v,e}, s^\epsilon_{\gamma,v,e'}, s^\epsilon_{\gamma,v,\tilde{e}}$ and the arcs $a^\epsilon_{\gamma,v,e,e'}, a^\epsilon_{\gamma,v,e',\tilde{e}}, a^\epsilon_{\gamma,v,\tilde{e},e}$ from which loops $\alpha^\epsilon(\gamma; v; e, e')$, etc. are built and triangulate the rest of σ arbitrarily. The ordered triple $e, e', \tilde{e}$ is such that their tangents at v, in this sequence, form a matrix of positive determinant.

Requirement (a) prevents the action of the Hamiltonian constraint operator from being trivial. Requirement (b) guarantees that the regulated operator $\hat{H}^\epsilon(N)$ is densely defined for each ϵ. Requirements (c), (d) and (e) specify the triangulation in the neighbourhood of each vertex of γ and leave it unspecified outside of them. The more detailed prescription of [438] that uses Puisseaux' theorem shows that triangulations satisfying all of these requirements always exist[4] and can also deal with degenerate situations, for example, how to construct a tetrahedron for a planar vertex. More specifically, what was done in [438] is to fix the routing or braiding of the analytical arcs through the 'forest' of the already present edges in such a way that it is invariant under semianalytic diffeomorphisms that leave γ invariant and the arcs semianalytic. Here we are more general than in [438] in that we just use the •••• • •• ••••••. That is, we only use that a choice function

$$a^\epsilon : \Gamma^\omega_0 \to \Gamma^\omega_0; \gamma \mapsto \{a^\epsilon_{\gamma,v,e,e'}\}_{v\in V(\gamma); \; e,e'\in E(\gamma); \; v\in\partial e\cap\partial e'} \qquad (10.4.1)$$

subject to requirements (a)–(e) always exists and leave it unspecified otherwise. The reason why those tetrahedra lying outside the neighbourhoods of the vertices described above are irrelevant rests crucially on the choice of ordering (10.3.13) with $[\hat{h}^{-1}_s, \hat{V}]$ on the rightmost and on our choice of the volume operator [427]: if f is a cylindrical function over γ and s has support outside the neighbourhood

---

[4] Basically one wants that the arcs intersect the graph only in their endpoints. Thus for sufficiently fine triangulations it is enough to avoid intersections with the edges $\tilde{e} \neq e, e'$ also incident at the vertex in question. One first shows that there always exists an adapted frame, that is, a frame such that $s_e, s_{e'}$ lie in the x, y plane for sufficiently short $s_e, s_{e'}$. Now one shows that for any other edge $\tilde{e}$ of the graph whose beginning segment is not aligned with either s_e or $s_{e'}$ there are only two possibilities. (A) Either for all adapted frames the beginning segment of $\tilde{e}$ lies above or below the x, y plane and whether it is above or below is independent of the adapted frame. (B) Or there exists an adapted frame such that the beginning segment $\tilde{e}$ lies above the x, y plane. This can be achieved simultaneously for all edges incident at the vertex in question. The natural prescription is then to let the arc $a_{e,e'}$ be the straight line in the selected frame connecting the endpoints of $s_e, s_{e'}$ at which it intersects transversally.

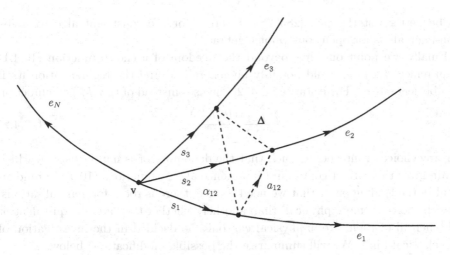

Figure 10.1 The meaning of tetrahedron, segments and arcs determined by a triple of edges meeting in a common vertex.

of any vertex of γ, then $V(\gamma \cup s) - V(\gamma)$ consists of planar at most four-valent vertices only so that $[\hat{h}_s^{-1}, \hat{V}]f = 0$. Notice, however, that [425] does not vanish on planar vertices and so $[\hat{h}_s^{-1}, \hat{V}]f$ would not vanish even on trivalent vertices in $V(\gamma \cup s) - V(\gamma)$ because it is not gauge-invariant. In other words, in the limit of small ϵ the operator would map us out of the space of cylindrical functions. Therefore the Hamiltonian constraint operator inherits from the volume operator a basic property: it annihilates all states cylindrical with respect to graphs with only co-planar vertices as can be understood from the fact that the volume operator enters the construction of both $\hat{H}_E^\epsilon(N), \hat{H}^\epsilon(N)$. In other words, the dynamics 'happens only at the vertices of a graph'. See Figure 10.1 for a sketch of these objects.

Notice that (a)–(e) are natural extensions to arbitrary graphs of what one does in lattice gauge theory [576] with one exception: what we will get is not an operator $\hat{H}^\epsilon(N)$ to begin with, but actually a family of operators $\hat{H}_\gamma^\epsilon(N)$, one for each graph γ. This happened because we adapted the triangulation to the graph of the state on which the operator acts. One must then worry that this does not define a linear operator any more, that is, it is not cylindrically consistently defined. Here we circumvent that problem as follows: we do not define the operator on functions cylindrical over graphs but cylindrical over •• •• ••• ••••••, that is, we define it on spin-network functions. The domain for the operator that we will choose is a finite linear combination of spin-network functions, hence this defines the operator uniquely as a linear operator. Any operator automatically becomes consistent if one defines it on a basis, the consistency condition simply drops out.

Moreover, the regulated operator $\hat{H}^\epsilon(N)$ is by construction background-independently defined for each ϵ but not symmetric which, as described in Chapter 30, is not a necessary requirement for a constraint operator and even argued to be better not the case [283, 284] in order for the constraint algebra to be non-anomalous for open constraint algebras.

Finally, we point out that beyond the freedom of a choice function (10.4.1) requirements (a)–(e) could possibly be generalised and the regularisation itself can be generalised. For instance in [452] one uses instead of $\mathrm{tr}(\tau_j h_\alpha)$ the function

$$\frac{\sum_{k=1}^{N} \mathrm{tr}\left(\tau_j h_\alpha^{n_k}\right)}{\sum_{k=1}^{N} n_k} \tag{10.4.2}$$

for any choice of integers n_k such that the denominator is non-vanishing, which again gives the correct continuum limit since all the functions (10.4.1) are identical in the leading order that we need. Hence, there is room for generalisations. Which choice is 'more physical' than another, whether they are all equivalent or whether all of them are unphysical can only be decided in the investigation of the classical limit. We will summarise the possible modifications below.

Let us then display the action of the Hamiltonian constraint on a spin-network function f_γ cylindrical with respect to a graph γ. It is given by

$$\left[\hat{H}_E^\epsilon(N)\right]^\dagger f_\gamma = \frac{32}{3i\kappa\ell_p^2} \sum_{v\in V(\gamma)} \frac{N(v)}{E(v)} \sum_{v(\Delta)=v} \epsilon^{IJK}$$

$$\times \mathrm{tr}\left(h_{\alpha_{IJ}(\Delta)} h_{e_K(\Delta)} \left[h_{e_K(\Delta)}^{-1}, \hat{V}(U_\epsilon(v))\right]\right) f_\gamma \tag{10.4.3}$$

$$\left[(\hat{H}^\epsilon - \hat{H}_E^\epsilon)(N)\right]^\dagger f_\gamma = \frac{128}{3\kappa\left(i\ell_p^2\right)^3} \sum_{v\in V(\gamma)} \frac{N(v)}{E(v)} \sum_{v(\Delta)=v} \epsilon^{IJK} \mathrm{tr}\left(h_{e_I(\Delta)} \left[h_{e_I(\Delta)}^{-1}, \hat{K}^\epsilon\right] h_{e_J(\Delta)}\right.$$

$$\times \left[h_{e_J(\Delta)}^{-1}, \hat{K}^\epsilon\right] h_{e_K(\Delta)} \left[h_{e_K(\Delta)}^{-1}, \hat{V}(U_\epsilon(v))\right]\right) f_\gamma \tag{10.4.4}$$

where $\hat{K}_\epsilon$ is defined by (10.3.17).

The reason for the adjoint operation is as follows: since H is classically real-valued it does not make any difference whether we quantise H or $\overline{H}$. We chose to quantise $\overline{H}$ resulting in $\hat{H}^\dagger$ in order to be able to easily use the definition of a dual operator as given in Chapter 30. Clearly, this does not make any difference if $\hat{H}$ is self-adjoint, however, as (10.4.3) and (10.4.4) stand the operator is not even symmetric. This actually is required for a first-class algebra with structure functions [283, 284] in order that it closes as we have shown already, see also Chapter 30. The difference between the symmetrised version of (10.4.3), (10.4.4) and (10.4.3), (10.4.4) itself is of course an $\hbar$ correction but the ordering (10.4.3), (10.4.4) turns out to be the only one in which (1) the constraint algebra closes and (2) the final operator is densely defined as we remove the regulator.[5]

---

[5] Basically, in case that the curvature term would be ordered to the right then the volume operator would contribute for all the interior points of an edge in the limit $\epsilon \to 0$, not only at vertices, because the volume operator does not vanish on 3- or 4- valent gauge-variant vertices. For the same reason it is required that the volume operator does not make contributions at planar vertices, which is why we must use version [427] rather than [425] as otherwise the retraced path holonomies h_s intersecting any interior point of an edge would contribute.

Let us explain the notation: the first sum is over all the vertices of a graph and the second sum over all ordered tetrahedra of the triangulation $T(\epsilon, \gamma)$ that saturate the vertex (the remaining tetrahedra drop out). The symbols $e_I(\Delta)$, etc. mean the same as in (10.3.12), (10.3.13) just that now the tetrahedra in question are the particular ones as specified in (a)–(e) above. Here the numerical factors $E(v) = \binom{n(v)}{3}$, where $n(v)$ is the valence of the vertex v, come about as follows.

Given a triple of edges (e, e', e'') incident at v with outgoing orientation consider the tetrahedron $\Delta^\epsilon(\gamma, v, e, e', e'')$ bounded by the three segments $s^\epsilon_{\gamma,v,e} \subset e$, $s^\epsilon_{\gamma,v,e'} \subset e'$, $s^\epsilon_{\gamma,v,e''} \subset e''$ incident at v and the three arcs $a^\epsilon_{\gamma,v,e,e'}, a^\epsilon_{\gamma,v,e',e''}, a^\epsilon_{\gamma,v,e'',e}$. We now define the 'mirror images' (see Figure 10.2)

$$s^\epsilon_{\gamma,v,\bar{p}}(t) := 2v - s^\epsilon_{\gamma,v,p}(t)$$
$$a^\epsilon_{\gamma,v,\bar{p},\bar{p}'}(t) := 2v - a^\epsilon_{\gamma,v,p,p'}(t)$$
$$a^\epsilon_{\gamma,v,\bar{p},p'}(t) := a^\epsilon_{\gamma,v,\bar{p},\bar{p}'}(t) - 2t\left[v - s^\epsilon_{\gamma,v,p'}(1)\right]$$
$$a^\epsilon_{\gamma,v,p,\bar{p}'}(t) := a^\epsilon_{\gamma,v,p,p'}(t) + 2t\left[v - s^\epsilon_{\gamma,v,p'}(1)\right] \tag{10.4.5}$$

where $p \neq p' \in \{e, e', e''\}$ and we have chosen some parametrisation of segments and arcs. Using the data (10.4.5) we build seven more 'virtual' tetrahedra bounded by these quantities so that we obtain altogether eight tetrahedra that saturate v and triangulate a neighbourhood $U^\epsilon_{\gamma,v,e,e',e''}$ of v. Let $U^\epsilon_{\gamma,v}$ be the union of these neighbourhoods as we vary the ordered triple of edges of γ incident at v. The $U^\epsilon_{\gamma,v}, v \in V(\gamma)$ were chosen to be mutually disjoint in point (d) above. Let now

$$\bar{U}^\epsilon_{\gamma,v,e,e',e''} := U^\epsilon_{\gamma,v} - U^\epsilon_{\gamma,v,e,e',e''}$$
$$\bar{U}^\epsilon_\gamma := \sigma - \bigcup_{v \in V(\gamma)} U^\epsilon_{\gamma,v} \tag{10.4.6}$$

then we may write any classical integral (symbolically) as

$$\int_\sigma = \int_{\bar{U}^\epsilon_\gamma} + \sum_{v \in V(\gamma)} \int_{U^\epsilon_{\gamma,v}}$$

$$= \int_{\bar{U}^\epsilon_\gamma} + \sum_{v \in V(\gamma)} \frac{1}{E(v)} \sum_{v=b(e) \cap b(e') \cap b(e'')} \left[\int_{U^\epsilon_{\gamma,v,e,e',e''}} + \int_{\bar{U}^\epsilon_{\gamma,v,e,e',e''}}\right]$$

$$\approx \int_{\bar{U}^\epsilon_\gamma} + \sum_{v \in V(\gamma)} \frac{1}{E(v)} \left[\sum_{v=b(e) \cap b(e') \cap b(e'')} 8 \int_{\Delta^\epsilon_{\gamma,v,e,e',e''}} + \int_{\bar{U}^\epsilon_{\gamma,v,e,e',e''}}\right] \tag{10.4.7}$$

where in the last step we have noticed that classically the integral over $U^\epsilon_{\gamma,v,e,e',e''}$ converges to eight times the integral over $\Delta^\epsilon_{\gamma,v,e,e',e''}$. Now when triangulating the regions of the integrals over $\bar{U}_{v,e,e',e''}$ and $\bar{U}^\epsilon_\gamma$ in (10.4.7), regularisation and quantisation gives operators that vanish on f_γ because the corresponding regions do not contain a non-planar vertex of γ.

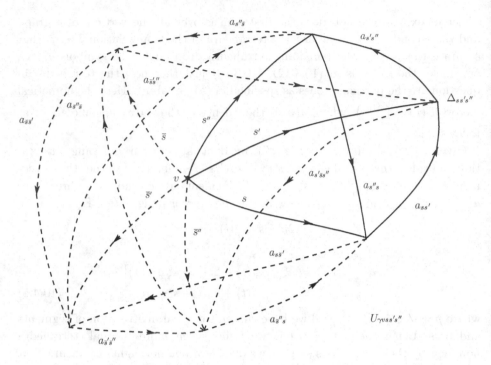

Figure 10.2 The construction of the mirror edges saturating a vertex.

Notice that (10.4.3) and (10.4.4) are •• ••• for each $\epsilon > 0$, that is, densely defined without that any renormalisation is necessary and with range in the smooth cylindrical functions again. Furthermore, the adjoints of the expressions (10.4.3) and (10.4.4) are densely defined on smooth cylindrical functions again so that we get in fact a consistently and densely defined family of closable operators on $\mathcal{H}^0$ (see below).

Let us check the dimensionalities: the volume operator in (10.4.3) is given by ℓ_p^3 times a dimension-free operator, hence (10.4.3) is given by $\ell_p/\kappa = m_p$ times a dimension-free operator. Hence the correct dimension of Planck mass $m_p = \sqrt{\hbar/\kappa}$ has popped out. Therefore, by inspection, (10.3.17) has dimension of $\ell_p^3 m_p/\hbar = \ell_p^2$ which is correct since $K(x) = \sqrt{\det(q)}(x)K_{ab}(x)q^{ab}(x)$ has dimension cm^{-1} so that $K = \int d^3x K(x)$ has dimension cm^2. Finally therefore (10.4.4) has the correct dimension of $(\ell_p^2)^2 \ell_p^3/\kappa \ell_p^6 = m_p$ again.

10.4.2 Operator limits

Basically there are two, technically equivalent viewpoints towards treating the limit $\epsilon \to 0$.

(A) •• ••••• •••••••• •••• ••••

The more radical proposal is •• •••• ••• ••••• •••• ϵ from all formulae. That is, take a choice function a once and for all. One gets a densely defined family of closed operators. One may object that on a given graph γ with only a few edges one does not get a quantisation of the full classical expressions (10.3.12), (10.3.13), however, that is only because the graph γ does not fill all of σ. In other words, the continuum limit of infinitely fine triangulation of the Riemann sum expressions (10.3.12), (10.3.13) in the classical theory is nothing else than taking the graphs, on which the operator is probed, finer and finer. This is a new viewpoint not previously reported in the literature and could be called the •• •••••• •••••••• •••• ••••• because on fine but not infinitely fine graphs the classical limit of the operator will only ••• •••••• ••• the exact classical expression in the same way as (10.3.12) and (10.3.13) only approximate (10.3.10) and (10.3.11). However, it may be that this •• ••• •• ••••• ••• •••••••• and classical physics is just an approximation to it. This way the UV regulator ϵ corresponding to the continuum limit is trivially removed and our family of operators is really defined on $\mathcal{H}^0$. Whether the operator $\hat{H}^\dagger$ that we then obtain has the correct classical limit cannot be decided at this stage but is again subject to a rigorous semiclassical analysis which requires new input, see Chapter 11.

(B) •• •• •••••••• •••• ••••• •

The challenge is to find an operator topology (see, e.g., Chapter 26)[6] in which the one-parameter family of operators $(\hat{H}^\epsilon)^\dagger$ converges. The operators (10.4.3) and (10.4.4) are easily seen to be unbounded (already the volume operator has this property). Thus, a convergence in the uniform topology is ruled out. Next, one may try the weak operator topology (matrix elements converge pointwise) but with respect to this topology the limit would be the zero operator (it is too weak): for instance, a matrix element between two spin-network states is non-zero for at most one value of ϵ. Since the weak operator topology is coarser than the strong, also the strong operator topology does not work. Finally, we try the weak* topology, that is, we must check whether $\Psi((\hat{H}^\epsilon(N))^\dagger f)$ converges for each $\Psi \in \mathcal{D}', f \in \mathcal{D}$ where $\mathcal{D} = C^\infty(\overline{\mathcal{A}})$ with its natural nuclear[7] topology is a dense domain and $\mathcal{D}'$ is its topological dual. It turns out that this topology is a little bit too strong, however, convergence holds with respect to a topology which we

---

[6] A net of bounded operators A_I is said to converge uniformly to a bounded operator A provided that $||A_I - A|| \to 0$ where $||A|| := \sup_{||\psi||=1} ||A\psi||$. A net of unbounded operators A_I is said to converge (1) strongly, (2) weakly or (3) in the weak *-topology to an unbounded operator A provided that all A_I, A have a common domain D and (1) $||(A_I - A)\psi|| \to 0$ for every $\psi \in D$ or (2) $< \psi, (A_I - A)\psi' > \to 0$ for every $\psi' \in D$ and all $\psi \in \mathcal{H}$ or (3) D is invariant under A, A_I and $l[(A_I - A)\psi] \to 0$ for every $\psi \in D$ and all $l \in D^*$ where D^* is some space of linear functionals on D. In case that D carries its own topology finer than that of $\mathcal{H}$, we may restrict D^* to the space of continuous linear functionals on D which in this case is larger than $\mathcal{H}$. The rate of convergence in the case of the strong, weak and weak *-topology may depend on ψ, ψ', l.

[7] This is the topology inherited from the nuclear topology on $C^\infty(G)$ generated by the Schwarz seminorms $||f||_{\alpha\beta} = \sup_{h \in G} |h_{\alpha_1} \ldots h_{\alpha_m} \partial^n f / \partial h_{\beta_1} \ldots \partial h_{\beta_n}|$ where $\alpha_k, \beta_l = (AB)$ run through the set of matrix indices.

might call • • •••• • •••••••••• • •••• • •••••••• (URST) in appreciation of the fact that Rovelli and Smolin first pointed out in [567] that, if instead of $\mathcal{D}'$ we consider the space $\mathcal{D}^*_{\text{diff}}$ of •••• ••• •••• ••• ••• ••••••• • •••••• •• ••••••• •••• • on $\mathcal{D}$, then objects of the form $\Psi((\hat{H}^\epsilon(N))^\dagger f)$ do not depend at all on the position or shape of the arcs $a^\epsilon_{\gamma,v,e,e'}$ alluded to above. In their original work [567] Rovelli and Smolin did not spell out this property in the context of $\mathcal{H}^0$ and also they did not have a well-defined constraint operator, but their observation applies to a huge class of operators, their only feature being an analogue of property (c) above. This is how one proceeded in [437–439].

Therefore, since all the triangulations $T(\gamma, \epsilon)$ restricted to each of the neighbourhoods $U^\epsilon_{\gamma,v}$ are diffeomorphic by property (c) above, the numbers $\Psi((\hat{H}^\epsilon(N))^\dagger f)$ are actually •••••• ••••••••••• •• ϵ! Accordingly, we have the striking result that with respect to the URST

$$(\hat{H}(N))^\dagger := \lim_{\epsilon \to 0}(\hat{H}^\epsilon(N))^\dagger = (\hat{H}^{\epsilon_0}(N))^\dagger \qquad (10.4.8)$$

where ϵ_0 is an arbitrary but fixed positive number. Notice that we require that for each $\delta > 0$ there exists an $\epsilon'(\delta) > 0$ such that for each $f \in \mathcal{D}, \Psi \in \mathcal{D}^*_{\text{diff}}$

$$|\Psi((\hat{H}^\epsilon(N))^\dagger f) - \Psi((\hat{H}^{\epsilon_0}(N))^\dagger f)| < \delta$$

for all $\epsilon < \epsilon'(\delta)$ where $\epsilon'(\delta)$ depends only on δ but not on f, Ψ. In other words, we have convergence •• •••••• in $\mathcal{D} \times \mathcal{D}^*_{\text{diff}}$ rather than pointwise. This will be important in what follows.

Notice that therefore the convergence in the URST is very similar to the effective operator viewpoint in the sense that it gives a topology in which it is allowed to drop the label ϵ from the choice function altogether. In particular we stress that •• ••• •••••• •• ••• •••• ••••• ••••• •• •••••••••• • • •••••• •••• ••• ••••••••• •••••• •• $\mathcal{H}^0$ ••• ••• •• ••• •••• •••••••• $\mathcal{D}^*_{\text{diff}} \subset \mathcal{D}^*$ •• •• •••••• •••• •••••••, precisely in the same sense as the limit of a family of operators which converges in the weak *-topology on $\mathcal{D}$ is still considered an operator on $\mathcal{D}$ and not a dual operator on $\mathcal{D}'$. In fact, the dual of $\hat{H}(N)$ cannot be defined on $\mathcal{D}^*_{\text{diff}}$ because that space is not left-invariant by $\hat{H}(N)'$ as we pointed out frequently, which is why the authors of [579, 580] have to take an extension to the so-called 'vertex smooth' distributions $\mathcal{D}^*_{\text{diff}} \subset \mathcal{D}^*_* \subset \mathcal{D}^*$ which is genuinely bigger than $\mathcal{D}^*_{\text{diff}}$ and therefore unphysical. Our viewpoint is completely different: • • •• ••••••••• •• ••••• $\hat{H}'(N)$ •• ••• •• •••• ••• $\mathcal{D}^*_{\text{diff}}$ ••• • •••• •• ••••• • ••••••!

On the other hand, the physical reason for testing convergence of the operator only on $\mathcal{D}^*_{\text{diff}}$ rather than on a bigger space is precisely because we are eventually going to look for the space of solutions to all constraints, which in turn must be a subspace $\mathcal{D}^*_{\text{phys}}$ of $\mathcal{D}^*_{\text{diff}}$, so in a sense we do not need stronger convergence. Notice that $\mathcal{D}^*_{phys}$ •• left-invariant by the dual action of $\hat{H}(N)$ (namely it is mapped to zero).

Again, whether the continuum operator thereby obtained has the correct classical limit must be decided in an additional step.

Which viewpoint one takes is a matter of taste, technically they are completely equivalent. The limit operator viewpoint has the advantage that it shows that many choice functions are going to be physically equivalent and this decreases (but does not remove) the degree of redundancy. In what follows we will therefore drop the label ϵ.

The limit (10.4.8) certainly only depends on the diffeomorphism-invariant characteristics of the particular triangulation $T(\gamma, \epsilon)$ that we chose. For instance, the limit would be different if we used arcs that intersect the graph tangentially or which are smooth rather than semianalytical. Other than that, there is no residual 'memory' of the triangulation. This is important for the following reason: by the axiom of choice, a choice function certainly exists and for each different choice function we get a different operator on the kinematical Hilbert space. Thus, we get a huge ambiguity at the kinematical level. This is, however, worrysome only if the physical states depend on that ambiguity. Fortunately, physical states are in particular spatially diffeomorphism-invariant and thus the dependence on the choice function up to its diffeomorphism class drops out completely. Hence, the amount of ambiguity is vastly removed at the level of the physical Hilbert space.

Let us show that the operator $(\hat{H}(N))^\dagger$ is not only densely defined on the finite linear span $\mathcal{D}$ of spin-network functions but that it is also closable, that is, its adjoint[8] $\hat{H}(N)$ is also densely defined. Recall that the domain $D(T^\dagger)$ of the adjoint $T^\dagger$ of an operator T densely defined on a domain $D(T)$ of a Hilbert space $\mathcal{H}$ is given by

$$D(T^\dagger) := \left\{ \psi \in \mathcal{H}; \quad \sup_{f \in D(T),\, \|f\|=1} |< \psi, Tf >| < \infty \right\} \qquad (10.4.9)$$

For $\psi \in D(T^\dagger)$ the linear form $f \mapsto <\psi, Tf>$ is therefore bounded and can be extended to all of $\mathcal{H}$ by the bounded linear functional theorem with the same bound. By the Riesz lemma this then defines a unique vector $T^\dagger\psi$ defined by $<T^\dagger\psi, f>:=<\psi, T f>$.

We now claim that $\mathcal{D} \subset D(\hat{H}(N))$. To see this it is enough to show that $T_s \in D(\hat{H}(N))$ for any spin-network $s \in \mathcal{S}$. Now for given s, by inspection of the explicit formulae (10.4.3), (10.4.4) the set $\mathcal{S}(s)$ of those $s' \in \mathcal{S}$ for which $<T_s, (\hat{H}(N))^\dagger T_{s'} > \neq 0$ is finite. Thus for arbitrary $f = \sum_{s' \in \mathcal{S}} z_{s'} T_{s'}$ with at

---

[8] To be precise we should write $((\hat{H}(N))^\dagger)^\dagger$ but we can drop the adjoints by passing to the closures. We will abuse the notation somewhat this way.

most finitely many $z_{s'} \neq 0$ and $\sum_{s'} |z_{s'}|^2 = 1$ we have

$$| <T_s, (\hat{H}(N))^\dagger f> | \leq \sum_{s' \in S(s)} | <T_s, (\hat{H}(N))^\dagger T_{s'} > | \qquad (10.4.10)$$

where we have used $|z_{s'}| \leq 1$. The right-hand side no longer depends on f, hence $T_s \in D(\hat{H}(N))$ for arbitrary s as claimed.

10.4.3 Commutator algebra

We now come to question III, whether the commutator between two Hamiltonian constraints and between Hamiltonian and diffeomorphism constraints exists and is free of anomalies.

1.

Recall that the infinitesimal generator of diffeomorphisms is ill-defined so that we must check the commutator algebra in terms of finite diffeomorphisms. The classical infinitesimal relation $\{\vec{H}(u), H(N)\} = -H(u[N])$ can be exponentiated and gives

$$e^{t\mathcal{L}_{\chi_{\vec{H}(u)}}} \cdot H(N) = H\left(\left(\left[\varphi_t^u\right]^{-1}\right)^* N\right)$$

where $\chi_{\vec{H}(u)}$ denotes the Hamiltonian vector field of $\vec{H}(u)$ on the classical continuum phase space $\mathcal{M}$ and φ_t^u the one-parameter family of diffeomorphisms generated by the integral curves of the vector field u. It tells us that $H(x)$ is a scalar density of weight one. Therefore we expect to have in quantum theory the relation

$$\hat{U}(\varphi)^{-1}(\hat{H}(N))^\dagger \hat{U}(\varphi) = (\hat{H}(\varphi^* N))^\dagger \qquad (10.4.11)$$

To check whether (10.4.11) is satisfied, we notice that for a spin-network function f_γ we have by the definition of the action of the diffeomorphism group $\hat{U}(\varphi)$ on $\mathcal{H}^0$ on the one hand

$$\hat{U}(\varphi)^{-1}(\hat{H}(N))^\dagger f_\gamma = \hat{U}(\varphi) \sum_{v \in V(\gamma)} N(v) \hat{H}^\dagger_{v,a(\gamma)} f_\gamma$$

$$= \sum_{v \in V(\gamma)} N(v) \hat{H}^\dagger_{\varphi^{-1}(v), \varphi^{-1}(a(\gamma))} f_{\varphi^{-1}(\gamma)}$$

$$= \sum_{v \in V(\gamma)} (\varphi^* N)(\varphi^{-1}(v)) \hat{H}^\dagger_{\varphi^{-1}(v), \varphi^{-1}(a(\gamma))} f_{\varphi^{-1}(\gamma)}$$

$$= [\hat{U}(\varphi)^{-1}(\hat{H}(N))^\dagger \hat{U}(\varphi)] \hat{U}(\varphi)^{-1} f_\gamma$$

$$= [\hat{U}(\varphi)^{-1}(\hat{H}(N))^\dagger \hat{U}(\varphi)] f_{\varphi^{-1}(\gamma)} \qquad (10.4.12)$$

and on the other hand

$$(\hat{H}(\varphi^*N))^\dagger f_{\varphi^{-1}(\gamma)} = \sum_{v\in V(\varphi^{-1}(\gamma))} (\varphi^*N)(v)\hat{H}^\dagger_{v,a(\varphi^{-1}(\gamma))} f_{\varphi^{-1}(\gamma)}$$

$$= \sum_{v\in V(\gamma)} (\varphi^*N)(\varphi^{-1}(v))\hat{H}^\dagger_{\varphi^{-1}(v),a(\varphi^{-1}(\gamma))} f_{\varphi^{-1}(\gamma)} \quad (10.4.13)$$

Here $\hat{H}^\dagger_{v,a(\gamma)}$ is the operator coefficient of $N(v)$ in (10.4.3), (10.4.4) which depends on the graph $a(\gamma)$ assigned to γ through the choice function a, that is, the segments $s_{\gamma,v,e}$ and arcs $a_{\gamma,v,e,e'}$. Comparing (10.4.12) and (10.4.13) we get equality provided that

$$\varphi \circ a = a \circ \varphi \;\forall \varphi \in \mathrm{Diff}^\omega_{sa}(\sigma) \quad (10.4.14)$$

This seems to burden us with the proof that such a choice function really exists and in fact we do not have a proof, although it would be very nice to have one since it would decrease the possible number of choice functions. However, we can avoid this by the observation that our choice function was constructed in such a way that the assignments $a(\gamma)$ and $a(\varphi(\gamma))$ are piecewise (semi)analytically diffeomorphic. In other words we always find a semianalytic diffeomorphism $\varphi'_{\varphi^{-1}(\gamma)}$ which preserves $\varphi^{-1}(\gamma)$ such that

$$\left[\hat{U}(\varphi)^{-1}(\hat{H}(N))^\dagger \hat{U}(\varphi)\right] f_{\varphi^{-1}(\gamma)} = \left[\hat{U}\left(\varphi'_{\varphi^{-1}(\gamma)}\right)(\hat{H}(\varphi^*N))^\dagger \hat{U}\left(\varphi'_{\varphi^{-1}(\gamma)}\right)^{-1}\right] f_{\varphi^{-1}(\gamma)}$$

$$(10.4.15)$$

for any γ and any $f_{\varphi^{-1}(\gamma)}$. Thus, while (10.4.11) is violated, it is violated in an allowed way because the 'anomaly' is a constraint operator again. Put differently, the 'anomaly' is not seen in the URST so that (10.4.11) is an ••••• •••••••• •••• ••••• in the URST.

In that sense then, $(\hat{H}(N))^\dagger$ is a diffeomorphism covariant, densely defined, closable operator on $\mathcal{H}^0$.

2. ••• •••••• ••• • ••• • •• ••••• ••• ••• •••• ••• •

There are three important properties of the operator $(\hat{H}(N))^\dagger$ that follow from our class of choice functions (properties (a)–(e)):

(A) First of all, we observe that $(\hat{H}(N))^\dagger$ has dense domain and range consisting of smooth (in the sense of $\mathcal{D}$) cylindrical functions. Therefore it makes sense to multiply operators and in particular to compute commutators.

(B) Secondly, it annihilates planar vertices.

(C) Thirdly, ••• • •• •••••• •••••• •• •••••• •• •• •••• ••• •••••••• •• ••• ••• ••••• •••• • ••• ••• •••••••• •• ••••••• ••• •• •• •••• ••••• ••• ••••• • •••••••• •• •••••

••••••• •••• ••• ••• •• •••••• is it true that in fact any finite product of operators $(\hat{H}^{\epsilon_1}(N_1))^\dagger \ldots (\hat{H}^{\epsilon_n}(N_n))^\dagger$ is independent of the parameters $\epsilon_1, \ldots, \epsilon_n$ in the URST.

The second and third properties do not hold for a more general class of operators considered in the papers [579, 580] so that there is no convergence in the URST – not even of the operators themselves, not to speak of their commutators. Since certainly none (of the duals) of these operators leaves the space $\mathcal{D}^*_{\text{diff}}$ invariant, in order to compute commutators these authors suggest introducing the larger, unphysical space $\mathcal{D}^*_\star$ already mentioned on which one can compute limits $\hat{H}'(N) = \lim_{\epsilon \to 0} (\hat{H}^\epsilon(N))'$ •••••• ••• in $\mathcal{D}^*_\star \times \mathcal{D}$ of their duals and products of these limits.

Let again f_γ be a spin-network function over some graph γ. Then we compute

$$[(\hat{H}(N))^\dagger, (\hat{H}(N'))^\dagger] f_\gamma = \sum_{v \in V(\gamma)} [N'(v)(\hat{H}(N))^\dagger - N(v)(\hat{H}(N'))^\dagger] \hat{H}^\dagger_{a(\gamma)|_v} f_\gamma$$

$$= \sum_{v \in V(\gamma)} \sum_{v' \in V(\gamma) \cup a(\gamma)|_v} [N'(v)N(v')$$

$$- N(v)N'(v')] \hat{H}^\dagger_{a(\gamma \cup a(\gamma)|_v)|_{v'}} \hat{H}^\dagger_{a(\gamma)|_v} f_\gamma \qquad (10.4.16)$$

where for clarity we have written $\hat{H}^\dagger_{a(\gamma)|_v} \equiv \hat{H}^\dagger_{v,a(\gamma)}$ in order to indicate that $\hat{H}^\dagger_{v,a(\gamma)}$ does not depend on all of $a(\gamma)$ but only on its restriction to the arcs and segments around v. We are abusing somewhat the notation in the second step because one should really expand $\hat{H}^\dagger_{a(\gamma)|_v} f_\gamma$ into spin-network functions over $\gamma \cup a(\gamma)|_v$ and then apply the second operator to that expansion into spin-network functions. In particular, $\hat{H}^\dagger_{a(\gamma)|_v} f_\gamma$ is really a finite linear combination of terms where each of them depends only on $\gamma \cup a(\gamma)|_{v,e,e'}$ for some edges e, e' incident at v and each of those should be expanded into spin-network functions. We will not write this explicitly because it is just a bookkeeping exercise and does not change anything in the subsequent argument. So either one writes out all the details or one just assumes for the sake of the argument that $\hat{H}^\dagger_{a(\gamma)|_v} f_\gamma$ is a spin-network function over $\gamma \cup a(\gamma)|_v$. Everything we say is more or less obvious for the Euclidean Hamiltonian constraint, but a careful analysis shows that it extends to the Lorentzian one as well [438].

Let us now analyse (10.4.16). The right-hand side surely vanishes for $v' = v$. We notice that any vertex $v' \in V(\gamma \cup a(\gamma)|_v) - V(\gamma)$ is planar and since $\hat{H}^\dagger_{v',a(\gamma \cup a(\gamma))}$ has an operator of the form $[h_s^{-1}, \hat{V}]$ to the outmost right-hand side where s is a segment, incident at v', of an edge incident at v', it follows that none of these vertices contributes. Here it was again crucial that we used the operator [427] rather than the operator [425]! Thus (10.4.16) reduces to

$$[(\hat{H}(N))^\dagger, (\hat{H}(N'))^\dagger] f_\gamma = \sum_{v \neq v' \in V(\gamma)} [N'(v)N(v') - N(v)N'(v')]$$

$$\times \hat{H}^\dagger_{a(\gamma \cup a(\gamma)|_v))|_{v'}} \hat{H}^\dagger_{a(\gamma)|_v} f_\gamma$$

$$= \frac{1}{2} \sum_{v \neq v' \in V(\gamma)} [N'(v)N(v') - N(v)N'(v')]$$

$$\times [\hat{H}^\dagger_{a(\gamma \cup a(\gamma)|_v)|_{v'}} \hat{H}^\dagger_{a(\gamma)|_v} - \hat{H}^\dagger_{a(\gamma \cup a(\gamma)|_{v'})|_v} \hat{H}^\dagger_{a(\gamma)|_{v'}}] f_\gamma$$

$$(10.4.17)$$

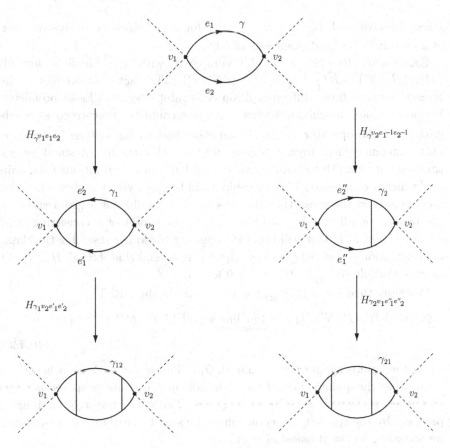

Figure 10.3 Vanishing of the commutator between two (Euclidean) Hamiltonian constraints up to a diffeomorphism.

where in the second step we used the antisymmetry of the expression $[N'(v)N(v') - N(v)N'(v')]$ in v, v'. Now the crucial point is that for $v \neq v' \in V(\gamma)$ the prescription of how to attach the arcs first around v and then around v' compared with the opposite may not be the same because our prescription depends explicitly on the graph to which we apply it, however, they are certainly analytically diffeomorphic (see Figure 10.3). Thus, there exist analytical diffeomorphisms $\varphi_{\gamma, v, v'}$ preserving $\gamma \cup a(\gamma)_{|v}$ such that

$$\hat{H}^\dagger_{a(\gamma \cup a(\gamma)_{|v})_{|v'}} \hat{H}^\dagger_{a(\gamma)_{|v}} f_\gamma = \hat{U}(\varphi_{\gamma, v, v'}) \hat{H}^\dagger_{a(\gamma)_{|v'}} \hat{H}^\dagger_{a(\gamma)_{|v}} f_\gamma \qquad (10.4.18)$$

for any $v \neq v' \in V(\gamma)$. It follows that

$$[(\hat{H}(N))^\dagger, (\hat{H}(N'))^\dagger] f_\gamma = \frac{1}{2} \sum_{v \neq v' \in V(\gamma)} [N'(v)N(v') - N(v)N'(v')]$$

$$\times [\hat{U}(\varphi_{\gamma, v, v'}) - \hat{U}(\varphi_{\gamma, v', v})] \hat{H}^\dagger_{a(\gamma)_{|v'}} \hat{H}^\dagger_{a(\gamma)_{|v}} f_\gamma$$

$$(10.4.19)$$

where we have used $[\hat{H}^\dagger_{a(\gamma)_{|v'}}, \hat{H}^\dagger_{a(\gamma)_{|v}}] = 0$ for $v \neq v'$ since the derivative operators involved act on disjoint sets of edges.

Expression (10.4.19) is to be compared with the classical formula $\{H(N), H(N'\} = \vec{H}(q^{-1}[(dN)N' - (dN')N])$. The fact that we get a difference between finite diffeomorphism constraint operators looks promising, because for next-neighbour vertices v, v' this could be interpreted as a substitute for the operator $\hat{\vec{H}}$ which somehow had to be written in terms of finite diffeomorphism anyway because we know that the infinitesimal generator does not exist. Unfortunately there could also be contributions from pairs v, v' which are far apart. This we could avoid by specifying the choice function more closely in the sense that the arcs $a_{\gamma,v,e,e'}$ should, for a given vertex v, not depend on all of γ but only on $\gamma_v \subset \gamma$, the subset of γ consisting of all edges incident at v. But still (10.4.19) does not obviously resemble the classical calculation too closely because there it is crucial that $\{H(x), H(x')\} \neq 0$ as $x \to x'$ while $[\hat{H}^\dagger_{a(\gamma)_{|v'}}, \hat{H}^\dagger_{a(\gamma)_{|v}}] = 0$ for any $v \neq v'$.

Certainly then for $\Psi \in \mathcal{D}^*_{\text{diff}}, f \in \mathcal{D}$ we have in the URST

$$\Psi([(\hat{H}(N))^\dagger, (\hat{H}(N'))^\dagger]f) := \lim_{\epsilon \to 0} \lim_{\epsilon' \to 0} \Psi([(\hat{H}^\epsilon(N))^\dagger, (\hat{H}^{\epsilon'}(N'))^\dagger]f) = 0$$

(10.4.20)

where the limit is again •• •••••• in both Ψ, f. This is a crucial result because it means that the quantisation of the Hamiltonian constraint proposed is • •••• •• •••••• ••• ••• •••••• •• ••••• •• •• •••• •••. The commutator is annihilated precisely by the spatially diffeomorphism-invariant distributions as it should be according to the classical algebra.

Yet, one would like to have a stronger result, namely that the right-hand side of (10.4.19) can be manifestly considered as a quantisation of $\vec{H}(q^{-1}(dNN' - dN'N))$. In the next section we will quantise the function $\vec{H}(q^{-1}(dN\ N' - dN'\ N))$ independently on $\mathcal{H}^0$ and see that the regulated net of operators converges in the URST and is annihilated by spatially diffeomorphism-invariant distributions. Hence, in the URST there is no difference between (10.4.19) and that operator. But that would be the case for all operators which are of the form $(\hat{U}(\varphi) - 1_{\mathcal{H}^0})\hat{O}$ where $\hat{O}$ is an arbitrary operator. This therefore does not show that on $\mathcal{H}^0$ these two operators have anything to do with each other.

Why is this so difficult to decide? If we recall that we needed pages of calculation in Section 1.5 in order to write $\{H(N), H(N')\}$ as $\vec{H}(q^{-1}(dN\ N' - dN'\ N))$ where we have used manipulations such as (1) integrations by parts, (2) reordering of terms, (3) differential geometric identities and (4) multiplying fractions by functions in both numerator and denominator then it is not surprising that one cannot simply see a relation between the two operators. Not only is the classical calculation already quite involved, but moreover the above-mentioned operations are difficult to perform at the quantum level due to the non-commutativity of operators and their possible non-invertabilty.

But maybe we are asking too much: after all, a classical identity must be reproduced in the semiclassical limit only. This is precisely the reason why one should test the correspondence between these two operators by using coherent states, because within the corresponding expectation values one can basically replace all operators by their corresponding classical functions (plus quantum corrections) and perform the above-mentioned calculations. Thus, the 'correctness' of any choice of Hamiltonian constraint can be answered maybe only in step V of the programme.

To summarise: the constraint algebra of the Hamiltonian constraints among each other is mathematically consistent but this does not yet prove that it has the correct classical limit.

The proposed quantisation of $H(N)$ has been criticised in the literature on several grounds. In order to avoid confusion and to clarify what really has been shown, let us briefly discuss them and show why these criticisms are inconclusive and sometimes simply false. See also the discussion in [577, 578].

(i) In [579, 580] the authors prove a statement similar to (10.4.20) on their space $\mathcal{D}_*^*$. The algebra of their dual constraint operators becomes • • •••• • for a large class of operators, which even classically do not need to be proportional to a diffeomorphism constraint. They then argue that the quantisation method proposed here cannot be correct because it either implies a departure from the classical calculation or, even worse, that the (dual of the) quantum metric operator $\hat{q}^{ab}$ vanishes identically.

We disagree with both conclusions for two reasons:
1. Their limit dual operators are defined by

$$[\hat{H}'(N)\Psi](f) := \lim_{\epsilon \to 0} \Psi((\hat{H}^\epsilon(N))^\dagger f) \qquad (10.4.21)$$

where convergence is only pointwise, that is, for any $\delta > 0$, $\Psi \in \mathcal{D}_*^*, f \in \mathcal{D}$ there exists $\epsilon(\delta, \Psi, f)$ such that

$$|[\hat{H}'(N)\Psi](f) - \Psi((\hat{H}^\epsilon(N))^\dagger f)| < \delta \qquad (10.4.22)$$

for any $\epsilon < \epsilon(\delta, \Psi, f)$. Thus, while they have blown up $\mathcal{D}_{\text{diff}}^*$ to $\mathcal{D}_*^*$, their convergence is weaker when restricted to $\mathcal{D}_{\text{diff}}^*$ so that it is not easy to compare the two operator topologies (notice that we can also define a dual operator via (10.4.21) restricted to $\mathcal{D}_{\text{diff}}^*$ considered as a subspace of $\mathcal{D}^*$, however this subspace is just not left-invariant so we cannot compute commutators of duals). However, it is clear that the subspace $\mathcal{D}_*^*$ is a sufficiently small extension of $\mathcal{D}_{\text{diff}}^*$ in order to make sure that a much wider class of operators converges in their topology than the class that we have in mind for our topology, since our topology roughly requires that $\Psi((\hat{H}^\epsilon(N))^\dagger f)$ is already independent of ϵ while their topology only requires that the ϵ dependence rests in the smearing functions N which are required to be smooth at vertices.

Therefore our first conclusion is that it is not surprising that in their topology more operators converge.

Next, let us turn to commutators. In our topology, what is required is that the expression $\Psi([(\hat{H}^\epsilon(N))^\dagger, (\hat{H}^{\epsilon'}(N'))^\dagger]f)$ •••• ••• •• •• •••• ••••• •••• •••• f •••••••• because we have identified $(\hat{H}^\epsilon(N))^\dagger$ with the continuum operator. In their topology what happens is that unless the operator $(\hat{H}(N))^\dagger$ also has the properties (B), (C) besides (A) then one gets for the commutator an expression of the form (10.4.17) on which one acts with an element $\Psi \in \mathcal{D}_*^*$, the result of which is that one gets

$$([\hat{H}(N')', \hat{H}(N)']\Psi)(f_\gamma) = \lim_{\epsilon \to 0} \lim_{\epsilon' \to 0} \sum_{v \in V(\gamma)} \sum_{v' \in V(\gamma \cup a^\epsilon(\gamma)_{|v}) - V(\gamma)} [N'(v)N(v')$$
$$- N(v)N'(v')]\Psi\left[\hat{H}^\dagger_{a^{\epsilon'}(\gamma \cup a^\epsilon(\gamma)_{|v})_{|v'}} \hat{H}^\dagger_{a^\epsilon(\gamma)_{|v}} f_\gamma\right]$$

$$(10.4.23)$$

For the same reason as for $\Psi \in \mathcal{D}^*_{\text{diff}}$ each evaluation of Ψ that appears on the right-hand side is already independent of ϵ, ϵ' for any $\Psi \in \mathcal{D}^*_*$ by definition of that space. Therefore the only ϵ, ϵ' dependence rests in the function $[N'(v)N(v') - N(v)N'(v')]$. Now, while each of the roughly $|V(\gamma)|$ Ψ-evaluations is non-vanishing, since we take the limit pointwise and the N, N' are smooth, the limit vanishes. If we had not taken pointwise convergence, then for each finite ϵ, ϵ' we could find f_γ, Ψ such that the right-hand side of (10.4.23) takes an arbitrarily large value. The reason why this happens is that since one of the conditions (B), (C) does not hold, now the vertices $V(\gamma \cup a^\epsilon(\gamma)_{|v}) - V(\gamma)$ in fact do contribute.

We conclude that their topology is too weak in order to detect even a mathematical inconsistency. In fact, the extension $\mathcal{D}_*^*$ is rather unphysical because it is not annihilated by the spatial diffeomorphism constraint. Since physical states will be in particular spatially diffeomorphism-invariant, nothing has been gained by considering $\mathcal{D}_*^*$, which is why this space should not be used at all. The Hamiltonian constraint • ••• •• •• • ••• •• ••• • ••••• •••••• • •••••• ••••• and there can be absolutely no debate about that.

2. Coming to their second conclusion, we will explicitly display in the next subsection a quantisation of $\vec{H}(q^{-1}[(dN)N' - (dN')N])$. Now in their topology, the dual of that operator again annihilates $\mathcal{D}_*^*$ but this is again only because one takes only pointwise rather than uniform limits. If one tests this operator on a finite graph then, again because there are finitely many contributions each of which is evidently proportional to a term of the form $[N'(v)N(v') - N(v)N'(v')]$, the limit must vanish pointwise, however, uniformly it blows up for operators not satisfying (B), (C).

Finally, it is •• ••• that in LQG $\hat{q}^{ab}$ is the zero operator. For instance, it was shown [581] that for any one-form ω the operator corresponding to $Q(\omega) := \int d^3x \sqrt{\det(q)} q^{ab} \omega_a \omega_b$ admits a well-defined quantisation on $\mathcal{H}^0$ and is non-trivial. By the same methods one can show that this holds with respect to $Q_1(\omega) := \int d^3x \sqrt{\det(q)} q^{ab} \omega_a \omega_b$. However, Q_1 is just the smeared volume operator with the 'smearing function' $q(\omega) := q^{ab}\omega_a\omega_b$. Hence, if $\hat{q}(\omega) = 0$ then $\hat{Q}_1(\omega) = 0$ which is not the case.

(ii) Unfortunately, the papers [579, 580] have led to a folklore knowledge in the field that the operators $\hat{H}(N)$ are (1) defined on $\mathcal{H}_{\text{diff}}$ and (2) mutually commuting. Both statements are •• •••: first of all, we have shown that the operators $(\hat{H}(N))^\dagger$ defined on $\mathcal{H}^0$ are ••• • • ••• ••• ••• • • ••••. Next, we have shown that their dual •• •••• •• •• •••••• •• $\mathcal{H}_{\text{diff}}$ due to the structure of the Dirac algebra $\mathfrak{D}$. What •• true is that the net of regulated dual operators can be defined on an unphysical extension $\mathcal{D}_*^*$ of $\mathcal{D}_{\text{diff}}^*$ and converge there, in a topology which we just have argued to be too weak, to mutually commuting dual operators.

(iii) In [582] we find the claim that the action of the Hamiltonian constraint is too local in order to allow for interesting critical points in the renormalisation flow of the theory and that therefore the Hamiltonian constraint must be changed drastically if possible at all.

Four comments are appropriate:

1. First of all the claim is not even technically correct, how non-local the operator $(\hat{H}(N))^\dagger$ is depends on our choice function a which builds a new graph around any vertex of a given graph γ and the details of that new graph around v may depend on an ••• •••• •••• •• ••• •••••••••••• •• v (where a neighbourhood of degree n can be background-independently defined as the set of edges that one can trace within γ if one performs a closed loop with endpoints v using at most n edges).

2. Second, it is unclear what role a renormalisation group should play in a diffeomorphism-invariant theory; after all, renormalisation group analysis has much to do with scale transformations (integrating out momentum degrees of freedom above a certain scale) which are difficult to deal with in absence of a background metric.

3. Third, suppose that we managed to write down a physically correct Hamiltonian operator of the type $(\hat{H}(N))^\dagger$. We could order it symmetrically and presumably find a self-adjoint extension. It would then be possible to diagonalise it and in the associated 'eigenbasis' the operator would act in an ultralocal way! Thus any non-local operator can be made ultralocal in an appropriate basis. A good example is given by the Laplace operator in $\mathbb{R}^n$ which is non-local in position space but ultralocal in momentum space. Of course the momentum eigenfunctions are not eigenfunctions but rather distributions and we must take an uncountably infinite linear combination of them (rather, an integral against a sufficiently nice function,

that is, a Fourier transform) in order to obtain an L_2 function on which the Laplacian looks rather non-local. Thus, non-locality is hidden in infinite linear combinations, which is the reason why we are working with $\mathcal{D}^*$ rather than with $\mathcal{D}$. This is precisely what happens in $2+1$ gravity [440].

4. Finally, the intuition on which [582] is based comes from lattice gauge theory. Thus, in renormalisation group language, we are right at the fix point and there is no renormalisation, no second-order phase transition necessary. Hence it is inappropriate to use intuition based on a discrete approximation to a continuum limit as a guideline, these are just two different theories. Another difference with lattice gauge theory is that there one works with an honest Hamiltonian, that is, one looks for eigenfunctions of $\mathbf{H} = \int d^3x H(x)$ in contrast to simultaneous zero eigenvectors of the $H(x)$. Obviously, $\mathbf{H}$ is more non-local than the $H(x)$, hence intuition based on Hamiltonians rather than Hamiltonian constraints might be very misleading.

(iv) The Hamiltonian constraint operators potentially suffer from a huge amount of ambiguities. There are several qualitatively different sources of ambiguities:

1. We have attached the loop in the spin $j = 1/2$ representation. However, the analysis in [452] shows that one can work with higher spin representations without affecting the naive semiclassical limit of the operator. Recently [583] it was shown that higher spin representations lead to spurious solutions of the Hamiltonian constraint in $2+1$ dimensions (where one knows the physical Hilbert space by independent methods). This result presumably extends to $3+1$ dimensions so that this kind of higher spin ambiguity is apparently absent.

   Notice also that this kind of ambiguity is also present in ordinary QFT: take a canonically quantised free scalar field theory and consider instead of the momentum operator-valued distribution $\pi(x)$ the quantity $\pi_F(x) = [F\pi(x)F^{-1} + \bar{F}^{-1}\pi(x)\bar{F}]/2$ where F is an arbitrary never-vanishing multiplication operator. Clearly classically $\pi(x) = \pi_F(x)$ but in quantum theory this substitution will generically lead to a different spectrum of the Hamiltonian so that the two theories are unitarily inequivalent. Of course, it is •••••••• to perform this substitution. To use higher spin representations is equally unnatural.

2. We have decided to order the dependence of the Hamiltonian constraint operator on the electric field to the right of the connection operators. Could we have chosen a different ordering? The answer is negative: a different ordering leads to an operator which is ill-defined on every spin-network state because it makes a contribution at every vertex of the triangulation, not only at the vertices of the graph. Thus it maps, as the triangulation is refined, a normalisable state to a non-normalisable 'state'. Thus, there is in fact no normal ordering ambiguity.

3. As discussed above, if one takes the habitat $\mathcal{D}_*^*$ point of view, then there are an infinite number of these spaces that one could consider as a domain of definition of the unregularised dual of the Hamiltonian constraint operators. Fortunately, the habitats are not only irrelevant as we have seen, they are unphysical. Hence this habitat ambiguity is absent.

4. The largest amount of ambiguity comes from the loop attachment. In the above discussion we have chosen to align the beginning and ending segments of the loop with the edges of the graph and we have chosen to let the additional edges intersect the graph transversally. This is not forced on us. We could for instance detach the beginning and ending segments slightly from the edges of the graph, we could let them wind an arbitrary number of times around those edges, we could let the additional edge not intersect the graph at all or in a $C^{(n)}$ fashion.

   The first observation is that these ambiguities are only bad if they affect the structure of the space of solutions. Since solutions are spatially diffeomorphism-invariant, only the (semianalytic) diffeomorphism-invariant characteristics of these different loop assignments are important. These are counted by a discrete number of possibilities, namely (A) aligned or not, (B) transversal additional edge or at least $C^{(1)}$ (all $C^{(n)}$, $n > 0$ possibilities can be reduced to $C^{(1)}$ by semianalytic diffeomorphisms), (C) the braiding of the additional edge through the present edges of the graph and (D) the winding number n. If we appeal to the notion of naturalness, then we can rule out the winding number $n > 0$ because this is not what one would do in lattice gauge theory. Also detachment would never be considered in lattice gauge theory and there is an additional argument given in [578] which comes from a different point splitting regularisation of the Hamiltonian constraint operators which speaks in favour of alignment. Finally, a natural choice of braiding, based on Puisseaux' theorem was shown to exist in [438].

Thus we conclude that the amount of ambiguity of the operator is not as bad as it first appears. In fact, the concrete proposal given above is free of ambiguities once we pass to the physical Hilbert space.

10.4.4 The quantum Dirac algebra

Recall from Section 10.4.3 that in the URST the commutator of two Hamiltonian constraints vanishes: the non-zero operator on $\mathcal{H}^0$ given by $[(\hat{H}(N))^\dagger, (\hat{H}(N'))^\dagger]$ is indistinguishable from the zero operator in the URST. We would like to know whether there exists an operator corresponding to $\vec{H}(q^{-1}[(dN)N' - (dN')N])$ and if it is also indistinguishable from the zero operator in the URST. If that were true, then we could equate the two operators in the URST. Notice that this is still not satisfactory because one cannot test the correctness of the algebraic form of an operator on its kernel, but it is still an important consistency check whether an operator corresponding to $\vec{H}(q^{-1}[(dN)N' - (dN')N])$ exists at all. More explicitly, we wish to study whether we can quantise

$$O(N, N') := \int d^3x (NN'_{,a} - N_{,a}N') q^{ab} H_b \qquad (10.4.24)$$

In [437] this question is answered affirmatively, that is, we manage to quantise a regulated operator $\hat{O}^\epsilon(N, N')$ corresponding to (10.4.24) and prove that it converges in the URST to an operator $\hat{O}(N, N')$. We will not derive the operator but merely give its final expression. However, let us point out once more that while H_a and q^{ab} are known not to have well-defined quantisations because the infinitesimal generator of diffeomorphisms does not exist (Section 13.6) and since q^{ab} has the wrong density weight (Section 10.4.1), the combination $\omega_a q^{ab} V_b$ is a scalar density of weight one and therefore has a chance to result in a well-defined operator for any co-vector field ω_a such as $\omega_a = NN'_{,a} - N_{,a}N'$.

Let γ be a graph, $V(\gamma)$ its set of vertices, $v \in V(\gamma)$ a vertex of γ, introduce the triangulation $T(\gamma)$ of Section 10.4 adapted to γ, let Δ be a tetrahedron of that triangulation such that $v(\Delta) = v$, let $\chi_{\epsilon,v}(x)$ be the smoothed out characteristic function of the neighbourhood $U(v)$ (using, for instance, a smooth partition of unity) and finally let $s_I(\Delta)$ be the endpoint of the edge $e_I(\Delta)$ of Δ incident at v. We define a vector field on σ of compact support by

$$\xi^a_{\epsilon,v,\Delta,I}(x) := \chi_{U(v)}(x) \frac{s^a_I(\Delta) - v^a}{\epsilon} \qquad (10.4.25)$$

where ϵ^3 is the coordinate volume of $U(v)$ and for any vector field ξ on σ let φ^ξ_t be the one-parameter group of diffeomorphisms that it generates. Let us also introduce the short-hand notation $\hat{V}(v) := \hat{V}(U(v))$. It was shown in [437] that there is a classical object $O_\gamma(N, N')$ which uses the triangulation $T(\gamma)$ and whose limit, as $\gamma \to \sigma$, in the topology of the phase space coincides with (10.4.24). The quantisations of these objects define densely defined operators $\hat{O}(N, N')$ with consistent cylindrical projections $\hat{O}_\gamma(N, N')$ given by their action on functions f_γ cylindrical over a graph γ. The explicit form of these projections is given by

$$\hat{O}(N, N') f_\gamma = -i \frac{16\epsilon_{ijk}\epsilon_{ilm}}{\hbar \ell_p^2} \sum_{v \in V(\gamma)} \sum_{v(\Delta) = v(\Delta') = v} [\hat{U}(\varphi^{\xi_{\epsilon,v,\Delta',R}}_\epsilon) - \mathrm{id}_\mathcal{H}]$$

$$\times \epsilon^{RST} \epsilon^{NPQ} [N(v)N'(s_N(\Delta)) - N(s_N(\Delta))N'(v)]$$

$$\times \mathrm{tr} \left(\tau_j h_{e_P(\Delta)} \left[h^{-1}_{e_P(\Delta)}, \sqrt{\hat{V}(v)} \right] \right) \mathrm{tr} \left(\tau_k h_{e_Q(\Delta)} \left[h^{-1}_{e_Q(\Delta)}, \sqrt{\hat{V}(v)} \right] \right)$$

$$\times \mathrm{tr} \left(\tau_l h_{e_S(\Delta')} \left[h^{-1}_{e_S(\Delta')}, \sqrt{\hat{V}(v)} \right] \right)$$

$$\times \mathrm{tr} \left(\tau_m h_{e_T(\Delta')} \left[h^{-1}_{e_T(\Delta')}, \sqrt{\hat{V}(v)} \right] \right) f_\gamma \qquad (10.4.26)$$

Basically, what happened in the quantisation step was that one had to introduce a point splitting which is why one has a double sum over tetrahedra and again factors of $1/\sqrt{\det(q)}$ got absorbed into Poisson brackets which then were replaced by commutators. Notice that in (10.4.26) the square root of the volume operator appears.

The fact that the combination $[\hat{U}(\varphi_\epsilon^{\xi_{\epsilon,v,\Delta',R}}) - \mathrm{id}_\mathcal{H}]$ stands •• ••• ••• shows that $\Psi(\hat{O}(M, N')f) = 0$ uniformly in $\Psi \in \mathcal{D}_{\mathrm{diff}}^*$ and $f \in \mathcal{D}$ for any N, N'.

10.5 The kernel of the Wheeler–DeWitt constraint operator

In [439] it was investigated to what extent one can solve the •• •• ••• • •• •• •••• ••• •••••• for $\Psi \in \mathcal{D}_{\mathrm{diff}}^*$

$$\Psi((\hat{H}(N))^\dagger f) = 0 \tag{10.5.1}$$

for all $N \in C^\infty(\sigma), f \in \mathcal{D}$. This section is devoted to an outline of an explicit construction of the complete and rigorous kernel of the proposed operator $(\hat{H}(N))^\dagger$. The methods that we display here will prove useful for all possible choices of $(\hat{H}(N))^\dagger$ of the type described. Notice that these solutions are rigorous solutions to the Wheeler–DeWitt constraint in full four-dimensional, Lorentzian quantum General Relativity in terms of connections compared with the calculations performed in [344–349]. Also, they are the first ones that have non-zero volume and which do not need non-zero cosmological constant.

We first want to give an intuitive picture of the way that the Hamiltonian constraint acts on cylindrical functions. When looking at (10.4.3) and (10.4.4) one realises the following: the Euclidean Hamiltonian constraint operator, when acting on, say, a spin-network state T over a graph γ, looks at each non-planar vertex v of γ and for each such vertex considers each triple of distinct edges $e, e', \tilde{e}$ incident at it. For each such triple, the constraint operator contains three terms labelled by the three possible pairs of edges that one can form from $\{e, e', \tilde{e}\}$. Let us look at one of them, say (neglecting numerical factors)

$$\mathrm{tr}\big([h_{\alpha(v;e,e')} - h_{\alpha(v;e,e')^{-1}}]h_{\tilde{s}}[h_{\tilde{s}}^{-1}, \hat{V}(U(v))]\big)T \tag{10.5.2}$$

The notation is as follows: $s, s', \tilde{s}$ are the segments of $e, e', \tilde{e}$ incident at v that end in the endpoints of the three arcs $a(v; e, e')$, etc., $\alpha(v; e, e')$ is the loop $s \circ a(v; e, e') \circ (s')^{-1}$ and $U(v)$ is any system of mutually disjoint neighbourhoods, one for each vertex v. For notational simplicity we have dropped the graph label. Let $j, j', \tilde{j}$ be the spins of the edges $e, e', \tilde{e}$ in T. First of all it is easy to see that the piece $h_{\tilde{s}}[h_{\tilde{s}}^{-1}, \hat{V}(U_{\epsilon_0}(v))]$ is invariant under a gauge transformation at the endpoint $\tilde{p}$ of $\tilde{s}$. Therefore the state (10.5.2) is also invariant at $\tilde{p}$ and since $\tilde{p}$ is a two-valent vertex this is only possible if the segments $\tilde{s}$ and $\tilde{e} - \tilde{s}$ of $\tilde{e}$ carry the same spin in the decomposition of (10.5.2) into spin-network states T'. But since $\tilde{e} - \tilde{s}$ carries still spin $\tilde{j}$ (no holonomy along $\tilde{e} - \tilde{s}$ appears in (10.5.2)) we conclude that the spin of $\tilde{e}$ is unchanged in T' compared with T.

However, the same is not true for e, e': the piece $[h_{\alpha(v;e,e')} - h_{\alpha(v;e,e')^{-1}}]$ is a multiplication operator and raises the spin of $a(v; e, e')$ from zero to $1/2$ and (10.5.2) decomposes into, in general, four spin-network states T' where the spins of the segments s, s' are raised or lowered in units of $1/2$ compared with T, that is, they are $j \pm 1/2, j' \pm 1/2$ respectively while the spins of the segments $e - s, e' - s'$ remain unchanged, namely j, j'. All this follows from basic Clebsch–Gordan decomposition theory for SU(2).

Next we look at the remaining piece $(\hat{H}(N))^\dagger + (\hat{H}_E(N))^\dagger$ of the Lorentzian Hamiltonian constraint. Its most important ingredient are the two factors of the operator $\hat{K}$ which, up to a numerical factor, equal $[\hat{V}(\sigma), (\hat{H}_E(1))^\dagger]$. Now as shown in [438], when inserting this operator into (10.4.4) what survives in the term corresponding to the vertex v of the graph is just $[\hat{V}(U(v)), (\hat{H}_E(U(v)))^\dagger]$. Thus, since the volume operator does not change any spins, the spin-changing ingredient of the action of the remaining piece of $(\hat{H}(N))^\dagger$ at v are two successive actions of $(\hat{H}_E(U(v)))^\dagger$ as just outlined.

In summary, the Hamiltonian constraint operator has an action similar to a fourth-order polynomial consisting of creation and annihilation operators. What is being created or annihilated are the spins of edges of a graph (notice that an edge with spin zero is the same as no edge at all).

Let us now look at this action in more detail. We will restrict attention only to the Euclidean piece, for the more complicated full action see [439].

Notice that the Euclidean constraint operator creates edges of a special kind, called •••••••• •• ••• •••••, namely the arcs $a = a(v; e, e')$. What is special about them is that they end in planar vertices which are either bi- or trivalent. If they are trivalent then, moreover, the vertex is the intersection of the two semianalytical edges a, e where a just ends on an interior point of e. Moreover, let e, e' be the edges on which a ends. Then there exist semianalytical extensions of e, e' which end in at least one point and the two possible earliest of these intersection points away from $a \cap e, a \cap e'$ are, together with these semianalytical extensions, non-planar vertices of γ. However, not only are these edges special, also the spin they carry is special, namely the arc a carries always spin $1/2$. We will continue to call this whole set of extraordinary structures an extraordinary edge.

The special nature of these edges allows us to classify the full set of labels $\mathcal{S}$ of spin-network states, called ••• •• •••, as follows. Denote by $\mathcal{S}_0 \subset \mathcal{S}$, called ••• ••••, the set of spin-nets, corresponding to graphs with no extraordinary edges at all.

From these sources one constructs iteratively derived sets $\mathcal{S}_n(s_0)$, $n = 0, 1, 2, \ldots$ for each source $s_0 \in \mathcal{S}_0$, called spin-nets of level n based on s_0. Put $\mathcal{S}_0(s_0) := \{s_0\}$ and define $\mathcal{S}_{n+1}(s_0)$ as follows: take each $s \in \mathcal{S}_n(s_0)$, compute $(\hat{H}_E(N))^\dagger T_s$ for all possible lapse functions N, decompose it into spin-network states and enter the appearing spin-nets into the set $\mathcal{S}_{n+1}(s_0)$.

In [439] it is shown that the sets $\mathcal{S}_n(s_0), \mathcal{S}_{n'}(s_0')$ are disjoint unless $s_0 = s_0'$ and $n = n'$. It is easy to see that the complement of the set of sources $\overline{\mathcal{S}_0} = \mathcal{S} - \mathcal{S}_0$

coincides with the set of derived spin-nets of level greater than zero. Moreover, for each $s \in \mathcal{S}$ there is a unique integer n and a unique source s_0 such that $s \in \mathcal{S}_n(s_0)$.

The purpose for doing all this is, of course, that this classification leads to a simple construction of all rigorous solutions of the Euclidean Hamiltonian constraint based on the observation that

$$(\hat{H}_{\mathrm{E}}(N))^{\dagger} \cdot \mathrm{span}\{T_s\}_{s \in \mathcal{S}_n(s_0)} \subset \mathrm{span}\{T_s\}_{s \in \mathcal{S}_{n+1}(s_0)} \tag{10.5.3}$$

Since a solution Ψ of (10.5.1) is a diffeomorphism-invariant distribution in $\mathcal{D}^*_{\mathrm{diff}}$ we define first $[\mathcal{S}_n(s_0)] := \{[s]\}_{s \in \mathcal{S}_n(s_0)}$ where $[s]$ is the label for the diffeomorphism-invariant distribution $T_{[s]}$ (recall Section 13.6). We can now make an Ansatz for a basic solution of the form

$$\Psi := \Psi_{[s_0], \vec{n}} := \sum_{k=1}^{N} \sum_{[s] \in [\mathcal{S}_{n_k}(s_0)]} c_{[s]} T_{[s]} \tag{10.5.4}$$

with complex coefficients $c_{[s]}$ which are to be determined from the quantum Einstein equations (10.5.1). Now from (10.5.4) it is clear that $\Psi_{[s_0], [\vec{n}]}((\hat{H}_{\mathrm{E}}(N))^{\dagger} T_s)$ can be non-vanishing if and only if $[s] \in [\mathcal{S}_{n_k-1}(s_0)]$ for some $k = 1, \ldots, n$, say $k = l$. Choose a representative $s \in [s]$ and let γ be the graph underlying s and $V(\gamma)$ its set of vertices. We then find, writing $(\hat{H}_{\mathrm{E}}(N))^{\dagger} = \sum_{v \in V(\gamma)} N(v) \hat{H}^{\dagger}_{\mathrm{E}}(v)$, that

$$\Psi_{[s_0], \vec{n}}((\hat{H}_{\mathrm{E}}(N))^{\dagger} T_s) = \sum_{[s'] \in [\mathcal{S}_{n_l}(s_0)]} c_{[s']} \sum_{v \in V(\gamma)} N(v) T_{[s']}(\hat{H}^{\dagger}_{\mathrm{E}}(v) T_s) \tag{10.5.5}$$

should vanish for any choice of lapse function $N(v)$. Since $N(v)$ can be any smooth function we find the condition that

$$\sum_{[s'] \in [\mathcal{S}_{n_l}(s_0)]} c_{[s']} T_{[s']}((\hat{H}_{\mathrm{E}}(v))^{\dagger} T_s) = 0 \tag{10.5.6}$$

should vanish for each choice of the finite number of vertices $v \in V(\gamma)$ and for each of the finite number of spin-nets $s \in \mathcal{S}_{n_l-1}(s_0)$. This follows from the fact that the numbers $T_{[s']}((\hat{H}_{\mathrm{E}}(v))^{\dagger} T_s)$ are diffeomorphism-invariant and therefore do not actually depend on v itself but only on the diffeomorphism-invariant information that is contained in the graph γ together with the vertex v singled out.

Therefore, (10.5.6) is a finite system of linear equations for the coefficients $c_{[s']}$. As the cardinality of the sets $\mathcal{S}_n(s_0)$ grows exponentially with n this system is far from being overdetermined and we arrive at an infinite number of solutions. The most general solution will be a linear combination of the elementary solutions (10.5.6). Qualitatively the same result holds for the Lorentzian constraint [439], however, it is more complicated because coefficients from different levels get coupled and so one gets solutions labelled also by the highest level that was used (possibly one has to allow all levels, that is, the highest level is always infinity).

Nevertheless it is remarkable how the solution of the quantum Einstein equations is reduced to an exercise in finite-dimensional linear algebra (although the computation of the coefficients $T_{[s']}((\hat{H}_E(v))^\dagger T_s)$ is far from easy, see, e.g., [451] which, although the authors restrict to trivalent graphs and $(\hat{H}_E(N))^\dagger$ only, is already rather involved). On the other hand, it is expected that physically interesting solutions will actually be infinite linear combinations of coupled solutions, that is, solutions of infinite level, an intuition coming from [440].

Notice that the solutions (10.5.6) are bona fide elements of $\mathcal{D}^*_{\text{diff}}$ and therefore give rigorously defined solutions to the diffeomorphism and the Hamiltonian constraint of full, four-dimensional Lorentzian quantum General Relativity in the continuum, subject to the reservation that we still have to prove that the classical limit of this theory is in fact General Relativity. One should now organise these solutions into a Hilbert space such that adjointness and canonical commutation relations of full Dirac observables are faithfully implemented. However, since group averaging does not work for open algebras, there is no good proposal at this point for how to do that. This is precisely one of the motivations for the Master Constraint Programme, to which we turn in the next section.

•• ••• •••: One could hope that the so-called Kodama state $\Psi_{\text{Kodama}}[A] := \exp(-\frac{2i}{\Lambda\hbar\kappa}S_{\text{CS}}[A])$ is an exact solution of all quantum constraints of vacuum loop quantum gravity (see, e.g., [584] and references therein for a recent review) with a cosmological constant Λ. Here

$$S_{\text{CS}} = \int_\sigma \text{Tr}\left(A \wedge F - \frac{1}{3}A \wedge A \wedge A\right)$$

is the Chern–Simons action for the connection A which has the property that, considered as a functional of smooth connections, we have $\delta S_{\text{CS}}[A]/\delta A^j_a = B^a_j, B^a_j = \frac{1}{2}\epsilon^{abc}F^j_{bc}$, that is, it is the generating functional for the magnetic field B of A. The hope could be based on the fact that the classical Hamiltonian constraint with a cosmological constant can be written in the form $\tilde{H} = \epsilon_{abc}\text{Tr}(E^a E^b[B^c - \Lambda E^c])$ and if we quantise as $\hat{E}^a_j(x) := i\hbar\kappa/2\delta/\delta A^j_a(x)$ then in this ordering for $\tilde{H}$ we formally have $\widehat{\tilde{H}}(x)\Psi_{\text{Kodama}} = 0$.

We will now show that this argument is at least very misleading and far from established for many reasons:

- The constraint $H = \tilde{H}/\sqrt{|\det(E)|}$ is the vacuum constraint for GR with a cosmological constant only if we use the complex-valued self-dual variables. However, in this case we do not even have a kinematical inner product for two reasons: First, the uniform measure, the analogue of the measure μ_0 that we have constructed, does not exist for non-compact gauge groups such as $SL(2, \mathbb{C})$ because the Haar measure is not normalisable and there does not exist a G-invariant mean for any gauge group which contains $SL(2, \mathbb{R})$ as a subgroup since $SL(2, \mathbb{R})$ is a non-amenable group [562,585]. The second problem is even worse: even if an analogue of the uniform measure existed, it would implement the wrong reality conditions. Namely, for the self-dual connection we would need to impose the operator analogue of $E - \bar{E} = 0$ and $A + \bar{A} = 2\Gamma[E]$ where Γ is the highly non-polynomial spin connection of E.

•••• ••• this can be done as follows: we consider an L_2 space holomorphic of holomorphic wave functions $\psi(A)$ of the form $L_2(\mathcal{A}, K(A, \bar{A})[DA \, D\bar{A}])$ where $\mathcal{A}$ is some unspecified space of $SL(2, \mathbb{C})$ connections, $[DA]$ denotes the formal infinite-dimensional Lebesgue measure and K is a kernel to be determined. The canonical brackets $\{E_j^a(x), A_b^k(y)\} = i\kappa\delta_b^a\delta_j^k\delta(x,y)$ motivate us to represent $A_a^j(x)$ as a multiplication operator and $E_j^a(x)$ as the functional derivative $-\ell_P^2 \delta/\delta A_a^j(x)$. Since wave functions are holomorphic, in order that $\hat{E}$ be symmetric[9] we need that $K(A, \bar{A}) = \rho(\Re(A))$ for some functional ρ. In order to satisfy the second reality condition we display ρ by its formal Fourier functional integral

$$\rho(\Re(A)) = \int_{\mathcal{E}} [dE] \exp\left(i \int_\sigma d^3x \, E_j^a \Re(A)_a^j \right) \rho(\tilde{E}) \qquad (10.5.7)$$

over some unspecified space of electric fields because the exponential function in (10.5.7) is a generalised eigenfunction of the functional derivative operator. The adjointness relation $\hat{A}^\dagger(x) + \hat{A}(x) = 2\Gamma(\hat{E}(x))$ is then equivalent[10] to the following system of functional differential equations

$$\left[\Re\big(A_a^j(x)\big) - \Gamma_a^j \left(\frac{\ell_P^2}{2} \delta/\delta\Re(A(x)) \right) \right] \rho(\Re(A)) = 0 \qquad (10.5.8)$$

which can be solved, using (10.5.7) by[11]

$$K(A, \bar{A}) = \int_{\mathcal{E}} dE \, e^{i \int_\sigma d^3x [\frac{A_a^j + \bar{A}_a^j}{2} - \Gamma_a^j] E_j^a} \qquad (10.5.9)$$

Unfortunately, this gauge-invariant kernel does not satisfy the decay assumptions under which it was derived. Moreover, the kernel is not manifestly positive so that it is unclear whether the corresponding formal inner product is positive (semi)definite.

In summary, for the self-dual connection there is presumably no Hilbert space representation •• •••. Therefore we do not even know whether (the smeared form of) $\widehat{\tilde{H}}$ is densely defined. Even worse, the fact that $\tilde{H}$ is a density of weight two makes this impossible to happen in any background-independent representation, as we showed in this chapter.

---

[9] When performing the corresponding functional integration by parts we must require that K decays sufficiently fast at infinity in $\mathcal{A}$. This is not possible for the holomorphic wave functions by Liouville's theorem.

[10] This only holds formally if we think of $E \mapsto \Gamma(E)$ as being approximated by a Taylor series in E.

[11] Use that Γ is a homogeneous rational function of degree zero and that Γ has the generating functional (4.2.24).

- Even when including the proper factor $1/\sqrt{|\det(E)|}$ which turns H into a density of weight one then we do not know how to quantise the volume operator associated with $\sqrt{|\det(E)|}$ because we would need to know whether the associated flux operators are self-adjoint, otherwise we do not know what operators are positive and we cannot define the absolute value $|A| := \sqrt{A^\dagger A}$ of an operator A nor the square root. We need the volume operator in order to define $1/\sqrt{|\det(E)|}$ via the Poisson bracket identity between the connection and the volume, otherwise the operator corresponding to H would not be densely defined in any representation due to the zero modes of the volume operator (e.g., constant functions, assuming that they are normalisable).

- We might think of $\widehat{\tilde{H}}$ as the dual of an operator where the dual is defined on some space of distributions over some space of functions of connections and then the above chosen ordering is formally the same as we have defined for the Hamiltonian constraint in that the connection dependence is to the right for the dual operator (to the left for the actual operator). However, none of these spaces can be specified and thus it is not possible to see what the actual operator should be, that is, we do not know whether $\widehat{\tilde{H}}$ is the dual quantisation of $\tilde{H}$. Since we do not know what the spaces involved are we cannot perform spectral analysis. We therefore cannot determine the physical inner product and cannot check whether Ψ_{Kodama} is normalisable and has non-zero norm.

- If we take the connection to be real-valued then all these mathematical notions are rigorously defined, however, then the Kodama state is no solution to the constraints any more because in this case the constraint $\tilde{H}$ acquires an extra term. Without this term we just have the Euclidean piece of the constraint. Even then it is not at all clear that Ψ_{Kodama} can be extended to a measurable function on $\overline{\mathcal{A}}$, but even if one could somehow define the Chern–Simons action as some limit of functions of holonomies then the resulting functional would not be an L_2 function any more, being an at least uncountably infinite linear combination of mutually orthogonal spin-network states. Hence at best it can be defined as a distribution on the finite linear span of spin-network functions. But then the dual of the Hamiltonian constraint does not annihilate it, not even the Euclidean piece, because in order to define the Euclidean Hamiltonian constraint we have to take the factor $1/\sqrt{|\det(E)|}$ into account and quantise it with the Poisson bracket identity. In that form then we get at each vertex an actual, smeared Euclidean Hamiltonian constraint of the form $\text{Tr}(h_\alpha h_e[h_e^{-1}, \hat{V}] - \Lambda \hat{V})$ whose dual does not annihilate Ψ_{Kodama} because there is no factor $B - \Lambda E$ any more.

- Finally, while S_{CS} is formally spatially diffeomorphism-invariant, it is well known not to be invariant under finite (rather than infinitesimal) gauge transformations. Hence it does not even solve the Gauß constraint.

Hence, if one wants to define a rigorously defined Kodama state as an element of $\mathcal{D}^*$ using all the machinery of $\overline{\mathcal{A}}$ then a lot more work is required.

10.6 The Master Constraint Programme

In this section we describe the Master Constraint Programme applied to GR. For an outline of the method for a general theory see Chapter 30.

10.6.1 Motivation for the Master Constraint Programme in General Relativity

There are three good reasons for setting up the Master Constraint Programme (MCP):

1. • • •• ••• • ••• • • • ••• • • ••• •• • • •• • •• • •

We have seen that spatial diffeomorphism-invariance plays a very important role in showing that the limit of the Hamiltonian constraint, as the regulator is removed, converges to a well-defined operator on $\mathcal{H}^0$. It is therefore natural to try to define $\hat{H}(N)$ (or rather the dual $\hat{H}'(N)$) directly on $\mathcal{H}_{\text{diff}}$. However, we simply cannot do that because the relation $\{\vec{H}, H\} \propto H$ of the Dirac algebra $\mathfrak{D}$ dictates that $\hat{H}'(N)$ must not preserve $\mathcal{H}_{\text{diff}}$. This is the reason for introducing a topology of the kind of the URST which allows us to take advantage of spatial diffeomorphism-invariance while $(\hat{H}(N))^\dagger$ is defined on the kinematical Hilbert space $\mathcal{H}^0$. However, it would be much cleaner to work directly on $\mathcal{H}_{\text{diff}}$ and to use a usual strong or weak operator topology. In particular, the axiom of choice would no longer be necessary because all choice functions are spatially diffeomorphically equivalent up to diffeomorphism-invariant characteristics.

2. • • •• ••• ••• •• •• ••• ••• •• ••• •• • •• ••• ••• • ••• ••• ••• ••• •

We have seen that one-parameter subgroups of the spatial diffeomorphism group are not weakly continuous, hence a self-adjoint generator corresponding to $\vec{H}(\vec{N})$ does not exist. Therefore it is not possible to implement the relation $\{H(N), H(N')\} = \propto \vec{H}(q^{-1}(dNN' - NdN'))$ in the quantum theory with infinitesimal $\vec{H}$, it is at most possible with finite diffeomorphisms $U(\varphi) - 1_{\mathcal{H}^0}$ and it seems that this is what actually happens. Notice that due to the uniqueness of $\mathcal{H}^0$ it is not possible to switch to another representation in which the infinitesimal generator does exist unless one changes the algebra $\mathfrak{P}$, that is, the very starting point of the quantisation programme. Hence, trouble with implementing the relation $\{H(N), H(N')\} = \propto \vec{H}(q^{-1}(dNN' - NdN'))$ is to be expected and it would be nice to circumvent that problem.

3. • • ••• •• •• •••• • •• • ••• ••

The fact that we get structure functions $q^{-1}(dNN' - NdN')$ rather than structure constants (with respect to the phase space) disqualifies $\mathfrak{D}$ as an honest Lie algebra and prohibits using group averaging techniques in order to construct solutions to the infinite number of Hamiltonian constraints, a physical inner product thereon and Dirac observables by RAQ methods, see Chapter 30. It would be important to develop methods for dealing with first-class algebras which are not Lie algebras.

These three observations motivate us to reformulate the constraint algebra $\mathfrak{D}$ such that the new algebra $\mathfrak{M}$, the Master Constraint algebra, is free from all three problems while still encoding the same reduced phase space, that is, the same constraint surface and the same (weak) Dirac observables. To circumvent problem 1 one would need to define the new Hamiltonian constraints directly as operators on $\mathcal{H}_{\text{diff}}$. This is clearly only possible if the corresponding classical Hamiltonian constraints are spatially diffeomorphism-invariant and hence would solve problem 2 as well. To solve problem 3 the new constraints would need to close with structure constants but without involving the infinitesimal spatial diffeomorphism constraints. Finally they would need to be spatial scalars with density weight one in order to have a chance of resulting in well-defined smeared operators. Hence, denoting the new Hamiltonian constraints with the symbol $M(x)$ and their smeared version with $M(N)$ we would need $\{\vec{H}(\vec{N}), M(N)\} = 0$ and $\{M(N), M(N')\} \propto M(N''(N, N'))$ where N'' is a phase space-independent functional of N, N' with $N''(N, N') = -N''(N', N)$. The latter condition implies that N'' cannot be ultralocal, that is, it must involve at least spatial derivatives of N, N'. But then it must depend on q^{ab} which is not allowed. Hence we conclude $N'' = 0$, that is, $\{M(N), M(N')\} = 0$.

Quite surprisingly one can classify all solutions to this Abelian condition [586–588]. The corresponding $M(x)$ are algebraic aggregates built from $H(x)$ and $q^{ab} H_a H_b$. Unfortunately, the only density one-valued solution is $M(x) = \sqrt{H^2 - q^{ab} H_a H_b}$ whose Hamiltonian vector field vanishes on the constraint surface and therefore does not result in a well-defined gauge flow. Apart from this classical problem the argument of the square root is indefinite and hence it is hard to give meaning to it in the quantum theory. However, the worst problem is that the $M(x)$, just as the $H(x)$, are scalar densities of weight one, therefore we have automatically $\{\vec{H}(\vec{N}), M(N)\} \propto M(\mathcal{L}_{\vec{N}} N)$ which therefore does not solve problem 1. The only spatially diffeomorphism-invariant quantity that can be built from $M(x)$ is $M(N)_{N=1} = \int d^3x M(x)$, however, this gives us not enough information because we need all the $M(N)$ for arbitrary N in order to conclude from $M(N) = 0$ for all N that $M(x) = 0$ for all x.

The way out is the $\bullet$ $\bullet\bullet\bullet\bullet\bullet$ $\bullet$ $\bullet\bullet\bullet\bullet\bullet\bullet\bullet$

$$= \frac{1}{2} \int_\sigma d^3x \, \frac{H(x)^2}{\sqrt{\det(q)(x)}} \tag{10.6.1}$$

Clearly the integrand is a density of weight one, therefore $\{\vec{H}(\vec{N}), \quad\} = 0$ which solves problem 1.[12] Since the integrand is positive, the $\bullet\bullet\bullet\bullet\bullet$ $\bullet$ $\bullet\bullet\bullet\bullet\bullet$ $\bullet$ $\bullet\bullet\bullet\bullet\bullet\bullet$

$= 0$ is in fact equivalent to the infinite number of constraints $H(x) = 0$, thus the problem just stated does not appear.[13] Also the integrand is differentiable on

---

[12] Since q_{ab} is classically non-degenerate the integrand is classically non-singular.

[13] Actually we can only conclude that $H(x) = 0$ a.e. with respect to d^3x. However, also $H(N) = 0$ for all N just means that $H(x) = 0$ for a.a. x. But since $x \mapsto H(x)$ is classically continuous, even smooth, we in fact may conclude $H(x) = 0$ for all $x \in \sigma$.

the constraint surface in contrast to $M(x)$. Now since we are proposing to take only one constraint   instead of infinitely many, the constraint algebra among the 'Master Constraints' trivialises $\{\ ,\ \} = 0$.

Thus we have managed to solve all three problems listed above. However, while we have verified that $= 0$ defines the same constraint surface as the $H(N)$, we must still check that   is able to detect the same (weak) Dirac observables as the $H(N)$ do. Recall that a spatially diffeomorphism-invariant function O is called a weak Dirac observable provided that $\{O, H(N)\}_{=0} = 0$ for all N. Now the analogous condition $\{O,\ \}_{=0} = 0$ is trivially satisfied for arbitrary O because

$$\{O,\ \} = \int_\sigma d^3x\ \frac{\{O, H(x)\}\, H(x) - \frac{1}{2}\{O, \ln(\sqrt{\det(q)}(x))\}H(x)^2}{\sqrt{\det(q)}(x)} \tag{10.6.2}$$

obviously vanishes at $= 0$ as long as O is differentiable. Thus it seems that   is not enough in order to detect Dirac observables. However, we notice that

$$\{O, \{O,\ \}\}_{=0} = \int_\sigma d^3x\ \frac{\left(\{O, H(x)\}_{=0}\right)^2}{\sqrt{\det(q)}(x)} \tag{10.6.3}$$

vanishes if and only if $\{O, H(N)\}_{=0} = 0$ for all N. Thus, again the infinite number of conditions $\{O, H(N)\}_{=0} = 0$ for all N is replaced by the $\cdots\cdots$ $\cdots\cdots\cdots\cdots$ $\{O, \{O,\ \}\}_{=0} = 0$, the only price we have to pay is that we need to work with double Poisson brackets instead of a single one.[14]

Since we now have managed to recast the constraint algebra $\mathfrak{D}$ into a much simpler form, namely the $\cdots\cdots\cdots\cdots\cdots\cdots\cdots$ $\mathfrak{M}$

$$\{\vec{H}(\vec{N}), \vec{H}(\vec{N}')\} = -\kappa \vec{H}(\mathcal{L}_{\vec{N}}\vec{N}')$$
$$\{\vec{H}(\vec{N}),\ \} = 0$$
$$\{\ ,\ \} = 0 \tag{10.6.4}$$

we propose to quantise   rather than the $H(N)$ directly on $\mathcal{H}_{\text{diff}}$ and to apply the direct integral method reviewed in Chapter 30 in order to solve the constraint, to define the physical inner product and to find Dirac observables. In fact, it is necessary to define the $\cdot\ \cdots\cdot\ \cdots\cdots\cdots\cdots\cdots$ $\widehat{\ }$ directly on $\mathcal{H}_{\text{diff}}$ because, as we have seen, graph-changing spatially diffeomorphism-invariant operators cannot be defined on $\mathcal{H}^0$. Thus we see that several facts work nicely together.

One may ask why we then took the effort to go through the construction of $(\hat{H}(N))^\dagger$ at all. The answer is that the same techniques that we used to define $(\hat{H}(N))^\dagger$ can be applied to define $\widehat{\ }$ because $\widehat{\ }$ is closely related to the square of the $H(x)$, modulo the important factor $1/\sqrt{\det(q)}$.

---

[14] In fact we could have used any even power of H, that is, $H^{2n}/\sqrt{\det(q)}^{2n-1}$ and then   would have to consider multiple Poisson brackets of order $2n$. However, $n = 1$ is the simplest choice.

10.6.2 Definition of the Master Constraint

The strategy to implement the Master Constraint is as follows. Let $T(\epsilon)$ be a triangulation of σ into tetrahedra Δ and denote by $\epsilon \to 0$ the limit in which the triangulation is infinitely refined. Then the classical Master Constraint is the limit of the Riemann sum

$$= \lim_{\epsilon \to 0} \sum_{\Delta \in T(\epsilon)} \frac{H(\Delta)^2}{V(\Delta)} \tag{10.6.5}$$

where $H(\Delta) = H(\chi_\Delta)$, $V(\Delta) = \int_\Delta d^3x \sqrt{\det(q)}$ and χ_Δ is the characteristic function of the set Δ. Now recall formulae (10.3.12), (10.3.13) and consider there the term for a given Δ. It is easy to see that in the limit $\epsilon \to 0$ this term coincides with $N(v(\Delta))H(\Delta)$ where $H(\Delta)$ is defined as above. Now $H(\Delta)$ is proportional to the Poisson bracket $\{h_{e_K(\Delta)}^{-1}, V(R_{v(\Delta)})\}$ where $V(R_{v(\Delta)})$ can be chosen to be identical to the $V(\Delta)$ used in the notation used in this section. We now write

$$\frac{H(\Delta)^2}{V(\Delta)} = \left(\frac{H(\Delta)}{\sqrt{V(\Delta)}} \right)^2 =: C(\Delta)^2 = \overline{C(\Delta)}C(\Delta) \tag{10.6.6}$$

where we used $\{., V(\Delta)\}/\sqrt{V(\Delta)} = 2\{., \sqrt{V(\Delta)}\}$ and defined $C(\Delta)$ to be the same as $H(\Delta)$ just that $\{h_{e_K(\Delta)}^{-1}, V(R_{v(\Delta)})\}$ is replaced by $2\{h_{e_K(\Delta)}^{-1}, \sqrt{V(R_{v(\Delta)})}\}$.

This is a huge simplification because the $C(\Delta)$ can be quantised precisely as the $H(\Delta)$ with this simple change in the power of the volume operator. All the qualitative features remain the same, only the numerical values of the matrix elements of the corresponding regularised $\hat{C}_\epsilon^\dagger(\Delta)$ change. We may therefore compute the corresponding, regularised dual operators $\hat{C}_\epsilon'(\Delta)$ on $\mathcal{D}^*$ and when restricted to $l \in \mathcal{D}_{\text{diff}}^*$ the dependence on ϵ in $\hat{C}_\epsilon'(\Delta)l$ actually drops out. However, as before $\hat{C}_\epsilon'(\Delta)$ does not preserve $\mathcal{H}_{\text{diff}}$. Now since we have managed to recast (10.6.5) into the form

$$= \lim_{\epsilon \to 0} \sum_{\Delta \in T(\epsilon)} \overline{C(\Delta)}C(\Delta) \tag{10.6.7}$$

and since we must implement $\widehat{}$ directly on $\mathcal{H}_{\text{diff}}$ we try to define the •••• •• ••• ••••

$$Q \ (l, l') := \lim_{\epsilon \to 0} \sum_{\Delta \in T(\epsilon)} < l, (\hat{C}_\epsilon'(\Delta))^* \hat{C}_\epsilon'(\Delta)l' >_{\text{diff}}$$

$$= \lim_{\epsilon \to 0} \sum_{\Delta \in T(\epsilon)} < \hat{C}_\epsilon'(\Delta)l, \hat{C}_\epsilon'(\Delta)l' >_{\text{diff}} \tag{10.6.8}$$

where $(.)^*$ denotes the adjoint operation on $\mathcal{H}_{\text{diff}}$. However, at least at finite ϵ equation (10.6.8) is ill-defined because we are using the scalar product on $\mathcal{H}_{\text{diff}}$ while $\hat{C}_\epsilon'(\Delta)l \notin \mathcal{H}_{\text{diff}}$. For the same reason the adjoint operation carried out in the second step is unjustified.

The hope is, of course, that (10.6.8) makes sense in the limit $\epsilon \to 0$ when the corresponding classical quantity becomes spatially diffeomorphism-invariant. The tool to arrive at this is to equip the space $\mathcal{D}^*$ with an inner product which reduces to the one on $\mathcal{H}_{\mathrm{diff}}$ when evaluated on $\mathcal{D}^*_{\mathrm{diff}}$. This can be done, formally, as follows: given a spin-network diffeomorphism equivalence class $[s]$ we define the non-standard number or • •• ••• • •••••

$$\aleph([s]) := |[s]| := |\{s' \in \mathcal{S};\ [s'] = [s]\}| \tag{10.6.9}$$

as the size of the orbit $[s]$. Now recall that the preferred elements of $\mathcal{D}^*_{\mathrm{diff}}$ were given by

$$l_{[s]} := \sum_{s' \in [s]} < T_{s'}, . >_{\mathrm{kin}}, \eta(T_s) = \eta_{[s]} l_{[s]} \tag{10.6.10}$$

with positive numbers $\eta_{[s]}$ and

$$< \eta(T_s), \eta(T_{s'}) >_{\mathrm{diff}} = \eta(T_{s'})[T_s] \tag{10.6.11}$$

An arbitrary element of $\mathcal{D}^*$ is of the form $l = \sum_{s \in \mathcal{S}} c_s < T_s, . >_{\mathrm{kin}}$. Formally, we may define an inner product $< ., >_*$ on $\mathcal{D}^*$ by

$$< l, l' >_* := \sum_{s,s'} \overline{c_s} c'_{s'} << T_s, . >_{\mathrm{kin}}, < T_{s'}, . >_{\mathrm{kin}} >_*$$

$$:= \sum_{s,s'} \overline{c_s} c'_{s'} < T_{s'}, T_s >_{\mathrm{kin}} \frac{\sqrt{\eta_{[s]} \eta_{[s']}}}{\sqrt{\aleph([s]) \aleph([s'])}} = \sum_s \overline{c_s} c'_s \frac{\eta_{[s]}}{\aleph([s])} \tag{10.6.12}$$

This reproduces the inner product between the $\eta_{[s]}$ which correspond to $c_{s'} = \chi_{[s]}(s')$. It also formally corresponds to formally extending (10.6.12) to $\mathcal{H}_{\mathrm{kin}}$ with

$$< T_s, T_{s'} >_* := < T_s, T_{s'} >_{\mathrm{kin}} \frac{\sqrt{\eta_{[s]} \eta_{[s']}}}{\sqrt{\aleph([s]) \aleph([s'])}} \tag{10.6.13}$$

but of course elements of $\mathcal{H}_{\mathrm{kin}}$ have zero norm in this inner product. Hence by far not all elements of $\mathcal{D}^*$ are normalisable in this inner product and many elements have zero norm with respect to it. By passing to the quotient by the null vectors and completing we may turn the normalisable elements of $\mathcal{D}^*$ into a Hilbert space $\mathcal{H}_* \subset \mathcal{D}^*$. Notice that (10.6.12) is the first inner product to be proposed on (a subset of) $\mathcal{D}^*$.

It is curious to note that we may formally define a partial isometry

$$V: \mathcal{H}_* \to \mathcal{H}_{\mathrm{kin}}; l = \sum_s c_s < T_s, . >_{\mathrm{kin}} \mapsto \tilde{l} = \sum_s c_s \sqrt{\frac{\eta_{[s]}}{\aleph([s])}} T_s \tag{10.6.14}$$

so that we may formally identify $< ., . >_*$ with the kinematical inner product $< ., . >_{\mathrm{kin}}$ under the map $l \mapsto \tilde{l}$.

The idea is then to use $< .,. >_*$ and its associated adjoint operation to define (10.6.8) properly, that is,

$$Q\ (l, l') := \lim_{\epsilon \to 0} \sum_{\Delta \in T(\epsilon)} < l, (\hat{C}'_\epsilon(\Delta))^* \ \hat{C}'_\epsilon(\Delta)\ l' >_*$$

$$= \lim_{\epsilon \to 0} \sum_{\Delta \in T(\epsilon)} < \hat{C}'_\epsilon(\Delta)\ l, \hat{C}'_\epsilon(\Delta)\ l' >_* \qquad (10.6.15)$$

which is now well-defined. To evaluate $< .,. >_*$ we write

$$\hat{C}'_\epsilon(\Delta) l = \sum_{s \in S} c^l_s(\Delta, \epsilon) < T_s, . >_{\text{kin}} \Rightarrow c^l_s(\Delta, \epsilon) = l(C^\dagger_\epsilon(\Delta) T_s) \quad (10.6.16)$$

where the dependence on ϵ is actually trivial. Hence (10.6.16) becomes

$$Q\ (l, l') = \lim_{\epsilon \to 0} \sum_{\Delta \in T(\epsilon)} \sum_s \overline{c^l_s(\Delta, \epsilon)}\ c^{l'}_s(\Delta, \epsilon)\ \frac{\eta_{[s]}}{\aleph([s])}$$

$$= \lim_{\epsilon \to 0} \sum_{\Delta \in T(\epsilon)} \sum_{[s]} \frac{\eta_{[s]}}{\aleph([s])} \sum_{s' \in [s]} \overline{c^l_{s'}(\Delta, \epsilon)}\ c^{l'}_{s'}(\Delta, \epsilon) \quad (10.6.17)$$

We notice that for given l, l' only a finite number of $[s]$ contribute to (10.6.17): namely, both l, l' are finite linear combinations of the $l_{[s_1]}$ in (10.6.10), hence it suffices to show that for any $[s_1], [s_2]$ the numbers

$$\overline{c^{l_{[s_1]}}_{s'}(\Delta, \epsilon)}\ c^{l_{[s_2]}}_{s'}(\Delta, \epsilon) \qquad (10.6.18)$$

are non-vanishing only when $s' \in [s]$ and $[s]$ ranges over a finite number of classes. In order that $c^{l_{[s_1]}}_{s'}(\Delta, \epsilon) \neq 0$ we must have that $\hat{C}^\dagger_\epsilon(\Delta) T_{s'}$ is a finite linear combination of spin-network states which involves at least one of the $T_{s'_1}$ with $s'_1 \in [s_1]$. But from the explicit action of $\hat{C}^\dagger_\epsilon(\Delta)$ it is clear that for each $s'_1 \in [s_1]$ there is only a finite set $S(s'_1)$ of s' with this property.[15] Moreover, for each $s'_1 \in [s_1]$ the number of elements of $S(s'_1)$ is the same and the classes of the elements of $S(s'_1)$ do not depend on the representative $s'_1 \in [s_1]$. Denote the finite set of these classes by $[S]([s_1])$.

The sum over $[s]$ in (10.6.17) is therefore only over the finite set $[S]([s_1]) \cap [S]([s_1])$ for $l = l_{[s_1]}$, $l' = l_{[s_2]}$, hence for any $l, l' \in \mathcal{H}_{\text{diff}}$ the sum over $[s]$ in (10.6.17) is finite. We may therefore interchange the sum $\sum_{[s]}$ with the $\sum_\Delta$ and the limit $\lim_{\epsilon \to 0}$ and arrive at

$$Q\ (l, l') = \sum_{[s]} \frac{\eta_{[s]}}{\aleph([s])} \lim_{\epsilon \to 0} \sum_{\Delta \in T(\epsilon)} \sum_{s' \in [s]} \overline{c^l_{s'}(\Delta, \epsilon)}\ c^{l'}_{s'}(\Delta, \epsilon) \quad (10.6.19)$$

---

[15] This is only true if we restrict attention to those s' such that $[s']$ is in a given, invariant θ equivalence class as discussed in Section 10.6.3. The reason is that $\hat{C}^\dagger_\epsilon(\Delta)$ may remove moduli within $T_{s'}$ that $T_{s'_1}$ does not know about. Hence, without this assumption we would need to sum over uncountably many $[s]$ with different moduli. Thus we will make this restriction here and in the next subsection which will be justified in Section 10.6.3.

Fix $s' \in [s]$ and consider $\hat{C}_\epsilon^\dagger(\Delta)T_{s'}$. From Section 10.4 we know that this can be written in the form

$$\hat{C}_\epsilon^\dagger(\Delta)T_{s'} = \sum_{v \in V(\gamma(s')) \cap \Delta} \hat{C}_{\epsilon,v}^\dagger T_{s'} \qquad (10.6.20)$$

For sufficiently small ϵ each Δ contains at most one vertex and the sum over Δ therefore reduces to the finite set $T(\epsilon, s')$ of those Δ's containing precisely one vertex of $\gamma(s')$. We may therefore interchange the sum $\sum_{s'}$ with the $\sum_\Delta$ and the limit $\epsilon \to 0$ and obtain

$$
\begin{aligned}
Q\ (l, l') &= \sum_{[s]} \frac{\eta_{[s]}}{\aleph([s])} \sum_{s' \in [s]} \lim_{\epsilon \to 0} \sum_{\Delta \in T(\epsilon, s')} \overline{c_{s'}^l(\Delta, \epsilon)}\, c_{s'}^{l'}(\Delta, \epsilon) \\
&= \sum_{[s]} \frac{\eta_{[s]}}{\aleph([s])} \sum_{s' \in [s]} \lim_{\epsilon \to 0} \sum_{v \in V(\gamma(s'))} \overline{c_{s'}^l(v, \epsilon)}\, c_{s'}^{l'}(v, \epsilon) \\
&= \sum_{[s]} \frac{\eta_{[s]}}{\aleph([s])} \sum_{s' \in [s]} \sum_{v \in V(\gamma(s'))} \overline{c_{s'}^l(v)}\, c_{s'}^{l'}(v) \qquad (10.6.21)
\end{aligned}
$$

where

$$c_{s'}^l(v, \epsilon) = l(\hat{C}_{\epsilon,v}^\dagger T_{s'}) = l(\hat{C}_{\epsilon_0,v}^\dagger T_{s'}) =: c_{s'}^l(v) \qquad (10.6.22)$$

for any choice ϵ_0 by spatial diffeomorphism-invariance of l. In the second step the sum over the contributing Δ could be replaced by the sum over vertices and since then nothing depends on ϵ any more the limit $\epsilon \to 0$ is trivial.

We now claim that

$$a(s') := \sum_{v \in V(\gamma(s'))} \overline{c_{s'}^l(v)}\, c_{s'}^{l'}(v) \qquad (10.6.23)$$

only depends on the class $[s]$ of s'. Indeed,

$$
\begin{aligned}
a(\varphi \cdot s') &= \sum_{v \in V(\gamma(\varphi \cdot s'))} \overline{c_{\varphi \cdot s'}^l(v)}\, c_{\varphi \cdot s'}^{l'}(v) \\
&= \sum_{v \in \varphi(V(\gamma(s')))} \overline{c_{\varphi \cdot s'}^l(v)}\, c_{\varphi \cdot s'}^{l'}(v) \\
&= \sum_{v \in V(\gamma(s'))} \overline{c_{\varphi \cdot s'}^l(\varphi(v))}\, c_{\varphi \cdot s'}^{l'}(\varphi(v)) \qquad (10.6.24)
\end{aligned}
$$

but

$$
\begin{aligned}
c_{\varphi \cdot s'}^l(\varphi(v)) &= l(\hat{C}_{\epsilon_0, \varphi(v)}^\dagger \hat{U}(\varphi) T_{s'}) = l(\hat{U}(\varphi)\hat{C}_{\epsilon_0,v}^\dagger T_{s'}) \\
&= l(\hat{C}_{\epsilon_0,v}^\dagger T_{s'}) = l(\hat{C}_{\epsilon_0,v}^\dagger T_{s'}) \\
&= c_{s'}^l(v) \qquad (10.6.25)
\end{aligned}
$$

where in the first step we used the fact that $\mathrm{Diff}_{sa}^\omega(\sigma)$ is unitarily implemented, in the second we have used the covariance relation (10.4.15) up to a diffeomorphism under which the choice ϵ_0 may change but two choices are related by a diffeo-morphism and in the last two steps we used diffeomorphism-invariance of l.

It follows that all the $\aleph([s])$ terms in the sum $\sum_{s'\in[s]}$ are identical. Let $s_0([s])$ be a representative of $[s]$ then we may finish our derivation and get the final result

$$Q\ (l,l') = \sum_{[s]} \eta_{[s]} \sum_{v\in V(\gamma(s_0[s]))} \overline{l\big(\hat{C}_v^\dagger T_{s_0([s])}\big)}\, l'\big(\hat{C}_v^\dagger T_{s_0([s])}\big) \qquad (10.6.26)$$

The explicit expression for $l(\hat{C}_v^\dagger T_{s_0([s])})$ is given by the evaluation of l on (10.4.3), (10.4.4) with $\hat{V}(U_{\epsilon_0}(v))$ replaced by $2\sqrt{\hat{V}(U_{\epsilon_0}(v))}$. We have dropped the irrele-vant label ϵ_0. Since we showed that the sum over $[s]$ collapses to a finite number of terms, (10.6.26) is well-defined. Readers who dislike the formal steps performed involving division by and summing over $\aleph([s])$ terms may take (10.6.26) as a definition.

However, we are not yet finished because $Q\ $ only defines a quadratic form. See Definition 26.8.1. In our case we can take as the domain $D(Q\ )$ the finite linear span of the $l_{[s]}$. Our $Q\ $ is manifestly positive and sesqui linear. It remains to show that it is closable. The problem that one might encounter is the following: the Hilbert space $\mathcal{H}_{\text{diff}}$ has the orthonormal basis $T_{[s]} := l_{[s]}/\sqrt{\eta_{[s]}}$ and we would like to define an operator $\hat{\ }$ densely on $D(Q\ )$ by

$$\hat{\ }\,T_{[s_2]} := \sum_{[s_1]} Q\ (T_{[s_1]}, T_{[s]})\, T_{[s_1]} \qquad (10.6.27)$$

However, the right-hand side should be an element of $\mathcal{H}_{\text{diff}}$, that is

$$\|\hat{\ }\,T_{[s]}\|^2 := \sum_{[s_1]} |Q\ (T_{[s_1]}, T_{[s_2]})|^2 < \infty \qquad (10.6.28)$$

Hence there is a convergence issue to be resolved.

Theorem 10.6.1

••• • •• •• ••••• •••• •••••••• •• •• Q •••••••• •• • •••••• ••• •• ••••• • • •• ••• •••••• •• ••••• •• •• •••••••• $\hat{\ }$ •• $\mathcal{H}_{\text{diff}}$•
•••• • •••• •• ••• • ••••• •••• •• ••• ••• ••• •••• • ••••••• •• $\hat{\ }$•

• ••••

(i) Since, given $[s_2]$ the 'matrix element' $Q\ (T_{[s_1]}, T_{[s_2]})$ is finite for every $[s_1]$ in order to prove convergence of (10.6.28) it will be sufficient to show that $Q\ (T_{[s_1]}, T_{[s_2]}) \neq 0$ for at most a finite number of $[s_1]$ only.
1. Let us fix $[s_1], [s_2]$ and consider the term corresponding to $[s]$ in (10.6.26). In order that it does not vanish, the expression

$$\sum_{v\in V(\gamma(s_0[s]))} \overline{T_{[s_1]}\big(\hat{C}_v^\dagger T_{s_0([s])}\big)}\, T_{[s_2]}\big(\hat{C}_v^\dagger T_{s_0([s])}\big) \qquad (10.6.29)$$

must be non-zero. Hence the spin-network decomposition of $\hat{C}_v^\dagger T_{s_0([s])}$ must contain a term diffeomorphic to T_{s_1} and a term diffeomorphic to T_{s_2} for at least one $v \in V(\gamma(s_0([s])))$. Let us estimate the number of $[s]$ for which this is possible. The action of $\hat{C}_v^\dagger$ on $T_{s_0([s])}$ consists of two terms.

The first term adds an arc in between any possible pair of edges with two possible orientations and changes the spin of the two corresponding adjacent segments by $\pm 1/2$. Therefore it adds two more vertices. Working at the gauge-variant level (there are more gauge-variant SNWFs than invariant ones) this also changes the magnetic quantum numbers at the endpoints of all three edges by $\pm 1/2$, which results in an additional factor of 4^3 at most. Hence per vertex of valence $n(v)$ we get this way no more than $4 \cdot 2 \cdot 4^3 n(v)(n(v) - 1)/2 = 4^4 n(v)(n(v) - 1)$ new spin-network states from the first term.

The second term is the square of the first term as far as the counting of new states is concerned. Hence we get $4^8 n(v)^2 (n(v) - 1)^2$ new spin network states from the second term depending on two more arcs and four more vertices.

Now in order that any of those is diffeomorphic to T_{s_1} the graph $\gamma(s_0([s]))$ must have one or two edges less than $\gamma(s_1)$ and two or four vertices less than $\gamma(s_1)$. Moreover, the spins of the segments of edges adjacent to the arcs must differ by $\pm 1/2$ and the magnetic quantum numbers of arcs and edges must differ by $\pm 1/2$. We conclude that if N_1 is the maximal valence of a vertex of $\gamma(s_1)$ then the number of $[s]$ that can contribute is bounded by $4^8 N_1^4 |V(\gamma(s_1))|$ which depends only on $[s_1]$. The same applies to s_2 of course. The actually contributing number of $[s]$ is certainly smaller than the maximum of $4^8 N_1^4 |V(\gamma(s_1))|, 4^8 N_2^4 |V(\gamma(s_2))|$.

2. Let us now fix $[s_2]$ and let $[s_1]$ run. There are only $4^8 N_2^4 |V(\gamma(s_2))|$ classes $[s]$ which can contribute no matter which $[s_1]$ we choose. By a similar argument, for each of those $[s]$ the number of $[s_1]$ which lead to a non-vanishing contribution is bounded by $4^8 N^4 |V(\gamma(s_0([s])))| + 2$ where N is the maximal vertex valence of $\gamma(s_0([s]))$. Since $N = N_2$ and $|V(\gamma(s_0([s])))| \leq |V(\gamma(s_2))|$ we conclude that $Q\ (T_{[s_1]}, T_{[s_2]})$ is non-vanishing for at most $4^{16} N_2^8 |V(\gamma(s_0([s_2])))|^2$ of the classes $[s_1]$.

We thus have shown that there is a positive symmetric operator with dense domain $\mathcal{D}_{\text{diff}}$, the finite linear span of the $T_{[s]}$, defined by (10.6.28) whose quadratic form coincides with $Q\ $ on the form domain $D(Q\ ) = \mathcal{D}_{\text{diff}}$. Hence by Theorem 26.8.2 (iii) $Q\ $ has a positive closure and induces a unique self-adjoint (Friedrich) extension of $\hat{\ }\ $ by Theorem 26.8.2 which we denote by $\ $ as well.

(ii) Notice that the construction of the solutions of $\hat{H}'(N)l = 0$ for all N (which produces zero $\cdots\cdots\cdots$, i.e., normalisable elements of $\mathcal{H}_{\mathrm{diff}}$) which we sketched in Section 10.5 can be directly transcribed to the construction of solutions to $\hat{}\, l = 0$. Namely, $\hat{}\, l = 0$ implies $Q\,(l,l) = 0$ which in turn enforces $l(\hat{C}_v^\dagger T_{s_0([s])}) = 0$ for all $[s]$ and all $v \in V(\gamma(s_0([s])))$. This is equivalent to $l(\hat{C}^\dagger(N)T_s) = 0$ for all s and all N where $\hat{C}^\dagger(N)$ is defined identically as $\hat{H}^\dagger(N)$ just that one of the volume operators is replaced by two times its square root. Thus, in particular $T_{[s]}$ where s has no extraordinary edges are normalisable solutions. $\qquad\square$

Hence the Master Constraint operator has a kernel as rich as the Hamiltonian constraint. Moreover, it gives us additional flexibility in the following sense: in order to have a consistent constraint algebra the action of the Hamiltonian constraint had to be trivial at the vertices that it creates itself. However, the Master Constraint does not have to satisfy any non-trivial constraint algebra, hence this restriction can be relaxed which is probably welcome to those [582] who believe that the action of the operator is too local. Whether such modifications lead to a sufficiently large semiclassical sector is, of course, not clear a priori and is subject to a detailed semiclassical analysis in step V.

10.6.3 Physical inner product and Dirac observables

Given the self-adjoint Master Constraint operator $\hat{}$ on $\mathcal{H}_{\mathrm{diff}}$ of the previous section one would now like to use the machinery of the direct integral decomposition of Chapter 30 in order to define the physical Hilbert space. However, there is one additional obstacle: while the spectral theorem holds also in non-separable Hilbert spaces, the direct integral decomposition can be performed only in the separable case for the reasons spelt out in Chapter 30. However, $\mathcal{H}_{\mathrm{diff}}$ is not separable unless, possibly, if we admit piecewise analytic diffeomorphisms, that is, bijections which are almost everywhere analytic but only homeomorphisms on certain lower-dimensional submanifolds, which remove the continuous moduli for vertices of valence five or higher. Now using homeomorphisms is forbidden because we must use the volume operator [427] rather than [425], which depends on a $C^{(1)}$ structure and which is absolutely crucial in order that $\hat{}$ or $\hat{H}'(N)$ be even densely defined. Thus, the direct integral method seems not to be applicable.

Fortunately, $\mathcal{H}_{\mathrm{diff}}$ can be decomposed as an uncountably infinite, almost direct, sum of separable, invariant Hilbert spaces as follows.

Definition 10.6.2. $\cdots\cdots\cdots\cdots\cdots\cdots$ γ_1, γ_2 $\cdots$
$\theta\cdots\cdots\cdots\cdots\cdots\cdots\cdots\cdots\cdots\cdots\cdots\cdots\cdots\cdots$
$\cdots\cdots\cdots\cdots\cdots\cdots$ b $\cdots$ σ $\cdots\cdots$

$\bullet\bullet$ $b(\gamma_1) = \gamma_2 \bullet$

$\bullet\bullet$ γ_1, γ_2 $\cdots$

$\cdots$

$\bullet\bullet\bullet$ $v \in V(\gamma_2)$ $\cdots$ $\cdots$ $\cdots$ $\cdots\bullet\bullet$ $e_1, e_2, e_3 \in E(\gamma_1)$ $\bullet\bullet$ $\bullet\bullet\bullet\bullet\bullet\bullet$ $\bullet\bullet$ $\bullet\bullet\bullet$

$\epsilon(e_1, e_2, e_3) = \epsilon(b(e_1), b(e_2), b(e_3)) \bullet$

The definition almost introduces piecewise analytic homeomorphisms but not quite because the graphs remain semianalytic (i.e., edges remain at least $C^{(1)}$). Also we do not know whether the inverse of a piecewise analytic homeomorphism is still piecewise semianalytic, that is, we do not know whether they form a group. However, definition (10.6.2) retains some tangential space structure in terms of the degeneracy type which is enough for the volume operator [426] to remain consistent. In fact we could have equivalently defined θ-equivalence by asking that for each triple of edges e_1, e_2, e_3 in $E(\gamma_1)$ there exists a semianalytic diffeomorphism φ_{e_1, e_2, e_3} such that $\varphi_{e_1, e_2, e_3}(e_I) \in E(\gamma_2)$ is consistently defined as we vary the triples, such that these maps altogether define a bijection $E(\gamma_1) \to E(\gamma_2)$ and such that $\epsilon(e_1, e_2, e_3) = \epsilon(\varphi_{e_1, e_2, e_3}(e_1), \varphi_{e_1, e_2, e_3}(e_2), \varphi_{e_1, e_2, e_3}(e_3))$.

Denote by $[\Gamma]$ the set of semianalytic diffeomorphism equivalence classes $[\gamma]$ of graphs $\gamma \in \Gamma$ and by (Γ) the set of θ-equivalence classes (γ) of graphs. Given (γ), let $\Theta'_{(\gamma)}$ be the set of moduli that are necessary to specify all the $[\gamma']$ with $(\gamma') = (\gamma)$. Hence any element $[\gamma] \in [\Gamma]$ is now uniquely specified by a pair $((\gamma), \theta') \in (\Gamma) \times \Theta'_{(\gamma)}$. Let

$$\Theta' := \times_{(\gamma) \in (\Gamma)} \Theta'_{(\gamma)} \ni \theta' = \{\theta'_{(\gamma)}\}_{(\gamma) \in (\Gamma)} \tag{10.6.30}$$

Then the direct sum of Hilbert spaces

$$\mathcal{H}_{\text{diff}} = \oplus_{[\gamma] \in [\Gamma]} \mathcal{H}_{\text{diff}}^{[\gamma]} \tag{10.6.31}$$

where $\mathcal{H}_{\text{diff}}^{[\gamma]}$ is the closure of the finite linear span of $T_{[s]}$ with non-trivial representations on all edges can be decomposed also as

$$\mathcal{H}_{\text{diff}} = \oplus_{(\gamma) \in (\Gamma)} \oplus_{\theta'_{(\gamma)} \in \Theta'_{(\gamma)}} \mathcal{H}_{\text{diff}}^{((\gamma), \theta'_{(\gamma)})} \tag{10.6.32}$$

However, it cannot be written as

$$\mathcal{H}_{\text{diff}} \neq \oplus_{\theta' \in \Theta'} \oplus_{(\gamma) \in (\Gamma)} \mathcal{H}_{\text{diff}}^{((\gamma), \theta'_{(\gamma)})} =: \oplus_{\theta' \in \Theta'} \mathcal{H}_{\text{diff}}^{\prime \theta'} \tag{10.6.33}$$

This is because two points $\theta'_1 \neq \theta'_2$ in Θ' are different whenever there is at least one entry $\theta'_{1,(\gamma)} \neq \theta'_{2,(\gamma)}$.

Thus the Hilbert spaces $\mathcal{H}_{\text{diff}}^{\prime \theta'}$ corresponding to a fixed choice $\theta' \in \Theta'$ are not mutually orthogonal. However, they are naturally isomorphic under the unitary operator that maps the basis of spin knot work functions over $((\gamma), \theta'_{1,(\gamma)})$ into the basis over $((\gamma), \theta'_{1,(\gamma)})$. Moreover, the Hilbert spaces $\mathcal{H}_{\text{diff}}^{\prime \theta'}$ are separable. Separability follows from the fact that at fixed θ' a spin-network label is completely specified by (1) the number of vertices and their connectivities, (2) the braiding and orientation of the corresponding edges, (3) the degeneracy type and (4) the spin and intertwining quantum numbers. Each of the four label sets is countable, hence it has at most cardinality $\mathbb{N}^4$, which is countable.

Unfortunately, the spaces $\mathcal{H}'^{\theta'}_{\text{diff}}$ are generically not left-invariant by $\hat{}$: this follows from the fact that $\hat{}$ can have non-vanishing matrix elements between $T_{[s]}, T_{[s']}$ where $\gamma(s), \gamma(s')$ differ by an arc and possibly in addition by one or two edges which are the beginning segments of edges within $\gamma(s)$ that connect the vertex in question to the arc in $\gamma(s')$ (this happens when one or two of the corresponding edges carries spin $1/2$). The attachment of the arc creates a new trivalent or bivalent vertex and thus does not require moduli. However, the moduli created by the annihilation of one or two edges at the given vertex in $\gamma(s')$ may differ from the value $\theta'_{(\gamma(s'))}$ which is assigned to the entry with label $(\gamma(s'))$ in θ', which labels $\mathcal{H}'^{\theta'}$.

Hence, if we do not want to identify all the $\mathcal{H}'^{\theta'}$, which is not what the formalism forces us to do, then in order to select an invariant element $\mathcal{H}'^{\theta'}$ we must proceed differently. We can combine the θ-moduli classification with the classification by sources $\mathcal{S}_0$ and derived spin nets $\mathcal{S}_n(s_0)$ of level n developed in Section 10.5 as follows.

Denote by $[\mathcal{S}_0]$ the set of diffeomorphism equivalence classes of sources. For any two representatives $s_1([s_0]), s_2([s_0]) \in \mathcal{S}_0$ the set of diffeomorphism equivalence classes of the members of $\mathcal{S}_n(s_1([s_0])), \mathcal{S}_n(s_2([s_0]))$ coincide, that is, they depend only on $[s_0]$. We will denote this set therefore by $[\mathcal{S}_n]([s_0])$. We notice that the moduli parameters of all the $[s] \in [\mathcal{S}_n]([s_0]), n = 0, 1 \ldots$ are completely determined by those of $[s_0]$. The completion of the finite linear span of these $T_{[s]}$ will be denoted $\mathcal{H}^{[s_0]}_{\text{diff}}$ and this Hilbert space is separable by construction. Now the following issue arises: the action of $\hat{}$ consists in adding and removing arcs to a graph and sometimes it reduces the valence of a vertex by one or two units. It therefore happens that given $[s_0] \neq [s'_0]$ with $(s_0) = (s'_0)$ the set $[\mathcal{S}_n]([s_0]) \cap [\mathcal{S}_n]([s'_0])$ is not empty. For instance, a five-valent vertex, which has moduli, could be turned into a three-valent one which does not have moduli. Hence it is almost but not quite true that $\mathcal{H}_{\text{diff}}$ is the uncountable direct sum of the $\mathcal{H}^{[s_0]}_{\text{diff}}, [s_0] \in [\mathcal{S}_0]$.

Let us write $[s_0] = ((s_0), \theta_{(s_0)} := \theta_{(\gamma(s_0))})$ where (s_0) is the θ-equivalence class of s_0 which is determined by the $(\gamma(s_0))$. Let $(\mathcal{S}_0)$ be the set of those (s_0) and let Θ' be the collection of the $\theta_{(s_0)}, (s_0) \in (\mathcal{S}_0)$. Then

$$\mathcal{H}_{\text{diff}} = \cup_{\theta \in \Theta} \mathcal{H}^{\theta}_{\text{diff}} := \cup_{(s_0) \in (\mathcal{S}_0)} \mathcal{H}^{((s_0), \theta_{(s_0)})}_{\text{diff}} \tag{10.6.34}$$

Notice that the unions are almost direct sums but not quite as just pointed out. However, each of the spaces $\mathcal{H}^{\theta}_{\text{diff}}$ is a separable and $\hat{}$-invariant subspace of $\mathcal{H}_{\text{diff}}$ and all of them are mutually isomorphic. Moreover, each of them contains information about all θ-equivalence classes of spin-network states and therefore all the physically relevant information.

Thus, while these are not sectors in the strict sense, we may just pick one of these subspaces and apply the direct integral decomposition method to it.

Theorem 10.6.3. •••••• •• • •• ••••• ••••••• V •••• •••• $V\mathcal{H}^\theta_{\text{diff}}$ •• ••• ••••••
•• ••••••• • •••••• •• •••

$$\mathcal{H}^\theta_{\text{diff}} \propto \int_{\mathbb{R}^+}^{\oplus} d\mu(\lambda) \, \mathcal{H}^\theta_{\text{diff}}(\lambda) \qquad (10.6.35)$$

•• •••• ••• • •• •••• •• ••• •• •• μ ••• ••• • •••••• •••••• $\mathcal{H}^\theta_{\text{diff}}(\lambda)$• •• • ••••• $V \,\widehat{\,}\, V^{-1}$
•••• •• • • •••• •••• ••• • •• λ• ••• • •• ••• •• •••••• •• •••
••• •••••••••• • •••••• ••••• •• ••••• ••[16] $\mathcal{H}^\theta_{\text{phys}} = \mathcal{H}^\theta_{\text{diff}}(0)$•

Thus, we see that we can complete the fourth step of the quantisation programme by supplying a physical inner product and a separable physical Hilbert space. Notice that there is an explicit construction behind Theorem 10.6.3 as outlined in Chapter 30, however, it is rather involved. Therefore, while Theorem 6.2.26 gives us existence and, possibly uniqueness, of $\mathcal{H}^\theta_{\text{phys}}$ once $< ., . >_{\text{diff}}$, θ, $\widehat{\,}$ have been specified, to be practically useful, approximation methods must be developed.

Dirac observables could now be constructed from spatially diffeomorphism-invariant operators which preserve any $\mathcal{H}^\theta_{\text{diff}}$, for example, by using the ergodic projection technique or the partial observable Ansatz of Chapter 30. Any spatially diffeomorphism-invariant operator regularised in the same fashion as the Hamiltonian constraint operator has the property of preserving each of the subspaces $\mathcal{H}^\theta_{\text{diff}}$ separately, hence this is no restriction.

10.6.4 Extended Master Constraint

We sketch here another possibility to implement the Master Constraint Programme [252, 589–591]: this is based on the fact that we may consider the
•••••••• • ••••• • •• •••• •• ••

$$\mathbf{E} = \int_\sigma d^3x \, \frac{C^2 + q^{ab} C_a C_b}{\sqrt{\det(q)}}$$

$$\mathbf{E}\mathbf{E} = \int_\sigma d^3x \, \frac{C^2 + q^{ab} C_a C_b + C_j C_j}{\sqrt{\det(q)}} \qquad (10.6.36)$$

where C, C_a, C_j denote Hamiltonian, spatial diffeomorphism and Gauß constraint respectively. Both constraints are spatially diffeomorphism-invariant.

---

[16] Actually the spaces $\mathcal{H}^\theta_{\text{diff}}(\lambda)$ are defined up to measure μ zero sets. See Chapter 30 for physical criteria to choose an appropriate candidate.

However, M_E allows us to implement both the Hamiltonian and the spatial diffeomorphism constraint on $\mathcal{H}_{\text{kin}}$ (and M_{EE} also the Gauß constraint in addition) provided we implement the corresponding operators in a non-graph-changing fashion. Both operators will implement spatial diffeomorphism-invariance in a completely different way than we have done in the main text because the operator corresponding to $q^{ab}C_a C_b / \sqrt{\det(q)}$, which exists as we will show in Section 12.3.4, is the 'weighted square' of the infinitesimal generator of spatial diffeomorphisms which does not exist for the representation $\varphi \mapsto U(\varphi)$ on $\mathcal{H}_{\text{kin}}$. This is possible because the weight is phase space-dependent and thus effectively turns the operator into something less singular than the square of the generators. Hence, the space $\mathcal{H}_{\text{diff}}$ will not be in the kernel of this weighted integral of squared spatial diffeomorphism constraint operators and thus we are actually forced to implement them on $\mathcal{H}_{\text{kin}}$. Incidentally, since the spatial diffeomorphism group is not implemented by its pull-back action as in Chapter 7, the uniqueness theorem of Chapter 8 is not available and one may look for new representations.

For the sake of this section, let us stick with the representation $\mathcal{H}_{\text{kin}}$ derived in the main text. In order to define the corresponding operator $\hat{M}_E$ in a non-graph-changing way we may proceed as in [252] where a diffeomorphism-invariant rule is prescribed for how to select a loop $\alpha_{\gamma,v,e,e'}$. Basically one calls a loop within γ with endpoint $v \in V(\gamma)$, starting and ending along $e \in E(\gamma)$ and $(e')^{-1}$ respectively, minimal if there is no other loop with the same properties and fewer edges of γ traversed. If there is more than one minimal loop then one averages over them. The operator $\hat{M}_E$ is similarly defined as $\hat{M}$, just that $\alpha_{\gamma,v,e,e'}$ is chosen as explained and no limits of triangulations are to be taken. This way one is able to define $\hat{M}_E$ on all graphs but only on sufficiently fine ones does one recover the semiclassical limit because only then are the loops attached sufficiently small.

As we have shown in previous sections, a non-graph-changing Hamiltonian constraint that effectively enters the construction of the extended Master Constraint is anomalous. This will be detected by the spectrum of the $\hat{M}_E$ which will presumably not include zero. As we have explained in Chapter 30, this can be dealt with by subtracting the minimum of the spectrum from the Master Constraint (assuming it to be finite and proportional to $\hbar$ such that the modified operator has the same classical limit as the original one) and thus poses no problem for the Master Constraint Programme. For instance, in lattice quantum gravity (see, e.g., [112] for a review) an obstacle is usually the spatial diffeomorphism group because the discrete generators are anomalous and the representation $\varphi \mapsto U(\varphi)$ does not preserve the lattice unless φ is a symmetry of the lattice. This is no obstacle any longer when using M_E instead since there is no non-trivial algebra to be checked. While that is true, it would still be desirable to work with non-anomalous constraints. A possibility for doing that is the concept of perfect actions [593], which is based on renormalisation group ideas and basically consists in replacing naive next-neighbour discretisations by more complicated ones with

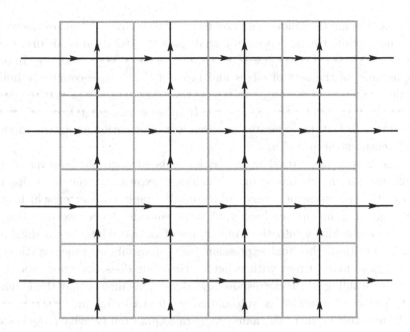

Figure 10.4 Countable subset of D families of congruences of curves and faces respectively for $D = 2$.

improved continuum limit properties. For instance, it is possible to construct a perfect, discretised Laplace operator which nevertheless has the spectrum of the continuum operator [594, 595].

By the methods of [589–591] one can indeed show that the extended Master Constraint of LQG has the correct classical limit at least when σ is compact and the graph-dependent semiclassical states chosen are sufficiently fine. This graph dependence of the semiclassical states is one of the motivations for a reformulation of LQG, called Algebraic Quantum Gravity (AQG), to which we turn now.

10.6.5 Algebraic Quantum Gravity (AQG)

In [589–591] the following observation was made: to consider all graphs γ and all surfaces S is a vast overcoordinatisation of the space of connections and electric fields respectively. It would be sufficient to consider three linearly independent congruences of curves and foliations of σ respectively and just to consider paths within the elements of the congruence and surfaces within the leaves of the foliations respectively. From these one can extract all possible holonomies and fluxes by limiting procedures. The set of paths and surfaces within those structures is still an uncountably infinite set. Let us consider a countable subset. This consists of a graph γ of cubic topology and a dual cell complex γ^* consisting of squares as faces. See Figure 10.4 for the situation in $D = 2$. Let us define $\hat{\;}_E$ by discretising it using the holonomies and fluxes allowed by γ, γ^* respectively. The point is now, while this is completely analogous to lattice gauge theory, •••• •• •••••••••• ••••••••••••• ••• •• ••••• ••• ••• ••• • •• ••• ••••• ••••• $\widehat{}$ $_E$. In other words, as long as the sets of edges and faces of γ, γ^* have countably infinite cardinality, ••• •••••• ••••••• ••• ••• •••••• •••••• ••• • •••••• •• •• ••••• ••••• •• •••••• •• • ••••• •• •• •••••• •• •••••••• In other words: ••• ••• ••• ••• ••• •• •• •••••• ••••• . In fact, the operator has even lost information about the topology and differential structure of σ.

The graph that we started from was an embedded graph, however, due to the diffeomorphism invariance of the classical expression, the final operator looks exactly the same on every cubic (dual) graph no matter which σ we started from and no matter how γ, γ^* were embedded because any two diffeomorphic embeddings into the same σ are of course related by a diffeomorphism. In addition, the final expression just knows about which vertices are connected how many times with other vertices but does not know about the knotting or braiding of the corresponding edges. This means that the extended Master Constraint operator is automatically lifted to such an •••••••• ••••• [527–529] and this is where the name Algebraic Quantum Gravity (AQG) stems from.

The proposal is then just to work with the countably infinite γ, γ^* indicated. Notice that since there is precisely one face of γ^* dual to a given edge of γ, eventually all fluxes are just labelled by the edges which the corresponding face intersects, so that the final algebra of holonomies and fluxes just depends on the algebraic abstraction of γ which we denote by α. All operators are labelled by α. There is just this one algebraic graph and therefore all graph dependence disappears. The algebraic graph acquires the meaning of a fundamental structure. The extended Master Constraint and all other operators live on α, which is why cylindrical consistency is automatically satisfied, there are no subgraphs to consider.

It is tempting to think that at this algebraic level spatial diffeomorphism invariance is automatically taken care of and that one should therefore work with the unextended Master Constraint, that is, without imposing spatial diffeomorphism invariance, but that was shown to be wrong in [592]. It would reduce an insufficient number of degrees of freedom only.

The information about the topology and differential structure of σ as well as the embedding of the algebraic graph and its dual is recovered in the semiclassical limit. The corresponding semiclassical states are similar to the cutoff (or graph-dependent) states as discussed in the next chapter. However, there is a crucial difference: in LQG these states are labelled by an embedded graph and they are linear combinations of spin-network functions over the embedded graphs with judiciously chosen coefficients. In AQG the states are also labelled by an embedded graph, however, they are linear combinations of spin-network states when embedded into compact[17] σ over the algebraic graph with the same coefficients. This means that the AQG semiclassical states always control the fluctuations of all present degrees of freedom because all of them live on the algebraic graph. In contrast, in LQG there are, for any given graph, zillions of degrees of freedom on other graphs whose fluctuations the given semiclassical state cannot control. This is what makes the AQG semiclassical analysis superior over that in LQG.

Precisely due to this fact it is possible to show that $\hat{}_{\mathrm{E}}$ does have the correct semiclassical limit [589–591]. Thus, the extended Master Constraint proposal together with the AQG framework could be an interesting alternative to the LQG Master Constraint Programme derived above, which is very close in spirit to lattice gauge theory and in a sense conceptually and technically much simpler.

As far as the computation of the physical inner product is concerned, the same construction (the direct integral decomposition) applies to both , $_{\mathrm{E}}$. The difference is, of course, that is defined on $\mathcal{H}_{\mathrm{diff}}$ while $_{\mathrm{E}}$ is defined on a space of the form $\mathcal{H}_{\gamma}^{0}$, however, γ is an infinite graph. Such infinite graph Hilbert spaces are not contained in the kinematical Hilbert space and we must pass to the infinite tensor product extension discussed in Section 11.2.6. Both spaces are not separable and we have dealt with the associated problems on the level of $\mathcal{H}_{\mathrm{diff}}$ in Section 10.6.3. As far as the infinite tensor product is concerned it turns out that $\mathcal{H}_{\gamma}^{0}$ decomposes into an uncountably infinite direct sum of separable spaces which are left-invariant by $\hat{}_{\mathrm{E}}$.[18] These separable subspaces are closures of finite linear combinations of states which can be obtained by exciting a given infinite tensor product state in finitely many factors. The operator $\hat{}_{\mathrm{E}}$ is a countable sum of operators which affect an infinite tensor product state in finitely many factors only, hence if the given state is in the domain of the operator at all then it preserves the given sector. Thus non-separability is no obstacle in defining the physical inner product by direct integral decomposition in AQG.

It is an interesting speculation that some of these separable subspaces of the infinite tensor product correspond to Fock-like spaces and thus maybe to certain QFTs on given background spacetimes because also Fock states are closed linear spans of finite excitations of a given vacuum vector. See especially the introduction of [589] for the conceptual framework of AQG and all the details which we do not want to incorporate here because the proposal is yet rather unexplored.

---

[17] If σ is spatially compact, this means that the embedded infinite graphs γ, γ^{*} have accumulation points of edges and faces. This does not lead to problems because we can leave the degrees of freedom associated with all but finitely many edges of the algebraic graph unexcited in the semiclassical state under consideration.

[18] All sectors on which $\hat{}_{\mathrm{E}}$ is not densely defined are dropped from the Hilbert space. As shown in [589–591], $_{\mathrm{E}}$ is densely defined on the semiclassical sectors.

10.7 $^+$ Further related results

We list here further results that are directly connected to the issues that we have touched upon in this section, which however can safely be skipped on a first reading.

10.7.1 The Wick transform

This section describes an idea for how to write the theory in complex variables, thus simplifying the Hamiltonian constraint, while being able to use the Hilbert space machinery developed in earlier chapters.

The Bargmann–Segal transform for quantum gravity discussed in [321] gives a rigorous construction of quantum kinematics on a space of complexified, distributional connections by means of key results obtained by Hall [318–320]. Since the transform depended on a background structure, it was clear that the associated scalar product did not implement the correct reality conditions. To fix this was the purpose of [315], where a general theory was developed of how to trivialise reality conditions while keeping the algebraic structure of a functional as simple as when complex variables are being used. The same idea proves very useful in order to obtain a very general class of coherent states, as we will see in Chapter 11. Moreover, as a side result, it is possible to improve the coherent state transform as defined by Hall in the following sense.

Notice that the prescription given by Hall turns out to establish indeed a unitary transformation but that it was 'pulled out of the hat', that is, it was guessed by an analogy consideration with the transform on $\mathbb{R}^n$ and turned out to work. It would be much more satisfactory to have a derivation of the transform $\hat{U}_t$ and the measure ν_t on the complexified configuration space from first principles, that is, one should be able to compute them just from the knowledge of the two polarisations of the phase space. We will first describe the general scheme in formal terms and then apply it to quantum gravity, following closely [315].

•• • •• •• •• • ••• ••• •••••• •

Consider an arbitrary phase space $\mathcal{M}$ of cotangential bundle type, finite or infinite, with local real canonical coordinates (p, q) where q is a configuration variable and p its conjugate momentum (we suppress all discrete and continuous indices in this subsection). Furthermore, we have a Hamiltonian (constraint) $H'(p, q)$ which unfortunately looks rather complicated in the variables p, q (the reason for the prime will become evident in a moment). Suppose that, however, we are able to perform a •••••• ••• transformation on $\mathcal{M}$ which leads to the complex canonical pair $(p_{\mathbb{C}}, q_{\mathbb{C}})$ such that the Hamiltonian becomes algebraically simple (e.g., a polynomial $H_{\mathbb{C}}$ in terms of $p_{\mathbb{C}}, q_{\mathbb{C}}$). That is, we have a complex symplectomorphism $(p_{\mathbb{C}}, q_{\mathbb{C}}) := W^{-1}(p, q)$ such that $H_{\mathbb{C}} = H' \circ W$ is algebraically simple. Notice that we are not complexifying the phase space, we just happen to find it convenient to coordinatise it by complex-valued coordinates. The reality conditions on $p_{\mathbb{C}}, q_{\mathbb{C}}$ are encoded in the map W.

We now wish to quantise the system. We choose two Hilbert spaces, the first one, $\mathcal{H}$, for which the q's become a maximal set of mutually commuting, diagonal operators and a second one, $\mathcal{H}_{\mathbb{C}}$, for which the $q_{\mathbb{C}}$'s become a maximal set of mutually commuting, diagonal operators. According to the canonical commutation relations we represent $\hat{p}, \hat{q}$ on $\psi \in \mathcal{H}$ by $(\hat{p}\psi)(x) = i\hbar\partial\psi(x)/\partial x$ and $(\hat{q}\psi)(x) = x\psi(x)$. Likewise, we represent $\hat{p}_{\mathbb{C}}, \hat{q}_{\mathbb{C}}$ on $\psi_{\mathbb{C}} \in \mathcal{H}_{\mathbb{C}}$ by $(\hat{p}_{\mathbb{C}}\psi_{\mathbb{C}})(z) = i\hbar\partial\psi_{\mathbb{C}}(z)/\partial z$ and $(\hat{q}_{\mathbb{C}}\psi)(x) = z\psi_{\mathbb{C}}(z)$. The fact that p, q are real-valued forces us to set $\mathcal{H} := L_2(\mathcal{C}, d\mu_0)$ where $\mathcal{C}$ is the quantum configuration space and μ_0 is the uniform (translation-invariant) measure on $\mathcal{C}$ in order that $\hat{p}$ be self-adjoint.

In order to see what the Hilbert space $\mathcal{H}_{\mathbb{C}}$ should be, we also represent the operators $\hat{p}_{\mathbb{C}}, \hat{q}_{\mathbb{C}}$ on $\mathcal{H}$ by choosing a particular ordering of the function W^{-1} and substituting p, q by $\hat{p}, \hat{q}$. In order to avoid confusion, we will write them as $(\hat{p}', \hat{q}') := W^{-1}(\hat{p}, \hat{q})$ where the prime means that the operators are defined on $\mathcal{H}$ but are also quantisations of the classical functions $p_{\mathbb{C}}, q_{\mathbb{C}}$. Now, the point is that the operators $\hat{p}', \hat{q}'$, possibly up to $\hbar$ corrections, automatically satisfy the correct adjointness relations on $\mathcal{H}$ declining from the reality conditions on $p_{\mathbb{C}}, q_{\mathbb{C}}$. This follows simply by expanding the function W^{-1} in terms of $\hat{p}, \hat{q}$, computing the adjoint and defining the result to be the quantisation of $\bar{p}_{\mathbb{C}}, \bar{q}_{\mathbb{C}}$ on $\mathcal{H}$ which equals any valid quantisation prescription up to $\hbar$ corrections. Thus, if we could find a unitary operator $\hat{U} : \mathcal{H} \to \mathcal{H}_{\mathbb{C}}$ such that

$$\hat{p}_{\mathbb{C}} = \hat{U}\hat{p}'\hat{U}^{-1} \text{ and } \hat{q}_{\mathbb{C}} = \hat{U}\hat{q}'\hat{U}^{-1} \qquad (10.7.1)$$

then we have automatically implemented the reality conditions on $\mathcal{H}_{\mathbb{C}}$ as well because by unitarity

$$(\hat{p}_{\mathbb{C}})^\dagger = \hat{U}(\hat{p}')^\dagger\hat{U}^{-1} \text{ and } (\hat{q}_{\mathbb{C}})^\dagger = \hat{U}(\hat{q}')^\dagger\hat{U}^{-1} \qquad (10.7.2)$$

where the $\dagger$ operations in (10.7.2) on the left- and right-hand side respectively are to be understood in terms of $\mathcal{H}_{\mathbb{C}}$ and $\mathcal{H}$ respectively. In other words, the adjoint of the operator on $\mathcal{H}_{\mathbb{C}}$ is the image of the correct adjoint of the operator on $\mathcal{H}$.

To see what $\hat{U}$ must be, let $\hat{K} : \mathcal{H} \cap \text{Ana}(\mathcal{C}) \to \mathcal{H}_{\mathbb{C}}$ be the operator of analytical extension of real analytical elements of $\mathcal{H}$ and likewise $\hat{K}^{-1}$ the operator that restricts the elements of $\mathcal{H}_{\mathbb{C}}$ (all of which are holomorphic) to real values. We then have the identities

$$\hat{p}_{\mathbb{C}} = \hat{K}\hat{p}\hat{K}^{-1} \text{ and } \hat{q}_{\mathbb{C}} = \hat{K}\hat{q}\hat{K}^{-1} \qquad (10.7.3)$$

We now exploit that W^{-1} was supposed to be a canonical transformation (an automorphism of the phase space that preserves the symplectic structure but not the reality structure). Let C be its infinitesimal generator, called the •• • •• •• •• ••, that is, for any function f on $\mathcal{M}$,

$$f(p_{\mathbb{C}}, q_{\mathbb{C}}) := f_{\mathbb{C}}(p, q) := ((W^{-1})^* f)(p, q) = \sum_{n=0}^{\infty} \frac{i^n}{n!} \{C, f\}_{(n)} \qquad (10.7.4)$$

where the multiple Poisson bracket is inductively defined by $\{C, f\}_{(0)} = f$ and $\{C, f\}_{(n+1)} = \{C, \{C, f\}_{(n)}\}$. Using the substitution rule that Poisson brackets become commutators times $1/(i\hbar)$ we can quantise (10.7.4) by

$$\hat{f}' := f_{\mathbb{C}}(\hat{p}, \hat{q}) := \sum_{n=0}^{\infty} \frac{1}{\hbar^n n!} [\hat{C}, \hat{f}]_{(n)} = (\hat{W}_t)^{-1} \hat{f} \hat{W}_t \qquad (10.7.5)$$

where we have defined the generalised 'heat kernel' operator

$$\hat{W}_t := e^{-t\hat{C}} \qquad (10.7.6)$$

and $t = 1/\hbar$. That is, the generator C motivates a natural ordering of $W^{-1}(p, q)$.

Substituting (10.7.6) into (10.7.4) we find

$$\hat{p}_{\mathbb{C}} = \hat{U}_t \hat{p}' \hat{U}_t^{-1} \text{ and } \hat{q}_{\mathbb{C}} = \hat{U}_t \hat{q}' \hat{U}_t^{-1} \qquad (10.7.7)$$

where we have defined the generalised coherent state or • ••• •• •• ••• • ••• •• ••• ••

$$\hat{U}_t := \hat{K} \hat{W}_t \qquad (10.7.8)$$

with $t = 1/\hbar$. The reason for the names we chose will become obvious in the next subsection.

It follows that if $\hat{C}, \hat{W}_t$ exist on real analytic functions and if we can then extend $\hat{U}_t$ to a unitary operator from $\mathcal{H}$ to $\mathcal{H}_{\mathbb{C}} := L_2(\mathcal{C}_{\mathbb{C}}, d\nu_t) \cap \text{Hol}(\mathcal{C}_{\mathbb{C}})$ where $\mathcal{C}_{\mathbb{C}}$ denotes the complexification of $\mathcal{C}$ then we have completed the programme.

Moreover, as a bonus we would have ••• • ••• •• ••• •••••••• •• ••••• of the operator that corresponds to the quantisation of H'.

First of all we define an unphysical Hamiltonian (constraint) operator $\hat{H}$ on $\mathcal{H}$ simply by choosing a suitable ordering of the function

$$H(p, q) := H_{\mathbb{C}}(p_{\mathbb{C}}, q_{\mathbb{C}})_{|p_{\mathbb{C}} \to p, q_{\mathbb{C}} \to q} = (K^{-1} \cdot H_{\mathbb{C}})(p, q) \qquad (10.7.9)$$

and substituting p, q by the operators $\hat{p}, \hat{q}$. Thus we obtain an operator $\hat{H}_{\mathbb{C}}$ on $\mathcal{H}_{\mathbb{C}}$ by $\hat{H}_{\mathbb{C}} := \hat{K} \hat{H} \hat{K}^{-1}$. It follows that if we ••••• the quantisation of the physical Hamiltonian (constraint) H' on $\mathcal{H}$ by $\hat{H}' := \hat{W}_t^{-1} \hat{H} \hat{W}_t$ then in fact $\hat{H}_{\mathbb{C}} = \hat{U}_t \hat{H}' \hat{U}_t^{-1}$ and since $\hat{U}_t$ is unitary the spectra of $\hat{H}'$ on $\mathcal{H}$ and of $\hat{H}_{\mathbb{C}}$ on $\mathcal{H}_{\mathbb{C}}$ ••• •• ••••. But since $\hat{H}_{\mathbb{C}}$ is an algebraically simple function of the elementary operators $\hat{p}_{\mathbb{C}}, \hat{q}_{\mathbb{C}}$ it follows that one has drastically simplified the spectral analysis of the complicated operator $\hat{H}'$! Finally, given a (generalised) eigenstate $\psi_{\mathbb{C}}$ of $\hat{H}_{\mathbb{C}}$, we obtain a (generalised) eigenstate $\psi := \hat{U}_t^{-1} \psi_{\mathbb{C}}$ of $\hat{H}'$ by the ••• •••• of the coherent state transform.

The crucial question then is whether we can actually make $\hat{U}_t$ unitary. In [315] the following formula for the unitarity implementing measure ν_t on $\mathcal{C}_{\mathbb{C}}$ was derived:

$$dv_t(z, \bar{z}) := \nu_t(z, \bar{z})d\mu_0^{\mathbb{C}}(z) \otimes d\bar{\mu}_0^{\mathbb{C}}(\bar{z})$$

$$\nu_t(z, \bar{z}) := (\hat{K}\overline{[[\hat{W}_t]^\dagger]\hat{K}^{-1}})^{-1}((\hat{K}\overline{[[\hat{W}_t]^\dagger]\hat{K}^{-1}}))^{-1}\delta(z, \bar{z}) \qquad (10.7.10)$$

The adjoint operation is meant in the sense of $\mathcal{H}$, $\hat{K}$ means analytical extension as before and the bar means complex conjugation of the expression of the operator (i.e., any appearance of multiplication or differentiation by z is replaced with multiplication or differentiation by $\bar{z}$ and vice versa, and, of course, also numerical coefficients are complex conjugated). Here $\mu_0^{\mathbb{C}}$ and $\bar{\mu}_0^{\mathbb{C}}$ are just the analytic and anti-analytic extensions of the measure μ_0 on $\mathcal{C}$ (they are just complex conjugates of each other thanks to the positivity of μ_0) and the distribution in the second line of (10.7.9) is defined by

$$\int_{\mathcal{C}_{\mathbb{C}}} d\mu_0^{\mathbb{C}}(z)d\bar{\mu}_0^{\mathbb{C}}(\bar{z})f(z, \bar{z})\delta(z, \bar{z}) = \int_{\mathcal{C}} d\mu_0(x)f(x, x) \qquad (10.7.11)$$

for any smooth function f on the complexified configuration space of rapid decrease with respect to μ_0.

Whenever (10.7.9) exists (it is straightforward to check that (10.7.9) does the job formally), the extension of $\hat{U}_t$ to a unitary operator (isometric, densely defined and surjective) in the sense above can be expected [315]. A concrete proof is model-dependent.

In summary, we have solved •• • •• ••••• • •• ••• ••••••: we have implemented the correct adjointness relations and we have simplified the Hamiltonian (constraint) operator.

A couple of remarks are in order:

- The method does not require that $\hat{C}$ is self-adjoint, positive, bounded or at least normal. All that is important is that $\hat{W}_t$ exists on real analytic functions in the sense of Nelson's analytic vector theorem, see [282 vol. 2].
- It reproduces the cases of the harmonic oscillator and the case considered by Hall [318]. But it also explains • •• it works the way it works, namely it answers the question of how to identify analytic continuation with a given complex polarisation of the phase space as is obvious from $\hat{K} = \hat{U}_t\hat{W}_t^{-1}$. The computation of ν_t via (10.7.9), (10.7.11) is considerably simpler. The harmonic oscillator corresponds to the complexifier $C = \frac{1}{2}p^2$.
- One might wonder why one should compute ν_t at all and bother with $\mathcal{H}_{\mathbb{C}}$ [317]? Could one not just forget about the analytic continuation and work only on $\mathcal{H}$ simply by studying the spectral analysis of the unphysical operator $\hat{H}$ and defining the physical operator by $\hat{H}' := \hat{W}_t^{-1}\hat{H}\hat{W}_t$? The problem is that, while it is true that restrictions to real arguments of (generalised) eigenvectors of $\hat{H}_{\mathbb{C}}$ are •••• •• eigenvectors of $\hat{H}$, these are typically not (generalised) eigenvectors in the sense of the topology of $\mathcal{H}$. Intuitively, what happens is that the measure ν_t provides for the necessary much stronger fall-off in order to turn the analytic

extension of the badly behaved formal eigenvectors $\hat{W}_t^{-1}\psi$ of $\hat{H}'$ into well-defined (generalised) eigenvectors $\hat{K}\psi$ of $\hat{H}_{\mathbb{C}}$. One can see this also from another point of view: by unitarity, whenever $\hat{H}_{\mathbb{C}}$ is self-adjoint, so is $\hat{H}'$ but in general $\hat{H}$ is not. Thus, one would not expect the spectra of $\hat{H}, \hat{H}'$ to coincide. See the appendix of [315] for a discussion of this point.

• There are also other applications of this transform, for example in Yang–Mills theory it can be used to turn the Hamiltonian from a fourth-order polynomial into a polynomial of order three only [315]!

This completes the outline of the general framework. We will now turn to the interesting case of quantum gravity.

•• •• •• • • ••• •••• ••• • ••• •• • •• • ••• • •••

As Barbero [310, 311] correctly pointed out, all the machinery that is associated with the quantum configuration space $\overline{\mathcal{A}}$ and the uniform measure μ_0 is actually also available for Lorentzian Quantum General Relativity if one chooses the Immirzi parameter β to be real. However, the Hamiltonian constraint then does not simplify at all compared with the ADM expression and so the virtue of the new variables would be lost. The coherent state transform as derived below in principle combines both advantages, namely a well-defined calculus on $\overline{\mathcal{A}}$ and a simple Wheeler–DeWitt constraint.

Let us then apply the framework of the previous section. The phase space of Lorentzian General Relativity can be given a real polarisation through the canonical pair $(A_a^j := \Gamma_a^j + K_a^j, E_j^a/\kappa)$ (the case considered by Barbero with $\beta = 1$) and a complex polarisation through the canonical pair $(({}^{\mathbb{C}}A_a^i) := \Gamma_a^j - iK_a^j, ({}^{\mathbb{C}}E_j^a) := iE_j^a/\kappa)$ (the case considered by Ashtekar). The rescaled Hamiltonian constraint looks very simple in the complex variables, namely

$$\tilde{H}_{\mathbb{C}}(A_{\mathbb{C}}, E_{\mathbb{C}}) = \epsilon_{ijk} ({}^{\mathbb{C}}F_{ab}^i)({}^{\mathbb{C}}E_j^a)({}^{\mathbb{C}}E_k^b) \tag{10.7.12}$$

but if we write $A_{\mathbb{C}}, E_{\mathbb{C}}$ in terms of A, E then the resulting Hamiltonian $\tilde{H}'(A, E)$ becomes extremely complicated. Let us compute the map W. We first of all see that we can go from (A, E) to $(A_{\mathbb{C}}, E_{\mathbb{C}})$ in a sequence of three canonical transformations given by

$$(A = \Gamma + K, E/\kappa) \rightarrow (K, E/\kappa) \rightarrow (-iK, iE/\kappa) \rightarrow (A_{\mathbb{C}} = \Gamma - iK, E_{\mathbb{C}} = iE/\kappa)$$

That the first and third step are indeed canonical transformations was already shown. The second step is a ••••• •••••• • ••• •• ••• •••• . Since (K, E) is a canonical pair it is trivial to see that we have

$$-iK = \sum_{n=0}^{\infty} \frac{i^n}{n!} \{C, K\}_{(n)} \text{ and } iE = \sum_{n=0}^{\infty} \frac{i^n}{n!} \{C, E\}_{(n)} \tag{10.7.13}$$

where the ••• •• •• •• •• or generator of the Wick transform is given by

$$C = -\frac{\pi}{2\kappa} \int_\sigma d^3x K^i_a E^a_i \qquad (10.7.14)$$

which is easily seen to be the integrated densitised trace of the extrinsic curvature. C generates infinitesimal constant scale transformations. It now seems that we need to compute the generator of the transform that adds and subtracts the spin-connection Γ. However, we have seen in Chapter 1 that the spin-connection in three dimensions is a homogeneous polynomial of degree zero in E and its derivatives, and since a •• •• •• •• scale factor is unaffected by derivatives we have $\{\Gamma, C\} = 0$. Thus in fact we have

$$A_{\mathbb{C}} = \sum_{n=0}^{\infty} \frac{i^n}{n!} \{C, A\}_{(n)} \text{ and } E_{\mathbb{C}} = \sum_{n=0}^{\infty} \frac{i^n}{n!} \{C, E\}_{(n)} \qquad (10.7.15)$$

The task left is to define the operator $\hat{C}$ and to compute the corresponding measure ν_t. This seems to be a very hard problem because $K^i_a = A^i_a - \Gamma^i_a$ and Γ^i_a is just a very complicated function to quantise. Nevertheless, it can be done as we have seen in this chapter.

We conclude this section with a few remarks:

1. The Wick transform is a phase space Wick rotation and has •• •••• •• •• •• •• •• •• ••• •• •• •• •••• •• •• ••• •••• •• ••• ••• • •• •• •• •••• $t!$ Mena Marugán [322, 323] has given a formal relation with the usual Wick rotation corresponding to an analytical continuation of time together with a complex conformal rescaling of the four-dimensional metric.

2. As we have seen in this chapter, one ••• construct a well-defined operator $\hat{C}$, whether its exponential makes any sense though is an open question. Notice that in principle one can dispense with the complex variables altogether because one can give meaning to the •• •••••• ••• • ••••• •• Hamiltonian constraint $H' = \tilde{H}'/\sqrt{\det(q)}$ in terms of the real variables (A, E) as we have seen. Still, although the complexifier C is then not used any more for the purpose of a Wick rotation, it still plays a crucial role in the quantisation scheme displayed there (in order to write the extrinsic curvature as a triple Poisson bracket between the Euclidean Hamiltonian constraint, the volume and the Wick rotator). It comes out rigorously quantised from that scheme. The corresponding operator $\hat{H}$ which we construct directly on the Hilbert space $\mathcal{H}^0$ is surprisingly not terribly complicated. Still, it may be important to construct a Wick transform one day because (a) it could simplify the construction of rigorous solutions and since (b) a coherent state transform always has a close connection with semiclassical physics which is important for the interpretation and the classical limit of the theory.

3. In order to define the exponential of C one could use the spectral theorem applied to a self-adjoint extension of the symmetrised version of C or Nelson's analytic vector theorem. Self-adjoint extensions of $C + C^\dagger$ exist by

•••• •• • ••• •• • ••• •• ••••• ••• •• •• ••••• •• ••• • •• ••••• ••• •• ••••••• •

von Neumann's theorem.[19] or Nelson's analytic vector theorem.[20] See [315] for more details.

4. Not surprisingly, the unphysical Hamiltonian $\tilde{H}(A, E) := \tilde{H}_{\mathbb{C}}(A_{\mathbb{C}} := A, E_{\mathbb{C}} := E)$ can be recognised as the Hamiltonian constraint that one obtains from the Hamiltonian formulation of Riemannian General Relativity (i.e., ordinary General Relativity just that one considers four-metrics of Euclidean signature).

5. The Wick transform derived in [315] is the first concrete proposal for a solution of the reality conditions for the ••• •••• connection variables. For a different proposal geared to a Minkowski space background, see [596].

10.7.2 Testing the new regularisation technique by models of quantum gravity

Presently there are two positive tests for the quantisation procedure that we applied to the Hamiltonian constraint, namely Euclidean $2 + 1$ gravity [440] and isotropic and homogeneous Bianchi cosmologies quantised in a non-standard fashion [497–499]. (For an introduction to quantum cosmology, see, e.g., the reviews [597–599] and Section 16.2.)

The first model is a dimensional reduction of $3 + 1$ gravity which one can formulate also as a quantum theory of SU(2) connections and SU(2) electric fluxes with precisely the same algebraic form of all constraints. Hence, one can introduce the full mathematical structure of $\overline{\mathcal{A}}, \mu_0, \mathcal{H}^0$ as well as the quantum constraints $G_j = \mathcal{D}_a E_j^a, V_a = F_{ab}^j E_j^b, H_E = F_{ab}^j E_k^a E_l^b \epsilon_{jkl} / \sqrt{\det(q)}$, the only difference with the Lorentzian $3 + 1$ theory being that now indices $a, b, c, \ldots = 1, 2$ have range in one dimension less and that there is only the Euclidean constraint.

The second model is $3 + 1$ Lorentzian gravity but all degrees of freedom except for finitely many are switched off by hand by performing the usual Killing reduction. However, instead of using a Schrödinger representation of the canonical commutation relations one uses an LQG type of representation, the Bohr representation, see Section 16.2 and Chapter 28.

In both models one then follows step by step the regularisation procedure outlined in Sections 10.3, 10.4. The outcomes in the $2 + 1$ theory are as follows (we devote Section 16.2 to the quantum cosmology models): The quantisation of $2 + 1$ general relativity is an exhaustively studied problem (see, e.g., [600–612] and [354, 355] as well as references in all of those). Several different quantisation

---

[19] This says that if a densely defined symmetric operator commutes on its domain with a conjugation operator that preserves its domain then there exist self-adjoint extensions. A conjugation operator is a bounded, anti-linear operator which squares to the identity. In our case the expression for $C + C^\dagger$ is real and an appropriate conjugation operator is just complex conjugation.

[20] This does not assume that C is even symmetric, however, one must show that there exists a dense set D of vectors on which the power expansion of the exponential converges absolutely pointwise, that is, $\sum_{n=0}^{\infty} t^n \|C^n \psi\| / (n!) < \infty$ for all $\psi \in D$ and for all $|t| \leq \hbar^{-1}$.

techniques have been applied and were shown to give consistent results. The reader might wonder why $2 + 1$ •• • •• ••• ••• quantum gravity should serve as a test model for $3 + 1$ •• ••• •• ••• quantum gravity. The reason for this is that, as pointed out in [609–612], the Hamiltonian formulation of $2 + 1$ gravity via connections leads to the non-compact gauge group $SU(1, 1)$ for three-metrics of Lorentzian signature while for three-metrics of Euclidean signature we have the same compact gauge group as in Lorentzian $3 + 1$ gravity, namely $SU(2)$. Thus, in order to maximally simulate the $3 + 1$ theory, we should consider Euclidean $2 + 1$ gravity.

However, in order to maximally test the new technique introduced in Sections 10.3, 10.4 and the constraints of the $3 + 1$ theory one has to develop techniques different from those that people normally employ in $2 + 1$ gravity which make [440] of interest by itself. In particular, it contains a full-fledged derivation of the $2 + 1$ volume operator. The reason is the following: pure $2 + 1$ gravity on a Riemann surface of some fixed genus is a topological field theory, that is, there are only finitely many degrees of freedom. This can easily be seen from the fact that we have six canonical pairs and six first-class constraints. When the metric q_{ab} is non-degenerate, the diffeomorphism and Hamiltonian constraint together are equivalent to the •• ••• ••• ••• ••••••• • $C^j := \epsilon^{ab} F^j_{ab} = 0$. Almost exclusively the theory is quantised using C_j rather than V_a, H, see in particular [602] and [354, 355]. But of course we must use V_a, H in order to test the $3 + 1$ theory appropriately and this is what has been done successfully in [440].

10.7.3 Quantum Poincaré algebra

In [442] an investigation was started in order to settle the question whether $\mathcal{H}^0$ supports the quantisation of the ADM energy surface integral

$$E_{\text{ADM}}(N) = -\frac{2}{\kappa} \int_{\partial\sigma} dS_a \frac{N}{\sqrt{\det(q)}} E^a_j \partial_b E^b_j \qquad (10.7.16)$$

for an asymptotically flat spacetime M (here $\partial\sigma$ corresponds to spatial infinity i^0 in the Penrose diagram describing the conformal completion of M). As we saw in Section 1.5.2, (10.7.16) is the value of the gravitational energy (at unit lapse $N = 1$) only when the constraints are satisfied, otherwise one has to add to (10.7.16) the Hamiltonian constraint $H(N)$. In particular one has to use $H_{\text{ADM}}(N) = H(N) + E_{\text{ADM}}(N)$ in order to compute the equations of motion. If N is, say, of rapid decrease, then $H_{\text{ADM}}(N) = H(N)$ generates gauge transformations (time reparametrisations), if it is asymptotically constant then it generates symmetries. There are nine more surface integrals of the type (10.7.16) and together they generate the asymptotic Poincaré algebra. They are the only ten global (in phase space) Dirac observables known for full, Lorentzian, asymptotically flat gravity in four dimensions. For a discussion of these and related issues, see, for example, [336] and references therein.

In [442] only time translations (10.7.16), spatial translations and spatial rotations were treated. Boosts, which are much harder to define, see Section 1.5.2, were not considered here but there is no principal problem to do so. We will focus here only on the quantisation of (10.7.16) for reasons of brevity. The methods of regularisation and quantisation completely parallel those displayed in Sections 10.3, 10.4 and will not be repeated here. The only new element that goes into the classical regularisation is the exploitation of the fall-off conditions on the classical fields, in particular that $A = O(1/r^2)$ in an asymptotic radial coordinate. This enables one to replace, effectively, $\partial_b E_j^b$ by the gauge-invariant quantity $G_j = \mathcal{D}_b E_j^b$ in (10.7.16), that is, the Gauß constraint. At first sight one is tempted to set it equal to zero. However, the detailed analysis of Section 1.5.2 shows that for the Gauß constraint to be functionally differentiable, its Lagrange multiplier must fall off as $1/r^2$, which means that the Gauß constraint does not need to hold at $\partial\sigma$ although the smeared constraint $\int d^3x \Lambda^j \hat{G}_j$ vanishes identically on states which are gauge-invariant at finite r but not at $r = \infty$. Thus, it would be physically incorrect to require $G_j = 0$ at $\partial\sigma$, in other words, quantum states do not need to be gauge-invariant at $\partial\sigma$ or, put differently, the motions generated by G_j at $\partial\sigma$ are not gauge transformations but symmetries.

The final answer is $(E_{\text{ADM}} = E_{\text{ADM}}(1))$

$$\hat{E}_{\text{ADM}} f_\gamma = -2m_p \sum_{v \in V(\gamma) \cap \partial\sigma} \frac{\ell_p^3}{\hat{V}_v} R_v^j R_v^j f_\gamma \tag{10.7.17}$$

where $R_v^j = \sum_{f(e)=v} R_e^j$, $\hat{V}_v = \lim_{R_v \to \{v\}} \hat{V}(R_v)$ and $x \mapsto R_x$ is an open region-valued function with $x \in R_x$. The operator (10.7.17) is defined actually on an extension of $\mathcal{H}^0$ which allows for edges that are not compactly supported. Moreover, we must require that (1) for each $v \in \gamma \cap \partial\sigma$ the eigenvalues of $\hat{V}_v$ are non-vanishing and (2) $e \cap \partial\sigma$ is a discrete set of points for every $e \in E(\gamma)$. We have assumed w.l.g. that all edges with $e \cap \partial\sigma \neq \emptyset$ are of the 'up' type with respect to the surface $\partial\sigma$.

Under these assumptions one can show the following:

(i) • •••••••• ••• ••••••••••• •••

   (10.7.17) defines a self-consistent family of essentially self-adjoint, positive semi-definite operators. This is like a quantum positivity of energy theorem but it rests heavily on the two assumptions (1) and (2) made above whose physical justification is unclear.

(ii) • ••• •••••• •• ••••••••••• ••• •

   Since the volume operator is gauge-invariant, it follows that it commutes with the Laplacian $\Delta_v = (R_v^j)^2$ and therefore we can simultaneously diagonalise these operators. It is clear that the eigenstates are certain linear combinations of spin-network states and the eigenvalues are of the form $j_v(j_v + 1)/\lambda_v$ (where λ_v is a volume eigenvalue) times m_p. Thus we can complete the intuitive picture that the Hamiltonian constraint gave us: while the constraint changes the spin quantum numbers, the energy is diagonal in very much the same way as the annihilation and creation operators of quantum mechanics change the occupation number of an energy eigenstate. • • • •• •••• •• •••••••• ••• •••• ••••••• ••• •••• •• ••••••••••• ••• •••• ••• ••••••••••• ••• •••• ••••••••••••••. In quantum field theory we label Fock states by occupation numbers n_k for momentum modes k. Here we have occupation numbers j_e for 'edge modes' e.

(iii) •••••••• •••••••••

The eigenvalues are discrete and unbounded from above but in contrast to the geometry operators there is no energy gap. Rather there is an accumulation point at zero because $[\Delta_v, \hat{V}_v] = 0$ (we can choose the state to be very close to being gauge-invariant but to have arbitrarily large volume). This is to be expected on physical grounds because we should be able to detect arbitrarily soft gravitons at spatial infinity.

(iv) •••••• • •••• •••••••••• ••• ••••••••••• ••• •••••••

(10.7.17) trivially commutes with all constraints (since diffeomorphisms φ and lapses N that generate gauge transformations are trivial (identity and zero) at $\partial\sigma$) and therefore represents a true quantum Dirac observable. In principle we can now solve 'the problem of time' since a physically meaningful time parameter is selected by the one-parameter unitary groups generated by $\hat{E}_{\mathrm{ADM}}$, in other words, we have a Schrödinger equation

$$-i\hbar\frac{\partial\Psi}{\partial t} = \hat{E}_{\mathrm{ADM}}\Psi \tag{10.7.18}$$

and thus have solved the quantum problem of time because (10.7.18) puts us into the conceptual situation of a standard canonical theory with a Hamiltonian. In fact, (10.7.18) is only correct if we impose the quantum analogue of $H(x) = 0$ also at $\partial\sigma$, which is not what the formalism tells us to do because $\int d^3x N(x)\hat{H}(x)$ vanishes on all states which only satisfy the constraint at finite r but are otherwise arbitrary at $r \to \infty$, since for gauge transformations N must vanish at $\partial\sigma$. Hence, in the general situation we must add to the right-hand side of (10.7.18) the term $\int_\sigma d^3x\hat{H}(x)$ which contains both matter and geometry contributions and explains why the matter Hamiltonian densities derived in Chapter 12, and which reduce to the usual energy densities of Minkowski space when we evaluate the gravitational field on Minkowski initial data, still determine energy also when coupled to gravity. Writing $\mathbf{H}_{\mathrm{matter}} = \int d^3x H_{\mathrm{matter}}(x)$ and $\mathbf{H}_{\mathrm{geometry}} = \int d^3x H_{\mathrm{geometry}}(x)$ we find for the total Hamiltonian $\mathbf{H} = E_{\mathrm{ADM}} + \mathbf{H}_{\mathrm{matter}} + \mathbf{H}_{\mathrm{geometry}}$. Physical states must then satisfy $[H_{\mathrm{matter}}(x) + H_{\mathrm{geometry}}(x)]\Psi = 0$ at finite r and evolve according to (10.7.18) with E_{ADM} replaced by $\mathbf{H}$. Since $\mathbf{H}_{\mathrm{matter}}$ is manifestly positive even when coupled to geometry, one way to ensure $\mathbf{H} \geq 0$ is by looking for physical states satisfying $[E_{\mathrm{ADM}} + \mathbf{H}_{\mathrm{geometry}}]\Psi = 0$, which is an equation that involves only the gravitational degrees of freedom. This implies in particular that $E_{\mathrm{ADM}} = \mathbf{H}_{\mathrm{matter}}$.

Actually in [442] concepts that go beyond $\mathcal{H}^0$ were needed and introduced heuristically. They go under the name 'Infinite Tensor Product Extension' and were properly defined only later in [479]. They will be discussed briefly in Section 11.2.

10.7.4 Vasiliev invariants and discrete quantum gravity

Recently, a second approach towards solving the Hamiltonian constraint has been proposed [613, 614] which is constructed on (almost) diffeomorphism-invariant distributions which are based on Vasiliev invariants. What is exciting about this is that one can define something like an area derivative [346] in this space and therefore the arc attachment which we described above becomes much less ambiguous.

Also recently a third approach has emerged [615–627] which starts with a fundamentally discrete formulation at the classical level. The discrete evolution equations then become inconsistent unless one solves for lapse and shift functions, which means that the discretisation acts like a gauge-fixing procedure. In this formalism there are therefore no constraints, which has technical and conceptual advantages. Moreover, the formalism was demonstrated to work in several models and leads to possibly observable effects such as quantum gravity-induced decoherence.

Since these developments are somewhat removed from the main thrust of the book, unfortunately we cannot review them here due to reasons of space, but must refer the reader to the literature.

11

Step V: semiclassical analysis

Normally, when constructing a perturbative quantum field theory on Minkowski space or any other background spacetime one never doubts that the resulting theory has the correct classical limit. One is satisfied with having found a Fock representation and a definition of the S-matrix, that is, matrix elements of powers of a normal ordered Hamiltonian operator. In fact it is clear from the outset that a theory written in terms of (an infinite number of) annihilation and creation operators has the correct classical limit because one can construct the usual coherent states for the underlying free field theory and then one knows that operators written in terms of the annihilation and creation operators have expectation values very close to the classical values that the corresponding classical function takes at the point in phase space where the coherent state is peaked.[1]

In a constrained, non-perturbative quantum field theory without background structure the question about the classical limit is much less trivial. First of all, since we are using a non-perturbative approach we cannot expand around a free field theory and hence cannot use Fock space (coherent state) techniques. Secondly, since we must work without a background spacetime we are forced to use completely new types of Hilbert spaces for which no semiclassical techniques have been developed so far. Thirdly, the theory is highly non-linear: for example, the constraint operators are simply not polynomials of the basic variables A, E for which one would hope to be able to construct semiclassical states which approximate those. Thus, not only is there no natural choice of operators to which one should adapt the coherent states, moreover, the calculations to be performed are going to be of a new type since we must deal with expectation values of square roots of powers of those variables. Finally, there is even the basic question of what we mean by coherent states in the presence of constraints: do we mean states that are completely kinematical or at least gauge-invariant and/or at least spatially diffeomorphism-invariant or do we mean physical coherent states?

As we mentioned several times in Chapter 10, the goals of the semiclassical analysis to be performed in step V are twofold:

---

[1] This does not guarantee, however, that the finite quantum time evolution is close to the classical one. Consider, for example, the anharmonic oscillator $H_\lambda = H_0 + \lambda q^4$ where $H_0 = (p^2 + q^2)/2$. Let ψ_{m_0} be the harmonic oscillator coherent state peaked at $m_0 = (q_0, p_0)$ for the free Hamiltonian H_0 and let m_t be the classical time evolution of the initial data m_0 with respect to H_λ. Then $\exp(it\hat{H}_\lambda)\psi_{m_0}$ is very different from ψ_{m_t} for large t, they match only for small t.

1. We must verify that the Hamiltonian constraint $\hat{H}(N)$ or the Master Constraint $\hat{}$ have the correct classical limit.
2. We must check that the physical Hilbert space determined by either of these two operators contains enough semiclassical states.

Concerning the first goal, since $\hat{H}(N)$, $\hat{}$ are defined on $\mathcal{H}_{kin}, \mathcal{H}_{diff}$ respectively, the semiclassical states that test them must be taken from $\mathcal{H}_{kin}, \mathcal{H}_{diff}$ respectively. In particular, it does not make sense to construct physical semiclassical states in order to test these operators since on $\mathcal{H}_{phys}$ they are identically zero by definition. Concerning the second goal, we must construct physical coherent states with respect to which Dirac observables have good semiclassical properties.

Hence, given a quantum field theory with constraints, the following questions arise:

(A) What are semiclassical states if there is no Hamiltonian which would suggest them?
(B) On which Hilbert space $\mathcal{H}$ do we want to construct semiclassical states, that is, before or after imposing all or some of the constraints?
(C) Which (kinematical) algebra $\mathfrak{A}$ of observables should be approximated especially well?
(D) How do we construct coherent states once $(\mathcal{H}, \mathfrak{A})$ have been chosen?
(E) What is the relation between kinematical semiclassical states and physical semiclassical states?

In the next sections we will address these questions in more detail. For now, let us briefly outline what we will do:

(A) First of all we will give a possible, reasonable definition for semiclassical or coherent states for a general theory.
(B) Next, as we have argued above, we need semiclassical states on all three Hilbert spaces $\mathcal{H}_{kin}, \mathcal{H}_{diff}, \mathcal{H}_{phys}$.
(C) For $\mathcal{H}_{kin}$ it is of course natural to assume that the appropriate algebra $\mathfrak{A}$ to approximate is the one that we have based the kinematical representation theory on. As we will see, due to the non-separability of $\mathcal{H}_{kin}$ this is not quite true: natural coherent states for $\mathfrak{A}$ are distributions Ψ_m, peaked at the field configuration $m = (A, E)$ of the classical phase space and hence are not normalisable in $\mathcal{H}_{kin}$. One can write them as an uncountable sum over all graphs $\Psi_{kin}^m = \sum_\gamma < \psi_\gamma^m, . >_{kin}$ where $\psi_\gamma^m \in \mathcal{H}_{kin}$ is a linear combination of spin-network states over the finite graph γ with only non-trivial spins of all edges. The ψ_γ^m are normalisable, however, these states do not approximate all elements of $\mathfrak{A}$, for example, only those holonomies along paths which are compositions of edges of γ. However, it turns out that they approximate well a restricted class of compound operators arising from classical functions of the form $O = \int_\sigma d^3x F(A, E)$ where F has density weight one. In

particular, O could be spatially diffeomorphism-invariant. The restriction consists in the requirement that $\hat{O}$ be non-graph-changing, as otherwise the edges that $\hat{O}$ adds to γ are not approximated in $<\psi_\gamma^m, \hat{O}\psi_\gamma^m>/||\psi_\gamma^m||^2$. It also does not help to replace this by $\Psi_{\rm kin}^m(\hat{O}\psi_\gamma)/||\psi_\gamma||^2$ as we will see. Thus, at the level of $\mathcal{H}_{\rm kin}$ the currently available techniques only suffice to approximate non-graph-changing operators, which excludes, for example, $\hat{H}(N)$ but includes the extended Master Constraint which also incorporates the spatial diffeomorphism constraint.

Things look better at the spatially diffeomorphism-invariant level because now the Hilbert space $\mathcal{H}_{\rm diff}$ is (or can be chosen to be) separable. What one does is to replace $\Psi_m^{\rm kin}$ by $\Psi_m^{\rm diff} = \sum_{(\gamma)} \psi_{m,(\gamma)}$ where the sum is over θ-equivalence classes of graphs and $\psi_{m,(\gamma)}$ arises from $\psi_{m,\gamma} = \sum_{\gamma(s)=\gamma} \psi_{m,s} T_s$ by replacing T_s by $T_{[s]}$ where we have made a definite choice $\theta^0_{(\gamma)}$ for the θ-moduli involved in the decomposition $[\gamma] = ((\gamma), \theta_{(\gamma)})$, see Section 10.6.3. The states $\Psi_m^{\rm diff}$ are now (potentially) normalisable. Notice that $\Psi_m^{\rm diff}$ is *not* labelled by spatial diffeomorphism equivalence classes $[m]$ of points m in the classical phase space but rather by points in the unconstrained phase space. However, if $\hat{O}$ is an operator on $\mathcal{H}_{\rm diff}$ which is well approximated by $\Psi_m^{\rm diff}$ in the sense that the expectation value is $O(m)$ to zeroth order in $\hbar$, then the choice of m in the class $[m]$ is irrelevant. Hence we see that the construction of $\Psi_m^{\rm kin}$, while not helpful for $\hat{H}(N)$, could still be of help for the knot class-changing operator $\hat{}$ in that $\Psi_m^{\rm kin}$ suggests natural candidates $\Psi_m^{\rm diff}$.

Finally, in order to construct a physical coherent state, a natural Ansatz is $\Psi_m^{\rm phys} := \Psi_m^{\rm diff}(0)$ where $\Psi_m^{\rm diff} = (\Psi_m^{\rm diff}(\lambda))_{\lambda \in \mathbb{R}^+}$ is the direct integral presentation of $\Psi_m^{\rm diff}$, see Section 30.2. Thus we see that natural Ansätze for coherent states on all three Hilbert spaces can be obtained by constructing (distributional) coherent states for the kinematical algebra $\mathfrak{A}$ on $\mathcal{H}_{\rm kin}$.

(D) This leaves us with the problem of constructing coherent states for $\mathfrak{A}$. It turns out that there is a general construction principle available, called the *complexifier* method, which is applicable if the underlying phase space is a cotangent bundle (which is the case for our unconstrained phase space). In fact, all coherent states for free field, background-dependent Wightman theories can be formulated in this language. The only input is a certain positive function on phase space, called the complexifier C, which generates these states. The choice of C is constrained by the desire to achieve an optimal approximation of a given Hamiltonian (constraint) and the time evolution it generates.

(E) Calculating at the level of $\mathcal{H}_{\rm phys}$ is of course very hard because $\mathcal{H}_{\rm phys}$ is only implicitly known. Hence the question arises whether one cannot approximate expectation value calculations in $\mathcal{H}_{\rm phys}$ by calculations in $\mathcal{H}_{\rm diff}$ or $\mathcal{H}_{\rm kin}$. The idea would be to compare, for Dirac observables $\hat{O}$, for instance, the expectation values $<\Psi_{\rm phys}^m, \hat{O}\Psi_{\rm phys}^m>_{\rm phys}$ with $<\Psi_{\rm diff}^m, \hat{O}\Psi_{\rm diff}^m>_{\rm diff}$ assuming that

$\hat{O}$ has a representation on $\mathcal{H}_{\text{diff}}$ as well. Under certain circumstances it happens that these numbers and the corresponding fluctuations are close to each other and work is in progress to systematically analyse sufficient criteria for this to happen.

Three proposals for semiclassical states have appeared in the literature so far: historically the first ones are the so-called 'geometrical weaves' [484,628–630,640, 641] which try to approximate kinematical geometric operators only, see Chapter 13. Also 'connection weaves' have been considered [631, 632] (see also [633] for a related proposal) which are geared to approximate kinematical holonomy operators on a given graph. Finally, one can get rid of a certain graph dependence of geometrical weaves through a clever statistical average [634–636] resulting in 'statistical weaves'.

The second proposal consists in the complexifier method [487] just mentioned above for any canonical quantum field theory whose underlying phase space is a cotangent bundle. Also the Wick transform [315, 316], see Section 10.7.1, is based on the complexifier method. This programme was applied to full nonlinear, non-Abelian Loop Quantum Gravity [485–490] for a specific choice of complexifier. As already said, these states are not normalisable but rather are distributions in $C^\infty(\overline{\mathcal{A}})^*$. However, their cylindrical projections, which we will call cutoff states, are normalisable as we outlined above. For these the desired properties like overcompleteness, saturation of the Heisenberg uncertainty relation, peakedness in *phase space* (thus both connection and electric flux are well approximated), construction of annihilation and creation operators and corresponding Ehrenfest theorems were confirmed. Given such cutoff coherent state, its excitations can be interpreted as the analogue of the usual graviton states [637, 638]. One can combine these methods with a statistical average of the kind considered above to eliminate some of the graph dependence of the cutoff states.

The complexifier method also encompasses an apparently different third proposal [491] which seems to be especially well-suited for the semiclasscal analysis of free Maxwell theory and linearised gravity. It was originally discovered by using a striking isomorphism between the usual Poisson algebra in terms of connections smeared in D dimensions and unsmeared electric fields on the one hand and the algebra obtained by one-dimensionally smeared connections and electric fields smeared in D dimensions on the other hand. Using this observation, which however does not carry over to the non-Abelian case, one can carry states on the usual Fock–Hilbert space into distributions over $C^\infty(\overline{\mathcal{A}})$ and drag the Fock inner product into a new inner product on the space of these distributions with respect to which they are normalisable. In [495] these observations were interpreted in a more abstract way, in particular, a measure-theoretic interpretation of the distributions constructed via the technique of [491–493] was given. In [495] the states that we refer to as cutoff states were called 'shadows' and in [549] a simple quantum mechanical model was studied using these.[2] In [639] it is shown that, for the Abelian case, the dragged Fock measure and the uniform measure are mutually singular with respect to each other and that the dragged Fock measure does not support an electric field operator smeared in $D - 1$ dimensions, which are essential to use in the non-Abelian case. This indicates that all the nice structure that comes with U(1) does not generalise to SU(2) [487] even if one allows background-dependent representations. In fact, under very mild assumptions one seems always to be back to the representation of $\mathfrak{A}$ on $\mathcal{H}_0$ [514, 515]. This shows that the representation theory of the holonomy flux algebra is rather robust. It is therefore rather certain that the only useful coherent states on $\mathcal{H}_{\mathrm{kin}}$ are normalisable linear combinations of cutoff states (and of their diffeomorphism-invariant images on $\mathcal{H}_{\mathrm{diff}}$).

In what follows we will mainly describe the complexifier method and specific realisations thereof as it seems to be the currently unifying framework.

11.1 + Weaves

Let us briefly summarise the early work on semiclassical states.

(a) *Geometric weaves*

The early geometric weaves [484] were constructed as follows: let q^0_{ab} be a background metric. Notice that we are not introducing some background dependence here, all states still belong to the background-independent Hilbert space $\mathcal{H}^0$, we are just looking for states that have low fluctuations around a given classical three-metric. Using that metric, sprinkle non-intersecting (but possibly linked), circular, smooth loops at random with mean separation ϵ and mean radius ϵ (as measured by q^0_{ab}). The union of these loops is a graph, more precisely a link, γ without intersections (see Figure 11.1 for an example). The random process used was, however, not specified in [484]. Consider the state given by the product of the traces of the holonomies along those loops. The reason for choosing non-intersecting loops was that such a state was formally annihilated by the Hamiltonian constraint. Consider any surface S. From our discussion in Chapter 13 it is clear that this state is an eigenstate of the area operator $\widehat{\mathrm{Ar}}(S)$ with eigenvalue $\ell_p^2 \sqrt{3} N(S, q^0, \epsilon)/4$ where $N(S, q^0, \epsilon)$ is the number of intersections of S with the link γ. If q^0 does not vary too much at the scale ϵ then this number

---

[2] The shadow framework is almost the same as the cutoff state framework, the only difference being that one considers, instead of the real-valued expectation values $< \psi_\gamma, \hat{O}\psi_\gamma >_{\mathcal{H}_{\mathrm{kin}}} /||\psi_\gamma||^2$ where ψ_γ is the cutoff state on γ of the distribution Ψ, the complex numbers $\Psi[\hat{O}\psi_\gamma]/\Psi[\psi_\gamma]$. For non-graph-changing operators these two numbers coincide, for graph-changing ones there is a small difference. However, as we will see, this more general prescription is also not able to reproduce the correct expectation values for graph-changing operators.

Figure 11.1 The Booromean rings – a simple example of a link. A complicated fabric of a vast number of such linked loops defines a weave.

is roughly given by $Ar_{q^0}(S)/\epsilon^2$. Notice that all of this was done still in the complex connection representation and therefore outside of a Hilbert space context. Yet, the eigenvalue equation $\ell_p^2 Ar_{q^0}(S)/\epsilon^2$ tells us *that canonical quantum gravity seems to have a built-in finiteness*: it does not make sense to take an arbitrarily fine graph $\epsilon \to 0$ since the eigenvalue would blow up. In order to get the correct eigenvalue one must take $\epsilon \approx \ell_p$, that is, the loops have to be sprinkled at Planck scale separation. This observation rests crucially on the fact that there is an area gap. These calculations were done for metrics q^0 that are close to being flat. In [640] weaves for Schwarzschild backgrounds were considered, which requires an adaption of the sprinkling process to the local curvature of q^0 in order that one obtains reasonable results.

Finally, in [641] the link γ was generalised to disjoint collections of triples of smooth multi-loops. Each triple intersects in one point with linearly independent tangents there. The motivation for this generalisation was that then the volume operator (which vanishes if there are no intersections) could also be approximated by the same technique.

(b) *Connection weaves*

For an element h of $SU(2)$ we have $\text{Tr}(h) \leq 2$ where equality is reached only for $h = 1$. Thus $h \mapsto 2 - \text{tr}(h)$ is a non-negative function. Let now α be one of the loops considered in [641] and let $A \in \overline{\mathcal{A}}$. Then $A \mapsto e^{-\beta[2-\text{tr}(A(\alpha))]}$ is sharply peaked at those $A \in \overline{\mathcal{A}}$ with $A(\alpha) = 0$, that is, at a flat connection (since the α are contractible). Arnsdorf [480] then considers the product of all those functions for loops which generate the fundamental group of a given graph γ (this function is precisely of the form of the exponential of the Wilson action employed in lattice gauge theory [576]).

Since [480] is written in the context of the Hilbert space $\mathcal{H}^0$ and since non-compact topologies of σ were considered, in contrast to [484] one had to deal with the case that the graph γ becomes infinite (the number of loops becomes infinite). Since such a state is not an element of $\mathcal{H}^0$, Arnsdorf constructed a positive linear functional on the algebra of local operators using that formal state and then used the GNS construction (see Chapter 29) in order to obtain a new Hilbert space in which one can now compute expectation values of various operators. Expectedly, holonomy operators along paths in l have expectation values close to their classical value at flat connections while the semiclassical behaviour of electric flux operators is not reproduced.

(c) *Statistical weaves*

In both the geometric and connection weave construction an arbitrary but fixed graph γ had to be singled out. This is unsatisfactory because it involves a huge amount of arbitrariness. Which graph should one take? Also, unless the graph γ is sufficiently random the expectation values, say of the area operator in a geometric weave for a flat background metric q^0, are not rotationally invariant.

To improve this, Ashtekar and Bombelli [634, 635] have employed the Dirichlet–Voronoi construction, often used in statistical mechanics [642], to the geometrical weave (see Figure 11.2). Roughly, this works as follows: given a background metric q^0, a *compact* hypersurface σ and a density parameter λ one can construct a subset $\Gamma(q^0, \lambda) \subset \Gamma_{sa}^\omega$ of semianalytic graphs each of which, in D spatial dimensions, is such that each of its vertices is $(D + 1)$-valent. A member $\gamma_{x_1,\ldots,x_N} \in \Gamma(q^0, \lambda)$ is labelled by $N \approx [\lambda \text{Vol}_{q^0}(\sigma)]$ points $x_k \in \sigma$ where $[.]$ denotes the Gauß bracket. The graph $\gamma_{x_1,\ldots,x_N}$ is obtained unambiguously from the set of points $x_1, \ldots, x_N$ and the metric q^0 (provided that it is close to being flat) by employing natural notions like minimal geodesic distances, etc. Next, given a spin label j and an intertwiner I we can construct a gauge-invariant spin-net $s_{x_1,\ldots,x_N}(j, I)$ by colouring each edge with the same spin and each vertex with the same intertwiner. From these data one can construct the 'density operator'

$$\hat{\rho}(q^0, \lambda, I, j) := \int_{\sigma^N} d\mu_{q^0}(x_1) \ldots d\mu_{q^0}(x_N) T_{s_{x_1,\ldots,x_N}(j,I)} < T_{s_{x_1,\ldots,x_N}(j,I)}, \cdot >$$

(11.1.1)

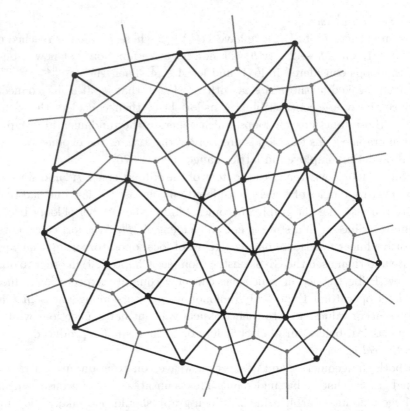

Figure 11.2 A Dirichlet–Voronoi graph in 2D: the random process creates the faces depicted in dark grey while the dual graph is depicted in light grey. In D dimensions, the valence of each vertex is precisely $D + 1$.

where

$$d\mu_{q^0}(x) := \frac{\sqrt{\det(q^0)(x)}d^D x}{\mathrm{Vol}_{q^0}(\sigma)} \tag{11.1.2}$$

is a probability measure (it is here where compactness of σ is important). The reason for the inverted commas in 'density operator' is that (11.1.1) actually is the zero operator. To see this, notice that for any spin-network state T_s we have $< T_{s_{x_1,\ldots,x_N}(j,I)}, T_s >= \delta_{s_{x_1,\ldots,x_N}(j,I),s}$ which in particular means that $\gamma_{x_1,\ldots,x_N} = \gamma(s)$. But the set of points satisfying this is certainly thin with respect to the measure (11.1.2). What happens is that although for any spin-network state T_s the one-dimensional projector $T_s < T_s. >$ is a trace class operator of unit trace, the trace operation does not commute with the integration in (11.1.1). However, one can then define a positive linear functional $\omega_{q^0,\lambda,I,j}$ on the algebra of linear operators on $\mathcal{H}^0$ by

$$\omega_{q^0,\lambda,j,I}(\hat{O}) := \int_{\sigma^N} d\mu_{q^0}(x_1)\ldots d\mu_{q^0}(x_N) < T_{s_{x_1,\ldots,x_N}(j,I)}, \hat{O} T_{s_{x_1,\ldots,x_N}(j,I)} > \tag{11.1.3}$$

which would equal $\text{Tr}(\hat{\rho}(q^0, \lambda, j, I)\hat{O})$ if integration and trace commuted. Via the GNS construction one can now define a new representation $\mathcal{H}^0_{q^0, \lambda, j, I}$ which depends on a background structure. The representations $\mathcal{H}^0$ and $\mathcal{H}^0_{q^0, \lambda, j, I}$ are certainly not (unitarily) inequivalent. The problem that (11.1.1) is the zero operator in LQG is avoided in Algebraic Quantum Gravity (AQG) because there the algebraic graph is not subject to the sprinkling process, see the discussion in [589].

What is interesting about (11.1.3) is that for an exactly flat background the expectation values of, say the area operator, are *Euclidean invariant*. In order to match the expectation values of $\widehat{\text{Ar}}(S)$ with the value $\text{Ar}_{q^0}(S)$ one must choose j according to $[\sqrt{j(j+1)}\ell_p^2 \beta \lambda^{2/3}/2] = 1$. A similar calculation for the volume operator presumably fixes the value I for the intertwiner.

11.2 Coherent states

Especially the statistical weave construction of the previous section looks like a promising starting point for semiclassical analysis. However, there are several drawbacks with weaves:

(i) *Phase space approximation*

All the weaves discussed above seem to approximate either the connection or the electric field appropriately, although the degree of their approximation has never been checked (are the fluctuations small?). However, what we really need are states which approximate the connection and the electric field simultaneously with small fluctuations.

(ii) *Arbitrariness of spins and intertwiners*

All weaves proposed somehow seem to arbitrarily single out special and uniform values for spin and intertwiners. Drawing an analogy with a system of uncoupled harmonic oscillators, it is like trying to build a semiclassical state by choosing an arbitrary but fixed occupation number (spin) for each mode (edge). However, we know that the preferred semiclassical states for the harmonic oscillator are coherent states which depend on all possible occupation numbers. As we will see, issues (i) and (ii) are closely related.

(iii) *Arbitrariness of graphs*

Even in the statistical weave construction we select arbitrarily only a certain subclass of graphs. Again, drawing an analogy with the harmonic oscillator picture, this is like selecting a certain subset of modes in order to build a semiclassical state. However, then not all modes can behave semiclassically.

(iv) *Missing construction principle*

The weave states constructed suffer from a missing enveloping construction principle that would guarantee from the outset that they possess desired semiclassical properties.

The aim of the series of papers [485–490] was to decrease this high level of arbitrariness, to look for a systematic construction principle and to make semiclassical states for quantum gravity look more similar to the semiclassical states for free Maxwell theory, which are in fact *coherent states* and have been extremely successful (see, e.g., [643, 644] and references therein).

11.2.1 Semiclassical states and coherent states

Recall that quantisation is, roughly speaking, an attempt to construct a * homomorphism

$$\bigwedge : (\mathcal{M}, \{.,.\}, \mathcal{O}, \overline{(.)}) \to \left(\mathcal{H}, \frac{[.,.]}{i\hbar}, \hat{\mathcal{O}}, (.)^\dagger\right) \tag{11.2.1}$$

from a subalgebra $\mathcal{O} \subset C^\infty(\mathcal{M})$ of the Poisson algebra of complex-valued functions on the symplectic manifold $(\mathcal{M}, \{.,.\})$ to a subalgebra $\hat{\mathcal{O}} \subset \mathcal{L}(\mathcal{H})$ of the algebra of linear operators on a Hilbert space $\mathcal{H}$ with inner product $< .,. >$ such that Poisson brackets turn into commutators and complex conjugation into the adjoint operation. Notice that the map cannot be extended to all of $C^\infty(\mathcal{M})$ (only up to quantum corrections) unless one dives into deformation quantisation (see, e.g., [645] and references therein), the subalgebra for which it holds is referred to as the algebra of elementary functions (operators). The algebra $\mathcal{O}$ should be sufficiently large in order that more complicated functions can be expressed in terms of elements of it so that they can be quantised by choosing a suitable factor ordering (mathematically speaking, $\mathcal{O}$ should separate the points of $\mathcal{M}$).

Dequantisation is the inverse of the map (11.2.1). A possible way to phrase this more precisely is:

Definition 11.2.1. *A system of states $\{\psi_m\}_{m\in\mathcal{M}} \in \mathcal{H}$ is said to be semiclassical for an operator subalgebra $\hat{\mathcal{O}} \subset \mathcal{L}(\mathcal{H})$ provided that for any $\hat{O}, \hat{O}' \in \hat{\mathcal{O}}$ and any generic point $m \in \mathcal{M}$*

1. *Expectation value property*

$$\left| \frac{< \psi_m, \hat{O}\psi_m >}{O(m)} - 1 \right| \ll 1 \tag{11.2.2}$$

2. *Infinitesimal Ehrenfest property*

$$\left| \frac{< \psi_m, [\hat{O}, \hat{O}]\psi_m >}{i\hbar\{O, O'\}(m)} - 1 \right| \ll 1 \tag{11.2.3}$$

3. *Small fluctuation property*

$$\left| \frac{< \psi_m, \hat{O}^2\psi_m >}{< \psi_m, \hat{O}\psi_m >^2} - 1 \right| \ll 1 \tag{11.2.4}$$

The quadruple $(\mathcal{M}, \{.,.\}, \mathcal{O}, \overline{(.)})$ is then called the classical limit of $(\mathcal{H}, \frac{[.,.]}{i\hbar}, \hat{\mathcal{O}}, (.)^\dagger)$.

Clearly Definition 11.2.1 makes sense only when none of the denominators displayed vanish, so they will hold at most at generic points m of the phase space (meaning a subset of $\mathcal{M}$ whose complement has Liouville measure comparable to a phase cell), which will be good enough for all practical applications. An alternative definition could depend on additional scale parameters s, one for each observable to be approximated, which says that, for example, (11.2.2) is replaced by $|<\psi_m, \hat{O}\psi_m> -O(m)| \ll s$.

Notice that if (1) holds for $\hat{O}$ then it holds for $\hat{O}^\dagger$ automatically. Condition (1) is for polynomial operators sometimes required in the stronger form that (11.2.2) should vanish exactly, which can always be achieved by suitable (normal) ordering prescriptions. Condition (2) ties the commutator to the Poisson bracket and makes sure that the infinitesimal quantum dynamics mirrors the infinitesimal classical dynamics. If the error in (2) vanishes then we have a finite Ehrenfest property, which in non-linear systems is very hard to achieve. Finally, (3) controls the quantum error, the fluctuation of the operator.

Coherent states have further properties which can be phrased roughly as follows:

Definition 11.2.2. *A system of states $\{\psi_m\}_{m\in\mathcal{M}} \in \mathcal{H}$ is said to be coherent for an operator subalgebra $\hat{O} \subset \mathcal{L}(\mathcal{H})$ provided that for any $\hat{O}, \hat{O}' \in \hat{O}$ and any generic point $m \in \mathcal{M}$ in addition to properties (1)–(3) we have*

4. *Overcompleteness property*
   There is a resolution of unity

$$1_\mathcal{H} = \int_\mathcal{M} d\nu(m)\psi_m <\psi_m, .> \qquad (11.2.5)$$

   for some measure ν on $\mathcal{M}$.

5. *Annihilation operator property*
   There exist elementary operators $\hat{g}$ (forming a complete system) such that

$$\hat{g}\psi_m = g(m)\psi_m \qquad (11.2.6)$$

6. *Minimal uncertainty property*
   For the self-adjoint operators $\hat{x} := (\hat{g} + \hat{g}^\dagger)/2$, $\hat{y} := (\hat{g} - \hat{g}^\dagger)/(2i)$ the (unquenched) Heisenberg uncertainty relation is saturated

$$<(\hat{x}- <\hat{x}>_m)^2 >_m=< (\hat{y}- <\hat{y}>_m)^2 >_m= \frac{1}{2}| <[\hat{x}, \hat{y}] >_m | \qquad (11.2.7)$$

7. *Peakedness property*
   For any $m \in \mathcal{M}$, the overlap function

$$m' \mapsto | <\psi_m, \psi_{m'} > |^2 \qquad (11.2.8)$$

is concentrated in a phase cell of Liouville volume $\frac{1}{2}| <[\hat{p}, \hat{h}] >_m |$ if $\hat{p}$ is a momentum operator and $\hat{h}$ a configuration operator.

These four conditions are not completely independent of each other, in particular, (5) implies (6) but altogether (1)–(7) comprises a fairly complete list of desirable properties for semiclassical (coherent) states.

We have phrased our definitions in terms of pure states for simplicity. But more generally we might need to consider families of positive linear functionals $m \mapsto \omega_m$ on the algebra of operators to be approximated and which do not need to be pure. Properties (1)–(3), (6) and (7) can then be phrased with $\omega_m(.)$ instead of $< \psi_m, \cdot \psi_m > /||\psi_m||^2$. However, conditions (4) and (5) are specific to pure states.

An additional property which is satisfied for the harmonic oscillator coherent states is that the quantum evolution $\psi_{m_0} \mapsto \psi_{m_0}^t := \exp(it\hat{H})\psi_{m_0}$ driven by a Hamiltonian operator $\hat{H}$ approximates the classical trajectory $\psi_{m_0} \mapsto \psi_{m_{m_0}(t)}$ where $m \mapsto m_{m_0}(t)$ is the solution of the classical equations of motion with initial condition $m_{m_0}(0) = m_0$. This condition is difficult, actually unknown, to meet even in non-linear systems with a finite number of degrees of freedom as simple as the anharmonic oscillator.

11.2.2 Construction principle: the complexifier method

Usually one introduces coherent states for the harmonic oscillator as eigenstates of the annihilation operator in terms of superpositions of energy eigenstates. This method has the disadvantage that one needs a preferred Hamiltonian, that is, dynamical input in order to define suitable annihilation operators. Even if one has a Hamiltonian, the construction of annihilation operators is no longer straightforward if we are dealing with a non-linear system. Since we neither have a Hamiltonian nor a linear system, and since for the time being we are anyway interested in kinematical coherent states, we have to look for a different constructive strategy.

A hint comes from a different avenue towards the harmonic oscillator coherent states. Let the Hamiltonian be given by

$$H := \frac{1}{2}[p^2/m + m\omega^2 x^2] = \omega \bar{z} z \quad \text{where} \quad z = \frac{\sqrt{m\omega} x - ip/\sqrt{m\omega}}{\sqrt{2}} \quad (11.2.9)$$

Define the *complexifier* function

$$C := \frac{p^2}{2m\omega} \quad (11.2.10)$$

then it is easy to see that

$$z = \sqrt{\frac{m\omega}{2}} \sum_{n=0}^{\infty} \frac{(-i)^n}{n!} \{C, x\}_n \quad (11.2.11)$$

(recall that in our terminology $\{p, x\} = 1$). Translating this equation into quantum theory we find

$$\hat{z} = \frac{\sqrt{m\omega}}{\sqrt{2}} \sum_{n=0}^{\infty} \frac{(-i)^n}{n!} \frac{[\hat{C}, \hat{x}]_n}{(i\hbar)^n} = e^{-t(-\Delta/2)} \frac{\hat{x}\sqrt{m\omega}}{\sqrt{2}} \left(e^{-t(-\Delta/2)}\right)^{-1} \qquad (11.2.12)$$

where the *classicality parameter*

$$t := \hbar/(m\omega) \qquad (11.2.13)$$

has naturally appeared and which for this system has dimension cm^2. The operator $\hat{z}$ is usually chosen by hand as the annihilation operator. Let us accept that coherent states ψ_z are eigenstates of $\hat{z}$. Given formula (11.2.13) we can trivially construct them as follows: let δ_x be the δ-distribution, supported at x, with respect to the Hilbert space measure dx. Define $\psi_x := e^{-t\hat{C}/\hbar^2}\delta_x$. Then formally

$$\hat{z}\psi_x = e^{-t\hat{C}/\hbar^2}\frac{\sqrt{m\omega}\hat{x}}{\sqrt{2}}\delta_x = \frac{x\sqrt{m\omega}}{\sqrt{2}}\psi_x \qquad (11.2.14)$$

because δ_x is an eigendistribution of the operator $\hat{x}$. The crucial point is now that ψ_x is an analytic function of x as one can see by using the Fourier representation for the δ-distribution $\delta_x = \int_{\mathbb{R}} dk/(2\pi)e^{ikx}$. We can therefore analytically extend ψ_x to the complex plane $x \to x - ip/(m\omega)$ and arrive with the trivial redefinition $\psi_{x - ip/(m\omega)} \mapsto \psi_z$ at

$$\hat{z}\psi_z = z\psi_z \qquad (11.2.15)$$

One can check that the state $\psi_z/\|\psi_z\|$ coincides with the usual harmonic oscillator coherent states up to a phase.

We see that the harmonic oscillator coherent states can be naturally put into the language of the Wick rotation transform of Section 10.7.1. This observation, stripping off the particulars of the harmonic oscillator, admits a generalisation that applies to any symplectic manifold $\mathcal{M}, \{.,.\}$ which is a cotangent bundle $\mathcal{M} = T^*\mathcal{C}$ where $\mathcal{C}$ is the configuration base space of $\mathcal{M}$. It is called the *complexifier method* and provides a *systematic construction mechanism*. The method has been introduced for the first time in [315] and is by now also appreciated by mathematicians (see [319, 320]). See [487] for a more detailed account and comparison with other proposals.

Let $(\mathcal{M}, \Omega)$ be a symplectic manifold with strong symplectic structure Ω (notice that $\mathcal{M}$ is allowed to be infinite-dimensional). We will assume that $\mathcal{M} = T^*\mathcal{C}$ is a cotangent bundle. Let us then choose a real polarisation of $\mathcal{M}$, that is, a real Lagrangian submanifold $\mathcal{C}$ which will play the role of our configuration space. Then a loose definition of a complexifier is as follows:

Definition 11.2.3. *A complexifier is a positive definite function C on $\mathcal{M}$ with the dimension of an action, which is smooth a.e. (with respect to the Liouville measure induced from Ω) and whose Hamiltonian vector field is everywhere nonvanishing on $\mathcal{C}$. Moreover, for each point $q \in \mathcal{C}$ the function $p \mapsto C_q(p) = C(q, p)$*

grows stronger than linearly with $\|p\|_q$ *where* p *is a local momentum coordinate and* $\|.\|_q$ *is a suitable norm on* $T_q^*(\mathcal{C})$.

In the course of our discussion we will motivate all of these requirements.

The reason for the name *complexifier* is that C enables us to generate a *complex polarisation* of $\mathcal{M}$ from C as follows: if we denote by q local coordinates of $\mathcal{C}$ (we do not display any discrete or continuous labels but we assume that local fields have been properly smeared with test functions) then

$$z(m) := \sum_{n=0}^{\infty} \frac{i^n}{n!} \{q, C\}_{(n)}(m) \tag{11.2.16}$$

define local complex coordinates of $\mathcal{M}$ provided we can invert $z, \bar{z}$ for $m := (q, p)$ where p are the fibre (momentum) coordinates of $\mathcal{M}$. This is granted at least locally by definition (11.2.3). Here the multiple Poisson bracket is inductively defined by $\{C, q\}_{(0)} = q$, $\{C, q\}_{(n+1)} = \{C, \{C, q\}_{(n)}\}$ and makes sense due to the required smoothness. What is interesting about (11.2.16) is that it implies the following bracket structure

$$\{z, z\} = \{\bar{z}, \bar{z}\} = 0 \tag{11.2.17}$$

while $\{z, \bar{z}\}$ is necessarily non-vanishing. The reason for this is that (11.2.16) may be written in the more compact form

$$z = e^{-i\mathcal{L}_{\chi_C}} q = \left([\varphi_{\chi_C}^t]^* q \right)_{t=-i} \tag{11.2.18}$$

where χ_C denotes the Hamiltonian vector field of C, unambiguously defined by $i_{\chi_C} \Omega + dC = 0$, $\mathcal{L}$ denotes the Lie derivative and $\varphi_{\chi_C}^t$ is the one-parameter family of symplectomorphisms generated by χ_C. Formula (11.2.18) displays the transformation (11.2.16) as the analytic extension to imaginary values of the one-parameter family of diffeomorphisms generated by χ_C and since the flow generated by Hamiltonian vector fields leaves Poisson brackets invariant, (11.2.17) follows from the definition of a Lagrangian submanifold. The fact that we have to continue to the negative imaginary axis rather than the positive one is important in what follows and has to do with the required positivity of C.

The importance of this observation is that either of $z, \bar{z}$ are coordinates of a Lagrangian submanifold of the complexification $\mathcal{M}^{\mathbb{C}}$, that is, a complex polarisation and thus may serve to define a Bargmann–Segal representation of the quantum theory (wave functions are holomorphic functions of z). The diffeomorphism $\mathcal{M} \to \mathcal{C}^{\mathbb{C}}$; $m \mapsto z(m)$ shows that we may think of $\mathcal{M}$ either as a symplectic manifold or as a complex manifold (complexification of the configuration space). Indeed, the polarisation is usually a positive Kähler polarisation with respect to the natural Ω-compatible complex structure on a cotangent bundle defined by local Darboux coordinates, if we choose the complexifier to be a function of p only. These facts make the associated Segal–Bargmann representation especially attractive. For a short account on symplectic and complex geometry respectively see Sections 19.3 and 19.4 respectively.

We now apply the rules of canonical quantisation: a suitable Poisson algebra $\mathcal{O}$ of functions O on $\mathcal{M}$ is promoted to an algebra $\hat{\mathcal{O}}$ of operators $\hat{O}$ on a Hilbert space $\mathcal{H}$ subject to the condition that Poisson brackets turn into commutators divided by $i\hbar$ and that reality conditions are reflected as adjointness relations, that is,

$$[\hat{O}, \hat{O}'] = i\hbar\widehat{\{O, O'\}} + O(\hbar), \quad \hat{O}^\dagger = \hat{\bar{O}} + O(\hbar) \qquad (11.2.19)$$

where quantum corrections are allowed (and in principle unavoidable except if we restrict $\mathcal{O}$, say, to functions linear in momenta). We will assume that the Hilbert space can be represented as a space of square integrable functions on (a distributional extension $\overline{\mathcal{C}}$ of) $\mathcal{C}$ with respect to a positive, faithful probability measure μ, that is, $\mathcal{H} = L_2(\overline{\mathcal{C}}, d\mu)$ as it is motivated by the real polarisation (cotangent bundle structure).

The fact that C is positive motivates to quantise it in such a way that it becomes a self-adjoint, positive definite operator. We will assume this to be the case in what follows. Applying then the quantisation rules to the functions z in (11.2.16) we arrive at

$$\hat{z} = \sum_{n=0}^{\infty} \frac{i^n}{n!} \frac{[\hat{q}, \hat{C}]_{(n)}}{(i\hbar)^n} = e^{-\hat{C}/\hbar} \hat{q} e^{\hat{C}/\hbar} \qquad (11.2.20)$$

The appearance of $1/\hbar$ in (11.2.20) justifies the requirement for $C/\hbar$ to be dimensionless in (1.1.1). We will call $\hat{z}$ the *annihilation operator* for reasons that will become obvious in a moment.

Let now $q \mapsto \delta_{q'}(q)$ be the δ-distribution with respect to μ with support at $q = q'$. (In more mathematical terms, consider the complex probability measure, denoted as $\delta_{q'} d\mu$, which is defined by $\int \delta_{q'} d\mu f = f(q')$ for measurable f.) Notice that since C is non-negative and necessarily depends non-trivially on momenta (which will turn into (functional) derivative operators in the quantum theory), the operator $e^{-\hat{C}/\hbar}$ is a *smoothing operator*. Therefore, although $\delta_{q'}$ is certainly not square integrable, the complex measure (which is probability if $\hat{C} \cdot 1 = 0$)

$$\psi_{q'} := e^{-\hat{C}/\hbar} \delta_{q'} \qquad (11.2.21)$$

has a chance to be an element of $\mathcal{H}$. Whether or not it is depends on the details of $\mathcal{M}, \Omega, C$. For instance, if C as a function of p at fixed q has flat directions, then the smoothing effect of $e^{-\hat{C}/\hbar}$ may be insufficient, so in order to avoid this we required that C is positive definite and not merely non-negative. If C was indefinite, then (11.2.21) has no chance to make sense as an L_2 function.

We will see in a moment that (11.2.21) qualifies as a candidate *coherent state* if we are able to analytically extend (1.1.6) to complex values z of q' where the label z in ψ_z will play the role of the point in $\mathcal{M}$ at which the coherent state is peaked. In order that this is possible (and in order that the extended function is

still square integrable), (11.2.21) should be entire analytic. Now $\delta_{q'}(q)$ roughly
has an integral kernel of the form $e^{ik[(q-q')]}$ (where k as a cotangential vector is
considered as a linear functional $k[.]$ on the space of tangential vectors) which is
analytic in q' but the integral over k, after applying $e^{-\hat{C}/\hbar}$, will produce an entire
analytic function only if there is a damping factor which decreases faster than
exponentially. This provides the intuitive explanation for the growth requirement
in Definition 11.2.3 . Notice that the ψ_z are not necessarily normalised.

Let us then assume that

$$q \mapsto \psi_m(q) := [\psi_{q'}(q)]_{q' \to z(m)} = \left[e^{-\hat{C}/\hbar}\delta_{q'}(q)\right]_{q' \to z(m)} \qquad (11.2.22)$$

is an entire L_2 function. Then ψ_m is automatically an *eigenfunction of the anni-*
hilation operator $\hat{z}$ with eigenvalue z since

$$\hat{z}\psi_m = \left[e^{-\hat{C}/\hbar}\hat{q}\delta_{q'}\right]_{q' \to z(m)} = \left[q'e^{-\hat{C}/\hbar}\delta_{q'}\right]_{q' \to z(m)} = z(m)\psi_m \qquad (11.2.23)$$

where in the second step we used the fact that the delta distribution is a generalised eigenfunction of the operator $\hat{q}$. But to be an eigenfunction of an annihilation operator *is one of the accepted definitions of coherent states!*

Next, let us verify that ψ_m indeed has a chance to be peaked at m. To see
this, let us consider the self-adjoint (modulo domain questions) combinations

$$\hat{x} := \frac{\hat{z} + \hat{z}^\dagger}{2}, \quad \hat{y} := \frac{\hat{z} - \hat{z}^\dagger}{2i} \qquad (11.2.24)$$

whose classical analogues provide real coordinates for $\mathcal{M}$. Then we have automatically from (1.1.8)

$$< \hat{x} >_m := \frac{< \psi_m, \hat{x}\psi_m >}{||\psi_m||^2} = \frac{z(m) + \bar{z}(m)}{2} =: x(m) \qquad (11.2.25)$$

and similar for y. Equation (11.2.25) tells us that the operator $\hat{z}$ should really
correspond to the function $m \mapsto z(m)$, $m \in \mathcal{M}$.

Now we compute by similar methods that

$$< [\delta\hat{x}]^2 >_m := \frac{< \psi_m, [\hat{x} - < \hat{x} >_m]^2\psi_m >}{||\psi_m||^2} = < [\delta\hat{y}]^2 >_m = \frac{1}{2}| < [\hat{x}, \hat{y}] >_m | \qquad (11.2.26)$$

so that the ψ_m are automatically *minimal uncertainty states for $\hat{x}, \hat{y}$*, moreover
the fluctuations are unquenched. This is the second motivation for calling the
ψ_m coherent states. Certainly one should not only check that the fluctuations
are minimal but also that they are small compared with the expectation value,
at least at generic points of the phase space, in order that the quantum errors
are small.

The *infinitesimal Ehrenfest* property

$$\frac{< [\hat{x}, \hat{y}] >_z}{i\hbar} = \{x, y\}(m) + O(\hbar) \qquad (11.2.27)$$

follows if we have properly implemented the canonical commutation relations
and adjointness relations. The size of the correction, however, does not follow

from these general considerations but the minimal uncertainty property makes small corrections plausible. Condition (11.2.27) supplies information about how well the symplectic structure is reproduced in the quantum theory.

For the same reason one expects that the peakedness property

$$\left| \frac{<\psi_m, \psi_{m'}>|^2}{||\psi_m||^2\,||\psi_{m'}||^2} \approx \chi_{K_m}(m') \right. \tag{11.2.28}$$

holds, where K_m is a phase cell with centre m and Liouville volume $\approx \sqrt{<[\delta\hat{x}]^2>_m<[\delta\hat{y}]^2>_m}$ and χ denotes the characteristic function of a set.

Finally one wants coherent states to be overcomplete in order that every state in $\mathcal{H}$ can be expanded in terms of them. This has to be checked on a case-by-case basis but by the fact that our complexifier coherent states are for real z, nothing else than regularised δ-distributions which in turn provide a (generalised) basis makes this property plausible to hold.

The reader should verify explicitly that the usual coherent states for the harmonic oscillator fall precisely into our scheme.

Remark: It is crucial to know the map $m \mapsto z(m)$. If we are just given some states ψ_z with $z \in \mathcal{C}^{\mathbb{C}}$ then we have no way of finding the point $m \in \mathcal{M}$ to which z corresponds (there are certainly infinitely many diffeomorphisms between $\mathcal{M}, \mathcal{C}^{\mathbb{C}}$) and the connection with the classical phase space is lost. Without this knowledge we cannot check, for instance, whether the infinitesimal Ehrenfest property holds. This is one of the nice things that the complexifier method automatically does for us. In order to know the function $z(m)$ we must know what the classical limit of $\hat{C}$ is, if we are just given some abstract operator without classical interpretation, then again we do not know $z(m)$. Of course, if we are given just some set of states ψ_z we could try to construct an appropriate map $m \mapsto z(m)$ as follows: find a (complete) set of basic operators $\hat{O}$ whose fluctuations are (close to) minimal and define a map $z \mapsto O'(z):=<\psi_z, \hat{O}\psi_z>/||\psi_z||^2$. Also define $\{O', \bar{O}'\}'(z):= \lim_{\hbar \to 0} <\psi_z, \frac{[\hat{O},\hat{O}^\dagger]}{i\hbar}\psi_z>/||\psi_z||^2$. Now construct $m \mapsto z(m)$ by asking that the pull-back functions $O(m):= O'(z(m))$ satisfy

$$\{O, \bar{O}\}(m) = \{O', \bar{O}'\}'(z(m)) \tag{11.2.29}$$

in other words, that the symplectic structure defined by $\{.,.\}'$ is the symplectomorphic image of the original symplectic structure $\{.,.\}$ under the canonical transformation $m \mapsto z(m)$. The reader will agree that this procedure is rather indirect and especially in field theory will be hard to carry out. Notice that by far not all symplectic structures are equivalent, so that even to find appropriate operators for given ψ_z such that at least one map $m \mapsto z(m)$ exists will be a non-trivial task. The complexifier method guarantees all of this to be the case from the outset, since the transformation (11.2.16) is a canonical transformation by construction.

11.2.3 Complexifier coherent states for diffeomorphism-invariant theories of connections

After having chosen a Hilbert space $\mathcal{H}_0 = L_2(\overline{\mathcal{A}}, d\mu_0)$, the only input required in the complexifier construction is the choice of a complexifier itself. We will restrict our class of choices to functions $C = C(E)$ which are gauge-invariant but not necessarily diffeomorphism-invariant (since we can use the (D-)metric to be approximated as a naturally available background metric) and only depend on the electric field to make life simple. We suppose that the associated operator $\hat{C}$ is a densely defined positive definite operator on $\mathcal{H}_0$ whose spectrum is pure point (discrete). The latter assumption is not really a restriction because operators which are constructed from (limits of) electric flux operators quite generically have this sort of spectrum, as we will see in Chapter 13. Let T_s, $s \in \mathcal{S}$ be the associated uncountably infinite orthonormal basis of eigenvectors. The labels $s = (\gamma, \vec{\pi}, \vec{I})$ are triples consisting of a semianalytic graph γ, an array of equivalence classes of non-trivial irreducible representations π_e, one for each edge e of γ and an array of intertwiners I_v, one for each vertex v of γ. The intertwiners are chosen in such a way that the T_s are not only gauge-invariant but also eigenfunctions of $\hat{C}$. The space of possible $\vec{I}$ at given $\vec{\pi}$ is always finite-dimensional and the operators of the form $\hat{C}$ which we consider here can never change $\vec{\pi}, \gamma$. Thus, the T_s are just suitable linear combinations of the usual spin-network functions.

Let λ_s be the corresponding eigenvalues. Then

$$\delta_{A'} = \sum_{s \in \mathcal{S}} T_s(A')\overline{T_s} \tag{11.2.30}$$

is a suitable representation of the δ distribution with respect to μ_0, that is,

$$\int_{\overline{\mathcal{A}}} d\mu_0(A)\delta_{A'}(A)f(A) = \sum_s T_s(A') < T_s, f >= f(A') \tag{11.2.31}$$

and our complexifier coherent states become explicitly

$$\psi_m = \sum_{s \in \mathcal{S}} e^{-\lambda_s/\hbar} T_s(Z(m))\overline{T_s} \tag{11.2.32}$$

where we have made use of the fact that the expression for $\hat{C}$ is real, $m = (A, E)$ is the point in $\mathcal{M}$ to be approximated and

$$[Z_a^j(m)](x) := [^{(\mathbb{C})}A_a^j(m)](x) := A_a^j(x) - i\kappa \frac{\delta C}{E_j^a(x)} = \sum_{n=0}^{\infty} \frac{i^n}{n!} \{A_a^j(x), C\}_{(n)} \tag{11.2.33}$$

is a *complex-valued G-connection* since C is supposed to be gauge-invariant.

Since there are more than countably many terms different from zero in (11.2.32) the states ψ_m are not elementsof $\mathcal{H}^0$. Rather, they define algebraic distributions in Cyl* defined by

$$\psi_m[f] := < 1, \psi_m \ f >$$

$$= < 1, \left[e^{-\hat{C}/\hbar} \delta_{A'}\right] f >_{A' \to Z(m)} = < \overline{f}, \left[e^{-\hat{C}/\hbar} \delta_{A'}\right] >_{A' \to Z(m)}$$

$$= < e^{-\hat{C}/\hbar} \overline{f}, \delta_{A'} >_{A' \to Z(m)} = < 1, \delta_{A'} \overline{e^{-\hat{C}/\hbar} \overline{f}} >_{A' \to Z(m)}$$

$$= \left(\delta_{A'} \left[\overline{e^{-\hat{C}/\hbar} \overline{f}}\right]\right)_{A' \to Z(m)} \tag{11.2.34}$$

For $f = T_s$ the right-hand side of (11.2.33) becomes $e^{-\lambda_s/\hbar} T_s(Z(m))$ and since C is supposed to depend on a sufficiently high power of E and since $|T_s(Z(m))|$ grows at most exponentially with the highest weight of $\vec{\pi}$, these numbers are actually bounded from above so that the distribution is well-defined. Equivalently, we can consider ψ_m as a complex probability measure (since the δ distribution is).

Consider for each semianalytic path e the *annihilation operators*

$$\hat{g}_e := e^{-\hat{C}/\hbar} \hat{A}(e) e^{\hat{C}/\hbar} \tag{11.2.35}$$

which are the quantum analogues of the classical functions $Z(m)(e) = h_e(^{(\mathbb{C})}A(m)) = g_e(m)$ where $h_e(A) = A(e)$ denotes the holonomy of A along e. Thus $g_e(m)$ is the holonomy along e of the complex connection $^{(\mathbb{C})}A$. The holonomy property can also be explicitly checked for the operators $\hat{g}_e$ themselves, since for a composition of paths $e = e_1 \circ e_2$ we have from the holonomy property for $\hat{A}$ that

$$\hat{g}_{e_1} \hat{g}_{e_2} = e^{-\hat{C}/\hbar} \hat{A}(e_1) \hat{A}(e_2) e^{\hat{C}/\hbar} = \hat{g}_e \text{ and } \hat{g}_{e^{-1}} = (\hat{g}_e)^{-1} \tag{11.2.36}$$

where product and inversion is that within $G^{\mathbb{C}}$.

As one can explicitly check, ψ_m is a simultaneous generalised eigenvector of all the $\hat{g}_e$, that is,

$$(\hat{g}_e \psi_m)[f] := < 1, [\hat{g}_e \psi_m] \ f >$$

$$= < \hat{g}_e^\dagger \overline{f}, \psi_m >=< 1, \psi_m \overline{\hat{g}_e^\dagger \overline{f}} >= \psi_m[\overline{\hat{g}_e^\dagger \overline{f}}]$$

$$= h_e(Z(m)) \psi_m[f] \tag{11.2.37}$$

The crucial point is now that although the ψ_m are not normalisable, we may be able to define a positive linear functional ω_m on our algebra of functions as expectation value functional

$$\omega_m(\hat{O}) := \frac{< \psi_m, \hat{O} \psi_m >}{||\psi_m||^2} \tag{11.2.38}$$

where we have *used the inner product on* $\mathcal{H}^0$ and no other additional inner product! This is conceptually appealing because, if we can give meaning to (11.2.38), then we arrive at a new representation of the canonical commutation relations *which is derived from* $\mathcal{H}^0$, whence $\mathcal{H}^0$ plays the role of the fundamental representation, very much in the same way as temperature representations in ordinary quantum field theory can be derived from the Fock representation by limits of the kind performed in (11.2.38). This idea has been made mathematically precise in [496].

Expression (11.2.38) is very formal in the sense that it is the quotient of two uncountably infinite series. However, notice that we can easily give meaning to it at least for *normal ordered functions of annihilation and creation operators* as

$$\omega_m(: \hat{O} :) = O(m) = O(\{g_e(m), \overline{g_e(m)}\}) \qquad (11.2.39)$$

which has no quantum corrections at all. Thus, if the functions $m \mapsto g_e(m)$ separate the points of $\mathcal{M}$ as e varies, then we may use them as the basic variables in the quantum theory and they, together with their adjoints, have the correct expectation values in the representation induced by ω_m via the GNS construction, moreover, that representation by construction also solves the adjointness and canonical commutation relations. Of course, (11.2.39) will be an interesting functional only if the normal ordering corrections of interesting operators are finite. This can only be decided in a case-by-case analysis.

As an illustrative example (see [485, 486] for more details) let $Q^{ab} := E_j^a E_k^b \delta^{jk}$ and consider the diffeomorphism-invariant complexifier (recall that E is a density of weight one)

$$C := \frac{1}{a\kappa} \int_\sigma d^D x (\sqrt{\det(Q)})^{1/(D-1)} \qquad (11.2.40)$$

where a is a parameter with units of $(\hbar\kappa)^{1/(D-1)}$. Our convention is that A has dimension of cm^{-1}, thus $\frac{1}{\kappa} \int_{\mathbb{R}} dt \int_\sigma d^D x \dot{A}_a^j E_j^a$, the kinetic term in the canonical action, must have dimension of an action, therefore $E/(\hbar\kappa)$ must have dimension cm$^{-(D-1)}$. Thus, in order that $C/\hbar$ be dimension-free, a must have the said dimension. For example, for general relativity in $D+1 = 4$ dimensions, $(\hbar\kappa)^{1/(D-1)} = \ell_p$ is the Planck length, (11.2.40) is essentially the volume functional V for σ and if we are interested in cosmological questions or scales, then $a = 1/\sqrt{\Lambda}$ would be a natural choice, where Λ is the cosmological constant. In that case the quantised complexifier would simply be given by

$$\hat{C}/\hbar = \frac{1}{a\ell_p^2} \hat{V} = \frac{\ell_p}{a} \hat{v} \qquad (11.2.41)$$

where $\hat{v} = \hat{V}/\ell_p^3$ is the dimension-free volume functional which has discrete spectrum (the eigenvalues of the volume itself are multiples of ℓ_p^3). Thus $\hat{C} = t\hat{v}$ where the tiny *classicality parameter*

$$t = \frac{\ell_p}{a} = \sqrt{\hbar\kappa\Lambda} \qquad (11.2.42)$$

has entered the stage (it equals 10^{-60} for the current value of Λ). We easily compute the complexified connection in this case as

$$^{(\mathbb{C})}A = A - ie/(2a) \qquad (11.2.43)$$

where e is the dimension-free co-triad. Thus, with the volume as the complexifier, the $g_e(m)$ indeed separate the points of $\mathcal{M}$!

However, in order to qualify as a good semiclassical state, at the very least the fluctuations of our basic operators with respect to ω_m should be small compared with the expectation values at generic points of the phase space, in particular, they should be finite. Whether or not this is the case has to be checked for the explicit choices for C.

It should be noted, however, that even if the fluctuations do not come out finite, then we can still produce graph-dependent coherent states, which we will call *cutoff* states, because the finite graph on which they are based serves as a cutoff in the number of degrees of freedom to be considered. In particular, they *are* elements of $\mathcal{H}^0$ defined as follows: given a graph γ, consider all of its subgraphs $\gamma' \subset \gamma$ obtained by removing edges in all possible ways. Given a label s we write $s = (\gamma(s), \vec{\pi}(s), \vec{I}(s))$ and define a graph-dependent δ-distribution

$$\delta_{A',\gamma}(A) := \sum_{\gamma' \subset \gamma} \sum_{s;\, \gamma(s)=\gamma'} T_s(A')\overline{T_s(A)} \tag{11.2.44}$$

It is easy to check that (11.2.44) is a δ-distribution restricted to those functions on $\overline{\mathcal{A}}$ which can be written in terms of the holonomies $A(p)$ where $p \subset \gamma$. In fact, (11.2.44) is the *cutoff* of (1.1.14) with the cutoff given by the graph γ since (11.2.44) is the restriction of the uncountably infinite series in (11.2.30) to the countably infinite one in (11.2.44) given by restricting the sum over $s \in \mathcal{S}$ to $s \in \mathcal{S}_\gamma$ where

$$\mathcal{S}_\gamma = \{s \in \mathcal{S};\ \gamma(s) \subset \gamma\} \tag{11.2.45}$$

In fact, we can consider the Hilbert space $\mathcal{H}^0_\gamma = L_2(\overline{\mathcal{A}}_\gamma, d\mu_{0,\gamma})$ where $\mu_{0,\gamma}$ is the push-forward of μ_0 to the space $\overline{\mathcal{A}}_\gamma$ which is the spectrum of holonomy algebra restricted to paths within γ. Then δ_γ is in fact the δ-distribution with respect to $\mu_{0\gamma}$. In other words, δ_γ is the cylindrical projection of the complex measure δ.

We now obtain normalisable, graph-dependent coherent states

$$\psi_{\gamma,m}(A) = \left(\left[e^{-\hat{C}/\hbar} \delta_{\gamma,A'} \right]_{A' \to {}^{(\mathbb{C})}A(m)} \right)(A) = \sum_{s \in \mathcal{S}_\gamma} e^{-\lambda_s/\hbar} T_s({}^{(\mathbb{C})}A(m))\overline{T_s(A)} \tag{11.2.46}$$

with norm

$$||\psi_{\gamma,m}||^2 = \sum_{s \in \mathcal{S}_\gamma} e^{-2\lambda_s/\hbar} |T_s({}^{(\mathbb{C})}A(m))|^2 \tag{11.2.47}$$

which converges due to our assumptions on the spectrum λ_s. Notice that these assumptions might not hold for the volume complexifier (the volume operator is only non-negative but not positive definite, the spectrum has flat directions and it would be crucial to know how generic these are, a problem very similar in nature (but much simpler) to the convergence proof of the partition function of Euclidean Yang–Mills theory). By arguments very similar to those from above it is easy to check that the $\psi_{\gamma,m}$ are still eigenstates of the operators $\hat{g}_e$ provided that the path e lies within γ. In other words, for normal ordered functions of some set of operators $\hat{g}_e, \hat{g}_e^\dagger$ it is unimportant whether we work with the complete state ψ_m or with the cutoff state $\psi_{\gamma,m}$, as far as expectation values are concerned, as long as γ contains all the paths e under consideration. However, the fluctuations will be significantly different in general since the square of a normal ordered operator is no longer normal ordered. As one might expect, it is the finiteness of the fluctuations which will force us to usually work with graph-dependent coherent states.

Thus, we arrive at a *coherent state family* $\{\psi_{\gamma,m}\}_{\gamma \in \Gamma_0^\omega}$ for each $m \in \mathcal{M}$ where Γ_0^ω denotes the set of semianalytic, compactly supported graphs embedded into σ. They define a complex probability measure μ_m through the consistent family of measures $d\mu_{\gamma,m} := \psi_{\gamma,m} d\mu_{0,\gamma}$. To see that this family of measures is automatically consistent we consider for $\gamma' \subset \gamma$ the projections $p_{\gamma'\gamma} : \overline{\mathcal{A}}_\gamma \to \overline{\mathcal{A}}_{\gamma'}$ defined by restricting connections from paths within γ to paths within γ'. Now the Hilbert space $\mathcal{H}^0$ is in fact the inductive limit of the Hilbert spaces $\mathcal{H}_\gamma^0$, that is, there exist isometric monomorphisms

$$\hat{U}_{\gamma'\gamma} : \mathcal{H}_{\gamma'}^0 \to \mathcal{H}_\gamma^0; \ f_{\gamma'} \mapsto p_{\gamma\gamma'}^* f_{\gamma'} \tag{11.2.48}$$

for all $\gamma' \subset \gamma$. These maps satisfy the consistency condition

$$\hat{U}_{\tilde{\gamma}\gamma} \hat{U}_{\gamma'\tilde{\gamma}} = \hat{U}_{\gamma'\gamma} \tag{11.2.49}$$

for all $\gamma' \subset \tilde{\gamma} \subset \gamma$. Recall that an operator $\hat{O}$ on $\mathcal{H}_0$ can be thought of as the inductive limit of a family of operators $\{\hat{O}_\gamma\}_{\gamma \in \Gamma}$, that is, $\hat{O}_\gamma$ is densely defined on $\mathcal{H}_\gamma$ subject to the consistency condition

$$\hat{O}_\gamma \hat{U}_{\gamma'\gamma} = \hat{U}_{\gamma\gamma'} \hat{O}_{\gamma'} \tag{11.2.50}$$

for all $\gamma' \subset \gamma$ (there is also a condition for the domains of definition which we skip here). Thus, in particular, the complexifier is a consistently defined operator family all of whose members are self-adjoint and positive on the respective $\mathcal{H}_\gamma^0$. Therefore, if $f_{\gamma'}$ depends only on connections restricted to paths within γ' we have

$$\int_{\mathcal{A}/\mathcal{G}_\gamma} d\mu_{\gamma,m} [p_{\gamma\gamma'}^* f_{\gamma'}] = \left(\int_{\mathcal{A}/\mathcal{G}_\gamma} d\mu_{\gamma,0} \delta_{A',\gamma} [e^{-\hat{C}_\gamma/\hbar} \hat{U}_{\gamma'\gamma} f_{\gamma'}] \right)_{A \to A^{(\mathbb{C})}}$$

$$= \left(\int_{\mathcal{A}/\mathcal{G}_\gamma} d\mu_{\gamma,0} \delta_{A',\gamma} [\hat{U}_{\gamma'\gamma} e^{-\hat{C}_{\gamma'}/\hbar} f_{\gamma'}] \right)_{A \to A^{(\mathbb{C})}}$$

$$= \left(\int_{\mathcal{A}/\mathcal{G}_{\gamma'}} d\mu_{\gamma',0} \delta_{A',\gamma'} [e^{-\hat{C}_{\gamma'}/\hbar} f_{\gamma'}] \right)_{A \to A^{(\mathbb{C})}}$$

$$= \int_{\mathcal{A}/\mathcal{G}_{\gamma'}} d\mu_{\gamma',m} f_{\gamma'} \tag{11.2.51}$$

The projective limit of these measures coincides with the measure $\psi_m d\mu_0$. The notation is abusing because it suggests that μ_m is absolutely continuous with respect to μ_0, which certainly is not the case because $\psi_m \notin L_1(\overline{\mathcal{A}}, d\mu_0)$.

11.2.4 Concrete example of complexifier

The first example of complexifier coherent states for the gauge group SU(2) was constructed in [488, 489]. Here we will exhibit an improved derivation [487] of those states starting from a gauge-invariant classical complexifier whose corresponding operator is densely and cylindrically consistently defined with explicitly known pure point spectrum. This works for arbitrary compact gauge groups. Moreover, the complexifier does not require the additional structure of the dual cell complex introduced in [488]. The cell complex is replaced by another structure which in turn defines dimensionless numbers l_e subject to $l_{eoe'} = l_e + l_{e'}$, $l_{e^{-1}} = l_e$ which are important for cylindrical consistency. The analysis of the semiclassical properties of these states can be reduced to that carried out in [488] as we will show, and will not be repeated here.

The clue for how to construct a complexifier with all of these properties comes from the observation that for non-Abelian gauge theories whose Hilbert space is based on holonomies the only known, well-defined and cylindrical momentum operators come from electric fluxes

$$E_j(S) = \int_S dS_a(x) E_j^a(x) \qquad (11.2.52)$$

These objects are not gauge-invariant, however, there are precisely two basic invariants that one can build from those, namely $E_j(S)E_k(S')\delta^{jk}$ and $E_j(S)E_k(S')E_l(S')\epsilon^{jkl}$ in the limit as the surfaces involved shrink to a single point. The operators on $\mathcal{H}^0$ for which this shrinking process converges to a well-defined operator are precisely the area operator on the one hand and volume and length operators on the other hand, as we will see in Chapter 13. We have already discussed the volume operator as a possible complexifier above and, in fact, it seems to be the more natural possibility because we do not need to introduce any other structure, however, since its spectrum is presently only poorly understood, we will turn to the area operator. By definition, the area operator is only supported on a given surface but we must obtain a complexifier which is supported everywhere in order that a damping factor is produced *for every graph*. Moreover, as we have shown in Section 11.2.2, we must use a power of the area operator which is greater than one in order to arrive at an entire analytic function (convergence) and since with an embedding $X : \check{S} \subset \mathbb{R}^2 \to S$

$$\mathrm{Ar}(S) = \int_{X^{-1}(S)} d^2u \sqrt{\det(X^*q)}(u) = \int_{X^{-1}(S)} d^2u \sqrt{\left[E_j^a(X(u)) n_a^S(u)\right]^2}$$

$$(11.2.53)$$

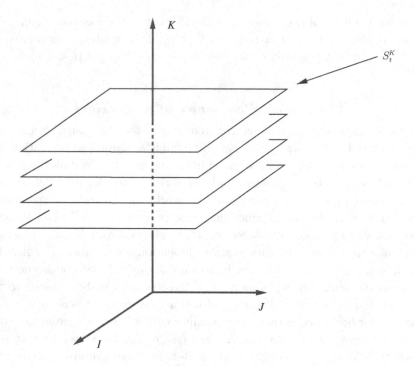

Figure 11.3 A foliation by surfaces.

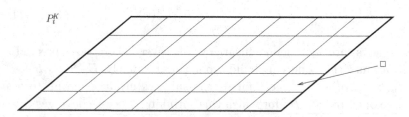

Figure 11.4 A parquet within a leaf of the foliation.

where $n_a^S(u) = \epsilon_{abc} X_{,u^1}^b X_{,u^2}^c$ we see easily that $\lim_{S \to x} [\mathrm{Ar}(S)]^2 / [E_j(S) E_j(S)] = 1$. Thus, the natural power, from the point of view of [488,489] which was built on a gauge-invariant version of objects of the type $E_j(S) E_j(S)$, is two. We will then approximate a Gaußian decay as closely as we can in the non-Abelian context.

How should we then construct a complexifier built from objects of the kind $[\mathrm{Ar}(S)]^2$ which is supported everywhere in σ? There are many possibilities and we will present just one of them based on the structure of a *foliation and parquet* (see Figure 11.4): let us introduce D linearly independent foliations X_t^I of σ, that is, for each $t \in \mathbb{R}$ we obtain an embedding of a $D-1$ surface[3] $X_t^I : \check{S}_t^I \subset \mathbb{R}^{D-1} \to \sigma$

---

[3] The topology of that surface will depend on t if σ is topologically non-trivial.

whose topology may vary with t, I and linear independence means that at each point $x \subset \sigma$ the D 'normal' co-vectors

$$n_a^I(x) := \epsilon_{aa_1\ldots a_{D-1}} \left[X_{t,u^1}^{Ia_1} \ldots X_{t,u^{D-1}}^{Ia_{D-1}} \right]_{X_t^I(u)=x} \tag{11.2.54}$$

or the D tangents $[(\partial X_t^I(u))/\partial t]_{X_t^I(u)=x}$ are linearly independent. Within each leaf of the foliation X_t^I fix a *parquet* P_t^I, that is, a partition into smaller $D-1$ surfaces of fixed (say simplicial) topology and we require that for each I the parquet varies smoothly with I. Notice that all of these structures do not refer to a background metric. The parquet is quite similar in nature to the polyhedronal decomposition dual to a graph defined in [488], but it is different because it is *graph-independently defined* so that the resulting complexifier can be defined already classically rather than only in quantum theory graph-wise. We then propose

$$C = \frac{1}{2a\kappa} \sum_{I=1}^{D} \int_{\mathbb{R}} dt \sum_{\square \in P_t^I} [\mathrm{Ar}(\square)]^2 \tag{11.2.55}$$

where a is again an appropriate dimensionful parameter. For instance, for Quantum General Relativity in $D = 3$, a would have dimension cm^2 if we take the parameter t dimension-free.

The corresponding complexified connection would be

$$A_a^{\mathbb{C}j}(x) = A_a^j(x) - i \sum_{I=1}^{D} \left(\frac{\mathrm{Ar}(\square_x^I)}{|\det (\partial X_t^I / \partial(t, u))|} \frac{E_j^b(x) n_b^{\square_x^I}(t, u)}{\sqrt{\left[E_j^c(x) n_c^{\square_x^I}(t, u)\right]^2}} n_a^{\square_x^I}(t, u) \right)_{X_t^I(u)=x} \tag{11.2.56}$$

where $\square_x^I \in P_{t^I(x)}^I, X_{t^I(x)}^I(u^I(x)) = x$ is the surface containing x. From (11.2.56) we see why we cannot do without the parquet since then we would have to work with the areas of the whole leaves, which would be an insufficiently local object. However, even (11.2.56) only allows us to reconstruct E from $A^{\mathbb{C}}$ with a precision that is defined by how fine the parquet is. Strictly speaking then, $A^{\mathbb{C}}$ does not separate the points of $\mathcal{M}$, but it does so with a precision that is sufficient for semiclassical purposes depending on how fine the parquet is.

The spectrum of the corresponding complexifier operator is essentially derived from the known spectrum of the area operator, together with an important key observation which is responsible for making this operator really leave all the Cyl$_\gamma$ separately invariant. As we will see in detail in Chapter 13, given an open, semianalytic, oriented surface S and a graph γ we can always subdivide its edges in such a way that any of them belongs to precisely one of the four disjoint subsets $E_{\mathrm{in}}, E_{\mathrm{out}}, E_{\mathrm{up}}, E_{\mathrm{down}}$ of edges of γ where $e \in E_{\mathrm{in}} \Rightarrow e \cap S = e$, $e \in E_{\mathrm{out}} \Rightarrow e \cap S = \emptyset$, $e \in E_{\mathrm{up}} \Rightarrow e \cap S = b(e)$ and e points up, $e \in E_{\mathrm{down}} \Rightarrow e \cap S = b(e)$ and e points down. Here 'up, down' means that there exists a neighbourhood U of $b(e)$ such that $e \cap U$ lies entirely within U^+, U^- respectively. Here U^+, U^- are the two disjoint halves into which S cuts U and U^+ is the half into which the co-normal of S points. Let $P(S, \gamma) = \{b(e), \; e \in E_{\mathrm{up}} \cup E_{\mathrm{down}}\}$ and given $p \in P(S, \gamma)$ let $X_{\mathrm{up}}^j(p) = \sum_{e \in E_{\mathrm{up}}(p)} X_e^j$, $X_{\mathrm{down}}^j(p) = \sum_{e \in E_{\mathrm{down}}(p)} X_e^j$. The operators $\Delta_{\mathrm{up}}(p) = (X_{\mathrm{up}}^j(p))^2$, $\Delta_{\mathrm{down}}(p) = (X_{\mathrm{down}}^j(p))^2$, $\Delta_{\mathrm{updown}}(p) = (X_{\mathrm{up}}^j(p) + X_{\mathrm{down}}^j(p))^2$ are simultaneously diagonisable with $(-2$ times$)$ total angular momentum spectrum. The area operator is given by

$$[\widehat{\mathrm{Ar}(S)}]_{\mathrm{Cyl}_\gamma} = \frac{\hbar\kappa}{8} \sum_{p \in P(S, \gamma)} \sqrt{-2\Delta_{\mathrm{up}}(p) - 2\Delta_{\mathrm{down}}(p) + \Delta_{\mathrm{updown}}(p)} \quad (11.2.57)$$

and its spectrum for $SU(2)$ reads explicitly

$$[\mathrm{Spec}(\widehat{\mathrm{Ar}(S)})]_{\mathrm{Cyl}_\gamma}$$
$$= \frac{\hbar\kappa}{4} \sum_{p \in P(S, \gamma)} \sqrt{2j_{\mathrm{u}}(p)(j_{\mathrm{u}}(p) + 1) + 2j_{\mathrm{d}}(p)(j_{\mathrm{d}}(p) + 1) - j_{\mathrm{ud}}(p)(j_{\mathrm{ud}}(p) + 1)}$$

$$(11.2.58)$$

with $j_{\mathrm{u}}(p) + j_{\mathrm{d}}(p) \geq j_{\mathrm{ud}}(p) \geq |j_{\mathrm{u}}(p) - j_{\mathrm{d}}(p)|$. We have set $\beta = 1$ for simplicity.

The key point is now that the subdivision of edges of γ into the classes $E_{\mathrm{in}}, E_{\mathrm{out}}, E_{\mathrm{up}}, E_{\mathrm{down}}$ *depends on the surface S!* That is, a given spin-network state T_s is not an eigenstate of a given operator $\widehat{\mathrm{Ar}(S)}$, rather we must subdivide the edges of $\gamma(s)$ adapted to S and then decompose the intertwiners $I(s)$ in such a way that we get eigenfunctions of $\Delta_{\mathrm{up}}(p), \Delta_{\mathrm{down}}(p), \Delta_{\mathrm{updown}}(p)$ for all vertices p of γ respectively. It follows that the function $\widehat{\mathrm{Ar}(S)}T_s$ depends, in the non-Abelian case, generally no longer only on the edges of γ but also on the subdivision of the edges of γ as adapted to S. This is dangerous because we are dealing with operators of the form $\int dt [\widehat{\mathrm{Ar}(S_t)}]^2$ for a foliation $t \mapsto S_t$ and the function $[\widehat{\mathrm{Ar}(S_t)}]^2 T_s$ therefore depends on the parameter t. If it depended on a graph γ_t where γ_t depends on a subdivision of edges according to S_t then the operator $\hat{C}$ would not exist since $[\widehat{\mathrm{Ar}(S_t)}]^2 T_s$ is not dt-measurable as we showed in Section 8.2.3. Fortunately this does not happen.

A point $p \in P(S_t, \gamma)$ falls into only one of the two categories: either it is a vertex of γ in which case the subdivision of edges does not change the graph or p is an *interior point of a single edge*. However, in the latter case a spin-network function is already an eigenfunction: if $e = e_u(t) \circ e_d^{-1}(t)$ denotes the adapted decomposition of the corresponding edge of γ with $p := S_t \cap e = b(e_u(t)) = b(e_d(t))$ then from (13.6.2) due to gauge-invariance at p

$$\sqrt{-2\Delta_{\mathrm{up}}(p) - 2\Delta_{\mathrm{down}}(p) + \Delta_{\mathrm{updown}}(p)} T_s = \hbar\kappa \sqrt{j_e(j_e + 1)} \quad (11.2.59)$$

is completely independent of t. We conclude that spin-network functions T_s are simultaneous eigenfunctions of all possible $\widehat{\mathrm{Ar}(S_t)}$ as long as S_t does not contain a vertex of $\gamma(s)$. However, for given T_s the number of vertices of γ is finite and the set $\{t \in \mathbb{R}; S_t \cap V(\gamma) \neq \emptyset\}$ is discrete and thus has dt measure zero.

The spectrum of our complexifier operator therefore can easily be computed as follows: we will assume that the graph $\gamma(s)$ is contained in a region such that each of the embedded surfaces $t \mapsto X_t^I$, $t \in [a, b]$ has topology independent of t with $\gamma \subset \cup_{t \in [a,b]} X_t^I$ for all I. The more general case including topology change just involves introducing more notation and does not lead to new insights and thus will be left to the reader. Our assumptions about the parquet imply then that, given I, we have a corresponding family of surfaces $S_{\square,t}^I$ with a discrete label $\square$. Fix $I, \square$, and a set of intersection numbers $n_e^{I,\square} = 0, 1, 2, \ldots$; $e \in E(\gamma)$ one for each edge of γ and denote by $t_\square^I(\gamma, \vec{n}^{\square,I})$ the dt-measure of the set $\{t \in [a, b]; |S_{\square,t}^I \cap e| = n_e^{\square,I} \; \forall \, e \in E(\gamma)\}$ (notice that we only count isolated intersection points). Then

$$
\begin{aligned}
\frac{\hat{C}}{\hbar} T_s &= \frac{\ell_p^2}{a} \left\{ \sum_{I,\square,\vec{n}^{\square,I}(s)} t_\square^I(\gamma(s), \vec{n}^{\square,I}(s)) \left[\sum_{e \in E(\gamma)} n_e^{\square,I} \sqrt{j_e(j_e + 1)} \right]^2 \right\} T_s \\
&= \frac{\ell_p^2}{a} \left\{ \sum_{e,e' \in E(\gamma)} \sqrt{j_e(j_e + 1)} \sqrt{j_{e'}(j_{e'} + 1)} \sum_{I,\square,\vec{n}^{\square,I}(s)} t_\square^I(\gamma(s), \vec{n}^{\square,I}(s)) n_e^{\square,I} n_{e'}^{\square,I} \right\} T_s \\
&=: \frac{\ell_p^2}{a} \left\{ \sum_{e,e' \in E(\gamma)} G_s^{e,e'} \sqrt{j_e(j_e + 1)} \sqrt{j_{e'}(j_{e'} + 1)} \right\} T_s \qquad (11.2.60)
\end{aligned}
$$

The last equality defines a non-Abelian generalisation of an edge metric [491] which is automatically consistent because the area operator is.

Interestingly, if the parquet is much finer than the graph then each of the surfaces $S_\square^I$ will typically intersect at most one edge $e_\square^I$ of the graph and if so then only once. Therefore, $t_\square^I(\gamma(s), \vec{n}^{\square,I}(s)) n_e^{\square,I} n_{e'}^{\square,I}$ vanishes unless $n_e^{\square,I} = \delta_{e,e_\square^I}$ up to small corrections in the vicinity of vertices. Thus the sum over edges reduces approximately to diagonal contributions and the sum over surfaces and their intersection numbers at given e reduces approximately to l_e^I, the dt-measure of the set $\{t \in [a, b]; |S_t^I \cap e| = 1\}$. This means that (13.6.4) is approximated by

$$
\frac{\hat{C}}{\hbar} T_s \approx \frac{\ell_p^2}{a} \sum_{e \in E(\gamma)} j_e(j_e + 1) \left[\sum_I l_e^I \right] T_s =: \frac{\ell_p^2}{a} \sum_{e \in E(\gamma)} j_e(j_e + 1) l_e T_s \qquad (11.2.61)
$$

which provides a concrete realisation and classical interpretation of the numbers l_e. In other words, at least for parquets much finer than a given graph, the function (11.2.55) provides a suitable continuum limit of the complexifier used in [488]! Of course, the exact operator has a non-diagonal edge metric and one has to take care of this when repeating all the estimates of [488,489] for this more general case, however, on graphs sufficiently coarse compared with the parquet the approximation given by (11.2.61) is quite good.

These are the general formulae. For the rest of this section we consider a specific situation in which these formulae simplify drastically and enable us to use the analytical results from [488, 489].

First of all, on sufficiently coarse graphs we can describe the results as follows (we fix $D = 3$ for simplicity): define

$$\delta_A^\gamma := \sum_{s \in \mathcal{S};\, \gamma(s)=\gamma} T_s(A) < T_s, . >$$

$$\delta_{\gamma,A} := \sum_{\gamma' \subset \gamma} \delta_A^{\gamma'} \tag{11.2.62}$$

We evidently have the identity

$$\delta_A = \sum_{\gamma \in \Gamma_0^\omega} \delta_A^\gamma \tag{11.2.63}$$

so that the second line in (11.2.62) is the 'δ-distribution cutoff at γ'. A simplification arises at the gauge-variant level since then evidently

$$\delta_{\gamma,A} = \prod_{e \in E(\gamma)} \delta_{e,A} \tag{11.2.64}$$

factorises. Now $\delta_{e,A} = \delta_{A(e)}$ where the latter distribution is with respect to the Haar measure. Due to the Peter and Weyl theorem

$$\delta_h(h') = \sum_{\pi \in \Pi} d_\pi \chi_\pi(h(h')^{-1}) \tag{11.2.65}$$

which demonstrates that with $\delta_A^e = \delta_{e,A} - 1$ we also have

$$\delta_A^\gamma = \prod_{e \in E(\gamma)} \delta_A^e \tag{11.2.66}$$

Let us now specify C_γ. First of all, if we set $A_e^j := \int_e A^j$ then we see that for sufficiently fine parquet (11.2.56) reduces to

$$A_e^{j\mathbb{C}} \approx A_e^j - \frac{i}{a^2} \sum_I \int dt \sum_{\square \in P_t^I} \epsilon(e, \square) E_j(\square) \tag{11.2.67}$$

where we have assumed that e intersects each of the $\square$ at most once transversally.[4] Here $\epsilon(e, \square)$ is the signed intersection number of $e, \square$ which by assumption takes only values ± 1 for transversal intersections and 0 for segments of e within $\square$. This formula can be simplified further if we assume that the graph under consideration is cubic and adapted to the parquet in the sense that the edges running in the direction of I (1) intersect the leaves of the Ith foliation transversally and such that $\sigma(e, \square) \geq 0$ and (2) intersect the leaves of the other foliations

---

[4] By the same computation as for the holonomy–flux algebra, contributions from segments lying entirely within the $\square$ drop out.

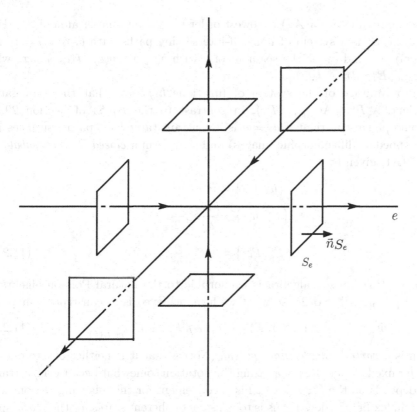

Figure 11.5 The contributing faces of the parquet in the approximate description of the coherent state. If the coarseness of graph and parquet are comparable, the contributing faces form a dual polyhedronal decomposition.

not at all or they lie entirely within the leaves of those other foliations. Then, for not too wild functions E and sufficiently small $\Box$ we may pick for each $e \in E(\gamma)$ one of the surfaces $\Box_e =: S_e$ which intersects e in an interior point of both e, S_e and then (11.2.67) simplifies further to

$$A_e^{j\mathbb{C}} \approx A_e^j - \frac{i l_e}{a^2} E_j(\Box_e) \tag{11.2.68}$$

where l_e is the dt-measure of the set $\{t \in \mathbb{R} : P_t^I \cap e \neq \emptyset\}$ for e in I direction. Notice that e is of type 'up' with respect to S_e by construction. Equation (11.2.68) holds to a good approximation only if the $\Box$ with $\Box \cap e \neq \emptyset$ vary little compared with $\Box_e$ over the range of e. In the situation described the $\Box_e$ chosen forms almost a complex dual to the graph if the graph is of the same fineness as the parquet, which is then very similar to the situation of [488] (see Figure 11.5). In [488, 489] instead, with $-\Im(A^{\mathbb{C}})$ one was dealing with the following functions

$$P_j^e(A, E) := -\frac{1}{2a_e^2} \operatorname{Tr}\left(\tau_j \int_{S_e} \operatorname{Ad}_{A(e_{x_e} \circ \rho_e(x))}(\ast E(x))\right) \tag{11.2.69}$$

which approximate $-\Im(A^{\mathbb{C}})$ to lowest order ϵ^2 (the parameter area of S_e). Here $\rho_e(x)$, $x \in S_e$ is a system of non-self-intersecting paths with $b(\rho_e(x)) = x_e$ and $f(\rho_e(x)) = x$ and e_{x_e} is the segment of e with $b(e_{x_e}) = b(e)$, $f(e_{x_e}) = x_e$ while $*E = \epsilon_{abc} E_j^a \tau_j dx^b \wedge dx^c$.

The advantage of the system of functions h_e, P^e is that they are gauge-covariant, $\lambda_g^* P^e = \mathrm{Ad}_{g(b(e))}(P^e)$, in contrast to the $E_j(S)$ of Section 29.1.1, diffeomorphism-covariant if $a_e = a$ is a constant (all edges, paths, surfaces just get mapped to diffeomorphic images) and they form a *closed Poisson subalgebra* of $C^\infty(\mathcal{M})$ given by

$$\{h_e, h_{e'}\} = 0$$

$$\{P_j^e, h_{e'}\} = \frac{\kappa}{a_e^2} \delta_{e'}^e \frac{\tau_j}{2} h_e$$

$$\{P_j^e, P_k^{e'}\} = -\delta^{ee'} \frac{\kappa}{a_e^2} \epsilon_{jkl} P_l^e \tag{11.2.70}$$

However, this Poisson algebra is isomorphic to the natural Poisson algebra on $\mathcal{M}_\gamma := \prod_{e \in E(\gamma)} T^*(\mathrm{SU}(2))$ so what we have achieved is to construct a map

$$\Phi_\gamma' : \mathcal{M} \to \mathcal{M}_\gamma; \; (A, E) \mapsto \big(h_e(A), P_j^e(A, E) \big)_{e \in E(\gamma)} \tag{11.2.71}$$

which is a *partial symplectomorphism*. (Notice that it is neither one to one nor onto for fixed γ. Here we are abusing the notation somewhat because Φ_γ' certainly also depends on the $S_e, \rho_e(x)$.) This is convenient for the following reason: what the complexifier $\hat{C}_\gamma$ does for us is to construct coherent states for the phase space $\mathcal{M}_\gamma := [T^*(\mathrm{SU}(2))]^{|E(\gamma)|}$ and since the Poisson structures of the phase spaces $\Phi_\gamma'(\mathcal{M})$ and $\mathcal{M}_\gamma$ coincide we automatically have proved the Ehrenfest property for $\Phi_\gamma'(\mathcal{M})$. Now, if γ gets sufficiently fine, we can approximate any function on $\mathcal{M}$ by functions in $\Phi_\gamma'(\mathcal{M})$ and in that sense we are constructing *approximate coherent states* for $\mathcal{M}$.

Next we describe the cylindrical projections C_γ. They really come from (11.2.55), (11.2.60) and (11.2.61) but the following description holds if γ is sufficiently coarse with respect to the parquet. One finds

$$C_\gamma := \frac{a}{2\kappa} \sum_{e \in E(\gamma)} l_e^{-1} (P_j^e)^2 \tag{11.2.72}$$

One may check that this leads to the complexification

$$g_e := \sum_{n=1}^\infty \frac{(-i)^n}{n!} \{C_\gamma, h_e\}_n = e^{-iP_j^e \tau_j/2} h_e \tag{11.2.73}$$

where the Poisson brackets are those of $\mathcal{M}$. In (11.2.73) we have stumbled naturally on the diffeomorphism

$$T^*(\mathrm{SU}(2)) \to \mathrm{SL}(2, \mathbb{C}); \; (h, P) \mapsto e^{-iP^j \tau_j/2} h \tag{11.2.74}$$

where the inverse of (11.2.74) is given by polar decomposition. Now, while the complexification of $\mathbb{R}$ is given by $\mathbb{C}$, the complexification of a Lie group G with

Lie algebra Lie(G) is given by the image under the exponential map of the complexification of its Lie algebra (that is, we allow arbitrary complex coefficients θ^j of the Lie algebra basis τ_j rather than only real ones) and (11.2.73) tells us precisely how this is induced by the complexifier. The map (11.2.74) allows us to identify $\mathcal{M}_\gamma$ with $SL(2, \mathbb{C})^{|E(\gamma)|}$ so that we have altogether a map

$$\Phi_\gamma : \mathcal{M} \to \mathcal{M}_\gamma; \ (A, E) \mapsto m_\gamma(A, E) := \left(g_e(A, E) := e^{-iP_j^e \tau_j/2} h_e\right)_{e \in E(\gamma)} \tag{11.2.75}$$

The Poisson algebra (11.2.70) is consistent with the quantisation $\hat{P}_j^e = it_e R_e^j/2$ on $\mathcal{H}_\gamma^0$ while $\hat{h}_e$ is a multiplication operator. Here the classicality parameters

$$t_e := l_e \frac{\ell_p^2}{a^2} \tag{11.2.76}$$

have naturally appeared and it follows that

$$\hat{C}_\gamma/\hbar = -\frac{1}{2} \sum_{e \in E(\gamma)} t_e \Delta_e \tag{11.2.77}$$

where $\Delta_e = (R_e^j)^2/4$. Our annihilation operators become

$$\hat{g}_e := e^{-\hat{C}_\gamma/\hbar} \hat{h}_e \left(e^{-\hat{C}_\gamma/\hbar}\right)^{-1} = e^{-t_e \tau_j^2/8} e^{-i\hat{P}_j^e \tau_j/2} \hat{h}_e \tag{11.2.78}$$

which up to a quantum correction is precisely the quantisation of (11.2.73). Then we can define abstract coherent states for $\mathcal{H}_\gamma$ by

$$\psi_{\gamma,m_\gamma} := \left[e^{-\hat{C}_\gamma/\hbar} \delta_{\gamma,h_\gamma}\right]_{h_\gamma \to m_\gamma}$$

$$= \prod_{e \in E(\gamma)} \left[e^{t_e \Delta_e/2} \delta_{h_e}\right]_{h_e \to g_e}$$

$$\psi_{m_\gamma}^\gamma := \left[e^{-\hat{C}_\gamma/\hbar} \delta_{h_\gamma}^\gamma\right]_{h_\gamma \to m_\gamma}$$

$$= \prod_{e \in E(\gamma)} \left[e^{t_e \Delta_e/2} \delta_{h_e} - 1\right]_{h_e \to g_e}$$

$$\psi_g := \left[e^{t\Delta/2} \delta_h\right]_{h \to g} = \sum_{j=0,1/2,1,3/2,\dots} (2j+1)e^{-tj(j+1)/2} \chi_j(gh^{-1}) \tag{11.2.79}$$

and coherent states on $\mathcal{H}^0$ by

$$\psi_{\gamma,m} := \hat{U}_\gamma \psi_{\gamma,\Phi_\gamma(m)} \text{ and } \psi_m^\gamma := \hat{U}_\gamma \psi_{\Phi_\gamma(m)}^\gamma \tag{11.2.80}$$

where $\hat{U}_\gamma : \mathcal{H}_\gamma^0 \to \mathcal{H}^0$ is the usual isometric monomorphism.

In [488, 489] peakedness, expectation value, small fluctuation and Ehrenfest properties for the gauge-variant states ψ_{γ,m_γ} and the algebra of operators $\mathcal{L}(H_\gamma^0)$ were proved. All proofs can be reduced to proving it for a single copy of SU(2). Overcompleteness follows from the results due to Hall [318] for the states ψ_g on $L_2(SU(2), d\mu_H)$. Annihilation and creation operators have been defined above and for those minimal uncertainty properties follow. See Figure 11.6 for a

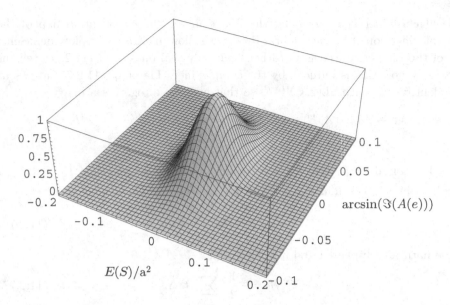

Figure 11.6 Resolution of the Gaußian-shaped peak in the overlap function of the coherent states (peakedness in phase space) for one edge. For illustrative purposes only two dimensions of the six-dimensional phase space are displayed, one electric flux component and one holonomy angle.

graphical illustration of these properties. For examples of semiclassical calculations, see Section 32.4.

11.2.5 Semiclassical limit of loop quantum gravity: graph-changing operators, shadows and diffeomorphism-invariant coherent states

One of the most important tasks of the semiclassical analysis is to verify that the quantum dynamics is implemented correctly. Hence one would need to show that the Master Constraint in either the graph-changing version on the level of $\mathcal{H}_{\text{diff}}$ or the non-graph-changing version (the extended Master Constraint) on the level of $\mathcal{H}_{\text{kin}}$ has the correct classical limit. Alternatively one can try to do this for the Hamiltonian constraint on the level of $\mathcal{H}_{\text{kin}}$. For the Hamiltonian constraint only the graph-changing version is available because the non-graph-changing version is anomalous as one immediately derives from the results of Section 10.4.3. A version on $\mathcal{H}_{\text{diff}}$ is also not possible since the Hamiltonian constraint is itself not spatially diffeomorphism-invariant.

In what follows we will describe the status of these programmes in some detail.

I. *Non-graph-changing approach*

Here one uses cutoff states for which in general the question arises how to choose the graph γ on which they are based. This question is analysed in detail in [490]. One possibility is to form a density matrix similar to the one we discussed before for the statistical weave but averaging only over a countable number of states (thus not leaving $\mathcal{H}^0$). Another would be to choose for γ a generic random graph which does not display any direction dependence on large scales. In any of these scenarios the picture that arises is the following: given γ, m we can extract from these two data two scales. The first is a *graph scale* ϵ given by the average edge length as measured by the metric determined by m. The second is a *curvature scale L* determined both by the mean curvature radius of the four-dimensional metric determined by m and the mean curvature of the induced metric on the embedded submanifolds $e, S_e, \rho_e(x)$ (so that even in the case that m are exactly flat initial data the scale L is not necessarily infinity). We then must decide which (kinematical) observables should behave maximally semi-classically. This is a choice that must be made and the choice of γ will depend largely on this physical input. In [490] we chose these observables to be electric and magnetic fluxes. When one then tries to minimise the fluctuations of these observables the parameters ϵ and a (the parameter that appears in $t_e = l_e \ell_p^2/a^2$) get locked at $a \approx L$ and $\epsilon = \ell_p^\alpha L^{1-\alpha}$ for some $0 < \alpha < 1$. These considerations suggest the following conclusions:

1. *Three scales*

   There are altogether three scales; the microscopic Planck scale ℓ_p, the mesoscopic scale ϵ and the macroscopic scale L. Since $\ell_p \ll L$ we have $\ell_p \ll \epsilon \ll L$ provided that (as in this case) α is not too close to the values $0, 1$.

2. *Geometric mean*

   The mesoscopic scale takes a *geometric mean* between the microscopic and macroscopic scales. In particular, it lies well above the microscopic scale ℓ_p in contrast to the geometric weave states. The reason for this is that not only electric fluxes had to be well approximated but also magnetic ones: the weave states are basically spin-network functions which in turn are very similar to momentum eigenfunctions. Since then electric fluxes are very sharply peaked, magnetic ones are not peaked at all due to the Heisenberg uncertainty relation. This can best be seen by the observation that $< T_s, (\hat{h}_p)_{AB} T_s >= 0$ for any spin-network state and any $A, B = 1, 2$ (and therefore also $\omega_{q^0, \lambda, j, I}(\hat{h}_p) = 0$ for the statistical weave) which is an unacceptable expectation value since $\hat{h}_p$ should be SU(2)-valued. In order to approximate holonomies *one must take an average over large numbers of spins*. This is precisely what our coherent states do. As a consequence, the elementary observables, those that are defined at the smallest scale which still allows semiclassical behaviour, are now defined at scales not smaller than $\epsilon \gg \ell_p$.

3. *Continuum limit*

   Notice that all our states and operators are defined in the continuum, therefore no continuum limit has to be taken. Yet, the scale ϵ could be associated with a measure for closeness to the continuum in which the graphs with which we probe operators tend to the continuum. The relation $\epsilon = \ell_p^\alpha L^{1-\alpha}$ reveals that not only can one not take $\epsilon \to 0$ at finite ℓ_p because fluctuations would blow up, but also the *'continuum limit'* $\epsilon \to 0$ *and the classical limit* $\ell_p \to 0$ *get synchronised.*

4. *Staircase problem*

The states displayed above have been criticised due to the following problem: if one computes the area operator expectation value for a surface which is not aligned with the surfaces dual to the graph by which the coherent state is labelled then it does not take the classical value. Even worse, if one computes the expectation value for a holonomy operator for a path not lying in the graph by which the coherent state is labelled, then one gets zero. One could take the point of view that therefore the states displayed above are not good. Unfortunately, averaging them does not help as shown in [487]. However, a natural way out is the following: electric flux and holonomy operators are simply not well approximated by these states but they approximate very well other observables of the theory which suffice to separate the points of the phase space. They are classically given as three-dimensional integrals over phase space functions rather than one- or two-dimensional ones, hence the direction dependence of holonomies, fluxes and areas disappears. The reason is that these functions have support everywhere and thus in particular along the graph where the coherent state is excited. Technically this leads to the fact, as we have seen for the example of the Master Constraint, that the operator corresponding to the three-dimensional integral is automatically adapted to the graph by which the coherent state is labelled, since it uses precisely the edges of that graph in its cylindrical projections. This happens due to background independence, however, it only holds for non-graph-changing operators.

We expect many of those properties to hold generically for any cutoff semiclassical states that one may want to build for canonical Quantum General Relativity and that the extensive proofs of their properties provided in [488, 489] will be useful for a whole class of states of this kind.

The machinery of [488, 489] has been applied already to the extended Master Constraint which is not graph-changing. The extended Master Constraint is exactly of the type of operators to which the cutoff states are especially well adapted: it derives from a classical three-dimensional integral of a density of weight one. The result is positive: using cutoff states on cubical graphs one can prove [589–591] that the extended Master Constraint has the correct classical limit. This result is very promising and one of the facts that makes the extended Master Constraint especially attractive. The result, however, does not yet establish that the physical Hilbert space that $_E$ constructs is large enough. The next step must therefore be to gain sufficient information about the corresponding space of physical states and the semiclassical behaviour of Dirac observables.

II. *Graph-changing approach*

So far the graph-changing approach is not very well developed. In what follows we will try to explain the underlying reasons and technical difficulties with graph-changing operators in the context of the current semiclassical framework.

1. *Spatially diffeomorphism-invariant approach*

Working with spatially diffeomorphism-invariant semiclassical states is appropriate for the unextended Master Constraint. Since $\mathcal{H}_{\text{diff}}$ is (or can be made) separable, cutoff states are not necessary here which is good because this removes the graph (or spatial diffeomorphism equivalence class of graphs) dependence while the state remains normalisable. On the other hand, since $\mathcal{H}_{\text{diff}}$ is not obviously of the form $L_2(\overline{\mathcal{A}}/\text{Diff}(\sigma), d\mu)$ we do not know how to define the δ-distribution with respect to the measure μ, if it exists at all. All we can do at the moment is to take, as an Ansatz for a spatially diffeomorphism-invariant semiclassical state

$$\Psi_m := \eta_{\text{diff}}[\psi_m]; \quad \psi_m = e^{-\hat{C}/\hbar} \sum_{[s]} T_{s_0([s])}(Z(m)) < T_{s_0([s])}, \cdot >_{\text{kin}}$$

(11.2.81)

Here the sum is over equivalence classes of spin-network labels, $s_0([s])$ is a representative of the class $[s]$ and $l_{[s]} = \eta(T_{s_0([s])})$ is an orthonormal basis of $\mathcal{H}_{\text{diff}}$. As before, η_{diff} denotes group averaging with respect to the spatial diffeomorphism group and $Z(m) = A^{\mathbb{C}}$ is the complexification of a connection induced by a complexifier. For instance, if $\hat{C}/\hbar$ is diagonal on spin-network functions T_s with eigenvalues λ_s then (11.2.81) becomes explicitly

$$\Psi_m = \sum_{[s]} T_{s_0([s])}(m) \, e^{-\lambda_{s_0([s])}} \, l_{[s]}$$

(11.2.82)

The resulting states are possibly normalisable elements of $\mathcal{H}_{\text{diff}}$ upon choosing suitable $\hat{C}$. They are simply 'projections' onto $\mathcal{H}_{\text{diff}}$ of certain elements of $\mathcal{H}_{\text{kin}}$ which are no longer of the cutoff form because the union of the graphs involved is no longer finite. However, at this point their semiclassical properties are unknown and further work is required. Notice that the graph dependence of cutoff states is replaced by the dependence on the representative s_0, or, equivalently, by the representative m because $T_s(\varphi^*(m)) = T_{\varphi \cdot s}(m)$. It is not clear how to get rid of this nor whether it is necessary.

2. *Kinematical approach*

Let us now turn to problems associated with the graph-changing nature of operators at the kinematical level, which is appropriate for the Hamiltonian constraint itself defined on $\mathcal{H}_{\text{kin}}$.

(A) *Cutoff states*

We consider first cutoff coherent states on a graph γ. The states are of the form $\psi_{\gamma,m} = \sum_{\gamma(s)=\gamma} c_s(m) T_s$ and we have

$$\hat{H}(N)\psi_{\gamma,m} = \sum_{v \in V(\gamma)} N(v)\, \hat{H}_{\gamma,v}\, \psi_{\gamma,m} \qquad (11.2.83)$$

Here, as before, $\hat{H}_{\gamma,v}$ is a linear combination of operators each of which changes the graph γ in the vicinity of the vertex v by adding one or more edges $a_{\gamma,v,e,e'}$, one for each pair of edges e, e' adjacent to v, in a representation with non-trivial spin. Notice that $a_{\gamma,v,e,e'}$ *never coincides with an edge in* $E(\gamma)$. It follows that trivially

$$< \hat{H}(N) >_{\gamma,m} := \frac{< \psi_{\gamma,m}, \hat{H}(N)\psi_{\gamma,m} >}{||\psi_{\gamma,m}||^2} = 0 \qquad (11.2.84)$$

because the decomposition of $\hat{H}(N)\psi_{\gamma,m}$ into spin-network states consists only of terms which depend with non-zero spin on the edge $a_{\gamma,v,e,e'}$. Hence, cutoff states are not suitable to establish the semiclassical limit of $\hat{H}(N)$.

(B) *Coherent states on fractals*

A possible solution could be states defined on *fractals*: remember that the Hamiltonian constraint changes a graph in the vicinity of its vertices. Choose a graph γ_0 and let $\Gamma_n(\gamma_0)$ be the set of all graphs on which the spin-networks in the decomposition of any $\hat{H}(N)^n T_s$; $\gamma(s) = \gamma_0$, $N \in C^\infty(\sigma)$ into spin-network states depend (see Figure 11.7). Define

$$\Psi^n_{\gamma_0,m} := \sum_{n'=0}^{n} \sum_{\gamma \in \Gamma_{n'}(\gamma_0)} \psi_{\gamma,m} \qquad (11.2.85)$$

The sum is orthogonal (we have suppressed a possible different weight for the individual terms) and now the analogue of (11.2.84) is certainly not trivial because by design for $\gamma \in \Gamma_{n'}$, $n' = 0, 1, \ldots, n$ the spin-network decomposition of $\hat{H}(N)\psi_{\gamma,m}$ produces terms contained in the $\psi_{\gamma',m}$, $\gamma' \in \Gamma_{n'+1}$. We call (11.2.85) a fractal state because around each vertex the structure of arcs attached is reproducing when looked at at ever finer resolutions (as $n \to \infty$). However, again the semiclassical properties of these states remain largely unexplored so far.

(C) *Shadows*

Another proposal is the *shadow* framework. Given a distributional state $\Psi_m = \sum_\gamma \psi_{\gamma,m}$ with cutoffs $\psi_{\gamma,m}$ we define the generalised 'expectation value'

$$< \hat{H}(N) >'_{\gamma,m} := \frac{\Psi[\hat{H}(N)\psi_{\gamma,m}]}{\Psi_m[\psi_{\gamma,m}]} \qquad (11.2.86)$$

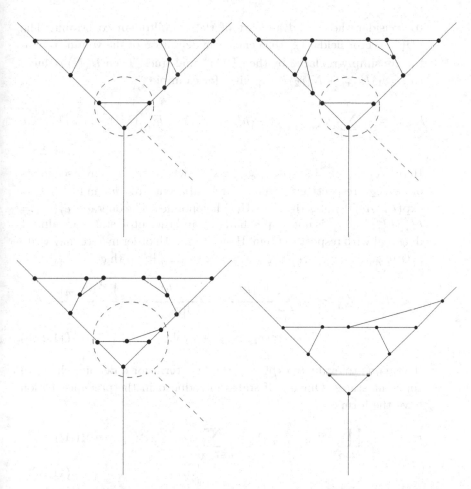

Figure 11.7 Solutions of the Hamiltonian constraint depend on a fractal graph because it acts in the vicinity of a vertex by adding arcs closer and closer to it. Hence, resolving the vertex at an ever finer scale unveils a self-similar structure. Here, for illustrative purposes, we have allowed the action of the operator to be non-trivial at the vertices it creates.

which is non-vanishing. For non-graph-changing operators $\hat{A}$, this prescription coincides with (11.2.84), however for graph-changing ones, even if self-adjoint or positive, the result is not necessarily real-valued or non-negative and thus does not qualify as an expectation value functional. We will now show that this functional still does not produce the correct semiclassical physics: we will do the calculation in the Abelian context, the non-Abelian one is technically more difficult but reveals the same effects. We just consider the simplest graph $\gamma = \{e_1, e_2, e_3\}$ consisting of three edges meeting in two vertices $\{v_1, v_2\} = e_1 \cap e_2 \cap e_3$ in their common endpoints. It will be sufficient to consider the Euclidean part of the Hamiltonian constraint. The SU(2) vector fields $X^j_{e_I}$ that enter the definition of the volume operator are simply replaced by their $U(1)^3$ analogues. Then $\hat{H}(N)$ reduces to $N(v_1)\hat{H}_{\gamma,v_1} + N(v_2)\hat{H}_{\gamma,v_2}$ with, for example,

$$\hat{H}_{\gamma,v_1} = -i\sum_{j=1}^{3}\left[h^j_{\alpha_{\gamma,v_1,e_1,e_2}} - \left(h^j_{\alpha_{\gamma,v_1,e_1,e_2}}\right)^{-1}\right]h^j_{e_3}\left[\left(h^j_{e_3}\right)^{-1}, V_{v_1}\right] + \text{cyclic}$$

(11.2.87)

Here $\alpha_{\gamma,v_1,e_1,e_2} = s_1 \circ a_{\gamma,v_1,e_1,e_2} \circ s_2^{-1}$ where s_1, s_2 are segments of e_1, e_2 respectively which are adjacent to v and $h^j_e(A) = \exp(i\int_e A^j)$ denotes the three $U(1)$ holonomies. The operator $e^j_I(v_1) := h^j_{e_I}\left[(h^j_{e_I})^{-1}, V_{v_1}\right]$ is not graph-changing and the cutoff states are almost diagonal with respect to them. Hence to zeroth order in $\hbar$ we may write $\hat{e}^j_I(v)\psi_{\gamma,m} = <\psi_{\gamma,m}, \hat{e}^j_I(v)\psi_{\gamma,m}>\psi_{\gamma,m}/||\psi_{\gamma,m}||^2$ so that

$$< \hat{H}_{\gamma,v_1} >'_{\gamma,m} \approx -i\sum_j \frac{\Psi\left[\left(h^j_{\alpha_{\gamma,v,e_1,e_2}} - \left(h^j_{\alpha_{\gamma,v,e_1,e_2}}\right)^{-1}\right)\psi_{\gamma,m}\right]}{||\psi_{\gamma,m}||^2}$$

$$\times < \hat{e}^j_3(v_1) >_{\gamma,m} + \text{cyclic} \qquad (11.2.88)$$

It remains to evaluate explicitly the first term for a definite choice of coherent states. Our cutoff states introduced in the previous section have the form

$$\psi_{\gamma,m} = \prod_{j,e\in E(\gamma)} \psi^j_{e,m}, \quad \psi^j_{e,m}(A) = \sum_{n\in\mathbb{Z}-\{0\}} e^{-t_e n^2/2}\left(g^j_e(m)\left(h^j_e(A)\right)^{-1}\right)^n$$

(11.2.89)

where $g^j_e(m) = h^j(A^{\mathbb{C}}) = \exp(i\int_e[A^j - i\{C, A^j\}])$ for a complexifier $C = C(E)$ which depends only on the electric fields. We can drop the index j since the calculation is the same for every j and everything factorises. We find to zeroth order in $\hbar$

$$-i\frac{\Psi\left[(h_\alpha - (h_\alpha)^{-1})\psi_{\gamma,m}\right]}{||\psi_{\gamma,m}||^2}$$

$$= -i\frac{<\psi_{s_1,m}\otimes\psi_{\bar{s}_1,m}, h_{s_1}\psi_{e_1,m}>}{||\psi_{e_1,m}||^2}$$

$$\times\frac{<\psi_{s_2,m}\otimes\psi_{\bar{s}_2,m}, h^{-1}_{s_2}\psi_{e_2,m}>}{||\psi_{e_2,m}||^2}<\psi_{a,m}, h_a 1>$$

$$+i\frac{<\psi_{s_1,m}\otimes\psi_{\bar{s}_1,m}, h^{-1}_{s_1}\psi_{e_1,m}>}{||\psi_{e_1,m}||^2}$$

$$\times\frac{<\psi_{s_2,m}\otimes\psi_{\bar{s}_2,m}, h_{s_2}\psi_{e_2,m}>}{||\psi_{e_2,m}||^2}<\psi_{a,m}, h^{-1}_a 1> \qquad (11.2.90)$$

where we have split the edges as $e_I = s_I \circ \bar{s}_I$. We now employ the Poisson resummation formula, Theorem 32.4.1 or [488, 646]

$$\sum_{n \in \mathbb{Z}} f(ns) = \frac{1}{s} \sum_{n \in \mathbb{Z}} \int_{\mathbb{R}} dx \, e^{2\pi i n x / s} f(x) \qquad (11.2.91)$$

whose conditions of applicability are satisfied because of the exponential damping factor $e^{-n^2 t / 2}$, $t = s^2$. Using the relations $h_e = h_s h_{\bar{s}}$, $g_e = g_s g_{\bar{s}}$, $g_e = e^{p_e} h_e$, $t_e = t_s + t_{\bar{s}}$ we find, to zeroth order in $\hbar \propto t_e, t_s, t_{\bar{s}}$ that (for examples of coherent state calculations see Section 32.4)

$$\frac{\Psi[(h_\alpha - (h_\alpha)^{-1})\psi_{\gamma,m}]}{||\psi_{\gamma,m}||^2}$$

$$= -ih_\alpha \exp\left(\frac{t_{s_1} p_{\bar{s}_1} - t_{\bar{s}_1} p_{s_1}}{t_{e_1}} - p_a - \frac{t_{s_2} p_{\bar{s}_2} - t_{\bar{s}_2} p_{s_2}}{t_{e_2}}\right)$$

$$+ ih_\alpha^{-1} \exp\left(-\left[\frac{t_{s_1} p_{\bar{s}_1} - t_{\bar{s}_1} p_{s_1}}{t_{e_1}} - p_a - \frac{t_{s_2} p_{\bar{s}_2} - t_{\bar{s}_2} p_{s_2}}{t_{e_2}}\right]\right)$$

$$(11.2.92)$$

where p_e denotes, approximately, the electric flux of a surface S_e divided by a_e^2, $t_e = \ell_P^2 / a_e^2$, which typically intersects only the edge e of γ (in an interior point) if the graph γ is sufficiently coarse compared with the parquet. The loop α encloses a surface S_α which has a parameter area of the same order of magnitude, say ϵ^2, as the surfaces S_e and we are precisely interested in that order ϵ^2 since the replacement of the magnetic field $B(S_\alpha) = \int_{S_\alpha} dA = \oint_\alpha A$ appearing in the classical Hamiltonian constraint by $[h_\alpha - h_\alpha^{-1}]/(2i)$ is correct precisely to that order of magnitude. Notice that the quotients t_{s_1}/t_{e_1} are of order zero in $\hbar$. It follows that to order ϵ^2 equation (11.2.92) equals

$$2\left[B(\alpha) - i\left\{\frac{t_{s_1} p_{\bar{s}_1} - t_{\bar{s}_1} p_{s_1}}{t_{e_1}} - p_a - \frac{t_{s_2} p_{\bar{s}_2} - t_{\bar{s}_2} p_{s_2}}{t_{e_2}}\right\}\right] \qquad (11.2.93)$$

We see that the shadow expectation value is *off* the expected result to the required order in ϵ unless the second term in (11.2.93) vanishes to order ϵ^2. Since this is generically not the case for general γ we conclude that more work is required in order to define semiclassical states for graph-changing operators. For instance, the 'expectation value' (11.2.93) becomes generically complex-valued and is in fact purely imaginary for flat space for which $A(x) = 0$, $E(x) = \text{const.}$ Moreover, when summing over vertices (corresponding to lapse functions of large support) the expectation value of the full Hamiltonian constraint blows up because $E(x)$ compared with $A(x)$ does not decay at infinity.

We close this section with some concluding remarks:

1. Let us come back once more to the issue of using kinematical rather than dynamical coherent states. Given the complicated structure of the Hamiltonian constraint or Master Constraint it is not likely that one will determine the physical Hilbert space exactly even after we have settled the issue of which Hamiltonian or Master Constraint to use. Thus, a more practical approach than to construct physical semiclassical states will be to consider kinematical coherent states ψ_m where m is a point on the constraint surface of the full phase space. The virtue of this is that the expectation value of full Dirac observables is *approximately gauge-invariant* since

$$\delta_N < \psi_m, \hat{O}\psi_m >=< \psi_m, \frac{[(\hat{H}(N))^\dagger, \hat{O}]}{i\hbar}\psi_m >\approx \{H(N), O\}(m) = 0$$

because O is a Dirac observable. Moreover

$$< \psi_m, \hat{O}\psi_m >\approx O(m) = O([m])$$

does not depend on the point m in the gauge orbit $[m]$ for the same reason. Thus, at least to zeroth order in $\hbar$ the expectation values of full Dirac observables and their infinitesimal dynamics should coincide whether we use kinematical or dynamical coherent states. This attitude is similar as in numerical classical gravity where one cannot just compute the time evolution of a given initial data set because for practical reasons one can only evolve approximately. The art is then to gain control on the error of these computations. Notice that Dirac observables themselves are difficult to construct even classically as shown in Section 2.2. However, the infinite series involved can be truncated close to the gauge cut defined by the clock variables after the first few terms and if we choose the point m on the gauge cut then we may use these approximate Dirac observables to very high accuracy. This will be discussed in more detail also in Section 16.1 in the context of *quantum gauge fixing*.

2. The classical starting point of Loop Quantum Gravity is a manifold diffeomorphic to $\mathbb{R} \times \sigma$ where σ has fixed topology. Thus, in the classical theory there is no topology change. Now let us look at typical (i.e., a dense set of) kinematical states: these are finite superpositions of spin-network states over finite graphs. The volume operator for sufficiently small regions R vanishes identically on such states. Now physically, if R has empty volume then R *does not exist!* In other words, typical quantum states do not describe σ, they describe σ with a large number of holes. However, a manifold with holes has a topology different from σ. Therefore, in Loop Quantum Gravity *dynamical topology change is already built in* because the Hamiltonian constraint changes the hole structure of σ in each time step. It is only when we pass to states which are excited everywhere that we will regain the topology of σ. This is

again the realm of semiclassical states: given a spatial metric q_0 to be approximated we can define a resolution length L with respect to q_0 up to which we do not want to have holes. Then we must take a coherent state in the completion of the Hilbert space whose underlying graph scale is smaller than L. Such states will then have non-zero volume 'everywhere' (up to L at least).

11.2.6 $^+$ The infinite tensor product extension

Quantum field theory on curved spacetimes is best understood if the spacetime is actually flat Minkowski space on the manifold $M = \mathbb{R}^4$. Thus, when one wants to compute the low-energy limit of canonical Quantum General Relativity to show that one gets the standard model (plus corrections) on a background metric one should do this first for the Minkowski background metric. Any classical metric is macroscopically non-degenerate. Since the quantum excitations of the gravitational field are concentrated on the edges of a graph, in order that, say, the expectation values of the volume operator for any macroscopic region is non-vanishing and changes smoothly as we vary the region, the graph must fill the initial value data slice densely enough, the mean separation between vertices of the graph must be much smaller than the size of the region (everything is measured by the three-metric, determined by the four-metric to be approximated, in this case the Euclidean one). Now $\mathbb{R}^4$ is spatially non-compact and therefore such a graph must necessarily have an at least *countably infinite* number of edges whose union has *non-compact* range.

However, the Hilbert spaces in use for Loop Quantum Gravity have as dense subspace the space of cylindrical functions labelled either by a semianalytic graph with a *finite* number of edges or by a so-called web, a piecewise smooth graph determined by the union of a *finite* number of smooth curves that intersect in a controlled way, albeit possibly a countably infinite number of times. Moreover, in both cases the edges or curves respectively are contained in *compact* subsets of the initial data hypersurface. These categories of graphs will be denoted by Γ_0^ω and Γ_0^∞ respectively where $\omega, \infty, 0$ stands for semianalytic, smooth and compactly supported respectively. Thus, the only way that the current Hilbert spaces can actually produce states depending on a countably infinite graph of non-compact range is by choosing elements in the closure of these spaces, that is, states that are countably infinite linear combinations of cylindrical functions.

The question is whether it is possible to produce semiclassical states of this form, that is, $\psi = \sum_n z_n \psi_{\gamma_n}$ where γ_n is either a finite semianalytic graph or a web, z_n is a complex number and we are summing over the integers. It is easy to see that this is not the case: Minkowski space has the Poincaré group as its symmetry group and thus we will have to construct a state which is at least invariant under (discrete) spatial translations. This forces the γ_n to be translations of some γ_0 and $z_n = z_0$. Moreover, the dependence of the state on each of the edges has to be the same and therefore the γ_n have to be mutually disjoint. It follows that the norm of the state is given by

$$||\psi||^2 = |z_0|^2 \left\{ 2\left[\sum_n 1\right] [1 - |<1, \psi_{\gamma_0}>|^2] + \left[\sum_n 1\right]^2 |<1, \psi_n>|^2 \right\}$$

where we assume without loss of generality that $||\psi_{\gamma_0}|| = 1$ and we use the diffeomorphism invariance of the measure and 1 is the normalised constant state. Decompose $\psi_n := \psi_n' + <1, \psi_n> 1$ to see this using that the states ψ_n' are mutually orthogonal and orthogonal to 1. By the Schwartz inequality the first term is non-negative and convergent only if $\psi_{\gamma_0} = 1$ while the second is non-negative and convergent only if $<1, \psi_{\gamma_0}> = 0$. Thus the norm diverges unless $z_0 = 0$.

This caveat points to its resolution: we notice that the formal state $\psi :=$ $\prod_n \psi_{\gamma_n}$ really depends on an infinite graph and has unit norm if we formally compute it by $\lim_{N \to \infty} || \prod_{n=-N}^N \psi_{\gamma_n}|| = \lim_{N \to \infty} \prod_{n=-N}^N ||\psi_{\gamma_n}|| = 1$ where the second identity follows from the disjointness of the γ_n. For instance with $c_n = <1, \psi_n>$

$$||\psi_1\psi_2||^2 = ||c_1c_2 + c_1\psi_2' + c_2\psi_1' + \psi_1'\psi_2'||^2$$
$$= |c_1|^2 |c_2|^2 + |c_1|^2 ||\psi_2'||^2 + |c_2|^2 ||\psi_1'||^2 + ||\psi_1'||^2 ||\psi_2'||^2 = ||\psi_1||^2 ||\psi_2||^2$$

The only problem is that this state is no longer in our Hilbert space, it is not the Cauchy limit of any state in the Hilbert space: defining $\psi_N := \prod_{n=-N}^N \psi_{\gamma_n}$ we find $| <\psi_N, \psi_M> | = | <1, \psi_{\gamma_0}> |^{2|N-M|}$ so that ψ_N is not a Cauchy sequence unless $\psi_{\gamma_0} = 1$. However, it turns out that it belongs to the *Infinite Tensor Product (ITP) extension* of the Hilbert space.

To construct this much larger Hilbert space [479] we must first describe the class of graphs that we want to consider. We will consider graphs of the category Γ_σ^ω where σ now stands for countably infinite. More precisely, an element of Γ_σ^ω is the union of a countably infinite number of semianalytic, mutually disjoint (except possibly for their endpoints) curves called edges of compact or non-compact range which have no accumulation points of edges or vertices. In other words, the restriction of the graph to any compact subset of the hypersurface looks like an element of Γ_0^ω. These are precisely the kinds of graphs that one would consider in the thermodynamic limit of lattice gauge theories and are therefore best suited for our semiclassical considerations since it will be on such graphs that one can write actions, Hamiltonians and the like.

The construction of the ITP of Hilbert spaces is due to von Neumann [647] and already more than 60 years old. We will try to outline briefly some of the notions involved, see [479] for a concise summary of all definitions and theorems involved.

Let for the time being I be any index set whose cardinality $|I| = \aleph$ takes values in the set of non-standard numbers (Cantor's alephs). Suppose that for each $e \in I$ we have a Hilbert space $\mathcal{H}_e$ with scalar product $< ., . >_e$ and norm $||.||_e$. For complex numbers z_e we say that $\prod_{e \in I} z_e$ *converges* to the number z provided that for each positive number $\delta > 0$ there exists a finite set $I_0(\delta) \subset I$ such that for any other finite J with $I_0(\delta) \subset J \subset I$ it holds that $|\prod_{e \in J} z_e - z| < \delta$. We say that $\prod_{e \in I} z_e$ is *quasi-convergent* if $\prod_{e \in I} |z_e|$ converges. If $\prod_{e \in I} z_e$ is quasi-convergent but not convergent we define $\prod_{e \in I} z_e := 0$. Next we say that for $f_e \in \mathcal{H}_e$ the ITP $\otimes_f := \otimes_e f_e$ is a C_0 vector (and $f = (f_e)$ a C_0 sequence) if $\| \otimes_f \| := \prod_{e \in I} \|f_e\|_e$ converges to a non-vanishing number. Two C_0 sequences f, f' are said to be strongly or weakly equivalent respectively provided that

$$\sum_e | < f_e, f'_e >_e - 1| \quad \text{resp.} \quad \sum_e || < f_e, f'_e >_e | - 1|$$

converges. The strong and weak equivalence class of f is denoted by $[f]$ and (f) respectively and the set of strong and weak equivalence classes by $\mathcal{S}$ and $\mathcal{W}$ respectively. We define the ITP Hilbert space $\mathcal{H}^\otimes := \otimes_e \mathcal{H}_e$ to be the closed linear span of all C_0 vectors. Likewise we define $\mathcal{H}^\otimes_{[f]}$ or $\mathcal{H}^\otimes_{(f)}$ to be the closed linear spans of only those C_0 vectors which lie in the same strong or weak equivalence class as f. The importance of these notions is that they determine much of the structure of $\mathcal{H}^\otimes$, namely:

1. All the $\mathcal{H}^\otimes_{[f]}$ are isomorphic and mutually orthogonal.
2. Every $\mathcal{H}^\otimes_{(f)}$ is the closed direct sum of all the $\mathcal{H}^\otimes_{[f']}$ with $[f'] \in \mathcal{S} \cap (f)$.
3. The ITP $\mathcal{H}^\otimes$ is the closed direct sum of all the $\mathcal{H}^\otimes_{(f)}$ with $(f) \in \mathcal{W}$.
4. Every $\mathcal{H}^\otimes_{[f]}$ has an explicitly known orthonormal von Neumann basis.
5. If s, s' are two different strong equivalence classes in the same weak one then there exists a unitary operator on $\mathcal{H}^\otimes$ that maps $\mathcal{H}^\otimes_s$ to $\mathcal{H}^\otimes_{s'}$, otherwise such an operator does not exist, the two Hilbert spaces are unitarily inequivalent subspaces of $\mathcal{H}^\otimes$.

Notice that two isomorphic Hilbert spaces can always be mapped into each other such that scalar products are preserved (just map some orthonormal bases) but here the question is whether this map can be extended unitarily to all of $\mathcal{H}^\otimes$. Intuitively then, strong classes within the same weak classes describe the same physics, those in different weak classes describe different physics such as an infinite difference in energy, magnetisation, volume, etc. See [648] and references therein for illustrative examples.

Next, given a (bounded) operator a_e on $\mathcal{H}_e$ we can extend it in the natural way to $\mathcal{H}^\otimes$ by defining $\hat{a}_e$ densely on C_0 vectors through $\hat{a}_e \otimes_f = \otimes_{f'}$ with $f'_{e'} = f_{e'}$ for $e' \neq e$ and $f'_e = a_e f_e$. It turns out that the algebra of these extended operators is automatically a von Neumann algebra [22, 167, 168, 535–537, 649–651] for $\mathcal{H}^\otimes$ (a weakly closed subalgebra of the algebra of bounded operators on a Hilbert space) and we will call the weak closure of all these algebras the von Neumann algebra $\mathcal{R}^\otimes$ of local operators. This way, adjointness relations and canonical commutation relations (Weyl algebra) are preserved.

Given these notions, the strong equivalence class Hilbert spaces can be characterised further as follows. First of all, for each $s \in \mathcal{S}$ one can find a representative

$\Omega^s \in s$ such that $\|\Omega^s\| = 1$. Moreover, one can show that $\mathcal{H}_s^\otimes$ is the closed linear span of those C_0 vectors $\otimes_{f'}$ such that $f'_e = \Omega_e^s$ for all but finitely many e. In other words, the strong equivalence class Hilbert spaces are irreducible subspaces for $\mathcal{R}^\otimes$, Ω^s is a cyclic vector in $\mathcal{H}_s^\otimes$ for $\mathcal{R}^\otimes$ on which the local operators annihilate and create local excitations and thus, if I is countable, $\mathcal{H}_s^\otimes$ is actually separable. We see that we naturally make contact with Fock space structures, von Neumann algebras and their factor type classification [167, 168], modular theory and algebraic quantum field theory [22]. The algebra of operators on the ITP which are not local (i.e., are not elements of $\mathcal{R}^\otimes$) do not have an immediate interpretation but it is challenging that they map between different weak equivalence classes and thus change the physics in a drastic way.

A number of warnings are in order:

1. Scalar multiplication is not multi-linear! That is, if f and $z \cdot f$ are C_0 sequences where $(z \cdot f)_e = z_e f_e$ for some complex numbers z_e then $\otimes_{z \cdot f} = (\prod_e z_e) \otimes_f$ is in general wrong, it is true if and only if $\prod_e z_e$ converges.

2. Unrestricted use of the associative law of tensor products is false! Let us subdivide the index set I into mutually disjoint index sets $I = \cup_\alpha I_\alpha$ where α runs over some other index set A. One can now form the different ITP $\mathcal{H}'^\otimes = \otimes_\alpha \mathcal{H}_\alpha^\otimes$, $\mathcal{H}_\alpha^\otimes = \otimes_{e \in I_\alpha} \mathcal{H}_e$. Unless the index set A is finite, a generic C_0 vector of $\mathcal{H}'^\otimes$ is orthogonal to all of $\mathcal{H}^\otimes$. This fact has implications for quantum gravity which we outline below.

Let us now come back to canonical Quantum General Relativity. In applying the above concepts we arrive at the following surprises:

(i) First of all, we fix an element $\gamma \in \Gamma_\sigma^\omega$ and choose the countably infinite index set $E(\gamma)$, the edge set of γ. If $|E(\gamma)|$ is finite then the ITP Hilbert space $\mathcal{H}_\gamma^\otimes := \otimes_{e \in E(\gamma)} \mathcal{H}_e$ is naturally isomorphic with the subspace $\mathcal{H}_\gamma^0$ of $\mathcal{H}^0$ obtained as the closed linear span of cylinder functions over γ. However, if $|E(\gamma)|$ is truly infinite then a generic C_0 vector of $\mathcal{H}_\gamma^\otimes$ is orthogonal to any possible $\mathcal{H}_{\gamma'}^0$, $\gamma' \in \Gamma_0^\omega$. Thus, even if we fix only one $\gamma \in \Gamma_\sigma^\omega$, the total $\mathcal{H}^0$ is orthogonal to almost every element of $\mathcal{H}_\gamma^\otimes$.

(ii) Does $\mathcal{H}_\gamma^\otimes$ have a measure-theoretic interpretation as an L_2 space? By the Kolmogorov theorem [532] the infinite product of probability measures is well-defined and thus one is tempted to identify $\mathcal{H}_\gamma^\otimes = \otimes_e L_2(\mathrm{SU}(2), d\mu_H)$ with $\mathcal{H}_\gamma^{0'} := L_2(\times_e \mathrm{SU}(2), \otimes_e d\mu_H)$. However, this cannot be the case, the ITP Hilbert space is non-separable (as soon as $\dim(\mathcal{H}_e) > 1$ for almost all e and $|E(\gamma)| = \infty$) while the latter Hilbert space is separable, in fact, it is the subspace of $\mathcal{H}^0$ consisting of the closed linear span of cylindrical functions over γ' with $\gamma' \in \Gamma_0^\omega \cap E(\gamma)$.

(iii) Yet, there is a relation between $\mathcal{H}_\gamma^\otimes$ and $\mathcal{H}^0$ through the inductive limit of Hilbert spaces: we can find a directed sequence of elements $\gamma_n \in \Gamma_0^\omega \cap E(\gamma)$, that is, $\gamma_m \subset \gamma_n$ for $m \leq n$, such that γ is its limit in Γ_σ^ω. The subspaces

$\mathcal{H}^0_{\gamma_n} \subset \mathcal{H}^0$ are isometric isomorphic with the subspaces of $\mathcal{H}^\otimes_\gamma$ given by the closed linear span of vectors of the form $\psi_{\gamma_n} \otimes [\otimes_{e \in E(\gamma - \gamma_n)} 1]$ where $\psi_{\gamma_n} \in \mathcal{H}^0_{\gamma_n} \equiv \mathcal{H}^\otimes_{\gamma_n}$, which provides the necessary isometric monomorphism to display the strong equivalence class $\mathcal{H}^\otimes_{\gamma,[1]}$ as the inductive limit of the $\mathcal{H}^0_{\gamma_n}$.

(iv) So far we have looked only at a specific $\gamma \in \Gamma^\omega_\sigma$. We now construct the total Hilbert space

$$\mathcal{H}^\otimes := \overline{\cup_{\gamma \in \Gamma^\omega_\sigma} \mathcal{H}^\otimes_\gamma}$$

equipped with the natural scalar product derived in [479]. This is to be compared with the Hilbert space

$$\mathcal{H}^0 := \overline{\cup_{\gamma \in \Gamma^\omega_\sigma} \mathcal{H}^0_\gamma} = \overline{\cup_{\gamma \in \Gamma^\omega_\sigma} \mathcal{H}^0_{\gamma,[1]}}$$

The identity in the last line enables us to specify the precise sense in which $\mathcal{H}^0 \subset \mathcal{H}^\otimes$: for any $\gamma \in \Gamma^\omega_\sigma$ the space $\mathcal{H}^{0\prime}_\gamma$ is isometric isomorphic as specified in (iii) with the strong equivalence class Hilbert subspace $\mathcal{H}^\otimes_{\gamma,[1]}$ where $1_e = 1$ is the constant function equal to one. Thus, the Hilbert space $\mathcal{H}^0$ describes the local excitations of the 'vacuum' Ω^0 with $\Omega^0_e = 1$ for any possible semianalytic path e.

Notice that both Hilbert spaces are non-separable, but there are two sources of non-separability: the Hilbert space $\mathcal{H}^0$ is non-separable because Γ^ω_0 has uncountable infinite cardinality. This is also true for the ITP Hilbert space but it has an additional character of non-separability: even for fixed γ the Hilbert space $\mathcal{H}^\otimes_\gamma$ splits into an uncountably infinite number of mutually orthogonal strong equivalence class Hilbert spaces and $\mathcal{H}^{0\prime}_\gamma$ is only one of them.

(v) Recall that spin-network states form a basis for $\mathcal{H}^0$. The result of (iv) states that they are no longer a basis for the ITP. The spin-network basis is in fact the von Neumann basis for the strong equivalence class Hilbert space determined by $[\Omega^0]$ but for the others we need uncountably infinitely many other bases, even for fixed γ. The technical reason for this is that, as remarked above, the unrestricted associativity law fails on the ITP.

We would now like to justify this huge blow-up of the original Hilbert space $\mathcal{H}^0$ from the point of view of physics. Clearly, there is a blow-up only when the initial data hypersurface is non-compact as otherwise $\Gamma^\omega_0 = \Gamma^\omega_\sigma$. Besides the fact that like $\mathcal{H}^0$ it is another solution to implementing the adjointness and canonical commutation relations, we have the following:

(a) Let us fix $\gamma \in \Gamma^\omega_\sigma$ in order to describe semiclassical physics on that graph in one of the cutoff schemes described in the previous section. Given a classical initial data set m we can construct a coherent state $\psi_{\gamma,m}$ which in fact is a C_0 vector $\otimes^\gamma_{\psi_m}$ for $\mathcal{H}^\otimes_\gamma$ of unit norm. This coherent state can be considered

as a 'vacuum' or 'background state' for quantum field theory on the associated spacetime. As remarked above, the corresponding strong equivalence
class Hilbert space $\mathcal{H}^{\otimes}_{\gamma,[\psi_m]}$ is obtained by acting on the 'vacuum' by local
operators, resulting in a space isomorphic with the familiar Fock spaces and
which is separable. In this sense, the fact that $\mathcal{H}^{\otimes}_{\gamma}$ is non-separable, being an
uncountably infinite direct sum of strong equivalence class Hilbert spaces,
could simply account for the fact that in quantum gravity *all vacua have to be
considered simultaneously, there is no distinguished vacuum as we otherwise
would introduce a background dependence into the theory.*

(b) The Fock space structure of the strong equivalence classes immediately suggests trying to identify suitable excitations of $\psi_{\gamma,m}$ as graviton states propagating on a spacetime fluctuating around the classical background determined by m [637,638]. Also, it is easy to check whether for different solutions
of Einstein's equations the associated strong equivalence classes lie in different weak classes and are thus physically different. For instance, preliminary
investigations indicate that Schwarzschild black hole spacetimes with different masses lie in the *same* weak class. Thus, *unitary* black hole evaporation
and formation seems not to be excluded from the outset.

(c) From the point of view of $\mathcal{H}^{0\prime}_{\gamma}$ the Minkowski coherent state is an everywhere excited state like a thermal state, the strong classes $[\Omega^0]$ and $[\psi_m]$ for
Minkowski data m are orthogonal and lie in different weak classes. The state
Ω^0 has no obvious semiclassical interpretation in terms of coherent states for
any classical spacetime.

(d) It is easy to see that the GNS Hilbert space used in [631, 632] is isometric
isomorphic with a strong equivalence class Hilbert space of our ITP construction. Thus, our ITP framework collects a huge class of representations
in the 'folium' [22] of the representation corresponding to the Hilbert space
$\mathcal{H}^0$ and embeds them isometrically into one huge Hilbert space $\mathcal{H}^{\otimes}$, thus
we have now *an inner product between different GNS Hilbert spaces*! This
demonstrates the power of this framework because inner products between
different GNS Hilbert spaces are normally not easy to motivate.

11.3 Graviton and photon Fock states from $L_2(\overline{\mathcal{A}}, d\mu_0)$

In [491–493] Varadarajan investigated the question of in which sense the tech-
niques of $\overline{\mathcal{A}}, \mu_0$, which in principle apply to any gauge field theory of connections
for compact gauge groups, can be used to describe the Fock states of Maxwell
theory and linearised gravity on a Minkowski background spacetime. Both theories are Abelian gauge theories.[5] This is not at all an academic question: while
we will explicitly couple Maxwell fields to gravity in a background-independent

---

[5] Linearised gravity can be described in terms of connections as well [652,653] where it
becomes effectively a U(1)^3 Abelian gauge theory just like Maxwell theory.

way in Chapter 12, in order to make contact with low-energy physics we must understand how to recover background-dependent ordinary QFT from this perspective.

This therefore seems a hard problem to solve because the languages that one uses in both frameworks are so different: the photon Fock space uses the background metric in many important ways while there is no room for this metric in LQG. The way Varadarajan partly solved this problem was by discovering a background-dependent representation of a (modified) holonomy–flux algebra. See [654, 655] for earlier work on this subject but where the relation with LQG was not clear. Thus the new representation is background-dependent but it is based on the (at least partly) background-independent modified algebra $\mathfrak{A}$ which is similar to the one employed in LQG. We should point out that this modified algebra exists only in the Abelian case and does not admit an immediate generalisation to the non-Abelian case. Nevertheless, these works lie somewhat halfway between what one wants to achieve. More precisely, Varadarajan succeeded in displaying Fock states within the framework of $\overline{\mathcal{A}}, \mu_0$ in a very precise way. We describe these results in some detail below for the Maxwell case, the linearised gravity case is completely analogous.

The crucial observation, unfortunately only valid if the gauge group is Abelian, is the following isomorphism between two different Poisson subalgebras of the Poisson algebra on $\mathcal{M}$. consider a one-parameter family of test functions of rapid decrease which are regularisations of the δ-distribution, for instance

$$f_r(x, y) = \frac{e^{-\frac{||x-y||^2}{2r^2}}}{(\sqrt{2\pi}r)^3} \tag{11.3.1}$$

where we have made use of the Euclidean spatial background metric. Given a path $p \in \mathcal{P}$ we denote its form factor by

$$X_p^a(x) := \int_0^1 dt \dot{p}^a(t)\delta(x, p(t)) \tag{11.3.2}$$

The smeared form factor is defined by

$$X_{p,r}^a(x) := \int d^3y f_r(x, y) X_p^a(y) = \int_0^1 dt \dot{p}^a(t) f_r(x, p(t)) \tag{11.3.3}$$

which is evidently a test function of rapid decrease. Notice that a U(1) holonomy can be written as

$$h_p(A) := e^{i \int d^3x X_p^a(x) A_a(x)} \tag{11.3.4}$$

and we can define a smeared holonomy by

$$h_{p,r}(A) := e^{i \int d^3x X_{p,r}^a(x) A_a(x)} \tag{11.3.5}$$

It is possible to show that the smeared holonomies (11.3.5) are algebraically independent and that the finite linear span of form factors (for the same value of r) is dense in the space of test functions of rapid decrease [491]. Likewise we may define smeared electric fields as

$$E_r^a(x) := \int d^3 y f_r(x,y) E^a(y) \qquad (11.3.6)$$

If we denote by q the electric charge (notice that in our notation $\alpha = \hbar q^2$ is the fine structure constant), then we obtain the following Poisson subalgebras: on the one hand we have smeared holonomies but unsmeared electric fields with

$$\{h_{p,r}, h_{p',r}\} = \{E^a(x), E^b(y)\} = 0, \quad \{E^a(x), h_{p,r}\} = iq^2 X_{p,r}^a(x) h_{p,r} \ (11.3.7)$$

and on the other hand we have unsmeared holonomies but smeared electric fields with

$$\{h_p, h_{p'}\} = \{E_r^a(x), E_r^b(y)\} = 0, \quad \{E_r^a(x), h_p\} = iq^2 X_{p,r}^a(x) h_p \quad (11.3.8)$$

Thus the two Poisson algebras are isomorphic and also the $*$ relations are isomorphic, both $E^a(x), E_r^a(x)$ are real-valued while both $h_p, h_{P,r}$ are U(1)-valued. Thus, as abstract $*$- Poisson algebras these two algebras are indistinguishable and we may ask if we can find different representations of it. Even better, notice that $h_{p,r} h_{p',r} = h_{pop',r}$, $h_{p,r}^{-1} = h_{p^{-1},r}$ so the smeared holonomy algebra is also isomorphic to the unsmeared one. It is crucial to point out that the right-hand side of both (11.3.7), (11.3.8) is a cylindrical function again only in the Abelian case. Therefore all that follows *is not true for* SU(2), see [487] for a proof.

Now we know that the unsmeared holonomy algebra is well represented on the Hilbert space $\mathcal{H}^0 = L_2(\overline{\mathcal{A}}, d\mu_0)$ while the smeared holonomy algebra is well represented on the Fock–Hilbert space $\mathcal{H}_F = L_2(\mathcal{S}', d\mu_F)$ where $\mathcal{S}'$ denotes the space of divergence-free, tempered distributions and μ_F is the Maxwell–Fock measure of the Gaußian type. These measures are completely characterised by their generating functional

$$\omega_F(\hat{h}_{p,r}) := \mu_F(h_{p,r}) = e^{-\frac{1}{4\alpha} \int d^3 x X_{p,r}^a(x) \sqrt{-\Delta}^{-1} X_{p,r}^b \delta_{ab}} \qquad (11.3.9)$$

since finite linear combinations of the $h_{p,r}$ are dense in $\mathcal{H}_F$ [491]. Here $\Delta = \delta^{ab} \partial_a \partial_b$ denotes the Laplacian and we have taken a loop p rather than an open path so that $X_{p,r}$ is transversal. Also unsmeared electric fields are represented through the Fock state ω_F by

$$\omega_F(\hat{h}_{p,r} \hat{E}^a(x) \hat{h}_{p',r}) = -\frac{\alpha}{2} [X_{p,r}^a(x) - X_{p',r}^a(x)] \omega_F(\hat{h}_{pop',r}) \qquad (11.3.10)$$

and any other expectation value follows from these and the commutation relations.

Since ω_F defines a positive linear functional we may define a new representation of the algebra h_p, E_r^a by

$$\omega_r(\hat{h}_p) := \omega_F(\hat{h}_{p,r}) \text{ and } \omega_r(\hat{h}_p \hat{E}_r^a(x) \hat{h}_{p'}) := \omega_F(\hat{h}_{p,r} \hat{E}^a(x) \hat{h}_{p',r}) \qquad (11.3.11)$$

called the r-Fock representation. In order to see whether there exists a measure μ_r on $\overline{\mathcal{A}}$ that represents ω_r in the sense of the Riesz representation theorem we must check that ω_r is a positive linear functional on $C(\overline{\mathcal{A}})$. This can be done [491]. In [639] Velhinho has computed explicitly the cylindrical projections of this measure and showed that the one-parameter family of measures μ_r are expectedly mutually singular with respect to each other and with respect to the uniform measure μ_0. Thus, none of these Hilbert spaces is contained in any other. The Hilbert space $\mathcal{H}_r$, for any r, is unitarily equivalent to $\mathcal{H}_F$ (and thus they are mutually unitarily equivalent for different r) by construction but certainly not to the kinematical Hilbert space $\mathcal{H}_0$ since $\mathcal{H}_F$ is separable while $\mathcal{H}_0$ is not.

In fact, we have a natural map

$$\Theta_r : \mathcal{S}' \to \overline{\mathcal{A}/\mathcal{G}}; \; A \mapsto \Theta_r(A) \text{ where } [\Theta_r(A)](p) := e^{i \int d^3 x X_{p,r}^a A_a(x)} \quad (11.3.12)$$

and Velhinho showed that $\mu_r = (\Theta_r)_* \mu_F$ is just the push-forward of the Fock measure.

Recall that the Fock vacuum Ω_F is defined to be the zero eigenvalue coherent state, that is, it is annihilated by the annihilation operators

$$\hat{a}(f) := \frac{1}{\sqrt{2\alpha}} \int d^3 x f^a [\sqrt[4]{-\Delta} \hat{A}_a - i(\sqrt[4]{-\Delta})^{-1} \hat{E}^a] \quad (11.3.13)$$

where f^a is any transversal smearing field. We then have in fact that $\omega_F(.) = <\Omega_F, .\Omega_F>_{\mathcal{H}_F}$, that is, Ω_F is the cyclic vector that is determined by ω_F through the GNS construction. The idea is now the following: from (11.3.11) we see that we can easily answer any question in the r-Fock representation which has a pre-image in the Fock representation, we just have to replace everywhere $h_{p,r}, E^a(x)$ by $h_p, E_r^a(x)$. Since in the r-Fock representations only exponentials of connections are defined, we should exponentiate the annihilation operators and select the Fock vacuum through the condition

$$e^{i\hat{a}(f)} \Omega_F = \Omega_F \quad (11.3.14)$$

In particular, choosing $f = \sqrt{2\alpha}(\sqrt[4]{-\Delta})^{-1} X_{p,r}$ for some loop p we get

$$e^{\int d^3 x X_{p,r}^a [i\hat{A}_a + (\sqrt[4]{-\Delta})^{-1} \hat{E}^a]} \Omega_F = \Omega_F \quad (11.3.15)$$

Using the commutation relations and the Baker–Campell–Hausdorff formula one can write (11.3.15) in terms of $\hat{h}_{p,r}$ and the exponential of the electric field appearing in (11.3.15) times a numerical factor. The resulting expression can then be translated into the r-Fock representation.

This was Varadarajan's idea. He found that in fact there is no state in $\mathcal{H}^0$ which satisfies the translated analogue of (11.3.15) but that there exists a distribution that does (we must translate (11.3.15) first into the dual action to compute that distribution). It is given (up to a constant) by

$$\Omega_r = \sum_s e^{-\frac{\alpha}{2} \sum_{e,e' \in E(\gamma(s))} G_{e,e'}^r n_e(s) n_{e'}(s)} T_s < T_s, . >_{\mathcal{H}^0} \quad (11.3.16)$$

where $s = (\gamma(s), \{n_e(s)\}_{e \in E(\gamma(s))})$ denotes a charge-network (the U(1) analogue of a spin-network) and

$$G_{e,e'}^r = \int d^3x X_{e,r}^a \sqrt{-\Delta}^{-1} X_{e',r}^b \delta_{ab}^T \tag{11.3.17}$$

where $\delta_{ab}^T = \delta_{ab} - \partial_a \Delta^{-1} \partial_b$ denotes the transverse projector.

Several remarks are in order concerning this result:

1. *Distributional Fock states*

   n-particle state excitations of the state Ω_F (and also coherent states [495]) can easily be translated into distributional n-particle states (coherent states) by using Varadarajan's prescription above. Thus, we get in fact a **Varadarajan map**

$$V : (\mathcal{H}_F, \mathcal{L}(\mathcal{H}_F)) \mapsto (\mathcal{D}^*, \mathcal{L}'(\mathcal{D})) \tag{11.3.18}$$

   Since none of the image states is normalisable with respect to μ_0, this raises the question of in which sense the kinematical Hilbert space is useful at all in order to do semiclassical analysis. One can *in this case define* a new scalar product on these distributions simply by

$$< V \cdot \psi, V \cdot \psi' >_r := < \psi, \psi' >_F \tag{11.3.19}$$

   In particular we obtain $< \Omega_r, . \Omega_r >_r = \omega_r$ so Ω_r can be interpreted as the GNS cyclic vector underlying ω_r. With respect to this inner product one can now perform semiclassical analysis. But how would one have guessed (11.3.19) from first principles? In [487] it is shown that one can arrive at the new representation (11.3.19) directly from the kinematical Hilbert space through a limiting procedure by using the complexifier machinery, thus one can take the point of view that the kinematical, background-independent representation is the fundamental one from which certain others, including the background-dependent one (11.3.19), can be derived (in this case motivated by providing a suitable Hamiltonian operator).

2. *Electric flux operators*

   In the non-Abelian theory it was crucial not to work with electrical fields smeared in D dimensions but rather with those smeared in $D-1$ dimensions. However, $(D-1)$-smeared electrical fields have no pre-image under V and in fact Velhinho showed that there is no electric flux operator in the r-Fock representation as to be expected. This seems to be an obstruction to transfer the Varadarajan map to the non-Abelian case.

3. *Comparison with complexifier coherent states*

   Formula (11.3.16) reminds us of the complexifier framework with complexifier chosen to be a quadratic function of the electric fields (see also [495]). We can write (11.3.16) more suggestively as

$$\Omega_r = \sum_s e^{\frac{\alpha}{2} \sum_{e,e' \in E(\gamma(s))} G_{e,e'}^r R_e R_{e'}} T_s < T_s, . >_{\mathcal{H}^0} \tag{11.3.20}$$

where R_e are right-invariant vector fields on U(1). This formula just asks to be analytically continued in order to arrive at a coherent state as in the complexifier framework. The deeper origin of this apparent coincidence is unravelled in [487] where it is shown that the Varadarajan coherent distributions are complexifier distributions generated by the complexifier

$$ C = \frac{\alpha}{2} \int_{\mathbb{R}^3} d^3x \, \delta_{ab} E_r^a \sqrt{-\Delta}^{-1} E_r^b $$

In [495] it is proposed that one should generalise (11.3.20) in the obvious way to the non-Abelian case by replacing charge nets by spin nets and $R_e R_{e'}$ by $R_e^j R_{e'}^j$ and using the associated cutoff states (called 'shadows' there) for semi-classical analysis. However, it is unclear whether these shadows have similarly nice properties as the cutoff states introduced in [485, 486, 488–490] because the metric $G_{ee'}^r$ is not diagonal. Also it is unclear how one should define the corresponding distributional non-Abelian Fock states since the Laplacians do not form a cylindrically consistent family. Finally it is not clear what the interpretation of the complexified connection should be because the Laplacians do not obviously come from a classical function. Progress on these questions has been made recently in [487] by using the complexifier approach and where instead of Laplacians one uses area operators as displayed in Section 11.2.6.

4. *Other operators*

One should not forget that important operators of Maxwell theory such as the Hamiltonian operator are expressed as polynomials of *non-exponentiated* annihilation and creation operators. However, such operators are not defined either in the r-Fock representation or in $\mathcal{H}^0$. In [637, 638] it is shown how to circumvent that problem.

In conclusion, Varadarajan's construction [491–493] is an important contribution to building the semiclassical sector of the theory and [447, 495] rest crucially on it. In particular, at least as far as free field theories on a (Minkowski) background spacetime are concerned, one can construct a polymer-like image of the usual Fock space starting from the background-independent kinematical loop quantum gravity Hilbert space through a limiting procedure. The open issue is of course the construction of semiclassical states for an interacting, background-independent theory which was started in [485–490]. Here the idea is that the background metric is supplied by a semiclassical state. Those states are still elements of a background-independent representation, however, they are peaked on (initial data of) a certain background spacetime. See in particular [637, 638] where the interacting Klein–Gordon–Maxwell–Einstein system was considered. Also one should understand how to describe gravitons and in particular the vacuum state from the point of view of the SU(2) theory rather than from the U(1)^3 theory. See [652, 653] for first steps in that direction.

Notice that the developments of Sections 11.2 (coherent states for interacting quantum field theories and cutoff states) and 11.3 (Varadarajan states for background-dependent linear quantum field theories and shadow states) started independently of each other. However, as we just said, the complexifier method spelt out initially in [486] is also the underlying framework for [447, 491–493, 495] as was demonstrated in [487] so that *there is actually one enveloping concept*: the complexifier concept. This is helpful to know as it unifies the recently started semiclassical programmes.

III

Physical applications

11
Physical applications

12

Extension to standard matter

The exposition of Chapter 10 would be incomplete if we could not extend the framework to matter also, at least to the matter of the standard model. This is straightforward for gauge field matter, however for fermionic and Higgs matter one must first develop a background-independent mathematical framework [443]. We will discuss the essential steps in the next section and then outline the quantisation of the matter parts of the total Hamiltonian constraint in the section after that, see [441] for details.

We should point out that these representations are geared towards a background-independent formulation. The matter Hamiltonian operator of the standard model in a background spacetime *is not carried* by these representations. They make sense *only if we couple quantum gravity*. Also, while we did not treat supersymmetric matter explicitly, the following exposition reveals that it is straightforward to extend the formalism to Rarita–Schwinger fields. We will follow closely [441, 443].

Before we start we comment on a frequently stated criticism: as we will see there is no obstacle in finding background-independent kinematical representations of standard matter quantum field theories and these support the matter contributions to the Hamiltonian constraint. Thus, it seems as if in LQG there is no restriction on the matter content of the world. However, that is a premature conclusion: the associated Master Constraint of geometry and matter could have zero in its spectrum depending on the type of matter coupled. Indeed, the reason why the spectrum of the Master Constraint could not contain zero is due to normal or factor ordering effects which are finite but similar in nature to the infinite vacuum energies of background-dependent quantum field theories. A well-known procedure for how to cancel these (infinite) vacuum energies is by supersymmetry: every positive contribution from a bosonic mode is cancelled by a negative contribution from a fermionic mode. Hence, in LQG, when it comes to the construction of the physical Hilbert space, restrictions on the matter content of the world might occur by a mechanism rather similar to those that lead to supersymmetry.

12.1 The classical standard model coupled to gravity

The Lagrangian of the standard model coupled to gravity can be reduced to a linear combination of actions of the following types

$$S_{\text{Einstein}} = \frac{1}{\kappa} \int_M d^4 X \sqrt{|\det(g)|} R$$

$$S_{\text{cosmo}} = \frac{\Lambda}{\kappa} \int_M d^4 X \sqrt{|\det(g)|}$$

$$S_{\text{YM}} = -\frac{1}{4Q^2} \int_M d^4 X \sqrt{|\det(g)|} g^{\mu\nu} g^{\rho\sigma} \underline{F}^I_{\mu\rho} \underline{F}^J_{\nu\sigma} \delta_{IJ}$$

$$S_{\text{Higgs}} = \frac{1}{2\lambda} \int_M d^4 X \sqrt{|\det(g)|} (g^{\mu\nu} [\nabla_\mu \phi_I][\nabla_\nu \phi_J] + V(\phi))$$

$$S_{\text{Dirac}} = \frac{i}{2} \int_M d^4 X \sqrt{|\det(g)|} \left(\left[\overline{\Psi}_r \gamma^\alpha \epsilon^\mu_\alpha \nabla_\mu \Psi_s - \overline{\nabla_\mu \Psi}_r \gamma^\alpha \epsilon^\mu_\alpha \nabla_\mu \Psi_s \right] \delta^{rs} - iJ(\overline{\Psi}, \Psi) \right)$$

$$(12.1.1)$$

The first two terms are the already familiar Einstein–Hilbert action and a cosmological term. The third is a Yang–Mills action for some compact gauge group G where $\underline{F}$ is the curvature of some G-connection $\underline{A}$ and Q is a coupling constant. For the standard model G = U(1) × SU(2) × SU(3). The fourth term is a scalar (Higgs) contribution with some potential V which for the standard model is a fourth-order polynomial which induces spontaneous symmetry breaking and λ is a coupling constant. For definiteness we have written a Higgs field which as $\underline{F}$ also transforms in the adjoint representation of Lie(G), however, that is not necessary. In fact, for the standard model the Higgs transforms in the fundamental representation of the electroweak SU(2). Finally, the last term is the fermionic contribution where for definiteness we have assumed that Ψ is a Dirac spinor. Here γ^α, $\alpha = 0, 1, 2, 3$ are Minkowski space Dirac matrices and e^μ_α are tetrads. As usual, $\overline{\Psi} = (\Psi^*)^T \gamma^0$ denotes the conjugate spinor.

In the form displayed the action is appropriate for the quarks. For the leptons we must insert additional appropriate chiral projectors $(1_4 \pm \gamma_5)/2$, $\gamma_5 = i\gamma_0 \ldots \gamma_3$ to isolate the left-handed and right-handed contributions (notice that by now it is fairly sure that neutrinos have masses leading to neutrino oscillations). In the terminology used in Section 15.1.4, Dirac spinors transform in the direct sum of the two fundamental representations of SL(2, ℂ) (unprimed and primed spinors) while the Weyl fermions corresponding to the chiral projections transform in one of the two fundamental representations (left-handed corresponds to unprimed). In addition, as is the case for the standard model, the fermions may transform in some irreducible representation of G, here indicated by the indices r, s. For quarks this is the tensor product of the defining representation of U(1) × SU(3) while for the leptons it is U(1) × SU(2) for the charged particles while only SU(2) for the neutrinos. We are describing here the standard model in the form before symmetry breakdown and thus refrain from introducing Higgs vacuum expectation values and Cabibbo–Kobayashi–Maskawa mass matrices (and similar ones for the neutrinos). These could be absorbed into the current J which is supposed to be a real-valued, bilinear, G-invariant scalar. Notice that all fermion fields are taken to be Grassman-valued.

Finally, ∇_μ denotes the $SL(2, \mathbb{C}) \times G$ covariant derivative. Thus it annihilates $g_{\mu\nu}$, e_α^μ, γ^α and takes the explicit form $\nabla\phi_I = d\phi_I + f_{IJK}\underline{A}_K\phi_K, \nabla\Psi_r = d\Psi_r + \Gamma\Psi_r + \underline{A}_I(\underline{\tau}_I)_{rs}\Psi_s$. Here f_{IJK} are the structure constants for the chosen basis $\underline{\tau}_I$ of Lie(G) with $(\underline{\tau}_J)_{IK} = f_{IJK}$, Γ is the spin connection of the tetrad acting on the corresponding spinor type, see Section 15.1.4, and $(\underline{\tau}_I)_{rs} = (d/dt)_{t=0}[\rho(\exp(t\underline{\tau}_I))]_{rs}$ where ρ is the G-representation in which the fermions transform. We are using the second-order formalism since we require that Γ is defined by e.

One can also consider any of the supersymmetric generalisations of the standard model, thus leading to 4D supergravity theories. The presentation below would not change, we would just need to introduce additional Rarita–Schwinger fields and superpartners in order to complete the super multiplets. The canonical formulations of such extensions can be found, for example, in [295, 296]. All the essentially new ideas can already be understood from the types of matter considered above, in fact, as we will show, the matter quantisation in LQG is universal and can deal with any type of matter, at least before solving the constraints. Restrictions on allowed matter couplings will arise, as already metioned, when solving the corresponding Master Constraint: adding or delcting matter modes may result in a shift of the minimum of the spectrum away from zero, which would mean that the space of solutions is empty. Hence we expect that certain types of matter are dynamically excluded.

We must now cast (12.1.1) into canonical form. Clearly the cosmological term just reduces to $\frac{\Lambda}{\kappa} \int_{\mathbb{R}} dt \, \sigma \, d^3 x N \sqrt{\det(q)}$ and does not need to be discussed any further, it thus contributes the volume operator to the Hamiltonian constraint.

12.1.1 Fermionic and Einstein contribution

We begin with the fermionic contribution and want to write this in terms of the $(A_a^i = \Gamma_a^i + \beta K_a^i, E_i^a)$ where for simplicity we set $\beta = 1$. Let us write the Dirac bi-spinor explicitly as $\Psi_r = (\psi_r, \eta_r)$ where $\psi_r = (\psi_r^A)$ and $\eta_r = (\eta_{A'r})$ transform according to the fundamental representations of $SL(2, \mathbb{C})$, the representation ρ of G and are scalars of density weight zero. We take as usual $M = \mathbb{R} \times \sigma$, let T^μ be the time foliation vector field of M and denote by n^μ the normal vector field of the time slices σ. Then the tetrad can be written $e_\alpha^\mu = e_\alpha^\mu - n^\mu n_\alpha$ with $e_\alpha^\mu n_\mu = e_\alpha^\mu n^\alpha = 0$ so that e_α^μ is a triad and $\eta^{\alpha\beta} n_\alpha n_\beta = -1$ is an internal unit timelike vector which we may choose to be $n_\alpha = -\delta_{\alpha,0}$ ($\eta = \text{diag}(-, +, +, +)$ is the Minkowski metric). Finally, inserting lapse and shift fields by $(\partial_t)^\mu = T^\mu = Nn^\mu + N^\mu$ with $N^\mu n_\mu = 0$ one sees that the action can be written, after lengthy computations, in terms of Weyl spinors as (using the Weyl representation for the Dirac matrices, for instance, to expand out various terms) and neglecting the current J and the indices r over which implicit summation is assumed

$$S_{\text{Dirac}} = \frac{i}{2} \int_{\mathbb{R}} dt \int_{\Sigma} d^3x N \sqrt{\det(q)} \left[\frac{T^\mu - N^\mu}{N} (\psi^\dagger \nabla^+_\mu \psi + \eta^\dagger \nabla^-_\mu \eta - c.c.) \right.$$

$$\left. + e^\mu_i (\psi^\dagger \sigma_i \nabla^+_\mu \psi - \eta^\dagger \sigma_i \nabla^-_\mu \eta - c.c.) \right] \tag{12.1.2}$$

where the *c.c.* in $(* + c.c.)$ stands for 'complex conjugate of $*$'. Here we have defined $e^\mu_\alpha = (0, e^\mu_i)$ and abused the notation in writing $e^\mu_i (= (e^t_i = 0, e^a_i)$, σ_i are the Pauli matrices, $\psi^\dagger := (\psi^*)^T$ and $\nabla^\pm_\mu$ is the self-dual respectively anti-self-dual part of ∇_μ in the Weyl representation. More precisely, $\nabla^\pm_\mu = \partial_\mu + \omega^\pm_\mu + \underline{A}_\mu$, $\omega^\pm_\mu = -i\sigma_j \omega^{j\pm}_\mu$, $\omega^{j+}_\mu = -1/2\epsilon_{jkl}\omega^{kl+}_\mu$, $\omega^{\alpha\beta+}_\mu = \frac{1}{2}(\omega^{\alpha\beta}_\mu - i\epsilon^{\alpha\beta}{}_{\gamma\delta}\omega^{\gamma\delta}_\mu)$, $\omega^{j-}_\mu = \overline{\omega^{j+}_\mu}$ and $\omega^{\alpha\beta}_\mu$ is the spin-connection of e^a_α. These formulae can be derived directly from Section 15.1.4.

It is easy to see that the spatial part of ω^{j+}_μ is just given by $\frac{1}{2}A^{j\mathbb{C}}_a$ where $A^{j\mathbb{C}}_a = \Gamma^j_a + iK^j_a$ is the complex-valued Ashtekar connection. We have already seen this in Chapter 4. Denoting $\mathcal{D}^{\mathbb{C}}_a \psi_r = (\partial_a + A^{j\mathbb{C}}_a \tau_j)\psi_r + \underline{A}^{rs}_a \psi_s$ and $\overline{\mathcal{D}}^{\mathbb{C}}_a \eta_r = (\partial_a + \overline{A}^{j\mathbb{C}}_a \tau_j)\eta_r + \underline{A}^{rs}_a \eta_s$ with $\tau_j = -\frac{i}{2}\sigma_j$ (Pauli matrices) and $A^{j\mathbb{C}}_t = T^\mu \omega^{j+}_\mu$, $\dot{\psi} = T^\mu \partial_\mu \psi$ we end up with

$$S_{\text{Dirac}} = \frac{i}{2} \int dt \int d^3x \sqrt{\det(q)} \left[(\psi^\dagger \dot{\psi} + \eta^\dagger \dot{\eta} - c.c.) \right.$$

$$- \left(-(A^{j\mathbb{C}}_t \psi^\dagger \tau_j \psi + \overline{A^{j\mathbb{C}}_t} \eta^\dagger \tau_j \eta - c.c.) + i\underline{A}^{rs}_t \delta^{AB}(\bar{\psi}_A \psi_B + \bar{\eta}_A \psi_B) \right.$$

$$\left. + N^a(\psi^\dagger \mathcal{D}^{\mathbb{C}}_a \psi + \eta^\dagger \overline{\mathcal{D}}^{\mathbb{C}}_a \eta - c.c.) + N e^a_i (-\psi^\dagger \sigma_i \mathcal{D}^{\mathbb{C}}_a \psi + \eta^\dagger \sigma_i \overline{\mathcal{D}}^{\mathbb{C}}_a \eta - c.c.)) \right] \tag{12.1.3}$$

Let us now introduce $D_a \psi = (\partial_a + \tau_j \Gamma^j_a)\psi$, $E^i_a = |\det(e^i_a)|e^a_i$, $A^j_t := \Re(A^{j\mathbb{C}}_t)$ then we see by explicitly evaluating *c.c.* that

$$S_{\text{Dirac}} = \frac{i}{2} \int dt \int d^3x \sqrt{\det(q)} \left[(\psi^\dagger \dot{\psi} + \eta^\dagger \dot{\eta} - c.c.) \right.$$

$$- (-2A^j_t(\psi^\dagger \tau_j \psi + \eta^\dagger \tau_j \eta) + N^a(\psi^\dagger D_a \psi + \eta^\dagger D_a \eta - c.c.)$$

$$+ i\underline{A}^{rs}_t \delta^{AB}(\bar{\psi}_A \psi_B + \bar{\eta}_A \psi_B) + N \frac{E^a_i}{\sqrt{\det(q)}}([-\psi^\dagger \sigma_i D_a \psi + \eta^\dagger \sigma_i D_a \eta - c.c.]$$

$$\left. + 2[K_a, E^a]^j (\psi^\dagger \tau_j \psi - \eta^\dagger \tau_j \eta))) \right] \tag{12.1.4}$$

This is the $3 + 1$ split Dirac action that we are going to combine with the $3 + 1$ split Einstein action to obtain the desired form in terms of (A^i_a, E^a_i).

We come to the Einstein action which now derives in terms of the new variables from an action principle following [295, 296]. One takes the Palatini Lagrangian in first-order form $L = F_{\alpha\beta}(\omega) \wedge *(e^\alpha \wedge e^\beta)$ where $*T^{\alpha\beta} = \frac{1}{2}\epsilon^{\alpha\beta\gamma\delta}T_{\gamma\delta}$ denotes the flat Hodge dual (see Section 15.1.4; all flat indices are moved with the Minkowski metric η), use the identity $T = T_+ + T_-$ where $T_\pm = (T \mp i * T)/2$ as well as $F_\pm(\omega) = F(\omega_\pm)$ and see that $L = i(L_+ - L_-)$ where $L_\pm = F^\pm_{\alpha\beta}(\omega) \wedge (e^\alpha \wedge e^\beta)$.

It follows that $S_{\text{Einstein}} = \Re(S^+_{\text{Einstein}})$ where S^+_{Einstein} is the self-dual part of S_{Einstein}, which in our notation is written as

$$S^+_{\text{Einstein}} = \frac{1}{\kappa} \int dt \int d^3x \left[-i\dot{A}^{j\mathbb{C}}_a E^a_j - \left(iA^{j\mathbb{C}}_t \mathcal{D}^{\mathbb{C}}_a E^a_j - iN^a \text{tr}(F^{\mathbb{C}}_{ab} E^b) \right. \right.$$

$$\left. \left. + \frac{N}{2\sqrt{\det(q)}} \text{tr}(F^{\mathbb{C}}_{ab}[E^a, E^b]) \right) \right] \tag{12.1.5}$$

where $F^{\mathbb{C}}$ denotes the curvature of $A^{\mathbb{C}}$ and κ the gravitational coupling constant. Computing the real part reveals

$$S_{\text{Einstein}} = \frac{1}{\kappa} \int dt \int d^3x \left[\dot{K}^j_a E^a_j - \left(-A^j_t[K_a, E^a]_j + 2N^a D_{[a} K^j_{b]} E^b_j \right. \right.$$

$$\left. \left. - \frac{N}{2\sqrt{\det(q)}} \text{tr}(([K_a, K_b] - R_{ab})[E^a, E^b]) \right) \right] \tag{12.1.6}$$

where R_{ab} is the curvature of e^i_a.

The point of this alternative derivation is that it gives a four-dimensional interpretation of the Lagrange multiplier $\Lambda^j = A^j_t$ of the gravitational Gauß constraint. Thus, putting both actions together, we find that the gravitational Gauß constraint is given by (no other matter contributes to it)

$$\mathcal{G}_j = \frac{1}{\kappa}[K_a, E^a]_j + i\sqrt{\det(q)}[\psi^\dagger \tau_j \psi + \eta^\dagger \tau_j \eta] \tag{12.1.7}$$

Here we have assumed that Ψ transforms trivially under G. Otherwise we have to sum over r for each species ψ_r, η_r.

We can now perform a canonical point transformation on the gravitational phase space given by $(K^i_a, E^a_i) \to (A^i_a, E^a_i)$ (the generator is $\int d^3x \Gamma^i_a E^a_i$ as we checked) and we must then express the constraints in terms of A^i_a. Let us therefore introduce the real-valued derivative $D_a\psi := (\partial_a + A^j_a \tau_j)\psi$ and denote by F_{ab} the curvature of A^i_a. Using that $D_a E^a_i = 0$ we can immediately write

$$\mathcal{G}_j = \frac{1}{\kappa} D_a E^a_j + i\sqrt{\det(q)}[\psi^\dagger \tau_j \psi + \eta^\dagger \tau_j \eta] \tag{12.1.8}$$

Next, we expand F_{ab} in terms of Γ_a, K_a, use the Bianchi identity $\text{tr}(R_{ab}E^b) = 0$ and find that the vector constraint H_a, the coefficient of N^a in $S_{\text{Dirac}} + S_{\text{Einstein}}$ is given, up to a term proportional to $\mathcal{G}_j$, by

$$H_a = \text{tr}(F_{ab}E^b) + \frac{i}{2}\sqrt{\det(q)}(\psi^\dagger D_a\psi + \eta^\dagger D_a\eta - c.c.) \tag{12.1.9}$$

Finally, let, as in the source-free case

$$H_E = \frac{1}{2\kappa} \text{tr}\left(F_{ab} \frac{[E^a, E^b]}{\sqrt{\det(q)}} \right) \tag{12.1.10}$$

which has the interpretation of the source-free Euclidean Hamiltonian constraint. Furthermore, let

$$H_G := -H_E + \frac{2}{2\kappa} \text{tr}\left([K_a, K_b] \frac{[E^a, E^b]}{\sqrt{\det(q)}}\right) \tag{12.1.11}$$

which in the source-free case would be the full Lorentzian Hamiltonian constraint. Then the Einstein contribution to the Hamiltonian constraint of $S_{\text{Dirac}} + S_{\text{Einstein}}$ is given by

$$H = H_G - \frac{2}{2\kappa} D_a \text{tr}\left([K_b, E^b] \frac{E^a}{\sqrt{\det(q)}}\right) =: H_G + T \tag{12.1.12}$$

Notice that in the source-free case the correction T of H to H_G is proportional to a Gauß constraint and therefore would vanish separately on the constraint surface. However, in our case, using the Gauß constraint (12.1.7) we find that

$$T = -\frac{1}{2}\left([K_a, E^a]^j - E_j^a \mathcal{D}_a\right) J_j \tag{12.1.13}$$

where we have defined the current $J_j := \psi^\dagger \sigma_j \psi + \eta^\dagger \sigma_j \eta$. On the other hand, writing also the Dirac contribution to the Hamiltonian constraint in terms of $\mathcal{D}_a$ rather than D_a and combining with H we find that the first term on the right-hand side of (12.1.13) cancels against a similar term. We end up with the contribution H from both the Einstein and Dirac sector to the Hamiltonian constraint which is given, up to a term proportional to the gravitational Gauß constraint, by

$$H = H_G + \frac{E_j^a}{2\sqrt{\det(q)}}\left(\mathcal{D}_a\left(\sqrt{\det(q)}J_j\right) + i\sqrt{\det(q)}[\psi^\dagger \sigma_j \mathcal{D}_a \psi - \eta^\dagger \sigma_j \mathcal{D}_a \eta - c.c.]\right.$$

$$\left. - K_a^j \sqrt{\det(q)}(\psi^\dagger \psi - \eta^\dagger \eta)\right) \tag{12.1.14}$$

In order to arrive at (12.1.14) one has to use the Pauli matrix algebra $\sigma_j \sigma_k = \delta_{jk} 1_{SU(2)} + i\epsilon_{jkl}\sigma_l$ at several stages when computing $c.c.$ Notice that we can write (12.1.14) also in terms of the half-densities $\xi = \sqrt[4]{\det(q)}\psi$, $\rho = \sqrt[4]{\det(q)}\eta$ by absorbing the $\sqrt{\det(q)}$ appropriately and using that $\mathcal{D}_a \det(q) = 0$. We find

$$H = H_G + \frac{E_j^a}{2\sqrt{\det(q)}}\left(\mathcal{D}_a(\xi^\dagger \sigma_j \xi + \rho^\dagger \sigma_j \rho)\right.$$

$$\left. + i[\xi^\dagger \sigma_j \mathcal{D}_a \xi - \rho^\dagger \sigma_j \mathcal{D}_a \rho - c.c.] - K_a^j(\xi^\dagger \xi - \rho^\dagger \rho)\right) \tag{12.1.15}$$

Note also that $i\sqrt{\det(q)}[\psi^\dagger \dot{\psi} - \dot{\psi}^\dagger \psi] = i[\xi^\dagger \dot{\xi} - \dot{\xi}^\dagger \xi]$ so that our change of variables is actually a symplectomorphism!

This is the form of the constraint that we have been looking for: up to K_a^i we have expressed everything in terms of real-valued quantities and the canonically conjugate pairs $(\xi, i\bar{\xi}), (\rho, i\bar{\rho})$. Now, let us in the source-free case denote by $V = \int_\sigma d^3x \sqrt{\det(q)}$ the total volume of σ and $H_E(1) = \int_\sigma d^3x H_E(x)$. Then it is still true that $K_a^j = -\{A_a^j, \{V, H_{\rm E}(1)\}\}$ and since $V, H_{\rm E}(1)$ admit well-defined quantisations as we saw in Chapter 9 we conclude that despite its complicated appearance (12.1.15) admits a well-defined quantisation as well. Note that if we did not work with half-densities ξ, ρ but with the ψ, η then, while $i\sqrt{\det(q)}\bar{\psi}$ is the momentum conjugate to ψ, the gravitational connection would get a correction proportional to $ie_a^i[\psi^\dagger\psi + \eta^\dagger\eta]$. Thus we would have had to admit a complex connection, which would be disastrous as our Hilbert space techniques would not be at our disposal. Hence we will use the half-densities.

12.1.2 Yang–Mills and Higgs contribution

By carrying out literally the same steps as for the fermionic contribution and using the explicit expression for the metric and its inverse in the ADM frame one finds after performing the Legendre transform

$$
S_{\rm YM} = \frac{1}{Q^2} \int_{\mathbb{R}} dt \int_\sigma d^3x \left\{ \dot{A}_a^I E_I^a - \left[-A_t^I D_a E_I^a + N^a F_{ab}^I E_I^b \right.\right.
$$
$$
\left.\left. + \frac{q_{ab}}{2\sqrt{\det(q)}} (E_I^a E_J^b + B_I^a B_J^b)\delta^{IJ} \right] \right\}
$$
$$
S_{\rm Higgs} = \frac{1}{\lambda} \int_{\mathbb{R}} dt \int_\sigma d^3x \left\{ \dot\phi_I \pi^I - \left[-A_t^I f_{IJK}\phi^J \pi^K + N^a \pi^I D_a \phi_I \right.\right.
$$
$$
\left.\left. + \frac{1}{2} \left(\left[\frac{\pi^I \pi^J}{\sqrt{\det(q)}} + \sqrt{\det(q)}\,[D_a\phi_I][D_a\phi_I] \right]\delta^{IJ} + \sqrt{\det(q)}V(\phi) \right) \right] \right\}
$$

$$(12.1.16)$$

Here D_a acts like the Levi–Civita connection on tensor indices and like A on indices connected with representations of G. For instance, $D_a E_I^a = \partial_a E_I^a + f_{IJK}A_a^J E_K^a$ and $D_a\phi_I = \partial_a\phi_I + f_{IJK}A_a^J\phi_K$. Of course, F is the pull-back to σ of the four-dimensional Yang–Mills curvature and $B^a = \epsilon^{abc}F_{bc}$ its magnetic field.

The Legendre transform which has given rise to (12.1.16) can be rediscovered by solving the equation of motion for A, ϕ respectively, which follow by imposing that (A, E), (ϕ, π) are canonical pairs, in other words by varying (12.1.16) with respect to E, π respectively. We find

$$
\pi_I = \sqrt{\det(q)}n^\mu D_\mu \phi_I, \quad E_I^a = \sqrt{\det(q)}q^{ab}n^\mu F_{\mu b} \qquad (12.1.17)
$$

where $Nn^\mu = T^\mu - N^\mu, T^\mu = X_{,t}^\mu, N^\mu = N^a X_{,a}^\mu$ in terms of the foliation $X(t,\sigma) = \Sigma_t$ of M.

From (12.1.16) we read off the contributions to the spatial diffeomorphism constraint and the Hamiltonian constraint. In particular, the Yang–Mills Gauß constraint becomes

$$\underline{G}_I = \underline{D}_a \underline{E}_I^a + f_{IJK} \phi^J \pi^K + i\sqrt{\det(q)} \delta^{AB} [\rho(\tau_I)]_{rs} \bar{\psi}_A^r \psi_B^s + [\rho(\tau_I)]_{rs} \bar{\eta}_A^r \eta_B^s$$

$$(12.1.18)$$

where we have taken care of the G-transformation behaviour of the fermions.

As in the matter-free case, the spatial diffeomorphism constraint and Gauß constraints generate the expected gauge transformations on the matter extended phase space and therefore we will quantise them precisely as for the pure geometry part. The non-trivial structure lies again with the Hamiltonian constraint on which we focus in what follows. Before we do that we must first introduce suitable background-independent representations for the matter degrees of freedom based on suitable kinematical algebras. With respect to the gauge fields we proceed exactly as with gravity, so this is already achieved because our considerations in the second part of the book did not impose any restriction on G. However, for fermions and scalars we must invest additional work. This is what we will do in the next section.

12.2 Kinematical Hilbert spaces for diffeomorphism-invariant theories of fermion and Higgs fields

First attempts to couple quantum field theories of fermions to Quantum General Relativity gravity were made in the pioneering work [444, 445]. However, these papers were still written in terms of the complex-valued Ashtekar variables for which the kinematical framework was missing. Later on [446] appeared, in which a kinematical Hilbert space for diffeomorphism-invariant theories for fermions was proposed, coupled to arbitrary gauge fields and real-valued Ashtekar variables using the kinematical framework developed in Part II. Also, the diffeomorphism constraint was solved in [446] but not the Hamiltonian constraint. However, that fermionic Hilbert space did implement the correct reality conditions for the fermionic degrees of freedom only for a subset of all kinematical observables. In [443] this problem was removed by introducing new fermionic variables, so-called *Grassmann-valued half-densities*, and the framework was extended to Higgs fields. This section is accordingly subdivided into one subsection each for the fermionic and the Higgs sector respectively and in the third subsection we collect results and define the most general gauge and diffeomorphism-invariant states of connections, fermions and Higgs fields by group averaging.

12.2.1 Fermionic sector

We will take the fermionic fields to be Grassmann-valued, see [235, 656] for a mathematical introduction into these concepts. Furthermore, the Grassmann field η_{Ar} is a scalar with respect to diffeomorphisms of σ which carries two indices, $A, B, C, \ldots = 1, 2$ and $r, s, t = 1, \ldots, \dim(G)$ corresponding to the fact that it transforms according to the fundamental representation of SU(2) and some irreducible representation of the compact, connected, unimodular gauge group G of a Yang–Mills gauge theory to which it may couple. This can be generalised to arbitrary representations of SU(2) × G but we refrain from doing that for the sake of concreteness. Notice that it is no loss of generality to restrict ourselves to only one helicity of the fermion as we can always perform a canonical transformation $(i\bar{\sigma}^{A'}, \sigma_{A'}) \rightarrow (i\epsilon^{AB'}\sigma_{B'}, \epsilon_{AB'}\bar{\sigma}^{B'}) =: (i\bar{\eta}^A, \eta_A)$. We will restrict to only one fermionic species in order not to clutter the formulae.

Recall the Hamiltonian form for any diffeomorphism-invariant theory of fermions from the preceding section

$$S_{\mathrm{F}} = \int_{\mathbb{R}} dt \int_{\sigma} d^3 x \left(\frac{i}{2} \sqrt{\det(q)} [\bar{\eta}^{Ar} \dot{\eta}_{Ar} - \dot{\bar{\eta}}^{Ar} \eta_{Ar}] - [\text{more}] \right) \qquad (12.2.1)$$

where summation over A, r is understood and where 'more' stands for various constraints and possibly a Hamiltonian and $\det(q)$ is the determinant of the gravitational three-metric which appears because in four spacetime dimensions one needs a metric to define a diffeomorphism-invariant theory of fermions. Notice that (12.2.1) is real-valued with respect to the usual involution $(\theta_1 \ldots \theta_n)^* = \bar{\theta}_n \ldots \bar{\theta}_1$ for Grassmann variables $\theta_1, \ldots, \theta_n$ since indices A, r are raised and lowered with the Kronecker symbol (the involution is just complex conjugation with respect to bosonic variables).

The immediate problem with (12.2.1) is that it is not obvious what the momentum $\pi^{A\mu}$ conjugate to η_{Ar} should be. One strategy would be to integrate the second term in (12.2.1) by parts (the corresponding boundary term being the generator of the associated canonical transformation) and to conclude that it is given by $i\sqrt{\det(q)}\bar{\eta}^{Ar}$. However, there is a second term from the integration by parts given by $i\dot{E}_i^a e_a^i \bar{\eta}^{Ar} \eta_{Ar}$ which after a further integration by parts combines with the symplectic potential of the real-valued Ashtekar variables to the effect that A_a^i is replaced by $({}^{\mathbb{C}}A_a^i) = A_a^i - ie_a^i \bar{\eta}^{Ar} \eta_{Ar}$ (recall that E_i^a is the momentum conjugate to A_a^i). This is bad because the connection is now complex-valued and the techniques from Part II do not apply any longer so that we are in fact forced to look for another method. The authors of [446] also noticed this subtlety in the following form. If one assumes that the connection is still real-valued while $\pi = i\sqrt{\det(q)}\eta$ is taken as the momentum conjugate to η then one discovers the following contradiction: by assumption we have the classical Poisson bracket $\{\pi(x), A(y)\} = 0$. Taking the involution of this equation results in $0 = -i\eta(x)\{\sqrt{\det(q)}(x), A(y)\} \neq 0$. If we, however, insert instead of A the above complex variable $({}^{\mathbb{C}}A)$ into these equations then in fact there is no contradiction as was shown in [443].

The idea of how to preserve the real-valuedness of A_a^i and to simplify the reality conditions on the fermions is as follows: notice that if we define the

Grassmann-valued half-density

$$\xi_{Ar} := \sqrt[4]{\det(q)}\eta_{Ar} \tag{12.2.2}$$

then (12.2.1) in fact equals

$$S_F = \int_{\mathbb{R}} dt \int_{\sigma} d^3x \left(\frac{i}{2}[\bar{\xi}^{Ar}\dot{\xi}_{Ar} - \dot{\bar{\xi}}^{Ar}\xi_{Ar}] - [\text{more}] \right) \tag{12.2.3}$$

without picking up a term proportional to $d\det(q)/dt$. Thus the momentum conjugate to ξ_{Ar} and the reality conditions respectively are simply given by

$$\pi^{Ar} = i\bar{\xi}_{Ar} \text{ and } (\xi)^* = -i\pi, \ (\pi)^* = -i\xi \tag{12.2.4}$$

The fact that ξ, π are half-densities may seem awkward at first sight but it does not cause any immediate problems. Also, recall that 'half-density-quantisation' is a standard procedure in the theory of geometric quantisation of phase spaces with real polarisations [218].

It is in fact possible to base the quantisation on the half-density ξ as a quantum configuration variable as far as the solution to the Gauß constraint is concerned. Namely, as has been pointed out by many (see, e.g., [444, 445]) an example for a natural, classical, gauge-invariant observable is given by

$$P_e(\xi, A, \underline{A}) := \xi_{Ar}(e(0))C_1^{Ar,Cs}(h_e(A))_{CD}(\pi(\underline{h}_e(\underline{A})))_{st}C_2^{Dt,Bu}\xi_{Bu}(e(1)) \tag{12.2.5}$$

where the notation is as follows: by $(A, h_e, \pi(h_e))$ and $(\underline{A}, \underline{h}_e, \pi(\underline{h}_e))$ respectively we denote (connection, holonomy along an edge e, irreducible representation evaluated at the holonomy) of the gravitational SU(2) and the Yang–Mills gauge group G respectively. The matrices $C^{Ar,Bt}$ are projectors on singlet representations of the decomposition into irreducibles of tensor product representations that appear under gauge transformations on both ends of the path $[0, 1] \ni t \to e(t)$ and the irreducible representation π has to be chosen in such a way that a singlet can occur. For example, if G = SU(N) then we can choose π to be the complex conjugate of the defining representation. In particular, if G = SU(2) as well we can take π to be the fundamental representation and $C_1^{Ar,Cs} = \epsilon^{AC}\epsilon^{rs}, C_2^{Dt,Bu} = \delta^{DB}\epsilon^{tu}$. For more general groups we may have to take more than one spinor field at each end of the path in order to satisfy gauge invariance.

All this works fine until it comes to diffeomorphism invariance: notice that the objects (12.2.5) behave strangely under a diffeomorphism φ, namely $\varphi \cdot P_e = P_{\varphi(e)}(J_\varphi(e(0))J_\varphi(e(1)))^{-1/2}$ where $J_\varphi(x) = |\det(\partial\varphi(x)/\partial x)|$ is the Jacobian. Since there are semianalytic diffeomorphisms which leave e invariant but such that, say, $J_\varphi(e(0))$ can take any positive value it follows that the average of P_e over diffeomorphisms is meaningless. We are therefore forced to adopt another strategy.

The new idea [443] is to 'dedensitise' ξ by means of the δ-distribution $\delta(x, y)$ which itself transforms as a density of weight one in one argument and as a scalar in the other. Let $\theta(x)$ be a smooth Grassmann-valued scalar (we drop the indices $A\mu$) and we *define* $\xi(x)$ not to be a smooth function but rather a distribution (already classically). Let $\delta_{x,y} = 1$ for $x = y$ and zero otherwise (a Kronecker δ, not a distribution). Then on the space of test functions of rapid decrease the distribution $\sqrt{\delta(x,y)\delta(z,y)}$ is well-defined and equals $\delta_{x,z}\delta(x,y)$ [443]. As shown in [443] the following transformations (and corresponding ones for the complex conjugate variables)

$$\theta(x) := \int_\sigma d^3y \sqrt{\delta(x,y)}\xi(y) \tag{12.2.6}$$

$$\xi(x) = \sum_{y\in\sigma} \sqrt{\delta(x,y)}\theta(y) \tag{12.2.7}$$

are canonical transformations between the symplectic structures defined by the symplectic potentials $i\int_\sigma d^3x\bar{\xi}(x)\dot{\xi}(x)$ and $i\sum_{x\in\sigma}\bar{\theta}(x)\dot{\theta}(x)$ respectively. Notice that (12.2.6) makes sense precisely when ξ is a distributional half-density and in fact one can show that $\xi = \eta\sqrt[4]{\det(q)}$ will precisely display such a behaviour (at least upon quantisation) since $\sqrt{\det(q)}$ becomes an operator-valued distribution proportional to the δ-distribution (recall the formula for the volume operator). The non-trivial anti-Poisson brackets in either case are given by

$$\{\xi(x), \bar{\xi}(y)\}_+ = -i\delta(x,y) \text{ and } \{\theta(x), \bar{\theta}(y)\}_+ = -i\delta_{x,y} \tag{12.2.8}$$

In summary, we conclude that we can base the quantisation of the fermionic degrees of freedom on θ as a configuration variable with conjugate momentum and reality structure given by

$$\pi^{Ar} = i\bar{\theta}_{Ar} \text{ and } (\theta)^* = -i\pi, \ (\pi)^* = -i\theta \tag{12.2.9}$$

Notice that only the half-densities ξ have a classical meaning, more precisely, local bilinear expressions (currents) constructed from them. Consider for instance the current $J(x) := \bar{\xi}(x)\xi(x)$. Using the formal identity $\sqrt{\delta(x,y)\delta(x,z)} = \delta(x,y)\delta_{y,z}$ we see that $J(B) := \int_B d^3x \, J(x) = \sum_{x\in B}\bar{\theta}(x)\theta(x)$. Since in matter Hamiltonians we encounter precisely limits of smeared currents such as $J(B)$ as B shrinks to a point, we see that the currents $j(x) = \bar{\theta}(x)\theta(x)$ appear naturally.

We now have to develop integration theory. This will be based, of course, on the Berezin 'integral' [235, 656]. Let $\mathcal{F}(x)$ be the superspace underlying the $2d$ fermionic configuration degrees of freedom $\theta_{A\mu}(x)$ for any $x \in \sigma$ where $d = 2\dim(G)$. Of course, all these spaces are just copies of a single space $\mathcal{F}$. This superspace can be turned into a trivial σ-algebra $\mathcal{B}(x)$ consisting of $\mathcal{F}(x)$ and the empty set. On $\mathcal{B}(x)$ one can define a probability 'measure' dm_x with the additional property that it is positive on 'holomorphic' functions (that is, those which depend on $\theta(x)$ only and not on $\bar{\theta}(x)$) in the sense that $\int_{\mathcal{F}} dm_x f(\theta(x))^* f(\theta(x)) \geq 0$ where equality holds if and only if $f = 0$. This measure is given by

$$dm(\bar{\theta}, \theta) = \prod_{Ar}(1 + \bar{\theta}^{Ar}\theta_{Ar})d\bar{\theta}^{Ar}d\theta_{Ar} \qquad (12.2.10)$$

and $dm_x = dm(\bar{\theta}(x), \theta(x))$.

Let now $\overline{\mathcal{F}} := \times_{x \in \sigma} \mathcal{F}_x$ be the fermionic quantum configuration space with σ-algebra $\mathcal{B}$ given by the *direct product* of the $\mathcal{B}(x)$. The Kolmogorov theorem [532] for uncountable direct products of probability measures ensures that

$$d\mu_F(\bar{\theta}, \theta) := \otimes_{x \in \sigma} dm_x \qquad (12.2.11)$$

is a rigorously defined probability measure on $\overline{\mathcal{F}}$. It can be recovered as the direct product limit (rather than projective limit) from its finite-dimensional joint distributions defined by cylindrical functions. Here a function F on $\overline{\mathcal{F}}$ is said to be cylindrical over a finite number of points $x_1, \ldots, x_n$ if it is a function only of the finite number of degrees of freedom $\theta(x_1), \ldots, \theta(x_n)$ and their complex conjugates, that is, $F(\theta) = f_{x_1,\ldots,x_n}(\bar{\theta}_1(x_1), \theta_1(x_1), \ldots, \bar{\theta}_n(x_n), \theta_n(x_n))$ where $f_{x_1,\ldots,x_n}$ is a function on $\mathcal{F}^n$. We then have

$$\int_{\overline{\mathcal{F}}} d\mu_F F = \int_{\mathcal{F}^n} dm(\bar{\theta}_1, \theta_1) \ldots dm(\bar{\theta}_n, \theta_n) f_{x_1,\ldots,x_n}(\bar{\theta}_1, \theta_1, \ldots, \bar{\theta}_n, \theta_n)$$

$$(12.2.12)$$

Basic cylindrical functions are the fermionic vertex functions. These are defined as follows: order the labels Ar from 1 to $2d$ and denote them by $i, j, k, \ldots$ (confusion with the SU(2) labels should not arise). Denote by I an array $1 \leq i_1 < \ldots < i_k \leq 2d$ and define $|I| = k$ in this case (confusion with the Lie(G) or spin-network labels should not arise). Then for each set of distinct points $v_1, \ldots, v_n$ we define

$$F_{\vec{v}, \vec{I}} = \prod_{l=1}^{n} F_{v_l, I_{v_l}}, \quad F_{v_l, I_{v_l}} = \prod_{j=1}^{|I_{v_l}|} \theta_{i_j(v_l)}(v_l) \qquad (12.2.13)$$

Is this the correct measure, that is, are the adjointness relations $\hat{\pi}^\dagger = -i\hat{\theta}$, $\hat{\theta}^\dagger = -i\hat{\pi}$ and the canonical anti-commutation relations $[\hat{\theta}_{Ar}(x), \hat{\pi}^{Bs}(y)]_+ = i\hbar\delta_A^B\delta_r^s\delta_{x,y}$ faithfully implemented? It is sufficient to check this to be the case on cylindrical subspaces if we represent $\hat{\theta}(x)$ as a multiplication operator and $\hat{\pi}(x)$ as $i\hbar\partial^l/\partial\theta(x)$ where the superscript stands for the left *ordinary* derivative (not a functional derivative). In fact, the measure $d\mu_F$ is uniquely selected by these relations given the representation just as in the case of the theory of distributional connections $\overline{\mathcal{A}}$. Also, it is trivially diffeomorphism invariant since the integrals of a function cylindrical over n points and of its diffeomorphic image coincide.

In summary, the correct kinematical fermion Hilbert space is therefore defined to be $\mathcal{H}_F := L_2(\overline{\mathcal{F}}, d\mu_F)$. It follows immediately from these considerations that the quantum fermion field *at a point* (i.e., totally unsmeared) becomes a densely defined operator. This seems astonishing at first sight but it is only a little bit more surprising than to assume that Wilson loop operators, the quantum connection being smeared in one direction only, are densely defined. When quantising diffeomorphism-invariant theories which lack a background structure one has to give up standard representations and construct new ones.

12.2.2 Higgs sector

It turns out that it is also not possible to combine the well-developed theory of Gaußian measures for scalar field theories with diffeomorphism invariance in order to obtain a kinematical framework for diffeomorphism-invariant theories of Higgs fields. The basic obstacle is that a Gaußian measure is completely defined by its covariance which, however, depends on a background structure (see [441] for a detailed discussion of this point). We are therefore again led to a new non-standard representation. We will describe two possible avenues. The first is very similar to the procedure adopted for the fermion field, the second is more similar to the procedure adopted for the gauge fields.

12.2.2.1 Diffeomorphism-invariant Fock representation

We have seen that the successful quantisation of the gauge fields was based on the fact that we had a canonical pair consisting of a $p-1$ form (the connection) and a $(D-p)$-pseudo-form (the electric field). These fields were then smeared in their natural dimensions. For scalar fields the situation is similar: we have a canonical pair (ϕ_I, π^I) consisting of a 0-form ϕ and a pseudo D-form π. Here we assume that we are considering a real Higgs field ϕ_I transforming in some irreducible representation of a compact Yang–Mills gauge group G. The case of a complex field can be reduced to that case by treating the real and complex parts separately, giving rise to two scalar species. Hence we are naturally led to the following object

$$\pi(f) := \int_\sigma d^D x \, f_I(x) \pi^I(x) \tag{12.2.14}$$

for some test field of compact support, for example, $f_I(x) = \chi_B(x) n_I(x)$ where B is an open region σ contained in a compact set and n_I is a s.a. function. The functions f_I transform in the same representation as the fields ϕ, π.

As we have seen, the canonical action of a scalar field takes the form

$$S = \frac{1}{\lambda} \int_{\mathbb{R}} dt \int_\sigma d^D x \, [\dot{\phi}_I \pi^I - \text{more}] \tag{12.2.15}$$

where 'more' contains constraints and possibly a Hamiltonian and λ is some coupling constant. The Hamiltonian constraint always has a term proportional to π^2 if the scalar field action is polynomial in the scalar field and is minimally coupled to the geometry. We will assume that ϕ is dimensionless so that $\pi \propto \dot{\phi}$ (according to the equations of motion) has dimension cm^{-1}. Hence $\hbar\lambda$ has dimension cm^{D-1}. From (12.2.15) we see that (ϕ, π) form a canonical pair and as just motivated we would like to base the quantisation on the closed Poisson algebra of the $\phi_I(x)$, $\pi(f)$, that is,

$$\{\phi_I(x), \phi_J(y)\} = \{\pi(f), \pi(f')\} = 0, \ \{\pi(f), \phi_I(x)\} = \lambda f_I(x) \quad (12.2.16)$$

Notice that the functions $\phi(x)$ are *unsmeared*. Now it would seem natural to base a quantisation on annihilation and creation operators whose classical counterpart has the form

$$a'_I(x) := \frac{1}{\sqrt{2\hbar\lambda L}}[\phi_I(x) - iL\pi_I(x)] \quad (12.2.17)$$

where L is some length scale needed in order to match the dimensions of the objects in the square brackets. However, that is not compatible with diffeomorphism covariance: the imaginary part of a'_I transforms as a scalar density while the real part transforms as a scalar. Thus a'_I does not make sense as a linear combination of tensors of different type.

From experience with the fermions we are thus led to the idea to 'dedensitise' π. Consider the formal object

$$p^I(x) := \int_\sigma d^D y \, \delta_{x,y} \, \pi^I(y) \quad (12.2.18)$$

where $\delta_{x,y}$ is the Kronecker δ, not the δ-distribution. Notice that p has dimension cm^{D-1}. Formally, the inversion of (12.2.18) is given by

$$\pi^I(x) = \sum_{y\in\sigma} \delta(x,y) \, p^I(y) \quad (12.2.19)$$

which one can check by interchanging the discrete sums with the integrals and using $\int d^D y \, \delta_{x,y} \delta(y,z) = \delta_{x,z}$ as well as $\sum_y \delta(x,y)\delta_{y,z} = \delta(x,z)$. Of course, for smooth π the quantity (12.2.18) vanishes identically and for smooth p the quantity (12.2.19) would blow up. Hence, these formulae will make sense only for the corresponding operator-valued distributions for which indeed π will have a δ-distribution-like singularity structure with support on a finite number of points while p will be discontinuous with support at a finite number of points.

Using these formulae one can check that formally (12.2.16) becomes

$$\{\phi_I(x), \phi_J(y)\} = \{p^I(x), p^J(y)\} = 0, \ \{p^I(x), \phi_J(y)\} = \lambda \delta^I_J \delta x, y$$
$$(12.2.20)$$

and that p transforms as an honest scalar so that the following object makes geometrical and dimensional sense

$$a_I(x) := \frac{1}{\sqrt{2\hbar\lambda L^{-(D-1)}}}[\phi_I(x) - iL^{-(D-1)}p_I(x)] \quad (12.2.21)$$

We now impose canonical commutation relations for the corresponding operators

$$[\hat{a}_I(x), \hat{a}_J(y)] = [\hat{a}_I^\dagger(x), \hat{a}_J^\dagger(y)] = 0, \; [\hat{a}_I(x), \hat{a}_J^\dagger(y)] = \delta_{IJ}\delta_{x,y} \quad (12.2.22)$$

Notice that these are really operators, not operator-valued distributions!

In seeking representations of the algebra (12.2.23) we use a Fock representation defined by declaring the existence of a ground state Ω satisfying $\hat{a}_I(x)\Omega = 0$ for all $x \in \sigma$, I. Then the expectation value functional $\omega := < \Omega, .\Omega >$ is positive and the corresponding Fock space is the GNS Hilbert space descending from ω. We may also describe this measure theoretically: we consider dimension-free functions $f_I : \sigma \to \mathbb{R}$ with the property that the support of f_I is compact and discrete, that is, it consists of a finite number of points. Let $\phi(f) := \sum_{x \in \sigma} f_I(x)\phi_I(x)$ and let us consider the C^*-algebra $\mathfrak{A}$ of functions of smooth ϕ generated by the 'holonomies' $h(f) := \exp(i\phi(f))$ and completed in the sup-norm. We define the positive linear functional

$$\Lambda(h(f)) := \omega\left(e^{i\hat{\phi}(f)}\right) = e^{-\frac{\hbar\lambda}{2L^{D-1}}||f||^2} \quad (12.2.23)$$

where we expressed $\hat{\phi}(f) := \sqrt{\frac{\hbar\lambda}{2L^{D-1}}}(\hat{a}(f) + \hat{a}^\dagger(f))$ and used the properties of Ω and the Baker–Campbell–Hausdorff formula. The norm squared of f is $||f||^2 = \sum_{x \in \sigma} f_I(x)^2$ which converges due to the assumptions about f. Formula (12.2.23) together with the Riesz–Markov theorem of Chapter 25 reveals that $\Lambda(.) = \int_{\overline{\Phi}} d\mu_S(.)(.)$ where μ_S is a white noise Gaußian measure with covariance $C(f) = ||f||^2\hbar\lambda/L^{D-1}$ and where $\overline{\Phi}$ is a space of distributional scalar fields, namely the spectrum of $\mathfrak{A}$. A more geometric characterisation of $\overline{\Phi}$ can be given as well. From the point of view of the scalar field theory, $L := \sqrt[D]{\hbar\lambda}$ is a natural choice since there is no other scale in the problem and we will adopt this choice for what follows.

In order to draw an analogy with the case of gauge fields we define cylindrical functions as functions of the form $F(\phi) = F_P(\{\phi(x)\}_{x \in S})$ where P is a finite set of points in σ and F_P is a complex-valued function of the arguments displayed. We call a cylindrical function smooth if F_P is smooth. For the case of $h(f)$ we see that $P = \text{supp}(f)$. The sets P form a directed set by inclusion so that we may apply the framework adopted for gauge fields. It follows that the cylindrical projections μ_S^P of the Gaußian measure are given by

$$d\mu_S^P(\phi) = \prod_{x \in S} d\mu_G(\phi(x)), \; d\mu_G(X) = \frac{1}{\sqrt{2\pi}^n} e^{-\sum_{I=1}^n X_I^2} d^n X \quad (12.2.24)$$

where n is the dimension of the representation in which ϕ transforms. Of course we only consider normalisable cylindrical functions.

Consider then the Hilbert space $\mathcal{H}_0 := L_2(\overline{\Phi}, d\mu_S)$. The GNS Hilbert space $\mathcal{H}_\omega$ can be realised as $\mathcal{H}_0$ such that $\Omega = 1$ (constant function) and in which the operators $\hat{\phi}_I(x)$ act by multiplication and the $\hat{p}^I(x)$ as follows: let F, F' be any

C^1 cylindrical functions, then

$$
\begin{aligned}
< F, \hat{p}_I(x)F' >_0 &:= \; < \Omega, \pi_\omega(\bar{F}p^I(x)F')\Omega >_\omega \\
&= \frac{i\hbar\lambda}{\sqrt{2}} < \Omega, \pi_\omega(\bar{F}(a^I(x) - \bar{a}_I(x))F')\Omega >_\omega \\
&= \frac{i\hbar\lambda}{\sqrt{2}} (< \pi_\omega(F)\Omega, [\pi_\omega(a_I(x)), \pi_\omega(F')]\Omega >_\omega \\
&\quad - < [\pi_\omega(a_I(x)), \pi_\omega(F)]\Omega, \pi_\omega(F')\Omega >_\omega) \\
&= \frac{1}{2} (< \pi_\omega(F)\Omega, [\pi_\omega(p_I(x)), \pi_\omega(F')]\Omega >_\omega \\
&\quad + < [\pi_\omega(p_I(x)), \pi_\omega(F)]\Omega, \pi_\omega(F')\Omega >_\omega) \\
&= \frac{i\hbar\lambda}{2} (< \pi_\omega(F)\Omega, \pi_\omega(Y_I(x) \cdot F')\Omega >_\omega \\
&\quad - < \pi_\omega(Y_I(x) \cdot F)\Omega, \pi_\omega(F')\Omega >_\omega) \\
&= \frac{i\hbar\lambda}{2} (< F, Y_I(x) \cdot F' >_0 - < Y_I(x) \cdot F, F' >_0) \qquad (12.2.25)
\end{aligned}
$$

where $Y_I(x) = \partial/\partial\phi_I(x)$ is the Hamiltonian vector field of $p_I(x)$. Using (12.2.24) it follows that

$$
\hat{p}_I(x) = i\hbar\lambda \left[Y_I(x) - \frac{1}{2}\phi_I(x) \right] \qquad (12.2.26)
$$

An orthonormal basis for $\mathcal{H}_0$ are functions of the form

$$
T_{S,\vec{n}}(\phi) = \prod_{x \in S, I} H_{n_{xI}}(\phi_I(x)) \qquad (12.2.27)
$$

where $\vec{n} = \{n_I(x)\}$, S runs through all finite point sets and H_n are Hermite functions.

We notice that the state ω is $\mathfrak{G}$ invariant where

$$
\alpha_g(a_I(x)) = \rho_{IJ}(g(x))a_J(x), \; \alpha_\varphi(a_I(x)) = a_I(\varphi(x)) \qquad (12.2.28)
$$

since the matrices $\rho(g)$ are unitary (ρ is the irreducible representation of G in which the Higgs field transforms). Hence $\mathfrak{G}$ is implemented unitarily. We also notice that we could have defined Weyl operators

$$
W(a, b) = \exp(i[\phi(a) + \pi(b)/(\hbar\lambda)]) \qquad (12.2.29)
$$

and that the representation $\mathcal{H}_0$ is weakly continuous in both a, b. This is different from the case of the gauge field. Of course, the representation $\mathcal{H}_\omega$ is far from unique: for instance, we could have used any other (white noise) covariance to define a Gaußian measure without breaking background independence. This is also different from the case of the gauge field.

We may ask whether the original operators corresponding to $\pi(f)$ are well-defined in this representation where f is a s.a. function. It is easy to see that this is not the case: using (12.2.19) we see that $\hat{\pi}(f) = \sum_{x \in \sigma} f^I(x)\hat{p}_I(x)$ so that

$||\hat{\pi}_I(f)\mathbf{1}||^2 \propto ||f||^2 = \infty$ is not even defined on the vacuum state because a s.a. function f does not have discrete support. Since the objects p have no classical interpretation in contrast to the π we may ask what has been gained. The answer is that, as we will see, it is the p's which are of relevance when we couple geometry to matter (recall that the representations used here are anyway of use only when coupling geometry to matter). Basically this happens because interesting, spatially diffeomorphism-invariant observables will come from functionals that depend on both geometry and matter. The geometry part of the corresponding operator will force the action to be non-trivial at the vertices of a graph so that one will be forced to take the limit of $\hat{\pi}(f)$ as the support of f shrinks to one point v. But that is precisely the definition of $\hat{p}(v)$.

We will define in the next subsection also representations for which the $\pi(f)$'s can be defined, however, in these representations part of the Weyl operators are again not weakly continuous and Fock representations are therefore not available.

12.2.2.2 Point holonomies

In the following we restrict ourselves to real-valued Higgs fields ϕ_I which transform according to the adjoint representation of G. Other cases can be treated by similar methods. This also covers the case of scalar fields (without internal degrees of freedom). Actually, we are not going to deal with ϕ_I itself but with the *point-holonomies*, which also play a crucial role in Bojowald's series [497–499]

$$U_x(\phi) := \exp(\phi_I(x)\tau_I) \tag{12.2.30}$$

where τ_I denotes a basis of the Lie algebra Lie(G) of the Yang–Mills gauge group. The name stems from the fact that under a gauge transformation $g(x)$ at x we have that $U(x) \to \mathrm{Ad}_{g(x)}(U(x))$ which is precisely the transformation behaviour of a holonomy $\underline{h}_e$ starting at x in the limit of vanishing edge length. In the case of a simple scalar field we define $U_x = e^{i\phi(x)}$. These variables play a role similar to the Wilson loop variables in lattice gauge theory [576] and it is understood that any action written in terms of ϕ_I should be rewritten in terms of the $U(x)$ in analogy to the replacement of the Yang–Mills action by the Wilson action.

Notice that while one can easily extract the connection from a path holonomy by considering the limit of paths shrinking to a point, one cannot do the same for point holonomies which are already labelled by points. Hence it seems that point holonomies do not separate the points of the classical configuration space because the map $\phi_I(x) \mapsto U(x)$ is many to one (the ϕ_I have non-compact range while the U have compact range). However, this is not the case: for instance in the case of U(1) (scalar field) we have $(dU(x))U^{-1}(x)/i = d\phi(x)$, for SU(2) we have $d\chi = -d\mathrm{Tr}(U(x))/\sqrt{4 - [\mathrm{Tr}(U(x))]^2}$ where we have parametrised $U = \cos(\chi) + \tau_j n_j \sin(\chi)$ with $\phi_j = \chi n_j$, $n_j^2 = 1$ so that $n_j = -\mathrm{Tr}(\tau_j U)/\sqrt{4 - [\mathrm{Tr}(U(x))]^2}$. For higher groups similar formulae hold. Thus we are able to construct the differential $d\phi_I(x)$ from knowledge of all the $U(x)$ so that $\phi_I(x) - \phi_I(x_0) = \int_{x_0}^x d\phi_I$. If σ is asymptotically flat we may choose $x_0 = \infty$ and then $\phi_I(x_0) = 0$. Of course

$\int_{x_0}^{x} d\phi_I[U]$ is not a cylindrical function but it can be approximated arbitrarily well by a Riemann sum which is a cylindrical function. An alternative possibility adopted in the application of LQG methods to string theory [205] is to actually use 'two-point' holonomies $U(p) = U(f(p))U(b(p))^{-1} = \exp(i \int_p d\phi)$ which however only works for U(1). Yet another possibility [447–450] is to use the generalised point holonomies $U_\lambda(x) = \exp(i\lambda\phi_I(x)\tau_I)$ where λ is any real number. These have the advantage of extracting ϕ_I more locally in the limit $\lambda \to 0$ for every single point x. One is naturally led to such a procedure in quantum cosmology where due to spatial homogeneity there is only one spatial point left so that the differentials used above are not available in order to separate the points, see Chapter 16, or for the polymer particle [549]. This procedure has recently been criticised in [657] because already a much smaller algebra of generalised point holonomies separates the points of the classical configuration space, see Chapter 16. However, one can combine the proposal of [447, 657] with the one advertised here because for spatially compact σ we can extract only differences $\phi(x) - \phi(x_0)$ since there is no distinguished point x_0 w.t. which ϕ takes a designated value. If we do not want to fix $x_0, \phi(x_0)$ by hand we must extract $\phi(x_0)$ in a local way for one single point. This can be done, for example, by considering a finite set $U_{\lambda_1}(x_0), \ldots, U_{\lambda_n}(x_0)$ where n depends on the gauge group (e.g., $\lambda_1 = 1, \lambda_2 = \sqrt{2}; n = 2$ for U(1), see Chapter 16). Summarising, we may extract ϕ_I from point holonomies and therefore can construct mass terms and potentials and the like albeit in a non-local way (derivative terms can be constructed locally).

This analogy with holonomies suggests a step-by-step repetition of the Ashtekar–Isham–Lewandowski framework of Part II [443], which we are going to consider below. The integrated quantity

$$\pi^I(B) := \int_B d^3x\,\pi^I(x) \qquad (12.2.31)$$

for any open region B in σ is diffeomorphism covariantly defined and the formal Poisson brackets $\{\pi^I(x), \phi_J(y)\} = \delta_J^I\delta(x, y)$ translate into

$$\{\pi^I(B), U_x\} = \chi_B(x)\frac{1}{2}[\tau_I U_x + U_x\tau_I] \qquad (12.2.32)$$

(in order to see this one must regularise U_x as in [443] and then remove the regulator. Only in the Abelian case this works without regularisation). The other elementary Poisson bracket is $\{U_x, U_y\} = 0$. Actually one has to generalise the Poisson algebra to the Lie algebra of functions on smooth ϕ_I's and vector fields thereon just as in the case of connections in order to obtain a true Lie algebra which one can quantise. Finally, the reality conditions are that $\pi^I(B)$ is real-valued and U_x is G-valued.

The construction of a quantum configuration space $\overline{\mathcal{U}}$ and a diffeomorphism-invariant measure $d\mu_{\mathcal{U}}$ thereon now proceeds just in analogy with Part II: a

Higgs vertex function $H_{\vec{v},\vec{\pi},\vec{\mu},\vec{\nu}}$ is just given by

$$H_{\vec{v},\vec{\pi},\vec{\mu},\vec{\nu}} = \prod_{k=1}^{n} \sqrt{d_{\pi_k}} (\pi_k(U(v_k)))_{\mu_k \nu_k} \tag{12.2.33}$$

where π_k are chosen from a complete set of irreducible, inequivalent representations of G and $v_1, \ldots, v_k$ are distinct points of σ. Consider the Abelian C*-algebra given by finite linear combinations of Higgs vertex functions and completed in the sup-norm over the set of smooth Higgs fields $\mathcal{U}$. Then $\overline{\mathcal{U}}$, the quantum configuration space of distributional Higgs fields, is the spectrum of that algebra equipped with the weak *-topology (Gel'fand topology).

The characterisation of the spectrum is as follows: points $\bar{\phi}$ in $\overline{\mathcal{U}}$ are in one-to-one correspondence with the set $\text{Fun}(\sigma, G)$ of G-valued functions on σ, the correspondence being given by $\bar{\phi} \leftrightarrow U_{\bar{\phi}}$ where $\sqrt{d_{\pi_0}} (U_{\bar{\phi}})_{\mu\nu}(v) = \bar{\phi}(H_{v,\pi_0,\mu,\nu})$ and π_0 is the fundamental representation of G.

Again, since the spectrum is a compact Hausdorff space one can define a regular Borel probability measure μ on it through positive, normalised, linear functionals Γ on the set of continuous functions f thereon, the correspondence being given by $\Gamma(f) = \int_{\overline{\mathcal{U}}} d\mu f$. We define the measure μ_U by

$$\Gamma_{\mu_U}(H_{\vec{v},\vec{\pi},\vec{\mu},\vec{\nu}}) = \begin{cases} 1 & H_{\vec{v},\vec{\pi},\vec{\mu},\vec{\nu}} = 1 \\ 0 & \text{otherwise} \end{cases} \tag{12.2.34}$$

and one easily sees that this measure is just the Haar measure on G^n for functions cylindrical over n distinct points. In particular, the Higgs vertex functions form a complete orthonormal basis by an appeal to the Peter and Weyl theorem. The measure μ_U can be shown [443] to be concentrated on nowhere continuous Higgs fields, in particular $\mu_U(\mathcal{U}) = 0$.

Finally, $\hat{U}(x)$ is just a multiplication operator on cylindrical functions and if we replace π^I by $-i\hbar\delta/\delta\phi_I$ then we find for a function $F = f_{\vec{v}}$ cylindrical over n points $\vec{v}$ that $\hat{\pi}^I(B)F = -i\hbar \sum_{k=1}^{n} \chi_B(v_k) X_{v_k}^I f_{\vec{v}}$ where $X_v^I = X^I(U(v))$, $X^I(g) = \frac{1}{2}[X_R^I(g) + X_L^I(g)]$ and X_L, X_R are, respectively, left- and right-invariant vector fields on G. The canonical commutation relations as well as the adjointness relations are then faithfully implemented and an appropriate kinematical Higgs field Hilbert space can be chosen to be $\mathcal{H}_U := L_2(\overline{\mathcal{U}}, d\mu_U)$.

12.2.3 Gauge and diffeomorphism-invariant subspace

We now put everything together to arrive at the complete solution to the Gauß and diffeomorphism constraint for quantum gravity coupled to gauge fields, Higgs fields and fermions. To be explicit, let us do this for the representation of Section 12.2.2.1.

We begin with the kinematical Hilbert space

$$\mathcal{H} = L_2(\overline{\mathcal{A}}_{SU(2)}, d\mu_0^{SU(2)}) \otimes L_2(\overline{\mathcal{A}}_G, d\mu_0^G) \otimes L_2(\overline{\mathcal{F}}, d\mu_F) \otimes L_2(\overline{\Phi}, d\mu_U) \tag{12.2.35}$$

and now consider its subspace consisting of gauge-invariant functions. A basis of such functions is labelled by a graph γ, a labelling of its edges e by spins j_e and colours c_e corresponding to irreducible representations of SU(2) and G respectively and a labelling of its vertices v by an array I_v, another colour C_v and two projectors p_v, q_v. The array I_v indicates a fermionic dependence at v by F_{v,I_v} and C_v stands for an irreducible representation of G which one forms out of tensor products of the representation ρ in which the Higgs field transforms. Finally, decompose the tensor product of irreducible representations of SU(2) given by the fundamental representations corresponding to F_{v,I_v} and the representations π_{j_e} for those edges e incident at v and project with p_v on a singlet that appears. Likewise, decompose the tensor product of irreducible representations of G given by the fundamental representations corresponding to F_{v,I_v}, the representations π_{c_e} for those edges e incident at v and the representation π_{C_v} and project with q_v on a singlet that appears.

The result is a gauge-invariant state $T_{\gamma,[\vec{j},\vec{I},\vec{p}],[\vec{c},\vec{C},\vec{q}]}$ called a *spin-colour-network state* extending the definition of a purely gravitational spin-network state. Consider the action $\overline{\mathcal{G}}$ of the gauge group SU(2) × G on all distributional fields. Then the spin-colour-network states contain the space of gauge-invariant functions, which is the same as the Hilbert space

$$\mathcal{H} = L_2\left([\overline{\mathcal{A}}_{\mathrm{SU}(2)} \times \overline{\mathcal{A}}_{\mathrm{G}} \times \overline{\mathcal{F}} \times \overline{\Phi}]/\overline{\mathcal{G}}, d\mu_0^{\mathrm{SU}(2)} \otimes d\mu_0^{\mathrm{G}} \otimes d\mu_{\mathrm{F}} \otimes d\mu_{\mathrm{U}}\right) \quad (12.2.36)$$

that is, the L_2 space on the moduli space.

To get the solution to the diffeomorphism constraint one considers the spaces $\mathcal{D}_{\mathrm{SU}(2)}, \mathcal{D}_{\mathrm{G}}, \mathcal{D}_{\mathrm{F}}, \mathcal{D}_S$ of smooth cylindrical functions (smooth in the sense of the nuclear topology of $\mathrm{SU}(2)^n, \mathrm{G}^n, \mathcal{F}^n, \mathrm{G}^n$ respectively) and their corresponding algebraic duals. Then we form the gauge-invariant subspaces of the spaces

$$\mathcal{D} := \mathcal{D}_{\mathrm{SU}(2)} \times \mathcal{D}_{\mathrm{G}} \times \mathcal{D}_{\mathrm{F}} \times \mathcal{D}_{\mathrm{U}} \text{ and } \mathcal{D}^* := \mathcal{D}_{\mathrm{SU}(2)}^* \times \mathcal{D}_{\mathrm{G}}^* \times \mathcal{D}_{\mathrm{F}}^* \times \mathcal{D}_{\mathrm{U}}^* \quad (12.2.37)$$

Now the spin-colour-network states span the invariant subspace of $\mathcal{D}$ and the diffeomorphism group acts unitarily by

$$\hat{U}(\varphi)T_{\gamma,[\vec{j},\vec{I},\vec{p}],[\vec{c},\vec{C},\vec{q}]} = T_{\varphi(\gamma),[\vec{j},\vec{I},\vec{p}],[\vec{c},\vec{C},\vec{q}]} \quad (12.2.38)$$

and similar as in the purely gravitational case we get diffeomorphism-invariant distributions in $\mathcal{D}^*$ by judiciously group averaging the action (12.2.38), that is, we take the continuous sum over all states contained in an orbit under the action (12.2.38).

12.3 Quantisation of matter Hamiltonian constraints

The quantisation of matter Hamiltonian constraints follows the same pattern as for the pure geometry Hamiltonian constraint so that we do not need to display all the details, which we leave to the ambitious reader. The details can be found in [441].

What we will find is that certain ultraviolet divergences, which appear when we consider matter fields propagating on a background spacetime, *disappear* when we let the spacetime metric fluctuate as well. The underlying reason is background independence as heuristically explained in Section 10.2. We do not claim that this proves finiteness of quantum gravity because, first, we must prove that the quantum theory constructed has General Relativity as its classical limit, and second, besides the Hamiltonian constraint we also must show that quantisations of classical observables of the theory are finite and, third, we must establish that those operators remain non-singular upon passing to the physical Hilbert space. However, at the very least, these are first promising indications for an UV finite theory.

In the next three subsections we explain the quantisation of various matter Hamiltonians. One can verify that the quantum Dirac algebra of the complete Hamiltonian constraint consisting of the sum of all matter and geometry contributions closes in a similar fashion as outlined in Section 10.5 and which is shown explicitly in [441]. We will not repeat this here because the mechanism is identical to the one for the case of pure geometry. In the last section of this chapter we will show that this is no coincidence: we will explain the general scheme, how coupling to gravity is able to regulate certain ultraviolet divergences in a consistent way for any background-independent matter coupling.

12.3.1 Quantisation of Einstein–Yang–Mills theory

The canonical pair coordinatising the Yang–Mills phase space is given by $(\underline{E}_I^a, \underline{A}_a^I)$ with symplectic structure formally given by

$$\{\underline{E}_I^a(x), \underline{A}_b^J(y)\} = Q^2 \delta_b^a \delta_I^J \delta(x, y) \tag{12.3.1}$$

where as before $I, J, K, \ldots = 1, \ldots, \dim(G)$ denote Lie(G) indices. The contribution of the Yang–Mills field to the Hamiltonian constraint is

$$H_{\mathrm{YM}} = \frac{q_{ab}}{2Q^2 \sqrt{\det(q)}} [\underline{E}_I^a \underline{E}_I^b + \underline{B}_I^a \underline{B}_I^b] \tag{12.3.2}$$

where Q is the Yang–Mills coupling constant, $\underline{B}_I^a := \frac{1}{2} \epsilon^{abc} F_{bc}^I$ the magnetic field of the connection $\underline{A}_a^I$ and F_{ab}^I its curvature. The integrated form is given by $H_{\mathrm{YM}}(N) = \int_\sigma d^3x N H_{\mathrm{YM}}$ where N is the lapse function.

We will focus first on the electric part of $H_{\mathrm{YM}}(N)$ which we write in the form

$$H_{\mathrm{YM,el}}(N) = \frac{1}{2Q^2} \int d^3x N \frac{[e_a^i \underline{E}_I^a]}{\sqrt{\det(q)}} [e_b^i \underline{E}_I^b] \tag{12.3.3}$$

where Q is the Yang–Mills coupling constant. Using the same notation as in Chapter 10 we can also write this as

$$H_{\text{YM,el}}(N) = \frac{1}{8\kappa^2 Q^2} \int d^3x\, N(x) \frac{[\{A_a^i(x), V(R_x)\} \underline{E}_I^a(x)]}{\sqrt{\det(q)}(x)} [\{A_b^i(x), V(R_x)\} \underline{E}_I^b(x)]$$

(12.3.4)

Since $\underline{E}_I^a = \frac{1}{2}\epsilon^{abc}\underline{e}_{bc}^I$ is Hodge dual to a two-form $\underline{e}^I$ we can also write this as

$$H_{\text{YM,el}}(N) = \frac{1}{8\kappa^2 Q^2} \int d^3x\, N(x) \frac{[\{A_a^i(x), V(R_x)\} \underline{E}_I^a(x)]}{\sqrt{\det(q)}(x)} [\{A_i(x), V(R_x)\} \wedge \underline{e}_I(x)]$$

(12.3.5)

which suggests approximating the integral by a Riemann sum utilising a triangulation of σ as in Section 10.3. Using the same notation as there we get

$$H_{\text{YM,el}}^\epsilon(N) = \frac{1}{8\kappa^2 Q^2} \sum_{\Delta \in T(\epsilon)} N(v(\Delta)) \frac{[\{A_a^i(v(\Delta)), V(R_{v(\Delta)})\} \underline{E}_I^a(v(\Delta))]}{\sqrt{\det(q)}(v(\Delta))}$$

$$\times \epsilon^{LMN}\big[\text{tr}\big(\tau_i h_{e_L(\Delta)}\{h_{e_L(\Delta)}^{-1}, V(R_{v(\Delta)})\}\big) E_I(S_{MN}(\Delta))\big] \quad (12.3.6)$$

where we have used that $S_{MN}(\Delta)$ is any oriented triangular surface with boundary $e_M(\Delta) \circ a_{MN}(\Delta) \circ e_N(\Delta)^{-1}$.

We now apply the same trick that we used already in previous sections: let $\chi_{\epsilon,x}(y)$ be the characteristic function of a box $U_\epsilon(x)$ with coordinate volume ϵ^3 and centre x. Then

$$V(U_\epsilon(x)) = \epsilon^3 \sqrt{\det(q)}(x) + o(\epsilon^4) \tag{12.3.7}$$

and

$$\int \chi_{\epsilon,x}(y) \frac{[\{A_i(y), V(R_y)\} \wedge \underline{e}_I(y)]}{\sqrt{V(U_\epsilon(y))}} = \epsilon^3 \frac{[\{A_a^i(x), V(R_x)\} \underline{E}_I^a(x)]}{\sqrt{V(U_\epsilon(x))}} + o(\epsilon^3)$$

(12.3.8)

which allows us to replace (12.3.6) by

$$H_{\text{YM,el}}^\epsilon(N) = \frac{1}{2\kappa^2 Q^2} \sum_{\Delta,\Delta' \in T(\epsilon)} N(v(\Delta)) \chi_{\epsilon,v(\Delta)}(v(\Delta')) \epsilon^{LMN} \epsilon^{RST}$$

$$\times \frac{\text{tr}\big(\tau_i h_{e_L(\Delta)}\{h_{e_L(\Delta)}^{-1}, V(R_{v(\Delta)})\}\big) E_I(S_{MN}(\Delta))}{2\sqrt{V(U_\epsilon(v(\Delta)))}}$$

$$\times \frac{\text{tr}\big(\tau_i h_{e_R(\Delta')}\{h_{e_R(\Delta')}^{-1}, V(R_{v(\Delta')})\}\big) E_I(S_{ST}(\Delta'))}{2\sqrt{V(U_\epsilon(v(\Delta')))}} \tag{12.3.9}$$

Again, the region-valued function $x \to R_x$ is completely arbitrary up to this point and if we choose $R_x = U_\epsilon(x)$ then we obtain the final formula

$$H^\epsilon_{\text{YM,el}}(N) = \frac{1}{2\kappa^2 Q^2} \sum_{\Delta,\Delta' \in T(\epsilon)} N(v(\Delta)) \chi_{\epsilon,v(\Delta)}(v(\Delta')) \epsilon^{LMN} \epsilon^{RST}$$

$$\times \left[\text{tr} \left(\tau_i h_{e_L(\Delta)} \left\{ h^{-1}_{e_L(\Delta)}, \sqrt{V(U_\epsilon(v(\Delta)))} \right\} \right) E_I(S_{MN}(\Delta)) \right]$$

$$\times \left[\text{tr} \left(\tau_i h_{e_R(\Delta')} \left\{ h^{-1}_{e_R(\Delta')}, \sqrt{V(U_\epsilon(v(\Delta')))} \right\} \right) E_I(S_{ST}(\Delta')) \right]$$

$$(12.3.10)$$

in which the $1/\sqrt{\det(q)}$ was removed from the denominator and so qualifies as the starting point for the quantisation. The pointwise limit of (12.3.10) on the phase space gives back (12.3.2) for *any* triangulation.

The theme repeats: in order to arrive at a well-defined result on a dense set of vectors given by functions cylindrical over graphs γ one must adapt the triangulation to the γ in question. The limit of (12.3.10) with respect to the so-obtained $T(\epsilon,\gamma)$ still gives back (12.3.2). The only new ingredient of the triangulation compared with the one outlined in Section 10.4 is that, at fixed ϵ, we deform the surfaces $S_{MN}(\Delta)$, controlled by a further parameter δ, to the effect that $\lim_{\delta \to 0} S_{MN}(\Delta,\delta) = S_{MN}(\Delta)$ and at finite δ the edge $e_L(\Delta)$, $\epsilon^{LMN} = 1$ is the only one that intersects $S_{MN}(\Delta,\delta)$ transversally. This can be achieved by detaching $S_{MN}(\Delta)$ slightly from $v(\Delta)$ and otherwise choosing the shape of $S_{MN}(\Delta)$ appropriately. After replacing Poisson brackets by commutators times $1/(i\hbar)$ and the Yang–Mills electric field by $-i\hbar Q^2$ times functional derivatives we first get a family of operators $(\hat{H}^{\epsilon,\delta}_{\text{YM,el}}(N)_\gamma)^\dagger$, the limit $\delta \to 0$ of which, in the topology of smooth connections, converges to a family of operators $(\hat{H}^\epsilon_{\text{YM,el}}(N)_\gamma)^\dagger$ which can be extended to all of $\overline{\mathcal{A}}$. One verifies that this family of operators, for sufficiently small ϵ depending on γ, qualifies as the set of cylindrical projections of an operator $(\hat{H}^\epsilon_{\text{YM,el}}(N))^\dagger$ and the limit $(\hat{H}_{\text{YM,el}}(N))^\dagger$ as $\epsilon \to 0$ in the URST exists and is given by $(\hat{H}^{\epsilon_0}_{\text{YM,el}}(N))^\dagger$ for any arbitrary but fixed $\epsilon_0 > 0$. We give the final result

$$(\hat{H}_{\text{YM,el}}(N))^\dagger f_\gamma = -\frac{m_p \alpha_Q}{2\ell^3_p} \sum_{v \in V(\gamma)} \sum_{b(e)=b(e')=v} N(v) \text{tr}\left(\tau_i h_e \left[h^{-1}_e, \sqrt{\hat{V}(U_{\epsilon_0}(v))} \right] \right)$$

$$\times \text{tr}\left(\tau_i h_{e'} \left[h^{-1}_{e'}, \sqrt{\hat{V}(U_{\epsilon_0}(v))} \right] \right) \underline{R}^I_e \underline{R}^I_{e'} f_\gamma \qquad (12.3.11)$$

where the Planck mass $m_p = \sqrt{\hbar/\kappa}$ and the dimensionless fine structure constant $\alpha_Q = \hbar Q^2$ have peeled out (in our notation, Q^2 has the dimension of $1/\hbar$) while the Planck volume ℓ^3_p in the denominator makes the rest of the expression dimensionless. As before, $\underline{R}^I_e = R^I(\underline{h}_e)$ and $\underline{R}^I(g)$ is the right-invariant vector field on G and $\underline{h}_e$ is the holonomy of $\underline{A}$ along e. Expression (12.3.11) is manifestly gauge-invariant and diffeomorphism-covariant. Notice that for ϵ_0 sufficiently small we can relace $\hat{V}(U_{\epsilon_0}(v))$ by the regulator-independent operator

$\hat{V}_v := \lim_{\epsilon_0 \to 0} \hat{V}(U_{\epsilon_0}(v))$ which exists and is ϵ_0 independent. As one can check, $\hat{V}_v$ is uniquely defined by $\hat{V}(R)f_\gamma = \hat{V}_v f_\gamma$ whenever $R \cap V(\gamma) = \{v\}$.

Notice that, expectedly, (12.3.11) resembles (minus) a Laplacian. Indeed, one can show [441] that $(\hat{H}_{\text{YM,el}}(N=1))^\dagger$ is an essentially self-adjoint, positive semidefinite operator on $\mathcal{H}$. In particular, (12.3.11) is densely defined and does not suffer from any singularities, it is *finite*! This extends to the magnetic part of the Yang–Mills Hamiltonian whose action on cylindrical functions is given by

$(\hat{H}_{\text{YM,mag}}(N))^\dagger f_\gamma$

$$
= -\frac{m_p}{2\alpha_Q (12N)^2 \ell_p^3} \sum_{v \in V(\gamma)} \sum_{v(\Delta) = v(\Delta') = v} N(v) \left(\frac{8}{E(v)}\right)^2 \epsilon^{LMN} \epsilon^{RST}
$$

$$
\times \text{tr}\left(\tau_i h_{e_L(\Delta)} \left[h_{e_L(\Delta)}^{-1}, \sqrt{\hat{V}_v}\right]\right)
$$

$$
\times \text{tr}\left(\tau_i h_{e_R(\Delta')} \left[h_{e_R(\Delta')}^{-1}, \sqrt{\hat{V}_v}\right]\right) \text{tr}\left(\underline{\tau}_I \underline{h}_{\alpha_{MN}(\Delta)}\right) \text{tr}\left(\underline{\tau}_I \underline{h}_{\alpha_{ST}(\Delta')}\right) f_\gamma
$$

$$
(12.3.12)
$$

(we use the convention $\text{tr}(\underline{\tau}_I \underline{\tau}_J) = -\delta_{IJ}/N$ for the normalisation of the generators of Lie(G)). Here the sum is over tetrahedra Δ adapted to the graph as defined for the gravitational constraint, that is, each Δ is defined by beginning segments $s_I(\Delta)$ of triples of edges e_I, $I = 1, 2, 3$ incident at $v = v(\Delta)$ and the corresponding arcs connecting the endpoints of the $s_I(\Delta)$ from which the loops $\alpha_{IJ}(\Delta)$ are formed. Notice the non-perturbative dependence of (12.3.12) on the fine structure constant. The regulator dependence on those choices of beginning segments and arcs drops out trivially in the URST upon choosing a loop attachment, just as for the gravitational term.

In summary, the Yang–Mills contribution to the Hamiltonian constraint can be densely defined on $\mathcal{H}$. We can see explicitly the regularising role that the gravitational quantum field has played in the quantisation process: the volume operator acts only at vertices of a graph and therefore also restricts the Yang–Mills Hamiltonian to an action at those points. Therefore, *the volume operator acts as an infrared cutoff*! Next, the divergent factor $1/\epsilon^3$ stemming from the point splitting of the two Yang–Mills electric fields was *absorbed* by the volume operator which *must happen* in order to preserve diffeomorphism covariance as the point-splitting volume should not be measured by the coordinate background metric but by the dynamical metric itself. Therefore, *the volume operator also acts as an ultraviolet cutoff*! The volume operator thus plays a key role in the quantisation process, which is why a more detailed knowledge about its spectrum would be highly desirable.

12.3.2 Fermionic sector

In this section we will only focus on the first term displayed in the expression for H_{Dirac}. The other two terms can be quantised similarly, for the quantisation of K^i_a we adopt a procedure identical to the one used for the quantisation of the Einstein contribution to the Hamiltonian constraint.

We begin by rewriting the classical constraint using that classically $E^a_i = \pm \frac{1}{2} \epsilon^{abc} \epsilon_{IJK} e^j_b e^k_c$. We find by an already familiar procedure that

$$H_{\text{Dirac}}(N) = -\frac{i}{2\kappa^2} \int d^3x N(x) \epsilon^{IJK} \epsilon^{abc} \frac{4\{A^i_a(x), V(x,\delta)\}\{A^j_b(x), V(x,\delta)\}}{\sqrt{\det(q)}(x)}$$

$$\times \left[(\tau_k \mathcal{D}_c \xi)_{A\mu}(x) \pi_{A\mu}(y) - c.c. \right] \tag{12.3.13}$$

where δ is an arbitrarily small but finite parameter and a possible sign was absorbed into N (we could also quantise the sign function as mentioned previously). The minus sign comes from moving the classical momentum variable to the right.

The first task is to rewrite (12.3.13) in terms of the quantities θ. To that end let f^a_i be a real-valued, $\text{Ad}_{SU(2)}$ transforming vector field and consider the discrete sum (we abbreviate A, r, etc. as I, etc.)

$$\sum_x f^a_i(x) (\tau_i \mathcal{D}_a \theta)_I(x) \bar{\theta}_I(x) \tag{12.3.14}$$

Recall that $\theta_I(x) := \int d^3y \sqrt{\delta(x,y)} \xi_I(y) := \lim_{\epsilon \to 0} \theta^\epsilon_I(x)$ where $\theta^\epsilon_I(x) = \int d^3y \frac{\chi_\epsilon(x,y)}{\sqrt{\epsilon^3}} \xi_I(y)$ and $\chi_\epsilon(x,y)$ denotes the characteristic function of a box with Lebesgue measure ϵ^3 and centre x. We define $(\partial_a \theta_I)(x) := \lim_{\epsilon \to 0} \partial_{x^a} \theta^\epsilon_I(x)$ and find

$$\partial_{x^a} \theta^\epsilon_I(x) = \int d^3y \frac{\partial_{x^a} \chi_\epsilon(x,y)}{\sqrt{\epsilon^3}} \xi_I(y)$$

$$= -\int d^3y \frac{\partial_{y^a} \chi_\epsilon(x,y)}{\sqrt{\epsilon^3}} \xi_I(y) = \int d^3y \frac{\chi_\epsilon(x,y)}{\sqrt{\epsilon^3}} \partial_{y^a} \xi_I(y)$$

since $\chi_\epsilon(x,y) = \chi_\epsilon(y,x)$ and there was no boundary term dropped in the integration by parts because χ_ϵ is of compact support. Let us partition σ by a countable number of boxes B_n of Lebesgue measure ϵ^3 and centre x_n and interpret (12.3.14) as the $\epsilon \to 0$ limit of

$$\sum_n f^a_i(x_n)(\tau_i \mathcal{D}_a \theta^\epsilon)_I(x_n) \bar{\theta}^\epsilon_I(x_n) \tag{12.3.15}$$

Substituting for θ^ϵ in terms of ξ, (12.3.15) becomes

$$\int d^3x \int d^3y \left[\sum_n f^a_i(x_n) \frac{\chi_\epsilon(x,x_n)\chi_\epsilon(y,x_n)}{\epsilon^3} \right] [(\tau_i \partial_a \xi_I(x) + (\omega_a(x_n)\xi(x))_I] \bar{\xi}_I(y)$$

$$\tag{12.3.16}$$

We have not written the Levi–Civita connection in (12.3.16) which is needed due to the density weight of ξ because it drops out in the final antisymmetric sum $i[(.) - (.)^\star] = i[(.) - c.c.]$ in (12.3.13). Now, as $\epsilon \to 0$ (the partition of σ becomes finer and finer) we can replace $\chi_\epsilon(x,x_n)$ by $\delta(x,x_n)$ and $\chi_\epsilon(y,x_n)$ by $\delta_{x_n,y}$ and

(12.3.16) becomes, upon performing the x-integral and the sum over x_n,

$$\int d^3x f_i^a(x)(\tau_i \mathcal{D}_a \xi_I)(x)\bar{\xi}_I(y) \tag{12.3.17}$$

which is precisely (12.3.13) with the proper interpretation of f_i^a. Expression (12.3.17) is written in a form that is well defined on the kinematical Hilbert space, which consists of functions of θ rather than ξ.

Now, in quantising expression (12.3.14) we keep the fermionic momenta to the right and replace $\bar{\theta}_{Ar}(x)$ by $\hbar\partial/\partial\theta_{Ar}$, which is the proper quantisation rule for the θ variables as derived in Section 12.2. Also, we multiply nominator and denominator by δ^3 and replace $\delta^3\sqrt{\det(q)}(x)$ by $V(x,\delta)$ in the denominator, which by the standard trick we can absorb into the Poisson bracket. Finally we replace the Poisson bracket by a commutator times $1/(i\hbar)$. Labelling the regulated operator with the parameter δ, we find a function f_γ cylindrical with respect to a graph γ with fermionic insertions $\theta_{A\mu}$ at the vertices $v \in V(\gamma)$

$$\hat{H}^\delta_{\mathrm{Dirac}}(N)f_\gamma = -\frac{\hbar}{2\ell_p^4}\sum_{v\in V(\gamma)}\sum_x N(x)\epsilon^{IJK}\epsilon^{abc}\delta^3\left[A_a^i(x),\sqrt{\hat{V}(x,\delta)}\right]$$

$$\times\left[A_b^j(x),\sqrt{\hat{V}(x,\delta)}\right]\left[(\tau_k\mathcal{D}_c\theta)_{Ar}(v)\frac{\partial}{\partial\theta_{Ar}(v)}\delta_{x,v}+h.c.\right]f_\gamma$$

$$\tag{12.3.18}$$

Notice that the sum over all $x \in \Sigma$ already collapses to a sum over the vertices of γ. Next we triangulate σ in adaption to γ. We have the expansion $h_s(0,\delta)\theta(s(\delta)) - \theta(s(0)) = \delta\dot{s}^a(0)(\mathcal{D}_a\theta)(s(0))$ for the holonomy. Therefore we just introduce as in the sections before a holonomy at various places to absorb the factor of δ^3 and replace $\hat{V}(v,\delta)$ by $\hat{V}_v$. Thus,

$$\hat{H}^\delta_{\mathrm{Dirac}}(N) = -\frac{m_p}{2\ell_p^3}\sum_{v\in V(\gamma)}N_v\frac{1}{E(v)}\sum_{v(\Delta)=v}\epsilon^{IJK}\epsilon^{mnp}\,\mathrm{tr}\left(\tau_i h_{s_m(\Delta)}\left[h^{-1}_{s_m(\Delta)},\sqrt{\hat{V}_v}\right]\right)$$

$$\times\,\mathrm{tr}\left(\tau_j h_{s_n(\Delta)}\left[h^{-1}_{s_n(\Delta)},\sqrt{\hat{V}_v}\right]\right)$$

$$\times\left[(\tau_k[H_{s_p(\Delta)}\theta(s_p(\Delta)(\delta)) - \theta(v)]_{Ar}\frac{\partial}{\partial\theta_{Ar}(v)}+h.c.\right]$$

$$= -\frac{m_p}{2\ell_p^3}\sum_{v\in V(\gamma)}N_v\sum_{v(\Delta)=v}\epsilon^{IJK}\epsilon^{mnp}\,\mathrm{tr}\left(\tau_i h_{s_m(\Delta)}\left[h^{-1}_{s_m(\Delta)},\sqrt{\hat{V}_v}\right]\right)$$

$$\times\,\mathrm{tr}\left(\tau_j h_{s_n(\Delta)}\left[h^{-1}_{s_n(\Delta)},\sqrt{\hat{V}_v}\right]\right)[(Y_k(s_p(\Delta))) - Y_k(v) + h.c.]$$

$$=: \hat{H}^T_{\mathrm{Dirac}} \tag{12.3.19}$$

where $E(v) = \binom{n_v}{3}$, n_v is the valence of v and where the label T reminds us of the triangulation dependence (we have naturally chosen the value of δ in such a way that (a) $e(\delta)$ coincides with the endpoint of the segment of e starting at $v = e(0)$ and (b) is part of the definition of the triangulation adapted to γ). We have defined

$$Y_i(e) := \text{tr}\left(\tau_i H_e \xi(e(1)) \frac{\partial}{\partial \xi(e(0))}\right) \quad \text{and} \quad Y_i(v) := Y_i(e = v)$$

and $e : [0, 1] \to \sigma$ is a suitable parametrisation of the edge e.

The Hermitian conjugation operation '*h.c.*' involved in (12.3.19) is meant with respect to the inner product on the Hilbert space and with respect to the operator of which the first term in (12.3.19) is the projection on the cylindrical subspace labelled by the graph γ. Again the sum is over tetrahedra Δ adapted to γ with beginning segments $s_I(\Delta)$ of all triples of edges e_I, $I = 1, 2, 3$ incident at $v = v(\Delta)$. Removing the triangulation dependence in the URST now simply corresponds, as before, to choosing the beginning segments of all edges $s_i(\Delta)$ once and for all for all graphs in (12.3.19).

Notice that the classical fermionic Hamiltonian constraint is a density of weight one and that the operator defined by (12.3.19) precisely respects this because the θ are scalar-valued and not density-valued. If we were dealing with the ξ instead of the θ we would run into conflict with diffeomorphism covariance at this point.

12.3.3 Higgs sector

We finally come to regularise the Higgs sector. Especially for this sector a general scheme will become evident of how to systematically *take advantage* of the factor ordering ambiguity in order to arrive at a densely defined operator.

The term in the scalar Hamiltonian constraint proportional to $(\pi^I)^2$ looks hopelessly divergent: even if we could manage to replace the denominator by the volume operator we end up with a singular, not densely defined operator because the volume operator has a huge kernel. We need a new trick as follows: we insert the number $1 = [\det(e_a^i)]^2/[\sqrt{\det(q)}]^2$ (one) into the kinetic term, which apparently makes the singularity even worse. However, consider the following regulated *four-fold* point splitting of the kinematic term (we set $\lambda = \kappa$ for simplicity which is dimensionally possible)

$$H^\epsilon_{\text{Higgs,kin}}(N) = \frac{1}{2\kappa} \int d^3x N(x)\pi^I(x) \int d^3y \, \pi^I(y) \int d^3u \left(\frac{\det\left(e_a^i\right)}{[\sqrt{V(u, \epsilon)}]^3}\right)(u)$$

$$\times \int d^3v \left(\frac{\det\left(e_a^i\right)}{[\sqrt{V(v, \epsilon)}]^3}\right)(v) \, \chi_\epsilon(x, y)\chi_\epsilon(u, x)\chi_\epsilon(v, y)$$

$$= \frac{1}{2\kappa} \frac{(-2)^2}{(3!)^2\kappa^6} \int d^3x N(x)\pi^I(x) \int d^3y \pi^I(y)$$

$$\times \int \text{tr}(\{A(u), \sqrt{V(u, \epsilon)}\} \wedge \{A(u), \sqrt{V(u, \epsilon)}\} \wedge \{A(u), \sqrt{V(u, \epsilon)}\})$$

$$\times \int \text{tr}(\{A(v), \sqrt{V(v, \epsilon)}\} \wedge \{A(v), \sqrt{V(v, \epsilon)}\} \wedge \{A(v), \sqrt{V(v, \epsilon)}\})$$

$$\times \chi_\epsilon(x, y)\chi_\epsilon(u, x)\chi_\epsilon(v, y) \tag{12.3.20}$$

Recall that $\int d^3x \det(e_a^i) = \frac{1}{3!} \int \epsilon_{IJK} e^i \wedge e^j \wedge e^k = -\frac{1}{3} \int \text{tr}(e \wedge e \wedge e)$ in order to see this.

Now we replace π^I by $-i\hbar(\kappa)\delta/\delta\phi^I$, replace the volume by its operator version and Poisson brackets by commutators times $1/(i\hbar)$ and find, when applying the operator to a cylindrical function f_γ, that

$$\hat{H}^\epsilon_{\text{Higgs,kin}}(N)f_\gamma$$

$$= \frac{(-i)^2}{i^6} \frac{\hbar^2\kappa^2}{18\hbar^6\kappa^7} \sum_{v,v' \in V(\gamma)} N(v)Y^I(v)Y^I(v')\chi_\epsilon(v,v')$$

$$\times \int \text{tr}\left(\left[A(x),\sqrt{\hat{V}(x,\epsilon)}\right] \wedge \left[A(x),\sqrt{\hat{V}(x,\epsilon)}\right] \wedge \left[A(x),\sqrt{\hat{V}(x,\epsilon)}\right]\right)$$

$$\times \int \text{tr}\left(\left[A(y),\sqrt{\hat{V}(y,\epsilon)}\right] \wedge \left[A(y),\sqrt{\hat{V}(y,\epsilon)}\right] \wedge \left[A(y),\sqrt{\hat{V}(y,\epsilon)}\right]\right)$$

$$\times f_\gamma \chi_\epsilon(x,v)\chi_\epsilon(y,v') \tag{12.3.21}$$

where $X_I(x) = Y_I(x) - \frac{1}{2}\phi_I(x)$, $Y^I(x) = \partial/\partial\phi_I(x)$, recall (12.2.26) for the Fock space quantisation. The expression for the point holonomy quantisation is similar, see [441].

Certainly we are now going to triangulate σ in adaption to γ in an already familiar fashion and write

$$\int_\Delta \text{tr}\left(\left[A(x),\sqrt{\hat{V}(x,\epsilon)}\right] \wedge \left[A(x),\sqrt{\hat{V}(x,\epsilon)}\right] \wedge \left[A(x),\sqrt{\hat{V}(x,\epsilon)}\right]\right)$$

$$\approx \frac{1}{6}\epsilon^{ijk}\text{tr}\left(h_{s_i(\Delta)}\left[h^{-1}_{s_i(\Delta)},\sqrt{\hat{V}(v(\Delta),\epsilon)}\right]\right) \text{tr}\left(h_{s_j(\Delta)}\left[h^{-1}_{s_j(\Delta)},\sqrt{\hat{V}(v(\Delta),\epsilon)}\right]\right)$$

$$\times \text{tr}\left(h_{s_k(\Delta)}\left[h^{-1}_{s_k(\Delta)},\sqrt{\hat{V}(v(\Delta),\epsilon)}\right]\right) \tag{12.3.22}$$

which results in

$$\hat{H}^\epsilon_{\text{Higgs,kin}}(N)f_\gamma = \frac{m_p}{18\ell_p^9}\frac{1}{36} \sum_{p,q,r,s \in V(\gamma)} N(p)X^I(p)X^I(q)\chi_\epsilon(p,q)\frac{8}{E(r)}\chi_\epsilon(r,p)$$

$$\times \sum_{v(\Delta)=r} \epsilon^{ijk}\frac{8}{E(s)}\chi_\epsilon(s,q) \sum_{v(\Delta')=s} \epsilon^{lmn}\text{tr}\left(h_{s_i(\Delta)}\left[h^{-1}_{s_i(\Delta)},\sqrt{\hat{V}(v(\Delta),\epsilon)}\right]\right)$$

$$\times \text{tr}\left(h_{s_j(\Delta)}\left[h^{-1}_{s_j(\Delta)},\sqrt{\hat{V}(v(\Delta),\epsilon)}\right]\right) \text{tr}\left(h_{s_k(\Delta)}\left[h^{-1}_{s_k(\Delta)},\sqrt{\hat{V}(v(\Delta),\epsilon)}\right]\right)$$

$$\times \text{tr}\left(h_{s_l(\Delta')}\left[h^{-1}_{s_l(\Delta')},\sqrt{\hat{V}(v(\Delta'),\epsilon)}\right]\right) \text{tr}\left(h_{s_m(\Delta')}\left[h^{-1}_{s_m(\Delta')},\sqrt{\hat{V}(v(\Delta'),\epsilon)}\right]\right)$$

$$\times \text{tr}\left(h_{s_n(\Delta')}\left[h^{-1}_{s_n(\Delta')},\sqrt{\hat{V}(v(\Delta'),\epsilon)}\right]\right) f_\gamma \tag{12.3.23}$$

since only tetrahedra based at vertices of γ contribute in the sum $\int_\Sigma = \sum_\Delta \int_\Delta$.

Now we just take ϵ to zero, realise that only terms with $v = p = q = r = s$ contribute and find that

$$\hat{H}_{\text{Higgs,kin}}(N)f_\gamma = \frac{8m_p}{9^2\ell_p^9} \sum_{v \in V(\gamma)} N(v)X^I(v)X^I(v)\frac{1}{E(v)^2} \sum_{v(\Delta)=v(\Delta')=v} \epsilon^{ijk}$$

$$\times \text{tr}\left(h_{s_i(\Delta)}\left[h_{s_i(\Delta)}^{-1}, \sqrt{\hat{V}_v}\right] \right) \text{tr}\left(h_{s_j(\Delta)}\left[h_{s_j(\Delta)}^{-1}, \sqrt{\hat{V}_v}\right] \right)$$

$$\times \text{tr}\left(h_{s_k(\Delta)}\left[h_{s_k(\Delta)}^{-1}, \sqrt{\hat{V}_v}\right] \right) \epsilon^{lmn}\text{tr}\left(h_{s_l(\Delta')}\left[h_{s_l(\Delta')}^{-1}, \sqrt{\hat{V}_v}\right] \right)$$

$$\times \text{tr}\left(h_{s_m(\Delta')}\left[h_{s_m(\Delta')}^{-1}, \sqrt{\hat{V}_v}\right] \right) \text{tr}\left(h_{s_n(\Delta')}\left[h_{s_n(\Delta')}^{-1}, \sqrt{\hat{V}_v}\right] \right) f_\gamma \quad (12.3.24)$$

The operator (12.3.24) is certainly quite complicated but it is densely defined!

Next we turn to the term containing the derivatives of the scalar field. We write

$$q^{ab}\sqrt{\det(q)} = \frac{E_i^a E_i^b}{\sqrt{\det(q)}} \quad \text{and} \quad E_i^a = \pm\epsilon^{acd}\epsilon_{ijk}\frac{e_c^j e_d^k}{2}$$

where the sign drops out when taking the square and regulate (again we could have chosen to replace only one of the E_i^a by the term quadratic in e_a^i and would still arrive at a well-defined result at the price of losing symmetry of the expression)

$$H_{\text{Higgs,der}}^\epsilon(N)$$

$$= \frac{1}{2\kappa}\int d^3x \int d^3y\, N(x)\chi_\epsilon(x,y)\epsilon^{ijk}\epsilon^{imn}\epsilon^{abc}\frac{(\mathcal{D}_a\phi_I e_b^j e_c^k)(x)}{\sqrt{V(x,\epsilon)}}\epsilon^{bef}\frac{(\mathcal{D}_b\phi_I e_e^m e_f^n)(y)}{\sqrt{V(y,\epsilon)}}$$

$$= \frac{1}{2\kappa^5}\left(\frac{2}{3}\right)^4\int N(x)\epsilon^{ijk}\mathcal{D}\phi_I(x) \wedge \{A^j(x), V(x,\epsilon)^{3/4}\} \wedge \{A^k(x), V(x,\epsilon)^{3/4}\}$$

$$\times \int \chi_\epsilon(x,y)\epsilon^{imn}\mathcal{D}\phi_I(y) \wedge \{A^m(x), V(y,\epsilon)^{3/4}\} \wedge \{A^n(y), V(y,\epsilon)^{3/4}\}$$

$$(12.3.25)$$

It is clear what we are driving at. We replace Poisson brackets by commutators times $1/i\hbar$ and V by its operator version. Furthermore we introduce the already familiar triangulation of σ and have, using that with $v = s(0)$ for some path s

$$\text{Ad}(\underline{h}_s(0,\delta t))[\phi(s(\delta t))] - U(v) = \underline{h}_s(0,\delta t)\phi(s(\delta t))\underline{h}_s(0,\delta t)^{-1} - \phi(v)$$

$$= \exp([1 + \delta t\dot{s}^a(0)\underline{A}_a][\phi(v) + \delta t\dot{s}^a(0)\partial_a\phi(v)]$$

$$\times [1 - \delta t\dot{s}^a(0)\underline{A}_a] + o((\delta t)^2)) - \phi(v)$$

$$= \exp(\delta t\dot{s}^a(0)(\partial_a\phi(v) + [\underline{A}_a, \phi(v)]) + o((\delta t)^2))$$

$$= -\phi(v)\delta t\dot{s}^a(0)\mathcal{D}_a\phi(v) + o((\delta t)^2) \quad (12.3.26)$$

and with $\text{tr}(\tau_i\tau_j) = -\delta_{ij}/2, \text{tr}(\mathcal{I}_I\mathcal{I}_J) = -d\delta_{IJ}$, d the dimension of the fundamental representation of G that

$$6\int_\Delta \mathcal{D}\phi_I(x) \wedge \{A^j(x), V(x,\epsilon)^{3/4}\} \wedge \{A^k(x), V(x,\epsilon)^{3/4}\}$$

$$\approx -\frac{4}{d}\epsilon^{mnp}\text{tr}(\mathcal{I}_I[\text{Ad}(\underline{h}_{s_m(\Delta)})[\phi(s_m(\Delta))] - \phi(v(\Delta))])$$

$$\times \text{tr}(\tau_j h_{s_n(\Delta)}\{h^{-1}_{s_n(\Delta)}, V(v(\Delta),\epsilon)^{3/4}\})\text{tr}(\tau_k h_{s_p(\Delta)}\{h^{-1}_{s_p(\Delta)}, V(v(\Delta),\epsilon)^{3/4}\})$$

$$(12.3.27)$$

Then we find on a cylindrical function

$$\hat{H}^\epsilon_{\text{Higgs,der}}(N)f_\gamma = \frac{1}{2\kappa^5\hbar^4}\left(\frac{2}{3}\right)^4\left(\frac{2}{3d}\right)^2 \sum_{v,v'\in V(\gamma)} N(v)\chi_\epsilon(v,v')\epsilon^{ijk}\epsilon^{ilm}$$

$$\times \sum_{v(\Delta)=v} \frac{8}{E(v)}\epsilon^{npq}\text{tr}(\mathcal{I}_I[\text{Ad}(\underline{h}_{s_n(\Delta)})[\phi(s_n(\Delta))] - \phi(v(\Delta))])$$

$$\times \text{tr}(\tau_j h_{s_p(\Delta)}[h^{-1}_{s_p(\Delta)}, \hat{V}^{3/4}_v])\text{tr}(\tau_k h_{s_q(\Delta)}[h^{-1}_{s_q(\Delta)}, \hat{V}^{3/4}_v])$$

$$\times \sum_{v(\Delta')=v'} \frac{8}{E(v')}\epsilon^{rst}\text{tr}(\mathcal{I}_I[\text{Ad}(\underline{h}_{s_r(\Delta')})[\phi(s_r(\Delta'))] - \phi(v(\Delta'))])$$

$$\times \text{tr}(\tau_l h_{s_s(\Delta')}[h^{-1}_{s_s(\Delta')}, \hat{V}^{3/4}_{v'}])\text{tr}(\tau_m h_{s_t(\Delta')}[h^{-1}_{s_t(\Delta')}, \hat{V}^{3/4}_{v'}])f_\gamma$$

$$(12.3.28)$$

since only tetrahedra with vertices as basepoints contribute. Thus we find in the limit $\epsilon \to 0$ in the URST (i.e., choose finite beginning segments of the edges at each vertex once and for all, identical with the choice for the other Hamiltonian constraint contributions) $s(\Delta)$

$$\hat{H}_{\text{Higgs,der}}(N)f_\gamma = \frac{4^6 m_p}{2\ell_p^9 d^2 3^6} \sum_{v\in V(\gamma)} N(v)\epsilon^{ijk}\epsilon^{ilm}$$

$$\times \sum_{v(\Delta)=v(\Delta')=v} \frac{1}{E(v)^2}\epsilon^{npq}\epsilon^{rst}\text{tr}(\mathcal{I}_I[\text{Ad}(\underline{h}_{s_n(\Delta)})[\phi(s_n(\Delta))] - \phi(v)])$$

$$\times \text{tr}(\tau_j h_{s_p(\Delta)}[h^{-1}_{s_p(\Delta)}, \hat{V}^{3/4}_v])\,\text{tr}(\tau_k h_{s_q(\Delta)}[h^{-1}_{s_q(\Delta)}, \hat{V}^{3/4}_v])$$

$$\times \text{tr}(\mathcal{I}_I[\text{Ad}(\underline{h}_{s_r(\Delta')})[\phi(s_r(\Delta'))] - \phi(v)])$$

$$\times \text{tr}(\tau_l h_{s_s(\Delta')}[h^{-1}_{s_s(\Delta')}, \hat{V}^{3/4}_v])\,\text{tr}(\tau_m h_{s_t(\Delta')}[h^{-1}_{s_t(\Delta')}, \hat{V}^{3/4}_v])f_\gamma$$

$$(12.3.29)$$

Again, despite its complicated appearance, (12.3.29) defines a densely defined operator. Finally the potential term, like the cosmological constant term, is trivial to quantise because ϕ is just a multiplication operator and the $\sqrt{\det(q)}$ plus the integral just becomes the sum over vertices times the volume operator at those times the potential evaluated at those. Hence this term becomes like the cosmological term

$$\hat{H}_{\mathrm{Higgs,pot}}(N)f_\gamma = \frac{m_p}{\ell_p^3} \sum_{v \in V(\gamma)} N_v V(\phi(v)) \hat{V}_v f_\gamma$$

$$\hat{H}_{\mathrm{cosmo}}(N)f_\gamma = \frac{m_p \lambda}{\ell_p^3} \sum_{v \in V(\gamma)} N_v \hat{V}_v f_\gamma \tag{12.3.30}$$

This furnishes the quantisation of the matter sector. Notice that all Hamiltonians have the same structure, namely an operator which carries out a discrete operation on a cylindrical function, like adding or subtracting lines, fermions or Higgs fields, multiplied by the Planck mass and divided by an appropriate power of the Planck length which compensates the power of the Planck length coming from the action of the volume operator. It follows that in this sense the matter Hamiltonians are quantised in multipla of the Planck mass when we go to the diffeomorphism-invariant sector.

12.3.4 A general quantisation scheme

Looking at what happened in Sections 10.4 and 12.3.1 it seems that one can quantise any Hamiltonian constraint which is a scalar density of weight one in such a way that it is densely defined. Indeed, in [441] a proof of this observation is given which we sketch below (we restrict ourselves here to non-fermionic matter and to $D = 3$ spatial dimensions for the sake of clarity). It applies to any field theory in any dimension $D \geq 2$ which is given in Hamiltonian form, that is, any generally covariant field theory deriving from a Lagrangian (for theories including higher derivatives as in higher derivative gravity [658] or as predicted by the effective action of string theory [45] one can apply the Ostrogradsky method [659] to bring it into Hamiltonian form).

Suppose then that we are given a scalar density $H(x)$ of weight one. Without loss of generality we can assume that all the momenta P of the theory are tensor densities of weight one and act by functional derivation with respect to the configuration variables Q which are associated dual tensor densities of weight zero. By contracting them with triad and co-triad fields we obtain new canonical variables without tensor indices but with su(2) indices. The corresponding canonical transformation is generated by a functional which changes the definition of the real-valued connection variable A_a^i but preserves its real-valuedness and thus does not spoil the kinematical Hilbert space of Part II. Spatial covariant derivatives are then with respect to A_a^i.

The general form of this density $H(x)$ is then a sum of homogeneous polynomials of the form (not displaying internal indices)

$$H_{m,n}(x) = [P(x)]^n E^{a_1}(x) \dots E^{a_m}(x) f_{m,n}[Q]_{a_1 \dots a_m}(x) \frac{1}{[\sqrt{\det(q)}(x)]^{m+n-1}}$$

$$(12.3.31)$$

where f is a local tensor depending only on configuration variables and their covariant derivatives with respect to A_a^i. In order to quantise (12.3.31) we must point split the momenta P, E^a. Multiply (12.3.31) by $1 = [\frac{|\det((e_a^i))|}{\sqrt{\det(q)}}]^k$ where $k = 0, 1, 2, \dots$ is an integer to be specified later on. Since up to a numerical constant $|\det((e_a^i))|$ equals $\epsilon^{abc} \epsilon_{ijk} \{A_a^i, V(R)\}\{A_b^j, V(R)\}\{A_c^k, V(R)\}$ for some appropriately chosen region we see that this factor is worth Dk volume functionals in the numerator and k factors of $\sqrt{\det(q)}$ in the denominator. We now introduce $m + n + k - 1$ point splittings by the point splitting functions $\chi_{\epsilon,x}(y)/\epsilon^D$ of the previous section to point split both the momenta and the factors of $|\det((e_a^i))|$. The factor $1/\epsilon^{D(m+n+k-1)}$ can be absorbed into the $\sqrt{\det(q)}$'s as before so that we get a power of $m + n + k - 1$ of volume functionals of the form $V(U_\epsilon(x))$ in the denominator. Now choose k large enough until $Dk > m + n + k - 1$ or $(D - 1)k > m + n - 1$. By suitably choosing the arguments in the process of point splitting and choosing $R = U_\epsilon(.)$ we can arrange, as in the previous section, that the only dependence of (12.3.31) on the volume functional is through Dk factors of the form

$$\frac{\{A_a^i, V(U_\epsilon)\}}{V(U_\epsilon)^{\frac{m+n+k-1}{Dk}}} = \frac{\{A_a^i, V(U_\epsilon)^{1 - \frac{m+n+k-1}{Dk}}\}}{1 - \frac{m+n+k-1}{Dk}}$$

$$(12.3.32)$$

so that the volume functional is removed from the denominator. The rest of the quantisation proceeds by choosing a triangulation of σ replacing connections by holonomies along its edges, Higgs fields by point holonomies at vertices or corresponding gauge-covariant polynomials, momenta by functional derivatives and Poisson brackets by commutators. By carefully choosing the factor ordering (momenta to the right-hand side) one always finds a densely defined operator whose limit (as the regulator is removed) exists in the URST and whose commutator algebra is non-anomalous.

The proof shows that the density weight of one for $H(x)$ was crucial: if it was lower than one then point splitting would result in a regulated operator whose limit is the zero operator and if it was higher than one then the limit diverges, as mentioned earlier. Notice that the final result suffers from factor ordering *ambiguities* but *not* from factor ordering *singularities*.

13
Kinematical geometrical operators

In this chapter we will describe the so-called kinematical geometrical operators of Loop Quantum Gravity. These are gauge-invariant operators which measure the length, area and volume respectively of coordinate curves, surfaces and volumes for $D = 3$. The area and volume operators were first considered by Smolin in [660] and then formalised by Rovelli and Smolin in the loop representation [425]. In [575] Loll discovered that the volume operator vanishes on gauge-invariant states with at most trivalent vertices and used area and volume operators in her lattice theoretic framework [661–663]. Ashtekar and Lewandowski [427] used the connection representation defined in previous chapters and could derive the full spectrum of the area operator, while their volume operator differs from that of Rovelli and Smolin on graphs with vertices of valence higher than three, which can be seen as the result of using different diffeomorphism classes of regularisations. In [664] de Pietri and Rovelli computed the matrix elements of the RS volume operator in the loop representation and de Pietri created a computer code for the actual case-by-case evaluation of the eigenvalues. In [559] the connection representation was used in order to obtain the complete set of matrix elements of the AL volume operator.

Area and volume operators could be quantised using only the known quantisations of the electric flux of Section 6.3 but the construction of the length operator [424] required the new quantisation technique of using Poisson brackets with the volume operator, which was first employed for the Hamiltonian constraint, see Chapter 10. To the same category of operators also belong the ADM energy surface integral [442], angle operators [429, 430] and other similar operators that test components of the three-metric tensor [581].

In D-dimensions we have analogous objects corresponding to d-dimensional submanifolds of σ with $1 \leq d \leq D$. To get an idea of the constructions involved we will start with the simplest operator, the so-called area operator which we construct in D dimensions and which measures the area of an open $(D-1)$-dimensional submanifold of σ. A common feature of all these operators is that they are essentially self-adjoint, positive semidefinite unbounded operators with pure point (discrete) spectrum which has a length, area, volume, ... gap respectively of the order of the Planck length, area, volume, etc. (that is, zero is not an accumulation point of the spectrum).

We call these operators kinematical because they do not (weakly) commute with the spatial diffeomorphism or Hamiltonian constraint operator. One may therefore ask what their physical significance should be. Apart from the fact that the kinematical volume operator plays a pivotal role for the very definition of the Hamiltonian constraint, as a partial answer we will sketch a proof that if the curves, surfaces and regions are not coordinate manifolds but are invariantly defined through matter, then they not only weakly commute with the spatial diffeomorphism constraint but also their spectrum remains unaffected. There is no such argument with respect to the Hamiltonian constraint yet, however. We will follow the treatment in [427, 559, 665].

13.1 Derivation of the area operator

Let S be an oriented, embedded, open, compactly supported, semianalytical surface and let $X : U_0 \to S$ be the associated embedding where U is an open submanifold of $\mathbb{R}^{D-1}$. The area functional $\mathrm{Ar}[S]$ of the D-metric tensor q_{ab} is the volume of $X^{-1}(S)$ in the induced $(D-1)$-metric

$$\mathrm{Ar}[S] := \int_{U_0} d^{D-1}u\sqrt{\det([X^*q](u))} \tag{13.1.1}$$

which coincides with the Nambu–Goto action for the bosonic Euclidean $(D-1)$-brane propagating in a D-dimensional target spacetime (σ, q_{ab}). Using the co-vector densities

$$n_a(u) := \epsilon_{aa_1\ldots a_{D-1}} \prod_{k=1}^{D-1} \frac{\partial X^{a_k}}{\partial u_k}(u) \tag{13.1.2}$$

familiar from Section 6.3 it is easy to see that we can write (13.1.1) in the form

$$\mathrm{Ar}[S] := \int_{U_0} d^{D-1}u\sqrt{n_a(u)n_b(u)E_j^a(X(u))E_j^b(X(u))} \tag{13.1.3}$$

Let now $U_0 = \bigcup_{U \in \mathcal{U}} U$ be a partition of U_0 by closed sets U with open interior and let $\mathcal{U}$ be the collection of these open sets. Then the Riemann integral (13.1.3) is the limit as $|\mathcal{U}| \to \infty$ of the Riemann sum

$$\mathrm{Ar}_{\mathcal{U}}[S] := \sum_{U \in \mathcal{U}} \sqrt{E_j(S_U)E_j(S_U)} \tag{13.1.4}$$

where $S_U = X(U)$ and $E_j(S_U)$ is the electric flux function of Section 6.3. The strategy for quantising (13.1.4) will be to use the known quantisation of $E_j(S_U)$, to plug it into (13.1.4), to apply it to cylindrical functions and to hope that in the limit $|\mathcal{U}| \to \infty$ we obtain a consistently defined family of positive semidefinite operators. Notice that the square root involved makes sense because its argument will be a sum of squares of (essentially) self-adjoint operators which has non-negative real spectrum and we may therefore define the square root by the spectral resolution of the operator.

Let then $l = l(\gamma)$ be any subgroupoid and $f_l \in C^2(X_l)$. Using the results of Section 6.3 we obtain for any surface S

$$\hat{E}_j(S)\hat{E}_j(S)p_l^* f_l = -p_{l_S}^* \frac{\ell_p^4 \beta^2}{64} \left\{ \sum_{e \in E(\gamma_S)} \epsilon(e, S)R_e^j \right\}^2 p_{l_S l}^* f_l \tag{13.1.5}$$

where $l_S = l(\gamma_S)$ is any adapted subgroupoid $l \prec l_S$.

When we now plug (13.1.5) into (13.1.4) we can exploit the following fact: since (13.1.4) classically approaches (13.1.3) for *any* uniform refinement of the partition $\mathcal{U}$, for given l and adapted l_S we can refine in such a way that for all $e \in E(\gamma)$ with $\epsilon(e, S) \neq 0$ (e is of the up or down type with respect to S) we have always that $e \cap S$ is an interior point of some $U \in \mathcal{U}$. Notice that then $\epsilon(e, S) = \epsilon(e, S_U)$ and $e \cap S = e \cap S_U$. If on the other hand $\epsilon(e, S) = 0$ but $S \cap e \neq \emptyset$ (e is of the inside type with respect to S) then for those U with $U \cap e \neq \emptyset$ we also have $\epsilon(e, S_U) = 0$. Clearly, if $e \cap S = \emptyset$ then $e \cap U = \emptyset$ for all $U \in \mathcal{U}$ so again $\epsilon(e, S) = \epsilon(e, S_U)$. We conclude that under such refinements the subgroupoid l_S stays adapted for all S_U. Let us denote an adapted partition and their refinements by $\mathcal{U}_l$. Then

$$\widehat{Ar}_{\mathcal{U}_l}[S]p_l^* f_l = \frac{\ell_p^2 \beta}{8} p_{l_S}^* \sum_{U \in \mathcal{U}_l} \sqrt{-\left\{ \sum_{e \in E(\gamma_S)} \epsilon(e, S_U)R_e^j \right\}^2} \, p_{l_S l}^* f_l \tag{13.1.6}$$

Let us introduce the set of isolated intersection points between γ and S

$$P_l(S) := \{e \cap S; \ \epsilon(e, S) \neq 0, \ e \in E(\gamma_S)\} \tag{13.1.7}$$

which is independent of the choice of γ_S of course. After sufficient refinement, every S_U will contain at most one point which is the common intersection point of edges of the up or down type respectively. Let then for each $x \in P_l(S)$ the surface that contains x be denoted by S_{U_x}. From our previous discussion we know that then $\epsilon(e, S) = \epsilon(e, S_{U_x})$ for any $e \in E(\gamma_S)$ with $x \in \partial e$. It follows that (13.1.6) simplifies after sufficient refinement to

$$\widehat{Ar}_{\mathcal{U}_l}[S]p_l^* f_l = \frac{\ell_p^2 \beta}{8} p_{l_S}^* \sum_{x \in P_l(S)} \sqrt{-\left\{ \sum_{e \in E(\gamma_S), x \in \partial e} \epsilon(e, S)R_e^j \right\}^2} \, p_{l_S l}^* f_l \tag{13.1.8}$$

Now the right-hand side no longer depends on the degree of the adapted refinement and hence the limit becomes trivial

$$\widehat{Ar}_l[S]p_l^* f_l = \frac{\ell_p^2 \beta}{8} p_{l_S}^* \sum_{x \in P_l(S)} \sqrt{-\left\{ \sum_{e \in E(\gamma_S), x \in \partial e} \epsilon(e, S)R_e^j \right\}^2} \, p_{l_S l}^* f_l \tag{13.1.9}$$

Thus, we have managed to derive a family of operators $\widehat{Ar}_l[S]$ with dense domain $\mathrm{Cyl}^2(\overline{\mathcal{A}})$. The independence of (13.1.9) of the adapted graph follows from that of the $\hat{E}_j(S)$. Here we have encountered again a common theme throughout the formalism: a state (or graph)-dependent regularisation. One must make sure therefore that the resulting family of operators is consistent.

13.2 Properties of the area operator

The following properties go through with minor modifications also for the length and volume operators.

1. *Consistency*

   We must show that for any $l \prec l'$ it holds that (a) $\hat{U}_{ll'} C^2(X_l) \subset C^2(X_{l'})$ and (b) $\hat{U}_{ll'} \widehat{\mathrm{Ar}}_l[S] = \widehat{\mathrm{Ar}}_{l'}[S] \hat{U}_{ll'}$ where $\hat{U}_{ll'} f_l = p_{l'l}^* f_l$. Since the $p_{ll'}^*$ are analytic, (a) is trivially satisfied. To verify (b) we notice that (13.1.9) can be written as

   $$\hat{U}_l \widehat{\mathrm{Ar}}_l[S] = \hat{U}_{l_S} \widehat{\mathrm{Ar}}_{l_S}[S] \hat{U}_{l l_S} \tag{13.2.1}$$

   where $\widehat{\mathrm{Ar}}_{l_S}[S]$ is simply the middle operator in (13.1.9) between the two pull-backs for the case that l is already adapted. First we must check that (13.2.1) is independent of the adapted subgroupoid $l \prec l_S$. Let $l \prec l_S'$ be another subgroupoid and take a third adapted subgroupoid with $l_S, l_S' \prec l_S''$. If we can show that for any adapted subgroupoids with $l_S \prec l_S''$ we have

   $$\widehat{\mathrm{Ar}}_{l_S''}[S] \hat{U}_{l_S l_S''} = \hat{U}_{l_S l_S''} \widehat{\mathrm{Ar}}_{l_S}[S] \tag{13.2.2}$$

   then we will be done. To verify (13.2.2) we must make a case-by-case analysis as in Section 6.5 for the electric flux operator. But since (13.1.9) is essentially the sum of square roots of the sum of squares of electric flux operators, the analysis is completely analogous and will not be repeated here.

   Finally, let $l \prec l'$. We find an adapted subgroupoid $l, l' \prec l_S$. Then

   $$\hat{U}_{l'} \widehat{\mathrm{Ar}}_{l'}[S] \hat{U}_{ll'} = \hat{U}_{l_S} \widehat{\mathrm{Ar}}_{l_S}[S] \hat{U}_{l'l_S} \hat{U}_{ll'} = \hat{U}_{l_S} \widehat{\mathrm{Ar}}_{l_S}[S] \hat{U}_{ll_S} = \hat{U}_l \widehat{\mathrm{Ar}}_l[S]$$
   $$= \hat{U}_{l'} \hat{U}_{ll'} \widehat{\mathrm{Ar}}_l[S] \tag{13.2.3}$$

   which is equivalent with consistency.

   That the operator exists at all is like a small miracle: not only did we multiply two functional derivatives $\hat{E}_j^a(x)$ at the same point, even worse, we took the square of it. Yet it is a densely defined, positive semidefinite operator without encountering any need for renormalisation after taking the regulator (here the fineness of the partition) away. The reason for the existence of the operator is the *payoff for having constructed a manifestly background-independent representation*. We will see more examples of this 'miracle' in the sequel.

2. *Essential self-adjointness*

To see that the area operator is symmetric, let $f_l \in C^2(X_l)$, $f_{l'} \in C^2(X_{l'})$. Then we find an adapted subgroupoid $l, l' \prec l_S$ whence

$$< p_l^* f_l, \widehat{\text{Ar}}[S] p_{l'}^* f_{l'} > = < p_{l_S l}^* f_l, \widehat{\text{Ar}}_{l_S}[S] p_{l_S l'}^* f_{l'} >_{L_2(X_{l_S}, d\mu_{0l_S})}$$

$$= < \widehat{\text{Ar}}_{l_S}[S] p_{l_S l}^* f_l, p_{l_S l'}^* f_{l'} >_{L_2(X_{l_S}, d\mu_{0l_S})}$$

$$= < \widehat{\text{Ar}}[S] p_l^* f_l, p_{l'}^* f_{l'} > \tag{13.2.4}$$

where in the second step we used the fact that $\widehat{\text{Ar}}_{l_S}[S]$ is symmetric on $L_2(X_{l_S}, d\mu_{0l_S})$ with $C^2(X_{l_S})$ as dense domain.

Thus, the area operator is certainly a symmetric, positive semidefinite operator. Therefore we know that it possesses at least one self-adjoint extension, the so-called Friedrich extension, see Theorem 26.8.1.

However, we can show that $\widehat{\text{Ar}}[S]$ is even essentially self-adjoint. The proof is quite similar to proving essential self-adjointness for the electric flux operator: let $\mathcal{H}^0_{\gamma,\vec{\pi}}$ be the finite-dimensional Hilbert subspace of $\mathcal{H}^0$ given by the closed linear span of spin-network functions over γ where all edges are labelled with the same irreducible representations given by $\vec{\pi}$. Then the Hilbert space may be written as

$$\mathcal{H}^0 = \overline{\oplus_{\gamma \in \Gamma_0^\omega, \vec{\pi}} \mathcal{H}^0_{\gamma,\vec{\pi}}} \tag{13.2.5}$$

Given a surface S we can without loss of generality restrict the sum over graphs to adapted ones because for $r(\gamma) = r(\gamma_S)$ we have $\mathcal{H}^0_{\gamma,\vec{\pi}} \subset \mathcal{H}^0_{\gamma_S,\vec{\pi}'}$ for the choice $\pi'_{e'} = \pi_e$ with $E(\gamma_S) \ni e' \subset e \in E(\gamma)$. Since then $\widehat{\text{Ar}}[S]$ preserves each $\mathcal{H}^0_{\gamma,\vec{\pi}}$ its restriction is a symmetric operator on a finite-dimensional Hilbert space, therefore it is self-adjoint. It follows that $\widehat{\text{Ar}}_{|\gamma,\vec{\pi}}[S] \pm i \cdot 1_{\gamma,\vec{\pi}}$ has dense range on $\mathcal{H}^0_{\gamma,\vec{\pi}} = C^\infty(X_{l(\gamma)})_{\vec{\pi}} = C^2(X_{l(\gamma)})_{\vec{\pi}}$. Therefore

$$[\widehat{\text{Ar}}[S] \pm i \cdot 1_{\mathcal{H}^0}] C^2(\overline{\mathcal{A}}) = \oplus_{\gamma,\vec{\pi}} [\widehat{\text{Ar}}_{|\gamma,\vec{\pi}}[S] \pm i \cdot 1_{\gamma,\vec{\pi}}] C^2(X_{l(\gamma)})_{\vec{\pi}}$$

$$= \oplus_{\gamma,\vec{\pi}} [\widehat{\text{Ar}}_{|\gamma,\vec{\pi}}[S] \pm i \cdot 1_{\gamma,\vec{\pi}}] \mathcal{H}^0_{\gamma,\vec{\pi}} = \oplus_{\gamma,\vec{\pi}} \mathcal{H}^0_{\gamma,\vec{\pi}} \tag{13.2.6}$$

is dense in $\mathcal{H}^0$, hence $\widehat{\text{Ar}}[S]$ is essentially self-adjoint. Here we have used the criterion of (essential) self-adjointness, Theorem 26.7.1.

3. *Spectral properties*

(i) *Discreteness*

Since $\widehat{\text{Ar}}[S]$ leaves the $\mathcal{H}^0_{\gamma,\vec{\pi}}$ invariant it is simply a self-adjoint matrix there with non-negative eigenvalues. Since

$$\mathcal{H}^0_\gamma = \overline{\oplus_{\vec{\pi}} \mathcal{H}^0_{\gamma,\vec{\pi}}}$$

and the set of $\vec{\pi}$ is countable it follows that $\mathcal{H}^0_\gamma$ has a countable basis of eigenvectors for $\widehat{\text{Ar}}[S]$ so that the spectrum is pure point (discrete), that is, it does not have a continuous part. Now, as we vary γ we get a non-separable Hilbert space, however, the spectrum of $\widehat{\text{Ar}}[S]$ depends only on (a) the number of intersection points with edges of the up and down type, (b) their respective number per such intersection point and (c) the irreducible representations they carry and not on any other intersection characteristics. These possibilities are countable, whence the entire spectrum is pure point and each eigenvalue comes with an uncountably infinite multiplicity.

(ii) *Complete spectrum*

It is even possible to compute the complete spectrum directly and to prove the discreteness from an explicit formula. Such a closed formula is unfortunately not yet available for the volume and length operators, while highly desirable for purposes in particular connected with quantum dynamics as we will see in the next chapter.

From the explicit formula (13.1.9) it is clear that we may compute the eigenvalues for each intersection point x of S with edges of γ_S of the up or down type separately. Let $E_{x,\star}(\gamma_S) = \{e \in E(\gamma_S); \ x = b(e); \ e = \star \text{ type}\}$ where $\star = u, d, i$ for 'up, down, inside' respectively and let $R^j_{x,\star} = \sum_{e \in E_{x,\star}(\gamma_S)} R^j_e$. Then we have

$$\left\{ \sum_{e \in E(\gamma_S), x \in \partial e} \epsilon(e, S) R^j_e \right\}^2 = \left[R^j_{x,u} - R^j_{x,d} \right]^2 = \left(R^j_{x,u} \right)^2 + \left(R^j_{x,d} \right)^2 - 2 R^j_u R^j_d$$

$$= 2 \left(R^j_{x,u} \right)^2 + 2 \left(R^j_{x,d} \right)^2 - \left(R^j_u + R^j_d \right)^2 \quad (13.2.7)$$

where we have used the fact that $[R^j_{x,u}, R^k_{x,d}] = 0$ (independent degrees of freedom). We check that $[R^j_{x,\star}, R^k_{x,\star}] = -2 f_{jk}{}^l R^j_{x,\star}$ so that also $[R^j_{x,u+d}, R^k_{x,u+d}] = -2 f_{jk}{}^l R^j_{x,u+d}$ with $R^j_{x,u+d} = R^j_u + R^j_d$. From this follows that $[R^k_\star, (R^j_u)^2] = [R^k_\star, (R^j_d)^2] = 0$ so that $\Delta_u = (R^j_{x,u})^2/4, \Delta_d = (R^j_{x,d})^2/4, \Delta_{u+d} = (R^j_{x,u+d})^2/4$ are mutually commuting operators and each of $R^j_{x,u}, R^j_{x,d}, R^j_{x,u+d}$ satisfies the Lie algebra of right-invariant vector fields. Thus their respective spectrum is given by the eigenvalues $-\lambda_\pi < 0$ of the Laplacian $4\Delta = (R^j)^2 = (L^j)^2$ on G in irreducible representations π for which all matrix element functions π_{mn} are simultaneous eigenfunctions with the same eigenvalue, see Chapter 31. It follows that

$$\text{Spec}(\widehat{\text{Ar}}[S]) = \left\{ \frac{\ell_p^2 \beta}{4} \sum_{n=1}^N \sqrt{2\lambda_{\pi_n^1} + 2\lambda_{\pi_n^1} - \lambda_{\pi_n^{12}}}; \ N \in \mathbb{N}, \ \pi_n^1, \pi_n^2, \pi_n^{12} \in \Pi; \right.$$

$$\left. \pi_n^{12} \in \pi_n^1 \otimes \pi_n^2 \right\} \quad (13.2.8)$$

where the last condition means that π_n^{12} is an irreducible representation that appears in the decomposition into irreducibles of the tensor product representation $\pi_n^1 \otimes \pi_n^2$. In case we are looking only at gauge-invariant states we actually have $R^j_{x,u+v} = -R^j_{x,i}$. The spectrum (13.2.8) is manifestly discrete by inspection. It is bounded from below by zero and is unbounded from above and depends explicitly on the Immirzi parameter.

(iii) *Area gap*

Let us discuss the spectrum more closely for $G = SU(2)$. Then per intersection point we have eigenvalues of the form

$$\lambda = \frac{\ell_p^2 \beta}{4} \sqrt{2j_1(j_1 + 1) + 2j_2(j_2 + 1) - j_{12}(j_{12} + 1)} \qquad (13.2.9)$$

where $|j_1 - j_2| \leq j_{12} \leq j_1 + j_2$ by recoupling theory, see Chapter 32. Recoupling theory [666], that is, coupling of N angular momenta also tells us how to build the corresponding eigenfunctions through an appropriate recoupling scheme. The lowest positive eigenvalue is given by the minimum of (13.2.9). At given j_1, j_2 the minimum is given at $j_{12} = j_1 + j_2$ which gives

$$\frac{\ell_p^2 \beta}{4} \sqrt{(j_1 - j_2)^2 + j_1 + j_2} = \frac{\ell_p^2 \beta}{4} \sqrt{(j_2 - (j_1 - 1/2))^2 + 2j_1 - 1/4}$$
$$(13.2.10)$$

Since (13.2.10) vanishes at $j_1 = j_2 = 0$ at least one of them must be greater than zero, say j_1. Then (13.2.10) is minimised at $j_2 = j_1 - 1/2 \geq 0$ and proportional to $\sqrt{2j_1 - 1/4}$ which takes its minimum at $j_1 = 1/2$. Thus we arrive at the *area gap*

$$\lambda_0 = \frac{\sqrt{3}\ell_p^2 \beta}{8} \qquad (13.2.11)$$

(iv) *Main series*

It is sometimes claimed [667] that the regularisation of the area operator is incorrect and that a different regularisation gives eigenvalues proportional to $\sqrt{j(j+1)}$ or $j + 1/2$ rather than (13.2.9). If that was the case then this would be of some significance for black hole physics, as we will see in Chapter 15. However, first of all regularisations in quantum field theory are never unique and may lead to different answers, the only important thing is that all of them give the same classical limit. Secondly, even if the regularisation performed in [667] is more aesthetic to some authors it is incomplete: in [667] one looks only at the so-called main series which results if we choose $j_1 = j_2 = j$, $j_{12} = 0$ and then just gives

$$\frac{\ell_p^2}{2}\beta\sqrt{j(j+1)}$$

(plus a quantum correction $j(j+1) \mapsto j(+1/2)^2$ due to the different regularisation which results in integral quantum numbers). However, the complete spectrum (13.2.9) is much richer, the side series have physical significance for the black hole spectrum as we will see and lead to a *correspondence principle*, that is, at large quantum numbers the spectrum approaches a continuum. To see this notice that at large eigenvalue, λ changes as

$$\frac{\delta\lambda}{\lambda} \approx \frac{2(2j_1+1)\delta j_1 + 2(2j_2+1)\delta j_2 - (2j_{12}+1)\delta j_{12}}{2[(j_1+1)j_1 + (j_2+1)j_2 - (j_{12}+1)j_{12}]} \tag{13.2.12}$$

Suppose we choose $j_1 = j_2 = j \gg 1$. Then $0 \le j_{12} \le 2j$ and we may choose $j_{12} = 0, \delta j_{12} = 1/2, \delta j_1 = \delta j_2 = 0$ (notice that such a transition is ignored if we do not discuss the side series). Then (13.2.12) can be written

$$\delta\lambda \approx -\frac{(\lambda_0)^2}{\lambda} \tag{13.2.13}$$

which becomes arbitrarily small at large j. The subsequent eigenvalues have been calculated numerically in [803], displaying a rapid transition to the continuum.

However, more is true: even if we just use the main series, the spectrum lies dense in $\mathbb{R}^+$ for large j. This happens when we take a large number of intersections into account as we will do for black hole physics in Chapter 15, in which the main series spectrum becomes $\ell_P^2/2 \sum_p \sqrt{j_p(j_p+1)}$. Due to the square roots involved the spectrum is not equally spaced as would happen if we replaced $\sqrt{j(j+1)}$ by $\sqrt{j(j+1)+1/4} = j+1/2$, as is favoured by some authors. However, this choice is not only physically unacceptable in view of the black body spectrum of the Hawking radiation as we will explain in Chapter 15, it is also mathematically incorrect as it leads to a cylindrically inconsistent operator, that is, to no operator at all [668].

(v) *Sensitivity to topology*

The eigenvalues (13.2.9) do detect some topological properties of σ as well. For instance, in the gauge-invariant sector the spectrum depends on whether $\partial S = \emptyset$ or not. Moreover, for $\partial S = \emptyset$ the spectrum depends on whether S divides σ into two disjoint regions or not.

13.3 Derivation of the volume operator

As we have seen, the volume operator is fundamental in order to even define the quantum constraints. We will now derive it using the point-splitting regularisation technique of [559]. We will set $\beta = 1$ for simplicity, otherwise multiply the final formula by $\beta^{3/2}$.

Let $R \subset \Sigma$ be an open, connected region of Σ. Since $E_j^a = \sqrt{\det(q)}e_j^a$ we have the identity

$$\frac{1}{3!}\epsilon_{abc}\epsilon^{ijk}E_i^a E_j^b E_k^c = \det\left((E_i^a)\right) = \text{sgn}(\det(E))\det((q_{ab})) \tag{13.3.1}$$

Notice that $\det(q) = [\det((e_a^i))]^2 \ge 0$. Since classically $\det(E) \ne 0$ we can write the volume of the region R as measured by the metric q_{ab} as follows

$$V(R) := \int_R d^3x \sqrt{\det(q)} = \sqrt{\left|\frac{1}{3!}\epsilon_{ijk}\epsilon^{abc}E_i^a E_j^b E_k^c\right|} \tag{13.3.2}$$

where the absolute values are necessary because $\det(E)$ is not positive. The next step is to smear the fields E_i^a. Let $\chi_\Delta(p, x)$ be the characteristic function in the coordinate x of a cube with centre p spanned by the three vectors $\vec{\Delta}_i = \Delta_i \vec{n}_i(\Delta)$ where $\vec{n}_i$ is a normal vector in the frame under consideration and which has coordinate volume $\mathrm{vol} = \Delta_1 \Delta_2 \Delta_3 \det(\vec{n}_1, \vec{n}_2, \vec{n}_3)$ (we assume the three normal vectors to be right-oriented). In other words, $\chi_\Delta(p, x) = \prod_{i=1}^3 \theta(\frac{\Delta_i}{2} - | < n_i, x - p > |)$ where $< ., . >$ is the standard Euclidean inner product and $\theta(y) = 1$ for $y > 0$ and zero otherwise.

We consider the smeared quantity

$$
E(p, \Delta, \Delta', \Delta'') := \frac{1}{\mathrm{vol}(\Delta)\mathrm{vol}(\Delta')\mathrm{vol}(\Delta'')} \int_\sigma d^3x \int_\sigma d^3y \int_\sigma d^3z \, \chi_\Delta(p, x) \chi_{\Delta'}(2p, x + y)
$$

$$
\times \chi_{\Delta''}(3p, x + y + z) \frac{1}{3!} \epsilon_{abc} \epsilon^{ijk} E_i^a(x) E_j^b(y) E_k^c(z) \qquad (13.3.3)
$$

Notice that if we take the limits $\Delta_i, \Delta_i', \Delta_i'' \to 0$ in any combination and in any rate with respect to each other then we get back to (13.3.1) evaluated at the point p. This holds for any choice of linearly independent normal vectors $\vec{n}_i, \vec{n}_i', \vec{n}_i''$. The strange arguments $x + y$, $x + y + z$ will turn out to be very crucial in obtaining a manifestly diffeomorphism-covariant result. We will see this in a moment.

Then it is easy to see that the classical identity

$$
V(R) = \lim_{\Delta \to 0} \lim_{\Delta' \to 0} \lim_{\Delta'' \to 0} \int_R d^3p \sqrt{|E(p, \Delta, \Delta', \Delta'')|} \qquad (13.3.4)
$$

holds. Observe that (13.3.3) is not gauge-invariant any longer in contrast to $V(R)$, however, we will be interested only in the limit of shrinking all Δ to points and, as it turns out, recover gauge invariance in that limit.

The virtue of introducing the quantities (13.3.3) is that they enable us to define operators corresponding to $V(R)$, in the limit that all Δ shrink to a point, and which have the dense domain $\mathrm{Cyl}^3(\overline{\mathcal{A}/\mathcal{G}})$. To do this we adopt the same strategy which led to the fundamental flux operators: according to the canonical brackets $\{A_a^i(x), E_j^b(y)\} = -\frac{\kappa}{2} \delta_a^b \delta_j^i \delta^{(3)}(x, y)$ we represent, just as for the flux operator, the operator corresponding to E_i^a by $\hat{E}_i^a(x) = i\frac{\ell_p^2}{2} \delta/\delta A_a^i(x)$ where $\ell_p = \sqrt{\hbar \kappa}$ is the Planck length (we set $\beta = 1$ for simplicity). The functional derivative makes sense only on functions of smooth connections. However, after removing the regulator we will see that the final formula can simply be lifted to functions on $\overline{\mathcal{A}}$. Hence the limit of vanishing regulator will map from a family of operators on the space of spin-network functions restricted to $\mathcal{A}$ to an honest operator on the kinematical Hilbert space of Loop Quantum Gravity.

Let a graph γ be given. In order to simplify the notation, we subdivide each edge e with endpoints v, v' which are vertices of γ into two segments s, s' where $e = s \circ (s')^{-1}$ and s has an orientation such that it is *outgoing* at v while s' has an orientation such that it is *outgoing* at v'. This introduces new vertices $s \cap s'$ which we will call pseudo-vertices because they are not points of non-semianalyticity of the graph. Let $E(\gamma)$ be the set of these segments of γ but $V(\gamma)$ the set of true (as opposed to pseudo) vertices of γ. Let us now evaluate the action of $\hat{E}_i^a(p, \Delta) := 1/\mathrm{vol}(\Delta) \int_\Sigma d^3 x \chi_\Delta(p, x) \hat{E}_i^a(x)$ on a function $f = p_\gamma^* f_\gamma$ cylindrical with respect to γ. We find ($e : [0,1] \to \sigma; \ t \to e(t)$ being a parametrisation of the edge e)

$$2\hat{E}_i^a(p, \Delta)f = \frac{i\ell_p^2}{\mathrm{vol}(\Delta)} \sum_{e \in E(\gamma)} \int_0^1 dt \chi_\Delta(p, e(t))\dot{e}^a(t)$$

$$\times \frac{1}{2}\mathrm{tr}\left([h_e(0,t)\tau_i h_e(t,1)]^T \frac{\partial}{\partial h_e(0,1)}\right) f_\gamma \qquad (13.3.5)$$

Here we have used (0) $\{E_j^a(x), A_b^k(y)\} = \kappa/2\delta(x,y)\delta_b^a\delta_j^k$, $\ell_P^2 = \hbar\kappa$, (1) the fact that a cylindrical function is already determined by its values on $\mathcal{A}/\mathcal{G}$ rather than $\overline{\mathcal{A}/\mathcal{G}}$ so that it makes sense to take the functional derivative, (2) the definition of the holonomy as the path-ordered exponential of $\int_e A$ with the smallest parameter value to the left, (3) $A = dx^a A_a^i \tau_i/2$ where SU(2) $\ni \tau_j = -i\sigma_j$, σ_j being the usual Pauli matrices, so that $[\tau_i/2, \tau_j/2] = \epsilon_{ijk}\tau_k/2$ and we have defined (4) $\mathrm{tr}(h^T \partial/\partial g) = h_{AB}\partial/\partial g_{AB}$, $A, B, C, \ldots$ being SU(2) indices. The state that appears on the right-hand side of (13.3.5) is actually well-defined, in the sense of functions of connections, only when A is smooth for otherwise the integral over t does not exist, see [418] and Section 8.2.3. However, as announced, we will be interested only in quantities constructed from operators of the form (13.3.5) and for which the limit of shrinking $\Delta \to 0$ to a point has a meaning in the sense of $\mathcal{H} = L_2(\overline{\mathcal{A}/\mathcal{G}}, d\mu_0)$ and therefore will not be concerned with the actual range of the operator (13.3.5) for the moment.

We now wish to evaluate the whole operator $\hat{E}(p, \Delta, \Delta', \Delta'')$ on f. It is clear that we obtain three types of terms, the first type comes from all three functional derivatives acting on f only, the second type comes from two functional derivatives acting on f and the remaining one acting on the trace appearing in (13.3.5) and finally the third type comes from only one derivative acting on f_γ and the remaining two acting on the trace. Explicitly we find (we mean by $\theta(t, t')$ the theta function which is unity if $0 < t < t' < 1$ and zero otherwise and likewise $\theta(t, t', t'')$ is 1 if $0 < t < t' < t'' < 1$ and zero otherwise)

$$8\hat{E}(p, \Delta, \Delta', \Delta'')f$$

$$= -\frac{i\ell_p^6}{8 \cdot 3! \mathrm{vol}(\Delta)\mathrm{vol}(\Delta')\mathrm{vol}(\Delta'')}\epsilon_{abc}\epsilon^{ijk}\int_{[0,1]^3} dt\, dt'\, dt''$$

$$\times \left\{ \sum_{e,e',e'' \in E(\gamma)} \dot{e}(t)^a \dot{e}'(t')^b \dot{e}''(t'')^c \chi_\Delta(p, e(t))\chi_{\Delta'}(2p, e(t) + e'(t')) \right.$$

$$\times \chi_{\Delta''}(3p, e(t) + e'(t') + e''(t''))\mathrm{tr}\left(h_{e''}(0,t'')\tau_k h_{e''}(t'',1)\frac{\partial}{\partial h_{e''}^T(0,1)}\right)$$

$$\times \mathrm{tr}\left(h_{e'}(0,t')\tau_j h_{e'}(t',1)\frac{\partial}{\partial h_{e'}^T(0,1)}\right)\mathrm{tr}\left(h_e(0,t)\tau_i h_e(t,1)\frac{\partial}{\partial h_e^T(0,1)}\right)$$

$$+ \sum_{e',e'' \in E(\gamma)} \dot{e}''(t)^a \dot{e}'(t')^b \dot{e}''(t'')^c \chi_\Delta(p, e''(t))\chi_{\Delta'}(2p, e''(t) + e'(t'))\chi_{\Delta''}(3p, e''(t))$$

$$+e'(t') + e''(t'')) \left[\theta(t,t'') \mathrm{tr} \left(h_{e''}(0,t) \tau_i h_{e''}(t,t'') \times \tau_k h_{e''}(t'',1) \frac{\partial}{\partial h_{e''}^T(0,1)} \right) \right.$$

$$\left. + \theta(t'',t) \mathrm{tr} \left(h_{e''}(0,t'') \tau_k h_{e''}(t'',t) \tau_i h_{e''}(t,1) \frac{\partial}{\partial h_{e''}^T(0,1)} \right) \right]$$

$$\times \mathrm{tr} \left(h_{e'}(0,t') \tau_j h_{e'}(t',1) \frac{\partial}{\partial h_{e'}^T(0,1)} \right) + \sum_{e',e'' \in E(\gamma)} \dot{e}'(t)^a \dot{e}'(t')^b \dot{e}''(t'')^c \chi_\Delta(p, e'(t)) \chi_{\Delta'}$$

$$\times (2p, e'(t) + e'(t')) \chi_{\Delta''}(3p, e'(t) + e'(t') + e''(t'')) \mathrm{tr} \left(h_{e''}(0,t'') \tau_k h_{e''}(t'',1) \frac{\partial}{\partial h_{e''}^T(0,1)} \right)$$

$$\times \left[\theta(t,t') \mathrm{tr} \left(h_{e'}(0,t) \tau_i h_{e'}(t,t') \tau_j h_{e'}(t',1) \frac{\partial}{\partial h_{e'}^T(0,1)} \right) + \theta(t',t) \mathrm{tr} \left(h_{e'}(0,t') \tau_j h_{e'} \right. \right.$$

$$\left. \left. \times (t',t) \tau_i h_{e'}(t,1) \frac{\partial}{\partial h_{e'}^T(0,1)} \right) \right] + \sum_{e,e'' \in E(\gamma)} \dot{e}(t)^a \dot{e}''(t')^b \dot{e}''(t'')^c \chi_\Delta(p, e(t)) \chi_{\Delta'}(2p, e(t) + e''(t'))$$

$$\times \chi_{\Delta''}(3p, e(t) + e''(t') + e''(t'')) \left[\theta(t',t'') \mathrm{tr} \left(h_{e''}(0,t') \tau_j h_{e''}(t',t'') \tau_k h_{e''}(t'',1) \frac{\partial}{\partial h_{e''}^T(0,1)} \right) \right.$$

$$\left. + \theta(t'',t') \mathrm{tr} \left(h_{e''}(0,t'') \tau_k h_{e''}(t'',t') \tau_j h_{e''}(t',1) \frac{\partial}{\partial h_{e''}^T(0,1)} \right) \right] \mathrm{tr} \left(h_e(0,t) \tau_i h_e(t,1) \frac{\partial}{\partial h_e^T(0,1)} \right)$$

$$+ \sum_{e'' \in E(\gamma)} \dot{e}''(t)^a \dot{e}''(t')^b \dot{e}''(t'')^c \chi_\Delta(p, e''(t)) \chi_{\Delta'}(2p, e''(t) + e''(t')) \chi_{\Delta''}(3p, e''(t)$$

$$+ e''(t') + e''(t'')) \left[\theta(t,t',t'') \mathrm{tr} \left(h_{e''}(0,t) \tau_i h_{e''}(t,t') \tau_j h_{e''}(t',t'') \tau_k h_{e''}(t'',1) \frac{\partial}{\partial h_{e''}^T(0,1)} \right) \right.$$

$$+ \theta(t,t'',t') \mathrm{tr} \left(h_{e''}(0,t) \tau_i h_{e''}(t,t'') \tau_k h_{e''}(t'',t') \tau_j h_{e''}(t',1) \frac{\partial}{\partial h_{e''}^T(0,1)} \right)$$

$$+ \theta(t',t'',t) \mathrm{tr} \left(h_{e''}(0,t') \tau_j h_{e''}(t',t'') \tau_k h_{e''}(t'',t) \tau_i h_{e''}(t,1) \frac{\partial}{\partial h_{e''}^T(0,1)} \right)$$

$$+ \theta(t',t,t'') \mathrm{tr} \left(h_{e''}(0,t') \tau_j h_{e''}(t',t) \tau_i h_{e''}(t,t'') \tau_k h_{e''}(t'',1) \frac{\partial}{\partial h_{e''}^T(0,1)} \right)$$

$$+ \theta(t'',t,t') \mathrm{tr} \left(h_{e''}(0,t'') \tau_k h_{e''}(t'',t) \tau_i h_{e''}(t,t') \tau_j h_{e''}(t',1) \frac{\partial}{\partial h_{e''}^T(0,1)} \right)$$

$$\left. + \theta(t'',t',t) \mathrm{tr} \left(h_{e''}(0,t'') \tau_k h_{e''}(t'',t') \tau_j h_{e''}(t',t) \tau_i h_{e''}(t,1) \frac{\partial}{\partial h_{e''}^T(0,1)} \right) \right] \right\} f_\gamma$$

$$=: [\hat{O}_{1,2,3} + \hat{O}_{2,31} + \hat{O}_{12,3} + \hat{O}_{1,23} + \hat{O}_{123}] f_\gamma \tag{13.3.6}$$

The fact that the integrand of the terms involved in $\hat{O}_{12,3}, \hat{O}_{1,23}, \hat{O}_{2,31}, \hat{O}_{123}$ vanishes if either of the cases $0 < t = t' < 1, 0 < t' = t'' < 1, 0 < t = t' = t'' < 1$ occurs is due to the fact that in this case in $\hat{O}_{12,3}, \hat{O}_{1,23}, \hat{O}_{2,31}, \hat{O}_{123}$ we get a trace which contains $\tau_{(i}\tau_{j)}, \tau_{(j}\tau_{k)}, \tau_{(k}\tau_{i)}$ contracted with ϵ^{ijk} which vanishes (to see this recall that the functional derivative is

$$\delta h_e(A)/\delta A_a^i(x) = \frac{1}{2} \int_0^1 dt \left[\frac{1}{2} \delta^{(3)}(e(t+), x) \dot{e}(t+)^a h_e(0,t) \tau_i h_e(t,1) \right.$$

$$\left. + \frac{1}{2} \delta^{(3)}(e(t-), x) \dot{e}(t-)^a h_e(0,t) \tau_i h_e(t,1) \right] \tag{13.3.7}$$

(one-sided derivatives and δ-distributions). This expression is also correct if x is an endpoint of e (in which case there is only one term which survives in (13.3.7): namely, in the case that we consider $h_{e_1} \tau_j h_{e_2}$ instead of h_e, $e = e_1 \circ e_2$ where $x = e_1 \cap e_2$ is a point of analyticity the result of (13.3.7) is a term involving $\tau_{(i}\tau_{j)})$.

Given a triple e, e', e'' of (not necessarily distinct) edges of γ, consider the functions

$$x_{ee'e''}(t, t', t'') := e(t) + e'(t') + e''(t'') \tag{13.3.8}$$

This function has the interesting property that the Jacobian is given by

$$\det\left(\frac{\partial\left(x^1_{ee'e''}, x^2_{ee'e''}, x^3_{ee'e''}\right)(t, t', t'')}{\partial(t, t', t'')}\right) = \epsilon_{abc}\dot{e}(t)^a \dot{e}'(t')^b \dot{e}''(t'')^a \tag{13.3.9}$$

which is precisely the form of the factor which enters all the integrals in (13.3.6). This is why we have introduced the strange argument $x + y + z$.

We now consider the limit $\Delta_i, \Delta'_i, \Delta''_i \to 0$. The idea is that all quantities in (13.3.6) are meaningful in the sense of functions on smooth connections and thus limits of functions as $\Delta \to 0$ are to be understood with respect to any Sobolov topology. The miracle is that the final function is again cylindrical and thus the operator that results in the limit has an extension to all of $\overline{A/\mathcal{G}}$.

Lemma 13.3.1. *For each triple of edges e, e', e'' there exists a choice of vectors $\vec{n}_i, \vec{n}'_i, \vec{n}''_i$ and a way to guide the limit $\Delta_i, \Delta'_i, \Delta''_i \to 0$ such that*

$$\int_{[0,1]^3} \det\left(\frac{\partial\left(x^a_{ee'e''}\right)}{\partial(t, t', t'')}\right) \chi_\Delta(p, e)\chi_{\Delta'}(2p, e + e')\chi_{\Delta''}(3p, e + e' + e'')\hat{O}_{ee'e''} \tag{13.3.10}$$

vanishes

(a) if e, e', e'' do not all intersect p or

(b) $\det\left(\frac{\partial(x^a_{ee'e''})}{\partial(t,t',t'')}\right)_p = 0$ (which is a diffeomorphism-invariant statement).

Otherwise it tends to $1/8 sgn\left(\det\left(\frac{\partial(x^a_{ee'e''})}{\partial(t,t',t'')}\right)\right)_p \hat{O}_{e,e',e''}(p) \prod_{i=1}^3 \Delta''_i$. Here we have denoted by $\hat{O}_{ee'e''}(t, t', t'')$ the trace(s) involved in the various terms of (13.3.6).

Remark: To adapt the regularisation to each triple of edges is justified by the fact that the classical expression does not depend on the way we regularise when we take the limit. This has been used already before for the Hamiltonian constraint.

Proof: If at least one of e, e', e'' does not intersect p then, if we choose Δ_i, etc. smaller than some finite number Δ_0, (13.3.10) vanishes identically since the support of the characteristic functions is in a neighbourhood around p which shrinks to zero with the Δ_i, etc. So let us assume that all of e, e', e'' intersect p at parameter value t_0, t'_0, t''_0 (this value is unique because the edges are not self-intersecting). Then we can write $e(t) = p + c(t - t_0)$ where c is analytic and vanishes at $\tau = l - l_0 = 0$. We have the case subdivision:

Case I: $\det \left(\frac{\partial(x^a_{ee'e''})}{\partial(t,t',t'')} \right)_{t_0} = 0$.

Case I(a): All of $\dot{c}(0), \dot{c}'(0), \dot{c}''(0)$ are co-linear.

Case I(b): Two of $\dot{c}(0), \dot{c}'(0), \dot{c}''(0)$ are co-linear and the third is linearly independent of them.

Case I(c): No two of $\dot{c}(0), \dot{c}'(0), \dot{c}''(0)$ are co-linear.

Case II: $\det \left(\frac{\partial(x^a_{ee'e''})}{\partial(t,t',t'')} \right)_{t_0} \neq 0$.

Notice that all vectors $\dot{c}(0), \dot{c}'(0), \dot{c}''(0)$ are non-vanishing by the definition of a curve.

We consider first case I. We exclude the trivial case that all three curves lie in a coordinate plane or line such that the determinant already vanishes for all finite values of the Δ's. Therefore there exist linearly independent unit vectors u, v, w (not necessarily orthogonal) in terms of which we may express c, c', c''.

In case I(a) we have an expansion of the form

$$
\begin{aligned}
c(t) &= au(t + o(t^2)) + bv(t^m + o(t^{m+1})) + cw(t^n + o(t^{n+1})) \\
c'(t) &= au(t + o(t^2)) + b'v(t^{m'} + o(t^{m'+1})) + c'w(t^{n'} + o(t^{n'+1})) \\
c''(t) &= a''u(t + o(t^2)) + b''v(t^{m''} + o(t^{m''+1})) + c''w(t^{n''} + o(t^{n''+1}))
\end{aligned}
$$

$$(13.3.11)$$

where $a, b, c, a', b', c', a'', b'', c''$ are real numbers with $aa'a'' \neq 0$ and at least one of the b's and c's being different from zero (also not for instance $b = c = b' = c' = 0$). Furthermore $m, m', m'', n, n', n'' \geq 2$. The characteristic functions have support in coordinate cubes spanned by the vectors $\vec{n}_i, \vec{n}'_i, \vec{n}''_i$. Now, since u, v, w are linearly independent we may simply *choose*, for instance, $\vec{n}_i := u, \vec{n}'_i := v, \vec{n}''_i := w$. It follows then and from the fact that $0 \leq \chi \leq 1$ that

$$
\begin{aligned}
&\chi_\Delta(p, e)\chi_{\Delta'}(2p, e + e')\chi_{\Delta''}(3p, e + e' + e'') \\
&= \tilde{\chi}_\Delta(0, c)\tilde{\chi}_{\Delta'}(0, c + c')\tilde{\chi}_{\Delta''}(0, c + c' + c'') \\
&\leq \theta_{\Delta_1}(< c, u >)\theta_{\Delta'_1}(< c + c', v >)\theta_{\Delta''_1}(< c + c' + c'', w >)
\end{aligned} \quad (13.3.12)
$$

From the explicit expansions of c, c', c'' we conclude that (13.3.12) has the bound

$$
\theta_{\delta_1 \Delta_1}(t)\theta_{\delta'_1 \Delta'_1}(t')\theta_{\delta''_1 \Delta''_1}(t'') \quad (13.3.13)
$$

for some sufficiently large numbers $\delta_1, \delta'_1, \delta''_1$. On the other hand we also see from the explicit expansion of $|\det \left(\frac{\partial(x^a_{ee'e''})}{\partial(t,t',t'')} \right)|$ around t_0 that it is bounded by $M(|t|^k + |t'|^k + |t''|^k)$ where M is a positive number and where $k = \min(m + n', m + n'', m' + n, m' + n'', m'' + n, m'' + n') - 2 \geq 2$.

The prescription of how to guide the limit in case I(a) is then to synchronise $\Delta_1 = \Delta_1' = \Delta_1'' = \Delta$ and to take the limit $\Delta \to 0$ first. The integral is at least of order Δ^5 while we divide only by an order of Δ^3 so that the result vanishes.

In case I(b) we have an expansion of the form (let w.l.g. c, c' have co-linear tangents)

$$c(t) = au(t + o(t^2)) + bv(t^m + o(t^{m+1})) + cw(t^n + o(t^{n+1}))$$
$$c'(t) = au(t + o(t^2)) + b'v(t^{m'} + o(t^{m'+1})) + c'w(t^{n'} + o(t^{n'+1}))$$
$$c''(t) = a''v(t + o(t^2)) + b''u(t^{m''} + o(t^{m''+1})) + c''w(t^{n''} + o(t^{n''+1})) \quad (13.3.14)$$

We now argue as above and find that the product of the characteristic functions can be estimated by

$$\theta_{\delta_1 \Delta_1}(t) \theta_{\delta_1' \Delta_1'}(t') \theta_{\delta_2'' \Delta_2''}(t'')$$

while the determinant can be estimated as above just that k is now given by $k = \min(m, m', m'', n, n', n'') - 1 \geq 1$.

The prescription is now $\Delta_1 = \Delta_1' = \Delta_2'' =: \Delta \to 0$ first and we conclude that the integral is at least of order Δ^4 while we divide again only by Δ^3 such that the limit vanishes.

In case I (c) finally we have an expansion of the form

$$c(t) = au(t + o(t^2)) + bv(t^m + o(t^{m+1})) + cw(t^n + o(t^{n+1}))$$
$$c'(t) = av(t + o(t^2)) + b'v(t^{m'} + o(t^{m'+1})) + c'w(t^{n'} + o(t^{n'+1}))$$
$$c''(t) = a''u(t + o(t^2)) + b''v(t + o(t^2)) + c''w(t^{n''} + o(t^{n''+1})) \quad (13.3.15)$$

This time we estimate the product of the characteristic functions for instance by

$$\theta_{\delta_1 \Delta_1}(t) \theta_{\delta_2' \Delta_2'}(t') \theta_{\delta_2'' \Delta_2''}(t'')$$

while the determinant can be estimated as above and k is given by $k = \min(m, m', n, n', n'') - 1 \geq 1$ so that we have actually the same situation as in case I(b) upon synchronising this time $\Delta_1 = \Delta_2' = \Delta_2'' =: \Delta \to 0$.

As for case II we observe that the non-vanishing of the functional determinant at p implies that the map $x_{ee'e''}$ is actually invertible in a neighbourhood of p by the inverse function theorem. In other words, there is only one point (t_0, t_0', t_0'') such that $x_{ee'e''}(t_0, t_0', t_0'') = p$. Moreover, since the determinant is non-vanishing at p, all three edges must be distinct from each other. It follows now from our choice of edges that p must be a vertex $v = e \cap e' \cap e''$ of γ in order that the result is non-vanishing and thus from the choice of parametrisation $t_0 = t_0' = t_0'' = 0$.

Therefore, if we take the limit $\Delta_i'' \to 0$ first in any order then the condition $\chi_{\Delta''}(p, x_{ee'e''}) = 1$ will actually imply $\chi_\Delta(p, e) = \chi_{\Delta'}(2p, e + e') = 1$ for small enough Δ_i'' so that we can take these characteristic functions out of the integral and replace them by 1 if p is a common vertex of all three edges. Also we can replace the operator $\hat{O}_{ee'e''}(t, t', t'')$ by $\hat{O}_{ee'e''}(v)$. This holds only if the triple intersects in p.

If not all of e, e', e'' intersect in p then the limit will vanish anyway if we take a suitable limit of the Δ_i as we have shown before. We can account for that case by replacing $\chi_\Delta(p, e), \chi_{\Delta'}(2p, e + e')$ by $\chi_\Delta(p, v)\chi_{\Delta'}(p, v)$. Here v is the common vertex at which the distinct e, e', e'' must be incident otherwise they could not even pass through a small enough neighbourhood of p. We can also assume that all three edges have linearly independent tangents at v and expand still around $t = 0$. The remaining integral divided by $\Delta_1'' \Delta_2'' \Delta_3''$ then tends to

$$\int_{[0,1]^3} d^3t \det\left(\frac{\partial x_{ee'e''}}{\partial t}\right) \delta^{(3)}(p, x_{ee'e''}) = s(e, e', e'') \int_{C_{ee'e''}} d^3 x \delta^{(3)}(p, x)$$

$$= \frac{1}{8} s(e, e', e'') \tag{13.3.16}$$

where

$$s(e, e', e'')_v := \operatorname{sgn}(\det(\dot{e}(0), \dot{e}'(0), \dot{e}''(0))) \tag{13.3.17}$$

The factor $1/8$ is due to the fact that in the limit $\Delta'' \to 0$ we obtain an integral over $C(e, e', e'')$, the cone based at p and spanned by $\dot{e}(0), \dot{e}'(0), \dot{e}''(0)$ where the orientation is taken to be positive. This integral just equals $\int_{\mathbb{R}^3_+} d^3 t \delta(0, t) \delta(0, t') \delta(0, t'') = 1/8$ as one can easily check. This furnishes the proof. □

We conclude that (13.3.6) reduces to (in particular, the operators $\hat{O}_{12,3}\hat{O}_{1,23}\hat{O}_{2,31}\hat{O}_{123}$ drop out)

$$\lim_{\Delta'' \to 0} \hat{E}(p, \Delta, \Delta', \Delta'')f = \sum_{e,e',e''} \frac{i\ell_p^6 s(e, e', e'')_v}{8^3 \cdot 3! \operatorname{vol}(\Delta)\operatorname{vol}(\Delta')}$$

$$\times \chi_\Delta(p, v)\chi_{\Delta'}(p, v)\hat{O}_{e,e',e''}(0, 0, 0)$$

where v on the right-hand side is the intersection point of the triple of edges and it is understood that we only sum over such triples of edges which are incident at a common vertex. There is no factor of 3^3 missing because it cancels against a similar factor in $\operatorname{vol}(\Delta'')$. Moreover,

$$\hat{O}_{e,e',e''}(0, 0, 0) = \epsilon_{ijk} X_{e''}^i X_{e'}^j X_e^k \text{ and } X_e^i := X^i(h_e(0, 1))$$

$$:= \operatorname{tr}\left((\tau_i h_e(0, 1))^T \frac{\partial}{\partial h_e(0, 1)}\right) \tag{13.3.18}$$

is a right-invariant vector field in the τ_i direction of SU(2), that is, $X(hg) = X(h)$. We have also extended the values of the sign function to include 0, which takes care of the possibility that one has triples of edges with linearly dependent tangents.

The final step consists in choosing $\Delta = \Delta'$ and taking the square root of the modulus. We replace the sum over all triples incident at a common vertex $\sum_{e,e',e''}$ by a sum over all vertices followed by a sum over all triples incident at the same vertex $\sum_{v \in V(\gamma)} \sum_{e \cap e' \cap e'' = v}$. Now, for small enough Δ and given p, at most one vertex contributes, that is, at most one of $\chi_\Delta(v, p) \neq 0$ because all vertices have finite separation. Then we can take the relevant $\chi_\Delta(p, v) = \chi_\Delta(p, v)^2$ out of the square root and take the limit, which results in

$$\hat{V}(R)_\gamma = \int_R d^3p \sqrt{\widehat{\det(q)(p)}}_\gamma = \int_R d^3p \hat{V}(p)_\gamma$$

$$\hat{V}(p)_\gamma = \left(\frac{\ell_p}{2}\right)^3 \sum_{v \in V(\gamma)} \delta^{(3)}(p, v) \hat{V}_{v,\gamma}$$

$$\hat{V}_{v,\gamma} = \sqrt{\left| \frac{i}{3! \cdot 8} \sum_{e,e',e'' \in E(\gamma), e \cap e' \cap e'' = v} s(e, e', e'') q_{ee'e''} \right|}$$

$$q_{ee'e''} = \epsilon_{ijk} X_e^i X_{e'}^j X_{e''}^k, \tag{13.3.19}$$

where we could switch the order of the X's because a triple contributes only if the corresponding edges are distinct and so the X's commute.

Expression (13.3.19) is the final expression for the volume operator and coincides precisely[1] with the expression found in [427]. Note that the final expression is manifestly diffeomorphism-covariant. Although the procedure of adapting the limiting to a given triple of edges is somewhat non-standard there is an argument in favour of such a procedure: the discussion in Lemma 13.3.1 reveals that any other regularisation which would result in a finite contribution for the case where $s(e, e', e'')$ is zero would necessarily depend on the higher-order intersection characteristics of a triple of edges. However, since such a quantity is not diffeomorphism-covariant, which is unacceptable, the dependence must be trivial.

As we will see, there are both *kinematical* and *dynamical* reasons to prefer the operator of [427] over [425]. The kinematical reason is that one can show that [425] is inconsistent with the flux operator on which the volume operator is based [573,574]. We will discuss this in more detail in Section 13.5. The dynamical reason is that the Hamiltonian constraint or Master Constraint would not even be densely defined if one used [425] in place of [427]. This is due to the fact that the volume operator of [425] does not annihilate coplanar at least trivalent and non-gauge-invariant vertices. Therefore, following the regularisation of the Hamiltonian constraint of Section 10.4 one realises that the resulting operator would not only act at the vertices of the graph of a spin-network function but (in the limit of infinite refinement) at all interior points of all edges unless one excludes such contributions by hand. However, even if one did that, the resulting operator would no longer be free of anomalies.

---

[1] In order to avoid confusion, in [427] one uses $Y_E^j = X_e^j/2$ and $\kappa' = \kappa/2$ so that $(\ell_p')^2 = \ell_p^2/2$. In terms of these quantities there is no factor $1/8$ in $\ell_p^3/8$.

13.4 Properties of the volume operator

This section is subdivided into three parts. First we prove that the family of operators derived in (13.3.19) defines a linear unbounded operator on $\mathcal{H}$. Next we show that the operator is symmetric, positive semidefinite and admits self-adjoint extensions (actually it is essentially self-adjoint) and finally we show that its spectrum is discrete and that the operator so defined is anomaly-free.

13.4.1 Cylindrical consistency

What we have obtained in (13.3.19) is a family of operators $(\hat{V}(R)_\gamma, D_\gamma)_{\gamma\in\Gamma}$. That is not enough to show that this family of cylindrical projections 'comes from' a linear operator on $\mathcal{H}$. As for the area operator, for this to be the case we need to check that whenever $\gamma \subset \gamma'$ then

1. $p^*_{\gamma\gamma'} D_\gamma \subset D_{\gamma'}$ where $p_{\gamma\gamma'}$ is the restriction from γ' to γ. This condition makes sure that the operator defined on bigger graphs can be applied to functions defined on smaller graphs.
2. $(\hat{V}(R)_{\gamma'})|_\gamma = \hat{V}(R)_\gamma$, this is the condition of cylindrical consistency and says that the operator on bigger graphs equals the operator on smaller graphs when restricted to functions thereon.

A graph $\gamma \subset \gamma'$ can be obtained from a bigger graph γ' by a finite series of steps consisting of the following basic ones:

(i) remove an edge from γ';
(ii) join two edges e', e'', such that $e' \cap e''$ is a point of analyticity, to a new edge $e - e' \circ (e'')^{-1}$;
(iii) reverse the orientation of an edge.

Clearly, a dense domain for $\hat{V}(R)_\gamma$ is given by $D_\gamma := \mathrm{Cyl}^3_\gamma(\overline{\mathcal{A}/\mathcal{G}})$. This choice trivially satisfies requirement (1) since functions which just do not depend on some arguments or only on special combinations $h_e = h_{e'}h_{e''}, h_{e'} = h_e^{-1}$ are still thrice continuously differentiable if the original function was (here we have used the fact that $SU(2)$ is a Lie group, that is, group multiplication and taking inverses is an analytic map).

Next, let us check cylindrical consistency. Consider first the case (i) that γ does not depend on an edge e on which γ' does. Then clearly $X^i_e f_\gamma = 0$ for any function cylindrical with respect to γ and so in the sum over triples over vertices in (13.3.19) the terms involving e drop out.

Next consider the case (ii). If $e = e' \circ (e'')^{-1}$ is an edge of γ and e', e'' are edges of γ' where $v := e' \cap e''$ is a point of analyticity for γ while for γ' it is not, then while v is a vertex for γ' it is only a pseudo-vertex for γ and so in $\hat{V}(D)_\gamma$ there is no term corresponding to v. On the other hand, since the vertex v is a pseudo-vertex for γ it is in particular only two-valent and so the corresponding term in $\hat{V}(D)_{\gamma'}$ drops out. Likewise, if v is a vertex for γ at which the outgoing edge e is incident, then from right invariance of the vector field we have $X_e = X_{e' \circ (e'')^{-1}} = X_{e'}$ and so at vertices that belong to both γ and γ' the corresponding vertex operators coincide.

Finally, case (iii) is actually excluded by our unambiguous choice of orientation.

We conclude that there exists an operator $(\hat{V}(R), D)$ on $\mathcal{H}$ which is densely defined on $D = \mathrm{Cyl}^3(\overline{\mathcal{A}/\mathcal{G}})$.

13.4.2 Symmetry, positivity and self-adjointness

Notice that the vector field iX_e is symmetric on $\mathcal{H}_\gamma$, the completion of $\mathrm{Cyl}^1_\gamma(\overline{\mathcal{A}/\mathcal{G}})$ with respect to $\mu_{0,\gamma}$, e an edge of γ, because the Haar measure is right-invariant. It follows from the explicit expression (13.3.19) in terms of the iX_e that all the projections $\hat{V}(R)_\gamma$ are symmetric. In this special case (namely, the volume operator leaves the space D_γ-invariant) this is enough to show that $\hat{V}(R)$ is symmetric on D.

Furthermore, all $\hat{V}(R)_\gamma$ are positive semidefinite by inspection so that $\hat{V}(R), D$ is a densely defined, positive semidefinite and symmetric operator. It follows that it has self-adjoint extensions, for instance its Friedrich extension. That this extension is actually the unique one follows from essential self-adjointness, which can be shown by the same method as applied to flux and area operators and which we leave to the reader.

13.4.3 Discreteness and anomaly-freeness

The operator $\hat{V}(R)$ has the important property that it leaves the dense subset $\mathrm{Cyl}^\infty_\gamma(\overline{\mathcal{A}/\mathcal{G}}) \subset \mathcal{H}$ invariant, separately for each $\gamma \in \Gamma$. Spin-network functions $T_{\gamma, \vec{j}, \vec{I}}$ are particular smooth functions of that sort. Notice that given $\gamma, \vec{j}$ there are only a finite number of linearly independent $\vec{I}$ compatible with $\gamma, \vec{j}$. Now it is obvious that the operator $\hat{V}(R)$ leaves the finite-dimensional vector space $U_{\gamma, \vec{j}}$ spanned by spin-network states compatible with $\gamma, \vec{j}$ invariant. The matrix

$$(V(R)_{\gamma \vec{j}})_{\vec{I}, \vec{I'}} := <T_{\gamma, \vec{j}, \vec{I}}|\hat{V}(R)|T_{\gamma, \vec{j}, \vec{I'}}> \tag{13.4.1}$$

is therefore finite-dimensional, positive semidefinite and symmetric. The task of computing its eigenvalues therefore becomes a problem in linear algebra!

Next, since from (13.3.19)

$$\hat{V}(R)_\gamma = \ell_p^3 \sum_{v \in V(\gamma) \cap R} \hat{V}_{v, \gamma} \tag{13.4.2}$$

and since $\hat{V}_{v, \gamma}$ involves only those $e \in E(\gamma)$ with $v \in e$, we find that $\hat{V}_{v, \gamma}$ can only change the entry I_v in $\vec{I}$. In other words, $[\hat{V}_{v, \gamma}, \hat{V}_{v', \gamma}] = 0$ and each $\hat{V}_{v, \gamma}$ can be diagonalised separately.

Finally, since the spins j_e only take discrete values it follows that $\mathcal{H}_\gamma$ has a countable basis and the spectrum that $\hat{V}(R)$ attains on D_γ is therefore pure point. Let us check whether this is the complete spectrum. Assume it were not and let $\hat{P}$ be the spectral projection on the rest of the spectrum (the existence of the spectral projections relies on the fact that $\hat{V}(R)$ is self-adjoint and not only symmetric). It follows that $u = \hat{P}v$ is orthogonal to D_γ where v is any vector in $\mathcal{H}_\gamma$. But D_γ is dense in $\mathcal{H}_\gamma$ and so we find for every $\epsilon > 0$ a $\phi \in D_\gamma$ with $||u - \phi|| < \epsilon$. Now we have from orthogonality $\epsilon^2 > ||u - \phi||^2 = ||u||^2 + ||\phi||^2 > ||u||^2$ and so $u = 0$. This shows that the complete spectrum is already attained on D_γ. It is purely discrete as well in the physical sense that it is attained on a countable basis so that the eigenvalues only comprise a countable set. In a mathematical sense one would need to check that there are no accumulation points and no eigenvalues of infinite multiplicity for a given graph. This is one possible future application of the explicit matrix element formulae which we derive in the next subsection.

Last, we wish to show that the volume operators are anomaly-free (given the fact that we have largely adapted our regularisation to a graph, this statement is far from trivial). By this we mean the following: given any two open sets $R_1, R_2 \subset \Sigma$ we have vanishing Poisson brackets $\{V(R_1), V(R_2)\} = 0$ because the functionals $V(R)$ depend on the momentum variable $E_i^a(x)$ only. Now, given a function f cylindrical with respect to a graph γ, it is not at all obvious any more that $[\hat{V}(R_1), \hat{V}(R_2)]f = 0$ for any such f. Fortunately, given the above characterisation of the spectrum, the commutator can easily be proved to vanish on cylindrical functions. To see this, note that the above results imply that if we choose any region $R(\gamma)$ such that $\gamma \subset R(\gamma)$ then there exists an eigenbasis of $\mathcal{H}_\gamma$ of $\hat{V}(R(\gamma))$. Now consider any region R. Since all regions are open by construction, all regions fall into equivalence classes with respect to γ: R, R' are equivalent if they contain the same vertices of γ (any vertex either has a neighbourhood which lies completely inside R or it lies outside). Therefore any two $\hat{V}(R), \hat{V}(R')$ differ at most by some of the $\hat{V}_{v,\gamma}$, all of which are contained in the expression for $\hat{V}(R(\gamma))$. Since the $\hat{V}_{v,\gamma}$ commute, the eigenbasis of $\hat{V}(R(\gamma))$ is a simultaneous eigenbasis of all $\hat{V}_{v,\gamma}$ for all $v \in V(\gamma)$ and so this eigenbasis is a simultaneous eigenbasis of all $\hat{V}(R)_\gamma$. Since all $\mathcal{H}_\gamma$ are orthogonal, we have a simultaneous eigenbasis for all $\hat{V}(R)$.

While it is in general not enough to verify that two self-adjoint, unbounded operators commute on a dense domain (rather, by definition, we have to check that the associated spectral projections commute) in our case we are done because the spectral projections *are* the projections on the various D_γ because the point spectrum is already the complete spectrum. Thus we have verified that the commutator algebra mirrors the classical Poisson algebra.

13.4.4 Matrix elements

In contrast to the area operator, the volume operator cannot be diagonalised in closed form. The reason for this is that the operator $Q_{v,\gamma}$, related to the volume operator by $Vf_\gamma = \sum_{v \in V(\gamma)} \sqrt{|Q_{v,\gamma}|}f_\gamma$ where f_γ is a function cylindrical over γ and $v \in V(\gamma)$, is a homogeneous polynomial of third order in the n_v right-invariant vector fields X_e^j where n_v is the valence of the vertex. While we can easily calculate matrix elements of $Q_{v,\gamma}$ in the spin-network basis using the quantum mechanics of n_v angular momentum operators by the technique displayed in Chapter 32 and which results in a finite-dimensional, antisymmetric and Hermitian matrix for each fixed choice of the spins j_e (because $Q_{v,\gamma}$ leaves the $\vec{j}$ invariant, it just changes the intertwiners $\vec{I}$ of the spin-network functions $T_{\gamma,\vec{j},\vec{I}}$), that matrix has no obvious special symmetries and hence its eigenvalues, for generic configurations of the j_e, cannot be calculated analytically any more by quadratures beyond rank nine. Hence, what needs to be done in order to compute matrix elements of the volume operator is to develop approximation methods which relate the matrix elements of V to the analytically available matrix elements of $Q^2 = V^4$. One such method is to use coherent states which we have discussed in Chapter 11. Essentially, coherent states are diagonal, within the limits of the Heisenberg uncertainty obstruction, for all operators, hence to zeroth order in $\hbar$ the expectation value of the volume operator can be replaced by its classical value at that point in phase space at which the coherent state is peaked. In order to compute the higher-order corrections we consider the Taylor expansion around the coherent state expectation value $<Q>$

$$V = \sqrt{|Q|} = \sqrt[4]{(<Q> + [Q - <Q>])^2}$$

$$= \sqrt{|<Q>|}\left\{1 + \frac{1}{4}\left[\left(\frac{Q - <Q>}{<Q>}\right)^2 - 2\frac{Q - <Q>}{<Q>} - \frac{3}{8}\left(\frac{Q - <Q>}{<Q>}\right)^2\right.\right.$$

$$\left.\left. + O\left(\left(\frac{Q - <Q>}{<Q>}\right)^3\right)\right]\right\}$$

Since the operator Q is unbounded while the radius of convergence of the Taylor expansion is bounded, the validity of this expansion must be established by independent means which is possible[2] by using properties of coherent states and semiclassical perturbation theory developed in [591]. The expectation values can then be computed with sufficient accuracy in $\hbar$ because expectation values of powers of Q can be computed analytically.

We see that we are left with computing matrix elements of Q with respect to spin-network states (coherent states are coherent superpositions of those). Furthermore, the matrix elements of Q are linear combinations of the matrix

---

[2] Basically, given a self-adjoint operator A and a function $f : \mathbb{R} \to \mathbb{R}$ one finds polynomial functions $f_\pm$ such that $f_- \leq f \leq f_+$. Then by positivity and the spectral theorem $<\hat{f}> \in [<\hat{f}_->, <\hat{f}_+>]$, where, for example, $\hat{f} = f(A)$, for the expectation values with respect to any states. For coherent states the range of the interval is indeed given by the fluctuations (to first order in $\hbar$) of the right-hand side of (13.4.3). All $\hbar$ corrections can be computed by this method using polynomials of sufficiently high degree.

elements of the operators (see Chapter 32 for the notation)

$$Q_{IJK} = -i\epsilon_{ijk}X_I^i X_J^j X_K^k = 8\epsilon_{ijk}Y_I^i Y_J^j Y_K^k = -8iY_I^i Y_J^j [Y_K^i, Y_K^j]$$
$$= -8i[Y_I^i Y_K^i, Y_J^j Y_K^j]$$

where $I < J < K$. The idea of computing the matrix elements of Q_{IJK} is to expand the spin-network states defined in Chapter 32 which are written in terms of the *standard recoupling scheme* $(j_{1...k-1}, j_k) \to j_{1...k}$, $k = 2, \ldots, n_v$ in terms of another basis of spin-network states which are adapted to the two operators $Y_I^i Y_K^i, Y_J^i Y_K^i$. Namely we define for $I < J$ the (I, J) recoupling scheme by $(j_I, j_J) \to j_{IJ}$, $(j_{IJ1...k-1}, j_k) \to j_{IJ1...k}$ for $k = 1, \ldots, I - 1$, $(j_{IJ1...I-1I+1...l-1}, j_l) \to j_{IJ1...I-1I+1...l}$ for $l = I + 1, \ldots, J - 1$, $(j_{IJ1...I-1I+1...J-1J+1...m-1}, j_m) \to j_{IJ1...I-1I+1...J-1J+1...m}$ for $m = J + 1$, $\ldots, n_v$. The purpose of doing this is of course that the operators $(Y_I^i + Y_J^i)^2$, $(Y_I^i + Y_J^i + Y_1^i + \cdots + Y_k^i)^2$, $(Y_I^i + Y_J^i + Y_1^i + \cdots + Y_{I-1}^i + Y_{I+1}^i + \cdots + Y_l^i)^2$, $(Y_I^i + Y_J^i + Y_1^i + \cdots + Y_{I-1}^i + Y_{I+1}^i + \cdots + Y_{J-1}^i + Y_{J+1}^i + \cdots + Y_m^i)^2$ respectively are diagonal in this basis with eigenvalues given by the recoupling angular momenta $j_*(j_* + 1)$. In particular, $(Y_{IJ}^i)^2, Y_{IJ}^i = Y_I^i + Y_J^i$ has eigenvalues $j_{IJ}(j_{IJ} + 1)$. Hence we compute

$$< (1, 2)|Q_{IJK}|(1, 2)' >$$
$$- \sum_{(I,K),(J,K)} [< (1, 2)|(I, K) > < (I, K)|(Y_{IK}^i)^2 (Y_{JK}^i)^2|(J, K) > < (J, K)|(1, 2)' >$$
$$- < (1, 2)|(J, K) > < (J, K)|(Y_{JK}^i)^2 (Y_{IK}^i)^2|(I, K) > < (I, K)|(1, 2)' >]$$
$$= \sum_{(I,K),(J,K)} j_{IK}(j_{IK} + 1) \; j_{JK}(j_{JK} + 1) < (I, K)|(J, K) >$$
$$\times [< (1, 2)|(I, K) > < (J, K)|(1, 2)' > - < (1, 2)|(J, K) > < (I, K)|(1, 2)' >]$$

where we are summing over all intermediate states of the adapted recoupling scheme. Here we have exploited that the coefficients $< (I, J)|(K, L) >$ are real-valued so that $< (I, J)|(K, L) > = < (K, L)|(I, J) >$. This follows from the fact that up to the unitary transformation W of Chapter 32 the coefficients $< (I, J)|(K, L) >$ are polynomials of Clebsch–Gordan coefficients, more precisely they are known as $3(n - 1) - j$ symbols for n degrees of freedom (n-valent vertex).

Most of the work in computing $< (1, 2)|Q_{IJK}|(1, 2)' >$ is devoted to computing $< (1, 2)|(I, J) >$ for which a closed but tedious expression was derived in [559]. That expression is a complicated polynomial of $6j$ symbols which are the coefficients of the unitary matrix, which for fixed j_1, j_2, j_3 mediates between the recoupling schemes $(j_1, j_2) \to j_{12}, (j_{12}, j_3) \to j_{123}$ and $(j_1, j_3) \to j_{13}, (j_{13}, j_2) \to j_{123}$. For the $6j$ symbols themselves a closed expression is available, the so-called Racah formula. However, that formula is again a complicated sum of fractions of large factorials which therefore even for numerical evaluations quickly becomes a challenge even for moderately large values of j_1, j_2, j_3. A tremendous simplification was achieved in [665] where by means of the Elliot–Biedenharn identity among $6j$ symbols the polynomials of $6j$ symbols could be eliminated. The end result is the quite simple expression which holds for $I > 1$, $J > I+1$ (the remaining cases require a tedious case-by-case analysis and can be found in [665]):

$$
< \vec{a} | Q_{IJK} | \vec{a}' >
$$

$$
= \frac{1}{4} (-1)^{j_K + j_I + a_{I-1} + a_K} (-1)^{a_I - a'_I} (-1)^{\sum_{n=I+1}^{J-1} j_n} (-1)^{-\sum_{p=J+1}^{K-1} j_p} X(j_I, j_J)^{\frac{1}{2}} X(j_J, j_K)^{\frac{1}{2}}
$$

$$
\times \sqrt{(2a_I + 1)(2a'_I + 1)} \sqrt{(2a_J + 1)(2a'_J + 1)}
$$

$$
\times \left\{ \begin{matrix} a_{I-1} & j_I & a_I \\ 1 & a'_I & j_I \end{matrix} \right\} \left[\prod_{n=I+1}^{J-1} \sqrt{(2a'_n + 1)(2a_n + 1)} (-1)^{a'_{n-1} + a_{n-1} + 1} \left\{ \begin{matrix} j_n & a'_{n-1} & a'_n \\ 1 & a_n & a_{n-1} \end{matrix} \right\} \right]
$$

$$
\times \left[\prod_{n=J+1}^{K-1} \sqrt{(2a'_n + 1)(2a_n + 1)} (-1)^{a'_{n-1} + a_{n-1} + 1} \left\{ \begin{matrix} j_n & a'_{n-1} & a'_n \\ 1 & a_n & a_{n-1} \end{matrix} \right\} \right] \left\{ \begin{matrix} a_K & j_K & a_{K-1} \\ 1 & a'_{K-1} & j_K \end{matrix} \right\}
$$

$$
\times \left[(-1)^{a'_J + a'_{J-1}} \left\{ \begin{matrix} a_J & j_J & a'_{J-1} \\ 1 & a_{J-1} & j_J \end{matrix} \right\} \left\{ \begin{matrix} a'_{J-1} & j_J & a'_J \\ 1 & a_J & j_J \end{matrix} \right\} \right.
$$

$$
\left. \times -(-1)^{a_J + a_{J-1}} \left\{ \begin{matrix} a'_J & j_J & a'_{J-1} \\ 1 & a_{J-1} & j_J \end{matrix} \right\} \left\{ \begin{matrix} a_{J-1} & j_J & a'_J \\ 1 & a_J & j_J \end{matrix} \right\} \right]
$$

$$
\times \prod_{n=2}^{I-1} \delta_{a_n a'_n} \prod_{n=K}^{N} \delta_{a_n a'_n} \tag{13.4.3}
$$

with $X(j_1, j_2) = 2j_1(2j_1 + 1)(2j_1 + 2)2j_2(2j_2 + 1)(2j_2 + 2)$ and we have abbreviated $a_k := j_{1...k}$. The result is written directly in the abstract angular momentum Hilbert space (the image of the map W displayed in Chapter 32) and we used $Q_{IJK} := [J_{IJ}^2, J_{JK}^2]$. Notice that all still appearing $6j$ symbols are just abbreviations for the following simple expressions in which no summations or products (factorials) need to be carried out any longer, for example (using $s = a+b+c$):

$$
\left\{ \begin{matrix} a & b & c \\ 1 & c & b \end{matrix} \right\} = (-1)^{s+1} \frac{2[b(b+1)c(c+1) - a(a+1)]}{[2b(2b+1)(2b+2)2c(2c+1)(2c+2)]^{\frac{1}{2}}} \tag{13.4.4}
$$

$$
\left\{ \begin{matrix} a & b & c \\ 1 & c-1 & b \end{matrix} \right\} = (-1)^s \left[\frac{2(s+1)(s-2a)(s-2b)(s-2c+1)}{2b(2b+1)(2b+2)(2c-1)2c(2c+1)} \right]^{\frac{1}{2}} \tag{13.4.5}
$$

$$
\left\{ \begin{matrix} a & b & c \\ 1 & c-1 & b-1 \end{matrix} \right\} = (-1)^s \left[\frac{s(s+1)(s-2a-1)(s-2a)}{(2b-1)2b(2b+1)(2c-1)2c(2c+1)} \right]^{\frac{1}{2}} \tag{13.4.6}
$$

$$
\left\{ \begin{matrix} a & b & c \\ 1 & c-1 & b+1 \end{matrix} \right\} = (-1)^s \left[\frac{(s-2b-1)(s-2b)(s-2c+1)(s-2c+2)}{(2b+1)(2b+2)(2b+3)(2c-1)2c(2c+1)} \right]^{\frac{1}{2}} \tag{13.4.7}
$$

We will not derive the final formula (13.4.3) here, the detailed proof can be found in [665]. For the case of a gauge-invariant four-vertex this result had been derived previously in [669] by graphical techniques. For this case we have $j_{123} = j_{12}$ in order that $J = j_{1234} = 0$ is possible so that the intertwiners are parametrised by j_{12} only. Furthermore, also due to gauge invariance we have $X_4^j = -(X_1^j + X_2^j + X_3^j)$ so that all Q_{IJK}, $I < J < K$ coincide with $\pm Q_{123}$ on gauge-invariant states. The non-vanishing matrix elements are then

$$
< j_{12}|Q_{123}|j_{12} - 1 >
$$

$$
= \frac{1}{\sqrt{(2j_{12} - 1)(2j_{12} + 1)}}[(j_1 + j_2 + j_{12} + 1)(-j_1 + j_2 + j_{12})(j_1 - j_2 + j_{12})
$$

$$
\times (j_1 + j_2 - j_{12} + 1)(j_3 + j_4 + j_{12} + 1)(-j_3 + j_4 + j_{12})(j_3 - j_4 + j_{12})
$$

$$
\times (j_3 + j_4 - j_{12} + 1)]^{\frac{1}{2}}
$$

$$
= - < j_{12} - 1|\hat{q}_{123}|j_{12} > \tag{13.4.8}
$$

Formula (13.4.3) holds for arbitrary valence and also for non-gauge-invariant states which is important in applications, for instance the Hamiltonian constraint or the length operator where non-gauge-invariant states appear in intermediate steps of the calculation since one writes triad operators, which are themselves not gauge-invariant but out of which gauge-invariant operators are composed, as commutators between non-gauge-invariant holonomies and the volume operator.

13.5 Uniqueness of the volume operator, consistency with the flux operator and pseudo-two-forms

The regularisation of the volume operator displayed in the previous section is quite involved and it is far from manifest that a different regularisation would have resulted in the same expression. Indeed, as we have said already, there exists an alternative regularisation due to Rovelli and Smolin [425] which does result in a qualitatively different operator while the operator derived here by a point-splitting regularisation agrees with the one derived by Ashtekar and Lewandowski by yet another (averaging) technique [427]. In terms of the operators $Q_{IJK} = \epsilon_{ijk} X_I^i X_J^j X_K^k$, defined for a given vertex v of a given graph γ, the difference between these two operators is roughly as follows

$$
8V_{\mathrm{RS},\gamma,v}/\ell_P^3 = c_{\mathrm{RS}} \sum_{I<J<K} \sqrt{|Q_{IJK}|}
$$

$$
8V_{\mathrm{AL},\gamma,v}/\ell_P^3 = c_{\mathrm{AL}} \sqrt{\left| \sum_{I<J<K} \sigma_{IJK}\, Q_{IJK} \right|}
$$

Apart from the overall factor $c_{\mathrm{RS}}, c_{\mathrm{AL}}$ which could depend on the details of the regularisation, we see two differences. First, the sum over ordered triples is outside the square root for RS and inside for AL. Second, the orientation factor $\sigma_{IJK} = \mathrm{sgn}(\det(\dot{e}_I(0)), \det(\dot{e}_J(0)), \det(\dot{e}_K(0)))$, $e_I(0) = v$ is absent for the

RS operator. Hence, the operator due to AL vanishes identically on vertices all of whose adjacent edges have co-planar tangents while the volume operator due to RS does not. In particular, the AL operator therefore depends on the differential structure of σ while the RS operator does not. One may like the latter property because when using diffeomorphisms which are only piecewise analytic homeomorphisms then the 'diffeomorphism-invariant' Hilbert space would be separable [557,558]. Notice, however, that homeomorphisms are not symmetries of the classical action and that the semianalytic (i.e., at least $C^{(1)}$) structure is indispensable in order to derive the uniqueness result on the kinematical representation of Loop Quantum Gravity. Also, piecewise analytic homeomorphisms are not known to form a group, which would be desirable in order to solve the diffeomorphism constraint by group averaging techniques. Furthermore, the quantum dynamics acts by adding or removing trivalent vertices which do not contain diffeomorphism-invariant information (θ moduli) and hence preserves each θ sector. Hence, if the Dirac observables share the same property, then all θ sectors are superselected and thus contain the same physical information. In other words, by just choosing one of these sectors we also arrive at a separable diffeomorphism-invariant Hilbert space even when only using semianalytic diffeomorphisms.

It is easy to see that both $V_{\mathrm{RS}}, V_{\mathrm{AL}}$ are cylindrically consistent and diffeomorphism-covariant and therefore a priori seem to be equally valid quantum volume operators. In [573,574] a consistency check on Loop Quantum Gravity was performed which could discriminate between $V_{\mathrm{RS}}, V_{\mathrm{AL}}$ and fix the regularisation constant. The idea is quite simple: as we have shown, the volume operator is derived from the known quantisation of the flux operator. Now we may in turn write the classical flux in terms of triads by using

$$E_j^a = \sqrt{\det(q)}e_j^a = \frac{1}{2}\mathrm{sgn}(\det(e))\epsilon^{abc}\epsilon_{jkl}e_b^k e_c^l$$

and the Poisson bracket identity $\{V(R), A_a^j(x)\} \propto e_a^j(x)$ for $x \in R$. At this point it is worthwhile mentioning that there is a classical canonical transformation $E \mapsto E' = \mathbf{S}E$, $A = \Gamma + \beta K \mapsto A' = \Gamma + \beta \mathbf{S}K$ where $\mathbf{S} = \mathrm{sgn}(\det(e)) = \mathrm{sgn}(\det(E))$ (this is a canonical transformation because $\mathbf{S} = \pm 1 = \mathrm{const.}$ classically). However, if one worked with E', A' rather than E, A then $\det(E') = \det(e)^2 \geq 0$ cannot take both signs. Thus, E' could not be represented as a self-adjoint (functional) derivative operator for the same reason that id/dx does not represent a classical momentum p on $\mathbb{R}$ subject to the anholonomic constraint $p \geq 0$ (this would be a contangent bundle over $\mathbb{R}^+$ as has been pointed out, e.g., in [670] and must be quantised differently). As a side result, the check performed in [573,574] could rule out that the Hilbert space representation of Loop Quantum Gravity is based on the classical variables E', A', hence E_j^a must be a true vector density rather than a pseudo-vector density (equivalently, $(*E^j)_{ab} = \epsilon_{abc}E_j^c$ is a pseudo-two-form rather than a two-form).

In order to quantise

$$\int_S *E \propto \int_S \mathbf{S}\ \{A, V\} \wedge \{A, V\}$$

written in terms of Poisson brackets with the volume functional, one partitions the surface S into plaquettes, replaces connections by holonomies along the boundaries of the plaquettes, the volume functional by the volume operator and Poisson brackets by commutators divided by $i\hbar$. Finally, one has to take the continuum limit of refining the partition. Notice that due to the appearance of the signum function $\mathbf{S}$ in that expression one had to quantise $\mathbf{S}$ as well in [573, 574].

The result is very simple: $\hat{\mathbf{S}}(x) = \mathrm{sgn}(\hat{Q}(x))$ where $\hat{V}_{\mathrm{AL}}(x) = \sqrt{|\hat{Q}(x)|}$. As displayed, it turns out that in order to define $\hat{\mathbf{S}}$ one necessarily has to use the regularisation of the volume operator due to Ashtekar and Lewandowski. Hence, already at this stage it would be strange to use the volume operator due to Rovelli and Smolin within the commutators because both $V, \mathbf{S}$ come from the same classical quantity $\det(E)$ and hence both should be regularised in the same fashion. However, even when doing this artificially, it turns out that the RS volume is inconsistent with the flux operator while the AL operator is fully consistent if and only if $c_{\mathrm{AL}} = 1/48$. This is even the case when taking factor ordering ambiguities into account. Notice that the flux is not a gauge-invariant operator and therefore in this calculation for the first time the non-trivial map W derived in Chapter 32 between the spin-network and abstract angular momentum Hilbert space was discovered. Interestingly, the very first, pioneering paper on the volume operator [660] contains a regularisation leading to a heuristic formula which, when made rigorous, would lead to [427] rather than [425].

One should not view this consistency check as a criticism of [425]. Rather, the fact that within Loop Quantum Gravity one can discriminate between the equally reasonable candidates by using mathematical consistency arguments is a strength of the theory.

13.6 Spatially diffeomorphism-invariant volume operator

We now sketch how to make the geometrical operators at least a weak observable with respect to spatial diffeomorphisms. This is easiest for the volume functional.

Let R be a coordinate region, that is, a D-dimensional submanifold of σ, then the volume functional is defined by

$$\mathrm{Vol}[R] := \int_R d^D x \sqrt{\det(q)} = \int_\sigma d^D x \chi_R \sqrt{\det(q)} \qquad (13.6.1)$$

where χ_R denotes the characteristic function of the set R. Suppose now that we couple gravity to matter (which is possible, see Chapter 12) and that ρ is a positive definite scalar density of any weight of the matter (and gravitational) degrees of freedom. Here by positive definite we mean that $\rho(x) = 0$ if and only if the matter field vanishes at x. For instance, if we have an electromagnetic field we could use the electromagnetic field energy density

$$\rho = \frac{q_{ab}}{2\sqrt{\det(q)}}[E^a E^b + B^a B^b]$$

Consider now the intrinsically defined region

$$R_\rho := \{x \in \sigma; \ \rho(x) > 0\} \tag{13.6.2}$$

Then

$$\text{Vol}[R_\rho] = \int_\sigma d^D x \tilde{\theta}(\rho)\sqrt{\det(q)} \tag{13.6.3}$$

where $\tilde{\theta}$ is the modified step function with $\tilde{\theta}(x) = 1$ if $x > 0$ and $\tilde{\theta}(x) = 0$ otherwise. We claim that (13.6.3) is in fact diffeomorphism-invariant. To see this, it is sufficient to show that $F_\rho(x) := \tilde{\theta}(\rho(x))$ is a scalar of density weight zero. Let ρ be of density weight n, then under a diffeomorphism

$$F_\rho(x) \mapsto \tilde{\theta}(|\det(\partial\varphi(x)/\partial x)|^n \rho(\varphi(x))) = \tilde{\theta}(\rho(\varphi(x))) = (\varphi^* F_\rho)(x)$$

since $\tilde{\theta}(cx) = \tilde{\theta}(x)$ for any $c > 0$.

The use of matter is not really essential, we could also have used a gravitational degree of freedom, say $\rho = \sqrt{\det(q)}R^2$ where R is the curvature scalar. The point is now that for scalar densities we can actually define $\hat{\rho}$ as an operator-valued distribution (see Chapter 10) *if and only if ρ has density weight one*. Let $\mathcal{U}$ be a partion of σ. If it is fine enough and $\rho(x) > 0$ then also $\rho[U] := \int_U d^D x \rho(x) > 0$ for $x \in U \in \mathcal{U}$, therefore (13.6.3) is approximated by

$$\text{Vol}_\mathcal{U}[R_\rho] = \sum_{U \in \mathcal{U}} \tilde{\theta}(\rho[U])\text{Vol}[U] \tag{13.6.4}$$

Now $\rho[U]$ can be turned into a densely defined positive definite operator and thus $\tilde{\theta}(\hat{\rho}[U])$ can be defined by the spectral theorem. Moreover, since $\tilde{\theta}(x)^2 = \tilde{\theta}(x)$ we can order (13.6.4) symmetrically and define

$$\widehat{\text{Vol}}_\mathcal{U}[\rho] = \sum_{U \in \mathcal{U}} \tilde{\theta}(\rho[\hat{U}])\widehat{\text{Vol}}[U]\tilde{\theta}(\rho[\hat{U}]) \tag{13.6.5}$$

One now has to refine the partition and show that the final operator $\widehat{\text{Vol}}[\rho]$, if it exists, is consistently defined. Since the spectrum of $\tilde{\theta}(\rho[\hat{U}])$ is given by $\{0, 1\}$, the spectra of that final operator and the coordinate volume operator should coincide and in that sense the discreteness of the spectrum is carried over to the diffeomorphism-invariant context. Of course there remain technical issues, for instance $\widehat{\text{Vol}}[U], \tilde{\theta}(\rho[\hat{U}])$ do not commute and cannot be diagonalised simultaneously, the existence of the limit is unclear, etc. The details will appear elsewhere [227].

What this sketch shows are three points:

1. Kinematical operators have a chance of becoming full Dirac observables by defining their coordinate regions invariantly through matter (for invariance under the Hamiltonian evolution, this requires them to be smeared over time intervals as well, see Section 2.2). Actually, this is physically the way that one defines regions!

2. The discreteness of the spectrum then has a chance of being an invariant property of the physical observables (depending on the choice of clock variable that one chooses with respect to the Hamiltonian constraint [227]).

3. If discreteness holds true also for the complete (Dirac) observables, then something amazing has happened: we started out with a semianalytic manifold σ and smooth area functions. Yet, their spectra are entirely discrete, hinting at a discrete Planck scale physics, quantum geometry is distributional rather than smooth. Hopefully, the semianalytic structure that we needed at the classical level everywhere can be lifted to a purely combinatorial structure in the final picture of the theory, as happened for $2 + 1$ gravity, see [355]. See also the Algebraic Quantum Gravity programme of Section 10.6.5 and [589].

14

Spin foam models

Spin foam models are an attempt at a fully covariant formulation of Loop Quantum Gravity. The subject took off when the Hamiltonian constraint of Chapter 10 was developed and one tried to use it in order to define a path integral formulation of its 'transition amplitudes'. The field has grown quite a bit since its incarnation and it almost deserves a book of its own. We will devote relatively little space to it because we focus on the most important aspect, namely its relation with the canonical formalism and the interpretation of spin foam models. For an introduction to spin foam models we recommend the really beautiful articles by Baez [671, 672] which contain an almost complete and up-to-date guide to the literature and the historical development of the subject. See also the articles by Barrett [673, 674] for the closely related subject of state sum models and the most updated review article by Perez [675] and the thesis by Oriti [676].

What follows is a structural overview of spin foam models which focuses on mediating the main ideas and the open problems in constructing spin foam models.

14.1 Heuristic motivation from the canonical framework

The prototype of spin foam models are state sum models that had been studied extensively [677–681] within the context of topological quantum field theories [682–691] long before spin foam models arose within quantum gravity. The concrete connection of state sum models with canonical quantum gravity was made by Reisenberger and Rovelli in their seminal paper [453], where they used the (Euclidean version of the) Hamiltonian constraint described in Chapter 10 in order to write down a path integral formulation of the theory.

A heuristic method of solving the Hamiltonian constraint is to take any kinematical state ψ and map it to $\delta(\hat{H}^\dagger)\psi$ where $\delta(\hat{H}^\dagger) = \prod_{x\in\sigma} \delta(\hat{H}^\dagger(x))$. Here one hopes to define $\delta(\hat{H}^\dagger(x)) := \int_{\mathbb{R}} dt \exp(it\hat{H}^\dagger(x))/(2\pi)$ formally by the functional calculus, see Chapter 29. It is clear that this proposal is not only mathematically formal due to the infinite product of δ-distributions but strictly speaking also ill-defined: the $\hat{H}^\dagger(x)$ are operator-valued distributions rather than operators and even when smearing them with test functions they are not normal, that is, they do not commute with their adjoint so that the spectral theorem cannot be used in order to define the exponential.

Notice also that we really must use $\psi \in \mathcal{H}_{\text{kin}}$ rather than $\mathcal{H}_{\text{diff}}$ because the Hamiltonian constraint does not preserve $\mathcal{H}_{\text{diff}}$. However, then not even formally does the infinite product define a projector because on $\mathcal{H}_{\text{kin}}$ the $\delta(\hat{H}^\dagger(x))$ do not commute, hence $\hat{H}^\dagger(y) \prod_x \delta(\hat{H}^\dagger(x)) \neq 0$. If we introduce some kind of lexicographic ordering among the points x then formally we have

$$\left[\hat{H}^\dagger(y), \prod_x \delta(\hat{H}^\dagger(x)) \right] = \sum_{z<y} \left[\prod_{x<z} \delta(\hat{H}^\dagger(x)) \right] [\hat{H}^\dagger(y), \delta(\hat{H}^\dagger(z))] \left[\prod_{z<x} \delta(\hat{H}^\dagger(x)) \right]$$

$$(14.1.1)$$

and one may hope that the infinite sum of commutators is proportional to a diffeomorphism constraint. However, even if that were the case, the diffeomorphism constraint again does not commute with the Hamiltonian constraints but rather the commutator is another Hamiltonian constraint so that (14.1.1) is not even annihilated by diffeomorphism-invariant states.

Proceeding formally anyway, we may use a path integral formulation of the δ-distribution. Neglecting an (infinite) constant as usual we obtain the functional integral

$$\delta(\hat{H}') = \int [dN] e^{i \int_\sigma d^3x N(x) \hat{H}'(x)} \qquad (14.1.2)$$

Interestingly, (14.1.2) is formally spatially diffeomorphism-invariant because under a diffeomorphism $\hat{H}'(N) \mapsto \hat{H}'(N \circ \varphi^{-1})$ and since φ is just a bijection on the set of points σ we have $[dN] = \prod_x dN(x) = \prod_x dN(\varphi^{-1}(x)) = \prod_{\varphi(x)} dN(x)$. Therefore (14.1.2) may be applied to spatially diffeomorphism-invariant states and this is why we have used the dual operator. However, even when neglecting the mathematical issues mentioned, the point raised above remains: (14.1.2) applied to a spatially diffeomorphism-invariant state is not annihilated by the $\hat{H}'(N)$ because, while $[\hat{H}'(N), \hat{H}'(N')]\psi = 0$ (as an element of $\mathcal{D}^*$) for $\psi \in \mathcal{H}_{\text{diff}}$ it is not true that $[[\hat{H}'(N), \hat{H}'(N')], \hat{H}'(N'')]\psi = 0$. To see this we simply compute for $f \in \mathcal{D}_{\text{kin}}$

$$([[\hat{H}'(N), \hat{H}'(N')], \hat{H}'(N'')]\psi)[f] = (\hat{H}'(N'')\psi)[[\hat{H}^\dagger(N'), \hat{H}^\dagger(N)]f] \quad (14.1.3)$$

Since $\hat{H}'(N'')\psi \notin \mathcal{D}^*_{\text{diff}}$ the result does not vanish. Unfortunately, such multiple commutators appear in the calculations of the kind (14.1.1).

Proceeding formally anyway, (14.1.2) looks like a group averaging operation and we may try to define a physical inner product between physical states $\psi_{\text{phys}} := \delta(\hat{H}')\psi$ as

$$< \psi_{\text{phys}}, \psi'_{\text{phys}} >_{\text{phys}} := < \psi, \delta(\hat{H}')\psi' >_{\text{diff}} = \int_{\mathcal{N}} [dN] < \psi, e^{i \int_\sigma d^3x N(x) \hat{H}'(x)} \psi' >_{\text{diff}}$$

$$(14.1.4)$$

where $\mathcal{N}$ is the set of all lapse functions on σ and the inner product in the last line is only formally defined because the exponential at fixed N is not spatially diffeomorphism-invariant.

In order to get time-dependent lapse functions $\bar{N}(x,t)$ consider the set of lapse functions $\overline{\mathcal{N}}_N$ on M with $\int_{-T}^{T} dt \bar{N}(x,t) = N(x)$ for some $T > 0$. Let also $\overline{\mathcal{N}}$ be the set of lapse functions over M. Then

$$\int_{\overline{\mathcal{N}}} [d\bar{N}] < \psi, e^{i \int_M d^4 x \bar{N}(x,t) \hat{H}'(x)} \psi' >_{\text{diff}}$$

$$= \lim_{T \to \infty} \int_{\overline{\mathcal{N}}} [d\bar{N}] < \psi, e^{i \int_{-T}^{T} dt \int_\sigma d^3 x \bar{N}(x,t) \hat{H}'(x)} \psi' >_{\text{diff}}$$

$$= \lim_{T \to \infty} \int_{\mathcal{N}} [dN] < \psi, e^{i \int_\sigma d^3 x N(x) \hat{H}(x)} \psi' >_{\text{diff}}$$

$$\times \left[\int_{\overline{\mathcal{N}}} [d\bar{N}] \prod_x \delta \left(\int_{-T}^{T} dt \bar{N}(x,t), N(x) \right) \right] \qquad (14.1.5)$$

where we used the fact that the operator $\hat{H}'(x)$ is not explicitly time-dependent. Consider the integral

$$I_N^T := \int_{\overline{\mathcal{N}}} [d\bar{N}] \prod_x \delta \left(\int_{-T}^{T} dt \bar{N}(x,t), N(x) \right) \qquad (14.1.6)$$

appearing in the square bracket in the last line of (14.1.5). We claim that it is actually independent of $N(x)$. This can be verified by introducing the constant shift $\bar{N}(x,t) \mapsto \bar{N}(x,t) + \frac{N'(x) - N(x)}{2T}$ so that $I_N^T = I_{N'}^T = \text{const.}$ We conclude that (14.1.5) and (14.1.4) are proportional to each other (by an infinite constant $\lim_{T \to \infty} I_N^T$). The formula (14.1.5) is then the starting point for formulating a path integral through the usual skeletonisation process.

In any case we can now formally expand the exponent in (14.1.4) and arrive at the following picture: given two spatially diffeomorphism-invariant spin-network functions $T_{[s]}, T_{[s']}$ we have

$$< T_{[s],\text{phys}}, T'_{[s'],\text{phys}} >_{\text{phys}} := \sum_{n=0}^{\infty} \frac{i^n}{n!} \int_{\mathcal{N}} [dN] < T_{[s]}, \hat{H}'(N)^n T_{[s']} >_{\text{diff}} \qquad (14.1.7)$$

If we pretend that $\hat{H}(N)$ is spatially diffeomorphism-invariant then we may *define* the last inner product by

$$< T_{[s]}, \hat{H}'(N)^n T_{[s']} >_{\text{diff}} := T_{[s']}([\hat{H}^\dagger(N)]^n T_s)$$

which is well-defined.

Since $\hat{H}^\dagger(N)$ is closed and densely defined on spin-network functions, the matrix elements of powers of the Hamiltonian constraint can be computed and since we integrate over all possible lapse functions the result is manifestly spatially diffeomorphism-invariant. Of course, the result is badly divergent, but

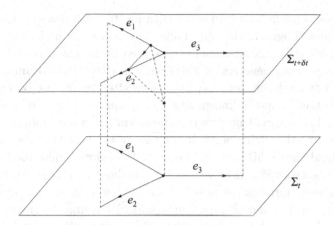

Figure 14.1 The action of the Euclidean piece of the Hamiltonian constraint can be interpreted as a discrete unphysical time evolution which builds a spin foam defined as the collection of branched surfaces defined by the dotted lines. The surfaces carry the same spin as the bounding edges.

cutting off the integral over N somehow the following picture emerges: the power of $[\hat{H}^\dagger(N)]^n$ corresponds to a discrete n time step evolution of an initial spin net s' to a final one s. At each step $\hat{H}^\dagger(N)$ changes the graph of the spin net s' according to the rules of Chapter 10. Let us associate a hypersurface with each time step and let the respective spin nets be embedded inside them. Connect the vertices of the spin nets in subsequent hypersurfaces by dotted lines. Since $\hat{H}(N)$ adds edges to a graph, one of these dotted lines branches up at some intermediate point into two additional dotted lines which connect with the two newly created vertices. We thus see that the quantum time evolution of edges become two-surfaces (bounded by one or two edges and two dotted lines), that is, *a spin foam* (see Figure 14.1). Such kind of 'transition amplitudes' are exactly of the form considered earlier by Reisenberger [692, 693].

Thus, the canonical theory seems to suggest a bubble evolution not unlike the worldsheet formulation of string theory, although spin foams define a background-independent string theory in which the worldsheet is not a smooth two-dimensional manifold but has necessarily (conical) singularities due to the fact that the Hamiltonian constraint acts non-trivially only at vertices in each time step.

Unfortunately, the concrete (Euclidean) Hamiltonian constraint constructed in Chapter 10 only generates what is known as a 0–3 move: as transpires from Figure 14.1, a vertex (which is one-dimensional) is transformed into a tetrahedron (in the figure one boundary triangle of that tetrahedron, composed of the beginning segments of edges adjacent to the vertex and the corresponding arcs, is displayed). However, if we think of a generic spin foam model which is defined in terms of a triangulation of M by four simplices as we will see, a triangulation by four simplices is such that they must be glued in all possible ways between two time slices. Then a p–$(3-p)$ move with $p = 0, 1, 2, 3$ is a situation in which the four-simplex intersects the initial slice in a p-simplex and the final slice in a $(3-p)$-simplex (the missing dimension is used up by the time evolution) and all these moves occur generically. However, the (Euclidean) Hamiltonian constraint does not have this property (not even after symmetric ordering, which also produces the 3–0 move) known as *crossing symmetry*. It is related to slicing independence because evidently we can transform the various moves among each other by changing the slicing of M. It is by no means clear that the Hamiltonian constraint should have this property because the evolution described above is an unphysical time evolution. However, at least without it, this evolution does not fit into the general formulation of spin foam models as described below.

To summarise: In order to give mathematical meaning to these amplitudes one obviously has to look for a better definition of the path integral. One way to proceed is by stripping off all the particulars of the specific theory that describes quantum gravity and considering very general spin foam models and searching for criteria when they converge and when they do not. Then, in a second step, one has to select among the converging ones the theory which describes quantum gravity (if any). This way one may discover an alternative route to the Hamiltonian constraint.

14.2 Spin foam models from BF theory

In this section we will explain the basic strategy employed in the construction of spin foam models. The descriptive discussion presented here will be complemented by a precise construction in the next section.

It turns out that a systematic starting point are the so-called BF topological field theories [682–691]. In $D + 1$ dimensions these are described by an action ($D \geq 2$)

$$S_{\mathrm{BF}} = \int_M \mathrm{Tr}(B \wedge F) \tag{14.2.1}$$

where B is a Lie(G)-valued $(D-1)$-form in a vector bundle associated with a principal G bundle P under the adjoint representation and F is the curvature of a connection A over P. The trace operation is with respect to the non-degenerate Cartan–Killing metric on Lie(G) (assuming G to be semi-simple), that is, basically the Kronecker symbol (up to normalisation). The equations of motion are given by $F = DB = 0$ where D is the covariant differential determined by A (see Chapter 21). Thus A is constrained to be flat. The action has a huge symmetry, namely it is gauge-invariant and invariant under $A \mapsto A$, $B \mapsto B + Df$ for any $(D-2)$-form f. Counting physical degrees of freedom it is easy to see that almost nothing is left, the theory has only a finite number of degrees of freedom, it is topological.

The connection with gravity is made through the Palatini (first-order) action (in this chapter we set $\kappa = 1$)

$$S_P = \int_M \mathrm{Tr}((*[e \wedge e]) \wedge F) \tag{14.2.2}$$

Here $e = (e^j_\mu)$ denotes the co-$(D+1)$-bein and $*$ denotes the Hodge dual with respect to the internal metric η_{ij}, which is just the Minkowski (Euclidean) metric for Lorentzian (Euclidean) General Relativity with gauge group $SO(D,1)$ ($SO(D+1)$). More specifically

$$(*[e \wedge e])_{ij} := \frac{1}{(D-1)!} \epsilon_{ijk_1 \ldots k_{D-1}} e^{k_1} \wedge \ldots \wedge e^{k_{D-1}} \tag{14.2.3}$$

and plugging this into (14.2.2) one easily sees that (14.2.2) equals the Einstein–Hilbert action for orientable M when A is the spin-connection of e (which is one of the equations of motion that one derives from (14.2.2)). Thus we see that gravity is a BF theory modulo the constraint that B is in this case not an arbitrary $(D-1)$-form but rather has to satisfy the so-called *simplicity* constraint

$$B = *[e \wedge e] \tag{14.2.4}$$

The idea for writing a path integral for General Relativity is then the following: a lot is known about the path integral quantisation of BF theory in three and four dimensions [677–681]. Thus, it seems advisable to consider General Relativity as a BF theory in which the sum over histories is constrained by (14.2.4). One might wonder how it can happen that a TQFT like BF theory with only a finite number of degrees of freedom plus additional constraints can give rise to a field theory like General Relativity with an infinite number of degrees of freedom. The answer is that (14.2.4) breaks a lot of the gauge invariance of BF theory, so that gauge degrees of freedom become physical degrees of freedom. In order to sum over histories of B's and A's with the constraint (14.2.4) we must first write it in a form in which only B's appear. The algebraic condition on B such that there exists e with (14.2.4) satisfied (up to a sign) has been systematically analysed by Freidel, Krasnov and Puzio in [694]. It can be written for $D \geq 3$ as

$$\epsilon^{ijklm_1 \ldots m_{D-3}} B^{\mu\nu}_{ij} B^{\rho\sigma}_{kl} = \epsilon^{\mu\nu\rho\sigma\lambda_1 \ldots \lambda_{D-3}} c^{m_1 \ldots m_{D-3}}_{\lambda_1 \ldots \lambda_{D-3}} \tag{14.2.5}$$

where c is any totally skew (in both sets of indices) tensor density and

$$B^{\mu\nu}_{ij} = \frac{1}{(D-1)!} \epsilon^{\mu\nu\rho_1 \ldots \rho_{D-1}} \eta_{ik} \eta_{jl} B^{kl}_{\rho_1 \ldots \rho_{D-1}} \tag{14.2.6}$$

Actually for $D = 3$ there is another solution to (14.2.5) besides (14.2.4) given by

$$B = \pm e \wedge e \tag{14.2.7}$$

but this solution gives rise again to a topological theory. The constraint (14.2.5) is enforced by adding to the BF action a term of the form (for $D = 3$)

$$\frac{1}{2}\int_M d^4x\,\Phi_{\mu\nu\rho\sigma}\epsilon^{ijkl}\left(\delta^\alpha_\mu\delta^\beta_\nu\delta^\gamma_\rho\delta^\delta_\sigma - \frac{1}{4!}\epsilon_{\mu\nu\rho\sigma}\epsilon^{\alpha\beta\gamma\delta}\right)B^{\alpha\beta}_{ij}B^{\gamma\delta}_{kl}$$

$$=: \frac{1}{2}\int_M \mathrm{tr}(B\wedge\Phi(B)) =: \int_M \Phi\cdot C \qquad (14.2.8)$$

where the Lagrange multiplier $\Phi^{\mu\nu\rho\sigma}$ has the symmetries $\Phi^{\mu\nu\rho\sigma} = -\Phi^{\nu\mu\rho\sigma} = -\Phi^{\mu\nu\sigma\rho} = \Phi^{\rho\sigma\mu\nu}$ and we have denoted the simplicity constraint by C. To see that this captures the right number of degrees of freedom in $D = 3$, notice that $B^{\mu\nu}_{ij}$ has $6^2 = 36$ degrees of freedom while e^i_μ has only $4^2 = 16$. Now Φ has $6\cdot 7/2 = 21$ independent components, however, the totally skew part is projected out in (14.2.8) which leads us to precisely the 20 independent constraints needed.

Now the 'partition function' for BF theory is given by

$$Z_{\mathrm{BF}} = \int [dA\,dB]e^{i\int_M \mathrm{tr}(B\wedge F)} \propto \int [dA]\delta(F) \qquad (14.2.9)$$

where for *either signature* the factor of i in front of the action has to be there in order to enforce the flatness constraint $\delta(F)$. That this defines the correct path integral (up to proper regularisation) has been verified by independent methods, see [677–691] and references therein. Since, from the point of view of BF theory, General Relativity is a 'perturbation' (with the role of the 'free' theory being played by BF theory) with interaction term (14.2.8), the partition function for General Relativity should be given by

$$Z_{\mathrm{P}} = \int [dA\,dB\,d\Phi]e^{i\int_M \mathrm{tr}(B\wedge[F+\frac{1}{2}\Phi(B)])} \propto \int [dA\,dB]\delta(C)e^{i\int_M \mathrm{tr}(B\wedge F)}$$

$$(14.2.10)$$

where the additional integral over the Lagrange multiplier enforces the simplicity constraint. Path integrals of the type (14.2.10) were studied by Freidel and Krasnov [695] in terms of a generating functional

$$Z[J] := \int [dA\,dB]e^{i\int_M \mathrm{tr}(B\wedge[F+J])} \qquad (14.2.11)$$

where J is a two-form current. It is easy to see that formally, by a trick familiar from ordinary quantum field theory,

$$Z_{\mathrm{P}} = \int [d\Phi]\{e^{i\frac{1}{2}\int_M \mathrm{tr}\left(\frac{\delta}{i\delta J}\Phi\left(\frac{\delta}{i\delta J}\right)\right])}Z[J]\}_{J=0} \qquad (14.2.12)$$

which could then be the starting point for perturbative expansions. Unfortunately, a truly systematic derivation of spin foam models for General Relativity starting directly from (14.2.12) is still missing, see below for the currently adopted substitute.

We see that in order to define the partition function for General Relativity we must first define the one for BF theory. Let us first consider the case that G is compact (Euclidean signature). Then the δ-distribution $\delta(F)$ in (14.2.9) can be interpreted as the condition that the holonomy of every contractible loop is trivial. Furthermore, in order to regularise the functional integral, we triangulate M, using some triangulation T and interpret the measure $[dA]$ as the uniform measure on $\overline{\mathcal{A}}$ restricted to T. The triangulation is considered as a topological triangulation, that is, the corresponding graph is an embedded graph modulo $\mathrm{Diff}(M)$. It is therefore often claimed that even at the triangulated level spin foam models already take care of four-dimensional diffeomorphism invariance. This has been demonstrated to be false already for $D = 2$ in [696, 697] and is expected to be false in $D = 3$ as well. There are still diffeomorphism symmetries in a given triangulation and these have to be gauge fixed when one sums over triangulations, see below.

The condition $F = 0$ amounts to saying that $h_\alpha = 1_G$ where α is any contractible loop within T. Let $\pi_1'(T)$ be a set of generators of the contractible subgroup of the fundamental group of T. Hence the regulated BF partition function becomes

$$Z_{\mathrm{BF}}(T) = \int_{\overline{\mathcal{A}}_T} d\mu_{0T}(A) \prod_{\alpha \in \pi_1'(T)} \delta(A(\alpha), 1_G) \qquad (14.2.13)$$

and we can use the Peter and Weyl theorem in order to write the δ-distribution as

$$\delta(h, 1_G) = \sum_{\pi \in \Pi} d_\pi \chi_\pi(h) \qquad (14.2.14)$$

Now magically the integral (14.2.13) is independent of the choice of triangulation which can be traced back to the fact that BF is a topological theory. The theory defined by (14.2.13) is known as the Turarev–Viro state sum model for $D = 2, G = \mathrm{SU}(2)$ and as the Turarev–Ooguri–Crane–Yetter model in $D = 3, G = \mathrm{SO}(4)$. Actually (14.2.13) is still divergent when one expands out the products of δ-distributions, but this can be taken care of by using a quantum group regularisation at a root of unity which cuts off the sum over representations at those of bounded dimension (see, e.g., [684]).

Let us now turn to Euclidean gravity for $D = 3$. We somehow must invoke the simplicity constraint into (14.2.13). The idea is to look at a canonical quantisation of BF theory with the additional simplicity constraint imposed. This analysis has been started by Barbieri [698, 699], leading to the consideration of *quantum tetrahedra* and was completed by Baez and Barrett [700]. The result is as follows: recall that $\mathrm{SO}(4)$ is homomorphic with $\mathrm{SU}(2) \times \mathrm{SU}(2)$, therefore its irreducible representations can be labelled by two spin quantum numbers (j, j') ('left-handed and right-handed'). This holds for both the representations on the links of spin-network states (the time evolution of which are faces) as well as the intertwiner representations of the vertices (the time evolution of which are edges). The simplicity constraint now amounts to the constraint $j = j'$ for both types of representations, explaining the word 'simplicity'. This motivates us to define, roughly speaking, the partition function for General Relativity by restricting the sum in

$$\delta(h, 1_{SO(4)}) = \sum_{j,j'} d_{\pi_{j,j'}} \chi_{\pi_{j,j'}}(h) \qquad (14.2.15)$$

to

$$\delta'(h, 1_{SO(4)}) = \sum_{j} d_{\pi_{j,j}} \chi_{\pi_{j,j}}(h) \qquad (14.2.16)$$

resulting in

$$Z_{\mathrm{P}}(T) = \int_{\overline{\mathcal{A}}_T} d\mu_{0T}(A) \prod_{\alpha \in \pi_1'(T)} \delta'(A(\alpha), 1_{\mathrm{G}}) \qquad (14.2.17)$$

(Some version of) (14.2.17) is referred to as the Barrett–Crane model [454]. The model has been improved in its degree of uniqueness by Reisenberger [701] and also by Yetter, Barrett, Barrett and Williams [702–704].

14.3 The Barrett–Crane model

In this section we will make the discussion of the previous section mathematically precise. The result is the most studied spin foam model to date in $D = 4$. It serves as a prototype for other spin foam models and we will learn about the various approximations that enter the derivation of the model from a path integral formulation. For pedagogical reasons we will restrict ourselves to the Euclidean case.

14.3.1 Plebanski action and simplicity constraints

The Plebanski action can be implicitly defined by

$$S_{\mathrm{Pl}} := \int_M B^{IJ} \wedge F_{IJ} + \lambda_{IJKL} B^{IJ} \wedge B^{KL} \qquad (14.3.1)$$

where λ is a Lagrange multiplier with the symmetries $\lambda_{IJKL} = -\lambda_{JIKL} = -\lambda_{IJLL} = \lambda_{KLIJ}$ and $\lambda_{IJKL} \epsilon^{IJKL} = 0$. Extremisation of (14.3.1) with respect to it imposes the *simplicity constraints*

$$B^{IJ} \wedge B^{KL} = \epsilon^{IJKL} \frac{1}{4!} \epsilon_{MNPQ} B^{MN} \wedge B^{PQ} \qquad (14.3.2)$$

The first result we need is the following not entirely trivial fact.

Theorem 14.3.1. *Suppose that (14.3.2) holds and that*

$$B := \frac{1}{4!} \epsilon_{MNPQ} B^{MN} \wedge B^{PQ} \qquad (14.3.3)$$

is non-vanishing. Then there exists a co-tetrad e^I such that either $B^{IJ} = \pm e^I \wedge e^J$ or $B^{IJ} = \pm\frac{1}{2}\epsilon_{IJKL}e^K \wedge e^L$.

Proof

Step I

The proof simplifies dramatically by making use of self-dual fields. Let T^{IJ} be an antisymmetric tensor. Its dual is defined by $(*T)^{IJ} := \frac{1}{2}\delta^{IK}\delta^{JL}\epsilon_{KLMN}T^{MN}$. The operator $*$ is called the Hodge operator with respect to the Euclidean metric δ_{IJ}. In what follows we will suppress it and no longer care about index positions. It is easy to see that $** = \mathrm{id}$. The (anti-)self-dual part of T is defined by $T_\pm = \frac{1}{2}[T \pm *T]$, which has the property that $*T_\pm = \pm T_*$. Defining $T_*^j := T_\pm^{0j}$ it follows that $T_\pm^{jk} = \pm\epsilon_{jkl}T_\pm^l$. Here we take $I, J, K, \ldots = 0, 1, 2, 3$ and $j, k, l, \ldots = 1, 2, 3$.

The antisymmetric tensors, considered as matrices, form the Lie algebra so(4) (the commutator of two antisymmetric matrices is again antisymmetric). Let A, B be two antisymmetric tensors, then it is an elementary exercise to show that $[A, ** B] = [*A, B] = *([A, B])$. From this it immediately follows that $[A_+, B-] = 0$ and $[A_\pm, B_\pm] = (A, B)_\pm$. Hence the (anti-)self-dual tensors form an ideal in so(4). These ideals are easily seen to be commuting copies of so(3), for instance by considering the basis of antisymmetric matrices P_{IJ}, $0 \leq I < J \leq 0$ with $P_{IJ} = [E_{IJ} - E_{JI}]/2$, $(E_{IJ})_{KL} := \delta_{IJ}\delta_{KL}$. Notice that $*P_{IJ} = \frac{1}{2}\epsilon_{IJKL}P_{KL}$. We then discover that $[P_\pm^j, P_\pm^k] = \pm\frac{1}{2}\epsilon_{jkl}P_\pm^l$. Hence so(4) $\cong$ so(3) $\oplus$ so(3). Thus locally SO(4) $\cong$ SO(3) $\times$ SO(3). Globally it turns out that SO(4)$/Z_2 \cong$ SO(3) $\times$ SO(3) and SU(2) $\times$ SU(2)$/Z_2 \cong$ SO(4) where the central and normal subgroup Z_2 is given by $Z_2 = \{1_4, -1_4\}$. Hence SU(2) $\times$ SU(2) is the universal covering[1] group of SO(4). Now take $(I, J) = (K, L)$ in (14.3.2). Then either $(I, J) = (0, j)$ or $(I, J) = (j, k)$ with $j < k$. These six conditions are then equivalent to

$$\left(B_+^j \pm B_-^j\right) \wedge \left(B_+^j \pm B_-^j\right) = 0 \tag{14.3.4}$$

(no summation over j). Next we take $I = K$ but I, J, L mutually different. Then either $(I, J) = (0, j)$, $(K, L) = (0, k)$ with $j \neq k$ or $(I, J) = (j, k)$, $(K, L) = (j, l)$ with j, k, l mutually different. This results in the 12 conditions

$$\left(B_+^j \pm B_-^j\right) \wedge \left(B_+^k \pm B_-^k\right) = 0 \tag{14.3.5}$$

for $j \neq k$. Finally we take the case that all indices are mutually different. The only independent equations result by taking, say, $(I, J) = (0, j)$, $(K, L) = (k, l)$ and results in

$$\left(B_+^j + B_-^j\right) \wedge \left(B_+^k - B_-^k\right) = \delta_{jk}B = \frac{\delta_{jk}}{3}\sum_l \left(B_+^l \wedge B_+^l - B_-^l \wedge B_-^l\right) \tag{14.3.6}$$

---

[1] We also have SU(2) $\times$ SU(2)$/Z_2' \cong$ SO(3) $\times$ SU(2), SU(2) $\times$ SU(2)$/Z_2'' \cong$ SU(2) $\times$ SO(3) and SU(2)$\times$ SU(2)$/Z_4 \cong$ SO(3) $\times$ SO(3) where $Z_2' = \{1_4, 1_2 \oplus (-1_2)\}$, $Z_2'' = \{1_4, (-1_2) \oplus 1_2\}$ and $\bar{Z}_2' = \{1_4, -1_4, 1_2 \oplus (-1_2), (-1_2) \oplus 1_2\}$ where $1_4 = 1_2 \oplus 1_2$.

Let us rewrite (14.3.4), (14.3.5), (14.3.6). Adding and subtracting the '+' part of (14.3.5) and (14.3.6) for $j \neq k$ gives $(B^j_+ + B^j_-) \wedge B^k_\pm = 0$. Adding and subtracting the '−' part of (14.3.5) and (14.3.6) for $j \neq k$ gives $(B^j_+ - B^j_-) \wedge B^k_\pm = 0$. It follows that

$$B^j_\epsilon \wedge B^k_\delta = 0 \tag{14.3.7}$$

for all $j \neq k$, $\epsilon, \delta = \pm$. Subtracting the '+' and '−' parts of (14.3.4) from each other we find

$$B^j_+ \wedge B^j_- = 0 \tag{14.3.8}$$

for all j. Adding the '+' and '−' parts of (14.3.4) and adding and subtracting from the resulting expression (14.3.6) for $j = k$ we find

$$B^j_+ \wedge B^j_+ = -B^j_- \wedge B^j_- = \frac{B}{2} \tag{14.3.9}$$

for all j. Now we can combine (14.3.7), (14.3.8), (14.3.9) into

$$0 = B^j_\pm \wedge B^k_\pm - \frac{1}{3} \sum_l B^l_\pm \wedge B^l_\pm$$

$$0 = B^j_+ \wedge B^k_-$$

$$0 = \sum_l (B^l_+ \wedge B^l_+ + B^l_- \wedge B^l_-) \tag{14.3.10}$$

for all j, k. Notice that the first set of equations in (14.3.10), which are symmetric in j, k, are only 10 conditions because taking the sum over $j = k$ results in two identities. The second set are nine conditions. Thus altogether we have 20 conditions as desired.

Step II

Suppose that B^j are three two-forms such that $B := \sum_l B^l \wedge B^l \neq 0$ and such that $B^j \wedge B^k = \frac{1}{3} \sum_l B^l \wedge B^l$. Then we will show that there are four independent one-forms e^I, $I = 0, 1, 2, 3$ and numbers ϵ, δ taking the values ± 1 such that $B^j = \frac{\epsilon}{2}(\omega^0 \wedge e^j + \frac{\delta}{2}\epsilon_{jkl}e^k \wedge e^l)$.

To see this, take any vector field $v \neq 0$ and define the one-forms ω^j by $\omega^j_\mu := v^\nu B^j_{\nu\mu}$. These are linearly independent for suppose they were not then there would be non-trivial real numbers z_j such that $\sum_j z_j \omega^j = 0$. It follows that the antisymmetric tensor field $A_{\mu\nu} := \sum_j z_j B^j_{\mu\nu}$ has a zero eigenvector v. Since A is antisymmetric, it therefore can have at most rank two and is thus of the form $A = \alpha \wedge \beta$ for some one-forms α, β. Thus $A \wedge A = [\sum_j z_j^2]B = 0$ hence $z_j = 0$. Now fix any-one form ω^0 such that $\omega^0_\mu v^\mu = 1$. Then the ω^I constitute a basis of one-forms and we can therefore expand

$$B^j = \alpha_{jk}\omega_0 \wedge \omega^k + \beta_{jkl}\omega^k \wedge \omega^l \tag{14.3.11}$$

where β_{jkl} is skew in k, l. Since $\omega_\mu^j v^\mu = 0$ we find $\alpha_{jk} = 2\delta_{jk}$. Now

$$B^j \wedge B^k = 4\omega^0 \wedge \omega^1 \wedge \omega^2 \wedge \omega^3 \epsilon_{mn(j}\beta_{k)mn} = \frac{\delta_{jk}}{3}B \qquad (14.3.12)$$

We conclude

$$\epsilon_{mn(j}\beta_{k)mn} = a\delta_{jk} + b^l \epsilon_{jkl} \qquad (14.3.13)$$

for certain numbers $a \neq 0, b^j$. It then follows from $\beta_{j(kl)} = 0$ that

$$\beta_{jkl} = \frac{1}{2}\epsilon_{ikl}\epsilon^{imn}\beta_{jmn} = \frac{1}{2}\left[a\epsilon_{jkl} + \delta^j_{[l}b_{k]}\right] \qquad (14.3.14)$$

Inserting (14.3.14) into (14.3.11) we obtain

$$B^j = 2a \left[\frac{1}{2}\frac{2\omega^0 - b_k\omega^k}{a} \wedge \omega^j + \frac{1}{4}\epsilon_{jkl}\omega^k \wedge \omega^l\right] \qquad (14.3.15)$$

Let us set $\delta = \mathrm{sgn}(a)$, $e^0 := \epsilon\, \mathrm{sgn}(a)\sqrt{2|a|}(2\omega^0 - b_k\omega^k)/a$ where $\epsilon = \pm 1$ is arbitrary and $e^j = \sqrt{2|a|}\omega^j$, then (14.3.15) takes the anticipated form.

Step III

Combining the first set of relations of (14.3.10) with the conclusions of step II we see that there are two bases of one-forms $e^I_\pm$ such that

$$B^j_\pm = s_\pm \left(\frac{1}{2}e^0_\pm \wedge e^j_\pm \pm \frac{1}{4}\epsilon_{jkl}e^k_\pm \wedge e^l_\pm\right) = s_\pm [P^j_\pm]_{IJ}e^I_\pm \wedge e^J_\pm \qquad (14.3.16)$$

for some $s_\pm \in \{+1, -1\}$ and $P^j_\pm$ was defined above. Since they are bases there exists $G \in GL(4, \mathbb{R})$ such that $e^I_- = G^I_J e^J_+$. Now from the last equation in (14.3.10) we obtain

$$0 = \sum_l (B^l_+ \wedge B^l_+ + B^l_- \wedge B^l_-) = \frac{3}{2}[e^0_+ \wedge e^1_+ \wedge e^2_+ \wedge e^3_+ - e^0_- \wedge e^1_- \wedge e^2_- \wedge e^3_-]$$

$$= \frac{3}{2}[e^0_+ \wedge e^1_+ \wedge e^2_+ \wedge e^3_+]\,[1 - \det(G)] \qquad (14.3.17)$$

hence $G \in SL(4, \mathbb{R})$.

Consider now the second set of conditions in (14.3.10)

$$0 = B^j_+ \wedge B^k_- = \pm [P^j_+]_{IJ}[P^k_-]_{KL}e^I_+ \wedge e^J_+ \wedge e^K_- \wedge e^L_-$$

$$= \pm ([P^j_+]_{IJ}[P^k_-]_{KL}G^K_M G^L_N \epsilon^{IJMN})e^0_+ \wedge e^1_+ \wedge e^2_+ \wedge e^3_+$$

$$= 2 \pm ([P^j_+]_{MN}[P^k_-]_{KL}G^K_M G^L_N)e^0_+ \wedge e^1_+ \wedge e^2_+ \wedge e^3_+ \qquad (14.3.18)$$

where self-duality was exploited. Equation (14.3.18) can be rewritten as

$$\mathrm{Tr}(GP^j_+ G^T P^k_-) = 0 \qquad (14.3.19)$$

Now recall that any non-degenerate matrix G can be written as $G = ODO'$ where D is positive definite and diagonal while $O, O' \in O(4)$. Since G is unimodular we must have $OO' \in SO(4)$ and since D is diagonal we may assume that

$O, O' \in SO(4)$. Moreover, since $SO(4)/Z_2 \cong SO(3) \times SO(3)$ we find $O_\pm$, $O'_\pm \in$ $SO(3)$ such that $O = O_+ O_- = O_- O_+$ and $O' = O'_+ O'_- = O'_- O'_+$ (up to a possible sign which drops out in the square of (14.3.16)). The two copies of $SO(3)$ that we are considering here have the algebra generated by the P^j_+ and P^j_- respectively as their Lie algebra. Therefore $[O_\pm, P^j_\mp] = [O'_\pm, P^j_\mp] = 0$. Moreover, $O_\pm P^j_\pm [O_\pm]^T = [\text{Ad}_{O_\pm}]_{jk} P^k_\pm$ defines the adjoint representation of $SO(3)$ on its Lie algebra (remember $O^T = O^{-1}$ for orthogonal matrices).

With these tools prepared we can now simplify (14.3.19) to

$$
\begin{aligned}
0 &= \text{Tr}\big(G P^j_+ G^T P^k_-\big) = \text{Tr}(O_+ O_- D O'_+ O'_- P^j_+ [O'_+]^T [O'_-]^T D [O_+]^T [O_-]^T P^k_-) \\
&= \text{Tr}(O_- D O'_+ P^j_+ [O'_+]^T D [O_-]^T P^k_-) \\
&= \big[\text{Ad}_{O'_+}\big]_{jm} \big[\text{Ad}_{(O_-)^1}\big]_{kn} \text{Tr}\big(D P^m_+ D P^n_-\big)
\end{aligned}
\tag{14.3.20}
$$

for all j, k. Since the representation matrices of the adjoint representation are non-singular, (14.3.20) is equivalent to

$$
\text{Tr}\big(D P^j_+ D P^k_-\big) = 2 D_0 D_j \delta_{jk} - \sum_{m,n} \epsilon_{jmn} \epsilon_{kmn} D_m D_n = 0
\tag{14.3.21}
$$

for all j, k. Here we have denoted the diagonal matrix elements of D by $D_{IJ} =: \delta_{IJ} D_I$. Equation (14.3.21) is an identity for $j \neq k$. For $j = k$ we obtain

$$
D_0 D_j = D_m D_n
\tag{14.3.22}
$$

where j, m, n are mutually distinct, that is, we obtain the three equations

$$
D_0 = \frac{D_1 D_2}{D_3} = \frac{D_2 D_3}{D_1} = \frac{D_3 D_1}{D_2}
\tag{14.3.23}
$$

Together with unimodularity $D_0 D_1 D_2 D_3 = 1$ from (14.3.17) and $D_I > 0$ the unique solution is $D_I = 1$, that is, $G \in SO(4)$.

Let us write $G = U_+ U_-$ with $U_\pm = O_\pm O'_\pm$ as before then

$$
\begin{aligned}
s_- B^j_- &= [P^j_-]_{IJ} e^I_- \wedge e^J_- \\
&= [P^j_-]_{IJ} [U_+ U_-]_{IK} [U_+ U_-]_{JL} e^K_+ \wedge e^L_+ \\
&= [P^j_- U_+]_{IN} [U_+ U_-]_{IK} [U_-]_{NL} e^K_+ \wedge e^L_+ \\
&= [U_+]_{IM} [P^j_-]_{MN} [U_+]_{IP} [U_-]_{PK} [U_-]_{NL} e^K_+ \wedge e^L_+ \\
&= [P^j_-]_{MN} [U_-]_{MK} [U_-]_{NL} e^K_+ \wedge e^L_+
\end{aligned}
\tag{14.3.24}
$$

while

$$
\begin{aligned}
s_+ B^j_+ &= [P^j_+]_{IJ} e^I_+ \wedge e^J_+ \\
&= [P^j_+]_{IJ} [U_-]_{MI} [U_-]_{MK} [U_-]_{NJ} [U_-]_{NL} e^K_+ \wedge e^L_+ \\
&= [U_- P^j_+]_{MJ} [U_-]_{MK} [U_-]_{NJ} [U_-]_{NL} e^K_+ \wedge e^L_+ \\
&= [P^j_+]_{MP} [U_-]_{PJ} [U_-]_{MK} [U_-]_{NJ} [U_-]_{NL} e^K_+ \wedge e^L_+ \\
&= [P^j_+]_{MN} [U_-]_{MK} [U_-]_{NL} e^K_+ \wedge e^L_+
\end{aligned}
\tag{14.3.25}
$$

Thus, if we define $e^I := [U_-]_{IJ}e^J_+ = [U^{-1}_+]_{IJ}e^J_-$ then

$$B^j_\pm = s_\pm [P^j_\pm]_{IJ} e^I \wedge e^j \qquad (14.3.26)$$

where all four sign combinations of s_+, s_- are possible. In case that $s_+ = s_- = s \in \{+1, -1\}$ we easily find

$$B^{IJ} = se^I \wedge e^J \qquad (14.3.27)$$

while for $s_+ = -s_- = s$ we find

$$B^{IJ} = s * (e^I \wedge e^J) := s\frac{1}{2}\epsilon_{IJKL}e^K \wedge e^L \qquad (14.3.28)$$

□

In the degenerate sector we have not only 20 conditions but in fact 21, given by $B^j_\epsilon \wedge B^k_\sigma = 0$ for all j, k and all $\epsilon, \sigma = \pm 1$. The degenerate sector does not have an interpretation as a theory of gravity and is described in more detail in [693]. Let us summarise our findings.

Corollary 14.3.2. *The simplicity constraints (14.3.2) allow for five different solution sectors*

$$B_{++} = e \wedge e, \ B_{+-} = -e \wedge e, \ B_{-+} = *(e \wedge e), \ B_{--} = - * (e \wedge e),$$
$$B_0 = \text{degenerate} \qquad (14.3.29)$$

Only the ++ sector alone reduces the Plebanski action to the Palatini action.

The fact that only one sector really corresponds to the Plebanski action must be taken care of in the path integral in order that one really quantises gravity and not a mixture of phases which altogether do not reduce to General Relativity in the semiclassical limit.

Before we close this section we notice that there is an equivalent formulation of (14.3.2) at least in the non-degenerate sector. Consider the quantity

$$\Sigma^{\mu\nu}_{IJ} = \frac{1}{4e}\epsilon^{\mu\nu\rho\sigma} \epsilon_{IJKL}B^{KL}_{\rho\sigma}, \quad e := \frac{1}{4!}\epsilon_{IJKL}\epsilon^{\mu\nu\rho\sigma} B^{IJ}_{\mu\nu}B^{KL}_{\rho\sigma} \qquad (14.3.30)$$

Then (14.3.2) is equivalent to

$$\Sigma^{\mu\nu}_{IJ}B^{KL}_{\mu\nu} = \delta^I_{[K}\delta^J_{L]} \qquad (14.3.31)$$

which says that Σ is a bi-vector inverse to the bi-co-vector B. Thus also

$$\Sigma^{\mu\nu}_{IJ}B^{IJ}_{\rho\sigma} = \delta^\mu_{[\rho}\delta^\nu_{\sigma]} \qquad (14.3.32)$$

which in turn is equivalent to

$$\epsilon_{IJKL}B^{IJ}_{\mu\nu}B^{KL}_{\rho\sigma} = e\epsilon_{\mu\nu\rho\sigma} \qquad (14.3.33)$$

We will use this form of the simplicity constraint in what follows.

14.3.2 Discretisation theory

As outlined in the previous section, the idea of the spin foam approach is to start from BF theory. Since we must regularise the theory by means of a discretisation and as BF theory is a topological theory, we want to use a discretisation which is compatible with the topological invariance of the theory. In order to do that we have to look for discrete analogues of the various operations that one can perform on p-forms such as the exterior product, exterior derivative and Hodge dual. This is what we will describe in the present subsection. The presentation is based on [705] and references therein.

Definition 14.3.3. *A p-simplex $\sigma^{(p)} = [v_0, \ldots, v_p]$ in $\mathbb{R}^D$ is given by the convex hull of $p + 1$ vectors, that is,*

$$\sigma^{(p)} := \left\{ \sum_{k=0}^{p} t_k v_k; \ t_k \geq 0, \ \sum_{k=0}^{p} t_k = 1 \right\} \tag{14.3.34}$$

which span a p-dimensional vector space.

Remarks

1. By solving the constraint $\sum_{k=0}^{p} t_k = 1$ for $0 \leq t_0 = 1 - \sum_{k=1}^{p} t_k$ we can also describe a p-simplex as the convex hull of the p vectors $v'_k = v_k - v_0$, $k = 1, \ldots, p$ which are linearly independent by assumption. However, we will not make use of this notation here.

2. Notice that p-simplices are oriented by the order in which the vertices v_k appear in the list $[v_0, \ldots, v_k]$. We say that for a permutation $\pi \in S_{p+1}$ the simplices $[v_0, \ldots, v_p]$ and $[v_{\pi(1)}, \ldots, v_{\pi(p+1)}]$ are equally oriented if π is an even permutation, otherwise they are oppositely oriented.

3. The boundary $\partial\sigma^{(p)}$ is defined as the set of points for which $t_k = 0$, $k = 0, \ldots, p$. These define $p + 1$ different $(p - 1)$-simplices $\sigma_k^{(p-1)} = [v_0, \ldots, \hat{v}_k, \ldots, v_p]$. Here the hat over a vertex denotes omission of that vertex. These are oriented equally relative to $[v_1, \ldots, v_p]$ if k is even and otherwise oppositely. This defines the induced orientation of these so-called faces of $\sigma^{(p)}$. It follows that (as sets, i.e., modulo orientation) $\partial\sigma^{(p)} = \cup_k \sigma_k^{(p-1)}$. By repeating this process we obtain all subsimplices of $\sigma^{(p)}$. It follows that a p-simplex has as many different k-simplices as subsimplices as there are possibilities (up to orientation) to omit $p - k$ from $p + 1$ points, that is, $\binom{p+1}{k+1}$ for $k = 0, \ldots, p$.

Definition 14.3.4. *The barycentre of a p-simplex $\sigma^{(p)} = [v_0, \ldots, v_p]$ is defined as the point*

$$\hat{\sigma}^{(p)} := \frac{\sum_{k=0}^{p} v_k}{p+1} \tag{14.3.35}$$

This is precisely the same formula known from mechanics for the barycentre of $p + 1$ points v_k with equal masses $m_k = m$.

Definition 14.3.5. *A simplicial complex K is a collection of simplices $\sigma_i^{(p)}$; $p = 0, \ldots, D$; $i = 1, \ldots, N_p$ with the following properties:*

(i) All the subsimplices of each $\sigma_i^{(p)}$ also belong to K.

(ii) Two simplices $\sigma_i^{(p)}$, $\sigma_j^{(q)}$ intersect at most in a common subsimplex, which has opposite orientation in both.

One can show that all differential manifolds M admit a simplicial complex (under the respective coordinate charts) as a partition, which is then called a triangulation. Of course, a triangulation need not be simplicial, that is, it can consist of more general (polyhedral) cells. This is generically the case for the so-called dual cell complex.

Definition 14.3.6. *Let $K = \{\sigma_i^{(p)}$; $p = 0, \ldots, D$; $i = 1, \ldots, N_p\}$ by a simplicial complex. Pick any $\sigma_{j_0}^{(p)} \in K$ and consider all possible $(D - p)$-tuples of simplices $\sigma_{j_k}^{(p+k)} \in K$ with $k = 1, \ldots, D - p$ and $1 \leq j_k \leq N_{p+k}$ subject to the following condition:*

For all $l = 0, \ldots, D - p - 1$ the simplex $\sigma_{j_l}^{(p+l)}$ is a face of $\sigma_{j_{l+1}}^{(p+l+1)}$ with the induced orientation.

For each such $(D - p)$-tuple of simplices construct the $(D - p)$-simplex $[\hat{\sigma}_{j_0}^{(p)}, \hat{\sigma}_{j_1}^{(p+1)}, \ldots, \hat{\sigma}_{j_{D-p}}^{(D)}]$ where we have used the barycentres of those simplices as defined in (14.3.35). The cell dual to $\sigma_{j_0}^{(p)}$ is then defined by

$$*_K\left[\sigma_{j_0}^{(p)}\right] := \bigcup_{\sigma_{j_l}^{(p+l)} \subset \partial\sigma_{j_{p+l+1}}^{(p+l+1)}; \, l=0,\ldots,D-p-1} [\hat{\sigma}_{j_0}^{(p)}, \hat{\sigma}_{j_1}^{(p+1)}, \ldots, \hat{\sigma}_{j_{D-p}}^{(D)}] \qquad (14.3.36)$$

The cell complex K^* dual to K is obtained by gluing dual cells along common subcells.

As is obvious from the construction, the p-simplices in K are in one-to-one correspondence with the $(D-p)$-cells in K^*. We can therefore define an operation $*_{K^*}$ on the p-cells of K^* by the inverse of $*_K$ (times $(-1)^{p(D-p)}$, see below).

Definition 14.3.7. *Let $K = \{\sigma_I^{(p)}$; $p = 0, \ldots, D$; $i = 0, \ldots, N_p\}$ be a simplicial complex*

(i) The vector space $C_p(K)$ of p-chains is defined as the formal real linear combination of the $\sigma_p^{(p)}$.

(ii) We turn $C_p(K)$ into a Hilbert space by defining the non-degenerate inner product

$$< \sigma_i^{(p)}, \sigma_j^{(p)} >_K := \delta_{ij} \qquad (14.3.37)$$

for all $i, j = 1, \ldots, N_p$, that is, the p-simplices of K provide an orthonormal basis. By means of this scalar product we can identify the dual space of

$C_p(K)$ *(the space $C^p(K)$ of linear forms on $C_p(K)$ called the space of p-cochains) with $C_p(K)$ itself.*

(iii) The boundary operator $\partial_K : C_p(K) \to C_{p-1}(K)$ is defined by

$$\partial_K \sigma_i^{(p)} := \sum_{k=0}^{p} (-1)^k [v_0, \ldots, \hat{v}_k, \ldots, v_p] \qquad (14.3.38)$$

for $v(p)_i = [v_0^i, \ldots, v_p^I]$. One easily verifies that $(\partial_K)^2 = 0$. The coboundary operator $d_K : C_p(K) \to C_{p+1}(K)$ is defined as the adjoint of ∂_K under the scalar product (14.3.37).

The operations $*_K$, ∂_K, d_K defined on p-chains in K as defined above are the analogues of the operations $*$, d, $*d*$ on the vector space $\Lambda^p(M)$ of p-forms as we will see in a moment. Here $*$ is the Hodge dual

$$(*\omega)_{\mu_1 \ldots \mu_{D-p}} := \frac{1}{p!} \sqrt{|\det(g)|} \epsilon_{\mu_1 \ldots \mu_{D-p} \nu_1 \ldots \nu_p} g^{\nu_1 \rho_1} \ldots g^{\nu_p \rho_p} \omega_{\rho_1 \ldots \rho_p} \qquad (14.3.39)$$

which needs a metric g. The normalisation here is such that $** = s(-1)^{p(D-p)}$id on p-forms where s is the signature of the metric used. The analogue of the scalar product on chains is given by $< \omega, \omega' >:= \int_M \omega \wedge *\omega'$.

In order to discretise actions on simplicial and dual complexes we must relate p-forms and p-chains. This will also serve to add the missing analogue of a wedge product. Consider the space $\Lambda^p(K)$ of p-forms restricted to the p-chains of K.

Definition 14.3.8. *Let K be a simplicial complex.*

(i) The Whitney map is defined by

$$W_K : C_p(K) \to \Lambda_p(K); \sigma^{(p)} = [v_0, \ldots, v_p] \mapsto p! \sum_{k=0}^{p} (-1)^k t_k dt_0 \wedge \ldots \wedge \widehat{dt_k} \wedge \ldots dt_p$$

$$(14.3.40)$$

(ii) The de Rham map is defined by

$$R_K : \Lambda_p(K) \mapsto C_p(K); \; < R_K(\omega), \sigma^{(p)} >_K := \int_{\sigma^{(p)}} \omega \qquad (14.3.41)$$

(iii) The wedge product on chains is defined by

$$\wedge_K : C_p(K) \times C_q(K) \to C_{p+q}(K); \; \sigma^{(p)} \wedge_K \sigma^{(q)} := R_K \left(W_K \left(\sigma^{(p)} \right) \wedge W_K \left(\sigma^{(q)} \right) \right)$$

$$(14.3.42)$$

The Whitney map is of course understood in the sense that the t_k with $\sum_{k=0}^{p} t_k = 1$ are local coordinates for $\sigma^{(p)}$. We will state without proof the following properties of the discrete wedge product, see [706] for more details.

Theorem 14.3.9. *The above operations obey the following relations*

$$\sigma^{(p)} \wedge_K \sigma^{(q)} = (-1)^{pq} \sigma^{(q)} \wedge_K \sigma^{(p)}$$

$$d_K\left(\sigma^{(p)} \wedge_K \sigma^{(q)}\right) = \left(d_K\sigma^{(p)}\right) \wedge_K \sigma^{(q)} + (-1)^p \sigma^{(p)} \wedge_K \left(d_K\sigma^{(q)}\right)$$

$$R_K \circ W_K = \mathrm{id}$$

$$d \circ W_k = W_K \circ d_K$$

$$d_K \circ R_k = R_K \circ d$$

$$\int_{\sigma^{(p)}} W_K\left(\sigma^{(p)\prime}\right) = <\sigma^{(p)}, \sigma^{(p)\prime}>_K \tag{14.3.43}$$

While the discrete wedge product is skew symmetric and obeys the Leibniz rule, it is not associative.

Notice that in the language introduced the map $*_K : C_p(K) \to C_{D-p}(K^*)$ cannot be iterated because $K \neq K^*$, in fact, K^* is no longer simplicial. To repair this we need the following.

Definition 14.3.10. *The barycentric subdivision of a p-simplex $\sigma^{(p)} = [v_0, \ldots, v_p]$ consists of $(p+1)!$ different p-simplices $\sigma_\pi^{(p)}$, one for each element $\pi \in S_{p+1}$ of the symmetric group, obtained as follows:*
Let for each $k = 0, \ldots, p$

$$\hat{\sigma}(k)_\pi := \frac{\sum_{l=0}^k v_{\pi(l)}}{k+1} \tag{14.3.44}$$

be the barycentre of the k-subsimplex $[v_{\pi(1)}, \ldots, v_{\pi(k)}]$ and set $\sigma_\pi^{(p)} := [\hat{\sigma}_\pi^{(0)}, \ldots, \hat{\sigma}_\pi^{(p)}]$. The collection of these $(p+1)!$ subdivisions for each p-simplex of K and for all $p = 0, \ldots, D$ defines the barycentric refinement $B(K)$ of K.

By construction, the p-cells of K^* are unions of the p-simplices of $B(K)$, hence we now have K, $K^* \subset B(K)$. Notice that $B(K)$ is simplicial again so that we can extend all operations from K to $B(K)$ as necessary. In particular we can extend all operations to K^* because $C_p(K^*)$ is a subspace of $C_p(B(K))$. Then the following crucial result holds [707].

Theorem 14.3.11. *Let $x \in C_p(K)$, $y \in C_{D-p}(K^*)$.*

(i) $$<*_K(x), y>_{K^*} = \frac{(D+1)!}{p! \, (D-p)!} \int_M W_{B(K)}(E(x)) \wedge W_{B(K)}(E(y))$$

$$<*_{K^*}(y), x>_K = \frac{(D+1)!}{p! \, (D-p)!} \int_M W_{B(K)}(E(y)) \wedge W_{B(K)}(E(x))$$

$$\tag{14.3.45}$$

where $E(x)$ is the linear combination (obeying compatibility of orientation) of $x \in C_p(K)$ in terms of elements of $C_p(B(K))$ and similarly for $E(y)$. The inner product $< .,. >_{K^*}$ on $C_{D-p}(K^*)$ is defined as $< .,. >_K$ on $C_p(K)$ by declaring dual cells as orthonormal.

(ii) $\partial_K = (-1)^{p(D-p)} *_{K^*} \circ d_{K^*} \circ *_K, \quad \partial_{K^*} = (-1)^{p(D-p)} *_K \circ d_K \circ *_{K^*}$

$$\text{(14.3.46)}$$

We will use this theorem in the next subsection in order to arrive at a discretisation of BF theory which is maximally topologically invariant.

14.3.3 Discretisation and quantisation of BF theory

The BF action is defined by

$$S[B, F] := \int_M \text{Tr}(B \wedge F) \tag{14.3.47}$$

where $B, F \in C_2(M)$ and the trace is with respect to the metric $\delta^I_{[K}\delta^J_{L]}$ on bi-covectors. Introduce a triangulation K of M. Using the relation $W_{B(K)} \circ R_{B(K)} = \text{id}$ between the Whitney and de Rham maps respectively we find

$$S[B, F] = \int_M \text{Tr}\big(W_{B(K)}\big(R_{B(K)}(B)\big) \wedge W_{B(K)}\big(R_{B(K)}(F)\big)\big)$$
$$= \text{Tr}\big(< *_{K^*}\big(R_{B(K)}(F)\big), R_{B(K)}(B) >_K\big) \tag{14.3.48}$$

where we have used the second relation in (14.3.45) and the skew symmetry of the exterior product. Using the orthonormal basis $\sigma^{(2)}$ of $C_2(K)$ we can write (14.3.48) as

$$S[B, F] = \sum_{\sigma^{(2)} \in C_2(K)} \text{Tr}\big(< *_{K^*}\big(R_{B(K)}(F)\big), \sigma^{(2)} >_K < \sigma^{(2)}, R_{B(K)}(B) >_K\big)$$

$$= \sum_{\sigma^{(2)} \in C_2(K)} \text{Tr}\left(< *_{K^*}\big(R_{B(K)}(F)\big), \sigma^{(2)} >_K \left[\int_{\sigma^{(2)}} B\right]\right)$$

$$= \sum_{\sigma^{(2)} \in C_2(K)} \text{Tr}\left(\left[\int_M W_{B(K)}\big(R_{B(K)}(F)\big) \wedge W_{B(K)}\big(E(\sigma^{(2)})\big)\right] \left[\int_{\sigma^{(2)}} B\right]\right)$$

$$= \sum_{\sigma^{(2)} \in C_2(K)} \text{Tr}\left(\left[\int_{*_K(\sigma^{(2)})} F\right]\left[\int_{\sigma^{(2)}} B\right]\right) \tag{14.3.49}$$

where we used the last relation in (14.3.43) and $W_{B(K)} \circ D_{B(K)} = \text{id}$ in the second step, in the third step we used the second relation of (14.3.45) again, in the fourth we used skew symmetry of the wedge product as well as the first relation in (14.3.45) and finally in the last step we used the last relation of (14.3.43) and $W_{B(K)} \circ D_{B(K)} = \text{id}$.

The result (14.3.49) is quite remarkable because it is *exact*, it is in particular independent of the chosen triangulation K. This expresses the toplogical nature of BF theory. In order to make further progress one now takes a further discretisation step which is no longer exact: let us sum over two cells $f \in C_2(K^*)$, called faces in what follows, and let $t(f) \in C_2(K)$ be the unique triangle to which it corresponds. Consider

$$S_{BF}(K^*) := \sum_{f \in C_2(K^*)} \text{Tr}(B_f U(\partial f)) \qquad (14.3.50)$$

where $B_f := \int_{t(f)} B$ and $U(\partial f)$ is the holonomy of the SO(4) connection along the loop ∂f. Then, since SO(4) is unimodular, the 1_4 term in the expansion $U(\partial f) = 1_4 + F(f) + \ldots, F(f) = \int_f F$ drops out of the trace in (14.3.50) so that (14.3.50) approximates (14.3.49). The partion function for BF theory is now defined, given K^*, by

$$Z_{BF}(K^*) := \int \prod_{e \in C_1(K^*)} d\mu_H(g_e) \prod_{f \in C_2(K^*)} d^6 B_f \exp(iS_{BF}(K^*)) \qquad (14.3.51)$$

where we have used the product Haar measure, one for each edge e of the dual complex.

The integral over the B field results in a product of δ-distributions on the real axis (up to a power of 2π)

$$Z_{BF}(K^*) = \int \prod_{e \in C_1(K^*)} d\mu_H(g_e) \prod_{f \in C_2(K^*)} \prod_{I<J} \delta_{\mathbb{R}}(\text{Tr}(P_{IJ} U(\partial f))) \qquad (14.3.52)$$

where P_{IJ} denote the generators of so(4) as in Section 14.3.1 and $B_f = \sum_{I<J} B_f^{IJ} P_{IJ}$. Since $[P_{IJ}]^{KL} = \delta_{[I}^K \delta_{J]}^L$, the support of the integrand on those elements g of SO(4) which satisfy $g = g^T$, that is $g^2 = 1_4$. There are several solutions $g \in$ SO(4), for instance $g = \pm 1_4$ or any other diagonal matrix with two entries ± 1 each.

What one now does is, strictly speaking, an *ad hoc manipulation which does not derive from first principles:* we exclude by hand all solutions $g \neq 1_4$. This is on the one hand well motivated physically because classically we know that the solutions of the equations of motion of BF theory are flat connections which therefore have trivial holonomy. On the other hand, the path integral is usually not concentrated precisely on the classical histories but contains non-classical paths. Hence it may not be entirely justified to cancel these configurations by hand. In any case, one now substitutes (14.3.52) by

$$Z_{BF}(K^*) = \int \prod_{e \in C_1(K^*)} d\mu_H(g_e) \prod_{f \in C_2(K^*)} \delta_{SO(4)}(U(\partial f)) \qquad (14.3.53)$$

where we have now replaced the product of six δ-distributions on the real axis by the δ-distribution on SO(4). We see now the advantage of having discretised the classical action by integrating the B field over triangles and the curvature over dual faces: there is precisely one triangle for each face, thus we get as many flatness conditions as there are holonomies. Had we discretised the classical action by integrating both F and B over triangles in K, then for each four-simplex Δ we would have obtained a contribution to the action, for instance of the form

$$\epsilon^{ijkl}\mathrm{Tr}(B(t_{ij}(\Delta))F(t_{kl}(\Delta))) \tag{14.3.54}$$

where $i, j, k, l \in \{1, 2, 3, 4\}$ and for each Δ we have singled out a vertex which is a common vertex of the six triangles $t_{ij}(\Delta)$ which are sub-two-simplices of Δ (one should probably average over all five vertices of Δ in order to use all 10 triangles of Δ). While this has the correct continuum limit, the corresponding model is not obviously equivalent to (14.3.53) because one and the same $B(t)$ appears in several four-simplices and thus gives rise to more complicated conditions on the holonomies than (14.3.53). This was the whole point of going through the previous section.

It remains to evaluate (14.3.53). In order to do so one uses the Peter and Weyl theorem proved in Section 31.2 in order to expand the δ-distributions in terms of characters on the group. Thus we obtain

$$Z_{\mathrm{BF}}(K^*) = \int \prod_{e \in C_1(K^*)} d\mu_H(g_e) \sum_{\{\rho_f\}} \prod_{f \in C_2(K^*)} d_f \,\mathrm{Tr}(\rho_f(U(\partial f))) \tag{14.3.55}$$

where d_f is the dimension of the irreducible representation ρ_f assigned to each dual face f. In order to carry out the integral over the edge holonomies we must collect all characters in which a given edge occurs. This is a book-keeping problem which can be handled by studying the relation between K and K^* in a bit more detail.

First of all we claim that each $e \in C_1(K^*)$ occurs in precisely four $f \in C_2(K^*)$. To see this notice that the dual of a four-simplex $\sigma^{(4)} \in C_4(K)$ is its barycentre. Furthermore, given a tetrahedron $\sigma^{(3)} \in C_3(K)$ in $\partial_K \sigma^{(4)} \cap \partial_K \sigma^{(4)'}$ (recall that two four-simplices are glued together precisely in one three-simplex by the definition of a simplicial complex) we have $e := *_K(\sigma^{(3)}) = [\hat{\sigma}^{(3)}, \hat{\sigma}^{(4)}] \cup [\hat{\sigma}^{(3)}, \hat{\sigma}^{(4)'}]$. Hence the dual edge e connects the barycentres of two neighbouring four-simplices through the barycentre of their common boundary tetrahedron. Since each four-simplex has five boundary tetrahedra, it follows that there are five dual edges starting in its barycentre.

Next we come to the dual faces. Consider a triangle $\sigma^{(2)} \in C_2(K)$ which is shared by two tetrahedra $\sigma^{(3)}$, $\sigma^{(3)'}$ in the boundary of a common four-simplex $\sigma^{(4)}$. For definiteness, let $\sigma^{(4)} = [v_0, v_1, v_2, v_3, v_4]$, $\sigma^{(3)} = [v_1, v_2, v_3, v_4]$, $\sigma^{(3)'} = [v_0, v_2, v_3, v_4]$ and $\sigma^{(2)} = [v_2, v_3, v_4]$. The face dual to $\sigma^{(2)}$ is given by

$$f := *_K(\sigma^{(2)})$$
$$= \left(\cup_{\sigma^{(3)} \in \partial_K \sigma^{(4)'}} [\hat{\sigma}^{(2)}, \hat{\sigma}^{(3)}, \hat{\sigma}^{(4)'}]\right) \cup \left(\cup_{\sigma^{(3)'} \in \partial_K \sigma^{(4)'}} [\hat{\sigma}^{(2)}, \hat{\sigma}^{(3)'}, \hat{\sigma}^{(4)'}]\right)$$

$$\tag{14.3.56}$$

Since $\sigma^{(3)}$, $\sigma^{(3)\prime}$ are common tetrahedra of our given $\sigma^{(4)}$ it follows that $\sigma^{(4)\prime} = \sigma^{(4)}$ is a possible choice in both terms in (14.3.55). It follows that $f = *_K(\sigma^{(2)})$ has the two edges that connect the barycentre of $\sigma^{(4)}$ through the barycentres of $\sigma^{(3)}$, $\sigma^{(3)\prime}$ respectively with the barycentres of other four-simplices in its boundary, and these are the only two edges in the boundary of f that meet in the barycentre of $\sigma^{(4)}$.

What all of this means is that the dual faces f can be labelled the barycentre v of a four-simplex of K with which it shares a corner and a pair of two distinct dual edges which are incident at v. Hence there are five dual edges and $\binom{5}{2} = 10$ dual faces incident at v. In particular, a given dual edge is in the boundary of as many dual faces f as there are possible pairs of edges of which e is a member, that is, four. For convenience, we will choose the orientation of the faces f in the following way: suppose that $e = f_1 \cap f_2 \cap f_3 \cap f_4$. Then the induced orientation on the boundary of two of these four faces coincides with that of e and for the other two it is opposite. This can be done consistently throughout K^*.

Given $f \in C_2(K^*)$ consider $E_f := \{e \in C_1(K^*); \ e \subset \partial_{K^*} f\}$ and define for $e \in E_f$ the orientation factors $\sigma(f, e) = +1$ if $e, \partial f$ are aligned and $\sigma(f, e) = -1$ otherwise. We also define $\sigma(f, e) = 0$ if e is not part of ∂f. We write for $\partial f = e_1^{\sigma_1} \circ \ldots \circ e_n^{\sigma_n}$ with $\sigma_k = \pm 1$

$$
\mathrm{Tr}(\rho_f(U_f)) = \sum_{\{M_e^f, N_e^f\}_{e \in E_f}} \delta_{N_{e_n}^f, M_{e_1}^f} \delta_{N_{e_1}^f, M_{e_2}^f} \cdots \delta_{N_{e_{n-1}}^f, M_{e_n}^f}
$$

$$
\times \left[\rho_f(g_{e_1}^{\sigma_1})\right]_{M_{e_1}^f, N_{e_1}^f} \cdots \left[\rho_f(g_{e_n}^{\sigma_n})\right]_{M_{e_n}^f, N_{e_n}^f}
$$

$$
=: \sum_{\{M_e^f, N_e^f\}_{e \in E_f}} C_{\rho_f}(\{M_e^f, N_e^f\}_{e \in E_f}) \prod_{e \in E_f} \left[\rho_f(g_e^{\sigma(f,e)})\right]_{M_e^f, N_e^f}
$$

$$
(14.3.57)
$$

where the superscript f in M_e^f, N_e^f is necessary because each e appears in precisely four of the f. Now (14.3.54) can be written as

$$
Z_{\mathrm{BF}}(K^*) = \sum_{\{\rho_f\}} \sum_{\{M_e^f, N_e^f\}} \left[\prod_{f \in C_2(K^*)} d_f \, C_{\rho_f}(\{M_e^f, N_e^f\}_{e \in E_f}) \right]
$$

$$
\times \prod_{e \in C_1(K^*)} \int d\mu_H(g_e) \prod_{e \in E_f} \left[\rho_f(g_e^{\sigma(f,e)})\right]_{M_e^f, N_e^f} \quad (14.3.58)
$$

What is left to do is compute the integral over four representation matrices in (14.3.58). This can be done explicitly as follows: using that fact that $SO(4) \cong SU(2) \times SU(2)/Z_2$ we know that the irreducible representations of $SO(4)$ are given by tensor products of two irreducible (fundamental) representations of $SU(2)$, that is, $\rho = \pi_{j^+} \otimes \pi_{j^-}$ where only integral spins $j^\pm$ appear and the dimension of this representation is $d_\rho = (2j^+ + 1)(2j^- + 1)$. The Haar measure on SO(4) is likewise the product Haar measure on two copies of SU(2) or SO(3) (up to a factor of two). It follows that the basic integral to be computed is of the form

$$
\int_{SO(4)} d\mu_H(g) \left[\prod_{k=1}^{4} [\rho_k(g_k^\sigma)] \right]_{M_k N_k} = \prod_{\epsilon=\pm} \left[\int_{SO(3)} d\mu_H(g) \prod_{k=1}^{4} [\pi_{j_k^\epsilon}(g_\epsilon^{\sigma_k})]_{m_k^\epsilon n_k^\epsilon} \right]
$$

(14.3.59)

where $\rho_k = (j_k^+, j_k^-)$, $M_k = (m_k^+, m_k^-)$, $N_k = (n_k^+, n_k^-)$. It is clear that (14.3.59) projects out the gauge-invariant piece of the decomposition into irreducibles of the tensor product of four representations in (14.3.59). To do this explicitly we must decide on a recoupling scheme. Let us assume that $\sigma_1 = \sigma_2 = -\sigma_3 = -\sigma_4 = 1$. We will choose to couple $j_1^\pm, j_2^\pm$ to $j_{12}^\pm$, then $j_3^\pm, j_4^\pm$ to $j_{34}^\pm$ and finally $j_{12}^\pm, j_{34}^\pm$ to $j_{1234}^\pm$. Now since only $j_{1234}^\pm = 0$ contributes to (14.3.59) we must necessarily have $j_{12}^\pm = j_{34}^\pm =: J^\pm$. It is possible to explicitly calculate the coefficients of the term corresponding to $J^\pm \in \{|j_1^\pm - j_2^\pm|, \ldots, j_1^\pm + j_2^\pm\} \cap \{|j_3^\pm - j_4^\pm|, \ldots, j_3^\pm + j_4^\pm\}$ by using formula (32.3.4).

Define

$$
C^{m_1,m_2,m_3,m_4}_{j_1,j_2,j_3,j_4;J} := \frac{1}{\sqrt{2J+1}} < Jm_1 + m_2 | j_1 m_1, j_2 m_2 >< Jm_3 + m_4 | j_3 m_3, j_4 m_4 >
$$

(14.3.60)

and

$$
C^{M_1,M_2,M_3,M_4}_{\rho_1,\rho_2,\rho_3,\rho_4;\rho} := \prod_{\epsilon=\pm} C^{m_1^\epsilon,m_2^\epsilon,m_3^\epsilon,m_4^\epsilon}_{j_1^\epsilon,j_2^\epsilon,j_3^\epsilon,j_4^\epsilon;J^\epsilon}
$$

(14.3.61)

where $\rho = (J^+, J^-)$ and similarly with (m_k, M_K) replaced by (n_k, N_k). Then it is easy to see that (14.3.60) turns into

$$
\int_{SO(4)} d\mu_H(g) \prod_{k=1}^{4} [\rho_k(g_k^\sigma)]_{M_k N_k}
$$

$$
= \sum_{\rho} C^{M_1,M_2,M_3,M_4}_{\rho_1,\rho_2,\rho_3,\rho_4;\rho} \overline{C^{N_1,N_2,N_3,N_4}_{\rho_1,\rho_2,\rho_3,\rho_4;\rho}} \delta_{M_1+M_2,N_3+N_4} \, \delta_{M_3+M_4,N_1+N_2} \quad (14.3.62)
$$

where, for example, $M_1 + M_2 = (m_1^+ + m_2^+, m_1^- + m_2^-)$, etc. We can now evaluate (14.3.58). Notice that at each dual vertex $v \in C_0(K^*)$ there are precisely five incident dual edges e. These edges are either ingoing ($v = t(e)$ is the terminal point of e) or outgoing ($v = b(e)$ is the beginning point of e) at v. For the outgoing edges we attribute the factors $C^{\{M_e^f\}_{e\in E_f}}_{\{\rho_f\}_{e\in E_f};\rho_e}$ to v and for the ingoing ones we assign the factor $\overline{C^{\{N_e^f\}_{e\in E_f}}_{\{\rho_f\}_{e\in E_f};\rho_e}}$ to v. Since there are altogether 10 such factors (because each of the five integrals of the form (14.3.62) produces two of them), to each vertex we can attribute five such factors. The final formula is therefore given by

$$
Z_{\mathrm{BF}}(K^*) = \sum_{\{\rho_f\}_{f \in C_2(K^*)}} \sum_{\{\rho_e\}_{e \in C_1(K^*)}} \sum_{\{M_e^f, N_e^f\}_{e \in C_1(K^*); \, \sigma(f,e) \neq 0}}
$$

$$
\times \prod_{f \in C_2(K^*)} d_f \, C_{\rho_f}\left(\{M_e^f, N_e^f\}_{e \in E_f}\right)
$$

$$
\times \prod_{e \in C_1(K^*)} \delta_{\sum_{\sigma(f,e)=1} M_e^f, \sum_{\sigma(f,e)=-1} N_e^f} \, \delta_{\sum_{\sigma(f,e)=1} N_e^f, \sum_{\sigma(f,e)=-1} M_e^f}
$$

$$
\times \prod_{v \in C_0(K^*)} \left[\prod_{b(e)=v} C^{\{M_e^f\}_{e \in E_f}}_{\{\rho_f\}_{e \in E_f}; \rho_e} \right] \left[\prod_{t(e)=v} \overline{C^{\{N_e^f\}_{e \in E_f}}_{\{\rho_f\}_{e \in E_f}; \rho_e}} \right]
$$

$$
=: \sum_{\{\rho_f\}} \sum_{\{\rho_e\}} \left[\prod_{f \in C_2(K^*)} A_f(\{\rho_f\}) \right] \left[\prod_{e \in C_1(K^*)} A_e(\{\rho_e\}) \right]
$$

$$
\times \left[\prod_{v \in C_0(K^*)} A_v(\{\rho_f\}, \{\rho_e\}) \right] \tag{14.3.63}
$$

Here the last line is a symbolic notation for the precise formula in which the summation over the magnetic quantum numbers $m_e^{\pm f}$, $n_e^{\pm f}$ has been suppressed. The factors (rather: tensors) A_f, A_e, A_v respectively are called face, edge and vertex amplitudes respectively. It turns out that a symbolic notation of this form is generic for all spin foam models: one assigns intertwiners ρ_e to edges, representations ρ_f to faces and sums over them with specific weights which are products of face, edge and vertex amplitudes depending on those representations and intertwiners.

In the literature quoted one does not find formula (14.3.63) but rather a graphical notation for the vertex amplitude which goes by the name *pentagon diagram*: for a given vertex v, list the intertwiner quantum numbers ρ_e associated with the edges e incident at v by $\rho_0, \ldots, \rho_4$. Each of the edges e_i, $i = 0, \ldots, 4$ is shared by four faces f_{ij}, $j \neq i$ whose associated loop ∂f_{ij} contributes the representation ρ_{ij}. If we now draw five points in a plane representing the edges and connect these points with each other in all possible ways we obtain a pentagon where the lines connecting points i, j are labelled by ρ_{ij} and the points themselves by ρ_i (see Figure 14.2 for an illustration). Notice that the 10 faces only intersect in the edges, giving rise to a three-dimensional projection of the four-dimensional situation which we draw here in two dimensions by suppressing one dimension. Quite remarkably, formula (14.3.63) is still invariant under change of the triangulation K, even when regularising the sums over representations by using quantum groups, that is, the model is topologically invariant at the quantum level.

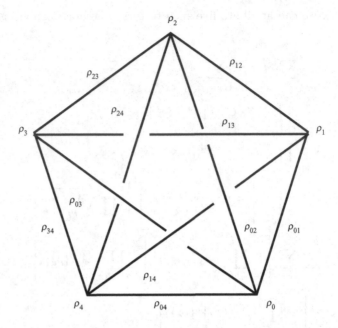

Figure 14.2 The pentagon diagram: the corners represent the five dual edges $e_i; i = 0, \ldots, 4$ incident at a dual vertex v and they are labelled by intertwiner representations ρ_i. The lines represent the 10 dual faces f_{ij} incident at v which contain the edges e_i, e_j; $0 \le i < j \le 4$ in its boundary and they are labelled by representations ρ_{ij}.

14.3.4 Imposing the simplicity constraints

The discretisation strategy for pure BF theory does not work for the Plebanski action because it would be awkward to have variables B labelled by both triangles t of K and dual faces f of K^* respectively: this would somehow mean doubling the number of degrees of freedom and one would not be able to write an expression that is bilinear in just $B(t)$ or just $B(f)$. Since gravity is not topological, there is fortunately no need to use a discretisation scheme which preserves topological invariance.

One way to discretise the simplicity part of the Plebanski action would be to just use K. For each four-simplex Δ consider its set of vertices $V(\Delta)$ and for each $v \in V(\Delta)$ consider the four edges $e_i^v(\Delta)$ outgoing from v. Consider the six triangles $t_{ij}^v(\Delta)$, $1 \le i < j \le 4$ incident at v whose boundary loop starts from v along $e_i^v(\Delta)$ and ends at v along $e_j^v(\Delta)^{-1}$. We take $t_{ij}^v(\Delta) = -t_{ji}^v(\Delta)$. Now it is easy to see that

$$\sum_{\Delta \in C_4(K^*)} \sum_{v \in V(\Delta)} \frac{\lambda_{IJKL}(v)}{5} \epsilon^{ijkl} B^{IJ}\left(t_{ij}^v(\Delta)\right) B^{KL}\left(t_{kl}^v(\Delta)\right) \qquad (14.3.64)$$

converges to

$$\int_M \lambda_{IJKL} \, B^{IJ} \wedge B^{KL} \tag{14.3.65}$$

as we refine the Riemann sum (14.3.64) in the limit $K \to M$ (up to a global factor). This is what one should do and integrate out the Lagrange multiplier field in the regularised path integral.

However, this approach has so far not been followed in the literature, there is no completely systematic construction of the partition function for the Plebanski theory starting from (14.3.65) yet available. Rather, what one does is to use the simplicity constraints in the form (14.3.33) which are now interpreted as follows in the discretised form:

$$\epsilon_{IJKL} B^{IJ}\big(t^v_{ij}(\Delta)\big) B^{KL}\big(t^v_{kl}(\Delta)\big) = \epsilon_{ijkl} \frac{1}{4!} \epsilon_{IJKL} \epsilon^{pqrs} B^{IJ}\big(t^v_{pq}(\Delta)\big) B^{KL}\big(t^v_{rs}(\Delta)\big)$$

$$\tag{14.3.66}$$

for all $\Delta, v \in V(\Delta)$, $i,j,k,l \in \{1,2,3,4\}$ where we have used the same notation for the triangles as in (14.3.64). To see that this reproduces the correct constraints in the continuum, use the parametrisation $(t_1,\ldots,t_4) \mapsto v + t_1 e_1 + \cdots + t_4 e_4$, $t_1 + \cdots + t_4 \le \epsilon, t_k \ge 0$ of Δ in a coordinate chart. Then $t^v_{ij}(\Delta)$ is parametrised by $(t_i, t_j) \mapsto v + t_i e_i + t_j e_j$, $t_i + t_j \le \epsilon$, $t_i, t_j \ge 0$. Now the lowest order in the ϵ expansion of (14.3.66) precisely reproduces the condition (14.3.33).

We can translate (14.3.66) more compactly into the condition that

$$\epsilon_{IJKL} B^{IJ}(t) B^{KL}(t') = 0 \text{ if } t = t' \text{ or } l \cap l' = e \tag{14.3.67}$$

that is, the triangles are equal (intersect in a face of K) or intersect in an edge of K while

$$\epsilon_{IJKL} B^{IJ}(t_{12}) B^{KL}(t_{34}) = \epsilon_{IJKL} B^{IJ}(t_{13}) B^{KL}(t_{42}) = \epsilon_{IJKL} B^{IJ}(t_{14}) B^{KL}(t_{23})$$

$$\tag{14.3.68}$$

when six triangles $t_{ij} = -t_{ji}$ only share a common vertex of a four-simplex with the orientation described above.

Let us denote the full set of simplicity constraints (14.3.66) by $C_\alpha(\{B_f\}_{f \in C_2(K^*)})$ where α runs through some set of labels and we have again made use of the one-to-one correspondence $C_2(K^*) \ni f \mapsto t(f) \in C_2(K)$ between dual faces and triangles and denoted $B_f := B(t(f))$. Then the idea is to introduce a δ-distribution for each α into the partition function of BF theory which one would obtain after integrating over a suitable set of Lagrange multipliers.

Hence one considers the *candidate* Plebanski partition function

$$Z'_P(K^*) := \int \left[\prod_{e \in C_1(K^*)} d\mu_H(g_e) \right] \left[\prod_{f \in C_2(K^*)} d^6 B_f \right] \left[\prod_\alpha \delta(C_\alpha(\{B_f\})) \right]$$

$$\times \exp\left(i \sum_f \mathrm{Tr}(B_f U(\partial f)) \right) \tag{14.3.69}$$

We say candidate because the actual partition function $Z_P(K^*)$ should arise from imposing the constraints that follow from (14.3.64). We indicated this by a first prime in (14.3.69). More approximations and modifications will come in what follows, which will be marked with an increasing number of primes.

Let $X_f^{IJ} := \mathrm{Tr}([P^{IJ}U_f]^T \partial/\partial U_f)$ be the right-invariant vector field on the copy of $SO(4)$ defined by $U_f := U(\partial f)$. Let $C_\alpha(\{X_f\})$ be the same as $C_\alpha(\{B_f\})$ just that B_f was replaced by X_f. Let $S = \sum_f \mathrm{Tr}(B_f U_f)$ then

$$\left[\prod_\alpha \delta(C_\alpha(\{X_f\})) \right] e^{iS} = e^{iS} \prod_\alpha \left[e^{-iS} \delta(C_\alpha(\{X_f\})) e^{iS} \right]$$

$$= e^{iS} \prod_\alpha \left[\delta(e^{-iS} C_\alpha(\{X_f\}) e^{iS}) \right] \tag{14.3.70}$$

where in the last step we made use of the representation $\delta(C) = \int dt/(2\pi) \exp(iC)$. Next we have

$$e^{-iS} \epsilon_{IJKL} X_f^{IJ} X_{f'}^{KL} e^{iS} = \epsilon_{IJKL} \left[e^{-iS} X_f^{IJ} e^{iS} \right] \left[e^{-iS} X_{f'}^{KL} e^{iS} \right]$$

$$= \epsilon_{IJKL} (X_f^{IJ} + i[X_f^{IJ}, S]) (X_{f'}^{KL} + i[X_{f'}^{KL}, S])$$

$$\tag{14.3.71}$$

But

$$[X_f^{IJ}, S] = [X^{IJ}, \mathrm{Tr}(B_f U_f)] = \mathrm{Tr}([P^{IJ}U_f]^T B_f) \tag{14.3.72}$$

Now if we assume that the measure in (14.3.69) is concentrated, also before integrating over the B_f, on flat connections then we can approximate $U_f \approx 1_4$ and can use $B_f = \sum_{I<J} P^{IJ} B_f^{IJ}$ as well as $\mathrm{Tr}(P^{IJ}P^{KL}) = -\delta_{[K}^I \delta_{L]}^J$ to conclude $[X_f^{IJ}, S] \approx B_f^{IJ}$. Of course, this assumption is not really justified since the flatness condition only arises *after* integrating over the B field. Only if one makes this assumption, however, can we deduce that

$$\left[\prod_\alpha \delta(C_\alpha(\{X_f\})) \right] e^{iS} \approx \left[\prod_\alpha \delta(C_\alpha(\{B_f\})) \right] e^{iS} \tag{14.3.73}$$

In particular we must really set $U_f = 1_4$ immediately after evaluation of $[X_f^{IJ}, S]$ in (14.3.72) as otherwise an infinite number of higher derivatives will act when working out the exponentials that are involved in $\prod_\alpha \int dt \exp(iC_\alpha)$ (notice that the same X_f appears in more than one C_α). Thus, the assumption made is really non-trivial and no real justification exists.

One has to pay a price for (14.3.73): the commuting constraints $C_\alpha(\{B_f\})$ were replaced by the non-commuting constraints $C_\alpha(\{X_f\})$. We will shortly see that they do not form a closed algebra, hence *they are anomalous*, and thus overconstrain the partition function. The advantage of the $C_\alpha(\{X_f\})$ is that we can now integrate over B_f in (14.3.69). Hence we start from a, strictly speaking, modified partition function

$$Z_P''(K^*) := \int \left[\prod_{e \in C_1(K^*)} d\mu_H(g_e) \right] \left[\prod_{f \in C_2(K^*)} d^6 B_f \right] \left[\prod_\alpha \delta(C_\alpha(\{X_f\})) \right]$$

$$\times \exp\left(i \sum_f \mathrm{Tr}(B_f U_f) \right)$$

$$= \int \left[\prod_{e \in C_1(K^*)} d\mu_H(g_e) \right] \left[\prod_\alpha \delta(C_\alpha(\{X_f\})) \right] \left[\prod_{f \in C_2(K^*)} \delta(U_f) \right]$$

$$\tag{14.3.74}$$

where we dropped a constant and made the same assumption about the support of the δ-distribution of U_f as for the BF partition function.

However, this is still not quite what one does. Rather than imposing (14.3.66) with respect to all triangles of K at once one imposes only a subset of those constraints, namely those that involve the triangles of a given four-simplex separately. One does not impose the constraints that involve triangles within different four-simplices, that is *one neglects interactions* between four-simplices. One does this in order to organise the products appearing in (14.3.74) in a form which resembles more BF theory, where one had a neat product of face, edge and vertex amplitudes. However, this clearly does not justify what follows.

One proceeds as follows: we have seen that edges e of K^* connect the baryonic centres $v = \hat{\sigma}^{(4)}$, $v' = \hat{\sigma}^{(4)\prime}$ (i.e., dual vertices) of neighbouring four-simplices $\sigma^{(4)}$, $\sigma^{(4)\prime}$ of K through the baryonic centre $\tilde{v} = \hat{\sigma}^{(3)}$ of the tetrahedron $\sigma^{(3)} = \sigma^{(4)} \cap \sigma^{(3)\prime}$ that they share. Let us split the edges into two halves at $\tilde{v}$, that is, $e = [v, v'] = [v, \tilde{v}] \circ [v', \tilde{v}]^{-1} =: e^v \circ (e^{v'})^{-1}$. The point of doing this is that the half-edges e^v, $e^{v'}$ can now be attributed to the vertex at which they start. We can label the half-edges attached to v by e_i^v, $i = 0, \ldots, 4$.

For the description of the geometrical situation, suppose that $v = \hat{\sigma}^{(4)}(v)$ with $\sigma^{(4)}(v) = [v_0, \ldots, v_4]$. Consider the boundary tetrahedra $\sigma_i^{(3)}(v) = [v_0, \ldots, \hat{v}_i, \ldots, v_4]$ with barycentre $b_i = \hat{\sigma}_i^{(3)}(v)$. Then $e_i^v = [v, b_i]$. Suppose that $i < j$ and consider the boundary triangle $\sigma_{ij}^{(2)}(v) = \sigma_i^{(3)}(v) \cap \sigma_j^{(3)}(v) = [v_0, \ldots, \hat{v}_i, \ldots, \hat{v}_j, \ldots, v_4]$ with barycentre $b_{ij} := \hat{\sigma}_{ij}^{(2)}(v)$. The wedge $w_{ij}^v(v)$ is a two-dimensional polyhedron bounded by the loop $[v, b_i] \circ [b_i, b_{ij}] \circ [b_{ij}, v_j] \circ [b_j, v]$ and composed out of the triangles $[v, b_i, b_{ij}] \cup [v, b_{ij}, b_j]$ which belong to the baryonic refinement $B(K)$ of K. The collection of wedges $w_{ij}(v)$, $i < j$ based at v is sometimes called a *fundamental atom* in the literature.

Now by definition, the face f dual to the triangle $\sigma_{ij}^{(2)}(v) = [v_0, \ldots, \hat{v}_i, \ldots, \hat{v}_j, \ldots, v_4]$ is the union of all triangles of the form $[b_{ij}, \hat{\sigma}^{(3)}, \hat{\sigma}^{(4)}]$ where $\sigma_{ij}^{(2)}(v) \subset \partial\sigma^{(3)}$, $\sigma^{(3)} \subset \partial\sigma^{(4)}$. Obviously, the possibilities are $\sigma^{(3)} = [v_0, \ldots, \hat{v}_i v_i', \ldots, \hat{v}_j, \ldots, v_4]$ or $\sigma^{(3)} = [v_0, \ldots, \hat{v}_i, \ldots, \hat{v}_j v_j', \ldots, v_4]$ for some v_i', $v_j' \in C_0(K)$ and $\sigma^{(4)} = [v_0, \ldots, \hat{v}_i v_i', \ldots, \hat{v}_j v_j'', \ldots, v_4]$ or $\sigma^{(4)} = [v_0, \ldots, \hat{v}_i v_i'', \ldots, \hat{v}_j v_j', \ldots, v_4]$ for some v_i'', $v_j'' \in C_0(K)$. The point is that all those triangles are of the form $[b_{ij}, b_i', v'']$ again where b_i' is the barycentre of a tetrahedron in the boundary of a four-simplex with barycentre v''. Hence, all dual faces are composed out of wedges, in particular, the face f dual to $\sigma_{ij}^{(2)}(v)$ is composed out of wedges which all have the point $b_{ij} = \hat{\sigma}_{ij}^{(2)}(v)$ in common.

Suppose that we are given a face $f \in C_2(K^*)$. Denote the barycentre of its dual triangle $t \in C_2(K)$ by b_f. Suppose that $\partial f = e_1 \circ \ldots \circ e_n$ is composed out of n dual edges $e_k = [v_k, v_{k+1}]$, $k = 1, \ldots, n$ with $v_{n+1} = v_1$. Let b_k be the barycentre of the tetrahedron which is shared by the four-simplices dual to v_k, v_{k+1} respectively. We consider the wedge w_k as the two-polyhedron bounded by the loop $[b_f, b_k] \circ [b_k, v_k] \circ [v_k, b_{k-1}] \circ [b_{k-1}, b_f]$ for $k = 1, \ldots, n$ with $b_0 = b_n$. Obviously $\partial w_n \circ \partial w_{n-1} \circ \ldots \circ \partial w_1 = \partial f$.

The purpose of introducing the wedges is that they allow for a useful reorganisation of the product of δ-distributions associated with the U_f. We compute with the abbreviations $p_k = [b_f, b_k]$, $e_k^1 = [v_k, b_k]$, $e_k^2 = [b_k, v_{k+1}]$

$$
\int \left[\prod_{k=1}^{n} d\mu_H(U(p_k)) \right] \left[\prod_{k=1}^{n} \delta(U(\partial w_k)) \right]
$$

$$
= \int \left[\prod_{k=1}^{n} d\mu_H(U(p_k)) \right] \left[\prod_{k=2}^{n} \delta(U(\partial w_k)) \right] \delta(U(\partial w_n) \ldots U(\partial w_1))
$$

$$
= \delta(U(\partial f)) \int \left[\prod_{k=1}^{n} d\mu_H(U(p_k)) \right] \left[\prod_{k=2}^{n} \delta\left(U(p_k) U\left(e_k^1\right)^{-1} U\left(e_{k-1}^2\right)^{-1} U(p_{k-1})^{-1}\right) \right]
$$

$$
\tag{14.3.75}
$$

The integration variable $U(p_1)$ appears only once in the last line in (14.3.75). Using translation invariance, we can drop the δ-distribution involving $U(p_1)$ from the integral as well as the integral over $U(p_1)$. After this the integrand only involves $U(p_2)$ once. Iterating we obtain finally an integral over $U(p_n)$ with no δ-distribution any more. That last integral equals unity due to the normalisation of the Haar measure.

Abstracting from the example, for $e \in E_f$ we introduce, as before, the beginning point $b(e)$, terminal point $t(e)$ and interior point $i(e)$. For the above example we have $b(e_k) = v_k$, $t(e_k) = v_{k+1}$ and $i(e_k) = b_k$. We also consider the paths $p_e^f = [b_f, i(e)]$, which in the above example would be $p_{e_k}^f = [b_f, b_k]$, and the wedges w_e^f, which in the above example would be $w_{e_k}^f = w_k$. Then we obtain the result

$$\delta(U(\partial f)) = \int \left[\prod_{e \in E_f} d\mu_H (U(p_e^f)) \right] \left[\prod_{e \in E_f} \delta(U(\partial w_e^f)) \right] \qquad (14.3.76)$$

Next, what we can do is write for the partition function of BF theory

$$Z_{\mathrm{BF}}(K^*) = \int \left[\prod_{e \in C_1(K^*)} d\mu_H(g_e) \right] \left[\prod_{f \in C_2(K^*)} \delta(U_f) \right]$$

$$= \int \left[\prod_{v \in C_0(K^*)} \prod_{i=0}^{4} d\mu_H(g_i^v) \right] \left[\prod_{f \in C_2(K^*)} \delta(U_f) \right] \qquad (14.3.77)$$

where we have denoted $g_i^v = U(e_i^v)$. This holds because the integrand only depends on the combinations $[b(e), i(e)] \circ [i(e), t(e)]$ of the half-edges, hence a change of variables from the e_i^v to the e and, say, the beginning segments of the e together with the normalisation and translation invariance of the Haar measure reveals the identity.

Now notice that each wedge can be attributed to precisely one dual face f or to precisely one dual vertex (by construction the wedges are adjacent to the baryonic centre of exactly one four-simplex of K). Thus, combining (14.3.76) and (14.3.77) and regrouping terms reveals

$$Z_{\mathrm{BF}}(K^*) = \int \left[\prod_{f \in C_2(K^*)} \prod_{e \in E_f} d\mu_H (U(p_e^f)) \right]$$

$$\times \prod_{v \in C_0(K^*)} \left\{ \int \prod_{i=0}^{4} d\mu_H(g_i^v) \left[\prod_{v \in w} \delta(U_w) \right] \right\} \qquad (14.3.78)$$

where the last product is over the wedges adjacent to v. In the literature one refers to the $U(p_e^f)$ as the *boundary data* of the fundamental atom. The curly bracket in expression (14.3.78) already takes the form of a vertex amplitude.

We can think of the product of δ-distributions appearing in the curly bracket of (14.3.78) as arising from the integral (up to a factor)

$$\prod_{v \in w} \delta(U_w) = \int \prod_{v \in w} d^6 B_w \, e^{i \mathrm{Tr}(B_w U_w)} \qquad (14.3.79)$$

Here the B_w should arise from the triangulation of the four-simplex corresponding to v. Indeed we could have first decomposed the classical integral as

$$\int_M \mathrm{Tr}(B \wedge F) = \sum_{v \in C_0(K^*)} \int_{\sigma^{(4)}(v)} \mathrm{Tr}(B \wedge F) \qquad (14.3.80)$$

and then we could triangulate each $\sigma^{(4)}(v)$ separately. Using the triangulation identities of the previous section we arrive precisely at

$$\int_{\sigma^{(4)}(v)} \mathrm{Tr}(B \wedge F) = \sum_{\sigma^{(2)}(v) \in C_2(\sigma^{(4)}(v))} \mathrm{Tr}\big(B(\sigma^{(2)}(v)) F\big(*_{\sigma^{(4)}(v)} \big(\sigma^{(2)}(v)\big)\big)\big)$$

(14.3.81)

But the $\sigma^{(2)}(v)$ are precisely the triangles $\sigma_{ij}^{(2)}(v)$ and the wedges are precisely their duals within $\sigma^{(4)}(v)$. To see the latter notice that the dual of $\sigma_{ij}^{(2)}(v)$ is the union of triangles $[b_{ij}, \hat{\sigma}^{(3)}, \hat{\sigma}^{(4)}]$ where b_{ij} is the barycentre of the triangle as before and $\sigma^{(3)} \subset \sigma^{(4)}(v)$ must contain the triangle in its boundary and $\sigma^{(4)} \subset \sigma^{(4)}(v)$ must contain $\sigma^{(3)}$ in its boundary. The only possibilities are, using the previous notation, $\sigma^{(3)} = \sigma_i^{(3)}(v)$, $\sigma_j^{(3)}(v)$ and $\sigma^{(4)} = \sigma^{(4)}(v)$ hence (neglecting orientation) $*_{\sigma^{(4)}(v)}(\sigma_{ij}^{(2)}(v)) = [b_{ij}, b_i, v] \cup [b_{ij}, b_j, v] = w_{ij}(v)$ as claimed.

The idea is now to impose the simplicity constraint only on the triangles of a given four-simplex individually. This is of course by no means justified because in principle one should also care about the relations among the $B_t, B_{t'}$ where the triangles t, t' belong to different four-simplices, that is, we should also care about the interaction between four-simplices which will not be the case in the analysis that now follows. In any case, one now replaces (14.3.79) by

$$\int \prod_{v \in w} d^6 B_w \prod_{\alpha_v} \delta(C_{\alpha_v}(\{B_w\})) \, e^{i\mathrm{Tr}(B_w U_w)}$$

$$\approx \int \prod_{v \in w} d^6 B_w \prod_{\alpha_v} \delta(C_{\alpha_v}(\{X_w\})) \, e^{i\mathrm{Tr}(B_w U_w)}$$

$$= \prod_{\alpha_v} \delta(C_{\alpha_v}(\{X_w\})) \prod_{v \in w} \delta(U_w))$$

(14.3.82)

where $X_w^{IJ} = X^{IJ}(U_w)$ is the right-invariant vector field associated with U_w and we have made the same (unjustified) assumption about setting $U_w = 1_4$ (assuming that the measure is concentrated on $U_w = 1_4$ before integrating over B_w) in the derivation of (14.3.82) as before. Here α_v is a label which runs through all the simplicity constraints that one imposes on the triangles contained in the four-simplex dual to v.

Notice that the δ-distribution is invariant under cyclic permutation, that is, $\delta(g_1 g_2) = \delta(g_2 g_1)$ because it involves only characters. Hence, if we define $c_{ij}(v) = [v_i(v), b_{ij}(v)] \circ [b_{ij}(v), v_j(v)]$ with $v_i(v) = t(e_i^v)$ then

$$\delta(U(\partial w_{ij}(v))) = \delta\big(U([b_{ij}(v), v_j(v)])(g_j^v)^{-1} g_i^v U([b_{ij}(v), v_i(v)])^{-1}\big)$$
$$= \delta\big(U([v_i(v), b_{ij}(v)] \circ [b_{ij}(v), v_j(v)])(g_j^v)^{-1} g_i^v\big)$$
$$= \delta\big(U(c_{ij}(v))(g_j^v)^{-1} g_i^v\big)$$

(14.3.83)

Thus when using $U_w = U(c_{ij}(v))(g_j^v)^{-1}g_i^v$ we can replace $X^{IJ}(U_w)$ by $X^{IJ}(U_c)$ in (14.3.82).

This allows us to write the yet once more modified definition of the partition function for the Plebanski theory in the form

$$Z_P'''(K^*) = \int \left[\prod_{f \in C_2(K^*)} \prod_{e \in E_f} d\mu_H(U(p_e^f)) \right] \prod_{v \in C_0(K^*)} \left[\prod_{\alpha_v} \delta(C_{\alpha_v}(\{X_c\})) \right]$$

$$\times \left\{ \int \prod_{i=0}^4 d\mu_H(g_i^v) \prod_{0 \le j \le k \le 4} \delta\big(h_{jk}^v (g_k^v)^{-1} g_i^v\big) \right\} \tag{14.3.84}$$

where we wrote $h_{ij}^v = U(c_{ij}(v))$ for simplicity.

We now perform the five integrals appearing in the curly bracket of (14.3.84). Dropping the label v we have

$$\int \prod_{i=0}^4 d\mu_H(g_i^v) \prod_{0 \le j \le k \le 4} \delta\big(h_{jk}^v (g_k^v)^{-1} g_i^v\big)$$

$$= \sum_{[\rho_{ij}]} \sum_{[L_{ij}],[M_{ij}],[N_{ij}]} \left[\prod_{i<j} d_{\rho_{ij}} [\rho_{ij}(h_{ij})]_{M_{ij} N_{ij}} \right]$$

$$\times \int \prod_k d\mu_H(g_k) \prod_{i<j} [\rho_{ij}(g_i)]_{L_{ij} M_{ij}} \overline{[\rho_{ij}(g_j)]_{L_{ij} N_{ij}}}$$

$$= \sum_{\{\rho_{ij}\}} \sum_{\{L_{ij}\},\{M_{ij}\},\{N_{ij}\}} \left[\prod_{i<j} d_{\rho_{ij}} [\rho_{ij}(h_{ij})]_{M_{ij} N_{ij}} \right]$$

$$\times \prod_i \int d\mu_H(g) \left[\prod_{j<i} [\rho_{ji}(g)]_{L_{ji} M_{ji}} \right] \left[\prod_{i<j} \overline{[\rho_{ij}(g)]_{L_{ij} N_{ij}}} \right]$$

$$= \sum_{\{\rho_{ij}\}} \sum_{\{L_{ij}\},\{M_{ij}\},\{N_{ij}\}} \left[\prod_{i<j} d_{\rho_{ij}} [\rho_{ij}(h_{ij})]_{M_{ij} N_{ij}} \right]$$

$$\times \sum_{\{\rho_i\}} \prod_i C^{\{L_{ji}\}_{j<i},\{L_{ij}\}_{i<j}}_{\{\rho_{ji}\}_{j<i},\{\rho_{ij}\}_{i<j};\rho_i} \overline{C^{\{N_{ji}\}_{j<i},\{M_{ij}\}_{i<j}}_{\{\rho_{ji}\}_{j<i},\{\rho_{ij}\}_{i<j};\rho_i}}$$

$$\times \delta_{\sum_{j<i} L_{ji} + \sum_{i<j} L_{ij},0} \, \delta_{\sum_{j<i} N_{ji} + \sum_{i<j} M_{ij},0} \tag{14.3.85}$$

where we have used the notation of (14.3.60), (14.3.61), (14.3.62), (14.3.63).

We can now impose the simplicity constraints for each four-simplex, which take the explicit form (again we drop the label v and write $X_{ij}^{IJ} := X^{IJ}(h_{ij})$; no summation over repeated indices i, j):

1. 10 triangle constraints

$$\epsilon_{IJKL} X_{ij}^{IJ} X_{ij}^{KL} = 0 \ \forall \, 0 \le i \le j \le 4 \tag{14.3.86}$$

2. 30 tetrahedron constraints

$$\epsilon_{IJKL} X_{ij}^{IJ} X_{ik}^{KL} = 0 \ \forall \, 0 \le i \le j \le 4, \ k \ne i, j \tag{14.3.87}$$

3. 10 four-simplex constraints

$$\epsilon_{IJKL} X_{ij}^{IJ} X_{kl}^{KL} = \epsilon_{IJKL} X_{ik}^{IJ} X_{lj}^{KL} = \epsilon_{IJKL} X_{il}^{IJ} X_{jk}^{KL} \ \forall \, 0 \le i < j < k < l \le 4 \tag{14.3.88}$$

Some of these constraints turn out to be redundant as we will see.

It is not difficult to show that for any pairs of indices $i < j$, $k < l$ we have

$$\epsilon_{IJKL} X_{ij}^{IJ} X_{kl}^{KL} = 8 \left[X_{ij}^{m+} X_{kl}^{m+} - X_{ij}^{m-} X_{kl}^{m-} \right] \tag{14.3.89}$$

where we have expanded $X = X^+ + X^-$ in terms of (anti-)self-dual right-invariant vector fields.

Triangle constraints
The triangle constraints now require that the self-dual and anti-self-dual Casimirs are equal for all $i < j$. Since $\rho_{ij} = \pi_{j_{ij}^+} \otimes \pi_{j_{ij}^-}$ and $[X_{ij}^{m\pm}]^2 \pi_{j_{ij}^\pm} = -j_{ij}^\pm (j_{ij}^\pm + 1) \pi_{j_{ij}^\pm}$ it follows that $j_{ij}^+ = j_{ij}^- =: j_{ij}$ for all $i < j$. This means that $\rho_{ij} = (j_{ij}, j_{ij})$ must be a *simple representation*.

Tetrahedron constraints
The tetrahedron constraints can be written as

$$\epsilon_{IJKL} X_{ij}^{IJ} X_{ik}^{KL} = 8 \left[\left(X_{ij}^{m+} + X_{ik}^{m+} \right)^2 - \left(X_{ij}^{m-} + X_{ik}^{m-} \right)^2 \right] = 0 \tag{14.3.90}$$

where we have exploited that the triangle constraints are already solved. To see what this means define $j_{ij} = j_{ji}$ and consider a particular index i. Then there are three possible recoupling schemes in order to couple the simple spins $j_1^i := j_{ii+1}, \ldots, j_4^i := j_{ii+4}$ with $j_{ij} = j_{ij-5}$ to zero, namely the ones based on coupling j_1^i, j_2^i or j_1^i, j_3^i or j_1^i, j_4^i first (the other three possibilities are equivalent because it does not matter whether we first couple say j_1^i, j_2^i then j_3^i, j_4^i or vice versa, which is why only get $5 \times 3 = 15$ rather than $5 \times 6 = 30$ independent tetrahedron constraints). Suppose we choose the $1, 2$ scheme. Then (14.3.90) says that the spins j_1^i, j_2^i for the self-dual part of the representation must couple to the same intertwining spin $J_i^+ = J_i$ as for the anti-self-dual copy, $J_i^- = J_i$. Thus the intertwiner representation $\rho_i = (J_i^+, J_i^-) = (J_i, J_i)$ must be simple in the $1, 2$ scheme. However, (14.3.90) asks that the intertwiner representation is simple in all three recoupling schemes.

To see what this amounts to we notice that since $X_{ii+1}^{IJ} + \cdots + X_{ii+4}^{IJ} = 0$ due to gauge invariance of the intertwiner we also have $X_{ii+1}^{m\pm} + \cdots + X_{ii+4}^{m\pm} = 0$.

Therefore we have for the $1, 4$ recoupling scheme (let us abbreviate $J_j^{(i)m\pm} = X_{ii+j}^{m\pm}$ and set $J_{jk}^{(i)\pm} := J_j^{(i)m\pm} J_k^{(i)m\pm}$)

$$J_{14}^{(i)+} - J_{14}^{(i)-} = -\left(J_{11}^{(i)+} - J_{11}^{(i)-}\right) - \left(J_{12}^{(i)+} - J_{12}^{(i)-}\right) - \left(J_{13}^{(i)+} - J_{13}^{(i)-}\right)$$

$$(14.3.91)$$

The first term on the right-hand side of (14.3.91) vanishes due to the triangle constraints. Thus the intertwiner simplicity in the $1, 4$ scheme holds automatically if it holds in both the $1, 2$ and $1, 3$ schemes. Therefore there are actually only 10 independent tetrahedron constraints.

However, these 10 tetrahedron constraints are inconsistent with each other, because we have the non-trivial 'integrability condition' that

$$\left[J_{12}^{(i)+} - J_{12}^{(i)-}, J_{13}^{(i)+} - J_{13}^{(i)-}\right] = \left(J_{123}^{(i)+} - J_{123}^{(i)-}\right) \qquad (14.3.92)$$

should vanish when the triangle and tetrahedron constraints hold. Here $J_{123}^{(i)\pm} = \epsilon_{jkl} J_1^{(i)j\pm} J_2^{(i)k\pm} J_3^{(i)l\pm}$ and we used $[J_j^{(i)m\epsilon}, J_k^{(i)n\epsilon}] = (\epsilon + \epsilon')\delta_{jk}\epsilon_{mnl} J_j^{(i)l\epsilon}/2$. The constraint (14.3.92) is a new independent constraint which arises for all $i = 0, \ldots, 4$. These new 'secondary' constraints could now lead to 'tertiary' constraints. We compute

$$\left[J_{12}^{(i)+} - J_{12}^{(i)-}, J_{123}^{(i)+} - J_{123}^{(i)-}\right]$$
$$= \left[J_{12}^{(i)+}, J_{123}^{(i)+}\right] + \left[J_{12}^{(i)-}, J_{123}^{(i)-}\right]$$
$$= \left[J_{11}^{(i)+} J_{23}^{(i)+} - J_{13}^{(i)+} J_{12}^{(i)+} + J_{12}^{(i)+} J_{23}^{(i)+} - J_{13}^{(i)+} J_{22}^{(i)+}\right]$$
$$\quad - \left[J_{11}^{(i)-} J_{23}^{(i)-} - J_{13}^{(i)-} J_{12}^{(i)-} + J_{12}^{(i)-} J_{23}^{(i)-} - J_{13}^{(i)-} J_{22}^{(i)-}\right]$$
$$= \left[J_{11}^{(i)+} + J_{12}^{(i)+}\right] J_{23}^{(i)+} - \left[J_{11}^{(i)-} + J_{12}^{(i)-}\right] J_{23}^{(i)-} \qquad (14.3.93)$$

where we used constraints already satisfied. Using the identity

$$J_{23}^{(i)\pm} = -J_{12}^{(i)\pm} - J_{13}^{(i)\pm} + \frac{1}{2}\left[J_{44}^{(i)\pm} - J_{11}^{(i)\pm} - J_{22}^{(i)\pm} - J_{33}^{(i)\pm}\right] \qquad (14.3.94)$$

where gauge invariance was used we see that (14.3.93) vanishes when the triangle, tetrahedron and secondary constraints hold. A similar calculation with the $1, 3$ constraint shows that there are no tertiary constraints.

The question is whether there are intertwiners which are simple in all recoupling schemes. Let us try to understand the integrability conditions (14.3.94): the recoupled states depending on the $h_{ij}(v)$ are mathematically equivalent to gauge-invariant spin-network states for $SO(4)$ with a four-valent vertex. The (anti-)self-dual volume operator is the modulus of the square root of a linear combination of the $J_{123}^{(i)\pm}, J_{234}^{(i)\pm}, J_{134}^{(i)\pm}, J_{124}^{(i)\pm}$ which can be reduced, using gauge invariance, to the modulus of the square root of $J_{123}^{(i)\pm}$. In order that the integrability conditions are met, each term in the sum over simple J_i must be an eigenstate of both volume

operators with the same eigenvalue because the intertwiners are linearly independent. However, the recoupling basis of spin-network states is not diagonal for the
volume operator as we explicitly saw, in particular for the four-vertex, in Section
13.4.4 whenever the space of possible gauge-invariant intertwiners is more than
one-dimensional. Hence the space of gauge-invariant intertwiners must be one-
dimensional. However, due to the antisymmetry of $J_{123}^{(i)\pm}$, in the one-dimensional
case the intertwiner state is automatically a zero-volume eigenstate.

Notice that the requirement that the space of gauge-invariant and simple
interwiners is one-dimensional is independent of the recoupling scheme because
a change of recoupling scheme corresponds to a unitary transformation which
preserves dimensionalities. The requirement on the one-dimensionality imposes
restrictions on the system of 10 spins j_{ij}, $i < j$: in order that there is a solution to
the simplicity constraint in the 1, 2 scheme the set $\{|j_1^{(i)} - j_2^{(i)}|, \ldots, j_1^{(i)} + j_2^{(i)}\} \cap$
$\{|j_3^{(i)} - j_4^{(i)}|, \ldots, j_3^{(i)} + j_4^{(i)}\}$ must be non-empty and in order that it consists of
one element only we must have

$$\max(|j_1^{(i)} - j_2^{(i)}|, |j_3^{(i)} - j_4^{(i)}|) = \min(j_1^{(i)} + j_2^{(i)}, j_3^{(i)} + j_4^{(i)}) \quad (14.3.95)$$

for all $i = 0, \ldots, 4$. Thus the sum over simple ρ_{ij} is further constrained by
(14.3.95). It admits non-trivial solutions. For instance the following:

$$j_{04} = j_{01} + j_{02} + j_{03}$$
$$j_{14} = j_{01} + j_{12} + j_{13}$$
$$j_{24} = j_{02} + j_{12} + j_{23}$$
$$j_{34} = j_{03} + j_{13} + j_{23}$$
$$j_{04} = j_{14} + j_{24} + j_{34}$$
$$\Rightarrow j_{12} = j_{13} = j_{23} = 0 \quad (14.3.96)$$

while j_{01}, j_{02}, j_{03} are unconstrained.

We conclude that the only solution to the tetrahedron constraints are those
for which the self-dual and anti-self-dual volume spanned by the triangles
$t_{ii+j}(v)$, $j = 1, 2, 3$ equals zero. In other words, the simplicity constraints completely over-constrain the system so that it only allows for degenerate three-
geometries. This means that the presented strategy to quantise the Plebanski
action does not quantise General Relativity [733].

One can evade this conclusion by an ad hoc modification of the intertwiner and
this results in the Barrett–Crane model: the intertwiner was given by (14.3.85),
which we display again as

$$\sum_{\{\rho_i\}} \prod_i C_{\{\rho_{ji}\}_{j<i},\{\rho_{ij}\}_{i<j};\rho_i}^{\{L_{ji}\}_{j<i},\{L_{ij}\}_{i<j}} \overline{C_{\{\rho_{ji}\}_{j<i},\{\rho_{ij}\}_{i<j};\rho_i}^{\{N_{ji}\}_{j<i},\{M_{ij}\}_{i<j}}}$$

$$\delta_{\sum_{j<i} L_{ji} + \sum_{i<j} L_{ij},0} \, \delta_{\sum_{j<i} N_{ji} + \sum_{i<j} M_{ij},0} \quad (14.3.97)$$

In terms of SO(3) representations or Clebsch–Gordan coefficients this becomes explicitly, using the simplicity in the various $1, 2$ schemes

$$\prod_i \left[\sum_{J(12)_i} \prod_{\epsilon=\pm} < \{j_{ij}\}_{j\neq i}, \{m_{ij}^\epsilon\}_{j\neq i} | J_i^{(12)} > \, < \{j_{ij}\}_{j\neq i}, \{n_{ij}^\epsilon\}_{j\neq i} | J_i^{(12)} > \right]$$

(14.3.98)

where we use the product of three CGCs

$$< \{j_{ij}\}_{j\neq i}, \{m_{ij}\}_{j\neq i} | J_i^{(12)} >: = \; < j_{i+1}m_{i+1}, j_{i+2}m_{i+2} | J_i^{(12)} m_{i+1} + m_{i+2} >$$
$$\times < j_{i+3}m_{i+3}, j_{i+4}m_{i+4} | J_i^{(12)} m_{i+3} + m_{i+4} >$$
$$\times < J_i^{(12)} m_{i+1} + m_{i+2}, J_i^{(12)} m_{i+3} + m_{i+4} | 00 >$$

and we have split L_{ij}, M_{ij}, N_{IJ} in the form (k^+, k^-) and denoted the entries by $m_{ij}^\pm$, $n_{ij}^\pm$. The superscript (12) in $J_i^{(12)}$ is to make it explicit that we have used the $1, 2$ scheme as before. The coefficients (14.3.99) are the expansion coefficients of the zero-momentum eigenstate $|J_i^{(12)} >$ with respect to the recoupling state $1, 2$ in the tensor product basis. To verify this formula, just iterate (32.3.4).

Now consider the following objects corresponding to a $1, 3$ scheme

$$< \{j_{ij}\}_{j\neq i}, \{m_{ij}\}_{j\neq i} | J_i^{(13)} >: = \; < j_{i+1}m_{i+1}, j_{i+3}m_{i+3} | J_i^{(12)} m_{i+1} + m_{i+3} >$$
$$\times < j_{i+3}m_{i+2}, j_{i+4}m_{i+4} | J_i^{(12)} m_{i+2} + m_{i+4} >$$
$$\times < J_i^{(12)} m_{i+1} + m_{i+3}, J_i^{(12)} m_{i+2} + m_{i+4} | 00 >$$

Since the states $|J_i^{(12)} >$, $J_i^{(13)} >$ form an orthonormal basis in the space of gauge-invariant states we can expand (14.3.98) in terms of the states of the $1, 3$ scheme

$$\prod_i \left[\sum_{J_i^{(13)\pm}, J_i^{(13)\pm\prime}} \prod_{\epsilon=\pm} < \{j_{ij}\}_{j\neq i}, \{m_{ij}^\epsilon\}_{j\neq i} | J_i^{(13)\epsilon} > \, < \{j_{ij}\}_{j\neq i}, \{n_{ij}^\epsilon\}_{j\neq i} | J_i^{(13)\epsilon\prime} > \right]$$

$$\times \left[\sum_{J(12)_i} < J_i^{(13)+} | J_i^{(12)} > \, < J_i^{(13)-} | J_i^{(12)} > \, < J_i^{(13)+\prime} | J_i^{(12)} > \, < J_i^{(13)-\prime} | J_i^{(12)} > \right]$$

(14.3.99)

We see what prevents (14.3.99) from reducing to a sum over a single label $J^{(13)}$: there are four factors in the last line of (14.3.99) involved. If there were only two then we could use the completeness relation $\sum_{J_i^{(12)}} |J_i^{(12)} >< J_i^{(12)}| = 1$ in order to achieve the goal. This motivates us to replace (14.3.98) by

$$\prod_i \left[\sum_{J_i^{(12)}, J_i^{(12)\prime}} \prod_{\epsilon=\pm} < \{j_{ij}\}_{j\neq i}, \{m_{ij}^\epsilon\}_{j\neq i} | J_i^{(12)} > \, < \{j_{ij}\}_{j\neq i}, \{n_{ij}^\epsilon\}_{j\neq i} | J_i^{(12)\prime} > \right]$$

(14.3.100)

or by

$$\prod_i \left[\sum_{J_i^{(12)+}, J_i^{(12)-}} \prod_{\epsilon=\pm} < \{j_{ij}\}_{j\neq i}, \{m_{ij}^\epsilon\}_{j\neq i} | J_i^{(12)\epsilon} > < \{j_{ij}\}_{j\neq i}, \{n_{ij}^\epsilon\}_{j\neq i} | J_i^{(12)\epsilon} > \right]$$

$$(14.3.101)$$

In both cases there are now two independent sums over $J^{(12)}$ recoupling spins so that the form of (14.3.100) or (14.3.101) is restored after introducing the $1, 3$ scheme. That these two substitutions are the only possible ones has been shown in [701], and one calls (14.3.100) or (14.3.101) a *simple intertwiner*. They lead to what is known as the Barrett–Crane A or B model respectively.

Four-simplex constraints
Neither the (more honest) degenerate model nor the Barrett–Crane substitution take the four-simplex constraints into account. They are not implied by the triangle and tetrahedron constraints because they involve new quadratic invariants of the X_{ij}^{IJ} with no repeated indices. In fact, it is a miracle that there is a solution to the triangle and tetrahedron constraints at all because including the secondary (consistency) conditions they themselves already comprise 25 rather than the wanted 20 conditions. The additional four-simplex conditions would add another 10 conditions, which are again mutually inconsistent and presumably would have no solutions.

14.3.5 Summary of the status of the Barrett–Crane model

As we have seen, at the moment there is no clear-cut derivation of the Barrett–Crane model from the Palatini or Plebanski action. The issues are summarised below, which we list here in order to make it explicit where improvements have to be made in the future.

(A) To begin with, the Palatini or Plebanski action are actions which give rise to second-class constraints in their canonical formulation. This implies that there is to be included into the measure a Jacobean associated with the corresponding Dirac bracket as is well known [263] and as will be explained in more detail below.

(B) The Plebanski action is not equivalent to the Palatini action. When solving the simplicity constraints one obtains five different sectors corresponding to plus or minus the Plebanski action, plus or minus a topological term or a theory which has no metrical interpretation at all. All of these terms could contribute with equal probability in the path integral. If that was the case, even if the path integral is dominated by classical configurations, its value would not be the exponential of the Palatini action.

(C) One does not really integrate over the Lagrange multiplier corresponding to the simplicity constraints as they appear in Plebanski's action. Rather, one implements the dual form of the constraints by hand.

(D) One substitutes the B field in those constraints by vector fields on the group. This can be justified at best when the path integral is dominated by flat connections. However, one does that obviously before integrating over the B field while the flatness property, as in BF theory, only arises after integrating over the B field. This is not justified.

(E) After integrating over the B field one obtains δ-distributions which are supported also on configurations for which the connection is not flat. These configurations are neglected by hand.

(F) One imposes the simplicity constraints only for triangles within every four-simplex, that is, for each four-simplex separately. Thus one neglects the constraints that arise from triangles belonging to different four-simplices. This means that one neglects interaction terms, the constraints are only imposed locally. The resulting constraints for a given four-vertex are then of three types: 10 triangle, 30 tetrahedron and 10 four-simplex constraints.

(G) Taken seriously, those constraints are inconsistent for every four-simplex. The theory is over-constrained even when only taking into account triangle and tetrahedron constraints, which give rise to 20 plus an additional 5 integrability constraints. The result is a model which describes degenerate three-geometries. More precisely, the triangle constraints impose simple representations on the dual wedges (faces), the tetrahedron constraints impose simple representations (intertwiners) on the dual edges. The integrability conditions impose a vanishing three-volume spanned by any three wedges.

(H) The remaining four-simplex constraints are not taken into account at all. If one did, the theory would presumably have a trivial partition function.

(I) One can modify the degenerate theory of item G by changing the definition of the simple intertwiner. The result is the Barrett–Crane model.

Despite these present shortcomings, the Barrett–Crane model has proved to be an important step in the development of spin foam models, with the aid of which many ideas have been tested. It is certainly not a candidate for quantum gravity, but rather a platform from which improved models can be constructed by modifying it.

14.4 Triangulation dependence and group field theory

Proceeding with our general description, in contrast to (14.2.13) the integral (14.2.17) is expectedly no longer independent of the triangulation T so that one has to sum over all triangulations in order to obtain triangulation independence. This amounts to defining

$$Z_{\mathrm{P}} = \sum_T w(T) Z_{\mathrm{P}}(T) \qquad (14.4.1)$$

Of course, the immediate question is how the weight factors $w(T)$ should be chosen.

A clue for how to do that comes from the matrix model approach to two-dimensional quantum gravity (see, e.g., [708] and references therein). Boulatov and Ooguri [463, 464] respectively have shown that a Feynman-like expansion of a certain field theory over a group manifold (rather than a space-time) gives rise to all possible triangulations of the Ponzano–Regge (or the Turarev–Viro) model in three dimensions with G = SU(2) and the Crane–Yetter model in four dimensions respectively [677–681] with G = SO(4). For a recent review of group field theory and its relation to spin foam models, see [709]. In [465] de Pietri, Freidel, Krasnov and Rovelli applied these ideas in order to recover the Barrett–Crane model from a field theory formulation. To see how this works, consider first the case of the BF theory in $D = 3$. Here one considers a real scalar field over SO(4)^4 which is right-invariant, that is, $\phi(h_1, h_2, h_3, h_4) = \phi(h_1 g, h_2 g, h_3 g, h_4 g)$ for any $g \in$ SO(4). One can always obtain such a ϕ from a non-invariant field ϕ' by $\phi = \int_{\mathrm{SU}(2)} d\mu_H(g) R_g^* \phi'$. The Boulatov–Ooguri action is then given by

$$
\begin{aligned}
S'_{\mathrm{BO}} = & \int_{\mathrm{SO}(4)^4} d\mu_H(h_1) d\mu_H(h_2) d\mu_H(h_3) d\mu_H(h_4) \phi^2(h_1, h_2, h_3, h_4) \\
& + \frac{\lambda}{5!} \int_{\mathrm{SO}(4)^{10}} d\mu_H(h_1) d\mu_H(h_2) d\mu_H(h_3) d\mu_H(h_4) d\mu_H(h_5) \\
& \times d\mu_H(h_6) d\mu_H(h_7) d\mu_H(h_8) d\mu_H(h_9) d\mu_H(h_{10}) \\
& \times \phi(h_1, h_2, h_3, h_4) \phi(h_5, h_6, h_7, h_8) \phi(h_7, h_3, h_8, h_9) \\
& \times \phi(h_9, h_6, h_2, h_{10}) \phi(h_{10}, h_8, h_5, h_1) \qquad (14.4.2)
\end{aligned}
$$

which looks almost like a $\lambda \phi^5$ theory. One can now develop the usual Feynman rules for this field theory, giving rise to propagators and vertex functions, and construct the perturbation theory as an expansion in powers of λ (see Figures 14.3 and 14.4). The result is

$$\int [d\phi] e^{-S_{\mathrm{BO}}(\phi)} = \sum_T w(T) Z_{\mathrm{BF}}(T) \qquad (14.4.3)$$

with specific weight factors $w(T) = \lambda^{N(T)}/S(T)$ where $N(T)$ is the number of Feynman graph vertices and $S(T)$ the usual combinatoric factor of Feynman graphs. Notice that the sum over triangulations is redundant for BF theory but not for General Relativity.

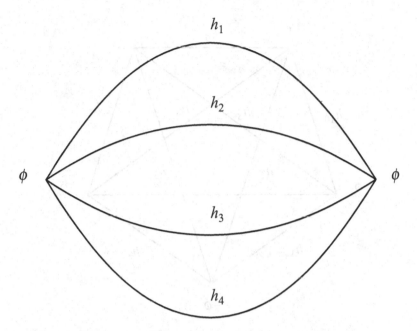

Figure 14.3 Kinetic term of the action of group field theory. The two field insertions each depend on the same quadruple of group variables.

The way in which Feynman diagrams produce triangulations is as follows: as in ordinary Euclidean QFT one defines n-point functions heuristically by

$$S(h_1, \ldots, h_n) := \frac{1}{Z} \int \cdots [d\phi] \ldots e^{-S(\phi)} \, \phi(h_1) \ldots \phi(h_n) \qquad (14.4.4)$$

which are the analogues of Schwinger functions. Here $h_j = (h_j^1, \ldots, h_j^4)$ is a quadruple of group elements. The two-point function is related to the propagator, while the five-point function is related to the vertex in the usual way. Due to the various translation invariances of the action S, the propagator only depends non-trivially on four group elements while the vertex only depends non-trivially on 10 group elements. This is like momentum conservation in ordinary QFT. Notice that as in usual QFT these facts are easiest to prove in Fourier space, that is, by summing over representations rather than integrating over the group. Hence a propagator can be pictured by a strand of four parallel lines, each line representing a group element or a representation while a vertex can be pictured by 10 lines or representations, where the ends of four of them bundle in a corner. In those corners a propagator can connect to the vertex (see Figures 14.5 and 14.6). The Feynman rules of the group field theory are now such that each of the lines in a strand can connect to the four lines in the corner of a vertex in several possible ways. Let us follow such a line as it runs through various propagators and vertices. When it reconnects to itself, we call such a line a cycle. We now define a map between the objects of the Feynman graph and the objects

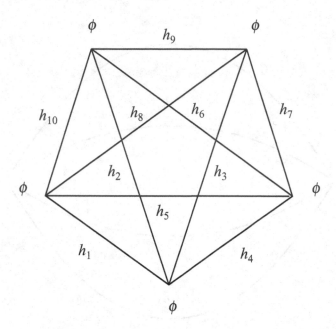

Figure 14.4 Potential term in the action of the group field theory. The five field insertions depend democratically on 10 group variables.

of a cell complex: each vertex of the Feynman graph corresponds to a vertex of the complex, each propagator of the Feynman graph to an edge of the complex and each cycle of the Feynman graph to a face of the complex. Notice that not all such abstract complexes obtained via Feynman graphs can be realised as a triangulation, hence the sum (14.4.3) also contains extra terms without such an interpretation. See [466] for a general analysis of when the 'fat Feynman graphs' that arise from this general type of tensor matrix model give rise to a triangulation of a manifold. The various projections in the models discussed below reduce the sums over representations to simple ones.

Given the fact that the Barrett–Crane model basically reduces the SO(4) $\cong$ SU(2)$_L \times$ SU(2)$_R/Z_2$ of the BF theory to SU(2) it was natural to try to reduce the Crane–Yetter model to the Barrett–Crane model by requiring separate right invariance under SU(2), that is, $\phi(g_1, g_2, g_3, g_4) = \phi(g_1 h_1, g_2 h_2, g_3 h_3, g_4 h_4)$ for any $h_1, \ldots, h_4 \in$ SU(2). Notice that such a field effectively only lives on SU(2)^4 precisely as wanted (more precisely, its Peter and Weyl expansion reduces to simple representations). This can be achieved by means of a projection

$$(P\phi)(g_1, \ldots, g_4)$$
$$= \int_{\mathrm{SU}(2)^4} d\mu_H(h_1) d\mu_H(h_2) d\mu_H(h_3) d\mu_H(h_4) \phi(g_1 h_1, g_2 h_2, g_3 h_3, g_4 h_4)$$

(14.4.5)

where we have chosen some internal direction in four-dimensional Euclidean space in order to write SO(4) in terms of two copies of SU(2) (to choose a

Figure 14.5 The propagator of the group field theory is represented by a strand composed of four lines. Each line is oriented and carries a representation label. The whole strand of lines corresponds to an edge in the simplicial complex interpretation of the group field theory Feynman diagram.

SU(2) subgroup of SO(4)). The field $P\phi$ is independent of that direction since it is invariant under simultaneous right action by SO(4) as well. The theory considered in [465] is given by (14.4.2) just that ϕ is replaced by $P\phi$, that is,

$$
\begin{aligned}
S'_{\text{BC}} = & \int_{\text{SO}(4)^4} d\mu_H(h_1) d\mu_H(h_2) d\mu_H(h_3) d\mu_H(h_4) (P\phi)^2(h_1, h_2, h_3, h_4) \\
& + \frac{\lambda}{5!} \int_{\text{SO}(4)^{10}} d\mu_H(h_1) d\mu_H(h_2) d\mu_H(h_3) d\mu_H(h_4) d\mu_H(h_5) \\
& \times d\mu_H(h_6) d\mu_H(h_7) d\mu_H(h_8) d\mu_H(h_9) d\mu_H(h_{10}) \\
& \times (P\phi)(h_1, h_2, h_3, h_4)(P\phi)(h_5, h_6, h_7, h_8)(P\phi)(h_7, h_3, h_8, h_9) \\
& \times (P\phi)(h_9, h_6, h_2, h_{10})(P\phi)(h_{10}, h_8, h_5, h_1)
\end{aligned}
\tag{14.4.6}
$$

It was shown that the resulting Feynman expansion indeed gives rise to a sum over triangulations of the Barrett–Crane model.

The individual terms of the resulting series, however, are still divergent. In [461] Perez and Rovelli suggested a slight modification of (14.4.6) by removing the projection in the quadratic term, that is,

$$
\begin{aligned}
S'_{\text{PR}} = & \int_{\text{SO}(4)^4} d\mu_H(h_1) d\mu_H(h_2) d\mu_H(h_3) d\mu_H(h_4) \phi^2(h_1, h_2, h_3, h_4) \\
& + \frac{\lambda}{5!} \int_{\text{SO}(4)^{10}} d\mu_H(h_1) d\mu_H(h_2) d\mu_H(h_3) d\mu_H(h_4) d\mu_H(h_5) \\
& \times d\mu_H(h_6) d\mu_H(h_7) d\mu_H(h_8) d\mu_H(h_9) d\mu_H(h_{10}) \\
& \times (P\phi)(h_1, h_2, h_3, h_4)(P\phi)(h_5, h_6, h_7, h_8)(P\phi)(h_7, h_3, h_8, h_9) \\
& \times (P\phi)(h_9, h_6, h_2, h_{10})(P\phi)(h_{10}, h_8, h_5, h_1)
\end{aligned}
\tag{14.4.7}
$$

Figure 14.6 The vertex of the group field theory is represented by a collection composed of 10 lines. Four of their ends terminate in five corners and this is where the strand of a propagator can connect. The orientation of the lines matches with the one on the lines within the propagators that connect. The representations on the lines are determined by the one on the cycles. The whole collection of lines corresponds to a vertex in the simplicial complex interpretation of the group field theory Feynman diagram.

which is free of certain bubble divergences in its Feynman expansion. In [462] Perez proved that the resulting model, which is only a slight variation of the Barrett–Crane model and which effectively only depends on simple representations, is actually *finite* order by order in perturbation theory (triangulation refinement). Of course, this does not show that the series converges but it is anyway a remarkable result that no renormalisation is necessary. Besides, in [710, 711] it was demonstrated that any Euclidean spin foam model can be written as a field theory over a compact group manifold. In a sense, group field theories are *dual* to spin foam models in the sense that the vertex amplitude of a spin foam model is in one-to-one correspondence to the interaction term of the group field theory (the kinetic term is universal).

Another virtue of the group field theory formulation is that it leads to a proposal for the physical inner product [709]: consider two spin-network functions T_{γ_i}, T_{γ_f} cylindrical over initial and final graphs γ_i, γ_f respectively which are four-valent. Now consider the set of all triangulations T appearing in (14.4.1) subject to two conditions: (1) the graphs γ_f, γ_i define the boundary of the dual $*T$; (2) T is a tree diagram in the sense of the group field theory (no closed

Feynman loops). It is not difficult to see that the set of these T is finite, hence (14.4.1) converges when restricting the sum to those.[2] Then the restricted partition function (14.4.1) with the spin-network states inserted is the definition of the physical inner product $< \eta(T_{\gamma_f}), \eta(T_{\gamma_i}) >_{\text{phys}}$ with a certain rigging map η. If one can show that this sesquilinear form is positive definite but does have a non-trivial kernel, a necessary requirement for a rigging map (different kinematical states get mapped to the same physical state), then it serves as a candidate for a physical inner product. At this stage this is just a proposal. It would be crucial to see in which sense if any the image of η satisfies (some version of the) Hamiltonian constraints. The positivity has been established for a certain class of vertex amplitudes including those of the Euclidean Barrett–Crane model in [735]. The positivity is based on the fact that $\text{Spin}(4) \cong \text{SU}(2) \times \text{SU}(2)$, which means that for simple representations the intertwiner always involves squares of higher j symbols which is non-negative. This idea of proof does not extend to the Lorentzian case in an obvious way, however, work on this is in progress.

So far we have only discussed the Euclidean theory. Can one also deal with the Lorentzian case? In [455] Barrett and Crane modified their Euclidean model to the Lorentzian case. One obstacle is that one now has to deal with the non-compact gauge group $\text{SO}(1,3)$ for which all non-trivial unitary representations are infinite-dimensional. The unitary representations of the universal covering group $\text{SL}(2,\mathbb{C})$ are labelled by a pair $(n,\rho) \in \mathbb{R}_0^+ \times \mathbb{N}_0^+$, quite similar to the case of the universal covering group $\text{SU}(2) \times \text{SU}(2)$ of $\text{SO}(4)$ which are labelled by a pair $(j,j') \in \mathbb{N}_0/2 \times \mathbb{N}_0/2$. For an exhaustive treatment see [712]. Following an analogous procedure that has led to the constraint $j = j'$ in the Euclidean case we now find that the simplicity constraint leads to $n\rho = 0$, that is, either $n = 0$ or $\rho = 0$. These representations pick an $\text{SL}(2,\mathbb{R})$ or $\text{SU}(2)$ subgroup within $\text{SL}(2,\mathbb{C})$ for $n = 0$ or $\rho = 0$ respectively. To see where this comes from, one notices that the B field of the BF theory essentially becomes, upon canonical quantisation, an angular momentum operator and the Casimir operators are given by $C_1 = L_{ij}L^{ij}$, $C_2 = L_{ij}(*L)^{ij}$, the simplicity constraint becomes $C_2 = 0$. In the Euclidean case the spectra are $C_1 = j(j+1) + j'(j'+1)$, $C_2 = j(j+1) - j'(j'+1)$ while in the Lorentzian case the spectra are $C_1 = [n^2 - \rho^2 - 4]/4$, $C_2 = n\rho/4$. We see that in the Euclidean case the simple representations are 'spacelike' representations $C_1 \geq 0$ while the simple representations with $n = 0, \rho = 0$ for the Lorentzian theory are timelike and spacelike respectively. The definition of the δ-distribution now becomes more complicated because there is no Peter and Weyl basis any longer. Rather, one has direct integrals and sums respectively for

---

[2] Since no loops in the Feynman diagrams are allowed, these diagrams only contain triangulations whose only faces are those dual to the triangles in the boundary spin-network graph. It is not clear whether this restriction captures enough dynamical information about the amplitude.

the simple continuous and discrete series of representations respectively and in order to evaluate the state sum amplitudes one must now perform complicated integrals also, rather than just discrete sums. In [713] Baez and Barrett proved that nevertheless a large class of these amplitudes are 'integrable'.

In [714] Perez and Rovelli managed to show that also (a variant of) the Lorentzian Barrett–Crane model can be defined as a field theory on a group manifold including the sum over triangulations again. Basically, what one does is to replace in (14.4.7) the group $SO(4)$ by $SL(2, \mathbb{C})$ while the projection P can now be performed with respect to any of the two subgroups $SL(2, \mathbb{R})$ and $SU(2)$ respectively while the field ϕ is now simultaneously $SL(2, \mathbb{C})$ right-invariant. In [714] the choice $SU(2)$ was made in order to define P, which is therefore given by (14.4.5) with $g_I \in SO(4)$ replaced by $g_I \in SL(2, \mathbb{C})$, $I = 1, 2, 3, 4$. Finally, in [715, 716] Crane, Perez and Rovelli succeeded in proving, using the results of [713], that the Lorentzian field theory [714] is finite order by order in perturbation theory at least on what they call 'regular' triangulations.

14.5 Discussion

(i) *Spin foams and canonical theory*

What is missing is an interpretation of these spin foam models. The interpretation of spin foam models *must go through the canonical theory*. In fact, spin foam models for GR are supposed to be path integral-like expressions for the physical inner product, which in turn is defined by the Hamiltonian constraint. This was precisely the starting point of Reisenberger and Rovelli as sketched at the beginning of this chapter. However, as we have seen, using the Hamiltonian constraint one presently cannot give any mathematical meaning to the associated model, moreover, due to the non-commutativity of the Hamiltonian constraint it is unlikely that it really defines a (generalised) projector. On the other hand, the current spin foam models start from a heuristic path integral Ansatz, the exponential of the Palatini action, but it is a priori unclear whether they have anything to do with the canonical theory. The reason for that is manifold:

 – From experience with field theory on Minkowski space one is used to the folklore knowledge that the transition amplitude with respect to a Hamiltonian is given, heuristically, by a path integral over configuration space histories weighted by the exponential of i times the Lagrangian action. This, however, implicitly assumes that the Hamiltonian is at most quadratic in the momenta. If that is not the case, then one can still write a path integral over histories in phase space weighted by i times the exponential of the Hamiltonian action, however, one cannot do the Gaußian integral over momentum space any longer so that the Lagrangian remains in its Hamiltonian, not manifestly covariant $D + 1$ split form. Hence for such more general theories, in particular highly

non-polynomial ones such as GR, it is in general not possible to arrive
at the Lagrangian, and hence four-dimensionally covariant form of the
action.

— The path integral for GR is *not a transition amplitude between initial and
final states at all*. It simply does not have this interpretation. What one
does is (1) to take kinematical states ψ, ψ' which are not solutions to the
Wheeler–DeWitt constraint, (2) to compute[3] the transition amplitude
$< \psi, \exp(i\hat{H}^\dagger(N))\psi >_{\mathrm{kin}}$ for *all the infinite number of 'Hamiltonians'*
$\hat{H}^\dagger(N)$ and (3) to functionally integrate over all lapse functions N with
some measure $[dN]$. There is no Hamiltonian here, therefore one should
never use the words 'transition amplitude'. The correct interpretation
is only that of a physical inner product between the formal solutions
$\int [dN] \exp(i\hat{H}'(N))\psi$.

— One might think that this still has a chance of giving rise to the Palatini
action or the Einstein–Hilbert action. Hence, given some states in the
kinematical Hilbert space, in order to get simultaneous solutions to all
constraints one formally hopes to get by the usual manipulations

$$\int [dN] [d\vec{N}] [d\Lambda] < \psi, \exp(i[\hat{H}^\dagger(N) + \widehat{\vec{H}}^\dagger(\vec{N}) + \hat{G}^\dagger(\Lambda)])\psi' >_{\mathrm{kin}}$$

$$= \int [dN] [d\vec{N}] [d\Lambda] [dE] [dA] \, \overline{\psi(A_0)}\psi'(A_1)$$

$$\times \exp\left(i \int_0^1 dt \int d^3x \{\dot{A}_j^a E_a^j - [-\Lambda^j G_j + N^a H_a + NH]\}\right) \quad (14.5.1)$$

where A_0, A_1 respectively are the field configurations of the spatial connection at $t = 0, t = 1$ respectively.

This is to be compared with the Palatini path integral Ansatz for the
partition function

$$\int [de] [d\omega] \, \exp\left(i \int_M \mathrm{Tr}(F(\omega) \wedge *(e \wedge e))\right) \quad (14.5.2)$$

with appropriate boundary conditions at $\partial M = \sigma_1 \cup \sigma_0$ imposed and
then integrated against $\overline{\psi(A_1)}\psi'(A_0)$. Apart from the fact that one
must show that this results in a positive semidefinite sesquilinear form,
even after dividing by the corresponding null space there is a striking
difference between (14.5.1) and (14.5.2): in (14.5.1) we integrate over
$1 + 3 + 3 + 3 \times 3 + 3 \times 3 = 25$ fields while in (14.5.2) we integrate over
$4 \times 4 + 4 \times 6 = 40$ fields. Indeed, in order to arrive at the canonical form
of the Palatini action which is the exponent of (14.5.6) one must write the
tetrad in the form $e_t^0 = N$, $e_t^j = e_a^j N^a$, $e_a^0 = 0$ where e_a^j is the triad. The
arbitrary choice $e_a^0 = 0$ is called the time gauge. The detailed canonical

---

[3] If it was possible, see however the discussion at the beginning of this chapter.

analysis of Chapter 24 reveals that this leads to 12 second-class constraints in addition to the seven first-class ones that appear in (14.5.6). Therefore, to match (14.5.1) and (14.5.2) one must take these second-class constraints into account, as has been pointed out in [736]. Implementing these second-class constraints reduces (14.5.2) to (14.5.1), however, it contains a non-trivial Jacobean from the determinant of the matrix of the Dirac brackets between the second-class constraints. This was demonstrated for the Plebanski action in [717]. This Jacobean makes it clear that there is a genuine difference between (14.5.1) and (14.5.2) even after taking the second-class constraints into account, whose significance is not very well understood.

– On general grounds it is almost clear that there cannot be an exact match between a covariant path integral and the physical inner product of the canonical theory. The reason is that the covariant path integral is invariant under the group $\text{Diff}(M)$ while the canonical theory is invariant under the dynamical Bergmann–Komar group $\text{BK}(M)$ introduced in Section 1.4. Thus, the two approaches a priori define different theories and what is important is that their respective semiclassical sectors agree with each other. This is not hopeless because precisely when the theory is on-shell, that is, when the equations of motion hold, then $\text{BK}(M)$ and $\text{Diff}(M)$ coincide.

Work towards establishing a link between the two Ansätze has been started in [718]. See also [456] where a sesquilinear form (but not a measure) is given for $\text{SL}(2, \mathbb{C})$ rather than $\text{SU}(2)$ and a connection is made with the canonical form of the Palatini action. The relation with the spin foam models so far constructed is, however, still veiled in 4D.

Recently, progress has been made in 3D: one can establish a precise relation between the canonical and spin foam approaches including Dirac observables, physical inner product and scattering amplitudes (when coupling to point particles) [719–729].

(ii) *Spin foam models from the Master Constraint*

The discussion above leads to the natural question whether one can develop a spin foam model with an exact match with the canonical theory. The natural starting point would seem to be the work of Reisenberger and Rovelli, but we have seen the many formal steps that go into their derivation. However, there is a possible alternative: we may use the Master Constraint. The Master Constraint is defined on the Hilbert space $\mathcal{H}_{\text{diff}}$, hence we would define physical states by[4]

$$\eta : \mathcal{D}_{\text{diff}} \to \mathcal{D}_{\text{phys}}; \ \psi \mapsto \int_{\mathbb{R}} \frac{dt}{2\pi} < \widehat{e^{it}} \ \psi, . >_{\text{diff}} \tag{14.5.3}$$

---

[4] This is a heuristic generalised projector. The mathematically precise projector is defined in Section 30.2.

whcrc $\mathcal{D}_{\text{diff}}$ is the dense subspace of $\mathcal{H}_{\text{diff}}$ consisting of the finite linear span of the $T_{[s]}$ and $\mathcal{D}_{\text{phys}}$ is the image of η which is dense in its completion $\mathcal{H}_{\text{phys}}$. Thc physical inner product is

$$< \eta(\psi), \eta(\psi') >_{\text{phys}} := (\eta(\psi'))[\psi] = \int_{\mathbb{R}} \frac{dt}{2\pi} < \psi', e^{-it\,\hat{\ }}\, \psi >_{\text{diff}} \quad (14.5.4)$$

Expression (14.5.4) looks like a transition amplitude if we formally interpret $\hat{\ }$ as a Hamiltonian (rather than a constraint) except for the integral over t. Notice the conceptual simplicity as compared with the Reisenberger–Rovelli proposal due to the fact that $\hat{\ }$ is a spatially diffeomorphism-invariant operator. But things get even better: we know that already in quantum mechanics the Feynman path integral formula for the transition amplitude $< \psi, e^{-it\hat{H}}\psi' >$ is ill-defined. What is well-defined is to consider that path integral formula for $< \psi, e^{-t\hat{H}}\psi' >$ for positive t *assuming that $\hat{H}$ is bounded from below* and then to analytically continue the result from t to it. The relation bctwccn the 'heat kernel' $< \psi, e^{-t\hat{H}}\psi' >$ and the associated path integral is known as the Feynman–Kac formula [282]. Now since $\hat{\ }$ is bounded from below (by zero), we may hope to give rigorous meaning to a path integral expression for the transition amplitude for the associated strongly continuous contraction semigroup $t \mapsto \exp(-t\,\hat{\ })$, $\|\exp(-t\,\hat{\ })\| \leq 1$. Moreover, the resulting expression *automatically involves a sum over triangulations*. To see this, notice that by the usual skeletonisation procedure

$$< \psi, e^{-t\,\hat{\ }}\, \psi' >_{\text{diff}} = \lim_{n \to \infty} \sum_{[s_1],\dots,[s_n]} < \psi, e^{-t\,\hat{\ }/n}T_{[s_1]} >_{\text{diff}}$$

$$\times < T_{[s_1]}, e^{-t\,\hat{\ }/n}T_{[s_2]} >_{\text{diff}} \dots < T_{[s_n]}, e^{-t\,\hat{\ }/n}\psi' >_{\text{diff}}$$

$$(14.5.5)$$

so that at each intermediate time step we can have any possible (diffeomorphism invariance class of) graph. In fact the expansion in terms of diffeomorphism equivalence classes of spin-network states is not the most convenient one, as has been pointed out by Klauder [643]. It is more practical to use diffeomorphism-invariant coherent states (yet to be developed) for which there is an associated resolution of unity. Notice that (14.5.5) automatically involves the sum over arbitrary triangulations.

(iii) *Semiclassical analysis*

The Perez–Rovelli variant of the Barrett–Crane model seems to be preferred at the moment but it is unclear whether the modification they performed changes the physics significantly or not. Moreover, as we have explained above, there is a chain of steps which are not fully justified in passing from the BF theory to General Relativity, in other words, while it is extremely convincing that one should pass to simple representations it would be nicer to start from the constrained BF theory partition function (14.2.10) and arrive at the Barrett–Crane model by integrating over the

Freidel–Krasnov–Puzio Lagrange multiplicator. Of course even then one has to make some guesses, like the choice of the measure $[dA\, dB\, d\Phi]$. So what one would like to have are some independent arguments that the models proposed have the correct classical limit, for instance by showing that they are a well-defined version of the Reisenberger–Rovelli projector (14.1.7). That this is actually the case without further modification is doubtful given the recent results [730–735] which indicate that the sum over spin configurations is largely dominated by zero or very low spins, which seems not to lead to a nice classical limit. The reason for this is presumably an inappropriate choice of the measure [736] or, in other words, of the precise coefficients in the sum over representation labels. It is also possible that this is related to the fact that classically the simplicity constraint has four solutions $B = \pm e \wedge e,\ \pm * e \wedge e$ which could all contribute to the path integral while only one of them gives the Palatini action.

More intuition concerning the semiclassical limit may come from coupling matter within the spin foam approach, see [719–729].

(iv) *Sum over triangulations*

While we seem to have finiteness proofs for the field theory formulation order by order ('triangulation by triangulation'), it would certainly be even better if one could establish that the sum over triangulations converges. However, that is not really necessary. The reason is that what we would really like to show is that

$$< O > := \frac{\int [d\phi] e^{-S[\phi]} O(\phi)}{Z} \qquad (14.5.6)$$

converges for a sufficiently large set of observables (how to express observables of General Relativity in terms of the field theory on the group manifold is another open question). This object should be regulated by cutting off the sum over triangulations and then one takes the regulator away. The objects (14.5.6) possibly define the finite moments of a rigorously defined measure on some field space on which the field ϕ lives. This is exactly how one usually performs constructive quantum field theory, see [99, 394, 399, 417]: even in free scalar quantum field theory none of the objects $[d\phi], e^{-S[\phi]}, Z$ makes sense separately, it is only the combination $\frac{[d\phi] e^{-S[\phi]}}{Z}$ which can be given a rigorous meaning. The most recent result is that the sum over topologies in three-dimensional Euclidean gravity seems to be uniquely Borel summable[5] [696, 697].

---

[5] Given a series $f(t) = \sum_{n=0}^{\infty} a_n t^{-n}$ its Borel transform $(Bf)(s) = \sum_{n=0}^{\infty} a_n s^{n-1}/(n-1)!$ is defined as the series formed from the term-by-term inverse Laplace transforms of the terms of the original series. If Bf has a non-zero radius of convergence, can be continued to the positive real line and grows at most exponentially along the positive real line then the Laplace transform of Bf exists and is called the Borel sum of f. Many divergent series have a convergent Borel sum which approximates the first few terms in the original series. This fact is used to define the divergent perturbation theory of QFT as an asymptotic series.

(v) *McDowell–Mansouri action*

A very interesting recent development, initiated by Starodubtsev [737–739], is a new type of spin foam model based on the McDowell–Mansouri action (so far in the Euclidean signature only). In this approach one takes a 4D BF theory based on the compact group SO(5) and adds to it a term which explicitly breaks the symmetry down to SO(4), specifically

$$S = \int_M \left(B^{IJ} \wedge F_{IJ} + \frac{v^M}{2} \epsilon_{IJKLM} B^{IJ} \wedge B^{KL} \right) \qquad (14.5.7)$$

where v is some fixed internal vector (we take B to have dimension cm^{-2} for simplicity so that (14.5.7) is dimension-free). Remarkably this action reproduces GR as follows: fix $v^I = \frac{\alpha}{2}\delta_5^I$ for some dimension-free constant α and decompose all the fields, similar to a Kaluza–Klein approach, into SO(4) tensors such as $\omega_{ij} = A_{ij}$, $i, j, \ldots = 1, \ldots, 4$ and additional fields such as A_{i5} where A is the connection underlying F. We identify $e_i := l A_{i5}$ with the tetrad where l is some length scale. The equation of motion for B^{i5} imposes $de + \omega \wedge e = 0$, hence it requires ω to be the spin connection derived from e. The equation of motion for B^{ij} can be solved for B^{ij} and leads to

$$S = \frac{1}{4\alpha} \int_M F^{ij} \wedge F^{kl} \epsilon_{ijkl} \qquad (14.5.8)$$

when reinserted into the action.[6] Decomposing $F^{ij} = R^{ij}(\omega) + e^i \wedge e^j / l^2$ where R is the curvature of ω one finds

$$S = \frac{1}{2\alpha} \int_M \left(F^{ij} \wedge e^k \wedge e^l / l^2 + \frac{1}{2} e^i \wedge e^j \wedge e^k \wedge e^l / l^4 \right) \epsilon_{ijkl}$$

$$+ \frac{1}{4\alpha} \int_M R^{ij} \wedge R^{kl} \epsilon_{ijkl} \qquad (14.5.9)$$

The second term has vanishing variation due to the Bianchi identity for ω, R; it is the Euler topological invariant. The first term is Palatini's action plus a positive cosmological constant term divided by $\hbar$ (recall that the action is dimension-free), provided we make the identification $2l^2\alpha = \hbar\kappa$, $4l^4\alpha = \hbar\kappa/\Lambda$, hence $2l^2\Lambda = 1$, $\alpha = \hbar\kappa\Lambda$.

This has the following interesting consequence: suppose we take the action (14.5.7) as the starting point of a spin foam model, that is, we use a path integral based on $\exp(iS)$. The first difference compared with the treatment in this chapter is that S is a deformed BF action, however, it is *unconstrained*. Next, interpreting the term proportional to α as a perturbation, the zeroth-order term is just a BF action in 4D for the gauge group SO(5). If we believe in the usual saddle point approximation, then

---

[6] Recall that the extrema of an action $S(x, y)$ with respect to both x, y can be found by determining the extremum $x(y)$ for x at fixed y and then determining the extremum of $S'(y) = S(x(y), y)$.

the classical limit of the path integral at $\alpha = 0$ should be proportional to $\exp(iS_{\text{extr}})$ where S_{extr} is the value of the action on-shell. The equations of motion of the BF action require the SO(5) curvature to vanish, in particular, $F^{ij}(A) = R^{ij}(\omega) + 2\Lambda e^i \wedge e^j = 0$. The unique solution is de Sitter space, which is in agreement with present observations. Furthermore, at the presently measured value of Λ, we find $\alpha \approx 10^{-120}$ which means that a perturbation expansion of the path integral around $\alpha = 0$ is *extremely rapidly converging*. It is a bit surprising that an unconstrained, topological BF theory with a symmetry-breaking term gives rise to an action for GR, however, the intuitive reason is that in each order of the perturbation expansion the symmetry-breaking term introduces more and more degrees of freedom as it breaks the symmetry more and more, thus transforming more and more gauge degrees of freedom into propagating ones. The full expansion is then entirely non-topological. More details about this can be found in [737, 738].

Clearly, this is a very interesting model and a complete analysis, which has just been started, is likely to yield valuable insights. Specifically, the fact that there is a rapidly converging expansion around the correct 'vacuum' is a feature that one would like to see incorporated, in some sense, also in the canonical approach. This could happen, for instance, in the sense of a semiclassical approximation by using excitations of a coherent state peaked on (the initial data of) de Sitter space.

(vi) *Other aspects of spin foam models*

In dealing with Lorentzian spin foams it is a valid question in which sense the corresponding quantum evolution is causal in any sense. These questions were first addressed in [740–745] by Markopoulou and Smolin. The idea is then to restrict the class of spin foams to be considered by allowing only those which are causal. See also [746–748].

A different question related to the issue of the classical limit is whether there is some notion of a renormalisation group within spin foam models, which then would answer the question in which sense they depend on the class of triangulations that we sum over or whether we are allowed to perform small changes in the 'initial field theory action' without changing the effective low-energy (semiclassical) theory, in other words whether there is a natural notion of universality classes and the like. A first pioneering work has recently been published by Markopoulou [749, 750] in which the Hopf algebra structure underlying renormalisation in ordinary field theory discovered by Connes and Kreimer [751–753] was applied to coarsening processes of the triangulations that underly spin foams. Related to this is recent work by Oeckl [754]. The idea here is that while GR is background-independent and thus the usual scaling transformations which lead to the renormalisation group are not available (because there is no background-independent notion of scale), one can still use Wilson's notion of the renormalisation group and effective fields theory arising by integrating out microscopic degrees of freedom (block spin transformations). The background-independent version of that should be to coarsen graphs or spin foams because this obviously leads to a reduction of the number of degrees of freedom. This will lead to modifications of the microscopic Hamiltonian constraint which (when using the equations of motion) can be rewritten as higher derivative terms at the effective field theory level. The usual running of the couplings and masses is also present in LQG. However, all the expressions are presumably finite while the physical screening effects, etc. that one observes in experiments are certainly there, it is just that one has to define things operationally (relationally: what is the coupling of field A at an energy level determined by some particle B).

The relation of spin foams with lattice gauge theory and state sum models was further analysed in [755–758]. The consequences of Diff(M)-invariance for spin foam models and its number of physical degrees of freedom were elaborated in [759, 760], which is also a nice collection of facts about smooth and piecewise linear structures on manifolds in various dimensions. The connection between spin foams, (2-)category theory and higher gauge theory was studied in [761, 762]. Matter coupled to spin foam models was investigated in [763–765]. Finally, various interesting aspects of spin foam models without particular category can be found in [766–770].

(vii) *Graviton propagator*

In order to make the connection with Minkowski spacetime, the general boundary formalism for spin foams was proposed in [771, 772]. This formalism was applied more recently in [773, 774] in order to define the graviton propagator from spin foam models: by definition, the spin foam partition function $Z(A_i, A_f)$ with fixed boundary connections A_i, A_f at an initial and final hypersurface σ_i, σ_f respectively, integrated against final and initial kinematical states $\psi_f(A_f), \overline{\psi_i(A_i)} \in \mathcal{H}^0$ respectively, is supposed to be the physical inner product

$$< \eta(\psi_f), \eta(\psi_i) >_{\text{phys}} = \eta(\psi_i)[\psi_f]$$
$$= \int_{\overline{A}} d\mu_0(A_f) \left[\int_{\overline{A}} d\mu_0(A_i) \, \overline{\psi_i(A_i)} Z(A_i, A_f) \right] \psi_f(A_f)$$

from which we read off the rigging map

$$\eta(\psi_i) = \int_{\overline{A}} d\mu_0(A_i) \, Z(A_i, .) \, \overline{\psi_i(A_i)}$$

The idea is now to choose ψ_i, ψ_f to be coherent states which are peaked on initial data m_0^i, m_0^f such that m_0^f is the gauge transform of m_0^i as described by the Einstein equations with respect to some choice of lapse and shift. The corresponding solution to Einstein's equations describes a spacetime background metric g_0 with pull-back to σ_i, σ_f given by q_0^i, q_0^f.

Next we consider the kinematical states $\psi_i' := [\hat{q}_{ab}(x_i) - q_{ab,0}^i(x_i)]\psi_i$, $\psi_f' :=$ $[\hat{q}_{ab}(x_f) - q_{ab,0}^i(x_f)]\psi_f$ (which should be properly smeared). When inserted into the physical inner product formula, one obtains an expression which depends on g_0 and the spatial points x_i, x_f. What is shown in [773, 774] is that the resulting expression, derived from some version of the group field theory definition of the Euclidean Barrett–Crane model, under various assumptions, becomes the correct graviton propagator of the linearisation of gravity around the background g_0! Notice that while the states used are background-dependent, they are still elements of the background-independent Hilbert space $\mathcal{H}^0$. This is because we can use background-dependent complex coefficients of background-dependent spin-network states in order to build a coherent state. The fact that Z is dominated by degenerate metrics therefore seems to be circumvented by choosing appropriate initial and final states. This is a curious result which deserves further exploration. For instance, one would like to understand how to define graviton annihilation and creation operators as Dirac observables and how this compares with the just-sketched heuristic calculation.

15

Quantum black hole physics

Any theory that claims to be a quantum theory of the gravitational field must give a microscopic explanation for the Bekenstein–Hawking entropy of a black hole [775–777] given by

$$S_{\text{BH}} = \frac{\text{Ar}(H)}{4\ell_p^2}$$

where $\text{Ar}(H)$ denotes the area of the event horizon H as measured by the metric that describes the corresponding black hole spacetime and in this chapter we set $\ell_p^2 = \hbar G_{\text{Newton}}$ instead of $\hbar\kappa = 16\pi\hbar G_{\text{Newton}}$.

Heuristically, the above formula arises as follows: Penrose and Hawking proved the famous area law theorems for black holes [207, 208] according to which there is no classical process that can decrease the area of a black hole. While mathematically not entirely trivial to prove, these theorems are physically not very surprising because by definition a black hole curves spacetime in such a way that not even light can escape. Even Newtonian physics tells us that compact massive bodies of mass m can have such a property, namely at best a photon can propagate on a circular orbit around the mass whose radius r is given by the formula $c^2/r = Gm/r^2$, provided of course that the body is so compact that its radius is smaller than $r = Gm/c^2$. General Relativity corrects the so-called Schwarzschild radius by a factor of two, that is, $r_S = 2Gm/c^2$. What happens is that the lightcones from the event horizon, defined by $r = r_S$, onwards are pointing into the interior of the black hole. Thus, no causal physics can prevent a body from falling inside once it has crossed $r = r_S$, which is why the mass of a black hole should only increase. Since the area of a black hole is thus proportional to m^2 it follows that the area can never decrease. This statement sounds familiar from thermodynamics, it reminds us of the second law according to which the entropy of a system in equilibrium can never decrease. This suggests we assume that the entropy of a black hole is proportional to the area of its event horizon. Since the entropy is a dimensionless quantity, from the only constants of nature available one would already guess that $S_{\text{BH}} \propto A/\ell_p^2$, which is precisely what Bekenstein did.

In order to obtain the constant of proportionality and a better physical explanation beyond dimensional analysis one has to go beyond classical physics. Let us consider an observer at rest somewhere in a Schwarzschild spacetime. By the equivalence principle, such an observer is in a situation not unlike a constantly accelerating observer in a Minkowski spacetime because both observers are not in geodesic motion. An accelerating observer in Minkowski space observes instantaneous rest frames which are changing with time. Effectively, the Hamiltonian associated with such an observer becomes time-dependent. In each rest frame one can define an instantaneous vacuum state (no particles) but the definition of vacuum or annihilation and creation operators changes with time and thus an initial vacuum state is no longer void of particles at a later time. This is the so-called *Unruh effect* [23, 24]. Its transcription to the curved spacetime case is called the *Hawking effect*.[1]

It predicts that black holes radiate, and the precise mathematics of QFT on curved spacetimes shows that the spectrum of this radiation is the Planck spectrum of a black body. More precisely, one forms a density matrix by taking the partial trace with respect to the degrees of freedom describing the interior of the black hole (the total Hilbert space is a tensor product of two Hilbert spaces describing the exterior and interior respectively) and that density matrix takes precisely the Gibbs–Planck form. The corresponding temperature is, not surprisingly, related to the peak of the spectrum at a wavelength $\lambda_S \approx r_S$, there is no other physical scale in the problem. It follows that the temperature of the black hole is given by Planck's relation $\hbar \omega_S \approx kT_S$, $\omega_S \approx c/\lambda_S$ where k is the Boltzmann constant. The energy of the black hole is given by its mass $E = mc^2$, thus its entropy is

$$S \approx E/(kT_S) \approx mc^2/(\hbar c/r_S) = r_S^2/(\hbar G/c^3) \qquad (15.0.1)$$

which is almost (15) except for the factor $1/4$ which only the precise calculation can provide.

Formula (15) gives rise to many puzzles:

1. *Microscopical explanation*

   Thermodynamics defines entropy as a measure for missing information. However, QFT on curved spacetimes cannot deliver this explanation because the framework breaks down in situations of extreme (diverging) curvature. Hawking's derivation was for a macroscopic, static Schwarzschild black hole and it is based on the construction of a density matrix which one obtains from a pure state in the total Hilbert space $\mathcal{H} = \mathcal{H}_{\text{out}} \times \mathcal{H}_{\text{in}}$ where out (in) denote degrees of freedoms associated with the outside or inside of the black hole by tracing over the degrees of freedom in $\mathcal{H}_{\text{in}}$. The derivation neglects backreaction effects and that one does not actually know how to describe the interior of the black hole quantum mechanically.

---

[1] Notice that both effects have absolutely nothing to do with the presence of matter and antimatter. We just mention this here because one often hears that the Hawking effect is due to matter–antimatter pair production in the vicinity of the event horizon, whereby magically the antimatter (negative energy) falls into the hole while matter (positive energy) escapes to infinity. The effect exists also for neutral matter and is simply due to the breakdown of the particle concept for accelerated observers.

2. *Information paradox*

   The entropy of the density matrix just described is somehow artificial because as long as the black hole has not completely evaporated due to Hawking radiation, the total system evolves unitarily and if the initial state is pure it remains so until the final stage of the radiation process. However, once the black hole has completely evaporated, the interior of the hole and thus $\mathcal{H}_{in}$ is gone. The system has evolved from a pure state (zero entropy) to a mixed state (non-zero entropy) and thus indicates a breakdown of unitarity.

3. *Redshift problem*

   Another detail that Hawking's derivation does not tell is who actually observes the black body spectrum. It is natural to assume that this is an observer located far away from the hole, perfectly at spatial infinity where the gravitational pull is negligible. However, then the immediate question is where the modes of frequency ω that reach spatial infinity have been created. It would be natural to assume that they were created close to the even horizon but then their frequency there could have been arbitrarily large due to the redshift effect that radiation encounters when climbing out of a gravitational well. However, modes of such large (trans-Planckian) energy must surely be properly described by quantum gravity.

So far, LQG can at best deliver the beginning of an answer to (1). This is what we will describe in detail in what follows.

In [467] Krasnov performed a bold computation: given any surface S with spherical topology, given some area A and an interval $[A - \Delta A, A + \Delta A]$, let us compute the number N of spin-network states T_s such that $< T_s, \widehat{Ar}(S)T_s > \in [A - \Delta A, A + \Delta A]$. Of course, N is infinite. But now let us mod out by the gauge motions generated by the constraints: most of the divergence of N stems from the fact that for a given number of punctures $S \cap \gamma(s)$ and fixed representations $\vec{\pi}(s)$, there are uncountably many different spin-network states with the same area expectation value because different positions of the punctures give different spin-network states. This is no longer the case after modding out by spatial diffeomorphisms. There is, however, still a source of divergence because what matters for the area eigenvalue is more or less only the number of punctures and the spins of the edges that intersect the surfaces S, what happens outside or inside the surface is irrelevant and certainly even after modding by spatial diffeomorphisms one still has $N = \infty$. Therefore Krasnov had to assume that this divergence would be taken care of after modding out the action of the Hamiltonian constraint. Hence, ignoring this final divergence his result for $\Delta \approx \ell_p^2$ was proportional to $Ar(S)/(4\ell_p^2)$. A similar computation by Rovelli [468] confirmed this value.

Of course, more work was necessary in order to make the derivation watertight: for instance, nothing in [467] could prevent one from performing the computation for *any* surface, not necessarily a black hole event horizon so that it was conceptually unclear what the computation showed. Somehow one had to invoke the information that H *is* an event horizon into the computation to get rid of the divergences that were just mentioned. Also, given the local nature of the area eigenvalue counting, it was desirable to localise the notion of an event horizon which can be determined only when one knows the entire spacetime (recall that an event horizon [207, 208] is the internal boundary of the portion of space-time that does not lie in the past of null future infinity), which is completely unphysical from an operational point of view because one would never know if a horizon is really an event horizon since the object under study could collide with a burnt out star in the late period of the universe when all life has deceased. Whether or not H is a horizon, one should be able to determine by performing local measurements in spacetime.

The physical requirement to have a more local notion of black hole horizons leads to the notion of *trapping horizons* [778–781], a special version of which are *isolated horizons* [782–785]. The subject deserves a volume of its own but we have here only space to summarise the main ingredients of the framework and to focus on the quantum aspects. Moreover, it is only for spherically symmetric isolated horizons for which a full quantum treatment is currently available, whence we will mostly treat this particular case.

A summary of the classical and quantum aspects of isolated horizons, that are used in black hole entropy calculations within LQG, can be found in [782]. For recent reviews and a comparison between dynamical, isolated and trapping horizons see [786, 787]. The pivotal papers that describe the details of the classical and quantum formulation respectively are [788–791] and [469] respectively.

15.1 Classical preparations

In order to motivate the notion of an isolated horizon we must recall some material from classical General Relativity.

15.1.1 Null geodesic congruences

A congruence through a region R of a spacetime (M, g) is a family of mutually non-intersecting curves, one through every point of R. The congruence is called null if the tangent vectors along all those curves are null. A null geodesic congruence is a null congruence consisting of geodesics. These are important notions and much of the classical singularity theorems and black hole area theorems of mathematical General Relativity employ techniques associated with them, in particular Raychaudhuri's equation. In the presence of curvature, the congruence has typically only a finite extension R because the curves tend to intersect each other as we will see.

We will denote by l the tangent vector field defined by the congruence. The fact that l is geodesic means that $\nabla_l l = \lambda l$ where λ is a function and

$\nabla g = 0$ is the covariant derivative compatible with g. The quantity λ will be called the acceleration of the congruence, in applications to black hole horizons also called the surface gravity. It is always possible to choose $\lambda = 0$ by means of a reparametrisation: consider the integral curves of l, that is, solve the system of ODEs $\dot{c}_p^l(t) = l(c_p^l(t))$, $c_p^l(0) = p$. Then the geodesic equation reads $D^2 c/dt^2 = (\lambda \circ c)\dot{c}$. Defining $\tilde{c}(t) := (c \circ f)(t)$ and requiring $D^2\tilde{c}/dt^2 = 0$ leads to the ODE $\lambda(f(t))\dot{f}(t)^2 + d^2 f(t)/dt^2 = 0$ which can be integrated by quadratures: integrating $F'(s) := \lambda(s)$ gives $F(f(t)) + \ln(\dot{f}(t)) = \text{const.}$ Integrating $G'(s) = \exp(F(s))$ gives $G(f(t)) = \text{const.}$, which can be inverted for f. Hence, if needed, we may always assume w.l.g. that $\nabla_l l = 0$ in a so-called affine parametrisation. This is convenient for timelike geodesic congruences because the norm l^2 is constant along an affinely parametrised geodesic. For null geodesics this is of course immaterial.

Abusing slightly the standard notation in General Relativity we will consider the following distributions (in the sense of Definition 19.3.3) of subspaces $\tilde{V}_p := \{u \in T_p(M); g(u,l) = 0\}$ and $\tilde{V}_p^* := \{\omega \in T_p^*(M); \omega(l) = 0\}$ for $p \in R$. We also define the equivalence classes $[u] := \{u + rl; r \in \mathbb{R}\}$, $[\omega] := \{\omega + rg(l,.); r \in \mathbb{R}\}$ and the corresponding spaces $\hat{V}_p$, $\hat{V}_p^*$. A tensor $T \in (T_b^a)_p(M)$, considered as a multilinear functional on $[\otimes^a T_p^*(M)] \otimes [\otimes^b T_p(M)]$, can be restricted to $[\otimes^a \tilde{V}_p^*(M)] \otimes [\otimes^b \tilde{V}_p(M)]$ and is then denoted by $\tilde{T}$. If and only if it vanishes when filling any of its entries with l or $g(l,.)$ and the remaining ones with general elements of $\tilde{V}_p$, $\tilde{V}_p^*$ can we define the tensor $\hat{T}$ on $[\otimes^a \hat{V}_p^*(M)] \otimes [\otimes^b \hat{V}_p(M)]$ by $\hat{T}([u_1], \ldots, [u_a]; [\omega_1], \ldots, [\omega_b]) := \tilde{T}(u_1, \ldots, u_a; \omega_1, \ldots, \omega_b)$. It is easy to show that $\hat{T} = [T] = \{T + S\}$ where S is of the form

$$S^{\mu_1 \ldots \mu_a}{}_{\nu_1 \ldots \nu_b} - \sum_{k=1}^{a} l^{\mu_k} S_k^{\mu_1 \ldots \hat{\mu}_k \ldots \mu_a}{}_{\nu_1 \ldots \nu_b} + \sum_{l=1}^{b} l_{\nu_l} S_{k+l}^{\mu_1 \ldots \mu_a}{}_{\nu_1 \ldots \hat{\nu}_l \ldots \nu_b}$$

and the tensors S_k are otherwise arbitrary.

An example for such a projectable tensor is the metric tensor $g_{\mu\nu}$ and its inverse $\hat{g}^{\mu\nu}$. Another is the tensor field $B_{\mu\nu} := \nabla_\mu l_\nu$ since for any $u \in \tilde{V}_p$ we have

$$u^\mu l^\nu B_{\mu\nu} = \frac{1}{2} u^\mu \nabla_\mu l^2 = 0 \text{ and } u^\nu l^\mu B_{\mu\nu} = u^\nu \nabla_l l_\nu = \lambda u^\nu l_\nu = 0 \quad (15.1.1)$$

where in the first equation we used that $l^2 = 0 = \text{const.}$ along the congruence, hence ∇l^2 is orthogonal (= tangential) to it while in the last equation we used the geodesic equation. In other words $l^\nu B_{\mu\nu} = \lambda' l_\mu 0$ and $l^\mu B_{\mu\nu} = \lambda l_\nu$. One now constructs $\hat{B}_{\mu\nu}$ and decomposes it into twist, shear and expansion

$$\hat{\omega}_{\mu\nu} := \hat{B}_{[\mu\nu]}, \ \hat{\sigma}_{\mu\nu} := \hat{B}_{(\mu\nu)} - \frac{1}{2}\theta\hat{g}_{\mu\nu}, \ \theta := \hat{g}^{\mu\nu}\hat{B}_{\mu\nu} \quad (15.1.2)$$

These notions come from an analogy with fluid dynamics by comparing the flow lines of the vector field l with the flow lines of a (generally relativistic) fluid. Notice that the restriction $\tilde{g}_p$ of g_p to $\tilde{V}_p$ is a degenerate 3D metric and that $\hat{g}_{\mu\nu}$ is a two-metric. We can display this more explicitly as follows: let e^I_μ be a co-tetrad for $g_{\mu\nu}$, that is, $g_{\mu\nu} = \eta_{IJ}e^I_\mu e^J_\nu$ where $\eta = \text{diag}(-1, 1, 1, 1)$ is the Minkowski metric. We may form the complex null co-tetrad

$$l_\mu := \frac{e^0_\mu - e^3_\mu}{\sqrt{2}}, \quad k_\mu := \frac{e^0_\mu + e^3_\mu}{\sqrt{2}}, \quad m_\mu := \frac{e^1_\mu + ie^2_\mu}{\sqrt{2}}, \quad \bar{m}_\mu := \frac{e^1_\mu - ie^2_\mu}{\sqrt{2}} \quad (15.1.3)$$

in which the metric takes the form

$$g_{\mu\nu} = -2l_{(\mu}k_{\nu)} + 2m_{(\mu}\bar{m}_{\nu)} \quad (15.1.4)$$

that is, $l^2 = k^2 = m^2 = \bar{m}^2 = l \cdot m = l \cdot \bar{m} = k \cdot m = k \cdot \bar{m} = 0$ and $-l \cdot k = m \cdot \bar{m} = 1$. Conversely, given a null vector field l we may always complete it to a null tetrad with these normalisations. It then follows that $\tilde{g}_{\mu\nu} = h_{\mu\nu} := 2m_{(\mu}\bar{m}_{\nu)}$ is a degenerate 3D metric with l as zero eigenvalue vector while $\hat{g}_{\mu\nu} = [h_{\mu\nu}]$ is a 2D metric of Euclidean signature.

We can now compute, using the definition of the Riemann tensor

$$\nabla_l B_{\mu\nu} = l^\rho([\nabla_\rho, \nabla_\mu] + \nabla_\mu\nabla_\rho)l_\nu$$
$$= l^\rho R_{\rho\mu\nu}{}^\sigma l_\sigma + \nabla_\mu\nabla_l l_\nu - B_\mu{}^\rho B_{\rho\nu} \quad (15.1.5)$$

where the term in the middle would vanish in an affine parametrisation. The tensor field $\nabla_l B_{\mu\nu}$ is projectable because for any u with $u \cdot l = 0$ we have

$$u^\mu l^\nu \nabla_l B_{\mu\nu} = -u^\mu (\nabla_l l^\nu) B_{\mu\nu} = -\lambda u^\mu l^\nu B_{\mu\nu} = 0$$
$$u^\nu l^\mu \nabla_l B_{\mu\nu} = -u^\nu (\nabla_l l^\mu) B_{\mu\nu} + u^\nu \nabla_l \lambda l_\nu = -\lambda u^\nu l^\mu B_{\mu\nu} + \lambda^2 u^\nu l_\nu = 0 \quad (15.1.6)$$

Likewise, all three terms on the right-hand side of (15.1.5) are separately projectable, which one can see from the symmetries of the Riemann tensor, the properties of B and the geodesic equation $\nabla_l l = \lambda l$.

We now derive Raychaudhuri's equation. On the one hand we have

$$h^{\mu\nu}\nabla_l B_{\mu\nu} = \nabla_l \theta - B_{\mu\nu}\nabla_l(l^\mu k^\nu + l^\nu k^\mu)$$
$$= \nabla_l \theta - B_{\mu\nu}(\lambda l^\mu k^\nu + l^\mu \nabla_l k^\nu + \lambda l^\nu k^\mu + l^\nu \nabla_l k^\mu)$$
$$= \nabla_l \theta - \lambda l_\nu (\lambda k^\nu + \nabla_l k^\nu)$$
$$= \nabla_l \theta - \lambda(-\lambda + \nabla_l k \cdot l - n^\nu \nabla_l l_\nu)$$
$$= \nabla_l \theta \quad (15.1.7)$$

where in the last step we used that $k \cdot l = -1$ is constant along the congruence. On the other hand we have, using the definition of the Ricci tensor and $h_{\mu\nu} = g_{\mu\nu} + 2l_{(\mu}k_{\nu)}$

$$h^{\mu\nu}\nabla_l B_{\mu\nu} = -R_{\mu\nu}l^\mu l^\nu - h^{\mu\nu}B_{\mu\rho}B^{\sigma\nu}g^{\rho\sigma} + h^{\mu\nu}(l_\nu \nabla_\mu \lambda + \lambda B_{\mu\nu})$$
$$= -R_{\mu\nu}l^\mu l^\nu - h^{\mu\nu}h^{\rho\sigma}B_{\mu\rho}B^{\sigma\nu} + h^{\mu\nu}(l^\rho n^\sigma + l^\sigma n^\rho)B_{\mu\rho}B^{\sigma\nu} + \lambda\theta$$
$$= -R_{\mu\nu}l^\mu l^\nu - h^{\mu\nu}h^{\rho\sigma}B_{\mu\rho}B^{\sigma\nu} + h^{\mu\nu}\lambda n^\rho B_{\mu\rho}l_\nu + \lambda\theta$$
$$= -R_{\mu\nu}l^\mu l^\nu - h^{\mu\nu}h^{\rho\sigma}B_{\mu\rho}B^{\sigma\nu} + \lambda\theta \quad (15.1.8)$$

Combining (15.1.7), (15.1.8) and using the decomposition (15.1.2) we find Raychaudhuri's equation

$$\nabla_l \theta = -R_{\mu\nu} l^\mu l^\nu + \hat{\omega}_{\mu\nu} \hat{\omega}^{\mu\nu} - \hat{\sigma}_{\mu\nu} \hat{\sigma}^{\mu\nu} - \frac{1}{2} \theta^2 + \lambda \theta \qquad (15.1.9)$$

where we used antisymmetry of the twist tensor. The last term vanishes in an affine parametrisation and it is that special case which is usually displayed in textbooks. Notice that the indices in the squared twist or shear terms are raised by h and thus we could replace ω, σ by $\hat{\omega}, \hat{\sigma}$.

The usefulness of (15.1.9) comes about when combining it with energy conditions on the energy momentum tensor $T_{\mu\nu}$ given by Einstein's equations, including a cosmological constant term

$$R_{\mu\nu} - \frac{1}{2} R g_{\mu\nu} + \Lambda g_{\mu\nu} = 8\pi G T_{\mu\nu} \qquad (15.1.10)$$

Definition 15.1.1. *We say that T satisfies the (i) weak, (ii) strong and (iii) dominant energy condition respectively provided that*

(i) $T_{\mu\nu} u^\mu u^\nu \geq 0$ *for all timelike u.*
(ii) $T_{\mu\nu} u^\mu u^\nu \geq -g^{\mu\nu} T_{\mu\nu}$ *for all unit timelike u.*
(iii) $-T^\mu_\nu u^\nu$ *is a causal (i.e., future directed and non-spacelike) vector for all future directed timelike u.*

Suppose then that the weak energy condition holds. By continuity, $R_{\mu\nu} l^\mu l^\nu \geq 0$ also for null l. Suppose furthermore that the distribution of the $\tilde{V}_p$ is integrable, that is, they are tangent to null surfaces. (A null surface, by definition, has a null normal. It is defined up to multiplication by a scalar function). One says that l is null surface orthogonal. By Frobenius' theorem, Theorem 19.3.4, this is equivalent to $\omega_{\mu\nu} := \nabla_{[\mu} l_{\nu]} = \alpha_{[\mu} l_{\nu]}$ for some one-form α. Then the squared twist term in (15.1.9) vanishes and we find, in an affine parametrisation

$$\nabla_l \theta + \frac{1}{2} \theta^2 = -R_{\mu\nu} l^\mu l^\nu - \hat{\sigma}_{\mu\nu} \hat{\sigma}^{\mu\nu} \leq 0 \qquad (15.1.11)$$

because, writing $\sigma_{\mu\nu} = \sigma_{IJ} e^I_\mu e^J_\nu$ in the tetrad basis we find $\hat{\sigma}^{\mu\nu} \hat{\sigma}_{\mu\nu} = \sigma_{I,J=1,2} \sigma_{ij} \sigma_{ji} = Tr(\sigma^T \sigma) \geq 0$. It follows that $\nabla_l \theta^{-1} \geq 1/2$ or with $l = \partial/\partial t$ that $\theta(t)^{-1} \geq \theta(0)^{-1} + t/2$. Hence, if $\theta(0) < 0$ then θ diverges in finite time. This indicates the breakdown of the congruence, that is, the emergence of caustics.

15.1.2 *Event horizons, trapped surfaces and apparent horizons*

We recall some important definitions from black hole physics. We will assume that (M, g) is asymptotically flat throughout this chapter.

Definition 15.1.2. *Given a globally hyperbolic spacetime (M, g) consider its Penrose diagram with future/past null infinity $\Upsilon^\pm$. Given any set S we denote by $J^\pm(S)$ its causal future/past (i.e., all points in M that can be connected to points in S by causal (=everywhere timelike or null) curves).*

(i) $B := M - J^-(\Upsilon^+)$ *is called the black hole region and* $H := B \cap J^-(\Upsilon^+)$ *is called the event horizon.*

(ii) *Given a spacelike hypersurface* Σ *we call* $T := B \cap \Sigma$ *the black hole at time* Σ.

(iii) *Let* Σ *be a spacelike hypersurface and* $S \subset \Sigma$ *be a compact, without boundary, 2D, smooth, spacelike submanifold* S *of* M. *Let* s *be the unit spacelike, outgoing normal of* S *within* Σ. *At each point* p *of* S, *there are two linearly independent, future oriented null vectors* l, k *orthogonal to* S *where, say,* $l \cdot s > 0$ *and* $k \cdot s < 0$. *Construct the two congruences of future directed null geodesics orthogonal to* S *and tangential to* l, k *respectively. We call them outgoing and ingoing respectively.*

We call S *a (1) trapped, (2) outer marginally trapped, (3) inner marginally trapped, or (4) marginally trapped surface if (1)* $\theta_l < 0$, $\theta_k < 0$, *(2)* $\theta_l \leq 0$, $\theta_k < 0$, *(3)* $\theta_l < 0$, $\theta_k \leq 0$, *or (4)* $\theta_l \leq 0$, $\theta_k \leq 0$.

(iv) *Let* Σ *be an asymptotically flat Cauchy surface which is spacelike at spatial infinity. A closed subset* $C \subset \Sigma$ *which is a 3D manifold with boundary and such that* $S = \partial C$ *is outer marginally trapped is called a trapped region within* Σ.

(v) *The closure of the union of all trapped regions within* Σ *is called the total trapped region* T. *Its boundary* $A := \partial T$ *within* Σ *is called an apparent horizon.*

Notice that an apparent horizon is an instantaneous (local in time) concept while an event horizon is a global concept and requires the knowledge of the entire spacetime. To avoid confusion, notice also that a trapped region need not be connected or compact and that a trapped surface is not necessarily the boundary of a trapped region. The usefulness of these definitions and the relation between apparent and event horizons is given by the following theorem.

Theorem 15.1.3. *Suppose that* (M, g) *is globally hyperbolic and that the weak energy condition holds.[2] Then:*

(i) *Any trapped surface is within the black hole region.*

(ii) *Any trapped region within an asymptotically flat Cauchy surface is contained in the black hole region. That is,* $C \subset \Sigma \cap B$, *in particular the total trapped region* $T \subset \Sigma \cap B$. *It follows that* $A = \partial T \subset \Sigma \cap \partial B = \Sigma \cap H$, *hence the apparent horizon lies within the event horizon at any time.*

---

[2] Global hyperbolicity, together with reasonable additional assumptions, implies that cosmic censorship conjecture holds in the sense that gravitational collapse always results in black holes rather than naked singularities, apart from initial singularities such as the big bang singularity. A singularity occurs when inextendible causal geodesics stop after finite parameter time. A singularity is called hidden if no causal curve starting at it can reach future null infinity Υ^+, that is, if it lies in the black hole region. Otherwise it is called naked. Thus, in the absence of naked singularities, observers outside of the black hole region cannot see the singularity.

(iii) *If the totally trapped region T within Σ is a 3D manifold with boundary then the apparent horizon is an outer marginally trapped surface.*

15.1.3 Trapping, dynamical, non-expanding and (weakly) isolated horizons

We would like to give an analytical criterion for the trapping condition which, as seen in the last section, plays a crucial role. Consider a surface S within a spacelike hypersurface Σ. The unit spacelike normal of S within Σ is denoted by s while the future oriented unit timelike normal of Σ within M is denoted by n. Let us normalise the outer and inner null normals to S such that $l \cdot k = -1$, that is $l = (n + s)/\sqrt{2}$, $k = (n - s)/\sqrt{2}$ with $l \cdot s > 0$, $k \cdot s < 0$ using $s^2 = -n^2 = 1$. Then the metric intrinsic to Σ is given by $q_{\mu\nu} = g_{\mu\nu} + n_\mu n_\nu$ while the metric intrinsic to S can be expressed as

$$h_{\mu\nu} = g_{\mu\nu} + 2l_{(\mu}k_{\nu)} = g_{\mu\nu} + n_\mu n_\nu - s_\mu s_\nu = q_{\mu\nu} - s_\mu s_\nu \qquad (15.1.12)$$

We conclude (notice that h is spatial, that is $h_{\mu\nu} = q_\mu^\rho q_\nu^\sigma h_{\rho\sigma}$ and $h_{\mu\nu}s^\nu = 0$, $s^2 = 1$)

$$\begin{aligned}
\sqrt{2}\theta_l &= h^{\mu\nu}\nabla_\mu(n_\nu + s_\nu) = (q^{\mu\nu} - s^\mu s^\nu)K_{\mu\nu} + q^{\mu\nu}\nabla_\mu s_\nu \\
&= K - s^\mu s^\nu K_{\mu\nu} + q^{\mu\nu}D_\mu s_\nu = s^\mu s^\nu(Kq_{\mu\nu} - K_{\mu\nu}) + q^{\mu\nu}D_\mu s_\nu \\
&= -s^\mu s^\nu \frac{P_{\mu\nu}}{\sqrt{\det(q)}} + D_\mu s^\mu
\end{aligned}$$

$$\sqrt{2}\theta_k = -s^\mu s^\nu \frac{P_{\mu\nu}}{\sqrt{\det(q)}} - D_\mu s^\mu \qquad (15.1.13)$$

where we have used the definition of the extrinsic curvature K of Σ, the momentum P conjugate to q and the torsion-free covariant differential D compatible with q. Equation (15.1.13) accomplishes the task of expressing the marginally outer trapping condition $\theta_l = 0$, $\theta_k < 0$ in terms of the ADM phase space variables.

We can now define:

Definition 15.1.4

(i) *A smooth 3D submanifold H of M is called a future, outer trapping horizon (FOTH), provided it can be foliated by closed 2D manifolds S such that (1) $\theta_l = 0$, (2) $\theta_k < 0$, (3) $\nabla_k \theta_l < 0$ where l, k are any two linearly independent future directed null normals to the leaves S.*

(ii) *A smooth 3D, spacelike submanifold H of M, possibly with boundary, is called a dynamical horizon (DH), provided it can be foliated by closed 2D manifolds S such that (1) $\theta_l = 0$, (2) $\theta_k < 0$ where the roles of l, n are as above.*

Notice that future directed null normals to a 2D surface are unique only up to $l \to fl, k \to gl$ where f, g are positive functions, however, the expansions $\theta_l = h^{\mu\nu}\nabla_\mu l_\nu$ and $\theta_k = h^{\mu\nu}\nabla_\mu k_\nu$ are independent of that scale freedom. Notice that in both definitions it is not required that H is a null surface, that is, l need not be tangential to H. In fact, for DHs this is excluded since H is supposed to be spacelike. The conditions (1) and (2) mean that FOTHs and DHs are foliated by outer marginally trapped surfaces. In addition, the condition $\nabla_k\theta_l < 0$ imposed for FOTH means that the surface becomes trapped when we move along the inward normal k. However, in contrast to the trapped surfaces of Definition 15.1.2 , the surfaces S do not refer to spacelike hypersurfaces or Cauchy surfaces Σ of M, which makes these notions much more local. The condition that H be spacelike for dynamical horizons seems strange at first, but it is actually not for the following reason: let t be tangential to H and orthogonal to the surfaces S. Then t is a linear combination of l, k, hence we find a function f such that $t = l - fk$. We may choose the normalisation of k, l such that $k \cdot l = -1$. Thus $t^2 = f$. Since by definition θ_l is constant along t we have $\nabla_t\theta_l = 0$, that is, by Raychaudhuri's equation

$$\nabla_l\theta_l = f\nabla_k\theta_l = -\hat{\sigma}_{\mu\nu}\hat{\sigma}^{\mu\nu} - R_{\mu\nu}l^\mu l^\nu \qquad (15.1.14)$$

where we have made use of the fact that the distribution of subspaces $D_p = \{u \in T_p;\ u \cdot l = u \cdot k = 0\}$ integrates to the surfaces S, hence by Frobenius' theorem, Theorem 19.3.4 $dl = \alpha \wedge l + \beta \wedge k$, $dk = \alpha' \wedge l + \beta' \wedge k$ for appropriate one-forms $\alpha, \beta, \alpha', \beta'$ so that the twists of both k, l vanish. If we use the weak energy condition and the physically motivated condition that typically $\nabla_k\theta_l < 0$ then we conclude that $f \geq 0$, that is, t is spacelike or null. It becomes null if and only if the shear of l vanishes, which means that the horizon becomes isolated, see below (i.e., non-dynamical). Hence, in truly dynamical situations, H should be spacelike. This should happen when there is energy flux across the horizon. When these processes stop and the horizon settles down to an equilibrium state, it becomes isolated. This is precisely the case we will be interested in when addressing the entropy of a black hole while the dynamical case would be of interest when addressing the issue of Hawking radiation, which is not worked out yet.

We will turn now to the notion of non-expanding and isolated horizons.

Definition 15.1.5. *A submanifold H of a spacetime (M, g) is said to be a non-expanding horizon (NEH) provided that*

1. *H is topologically $\mathbb{R} \times S^2$ and null.*
2. *Any null normal l of H has vanishing expansion θ_l.*
3. *All equations of motion hold at H and $-T^\mu_\nu l^\nu$ is a future directed causal vector for any future directed null normal l.*

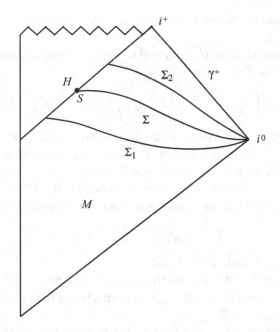

Figure 15.1 The portion of M in this Penrose diagram bounded by the isolated horizon H and the Cauchy surfaces Σ_1, Σ_2 describes a black hole in equilibrium. The intersection S of H with an intermediate Cauchy surface has spherical topology.

Two null normals to a NEH are said to be in the same equivalence class $[l] = [l']$ provided that $l' = cl$ for some positive constant $c > 0$.

Let us motivate these conditions and then draw some conclusions from them which will be of importance when we turn to the quantum theory. See also Figure 15.1 for a sketch of the situation. The last condition follows from the much stronger dominant energy condition and is thus not a restriction. The first condition is imposed for definiteness, it holds for all event horizons of physical interest. The key condition is the second one and means that the surfaces $S \cong S^2$ are marginally trapped. To see the implications of these conditions, first of all the twist of l vanishes when restricted to vectors tangent to H, that is, due to $\nabla_{[\mu} l_{\nu]} = \omega_{[\mu} l_{\nu]}$ for some one-form ω by Frobenius' theorem (H is integral manifold of l), we have $u^\mu v^\nu \nabla_{[\mu} l_{\nu]} = 0$ for all u, v such that $u \cdot l = v \cdot l = 0$. In particular we have

$$2 l^\mu \nabla_{[\mu} l_{\nu]} = \nabla_l l_\nu - \frac{1}{2} \nabla_\nu l^2 = \nabla_l l_\nu - \rho l_\nu = (\omega_\mu l^\mu) l_\nu \qquad (15.1.15)$$

where we used the fact that $l^2 = 0$ is constant along the integral curves of l so that its differential is orthogonal (i.e., tangential), that is, $\nabla l^2 = 2\rho l$ for some function ρ. It follows that l is geodesic with acceleration $\lambda_l := \rho + \omega_\mu^l l^\mu$. The acceleration parameter changes under the rescaling freedom $l \to l' := fl$, $f > 0$ as $\lambda^{l'} = f\lambda^l + \nabla_l f$.

Next, since θ_l vanishes on H, $\nabla_l \theta_l = 0$ so by Raychaudhuri's equation $\hat{\sigma}_{\mu\nu}\hat{\sigma} + R_{\mu\nu}l^\mu l^\nu = 0$. By the Einstein equations

$$R_{\mu\nu}l^\mu l^\nu = \left(R_{\mu\nu} + \left(\Lambda - \frac{1}{2}R\right)g_{\mu\nu}\right)l^\mu l^\nu = -8\pi G\left[-T_\nu^\mu l^\nu\right]l_\mu \quad (15.1.16)$$

Since $v^\mu := -T_\nu^\mu l^\nu$ is future-oriented and causal we have in tetrad components $-(t^0)^2 + \sum_{j=1}^3 (t^j)^2 \leq 0, t^0 > 0$. Choosing a Lorentz frame in which $l = l^0(1, 0, 0, \pm 1)$ with $l^0 > 0$ we get $l \cdot v = l^0(-v^0 \pm v^3) \leq 0$, thus $R_{\mu\nu}l^\mu l^\nu \geq 0$. Together with Raychaudhuri's equation we conclude $R_{\mu\nu}l^\mu l^\nu = 0$ and $\hat{\sigma}_{\mu\nu} = 0$. Thus, altogether, l is twist-free, expansion-free and shear-free. Hence

$$\nabla_\mu l_\nu = \omega_\mu^l l_\nu + \omega_\nu' l_\mu + f l_\mu l_\nu \quad (15.1.17)$$

for certain ω_μ^l, ω_μ', f with $\omega^l = \omega + \omega'$.

Next, from $R_{\mu\nu}l^\mu l^\nu = 0$ it follows that $T_{\mu\nu}l^\mu l^\nu = 0$. Since $-T_\nu^\mu l^\nu$ is future oriented and causal we conclude that $R_{\mu\nu}l^\nu$ is proportional to l_μ and so $R_{\mu\nu}l^\nu u^\mu = 0$ for all u such that $u \cdot l = 0$.

Furthermore, for any u, v such that $u \cdot l = v \cdot l = 0$ (notice that $\mathcal{L}_l u_\mu$ is tangential if u is because $u \cdot l$ is constant on H so that $l^\mu \mathcal{L}_l u_\mu = -u_\mu \mathcal{L}_l l^\mu = 0$)

$$\begin{aligned}\nabla_l(u \cdot v) &= (\mathcal{L}_l \tilde{g}_{\mu\nu})u^\mu v^\nu + [(\mathcal{L}_l u^\mu)v^\nu + (\mathcal{L}_l v^\nu)u^\mu]\tilde{g}_{\mu\nu}\\ &= (\mathcal{L}_l \tilde{g}_{\mu\nu})u^\mu v^\nu + [(\mathcal{L}_l u^\mu)v^\nu + (\mathcal{L}_l v^\nu)u^\mu]g_{\mu\nu}\\ &= (\mathcal{L}_l \tilde{g}_{\mu\nu})u^\mu v^\nu + [(\nabla_l u^\mu - \nabla_u l^\mu)v^\nu + (\nabla_l v^\nu - \nabla_v l^\nu)u^\mu]g_{\mu\nu}\\ &= (\mathcal{L}_l \tilde{g}_{\mu\nu})u^\mu v^\nu \nabla_l(u \cdot v)\end{aligned} \quad (15.1.18)$$

because $u^\mu v^\nu \nabla_\mu l_\nu = u^\mu v^\nu B_{\mu\nu} = 0$ (remember that l is twist-, shear- and expansion-free). It follows for the restriction $\tilde{g}$ of g to H that it is Lie dragged along l, that is,

$$\mathcal{L}_l \tilde{g}_{\mu\nu} = 0 \quad (15.1.19)$$

it is 'constant' along H or l is a Killing field of $\tilde{g}$. It follows that in particular the area of the spheres S of the foliation is constant. To see this, consider an embedding $Y : S^2 \to S$, then the pull-back of $\tilde{g}$ to S^2 is given by

$$h_{\alpha\beta} = Y_{,\alpha}^\mu Y_{,\beta}^\nu h_{\mu\nu} = 2m_{(\alpha}\bar{m}_{\beta)} \quad (15.1.20)$$

where $m_\alpha = Y_{,\alpha}^\mu m_\mu$ and the coordinates on S^2 are denoted by y^α, $\alpha = 1, 2$. Hence

$$\det(Y^*\tilde{g}) = \frac{1}{2}\epsilon^{\alpha\beta}\epsilon^{\gamma\delta}h_{\alpha\gamma}h_{\beta\delta} = -[\epsilon^{\alpha\beta}m_\alpha \bar{m}_\beta]^2 \quad (15.1.21)$$

so that

$$\mathrm{Ar}(S) = \int_{S^2} d^2y \sqrt{\det(Y^*\tilde{g})} = i \int_{S^2} d^2y\, \epsilon^{\alpha\beta}m_\alpha \bar{m}_\beta = i \int_S m \wedge \bar{m} \quad (15.1.22)$$

Here the area two-form

$$\eta := im \wedge \bar{m} \tag{15.1.23}$$

has appeared naturally.

To show that this quantity is constant along the integral curves of l it is sufficient to show that

$$u^\mu v^\nu \mathcal{L}_l m_{[\mu} \bar{m}_{\nu]} = 0 \tag{15.1.24}$$

for all u, v with $u \cdot l = v \cdot l = 0$. Now

$$u^\mu \mathcal{L}_l m_\mu = u^\mu \nabla_l m_\mu + (\nabla_u l^\mu) m_\mu = u^\mu \nabla_l m_\mu = u^\mu l^\nu (\nabla_\nu m_\mu - \nabla_\mu m_\nu) \tag{15.1.25}$$

for any u tangential to H because $u^\mu l^\nu \nabla_\mu m_\nu = -u^\mu m_\nu \nabla_\mu l^\nu = 0$. Now for any u, v tangential to H we have the general Ansatz (all terms proportional to l vanish)

$$u^\mu v^\nu \nabla_{[\mu} m_{\nu]} = iu^\mu v^\nu \left[(\bar{a} \bar{m}_{[\mu} + b n_{[\mu}) m_{\nu]} + c \bar{m}_{[\mu} n_{\nu]} \right] \tag{15.1.26}$$

We want to show that $c = 0$ and that b is imaginary. We have

$$-ic = 2m^\mu l^\nu \nabla_{[\mu} m_{\nu]} = l^\nu \nabla_m m_\nu - \frac{1}{2} \nabla_l m^2 = -m_\mu \nabla_m l^\mu = 0 \tag{15.1.27}$$

Hence, by a similar calculation

$$u^\mu v^\nu \left(\nabla_{[\mu} m_{\nu]} - i W_{[\mu} m_{\nu]} \right) = u^\mu v^\nu \left(\nabla_{[\mu} \bar{m}_{\nu]} + i W_{[\mu} \bar{m}_{\nu]} \right) = 0 \tag{15.1.28}$$

for a certain one-form W_μ and for all u, v tangential to H. Inserting this into (15.1.25) we find

$$u^\mu \mathcal{L}_l m_\mu = -i(u \cdot m)(l \cdot W) \tag{15.1.29}$$

and therefore from (15.1.19)

$$u^\mu v^\nu \mathcal{L}_l h_{\mu\nu} = i[(u \cdot m)(v \cdot \bar{m}) + (u \cdot \bar{m})(v \cdot m)](l \cdot [W - \bar{W}]) = 0 \tag{15.1.30}$$

Choosing $u = m, v = \bar{m}$ proves that b is real. Moreover, in (15.1.26) we may always add the term am to $\bar{a}\bar{m}$. We thus conclude

$$dm = iW \wedge m, \quad d\bar{m} = -iW \wedge \bar{m}, \quad W = am + \bar{a}\bar{m} + bn \tag{15.1.31}$$

when restricted to H and where b is real so that W is a real-valued one-form. Now finally

$$u^\mu v^\nu \mathcal{L}_l m_{[\mu} \bar{m}_{\nu]} = i[(u \cdot m)(v \cdot \bar{m}) - (u \cdot \bar{m})(v \cdot m)](l \cdot [W - \bar{W}]) = 0 \tag{15.1.32}$$

which proves that $\text{Ar}(S)$ is constant along l.

Just as in the case of spacelike hypersurfaces we want to define a covariant differential D on H which is torsion-free and compatible with $\tilde{g}$. It should map tangential tensors to tangential tensors and should be induced by ∇. Let u, v be tangential to H, then we set

$$D_u v^\mu := \nabla_u v^\mu \tag{15.1.33}$$

This is tangential because $l_\mu D_u v^\mu = -v^\mu \nabla_u l_\mu = 0$. This defines D on tangent vectors. To define it on tangential one-forms $\omega(l) = 0$ we set

$$v^\mu D_u \omega_\mu := \nabla_u(\omega(v)) - \omega_\mu D_u v^\mu \tag{15.1.34}$$

It follows that

$$([D_u, D_v] - [u, v])f = ([\nabla_u, \nabla_v] - [u, v])f = 0$$
$$D_u \tilde{g}(v, w) = \nabla_u \tilde{g}(v, w) = \nabla_u g(v, w) = g(\nabla_u v, w) + g(v, \nabla_u w)$$
$$= \tilde{g}(\nabla_u v, w) + \tilde{g}(v, \nabla_u w) \tag{15.1.35}$$

where we used metric compatibility of ∇. Hence D is torsion-free and $\tilde{g}$ compatible.

By definition of the Riemann tensor and due to $D_u l^\mu = \nabla_u l^\mu = \omega^l(u) l^\mu$ for any u, v tangential to H

$$([\nabla_u, \nabla_v] - \nabla_{[u,v]})l^\rho = \nabla_u(\omega^l(v) l^\mu) - \nabla_v(\omega^l(u) l^\mu) - \omega^l([u, v]) l^\rho$$
$$= [(\nabla_u \omega^l_\nu) v^\nu - (\nabla_v \omega^l_\mu) u^\mu] l^\rho = 2 u^\mu v^\nu l^\rho D_{[\mu} \omega^l_{\nu]}$$
$$= R_{\mu\nu}{}^\rho{}_\sigma l^\sigma \tag{15.1.36}$$

Using the definition of the Weyl tensor

$$C_{\mu\nu\rho\sigma} = R_{\mu\nu\rho\sigma} - (g_{\mu[\rho} R_{\sigma]\nu} - g_{\nu[\rho} R_{\sigma]\mu}) - \frac{1}{3} R g_{\mu[\rho} g_{\sigma]\nu} \tag{15.1.37}$$

we find

$$2 u^\mu v^\nu l_\rho D_{[\mu} \omega^l_{\nu]} = u^\mu v^\nu C_{\mu\nu\rho\sigma} l^\sigma \tag{15.1.38}$$

This implies

$$C_{\mu\nu\rho\sigma} u^\mu v^\nu w^\rho l^\sigma = 0 \tag{15.1.39}$$

for all tangential u, v, w. Specialising to $m, \bar{m}, l$ we find in particular that

$$\Psi_0 := C_{\mu\nu\rho\sigma} l^\mu m^\nu l^\rho m^\sigma = 0, \quad \Psi_1 := C_{\mu\nu\rho\sigma} l^\mu m^\nu l^\rho n^\sigma = 0 \tag{15.1.40}$$

where for the second coefficient we had to use $g = -l \otimes n - n \otimes l + m \otimes \bar{m} + \bar{m} \otimes m$. Ψ_0, Ψ_1 are so-called Newman–Penrose coefficients of the Weyl tensor, about which more will be said later. The result (15.1.40) implies that there is no flux of gravitational radiation across H and that the Weyl tensor is algebraically special of Petrov type II [207].

Contracting (15.1.38) with n_ρ we obtain

$$2 u^\mu v^\nu D_{[\mu} \omega^l_{\nu]} = u^\mu v^\nu C_{\mu\nu\rho\sigma} l^\rho n^\sigma \tag{15.1.41}$$

Using the null tetrad expansion for tangential u

$$u^\mu = -(u \cdot n)l^\mu + (u \cdot m)\bar{m}^\mu + (u \cdot \bar{m})m^\mu \qquad (15.1.42)$$

we obtain, using the fact that $\Psi_1 = 0$ and that the Weyl tensor has the same symmetries as the Riemann tensor (in particular $C_{\mu[\nu\rho\sigma]} = 0$)

$$
\begin{aligned}
2u^\mu v^\nu D_{[\mu}\omega^l_{\nu]} &= C_{\mu\nu\rho\sigma} \, m^\mu \bar{m}^\nu l^\rho n^\sigma [(u \cdot \bar{m})(v \cdot m) - (v \cdot \bar{m})(u \cdot m)] \\
&= [C_{\rho\mu\nu\sigma} - C_{\rho\nu\mu\sigma}] \, m^\mu \bar{m}^\nu l^\rho n^\sigma \, [(u \cdot \bar{m})(v \cdot m) - (v \cdot \bar{m})(u \cdot m)] \\
&= 4i\Im(\Psi_2) m_{[\mu}\bar{m}_{\nu]}u^\mu v^\nu \qquad (15.1.43)
\end{aligned}
$$

where we have defined the Newman–Penrose coefficient

$$\Psi_2 := C_{\rho\mu\nu\sigma} \, m^\mu \bar{m}^\nu l^\rho n^\sigma \qquad (15.1.44)$$

We may write equation (15.1.41) in the compact form

$$d\omega^l = \Im(\Psi_2)\eta \qquad (15.1.45)$$

which is to be understood in the sense of being true when restricted to H.

We now specialise non-expanding horizons further: for non-expanding horizons we have seen that the induced (degenerate) three-metric $\tilde{g}$ on H is Lie dragged, $\mathcal{L}_l\tilde{g} = 0$. One would, in analogy to the initial value formulation, expect that for an equilibrium system the extrinsic curvature is also Lie dragged. However, the definition of the extrinsic curvature $K_{\mu\nu} = q^\rho_\mu q^\sigma_\nu \nabla_\rho n_\sigma$ for spacelike hypersurfaces Σ, where n is the unit normal and q the induced metric, does not extend to null surfaces with n replaced by l because $\tilde{g}$ is degenerate on H (in contrast to q). However, notice that $u^\mu K^\nu_\mu = \nabla_u n^\nu$ for tangential u in the case of spacelike Σ. This suggests defining

$$u^\mu K^\nu_\mu := D_u l^\nu \qquad (15.1.46)$$

The quantity K^ν_μ is called the Weingarten map. We will thus impose the condition

$$u^\mu \mathcal{L}_l K^\nu_\mu = 0 \Leftrightarrow \mathcal{L}_l D_u l^\mu - D_{[l,u]}l^\mu = 0 \qquad (15.1.47)$$

for all tangential u. However, due to $D_u l = \omega^l(u)l$ this is equivalent to

$$(\mathcal{L}_l\omega^l)(u) = 0 \qquad (15.1.48)$$

for all tangential u. Notice that (15.1.47) can also be written

$$u^\mu(\mathcal{L}_l D_\mu - D_\mu \mathcal{L}_l)l^\nu = 0 \qquad (15.1.49)$$

for all tangential u. Hence (15.1.49) captures only restrictions on part of the connection D. In order to capture restrictions on the full connection one would impose

$$u^\mu(\mathcal{L}_l D_\mu - D_\mu \mathcal{L}_l)v^\nu = 0 \qquad (15.1.50)$$

for all u, v tangential to H. These considerations motivate the following:

Definition 15.1.6

   (i) *A weakly isolated horizon H (WIH) is a NEH such that $(\mathcal{L}_l \omega^l)(u) = 0$ for all u tangential to H and where the restriction to H of ω^l is determined by $\nabla_u l =: \omega^l(u)l$ for all u tangential to H.*

   (ii) *An isolated horizon (IH) is a NEH such that $(\mathcal{L}_l D_u - D_u \mathcal{L}_l - D_{[l,u]})v = 0$ for all u, v tangential to H.*

  (iii) *A non-rotating horizon is a NEH such that $\Im(\Psi_2) = 0$.*

Recall that a Killing horizon is a null surface with a null normal which is also a Killing vector field of g. Since l^2 is constant along H we must have that $\nabla_\mu l^2$ is normal to H, hence $\nabla_\mu l^2 =: -2\lambda^l l_\mu$. One can show that due to the Killing equation $\nabla_{(\mu} l_{\nu)} = 0$ the null normal is automatically geodesic along H with $\nabla_l l = \lambda^l l$ and that the surface gravity λ^l is constant along H. The latter result is known as the zeroth law of black hole thermodynamics. We will now establish a similar result for WIHs for the acceleration $\lambda^l := \omega^l(l)$ of a WIH. We have by definition

$$
\begin{aligned}
0 = [\mathcal{L}_l \omega^l](u) &= u^\mu \left[\nabla_l \omega^l_\mu + \nabla_\mu (l^\nu \omega^l_\nu) - l^\nu \nabla_\mu \omega^l_\nu \right] \\
&= -2u^\mu l^\nu D_{[\mu} \omega^l_{\nu]} + \nabla_u \lambda^l \\
&= -2u^\mu l^\nu \Im(\Psi_2) \eta_{\mu\nu} + \nabla_u \lambda^l \\
&= \nabla_u \lambda^l
\end{aligned}
\tag{15.1.51}
$$

Hence λ^l is constant on H for any WIH. In fact, obviously a NEH is a WIH if and only if $\lambda^l = $ const. on H.

We now show that we can use the gauge freedom $l \to fl$, $f > 0$ for a NEH to always arrange that it becomes a WIH. Under such a gauge transformation we have $\lambda^l \to f\lambda^l + \nabla_l f$ and we want to arrange that $f\lambda^l + \nabla_l f = k$. Introducing the parameter v along the integral curves of l we obtain the linear ODE $f\lambda^l + \partial f/\partial v = k$ which one can solve by the method of variation of constant. It is easy to show that for two different constants k, k' there is still gauge freedom left, because if l, l' led to k, k' respectively, then any $f > 0$ of the form $f = g \exp(-kv) + k'/k$, $\partial g/\partial v = 0$ mediates between the two, that is $l' = fl$. Hence the condition $\lambda^l = $ const. does not fix the equivalence class $[l]$ of a NEH. On the other hand, the condition for an IH is much stronger in the sense that not every NEH admits a gauge in which it is isolated. Finally, one can show that every Killing horizon is a WIH, however, notice that a WIH does not necessarily admit any Killing field of H and in this sense is much more general.

15.1.4 Spherically symmetric isolated horizons

The condition of non-rotation in Definition 15.1.6 comes from intuition of the Newman–Penrose formalism where $\Im(\Psi_2)$ indeed encodes angular momentum. We will now turn to the definition of a subclass of IHs, namely the spherically symmetric ones. These are precisely those for which a quantum framework has been developed so far. It turns out that these are automatically non-rotating. It is most convenient to define a spherically symmetric isolated horizon (SSIH) in spinorial language, to which we will give a brief introduction below.

Recall that the group $SL(2, \mathbb{C})$ has two fundamental representations. The corresponding representation spaces are two-dimensional and complex conjugates of each other. The elements of these spaces are often denoted by Weyl spinors and their complex conjugates. It is common to denote the components of these spinors by ψ^A, $\bar{\psi}^{A'}$ where $A, A' = 1, 2$; we will also call the corresponding spaces the spaces of (un)primed spinors. The components of the corresponding dual spaces are denoted by ψ_A, $\bar{\psi}_{A'}$ respectively. We define the entries of the spinor metric ϵ_{AB}, $\epsilon_{A'B'}$ to be the completely skew tensor in two dimensions, that is, $\epsilon_{12} = -\epsilon_{21} = 1$. The same holds for ϵ^{AB}, $\epsilon^{A'B'}$. Notice that $-\epsilon^{AB}$ is the inverse of ϵ_{AB}, that is, $\epsilon^{AC}\epsilon_{CB} = -\delta^A_B$. It allows us to identify, as usual, the dual spaces via $\psi^A := \epsilon^{AB}\psi_B$, $\psi_A = \epsilon_{BA}\psi^B$. Obviously $\psi^A\psi_A = 0$. The 'inner product' $\psi^A\xi^B\epsilon_{AB}$ is $SL(2, \mathbb{C})$-invariant, which is why the spinor metric is natural. The complex conjugate of a spinor is denoted as $\bar{\psi}^{A'} := \overline{[\psi^A]}_{A=A'}$.

By a spinor dyad we mean a pair (ι, o) normalised such that $o_A \iota^A = 1$. It is easy to show that $\epsilon_{AB} = 2o_{[A}\iota_{B]}$. Given a complex null tetrad $(l, k, m, \bar{m})$ subject to $-l \cdot k = m \cdot \bar{m} = 1$, all other inner products vanishing, we may define the soldering form

$$\sigma^\mu_{AA'} := -i[l^\mu \iota_A \bar{\iota}_{A'} + k^\mu o_A \bar{o}_{A'} - m^\mu \iota_A \bar{o}_{A'} - \bar{m}^\mu o_A \bar{\iota}_{A'}] \qquad (15.1.52)$$

We verify that

$$\sigma^\mu_{AA'}\sigma_\nu^{AA'} = g^\mu_\nu, \quad \sigma^\mu_{AA'}\sigma_\mu^{BB'} = \delta^A_B \delta^{A'}_{B'} \qquad (15.1.53)$$

where indices are pulled with g, ϵ. The soldering forms are anti-Hermitian, that is, $\overline{\sigma^\mu_{AA'}} = -\sigma^\mu_{A'A}$. Their purpose is to transform real vectors into anti-Hermitian spinors and vice versa via $v_{AA'} = v_\mu \sigma^\mu_{AA'}$, $v_\mu = v_{AA'}\sigma_\mu^{AA'}$. This also defines 'flat' soldering forms using a tetrad e^I_μ via $\sigma^I_{AA'} := e^I_\mu \sigma^\mu_{AA'}$, while the $\sigma^\mu_{AA'}$ are referred to as 'curved'. It is not difficult to check that the $\sigma^I_{AA'}$ can be chosen, up to numerical constants, to be $-i$ times the unit matrix for $I = 0$ and $-i$ times the Pauli matrix for $I = 1, 2, 3$. We note that

$$l_{AA'} = io_A \bar{o}_{A'}, \quad k_{AA'} = i\iota_A \bar{\iota}_{A'}, \quad m_{AA'} = io_A \bar{\iota}_{A'}, \quad \bar{m}_{AA'} = i\iota_A \bar{o}_{A'} \qquad (15.1.54)$$

and

$$l^\mu = i\sigma^\mu_{AA'} o^A \bar{o}^{A'}, \quad n^\mu = i\sigma^\mu_{AA'} \iota^A \bar{\iota}^{A'}, \quad m^\mu = i\sigma^\mu_{AA'} o^A \bar{\iota}^{A'}, \quad \bar{m}^\mu = i\sigma^\mu_{AA'} \iota^A \bar{o}^{A'}$$

$$(15.1.55)$$

Notice that the factors of i were introduced compared with other treatments in order to maintain the signature $(-, +, +, +)$.

We turn to antisymmetric tensors $T^{\mu\nu} = -T^{\nu\mu}$ whose spinorial equivalent is given by $T^{AA'BB'} = T^{\mu\nu} \sigma_\mu^{AA'} \sigma_\nu^{BB'}$. We have $T^{AA'BB'} = T^{(AA'B)B'} + T^{[AA'B]B'}$ and

$$T^{(AA'B)B'} = T^{(BA'A)B'} = -T^{(AB'B)A'} = T^{(A[A'B)B']},$$
$$T^{[AA'B]B'} = -T^{[BA'A]B'} = T^{[AB'B]A'} = T^{[A(A'B]B')} \qquad (15.1.56)$$

hence

$$T^{(AA'B)B'} = \epsilon^{A'B'} \left[\frac{1}{2} \epsilon_{C'D'} T^{AC'BD'} \right] =: \epsilon^{A'B'} T^{AB},$$

$$T^{[AA'B]B'} = \epsilon^{AB} \left[\frac{1}{2} \epsilon_{CD} T^{A'CB'D} \right] =: \epsilon^{AB} \bar{T}^{A'B'} \qquad (15.1.57)$$

Using hermiticity $\overline{T^{AA'BB'}} = T^{A'AB'B}$ we conclude $\overline{T^{AB}} = \bar{T}^{AB}$, $\overline{\bar{T}^{A'B'}} = T^{A'B'}$. The symmetric spinors $T_+^{AB} := T^{AB}$, $T_-^{A'B'} := \bar{T}^{A'B'}$ are called the self-dual and anti-self-dual parts of the spinor $T^{AA'BB'}$ respectively. In terms of flat tensorial indices one defines $T_\pm^{IJ} := (T^{IJ} \mp i\epsilon^{IJ}{}_{KL} T^{KL})/2$ where all indices are pulled with the Minkowski metric η_{IJ} and $\epsilon_{0123} = 1$. One can then verify explicitly that (15.1.57) is indeed the spinorial equivalent of $T_\pm^{IJ}$.

The covariant differential ∇ is extended to spinors by

$$\nabla \psi^A =: d\psi^A + \Gamma^A{}_B \psi^B, \quad \nabla \bar{\psi}^{A'} =: d\bar{\psi}^{A'} + \bar{\Gamma}^{A'}{}_{B'} \bar{\psi}^{B'} \qquad (15.1.58)$$

The Leibniz rule reveals that on mixed spinors the connection is given by

$$\nabla(\psi^A \bar{\xi}^{A'}) - d(\psi^A \bar{\xi}^{A'}) = \left[\Gamma^A{}_B \delta^{A'}_{B'} + \bar{\Gamma}^{A'}{}_{B'} \delta^A_B \right] (\psi^B \psi^{B'}) \qquad (15.1.59)$$

so that Γ^{AB}, $\bar{\Gamma}^{A'B'}$ are respectively the self- or anti-self-dual parts of the connection $\Gamma^{AA'BB'}$ acting on mixed spinors. Specialising to the anti-Hermitian spinor $i\psi^A \bar{\psi}^{A'}$ so that $k^\mu = i\psi^A \bar{\psi}^{A'} \sigma^\mu_{AA'}$ is a real null vector and requiring that, in analogy to the tetrad, the soldering form is covariantly constant, $\nabla \sigma_\mu^{AA'} = 0$, we find that $\Gamma^{AA'BB'}$ is the spinorial expression for the Lorentz spin connection $\Gamma^{IJ} = -\Gamma^{IJ}$ on all internal null vectors and thus on all vectors by linearity. Notice that due to the symmetry of the self-dual connection $\nabla \epsilon_{AB} = 0$.

We can now define a spherically symmetric isolated horizon. We first give it in tensorial form and then translate its key conditions into spinorial language, from which conclusions are easier to draw.

Definition 15.1.7. *A spherically symmetric isolated horizon (SSIH) is a submanifold H of (M, g) subject to the following conditions:*

1. *H is foliated by two spheres, that is, it is topologically $\mathbb{R} \times S^2$.*
2. *H is a null surface. If l is a future oriented null normal of H (which is tangential to H but normal to the two-sphere cross-sections) and k is the other future oriented null vector field normal to the two-sphere cross-sections and transversal to H then we require that the pull-back to H of k is closed. We*

extend k uniquely to M at points of H by requiring that it is null. Furthermore, we fix the relative normalisation by requiring that $l \cdot k = -1$.

3. *(a) l is expansion-free.*

   (b) k is shear-free with nowhere vanishing spherically symmetric expansion and vanishing Newman–Penrose coefficient $l^\mu \bar{m}^\nu \nabla_\mu k_\nu$ on H.

4. *All field equations hold at H.*

5. *$-T^\mu_\nu l^\nu$ is a causal vector and $T_{\mu\nu} l^\mu k^\nu$ is spherically symmetric at H.*

By 'spherically symmetric' is meant the following: since k is closed on H and $\mathbb{R} \times S^2$ is simply connected it follows that $k = -dv$ on H for some function v on H. Obviously then the null geodesic congruence generated by k is twist-free at H. Since k is orthogonal to the two-sphere cross-sections it follows that each leaf $S \cong S^2$ of the foliation is characterised by $v = $ const. The condition $l \cdot k = -1$ now implies that $l = \partial/\partial v$ coincides with the foliation vector field of H. This also fixes the scaling freedom of the null pair $(l, k) \to (fl, Fk)$ with $f, F > 0$ by $f = F^{-1}$ and $dF \wedge dv = 0$, that is, $F = F(v)$ only depends on v. This is what we mean by spherically symmetric.

We may draw some general conclusions about the form of the curvature already: since the definition of a SSIH implies that it is also a NEH we conclude as before that $R_{\mu\nu} l^\mu l^\nu = 0 = T_{\mu\nu} l^\mu l^\nu = 0$. Expanding $t^\mu = -T^\mu_\nu l^\nu$ into the null tetrad basis reveals that it can have no k component and thus is tangential, that is, it is of the form $t = el + am + \bar{a}\bar{m}$ so that $t^2 = 2|a|^2 \geq 0$. Since it is supposed to be causal, we must have $t^\mu = fl^\mu$ and we find $T^\mu_\nu l^\nu = -el^\mu$ with positive, spherically symmetric e. As for NEH we find $\Phi_{00} = \Phi_{01} = 0$. Contracting the field equations $G_{\mu\nu} + \Lambda g_{\mu\nu} = 8\pi G T_{\mu\nu}$ with $(l^\mu n^\nu + m^\mu \bar{m}^\nu)$ and using $\Phi_{11} := R_{\mu\nu}(l^\mu n^\nu + m^\mu \bar{m}^\nu)/4$ we conclude that

$$\Phi_{11} + \frac{R}{8} = \Lambda + 2\pi Ge \tag{15.1.60}$$

is spherically symmetric.

Choose a spinor dyad and complete l, k to a complex null tetrad where we consider the soldering form as variable while the spinor is fixed and constant along H. Consider now the set of equations imposed at points of H

$$o^A \nabla_u o_A = 0, \quad \iota^A \nabla_u \iota_A = g(\bar{m} \cdot u) \tag{15.1.61}$$

for all u tangential to H and where g is spherically symmetric, nowhere vanishing and real.

We want to translate (15.1.61) into tensorial language. Using the decomposition $\nabla_u o_A = a o_A + b \iota_A$, $\nabla_u \iota_A = c o_A + d \iota_A$ we find $b = 0$ and $c = g(\bar{m} \cdot u)$. This implies

$$\nabla_u l_{AA'} = i \nabla_u (o_A \bar{o}_{A'}) = (a + \bar{a}) l_{AA'} \tag{15.1.62}$$

which is the spinorial equivalent of the equation $\nabla_u l = \omega^l(u)l$. Contracting this equation with any tangential v we see that l is geodesic, twist-free, expansion-free and shear-free. Furthermore,

$$
\begin{aligned}
\nabla_u k_{AA'} = i\nabla_u(\iota_A \bar{\iota}_{A'}) &= i([co_A + d\iota_A]\bar{\iota}_{A'} + \iota_A[\bar{c}\bar{o}_{A'} + \bar{d}\bar{\iota}_{A'}]) \\
&= (d + \bar{d})k_{AA'} + cm_{AA'} + \bar{c}\bar{m}_{AA'}
\end{aligned}
\tag{15.1.63}
$$

Hence $\nabla_u n = fn + g[(u \cdot \bar{m})m + (u \cdot m)\bar{m}]$. Since $l^\mu \nabla_u k_\mu = -f = -k_\mu \nabla_u l^\mu = \omega^l(u)$ we conclude that

$$
u^\mu(\nabla_\mu k_\nu - [-\omega^l_\mu k_\nu + gh_{\mu\nu}]) = 0
\tag{15.1.64}
$$

for all u tangential to H. Thus $\theta_k := h^{\mu\nu}\nabla_\mu k_\nu = 2g$ is the expansion of k. Since the twist of k vanishes by definition, we conclude that $\omega^l_\mu = fk_\mu$ for some f. From the definitions $\nabla_u l = \omega^l(u)l$ and $\nabla_l l = \lambda^l l$ we infer $\lambda^l = \omega^l(l) = -f$. Hence $\omega^l = -\lambda^l k$. It follows that $\nabla_\mu k_\nu = \nabla_\nu k_\mu$. Hence for any u, v tangential to the two-sphere cross-sections $u^\mu v^\nu(\nabla_{(\mu} k_{\nu)} - \theta_k/2h_{\mu\nu}) = 0$, that is, k is shear-free. In summary we have

$$
\nabla_u l = \omega^l(u)l, \quad \nabla_u k = -\omega^l(u)k + \frac{1}{2}\theta_k h(u, .), \quad \omega^l = -\lambda^l k
\tag{15.1.65}
$$

with $\theta_k \neq 0$ spherically symmetric and with $a + \bar{a} = \omega^l(u) = -(d + \bar{d})$, $c = \theta_k(\bar{m} \cdot u)/2$. In particular, the Newman–Penrose coefficient $\bar{m}^\mu \nabla_l k_\mu = 0$ vanishes. Notice that $a = \iota^A \nabla_u o_A = -o_A \nabla_u \iota^A = o^A \nabla_u \iota_A = -d$.

Next

$$
\nabla_u \bar{m}_{AA'} = i\nabla_u(o_A \bar{\iota}_{A'}) = i(ao_A \iota_{A'} + o_A[\bar{c}\bar{o}_{A'} + \bar{d}\bar{\iota}_{A'}]) = (a - \bar{a})\bar{m}_{AA'} + \bar{c}l_{AA'}
\tag{15.1.66}
$$

In particular

$$
u^\mu v^\nu \nabla_{[\mu} m_{\nu]} = iW_{[\mu} m_{\nu]}
\tag{15.1.67}
$$

for some real-valued one-form W with $-iW(u) = a - \bar{a}$ which we derived in (15.1.31) for NEHs already. Thus $a = (\omega^l(u) - iW(u))/2$.

Having verified that (15.1.51) implies the conditions of Definition 15.1.7 we turn to the conditions imposed on the connection. Using the completeness relation and $do = d\iota = 0$

$$
\delta^B_A = \epsilon_{AC}\epsilon^{BC} = \epsilon^{BC}(o_A \iota_C - o_C \iota_A) = o_A \iota^B - o^B \iota_A
\tag{15.1.68}
$$

we find

$$
\nabla_u o_A = \Gamma_u{}_A{}^B o_B = ao_A, \quad \nabla_u \iota_A = \Gamma_u{}_A{}^B \iota_B = (co_A - a\iota_A)
\tag{15.1.69}
$$

hence

$$
\begin{aligned}
\Gamma_u{}_A{}^B = \Gamma_u{}^C_A(o_C \iota^B - \iota_C o^B) &= ao_A \iota^B - (co_A - a\iota_A)o^B \\
&= a(o_A \iota^B + \iota_A o^B) - co_A o^B
\end{aligned}
\tag{15.1.70}
$$

or

$$\Gamma_u^{AB} = (\omega^l(u) - iW(u))o^{(A}\iota^{B)} - \frac{1}{2}\theta_k\bar{m}(u)o^A o^B \qquad (15.1.71)$$

With this result we may now compute the curvature using

$$F_{uv\ A}{}^{B} = F_{uv\ A}{}^{C}(o_C\iota^B - o^B\iota_C)$$
$$= ([[\nabla_u,\nabla_v] - \nabla_{[u,v]}]o_A)\iota^B - ([[\nabla_u,\nabla_v] - \nabla_{[u,v]}]\iota_A)o^B \qquad (15.1.72)$$

and

$$([\nabla_u,\nabla_v] - \nabla_{[u,v]})\psi^A = u^\mu v^\nu(2\partial_{[\mu}\Gamma_{\nu]}^{AB} + \Gamma_\mu^{AC}\Gamma_{\nu\ C}{}^{B} - \Gamma_\nu^{AC}\Gamma_{\mu\ C}{}^{B})\psi_B$$
$$\qquad (15.1.73)$$

Combining (15.1.71), (15.1.72) and (15.1.73) yields

$$F^{AB} = -[d\lambda^l \wedge k + idW]o^{(A}\iota^{B)} - \frac{1}{2}[\nabla_l\theta_k + \lambda^l\theta_k]k \wedge \bar{m}o^A o^B \qquad (15.1.74)$$

to be understood to be true when contracted with vectors tangential to H. Here we have used $\omega^l = -\lambda^l k$, $d\theta_k = -k\nabla_l\theta_k$.

Next we use the fact that the self-dual part of the curvature of the Lorentz connection

$$F^{AA'BB'} = d\Gamma^{AA'BB'} + \Gamma^{AA'CC'} \wedge \Gamma_{CC'}{}^{BB'} \qquad (15.1.75)$$

equals the curvature of the self-dual part of the connection

$$F^{AB} := \frac{1}{2}F^{AA'BB'}\epsilon_{BB'} = d\Gamma^{AB} + \frac{1}{2}\epsilon_{BB'}(\Gamma^{AC}\epsilon^{A'C'} + \bar{\Gamma}^{A'C'}\epsilon^{AC})$$
$$\wedge(\Gamma_C^B\epsilon_{C'}^{B'} + \bar{\Gamma}_{C'}^{B'}\epsilon_C^B) = d\Gamma^{AB} + \Gamma_A^C \wedge \Gamma_C^B \qquad (15.1.76)$$

Since the (Palatini) equations of motion hold at H we know that the curvature of the spin connection F is related to the curvature of g by

$$F_{\mu\nu IJ} = R_{\mu\nu\rho\sigma}e_I^\rho e_J^\sigma \qquad (15.1.77)$$

This can also be derived from the covariant constance of the tetrad: if we denote by ∇' the covariant differential acting on tensor indices only and by ∇ the one which acts on Lorentz and spinor indices as well then $\nabla_\mu e_\nu^I = 0 = \nabla'_\mu e_\mu^I + \Gamma_\mu{}^I{}_J e_\nu^J = 0$. Thus

$$[\nabla'_\mu,\nabla'_\nu]e_\rho^I = R_{\mu\nu\rho}{}^\sigma e_\sigma^I$$
$$= -\nabla'_\mu(\Gamma_\nu{}^I{}_J e_\rho^J) + \nabla'_\nu(\Gamma_\mu{}^I{}_J e_\rho^J)$$
$$= -2(\nabla'_{[\mu}\Gamma_{\nu]}{}^I{}_J + \Gamma_{[\mu}{}^I{}_K\Gamma_{\nu]}{}^K{}_J)e_\rho^J$$
$$= -F_{\mu\nu}{}^I{}_J e_\rho^J \qquad (15.1.78)$$

The spinorial equivalent of (15.1.77) is given by

$$F_{\mu\nu}{}^{AA'BB'} = F_{\mu\nu}{}^{IJ}\sigma_I^{AA'}\sigma_J^{BB'} = R_{\mu\nu}{}^{\rho\sigma}\sigma_\rho^{AA'}\sigma_\sigma^{BB'} = R^{CC'DD'AA'BB'}\sigma_{\mu CC'}\sigma_{\nu DD'}$$
$$\qquad (15.1.79)$$

Using the algebraic symmetries of the Riemann tensor one can show (see, e.g., [207]) that the spinorial Riemann tensor allows for the following decomposition

$$R^{AA'BB'CC'DD'} = \epsilon_{A'B'}\epsilon_{CD}\Phi^{ABC'D'} + \epsilon_{A'B'}\epsilon_{CD}$$

$$\times \left[\Psi^{ABCD} - \frac{R}{12}\epsilon_{(A(C}\epsilon_{D)B)}\right] + c.c. \qquad (15.1.80)$$

where Φ corresponds to the trace-free piece of the Ricci tensor $R_{\mu\nu} - Rg_{\mu\nu}/4$ and Ψ to the Weyl tensor. The spinors $\Phi_{ABC'D'}$, Ψ_{ABCD} are totally symmetric in all indices of equal type and $\overline{\Phi}_{ABA'B'} = \Phi_{A'B'AB}$. Notice that for the spacetime metric we have $g_{AA'BB'} = \epsilon_{AB}\epsilon_{A'B'}$. Next one defines

$$\Psi_{m_1+m_2+m_3+m_4} := \Psi_{ABCD}\xi^A_{m_1}\xi^B_{m_2}\xi^C_{m_3}\xi^D_{m_4}$$

$$\Phi_{m_1+m_2,m_3+m_4} := \Phi_{ABA'B'}\xi^A_{m_1}\xi^B_{m_2}\bar{\xi}^{A'}_{m_3}\bar{\xi}^{B'}_{m_4} \qquad (15.1.81)$$

where $m_k = 0, 1$, $\xi_0 := o$, $\xi_1 := \iota$. Notice that $\overline{\Phi_{mn}} = \Phi_{nm}$. It is tedious but straightforward to show that the spinorial definition (15.1.81) is consistent with various tensorial definitions of Newman–Penrose coefficients that we have made before.

Let us denote the spinor equivalent of $e^I \wedge e^J$ by

$$\Sigma^{AA'BB'}_{\mu\nu} := \sigma^{AA'}_{[\mu}\sigma^{BB'}_{\nu]} \qquad (15.1.82)$$

Its (anti-)self-dual part is given by

$$\Sigma^{AB}_{\mu\nu} = \frac{1}{2}\epsilon_{A'B'}\sigma^{AA'}_{[\mu}\sigma^{BB'}_{\nu]}, \quad \Sigma^{A'B'}_{\mu\nu} = \frac{1}{2}\epsilon_{AB}\sigma^{AA'}_{[\mu}\sigma^{BB'}_{\nu]} \qquad (15.1.83)$$

Using $\Sigma^{AB} = \Sigma^{CD}(o_C\iota^A - o^A\iota_C)(o_D\iota^B - o^B\iota_D)$ it is not difficult to show that

$$\Sigma^{AB} = k \wedge \bar{m}\, o^A o^B - (m \wedge \bar{m} - l \wedge k)\, o^{(A}\iota^{B)} - l \wedge m\, \iota^A\iota^B \qquad (15.1.84)$$

Writing $F_{AA'BB'} = R_{AA'BB'CC'DD'}\Sigma^{CC'DD'}$ and employing (15.1.80) one finds when contracting with $\epsilon^{A'B'}/2$

$$F_{AB} = \Phi_{ABC'D'}\Sigma^{C'D'} + \left[\Psi_{ABCD} - \frac{R}{12}\epsilon_{(A(C}\epsilon_{D)B)}\right]\Sigma^{CD} \qquad (15.1.85)$$

Again using $F_{AB} = F_{CD}(o^C \iota_A - o_A \iota^C)(o^D \iota_B - o_B \iota^D)$ and using the definitions (15.1.81) one arrives, after tedious but straightforward calculations, at

$$
F_{AB} = \Big[(\Psi_3 + \Phi_{21}) l \wedge k - \Psi_4 l \wedge m - \Phi_{22} l \wedge \bar{m} + \Phi_{20} k \wedge m
$$
$$
+ \Big(\Psi_2 + \frac{R}{12} \Big) k \wedge \bar{m} - (\Psi_3 - \Phi_{21}) m \wedge \bar{m} \Big] o_A o_B
$$
$$
+ 2 \Big[\Big(\Psi_2 + \Phi_{11} - \frac{R}{24} \Big) l \wedge k - \Psi_3 l \wedge m - \Phi_{12} l \wedge \bar{m}
$$
$$
+ \Phi_{10} k \wedge m + \Psi_1 k \wedge \bar{m} - \Big(\Psi_2 - \Phi_{11} - \frac{R}{24} \Big) m \wedge \bar{m} \Big] o_{(A} \iota_{B)}
$$
$$
+ \Big[(\Psi_1 + \Phi_{01}) l \wedge k - \Big(\Psi_2 + \frac{R}{12} \Big) l \wedge m - \Phi_{02} l \wedge \bar{m}
$$
$$
+ \Phi_{00} k \wedge m + \Psi_0 k \wedge \bar{m} - (\Psi_1 - \Phi_{01}) m \wedge \bar{m} \Big] \iota_A \iota_B \qquad (15.1.86)
$$

Hence when restricted to vectors tangent to H

$$
F_{AB} = \Big[\Phi_{20} k \wedge m + \Big(\Psi_2 + \frac{R}{12} \Big) k \wedge \bar{m} - (\Psi_3 - \Phi_{21}) m \wedge \bar{m} \Big] o_A o_B
$$
$$
+ 2 \Big[\Phi_{10} k \wedge m + \Psi_1 k \wedge \bar{m} - \Big(\Psi_2 - \Phi_{11} - \frac{R}{24} \Big) m \wedge \bar{m} \Big] o_{(A} \iota_{B)}
$$
$$
+ [\Phi_{00} n \wedge m + \Psi_0 n \wedge \bar{m} - (\Psi_1 - \Phi_{01}) m \wedge \bar{m}] \iota_A \iota_B \qquad (15.1.87)
$$

We are now able to compare coefficients in (15.1.87) with (15.1.74): first of all, since (15.1.74) has no $\iota_A \iota_B$ term it follows that

$$
\Phi_{00} = \Psi_0 = 0, \quad \Psi_1 = \Phi_{01} = 0 \qquad (15.1.88)
$$

where $\Phi_{01} = 0$ holds for NEHs already. Next equating the $o_{(A} \iota_{B)}$ terms we find that

$$
\Big(\Psi_2 - \Phi_{11} - \frac{R}{24} \Big) m \wedge \bar{m} = d\lambda^l \wedge k + i dW \qquad (15.1.89)
$$

which means that λ^l is spherically symmetric and $i dW = (\Psi_2 - \Phi_{11} - \frac{R}{24}) m \wedge \bar{m}$. Since $m \wedge \bar{m}$ is imaginary and Φ_{11} is real, it follows that $\Im(\Psi_2) = 0$. Hence SSIHs are non-rotating. Finally, equating the $o_A o_B$ terms reveals that

$$
\Phi_{20} = \Psi_3 - \Phi_{21} = 0, \quad -\frac{1}{2}[\nabla_l \theta_k + \lambda^l \theta_k] = \Psi_2 + \frac{R}{12} \qquad (15.1.90)
$$

hence $\Psi_2 + R/12$ is spherically symmetric.

On the other hand, combining (15.1.84) and (15.1.85) one finds

$$
F_{AB} = \Big[\Big(\Psi_2 - \Phi_{11} - \frac{R}{24} \Big) \delta_A^C \delta_B^D - \Big(\frac{3}{2} \Psi_3 - \Phi_{11} \Big) o_A o_B \iota^C \iota^D \Big] \Sigma_{CD}
$$
$$
\qquad (15.1.91)
$$

again to be understood in the sense of being true when restricted to vectors tangent to H. In particular

$$m^\mu \bar{m}^\nu F_{\mu\nu AB} = \left(\Psi_2 - \Phi_{11} - \frac{R}{24}\right) \Sigma_{\mu\nu AB} m^\mu \bar{m}^\nu = \left(\Psi_2 - \Phi_{11} - \frac{R}{24}\right) o_{(A}\iota_{B)}$$
$$= -i\partial_{[\mu} W_{\nu]} m^\mu \bar{m}^\nu o_{(A}\iota_{B)} \qquad (15.1.92)$$

where the last equality follows from (15.1.74). Hence

$$\partial_{[\mu} W_{\nu]} m^\mu \bar{m}^\nu = i\left(\Psi_2 - \Phi_{11} - \frac{R}{24}\right) \qquad (15.1.93)$$

The pull-back of dW to the two spheres is proportional to $m \wedge \bar{m}$, thus it is given by

$$dW = -i\left(\Psi_2 - \Phi_{11} - \frac{R}{24}\right) m \wedge \bar{m} = -\left(\Psi_2 - \Phi_{11} - \frac{R}{24}\right)\eta \quad (15.1.94)$$

We have already derived that $\Psi_2 + R/12$, $\Phi_{11} + R/8$ are separately spherically symmetric. Hence $\Psi_2 - \Phi_{11} - R/24 = (\Psi_2 + R/12) - (\Phi_{11} + R/8)$ is also spherically symmetric. Thus, integrating (15.1.94) over a two-sphere cross-section we can pull this factor out of the integral and obtain

$$\int_S dW = -\left(\Psi_2 - \Phi_{11} - \frac{R}{24}\right)\int_S \eta = -\left(\Psi_2 - \Phi_{11} - \frac{R}{24}\right) \mathrm{Ar}(S) \quad (15.1.95)$$

Let us now interpret the one-form W: from (15.1.31) we have $dm = iW \wedge m$ when restricted to H. Now recall that $m = (e^1 + ie^2)/\sqrt{2}$ in terms of tetrads. We may consider e^1, e^2 as a Zweibein on S and can define the spin connection of the associated SO(2) bundle via the torsion-free equation

$$de^I + (\Gamma^{(S)})^I{}_J \wedge e^J = 0 \qquad (15.1.96)$$

If we set $(\Gamma^{(S)})_{IJ} = \Gamma^{(S)}\epsilon_{IJ}$ we obtain

$$de^1 + \Gamma^{(S)} \wedge e^2 = 0 = de^2 - \Gamma^{(S)} \wedge e^1 \quad \Leftrightarrow \quad dm = i\Gamma^{(S)} \wedge m \quad (15.1.97)$$

Since (15.1.96) defines $\Gamma^{(S)}$ uniquely it follows that $W_{|S} = \Gamma^{(S)}$ is the spin connection of the SO(2) bundle. By the Gauß–Bonnet theorem [234]

$$\int_S \sqrt{\det(h)} R^2 = 2\int_S d\Gamma^{(S)} = -2\pi\chi(S) \qquad (15.1.98)$$

where $\chi(S) = 2(1 - g(S))$ is the Euler characteristic of the compact Riemann surface S and $g(S)$ its genus. The first equality follows by a short computation, which relates the 2D Palatini action to the 2D Einstein–Hilbert action. Equivalently, $[D_a, D_b]v_i = R_{abij}v_j = 2\partial_{[a}\Gamma^{ij}_{b]}v_j$. For a sphere S we have $g(S) = 0$, hence combining (15.1.95) and (15.1.98) we conclude that

$$-c := \Psi_2 - \Phi_{11} - \frac{R}{24} = \frac{2\pi}{\mathrm{Ar}(S)} \qquad (15.1.99)$$

One can also check by a direct computation that $R^{(2)}(h) = R_{\mu\nu\rho\sigma}(g)h^{\mu\rho}h^{\nu\sigma} \propto \Psi_2 - \Phi_{11} - \frac{R}{24}$ and immediately consult the Gauß–Bonnet theorem. Formula (15.1.99) will play a crucial role for what follows.

We leave it to the reader to verify explicitly that the formulae derived imply, in particular, that a SSIH is a NEH.

15.1.5 Boundary symplectic structure for SSIHs

In Sections 4.2.1, 4.2.2 we derived the canonical transformation from the ADM phase space to the phase space of connections and electric fields. This was done by adding an exact one-form to the canonical action, which is thus closed and therefore does not change the symplectic structure. However, there we have assumed that there is only a boundary at spatial infinity, not an interior boundary such as in the case of a NEH. As we will see now, the form added in Section 4.2.2 is no longer exact in the presence of an interior boundary and needs to be altered.

We begin with the variation of the functional $F = \int_\sigma d^3x E_j^a \Gamma_a^j$ which would be the correct one if there were no boundaries at all. We assume that the timelike deformation vector field $T = \partial/\partial t$ becomes null at H and coincides with $l = \partial/\partial v$ at H. Then the internal boundary contribution to the variation of F is given by

$$\delta_{|S}F = \frac{1}{2}\int_S (\delta e^j) \wedge e^j \operatorname{sgn}(\det(e)) \qquad (15.1.100)$$

where the change of sign relative to (4.2.27) results from the fact that the two-sphere cross-sections at H are inner boundaries. Let us relate the e_a^j appearing in (15.1.100) to the tetrad. Comparing coefficients in the identity $g_{\mu\nu} = \eta_{IJ}e_\mu^I e_\nu^J$ and pulling back by the embeddings $X(t,x)$ reveals that

$$-N^2 + q_{ab}N^a N^b = -\left(e_t^0\right)^2 + e_t^j e_t^j, \; q_{ab}N^b$$
$$= -e_t^0 e_a^0 + e_t^j e_a^j, \; q_{ab} = -e_a^0 e_b^0 + e_a^j e_b^j \qquad (15.1.101)$$

hence, if $q_{ab} = e_a^j e_b^j$, we find $e_a^0 = 0$, $e_t^j = N^a e_a^j$, $e_t^0 = N$ where we used $N > 0$ and that e^0 is future oriented. Now

$$X_{,t}^\mu = e_I^\mu e_\nu^I X_{,t}^\mu = e_0^\mu e_t^0 + e_j^\mu e_t^j = Ne_0^\mu + N^a e_a^j e_j^\mu = Nn^\mu + N^a X_{,a}^\mu \qquad (15.1.102)$$

so that $n^\mu = e_0^\mu$, $e_a^j e_j^\mu = X_{,a}^\mu$. We conclude

$$m_a = \left(e_\mu^1 + ie_\mu^2\right)X_{,a}^\mu/\sqrt{2} = \left(e_a^1 + ie_a^2\right)/\sqrt{2}, \; m_t = \left(e_\mu^1 + ie_\mu^2\right)X_{,t}^\mu/\sqrt{2} = N^a m_a \qquad (15.1.103)$$

and using $g^{tt} = -1/N^2$, $g^{ta} = N^a/N^2$, $g^{ab} = q^{ab} - N^a N^b/N^2$ we find

$$m^t = g^{t\mu}m_\mu = 0, \; m^a = g^{a\mu}m_\mu = q^{ab}m_b \qquad (15.1.104)$$

The intersection of the embedded hypersurface Σ_t with the horizon is a two-sphere S_t. Since $\Sigma_t = X_t(\sigma)$ we obtain with an embedding $Y : S^2 \to \sigma$ the two-sphere $S_t = X_t(Y(S^2)) =: Y_t(S^2)$. Now the tangents $Y_{t,\alpha} = X_{t,a}Y^a_{,\alpha}$, $\alpha = 1, 2$ lie in the linear span of $m^\mu, \bar{m}^\mu$ and thus

$$\int_{S_t} (\delta e^j) \wedge e^j = \int_{S^2} d^2 y \delta (Y^\mu_{t,\alpha} e^j_\mu)(Y^\nu_{t,\beta} e^j_\nu)\epsilon^{\alpha\beta} = \sum_{j=1,2} \int_{S^2} d^2 y \delta (Y^a_{,\alpha} e^j_a)(Y^b_{,\beta} e^j_b)\epsilon^{\alpha\beta}$$

$$=: \sum_{I=1,2} \int_{S^2} d^2 y \delta (e^I_\alpha)(e^I_\beta)\epsilon^{\alpha\beta} \tag{15.1.105}$$

where we could restrict the sum in the third step due to the tangential properties of $Y_{,\alpha}$ and in the fourth we have denoted the pull-back to S^2 of e^j_a by e^j_α. Now we express e^I_α in terms of m_α, $\bar{m}_\alpha$ and find, assuming that $\text{sgn}(\det(e)) = 1$

$$\delta_{|S}F = \frac{1}{2} \int_{S^2} \delta m \wedge \bar{m} + \frac{1}{2} \int_{S^2} \delta \bar{m} \wedge m = \int_{S^2} \delta m \wedge \bar{m} - \delta \frac{1}{2} \int_{S^2} m \wedge \bar{m}$$

$$= \int_{S^2} \delta m \wedge \bar{m} + i\delta \frac{1}{2} \text{Ar}(S_t) \tag{15.1.106}$$

We have derived already that $\text{Ar}(S_t)$ is actually t-independent. We will now make the additional assumption that the single number $\text{Ar}(S)$ is moreover fixed (as a phase space degree of freedom, i.e., it is not only a constant of motion but furthermore takes only a single value). Then we see that the variation of F acquires the first term in (15.1.106) as a boundary term. We will compensate this term by adding a closed one-form (in field space), thereby obtaining a different boundary term which has the advantage of having a more familiar geometrical interpretation.

To do this, let us recall which boundary variations are allowed. By the boundary conditions we have for the pull-backs to S of the one-forms m, W that $dm = iW \wedge m$ and $dW = icm \wedge \bar{m}$ where $c = -2\pi/\text{Ar}(S)$. The allowed variations of m, W that preserve these conditions are $\delta m = -i\lambda m + \mathcal{L}_\xi m$ and $\delta W = -d\lambda + \mathcal{L}_\xi W$ where ξ is any vector field tangential to S (use $\delta c = 0$). These are precisely U(1) (or SO(2)) gauge transformations and diffeomorphisms of S^2. As one can check, these are precisely the gauge transformations that follow from m as a complex null dyad at S and from W as the associated spin connection. Let

$$\delta = \int_{S^2} d^2 y \left[\delta W_\alpha(y) \frac{\delta}{\delta W_\alpha(y)} + \delta m_\alpha(y) \frac{\delta}{\delta m_\alpha(y)} + \delta \bar{m}_\alpha(y) \frac{\delta}{\delta \bar{m}_\alpha(y)} \right] \tag{15.1.107}$$

be a general vector field on the space of fields $W, m, \bar{m}$ where $\delta W, \delta m, \delta \bar{m}$ are restricted to be of the form displayed above.[3] Notice that before the variation the fields $m, \bar{m}, W$ must be varied independently, only after the variation may the boundary conditions be used.

---

[3] These are considered to be the components of the vector field δ in the 'coordinate basis' $W_\alpha(y)$, $m_\alpha(y)$, $\bar{m}_\alpha(y)$.

Lemma 15.1.8. *The symplectic potential*

$$\Theta_S(\delta) := \int_S \delta m \wedge \bar{m} + \frac{1}{2c} \int_S \delta W \wedge W \qquad (15.1.108)$$

is closed.[4]

Proof: We have

$$i_\delta i_{\delta'} d\Theta_S = \delta'[\Theta_S(\delta)] - \delta[\Theta_S(\delta')] - \Theta_S([\delta', \delta])$$

$$= \int_S (\delta m \wedge \delta' \bar{m} - \delta' m \wedge \delta \bar{m}) + \frac{1}{c} \int_S \delta W \wedge \delta' W \qquad (15.1.109)$$

where $d\Theta_S$ denotes the exterior differential on field space. Inserting the explicit expressions for δ, δ' parametrised by λ, ξ and λ', ξ' respectively, the identities, $d\mathcal{L}_\xi = \mathcal{L}_\xi d$, $\mathcal{L}_\xi = di_\xi + i_\xi d$, the fact that S^2 is closed, Stokes' theorem (in particular, the Lie derivative of any two-form in 2D is exact) and the boundary condition $dW = icm \wedge \bar{m}$ one finds

$$\frac{1}{c} \int_S \delta W \wedge \delta' W = \int_S \left\{ i(\lambda \mathcal{L}_{\xi'} - \lambda' \mathcal{L}_\xi)(m \wedge \bar{m}) + \frac{1}{2c}(\mathcal{L}_{[\xi, \xi']} W) \wedge W \right\}$$

$$(15.1.110)$$

Similarly, by just inserting the definitions of δ, δ'

$$\int_S (\delta m \wedge \delta' \bar{m} - \delta' m \wedge \delta \bar{m}) = -\int_S \{ i(\lambda \mathcal{L}_{\zeta'} - \lambda' \mathcal{L}_\zeta)(m \wedge \bar{m}) - (\mathcal{L}_{[\zeta, \zeta']} m) \wedge \bar{m} \}$$

$$(15.1.111)$$

The first two terms in (15.1.110), (15.1.111) cancel each other in (15.1.109). Set $u = [\xi, \xi']$. Then, using repeatedly the boundary conditions $dW = icm \wedge \bar{m}, dm = iW \wedge m, d\bar{m} = -iW \wedge \bar{m}$

$$\int (\mathcal{L}_u W) \wedge W = \int (ici_u(m \wedge \bar{m}) \wedge W - (i_u W)dW)$$

$$= ic \int (-[i_u m]W \wedge \bar{m} + [i_u \bar{m}]W \wedge m - [i_u W]m \wedge \bar{m})$$

$$= ic \int (-i[i_u m]d\bar{m} - i[i_u \bar{m}]dm - [i_u(W \wedge m)] \wedge \bar{m} - [i_u m]W \wedge \bar{m})$$

$$= ic \int (-2i[i_u m]d\bar{m} - i[i_u \bar{m}]dm + i[i_u dm] \wedge \bar{m})$$

$$= -c \int (2[di_u m] \wedge \bar{m} + [di_u \bar{m}] \wedge m + [i_u dm] \wedge \bar{m})$$

$$= -c \int ([\mathcal{L}_u m] \wedge \bar{m} + [di_u m] \wedge \bar{m} - m \wedge [di_u \bar{m}]) \qquad (15.1.112)$$

---

[4] Let $\phi_I(y)$ be some space Φ of fields on some D-dimensional manifold σ with coordinates y. A vector field on Φ takes the general form $v = \int_\sigma d^D y \, v_I(y) \, \delta/\delta\phi_I(y)$ and a one-form on Φ takes the general form $\omega = \int_\sigma d^D y \, \omega^I(y) \, D\phi_I(y)$. Here the field theoretic analogue of the finite-dimensional coordinate basis relations $dx^a[\partial/\partial x^b] = \partial x^a/\partial x^b = \delta_b^a$ is given by the functional derivative $D\phi_I(y)[\delta/\delta\phi_J(y')] = \delta\phi_I(y)/\delta\phi_J(y') = \delta_J^I \delta(y, y')$ so that $\omega[v] = \int_\sigma d^d y \, \omega^I(y) \, v_I(y)$.

Now

$$\int ([di_u m] \wedge \bar{m} - m \wedge [di_u \bar{m}])$$

$$= -\int ([i_u m] d\bar{m} + [i_u \bar{m}] dm)$$

$$= -\int (m \wedge i_u d\bar{m} + \bar{m} \wedge i_u dm) = \int (i_u dm \wedge \bar{m} - m \wedge i_u d\bar{m})$$

$$= \frac{1}{2} \int ([\mathcal{L}_u m] \wedge \bar{m} - m \wedge [\mathcal{L}_u \bar{m}]) = \int [\mathcal{L}_u m] \wedge \bar{m} \qquad (15.1.113)$$

Thus

$$\int [\mathcal{L}_u W] \wedge W = -2c \int [\mathcal{L}_u m] \wedge \bar{m} \qquad (15.1.114)$$

and the second term in (15.1.110) cancels the second one in (15.1.111) when inserted into (15.1.109). □

Our strategy will be to add to the symplectic potential of the canonical action, given in Sections 4.2.1, 4.2.2 in terms of the canonically conjugate fields K^j_a, E^a_j, a closed one-form as follows (we will keep track of the Immirzi parameter β and drop the boundary term at spatial infinity for convenience)

$$\frac{2}{\kappa} \left(-\int_\sigma d^3x \, \delta K^j_a \delta E^a_j - \frac{1}{\beta} \delta \int_\sigma d^3x \, \Gamma^j_a \, E^a_j + \frac{1}{\beta} \int_S \delta e^j \wedge e^j + \frac{1}{c\beta} \int_S \delta W \wedge W \right)$$

$$= -\frac{2}{\kappa} \int_\sigma d^3x \,^{(\beta)}A^j_a \delta \,^{(\beta)}E^a_j + \frac{2}{c\beta\kappa} \int_S \delta W \wedge W \qquad (15.1.115)$$

Up to another exact one-form on field space, the first term is the symplectic potential used before and the second term reads explicitly

$$-\frac{2}{8\pi G\beta} \frac{\text{Ar}(S)}{4\pi} \int_S \delta W \wedge W \qquad (15.1.116)$$

This is the Chern–Simons contribution to the symplectic structure. That internal boundaries in spacetime naturally lead to Chern–Simons theory has been pointed out for the first time in [792].

Remarks

1. It might be confusing that the internal gauge freedom at H has been reduced from $SL(2, \mathbb{C})$ to $U(1)$, however, this is explained by the boundary conditions: we have fixed l, k up to spherically symmetric and mutually inverse scaling. It is equivalent, as far as $SL(2, \mathbb{C})$ transformations are concerned, to think of the soldering form as fixed and the spinor dyad as variable instead of the other way around as we did before. Since $l_{AA'} \propto o_A \bar{o}_{A'}$, $k_{AA'} \propto \iota_A \bar{\iota}_{A'}$ this fixes the transformation freedom to $o \mapsto \exp(\theta + i\varphi)o$, $\iota \mapsto \exp(-(\theta + i\varphi))\iota$ in order to preserve the normalisation condition $o_A \iota^A = 1$. Here θ must be spherically symmetric while φ may be a general function on H. This means that $m_{AA'} \propto \iota_A \bar{o}_{A'}$, $\bar{m}_{AA'} \propto \iota_A \bar{o}_{A'}$ really are reduced to $U(1)$ transformations. It is easy to see that $\theta_k \to \theta_k e^{2\theta}$, $\lambda^l \to e^{-2\theta}(\lambda^l - 2\nabla_l \theta_k)$ stay spherically symmetric

but are not invariant. In order to fix θ and therefore to meaningfully speak of the acceleration of SSIHs one now proceeds as follows: in spherically symmetric Reissner–Nordstrøm solutions $\theta_k = -2/r_S$, $\text{Ar}(S) =: 4\pi r_S^2$ irrespective of electric or magnetic charges. Thus requiring θ_k to have this value also for SSIHs (i) agrees with the Reissner–Nordstrøm case, (ii) completely exhausts the θ freedom and (iii) is a gauge which is always attainable.

2. Consider the U(1) Chern–Simons action

$$S_{\text{CS}} := \int_H dW \wedge W \tag{15.1.117}$$

which can be thought of as the generating functional of the curvature dW of the U(1) connection W. Its $2 + 1$ split is (recall that we identify the foliation vector fields of H and M at H and set $v = l$)

$$S_{\text{CS}} = \int_{\mathbb{R}} dt \int_S d^2 x\, \epsilon^{\alpha\beta} (\dot{W}_\alpha W_\beta + 2W_t \partial_\alpha W_\beta) \tag{15.1.118}$$

and we see that its symplectic potential is precisely the term that we have encountered above. Notice, however, that we are not adding Chern–Simons degrees of freedom to the classical phase space, in particular we do not require that W is closed. Rather, due to the boundary conditions that tie the bulk degrees of freedom $m, \bar{m}$ to the Chern–Simons surface degrees of freedom W, we have to change the symplectic potential. It is only in the quantum theory that we promote these surface degrees of freedom to additional dynamical degrees of freedom and remove them by imposing the boundary conditions as quantum conditions.

3. The Gauß constraint at H deserves special attention: in the bulk on $\Sigma = X(\sigma)$ it is given by $C_j = \partial_a E_j^a + \epsilon_{jkl} A_a^k E_l^a$. The bulk Gauß constraint is smeared with test functions vanishing at H and hence it is identically satisfied at H. As we move to $S = \Sigma \cap H$, the gauge group is reduced to U(1) as we just saw and thus we impose the U(1) Gauß constraint on the surface degrees of freedom $m, \bar{m}$.

4. Incidentally we notice that the boundary condition $dW = icm \wedge \bar{m}$, $c = -2\pi/\text{Ar}(S)$ can be expressed in the following way in terms of the canonical coordinates adapted to the $3 + 1$ decomposition: if $Y : S^2 \mapsto W$ is the embedding defined before then we really have $Y^* dW = icY^* m \wedge \bar{m}$. On the other hand, we had $Y^* F_{AB} = -iY^* dW o_{(A} l_{B)}$, $Y^* \Sigma_{AB} = m \wedge \bar{m} o_{(A} l_{B)}$ which means that

$$Y^*[F_{AB} - c\Sigma_{AB}] = 0 \tag{15.1.119}$$

Translating the expression for F_{AB} into Lorentz indices we get

$$Y^* F_{IJ}^+ = \Omega \sigma_I^{AA'} \sigma_J^{BB'} o_{(A} l_{B)} \epsilon_{A'B'} = -\Omega (l_{[I} k_{J]} - \bar{m}_{[I} m_{J]}) \tag{15.1.120}$$

where $\Omega = Y^*(-idW)$. Hence

$$Y^* F_{ij}^+ = Y^*(dW) \delta_{[i}^1 \delta_{j]}^2 \tag{15.1.121}$$

Comparing with the general expression for F_{ij}^+ as the curvature of the self-dual connection $A_{ij}^+ = (\Gamma_{ij} - iK_l \epsilon_{jkl})/2$ (recall that $^{(\beta=-i)}A = 2A^+$ is *twice* the pull-back to σ of the self-dual part of the spin connection of the tetrad) and pulling it back to S^2 we see that $Y^*K_j = 0$ because (15.1.121) has no imaginary part. Therefore also $Y^*(^{(\beta)}A_{ij}) = Y^*(\Gamma_{ij} + \beta K_l\epsilon_{ijl}) = Y^*\Gamma_{ij}$ and in particular $Y^*(^{(\beta)}A_j) = -Y^*((^{(\beta)}A_{kl})\epsilon_{jkl}/2) = (Y^*W)\delta_j^3$. Therefore our boundary condition can be formulated in terms of the canonical variables $(^{(\beta)}A, \, ^{(\beta)}E)$ as

$$-Y^*(^{(\beta)}F^j) = Y^*(dW)\delta_3^j = icY^*(m \wedge \bar{m})\delta_3^j = cY^*(e^1 \wedge e^2)\delta_3^j$$
$$= \frac{c}{2}Y^*(e^k \wedge e^l)\epsilon_{jkl}\delta_3^j = cY^*(*E)_j\delta_3^j \qquad (15.1.122)$$

where we have defined the two-form $(*E_j)_{ab} = \epsilon_{abc}E_j^c = \epsilon^{jkl}e_a^k e_b^l$ and no summation over j is assumed. Notice that by definition $*E^j = \frac{1}{2}(*E^j)_{ab}dx^a \wedge dx^b = \epsilon_{jkl}e^k \wedge e^l/2$ includes the factor $1/2$ involved in (15.1.122). Equation (15.1.122) is equivalent to

$$Y^*\left(^{(\beta)}F^j\right)\delta_j^3 = Y^*dW = \beta cY^*\left(* \; ^{(\beta)}E\right)_j\delta_3^j \qquad (15.1.123)$$

where summation over j is now assumed and $c = -2\pi/\text{Ar}(S)$ as before. Instead of the internal unit vector δ_j^3 we could have assumed any other bijection $S^2 \to S^2$ and U(1) would be the subgroup of SU(2) which preserves that internal vector. Indeed, we have reduced the SL(2, $\mathbb{C}$) gauge freedom to rotations in the $m, \bar{m}$ plane.

Equation (15.1.123) is the desired boundary equation that we want to impose in the quantum theory.

15.2 Quantisation of the surface degrees of freedom

In order to quantise (15.1.123) we must first quantise the phase space associated with the Chern–Simons degrees of freedom. See [793] for an exhaustive treatment of Chern–Simons theory quantisation. Basically, the idea is that this equation connects surface degrees of freedom with bulk degrees of freedom. We will start from a kinematical Hilbert space of the form $\mathcal{H}^0 = \mathcal{H}_B^0 \otimes \mathcal{H}_S^0$ where the first factor describes the kinematical bulk sector and $\mathcal{H}_S^0$ the kinematical surface sector. In the quantum theory, bulk degrees of freedom will be represented by kinematical states which can be thought of as finite linear combinations of spin-network states, each of whose underlying graphs intersects S or S^2 in a set $\mathcal{P}$ of points which we will call punctures. Let $D \subset S^2$ be a subset then the quantum boundary condition becomes, heuristically speaking,

$$\psi_B \otimes \left[\int_D Y^*\widehat{dW}\right]\psi_S = \beta c\hat{E}_3(D)\psi_B \otimes \psi_S \qquad (15.2.1)$$

where $E_3(D)$ is the flux operator as defined in previous chapters and the operator $\hat{W}$ will have to be specified in more detail later. Here we see already an essential feature: if the graph underlying the state ψ_B does not intersect D, that is, if $\mathcal{P} \cap D = \emptyset$ then the right-hand side vanishes and so must the left-hand side.

Since D can be arbitrarily small, it follows that the *quantum curvature* $\widehat{dW}$ *is flat except at the punctures*. Thus, while in the classical theory the flatness condition familiar from Chern–Simons theory did not follow from the analysis, in the quantum theory we are forced to consider the quantisation of a Chern–Simons theory with punctures. Now the bulk states associated with different sets $\mathcal{P}$ are mutually orthogonal (since the underlying graphs are necessarily different) and therefore we may perform the split

$$\mathcal{H}^0 = \oplus_\mathcal{P} \mathcal{H}^0_{B,\mathcal{P}} \otimes \mathcal{H}^0_{S,\mathcal{P}} \qquad (15.2.2)$$

where the sum runs over all finite subsets of points of S^2 and $\mathcal{H}^0_{B,\mathcal{P}}$ corresponds to the closed linear span of bulk spin-network states which intersect S precisely in $\mathcal{P}$ while $\mathcal{H}^0_{S,\mathcal{P}}$ denotes a kinematical Chern–Simons Hilbert space with punctures $\mathcal{P}$, which we will now construct.

15.2.1 *Quantum U(1) Chern–Simons theory with punctures*

The kinetic term of our U(1) Chern–Simons theory is

$$\frac{2}{\beta\kappa c} \int_S \epsilon^{\alpha\beta} \dot{W}_\alpha W_\beta = 2\frac{2}{\beta\kappa c} \int_S \dot{W}_1 W_2 - \frac{d}{dt}\frac{1}{\beta\kappa c} \int_S W_1 W_2 \qquad (15.2.3)$$

The second term is a total differential and does not contribute to the symplectic structure, hence

$$\{W_\alpha(y), W_\beta(y')\} = -\kappa' \epsilon_{\alpha\beta} \delta^{(2)}_{S^2}(y, y') \qquad (15.2.4)$$

where $2\kappa' = -\beta\kappa c/2 = \pi\beta\kappa/\text{Ar}(S)$, hence similar to the situation with fermions, Chern–Simons connections contain configuration and momentum degrees of freedom. As just mentioned, the quantum theory constrains the connection W to be flat everywhere except at $\mathcal{P} \subset S^2$. Let $\mathcal{A}^\mathcal{P}$ be the space of (generalised) U(1) connections which are flat except at $\mathcal{P}$, the group $\mathcal{G}^\mathcal{P}$ of local U(1) gauge transformations which reduce to the identity at $\mathcal{P}$ and the set $\text{Diff}^\mathcal{P}(S^2)$ of semianalytic diffeomorphisms which fix the points of $\mathcal{P}$. We will remove the restrictions of these transformations to be trivial at $\mathcal{P}$ later on.

Our first task is to coordinatise the phase space $\mathcal{A}^\mathcal{P}$ by suitable functions of W. To that end, notice that holonomies along arbitrary paths in S^2 separate the points of the space of *all* (classical) connections. Next, the $\mathcal{G}^\mathcal{P}$-invariant holonomies are those along closed loops in S^2 and any open paths connecting points of $\mathcal{P}$. However, since each $W \in \mathcal{A}^\mathcal{P}$ is flat except at $\mathcal{P}$, holonomies along closed loops not containing points of $\mathcal{P}$ are trivial. More generally, we are only interested in the homotopy type of paths because paths between the same endpoints such that the corresponding loop is contractable will have trivial holonomy (use $W(\alpha) = \exp(i \int_{D_\alpha} dW) = 1$ where $D_\alpha \subset S^2$ is the domain such that $\partial D_\alpha = \alpha$ to see that). Notice that the group $\text{Diff}^\mathcal{P}(S^2)$ does not change the homotopy type of paths because it preserves $\mathcal{P}$, hence one cannot detach an open

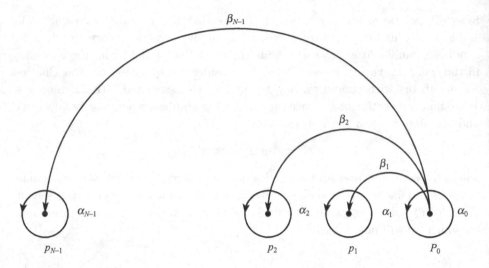

Figure 15.2 The surface structure of arcs and loops defined by the punctures of the boundary Chern–Simons theory.

path from a point of $\mathcal{P}$ and one cannot drag a loop across a point of $\mathcal{P}$. It follows that it is sufficient to fix $N-1$ mutually disjoint loops α_I and $N-1$ open paths β_I where α_I encircles the puncture p_I and β_I connects the puncture p_I with the puncture p_0. Here N is the number of punctures of $\mathcal{P} = \{p_0, p_1, \ldots, p_{N-1}\}$. The paths β_I are also mutually disjoint except for p_0 where we arrange by a homotopy that they intersect there in a C^∞ fashion (see Figure 15.2). One might think that we should also have a loop α_0 encircling p_0, however, let c_I, $I = 1, \ldots, N-1$ be any paths connecting a boundary point of α_{I-1} with a boundary point of α_I and otherwise not intersecting any of the α_J. Thereby each loop is split into two segments $\alpha_I = s_I \circ s_I'$. Then the path

$$\alpha_0 \circ \ldots \circ \alpha_{N-1}$$
$$= s_0 \circ c_1 \circ s_1 \circ c_2 \circ \ldots \circ c_{N-1} \circ s_{N-1} \circ s_{N-1}' \circ c_{N-1}^{-1} \circ \ldots \circ c_1^{-1} \circ s_0' \quad (15.2.5)$$

is contractible to a point 'over the back of the sphere' because it encircles all punctures and thus has trivial holonomy. Hence holonomies of flat connections along α_0 can be expressed in terms of the α_I, $I = 1, \ldots, N-1$.

To summarise, the phase space

$$\mathcal{A}^{\mathcal{P}}/(\mathcal{G}^{\mathcal{P}} \times \mathrm{Diff}^{\mathcal{P}}(S^2)) \quad (15.2.6)$$

that we are considering is topologically $U(1)^{N-1} \times U(1)^{N-1}$ since each of the holonomies $W(\alpha_I)$, $W(\beta_I)$, $I = 1, \ldots, N-1$ takes values in $U(1)$. This phase space is compact and far from being a cotangent bundle. Therefore we will apply geometric quantisation because it can deal with topologically non-trivial phase spaces. Before we do that we must compute the induced symplectic structure among the $W(\alpha_I)$, $W(\beta_I)$. First of all we have for any two curves c, c'

$$\{W(c), W(c')\} = -W(c)W(c')\left\{\int_c W, \int_{c'} W\right\}$$

$$= -\kappa' W(c)W(c') \int_0^1 dt \int_0^1 dt' \; \dot{c}^\alpha(t)\dot{c}'^\beta(t')\epsilon_{\beta\alpha}\delta(c(t), c'(t'))$$

$$= \kappa' W(c)W(c') \sum_{p \in c \cap c'} \text{sgn}(\det(\dot{c}(t), \dot{c}'(t')))_{c(t)=c'(t')=p} \quad (15.2.7)$$

if all intersections are interior points of both paths (there is an additional factor $1/2$ or $1/4$ if p is an endpoint of one or both paths respectively, which we do not display). It follows that

$$\{W(\alpha_I), W(\alpha_J)\} = \{W(\beta_I), W(\beta_J)\} = 0, \quad \{W(\beta_I), W(\alpha_J)\}$$
$$= \kappa'\delta_{IJ}W(\beta_I)W(\alpha_J) \quad (15.2.8)$$

The first two equalities follow from the fact that the sets α_I, α_J are mutually disjoint as are the sets β_I, β_J (at p_0 the signed intersection number in (15.2.7) vanishes due to the intersection properties of the β_I). The third equality follows from the fact that only α_I, β_I intersect and they do so in precisely one point.

We will now describe the phase space (15.2.7) in an equivalent but mathematically more convenient way. Consider the phase space $\mathbb{R}^{2(N-1)}$ with canonical brackets $\{y_I, x_J\} = \delta_{IJ}\kappa'$, $\{x_I, x_J\} = \{y_I, y_J\} = 0$. In order to obtain the torus $U(1)^{2(N-1)}$ we divide $\mathbb{R}^{2(N-1)}$ by the action of the discrete translation group or lattice $\Lambda_N = (2\pi\mathbb{Z})^{2(N-1)}$, that is, we identify points x_I, y_I up to translations by integer multiples of 2π. We can then make the identifications $W(\alpha_I) = \exp(iy_I)$, $W(\beta_I) = \exp(iy_I)$. The symplectic structure on this phase space which leads to these brackets is given by

$$\Omega = \frac{1}{\kappa'} \sum_{I=1}^{N-1} dy^I \wedge dx^I \quad (15.2.9)$$

As we want to quantise by means of geometric quantisation (see Chapter 23), in the prequantisation step we must ensure that Weil's integrality criterion is satisfied, see Theorem 23.1.4 or Corollary 23.1.5. The closed two-surfaces on the torus $T^{2(N-1)}$ are tori T^2_{IJ} wrapping around the x_I, y_J directions and the non-trivial restriction arises from choosing $I = J$:

$$\int_{T^2_{II}} \frac{\Omega}{2\pi\hbar} = \frac{(2\pi)^2}{2\pi\hbar\kappa'} = 2\frac{\text{Ar}(S)}{8\pi G\hbar\beta} =: K \quad (15.2.10)$$

Hence the number K, called the *level* of the Chern–Simons theory, must be integral.

Assuming this to be the case, the prequantisation condition is satisfied and we may proceed to choose a polarisation. We have

$$\Omega = \frac{\hbar K}{2\pi} dy^I \wedge dx^I \quad (15.2.11)$$

Let us set $z^I := x^I + iy^I$ which defines the usual positive Kähler polarisation on $\mathbb{R}^{2(N-1)}$, which restricts to the torus $T^{2(N-1)}$. We will first define the quantum theory for $\mathbb{R}^{2(N-1)}$ and then pass to the torus. The one-form $\Theta := -i\frac{\hbar K}{2\pi} z^I dx^I$ is a symplectic potential for Ω. Then $d\ln(\rho) := -2\Im(\Theta)/\hbar = d(x^2 K/(2\pi))$ defines the fibre metric of the associated complex line bundle. The Hamiltonian vector fields are $\chi_{x^I} = -\frac{2\pi}{\hbar K}\partial/\partial y^I$, $\chi_{y^I} = \frac{2\pi}{\hbar K}\partial/\partial x^I$ with our conventions for symplectic geometry $i_{\chi_f}\Omega + df = 0$, $\{f,g\} = i_{\chi_f}dg$, see Section 19.3. Hence the prequantum operators become

$$\hat{x}^I = i\hbar\chi_{x^I} - \Theta(\chi_{x^I}) + x^I = -i\frac{2\pi}{K}\partial/\partial y^I + x^I$$

$$\hat{y}^I = i\hbar\chi_{y^I} - \Theta(\chi_{y^I}) + y^I = i\frac{2\pi}{K}\partial/\partial y^I + ix^I \qquad (15.2.12)$$

Moreover

$$\nabla_{\partial/\partial\bar{z}^I} = \partial/\partial\bar{z}^I - \frac{1}{i\hbar}\Theta(\partial/\partial\bar{z}^I) = \partial/\partial\bar{z}^I + \frac{K}{4\pi}z^I \qquad (15.2.13)$$

Therefore polarised states are of the form

$$[\nabla_{\partial/\partial\bar{z}^I}\Psi](z,\bar{z}) = 0 \quad\Leftrightarrow\quad \Psi(z,\bar{z}) = e^{-K\bar{z}^I z^I/(4\pi)}\psi(z) \qquad (15.2.14)$$

We compute

$$\hat{x}^I\Psi = e^{-K\bar{z}^I z^I/(4\pi)}\left(-\frac{2\pi i}{K}\partial/\partial y^I + z^I\right) =: e^{-K\bar{z}^I z^I/(4\pi)}\hat{x}'_I\psi$$

$$\hat{y}^I\Psi = e^{-K\bar{z}^I z^I/(4\pi)}\frac{2\pi i}{K}\partial/\partial x^I\psi =: e^{-K\bar{z}^I z^I/(4\pi)}\hat{y}'_I\psi \qquad (15.2.15)$$

We therefore find

$$<\Psi, \hat{x}^I\Psi'> = \int_{\mathbb{R}^{2(N-1)}} d^{(N-1)}x\, d^{(N-1)}y\, e^{\frac{K}{2\pi}(x^I)^2}\left(e^{-K\bar{z}^I z^I/(4\pi)}\right)^2\bar{\psi}\hat{x}'_I\psi'$$

$$= \int_{\mathbb{R}^{2(N-1)}} d^{(N-1)}x\, d^{(N-1)}y\, e^{-\frac{K}{2\pi}(y^I)^2}\bar{\psi}\hat{x}'_I\psi' =: <\psi, \hat{x}'_I\psi'>$$

$$<\Psi, \hat{y}^I\Psi'> = <\psi, \hat{y}'_I\psi'> \qquad (15.2.16)$$

which is a representation of x, y by $\hat{x}', \hat{y}'$ on holomorphic functions of $\mathbb{C}^{N-1}$ respectively.

We must now deal with the fact that we actually want to quantise the torus. Hence one would naively ask that the wave functions are periodic. However, such functions do not exist, because otherwise the value of that function at arbitrarily large $|z^I|$ would be determined by its value in the compact set $\{x + iy,\ 0 \leq x, y \leq 2\pi\}$ and hence would be bounded, which is only possible for the constant function by Liouville's theorem. The way to proceed is to require that there is a unitary representation of the translation group defined by the lattice $(2\pi\mathbb{Z})^{2(N-1)}$ and to define physical states to be translation-invariant. To that end we define for

a_I, $b_I \in 2\pi\mathbf{Z}$ the unitary Weyl operators

$$V(b) := \exp\left(i\frac{K}{2\pi}\hat{x}'_I b_I\right), \quad U(a) := \exp\left(i\frac{K}{2\pi}\hat{y}'_I a_I\right) \qquad (15.2.17)$$

We compute, using the well-known formula $\exp(A + B) = \exp(A)\exp(B)\times$ $\exp(-[A, B]/2)$ applicable when $[A, B] = $ const.

$$(V(b)\psi)(z) = \exp\left(\frac{K}{2\pi}[ib \cdot z - ||b||^2/2]\right)\psi(z + ib), \quad (U(a)\psi)(z) = \psi(z + a)$$

$$(15.2.18)$$

Setting $c = a + ib$, $W(c) := U(a)V(b)$ we find

$$W(c)W(c') = \exp\left(-i\frac{K}{2\pi}b \cdot a'\right)W(c + c') = W(c + c') \qquad (15.2.19)$$

because K is integral and $b \cdot a'$ is an integer multiple of $(2\pi)^2$.

The transformations (15.2.18) are only translations up to a factor, however, this still defines a unitary representation of the translation group of the lattice on holomorphic functions. It is precisely due to this prefactor that the admissible state condition

$$W(c)\psi = \psi \;\; \forall \; c \in (2\pi\mathbf{Z})^{2(N-1)} \qquad (15.2.20)$$

will have non-trivial solutions. We can actually compute them here by elementary methods: since $U(a)$ generates actual translations into the real directions, the functions are real periodic and thus of the form

$$\psi(z) = \sum_{l \in \mathbf{Z}^{N-1}} \psi_l \, e^{il \cdot z} \qquad (15.2.21)$$

Applying $V(b)$ to $\psi(z)$ and $\exp(il \cdot z)$, using the definition (15.2.18) and comparing coefficients leads to the recursion relation

$$\psi_l = \psi_{l - Kb/(2\pi)} \, e^{-l \cdot b} \, e^{\frac{1}{2}\frac{K}{2\pi}b \cdot b} \qquad (15.2.22)$$

for all $b \in (2\pi\mathbf{Z})^{N-1}$. Setting $b_I = \delta_{IJ}$ for arbitrary $J = 1, \ldots, N-1$ reveals that ψ_l is determined already by those ψ_l with $l_I = 1, \ldots, K$ for all I. Let us call this the domain D_K. Hence the Hilbert space is at most K^{N-1}-dimensional. It is also easy to solve the recursion, with the result

$$\psi_{l+nK} = \psi_l \, e^{-2\pi l \cdot n} \, e^{-2\pi\frac{K}{2}n \cdot n} \qquad (15.2.23)$$

for all $l \in D_K$, $n \in (2\pi\mathbf{Z})^{N-1}$. Hence the general solution is

$$\psi(z) = \sum_{l \in D_K} \psi_l \vartheta_l^{K,\mathcal{P}}(z), \quad \vartheta_l^{K,\mathcal{P}}(z)$$

$$= \sum_{n \in \mathbf{Z}^{N-1}} e^{-2\pi l \cdot n} \, e^{-2\pi\frac{K}{2}n \cdot n} \exp\left(i\left(l + n\frac{K}{2\pi}\right) \cdot z\right) \qquad (15.2.24)$$

The functions $\vartheta_l^{K,\mathcal{P}}(z)$, $l \in D_K$ are linearly independent by inspection and are related to the standard Riemann ϑ function in one variable

$$\vartheta(z,\tau) := \sum_{n\in\mathbf{Z}} \exp(i\tau n^2 + 2\pi i n z) \tag{15.2.25}$$

with $\Im(\tau) > 0$ by the following formula (here reduced to $N = 2$, otherwise we get a product of ϑ functions)

$$\sum_{l=1}^{K} e^{-\pi l^2/K} \vartheta_l^{K,\mathcal{P}}(2\pi z) = \vartheta(z, \tau = i/K) \tag{15.2.26}$$

For more information on ϑ functions see [794].

To complete the quantisation we must restrict the integration domain in (15.2.16) with respect to the x_I variables to the interval $[0, 2\pi]$ as otherwise the integral would diverge due to the exact periodicity in the real directions. It is easy to check that in this inner product the functions $\vartheta_l^{K,\mathcal{P}}$ are mutually orthogonal for $l \in D_K$.

15.3 Implementing the quantum boundary condition

Now we are in a position to precisely state and solve the quantum boundary condition (15.1.123). For the Hilbert space labelled by $\mathcal{P}$ consider $p \in \mathcal{P}$. From the bulk an arbitrary but finite number of edges may intersect S in p *transversally*. The transversality restriction is here due to the fact that the edges are associated with bulk degrees of freedom and thus must extend into the bulk, that is, they cannot lie within S. Hence all the bulk edges are of type 'up' or 'down' with respect to S. For definiteness, let them be of type 'up'. Notice that since H is actually a boundary of M, these edges actually stop at S while it would be physically more reasonable to let them extend into the interior of the black hole. If that was the case, the eigenvalues of fluxes and areas of portions of S would be twice what they would be if we had edges only on one side of S, see the explicit formulae.[5] In order to capture at least an aspect of this more physical situation we will take $\hat{E}_3(D_p)$ as the limit of a one-parameter family $\hat{E}_3(D_p^\epsilon)$ where $\lim_{\epsilon\to 0} D_p^\epsilon = D_p$ but $D_p^\epsilon \cap S = \emptyset$ for $\epsilon > 0$, that is, D_p^ϵ lies in the bulk for $\epsilon > 0$. This will produce the wanted factor of two. We will choose a spin-network basis and so may assume that at each p the spins of the edges of the bulk states adjacent to p couple to some definite total angular momentum J_p. Consider a disk D_p containing p but no other element of $\mathcal{P}$. Then for such a spin-network state T_s we get for the corresponding flux

---

[5] This is a subtle point: if the surface cuts the edge in an interior point then $\hat{E}_3(D_p^\epsilon)$ is actually not diagonal, it is only in the limit $\epsilon = 0$.

operator

$$\hat{E}_3(D_p)T_s = 2\frac{i}{8}\hbar\kappa\beta \left[\sum_{e\in E(\gamma(s)),\ f(e)=p} \epsilon(e,S)R_3^e \right] T_s$$

$$= -2\frac{1}{4}\hbar\kappa\beta \left[\sum_{e\in E(\gamma(s)),\ f(e)=p} m_e \right] T_s = -\frac{\hbar\kappa}{2}\beta M_p T_s \quad (15.3.1)$$

where we have used that $-iR_3^e/2$ is diagonal with eigenvalue given by the magnetic quantum number m_e and their sum is the total magnetic quantum number M_p. The factor of two in the first equality is due to the limiting procedure performed. Let us label spin- network states in $\mathcal{H}_{B,\mathcal{P}}^0$ corresponding to total spin quantum numbers (J_p, M_p) at the punctures p by $T_{\{J\},\{M\},\alpha}^{\mathcal{P}}$ where α are other quantum numbers that are necessary to label that basis.

In order to determine the operator $\widehat{Y^*dW}$ corresponding to (15.1.123) we use that classically $W(\partial D_p) = \exp(i\int_{D_p} dW)$, so classically, according to (15.1.123)

$$W(\partial D_p) = \exp(i\beta c E_3(D_p)) \quad (15.3.2)$$

Now quantum mechanically $W(\partial D_p) = W(\alpha_I)$ for some I which we identified with $\exp(i\hat{y}_I')$ on the Chern–Simons Hilbert space of holomorphic functions with $\hat{y}_I' = \frac{2\pi i}{K}\partial/\partial u'$. Its action on the functions ϑ_l^K is given by

$$\widehat{W}(\alpha_I)\vartheta_l^{K,\mathcal{P}}(z) = \sum_{n\in\mathbb{Z}^{N-1}} e^{-2\pi l\cdot n}\, e^{-2\pi\frac{K}{2}n\cdot n} \exp\left(i\left(l+n\frac{K}{2\pi}\right)\vartheta_l(z) \right)$$

$$= \sum_{n\in\mathbb{Z}^{N-1}} e^{-2\pi l\cdot n}\, e^{-2\pi\frac{K}{2}n\cdot n} \exp\left(\left(i\left(l+n\frac{K}{2\pi}\right)\cdot z\right)\cdot z \right)$$

$$\times \exp\left(-i\frac{2\pi}{K}\left(l_I+n_I\frac{K}{2\pi}\right) \right)$$

$$= e^{-i\frac{2\pi l_I}{K}}\,\vartheta_l^{K,\mathcal{P}}(z) \quad (15.3.3)$$

and hence is diagonal in the basis ϑ_l^K. This is fortunate because a basis of $\mathcal{H}_{\mathcal{P}}^0 = \mathcal{H}_{B,\mathcal{P}}^0 \otimes \mathcal{H}_{S,\mathcal{P}}^0$ is given by the states $\vartheta_l^{K,\mathcal{P}} \otimes T_{\{J\},\{M\},\alpha}^{\mathcal{P}}$ and the quantum boundary condition becomes the following restriction on the quantum numbers l_p, M_p, $p\in\mathcal{P}$

$$e^{-i\frac{2\pi l_p}{K}} = e^{-i(\beta c)(\hbar\kappa M_p/2)} = e^{i\frac{2\pi(2M_p)}{K}} \quad (15.3.4)$$

that is

$$l_p + 2M_p = 0 \mod K \quad (15.3.5)$$

At this point we should recall that the loop α_0 could be expressed as the inverse of the composition of the other $N-1$ loops. Hence we automatically get the constraint

$$\sum_p l_p = 0 \mod K \quad \Rightarrow \quad 2\sum_p M_p = 0 \mod K \qquad (15.3.6)$$

which will be implicitly understood in what follows.

The effect of the boundary condition is thus to reduce the Hilbert space associated with $\mathcal{P}$ to the closed linear span of the mutually orthogonal states $T^{\mathcal{P}}_{J,M,\alpha} \otimes \vartheta_l^{K,\mathcal{P}}$ where M is constrained by $l + 2M \equiv 0 \ (K)$, J should be compatible with M in the sense that $|M_p| \leq J_p$ and finally α should be compatible with these data. We will loosely write

$$\mathcal{H}^0 = \oplus_{\mathcal{P},l,M:\, 2M+l\equiv\, (K)} \mathcal{H}^0_{\mathcal{P},M} \otimes \mathcal{H}^0_{\mathcal{P},l} \qquad (15.3.7)$$

15.4 Implementation of the quantum constraints

In the bulk we have to impose the SU(2) Gauß constraint, the spatial diffeomorphism constraint and the Hamiltonian constraint. In order to preserve the surface structure, obviously the SU(2) gauge transformations must reduce to U(1) gauge transformations, the group $\text{Diff}^\omega(\sigma)$ must reduce to $\text{Diff}^\omega(S)$ (where the two groups match semianalytically). Thus, together, these two reduced groups comprise exactly the symmetry group of the boundary theory, that is, the Chern–Simons theory. As far as the Hamiltonian constraint is concerned, if the lapse did not vanish at S as far as gauge transformations generated by it are concerned,[6] then the surface structure would not be preserved under Poisson brackets. Thus, there is no gauge transformation at S corresponding to the Hamiltonian constraint at S. Since on $\mathcal{H}^0_\mathcal{P}$ we have already solved U(1) gauge invariance and $\text{Diff}(S)$ invariance at S away from $\mathcal{P}$, it remains to impose both at $\mathcal{P}$ in order to reduce by the full symmetry of the surface theory, that is, the Chern–Simons theory. Yet, the bulk Hamiltonian constraint does have an effect on the surface quantum degrees of freedom. Namely, suppose that $T^\mathcal{B}_{J,M,\alpha}$ is a solution to all bulk constraints where α is now a reduced label (including spins on edges and intertwiners on vertices) corresponding to the fact that the bulk constraints have been imposed and taking into account compatibility with the surface data J, M given below. Such a compatibility might arise as follows: the Hamiltonian constraint acts non-trivially at vertices v away from S and imposes restrictions on the corresponding intertwiners and the spins of the edges outgoing from v. Some of those edges, say n, may intersect S in $p \in \mathcal{P}$ and may have only restricted range of spins $j_1, \ldots, j_n$ depending on α, whence also the coupling data J, M may not be freely choosable. Thus, although the Hamiltonian constraint does not act at $\mathcal{P}$, it still leaves its footprint on the data J, M. The same applies to the bulk SU(2) constraint. (This is not the case

---

[6] Recall the important distinction between symmetries and gauge from Section 1.5.

for the bulk diffeomorphism constraint which has a geometrical action on the graphs of the associated spin-network states and does not impose restrictions on spins and intertwiners.) We will come back to this issue when we compute the entropy.

15.4.1 Remaining U(1) gauge transformations

We begin with the U(1) Gauß constraint at S. As for the bulk degrees of freedom, recall from Chapter 9 that the self-adjoint generator $\hat{G}_j(p)$ of local gauge transformations at a vertex such as a puncture $p \in \mathcal{P}$ is proportional to $\sum_{f(e)=p} R_j^e$ when, as in our case, all edges are incoming to p. At S, the SU(2) gauge transformations reduce to U(1) generated by $\hat{G}_3(p)$. Thus we see that $\hat{C}_3(p)$ acts precisely as the operator $\hat{E}_3(D_p)$ where D_p was introduced above. The operator $\hat{C}_3(p)$ is thus diagonal with eigenvalues proportional to $2M_p$. Next we turn to the generator of gauge transformations for the surface degrees of freedom. Classically we have

$$\left\{ \int_S \lambda dW, W_\alpha(x) \right\} = -\left\{ \int_S d\lambda \wedge W, W_\alpha(x) \right\} = \kappa' \lambda_{,\alpha}(x) \qquad (15.4.1)$$

hence dW is essentially the generator of infinitesimal gauge transformations on the Chern–Simons degrees of freedom. It follows that in order to construct states which are invariant under U(1) gauge transformations at the points $p \in \mathcal{P}$ as well, we must compensate gauge transformations at each p with respect to the surface degrees of freedom by those of the bulk degrees of freedom. Therefore we must quantise an equation of the form $dW = f(*E_3)$ for some constant f. In order to be compatible with the boundary conditions (15.1.123) the constants must agree, that is, $f = \beta c$. Therefore we would like to impose $\int_S \lambda \, dW = \beta c \int_S \lambda * E_3$ for arbitrary smearing functions λ.

However, this is not possible because the operator dW does not exist, only holonomy operators exist in our chosen representation of the surface degrees of freedom. Therefore, we have to use the gauge-invariance condition in exponentiated form and in order to express the exponential as a holonomy, it follows that we cannot use arbitrary smearing functions λ but in fact only those which are integer-valued and constant on their support. However, then the invariance condition under U(1) gauge transformations at $\mathcal{P}$ becomes totally equivalent to the quantum boundary conditions and thus is already solved. It follows, in particular, that the gauge group at each puncture has been effectively reduced to the roots of unity determined by K, that is, all complex numbers z solving $z^K = 1$. This quantisation of the gauge group is a common theme in all Chern–Simons theories and corresponds to the substitution of the classical group U(1) by a quantum group $U_K(1)$.

15.4.2 Remaining surface diffeomorphism transformations

The surface diffeomorphism group $\mathrm{Diff}(S^2)$ is semianalytically matched at S to bulk diffeomorphisms. Hence this group evidently maps the structure labelled by the distinguished loops α_I, β_I to any diffeomorphic image and carries along with it the bulk edges in the vicinity of S. Therefore, what matters for the description of the physical Hilbert space is not the location $\mathcal{P}$ of the punctures p but only their sequence. In other words, by a diffeomorphism it is possible to move the punctures around on S^2 but it is not possible to interchange the role of p_I, p_J because one would have to cross the lines β_I, β_J, which is not something a diffeomorphism can do. In other words, the space $\mathcal{H}_{\mathcal{P}}^0$ becomes $\mathcal{H}_n^{\mathrm{phys}}$ labelled only by the number n of punctures but where the punctures themselves are in fact distinguishable, one cannot permute them.

15.4.3 Final physical Hilbert space

The physical Hilbert space is easiest described by selecting for each n a representative $\mathcal{P}$ together with the surface structure α_I, β_I and denoting the associated states by $T_{n,\{J\},\{M\},\alpha} \otimes \vartheta_l^{K,n}$; $2M \equiv l(K)$ and where we keep the dependence of J, M on α and the conditions $\sum_p l_p \equiv 0\ (K)$, $|M_p| \leq J_p$ in mind. We will denote it by

$$\mathcal{H}^{\mathrm{phys}} = \oplus_{n,l,2M\equiv l\,(K),J}\ \mathcal{H}_{S,n,l}^{\mathrm{phys}} \otimes \mathcal{H}_{B,J,M}^{\mathrm{phys}} \qquad (15.4.2)$$

Notice that $\mathcal{H}_{S,n,l}^{\mathrm{phys}}$ is the one-dimensional span of the vector $\vartheta_l^{K,n}$.

15.5 Entropy counting

The idea in entropy counting is to count *surface states*. This will then be the origin of a density matrix for the surface Hilbert space because we will form a partial trace the bulk Hilbert space. That is, given a value of $\mathrm{Ar}(S)$ and thus a value of K we want to count the number of states $\vartheta_l^{K,n}$ with variable n, l such that the corresponding area eigenvalue lies in the interval $[\mathrm{Ar}(S) - \ell_P^2, \mathrm{Ar}(S) + \ell_P^2]$ where for the purpose of this chapter we set $\ell_P^2 := \hbar G$. Notice that the area of S is a Dirac observable: it is left-invariant by the bulk symmetries, it is manifestly $SU(2)$-invariant and invariant under the diffeomorphisms of S. From Chapter 13 and Figure 15.3 we may read off the eigenvalues of $\widehat{\mathrm{Ar}}(S)$ on the states $T_{n,\{J\},\{M\},\alpha}$ to be

$$\lambda(n,\{J\}) = \frac{\hbar\kappa\beta}{4} \sum_{p=1}^{n} \sqrt{2J_p^{\mathrm{u}}(J_p^{\mathrm{u}}+1) + 2J_p^{\mathrm{d}}(J_p^{\mathrm{d}}+1) - J_p^{\mathrm{ud}}(J_p^{\mathrm{ud}}+1)} \qquad (15.5.1)$$

where the superscript stands for total up, down or intertwining spin at p respectively. Similar to the flux, for physical reasons we define the eigenvalue as the limit of those of $\widehat{\mathrm{Ar}}(S_\epsilon)$ where $\lim_{\epsilon\to 0} = S$ and $S \cap S_\epsilon = \emptyset$ for $\epsilon > 0$, that is, S_ϵ lies

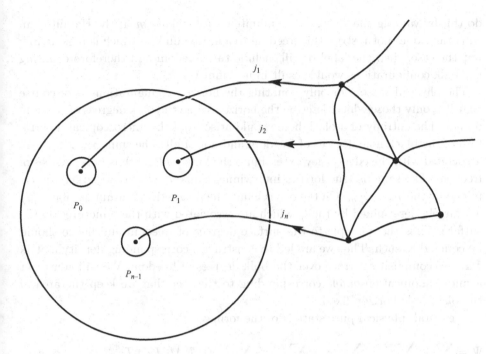

Figure 15.3 Entropy counting: how many surface states exist such that the corresponding bulk states are compatible with a given area eigenvalue?

in the bulk for $\epsilon > 0$. This will again produce a factor of two[7] compared with the situation of just using $\widehat{\mathrm{Ar}}(S)$ because at $\epsilon > 0$ we have $J_p^{\mathrm{u}} = J_p^{\mathrm{d}} = J_p$, $J_p^{\mathrm{ud}} = 0$ while at $\epsilon = 0$ we would have $J_p^{\mathrm{d}} = J_p^{\mathrm{ud}} = J_p$, $J_p^{\mathrm{u}} = 0$. Hence

$$\lambda(n, \{J\}) = \frac{\hbar\kappa\beta}{2} \sum_{p=1}^{n} \sqrt{J_p(J_p + 1)} \qquad (15.5.2)$$

We now count surface configurations n, l such that the associated bulk configurations $(n, \{J\}, \{M\}, \alpha)$ subject to $2M + l \equiv 0\ (K)$, $|M_p| \leq J_p$ and $2\sum_p M_p \equiv 0\ (K)$ satisfy $\lambda(n, \{J\}, \{M\}) \in [\mathrm{Ar}(S) - \ell_P^2, \mathrm{Ar}(S) + \ell_P^2]$. We will denote this number by $N(l, n)$. It is crucial that we only count the surface configurations l, n. If we counted the bulk configurations $(n, \{J\}, \{M\}, \alpha)$ compatible with $\lambda(n, \{J\}, \{M\}) \in [\mathrm{Ar}(S) - \ell_P^2, \mathrm{Ar}(S) + \ell_P^2]$ then this number would probably be just infinite. The reason is that for given J_p there is probably an infinite number of α such that the value J_p can be attained. Namely, we could have an arbitrary number, say m, of edges intersecting S in p carrying the spins $j_1, \ldots, j_m$ and the only requirement is that they couple to total spin J_p. The number of ways to

---

[7] This point is again subtle: the spins J_p are intertwining spins on vertices and not spins on edges. Now if we assume that at p say N spins $j_1, \ldots, j_N$ couple to J_p, then if we move S slightly into the bulk into S^ϵ the number of punctures and the total spins jump. Hence in order to make the argument here we must assume that for S^ϵ also the state changes in that there are still those N edges at the moved puncture p^ϵ which continue through the puncture and end at S. As $\epsilon \to 0$ the old situation is restored.

do this by varying the $j_1, \ldots, j_m$ is infinite even for fixed m. If the Hamiltonian constraint does not restrict this freedom to a finite number, which is presumably not the case, then the label α will include this freedom and therefore counting the bulk configurations would result in an infinity.

The physical reason for only counting the horizon configurations is of course that it is only those which describe the horizon while the bulk degrees of freedom do not. The entropy of a black hole should arise from the microscopical description of the black hole degrees of freedom and not of all. The spins $j_1, \ldots, j_m$ are associated with the edges, they extend into the bulk and are thus bulk degrees of freedom. The same is true for the interwining spins J_p which are largely determined by the $j_1, \ldots, j_m$. On the other hand, the magnetic quantum numbers M_p are largely determined by the l_p which are associated with the punctures on the surface. Thus these couple to the surface degrees of freedom and hence should be counted as such. Thus we are led to construct a corresponding density matrix which accomplishes a trace over the bulk degrees of freedom. We will construct a microcanonical ensemble corresponding to the fact that we keep the area of the black hole horizon fixed.

A general, physical pure state is of the form

$$
\Psi = \sum_{n=0}^{\infty} \sum_{l_1,\ldots,l_n=1}^{K} \sum_{M \in S(n,l)} \sum_{J \in S(n,M)} \sum_{\alpha \in S(n,J)} c(n,l,M,J,\alpha)\, T^{\text{phys}}_{n,\{J\},\{M\},\alpha} \otimes \vartheta_l^{n,K}
$$

(15.5.3)

where $S(n,l) = \{M : 2M + l \equiv 0\ (K),\ \sum_{p=1}^{n} M_p \equiv 0\ (K)\}$, $S(n,M) = \{J : |M_p| \le J_p,\ p = 1, \ldots, n\}$, and $S(n,J)$ is the set of α compatible with n, J. The expectation value of the area operator of the horizon is

$$
< \Psi, \widehat{\text{Ar}}(S)\Psi > = \sum_{n,l,M \in S(n,l), J \in S(n,M), \alpha \in S(n,J)} |c(n,l,M,J,\alpha)|^2\ \lambda(n,J)
$$

(15.5.4)

Let us introduce the set $S(n,\lambda) = \{J : \lambda(n,J) = \lambda\}$, then (15.5.3) can be written as

$$
< \Psi, \widehat{\text{Ar}}(S)\Psi >
$$
$$
= \sum_{\lambda \in \sigma(\widehat{\text{Ar}}(S))} \lambda \sum_{n,l} \sum_{M \in S(n,l), J \in S(n,M) \cap S(n,\lambda), \alpha \in S(n,J)} |c(n,l,M,J,\alpha)|^2
$$
$$
=: \sum_{\lambda \in \sigma(\widehat{\text{Ar}}(S))} \lambda \sum_{n,l} w(n,l,\lambda)
$$

(15.5.5)

where $w(n,l,\lambda)$ is the probability of finding an eigenstate of the horizon area operator in the state Ψ with eigenvalue λ and with surface configuration (n,l). We may write (15.5.6) as

$$
< \Psi, \widehat{\text{Ar}}(S)\Psi > = \text{Tr}(\hat{\rho}_{\text{BH}}\widehat{\text{Ar}}'(S)), \quad \hat{\rho}_{\text{BH}} = \sum_{\lambda \in \sigma(\widehat{\text{Ar}}(S))} \sum_{n,l} w(\lambda,n,l)\hat{P}_{\vartheta_l^{n,K}}
$$

(15.5.6)

where $\widehat{\mathrm{Ar}}'(S)$ is a fiducial operator in $\mathcal{H}_S^{\mathrm{phys}}$ which is diagonal, with eigenvalue λ, on the linear span of those $\vartheta_l^{n,K}$ compatible with that eigenvalue in the sense that there exists $\alpha \subset S(n, J)$ for some $J \in S(n, M) \cap S(n, \lambda)$ for some $M \in S(n, l)$. $\hat{P}_\psi$ denotes the projection onto the span of the state ψ. Hence the trace is taken in the surface Hilbert space and we have constructed a corresponding density matrix by taking the partial trace over the bulk degrees of freedom.

If we want to construct a state that describes a black hole with horizon area approximately equal to the expectation value of some classical given value A_0 determining K then $w(n, l, \lambda) = 0$ for $\lambda \notin [A_0 - 2\delta, A_0]$. Let $S(A_0, \delta)$ be the set (n, l) compatible with eigenvalues λ in that interval and let $N(A_0, \delta)$ be its cardinality. Then, as is well known, the entropy $S_{\mathrm{BH}} := -\mathrm{Tr}(\hat{\rho}_{\mathrm{BH}} \ln(\rho_{\mathrm{BH}}))$ of that state is extremised if the probabilities $w(\lambda, n, l)$ for all $(n, l) \in S(A_0, \delta)$ are equal to each other and thus must equal $w(\lambda, n, l) = N(A_0, \delta)^{-1}$, whence

$$S_{\mathrm{BH}} = \ln(N(A_0, \delta)) \tag{15.5.7}$$

To determine $N(A_0, \delta)$ we must make the following assumption: for each (n, l) such that there exist $\lambda \in \sigma(\widehat{\mathrm{Ar}}(S)) \cap [A_0 - 2\delta, A_0]$ and $J \in S(n, \lambda) \cap S(n, M)$ for some $M \in S(n, l)$ there exists at least one $\alpha \in S(n, J)$. This condition has not been verified to be true for the current version of the Hamiltonian constraint, however, it is likely to hold and if not would at most reduce $N(A_0, \delta)$. In that situation we are now able to estimate $N(A_0, \delta)$ from above and below as follows: we will count the number $\tilde{N}(A_0)$ of surface states which are compatible with areas below the area determined by the classical K. From this we can then determine $N(A_0, \delta) = \tilde{N}(A_0) - \tilde{N}(A_0 - 2\delta)$. Given then $M \in S(n, l)$, $J \in S(n, M) \cap S(n, \lambda)$, $\lambda \le A_0$ we have

$$A_0 \ge 8\pi G\hbar\beta \sum_{p=1}^{n} \sqrt{J_p(J_p + 1)} \ge 8\pi G\hbar\beta \sum_{p=1}^{n} \sqrt{|M|_p(|M|_p + 1)}$$

$$> 8\pi G\hbar\beta \sum_{p=1}^{n} |M|_p \ge 8\pi G\hbar\beta \sum_{p=1}^{n} M_p$$

$$= 4\pi G\hbar\beta \left(2 \sum_{p=1}^{n} M_p \right) = A_0 \frac{2\sum_{p=1}^{n} M_p}{K} \tag{15.5.8}$$

Since $M \in S(n, l)$ we have $2\sum_{p=1}^{n} M_p = mK$ for some integer m and thus by (15.5.8) we must have $\sum_{p=1}^{n} M_p = 0$ exactly. Next we also get from (15.5.8) that for each p necessarily $2|M_p| < K$ and since $l_p + 2M_p = 0$ modulo K we actually must have $2M_p = -l_p$ exactly. It follows that it is actually equivalent to count the M_p rather than the l_p and all we need to do is to ensure that $\sum_{p=1}^{n} M_p = 0$. As follows from (15.5.8), given M, to make sure that there exists J with $\lambda(n, J) \le A_0$ and $|M_p| \le J_p$ it is necessary that $\sum_p \sqrt{|M_p|(|M_p| + 1)} \le K/2$. Conversely, given M with $\sum_p \sqrt{|M_p|(|M_p| + 1)} \le K/2$ there exists $J_p := |M_p|$ satisfying this requirement. Notice that the number of possible J accomplishing this is not of interest to us because, according to our philosophy, to count surface degrees of freedom we count the l or M and not the J.

Hence our counting problem is reduced to considering all lists $(M_1, \ldots, M_n)$, $n = 0, 1, 2, \ldots$ (where $n = 0$ corresponds to the empty list) of non-zero half-integers subject to (1) $\sum_P M_p = 0$ and (2) $\sum_p \sqrt{|M_p|(|M_p| + 1)} \leq K/2$. The reason for requiring $M_p \neq 0$ for all p and for $n > 0$ is that $M_p = 0$ implies $l_p = 0$, which we excluded by our choice of basis of $\vartheta_l^{n,K}$, $l_p = 1, \ldots, D_K$. If we had instead allowed $l_p = 0, 1, \ldots, K - 1$ the lists with countably infinitely many zero entries and finitely many non-zero entries would contribute and the entropy would be infinite.

Let L_K be the set of those lists and $\tilde{N}(K)$ be the number of its elements. For $M \in L_K$ we certainly have $\sum_{p=1}^{n} |M_p| \leq K/2$ so that the set

$$L_K^+ := \emptyset \cup \left\{ (M_1, \ldots, M_n) : \ M_p \neq 0, \ n = 1, 2, \ldots; \ \sum_{p=1}^{n} |M_p| \leq K/2 \right\} \quad (15.5.9)$$

certainly includes L_K because we have also removed the first condition $\sum_{p=1}^{N} M_p = 0$. Denote the number of elements of L_K^+ by $\tilde{N}_K^+$.

Now notice that

$$L_K = \emptyset \cup \left\{ (M_1, \ldots, M_n) : \ n = 1, 2, \ldots; \ M_1, \ldots, M_n, \neq 0, \right.$$

$$\left. \sum_{p=1}^{n} M_p = 0, \ \sum_{p=1}^{n} \sqrt{|M_p|(|M_p| + 1)} \leq K/2 \right\}$$

$$= \emptyset \cup \left\{ (M_1, \ldots, M_n) : \ n = 2, 3, \ldots; \ M_1, \ldots, M_n \neq 0, \right.$$

$$\left. \sum_{p=1}^{n} M_p = 0, \ \sum_{p=1}^{n} \sqrt{|M_p|(|M_p| + 1)} \leq K/2 \right\}$$

$$= \emptyset \cup \left\{ (M_1, M_2, \ldots, M_n) : \ n = 2, 3, \ldots; \ M_2, \ldots, M_n \neq 0, \right.$$

$$\left. \sum_{p=1}^{n} M_p = 0, \ \sum_{p=1}^{n} \sqrt{|M_p|(|M_p| + 1)} \leq K/2 \right\}$$

$$- \left\{ (0, M_2, \ldots, M_n) : \ n = 2, 3, \ldots; \ M_2, \ldots, M_n \neq 0, \right.$$

$$\left. \sum_{p=2}^{n} M_p = 0, \ \sum_{p=2}^{n} \sqrt{|M_p|(|M_p| + 1)} \leq K/2 \right\}$$

$$=: \emptyset \cup L_K' - L_K'' \quad (15.5.10)$$

where in the step before the last one we have given up the restriction $M_1 \neq 0$ in L'_K and corrected this by subtracting the additional lists collected in L''_K. The point is now that $|L''_K| = |L_K| - 1$ so that $\tilde{N}_K = |L_K| = |L'_K|/2 + 1$. To estimate $|L'_K|$ we write more explicitly

$$L'_K = \left\{ (M_1, M_2, \ldots, M_n) : n = 2, 3, \ldots; \ M_2, \ldots, M_n \neq 0, \right.$$

$$\left. \sum_{p=1}^n M_p = 0, \ \sum_{p=1}^n \sqrt{|M_p|(|M_p|+1)} \leq K/2 \right\}$$

$$= \left\{ \left(-\sum_{p=2}^n M_p, M_2, \ldots, M_n \right) : n = 2, 3, \ldots; \ M_2, \ldots, M_n \neq 0, \right.$$

$$\left. \sum_{p=2}^n \sqrt{|M_p|(|M_p|+1)} + \sqrt{\left| \sum_{p=2}^n M_p \right| \left(\left| \sum_{p=2}^n M_p \right| + 1 \right)} \leq K/2 \right\} \quad (15.5.11)$$

where we could eliminate M_1 since the restriction $M_1 \neq 0$ was deleted. Obviously, $|L'_K| = |\tilde{L}_K|$ where

$$\tilde{L}_K = \left\{ (M_1, \ldots, M_n) : n = 1, 2, \ldots; \ M_1, \ldots, M_n \neq 0, \right.$$

$$\left. \sum_{p=1}^n \sqrt{|M_p|(|M_p|+1)} + \sqrt{\left| \sum_{p=1}^n M_p \right| \left(\left| \sum_{p=1}^n M_p \right| + 1 \right)} \leq K/2 \right\} \quad (15.5.12)$$

The advantage of $\tilde{L}_K$ is that the condition $\sum_p M_p = 0$ is deleted. We have

$$\sum_{p=1}^n \sqrt{|M_p|(|M_p|+1)} + \sqrt{\left| \sum_{p=1}^n M_p \right| \left(\left| \sum_{p=1}^n M_p \right| + 1 \right)}$$

$$\leq \sum_{p=1}^n \left(|M_p| + \frac{1}{2} \right) + \left(\left| \sum_{p=1}^n M_p \right| + \frac{1}{2} \right) \leq \frac{n+1}{2} + 2 \sum_{p=2}^n |M_p| \quad (15.5.13)$$

Hence the set

$$\tilde{L}_K^- := \emptyset \cup \left\{ (M_1, \ldots, M_n) : n = 1, 2, \ldots; \ M_p \neq 0, \right.$$

$$\left. \frac{n+1}{2} + 2 \sum_{p=1}^n \sqrt{|M_p|(|M_p|+1)} \leq K/2 \right\} \quad (15.5.14)$$

is certainly included in $\tilde{L}_K$ except for the empty list so that $|\tilde{L}_K^-| - 1 \leq |\tilde{L}_K| = |L'_K| = 2\tilde{N}_K - 2$. Let $|\tilde{L}_K^-| =: \tilde{N}_K^-$. Then we certainly have $(1 + \tilde{N}_K^-)/2 \leq \tilde{N}_K \leq \tilde{N}_K^+$.

We will now derive and solve recursion relations for $\tilde{N}_K^\pm$. Notice that $(M_1, \ldots, M_n) = (M_1', \ldots, M_{n'}')$ if and only if $n = n'$ and $M_p = M_p'$ for $p = 1, \ldots, n$ since the punctures are distinguishable. Suppose that $(M_1, \ldots, M_n) \in L_K^+$, then $l := 2|M_1| = 1, 2, \ldots, K$ and $(M_2, \ldots, M_n) \in L_{K-l}^+$. Moreover, any element of L_K^+ can be obtained from all the elements of L_{K-l}^+ by adjoining to the lists in L_{K-l}^+ (including the empty one) another first entry $M_1 = \pm l/2$, except for the empty list. We thus obtain the recursion relation

$$\tilde{N}_K^+ = 1 + 2 \sum_{l=1}^{K} \tilde{N}_{K-l}^+ = 1 + \sum_{l=0}^{K-1} \tilde{N}_l^+; \quad N_0^+ = 1 \tag{15.5.15}$$

It follows that

$$\tilde{N}_K^+ - \tilde{N}_{K-1}^+ = 2N_{K-1}^+ \quad \Rightarrow \quad \tilde{N}_K^+ = 3^K \tag{15.5.16}$$

Next, if $(M_1, \ldots, M_m) \in L_K^-$ then the largest value that M_1 can take is obtained for $n = 1$ so that $2|M_1| \leq (K-2)/2$. Hence, $l = 2|M_1| = 1, \ldots, (K-2)/2$. Now for $M_1 = \pm l/2$ and $(M_1, \ldots, M_n) \in L_K^-$ we have

$$\frac{n+1}{2} + l + 2 \sum_{p=2}^{n} |M_p| \leq K/2 \quad \Leftrightarrow \quad \frac{n}{2} + 2 \sum_{p=2}^{n} |M_p| \leq (K-1-2l)/2 \tag{15.5.17}$$

so that $(M_2, \ldots, M_n) \in L_{K-1-2l}^-$. Thus, by the same reasoning as above we obtain the recursion relation

$$\tilde{N}_K^- = 1 + 2 \sum_{l=1}^{[(K-2)/2]} \tilde{N}_{K-1-2l}^- \tag{15.5.18}$$

where [.] denotes the Gauß bracket. It follows that

$$\tilde{N}_K = 1 + 2 \begin{cases} \sum_{l=1,\, l \text{ odd}}^{K-3} N_l^- & K \text{ even} \\ \sum_{l=2,\, l \text{ even}}^{K-3} N_l^- & K \text{ odd} \end{cases} \tag{15.5.19}$$

whence

$$\tilde{N}_K^- - \tilde{N}_{K-2}^- = 2\tilde{N}_{K-3}^- \tag{15.5.20}$$

This is a linear three-step recursion with constant coefficients which is solved by $\tilde{N}_K^- = \sum_{I=0}^{3} c_I q_I^K$ where c_I are constants and q_I are the three roots of the cubic equation

$$q^3 - q = 2 \tag{15.5.21}$$

These can be found by the formulae due to Cardano and one finds

$$q_I = \frac{e^{-i\varphi_I}}{3r} + r e^{i\varphi_I}, \quad r = \sqrt[3]{1 + \sqrt{1 - \frac{1}{27}}}, \quad \varphi_I = 2I\pi/3, \quad I = 0, 1, 2 \tag{15.5.22}$$

The square of the modulus of the real root is the larger by $1/3$ than that of the complex ones. The coefficient c_0 can be determined from $\tilde{N}_{0,1,2}^-$ and is non-vanishing. Hence

$$\tilde{N}_K^- \propto \left(r + \frac{1}{3r}\right)^K , \quad r = \sqrt[3]{1 + \sqrt{1 - \frac{1}{27}}} \qquad (15.5.23)$$

Hence, we get an estimate of the form $c_1 q_1^K \le \tilde{N}_K \le c_2 q_2^K$, thus by varying K according to $\delta K = 2\delta/(4\pi\beta\ell_P^2)$ also $c_2 q_1^K \le N_K \le c_2 q_2^K$. The entropy of the black hole is therefore up to a constant estimated by

$$K \ln\left(r + \frac{1}{3r}\right) \le S_{BH} \le K \ln(3) \qquad (15.5.24)$$

This estimate shows already that the dominant contribution is indeed proportional to the area. Notice that even the lower bound is higher than the upper bound $\ln(2)/\sqrt{3}K$ reported in [469], which is due to an error made in that paper. The correct and exact value can be obtained using a more elaborate technique and one finds (remember $\ell_P^2 = \hbar G$ in this chapter) [795, 796]

$$S_{BH} = \frac{\beta_0}{\beta}\frac{A_0}{4\ell_P^2} - \frac{1}{2}\ln\left(\frac{A_0}{4\ell_P^2}\right) + O(1) \qquad (15.5.25)$$

plus subleading terms where β_0 is the solution to the equation[8]

$$1 = 2\sum_{k=1}^{\infty} e^{-\pi\beta_0\sqrt{k(k+1)}} , \quad \beta_0 = 0.23753295796592\ldots \qquad (15.5.26)$$

This number had been obtained before quite independently from holographic considerations in [802].

15.6 Discussion

In order to match the result (15.5.25) to the Hawking–Bekenstein value we must fix the Immirzi parameter to equal $\beta = \beta_0$. This strategy would be worthless if it was not the *same* value that one had to match for various kinds of black holes, not only the vacuum black holes that we have treated so far. However, as one can show [470–472] even for dilatonic and Yang–Mills hair black holes the same value works. This relies on the following facts: (a) the presence of this bosonic matter does not change the isolated horizon boundary conditions, (b) the matter fields are determined through W at S and therefore (c) matter has no independent

---

[8] This constant changes slightly if one re-attributes bulk and surface degrees of freedom. For instance one could argue that the total area spin quantum numbers J_p at the punctures should be counted as surface degrees of freedom [797–800], which is not unreasonable. See [801] for a discussion. We prefer the attribution as displayed here because there is a clear mathematical distinction between the quantum surface and bulk degrees of freedom respectively.

surface degrees of freedom. It should be pointed out that all of this works for astrophysically realistic (Schwarzschild), four-dimensional, non-supersymmetric black holes, which should be contrasted with the situation in string theory [56].

A couple of remarks are in order:

(i) *Non-triviality*

The derivation of this result is highly non-trivial: it involves an extensive list of physical arguments and one could not have expected from the outset that there would be a harmonic interplay between classical General Relativity (isolated horizon boundary conditions), quantum gravity (discrete eigenvalues of the area operator) and quantum Chern–Simons theory (horizon degrees of freedom).

Next, recall that there has been established a precise dictionary between the four laws of usual thermodynamics and black hole thermodynamics for *event horizons*. It turns out that one can write another dictionary for *isolated horizons* [790, 791]. Also, cosmological horizons can be described by isolated horizon methods.

(ii) *Emission spectrum*

The eigenvalues $Ar(n, \vec{j})$ are, luckily, not evenly spaced. In particular, one can show [816] that the number of eigenvalues in the interval $[A_0 - \ell_p^2, A_0 + \ell_p^2]$ grows as $e^{\sqrt{A_0}/\ell_p}$, which would not be the case for even spacing as seems to be favoured by the authors of [804]. Even spacing would have huge observational consequences: the peak of the black body Hawking spectrum from the black hole is at frequencies $\omega_0 \approx 1/r_0$ where $r_0 \approx GM$ is the Schwarzschild radius of the black hole (we neglect numerical constants and set $c = 1$). Now $A_0 = 4\pi r_0^2$ and since energy emission of the black hole is due to 'area transitions' we obtain spectral lines at $\hbar\omega \approx (\Delta M) \approx \Delta(\sqrt{A}/G) \approx (\Delta A)/(Gr_0) \approx \omega_0 \Delta A/G$. We see that if the spectrum was evenly spaced at $\Delta A \approx \hbar G$ then $\omega \approx n\omega_0$, so we would not get a black body spectrum at all, every line would be at a multiple of the peak line.

Yet, there seems to be a surprising reappearence [805] of the ad hoc quantisation condition proposed in [804]: if one plots $\ln(N(A_0, \delta))$ as a function of A_0 at fixed, generic value of δ one finds that this number displays oscillations with a fixed period given empirically by $\Delta = 8\beta\ell_P^2 \ln(3)$, which depends on the Immirzi parameter but not on δ. Even more interesting, if one sets $2\delta = \Delta$ then the oscillations of $\ln(N(A_0, \delta))$ disappear and the graph of that function becomes a staircase with step size given empirically by $\Delta S = 2\beta_0 \ln(3)$ where β_0 is the value displayed in (15.5.26). Thus, while the area spectrum is quasicontinuous for large spin quantum numbers, at that particular value of δ the range of the entropy is discrete, that is, entropy is quantised with even spacing but not area. Hence, while entropy is proportional to area, the proportionality is not exact but takes the form $S = [\text{Ar}(s)/4\ell_P^2]$ where $[.]$ is the Gauß bracket. This very interesting behaviour asks for a physically more intuitive explanation.

(iii) *Quasinormal modes*

An interesting and puzzling issue which has ignited a lively debate is the following: the classical perturbation theory of a Schwarzschild spacetime (linearisation of the field equations around an exact Schwarzschild metric) reveals that the frequencies of the corresponding Fourier modes are complex-valued, which means that the perturbations get damped by the energy loss due to the corresponding gravitational radiation. The spectrum of these ringing or quasinormal modes (see, e.g., [806] for a beautiful introduction) was determined first numerically in [807] to be $GM\omega_n/c^3 = 0.04371235 + \frac{i}{4}(n + 1/2)$ for the large damping limit. Interestingly, the real part asymptotes to a constant which equals *with eight digits precision* the number $\ln(3)/(8\pi)$ as observed first in [808]. This was later confirmed analytically in [809, 810]. The origin of the logarithm is awkward and suggests a connection with the Bekenstein–Hawking entropy as proposed by Dreyer in [811].

According to the LQG picture, the horizon area is quantised and given by $\text{Ar}(n, \vec{j}) = 8\pi\beta\ell_P^2 \sum_{p=1}^n \sqrt{j_p(j_p + 1)}$ with $\ell_P^2 = \hbar G/c^3$. If the black hole loses mass due to Hawking radiation then by the Schwarzschild radius area relation $A_S = 4\pi r_S^2 = 16\pi G^2 M^2/c^4$ we have $\delta A_s = 32\pi G^2 M\delta M/c^4$, hence if $c^2\delta M$ is a spectral line $\hbar\omega$ of the Hawking radiation then $GM\omega/c^3 = \delta A_S/(32\pi\ell_P^2)$. Let us write $\delta A_S = 8\pi\beta\ell_P^2\delta\lambda$ then $GM\omega/c^3 = \beta\delta\lambda/4$. Consider a transition due to disappearance of a puncture, that is, $\delta\lambda = \sqrt{j_m(j_m + 1)}$ where j_m is the minimal spin allowed to disappear in such a process. Assume also that most of the entropy comes from such j_m punctures by some selection principle which somehow is a trace from the non-trivial dynamics in the bulk not yet reflected in the current black hole calculations. It is easy to see that then the analysis made above changes in that the entropy would be $S = \ln(2j_m + 1)N$ where $N = A_0/(8\pi\beta\sqrt{j_m(j_m + 1)}\ell_p^2)$. In order that this be $A_0/(4\ell_P^2)$ we must have $\beta = \ln(2j_m + 1)/(2\pi\sqrt{j_m(j_m + 1)})$. Putting things together we find $GM\omega/c^3 = \ln(2j_m + 1)/(8\pi)$. If we now conjecture that the Hawking radiation spectrum has something to do with the quasinormal mode spectrum, then we find that $j_m = 1$.

This can be interpreted in many ways. First of all it could be that the conjecture is totally wrong, hence the whole analysis would be a pure coincidence. This viewpoint is supported by the fact that for charged black holes the real part of the quasinormal mode spectrum does not asymptote but rather oscillates wildly for large n and the amplitude of the oscillations does not decay in the limit of small charges. Hence the uncharged case is a rather special case hinting at a mere coincidence. Another argument in favour of this viewpoint is that the conjecture itself is very unnatural in that it is not clear what a microscopic quantum transition should have to do with a macroscopic, collective phenomenon. Another interpretation is that although the analysis only applies to the Schwarzschild family it still teaches us something about the dynamics,[9] for example, that only punctures with $j_m \geq 1$ are allowed due to some trace of the bulk dynamics on the horizon. This might be the case because, as mentioned, our calculation is based on the assumption that for each admissible entry in our list $L(K)$ there is a compatible bulk solution to the Hamiltonian constraint for which there is no argument at present.

No matter which viewpoint one takes, the subject is interesting and underlines once more that we need to understand much better the quantum dynamics of LQG. Notice, however, that if indeed all spins contribute to the entropy as was assumed in the main text and are not dynamically forbidden, then the entropy is not accounted for by 'Boolean degrees of freedom' [67].

(iv) *Open problems*

The case that we have treated above was for a static (spherically symmetric) isolated horizon. Rotating isolated horizons can be treated classically [812]. An effective way to describe them is in terms of classical multipole moments defined in [813]. It turns out that the corresponding multipole operators are simple functions of the area operator. Hence, if we construct a microcanonical ensemble corresponding to fixed area eigenvalues as in the spherically symmetric case, then this automatically fixes the multipole moment eigenvalues as well. The entropy counting therefore proceeds *exactly* as in the spherically symmetric case and gives the Bekenstein–Hawking result for the same value of the Immirzi parameter [814].

A different question is whether one can also treat Hawking radiation with the present framework and a pioneering Ansatz was made in [815]. Also, it has been conjectured that the Bekenstein–Hawking entropy is an inevitable, universal property of any kind of quantum gravity theory and a proof of that conjecture was begun in [817–819]. However, this calculation was shown not to apply in the present context [820]. A recent modification of that calculation, however, seems to fix the problem [821]. Finally, a better understanding of the role of the Immirzi parameter and whether or not it should be fixed as displayed here would be desirable. For instance, one could argue that renormalisation (quantum field theoretical screening effects of matter) affects Newton's constant, which then could be reabsorbed into the Immirzi parameter. The value of the Immirzi parameter could also be changed drastically if it turned out that due to the specific structure of the Hamiltonian constraint not all the states that we counted above are allowed, thus reducing the entropy and the Immirzi parameter (as long as the dominant term is

---

[9] Some speculate that $j_m = 1$ means that only integral values of j are allowed so that the gauge group is SO(3) rather than SU(2). This is impossible if we want to couple fermions.

linear in the area). This is due to the fact that the Hamiltonian constraint, while vanishing at S since the lapse vanishes there, does not vanish in the bulk and there does have an impact on the set of spins of the edges that puncture S because these edges must intersect in bulk vertices at which the Hamiltonian constraint restricts the space of possible intertwiners.

(v) *First principle calculation*

The isolated horizon description is an effective one (not from first principles) because the presence of an isolated horizon was put in at the classical level. It would be far more desirable to begin with the full quantum theory and to have *quantum criteria* at one's disposal for when a given state represents a quantum black hole. At this point the semiclassical analysis discussed in Section 11.2 could be of some help. For pioneering steps in that direction, in particular the resolution of the big bang singularity within the Schwarzschild minisuperspace model, see [822–824]. For more advanced results see [501, 825–828] where one quantises the condition for the formation of a trapping horizon in a spherically symmetric context using the more traditional ADM variables and also finds a resolution of the black hole singularity (see also the next chapter). This is also the reason why we have dealt with trapping and dynamical horizons in this book, because it seems to be the natural classical platform from which one can fully define quantum black holes.

For results within the full theory see [829–831]. For a conceptual framework concerning singularity avoidance see [828, 832, 834, 835].

16

Applications to particle physics and quantum cosmology

16.1 Quantum gauge fixing

Applications of Loop Quantum Gravity to Particle Physics (see A. [637,638] in the canonical framework where scalar, electromagnetic and fermionic-free matter propagation on fluctuating quantum spacetimes is studied, B. [773,774,836–839] in the 4D spin foam framework where graviton propagators are studied and C. [724, 840, 841] in the 3D spin foam framework where the relation with Feynman diagrams is studied) and Quantum Cosmology (see [834,835] for the full theory where the homogeneous sector has been studied; for homogeneous minisuperspace models see the next section) have just begun. This important research area is so far little explored because ideally one would need to have sufficient control over the physical Hilbert space. Since this is not yet the case, one must think about approximation schemes in order to make progress. As a possible starting point or approximation scheme one could use the kinematical, semiclassical framework developed in Chapter 11: namely, if we use states which are peaked on points in the phase space that solve the constraints, then the expectation value of the constraints in these states is either exactly or close to zero and the fluctuations are small in a suitable sense. Hence, while the states are not physical states, even the norm of the constraints on those states is small. They are therefore *approximately physical* states.

Notice that kinematical semiclassical states are labelled by a point on the constraint surface contained in some gauge orbit but not by the gauge orbit itself. In other words, we must choose a classical gauge fixing, that is, a section of the bundle whose total space is the constraint surface, whose fibres are the gauge orbits and whose base space is the reduced phase space. By definition, the constraints are exactly satisfied in this classical gauge fixing. However, the semiclassical states can fluctuate around the constraint surface: in contrast to reduced phase space quantisation or gauge fixing quantisation, we have not switched off the unphysical degrees of freedom (neither those in the orbits nor off the constraint surface), all degrees of freedom are still quantised. Hence, what the semiclassical framework provides is a *quantum gauge fixing* as advertised for the first time in [479, 485, 486, 488–490]. See [842, 843] for first tests of this idea in simple toy models.

Next, ideally one would try to analyse Dirac observables which, as we have seen, are hard to compute classically and even more so in the quantum theory. However, since by construction the semiclassical states are approximately annihilated by the Master Constraint operator $\hat{\;}$, the expectation value of any kinematical operator $\hat{O}$ and of its gauge transform $\exp(it\hat{\;})\hat{O}\exp(-it\hat{\;})$ are approximately identical, which can be interpreted as saying that we describe $\hat{O}$ in the gauge as chosen by the semiclassical state upon which it becomes, by definition, an observable. Indeed, the commutator of its expectation value is approximately zero, while about its fluctuation nothing general can be said. To improve on this, one can try to construct evolving constants as described in Chapter 12 on points of the phase space close to the gauge cut, so that the series involved can be terminated after a few terms because the series is an expansion around the points of the gauge cut.

Finally notice that quantum gauge fixing solves another mathematical problem with complete observables: complete observables for sufficiently complicated systems tend to be only locally defined on phase space. This is no problem in the classical theory, but we only know how to do quantum mechanics with densely defined operators. How should one implement into an operator the information that it is defined only on states with respect to which it has expectation values close to the classically allowed range? Quantum gauge fixing avoids these complications because we are working with kinematical but gobally defined quantum objects while the states are peaked on the classically allowed region only.

Thus, the semiclassical framework together with the partial observable framework provides a promising tool in order to circumvent solving the theory exactly and to develop suitable approximation methods. At the moment, little can be said about how close those approximations are in comparison to the exact theory, however, the subject is under close investigation.

More details about this approach can be found in [834, 835]. In what follows we will describe another development which is an approximation of a different kind: one studies a restricted class of spacetimes in the classical theory, so-called symmetry reduced mini-or midisuperspace models, which have fewer physical degrees of freedom than the full theory. The advantage is that these models are mathematically more tractable, however, they are also unreliable concerning their predictive power, because physical degrees of freedom, which in full theory are allowed to fluctuate and thus potentially destroy the imposed symmetry, are switched off by hand. In other words, it is presently not clear whether such models are stable under the mentioned fluctuations and whether they really capture all the essential information about the full theory.

16.2 Loop Quantum Cosmology

Loop Quantum Cosmology (LQC), to be precise, is *not* the cosmological sector of LQG. It is a toy model, namely Bianchi type models of full GR, quantised by using some of the methods of LQG. There are only a finite number of degrees of freedom involved. This is why these models, even as quantum theories, can be solved exactly. However, LQC is also *not* the usual quantisation of these minisuperspace models, which we refer to as Ordinary Quantum Cosmology (OQC). In both LQC and OQC one starts from the same classical model, the homogeneous (possibly isotropic) Bianchi type symmetry reduction of full classical GR. However, in OQC these models are quantised in the usual Schrödinger representation. In this representation the Weyl operators corresponding to the remaining degrees of freedom are weakly continuous. The idea, due to Bojowald [497–499,597–599,844–864], behind LQC is that this representation is not a good model for the representation used in full LQG, which is weakly continuous with respect to the fluxes but not with respect to the holonomies. Thus, to model this crucial aspect of LQG one uses a weakly discontinuous representation of the type described in Chapter 8 just after formula (8.1.1).

In this short chapter we will focus only on the qualitative aspects of LQC, for the details we refer the reader to the literature: the general reduction framework is developed in [497, 844]. Isotropic and flat models are studied in detail in [849–851] while general isotropic models are considered in [852–854]. Some homogeneous but non-isotropic models are analysed in [855], most importantly the classically chaotic Bianchi IX model in [861,862]. In all these models the big bang singularity is quantum mechanically absent in two different senses. First, there is no *curvature singularity* as zero-volume eigenstates are simultaneously eigenstates for the inverse scale factor operator (or generalised co-triad operator with finite eigenvalue). *This property is inherited from the full theory and thus can be viewed as a success of the techniqes of Chapter 10.* Second, one can extend the 'time' evolution to negative times. That is, in all these models the Hamiltonian constraint equation can be interpreted as a discrete time evolution equation, that is, a difference equation with respect to a partial clock observable (e.g., a component of a co-triad operator) and this equation can be uniquely continued through the classical singularity. This property is a particular feature of the simplified analytical appearence of the model and is not necessarily shared by the full theory. In any case, both properties together could be taken to mean that, within the model, there is no *initial singularity*. While there is this radical departure from OQC, consistency with the usual Wheeler–DeWitt differential equation at large volume is verified in [846], which also leads to initial conditions that are dynamically prescribed rather than by hand if one imposes semiclassical behaviour at late times (large volume). Furthermore, in the model, as in the full theory, quantisation ambiguities arise which can influence the length of a possible inflationary phase [860]. Finally, recent reviews are available in [863,864].

To be specific and in order to simplify the discussion, let us consider the simplest case, the spatially homogeneous and isotropic Friedman–Robertson–Walker (FRW) models which correspond to the line element

$$ds^2 = -dt^2 + a(t)^2 \left[\frac{dr^2}{1 - kr^2} + r^2(d\theta^2 + \sin(\theta)^2 d\varphi^2) \right] \qquad (16.2.1)$$

Here $k = 1, 0, -1$ corresponds to a spatially closed, flat and open universe respectively and the only dynamical degree of freedom is the scale factor $a(t) \geq 0$.

We will consider only the flat case $k = 0$ for simplicity, in which case $\sigma = \mathbb{R}^3$. Evidently the intrinsic metric is given by $q_{ab} = a^2 q_{ab}^0$ where $q_{ab}^0 = \delta_{ab}$ is not dynamical. Upon choosing a non-dynamical triad e_a^{0j} compatible with q_{ab}^0 we may define the physical triad by $e_a^j = a e_a^{0j}$. Then $E_j^a = \sqrt{\det(q)} e_j^a / \beta = a^2 E_j^{0a} / \beta$. Next, since the spin connection is a scale-invariant functional we have $\Gamma(e) = \Gamma(e^0)$ and we may choose the SU(2) gauge $e_a^{0j} = \delta_a^j$ so that $\Gamma(e) = 0$. Then $A_a^j = \Gamma_a^j + \beta K_{ab} e_j^b = \beta K_{ab} e_j^b$ modulo the Gauß constraint. Since $K_{ab} = (\dot{q}_{ab} + 2D_{(a} N_{b)})/(2N) = \dot{a} \delta_{ab} / (2N)$ due to spatial homogeneity we know that $K_{ab} = ac\delta_{ab}$ for some c. Thus $A_a^j = \beta(ca^{-1})\delta_{ab}(a^{-1}e_j^{0a}) = \beta c e_a^{0j}$. Finally with $E_j^a = p E_j^{0a} / \beta$ we conclude $\dot{A}_a^j E_j^a = 3\beta \dot{c} p$ where $a = \sqrt{|p|}$. In what follows we fix $\beta = 1$ for simplicity.

When we insert this Ansatz into the Einstein–Hilbert action the result diverges because all the fields are spatially constant and so the whole action is multiplied by $V_0 = \int_{\mathbb{R}^3} d^3x$. As usual, one can get a sensible result after dividing by that factor. Then evidently the symplectic structure becomes $\{p, c\} = \frac{3}{2}\kappa$, all others vanishing. Since the Gauß constraint has been solved already and the diffeomorphism constraint has been fixed by imposing spatial homogeneity and isotropy, the only constraint left is the Hamiltonian constraint. Due to spatial homogeneity the spatial curvature scalar of q vanishes and hence the constraint must be proportional to $\text{Tr}([A_a, A_b][E^a, E^b]) / \sqrt{\det(q)} \propto c^2 p^2 / a^3 = c^2 \sqrt{|p|}$.

Now we consider the associated holonomies and electric fields. We do not need to consider all $A(e)$, $E_n(S)$, only a sufficient number which are enough to separate the points in the symmetry reduced phase space. Let n be constant and S any surface, then $E_n(S) = p \int_S dS_a n^a$, hence we may fix any $S = S_0$ and any $n = n_0$ such that $\int_{S_0} dS_a n_0^a = 1$ and just consider $E_{n_0}(S_0) = p$. Likewise, take any curve $k_{0,r}$ which admits a parametrisation $k_{0,l}^a(t) = t2rk_0^a$ with $k_0^a k_0^b \delta_{ab} = 1$ then $A(k_{0,r}) = \cos(cr) + \sin(cr)k_0^j \tau_j$. Hence $c = -2 \lim_{l \to 0} \text{Tr}(A(k_{0,r})k_0^j \tau_j)/(2r)$ so these holonomies clearly separate the points.

The algebra of cylindrical functions Cyl is therefore equivalent to the finite linear span of the functions e^{ilc} where $l \in \mathbb{R}$. This is the algebra of almost or quasiperiodic functions also studied in Chapter 28 as an application of Gel'fand spectrum techniques. However, notice [657] that this algebra is vastly overcomplete: in fact, take any two real numbers $r_1, r_2 \neq 0$ which are not rationally dependent, that is, $r_1/r_2 \notin \mathbb{Q}$. Then the map $\mathbb{R} \to T^2$; $c \mapsto (e^{ir_1 c}, e^{ir_2 c})$ is a bijection between $\mathbb{R}$ and its image which is dense on the torus. Thus, for example, $r_1 = 1$, $r_2 = \sqrt{2}$ already suffices to separate the points. We will come back to this in a moment.

Proceeding as in the full theory we consider the C^*-algebra $\overline{\text{Cyl}}$ and its spectrum $\overline{\mathbb{R}}$. We then construct the analogue of the Hilbert space $\mathcal{H}_0 = L_2(\overline{\mathcal{A}}, d\mu_0)$. The result is constructed explicitly in Chapter 28 and is given by the Bohr compactification of the real line with associated uniform (translation-invariant) measure. In this Hilbert space the functions $T_r(c) = e^{irc}$ form an orthonormal system and hence the Hilbert space is not separable. The operator $\hat{c}$ therefore does not exist since the Weyl operators W_r corresponding to $\exp(irc)$ are represented weakly discontinuously as $W_r T_{r'} = T_{l+l'}$. The operator $\hat{p}$ on the other hand is the self-adjoint generator of the weakly continuous family W_t corresponding to $\exp(-itp)$ defined densely by $W_t T_r = \exp(irt\ell_p^2)T_l$, hence $\hat{p}T_r = -r\ell_p^2 T_l$. The functions T_r are analogous to the spin-network functions T_s of the full theory, the difference being that $s = (\gamma, \vec{j}, \vec{m}, \vec{n})$ is a mixture of continuous and discrete labels while r is a single continuous label. To make the analogy closer, as explained in Chapter 28, a graph can now be thought of as a finite collection of rationally independent numbers $\gamma = (r_1, \ldots, r_N)$ and spin quantum numbers are simply integers $(n_1, \ldots, n_N)$ which label the 'edges' r_k. They then generate the 'lattice' of real numbers $n_1 r_1 + \cdots + n_N r_N$ in $\mathbb{R}$.

We must now implement the Hamiltonian constraint $H = c^2 \sqrt{|p|} + H_{\text{matter}}$ where for cosmological applications, especially the very early universe, a spatially homogeneous inflaton scalar field ϕ with conjugate momentum π is of interest. Its contribution to the Hamiltonian constraint in the full theory is of the form (see Chapter 12)

$$H_{\text{matter}} = \frac{1}{2} \int d^3x \left[\frac{\pi^2}{\sqrt{\det(q)}} + \sqrt{\det(q)}(q^{ab}\phi_{,a}\phi_{,b} + V(\phi)) \right] \qquad (16.2.2)$$

with some self-interaction potential V. In a spatially homogeneous situation the derivative terms drop out and after dividing by the infinite coordinate volume as above one ends up with $H_{\text{matter}} = \frac{1}{2}[\pi^2/a^3 + a^3 V(\phi)]$. Using $a = \sqrt{|p|}$ this can also be written $H_{\text{matter}} = \frac{1}{2}[\pi^2/\sqrt{|p|}^3 + \sqrt{|p|}^3 V(\phi)]$. We see that while the geometry contribution $H_{\text{geometry}} = c^2\sqrt{|p|}$ together with the potential energy contribution is completely regular at the classical singularity $a = p = 0$, the kinetic energy term of the scalar field diverges there.

Now we want to quantise these expressions in close analogy to the full theory developed in Chapter 10. Thus, in particular we must use the volume and write co-triads as Poisson brackets with holonomies. The volume is evidently given by $V = \sqrt{|p|}^3$ (after dividing by the infinite volume factor), thus the triad is $\{V, c\} \propto \text{sgn}(p)\sqrt{|p|}$. In the classical theory $\text{sgn}(p) = 1 = \text{const.}$, thus we may multiply the classical Hamiltonian constraint with that factor. Alternatively we may declare the co-triad to be multiplied with the sign of p because for non-degenerate metrics $q \propto e^2$ is independent of that factor. It is also possible to keep the sign and quantise it the same way as it is possible in the full theory [573, 574]. Hence $e := \text{sgn}(p)\sqrt{|p|} \propto W_r\{W_r^{-1}, V\}/(ir\kappa)$. Now we use the same technique as for the quantisation of the matter Hamiltonian constraints of Chapter 12 in order to construct an expression for $\pi^2/\sqrt{|p|}^3$ which will be densely defined.[1] We have $2W_r\{W_r^{-1}, V^{1/3}\}/(ir\kappa) = \mathrm{sgn}(p)\sqrt{|p|}^{-1} = e^{-1}$, which is the triad. Therefore $\mathrm{sgn}(p)\sqrt{|p|}^{-3} \propto e^{-3}$. Hence the geometry factor in the scalar kinetic energy term can be made well-defined this way. The triad operator is given the symmetric ordering of the classical expression with Poisson brackets replaced by commutators divided by $i\hbar$, that is

$$\widehat{e^{-1}} = \left(W_r[W_r^{-1}, \hat{V}^{1/3}] - W_r^{-1}[W_r, \hat{V}^{1/3}]\right)/((ir\kappa)(i\hbar))$$
$$= \left(W_r\hat{V}^{1/3}W_r^{-1} - W_r^{-1}\hat{V}^{1/3}W_r\right)/(r\ell_p^2) \tag{16.2.3}$$

Its eigenvalue on $T_{r'}$ is thus up to a numerical factor equal to

$$[\sqrt{|r'+r|} - \sqrt{|r'-r|}]/(r\ell_p) = \frac{1}{r\ell_p} \frac{|r'+r| - |r'-r|}{\sqrt{|r'+r|} + \sqrt{|r'-r|}} \tag{16.2.4}$$

For large r' this becomes $(\mathrm{sgn}(r')\sqrt{r'}\ell_p)^{-1}$ which is precisely the inverse of the eigenvalue of the co-triad $\mathrm{sgn}(p)\sqrt{|p|}$ as it should be. At the classical singularity $r' = 0$ the eigenvalue is actually zero! Its maximum is taken at $r' = r$ and is given by $\sqrt{2/r}\ell_p^{-1}$, which is the maximal value of the inverse scale factor. Hence the operator corresponding to the inverse scale factor is bounded from above in this model as long as the arbitrary number r is kept finite. Moreover, the spectrum deviates from the effective one only for $|r'| \leq r$. Here we encounter the first aspect of the full theory not modelled by LQC: the inverse scale factor, or rather the geometry factor of the operator corresponding to $\int d^3x\pi^2/\sqrt{\det(q)}$ which was explicitly constructed in Chapter 12, *is not bounded from above* by inspection of the results of Chapter 12. Now what is important for the absence of the curvature singularity is that $\widehat{1/a}$ has finite eigenvalue when the eigenvalue of $\hat{a}$ vanishes. However, in the full theory [834, 835] the volume spectrum *is also not bounded at points of the volume spectrum where the volume vanishes, even when restricted to states which are homogeneous and isotropic on macroscopic scales.* The reason for this is that the spectrum of the volume operator of the full theory contains many flat directions. In other words, the full spectrum contains valleys of zero eigenvalues even if the spins j of the corresponding edges adjacent to the given vertex are arbitrarily large. The walls of these valleys become steeper the larger j. Now (16.2.4) is essentially a discrete derivative of eigenvalues of the third root of the volume operator. Thus in the full theory this discrete derivative is finite but unbounded even on states of zero volume. See [665] for the state of the art concerning the properties of the volume operator.

The number r can be interpreted as a regulator which in the model cannot be removed. Namely, in the expression for the gravitational Hamiltonian constraint we must replace the function c^2 by a holonomy, say $[(W_r - W_r^{-1})/(2ir)]^2$ (the actual function that is chosen differs from this one slightly and can be motivated

---

[1] Since the spectrum of $\hat{p}$ is discrete and contains the point zero, inverse powers of $\hat{p}$ are not densely defined.

by using the full theory). In the full theory the regulator had to do with the 'size' of the loop $\alpha_{\gamma,v,e,e'}$ to be attached to a given graph and the representation j of SU(2) used for the corresponding holonomy, which we pointed out in Section 10.3. The regulator could be removed in the full theory with respect to the loop size using spatial diffeomorphism-invariance while the representation j remains as a diffeomorphism invariant remnant after the regulator is removed. Thus there is a discrete ambiguity. In the model the regulator cannot be removed because spatial diffeomorphism invariance is fixed by imposing spatial homogeneity and thus there remains a continuous ambiguity. This is the second aspect of LQG not modelled by LQC.

The geometry part of the Hamiltonian constraint is therefore given by an expression of the form

$$\hat{H}^\dagger_{\text{geometry}} \propto \left[(W_r - W_r^{-1})/(2ir)\right]^2 \left[W_r \hat{V} W_r^{-1} - W_r^{-1} \hat{V} W_r\right]/(r\ell_p^2) \quad (16.2.5)$$

where we have used the ordering that is dictated in the full theory by asking that the operator be densely defined. We immediately see a third aspect of LQC which does not model the behaviour of full LQG: since the value r is arbitrary but fixed once and for all, the kinematical Hilbert space can be written as

$$\mathcal{H}_0 = \oplus_{0 \leq r' < r} \mathcal{H}_{0,r'} \quad (16.2.6)$$

where $\mathcal{H}_{0,r'}$ is the separable Hilbert space defined by the closed linear span of the $T_{r'+nr}$, $n \in \mathbf{Z}$. It is easy to see that $\mathcal{H}_{0,r}$ is an invariant subspace for (16.2.5). This means that in the model the Hamiltonian constraint is *not graph-changing*. This is a radical departure from the full theory if indeed the graph-changing Hamiltonian or Master Constraint is the correct one to use. As we have seen, the current semiclassical framework is only well-developed for non-graph-changing operators and therefore it is not surprising that one can use it in order to show that (16.2.6) has a good semiclassical behaviour for coherent states defined in any of the sectors $\mathcal{H}_{0,r'}$. However, this does not prove anything about the full theory where the graph-changing aspect could be crucial and actually might be the decisive feature that makes LQG a continuum theory rather than a lattice theory.

So far we have only focused on one aspect of the classical singularity, namely that the classical evolution equations become ill-defined at $a = 0$. In the quantum theory the evolution equation is replaced by the Wheeler–DeWitt equation $l[\hat{H}^\dagger f] = 0$ for all $f \in \mathcal{D}$ where $\mathcal{D}$ is the finite linear span of the $T_{r'}$. The solutions are distributions and one can solve the equation over every sector separately. One can interpret them as difference equations with respect to the 'clock' variable $\hat{p}$ so that the unphysical time becomes discrete with time steps of size $r\ell_p^2$ in each sector.[2] This is exactly the same thing that happens in the full theory if we

---

[2] In the full theory that is also possible by using the flux operator instead: given a surface S and a path p of type 'up' with respect to S one can construct right- and left-invariant

use the total volume as a clock variable. We have displayed the corresponding difference equations explicitly in Section 10.5. Now that given in the quantum theory the point $a = 0$ does not pose any mathematical problems, one can pose the more sophisticated question whether the quantum evolution equation can be solved 'through the classical singularity'. In other words, are all the coefficients in the Ansatz $l = \sum_n c_n(\phi) < T_{r'+nr}, . >$ completely determined if we specify, say, $c_{n_0}(\phi)$? Here $c_n(\phi)$ is a coefficient depending on the matter degrees of freedom. The answer turns out to be affirmative: the quantum evolution does not break down at $n = 0$ even in the sector $r' = 0$. This happens because the matter part of the full Hamiltonian actually vanishes at $n = 0, r' = 0$ as we have seen above. Of course, since triads are not bounded at zero volume in the full theory, this must be revisited in the full theory, see below.

Hence, in LQC one can explicitly construct the space of physical states. Unfortunately, this space is still not equipped with a physical inner product. One could use the spectral analysis of the Master Constraint Programme to construct it and to represent the corresponding Dirac observables, which is mathematically not trivial even in this model.[3] However, even without it, it was possible to show that the ambiguity in the choice of the initial conditions of the universe is almost completely removed if one asks that the solution behaves semiclassically at large volume (scale factor). Further insight is presumably gained after we have derived the physical inner product.

Let us come back to the issue raised above [657]: if we use the algebra of operators generated by $\hat{p}, W_1, W_{\sqrt{2}}$ which, regarded as classical functions, separate the points of the classical phase space then the Hilbert space $\mathcal{H}_0$ is hugely reducible. It decomposes into irreducible, separable subspaces. Thus, from the point of view of that minimal algebra, one should therefore just focus on one of its irreducible sectors. This is one more feature of LQC which is not shared by LQG and has to do with its high symmetry. It just expresses the fact that the model cannot

---

vector fields R_p^3, L_p^3 and a Laplacian $(R_p^j)^2 = (L_p^j)^2$ with eigenvalues $\propto (m_p, n_p, j_p(j_p + 1))$ on spin-network states $\propto [\pi_{j_p}(A(p))]_{m_p n_p}$ [488, 489]. Since the spectrum of the flux is unbounded from below and above one can use these operators to define a discrete time with range in a subset of the entire real line. This encodes also the spectrum of the sign operator constructed in [573, 574] whose eigenvalues change sign as we pass through the zero-volume eigenstates of the full theory.

[3] In [865] this was done for the pure gravity sector of LQC using coherent state techniques with the surprising result that the physical Hilbert space is non-separable even though the classical reduced phase space is zero-dimensional. Technically this is due to the fact that the finite constant r had to be introduced. Namely, it leads to a split of the kinematical Hilbert space into an uncountably infinite direct sum $\mathcal{H} = \oplus_{\delta \in [0,r)} \mathcal{H}_r$ of separable subspaces $\mathcal{H}_\delta$ which are invariant under the Master Constraint corresponding to LQC and, not surprisingly, direct integral decomposition leads to a physical Hilbert space of the form $\mathcal{H}_{phys} = \oplus_{\delta \in [0,r)} \mathbb{C}$. If each of the uncountably infinite copies of $\mathbb{C}$ was superselected (no Dirac observables switch between them) then this would be okay. However, in this model the physical Hilbert space turns out to be irreducible for the (weak) Dirac observables. We interpret this as an artefact of the model because the constant r is absent in LQG where diffeomorphism invariance removes the potential r dependence.

capture all the aspects of the full theory and therefore should not be considered as a bad feature.

Further results within LQC are:

1. Using the ambiguity parameter r and further ordering choices it is possible to tune the duration of the inflationary phase which might lead to observational signatures in the WMAP measurement of the anisotropies in the cosmological background radiation [500, 856–860].
2. According to the BKL scenario (see, e.g., [866] and references therein) the singularity structure of a given spacetime can be very well analysed assuming spatial homogeneity. This leads to the Bianchi IX model which is believed to behave chaotically (ergodically) [867]. In LQC this chaotic behaviour has been shown to be quantum mechanically absent [861, 862].

To summarise, while there are some features of LQG which are not very well modelled by LQC as we pointed out so that there can certainly be no claim that LQC is a reliable test of LQG which makes robust predictions, these results are very promising and hopefully extend to LQG, likely by a technically different incarnation. However, at least one aspect is common to both theories: the discreteness of the spectrum of volume and (co-)triad-like operators and the way to define them densely on the Hilbert space. Since these operators are not graph-changing in both theories, they can be analysed by similar coherent states. The fact that these operators have very good semiclassical behaviour in LQC therefore could be argued to be a strong piece of evidence for a similar behaviour of the (analogue of) the inverse scale factor in LQG.

A detailed corresponding comparison between LQG and LQC has been started in [834, 835]. The outcome is that, as already mentioned, in LQG the (analogue of) the inverse scale factor is *not* bounded from above, not even on zero-volume states and not even on any states of any kind of symmetry (such as homogeneous, spherically symmetric or axisymmetric, etc.). This is a robust result for any kind of reduced model and shows, in a drastic way, how much model and full theory can differ from each other in the details. On the other hand, it could be shown that the expectation value of the inverse scale factor with respect to semiclassical, kinematical states which are peaked on a classically singular trajectory (as we evolve the initial data backwards in unphysical time towards the big bang) remains *bounded*. This holds for the isotropic as well as more general homogeneous models (e.g., Kasner). For details see [834, 835].

What this means is the following: in isotropic LQC the inverse scale factor is a bounded operator (in anisotropic LQC models at least on zero-volume eigenstates). This implies that its expectation value with respect to *any* state is bounded as well. Now suppose that we take a state which is semiclassical at large volume and evolve it with respect to some physical Hamiltonian operator towards the classical singularity. Since time evolution is unitary (i.e., states remain normalisable), the expectation value remains bounded all the time. Hence, without specifying a physical Hamiltonian, we get boundedness. This is no longer true if, as in LQG, the inverse scale factor is not bounded on zero-volume eigenstates. Then it really matters what the states are which describe a collapsing universe. A candidate for such states are the semiclassical states peaked on the classical trajectory that we just described. The boundedness result just quoted in LQG is therefore by a completely different mechanism than in LQC.

However, these results are just preliminary because all that was shown so far is that boundedness is achieved when we peak the semiclassical state along the classical trajectory. The real question is what happens under the quantum time evolution of a physical Hamiltonian. In [868–870] this question could be analysed analytically and numerically in LQC for a scalar field coupled to the isotropic and homogeneous $k = 0$ model. For this model one could carry out the programme spelt out for the first time in [834], that is, compute physical observables, the physical Hilbert space and a physical Hamiltonian. The physical evolution is indeed deterministic and before and after the would-be big bang the universe behaves semiclassically. These are promising indications that something similar might happen in full LQG.

These and related questions will be investigated in detail in the future. Of particular importance is the computation of possibly observable effects of the inhomogeneities of full LQG in the WMAP and PLANCK spectrum [502] and the repetition of the beautiful analysis of [871].

17

Loop Quantum Gravity phenomenology

Beyond merely checking whether we have a quantum theory of the correct classical theory, namely General Relativity coupled to all known matter, quantum gravity has certainly a huge impact on the whole structure of physics. For instance, if the picture drawn in Chapter 12 is correct, then one must do quantum field theory on one-dimensional polymer-like structures rather than in a higher-dimensional manifold, presumably the ultraviolet divergences disappear and while there are still bare and renormalised charges, masses, etc., the bare charges will presumably be finite while the renormalised charges should better be called effective charges because they simply take into account physical screening effects.

Quantum gravity effects are notoriously difficult to measure because the Planck length is so incredibly tiny. It may therefore come as a surprise that recently physicists have started to seriously discuss the possibility of measuring quantum gravity effects, mostly from astrophysical data and gravitational wave detectors [503–506]. See also the discussion in the extremely beautiful review by Carlip [9] and references therein. Those who laugh at these ideas are recommended to have a look at the historical remarks in [872], which draws an analogy with the situation at the end of the nineteenth century when it was widely believed that it would never be possible to detect atomic effects. Einstein showed that the atomic structure of matter was not directly, but indirectly, visible through collective effects, in this case Brownian motion, and what we are about to describe goes in the same direction.

The challenge is of course to compute quantum gravity effects within Quantum General Relativity or more specifically LQG. First pioneering steps towards the computation of the so-called γ-ray burst effect have been made, to date mostly at a phenomenological level, in [508,509] for photons and [510,511] for neutrinos. A more detailed analysis based on the coherent states proposed in [485,486,488,489] is given in [637,638].

Due to reasons of space we cannot give a full-fledged account of these developments so we will restrict ourselves to presenting the main ideas for the γ-ray burst effect and otherwise point out further directions.

A γ-ray burst is a light signal of extremely high energetic photons (up to 1 TeV!) that travelled over cosmological distances (say 10^9 years). What is interesting about them is that the signal is like a flash, that is, the intensity decays on time scales as short as 10^{-3} s. The astrophysical origin of these bursts is still under debate (see the references in [510, 511]) and we will have nothing to add on this debate here. What is important though is that these photons probe the discrete (polymer) structure of spacetime more, the more energy they have, which should lead to an energy-dependent velocity of light (dispersion) very similar to the propagation of light in crystals. More specifically, if one plots the time signal of events as measured by an atmospheric Cerenkov light detector [873] within two disjoint energy channels $[E_1 - \Delta E, E_1 + \Delta E]$ and $[E_2 - \Delta E, E_2 + \Delta E]$ then one expects a time difference in the peak of these signals given by $t_2 - t_1 = \xi \frac{L}{c(0)}[(E_2/E_P)^\alpha - (E_1/E_P)^\alpha]$ where L is the difference from the source (measured by the red shift of the galaxy), $c(0)$ is the vacuum speed of light, E_P is the effective Planck scale energy of the order of $m_P c^2$ and α, ξ are theory-dependent constants of order unity. If $\alpha = \xi = 1$, $E_P = m_P c^2$ and $E_2 - E_1 = 1$ TeV then for $L = 10^9$ lightyears we get travel time differences of the order of 10^2 s which is much larger than the duration of the peak. At present, the sensitivity of available detectors is way below such a resolution mainly because no detectors have been built for this specific purpose, but the construction of better detectors such as GLAST is on its way [510, 511].

One may object that (1) quantum field theory effects from other interactions should be much stronger than quantum gravity effects so that this effect would not test so much quantum gravity but rather quantum field theory on Minkowski space, (2) there are many possible astrophysical disturbances that can cause dispersion such as interstellar dust and (3) it is not clear that the photons of different energies have been emitted simultaneously.

The answer to these objections is as follows: (1) is excluded by definition of quantum field theory on Minkowski space. Such a theory is Poincaré invariant by construction while an energy-dependent dispersion breaks Lorentz invariance. We see that the effect is *non-perturbative* because in any perturbative approach to quantum gravity one treats gravity like the other interactions as a quantum field theory on a Minkowski background. (2) is excluded by the fact that the effect gets stronger with higher energy while diffraction at dust gets weaker. The scale of dust or gas molecules is transparent for such highly energetic photons. (3) is apparently excluded by model computations in astrophysics [873] for the known scenarios that lead to the γ-ray burst effect.

How would one then compute the effect within LQG? Basically, one would look at quantum Einstein–Maxwell theory and consider states of the form $\psi_E \otimes \psi_M$ where ψ_E is a fixed coherent state for the gravitational degrees of freedom, peaked at Minkowski initial data and ψ_M is a quantum state for the Maxwell field. Given the Einstein–Maxwell Hamiltonian

$$H_{EM} = \frac{1}{2e^2} \int d^3x \frac{q_{ab}}{\sqrt{\det(q)}} [E^a E^b + B^a B^b]$$

one would quantise it as described in Chapter 12 and then define an effective Maxwell Hamiltonian by

$$< \psi_{\mathrm{M}}, \hat{H}_{\mathrm{M}}^{\mathrm{eff}} \psi_{\mathrm{M}}' >_{\mathcal{H}_{\mathrm{M}}} := < \psi_{\mathrm{E}} \otimes \psi_{\mathrm{M}}, \hat{H}_{\mathrm{EM}} \psi_{\mathrm{E}} \otimes \psi_{\mathrm{M}} >_{\mathcal{H}_{\mathrm{E}} \otimes \mathcal{H}_{\mathrm{M}}}$$

At the moment we can do this computation only at the kinematical level but as outlined in Section 11.2 this should approximate the full dynamical computation and at least gives an idea for the size of the effect.

Whatever technique is finally being used to carry out this computation, the effect, if it exists, is specific to background-independent approaches to quantum gravity. In fact, the technical reason for existence of the effect would be a corollary from the Heisenberg uncertainty relation: the quantum metric operators form a non-commuting set of operators (they depend both on magnetic and electric degrees of freedom) so that it is not possible to diagonalise them simultaneously. The best one can do is to construct an approximate eigenstate for all of them (namely a coherent state), but that state can then not be exactly Poincaré-invariant, only approximately.

A somewhat different research direction within Quantum Gravity Phenomenology is Doubly Special Relativity (DSR) discovered in [513]. See [874] for a recent review and references therein. Here one postulates that some fundamental quantum theory of gravity exists which gives rise to two invariant scales: the speed of light and the Planck energy. Mathematically, DSR is related to a so-called κ-deformation of the Poincaré Lie algebra [875, 876], to a Hopf algebra (or quantum group) or equivalently to a non-commutative version of Minkowski space defined in [513] as shown in [877]. The physical interpretation and the rule for addition of momenta in this theory is somewhat unclear at the moment, however, a possible DSR interpretation of $2 + 1$ gravity coupled to point particles has been proposed in [878]. For phenomenological consequences of DSR theories and a general review of Quantum Gravity Phenomenology based on some kind of modification of Lorentz invariance see [512, 879, 880]. For possible connections between DSR theories, spin foams and non-commutative geometry see [881–884].

IV

Mathematical tools and their connection to physics

In this last part of the book we collect some mathematical background material which is heavily used in the physics part of the book. There are several reasons for doing this: First of all, it makes the book almost self-contained. Secondly, some of this material is not covered by the obligatory courses in mathematics for physicists. Thirdly, while the material is covered in some mathematics courses, it is often presented in such a way that a physicist does not recognise it any more or it is not given sufficient attention. Clearly we can mostly give definitions and state theorems, proofs are often omitted for reasons of space. However, we try to motivate the mathematical theory from a physicists' point of view, explain how the various theorems fit together and indicate their various applications. We thus hope that the ambitious reader feels encouraged to study the mathematical theory in appropriate depth, going through the proofs by himself.

The material is presented in logical order, not in the order as it is applied in the physics part of the book. For instance, topology is needed before one speaks about differential geometry, measure theory and (functional) analysis.

18

Tools from general topology

We collect and prove here some important results from general topology needed in the main text. For more details, see, for example [533].

18.1 Generalities

Definition 18.1.1

I. (i) *Let X be a set and $\mathcal{U}$ a collection of subsets of X. We call X a topological space provided that*

1. *$\emptyset, X \in \mathcal{U}$.*
2. *$\mathcal{U}$ is closed under finite intersections: $U_1, \ldots, U_N \in \mathcal{U}, N \in \mathbb{N} \Rightarrow \bigcap_{k=1}^{N} U_k \in \mathcal{U}$.*
3. *$\mathcal{U}$ is closed under arbitrary (possibly uncountably infinite) unions. $U_\alpha \in \mathcal{U}, \alpha \in A \Rightarrow \bigcup_{\alpha \in A} U_\alpha \in \mathcal{U}$.*

   The sets $U \in \mathcal{U}$ are called open, their complements $X - U$ closed in X. A base $\mathcal{B}$ for $\mathcal{U}$ is such that any $O \in \mathcal{U}$ is an arbitrary union of elements $B \in \mathcal{B}$. A subset $N \subset X$ is called a neighbourhood of $x \in X$ if there is an open set O with $x \in O \subset N$. A neighbourhood base at x is a family $\mathcal{N}$ of neighbourhoods of x such that for any neighbourhood M of x we find $N \in \mathcal{N}$ with $N \subset M$. For example, if $\mathcal{B}$ is a base then $\{N \in \mathcal{B}; \ x \in N\}$ is a neighbourhood base at x. A topology $\mathcal{U}$ is called stronger (finer) than a topology $\mathcal{U}'$, which is then weaker (coarser) if $\mathcal{U}' \subset \mathcal{U}$.

   (ii) *Let $(X, \mathcal{U}), (Y, \mathcal{V})$ be topological spaces such that $Y \subset X$. The relative or subspace topology $\mathcal{U}_Y$ induced on Y is given by defining the sets $U \cap Y; \ U \in \mathcal{U}$ to be open. We say that we have a topological inclusion, denoted $Y \hookrightarrow X$, provided that the intrinsic topology is stronger than the relative one, that is, $\mathcal{U}_Y \subset \mathcal{V}$.*

II. (i) *A function $f : X \to Y$ between topological spaces X, Y is said to be continuous provided that the pre-image $f^{-1}(V)$ of any set $V \subset Y$ that is open in Y is open in X. (The pre-image is defined by $f^{-1}(V) = \{x \in X; \ f(x) \in V\}$ and despite the notation does not require f to be either an injection or a surjection.) One easily shows that f is continuous if it is continuous at each point $x \in X$. Here f is continuous at $x \in X$ if for any open neighbourhood V of $y = f(x)$ there exists an open neighbourhood U of x such that $f(x') \in V$ for all $x' \in U$ (i.e., $f(U) \subset V$).*

(ii) *If f is a continuous bijection and also f^{-1} is continuous then f is called a homeomorphism or a topological isomorphism.*

We see that a topology on a set X is simply defined by saying which sets are open, or equivalently, which functions are continuous. The importance of homeomorphisms f for topology is that not only can the spaces X, Y be identified set theoretically but also topologically, that is, open sets can be identified with each other.

In order to get more topological spaces with more structure one must add separation, and compactness, properties. The one we need here is the following.

Definition 18.1.2

1. *A topological space $(X, \mathcal{U})$ is called*
   (i) *T_1 iff for all $x, y \in X$; $x \neq y$ there exists $O \in \mathcal{U}$ with $x \notin O$, $y \in O$. Equivalently, T_1 spaces are such that all one-point sets $\{x\}$ are closed.*
   (ii) *Hausdorff (or T_2) iff for any two of its points $x \neq y$ there exist open neighbourhoods U, V of x, y respectively which are disjoint.*
   (iii) *Regular (or T_3) iff it is T_1 and if for all closed C and all points $x \notin C$ there exist open sets O_1, O_2 such that $x \in O_1$, $C \subset O_2$, $O_1 \cap O_2 = \emptyset$. Equivalently, T_3 spaces are such that the closed sets form a neighbourhood base.*
   (iv) *Normal (or T_4) iff it is T_1 and if for any closed C_1, C_2, $C_1 \cap C_2 = \emptyset$ we find open O_1, O_2 with $C_1 \subset O_1$, $C_2 \subset O_2$ such that $O_1 \cap O_2 = \emptyset$.*
   One can show that $T_4 \Rightarrow T_2 \Rightarrow T_3 \Rightarrow T_1$.
2. *A topological space is called*
   (i) *Separable iff it contains a countable set S of points which are dense in X (every neighbourhood of any point contains an element of S).*
   (ii) *First countable iff every point has a countable neighbourhood base.*
   (iii) *Second countable if X has a countable base.*
   We remark that metric spaces (those for which the open balls $B_\epsilon(x) = \{y \in X; d(x, y) < \epsilon\}$ for $\epsilon \in \mathbb{R}^+$ form a neighbourhood base) are (1) always first countable, (2) second countable if separable. Moreover, every second countable topological space is separable.
3. *A topological space X is called compact if every open cover $\mathcal{V}$ of X (a collection of open sets of X whose union is all of X) has a finite subcover.*

We remark that a topological space is called disconnected if it is the disjoint union of at least two non-empty closed sets.

Definition 18.1.3

(i) *A net (x^α) in a topological space X is a map $\alpha \to x^\alpha$ from a partially ordered and directed[1] index set A (relation $\geq$) to X.*

---

[1] For the definition of partially ordered and directed see Definition 6.2.11(i), (ii).

(ii) A net (x^α) converges to x, denoted $\lim_\alpha x^\alpha = x$ if for every open neighbourhood $U \subset X$ of x there exists $\alpha(U) \in A$ such that $x^\alpha \in U$ for every
$\alpha > \alpha(U)$ (one says that (x^α) is eventually in U).

(iii) A subnet $(x^{\alpha(\beta)})$ of a net (x^α) is defined through a map $B \to A$; $\beta \mapsto \alpha(\beta)$
between partially ordered and directed index sets such that for any $\alpha_0 \in A$
there exists $\beta(\alpha_0) \in B$ with $\alpha(\beta) \geq \alpha_0$ for any $\beta \geq \beta(\alpha_0)$ (one says that B
is co-final for A).

(iv) A net (x^α) in a topological space X is called universal if for any subset
$Y \subset X$ the net (x^α) is eventually either only in Y or only in $X - Y$.

Notice that for a subnet there is no relation between the index sets A, B except
that $\alpha(B) \subset A$ so that in particular the subnet of a sequence $(A = \mathbb{N})$ may not be
a sequence any longer. The notions of closedness, continuity and compactness can
be formulated in terms of nets. The fact that one uses nets instead of sequences
is that Lemma 18.1.4 is no longer true when $A = \mathbb{N}$ unless we are dealing with
metric spaces.

Lemma 18.1.4

(i) A subset Y of a toplogical space X is closed if for every convergent net (x^α)
in X with $x^\alpha \in Y \; \forall \alpha$ the limit actually lies in Y.

(ii) A function $f : X \to Y$ between topological spaces is continuous if for every
convergent net (x^α) in X, the net $(f(x^\alpha))$ is convergent in Y.

(iii) A topological space X is compact if every net has a convergent subnet
(Bolzano–Weierstrass theorem). The limit point of the convergent subnet
is called a cluster (accumulation) point of the original net.

The proof is standard and will be omitted. One easily sees that if a net converges (a function is continuous) in a certain topology, then it does so in any
weaker (stronger) topology. We warn the reader that in infinite-dimensional metric spaces such as Banach spaces the Heine–Borel theorem (compactness is equivalent to closure and boundedness) is false.

In our applications direct products of topological spaces are of fundamental
importance.

Definition 18.1.5. The Tychonov topology on the direct product $X_\infty = \prod_{l \in \mathcal{L}} X_l$ of topological spaces X_l, $\mathcal{L}$ any index set, is the weakest topology such
that all the projections

$$p_l : X_\infty \to X_l; \; (x_{l'})_{l' \in \mathcal{L}} \mapsto x_l \tag{18.1.1}$$

are continuous, that is, a net $x^\alpha = (x_l^\alpha)_{l \in \mathcal{L}}$ converges to $x = (x_l)_{l \in \mathcal{L}}$ iff $x_l^\alpha \to x_l$ for every $l \in \mathcal{L}$ pointwise (not necessarily uniformly) in $\mathcal{L}$. Equivalently, the
sets $p_l^{-1}(U_l) = [\prod_{l' \neq l} X_{l'}] \times U_l$ are defined to be open and form a base for the
topology of X_∞ (any open set can be obtained from those by finite intersections
and arbitrary unions).

The definition of this topology is motivated by the following theorem.

Theorem 18.1.6 (Tychonov). *Let $\mathcal{L}$ be an index set of arbitrary cardinality and suppose that for each $l \in \mathcal{L}$ a compact topological space X_l is given. Then the direct product space $X_\infty = \prod_{l \in \mathcal{L}} X_l$ is a compact topological space in the Tychonov topology.*

We will give an elegant proof of the Tychonov theorem using the notion of a universal net.

Lemma 18.1.7

 (i) *A universal net has at most one cluster point to which it then converges.*
 (ii) *For any map $f : X \to Y$ between topological spaces the net $f(x^\alpha)$ in Y is universal whenever (x^α) is universal in x with no restrictions on f.*
(iii) *Any net has a universal subnet.*

Proof

 (i) Suppose that x is a cluster point of a universal net (x^α) and that the subnet $x^{\alpha(\beta)}$ converges to it. Thus for any neighbourhood U of x the subnet is eventually in U, that is, there exists $\beta(U)$ such that $x^{\alpha(\beta)} \in U$ for any $\beta \geq \beta(U)$. Since (x^α) is universal it must eventually be in either U or $X - U$. Suppose there was α_0 such that $x^\alpha \in X - U$ for any $\alpha \geq \alpha_0$. By definition of a subnet we find $\beta(\alpha_0)$ such that $\alpha(\beta) \geq \alpha_0$ for any $\beta \geq \beta(\alpha_0)$. Without loss of generality we may choose $\beta(\alpha_0) \geq \beta(U)$. But then we know already that the $x^{\alpha(\beta)}$, $\beta \geq \beta(\alpha_0)$ are in U, which is a contradiction. Thus x^α is eventually in U. Since U was an arbitrary neighbourhood of x, it follows that (x^α) actually converges to x.
 (ii) Obviously $f(x^\alpha)$ is eventually in $f(X)$ so we must show that for any $V \subset f(X)$ we have $f(x^\alpha)$ eventually in V or $f(X) - V$. Let $U = f^{-1}(V)$ be the pre-image of V, then $f(X - U) = f(X) - V$. Since (x^α) is eventually in U or $X - U$, the claim follows.
(iii) The proof can be found in exercise 2J(d) together with theorem 2.5 in [885].

$\square$

Corollary 18.1.8. *A topological space X is compact iff every universal net converges.*

Proof
$\Rightarrow$: Take any universal net (x^α). Since X is compact it has a cluster point to which it actually converges by Lemma 18.1.4 (i).
$\Leftarrow$: Take any net (x^α). Then by Lemma 18.1.4 (iii) it has a universal subnet $x^{\alpha(\beta)}$ which converges by assumption. Thus, X is compact. $\square$

Proof of Theorem 18.1.6. Let $(x^\alpha) = (x_l^\alpha)_{l \in \mathcal{L}}$ be any universal net in $X_\infty = \prod_{l \in \mathcal{L}} X_l$. By Lemma 18.1.4 (ii) the net $p_l((x^\alpha)) = (x_l^\alpha)$ is universal in X_l.

Since X_l is compact, it converges to some x_l. Define $x := (x_l)_{l \in \mathcal{L}}$. By definition of the Tychonov topology, $x^\alpha \to x$ iff $x_l^\alpha \to x_l$ for any $l \in \mathcal{L}$, whence (x^α) converges. $\qquad\square$

This proof of the Tychonov theorem is shorter than the usual one in terms of the (in)finite intersection property, and technically clearer.

Definition 18.1.9. *Let Y be a subset of a topological space X. The subset topology induced by X on Y is defined through the collection of open sets $\mathcal{V} := \{U \cap Y; \ U \in \mathcal{U}\}$ where $\mathcal{U}$ defines the topology of X.*

Lemma 18.1.10. *A closed subset Y of a compact topological space X is compact in the subspace topology.*

Proof: Let $\mathcal{V}$ be any open cover for Y. Since Y is closed in X, $X - Y$ is open in X, whence $\mathcal{U} = \mathcal{V} \cup \{X - Y\}$ is an open cover for X. Since X is compact, it has a finite open subcover $\{U_k\}_{k=1}^N \cup \{X - Y\}$ for some $N < \infty$ where U_k is open in X. By definition of the subspace topology, $U_k \cap Y$ is open in Y so that $\{U_k \cap Y\}_{k=1}^N$ is a finite open subcover of $\mathcal{V}$. $\qquad\square$

We collect a number of rather important results which connect the notions of separability and compactness.

Theorem 18.1.11

(i) *Any compact Hausdorff space X is normal.*

(ii) *Let C_0, C_1 be closed disjoint sets in a normal space X. Then there exists a continuous function $f : X \to [0,1]$ with $f_{|C_0} = 0$, $f_{|C_1} = 1$. This is known as Urysohn's lemma.*

(iii) *Denote by $C_\mathbb{R}(X)$, $C(X)$ respectively the Banach algebras of real-valued and complex-valued functions on a compact Hausdorff space, complete in the sup-norm $\|f\| := \sup_{x \in X} |f(x)|$. We say that a collection B of functions on X separates the points of X if for each pair of points $x \neq y$ we find $f \in B$ such that $f(x) \neq f(y)$.*

If either (a) $B \subset C_\mathbb{R}(X)$ is a closed subalgebra or (b) $B \subset C(X)$ is a closed subalgebra also closed under complex conjugation and B separates the points of X then either $B = C_\mathbb{R}(X)$ or $B = C(X)$ respectively (e.g., if $1 \in B$) or there exists $x_0 \in X$ such that $B = \{f \in C_\mathbb{R}(X); \ f(x_0) = 0\}$ or $B = \{f \in C(X); \ f(x_0) = 0\}$ respectively. This is known as the real (respectively complex) Stone–Weierstrass theorem.

18.2 Specific results

In our discussion of the gauge orbit of connections we will deal with the quotient of connections by the set of gauge transformations, which is a topological space again. The resulting quotient space carries a natural topology, the quotient topology.

Definition 18.2.1

(i) *Let X, Y be topological spaces and $p : X \to Y$ a surjection. The map p is said to be a quotient map provided that $V \subset Y$ is open in Y if and only if $p^{-1}(V)$ is open in X.*

(ii) *If X is a topological space, Y a set and $p : X \to Y$ a surjection then there exists a unique topology on Y with respect to which p is a quotient map.*

(iii) *Let X be a topological space and let $[X]$ be a partition of X (i.e., a collection of mutually disjoint subsets of X whose union is X). Denote by $[x]$, $x \in X$ the subset of X in that partition of X which contains x. Equip $[X]$ with the quotient topology induced by the map $[] : X \to [X]$; $x \mapsto [x]$. Then $[X]$ is called the quotient space of X.*

Notice that the requirement for p to be a quotient map is stronger than that it be continuous, which would only require that $p^{-1}(V)$ is open in X whenever V is open in Y (but not vice versa). Clearly in (ii) we define the topology on the set Y to be those subsets V for which the pre-image $p^{-1}(V)$ is open in X and it is an elementary exercise in the theory of mappings of sets to verify that the collection of subsets of Y so defined satisfies the axioms of a topology of Definition 18.1.1.

Quotient spaces arise naturally if we have a group action $\lambda : G \times X \to X$; $(g, x) \to \lambda_g(x) := \lambda(g, x)$ on a topological space X and define $[x] := \{\lambda_g(x); \, g \in G\}$ to be the orbit of x. The orbits clearly define a partition of X.

Lemma 18.2.2. *Let X be a compact topological space, Y a set and $p : X \to Y$ a surjection. Then Y is compact in the quotient topology.*

Proof: First of all, consider any subsets V_1, V_2 of Y. On the one hand, suppose $x \in p^{-1}(V_1) \cap p^{-1}(V_2)$. Then there exist $y_1 \in V_1, y_2 \in V_2$ such that $y_1 = p(x) = y_2$, that is, $y_1 = y_2 \in V_1 \cap V_2$ so that actually $x \in p^{-1}(V_1 \cap V_2)$. We conclude $p^{-1}(V_1) \cap p^{-1}(V_2) \subset p^{-1}(V_1 \cap V_2)$.

On the other hand, let $x \in p^{-1}(V_1 \cap V_2)$, then there exists $y \in V_1 \cap V_2$ such that $x \in p^{-1}(y)$. Since $y \in V_1 \cap V_2$ we have $p^{-1}(y) \in p^{-1}(V_1)$ and $p^{-1}(y) \in p^{-1}(V_2)$, thus $x \in p^{-1}(V_1) \cap p^{-1}(V_2)$. We conclude $p^{-1}(V_1 \cap V_2) \subset p^{-1}(V_1) \cap p^{-1}(V_2)$.

Thus, altogether $p^{-1}(V_1) \cap p^{-1}(V_2) = p^{-1}(V_1 \cap V_2)$ and $p^{-1}(V_1) \cup p^{-1}(V_2) = p^{-1}(V_1 \cup V_2)$ by taking complements.

Next, let $\mathcal{V}$ be an open cover of Y. Then, by definition of the quotient topology, $p^{-1}(V)$ is open in X and $\mathcal{U} := \{p^{-1}(V); \, V \in \mathcal{V}\}$ covers X because $\bigcup_{U \in \mathcal{U}} U = \bigcup_{V \in \mathcal{V}} p^{-1}(V) = p^{-1}(\bigcup_{V \in \mathcal{V}} V) = p^{-1}(Y) = X$ since p is a surjection and $\mathcal{V}$ covers Y. We conclude that $\mathcal{U}$ is an open cover of X.

Since X is compact, we find a finite, open subcover $\{p^{-1}(V_k)\}_{k=1}^N$ of X so that $X = \bigcup_{k=1}^N p^{-1}(V_k) = p^{-1}(\bigcup_{k=1}^N V_k) = p^{-1}(Y)$, whence $Y = \bigcup_{k=1}^N V_k$, that is, $\{V_k\}_{k=1}^N$ is a finite open subcover of $\mathcal{V}$ and Y is compact. $\qquad\square$

Lemma 18.2.3. *Let X be a Hausdorff space and $\lambda : G \times X \to X$ a continuous group action on X (i.e., λ_g defined by $\lambda_g(x) := \lambda(g, x)$ is continuous for any $g \in G$). Then the quotient space $X/G := \{[x]; \ x \in X\}$ defined by the orbits $[x] = \{\lambda_g(x); \ g \in G\}$ is Hausdorff in the quotient topology.*

Proof: Let $[x] \neq [x']$, then certainly $x \neq x'$ since orbits are disjoint. Since X is Hausdorff we find disjoint open neighbourhoods U, U' of x, x' respectively. We want to show that U, U' can be chosen in such a way that

$$[U] := \{[y]; \ y \in U\}, \ [U'] := \{[y']; \ y' \in U'\} \tag{18.2.1}$$

are disjoint. First of all we notice that (p the projection map)

$$p^{-1}([U]) = \bigcup_{y \in U} p^{-1}([y]) = \{\lambda(g, y); \ y \in U, g \in G\} = \bigcup_{g \in G} \lambda_g(U) = \bigcup_{g \in G} \lambda_{g^{-1}}(U)$$

$$= \bigcup_{g \in G} (\lambda_g)^{-1}(U) \tag{18.2.2}$$

where we have made use of $\lambda_{g^{-1}} = (\lambda_g)^{-1}$. Since U is open in X and λ_g is continuous by assumption, we have that $\lambda_g^{-1}(U)$ is open in X. Since arbitrary unions of open sets are open it follows that $p^{-1}([U])$ is open in X, thus by the definition of the quotient topology we have $[U], [U']$ open in X/G. Next, obviously $[x] \in [U], [x'] \in [U']$ whence $[U], [U']$ are open neighbourhoods of $[x], [x']$ in X/G respectively.

Let us now choose V, V' to be open, disjoint neighbourhoods of the orbits $p^{-1}([x]) = \lambda_G(x), p^{-1}([x'])$ respectively. (This is certainly possible as otherwise there exists $g \in G$ such that $\lambda_g(x), x'$ have no disjoint neighbourhoods, which is impossible because $\lambda_g(x) \neq x'$ (otherwise $[x] = [x']$) and X is Hausdorff.) We claim that we can choose U, U' in such a way that $p^{-1}[U] := \bigcup_{g \in G} \lambda_g(U) \subset V$ and $p^{-1}[U'] := \bigcup_{g \in G} \lambda_g(U') \subset V'$.

Suppose that were not the case. Then for any neighbourhood U of x we find $z \in U$ and $g_0 \in G$ such that $\lambda_{g_0}(z) \notin V$. Since by construction of V we have that V is a common open neighbourhood of any $\lambda_g(x), g \in G$ we have in particular $y := \lambda_{g_0}(x) \in V$. It follows that we have found an open neighbourhood V of $y = \lambda_{g_0}(x)$ such that for any open neighbourhood U of x there exists $z \in U$ with $\lambda_{g_0}(z) \notin V$. This means that the map λ_{g_0} is not continuous at x, in contradiction to our assumption that λ_g is everywhere continuous for any $g \in G$.

Therefore $p^{-1}([U]) \cap p^{-1}([U']) = p^{-1}([U] \cap [U']) = \emptyset$, whence $[U] \cap [U'] = \emptyset$, thus X/G is Hausdorff. $\square$

Theorem 18.2.4. *Let X, Y be topological spaces and let G be a group acting (not necessarily continuously) on them via λ, λ' respectively. If $f : X \to Y$ is a homeomorphism with respect to which the actions λ, λ' are equivariant then f extends as a homeomorphism to the quotient spaces $X/G, Y/G$ in their respective quotient topologies.*

Proof: Equivariance means that $f \circ \lambda_g = \lambda'_g \circ f$ for all $g \in G$ and since f is a bijection, equivariance implies also $\lambda_g \circ f^{-1} = f^{-1} \circ \lambda'_g$. Consider the corresponding quotient maps

$$p: X \to X/G; \; x \mapsto [x]_\lambda = \{\lambda_g(x); \; g \in G\} \quad \text{and} \quad p': Y \to Y/G; \; y \mapsto [y]_{\lambda'}$$
$$= \{\lambda'_g(y); \; g \in G\} \tag{18.2.3}$$

Then due to equivariance

$$f([x]_\lambda) = \{f(\lambda_g(x)); \; g \in G\} = \{\lambda'_g(f(x)); \; g \in G\} = [f(x)]_{\lambda'} \tag{18.2.4}$$

and similarly $f^{-1}([y]_{\lambda'}) = [f^{-1}(y)]_\lambda$ so that f extends to a bijection between the corresponding equivalence classes.

Next we notice that $p^{-1}([x]_\lambda) = \{\lambda_g(x); \; g \in G\}$ whence by (18.2.4) we have $f(p^{-1}([x]_\lambda)) = (p')^{-1}([f(x)]_{\lambda'})$ for all $[x]_\lambda \in X/G$. This shows that equivariance also implies

$$f \circ p^{-1} = (p')^{-1} \circ f \; \Rightarrow \; f^{-1} \circ (p')^{-1} = p^{-1} \circ f^{-1} \tag{18.2.5}$$

Let then B be open in Y/G, thus $(p')^{-1}(B)$ is open in Y by definition of the quotient topology in Y/G, thus $(f^{-1} \circ (p')^{-1})(B) = (p^{-1} \circ f^{-1})(B)$ is open in X since f is continuous, thus $f^{-1}(B)$ is open in X/G by definition of the quotient topology in X/G. Likewise we see that A open in X/G implies $f(A)$ open in Y/G since f^{-1} is continuous. It follows that f, f^{-1} are continuous as maps between $X/G, Y/G$. $\qquad\square$

19

Differential, Riemannian, symplectic and complex geometry

In this chapter we collect the basic notions from differential geometry and its application to Riemannian, symplectic and complex manifolds. We restrict ourselves to finite-dimensional manifolds, the generalisation to infinite-dimensional manifolds is briefly sketched in Chapter 33 and can be found, for example, in [220, 900]. There are many excellent textbooks on differential geometry, for example, [234, 337, 887].

19.1 Differential geometry

Even without a Riemannian or symplectic structure the notion of a manifold enables us to generalise differential and integral calculus familiar from $\mathbb{R}^m$.

19.1.1 Manifolds

Definition 19.1.1

(i) *A topological space M is called an m-dimensional C^k manifold provided there is a family of pairs $(U_I, x_I)_{I \in \mathcal{I}}$ consisting of an open cover of M, that is, $M = \cup_{I \in \mathcal{I}} U_I$ and homeomorphisms $x_I : U_I \to x_I(U_I) \subset \mathbb{R}^m$; $p \mapsto x_I(p)$ such that for all $I, J \in \mathcal{I}$ with $U_I \cap U_J \neq \emptyset$ the map $\varphi_{IJ} := x_J \cap x_I^{-1} : x_I(U_I \cap U_J) \to x_J(U_I \cap U_J)$ is a C^k map between open subsets of $\mathbb{R}^m$.*

(ii) *The sets U_I are called charts, the functions x_I coordinates and the family of charts and coordinates comprises an atlas. The number m is called the dimension of M. Two atlases $(U_I, x_I)_{I \in \mathcal{I}}$, $(V_I, x_J)_{J \in \mathcal{J}}$ for a topological space M are said to be compatible if their union is again an atlas. Compatibility of atlases is an equivalence relation and an equivalence class is called a differentiable C^k structure.*

(iii) *A topological space M is said to be a manifold with a boundary ∂M provided each of the U_I is homeomorphic to an open subset of the negative half-space $H_- = \{x \in \mathbb{R}^m; \ x^1 \leq 0\}$. The smoothness condition on the coordinate functions is now applied as before, just that one asks that the φ_{IJ} are C^k on open subsets of $\mathbb{R}^m$ containing $x_I(U_I \cap U_J)$. The boundary points have coordinates $x^1 = 0$, that is, they lie in $\partial H_- = \{x \in \mathbb{R}^m; \ x^1 = 0\}$.*

(iv) *A map $\psi : M \to N$ between C^k manifolds M, N is called C^k if for all pairs of charts U_I, V_J of atlases for M, N respectively such that $\psi(U_I) \cap V_J \neq \emptyset$ the maps (where defined) $\psi_{IJ} := x_J \circ \psi \circ x_I^{-1} : x_I(U_I) \to x_J(V_J)$ are C^k*

maps between open subsets of $\mathbb{R}^m, \mathbb{R}^n$ *respectively. If all the* ψ_{IJ} *are invertible and also the inverses are* C^k *then* ψ *is called a* C^k *diffeomorphism.*

The diffeomorphisms of a manifold form a group which is denoted $Diff(M)$.

(v) An atlas (U_I, x_I) is said to be *locally finite* provided that every $p \in M$ has an open neighbourhood in M intersecting only a finite number of the charts. A manifold M is called *paracompact* if each atlas (U_I, x_I) admits a locally finite refinement (V_J, y_J) where each V_J is contained in some U_I.

(vi) Let N be a subset of an m-dimensional manifold M. We can equip N with the structure of a manifold provided the following condition holds: N naturally carries the induced (subspace) topology of M (i.e., the open sets are given by $N \cap U$ where U is open in M). Next we try to define an induced (subspace) differentiable structure, given an atlas (U_I, x_I) for M, by the atlas $(V_I = N \cap U_I, y_I = (x_I)_{|V_I})$ for N. This defines a differentiable structure only if the maps $\varphi_{IJ} = y_J \circ y_I^{-1}$ for $V_I \cap V_J \neq \emptyset$ have constant rank n.

Conversely, suppose that N is an n-dimensional manifold and that $\psi : N \to M$ is a C^k map. ψ is said to be a *local immersion* if each $q \in N$ has an open neighbourhood V such that $V \to \psi(V)$ is an injection. If ψ is a *global immersion*, that is, $N \to \psi(N)$ is an injection (the image of N in M does not intersect itself), then ψ is called an *embedding*. If moreover for each V open in N the set $\psi(V)$ is open in the subset topology induced from M, that is, it is of the form $U \cap \psi(V)$ for some open subset of M then ψ is called a *regular embedding* (the image of N does not come arbitrarily close to itself in M without ever self-intersecting). In the latter case we will say that N is an *embedded submanifold* of M.

An embedded submanifold of dimension $n = m - 1$ is called a *hypersurface*.

(vii) A manifold M is said to be *orientable* if it admits an atlas such that $\det(\partial x_J(p)/\partial x_I(p)) > 0$ for all $p \in U_I \cap U_J$. If M has a boundary then M induces an orientation on ∂M as follows: ∂M is a submanifold of M with atlas $(V_I = U_I \cap \partial M, y_I = (x_I)_{|V_I})$. By definition, if $U_I \cap \partial M \neq \emptyset$ then $x_I(U_I) \subset H_-$, $y_I(V_I) \subset \partial H_-$. Now $V_I \cap V_J \neq \emptyset$ requires $U_I \cap U_J \neq \emptyset$. By assumption the sign of $x_J^1(x_I)$ equals that of x_I^1, hence $[\partial x_J^1/\partial x_I^\mu](x_I^1 = 0, x_I^2, \ldots, x_I^m) = c\delta_\mu^1$ with $c > 0$ due to continuity. Since $\det(\partial x_J/\partial x_I) > 0$ in U_I, taking $x_I^1 \to 0$ shows that the coordinates $y_I = (x_I^2, \ldots, x_I^m)$ provide an orientation of ∂M.

(viii) A manifold is called *smooth* if it is C^∞. A manifold is called *real analytic* or C^ω if the maps φ_{IJ} are real analytic, that is, they have a convergent Taylor expansion in a neighbourhood of each point. A manifold of real dimension $2m$ is called *complex analytic* or a *holomorphic manifold* of complex dimension m provided that the maps $\varphi_{IJ} = z_J \circ z_I^{-1} : \mathbb{C}^m \to \mathbb{C}^m$ satisfy the Cauchy–Riemann equations and $(x_I, y_I) \to z_I = x_I + iy_I$ is the standard isomorphism between $\mathbb{R}^{2m}$ and $\mathbb{C}^m$.

Notice that if M, N are diffeomorphic then automatically $m := \dim(M) = \dim(N) =: n$. We will identify C^k manifolds whose differentiable structures are diffeomorphic. Hence diffeomorphisms classify differentiable manifolds into classes. Notice that the Möbius strip, defined as the topological space derived from the two-dimensional plane by identifying the points (x, y) and $(x + 2\pi, -y)$, does not admit an orientable atlas.

It is not easy to construct homeomorphisms of a topological space which are not simultaneously smooth diffeomorphisms as well. One can show that for $m < 4$ all homeomorphisms are also diffeomorphisms. For $m \geq 4$ things become more interesting. It has only relatively recently been shown that S^7 admits precisely 28 distinct differentiable structures and that $\mathbb{R}^4$ has an infinite number of distinct differentiable structures. On the other hand, one can show that any smooth, paracompact manifold admits an analytic structure which, however, is unique only up to smooth diffeomorphisms [525].

One can show that a connected, finite-dimensional, Hausdorff manifold is paracompact if and only if it has a countable base, that is, there is a countable family of open subsets of M such that any other open set can be written as the union of members of this family. (Recall that a topological space is called disconnected if and only if it is the union of at least two disjoint closed sets, otherwise it is called connected.) Unless otherwise stated, in what follows we will assume that M is a connected, Hausdorff, paracompact C^∞ manifold without boundary.

The importance of the concept of paracompactness is that it allows a practically useful theory of integration on manifolds. An important tool for this will be the concept of a partition of unity: let $(U_I, x_I)_{I \in \mathcal{I}}$ be a locally finite atlas of a paracompact C^k, $k \leq \infty$ manifold. Then one can always find [887] a system of C^k functions e_I, $I \in \mathcal{I}$ on M (f is said to be C^l, $l \leq k$ if $f \circ x_I^{-1}$ is C^l on $x_I(U_I)$ for all I) such that

1. $0 \leq e_I \leq 1$ on M.
2. The closure of the support $\text{supp}(e_I) := \{p \in M;\ e_I(p) \neq 0\}$ of e_I is contained in U_I.
3. $\sum_{I \in \mathcal{I}} e_I = 1$.

19.1.2 Passive and active diffeomorphisms

In physics one often talks about active and passive diffeomorphisms. An active diffeomorphism is simply a diffeomorphism as just defined which is different from the identity map. Hence it maps a point $p \in M$ in general to a different point $\psi(p) = q \in M$. On the other hand, if one and the same point p lies in the domain of two distinct charts U_I, U_J of the same atlas then its coordinates $x_I(p), x_J(p)$ will in general be distinct. However, by definition there is a diffeomorphism φ_{IJ} of $\mathbb{R}^m$ which maps between these two points. By a passive diffeomorphism of M one simply understands the diffeomorphisms of $\mathbb{R}^m$ between the various domains of parametrisations (coordinate systems) of the points of M. In physics, when we say that, for example, an action is diffeomorphism-invariant we really mean passive diffeomorphisms, that is, reparametrisations, because the action is an integral over $x(M)$ using specific coordinates. Diffeomorphism invariance hence means that smooth changes of coordinates do not affect the value of the action functional.

The notions of active and passive diffeomorphisms are connected as follows: given an active diffeomorphism ψ and an atlas (U_I, x_I) we can construct a new atlas $(V_I = \psi^{-1}(U_I), y_I = x_I \circ \psi)$. The compatibility criterion that $\varphi_{IJ}^\psi := y_J \circ x_I^{-1} = x_J \circ \psi \circ x_I^{-1}$ be differentiable on $x_I(U_I \cap V_J)$ coincides with the definition of differentiability of ψ, hence active diffeomorphisms simply produce compatible atlases and do not change the differentiable structure. On the other hand, they induce the passive diffeomorphisms φ_{IJ}^ψ. It follows that a reparametrisation-invariant functional is also invariant under active diffeomorphisms in this sense.

The notion of active and passive diffeomorphisms sometimes produces much confusion for the beginner for the following reason: as we have just seen, we can always trade an active diffeomorphism for a passive one, thus both are to be seen as gauge transformations in diffeomorphism-invariant physical theories. On the other hand, the real world is diffeomorphism-invariant and still objects at different spacetime locations are physically distinct, they should therefore not be gauge-equivalent. The way out of the apparent contradiction is the physical meaning that we associate with the points $p \in M$: so far we have used M as a purely mathematical object, the points $p \in M$ have no a priori physical meaning. In order to give meaning to them we have to label them, not only with a coordinate system but also with a physical measurement. To see the difference between the two, consider the example of the spatial volume

$$V_R[q] = \int_{x(R)} d^3x \, \sqrt{\det(q)}$$

of a submanifold R of $M = \mathbb{R}^3$ where we have used a specific global coordinate system x to parametrise it once and for all and q is a Riemannian metric on M. This is a functional of the field q and the question is whether it is invariant if we replace q by its diffeomorphic image. We will see later that under a change of coordinates (passive diffeomorphism) the metric tensor maps to the pull-back $q \mapsto \varphi^* q$. Hence $V_R[q] \mapsto V_R[\varphi^* q] = V_{R_\varphi}[q]$ where $x(R_\varphi) := \varphi(x(R))$. Thus $V_R[q]$ is not diffeomorphism-invariant. We see that the reason for this non-invariance is that the region R is a coordinate region, it is not attached to any physical process. Now we do something else: let ρ be some scalar built from the metric and/or some matter field, say $\rho = R$ where R is the Ricci scalar of q or maybe the electromagnetic field energy (divided by $\det(q)$). Let us now construct the volume of the region where ρ is not vanishing. This is mathematically described by the functional

$$V_\rho^{(3)}[q] := \int_{x(M)} d^3x [1 - \theta(-|\rho|)] \sqrt{\det(q)}$$

where θ is the Heaviside step function. Notice that the region where $\rho \neq 0$ is now dynamically determined and not abstractly prescribed in terms of coordinates. Under a passive diffeomorphism $\rho \mapsto \varphi^* \rho$, hence the region where $\rho \neq 0$ also gets transformed and thus the functional remains altogether invariant since $\varphi(x(M)) = x(M)$. The value of $V_\rho[g]$ is thus invariant under a passive diffeomorphism because it acts on both q, ρ in the same way. This corresponds to our everyday experience that we do not need to use a coordinate system in order to observe physical objects and what we have just constructed is an example of an invariant which uses the relational point of view as discussed in detail in Section 2.2.

The next confusion that arises is when it comes to dynamics: what we have just described are spatially diffeomorphism-invariant objects. The Einstein–Hilbert action, however, is spacetime diffeomorphism-invariant. Hence, the intuition would be that one should construct spacetime diffeomorphism-invariant objects. An example would be

$$V_\rho^{(4)}[g; a, b] := \int_{x(M)} d^4x \chi_{[a-b,a+b]}(\rho) \sqrt{|\det(g)|}$$

where now we have chosen $M = \mathbb{R}^4$, $\chi_{[a,b]}$ denotes the characteristic function of the interval $[a, b]$ and ρ, g are spacetime tensors. This functional is spacetime diffeomorphism invariant. However, now we are confronted with the following problem: in physics we are used to the fact that the dynamics is induced by time translations. Time translations are coordinate transformations, hence the above functional is time translation invariant. Hence it seems that in spacetime diffeomorphism-invariant theories the observables *do not evolve*, in clear contradiction to what we observe. The resolution of this contradiction is, among other things, the subject of Section 1.1.7, however, to sketch[1] what happens notice that the above functional depends on the two additional parameters a, b. Now roughly speaking one has to distinguish between the *unphysical* time reparametrisations of the coordinate x^0 under which $V_\rho^{(4)}[g, a, b]$ is truly invariant and *physical* time reparametrisations. By these we mean the selection of certain objects, in this case ρ, as *clocks*. Clocks are themselves not diffeomorphism-invariant but they are dynamical fields (i.e., not externally prescribed). The physical meaning of $V_\rho^{(4)}[g, a, b]$ is the spacetime volume of the spacetime region where the clock assumes values in $[a - b, a + b]$. If we fix b then $V_\rho^{(4)}[g, a, b]$ does evolve as we vary a because it describes different, diffeomorphism-invariant objects. This is now a physical time evolution because it is associated with the dynamical field ρ. Of course, one should show that it is generated by a spacetime diffeomorphism-invariant Hamiltonian. This is indeed the case if we follow the more complete construction of Chapter 2.

---

[1] The actual resolution is technically somewhat more complicated because of the mixture of gauge and dynamics in General Relativity.

19.1.3 Differential calculus

(i) *Functions*

A smooth function on M is a map $f : M \to \mathbb{C}$ such that $f \circ x_I^{-1}$ is smooth on $x_I(U_I) \subset \mathbb{R}^m$. The set of smooth functions $C^\infty(M)$ forms an Abelian *-algebra where operations are defined pointwise and the involution is given by complex conjugation.

(ii) *Vector fields*

A smooth vector field on M is a derivation on $C^\infty(M)$. That is, it is a linear map

$$v : \; C^\infty(M) \to C^\infty(M); \; f \mapsto v[f] \tag{19.1.1}$$

which obeys the Leibniz rule $v[fg] = v[f] \cdot g + f \cdot v[g]$ and annihilates constants, that is, $v[c] = 0$ if c is a constant function on M. If $f \in C^\infty(M)$ then fv is the vector field defined by $(fv)[f'] = f \cdot v[f']$. Given an atlas (U_I, x_I) we may define special vector fields ∂_μ^I on U_I defined by the condition

$$\left(\partial_\mu^I [x_I^\nu] \right)(p) = \delta_\mu^\nu \tag{19.1.2}$$

for $p \in U_I$ where $x(p) = (x^1(p), \ldots, x^m(p)) \in \mathbb{R}^m$ denote the components of $x(p)$. Given a vector field v, define $v_I^\mu(x_I(p)) := (v[x_I^\mu])(p)$. We claim that $v(p) = v_I^\mu(x_I(p)) \partial_\mu^I(p)$ and that this way of expanding v is independent of the chart in use. To see the first statement, one verifies that the formula reproduces v on polynomials of the $x_I^\mu(p)$ and that any continuous function can be approximated on a compact neighbourhood of p to arbitrary precision by the Weierstrass theorem (we assume here that every point has an open neighbourhood with compact closure, i.e., that M is locally compact. This is actually always the case for finite-dimensional M that we consider here). To see the second statement, notice that the Leibniz rule implies the chain rule to verify that

$$v_I^\mu(x_I(p)) \partial_\mu^I(p) = v_J^\mu(x_J(p)) \partial_\mu^J(p) \tag{19.1.3}$$

if $p \in U_I \cap U_J$, $x_J(p) = \varphi_{IJ}(x_I(p))$. It is now clear that

$$v[f] = v_I^\mu(x_I(p)) [\partial f_I(x)/\partial x^\mu]_{x = x_I(p)} \tag{19.1.4}$$

where $f_I = f \circ x_I^{-1}$. It follows from the definitions that

$$\partial_\mu^I(p) = \left[\partial \varphi_{IJ}^\nu(x_I(p))/\partial x_I^\mu(p) \right] \partial_\nu^J(p) \tag{19.1.5}$$

for $p \in U_I \cap U_J$, which explains the notation ∂_μ^I.

The space of smooth vector fields on M will be denoted by $T^1(M)$. It forms a Lie algebra where the Lie bracket is defined as

$$[.,.] : \; T^1(M) \times T^1(M) \to T^1(M); \; ([u, v])[f] := u[v[f]] - v[u[f]] \tag{19.1.6}$$

The antisymmetry and the Jacobi identity for (19.1.6) are easily verified.

(iii) *One-forms*

A smooth one-form is a linear map

$$\omega : T^1(M) \to C^\infty(M) \tag{19.1.7}$$

that is, for any $f, f' \in C^\infty(M)$ and $v, v' \in T^1(M)$ we have $\omega[fv + f'v'] = f\omega[v] + f'\omega[v']$. Given $f \in C^\infty(M)$ we may define an associated one-form df by the rule $df[v] := v[f]$. Applied to the coordinate functions x_I we find $(dx_I^\mu[\partial_\nu^I])(p) = \delta_\nu^\mu$, that is, the $dx_I^\mu(p)$, $\mu = 1, \ldots, m$; $p \in U_I$ form a local, dual coordinate basis. If follows that ω can be written as $\omega(p) = \omega_\mu^I(x_I(p))dx_I^\mu(p)$ where $\omega_\mu^I(x_I(p)) = (\omega[\partial_\mu^I])(p)$ and this way of writing ω is coordinate-independent again. In particular we find

$$(df)(p) = (\partial f_I(x)/\partial x^\mu)_{x=x_I(p)} \, dx_I^\mu(p) \tag{19.1.8}$$

It follows from the definitions that

$$dx_J^\mu(p) = [\partial \varphi_{IJ}^\mu(x_I(p))/\partial x_I^\nu(p)] dx_I^\nu(p) \tag{19.1.9}$$

for $p \in U_I \cap U_J$, explaining the notation d.

The space of smooth one-forms will be denoted by $T_1(M)$.

(iv) *Tensor fields*

A smooth tensor field of type (a, b) (called a-times contravariant and b-times covariant) is a multilinear functional (i.e., linear in each entry separately)

$$t : \left[\times_{r=1}^a T_1(M) \right] \times \left[\times_{s=1}^b T^1(M) \right] \to C^\infty(M) \tag{19.1.10}$$

It is clear that each such t is completely determined in terms of the component functions (which are smooth by definition)

$$(t_I)_{\nu_1 \ldots \nu_b}^{\mu_1 \ldots \mu_a}(x_I(p)) = \left(t[dx_I^{\mu_1}, \ldots, dx_I^{\mu_a}; \partial_{\nu_1}^I, \ldots, \partial_{\nu_b}^I]\right)(p) \tag{19.1.11}$$

and one writes t as the tensor product

$$t(p) = (t_I)_{\nu_1 \ldots \nu_b}^{\mu_1 \ldots \mu_a}(x_I(p)) \, \partial_{\mu_1}^I(p) \otimes \ldots \otimes \partial_{\mu_a}^I(p) \otimes dx_I^{\nu_1}(p) \ldots \otimes dx_I^{\nu_b}(p) \tag{19.1.12}$$

which is independent of the choice of chart.

The vector space of tensor fields of type (a, b) is denoted by $T_b^a(M)$. We use the notations $T^1(M) = T_0^1(M)$, $T_1(M) = T_1^0(M)$, $C^\infty(M) = T_0^0(M)$. It is invariant under multiplication by elements of $C^\infty(M)$. We can also define the tensor product of $t \in T_b^a(M)$, $t' \in T_{b'}^{a'}(M)$ as the element of $T_{b+b'}^{a+b}(M)$ defined by

$$(t \otimes t')[\omega_1, \ldots, \omega_{a+a'}; v_1, \ldots, v_{b+b'}] = t[\omega_1, \ldots, \omega_a; v_1, \ldots, v_b]$$

$$\times t'[\omega_{a+1}, \ldots, \omega_{a+a'}; v_{b+1}, \ldots, v_{b+b'}] \tag{19.1.13}$$

We may then form the direct sum of tensor fields

$$T(M) = \oplus_{a,b=0}^{\infty} T_b^a(M) \tag{19.1.14}$$

of formal sums $(t) = \oplus_{a,b=0}^{\infty} t_b^a$, $t_b^a \in T_b^a(M)$ with $t_b^a \neq 0$ for at most finitely many (a,b). With respect to the tensor product this is an algebra over $C^{\infty}(M)$, called the algebra of tensor fields with the operations

$$f \cdot (t) + f' \cdot (t') = \oplus_{a,b=0}^{\infty} \left[f t_b^a + f' t_b'^a \right]$$

$$(t) \otimes (t') = \oplus_{a,b=0}^{\infty} \left[\sum_{a'=0}^{a} \sum_{b'=0}^{b} t_{b'}^{a'} \otimes t_{b-b'}^{a-a'} \right] \tag{19.1.15}$$

Given a tensor field $t \in T_b^a(M)$, a vector field $v \in T^1(M)$ and a one-form ω we define contractions $i_v^k \cdot t \in T_b^{a-1}(M)$ for $1 \leq k \leq a$ and $i_\omega^k \cdot t \in T_{b-1}^a(M)$ for $1 \leq k \leq b$ by

$$\left(i_v^k \cdot t \right)[\omega_1, \ldots, \omega_a; v_1, \ldots, v_{b-1}] = t[\omega_1, \ldots, \omega_a; v_1, \ldots, v_{k-1}, v, v_k, \ldots, v_{b-1}]$$

$$\left(i_\omega^k \cdot t \right)[\omega_1, \ldots, \omega_{a-1}; v_1, \ldots, v_b] = t[\omega_1, \ldots, \omega_{k-1}, \omega, \omega_k, \ldots, \omega_{a-1}; v_1, \ldots, v_b]$$

$$\tag{19.1.16}$$

(v) *Tangent spaces and tensor bundles*

We can form a vector bundle $E_b^a(M)$ with base manifold M, typical fibre $T_b^a = \mathbb{R}^{(a+b)m}$ and structure group $GL(\mathbb{R}, m)^{a+b}$ as follows (refer to the next chapter for the bundle-theoretic terminology): form the product manifold $\tilde{E}_b^a(M) := \cup_{I \in \mathcal{I}} U_I \times T_b^a$ and consider the equivalence relation between $(p, t_I) \in U_I \times T_b^a$ and $(p', t_J) \in U_J \times T_b^a$ defined by $(p, f) \sim (p', f')$ iff $p = p'$ and

$$(t_J)_{\nu_1 \ldots \nu_b}^{\mu_1 \ldots \mu_a} = \left[\prod_{k=1}^{a} \left(\partial \varphi_{IJ}^{\mu_k}(x) / \partial x^{\mu'_k} \right)_{x = x_I(p)} \right] \left[\prod_{k=1}^{b} \left(\partial \varphi_{JI}^{\nu'_k}(x) / \partial x^{\nu_k} \right)_{x = x_J(p)} \right] (t_I)_{\nu'_1 \ldots \nu'_b}^{\mu'_1 \ldots \mu'_a}$$

$$\tag{19.1.17}$$

We will write the shorthand $t_J = h_{IJ}(p) \cdot t_I$ for (19.1.17) with $h_{IJ}(p) \in GL(\mathbb{R}, m)^{a+b}$. Let $E_b^a(M) = \tilde{E}_b^a(M) / \sim$ be the set of equivalence classes with local trivialisations $\phi_I(p, t) := [(p, t)]$ for $p \in U_I$ where $[(p, t)]$ is the class of (p, t) and canonical projection $\pi([p, t]) = p$.

The spaces $(T_b^a)_p(M) := \pi^{-1}(p)$ are called the tangent spaces over p and they are all isomorphic to T_b^a because for each $p \in M$ there are only a finite number of structure functions $h_{IJ}(p)$ if M is paracompact. A tensor field $t \in T_b^a(M)$ is then a cross-section in $E_b^a(M)$, that is, a global, smooth map $t : M \rightarrow E_b^a(M)$; $p \mapsto t(p)$ as follows: for any $p \in M$ choose an index $I(p) \in \mathcal{I}$ such that $p \in U_{I(p)}$. Then define $t(p) := [(p, t_{I(p)}(p))]$ where $t_I(p)$ denotes the collection of component functions (19.1.11) of t in the chart U_I. To see that this is well-defined, that is, independent of the choice $p \mapsto I(p)$ and hence really smooth, we notice that for $p \in U_I \cap U_J$ we have $t_J(p) =$

$h_{IJ}(p) \cdot t_I(p)$ so that different choices are identified under the equivalence relation just defined.

Notice that although the $(T_b^a)_p(M)$ are all isomorphic to T_b^a there is no natural way to compare these tangent spaces defined over different points simply because they are not subspaces of one and the same space. This is best illustrated by the sphere $M = S^2$ embedded in $\mathbb{R}^3$. The tangent spaces spanned by the vector fields $\partial_\theta, \partial_\varphi$ where θ, φ are polar coordinates can be expanded in terms of the Cartesian basis $\partial\mu$ of $\mathbb{R}^3$ and one sees that the tangent spaces are simply the planes in $\mathbb{R}^3$ tangent to the points of S^2. Evidently, these spaces are all different 2-dimensional subspaces in $\mathbb{R}^3$, all of which are isomorphic to $\mathbb{R}^2$. This is to be contrasted with the situation for $M = \mathbb{R}^2$ where all the tangent spaces coincide, namely they can be identified with M itself.

(vi) *Abstract index notation*

It is tedious to work with the symbols t and having to always state separately on which components certain operations have to be performed. We thus will frequently use the notation $t^{\mu_1\ldots\mu_a}_{\nu_1\ldots\nu_b}(p)$ for $t \in T_b^a(M)$. This is, just as t, a globally defined object, in fact it is the same as t, just that we display the index structure that t would acquire in any given coordinate basis. To distinguish this globally defined object from the locally defined component functions $(t_I)^{\mu_1\ldots\mu_a}_{\nu_1\ldots\nu_b}(x_I(p))$ we will drop the index I.

(vii) *n-forms*

An n-form is simply a tensor field in $T_n^0(M)$ whose component functions are totally skew (this is a coordinate-independent statement). It amounts to the statement that

$$\omega[v_1, \ldots, v_n] = \mathrm{sgn}(\pi)\omega[v_{\pi(1)}, \ldots, v_{\pi(n)}] \tag{19.1.18}$$

where $\pi \in S_n$ is a permutation and $\mathrm{sgn}(\pi)$ its sign. It follows that $n \leq m$. At this point it is convenient to introduce the total (anti)symmetriser on n symbols, for example,

$$\omega\big(v_{[1}, \ldots, v_{n]}\big) := \frac{1}{n!} \sum_{\pi \in S_n} \mathrm{sgn}(\pi)\, \omega\big(v_{\pi(1)}, \ldots, v_{\pi(n)}\big)$$

$$t_{(\mu_1\ldots\mu_n)} := \frac{1}{n!} \sum_{\pi \in S_n} t_{\mu_{\pi(1)}\ldots\mu_{\pi(n)}} \tag{19.1.19}$$

Clearly $\omega[v_1, \ldots, v_n] = \omega[v_{[1}, \ldots, v_{n]}]$.

The vector space of n-forms is denoted as $\Lambda_n(M)$. Three natural operations are defined on n-forms. The first is the exterior product (also called the wedge product)

$$\wedge : \Lambda_k(M) \times \Lambda_l(M) \to \Lambda_{k+l}(M)$$

$$(\omega \wedge \sigma)[v_1, \ldots, v_{k+l}] := \frac{1}{k!\, l!} \sum_{\pi \in S_{k+l}} \mathrm{sgn}(\pi)\omega\big[v_{\pi(1)}, \ldots, v_{\pi(k)}\big]$$

$$\times \sigma\left[v_{\pi(k+1)}, \ldots, v_{\pi(k+l)}\right]$$

$$= \binom{k+l}{k} \omega[v_{[1}, \ldots, v_k] \, \sigma[v_{k+1}, \ldots, v_{k+l]}] \tag{19.1.20}$$

It is non-commutative $\omega \wedge \sigma = (-1)^{kl}\sigma \wedge \omega$ but associative $\omega \wedge (\sigma \wedge \lambda) = (\omega \wedge \sigma) \wedge \lambda$. Use $[[\mu_1 \ldots \mu_k]\mu_{k+1} \ldots \mu_{k+l}] = [\mu_1 \ldots \mu_{k+l}]$ to see that. We may then form the finite-dimensional Grassmann algebra of forms as

$$\Lambda(M) = \oplus_{n=0}^m \Lambda_n(M) \tag{19.1.21}$$

with $\Lambda_0(M) = C^\infty(M)$, $\Lambda_1(M) = T_1(M)$.

The second operation is exterior derivation

$$d : \Lambda_n(M) \to \Lambda_{n+1}(M)$$

$$d\omega(v_0, \ldots, v_n) = \sum_{k=0}^n (-1)^k \, v_k[\omega[v_0, \ldots, \hat{v}_k, \ldots, v_n]]$$

$$+ \sum_{0 \le k < l \le n} (-1)^{k+l}\omega[[v_k, v_l], v_0, \ldots, \hat{v}_k, \ldots, \hat{v}_l, \ldots, v_n] \tag{19.1.22}$$

where the hat means omission of the argument. It is easy to see that $d(\omega \wedge \sigma) = d\omega \wedge \sigma + (-1)^k\omega \wedge d\sigma$, $d\Lambda_n(M) = 0$ and $d^2 = 0$.

Finally, we define the interior product of a k-form with a vector field as

$$i_v : \Lambda_n(M) \to \Lambda_{n-1}(M)$$
$$(i_v\omega)[v_1, \ldots, v_{n-1}] = \omega[v, v_1, \ldots, v_{n-1}] \tag{19.1.23}$$

with $i_v f := 0$ for $f \in C^\infty(M)$. Notice the relations $i_v^2 = 0$ and $i_v df = v[f]$.

Using the abstract index calculus one can rewrite these formulae much more compactly as follows. Consider the special n-forms corresponding to the n-fold wedge product of the coordinate one-forms

$$dx^{\mu_1} \wedge \ldots \wedge dx^{\mu_n} := \sum_{\pi \in S_n} dx^{\mu_{\pi(1)}} \otimes \ldots \otimes dx^{\mu_{\pi(n)}} \tag{19.1.24}$$

Then, since the component functions of an n-form are totally skew we have

$$\omega := \omega_{\mu_1 \ldots \mu_n} \, dx^{\mu_1} \otimes \ldots \otimes dx^{\mu_n} = \frac{1}{n!}\omega_{\mu_1 \ldots \mu_n} \, dx^{\mu_1} \wedge \ldots \wedge dx^{\mu_n} \tag{19.1.25}$$

Thus

$$\omega \wedge \sigma = \frac{1}{k! \, l!}\omega_{\mu_1 \ldots \mu_k} \, \sigma_{\mu_{k+1} \ldots \mu_{k+l}} \, dx^{\mu_1} \wedge \ldots \wedge dx^{\mu_{k+l}}$$

$$d\omega = \frac{1}{n!}\partial_{[\mu_0}\omega_{\mu_1 \ldots \mu_n]} \, dx^{\mu_0} \wedge \ldots \wedge dx^{\mu_n}$$

$$i_v[\omega] = \frac{1}{(n-1)!}v^{\mu_1}\omega_{\mu_1 \ldots \mu_n} \, dx^{\mu_2} \wedge \ldots \wedge dx^{\mu_n} \tag{19.1.26}$$

(viii) *Tensor transformation laws*

Passive diffeomorphisms: we had already seen that the expression

$$l(p) = (t_I)^{\mu_1 \cdots \mu_a}_{\nu_1 \cdots \nu_b}(x_I(p)) \partial^I_{\mu_1}(p) \otimes \ldots \otimes \partial^I_{\mu_a}(p) \otimes dx^{\nu_1}_I(p) \otimes \ldots dx^{\nu_b}_I(p)$$

(19.1.27)

depends only on the point p but not on the coordinate system. Comparing components in two different coordinate systems $x_I, x_J = \varphi_{IJ} \circ x_I$ gives

$$(t_I)^{\mu_1 \cdots \mu_a}_{\nu_1 \cdots \nu_b}(x_I(p)) = (t_J)^{\mu'_1 \cdots \mu'_a}_{\nu'_1 \cdots \nu'_b}(x_J(p)) \left[\prod_{k=1}^{a} \frac{\partial x^{\mu_k}_I(p)}{\partial x^{\mu'_k}_J(p)}\right] \left[\prod_{l=1}^{b} \frac{\partial x^{\nu'_l}_J}{\partial x^{\nu_l}_I(p)}\right]$$

$$= (t_J)^{\mu'_1 \cdots \mu'_a}_{\nu'_1 \cdots \nu'_b}(\varphi_{IJ}(x))_{x=x_I(p)} \left[\prod_{k=1}^{a} \frac{\partial (\varphi_{IJ}^{-1})^{\mu_k}(x)}{\partial x^{\mu'_k}}\right]_{x=\varphi_{IJ}(x_I(p))}$$

$$\times \left[\prod_{l=1}^{b} \frac{\partial \varphi^{\nu'_l}_{IJ}(x)}{\partial x^{\nu_l}}\right]_{x=x_I(p)}$$

$$=: (\varphi^*_{IJ} t_J)^{\mu_1 \cdots \mu_a}_{\nu_1 \cdots \nu_b}(x_I(p))$$

(19.1.28)

Abstracting from the coordinate systems x_I, x_J of the atlas under consideration and using a general passive diffeomorphism we call

$$(\varphi^* t)^{\mu_1 \cdots \mu_a}_{\nu_1 \cdots \nu_b}(x) := t^{\mu'_1 \cdots \mu'_a}_{\nu'_1 \cdots \nu'_b}(\varphi(x)) \left[\prod_{k=1}^{a} \frac{\partial (\varphi^{-1})^{\mu_k}(y)}{\partial y^{\mu'_k}}\right]_{y=\varphi(x)} \left[\prod_{l=1}^{b} \frac{\partial \varphi^{\nu'_l}(x)}{\partial x^{\nu_l}}\right]$$

(19.1.29)

the (components of) the pull-back tensor. Inverting (19.1.28) for t_J in terms of t_I one arrives likewise at the (components of) the push-forward tensor

$$(\varphi_* t)^{\mu_1 \cdots \mu_a}_{\nu_1 \cdots \nu_b}(\varphi(x)) := t^{\mu'_1 \cdots \mu'_a}_{\nu'_1 \cdots \nu'_b}(x) \left[\prod_{k=1}^{a} \frac{\partial \varphi^{\mu_k}(x)}{\partial x^{\mu'_k}}\right] \left[\prod_{l=1}^{b} \frac{\partial (\varphi^{-1})^{\nu'_l}(y)}{\partial y^{\nu_l}}\right]_{y=\varphi(x)}$$

(19.1.30)

Active diffeomorphisms: given an active diffeomorphism $\psi: M \to M$ and $f \in C^\infty(M)$ we can define the pull-back function

$$(\psi^* f)(p) := (f \circ \psi)(p) = f(\psi(p))$$

(19.1.31)

Given a vector field v on M we can define its push-forward by

$$((\psi_* v)[f])(\psi(p)) := (v[\psi^* f])(p)$$

(19.1.32)

for all $f \in C^\infty(M)$. Given a one-form ω on M we define its pull-back as

$$((\psi^* \omega)[v])(p) := (\omega[\psi_* v])(\psi(p))$$

(19.1.33)

for all $v \in T^1(M)$.

In these transformations we never need the inverse of ψ and thus the maps ψ^*, ψ_* respectively can be defined for all tensor fields of the type

$T_b^0(M)$, $T_0^a(M)$ by duality, that is, $(\psi^* t)[v_1, \ldots, v_b] := t[\psi_* v_1, \ldots, \psi_* v_b]$ and $(\psi_* t)[\omega_1, \ldots, \omega_a] := t[\psi^* \omega_1, \ldots, \psi^* \omega_a]$ respectively. However, in order to define these maps for general tensor fields of type $T_b^a(M)$ we must use invertible smooth maps, that is, diffeomorphisms. We define pull-backs and push-forwards respectively by

$$((\psi^* t)[\omega_1, \ldots, \omega_a, v_1, \ldots, v_b])(p)$$
$$:= (t[(\psi^{-1})^* \omega_1, \ldots, (\psi^{-1})^* \omega_a, \psi_* v_1, \ldots, \psi_* v_b])(\psi(p))$$
$$((\psi_* t)[\omega_1, \ldots, \omega_a, v_1, \ldots, v_b])(\psi(p))$$
$$:= (t[\psi^* \omega_1, \ldots, \psi^* \omega_a, (\psi^{-1})_* v_1, \ldots, (\psi^{-1})_* v_b])(p) \qquad (19.1.34)$$

This makes sense because $(\psi^{-1})^*$ pulls back one-forms from $\psi(p)$ to p and $(\psi^{-1})_*$ pushes forward vector fields from $\psi(p)$ to p. It follows that

$$((\psi_* \psi^* t)[\omega_1, \ldots, \omega_a, v_1, \ldots, v_b])(\psi(p))$$
$$= ((\psi^* t)[\psi^* \omega_1, \ldots, \psi^* \omega_a, (\psi^{-1})_* v_1, \ldots, (\psi^{-1})_* v_b])(p)$$
$$= (t[(\psi^{-1})^* \psi^* \omega_1, \ldots, (\psi^{-1})^* \psi^* \omega_a, \psi_* (\psi^{-1})_* v_1, \ldots, \psi_* (\psi^{-1})_* v_b])(\psi(p))$$
$$= (t[\omega_1, \ldots, \omega_a, v_1, \ldots, v_b])(\psi(p)) \qquad (19.1.35)$$

where we have used the fact that $\psi^* \circ (\psi')^* = (\psi' \circ \psi)^*$ and $\psi_* \circ (\psi')_* = (\psi \circ \psi')_*$. It follows that $\psi_* = (\psi^*)^{-1} = (\psi^{-1})^*$ so it is sufficient to consider pull-backs only. Notice the relation

$$\psi^* (t_1 \otimes t_2) = (\psi^* t_1) \otimes (\psi^* t_2) \qquad (19.1.36)$$

We can of course display (19.1.34) also in coordinates: given an atlas (U_I, x_I), pick $p \in M$ and choose neighbourhoods U_I, U_J respectively containing $p, \psi(p)$ respectively. Consider the neighbourhood $V_J := \psi^{-1}(U_J)$ containing p with coordinates $y_J := x_J \circ \psi$. Consider the passive diffeomorphism $\psi_{IJ} := y_J \circ x_I^{-1}$ defined on $x_I(U_I) \cap y_J(V_J)$. Then by definition

$$(\psi^* f)(p) = (\psi^* f)_I(x_I(p)) = f(\psi(p)) = f_J(y_J(p)) = f_J(\varphi_{IJ}(x_I(p)))$$
$$\Rightarrow (\psi^* f)_I(x) = f_J(\psi_{IJ}(x))$$
$$(\psi_* v)[f](\psi(p)) = (\psi_* v)_J^\mu(y_J(p))(\partial_\mu^J f_J)(y_J(p))$$
$$= (v[\psi^* f])(p) = v_I^\nu(x_I(p))(\partial_\nu^I(\psi^* f))_I(x_I(p))$$
$$= v_I^\nu(x_I(p)) \left(\frac{\partial \psi_{IJ}^\mu(x)}{\partial x^\nu} \right)_{x = x_I(p)} (\partial_\mu^J f_J)_I(y_J(p))$$
$$\Rightarrow (\psi_* v)_J^\mu(\psi_{IJ}(x)) = v_I^\nu(x) \frac{\partial \psi_{IJ}^\mu(x)}{\partial x^\nu}$$
$$(\psi^* \omega)[v](p) = (\psi^* \omega)_\mu^I(x_I(p)) v_I^\mu(x_I(p))$$
$$= \omega_\nu^J(\psi_{IJ}(x_I(p)))(\varphi_* v)_J^\nu(\psi_{IJ}(x_I(p))))$$
$$\Rightarrow (\psi^* \omega)_\mu^I(x) = \omega_\nu^J(\psi_{IJ}(x)) \frac{\partial \psi_{IJ}^\nu(x)}{\partial x^\mu} \qquad (19.1.37)$$

From this and (19.1.34) one deduces the coordinate expressions for $\psi^* t$, $\psi_* t$ which reproduce precisely (19.1.29) and (19.1.30) except that $\varphi_{IJ} \to \psi_{IJ}$. Hence, as we said before, there is no difference between passive and active diffeomorphisms from the analytical point of view, although they are philosophically very different.

(ix) *Lie derivative*

A smooth curve in M is a C^∞ map $t \mapsto c(t)$ from an interval of $\mathbb{R}$ into M, that is, the coordinate maps $t \mapsto x_I(c(t))$ are smooth maps from $\mathbb{R}$ into $\mathbb{R}^m$. To each curve we may assign a tangent vector field T_c along c by the rule

$$(T_c[f])(c(t)) := \frac{d}{dt} f(c(t)) = \frac{dx_I^\mu(t)}{dt} (\partial_\mu^I[f_I])(x_I(c(t))) \qquad (19.1.38)$$

where in the last equality we have assumed that $c(t) \in U_I$.

Conversely, given a vector field v in M and a point $p \in M$ an integral curve of v through p is a maximal curve $t \mapsto c_p^v(t)$ in M starting in p whose tangential vector field at each point of the curve coincides with v at each of those points. Mathematically this is described by the condition

$$c_p^v(0) = p, \quad (T_{c_p^v} - v)(c_p^v(t)) = 0 \qquad (19.1.39)$$

If $p \in U_I$ then also a segment of $c_p(t)$ lies in U_I and for that segment (19.1.32) can be written as

$$x_I(c_p^v(0)) = x_I(p), \quad \frac{d}{dt} x_I^\mu(c_p^v(t)) = v_I^\mu(x_I(c_p^v(t))) \qquad (19.1.40)$$

Notice that (19.1.33) is covariant under change of the coordinate system. This is a system of m first-order ordinary differential equations which, by well-known existence and uniqueness theorems has a unique maximal solution, hence maximal integral curves through any point always exist and are unique. The collection of integral curves generated by a vector field v is called the flow of v.

We can now turn the logic around and assign to each vector field v a one-parameter group of active diffeomorphisms by

$$\psi_t^v(p) := c_p^v(t) \qquad (19.1.41)$$

Notice that $\psi_t^v \circ \psi_s^v = \psi_{s+t}^v$ as follows directly from the definition of an integral curve.

With the help of the flow of a vector field we may define the Lie derivative of any tensor field $t \in T(M)$ along v by

$$(\mathcal{L}_v t)(p) := \left(\frac{d}{ds}\right)_{s=0} ((\psi_s^v)^* t)(p) \qquad (19.1.42)$$

Using coordinates and the coordinate representation of the diffeomorphism $\psi_I(x) := \psi_{II}(x) = x_I \circ \psi \circ x_I^{-1}$ in a chart U_I we deduce for $t \in T_b^a(M)$

that

$$((\mathcal{L}_v t)_I)^{\mu_1\dots\mu_a}_{\nu_1\dots\nu_b}(x) = v_I^\rho(x)\big(\partial_\rho^I (t_I)^{\mu_1\dots\mu_a}_{\nu_1\dots\nu_b}\big)(x) + \sum_{l=1}^{b}\big(\partial_{\nu_l}^I v_I^\rho\big)(x)(t_I)^{\mu_1\dots\mu_a}_{\nu_1\dots\hat{\nu}_l\rho\dots\nu_b}(x)$$

$$-\sum_{k=1}^{a}\big(\partial_\rho^I v_I^{\mu_k}\big)(x)(t_I)^{\mu_1\dots\hat{\mu}_k\rho\dots\mu_a}_{\nu_1\dots\nu_b}(x) \qquad (19.1.43)$$

A tensor field is said to be symmetric under a diffeomorphism if $\psi^* t = t$. A tensor field is symmetric under the flow of v if and only if $\mathcal{L}_v t = 0$.

The Lie derivative is a derivation on $T(M)$, that is, it satisfies the Leibniz rule

$$\mathcal{L}_v(t_1 \otimes t_2) = \left(\frac{d}{ds}\right)_{s=0}(\psi_s^v)^*(t_1 \otimes t_2) = \left(\frac{d}{ds}\right)_{s=0}\big[(\psi_s^v)^* t_1\big] \otimes \big[(\psi_s^v)^* t_2\big]$$

$$= (\mathcal{L}_v t_1) \otimes t_2 + t_1 \otimes (\mathcal{L}_v t_2) \qquad (19.1.44)$$

(x) *Derivations on the Grassmann algebra*

There is an interesting interplay between the objects $\mathcal{L}_v, d, i_v$ defined on the subset $\Lambda(M)$.

Theorem 19.1.2. *The following relations among* $d, i_v, \mathcal{L}_v$ *hold on n-forms*

$$\mathcal{L}_v = i_v \circ d + d \circ i_v$$
$$[\mathcal{L}_v, i_w] = i_{[v,w]}$$
$$[\mathcal{L}_v, \mathcal{L}_w] = \mathcal{L}_{[v,w]} \qquad (19.1.45)$$

To prove this theorem we notice the relation

$$d \circ \psi^* = \psi^* \circ d \qquad (19.1.46)$$

on $\Lambda(M)$. To see (19.1.46) we compute

$$(\psi^* df)[v] = df[\psi_* v] = (\psi_* v)[f] = v[\psi^* f] = (d\psi^* f)[v] \qquad (19.1.47)$$

so (19.1.46) holds on zero forms. Now any k-form is a linear combination of k-forms of the form $f_0 df_1 \wedge \dots \wedge df_k$. It follows

$$d[\varphi^* f_0 df_1 \wedge \dots \wedge df_k)]$$
$$= d[(\varphi^* f_0)d(\varphi^* f_1) \wedge \dots \wedge d(\varphi^* f_k)]$$
$$= d(\varphi^* f_0)d(\varphi^* f_1) \wedge \dots \wedge d(\varphi^* f_k)$$
$$= \varphi^*[df_0 df_1 \wedge \dots \wedge df_k] = \varphi^* d[f_0 df_1 \wedge \dots \wedge df_k] \qquad (19.1.48)$$

where we have made use of $d^2 = 0$. To establish (19.1.45) we use the following lemma:

Definition 19.1.3

(i) *A linear operator* $D : \Lambda(M) \to \Lambda(M)$, $D(z_1\omega_1 + z_2\omega_2) = z_1 D\omega_1 + z_2 D\omega_2$; $z_1, z_2 \in \mathbb{C}$ *is said to be of degree* $d \in \mathbb{Z}$ *if* $D : \Lambda^n(M) \to \Lambda^{n+d}(M)$.

(ii) *D is said to be an (anti-)derivation if its degree is even (odd) and obeys the (anti-)Leibniz rule*

$$D(\omega \wedge \sigma) = (D\omega) \wedge \sigma + (-1)^{dk}\omega \wedge D\sigma \qquad (19.1.49)$$

where k is the degree of D.

(iii) *D is said to be local if* $(D\omega)_{|U}$ *depends only on* $\omega_{|U}$ *for any open set* U.

It is clear that d, i_v are local anti-derivations of degree $+1, -1$ respectively.

Lemma 19.1.4

(i) *If D_j, A_j; $j = 1, 2$ are derivations and antiderivations respectively then* $[D_1, D_2] = D_1 D_2 - D_2 D_1$ *and* $\{A_1, A_2\} = A_1 A_2 + A_2 A_1$ *are derivations and $[D_j, A_k]$ is an antiderivation.*

(ii) *Two (anti-)derivations are equal if they coincide on 0-forms and 1-forms.*

(iii) *If D is a local (anti-)derivation which commutes with d then it is fully determined by its action on 0-forms.*

Proof

(i) is an elementary algebraic computation.

(ii) follows from the Leibniz rule and the fact that any k-form can be written as linear combinations of the form $\omega_1 \wedge \ldots \wedge \omega_k$.

(iii) follows from (ii) and from the fact that any 1-form can be written as $g \cdot df$ for $fg \in C^\infty(M)$, hence

$$D(g \cdot df) = Dg \wedge df + gDdf = Dg \wedge df + gdDf \qquad (19.1.50)$$

$\square$

Proof of Theorem 19.1.2

Since $d\psi^* = \psi^* d$ it follows that $[\mathcal{L}_v, d] = 0$. Hence by the lemma it is sufficient to establish the first relation in (19.1.45) on functions. Since $i_v f = 0$ we have

$$((di_v + i_v d)f)(p) = (i_v df)(p) = (df[v])(p) = (v[f])(p)$$

$$= \left(\frac{d}{ds}\right)_{s=0} ((\psi_s^v)^a st f)(p)$$

$$= \left(\frac{d}{ds}\right)_{s=0} f(c_p^v(s)) = (\mathcal{L}_v f)(p) \qquad (19.1.51)$$

as claimed.

The second relation in (19.1.45) merely has to be checked on 0-forms and 1-forms:

$$[\mathcal{L}_v, i_w]f = 0 = i_{[v,w]}f = 0 \tag{19.1.52}$$

since $i_u f = 0$ and

$$
\begin{aligned}
[\mathcal{L}_v, i_w]g\, df &= \mathcal{L}_v(gw[f]) - i_w(di_v + i_v d)g\, df \\
&= v[gw[f]] - i_w(d(gv[f]) + i_v dg \wedge df) \\
&= v[g]w[f] + gv[w[f]] - i_w(v[f]dg + gdv[f] \\
&\quad + v[g]df - dgv[f]) \\
&= v[g]w[f] + gv[w[f]] - gw[v[f]] - v[g]w[f] \\
&= v[g]w[f] + gv[w[f]] - gw[v[f]] - v[g]w[f] \\
&= g([v,w])[f] = i_{[v,w]}(g\, df) \tag{19.1.53}
\end{aligned}
$$

Finally the third relation only has to be confirmed on 0-forms since $\mathcal{L}_v$ and hence $[\mathcal{L}_v, \mathcal{L}_w]$ commute with d

$$[\mathcal{L}_v, \mathcal{L}_w]f = \mathcal{L}_v w[f] - \mathcal{L}_w v[f] = ([v,w])[f] = \mathcal{L}_{[v,w]}f \tag{19.1.54}$$

$\square$

(xi) *Integration of forms, Stokes' theorem, Poincaré lemma and de Rham cohomology*

A subset S of M is said to be of measure zero if all the sets $x_I(U_I \cap S)$ have Lebesgue measure zero. Notice that an m-form has only one independent component. An m-form is said to be measurable if its component under all coordinate charts is Lebesgue measurable. See Chapter 25 for more details on measure theory.

Let ω be an m-form on M. Since M is supposed to be paracompact we may choose a partition of unity e_I subordinate to M and define the m-forms $e_I \omega$ which have compact support in U_I. Then we have, for ω of compact support so that we may interchange summation and integration,

$$
\begin{aligned}
\int_M \omega &= \int_M \left[\sum_I e_I \omega \right] = \sum_I \int_M e_I \omega = \sum_I \int_{U_I} e_I \omega \\
&= \frac{1}{m!} \sum_I \int_{x_I(U_I)} (e_I)_I(x)(\omega_I(x))_{\mu_1 \dots \mu_m} dx^{\mu_1} \wedge \dots \wedge dx^{\mu_m} \\
&= \sum_I \int_{x_I(U_I)} (e_I)_I(x)(\omega_I(x))_{1 \dots m} dx^1 \wedge \dots \wedge dx^m \\
&:= \sum_I \int_{x_I(U_I)} d^m x (e_I)_I(x)(\omega_I(x))_{1 \dots m} \tag{19.1.55}
\end{aligned}
$$

where the last integral is an ordinary integral over $V_I \subset \mathbb{R}^m$. Notice that the Lebesgue measure $d^m x$ is independent of the sequence of integrations. This is taken into account in (19.1.55) by the fact that the sign from the wedge product under permutation of coordinates is absorbed by a corresponding minus sign in the totally skew components of ω. The m-form

$$dx^1 \wedge \ldots \wedge dx^m = \frac{1}{m!} \epsilon_{\mu_1 \ldots \mu_m} dx^{\mu_1} \wedge \ldots \wedge dx^{\mu_m} \qquad (19.1.56)$$

where $\epsilon_{\mu_1 \ldots \mu_m}$ is the totally skew symbol of m elements with $\epsilon_{1 \ldots m} = 1$, has the property $(dx^1 \wedge \ldots \wedge dx^m)[v_1, \ldots, v_m](p) = \det(v_1, \ldots, v_m)(p)$ and hence measures what one would intuitively call the 'volume' spanned by the tangent vectors $v_1(p), \ldots, v_m(p)$, hence one effectively replaces (19.1.56) by the Lebesgue measure $d^m x$.

In order that this definition makes sense we must show that it is independent of the partition of unity and of the choice of coordinate system. To that end, consider any other atlas (V_J, y_J) and partition of unity f_I subordinate to it. Then (19.1.55) would be replaced by

$$\int_M \omega = \sum_J \int_{y_J(V_J)} d^m y (f_J)_J(y)(\omega_J(y))_{1 \ldots m} \qquad (19.1.57)$$

Exploiting the fact that $\sum_I e_I(p) = \sum_J f_J(p) = 1$, that $e_I(p) = (e_I)_J(y_J(p))$ for $p \in V_J$, that e_I has support in U_I implies that $(e_I)_J$ has support in $y_J(U_I)$ and that $y_J(U_I) \cap y_J(V_J) = y_J(U_I \cap V_J)$, equality of (19.1.55) and (19.1.57) requires that

$$\sum_{I,J} \int_{x_I(U_I \cap V_J)} d^m x \, (e_I)_I(x) \, (f_J)_I(x)(\omega_I(x))_{1 \ldots m}$$

$$= \sum_{I,J} \int_{y_J(U_I \cap V_J)} d^m y \, (e_I)_J(y) \, (f_J)_J(y)(\omega_J(y))_{1 \ldots m} \qquad (19.1.58)$$

Introducing $\varphi_{IJ} = y_J \circ x_I^{-1} : x_I(U_I \cap V_J) \to y_J(U_I \cap V_J)$ we have by the transformation law of the Lebesgue measure on $\mathbb{R}^m$ that

$$d^m y (\omega_J(y))_{1 \ldots m} = \left| \det \left(\frac{\partial \varphi_{IJ}(x)}{\partial x} \right) \right| d^m x \omega_J(\varphi_{IJ}(x))_{1 \ldots m}$$

$$= \mathrm{sgn} \left(\det \left(\frac{\partial \varphi_{IJ}(x)}{\partial x} \right) \right) d^m x \, \omega_I(x)_{1 \ldots m} \qquad (19.1.59)$$

where we have used the transformation law of m-forms. Moreover $(e_I)_J(\varphi_{IJ}(x)) = (e_I)_I(x)$ and $(f_J)_J(\varphi_{IJ}(x)) = (f_J)_I(x)$. Hence (19.1.58) becomes

$$\sum_{I,J} \int_{x_I(U_I \cap V_J)} d^m x \ (e_I)_I(x) \ (f_J)_I(x)(\omega_I(x))_{1...m}$$

$$= \sum_{I,J} \int_{\varphi_{IJ}(x_I(U_I \cap V_J))} d^m y \ (e_I)_J(y) \ (f_J)_J(y)(\omega_J(y))_{1...m}$$

$$= \sum_{I,J} \int_{x_I(U_I \cap V_J)} d^m x \ (e_I)_I(x) \ (f_J)_I(x)(\omega_I(x))_{1...m} \mathrm{sgn}\left(\det\left(\frac{\partial \varphi_{IJ}(x)}{\partial x}\right)\right)$$

$$(19.1.60)$$

It is at this point that we must assume that M is orientable: without the assumption that M admits an atlas for which all the $\det(\frac{\partial \varphi_{IJ}(x)}{\partial x})$ are everywhere positive, the first and the last line in (19.1.60) may differ locally by a sign and the definition of the integral of n-forms becomes atlas- and partition-dependent.

It is possible to extend the definition of the integral of m-forms to non-orientable manifolds by using scalar densities of weight one rather than m-forms, which we postpone until later. Hence in what follows we assume that M is orientable and that ∂M is given the induced orientation with respect to an oriented atlas.

Notice that by the same calculation the integral of m-forms is invariant under active (orientation-preserving) diffeomorphisms

$$\int_M \omega = \int_{\psi M} \psi^* \omega = \int_M \psi^* \omega \qquad (19.1.61)$$

if $\psi(M) = M$, that is, $\psi \in \mathrm{Diff}(M)$.

Of fundamental importance is the following theorem which is the generalisation to forms of the fundamental theorem of calculus.

Theorem 19.1.5 (Stokes' theorem). *Let ω be an $(m-1)$-form of compact support. Then*

$$\int_M d\omega = \int_{\partial M} \omega \qquad (19.1.62)$$

Warning: It is an assumption of the theorem that ω is everywhere smooth on M. If that is not the case, the theorem fails to hold as is demonstrated by the example $\omega_\mu = -\epsilon_{\mu\nu} dx^\nu [(x^1)^2 + (x^2)^2]^{-1}$ on $\mathbb{R}^2 - \{0\}$. We have $d\omega = 0$ except at $x = 0$ where it is not only not differentiable but also not defined. Integrating ω over the circle and falsely applying Stokes' theorem gives the contradiction $2\pi = 0$. To give meaning to $d\omega$ one declares it as a distribution and smears it against test functions f of compact support.

Then

$$\int_{\mathbb{R}^2} f\, d\omega = \lim_{R\to 0} \int_{D_R} f\, d\omega := \lim_{R\to 0}\left[\int_{C_R} f\omega - \int_{D_R} df \wedge \omega\right]$$

$$= \lim_{R\to 0}\left[\int_0^{2\pi} d\varphi\left[f(R,\varphi) - \int_0^R dr\partial_r f(r,\varphi)\right]\right] = 2\pi f(0)$$

$$(19.1.63)$$

where we have used the fact that $d\omega = 0$ except at the origin so that integration could be reduced to the disk D_R with $r \le R$ and boundary C_R. This shows that $d\omega(x) = 2\pi\delta(x)dx^1 \wedge dx^2$.

Proof: We will only sketch the proof. One first realises that M can be written as $M = \cup_\Box \Box$ where the $\Box$ are mutually disjoint, up to common faces, embedded cubes. This is called a triangulation and we will assume that it is fine enough so that each $\Box$ lies in the domain of a chart $U_\Box$. We can find coordinates $x_\Box$ such that $x_\Box(\Box) = [-1,1]^m$. Then

$$\int_M d\omega = \sum_\Box d\omega = \frac{1}{(m-1)!}\sum_\Box \int_{[-1,1]^m}\partial_{\mu_1}(\omega_\Box)_{\mu_2...\mu_m}(x)dx^{\mu_1}\wedge\ldots\wedge dx^{\mu_m}$$

$$= \frac{1}{(m-1)!}\sum_\Box \int_{[-1,1]^m} d^m x \epsilon^{\mu_1\cdots\mu_m}\partial_{\mu_1}(\omega_\Box)_{\mu_2...\mu_m}(x)$$

$$= \frac{1}{(m-1)!}\sum_\Box\sum_{\mu=1}^m \int_{[-1,1]^{m-1}} d^{m-1}x \epsilon^{\mu\mu_2\cdots\mu_m}$$

$$\times \{[(\omega_\Box)_{\mu_2...\mu_m}]_{x^\mu=1} - [(\omega_\Box)_{\mu_2...\mu_m}]_{x^\mu=-1}\}(x)$$

$$= \sum_\Box\sum_{\mu=1}^m \int_{[-1,1]^{m-1}} d^{m-1}x(-1)^{\mu-1}\{[(\omega_\Box)_{1...\hat{\mu}...m}]_{x^\mu=1}$$

$$-[(\omega_\Box)_{1...\hat{\mu}...m}]_{x^\mu=-1}\}(x)$$

$$= \sum_\Box\sum_{\mu=1}^m \int_{[-1,1]^{m-1}} d^{m-1}x\{[(\omega_\Box^\mu)_{2...m}]_{x^1=1} - [(\omega_\Box^\mu)_{2...m}]_{x^1=-1}\}(x)$$

$$(19.1.64)$$

where in the last step we have relabelled $1 \leftrightarrow \mu$ and $[\omega_\Box^\mu]_{2...m}(x) = [\omega_\Box]_{2...m}$ $(x^\mu, x^2,\ldots, \hat{x}^\mu, x^1,\ldots, x^m)$. The sets $H_- = \{x \in \mathbb{R}^m; x^1 \le 1\}$, $H_+ = \{x \in \mathbb{R}^m; -x^1 \le 1\}$ respectively are such that the coordinates $(x^1,\ldots, x^m)$ and $(-x^1, -x^2, x^3,\ldots, x^m)$ respectively are positively oriented. The induced orientation on $\partial H_\mp$ is therefore such that $(\pm x^2, x^3,\ldots, x^m)$ is positively oriented. Defining $[\omega_\Box^{\mu\pm}]_{2...m}(x) = \mp[\omega_\Box]_{2...m}(\mp x^1, \mp x^2, x^3,\ldots x^m)$

we may thus write (19.1.64) as

$$\int_M d\omega = = \sum_{\square} \sum_{\mu=1}^m \int_{[-1,1]^{m-1}} d^{m-1}x\big\{[\omega_\square^{\mu-}]_{2...m} + [\omega_\square^{\mu+}]_{2...m}\big\}(x)_{x^1=1}$$

$$= \sum_{\square} \int_{\partial\square} \omega$$

$$= \int_{\partial M} \omega \qquad (19.1.65)$$

where in the last step we realised that all the contributions from the faces of the $\square$ in the interior of M cancel each other, leaving only those on ∂M with outward orientation. $\qquad\square$

Stokes' theorem generalises from M to any n-dimensional embedded submanifold N. Let $\psi : N \to M$ be the associated embedding, then we have for any $(n-1)$-form on M

$$\int_{\psi(N)} d\omega = \int_N \psi^* d\omega = \int_N d\psi^*\omega = \int_{\partial N} \psi^*\omega$$

$$= \int_{\psi(\partial N)} \omega = \int_{\partial\psi(N)} \omega \qquad (19.1.66)$$

where we have used Stokes' theorem applied to N in an intermediate step. One can also generalise the theorem to forms without compact support but with appropriate fall-off properties.

Definition 19.1.6

(i) *An n-form is called a cocycle if it is closed, that is, $d\omega = 0$. It is called a coboundary if it is exact, that is, if there is an $(n-1)$-form σ such that $\omega = d\sigma$. The vector spaces of closed (exact) n-forms are called $Z^n(M)$ and $B^n(M)$ respectively. The de Rham cohomology group (under addition) is given by $H^n(M) := Z^n(M)/B^n(M)$.*

(ii) *Let $\Lambda_n(M)$ be the formal real vector space of n-dimensional oriented submanifolds of M (actually slightly more general: one allows embedded n-simplices, that is, sets of the form $\{(t_1,\ldots,t_n);\ 0 \le t_j \le 1;\ t_1 + \cdots + t_n \le 1\}$). The elements of $\Lambda_n(M)$ are called chains. A chain C is called a cycle provided that $\partial C = \emptyset$ and a boundary if there is an $(n+1)$-chain C' such that $C = \partial C'$. Notice that $\partial : \Lambda_n(M) \to \Lambda_{n-1}(M)$ and that $\partial^2 = 0$. The vector spaces of cycles and boundaries respectively are denoted by $Z_n(M)$ and $B_n(M)$ respectively and $H_n(M) := Z_n(M)/B_n(M)$ is called the homology group.*

Obviously every exact form is also closed but not vice versa. Similarly, every boundary is a cycle but not vice versa. Stokes' theorem now establishes a precise duality between $H^n(M)$ and $H_n(M)$ via the so-called period integral

$$< C, d\omega > := \int_C \omega \qquad (19.1.67)$$

Thus Stokes' theorem takes the compact form

$$< \partial C, \omega > = < C, d\omega > \qquad (19.1.68)$$

Let $[C], [\omega]$ denote the classes of C, ω respectively, then it follows immediately that the bilinear form on $H_n(M) \times H^n(M)$ defined by

$$< [C], [\omega] > := < C, \omega > \qquad (19.1.69)$$

is well-defined, that is, independent of the representative.

Theorem 19.1.7 (de Rham). *If M is compact, then (19.1.69) is non-degenerate and $H_n(M), H^n(M)$ are finite-dimensional. Their common dimension $b_n(M) = b^n(M)$ is called the nth Betti number of M.*

The Betti numbers are connected with a well-known topological invariant, displaying a beautiful connection between topology and analysis.

Theorem 19.1.8 (Euler–Poincaré)

(i) *Let S be any simplicial decomposition of a compact manifold M (i.e., a partition into embedded m-simplices with $(m-1)$-simplices as boundaries, whose boundaries are $(m-2)$-simplices, etc.). Let $N(S, n)$ be the number of n-simplices in S. Then the Euler characteristic is defined as*

$$\chi(M) = \sum_{n=0}^{m} (-1)^n N(S, n) \qquad (19.1.70)$$

and is independent of S, that is, a topological invariant (under homeomorphisms).

(ii) *The topological invariant $\chi(M)$ is related to the diffeomorphism invariants $b_n(M)$ by*

$$\chi(M) = \sum_{n=0}^{m} (-1)^n b_n(M) \qquad (19.1.71)$$

In order that a form be exact it has to be closed, this is a necessary integrability criterion which underlies also the theory of exterior differential systems and Frobenius' theorem. A sufficient criterion for exactness is given by the following.

Theorem 19.1.9 (Poincaré's lemma). *An m-dimensional submanifold U of M is said to be contractible to a point $p_0 \in U$ if there is a smooth* map $F : U \times [0, 1] \to U$ such that $F(p, 1) = p$, $F(p, 0) = p_0$. In this case, any closed n-form on U is also exact.

Proof (sketch). We assume $n \geq 1$ since for $n = 0$ there is nothing to prove. We may assume that U lies in the domain of a chart, otherwise subdivide U. We may choose coordinates such that $x(p_0) = 0$ and use $F(x, t) := tx$. We claim that

$$\sigma(p) := \int_0^1 dt \, \frac{t^{n-1}}{(n-1)!} \, x^\nu \omega_{\nu\mu_2\ldots\mu_n}(tx(p)) \, dx^{\mu_2}(p) \wedge \ldots \wedge dx^{\mu_n}(p)$$

$$(19.1.72)$$

satisfies $d\sigma = \omega$. We compute

$$d\sigma(p) = \int_0^1 dt \, \frac{t^{n-1}}{(n-1)!} \, [\omega_{\mu_1\ldots\mu_n}(tx(p)) + tx^\nu(p)(\partial_{\mu_1}\omega_{\nu\mu_2\ldots\mu_n})(tx(p))]$$
$$\times \, dx^{\mu_1}(p) \wedge \ldots \wedge dx^{\mu_n}(p) \qquad (19.1.73)$$

Using $\partial_{[\nu}\omega_{\mu_1\ldots\mu_n]} = 0$ we see that

$$[n\partial_{\mu_1}\omega_{\nu\mu_2\ldots\mu_n} - \partial_\nu\omega_{\mu_1\ldots\mu_n}] \, dx^{\mu_1}(p) \wedge \ldots \wedge dx^{\mu_n}(p) = 0 \qquad (19.1.74)$$

Hence (19.1.72) simplifies to

$$d\sigma(p) = \int_0^1 dt \, \frac{1}{n!} \, [nt^{n-1}\omega_{\mu_1\ldots\mu_n}(tx(p)) + t^n x^\nu(p)(\partial_\nu\omega_{\mu_1\ldots\mu_n})(tx(p))]$$
$$\times \, dx^{\mu_1}(p) \wedge \ldots \wedge dx^{\mu_n}(p)$$

$$= \left[\int_0^1 dt \, \frac{1}{n!} \frac{d}{dt} \, [t^n \omega_{\mu_1\ldots\mu_n}(tx(p))] \right] dx^{\mu_1}(p) \wedge \ldots \wedge dx^{\mu_n}(p)$$

$$= \omega(p) \qquad (19.1.75)$$

$\square$

The theorem shows that every closed form is locally exact, hence the cohomology captures global properties about M. For instance, if M is not simply connected (not all loops can be contracted to a point, as for instance on the circle or the torus) then the cohomology will be non-trivial.

19.2 Riemannian geometry

As we have seen, in a general manifold there is no natural way to relate elements of tangent spaces at different points. In order to do that what is needed is a notion of transport of tensors from one point to another along curves connecting these points. One would think that this is accomplished by the pull-back of tensors under active diffeomorphisms generated by the flow of a vector field v. However, as one sees explicitly from the expression for the Lie derivative, the Lie-transported tensor not only depends on the points of the curve c^v generated by a vector field but also on the derivatives of v, which requires additional information. This is the point where we need a structure additional to differential geometry.

(i) *Covariant derivative*

Definition 19.2.1. *An affine connection or covariant differential is an operator $\nabla : T_b^a(M) \to T_{b+1}^a(M)$ satisfying the following axioms:*
1. *Linearity: $\nabla(z_1 t_1 + z_2 t_2) = z_1 \nabla t_1 + z_2 \nabla t_2$ for all $z_1, z_2 \in \mathbb{C}$.*
2. *Leibniz rule: $\nabla\, t_1 \otimes t_2 = (\nabla t_1) \otimes t_2 + t_1 \otimes (\nabla t_2)$.*
3. *Commutes with contractions:*

$$\nabla_u(t[\ldots,\omega,\ldots,v,\ldots]) = (\nabla_u t)[\ldots,\omega,\ldots,v,\ldots] + t[\ldots,\nabla_u\omega,\ldots,v,\ldots]$$
$$+ t[\ldots,\omega,\ldots,\nabla_u v,\ldots] \tag{19.2.1}$$

where $\nabla_u = i_u \circ \nabla : T_b^a(M) \to T_b^a(M)$ is called the covariant derivative with respect to ∇ in direction of the vector field u.
4. *$\nabla f = df$ for $f \in C^\infty(M)$.*

Notice that by definition $\nabla_{f_1 v_1 + f_2 v_2} = f_1 \nabla_{v_1} + f_2 \nabla_{v_2}$ for $f_1, f_2 \in C^\infty(M)$, so ∇_v depends only on v and not on its derivatives, in contrast to the Lie derivative.

(ii) *Parallel transport and geodesics*

Let us abbreviate $\nabla_\mu := \nabla_{\partial_\mu}$ in abstract index notation. Then we define the connection components as

$$\nabla_\mu \partial_\nu =: \Gamma^\rho_{\mu\nu} \partial_\rho \tag{19.2.2}$$

From the axioms we derive due to $\nabla \partial_\mu[x^\nu] = \nabla dx^\nu[\partial_\mu] = 0$

$$\nabla_\mu dx^\nu = -\Gamma^\nu_{\mu\rho} dx^\rho \tag{19.2.3}$$

From (19.2.2) and (19.2.3) we derive the covariant derivative of tensors of arbitrary type in abstract index notation

$$\nabla_\mu t^{\mu_1 \ldots \mu_a}_{\nu_1 \ldots \nu_b} := (\nabla_\mu t)^{\mu_1 \ldots \mu_a}_{\nu_1 \ldots \nu_b}$$
$$= \partial_\mu\left[t^{\mu_1 \ldots \mu_a}_{\nu_1 \ldots \nu_b}\right] + \sum_{k=1}^{a} \Gamma^{\mu_k}_{\mu\rho}\, t^{\mu_1 \ldots \hat{\mu}_k \rho \ldots \mu_a}_{\nu_1 \ldots \nu_b} - \sum_{l=1}^{b} \Gamma^\rho_{\mu\nu_l}\, t^{\mu_1 \ldots \mu_a}_{\nu_1 \ldots \hat{\nu}_l \rho \ldots \nu_b}$$

$$\tag{19.2.4}$$

The geometrical meaning of $(\nabla_v t)(p)$ is as follows: consider a curve along v starting in p. To first order in the curve parameter s this is given by $p_s = p + sv(p)$. Now take the components of $t(p)$ and transport them without change of the argument p of $t_{\cdots}(p)$ to p_s, however, since the bases ∂_μ, dx^μ are different we must express the $t_{\cdots}(p)$ in the new basis. This infinitesimal change of basis is precisely captured in (19.2.2), (19.2.3) and is what *defines* ∇: a covariant differential is simply a rule for how bases change when we change points along curves. Denote the so-transported tensor at p_s by $\tilde{t}(p_s)$. Then $(\nabla_v t)(p) := \lim_{s \to 0}[t - \tilde{t}](p_s)/s$. A tensor is said to be parallel transported along v provided its transport from p to p_s is proportional to the tensor defined there, that is, $\tilde{t}(p_s) \propto t(p_s)$. Taking $s \to 0$ gives the parallel transport equation $\nabla_v t \propto t$. Taking $v = \partial_s$ to be the tangential vector field along some *given* curve we get a first-order ordinary differential equation which always has a maximal solution. A particular case is obtained if we take $t = v$, that is, we consider the parallel transport of the tangential vector along a curve (to be determined) and ask that the tangential vector stays parallel to itself. This leads to the geodesic equation

$$(\nabla_v v)(c_v(s)) = \gamma(s)v(c_v(s)) \tag{19.2.5}$$

with $\dot{c}_v(s) = v(c_v(s))$ and $\gamma(s)$ is any function of s. One can show that upon choosing a so-called affine parameter $\tilde{s}(s)$ we can always choose $\gamma = 0$. From its definition, a geodesic is a curve which is 'as straight as possible'. (19.2.5) is a second-order ordinary differential equation which always has a unique and maximal solution again. Given initial data $p, v(p)$ we denote the solution by $\exp_s(v)$. This exponential map is defined for all p, v but the allowed range of s may vary. If it coincides with $\mathbb{R}$ for all $T_p(M)$ then M is said to be geodesically complete.

(iii) *Torsion and curvature*

Two tensor fields can be naturally associated with a covariant differential. The first is the torsion tensor field $T \in T_2^1(M)$

$$T[.; u, v] := \nabla_u v - \nabla_v u - [u, v] \tag{19.2.6}$$

with components

$$T^\rho_{\mu\nu} = T[dx^\rho; \partial_\mu, \partial_\nu] = 2\Gamma^\rho_{[\mu\nu]} \tag{19.2.7}$$

If ∇ is torsion-free then in the expressions for the Lie derivative of tensors and the exterior derivative for forms we may replace the partial derivatives ∂_μ by covariant derivatives ∇_μ. The geometric meaning of torsion is as follows: take $u, v \in T^1(M)$ and $p \in M$ and consider the points $c_p^u(r), c_p^v(s)$ of their integral curves through p corresponding to the parameter values r, s. Now parallel transport $u(p), v(p)$ respectively along c_p^v, c_p^u respectively to obtain $\tilde{u}(c_p^v(s)), \tilde{v}(c_p^u(r))$. Finally, construct the integral curves of these new vector fields through $c_p^v(s), c_p^u(r)$ respectively, that is, $c^{\tilde{u}}_{c_p^v(s)}(r), c^{\tilde{v}}_{c_p^u(r)}(s)$. The torsion now is directly proportional to the coefficient of rs of $x(c^{\tilde{u}}_{c_p^v(s)}(r)) - x(c^{\tilde{v}}_{c_p^u(r)}(s))$ and measures the deviation from the corresponding parallelism to close.

The second natural tensor field is the curvature tensor field $R \in T_3^1(M)$

$$R[.; w, u, v] := ([\nabla_u, \nabla_v] - \nabla_{[u,v]})w \tag{19.2.8}$$

For both T, R it is crucial to notice that (19.2.6), (19.2.8) evaluated at p depend only on the values of u, v, w at p, otherwise T, R would not define tensors of the indicated type. The components of R are

$$R^\rho{}_{\sigma\mu\nu} = T[dx^\rho; \partial_\sigma, \partial_\mu, \partial_\nu] = \partial_\mu \Gamma^\rho_{\nu\sigma} - \partial_\nu \Gamma^\rho_{\mu\sigma} + \Gamma^\rho_{\mu\lambda} \Gamma^\lambda_{\nu\sigma} - \Gamma^\rho_{\nu\lambda} \Gamma^\lambda_{\mu\sigma} \quad (19.2.9)$$

It is straightforward to verify that the curvature tensor enjoys the following symmetries

$$R^\rho{}_{\sigma(\mu\nu)} = 0, \quad R^\rho{}_{[\sigma\mu\nu]} = 0, \quad R^\rho{}_{\sigma[\mu\nu; \lambda]} = 0 \quad (19.2.10)$$

where $(.)_{; \lambda} := \nabla_\lambda(.)$. The last identity in (19.2.10) is called Bïanchi's identity. Finally we construct the Ricci tensor field $\mathrm{Ric} \in T^0_2(M)$ as

$$\mathrm{Ric}[u, v] := R[dx^\mu; u, \partial_\mu, v]; \quad R_{\mu\nu} = \mathrm{Ric}_{\mu\nu} = R^\rho{}_{\mu\rho\nu} \quad (19.2.11)$$

The geometric interpretation of curvature is as follows: consider two vector fields u, v and a point $p \in M$. Construct the integral curves $c^u_p, c^v_p, c^u_{c^v_p(s)}, c^v_{c^u_p(r)}$. To order rs we have $q = c^u_{c^v_p(s)}(r) = c^v_{c^u_p(r)}(s)$. Now take a third vector field w and construct its parallel transport from p to q (1) along the curve $c^u_p(r) \circ c^v_{c^u_p(r)}(s)$ resulting in $\tilde{w}_1(q)$ and (2) along the curve $c^v_p(s) \circ c^u_{c^v_p(s)}(r)$ resulting in $\tilde{w}_2(q)$. One can verify that the coefficient of rs of $\tilde{w}_1(q) - \tilde{w}_2(q)$ is precisely $R[.; w, u, v]$. Hence the vanishing of the curvature tensor is the integrability condition for the parallel transport of vectors along closed curves (loops) to coincide with the original vector, in other words, that the holonomy of the connection be trivial. If that is the case, we say that the connection ∇ is flat. See Chapter 21 for the bundle-theoretic language.

(iv) *Metric tensor field and Levi–Civita connection*

Definition 19.2.2. *A metric tensor field $g \in T^0_2(M)$ is a symmetric, non-degenerate two-times covariant tensor field of second rank. Due to continuity and non-degeneracy its signature $(n, m - n)$, given by the number of negative and positive eigenvalues of its component matrix, is constant on M. If $n = 0, 1$ respectively, g is said to be of Euclidean or Lorentzian signature respectively. The pair (M, g) consisting of a manifold and a metric is called a spacetime or (pseudo-)Riemannian space.*

One can show that every paracompact manifold admits Euclidean metrics while Lorentzian metrics can only be defined for non-compact, paracompact manifolds.

Lemma 19.2.3. *Given a spacetime (M, g), there is a unique torsion-free, metric-compatible covariant derivative ∇, that is, $\nabla g = 0$. The associated connection is called the Levi–Civita connection and its components are called Christoffel symbols.*

Proof: Write out the three conditions $\nabla_\mu g_{\nu\rho} = \nabla_\nu g_{\rho\mu} = \nabla_\rho g_{\mu\nu} = 0$ in terms of the $\Gamma^\rho_{\mu\nu}$. Using $\Gamma^\rho_{[\mu\nu]} = 0$, solve this system of linear equations for $\Gamma^\rho_{\mu\nu}$. One finds

$$\Gamma^\rho_{\mu\nu} = \frac{1}{2} g^{\rho\sigma} [g_{\sigma\mu,\nu} + g_{\sigma\nu,\mu} - g_{\mu\nu,\sigma}] \qquad (19.2.12)$$

where $g^{\mu\nu}$ are the components of the inverse metric tensor. $\qquad \square$

The Levi–Civita connection is therefore such that the 'scalar products' $g[u, v]$ of parallel transported vectors u, v remain constant.

The curvature tensor defined by a metric through the Levi–Civita connection is called the Riemann tensor. It has the additional symmetry

$$R_{\mu\nu\rho\sigma} = R_{\rho\sigma\mu\nu}, \quad R_{\mu\nu\rho\sigma} = g_{\mu\lambda} R^\lambda{}_{\nu\rho\sigma} \qquad (19.2.13)$$

The associated Ricci tensor is then symmetric and we can define the curvature scalar as $R = g^{\mu\nu} R_{\mu\nu}$.

(v) *Weyl tensor field*

Taking into account all the algebraic (non-differential) symmetries of the Riemann tensor we find that the pairs $(\mu\nu)$ and $(\rho\sigma)$ can take only $N = m(m-1)/2$ values due to antisymmetry. Moreover, the Riemann tensor is symmetric under exchange of these pairs, leaving us with $N(N+1)/2$ independent components. Finally we have to take into account the third condition in (19.2.10) which, using the other symmetries, can be written in the form

$$R_{\mu[\nu\rho\sigma]} = R_{[\nu\rho\sigma]\mu} = -R_{\nu[\mu\rho\sigma]} = -R_{[\mu\nu\rho]\sigma} = 0 \qquad (19.2.14)$$

Thus, since $\mu \neq \nu$ we get one non-trivial condition for one four-tuple of pairwise different indices, that is, we get $\binom{m}{4}$ additional conditions reducing the number of independent components to

$$N(N+1) - \binom{m}{4} = \frac{m^2(m^2-1)}{12} \qquad (19.2.15)$$

For $m = 1$ there can be no curvature, for $m = 2$ the curvature tensor is already determined by the curvature scalar, for $m = 3$ the curvature scalar is already determined by the Ricci tensor. It is only for $m \geq 4$ that the curvature tensor has more than the $m(m+1)/2$ components of the Ricci tensor. These can be extracted by removing the trace from the curvature tensor and results in Weyl's conformal tensor

$$C_{\mu\nu\rho\sigma} = R_{\mu\nu\rho\sigma} + \frac{2}{m-2}(g_{\mu[\sigma}R_{\rho]\nu} - g_{\nu[\sigma}R_{\rho]\mu}) + \frac{2}{(m-1)(m-2)}R g_{\mu[\rho}g_{\sigma]\nu} \qquad (19.2.16)$$

Another characterisation of C is that it is the unique linear combination of curvature tensor, Ricci tensor and curvature scalar such that $C^\mu{}_{\nu\rho\sigma}[\Omega^2 g] = C^\mu{}_{\nu\rho\sigma}[g]$, that is, the Weyl tensor is invariant under the

Weyl rescalings $g \mapsto \Omega^2 g$ for any everywhere non-vanishing function $\Omega \in C^\infty(M)$.

(vi) *Killing vector fields*

Let $t \mapsto \psi_t^v$ be the one-parameter family of diffeomorphisms generated by $v \in T^1(M)$. v is called a Killing vector field or an *isometry* provided that $(\varphi_t^v)^* g = g$. Equivalently, $\mathcal{L}_v g = 0$ by definition of the Lie derivative. v is called a conformal isometry if $(\varphi_t^v)^* g = (\Omega_t^v)^2 g$ for some nowhere vanishing function $\Omega_t^v \in C^\infty(M)$ or equivalently $\mathcal{L}_v g = fg$ for some $f \in C^\infty(M)$. Most metrics have no (conformal) isometries. They play a crucial role in quantum field theory on curved background spacetimes.

(vii) *Densities and volume forms*

In the previous section we have defined the integral of n-forms for orientable (sub-)manifolds. In the presence of a metric tensor field g we can now define such integrals for non-orientable submanifolds. Let $\psi : N \to M$ be an embedded n-dimensional submanifold and define the induced metric on N by $q := \psi^* g$. Let $f \in C^\infty(M)$. Then we define the integral of f over N by

$$\int_N \psi^* f := \sum_I \int_{x_I(V_I)} d^n y \, e_I \, \sqrt{|\det(q_I)|} (\psi^* f)_I \qquad (19.2.17)$$

where we have used a partition of unity and the induced atlas $(V_I = \psi^{-1}(U_I \cap \psi(N))$, $y_I = x_I \circ \psi)$ again. It is easy to see that (19.2.17) is independent of the choice of coordinates and the partition of unity even when N is not orientable. The reason for this is that the quantity

$$d^n y_I \sqrt{|\det(q_I(y_I))|} \qquad (19.2.18)$$

is an invariant measure on $\mathbb{R}^n$ rather than an n-form. Indeed if $y_J = \varphi_{IJ} \circ y_I$ then

$$\begin{aligned} d^n y_J \sqrt{\det(q_J(y_J))} &= d^n y_I \, |\det(\partial \varphi_{IJ}(y_I)/\partial y_I)| \sqrt{|\det(q_J(\varphi_{IJ}(y_I)))|} \\ &= d^n y_I \, \sqrt{|[\det(\partial \varphi_{IJ}(y_I)/\partial y_I)]^2 \det(q_J(\varphi_{IJ}(y_I)))|} \\ &= d^n y_I \, \sqrt{|\det((\varphi_{IJ}^* q_J)(y_I))|} = d^n y_I \, \sqrt{|\det(q_I(y_I))|} \end{aligned} \qquad (19.2.19)$$

We may define an associated volume n-form as follows. The totally skew, metric-independent Levi–Civita symbol in n dimensions is defined as

$$\epsilon_{a_1 \ldots a_n} := (n!) \delta^1_{[a_1} \ldots \delta^n_{a_n]}, \quad \epsilon^{a_1 \ldots a_n} := (n!) \delta^{[a_1}_1 \ldots \delta^{a_n]}_n \qquad (19.2.20)$$

where δ is the Kronecker symbol and a_k takes a range in $\{1, \ldots, n\}$. We define the pseudo[2] skew tensor

$$\eta_{a_1 \ldots a_n} = \sqrt{|\det(q)|}\, \epsilon_{a_1 \ldots a_n} \qquad (19.2.21)$$

Notice that $\delta_b^a = q^{ac} q_{cb}$ is an invariant tensor field and hence $\epsilon_{a_1 \ldots a_n}$ is as well. It follows that (19.2.21) indeed transforms as a pseudo n-form on N, that is, as an ordinary n-form provided N is orientable and under orientation-preserving passive diffeomorphisms. As shown, the function $\sqrt{|\det(q)|}$ is not a scalar but a so-called scalar density of weight one. An element $t \in T_b^a(N)$ is then also called a tensor field of density weight zero while a tensor field with the same transformation behaviour as $(\sqrt{|\det(q)|})^w t$ would be called a tensor of type $T_b^a(N)$ of density weight $w \in \mathbb{R}$. In particular, $\epsilon_{a_1 \ldots a_n}$, $\epsilon^{a_1 \ldots a_n}$ are pseudo tensor fields of type $T_n^0(N)$, $T_0^n(N)$ respectively of weight $-1, +1$ respectively. Notice that

$$\eta^{a_1 \ldots a_n} := q^{a_1 b_1} \ldots q^{a_n \ldots b_n} \eta_{b_1 \ldots b_n}$$
$$= \det(q)^{-1} \epsilon^{a_1 \ldots a_n} / \sqrt{|\det(q)|} = s \sqrt{\det(q)}^{-1} \epsilon^{a_1 \ldots a_n} \quad (19.2.22)$$

where s is the signature of q.

If $M = N$ and ∇ is the g-compatible covariant differential we must define the action of ∇ on tensor densities. Since $\eta_{\mu_1 \ldots \mu_n}$ is an ordinary tensor and since $\eta^{\mu_1 \ldots \mu_m} \eta_{\mu_1 \ldots \mu_m} = s(m!)$ we derive with $\nabla g^{-1} = 0$ (which follows from $\nabla_\mu \delta_\nu^\rho = 0$) that

$$\eta^{\mu_1 \ldots \mu_n} \nabla_\mu \eta_{\mu_1 \ldots \mu_n} = 0 \qquad (19.2.23)$$

Since $\nabla_\mu \eta_{\mu_1 \ldots \mu_n} = f_\mu \eta_{\mu_1 \ldots \mu_n}$ (there is only one totally skew tensor field of rank m in m dimensions up to multiplication with a scalar function) we conclude $\nabla_\mu \eta_{\mu_1 \ldots \mu_m} = 0$. Since $\epsilon_{\mu_1 \ldots \mu_m}$ is a linear combination of tensor products of components of g_ν^μ we conclude $\nabla \sqrt{|\det(g)|} = 0$. We can therefore extend the definition of ∇ to tensor densities of weight w by

$$\nabla \tilde{t} := (\sqrt{|\det(g)|})^w \, \nabla \frac{\tilde{t}}{(\sqrt{|\det(g)|})^w} \qquad (19.2.24)$$

where on the left-hand side the covariant differential now acts on an ordinary tensor field of weight zero in the usual way.

(viii) *Connection with fibre bundle theory and orthonormal frames*

As we will see in the next chapter, a connection in a fibre bundle can be defined as an assignment of a horizontal subspace in the fibre above each point of M. Equivalently, it is defined via a connection one-form in a

---

[2] A pseudo tensor field of certain index structure transforms in the same way as a tensor with the same index structure up to a sign difference under orientation-reversing diffeomorphisms.

principal fibre bundle over M. We want to make the connection between that definition and the definition of ∇ presented here.

The relevant principal fibre bundle turns out to be the bundle $L(M)$ of linear frames with structure group $G = GL(m, \mathbb{R})$. It is obtained as the equivalence class of smooth assignments of linearly independent bases $(e_\alpha)_{\alpha=1}^m$ modulo coordinate transformations in the fibre above each point. Given a standard frame any other frame is related to it by a local $GL(m, \mathbb{R})$ transformation, hence we can think of $L(M)$ as a principal $GL(m, \mathbb{R})$-bundle. A connection one-form in that bundle is given by the general formula $\omega = Ad_{h_I^{-1}}\pi^*A_I + h_I^{-1}dh_I$ where $\phi_I(p, h_I) = u$ are the local trivialisations of points $u \in \pi^{-1}(p)$, $h_I \in GL(m, \mathbb{R})$. The horizontal vector fields are annihilated by ω and hence are spanned by the vector fields

$$\partial/\partial x_I^\mu - (A_\mu^I)_\alpha^\gamma (h_I)_\gamma^\beta \partial/\partial_\alpha^\beta \qquad (19.2.25)$$

where $A_I = s_I^*\omega$ is the pull-back to M of ω by the local section $s_I(p) = \phi_I(p, 1_G)$. In particular, the horizontal lift of a curve $p(s)$ is given by $\tilde{p}(s) = \phi_I(p(s), h_I(s))$ where $h_I(s)$ is chosen such that $\omega[\dot{\tilde{p}}(s)] = 0$. This gives the holonomy equation

$$(\dot{h}_I(s))_\alpha^\beta = \dot{x}_I^\mu(s)(A_\mu^I(x_I(s)))_\alpha^\gamma (h^I)_\gamma^\beta(s) \qquad (19.2.26)$$

and it is easy to see that $\partial/\partial s = \dot{x}_I^\mu \partial/\partial x_I^\mu + (\dot{h}_I)_\alpha^\beta \partial/\partial(h_I)_\alpha^\beta$ is horizontal.

The bundles $E_b^a(M)$ are now associated with $L(M)$ via the corresponding tensor product representations of $GL(m, \mathbb{R})$. Namely, let (e_α) be a frame on which $GL(m, \mathbb{R})$ acts as $e_\alpha \mapsto h_\alpha{}^\beta e_\beta$. Let e^α be a dual frame of one-forms, that is, $e^\alpha[e_\beta] = \delta_\beta^\alpha$ from which we infer that $e^\alpha \mapsto (h^{-1})_\beta{}^\alpha e^\beta$. Let $t \in E_b^a(M)$ and express it in terms of these bases, that is

$$t = t_{\beta_1...\beta_b}^{\alpha_1...\alpha_a} e_{\alpha_1} \otimes ... \otimes e_{\alpha_a} \otimes e^{\beta_1} \otimes ... \otimes e^{\beta_b} \qquad (19.2.27)$$

It follows that t transforms in the representation τ where

$$t_{\beta_1...\beta_b}^{\alpha_1...\alpha_a} \mapsto (\tau(h))_{\gamma_1...\gamma_a}^{\alpha_1...\alpha_a} {}_{\beta_1...\beta_b}^{\delta_1...\delta_b} t_{\delta_1...\delta_b}^{\gamma_1...\gamma_a}$$

$$(\tau(h))_{\gamma_1...\gamma_a}^{\alpha_1...\alpha_a} {}_{\beta_1...\beta_b}^{\delta_1...\delta_b} = \overset{a}{\underset{k=1}{\prod}} h_{\gamma_k}{}^{\alpha_k} \overset{b}{\underset{l=1}{\prod}} (h^{-1})_{\beta_k}{}^{\delta_k} \qquad (19.2.28)$$

By definition the covariant derivative in the associated bundle is given by

$$\nabla_\mu t_{\beta_1...\beta_b}^{\alpha_1...\alpha_a} = \partial_\mu t_{\beta_1...\beta_b}^{\alpha_1...\alpha_a} + (A_\mu)_\alpha{}^\beta \frac{\partial(\tau(h))_{\gamma_1...\gamma_a}^{\alpha_1...\alpha_a} {}_{\beta_1...\beta_b}^{\delta_1...\delta_b}}{\partial h_\alpha{}^\beta} t_{\delta_1...\delta_b}^{\gamma_1...\gamma_a}$$

$$= \partial_\mu t_{\beta_1...\beta_b}^{\alpha_1...\alpha_a} + \overset{a}{\underset{k=1}{\sum}} (A_\mu)_\gamma^{\alpha_k} t_{\beta_1...\beta_b}^{\alpha_1...\hat{\alpha}_k\gamma...\alpha_a} - \overset{b}{\underset{l=1}{\sum}} (A_\mu)_{\beta_l}^\gamma t_{\beta_1...\hat{\beta}_l\gamma...\beta_b}^{\alpha_1...\alpha_a}$$

$$(19.2.29)$$

In (19.2.29) the covariant derivative acts on the $GL(m, \mathbb{R})$ indices $\alpha, \beta, \ldots$ while before it was acting on the tangent space indices $\mu, \nu, \ldots$. Comparing (19.2.29) with (19.2.4) and using

$$t^{\alpha_1 \ldots \alpha_a}_{\beta_1 \ldots \beta_b} = \prod_{k=1}^{a} e^{\alpha_k}_{\mu_k} \prod_{l=1}^{a} e^{\nu_l}_{\beta_l} \qquad (19.2.30)$$

we infer from the Leibniz rule the compatibility condition

$$\nabla_\mu e^\alpha_\nu = \partial_\mu e^\alpha_\nu - \Gamma^\rho_{\mu\nu} e^\alpha_\rho + (A_\mu)_\beta{}^\alpha e^\beta_\nu$$
$$\nabla_\mu e^\nu_\alpha = \partial_\mu e^\alpha_\nu + \Gamma^\nu_{\mu\rho} e^\rho_\alpha - (A_\mu)_\alpha{}^\beta e^\nu_\beta \qquad (19.2.31)$$

From this we derive the horizontal subspaces in the associated bundle $E^a_b(M)$ to be spanned by the vector fields given in local coordinates by

$$\frac{\partial}{\partial x^\mu} - \left[\sum_{k=1}^{a} \Gamma^{\mu_k}_{\mu\rho} t^{\mu_1 \ldots \hat\mu_k \rho \ldots \mu_a}_{\nu_1 \ldots \nu_l} - \sum_{l=1}^{b} \Gamma^\rho_{\mu\nu_l} t^{\mu_1 \ldots \mu_a}_{\nu_1 \ldots \hat\nu_l \rho \ldots \nu_b} \right] \frac{\partial}{\partial t^{\mu_1 \ldots \mu_a}_{\nu_1 \ldots \nu_b}} \qquad (19.2.32)$$

Suppose now that we reduce the structure group from $GL(m, \mathbb{R})$ to $O(n, m - p, \mathbb{R})$ and that the frames e_α are orthonormal, that is, $g[e_\alpha, e_\beta] = \eta_{\alpha\beta}$ where η is the diagonal metric with n entries of -1 and $m - n$ entries of $+1$. In this case the e_α are called m-Beine and the e^α co-m-Beine. Then from (19.2.29) we obtain, due to the definition $h_\alpha{}^\gamma h_\beta{}^\delta \eta_{\gamma\delta} = \eta_{\alpha\beta}$, that

$$\nabla_\mu \eta_{\alpha\beta} = 0 = (A_\mu)_{\alpha\beta} + (A_\mu)_{\beta\alpha} \qquad (19.2.33)$$

where $(A_\mu)_{\alpha\beta} = \eta_{\beta\gamma}(A_\mu)_\alpha{}^\gamma$. It follows that A_μ is an $o(n, m - n, \mathbb{R})$-valued one-form for a $O(n, m - n, \mathbb{R})$ gauge theory.

So far A_μ and the Levi–Civita connection have nothing to do with each other. However, the relations $g_{\mu\nu} = \eta_{\alpha\beta} e^\alpha_\mu e^\beta_\nu$, $\eta_{\alpha\beta} = g_{\mu\nu} e^\mu_\alpha e^\nu_\beta$ together with the covariant constancy of g, η motivate us to impose the restriction

$$\nabla_\mu e^\rho_\nu = 0 \qquad (19.2.34)$$

These equations are sufficient in order to express A_μ in terms of e^μ_α and its first derivatives and A_μ determined that way is called *spin connection*.

19.3 Symplectic manifolds

19.3.1 Symplectic geometry

Definition 19.3.1. *A symplectic structure for a differential manifold M is a non-degenerate, closed two-form ω. The pair (M, ω) is called a symplectic manifold.*

It follows that $2m = \dim(M)$ is even. By Poincaré's lemma, locally we can always find a symplectic potential, that is, a one-form θ such that $\omega = d\theta$.

Theorem 19.3.2 (Darboux). *Let (M, ω) be a symplectic manifold. Then for a neighbourhood Z of each point p one can choose so-called canonical coordinates $(x^\mu)^{2m}_{\mu=1} = (q^a, p_a)^m_{a=1}$ such that $\omega = dp_a \wedge dq^a$. The coordinates q, p are called configuration and momentum variables respectively.*

Proof: The coordinate components $\omega_{\mu\nu}(x)$ in a chart U form an antisymmetric $2m \times 2m$ matrix and thus can be brought into the standard form $\omega^0 = \frac{1}{2}\epsilon \otimes 1_m$ with $\epsilon_{IJ} = -\epsilon_{JI}$, $\epsilon_{12} = 1$ by a non-degenerate matrix $S(x)$, that is, $\omega_{\mu\nu}(x) = \omega^0_{\rho\sigma}S^\rho_\mu(x)S^\sigma_\nu(x)$. Choose p in U with coordinate x_0, let $\varphi^\mu_0(x) := S^\mu_\nu(x_0)x^\nu$ and set $\sigma := \varphi^*_0\omega_0$. Clearly $\sigma(x_0) = \omega(x_0)$ and $d(\sigma - \omega) = 0$ since both ω, ω_0 are closed. By Poincaré's lemma we find an open neighbourhood $V \subset U$ of x_0 such that $\sigma - \omega = d\alpha$ for some one-form α. Without loss of generality we may assume $\alpha(x_0) = 0$ (subtract df for some f if necessary).

Now consider on $[0, 1] \times V$ the two-form

$$\Omega := dt \wedge \alpha + \omega_t, \quad \omega_t = \omega + t(\sigma - \omega) \tag{19.3.1}$$

ω_t interpolates between ω, σ. We have $d\Omega = dt \wedge [\sigma - \omega - d\alpha] = 0$ and $\omega_t(x_0) = \omega(x_0)$ is non-degenerate at x_0, hence it is non-degenerate also in a neighbourhood $W \subset V$ of x_0 by continuity.

Ω is a two-form in $2m + 1$ dimensions and thus its antisymmetric component matrix is degenerate. It follows that the equation $i_Y\Omega = 0$ has a non-trivial solution $Y \in T^1(W)$. Since ω_t is non-degenerate in W, Y has the form $Y = \partial/\partial_t + v^\mu(x, t)\partial_\mu$ up to multiplication by a scalar function. We have explicitly $i_Y\Omega = i_v\omega_t + \alpha - [i_v\alpha]dt = 0$, from which we infer in particular that $i_v\alpha = 0$. Let $C^Y_x(s)$ be the integral curve of Y through $(0, x)$ in $[0, 1] \times W$ with $s = t \in [0, 1]$ as curve parameter. Hence $C^Y_x(t) = (s, c^v_x(t))$ with $\dot{c}^v_x(t) = v(t, x)$, $c^v_x(0) = x$. Since $\alpha(x_0) = 0$ we have $(i_Y\Omega)(t, x_0) = (i_v\omega_t)(t, x_0) = 0$. Since $\omega_t(x_0)$ is non-degenerate this means that $v(t, x_0) = 0$. This means that $c^v_{x_0}(t) = x_0$. We may now consider the one-parameter family of diffeomorphisms $t \mapsto \psi^v_t(x) := c^v_x(t)$ and since $\psi^v_t(x_0) = x_0$ is a fixed point, it follows from continuity that ψ^v_t has range in W for some open neighbourhood Z of x_0.

Let Ψ^Y_t be the corresponding one-parameter family of diffeomorphisms defined on $[0, 1] \times Z$, then $(\Psi^Y_t)^*\Omega = \Omega$ for all $t \in [0, 1]$ because $\mathcal{L}_Y\Omega = [d\, i_Y + i_Y\, d]\Omega = 0$. Defining $\psi = \psi^v_1$ we conclude

$$((\Psi^Y_1)^*\Omega)(0, x) = ((\psi^v_1)^*[\omega_t + dt \wedge \alpha]_{t=1})(x) = \psi^*\sigma$$

$$= \Omega(0, x) = ([\omega_t + dt \wedge \alpha]_{t=0})(x) = \omega(x) \tag{19.3.2}$$

hence $\omega = (\varphi_0 \circ \psi)^*\omega_0$. $\qquad\square$

We notice that every symplectic manifold is automatically orientable because the *Liouville form*

$$\Omega := \frac{(-1)^{m(m-1)}}{m!} \omega \wedge \dots \wedge \omega \tag{19.3.3}$$

given by the m-fold exterior product of ω is nowhere vanishing. It is easy to see that $\int_M f\Omega = \int_M f d^m p\, d^m q$ in local canonical coordinates.

Given $f \in C^\infty(M)$ one defines a unique Hamiltonian vector field $\chi_f \in T^1(M)$ by the equation

$$i_{\chi_f}\omega + df = 0 \qquad (19.3.4)$$

A diffeomorphism $\psi \in \mathrm{Diff}\,(M)$ is said to be a symplectic isometry or *canonical transformation* iff

$$\psi^*\omega = \omega \qquad (19.3.5)$$

The diffeomorphisms generated by the flow of χ_f are symplectic isometries because $\mathcal{L}_{\chi_f}\omega = [i_{\chi_f}\,d + d\,i_{\chi_f}]\omega = -d\,df = 0$. Conversely, if $\mathcal{L}_v\omega = d\,i_v\omega = 0$ then by Poincaré's lemma we find locally f_v with $i_v\omega = -df_v$. Hence every generator of a symplectic isometry is locally a Hamiltonian vector field. The locally Hamiltonian vector fields form a sub-Lie algebra of $T^1(M)$ since $\mathcal{L}_{[u,v]}\omega = [\mathcal{L}_u, \mathcal{L}_v]\omega = 0$. In particular,

$$i_{[u,v]}\omega = i_{\mathcal{L}_u v}\omega = \mathcal{L}_u(i_v\omega) - i_v(\mathcal{L}_u\omega) = [i_u\,d + d\,i_u]i_v\omega$$
$$= i_u[d\,i_v + i_v\,d] + d(i_u i_v\omega) = -d(i_v i_u\omega) \qquad (19.3.6)$$

hence $[u, v] = \chi_{i_v i_u\omega}$ so the Hamiltonian vector fields form a Lie ideal in the space of locally Hamiltonian vector fields.

The Poisson bracket is defined by

$$\{f, f'\} := -i_{\chi_f}\, i_{\chi_{f'}}\omega = \chi_f[f'] \qquad (19.3.7)$$

It is antisymmetric by inspection and from (19.3.6) we see that $[\chi_f, \chi_{f'}] = \chi_{\{f,f'\}}$. Next

$$0 = i_{\chi_f}\, i_{\chi_g}\, i_{\chi_h}\, d\omega$$
$$= i_{\chi_f}\, i_{\chi_g}\, [\mathcal{L}_{\chi_h} - d\,i_{\chi_h}]\omega = -i_{\chi_f}\, i_{\chi_g}\, d\,(i_{\chi_h})\omega)$$
$$= -\chi_f[i_{\chi_g} i_{\chi_h}\omega] + \chi_g[i_{\chi_f} i_{\chi_h}\omega] + i_{[\chi_f, \chi_g]}(i_{\chi_h}\omega)$$
$$= -\chi_f[\{h, g\}] + \chi_g[\{h, f\}] + i_{\chi_{\{f,g\}}}(i_{\chi_h}\omega)$$
$$= \{f, \{g, h\}\} + \{g, \{h, f\}\} + \{h, \{\{f, g\}\} \qquad (19.3.8)$$

Hence the Poisson bracket satisfies the Jacobi identity as a consequence of $d\omega = 0$. Conversely one can show that the Jacobi identity implies $d\omega = 0$ if ω is non-degenerate. Notice that ψ is a symplectic isometry if and only if $\psi^*\{f, f'\} = \{\psi^*, \psi^* f'\}$ for all $f, f' \in C^\infty(M)$.

19.3.2 Symplectic reduction

Definition 19.3.3

(i) *Let M be a smooth manifold. A distribution $D : M \mapsto E_0^1(M)$ is an assignment of a subspace $D_p(M) \subset T_p(M)$ of the tangent space for each point*

$p \in M$ *such that (1)* $\dim(D_p(M)) = n = const.$ *and (2) for each* $p \in M$
there is a neighbourhood U *of* p *and* n *vector fields* $v_k \in T^1(M)$, $k = 1, \ldots, n$
such that $D_q(M)$ *is spanned by them for each* $q \in U$.

(ii) A submanifold $L \subset M$ *is called an integral manifold of the distribution* D
provided that $T_p(L) = D_p(M)$ *for all* $p \in L$. *The distribution* D *is said to*
be integrable provided that the subspace $T^1(M, D)$ *of vector fields which*
are everywhere tangential to D *is a subalgebra of* $T^1(M)$. *An integrable*
distribution is called a foliation. By Frobenius' theorem, integral manifolds
exist if and only if D *is integrable. Maximal integral manifolds of a foliation*
are called leaves.

(iii) A foliation is called reducible provided that the space of leaves $M/D =$
$\{[p];\ p \in M\}$, $[p] = \{p' \in M; p, p'\ lie\ in\ the\ same\ leaf\}$ *is a Hausdorff man-*
ifold with smooth projection $\pi:\ M \to M/D$.

Notice that an integrable distribution is not necessarily reducible.

Theorem 19.3.4 (Frobenius). *A distribution of* n-*dimensional tangent*
spaces can be equivalently described by the specification of n *vector fields* v_k
which are everywhere tangent to D *or by* $m - n$ *one-forms* θ_α *which satisfy*
$(\theta^\alpha[v])(p) = 0$ *for all* v *tangent to* D. *A necessary and sufficient condition for a*
distribution D *to be integrable is one of the following two equivalent criteria:*

1. *The* v_k *form a subalgebra of* $T^1(M)$, *that is,* $[v_j, v_k] = f_{kl}{}^l v_l$ *for some func-*
   tions $f_{jk}{}^l$.
2. *The* θ^α *form a closed Pfaff system, that is,* $d\theta^\alpha = \omega^\alpha_\beta \wedge \theta_\beta$ *for some one-forms*
   ω^α_β.

Proof: We first show that the two criteria are equivalent. We have

$$i_{v_j} i_{v_k}(d\theta^\alpha) = v_j[\theta^\alpha[v_k]] - v_k[\theta^\alpha[v_j]] - \theta^\alpha[[v_j, v_k]] = -\theta^\alpha[[v_j, v_k]] \qquad (19.3.9)$$

since $\theta^\alpha[v_k] = 0$ by definition. Now criterion (1) implies $i_{v_j} i_{v_k}(d\theta^\alpha) = 0$ which
means that the two-form has the form $d\theta^\alpha = \omega^\alpha_\beta \wedge \theta_\beta$ that is (2). Conversely, if
(2) is satisfied then $\theta^\alpha[[v_j, v_k]] = 0$ which means that the commutator has the
form $[v_j, v_k] = f_{kl}{}^l v_l$. Hence it suffices to demonstrate (2).

If the distribution is integrable then there are local coordinates
$(x^\mu) = (y^j, z^\alpha)$, $\mu = 1, \ldots, m$, $j = 1, \ldots, n$, $\alpha = 1, \ldots, m - n$ such that $\theta^\alpha(x) =$
$\theta^\alpha_\beta(x)dz^\beta$ or equivalently $v_j(x) = v_j^k(x)\partial/\partial y^k$ where both θ^α_β, v_j^k are invertible
matrices. The y are then coordinates for the leaves while the z parametrise the
space of leaves. If the θ^α or v_j take this form then clearly the Frobenius criterion
is satisfied. We now show the converse.

We will prove by induction over m for fixed rank $r = m - n$ of the Pfaff system.
For $m = r$ there is nothing to show. For $m > r$ let us write $\theta^\alpha = \tilde{\theta}^\alpha + f^\alpha dx^m$

where $\tilde{\theta}^\alpha(x) = \sum_{\mu=1}^{m-1} \theta_\mu^\alpha(x) dx^\mu$ and $f^\alpha(x)$ is some function on M. Then

$$d\theta^\alpha = \left[\sum_{\mu,\nu \le m-1} (\partial_\mu \tilde{\theta}_\nu^\alpha) dx^\mu \wedge dx^\nu \right] + \left[\sum_{\mu=1}^{m-1} (-\partial_m \tilde{\theta}_\mu^\alpha + \partial_\mu f^\alpha) dx^\mu \right] \wedge dx^m$$

$$=: \tilde{d}\tilde{\theta}^\alpha + \xi \wedge dx^m$$

$$= \omega_\beta^\alpha \wedge \theta^\beta = [\omega_\beta^\alpha \wedge \tilde{\theta}^\beta] + \omega_\beta^\alpha f^\beta \wedge dx^m \qquad (19.3.10)$$

We conclude $\tilde{d}\tilde{\theta}^\alpha = \omega_\beta^\alpha \wedge \tilde{\theta}^\beta$ where $\tilde{d}$ is the restriction of d to the first $m-1$ coordinates. By induction assumption we conclude that $\tilde{\theta}^\alpha(x) = \tilde{\theta}_\beta^\alpha(x) d\tilde{z}^\beta$ with $\tilde{\theta}_\beta^\alpha(x)$ invertible and $\tilde{z}^\alpha$ independent of x^m since this is a Pfaff system in the manifold with local coordinates $x^1, \ldots x^{m-1}$. We can now construct the equivalent Pfaff system

$$\theta^{\alpha'} := (\tilde{\theta}^{-1})_\beta^\alpha \theta^\beta = d\tilde{z}^\alpha + (\tilde{\theta}^{-1})_\beta^\alpha f^\beta dx^m =: d\tilde{z}^\alpha + f^{\alpha'} dx^m \qquad (19.3.11)$$

This is not yet of the required form because we must write $\theta^{\alpha'}$ as $\theta_\beta^{\alpha'} dz^{\beta'}$. To achieve this, notice that (19.3.11) is still closed

$$d\theta^{\alpha'} = [d(\tilde{\theta}^{-1})_\gamma^\alpha + (\tilde{\theta}^{-1})_\delta^\alpha \omega_\gamma^\delta] \tilde{\theta}_\beta^\gamma \wedge \theta^{\beta'} =: \omega_\beta^{\alpha'} \wedge \theta^{\beta'} \qquad (19.3.12)$$

Thus

$$d\theta^{\alpha'} = df^{\alpha'} \wedge dx^m = \omega_\beta^{\alpha'} \wedge [d\tilde{z}^\beta + f^{\beta'} dx^m] \qquad (19.3.13)$$

If we note coordinates as $x^\mu = (y^1, \ldots, y^{m-r-1}, \tilde{z}^1, \ldots, \tilde{z}^r, x^m)$ and compare coefficients in (19.3.12) then we find $\partial f^{\alpha'}/\partial y^j = 0$, $j = 1, \ldots, m-r-1$, thus $\theta^{\alpha'}(x) = d\tilde{z}^\alpha + f^{\alpha'}(\tilde{z}, x^m) dx^m$. Define the 'time' x^m-dependent 'Hamiltonian'

$$H(\{\tilde{z}^\beta, p_\beta\}_{\beta=1}^r; x^m) := -f^{\alpha'}(\{\tilde{z}^\beta\}, x^m) p_\alpha \qquad (19.3.14)$$

and solve the associated Hamilton–Jacobi equation

$$\frac{\partial S}{\partial x^m}(\tilde{z}, x^m) + H\left(\left\{\tilde{z}^\beta, p_\beta = \frac{\partial S}{\partial \tilde{z}^\beta}\right\}(\tilde{z}, x^m); x^m\right) = 0 \qquad (19.3.15)$$

Since, as is well known, the Hamilton–Jacobi equation is equivalent to the system of Hamilton's $2r$ ordinary differential equations, a maximal solution, called the complete integral, always exists and depends on $r+1$ free parameters C, c^α, $S(\tilde{z}, x^m; C, c)$. The general integral is obtained from the complete integral as follows: prescribe an arbitrary function $C = F(c^\alpha)$ and solve the system of algebraic equations

$$\frac{\partial S}{\partial c^\alpha} + \left(\frac{\partial S}{\partial C}\right)_{C=F(c)} \frac{\partial C}{\partial c^\alpha} = 0 \qquad (19.3.16)$$

for $c^\alpha = f_C^\alpha(\tilde{z}, x^m)$, which is always possible locally by the implicit function theorem. Specialise to the case that $F = c^\alpha =: F^\alpha(c)$ and define

$$z^{\alpha'}(\tilde{z}, x^m) = S(\tilde{z}, x^m; C = F^\alpha(c(\tilde{z}, x^m)), c = f_{F^\alpha}(\tilde{z}, x^m)) \qquad (19.3.17)$$

It is easy to check that (19.3.17) still solves the Hamilton–Jacobi equation and that these r solutions are algebraically independent. Hence

$$dz^{\alpha\prime} = \frac{\partial z^{\alpha\prime}}{\partial \tilde{z}^\beta}(\tilde{z}, x^m)\theta^{\beta\prime} \qquad (19.3.18)$$

accomplishes our task since $\partial z'/\partial \tilde{z}$ is invertible. $\qquad\square$

Definition 19.3.5. *Let N be a submanifold of (M, ω). Given a closed two-form σ on N we call (N, σ) a presymplectic (or Poisson) submanifold. If σ is degenerate we call $K : N \to T(N)$; $K_p(N) = \{v \in T_p(N); i_v\sigma = 0\}$ the characteristic distribution of (N, σ). N is then said to be reducible if K is reducible.*

Lemma 19.3.6. *Every presymplectic manifold is integrable. If N is reducible then N/K carries a natural symplectic structure.*

Proof: Let $\theta^1, \ldots, \theta^r$ be the one-forms which determine the characteristic distribution. Then obviously $\sigma = \sigma_{\alpha\beta}\theta^\alpha \wedge \theta^\beta$ and since σ is closed we conclude

$$d\sigma_{\alpha\beta} \wedge \theta^\alpha \wedge \theta^\beta + 2\sigma_{\alpha\beta}d\theta^\alpha \wedge \theta^\beta = 0 \qquad (19.3.19)$$

hence $d\theta^\alpha = \omega^\alpha_\beta \wedge \theta^\beta$ so integrability follows from Frobenius' theorem.

Next let $\rho \in \Lambda^k(N)$. We claim that there exists $\tau \in \Lambda^k(N/K)$ with $\rho = \pi^*\tau$, $\pi : N \to N/K$ if and only if $i_v\rho = i_v d\rho = 0$ for all v tangential to K. To see this, denote local coordinates of N by $(x^\mu) = (y^a, z^\alpha)$ and vector fields in N tangential to K by $v = v^a(y, z)\partial/\partial y^a$. The projection is given by $\pi(x) = z$. We have in general $\rho(x) = \rho_{\mu_1\ldots\mu_k}(x)dx^{\mu_1} \wedge \ldots \wedge dx^{\mu_k}$ and $d\rho(x) = (\partial_{\mu_0}\rho_{\mu_1\ldots\mu_k})(x)dx^{\mu_0} \wedge \ldots \wedge dx^{\mu_k}$. Now $i_v\rho = 0$ means that $\rho_{\mu_1\ldots\mu_k} = 0$ whenever at least one of the $\mu_1, \ldots, \mu_k$ takes the value a. It follows then from $i_v d\rho = 0$ that $\partial_a\rho_{\alpha_1\ldots\alpha_k} = 0$. Hence

$$\rho(x) = \rho_{\alpha_1\ldots\alpha_k}(z)dz^{\alpha_1} \wedge \ldots \wedge dz^{\alpha_k}$$
$$= \rho_{\alpha_1\ldots\alpha_k}(\pi(x))d\pi^{\alpha_1}(x) \wedge \ldots \wedge d\pi^{\alpha_k}(x) = (\pi^*\tau)(x) \qquad (19.3.20)$$

with $\tau_{\alpha_1\ldots\alpha_k}(z) = \rho_{\alpha_1\ldots\alpha_k}((0, z))$ as claimed.

Applied to σ we have $i_v\sigma = 0$ by definition and $i_v d\sigma = 0$ because σ is closed, thus $\sigma = \pi^*\tau$ for some two-form τ on N/K. Clearly $d\sigma = \pi^* d\tau = 0$ so τ is closed. For $v \in T^1(N/K)$ we find $V \in T^1(N)$ such that $\pi_*V = v$ on N/K simply by choosing $V^\alpha((0, z)) = v^\alpha(z)$. Then $i_v\tau = i_{\pi_*V}\tau = i_V\pi^*\tau = i_V\sigma = 0$ implies that V is tangential to K, hence $v = \pi_*V = 0$. Hence τ is non-degenerate. $\qquad\square$

Definition 19.3.7

(i) Let (V, ω) be a symplectic vector space (i.e., $M = V$ is a vector space) and let F be a subspace of V. Then $F^\perp := \{u \in V; \omega[u, v] = 0 \,\forall\, v \in V\}$ is called the annihilator of F.

(ii) F is called (a) isotropic if $F \subset F^\perp$, (b) co-isotropic if $F^\perp \subset F$, (c) Lagrangian if $F = F^\perp$ and (d) symplectic if $F \cap F^\perp = \{0\}$.

(iii) A submanifold N of a symplectic manifold (M, ω) is called (a)–(d) if $T_p(N)$ is (a)–(d) at each point $p \in N$.

We verify immediately that if F, G are subspaces of V then (1) $F \subset G$ implies $G^\perp \subset F^\perp$, (2) $(F^\perp)^\perp = F$, (3) $(F + G)^\perp = F^\perp \cap G^\perp$ and (4) $(F \cap G)^\perp = F^\perp + G^\perp$. Furthermore, if $\dim(V) = 2m$, $\dim(F) = k$, that is, $\dim(F^\perp) = 2m - k$ then (a) isotropic, (b) co-isotropic, (c) Lagrangian and (d) symplectic respectively implies (a) $k \leq m$, (b) $k \geq m$, (c) $k = m$, (d) $k = 2n$ is even. Notice that every symplectic vector space has subspaces of either category but that the categories (a)–(d) are not exhaustive, for example, $F \cap F^\perp \neq \{0\}$ is possible while neither $F \subset F^\perp$ nor $F^\perp \subset F$.

The connection of this terminology with Dirac's formalism for constraints (see Chapter 24) is as follows: if (M, ω) is a given unconstrained phase space with a system of constraints C_I, $I = 1, \ldots, r$ then we define the constraint submanifold as $N = \{p \in M; \ C_I(p) = 0\}$ with projection $\rho : M \to N$ which is obtained by using the C_I as local coordinates (owing to the implicit function theorem) and defining ρ to be the map that sets $C_I = 0$. N is then identified with the manifold whose local coordinates are the remaining ones and we have an injection $j : N \to M$. The presymplectic structure on N is then defined as $\sigma = j^*\omega$ which is automatically closed. If K is the characteristic distribution of (N, σ) then N/K with its symplectic structure τ is called the reduced phase space. For any v tangential to N we have on N

$$v[C_I] = dC_I[v] = -i_v i_{\chi_{C_I}} \omega = 0 \qquad (19.3.21)$$

hence $\chi_{C_I} \in (T_p(N))^\perp$ for all $p \in N$ where χ_{C_I} denotes the Hamiltonian constraint vector fields with respect to ω. By definition the characteristic distribution is given by

$$\begin{aligned} K_p &= \{u \in T_p(N); \ \sigma[u, v] = 0 \ \forall \ v \in T_p(N)\} \\ &= \{u \in T_p(N); \ \omega[j_*u, j_*v] = 0 \ \forall \ v \in T_p(N)\} \\ &= \{u \in T_p(N); \ \omega[u, v] = 0 \ \forall \ v \in T_p(N)\} = \rho_*(T_p(N)^\perp) \quad (19.3.22) \end{aligned}$$

since we identify $T_p(N)$ and $j_*(T_p(N))$.

The classification of the constraints is now as follows:

(a) isotropic $T_p(N) \subset (T_p(N))^\perp$
   Then $T_p(N) = \rho_*(T_p(N))^\perp = K_p$, hence N/K is a single point. This includes the Lagrangian case.

(b) co-isotropic $(T_p(N))^\perp \subset T_p(N)$
   Then $(T_p(N))^\perp = \rho_*(T_p(N))^\perp = K_p$. Since we can always find a basis of $T_p(M)$ consisting of Hamiltonian vector fields because ω is non-degenerate we can always find a basis of K_p consisting of Hamiltonian vector fields χ_f. Now $i_{\chi_f} i_v \sigma = i_{\chi_f} i_v \omega = v[df] = 0$ for all $v \in T_p(N)$ holds precisely when $f_{|N} = 0$, hence f is a linear combination of constraints, hence the χ_{C_I} span K_p. It follows that $\dim(N/K) = 2(m - r)$ and by the Frobenius theorem the

χ_{C_I} are in involution on N, hence

$$[\chi_{C_I}, \chi_{C_J}] = f_{IJ}{}^K \chi_{C_K} = \chi_{f_{IJ}{}^K C_K} - \chi_{f_{IJ}{}^K C_K} = \chi_{f_{IJ}{}^K C_K} = \chi_{\{C_I, C_J\}}$$

(19.3.23)

on N. Thus in Dirac's terminology, the constraints are first class, the Hamiltonian constraint vector fields are tangential to the constraint surface and belong to the characteristic distribution.

(c) symplectic $T_p(N) \cap (T_p(N))^\perp = \{0\}$
Hence $K_p = \rho_*(T_p(N))^\perp = \{0\}$. Hence $N/K = N$ is already symplectic with non-degenerate symplectic two-form $\sigma = j^* \omega$ defining the Dirac bracket.

19.3.3 Symplectic group actions

What follows plays a crucial role in geometric quantisation [218] and group theoretical quantisation [281].

Definition 19.3.8

(i) *A Lie group G is said to have a smooth right action on a manifold M provided that there is a smooth map $\rho: G \times M \to M$; $(g, p) \mapsto \rho_g(p)$ such that $\rho_g \in \mathrm{Diff}(M)$ for all $g \in G$ and $\rho_g \circ \rho_{g'} = \rho_{g'g}$, $\rho_{1_G} = \mathrm{id}_M$. The group action is said to be (a) transitive, (b) effective or (c) free respectively provided that*
(a) For all $p, p' \in M$ there exists $g \in G$ such that $\rho_g(p) = p'$.
(b) $\rho_g = \mathrm{id}_M$ implies $g = 1_G$.
(c) $\rho_g(p) = p$ for all $g \in G$ has no solution in M (no fixed points).
(ii) *A Lie group G is said to be a canonical group for a symplectic manifold (M, ω) provided ρ_G is a subgroup of the isometry group of ω.*

Given an element of the Lie algebra $A \in \mathrm{Lie}(G)$ we have a one-parameter group of diffeomorphisms $t \mapsto \rho_{\exp(tA)}$ where $\exp: \mathrm{Lie}(G) \to G$ is the exponential map. These define vector fields v_A via

$$v_A[f] := \left(\frac{d}{dt}\right)_{t=0} \rho^*_{\exp(tA)} f$$

(19.3.24)

The map $v: \mathrm{Lie}(G) \to T^1(M)$; $A \mapsto v_A$ is a homomorphism of Lie algebras, that is, $[v_A, v_B] = v_{[A,B]}$ as one can check immediately by using the Baker–Campbell–Hausdorff formula. Now, if $H^1(M) = \{0\}$ we have seen that $v_A = \chi_{f_A}$ for some $f_A \in C^\infty(M)$, which is uniquely determined up to a constant $f_A \to f_A + c(A)$. This is because the map $\chi: C^\infty(M) \to T^1(M)$; $f \mapsto \chi_f$ is only a homomorphism of Lie algebras, $\chi_{\{f,g\}} = [\chi_f, \chi_g]$, since $\mathrm{Ker}(\chi) = \mathbb{R}$, that is, χ has the constant functions as kernel. We have for $z \in \mathbb{R}$

$$v_{A+zB} = \chi_{f_{A+zB}} = v_A + z v_B = \chi_{f_A} + z \chi_{f_B} = \chi_{f_A + z f_B}$$ (19.3.25)

$$[v_A, v_B] = v_{[A,B]} = \chi_{f_{[A,B]}} = [\chi_{f_A}, \chi_{f_B}] = \chi_{\{f_A, f_B\}}$$ (19.3.26)

hence we conclude that $f_{A+zB} = f_A + z f_B$, $f_{[A,B]} = \{f_A, f_B\}$ up to a constant.

Definition 19.3.9

(i) *A linear map $f : Lie(G) \to C^\infty(M)$; $A \mapsto f_A$ is said to be Hamiltonian provided that it is a homomorphism of Lie algebras, that is, $f_{[A,B]} = \{f_A, f_B\}$. The dual map $\mu : M \to Lie(G)^*$; $\mu(p)[A] := f_A(p)$ where $Lie(G)^*$ is the space of linear forms on $Lie(G)$ is called a momentum map.*

(ii) *An n-cochain on $Lie(G)$ is a completely skew and multilinear map $c : Lie(G)^n \to \mathbb{R}$. The coboundary operator maps n-cochains to $(n+1)$-cochains defined by $[\delta c](A_0, \ldots A_n) := c(A_{[0}, A_1, \ldots, A_{n]})$. Clearly $\delta^2 = 0$. An n-cochain is called a cocycle or coboundary respectively iff $\delta c = 0$ or $c = \delta c'$ for some $(n-1)$-cochain c' respectively. The cocycles modulo the coboundaries determine the cohomology on n-cochains.*

(iii) *A central extension of a Lie algebra $Lie(G)$ is a Lie algebra E together with a homomorphism $\pi : E \to Lie(G)$ such that $Ker(\pi) \subset Z(E)$ where $Z(E) = \{A \in E; [A,B] = 0 \,\forall\, B \in E\}$ is the centre of E.*

We may always choose a linear map $A \mapsto f_A$ simply by defining f_{τ_I} and then $f_{r^I \tau_I} := r^I f_{\tau_I}$ where τ_I, $I = 1, \ldots, \dim(G)$ is a basis of $Lie(G)$. Define the antisymmetric bilinear form

$$c : Lie(G)^2 \to \mathbb{R}; \ (A, B) \mapsto \{f_A, f_B\} - f_{[A,B]} \tag{19.3.27}$$

One verifies immediately from the Jacobi identity for $C^\infty(M)$ and $Lie(G)$ that

$$c([A, B], C) + c([B, C], A) + c([C, A], B) = 0 \tag{19.3.28}$$

Hence c defines a 2-cocycle, $\delta c = 0$, on the dual of $Lie(G)$. If c is a coboundary $c(A, B) = c'([A, B])$ define

$$A \mapsto \tilde{f}_A = f_A + c'(A) \tag{19.3.29}$$

Then

$$\{\tilde{f}_A, \tilde{f}_B\} - \tilde{f}_{[A,B]} = \{f_A, f_B\} - f_{[A,B]} - c'([A, B]) = c(A, B) - c'([A, B]) = 0 \tag{19.3.30}$$

so $\tilde{f}$ is a Hamiltonian map. Conversely, if f is Hamiltonian then $c = 0$. Thus the group action (G, ρ) has a moment on (M, ω) if and only if the cohomology of the obstruction cocycle c is trivial. That cohomology does not depend on the initial choice of f because for any other choice f' we have

$$c'(A, B) - c(A, B) = f_{[A,B]} - f'_{[A,B]} =: d([A, B]) = (\delta d)(A, B) \tag{19.3.31}$$

so c, c' lie in the same cohomology class.

If the class of the obstruction cocycle does not vanish we proceed as follows: construct a central extension E of $Lie(G)$ with basis $(\hat{\tau}_I)_{I=0}^{\dim(G)}$ and Lie algebra

$(I, J, K = 1, \ldots, \dim(G))$

$$[\hat{\tau}_0, \hat{\tau}_I] = 0, \quad [\hat{\tau}_I, \hat{\tau}_J] = f_{IJ}{}^K \hat{\tau}_K + c(\tau_I, \tau_J)\hat{\tau}_0 \qquad (19.3.32)$$

where $[\tau_I, \tau_J] = f_{IJ}{}^K \tau_K$ are the structure constants of Lie(G). The required homomorphism is given by $\pi(\hat{\tau}_0) = 0$, $\pi(\hat{\tau}_I) = \tau_I$. We define a group action of $\hat{G}$ generated by E by $\hat{\rho}_{\exp(\hat{A})} := \rho_{\exp(\pi(\hat{A}))}$. Finally we define a map $\hat{f}_{\hat{A}} := f_{\pi(\hat{A})} + z(\hat{A})$ where $z(\hat{A}) = A^0$ if $\hat{A} = \sum_{I=0}^{\dim(G)} A^I \hat{\tau}_I$. We immediately verify that $\hat{v}_{\hat{A}} = v_{\pi(\hat{A})}$. Hence $\hat{v}_{\hat{A}} = \chi_{f_{\pi(\hat{A})}} = \chi_{\hat{f}_{\hat{A}}}$. Now with the abbreviation $A = \pi(\hat{A})$

$$\hat{f}_{[\hat{A},\hat{B}]} = f_{\pi([\hat{A},\hat{B}])} + z([\hat{A}, \hat{B}]) = f_{[A,B]} + ([\hat{A}, \hat{B}])^0$$
$$= \{f_A, f_B\} - c(A, B) + c(\pi(\hat{A}), \pi(\hat{B})) = \{\hat{f}_{\hat{A}}, \hat{f}_{\hat{B}}\} \qquad (19.3.33)$$

hence $(\hat{G}, \hat{\rho})$ has a trivial obstruction cocycle.

19.4 Complex, Hermitian and Kähler manifolds

Recall that a map $f : \mathbb{C}^m \to \mathbb{C}^n$ is called holomorphic if the Cauchy–Riemann equations hold for all component functions, that is

$$f^\nu(z) = u^\nu(x, y) + iv^\nu(x, y), \quad z^\mu = x^\mu + iy^\mu$$
$$\Rightarrow \frac{\partial u^\nu}{\partial x^\mu} - \frac{\partial v^\nu}{\partial y^\mu} = \frac{\partial v^\nu}{\partial x^\mu} + \frac{\partial u^\nu}{\partial y^\mu} = 0, \quad \nu = 1, \ldots n, \ \mu = 1, \ldots, m \qquad (19.4.1)$$

Definition 19.4.1. *A complex manifold is a topological space together with an atlas (U_I, z_I) consisting of an open cover of M with sets U_I and local charts $z_I : U_I \to \mathbb{C}^m$ which are homeomorphisms and are subject to the following condition: if $U_I \cap U_J \neq \emptyset$ then $\varphi_{IJ} = z_J \circ z_I^{-1} : z_I(U_I \cap U_J) \to z_J(U_I \cap U_J)$ is a holomorphic map.*

The number m is called the complex dimension of M. Each complex manifold is also a smooth real manifold of real dimension $2m$. However, a complex structure (holomorphicity) is much stronger than a differentiable structure (smoothness).

A complex manifold with local complex coordinates $z_I^\mu = x_I^\mu + iy_I^\mu$ over U_I has the local coordinate vector fields $\partial/\partial x_I^\mu, \partial/\partial y_I^\mu$ and local coordinate one-forms dx_I^μ, dy_I^μ as local real basis and co-basis respectively. Since we are allowed to take complex linear combinations in a complex manifold we may alternatively use the complex basis and co-basis

$$\partial/\partial z_I^\mu = \frac{1}{2}[\partial/\partial x_I^\mu - i\partial/\partial y_I^\mu], \quad \partial/\partial \bar{z}_I^\mu = \frac{1}{2}[\partial/\partial x_I^\mu + i\partial/\partial y_I^\mu]$$
$$dz_I^\mu = dx_I^\mu + idy_I^\mu, \quad d\bar{z}_I^\mu = dx_I^\mu - idy_I^\mu \qquad (19.4.2)$$

where $\bar{z} = x - iy$. Notice that M considered as a real manifold would not admit the (co)basis (19.4.2) since one would only be allowed to take real linear combinations. In order to distinguish tensor fields on M when M is considered as a real $(2m)$-dimensional real manifold from those when M is considered as a complex m-dimensional complex manifold we introduce the notation $T^a_b(M)$ to mean, as before, tensor fields spanned by tensor products of the $\partial_x, \partial_y, dx, dy$ with *real-valued* component functions while $T^a_b(M)_{\mathbb{C}}$ to mean tensor fields with complex-valued component functions. Only for $T^a_b(M)_{\mathbb{C}}$ we may also use tensor products of the $\partial_z, \partial_{\bar{z}}, dz, d\bar{z}$ as tensor basis elements.

We define a tensor field $J_0 \in T^1_1(M) :\ T^1(M) \to T^1(M)$ locally by

$$(J_I[\partial/\partial x^{\mu}_I])(x_I(p)) = \partial/\partial y^{\mu}_I, \quad (J_I[\partial/\partial y^{\mu}_I])(x_I(p)) = -\partial/\partial x^{\mu}_I \qquad (19.4.3)$$

We notice immediately $J^2_I(p) = -\mathrm{id}_{T^1_p(M)}$. We claim that J is actually globally defined. To see this, assume that $U_I \cap U_K \neq \emptyset$ and denote $z_K = \varphi_{IK}(z_I)$. Then, dropping tensor indices

$$\begin{aligned}
J_K[\partial/\partial x_I] &= \frac{\partial x_K}{\partial x_I} J_K[\partial/\partial x_K] + \frac{\partial y_K}{\partial x_I} J_K[\partial/\partial y_K] \\
&= \frac{\partial x_K}{\partial x_I} \partial/\partial y_K - \frac{\partial y_K}{\partial x_I} \partial/\partial x_K \\
&= \frac{\partial y_K}{\partial y_I} \partial/\partial y_K - \frac{\partial x_K}{\partial y_I} \partial/\partial x_K \\
&= \partial/\partial y_I = J_I[\partial/\partial x_I]
\end{aligned} \qquad (19.4.4)$$

and similarly for $J_K[\partial/\partial y_K]$ where in the third step we have used the Cauchy–Riemann equations. It follows that $J_0 := J_I$ is a smooth, globally defined tensor field with constant component matrix $\epsilon \otimes 1_m$ where $\epsilon_{12} = 1$, $\epsilon_{(AB)} = 0$ in the chosen coordinates.

Conversely, if M is a real $(2m)$-dimensional manifold which admits the globally defined tensor field J_0 then the diffeomorphisms between overlapping charts obey the Cauchy–Riemann equations. Thus the existence of J_0 is equivalent to the existence of a complex structure. We thus arrive at the equivalent definition:

Definition 19.4.2

(i) *A $(2m)$-dimensional real manifold M admits a complex structure if and only if it admits a smooth tensor field $J_0 \in T^1_1(M)$ with $J^2_0(p) = -\mathrm{id}_{T_p(M)}$ which in suitable coordinates has canonical component matrix $\epsilon \otimes 1_m$. We then call M a complex m-dimensional manifold with complex structure J_0.*

(ii) *An m-dimensional real manifold M with smooth tensor field $J \in T^1_1(M)$ such that $J^2(p) = -\mathrm{id}_{T_p(M)}$ is called an almost complex manifold with almost complex structure J.*

Notice that $\det(J^2(p)) = (-1)^m = [\det(J(p))]^2 > 0$, hence almost complex manifolds have even-dimension. Not every even dimensional manifold admits an almost complex structure (e.g., S^4) and not every almost complex manifold (e.g., S^6) admits a complex structure (no coordinates exist such that $J = J_0$).

Theorem 19.4.3 (Newlander and Nirenberg). *Let the Nijenhuis tensor field $N \in T^1_2(M)$ on an almost complex manifold M be defined by*

$$N[u, v] := [u, v] + J[[J[u], v]] + J[[u, J[v]]] - [J[u], J[v]] \qquad (19.4.5)$$

Then M admits a complex structure if and only if $N = 0$.

We omit the proof of this deep theorem.

Remark: If (M, J) is a complex manifold then the complex structure realises on M, viewed as a real manifold, multiplication by the imaginary unit on M, viewed as a complex manifold, as follows: every vector field $u \in T^1(M)$ can be uniquely written as $u_x \oplus u_y := (u_x, u_y)$ where u_x, u_y is spanned by ∂_x, ∂_y respectively. Let us define the bijection $I : T^1(M) \to T^1(M)_{\mathbb{C}}$ by $I(u) := u_1 + iu_2$. Then with $z = a + ib$ we have

$$zI(u) = (au_1 - bu_2) + i(au_2 + bu_1) = I((au_1 - bu_2) \oplus (au_2 + bu_1))$$
$$= I((a \, \mathrm{id} + bJ)[u_1 \oplus u_2]) = I((a \, \mathrm{id} + bJ)[u]) \qquad (19.4.6)$$

hence

$$a + ib = I \circ (a \, \mathrm{id} + bJ) \circ I^{-1} \qquad (19.4.7)$$

Definition 19.4.4

(i) *Let (M, J) be a complex manifold of complex dimension m which at the same time is a Riemannian $(2m)$-dimensional real manifold with Riemannian structure g. Then (M, J, g) is called a Hermitian manifold provided that*

$$g[J[u], J[v]] = g[u, v] \ \forall \ u, v \in T^1(M) \qquad (19.4.8)$$

Then g is called a Hermitian structure and is said to be J-compatible.

(ii) *Let (M, J, g) be a Hermitian manifold. The so-called Kähler two-form is defined by*

$$\omega[u, v] := g[J[u], v] \ \forall \ u, v \in T^1(M) \qquad (19.4.9)$$

Notice that $\omega[u, v] = -\omega[v, u]$ due to $J^2(p) = -\mathrm{id}_{T_p(M)}$ and that $\omega[J[u], J[v]] = \omega[u, v]$ so that ω is also J-compatible.

(iii) *A Kähler manifold is a Hermitian manifold (M, J, g) such that the corresponding Kähler two-form is closed. Equivalently, a Kähler manifold (M, J, ω) is a complex manifold which also carries a J-compatible symplectic structure ω.*

Hence a Kähler manifold connects the notions of complex and symplectic manifolds.

The Kähler two-form turns out to be of rather special type. In local coordinates we have with the notation $z^{\bar{\mu}} = \bar{z}^\mu$ and similarly for $dz^{\bar{\mu}}$, $\partial_{\bar{\mu}}$

$$\omega = \omega_{\mu\nu} dz^\mu \wedge dz^\nu + \omega_{\bar{\mu}\nu} d\bar{z}^\mu \wedge dz^\nu + \omega_{\mu\bar{\nu}} dz^\mu \wedge d\bar{z}^\nu + \omega_{\bar{\mu}\bar{\nu}} d\bar{z}^\mu \wedge d\bar{z}^\nu \qquad (19.4.10)$$

We have $J[\partial_\mu] = i\partial_\mu, J[\partial_{\bar\mu}] = -i\partial_{\bar\mu}$ and $2\omega_{\alpha\beta} = i\partial_\beta i\partial_\alpha\omega$ for $\alpha, \beta \in \{\mu, \nu, \bar\mu, \bar\nu\}$. From the compatibility criterion we infer $\omega_{\mu\nu} = \omega_{\bar\mu\bar\nu} = 0$. Hence (19.4.6) simplifies to

$$\omega = [\omega_{\mu\bar\nu} - \omega_{\bar\nu\mu}] \, dz^\mu \wedge d\bar z^\nu =: \Omega_{\mu\bar\nu} \, dz^\mu \wedge d\bar z^\nu \qquad (19.4.11)$$

Clearly $\Omega_{\bar\nu\mu} = -\Omega_{\mu\bar\nu}$. Reality $\omega = \bar\omega$ implies

$$\overline{\Omega_{\mu\bar\nu}} = -\Omega_{\nu\bar\mu} \qquad (19.4.12)$$

Closure $\partial_{[\alpha}\omega_{\beta\gamma]} = 0$ implies

$$\partial_{[\mu}\Omega_{\nu]\bar\rho} = \partial_{[\bar\mu}\Omega_{\bar\nu]\rho} = 0 \qquad (19.4.13)$$

By Poincaré's lemma this implies locally $\Omega_{\mu\bar\nu} = \partial_\mu f_{\bar\nu} = -\partial_{\bar\nu}g_\mu$ for some $f_{\bar\nu}, g_\mu$. Applying closure again and using holomorphicity $\partial_\mu z^\nu = \partial_{\bar\nu}z^\mu = 0$ we infer again from Poincaré's lemma that

$$\Omega_{\mu\bar\nu}(z, \bar z) = i\partial_\mu\partial_{\bar\nu}K(z, \bar z) \;\Rightarrow\; \omega = id \wedge \bar dK \qquad (19.4.14)$$

where K is called the local Kähler potential for ω. From reality we infer that K is a real-valued function which is uniquely determined by ω up to $K(z, \bar z) \to K(z, \bar z) + f(z) + g(\bar z)$ where f, g are holomorphic and antiholomorphic functions respectively.

Notice that by definition

$$g_{\mu\bar\nu} = \omega[\partial_\mu, J[\partial_{\bar\nu}]] = \partial_\mu\partial_{\bar\nu}K \qquad (19.4.15)$$

are the components of the Kähler metric. It satisfies $\overline{g_{\mu\bar\nu}} = g_{\bar\mu\nu} = g_{\nu\bar\mu}$. We can now compute the curvature tensor of g and its associated Ricci tensor whose non-vanishing components turn out to be

$$R_{\mu\bar\nu} = -\partial_\mu\partial_{\bar\nu}\ln(\det(g)) = R_{\bar\nu\mu} \qquad (19.4.16)$$

(this is independent of the branch of the logarithm chosen due to the derivatives). The associated real Ricci form is defined by $\rho = id \wedge \bar d\ln(\det(g))$. It is closed by inspection but not exact because $\det(g)$ is not a scalar. In fact, $\rho \in H^2(M)$ has non-trivial cohomology class in general, called the first Chern class. A compact Kähler manifold (M, J, ω) whose first Chern class vanishes is called a *Calabi–Yau manifold*. These manifolds play a crucial role in string theory compactifications and we cite one of the most important results needed for those.

Theorem 19.4.5 (Calabi and Yau). *A Kähler manifold (M, J, ω) which admits a Ricci flat metric h has vanishing first Chern class (of g). If M is compact and the first Chern class (of g) vanishes then M admits a Ricci flat metric h.*

20

Semianalytic category

In this chapter we define semianalytic structures and draw conclusions from those which are important for the uniqueness of the kinematical representation of LQG. Semianalytic structures are intuitively the same thing as piecewise analytic structures, that is, objects such as paths or surfaces are analytic on generic subsets but analyticity may be violated on lower-dimensional subsets. On those subsets there is again a notion of semianalyticity. This enables one to take advantage of analyticity while making the constructions local: for instance, strictly analytical paths are determined everywhere on their analytic extension once they are known on an open set, thus making them very non-local. If we make it semianalytic then these data only determine the path up to the next point where analyticity is reduced to C^n, $n > 0$. This is important because we need to make sure that certain local constructions do not have an impact on regions far away from the region of interest. We will see this explicitly in the uniqueness proof.

We will now develop elements of semianalytic differential geometry in analogy to Chapter 19. We begin with $\mathbb{R}^n$ with its canonical analytic structure. For general manifolds M we will assume that they are differential manifolds with given smooth structure and that a compatible analytic structure has been fixed. An introduction to semianalytical notions can be found in [888].

20.1 Semianalytic structures on $\mathbb{R}^n$

Definition 20.1.1. *Let $U \subset \mathbb{R}^n$ be open and $h := \{h_1, \ldots, h_N\}$ be a finite system of real-valued analytic functions h_k defined on a neighbourhood of U. Furthermore, let $\sigma = \{\sigma_1, \ldots, \sigma_N\}$ be a corresponding tuple of relators taking values in $\sigma_k \in \{=, >, <\}$. We will denote the set of those 3^N possible tuples by $\Sigma(h)$. Define*

$$U_{h,\sigma} := \{x \in U : h_k(x)\sigma_k 0; \ k = 1, \ldots, N\} \tag{20.1.1}$$

The semianalytic partition of U subordinate to h, denoted by $P(U, h)$, is the collection of all the subsets $U_{h,\sigma} \subset U$ as σ ranges through $\Sigma(h)$. We will often write $h(x)\sigma 0$ to mean $h_k(x)\sigma_k 0$ for all k and $\sigma_k := \sigma_{|h_k}$ when considered as a map $\sigma : h \to \{=, >, <\}$.

Notice that precisely one of $h_k(x) = 0$, $h_k(x) > 0$, $h_k(x) < 0$ always holds, hence $U = \cup_{\sigma \in \Sigma(h)} U_{h,\sigma}$ and $U_{h,\sigma} \cap U_{h,\sigma'} = \emptyset$ for $\sigma \neq \sigma'$. Hence we really have defined a partition. Notice also that some of the $U_{h,\sigma}$ may be empty.

Definition 20.1.2

(i) A function $f : U \subset \mathbb{R}^n \to \mathbb{R}^m$ (U open) is said to be semianalytic (s.a.) provided that for each $x \in U$ there exists an open neighbourhood V equipped with some semianalytic partition $P(V, h)$ and for each $\sigma \in \Sigma(h)$ such that $V_{h,\sigma} \neq \emptyset$ there exists an analytic function $f_\sigma : V \to \mathbb{R}^m$ such that $f_\sigma = f$ on $V_{h,\sigma}$.

(ii) If f is s.a. and $x \in \text{dom}(f)$ then a corresponding s.a. partition $P(V, h)$; $x \in V$ is called compatible with f at x.

It is instructive to show that a real-valued function on the real axis which is analytic on a partition of $\mathbb{R}$ by closed intervals $I_n = [n, n+1]$; $n \in \mathbb{Z}$ is s.a.

Lemma 20.1.3. *Suppose that* $f_1 : U \to \mathbb{R}$, $f_2 : U \to \mathbb{R}^m$ *are s.a. on open* $U \subset \mathbb{R}^n$. *Then* $f_1 \cdot f_2 : U \to \mathbb{R}^m$ *and* $f_1 \times f_2 : U \to \mathbb{R}^{m+1}$ *are s.a.*

Proof: Let $P(V^j, h^j)$ be s.a. partitions compatible with f_j at $x \in U$, $j = 1, 2$. Set $V := V_1 \cap V_2$ and $h := h^1 \cup h^2$. Then obviously $V_{h,\sigma} = V \cap V^1_{h^1,\sigma^1} \cap V^2_{h^2,\sigma^2}$ for any $\sigma = (\sigma^1, \sigma^2)$. Hence for $x \in V_{h,\sigma}$ we have $(f_1 \cdot f_2)(x) = f_{1,\sigma^1}(x) f_{2,\sigma^2}(x)$, and $(f_1 \cdot f_2)_\sigma := f_{1,\sigma^1} \cdot f_{2,\sigma^2}$ (and similar for $f_1 \times f_2$) does the job. $\square$

It follows that if $f \neq 0$ on U and f is s.a. then $1/f$ is also s.a. on U with $(1/f)_\sigma = 1/f_\sigma$.

Theorem 20.1.4. *Let* $U \subset \mathbb{R}^n$, $U' \subset \mathbb{R}^{n'}$ *be open and* $f' : U' \to \mathbb{R}^m$, $\varphi : U \to U'$ *s.a. Then* $f := f' \circ \varphi : U \to \mathbb{R}^m$ *is s.a.*

Proof: By assumption, for any $x \in U$ we find an open neighbourhood V and s.a. partition $P(V, h)$ and analytic functions φ_σ on V such that $\varphi_\sigma = \varphi$ on $V_{h,\sigma}$ for all $\sigma \in \Sigma(h)$. Likewise, for any $x' \in U'$ we find an open neighbourhood V' and s.a. partition $P(V', h')$ and analytic functions $\varphi_{\sigma'}$ on V' such that $f_{\sigma'} = f'$ on $V'_{h',\sigma'}$ for all $\sigma' \in \Sigma(h')$.

Choosing $y := \varphi(x)$, due to analyticity we may choose V so small that $\varphi_\sigma(V) \subset V'$ for all $\sigma \in \Sigma(h)$. To be explicit, suppose that $h = \{h_1, \ldots, h_N\}$, $h' = \{h'_1, \ldots, h'_{N'}\}$. We define $\tilde{h}_{k\sigma} := h'_k \circ \varphi_\sigma$, $k = 1, \ldots, N'$, $\sigma \in \Sigma(h)$ on V and $\tilde{h}_k := h_k$, $k = 1, \ldots, N$ on V. Let $\tilde{h} := \{H_{k,\sigma}\} \cup \{H_k\}$. An element $\tilde{\sigma} \in \Sigma(\tilde{h})$ is then of the form $\tilde{\sigma} = (\sigma^*(\tilde{\sigma}), \sigma(\tilde{\sigma}))$ where

$$\sigma^*(\tilde{\sigma}) = \{\tilde{\sigma}_{k,\sigma_1}\}_{k=1,\ldots,N'; \, \sigma_1 \in \Sigma(h)}, \quad \sigma(\tilde{\sigma}) = \{\tilde{\sigma}_k\}_{k=1,\ldots,N} \qquad (20.1.2)$$

We also set

$$\sigma'(\tilde{\sigma})_k := \tilde{\sigma}_{k,\sigma(\tilde{\sigma})} \qquad (20.1.3)$$

Then

$$V_{\tilde{h},\tilde{\sigma}} = \{x \in V : \{\tilde{h}_{k\sigma_1}(x)\sigma^*(\tilde{\sigma})_{k,\sigma_1} 0\}, \, \{\tilde{h}_k(x)\sigma(\tilde{\sigma})_k 0\}\}$$
$$\subset \{x \in V : \{\tilde{h}_k(x)\sigma(\tilde{\sigma})_k 0\}\} = \{x \in V : \{h_k(x)\sigma(\tilde{\sigma})_k 0\}\}$$
$$= V_{h,\sigma(\tilde{\sigma})} \qquad (20.1.4)$$

Next, using the fact that $\varphi_\sigma(V) \subset V'$ for all $\sigma \in \Sigma(h)$, in particular for $\sigma(\tilde{\sigma})$

$$\varphi_{\sigma(\tilde{\sigma})}(V_{\tilde{h},\tilde{\sigma}}) = \{\varphi_{\sigma(\tilde{\sigma})}(x) \subset V' : \{\tilde{h}_{k\sigma_1}(x)\sigma^*(\tilde{\sigma})_{k,\sigma_1}0\}, \ \{\tilde{h}_k(x)\sigma(\tilde{\sigma})_k 0\}\}$$
$$\subset \{\varphi_{\sigma(\tilde{\sigma})}(x) \in V' : \tilde{h}_{k\sigma_1}(x)\sigma^*(\tilde{\sigma})_{k,\sigma_1}0 \ \forall k = 1,\ldots,N'; \ \sigma_1 \in \Sigma(h)\}$$
$$\subset \{\varphi_{\sigma(\tilde{\sigma})}(x) \in V' : \tilde{h}_{k\sigma(\tilde{\sigma})}(x)\sigma^*(\tilde{\sigma})_{k,\sigma(\tilde{\sigma})}0 \ \forall k = 1,\ldots,N'\}$$
$$= \{\varphi_{\sigma(\tilde{\sigma})}(x) \in V' : h'_k(\varphi_{\sigma(\tilde{\sigma})}(x))\sigma'(\tilde{\sigma})_k 0 \ \forall k = 1,\ldots,N'\}$$
$$\subset \{y \in V' : h'_k(y)\sigma'(\tilde{\sigma})_k 0 \ \forall k = 1,\ldots,N'\} = V'_{h',\sigma'(\tilde{\sigma})} \qquad (20.1.5)$$

where in the third step we reduced the number of restrictions from all $\sigma_1 \in \Sigma(h)$ to only one, namely $\sigma_1 := \sigma(\tilde{\sigma})$ and in the last step we used again that $\varphi_\sigma(V) \subset V'$. We conclude that

$$\varphi^{-1}(V'_{h',\sigma'(\tilde{\sigma})}) = \{x \in U : \varphi(x) \in V'_{h',\sigma'(\tilde{\sigma})}\}$$
$$\supset \{x \in U : \varphi(x) \in \varphi_{\sigma(\tilde{\sigma})}(V_{\tilde{h},\tilde{\sigma}})\}$$
$$= \{x \in U : \varphi(x) \in \varphi(V_{\tilde{h},\tilde{\sigma}})\} = V_{\tilde{h},\tilde{\sigma}} \qquad (20.1.6)$$

where in the second step we used the inclusion (20.1.5) while in the third we used the inclusion (20.1.4) and the fact that $\varphi_{\sigma(\tilde{\sigma})} = \varphi$ on $V_{h,\sigma(\tilde{\sigma})}$.

We now want to show that the s.a. partition $P(V,\tilde{h})$ is compatible with f at x. We set

$$f_{\tilde{\sigma}} :- f'_{\sigma'(\sigma)} \circ \varphi_{\sigma(\tilde{\sigma})} \qquad (20.1.7)$$

on V. This is analytic as a composition of analytic maps and makes sense due to $\varphi_\sigma(V) \subset V'$. Then for $x \in V_{\tilde{h},\tilde{\sigma}}$

$$f_{\tilde{\sigma}}(x) = f'_{\sigma'(\tilde{\sigma})}(\varphi_{\tilde{\sigma}}(x)) = f'_{\sigma'(\tilde{\sigma})}(\varphi(x)) = f'((\varphi(x)) = f(x) \qquad (20.1.8)$$

where we used the inclusion (20.1.4) in the second step (i.e., also $x \in V_{h,\sigma(\tilde{\sigma})}$) and in the third we used the inclusion (20.1.7) (i.e., also $\varphi(x) \in V'_{h',\sigma'(\tilde{\sigma})}$). $\qquad \square$

Theorem 20.1.5. *Suppose that $\varphi : U \to U'$ with $U, U' \subset \mathbb{R}^n$ is a s.a. bijection and that for all $x \in U$ there is a s.a. partition $P(V,h)$ compatible with φ at x subject to the condition that the functions φ_σ on V are analytic injections $\varphi_\sigma : V \to U'$ with analytic inverse $\varphi_\sigma^{-1} : \varphi_\sigma(V) \to V$. Then φ^{-1} is s.a.*

Proof: Suppose $y = \varphi(x)$ and $P(V,h)$ is a s.a. partition compatible with φ at x according to the assumptions. Set $V' := \cap_{\sigma \in \Sigma(h)}\varphi_\sigma(V)$, then all the φ_σ^{-1} are well-defined on V'. Let us partition V' by the sets $\varphi(V_{h,\sigma}) \cap V'$, $\sigma \in \Sigma(h)$. Then $\varphi(V_{h,\sigma}) \cap V' \subset \varphi(V_{h,\sigma}) = \varphi_\sigma(V_{h,\sigma}) \subset \varphi_\sigma(V)$ and thus by assumption $\varphi^{-1} = \varphi_\sigma^{-1}$ on $\varphi(V_{h,\sigma}) \cap V'$.

It remains to show that the $\varphi(V_{h,\sigma}) \cap V'$ define (or can be refined to be) a s.a. partition. To that end consider the functions $h'_{k,\sigma} := h_k \circ \varphi_\sigma^{-1} : V' \to V; \ \sigma \in \Sigma(h)$ and $h = \{h_1,\ldots,h_N\}$. The functions $h'_{k,\sigma}$ are s.a. by assumption and

$$\varphi(V_{h,\sigma}) \cap V' = \{\varphi(x) \in V'; h_k(x)\sigma_k 0; \ k = 1,\ldots,N\}$$
$$\subset \{\varphi_\sigma(x) \in V'; h'_{k\sigma}(\varphi_\sigma(x))\sigma_k 0; \ k = 1,\ldots,N\} = V'_{h\circ\varphi_\sigma^{-1},\sigma} \qquad (20.1.9)$$

where in the second step we used $V' \subset \varphi_\sigma(V)$ for all $\sigma \in \Sigma(h)$. Now consider the s.a. partition $P(V', h')$. Then (20.1.9) reveals that $\varphi(V_{h,\sigma}) \cap V'$ is a union of the $V'_{h',\sigma'}$ and we have $\varphi^{-1} = (\varphi^{-1})_{\sigma'}$ on $V'_{h',\sigma'}$ where $(\varphi^{-1})_{\sigma'} := \varphi_\sigma^{-1}$ whenever $V'_{h',\sigma'} \subset \varphi(V_{h,\sigma}) \cap V'$. $\square$

Corollary 20.1.6. *If $\varphi : U \to U'$ is a s.a. C^m, $m > 0$ diffeomorphism between $U, U' \subset \mathbb{R}^n$ then φ^{-1} is s.a.*

Proof: Since φ is an injection we know that its differential $D\varphi$ is nowhere degenerate on U. Given a s.a. partition $P(V, h)$ compatible with φ at x, $D\varphi$ is nowhere degenerate on V and thus on every $V_{h,\sigma}$. Since φ_σ is analytic on V and coincides with φ on $V_{h,\sigma}$ we see that $D\varphi_\sigma$ is nowhere degenerate at least on $V_{h,\sigma}$ (and thus a neighbourhood thereof) and therefore is injective and has an analytic inverse there by the inverse function theorem. However, the $V_{h,\sigma}$ do not necessarily coincide with V and some of the $V_{h,\sigma}$ may not even contain the point x. Hence we cannot apply Theorem 20.1.5, which requires that all the $D\varphi_\sigma$ are non-degenerate on all of V.

Consider all those σ such that the closure of $V(H, \sigma)$ contains x. For those σ let S_σ be the set of points such that $D\varphi_\sigma$ is non-degenerate and let V' be the intersection of these S. Then V' contains x and for the chosen σ we have that $D\varphi_\sigma$ is non-degenerate. Moreover, the sets $V'_{h,\sigma'} = V_{h,\sigma'} \cap V'$ are empty when σ' does not belong to those σ chosen and on $V'_{h,\sigma}$ we still have $\varphi = \varphi_\sigma$ otherwise. We may now apply Theorem 20.1.5. $\square$

Definition 20.1.7. *A semi-semianalytic (s.s.a.) partition $P(U, h)$ of an open set U is analogous to a s.a. partition, just that the functions h are not required to be analytic, they just have to be s.a.*

Of course, in order to define a s.s.a. partition, one needs s.a. partitions in order to define the s.a. functions h entering $P(U, h)$. Recall that a partition is called finer than another one if every element of the coarser partition is a *finite* union of elements of the finer partition.

Lemma 20.1.8. *Let a s.s.a. partition $P(U, h)$ of open $U \subset \mathbb{R}^n$ be given. Then each $x \in U$ has a neighbourhood V admitting a s.a. partition which is finer than the restriction $P(V, h)$ of $P(U, h)$.*

Proof: Let $h = \{h_1, \ldots, h_N\}$. Since h_k is s.a. we find, for each $x \in U$, a s.a. partition $P(V_k, H^{(k)})$ compatible with h_k at x. We set $V := \cap_K V_k$, $H := \cup_k h^{(k)}$. Then $P(V, H)$ is a s.a. partition compatible with all the h_k at x. In particular, for $\sigma = (\sigma_1, \ldots, \sigma_n) \in \Sigma(H)$ with $\sigma_k \in \Sigma(H^{(k)})$ we find $V_{H,\sigma} = \cap_k V_{H^{(k)},\sigma_k}$, hence $(h_k)_\sigma = (h_k)_{\sigma_k} = h_k$ on $V_{H,\sigma}$.
   Let

$$h' := H \cup_{\sigma \in \Sigma(H)} h_\sigma, \quad h_\sigma := \{(h_k)_\sigma, \; k = 1, \ldots, N\} \qquad (20.1.10)$$

Given $\sigma' \in \Sigma(h')$ set $\sigma(\sigma') := \sigma'_{|H} \in \Sigma(H)$ and $\sigma^*_{\sigma_1} = \sigma'_{|h_{\sigma_1}} \in \Sigma(h_{\sigma_1})$ for $\sigma_1 \in \Sigma(H)$. Furthermore, we define $\tilde{\sigma}(\sigma') := \sigma^*(\sigma(\sigma')) \in \Sigma(h)$. Then

$$
\begin{aligned}
V_{h',\sigma'} &= \{x \in V : H(x)\sigma(\sigma')0,\; h_{\sigma_1}(x)\sigma^*_{\sigma_1}0 \;\forall \sigma_1 \in \Sigma(H)\} \\
&= V_{H,\sigma(\sigma')} \cap \{x \in V : h_{\sigma_1}(x)\sigma^*_{\sigma_1}0 \;\forall \sigma_1 \in \Sigma(H)\} \\
&\subset V_{H,\sigma(\sigma')} \cap \{x \in V : h_{\sigma(\sigma')}(x)\sigma^*_{\sigma(\sigma')}0\} \\
&= V_{H,\sigma(\sigma')} \cap \{x \in V : h(x)\sigma^*_{\sigma(\sigma')}0\} \\
&\subset \{x \in V : h(x)\tilde{\sigma}(\sigma')0\} = V_{h,\tilde{\sigma}(\sigma')}
\end{aligned}
\qquad (20.1.11)
$$

where in the third step we dropped the conditions involving all σ_1 other than $\sigma_1 = \sigma(\sigma')$, in the fourth we used the fact that $h_\sigma = h$ on $V_{H,\sigma}$ and in the fifth the definition of $\tilde{\sigma}(\sigma')$. It follows that every element of $P(V,h')$ is contained in an element of $P(V,h)$. $\qquad\square$

Definition 20.1.9. *A s.a. partition is called an analytic partition provided that every element of the partition is a connected, analytic submanifold.*

We now state without proof a deeper result about s.a. partitions which will be crucial for the subsequent considerations. See, for example, proposition 2.10 in [889] for a proof.

Theorem 20.1.10. *For every s.a. partition $P(U,h)$ of an open $U \subset \mathbb{R}^n$ and every $x \in U$ there exists an open neighbourhood V of x which admits an analytic partition finer than $P(V,h)$.*

20.2 Semianalytic manifolds and submanifolds

Let there be given a differential manifold M of class C^m, $m > 0$ and fix a compatible analytic structure. Recall that an atlas of M consists of a system (U_I, x_I) of charts where the U_I define a locally finite, open cover of M and $x_I : U_I \to \mathbb{R}^m$ is a homeomorphism.

Definition 20.2.1

(i) An atlas (U_I, x_I) is called s.a. provided that $x_J \circ x_I^{-1} : x_I(U_I \cap U_J) \subset \mathbb{R}^n \to x_J(U_I \cap U_J) \subset \mathbb{R}^n$ is s.a. in the sense of the previous section for all I, J with $U_I \cap U_J \neq \emptyset$.

(ii) Two s.a. atlases are called compatible if their union is again s.a. A s.a. structure on M is an equivalence class of compatible s.a. atlases. A s.a. manifold is a differential manifold of class C^m, $m > 0$ with a s.a. structure.

(iii) A map $f : M \to M'$ between s.a. manifolds is called s.a. if $x'_{I'} \circ f \circ x_I^{-1}$ is s.a. for all pairs of indices I, I' for which the composition is defined. In particular, if $M' = \mathbb{R}^{n'}$ with its natural s.a. structure then we say that f is a s.a. function on M.

(iv) An n'-dimensional s.a. submanifold of M, possibly with boundary, is a subset S such that for all $x \in S$ there exists a s.a. chart (U_I, x_I) of M with $x \in U_I$ such that

$$x_I(S \cap U_I) = \{(x^1, \dots, x^n) \in \mathbb{R}^n : x^1 = \dots = x^{n-n'}$$
$$= 0; \ 0\sigma_1 x^{n-n'+1} < 1, \dots, 0\sigma_{n'} x^n < 1\} \qquad (20.2.1)$$

where $\sigma_k \in \{<\}$ if S has no boundary and $\sigma_k \in \{<, \leq\}$ if it does have a boundary.

All of these definitions and results were to prepare for the following key result, which will be crucial in our applications.

Theorem 20.2.2. *Let S_1, S_2 be two s.a. submanifolds of a s.a. manifold M. Then, for every $p \in S_1 \cap S_2$, there exists an open neighbourhood V of x in M such that $V \cap S_1 \cap S_2$ is a finite, disjoint union of connected s.a. submanifolds.*

Proof: Let $p \in S_1 \cap S_2$. Since S_1, S_2 are s.a. submanifolds of M we find charts (U_j, x_j), $j = 1, 2$ of M with $p \in U_j$ such that $x_j(S_j \cap U_j)$ has the form (20.2.1). Notice that the set $x_j(S_j \cap U_j)$ has precisely the form V_{h_j, σ_j} for some open subset of $\mathbb{R}^n$ containing the cube $[0, 1]^n$, for certain collections h_j of analytic functions which are just the x^k, $k = 1, \dots, n$ and certain $\sigma_j \in \Sigma(h_j)$ displayed in (20.2.1). Let $U = U_1 \cap U_2$ then

$$x_2(S_1 \cap S_2 \cap U) = x_2((S_1 \cap U_1) \cap (S_2 \cap U_2)) = x_2((S_1 \cap U_1)) \cap x_2(S_2 \cap U_2)$$
$$= [x_2 \circ x_1^{-1}](x_1(S_1 \cap U_1)) \cap x_2(S_2 \cap U_2) \qquad (20.2.2)$$

where in the first step we used the fact that $x_2(A \cap B) = x_2(A) \cap x_2(B)$ since x_2 is a bijection.

Since M is s.a. the map $\varphi := x_2 \circ x_1^{-1}$ is a s.a. map $\mathbb{R}^n \to \mathbb{R}^n$ and so is its inverse by definition of a s.a. manifold. Thus $x_2(S_1 \cap S_2 \cap U)$ is of the form

$$\varphi(V_{h_1, \sigma_1}) \cap V_{h_2, \sigma_2} = V'_{h_1 \circ \varphi^{-1}, \sigma_1} \cap V'_{h_2, \sigma_2} = V'_{(h_1 \circ \varphi^{-1}, h_2), (\sigma_1, \sigma_2)} =: V'_{h, \sigma} \quad (20.2.3)$$

where V' is any open set containing both $V, \varphi(V)$.

Since φ is only s.a., the functions h defined in (20.2.3) are only s.a. and thus only define a s.s.a. partition $P(V', h)$ of V'. However, due to Lemma 20.1.8, for each $x \in V'$ we find a neighbourhood $\tilde{V}_x$ of x in $\mathbb{R}^n$ which admits a s.a. partition finer than $P(\tilde{V}_x, h)$. Choosing $x = x_2(p)$ and replacing V' by $\tilde{V}_x$ we see that V' admits a s.a. partition $P(V', h')$ for certain analytic functions h'. Next, due to Theorem 20.1.10, for every $x \in V'$ we find a neighbourhood V'_x which is a finite partition of V'_x by connected, analytic submanifolds of $\mathbb{R}^n$ and which is finer than $P(V'_x, h')$. Restricting V' to V'_x where $x = x_2(p)$ it follows that all of V' admits a finite partition by analytic, connected submanifolds of $\mathbb{R}^n$, that is, they are of the form (20.2.1) again. The inverse image of the final V' by x_2 then shows that $S_1 \cap S_2 \cap W$, $W := x_2^{-1}(V')$ admits a finite partition by connected and s.a. submanifolds of M. $\qquad \Box$

Notice that if S_1 and S_2 are contained in a compact subset of M then so is $S_1 \cap S_2$. Thus, using an open covering of $S_1 \cap S_2$ by the sets $S_1 \cap S_2 \cap W$ there must be a finite subcover and the result of the theorem applies to all of $S_1 \cap S_2$.

Two particular cases of s.a. submanifolds play a special role.

Definition 20.2.3

(i) *An edge e is a connected, oriented, one-dimensional s.a. submanifold of M with a two-point boundary.*

(ii) *A face S is a connected s.a. submanifold of M of co-dimension one without boundary whose normal bundle is equipped with an orientation.*

The normal bundle of a submanifold N of M is the bundle with base N and fibres above $p \in N$ given by the quotient spaces $T_p(M)/T_p(N)$. The latter condition just means that at each $p \in S$ the one-dimensional quotient space $T_p(M)/T_p(S)$ can be equipped with a smooth (i.e., constant) assignment of a direction ('up' or 'down'). Theorem 20.2.2 applied to edges and faces contained in compact sets shows that the intersection of edges is a finite collection of edges and isolated points, the intersection of edges and faces is a finite collection of edges and isolated points and the intersection of faces is a finite collection of faces, s.a. submanifolds of co-dimension two, ..., edges and isolated points. This is what is crucial for our applications.

Theorem 20.2.4. *Let $K \subset M$ be a compact subset of a s.a. manifold M. By compactness there exists a finite open cover $U_1, \ldots, U_N$ of K. For every such cover there are s.a. functions $e_I : M \to \mathbb{R}$ with $\mathrm{supp}(e_I) \subset U_I$ and such that $\sum_{I=1}^{N} e_I = 1$ on K. In other words, every finite open cover has a subordinate s.a. partition of unity.*

The proof follows almost that for the case of a C^m manifold: using charts, one employs smooth functions of the form $\exp(-/||x||^2)$ where $||x||$ is the Euclidean norm on $\mathbb{R}^n$. These functions are actually analytic except at the point $x = 0$ where they are C^∞. The rest is standard (see, e.g., [234]) and will be omitted.

Elements of fibre bundle theory

This chapter recalls the most important structural elements of the theory of connections and holonomies on principal fibre bundles and follows closely the excellent exposition in [337] to which the reader is referred for more details. We will abuse slightly the notation in this chapter compared with Chapter 19 in that x will denote a point in the base manifold σ and *not* its coordinates, while p denotes a point in the total space P of the fibre bundle.

21.1 General fibre bundles and principal fibre bundles

Definition 21.1.1. *A fibre bundle over a differential manifold σ with atlas $\{U_I, \varphi_I\}$ is a quintuple (P, σ, π, F, G) consisting of a differentiable manifold P (called the total space), a differentiable manifold σ (called the base space), a differentiable surjection $\pi : P \to \sigma$, a differentiable manifold F (called the typical fibre) which is diffeomorphic to every fibre $\pi^{-1}(x)$, $x \in \sigma$ and a Lie group G (called the structure group) which acts on F on the left, $\lambda : G \times F \to F$; $(h, f) \mapsto \lambda(h, f) =: \lambda_h(f)$, $\lambda_h \circ \lambda_{h'} = \lambda_{hh'}$, $\lambda_{h^{-1}} = (\lambda_h)^{-1}$. Furthermore, for every U_I there exist diffeomorphisms $\phi_I : U_I \times F \to \pi^{-1}(U_I)$, called local trivialisations, such that $\phi_{Ix} : F \to F_x := \pi^{-1}(x)$; $f \mapsto \phi_{Ix}(f) := \phi_I(x, f)$ is a diffeomorphism for every $x \in U_I$. Finally, we require that there exist maps $h_{IJ} : U_I \cap U_J \neq \emptyset \to G$, called transition functions, such that for every $x \in U_I \cap U_J \neq \emptyset$ we have $\phi_{Jx} = \phi_{Ix} \circ \lambda_{h_{IJ}(x)}$.*

Conversely, given σ, F, G and the structure functions $h_{IJ}(x)$ with given left action λ on F we can reconstruct P, π, ϕ_I as follows: define $P' = \cup_I U_I \times F$ and introduce an equivalence relation $\sim$ by saying that $(x, f) \in U_I \times F$ and $(x', f') \in U_J \times F$ for $U_I \cap U_J \neq \emptyset$ are equivalent iff $x' = x$ and $f' = \lambda_{h_{IJ}(x)}(f)$. Then $P = P'/\sim$ is the set of equivalence classes $[(x, f)]$ with respect to this equivalence relation with bundle projection $\pi([(x, f)]) := x$ and local trivialisations $\phi_I(x, f) := [(x, f)]$.

Definition 21.1.2. *Two bundles defined by the collections of tuples $\{(U_I, \phi_I)\}_I$ and $\{(U'_J, \phi'_J)\}_J$ respectively are said to be equivalent if the combined collection of tuples $\{(U_I, \phi_I), (U'_J, \phi'_J)\}_{I,J}$ defines a bundle again. A bundle automorphism is a diffeomorphism of P that maps whole fibres to whole fibres. Equivalently then, two bundles are equivalent if there exists a bundle automorphism which* *reduces to the identity on the base space. A bundle is really an equivalence class of bundles.*

Notice that the transition functions satisfy the cocycle condition $h_{IJ}h_{JK}h_{KI} = 1_G$ over $U_I \cap U_J \cap U_K$ and $h_{IJ} = h_{JI}^{-1}$ over $U_I \cap U_J$. It is crucial to realise that in general h_{IJ} is not a coboundary, that is, there are in general no maps $h_I : U_I \to G$ such that $h_{IJ}(x) = h_I(x)^{-1}h_J(x)$.

Definition 21.1.3. *A fibre bundle is called trivial if its transition function cocycle is a coboundary.*

The reason for this notation is that trivial bundles are equivalent to direct product bundles $\sigma \times F$: given transition functions ϕ_I, it may be checked that the transition functions $\phi_I'(x, f) := \phi_I(x, \lambda_{h_I(x)^{-1}}(f))$ are actually independent of the label I and thus there is only one of them. Therefore the bundle is diffeomorphic with $\sigma \times F$.

Definition 21.1.4. *A local section of P is a smooth map $s_I : U_I \to P$ such that $\pi \circ s_I = id_{U_I}$. A cross-section is a global section, that is, defined everywhere on σ.*

Definition 21.1.5. *A principal G bundle is a fibre bundle where typical fibre and structure group coincide with G. On a principal fibre bundle we may define a right action $\rho : G \times P \to P$; $\rho_h(p) := \phi_I(\pi(p), h_I(p)h)$ for $p \in \pi^{-1}(U_I)$ where $h_I : P \to G$ is uniquely defined by $(\pi(p) = x_I(p), h_I(p)) := \phi_I^{-1}(p)$. Since G acts transitively on itself from the right, this right action is obviously transitive in every fibre and fibre-preserving. $s_I^\phi(x) := \phi_I(x, 1_G)$ is called the canonical local section. Conversely, given a system of local sections s_I one can construct local trivialisations $\phi_I^s(x, h) := \rho_h(s_I(x))$, called canonical local trivialisations.*

The right action on a principal bundle is globally defined since $\phi_I(\pi(p), h_I(p)h) = \phi_J(\pi(p), h_J(p)h)$ for any $\pi(p) \in U_I \cap U_J$. Notice the identity $p = \rho_{h_I(p)}(s_I^\phi(\pi(p))) = \phi_I(\pi(p), h_I(p)) = \phi_{I\pi(p)}(h_I(p))$ for any $p \in \pi^{-1}(U_I)$. If $U_I \cap U_J \neq \emptyset$ and $p \in \pi^{-1}(U_I \cap U_J)$ this leads to $\rho_{h_I(p)}(s_I^\phi(\pi(p))) = \rho_{h_J(p)}^\phi(s_J^\phi(\pi(p)))$. Using the fact that ρ is a right action we conclude $s_J^\phi(\pi(p)) = \rho_{h_I(p)h_J(p)^{-1}}(s_I^\phi(\pi(p)))$. Since the left-hand side does not depend any longer on the point p in the fibre above $x = \pi(p)$ we conclude that we have a G-valued function $h_{IJ} : U_I \cap U_J \to G$, $x \mapsto [h_J(p)^{-1}h_I(p)]_{p \in \pi^{-1}(x)}$ where the right-hand side is independent of the point in the fibre. The functions h_{IJ} are actually the structure functions of P: by definition we have $\phi_{Ix}(h_I(p)) = \phi_{Jx}(h_J(p))$, thus $h_I(p) = (\phi_{Ix}^{-1} \circ \phi_{Jx})(h_J(p))\lambda_{h_{IJ}(x)}(h_J(p)) = h_{IJ}(x)h_J(p)$, which also shows that the left action in P reduces to left translation in the fibre coordinate.

It is central to all of fibre bundle theory that the transition functions $h_{IJ}(x)$ can in general not be written as a cocycle $h_J(x)^{-1}h_I(x)$ for functions $h_I : U_I \to G$ unless the bundle is trivial, while it is true that there exist functions $h_I : P \to G$, which we denote by the same label for simplicity, such that

$h_{IJ}(x) = [h_J(p)^{-1}h_I(p)]_{p \in \pi^{-1}(x)}$ is independent of the point $p \in \pi^{-1}(x)$ in the fibre above x. One might think that one could just define $h_I(x) := h_I(s_I(x))$ using a local section. However, this would lead with $p = s_I(x) \in \pi^{-1}(x)$ to $h_{IJ}(x) = h_J(s_I(x))^{-1}h_I(s_I(x)) \neq h_J(s_J(x))^{-1}h_I(s_I(x))$, which is not of the required form.

In a principal G-bundle it is easy to see, using transitivity of the right action of G, that triviality is equivalent with the existence of a global section. This is not the case for vector bundles which always have the global section $s_I(x) = \phi_I(x, 0)$ but may have non-trivial transition functions.

Definition 21.1.6. *A vector bundle E is a fibre bundle whose typical fibre F is a vector space. The vector bundle associated with a principal G-bundle P (where G is the structure group of E) under the left representation τ of G on F, denoted $E = P \times_\tau F$, is given by the set of equivalence classes $[(p, f)] = \{(\rho_h(p), \tau(h^{-1})f); \ h \in G\}$ for $(p, f) \in P \times F$. The projection is given by $\pi_E([(p, f)]) := \pi(p)$ and local trivialisations are given by $\psi_I(x, f) = [(s_I(x), f)]$ since $[(\rho_h(s_I(x)), f)] = [(s_I(x), \tau(h)f)] = [(s_I(x), f')]$. Transition functions result from $u = [s_J(\pi(u)), f_J(u)] = [\rho_{h_{IJ}(\pi(u))}(s_I(\pi(u)), f_J(u))] = [(s_I(\pi(u)), \tau(h_{IJ}(\pi(u)))f_J(u))] = [(s_I(\pi(u)), f_I(u))]$ and are thus given by $\tau(\rho_{IJ}(x))$.*

Conversely, given any vector bundle E we can construct a principal G-bundle P such that E is associated with it by going through the above-mentioned reconstruction process and by using the same structure group (with τ as the defining representation) acting on the fibre G by left translations and the same transition functions. A vector bundle is then called trivial if its associated principal fibre bundle is trivial. Notice that the equivalence classes in Definition 21.1.6 are with respect to the whole group G over each point $x \in \sigma$ while the transition functions used in the reconstruction process generically comprise only a discrete subset of G over each point in σ, namely at most as many as there are pairs I, J such that $x \in U_I \cap U_J$.

21.2 Connections on principal fibre bundles

Every principal fibre bundle P is naturally equipped with a vertical distribution, that is, an assignment of a subspace $V_p(P)$ of the tangent space $T_p(P)$ at each point p of P that is tangent to the fibre above $\pi(p)$. (Notice that distributions are not necessarily integrable, i.e., they do not form the tangent spaces of a submanifold of P.) These vertical distributions are generated by the fundamental vector fields v_Y associated with an element $Y \in \mathrm{Lie}(G)$ of the Lie algebra of G, which are defined through their action on functions $f \in C^\infty(P)$:

$$(v_Y[f])(p) := \left(\frac{d}{dt}\right)_{=0} f(\rho_{\exp(tY)}(p)) \qquad (21.2.1)$$

where $\exp : \mathrm{Lie}(G) \to G$ denotes the exponential map. The map $v : \mathrm{Lie}(G) \to V_p(P); Y \to v_Y$ is a Lie agebra homomorphism by construction.

The complement $H_p(P)$ of $V_p(P)$ in $T_p(P)$ is called the horizontal distribution and is one way to define a connection on P. More precisely

Definition 21.2.1. *A connection on a principal G-bundle P is a distribution of horizontal subspaces $H_p(P)$ of $T_p(P)$ such that*

(a) $H_p(P) \oplus V_p(P) = T_p(P)$ *(i.e.,* $H_p(P) \cap V_p(P) = \{0\}$, $H_p(P) \cup V_p(P) = T_p(P)$*).*

(b) *If $v(p) = v^H(p) + v^V(p)$ denotes the unique split of a smooth vector field into its horizontal and vertical components respectively, then the components are smooth vector fields again.*

(c) $H_{\rho_h(p)}(P) = (\rho_h)_* H_p(P).$

Condition (c) tells us how horizontal subspaces in the same fibre are related. Here $((\rho_h)_* v)[f] = v[(\rho_h)^* f]$ denotes the push-forward of a vector field and $(\rho_h)^* f = f \circ \rho_h$ the pull-back of a function.

An equivalent, less geometrical definition of a connection consistent with Definition 21.2.1 is as follows:

Definition 21.2.2. *A connection on a principal G-bundle P is a Lie algebra-valued one-form ω on P which projects $T_p(P)$ into $V_p(P)$, that is*

(a) $\omega(v_Y) = Y$

(b) $(\rho_h)^* \omega = \mathrm{Ad}_{h^{-1}}(\omega)$

(c) $H_p(P) = \{v \in T_p(P); i_v \omega = 0\}.$

Here $\mathrm{Ad} : G \times \mathrm{Lie}(G) \to \mathrm{Lie}(G); (x, Y) \mapsto hYh^{-1}$ denotes the adjoint action (or adjoint representation) of G on its own Lie algebra and i_v denotes the contraction of vector fields with forms. To see that both definitions are equivalent we notice that

$$((\rho_h)^* \omega)_p(v_p) = (\omega)_{\rho_h(p)}((\rho_h)_* v_p) = (\mathrm{Ad}_{h^{-1}} \omega)_p(v_p) = h^{-1} \omega_p(v_p)h \quad (21.2.2)$$

so that $v_p \in H_p(P)$ implies $(\rho_h)_* v_p \in H_{\rho_h(p)}(P)$ indeed, demonstrating that conditions (b), (c) of Definition 21.2.2 imply condition (c) of Definition 21.2.1 . Condition (a) is an additional requirement fixing an otherwise free constant factor in ω.

For practical applications it is important to have a coordinate expression for ω. To that end, let us express ω in a local trivialisation $p = \phi_I(x, h)$. Introducing matrix element indices $A, B, C, \ldots$ for group elements $h = (h_{AB})$ we have

$$v_Y^\mu(p) = \left(\frac{\partial \phi_I^\mu(x, h)}{\partial h_{AB}} (hY)_{AB} \right)_{\phi_I(x,h)=p} \quad (21.2.3)$$

where p^μ denotes the coordinates of p. Recalling the definition $(x_I(p) = \pi(p), h_I(p)) := \phi_I^{-1}(p)$ we claim that

$$(\omega_I(p))_{AB} = \mathrm{Ad}_{h_I(p)^{-1}}(\pi^* A_I)(p)_{AB} + \left(h_I(p)^{-1} \right)_{AC} dh_I(p)_{CB} \quad (21.2.4)$$

where $A_I(x)$ is a Lie(G)-valued one-form on U_I. Let us check that properties (a), (b) and (c) are satisfied.

(a) We have $(\pi^* A_I)(v_Y)_p = A_I(\pi_* v_Y)_{\pi(p)}$ but $(\pi_* v_Y)^\mu(x) = [\partial \pi^\mu(\phi_I(x,h))/\partial h_{AB}](hY)_{AB} = 0$ since $\pi(\phi_I(x,h)) = x$ is independent of the fibre coordinate h. On the other hand

$$(h_I(p)^{-1} dh_I [v_Y]_p)_{AB}$$
$$= h_I(p)_{AC}^{-1} [\partial h_I(p)_{CB}/\partial p^\mu][\partial \phi^\mu(x,h)/\partial h_{DE}(hY)_{DE}]_{p=\phi_I(x,h)}$$
$$= h_I(p)_{AD}^{-1}(h_I(p)Y)_{DB} = Y_{AB} \qquad (21.2.5)$$

where the h_{AB}, $A, B = 1, \ldots, \dim(G)$ could be treated as independent coordinates (although, depending on the group, this may not be the case) because of the chain rule. More precisely,

$$h_I(p)^{-1} dh_I [v_Y]_p = h_I(p)^{-1} [\partial h_I(p)/\partial p^\mu] \left[\left(\frac{d}{dt} \right)_{t=0} \phi^\mu(x, he^{tY}) \right]_{=\phi_I(x,h)}$$
$$= h_I(p)^{-1} \left(\frac{d}{dt} \right)_{t=0} h_I(p) e^{tY} = Y \qquad (21.2.6)$$

(b) We have $\rho_h(p) = \phi_I(\pi(p), h_I(p)h) = \phi_I(\pi(p), h_I(\rho_h(p)))$ since ρ is fibre-preserving, whence $h_I(\rho_h(p)) = h_I(p)h$. Since $(\pi^* A_I)$ depends only on $\pi(p)$ we have $(\pi^* A_I)(\rho_h(p)) = (\pi^* A_I)(p)$. Finally, since $\rho^* d = d\rho^*$ we easily find

$$(\rho_h^* \omega)(p) = \mathrm{Ad}_{h_I(p)h}(\pi^* A_I)(p) + (h_I(p)h)^{-1} dh_I(p)h = \omega(\rho_h(p)) = \mathrm{Ad}_{h^{-1}}(\omega(p)) \qquad (21.2.7)$$

as claimed.

(c) Was already checked above.

Remark: It is easy to check from the above formulae that in local coordinates a horizontal vector field has the form

$$((\phi_I)_* v_H)(x, h) = v^a(x) \left[\frac{\partial}{\partial x^a} - [A_{aI}(x)h]_{AB} \frac{\partial}{\partial h_{AB}} \right]$$

where v is a vector field on σ. This gives rise to the notion of parallel transport, see below.

Consider the pull-back of ω to σ by the canonical local section $s_I^\phi(x) = \phi_I(x, 1_G)$. Obviously $h_I(s^\phi(x)) = 1_G$ whence $((s_I^\phi)^* dh_I)(x) = d1_G = 0$ and $((s_I^\phi)^* \pi^* A_I)(x) = ((\pi \circ s_I^\phi)^* A_I)(x) = A_I(x)$ since $\pi \circ s_I = \mathrm{id}_\sigma$ for any section. We conclude

Definition 21.2.3. *The so-called connection potentials*

$$A_I = \left(s_I^\phi \right)^* \omega \qquad (21.2.8)$$

are nothing else than the pull-back of the connection by local sections.

By its very defintion, the connection ω is globally defined, therefore the above coordinate formula must be independent of the trivialisation. This implies the following identity between the potentials $A_I(x)$

$$\pi^* A_I = \pi^* \left[\mathrm{Ad}_{h_{IJ}} A_J - dh_{IJ} h_{IJ}^{-1}\right] \tag{21.2.9}$$

as one can easily verify using $(\pi^* h)_{IJ}(p) = h_I(p) h_J(p)^{-1}$. We can also pull this identity back to σ and obtain

$$A_I = \mathrm{Ad}_{h_{IJ}}(A_J) - dh_{IJ} h_{IJ}^{-1} \tag{21.2.10}$$

which is called the transformation behaviour of the connection potentials under a change of section (or trivialisation or gauge). Since the bundle P can be reconstructed from G, σ and the transition functions $h_{IJ}(x)$, we conclude that a connection can be defined uniquely by a system of pairs consisting of connection potentials and local sections (A_I, s_I) respectively, subject to the above transformation behaviour.

Definition 21.2.4. *Given a principal G-bundle P over σ and a curve c in σ we define a curve $\tilde{c}$ to be the horizontal lift of c provided that*

(i) $\pi(\tilde{c}) = c$.
(ii) $d\tilde{c}(t)/dt \in H_{\tilde{c}(t)}(P)$ for any t in the domain $[0,1]$ of the parametrisation of c.

We now show that the lift is actually unique: we know that $\tilde{c}(t) - \phi_I(c(t), h_{cI}(t)^{-1}) = \rho_{h_{cI}(t)^{-1}}(s_I^\phi(c(t)))$ for some function $h_{cI}(t)$ (to be solved for) when $c(t)$ lies in the chart U_I. It follows that

$$d\tilde{c}(t)/dt = [\partial\phi_I/\partial x^a \dot{c}^a(t) + \partial\phi_I/\partial h_{AB}(\dot{h}_{cI}(t)^{-1})_{AB}](\phi_I(x,h)) = c(\tilde{t}) \tag{21.2.11}$$

That this vector is horizontal along $\tilde{c}(t)$ means that $\omega[\dot{\tilde{c}}]_{\tilde{c}(t)} = 0$. Using $\omega = \mathrm{Ad}_{h_I^{-1}}(\pi^* A_I) + h_I^{-1} dh_I$ we find

$$\omega[\dot{\tilde{c}}]_{\tilde{c}(t)} = h_{cI}(t)\left[A_{Ia}(c(t)) h_{cI}(t)^{-1} \dot{c}^a(t) + \frac{d}{dt}(h_{cI}(t)^{-1})\right] \tag{21.2.12}$$

implying the so-called parallel transport equation (dropping the index I)

$$\dot{h}_{cI}(t) = h_{cI}(t) A_{Ia}(c(t)) \dot{c}^a(t) \tag{21.2.13}$$

which is an ordinary differential equation of first order and therefore has a unique solution by the usual existence and uniqueness theorems if we provide an initial datum $\tilde{c}(0)$. The point $\tilde{c}(1)$ is called the parallel transport of $\tilde{c}(0)$. Since the point $c(1)$ in the base is already known, the essential information is contained in the group element $h_{cI} = h_{cI}(1)$, which we will also refer to as the holonomy of A_I along c. It should be noted, however, that while $\tilde{c}(1)$ is globally defined, h_{cI} depends on the choice of the local trivialisation. In fact, under a change of trivialisation $A_I(x) \mapsto A_J(x) = -dh_{JI}(x) h_{JI}^{-1}(x) + \mathrm{Ad}_{h_{JI}(x)}(A_I(x))$ we obtain $h_{cJ} = h_{JI}(c(0)) h_{cI} h_{JI}(c(1))^{-1}$ which may be checked by inserting these formulae into the parallel transport equation with $x, c(1)$ replaced by $c(t)$

and relying on the uniqueness property for solutions of ordinary differential equations. It is easy to check that if c is within the domain of a chart, then an analytic formula for $h_c(A)$ is given by

$$h_c(A) = \mathcal{P}e^{\int_c A} = 1 + \sum_{n=1}^{\infty} \int_0^1 dt_n \int_0^{t_n} dt_{n-1} \dots \int_0^{t_2} dt_1 A(t_1) \dots A(t_n)$$

$$(21.2.14)$$

where $A(t) = A_a^j(c(t))\dot{c}^a(t)\tau_j/2$, $\tau_j/2$ is a Lie algebra basis and $\mathcal{P}$ denotes the path ordering symbol (the smallest path parameter to the left).

Definition 21.2.5. *Let V be a vector space and $\psi \in \bigwedge^n(P) \otimes V$ be a vector-valued n-form on P. The covariant derivative $\nabla\psi$ of ψ is the element of $\bigwedge^{n+1}(P) \otimes V$ defined uniquely by*

$$(\nabla\psi)_p[v_1, \dots, v_{n+1}] := d\psi_p[v_1^H, \dots, v_{n+1}^H] \qquad (21.2.15)$$

where $v_k \in T_p(P)$, v_k^H is its horizontal component and d is the ordinary exterior derivative.

This definition can be applied to the connection one-form where the vector space is given by $V = \text{Lie}(G)$.

Definition 21.2.6. *The covariant derivative of the connection one-form $\omega \in \bigwedge^1(P) \otimes Lie(P)$ is called the curvature two-form $\Omega = \nabla\omega$ of ω.*

The curvature inherits from ω the property

$$\rho_h^* \Omega = \text{Ad}_{h^{-1}}(\Omega) \qquad (21.2.16)$$

To see this, notice that the property $(\rho_h)_* H_p(P) = H_{\rho_h(p)}(P)$ of the horizontal subspaces means that $(\rho_h)_* v_p^H \in H_{\rho_h(p)}(P)$ for any $v \in T_p(P)$. Since every element of $H_{\rho_h(p)}(P)$ can be obtained this way and $(\rho_h)_*$ is a bijection we conclude $[(\rho_h)_* v_p]^H = (\rho_h)_* v_p^H$. Thus

$$(\rho_h^*\Omega)_p(u_p, v_p) = \Omega_{\rho_h(p)}((\rho_h)_* u_p, (\rho_h)_* v_p) = d\omega_{\rho_h(p)}([[(\rho_h)_* u_p]^H, [(\rho_h)_* v_p]^H)$$

$$= d\omega_{\rho_h(p)}((\rho_h)_* u_p^H, (\rho_h)_* v_p^H) = (d\rho_h^*\omega)_p(u_p^H, v_p^H)$$

$$= \text{Ad}_{h^{-1}}(d\omega_p)(u_p^H, v_p^H) = \text{Ad}_{h^{-1}}(\Omega_p)(u_p, v_p) \qquad (21.2.17)$$

Definition 21.2.7. *An element $\psi \in \bigwedge^n(P) \otimes F$ is said to be of type (τ, F) (or equivariant under ρ) for some representation τ of G on F iff $\rho_h^*\psi = \tau(h)\psi$.*

It follows that the curvature Ω is of type $(\text{Ad}, \text{Lie}(G))$.

Definition 21.2.8. *Let $\psi \in \bigwedge^m(P) \otimes Lie(G), \xi \in \bigwedge^n(P) \otimes Lie(G)$ then*

$$[\psi, \xi] := \psi \wedge \xi - (-1)^{mn} \xi \wedge \psi = \psi^j \wedge \xi^k [\tau_j, \tau_k] \in \overset{m+n}{\bigwedge} (P) \otimes Lie(G) \quad (21.2.18)$$

where τ_j is some basis of the Lie algebra of G.

Theorem 21.2.9 (Cartan structure equation)

$$\Omega = d\omega + \omega \wedge \omega \qquad (21.2.19)$$

Proof: Using the split $u = u^H + u^V$ it is clear that $\omega \wedge \omega(u,v) = \omega \wedge \omega(u^V, v^V)$ because ω_p annihilates $H_p(P)$. Notice that $[\omega, \omega] = 2\omega \wedge \omega$.

Likewise we write

$$d\omega(u,v) = d\omega(u^H, v^H) + d\omega(u^H, v^V) + d\omega(u^V, v^H) + d\omega(u^V, v^V) \qquad (21.2.20)$$

and use the differential geometric identity $d\omega(u,v) = u[i_v\omega] - v[i_u\omega] - i_{[u,v]}\omega$ with $(i_u\psi)(v_1, \ldots, v_{n-1}) := \sum_{k=1}^{n}(-1)^{k+1}\psi(v_1, \ldots, v_{k-1}, u, v_{k+1}, \ldots, v_n)$ for the contraction of an n-form with a vector field.

To evaluate these four terms in (21.2.20) we need two preliminary results:

1. We can always find $X, Y \in \text{Lie}(G)$ such that $u^V = v_X, v^V = v_Y$ are displayed as fundamental vector fields. It is easy to verify that $[u^V, v^V] = [v_X, v_Y] = v_{[X,Y]} \in V_p(P)$ is a Lie algebra homomorphism. We will exploit that $\omega(u^V) = X$, etc. is a constant.
2. By definition of the Lie bracket of vector fields $[u^V, v^H] = (d/dt)_{t=0}[\rho_{h^{u^V}(t)}]_* v^V \in H_p(P)$ since the push-forward by the right action preserves horizontal vector fields ($h^{u^V}(t)$ denotes the integral curve of u^V). We will exploit that $\omega(w^H) = 0$ for any horizontal vector field w^H.

Using these two properties it is immediate that $d\omega(u^H, v^V) = d\omega(u^V, v^H) = 0$ and that $d\omega(u^V, v^V) = -\omega([v_X, v_Y]) = -[X, Y]$. On the other hand

$$\omega \wedge \omega(u^V, v^V) = i_v v\, i_u v\, \omega \wedge \omega\, i_v v\, [\omega(u_v)\omega - \omega\omega(u_v)] = [\omega(v_X), \omega(v_Y)] = +[X, Y]$$

Therefore we are left with

$$[d\omega + \omega \wedge \omega](u,v) = d\omega(u^H, v^H) = \Omega(u,v) \qquad (21.2.21)$$

$\square$

Corollary 21.2.10 (Bianchi identity)

$$\nabla \Omega = 0 \qquad (21.2.22)$$

To prove this, use the Cartan structure equation to infer $d\Omega = d\omega \wedge \omega - \omega \wedge d\omega = \Omega \wedge \omega - \omega \wedge \Omega$ and use $\omega(u^H) = 0$ again.

Definition 21.2.11. *The local field strength* $F_I := 2s_I^*\Omega = 2[dA_I + A_I \wedge A_I]$ *is twice the pull-back by local sections of the curvature two-form.*

Using the transformation behaviour of the connection potential under a change of trivialisation it is easy to verify the corresponding change of the field strength is given by

$$F_J(x) = \text{Ad}_{h_{JI}(x)}(F_I(x)) \qquad (21.2.23)$$

whence traces of polynomials in the field strength, used in classical action principles of gauge field theories, are globally defined (gauge-invariant).

Definition 21.2.12. *Let $E = P \times_\tau F$ be a vector bundle associated with P, c a curve in σ and $\tilde{c}$ its horizontal lift which we display as above as $\tilde{c}(t) = \rho_{h_{cI}(t)^{-1}}(s_I^\phi(c(t)))$. A local section of E is then given by $S_I(x) = [(s_I^\phi(x), f_I(x))]$ where $f_I(x)$ is called the fibre section, whence*

$$S_I(c(t)) = [(s_I^\phi(c(t)), f_I(c(t)))] = [\rho_{h_{cI}(t)^{-1}}(s_I^\phi(c(t)), \tau(h_{cI}(t))f_I(c(t)))]$$
$$= [\tilde{c}(t), \tau(h_{cI}(t))f_I(c(t))] \tag{21.2.24}$$

The covariant differential of S_I along $v := \dot{c}(0)$ at $x = c(0)$ is defined by

$$(\nabla_v S_I)_x := \left[\tilde{c}(0), \left(\frac{d}{dt}\right)_{t=0} \tau(h_{cI}(t))f_I(c(t))\right] \tag{21.2.25}$$

It is easy to see, using the equivalence relation in the definition of E and the definition of the horizontal lift, that (21.2.25) is actually independent of the initial datum for $\tilde{c}$ or, equivalently, the group element h_0 in $\tilde{c}(0) = \rho_{h_0}(s_I(x), 1_G)$. Notice that multiplication of sections by scalar functions is defined by $f(x)S_I(x) = [(s_I^\phi(x), f(x)f_I(x))]$ so that the covariant differential ∇ satisfies the usual axioms for a covariant differential (Leibniz rule).

As usual, one is interested for practical calculations in coordinate expressions. To that end, consider a constant basis e_α in F and consider the special sections $S_{I\alpha}(x) := [(s_I^\phi(x), e_\alpha)]$. From the differential equation for the holonomy (21.2.13) with initial condition $h_{cI}(0) = 1_G$ we conclude

$$(\nabla_v S_{I\alpha})(x) = \left[\left(s_I^\phi(x), \left(\frac{\partial \tau(h)}{\partial h_{AB}}\right)_{h=1_G}(A_{Ia}(x))_{AB}v^a e^\alpha\right)\right]$$
$$= v^a A_{Ia}^j(x)\left[\left(s_I^\phi(x), \left(\frac{d\tau(\exp(t\tau_j))}{dt}\right)_{t=0} e_\alpha\right)\right] = v^a A_{Ia}^j(x)\tau_j^\tau S_{I\alpha}(x) \tag{21.2.26}$$

where we have abbreviated by $\tau_j^\tau = (\frac{d\tau(\exp(t\tau_j))}{dt})_{t=0}$ a basis of Lie(G) in the representation τ and have expanded $A_I = A_I^j \tau_j$ correspondingly. Using the Leibniz rule and the fact that a general section may be written as $S_I(x) = f_I^\alpha(x)S_{I\alpha}(x)$ we find

$$\nabla_v S_I = i_v\left[df_I^\alpha S_{I\alpha} + f_I^\alpha A_I^j \tau_j^\tau S_{I\alpha}\right] \tag{21.2.27}$$

This expression becomes especially familiar if we use the standard basis $(e_\alpha)^\beta = \delta_\alpha^\beta$ whence $f_I^\alpha(M e_\alpha) = M_\alpha^\beta f_I^\alpha e_\beta = (Mf_I)^\alpha e_\alpha$ for any matrix M so that

$$\nabla_v S_I = i_v\left[df_I + A_I^j \tau_j^\tau f_I\right]^\alpha S_{I\alpha} =: \left[i_v(\nabla f_I)^\alpha\right]S_{I\alpha} \tag{21.2.28}$$

We now require that $S_I = S$ is actually globally defined, which will require a certain transformation behaviour of $f_I(x)$ under a change of section. We have

$p = \rho_{h_I(p)}(s_I^\phi(x)) = \rho_{h_J(p)}(s_J^\phi(x))$ so that $s_J^\phi(x) = \rho_{h_{IJ}(x)}(s_I^\phi(x))$, thus $S_J(x) = [(s_I^\phi(x), \tau(h_{IJ}(x))f_J(x))] = S_I(x)$ requires that the fibre section transforms as

$$f_J(x) = \tau(h_{JI}(x))f_I(x) \tag{21.2.29}$$

This leads to the following covariant transformation property of its covariant derivative $(c(0) = x, \dot{c}(0) = v)$:

$$(\nabla_v f_J)(x) = i_v(df_J)_x + \left(\frac{d}{dt}\right)_{t=0} \tau(h_{cJ}(t))f_J(x)$$

$$= \tau(h_{JI}(x))[i_v(df_I)_x + \tau(h_{JI}(x))^{-1}[i_v d\tau(h_{JI})](x)f_I(x)$$

$$+ \tau(h_{JI}(x))^{-1}\left(\frac{d}{dt}\right)_{t=0} \tau(h_{JI}(x)h_{cI}(t)h_{JI}(c(t))^{-1})\tau(h_{JI}(x))f_I(x)]$$

$$= \tau(h_{JI}(x))[i_v(df_I)_x + \tau(h_{JI}(x))^{-1}[i_v d\tau(h_{JI})](x)f_I(x)$$

$$+ \left(\frac{d}{dt}\right)_{t=0} \tau(h_{cI}(t)h_{JI}(c(t))^{-1})f_I(x)$$

$$+ \left(\frac{d}{dt}\right)_{t=0} \tau(h_{JI}(c(t))^{-1})\tau(h_{JI}(x))f_I(x)]$$

$$= \tau(h_{JI}(x))[(\nabla_v f_I)(x) + \{\tau(h_{JI}(x))^{-1}[i_v d\tau(h_{JI})](x)f_I(x)$$

$$+ [i_v d\tau(h_{JI})^{-1}](x)\tau(h_{JI}(x))\}f_I(x)]$$

$$= \tau(h_{JI}(x))(\nabla_v f_I)(x) \tag{21.2.30}$$

which implies that the cross-section S has a globally defined covariant differential.

Definition 21.2.13. *A cross-section S in $E = P \times_\tau F$ is said to be parallel transported along a curve c in σ iff $(\nabla_{\dot{c}(t)} S)(c(t)) = 0$ for all $t \in [0,1]$.*

Notice that we may consider the covariant differential as a map $\nabla : \mathcal{S}(E) \to \mathcal{S}(E) \otimes \bigwedge^1(\sigma)$ where $\mathcal{S}(E)$ denotes the space of sections of E. We extend this definition to $\nabla : \mathcal{S}(E) \otimes \bigwedge^n(\sigma) \to \mathcal{S}(E) \otimes \bigwedge^{n+1}(\sigma)$ through the 'Leibniz rule'

$$\nabla(S \otimes \psi) := (\nabla S) \wedge \psi + S \otimes d\psi \tag{21.2.31}$$

In this way we can rediscover the field strength through the square of the covariant differential:

$$\nabla^2 S = \nabla^2 S_\alpha \otimes f^\alpha = \nabla[\nabla S_\alpha \otimes f^\alpha + S_\alpha \otimes df^\alpha]$$

$$= \nabla S_\alpha \otimes [df^\alpha + A^\alpha_\beta f^\beta] = S_\alpha \otimes \{A^\alpha_\gamma \wedge [df^\gamma + A^\gamma_\beta f^\beta] + d(A^\alpha_\beta f^\beta)\}$$

$$= S_\alpha \otimes [dA^\alpha_\beta + A^\alpha_\gamma \wedge A^\gamma_\beta]f^\beta = \frac{1}{2}S_\alpha \otimes F^\alpha_\beta f^\beta \tag{21.2.32}$$

22

Holonomies on non-trivial fibre bundles

The formula for the holonomy element given in Chapter 21 is correct only if the bundle is trivial or if the curve c is contained in the domain of a chart. Only the horizontal lift is globally defined in any bundle however, in non-trivial bundles the horizontal lifts are not naturally identified with elements of G. Since, however, the definition and the topology on $\overline{A}$ depend crucially on such an identification, we must generalise the definition of $\overline{A} = \text{Hom}(\mathcal{P}, G)$ to non-trivial bundles and provide such an identification of horizontal lifts with elements of G. In what follows we will describe two complementary ways for doing this.

22.1 The groupoid of equivariant maps

We will describe here a possibility that avoids the local connection potentials, following the elegant description in [482,483]. Let ω be a connection in a principal G-bundle P over σ. For any path $c \in \mathcal{P}$ we can construct the horizontal lift $\tilde{c}_u^\omega : [0,1] \to P$ with initial condition $\tilde{c}_u^\omega(0) = u$ as in Chapter 21. We thus obtain a map

$$F_c^\omega : P_{b(c)} \to P_{f(c)}; \quad u \mapsto \tilde{c}_u^\omega(1) \tag{22.1.1}$$

where $b(c) = c(0)$, $f(c) = c(1)$ denote the beginning and final point of c. We claim that the map (22.1.1) is equivariant with respect to the right action ρ in P, that is,

$$\rho_g \circ F_c^\omega = F_c^\omega \circ \rho_g \tag{22.1.2}$$

for all $g \in G$. To see this, we will show that actually $\rho_g(\tilde{c}_u^\omega(t)) = \tilde{c}_{\rho_g(u)}^\omega(t)$ for all $t \in [0,1]$, from which (22.1.2) follows for $t = 1$. Clearly both curves start at $\rho_g(u)$ and both are lifts of the base curve c since $\pi \circ \rho_g = \pi$ is fibre-preserving. Thus, we just have to check that $\rho_g(\tilde{c}_u^\omega)$ is horizontal. This follows from the simple calculation

$$\omega \left[\frac{d}{dt} \rho_g(\tilde{c}_u^\omega(t)) \right] = \omega \left[(\rho_g)_* \frac{d}{dt} \tilde{c}_u^\omega(t) \right] = (\rho_g)^* \omega \left[\frac{d}{dt} \tilde{c}_u^\omega(t) \right]$$

$$= \text{Ad}_{g^{-1}} \left[\omega \left[\frac{d}{dt} \tilde{c}_u^\omega(t) \right] \right] = 0 \tag{22.1.3}$$

where in the last step we have used property (b) of Definition 21.2.2 and that $\tilde{c}_u^\omega$ is horizontal. Since the horizontal lift is unique, the claim follows.

Definition 22.1.1. *By Eq(x, y) we denote the set of ρ-equivariant maps F :* $\pi^{-1}(x) \to \pi^{-1}(y)$.

Let us now choose a reference set of points $\mathcal{R} = \{u_x^0\}_{x \in \sigma}$ where $u_x^0 \in \pi^{-1}(x)$. Using this reference set, we set up a bijection between Eq(x, y) and G as follows:

1. Given $F \in$ Eq(x, y) and $u_x \in \pi^{-1}(x)$ we find unique elements $g(u_x), g_F \in$ G such that $u_x = \rho_{g(u_x)}(u_x^0)$ and $F(u_x^0) = \rho_{g_F^{-1}}(u_y^0)$. Then, since F is equivariant

$$F(u_x) = \rho_{g(u_x)}(F(u_x^0)) = \rho_{g_F^{-1}g(u_x)}(u_y^0) \qquad (22.1.4)$$

so that F is completely characterised by $g_F \in$ G (the map $u_x \to g(u_x)$ is the same for any F).

2. Conversely, given $g \in$ G we define $F_g \in$ Eq(x, y) by $F_g(u_x) := \rho_{g^{-1}g(u_x)}(u_y^0)$. To see that F_g is equivariant we notice that by definition $\rho_{g(\rho_h(u_x))}(u_x^0) = \rho_h(u_x)$, hence $\rho_{g(\rho_h(u_x))h^{-1}}(u_x^0) = u_x = \rho_{g(u_x)}(u_x^0)$ so that $g(\rho_h(u_x)) = g(u_x)h$. Therefore

$$F_g(\rho_h(u_x)) = \rho_{g^{-1}g(\rho_h(u_x))}(u_y^0) = \rho_{g^{-1}g(u_x)h}(u_y^0)$$
$$= \rho_h(\rho_{g^{-1}g(u_x)}(u_y^0)) = \rho_h(F_g(u_x)) \qquad (22.1.5)$$

As an immediate consequence of this bijection we see that equivariant maps are invertible (since all elements of G are). Thus, the reference set $\mathcal{R}$ enables us to identify Eq(x, y) with G. If we import the topology from G to Eq(x, y) then Eq(x, y) is a compact Hausdorff space and we get a homeomorphism

$$\phi_{\mathcal{R}} : \text{Eq}(x, y) \to \text{G}; \ F \mapsto g_F \qquad (22.1.6)$$

Notice that for $F = F_h \in$ Eq(x, y), $F' = F_{h'} \in$ Eq(y, z) we have (notice that $y(u_x^0) = 1$ and $y(\rho_h(u_x)) = y(u_x)h$)

$$F_{h'} \circ F_h(u_x) = \rho_{(h')^{-1}y(F_h'(u_x))}(u_z^0) = \rho_{(h')^{-1}g(\rho_{h^{-1}g(u_x)}(u_y^0))}(u_z^0)$$
$$= \rho_{(h')^{-1}g(u_y^0)h^{-1}g(u_x)}(u_z^0) = \rho_{(hh')^{-1}g(u_x)}(u_z^0) = F_{hh'}(u_x) \qquad (22.1.7)$$

Next, if F is equivariant so is F^{-1}, since inversion of $F \circ \rho_h = \rho_h \circ F$ gives $F^{-1} \circ \rho_{h^{-1}} = \rho_{h^{-1}} \circ F^{-1}$ for all $h \in$ G. Thus F_h^{-1} may be written in the form $F_{h'}$ for some h'. To compute h' we use the definition $F_h^{-1}(u_y) = F_{h'}(u_y) = \rho_{(h')^{-1}g(u_y)}(u_x^0)$ and evaluate it at $u_y = F_h(u_x)$. Noticing that $g(F_h(u_x)) = h^{-1}g(u_x)$ we find

$$u_x = \rho_{g(u_x)}(u_x^0) = \rho_{(h')^{-1}h^{-1}g(u_x)}(u_x^0) \qquad (22.1.8)$$

whence $h' = h^{-1}$. Summarising

$$F_h^{-1} = F_{h^{-1}} \text{ and } F_{h'} \circ F_h = F_{hh'} \qquad (22.1.9)$$

so inversion and composition of equivariant maps corresponds to inversion and composition of the corresponding group elements. This is precisely the generalisation of the algebraic characterisation of holonomies.

Let us study the effect of a change of the reference set on this identification: consider a different reference set $\widehat{\mathcal{R}} = \{\hat{u}_x^0\}_{x \in \sigma}$. There is a unique collection of group elements $\{g_x\}_{x \in \sigma}$ such that $\rho_{g_x^{-1}}(u_x^0) = \hat{u}_x^0$. Then for any $F \in \mathrm{Eq}(x, y)$ we have

$$F(\hat{u}_x^0) = \rho_{g_x^{-1}}(F(u_x^0)) = \rho_{g_x^{-1}g_x^{-1}}(u_y^0) = \rho_{g_F^{-1}g_x^{-1}}(\rho_{g_y}(\hat{u}_y^0))$$

$$= \rho_{g_y g_F^{-1} g_x^{-1}}(\hat{u}_y^0) =: \rho_{\hat{g}_F^{-1}}(\hat{u}_y^0) \tag{22.1.10}$$

from which we infer $\hat{g}_F = g_x g_F g_y^{-1}$. In other words, $\phi_{\widehat{\mathcal{R}}} = g_x \phi_{\mathcal{R}} g_y^{-1}$, however, this is a homeomorphism and does not change the topology of $\mathrm{Eq}(x, y)$. Relation (22.1.10) corresponds precisely to a gauge transformation induced by a change of reference set $\mathcal{R}$.

It is important to notice that the set $\mathrm{Eq}(x, y)$ no longer depends on P but just on G because any fibre $\pi^{-1}(x)$ is naturally identified with G by means of the reference point u_x^0. This is also easy to see from the fact that the transition functions of P do not play any role in the definition of $\mathrm{Eq}(x, y)$.

Definition 22.1.2. *We define the disjoint union*

$$\mathrm{Eq}(\sigma) := \cup_{x, y \in \sigma} \mathrm{Eq}(x, y) \tag{22.1.11}$$

The quantity $\mathrm{Eq}(\sigma)$ carries naturally the structure of a groupoid: the objects of this category are the fibres $\pi^{-1}(x)$, $x \in \sigma$ and its morphisms are the elements of $\hom(\pi^{-1}(x), \pi^{-1}(y)) := \mathrm{Eq}(x, y)$. Composition of morphisms is just defined by composition of equivariant maps $F \in \mathrm{Eq}(x, y)$, $F' \in \mathrm{Eq}(y, z)$, which gives indeed an element $F' \circ F \in \mathrm{Eq}(x, z)$ since $F' \circ F \circ \rho_g = F' \circ \rho_g \circ F = \rho_g \circ F' \circ F$. Associativity follows from associativity of composition of maps. Every morphism is invertible since every element of G is, that is, we just use the correspondence $F \leftrightarrow g_F$, $g_F^{-1} \leftrightarrow F^{-1}$. Finally identities $\mathrm{id}_{\pi^{-1}(x)} \in \mathrm{Eq}(x, x)$ are provided by the identical maps in the fibres.

The crucial observation is now that every connection ω in any principal G-bundle P defines a groupoid homomorphism $F^\omega \in \mathrm{Hom}(\mathcal{P}, \mathrm{Eq}(\sigma))$ through (22.1.1), that is,

$$F^\omega(c) := F_c^\omega \in \mathrm{Eq}(b(c), f(c)) \tag{22.1.12}$$

This is clear from

$$[F^\omega(c \circ c')](u_{b(c)}) = \widetilde{c \circ c'}_{u_{b(c)}}^\omega(1) = \tilde{c}_{\tilde{c}'_{u_{b(c)}}^\omega(1)}^\omega(1) = F_c^\omega(F_{c'}^\omega(u_{b(c)}))$$

$$= [F^\omega(c') \circ F^\omega(c)](u_{b(c)})$$

$$[F^\omega(c^{-1})](u_{b(c^{-1})}) = \widetilde{c^{-1}}_{u_{b(c^{-1})}}^\omega(1) = [F^\omega(c)]^{-1}(u_{b(c^{-1})}) \tag{22.1.13}$$

since $u_{b(c^{-1})} = F_c^\omega(\widetilde{c^{-1}}_{u_{b(c^{-1})}}^\omega(1))$. We may then define the holonomy group element by $F^\omega(c) = F_c^\omega =: F_{h_c(\omega)}$ which then from (22.1.9) evidently satisfies

$$h_c(\omega) h_{c'}(\omega) = h_{c \circ c'}(\omega), \quad h_{c^{-1}}(\omega) = h_c(\omega)^{-1} \tag{22.1.14}$$

Thus, the space of connections from any bundle P is contained in the universal set

$$\overline{\mathcal{A}} := \mathrm{Hom}(\mathcal{P}, \mathrm{Eq}(\sigma)) \qquad (22.1.15)$$

which is a closed subset of the compact Hausdorff space (in the Tychonov topology) $\prod_{c \in \mathcal{P}} \mathrm{Eq}(b(c), f(c))$. From here on the constructions of Section 8.2 proceed in exactly the way depicted there, just that G is replaced by $\mathrm{Eq}(\sigma)$. Alternatively, observation (22.1.14) provides a canonical way to identify $\mathrm{Eq}(\sigma)$ with G.

22.2 Holonomies and transition functions

The description of the holonomy group elements in Section 22.1 does not use local connection potentials, which makes it more elegant. However, in order to use the definition of the Poisson bracket on the space of connections $\mathcal{A}$ we must know how to express the holonomy as a function of the connection potentials. This can easily be done if we take the transition functions of the bundle P into account.

We start again from the definition of the horizontal lift: suppose the path c is not contained in the domain of a single chart. Then we find $N < \infty$ labels $I_1, \ldots, I_N$ and a breaking of c into segments $c = c_1 \circ \ldots \circ c_N$ such that $c_k \subset U_{I_k}$, $k = 1, \ldots, N$. Then, using the local trivialisations ϕ_{I_k}, we may write the horizontal lift of c with initial condition $u \in \pi^{-1}(b(c))$ over U_{I_k} in the form

$$\tilde{c}_u^w(t) = \phi_{I_k}\big(c(t), H_{I_k,u,c}^{-1}(t)\big) \qquad (22.2.1)$$

where w.l.g. we assume that $c : [0, N] \to \sigma$, $c_k : [k-1, k] \to \sigma$. Now the breakpoint $c(k) = f(c_k) = b(c_{k+1})$, $k = 1, \ldots, N-1$ lies in $U_{I_k} \cup U_{I_{k+1}}$ and using $s_I^\phi(x) := \phi_I(x, 1)$, $\phi_I(x, h) = \rho_h(s_I^\phi(x))$ we conclude

$$\phi_{I_k}\big(c(k), H_{I_k,u,c}^{-1}(k)\big) = \rho_{H_{I_k,u,c}^{-1}(k)}\big(s_{I_k}^\phi(c(k))\big) = \phi_{I_{k+1}}\big(c(k), H_{I_{k+1},u,c}^{-1}(k)\big)$$

$$= \rho_{H_{I_{k+1},u,c}^{-1}(k)}\big(s_{I_{k+1}}^\phi(c(k))\big) \qquad (22.2.2)$$

Now the transition functions of P over $U_I \cup U_J$ were defined by

$$u = \phi_I(\pi(u), g_I(u)) = \rho_{g_I(u)}\big(s_I^\phi(\pi(u))\big) = \phi_J(\pi(u), g_J(u)) = \rho_{g_J(u)}\big(s_J^\phi(\pi(u))\big)$$

$$\Rightarrow s_J^\phi(\pi(u)) = \rho_{g_I(u)g_J(u)^{-1}}\big(s_I^\phi(\pi(u))\big) =: \rho_{g_{IJ}(\pi(u))}\big(s_I^\phi(\pi(u))\big) \qquad (22.2.3)$$

We conclude the important relation

$$g_{I_k I_{k+1}}(c(k)) = H_{I_k,u,c}^{-1}(k) H_{I_{k+1},u,c}^{-1}(k) \qquad (22.2.4)$$

Now the $H_{I_k,u,c}(t)$ satisfy the ordinary differential equation

$$\frac{d}{dt} H_{I_k,u,c}(t) = H_{I_k,u,c}(t) A_{I_k a}(c(t)) \dot{c}^a(t) \qquad (22.2.5)$$

where $A_I = (s_I^\phi)^* \omega$ is the pull-back of ω by the local sections and the initial value $H_{I_k,u,c}(k)$ is defined by $\tilde{c}_u^\omega(k) = \phi_{I_k}(c(k), H_{I_k,u,c}(k))$. Thus we write the solution to (22.2.5) in the form

$$H_{I_k,u,c}(t) = H_{I_k,u,c}(k) h_{I_k,c}(t) \tag{22.2.6}$$

where $h_{I_k,c}(t) = \mathcal{P} \exp(\int_k^t ds A_{I_k a}(c(t)) \dot{c}^a(t))$. We can now define the holonomy of the connection ω along the whole curve c as the group element

$$h_c^{\{I\}}(\omega) := H_{I_1,u,c}(0)^{-1} H_{I_N,u,c}(N) \tag{22.2.7}$$

where the superscript $\{I\}$ is to denote the dependence of this group element on the chosen trivialisation. Notice that by construction (22.2.7) no longer depends on $u \in \pi^{-1}(b(c))$. Using the relations (22.2.4) and (22.2.6) we can write the holonomy in terms of the local holonomies $h_{c_k}(A_{I_k}) := h_{I_k,c}(1)$ and the transition functions

$$\begin{aligned}
h_c^{\{I\}}(\omega) &= H_{I_1,u,c}(0)^{-1} H_{I_N,u,c}(N-1) h_{c_N}(A_{I_N}) \\
&= H_{I_1,u,c}(0)^{-1} H_{I_{N-1},u,c}(N-1) [H_{I_{N-1},u,c}^{-1}(N-1) H_{I_N,u,c}(N-1)] h_{c_N}(A_{I_N}) \\
&= H_{I_1,u,c}(0)^{-1} H_{I_{N-1},u,c}(N-1) g_{I_{N-1},I_N}(c(N-1)) h_{c_N}(A_{I_N}) \\
&= H_{I_1,u,c}(0)^{-1} H_{I_{N-1},u,c}(N-2) h_{c_{N-1}}(A_{I_{N-1}}) g_{I_{N-1},I_N}(c(N-1)) h_{c_N}(A_{I_N}) \\
&= h_{c_1}(A_{I_1}) g_{I_1 I_2}(f(c_1)) h_{c_2}(A_{I_2}) g_{I_2 I_3}(f(c_2)) \cdots g_{I_{N-1} I_N}(f(c_{N-1})) h_{c_N}(A_{I_N})
\end{aligned} \tag{22.2.8}$$

We should check that (22.2.9) is invariant, up to gauge transformations, under a change of breaking the curve into segments according to a different choice of charts. Thus, consider the same curve broken up into N' segments c_l', $l = 1, \ldots, N'$, that is, $c = c_1' \circ \ldots \circ c_{N'}'$ where $c_l' \subset U_{I_l'}$. Then

$$h_c^{\{I'\}}(\omega) = h_{c_1'}(A_{I_1'}) g_{I_1' I_2'}(f(c_1')) h_{c_2'}(A_{I_2'}) g_{I_2' I_3'}(f(c_2')) \cdots g_{I_{N'-1}' I_{N'}'}(f(c_{N'-1}')) h_{c_{N'}'}(A_{I_{N'}'}) \tag{22.2.9}$$

In order to compare (22.2.8), (22.2.9) we introduce the common split of c into $c = \tilde{c}_1 \circ \ldots \circ \tilde{c}_{\tilde{N}}$, $\tilde{N} \geq \max(N, N')$ where for each $k = 1, \ldots, N$ and $l = 1, \ldots, N'$ we have unique compositions $c_k = \tilde{c}_{n_{k-1}+1} \circ \ldots \circ \tilde{c}_{n_k}$ and $c_l' = \tilde{c}_{n_{l-1}'+1} \circ \ldots \circ \tilde{c}_{n_l'}$ with $n_0 = n_0' = 0$ and $n_N = n_{N'}' = \tilde{N}$. Notice that each of the $\tilde{c}_j$ is contained in a set of the form $U_{I_k} \cap U_{I_l'}$. Let us first of all write both (22.2.8) and (22.2.9) in terms of the $h_{\tilde{c}_j}(A_{I_k})$ and $h_{\tilde{c}_j}(A_{I_l'})$ respectively, namely

$$\begin{aligned}
h_c^{\{I\}}(\omega) &= \left[h_{\tilde{c}_1}(A_{I_1}) \ldots h_{\tilde{c}_{n_1}}(A_{I_1}) \right] g_{I_1 I_2}(f(\tilde{c}_{n_1})) \cdots g_{I_{N-1} I_N}(f(\tilde{c}_{n_{N-1}})) \\
&\quad \times \left[h_{\tilde{c}_{n_{N-1}+1}}(A_{I_N}) \ldots h_{\tilde{c}_{n_N}}(A_{I_N}) \right] \\
h_c^{\{I'\}}(\omega) &= \left[h_{\tilde{c}_1}(A_{I_1'}) \ldots h_{\tilde{c}_{n_1'}}(A_{I_1'}) \right] g_{I_1' I_2'}(f(\tilde{c}_{n_1'})) \cdots g_{I_{N'-1}' I_{N'}'}(f(\tilde{c}_{n_{N'-1}'})) \\
&\quad \times \left[h_{\tilde{c}_{n_{N'-1}'+1}}(A_{I_{N'}'}) \ldots h_{\tilde{c}_{n_{N'}'}}(A_{I_{N'}'}) \right]
\end{aligned} \tag{22.2.10}$$

Now we exploit the gauge transformation behaviour $h_{\tilde{c}}(A_{I'}) = g_{I'I}(b(\tilde{c}))h_{\tilde{c}}(A_I)$ $g_{I'I}(f(\tilde{c})^{-1})$ for $\tilde{c} \subset U_I \cap U_{I'}$ and the cocycle conditions $g_{IJ}(x)g_{JK}(x)g_{KI}(x) = 1, x \in U_I \cap U_J \cap U_K; g_{IJ}(x)g_{JI}(x) = 1, \ x \in U_I \cap U_J$ in order to replace the $A_{I'}$ by the A_I. The general case requires a tedious case-by-case analysis but the typical situation is already captured by the following simple example: $N = N' = 2, \ c_1 = \tilde{c}_1, c_2 = \tilde{c}_2 \circ \tilde{c}_3, c'_1 = \tilde{c}_1 \circ \tilde{c}_2, c'_2 = \tilde{c}_3$. Thus $\tilde{c}_1 \subset U_{I_1} \cap U_{I'_1}$, $\tilde{c}_2 \subset U_{I_2} \cap U_{I'_1}, \tilde{c}_3 \subset U_{I_2} \cap U_{I'_2}$. Then

$$h_c^{\{I'\}}(\omega) = h_{c'_1}(A_{I'_1})g_{I'_1 I'_2}(f(c'_1))h_{c'_2}(A_{I'_2})$$

$$= h_{\tilde{c}_1}(A_{I'_1})h_{\tilde{c}_2}(A_{I'_1})g_{I'_1 I'_2}(f(\tilde{c}_2))h_{\tilde{c}_3}(A_{I'_2})$$

$$= [g_{I'_1 I_1}(b(\tilde{c}_1))h_{\tilde{c}_1}(A_{I_1})g_{I'_1 I_1}(f(\tilde{c}_1))^{-1}][g_{I'_1 I_2}(b(\tilde{c}_2))h_{\tilde{c}_2}(A_{I_2})g_{I'_1 I_2}(f(\tilde{c}_2))^{-1}]$$

$$\times g_{I'_1 I'_2}(f(\tilde{c}_2))[g_{I'_2 I_2}(b(\tilde{c}_3))h_{\tilde{c}_3}(A_{I_2})g_{I'_2 I_2}(f(\tilde{c}_3))^{-1}]$$

$$= g_{I'_1 I_1}(b(c))h_{\tilde{c}_1}(A_{I_1})[g_{I'_1 I_1}(f(\tilde{c}_1))^{-1}g_{I'_1 I_2}(b(\tilde{c}_2))]h_{\tilde{c}_2}(A_{I_2})$$

$$\times [g_{I'_1 I_2}(f(\tilde{c}_2))^{-1}g_{I'_1 I'_2}(f(\tilde{c}_2))g_{I'_2 I_2}(b(\tilde{c}_3))]h_{\tilde{c}_3}(A_{I_2})g_{I'_2 I_2}(f(c))^{-1}$$

$$= g_{I'_1 I_1}(b(c))h_{\tilde{c}_1}(A_{I_1})[g_{I_2 I_1}(f(\tilde{c}_1))g_{I'_1 I_2}(f(\tilde{c}_1))g_{I_2 I_1}(f(\tilde{c}_1))]g_{I_1 I_2}(f(\tilde{c}_1))h_{\tilde{c}_2}(A_{I_2})$$

$$\times [g_{I_2 I'_1}(f(\tilde{c}_2))g_{I'_1 I'_2}(f(\tilde{c}_2))g_{I'_2 I_2}(f(\tilde{c}_2))]h_{\tilde{c}_3}(A_{I_2})g_{I'_2 I_2}(f(c))^{-1}$$

$$= g_{I'_1 I_1}(b(c))h_{\tilde{c}_1}(A_{I_1})g_{I_1 I_2}(f(\tilde{c}_1))h_{\tilde{c}_2}(A_{I_2})h_{\tilde{c}_3}(A_{I_2})g_{I'_2 I_2}(f(c))^{-1}$$

$$= g_{I'_1 I_1}(b(c))h_{c_1}(A_{I_1})g_{I_1 I_2}(f(\tilde{c}_1))h_{c_2}(A_{I_2})g_{I'_2 I_2}(f(c))^{-1}$$

$$= g_{I'_1 I_1}(b(c))h_c^{\{I\}}(\omega)g_{I'_2 I_2}(f(c))^{-1} \tag{22.2.11}$$

Thus, indeed the holonomy as defined in (22.2.7) on a non-trivial bundle transforms under a change of trivialisation by a gauge transformation at the endpoints of the path. These local gauge transformations, moreover, depend only on the charts in which the endpoints lie, so that $h_c^{\{I\}}(\omega)$ does not depend on all of the ϕ_I but only on the particular ϕ_I which have been chosen to trivialise over regions containing the endpoints of the path.

In order to remove this dependence of the gauge transformations on the chosen charts of an atlas in which the endpoints of the curve lie, one can proceed as follows: if $\mathcal{I}$ denotes the set of indices I which label the charts of the atlas we use the axiom of choice[1] in order to choose a map $I : \sigma \to \mathcal{I}; x \mapsto I(x)$. Now one proceeds as above, just that one subdivides the curve into the segments c_k for which $I(x) = $ const. for all $x \in c_k$. In fact, one can partition σ by sets V_I defined by the condition $I(x) = I \ \forall \ x \in V_I$.

One way of providing such a choice function is as follows. For a given atlas $\mathcal{U}$ of open sets covering σ we choose a partition of σ subordinate to $\mathcal{U}$ as follows: choose some I and define V_I to be the closure of U_I. Redefine σ by $\sigma - V_I$, for

---

[1] The axiom of choice states that given any collection of sets $S_I \neq \emptyset, \ I \in \mathcal{I}$ where $\mathcal{I}$ is an index set of arbitrary cardinality, then there exists a choice function $c : \mathcal{I} \to \prod_{I \in \mathcal{I}} S_I$, that is, it is possible to choose an element $c(I)$ from each S_I. The axiom of choice is equivalent to Zorn's lemma.

$J \neq I$ redefine U_J by $U_J - V_I$ and redefine $\mathcal{U}$ by the collection of the redefined U_J. Then iterate (notice that the closure of the redefined U_J in the redefined σ cannot contain points of V_I any more). The end result is now a partition $\mathcal{V}$ of σ by sets V_I which have the property that they have empty intersection and such that V_I equals a certain closure in σ of a subset of a non-empty U_I. Thus for each $x \in \sigma$ there is a unique index $I = I(x)$ such that $x \in V_I$. Then each curve can be broken uniquely into the segments $c \cap V_I$. Formula (22.2.8) can then be applied with the corresponding choice of I. This being understood, we can now drop the superscript $\{I\}$ from $h_c^{\{I\}}(w)$.

Let us check the algebraic properties of (22.2.7). We have

$$h_{c^{-1}}(w) = h_{c_N^{-1}}(A_{I_N}) g_{I_N I_{N-1}}(f(c_N^{-1})) \cdots g_{I_2 I_1}(f(c_2^{-1})) h_{c_1^{-1}}(A_{I_1})$$

$$= h_{c_N}(A_{I_N})^{-1} g_{I_{N-1} I_N}(f(c_{N-1}))^{-1} \cdots g_{I_1 I_2}(f(c_1)) h_{c_1}(A_{I_1})^{-1}$$

$$= (h_c(w))^{-1} \tag{22.2.12}$$

Next, let $c = c_1 \circ c_2$. Then $x = f(c_1) = b(c_2)$ lies in a unique V_I. There are three cases to consider: (1) x is an interior point of V_I, (2) x is a boundary point of V_I and $c_2 \cap V_I = \{x\}$ and (3) x is a boundary point of V_I and $c_1 \cap V_I = \{x\}$.

Let $\tilde{c}_1 = c_1 \cap V_I, \tilde{c}_2 = c_2 \cap V_I, \tilde{c} = c \cap V_I = \tilde{c}_1 \circ \tilde{c}_2$. We have in general $h_{c_1}(w) = h'_{c_1}(w) g_{I_1 I}(b(\tilde{c}_1)) h_{\tilde{c}_1}(A_I), h_{c_2}(w) = h_{\tilde{c}_2}(A_I) g_{I I_2}(f(\tilde{c}_2)) h'_{c_2}(w)$ where the pieces $h'_{c_1}(w), h'_{c_1}(w)$ do not involve transition functions or holonomies that depend on I.

1. $h_{c_1}(w) h_{c_2}(w) = h'_{c_1}(w) g_{I_1 I}(b(\tilde{c}_1)) [h_{\tilde{c}_1}(A_I) h_{\tilde{c}_2}(A_I)] g_{I I_2}(f(\tilde{c}_2)) h'_{c_2}(w)$

$$= h'_{c_1}(w) g_{I_1 I}(b(\tilde{c}_1)) h_{\tilde{c}}(A_I) g_{I I_2}(f(\tilde{c}_2)) h'_{c_2}(w)$$

$$= h_c(w) \tag{22.2.13}$$

2. $h_{c_1}(w) h_{c_2}(w) = h'_{c_1}(w) g_{I_1 I}(b(\tilde{c}_1)) [h_{\tilde{c}_1}(A_I)] g_{I I_2}(f(\tilde{c}_2)) h'_{c_2}(w)$

$$= h'_{c_1}(w) g_{I_1 I}(b(\tilde{c}_1)) h_{\tilde{c}}(A_I) g_{I I_2}(f(\tilde{c}_2)) h'_{c_2}(w)$$

$$= h_c(w) \tag{22.2.14}$$

since $h_{\{x\}}(A_I) = 1$ and $\tilde{c} = \tilde{c}_1$.

3. $h_{c_1}(w) h_{c_2}(w) = h'_{c_1}(w) g_{I_1 I}(b(\tilde{c}_1)) [h_{\tilde{c}_2}(A_I)] g_{I I_2}(f(\tilde{c}_2)) h'_{c_2}(w)$

$$= h'_{c_1}(w) g_{I_1 I}(b(\tilde{c}_1)) h_{\tilde{c}}(A_I) g_{I I_2}(f(\tilde{c}_2)) h'_{c_2}(w)$$

$$= h_c(w) \tag{22.2.15}$$

since $\tilde{c} = \tilde{c}_2$.

We conclude that the algebraic properties of the holonomy as defined in (22.2.7) equal those of the case of a trivial bundle.

Finally, consider the case of a change of choice function $x \mapsto I(x)$. Let the corresponding partitions be denoted by V_I and V'_J respectively. Then each $x \in \sigma$ lies in a unique $V_{I(x)}$ and in a unique $V'_{J(x)}$. It follows that the transition functions appearing in (22.2.11) can be denoted as $g(x) := g_{I(x), J(x)}(x)$ for $x = b(c), f(c)$

and now depend only on x. Thus, as we vary the choice functions, we vary the $g(x)$.

The relation between the holonomy group elements defined in this section and Section 22.1 should be clear: a choice function $x \mapsto I(x)$ is equivalent to a choice $\mathcal{R}$ of basepoints u_x^0 of the previous section via $u_x^0 := \phi_{I(x)}(x, 1)$. We claim that then the holonomies $h_c(\omega)$ that we defined in both sections coincide (any other relation between the choice map $x \mapsto I(x)$ and the map $x \mapsto u_x^0$ results in an identification of the holonomy group elements up to a gauge transformation): in order to see this, consider a curve c with $b(c) = x$, $f(c) = y$. On the one hand, by the definition of the horizontal lift according to the previous section

$$\tilde{c}_{u_x}^\omega(1) = F_c^\omega(u_x) = \rho_{h_c(\omega)^{-1}g(u_x)}\left(u_y^0\right) = \phi_{I(y)}(y, h_c(\omega)^{-1}g(u_x)) \quad (22.2.16)$$

where we have used $u_y^0 = \phi_{I(y)}(y, 1)$ and $h_c(\omega)$ is the group element defined by the equivariant map F_c^ω. Now by definition

$$u_x = \rho_{g(u_x)}\left(u_x^0\right) = \phi_{I(x)}(x, g(u_x)) \quad (22.2.17)$$

Comparing with (22.2.7) we notice that $I(x) = I_1, I(y) = I_N, c(0) = x, c(N) = y$, $u_x = u$ so that $g(u_x) = H_{I_1,u,c}(0)$ in that formula. Thus $h_c(\omega)$ defined in (22.2.7) coincides with that defined in (22.2.16).

23

Geometric quantisation

As an application of the concepts of Chapters 19, 21 and in order to see their interplay in a concrete physical application, we sketch the main ideas of geometric quantisation. This will also provide the necessary background material for the treatment of quantum black holes in LQG.

Geometric quantisation concerns the quantisation of an arbitrary symplectic manifold (M, ω) using only natural symplectic structures during the quantisation process. It consists of three steps: (1) prequantisation, (2) polarisation and (3) quantisation. In the first step one is able to quantise *every* function on phase space in a natural representation, provided that a certain topological condition, Weil's integrality criterion, is satisfied. The famous Groenwald–van Hove theorem is evaded because that representation is highly reducible. In order to obtain an irreducible representation one has to invoke the polarisation step which selects a subspace of the Hilbert space. The final step then consists of finding the induced subrepresentation of the operators.

The strength of geometric quantisation is that it applies to the case when M is not a cotangent bundle, for example, when M is compact. Its weakness is that only a limited number of functions on phase space survive the final quantisation step because they are supposed to preserve the subrepresentation. This is in particular a problem for Hamiltonians and/or constraints which are polynomials of high degree in the momenta, which is why one can usually apply geometric quantisation in its strict form (i.e., without introducing factor ordering ambiguities) only on the reduced phase space constructed in Section 19.3. While there are proposals to remove these limitations in a fourth step called metaplectic correction, the associated theory becomes quite complicated and has not yet been generalised to an infinite number of degrees of freedom.

23.1 Prequantisation

As we have seen, every symplectic manifold (M, ω) is orientable and carries a natural measure, the Liouville form $\Omega = \omega \wedge \ldots \wedge \omega / \hbar^m$, which we have made dimensionless with the help of Planck's constant. Thus it is very natural to define a so-called prequantum Hilbert space as

$$\mathcal{H}_{\mathrm{P}} := L_2(M, \Omega) \tag{23.1.1}$$

consisting of Ω-square integrable functions which are scalars on M. A dense subspace will be the functions $C_0^\infty(M)$ of compact support and we wish to define symmetric operators $\hat{f}$ corresponding to arbitrary real-valued functions $f \in C^\infty(M)$ on that subspace as dense domain. By definition, if χ_f denotes the Hamiltonian vector field of f then $\mathcal{L}_{\chi_f}\omega = 0$, hence $\mathcal{L}_{\chi_f}\Omega = 0$ and therefore the following definition indeed defines a symmetric operator on $\mathcal{H}_P$

$$\hat{f}'\psi := i\hbar\mathcal{L}_{\chi_f}\psi = i\hbar\chi_f[\psi] \qquad (23.1.2)$$

From the properties of the Lie derivative $[\mathcal{L}_u, \mathcal{L}_v] = \mathcal{L}_{[u,v]}$ and from $[\chi_f, \chi_{f'}] = \chi_{\{f,f'\}}$ we easily verify that the map $f \mapsto \hat{f}'$ defines a homomorphism between the *entire* Poisson algebra and the algebra $\mathcal{L}(\mathcal{H}_P)$ of linear operators on $\mathcal{H}_P$, that is,

$$[\hat{f}'_1, \hat{f}'_2] = i\hbar\widehat{\{f_1, f_2\}}' \qquad (23.1.3)$$

However, this homomorphism is unsatisfactory because the map $f \mapsto \chi_f$ has the constant functions as kernel and hence position and momentum would become commuting operators on $(M = \mathbb{R}^2, \omega = dp \wedge dq)$, violating the uncertainty relation.

Thus we must look for a generalisation of (23.1.2). The most general Ansatz which produces a linear (in f) and symmetric operator without introducing extra structure beyond symplectic geometry is given by (23.1.2) supplemented by a real-valued multiplication operator

$$\hat{f}\psi := [i\hbar\mathcal{L}_{\chi_f} + a\theta[\chi_f] + bf]\psi \qquad (23.1.4)$$

where a, b are real parameters to be determined and θ is a symplectic potential for ω, $\omega = d\theta$, which always exists locally by Poincaré's lemma and which is unique up to $\theta \mapsto \theta + d\lambda$, $\lambda \in C^\infty(M)$. The requirements $[\hat{f}_1, \hat{f}_2] - i\hbar\widehat{\{f_1, f_2\}}$, $\hat{c} = c$ for $c = $ const. uniquely fixes $b = -a = 1$ as one can verify by using $i_u i_v d\theta = i_u(\mathcal{L}_v\theta - div_v\theta) = v[\theta[u]] - u[\theta[v]] - i_{[v,u]}\theta$. Hence (23.1.4) becomes

$$\hat{f} = i\hbar i_{\chi_f} \circ \left[d - \frac{\theta}{i\hbar}\right] + f =: i\hbar i_{\chi_f}\nabla + f \qquad (23.1.5)$$

The problem is, of course, that θ is neither necessarily globally defined nor unique. Here we can use our knowledge of fibre bundle theory in order to solve the problem: the operator ∇ looks like a covariant differential in a U(1) gauge theory. Hence the idea would be to interpret the local one-form θ_I as the pullback under local sections of a globally defined connection A in a U(1) principal fibre bundle B over M. The states ψ_I would then be local sections in an, under the defining representation of U(1), associated vector bundle, called a complex line bundle, because the fibres, being isomorphic to $\mathbb{C}$, are one-dimensional. If U_I dentoes the local charts of M with local trivialisations $\phi_I : M \times U(1) \to P$ and $g_{IJ}(p) : \pi_I^{-1}(p) \to \pi_J^{-1}(p)$ for $U_I \cap U_J \neq \emptyset$ are the local gauge transformations which act as $\theta_J/(i\hbar) = \theta_I/(i\hbar) - dg_{IJ}g_{IJ}^{-1}$ (since U(1) is Abelian) and $\psi_J = g_{IJ}\psi_I$ then $\nabla_J\psi_J = g_{IJ}\nabla_I\psi_I$ is gauge-covariant and hence in a scalar product the combination $\overline{\psi}_I \nabla_I \psi'_I$ is independent of the local trivialisation because the phase g_{IJ} drops out.

This elegant solution removes both the non-uniqueness of θ, which is reabsorbed in a gauge transformation, and makes the whole construction globally defined. We may actually slightly generalise the construction further, as follows: as we know from the construction of coherent states or Fock states, it is often convenient to deal with complex coordinates for M. In such a situation neither the symplectic potential θ nor the non-unique exact differential $d\lambda$ which we can always add will be real. The effect of the latter is that the gauge transformations now become complex-valued, hence the gauge group is now shifted from $U(1)$ to $\mathbb{C} - \{0\}$. The effect of the former is that the operators (23.1.5) are no longer symmetric with respect to the measure Ω. To compensate this we introduce the measure $\mu = \rho\Omega$ with positive ρ and impose symmetry of $\hat{f}$ for real-valued f. We find

$$< \hat{f}\psi, \psi' > - < \psi, \hat{f}\psi' >$$
$$= \int_M \rho\omega \left[(-i\hbar\overline{\chi_f[\psi]} - \overline{\theta[\chi_f]\psi})\psi' - \overline{\psi}(i\hbar\chi_f[\psi'] - \theta[\chi_f]\psi') \right]$$
$$= \int_M \omega \left(i\hbar\chi_f[\rho] - \rho(\overline{\theta} - \theta)[\chi_f] \right) \overline{\psi} \, \psi'$$
$$= i\hbar \int_M \rho\,\omega \left(i_{\chi_f} \left[d\ln(\rho) + \frac{2}{\hbar}\Im(\theta) \right] \right) \overline{\psi} \, \psi' \tag{23.1.6}$$

for all ψ, ψ', f hence

$$d\ln(\rho) = -\frac{2}{\hbar}\Im(\theta) \tag{23.1.7}$$

is uniquely determined by θ if the bundle P exists. Notice that since ω is real we have $\omega = d\theta = d\Re(\theta)$, hence $d\Im(\theta) = 0$, so $\Im(\theta)$ is closed while by (23.1.7) ρ only exists if $\Im(\theta)$ is exact.

We notice that under a $\mathbb{C}$-gauge transformation $g = e^{s+i\lambda}$ with real-valued functions s, λ we have $-\theta/(i\hbar) \mapsto -\theta/(i\hbar) - dgg^{-1} = -\theta/(i\hbar) - ds - id\lambda$ hence $\Im(\theta)/\hbar \mapsto \Im(\theta)/\hbar + ds$ and so $\rho \mapsto \rho e^{-2s}$. Since on the other hand $\psi \mapsto g\psi$, it follows that the combination $\rho\overline{\psi}\nabla\psi'$ remains invariant under gauge transformations as does the curvature ω of θ and hence the measure Ω. This motivates us to call ρ a (one-dimensional) fibre metric which defines the fibre inner product between sections

$$\rho[\psi, \psi'] = \rho\overline{\psi}\psi' \tag{23.1.8}$$

By (23.1.7) the fibre metric is covariantly constant with respect to the covariant differential ∇

$$\rho[\nabla_u\psi, \psi'] + \rho[\nabla_u\psi, \psi'] = u[\rho[\psi, \psi']] \tag{23.1.9}$$

and the inner product between states is given by

$$< \psi, \psi' >= \int_M \Omega \, \rho[\psi, \psi'] \qquad (23.1.10)$$

In summary, in order to make this work, given a symplectic manifold (M, ω) we need a *prequantum bundle*, that is:

1. A principal $(\mathbb{C} - \{0\})$ bundle B over M with globally defined connection A whose local sections θ have $\omega = d\theta$ as globally defined curvature.
2. A vector bundle E over M, associated with P under the defining representation of $\mathbb{C} - \{0\}$ with typical fibre $\mathbb{C}$ and local sections ψ.
3. A ∇-compatible fibre metric ρ.

We will now state a necessary and sufficient criterion for the existence of these structures. First we need some preparations.

Definition 23.1.1. *Let M be a manifold with open cover $\mathcal{U} = (U_I)_{I \in \mathcal{I}}$ subordinate to an atlas of M.*

(i) *An n-cochain $\{g\} \in C^n(\mathcal{U})$ is a system of functions $g_{I_1 \ldots I_{n+1}} : U_{I_1} \cap \ldots \cap U_{I_{n+1}} \to \mathbb{C} - \{0\}$ of a definitive type (e.g., smooth, locally constant, ...), precisely one for any $I_1, \ldots, I_{n+1}$ with $U_{I_1} \cap \ldots \cap U_{I_{n+1}} \neq \emptyset$ such that for any $\pi \in S_{n+1}$*

$$g_{I_{\pi(1)} \ldots I_{\pi(n+1)}} = (g_{I_1 \ldots I_{n+1}})^{sgn(\pi)} \qquad (23.1.11)$$

The n-cochains form a group under pointwise multiplication for each multi-index.

(ii) *The coboundary operator $\delta : C^n(\mathcal{U}) \to C^{n+1}(\mathcal{U})$ is defined by*

$$(\delta g)_{I_1 \ldots I_{n+2}} = \prod_{k=1}^{n+2} (g_{I_1 \ldots \hat{I}_k \ldots I_{n+2}})^{(-1)^{k-1}} \qquad (23.1.12)$$

One shows that $\delta^2\{g\} = \{1\}$ where $\{1\}$ is the constant cochain taking the unit value for all index combinations.

(iii) *We call a cochain $\{g\} \in C^n(\mathcal{U})$ closed (a cocycle) or exact (a coboundary) respectively if $\delta\{g\} = \{1\}$ or $\{g\} = \{\delta h\}$ for some $\{h\} \in C^{n-1}(\mathcal{U})$ respectively and write $\{g\} \in Z^n(\mathcal{U})$ or $\{g\} \in B^n(\mathcal{U})$ respectively. The group $H^n(\mathcal{U}) := Z^n(\mathcal{U})/B^n(\mathcal{U})$ is called the n-th Čech cohomology group.*

(iv) *Instead of a multiplicative notation we can use an additive one by writing*

$$g_{I_1 \ldots I_{n+1}} = \exp(f_{I_1 \ldots I_{n+1}}), \ f_{I_{\pi(1)} \ldots I_{\pi(n+1)}} = sgn(\pi) f_{I_1 \ldots I_{n+1}},$$
$$(\delta f)_{I_1 \ldots I_{n+2}} = (n+2)\chi_{[I_1} f_{I_2 \ldots I_{n+2}]} \qquad (23.1.13)$$

where $\chi_I = \chi_{U_I}$ is the characteristic function of U_I. We use $\{f\}$ instead of $\{g\}$ but otherwise use the notation C^n, Z^n, B^n for the corresponding vector spaces.

The Čech cohomology seems to depend explicitly on the atlas $\mathcal{U}$. This dependence can be removed by taking an infinite refinement limit. In the cases of interest (M paracompact so that we can choose a locally finite, contractible cover) we have automatically a so-called Leray cover [218] for which the cohomology is already independent of the cover. We use the notation $H^n(\mathcal{U})$ only in order to distinguish it from the de Rham cohomology $H^n(M)$ of forms.

Čech cohomology appears naturally in principal fibre bundle theory for Abelian gauge groups G: the transition functions $g_{IJ} : U_I \cap U_J \to G$ satisfy $g_{IJ}g_{JI} = 1$, $g_{IJ}g_{JK}g_{KI} = (\delta g)_{IJK} = 1$ and so define a cocycle. In what follows we will only consider the Čech cohomology defined by locally constant functions.

Definition 23.1.2. *Let M be paracompact and $\mathcal{U}$ a locally finite, contractible open cover (any $p \in M$ is only in finitely many U_I and every U_I is contractible to a point). Let (e_I), $0 \le e_I \le 1$ be a partition of unity subordinate to (U_I) with compact support $\mathrm{supp}(e_I) \subset U_I$ in U_I, that is, $\sum_I e_I = 1$. Let $\{f\} \in C^n(\mathcal{U})$ be a locally constant n-cochain (i.e., each $f_{I_1 \ldots I_{n+1}}$ takes a constant value on each connected component of $U_{I_1} \cap \ldots \cap U_{I_n}$, possibly a different one on each component). We define*

$$\alpha : C^n(\mathcal{U}) \to C^n(M); \; \{f\} \mapsto \alpha_{\{f\}}(p)$$
$$:= f_{II_1 \ldots I_n}(p) \, e_I(p) \, de_{I_1}(p) \wedge \ldots \wedge de_{I_n}(p)$$
$$:= \sum_{I,I_1,\ldots,I_n \in \mathcal{I}} f_{II_1 \ldots I_n}(p) \, e_I(p) \, de_{I_1}(p) \wedge \ldots \wedge de_{I_n}(p) \qquad (23.1.14)$$

where the summation convention is applied.

The n-form $\alpha_{\{f\}}$ is everywhere defined because even though the $f_{I_1 \ldots I_{n+1}}$ are only defined on $U_{I_1} \cap \ldots \cap U_{I_{n+1}}$, the n-form $e_{I_1} \, de_{I_2} \wedge \ldots \wedge de_{I_{n+1}}$ vanishes outside that region anyway. Furthermore, the sum in (23.1.14) is finite for every $p \in M$ due to local finiteness of the cover.

Theorem 23.1.3 (de Rham isomorphism). *We have $d\alpha_{\{f\}} = \alpha_{\{\delta f\}}$ and α defines an isomorphism $H^n(\mathcal{U}) \to H^n(M)$.*

Proof: Notice that $df_{I_1 \ldots I_{n+1}} = 0$ in the compact support of $e_{I_1} \, de_{I_2} \wedge \ldots \wedge de_{I_{n+1}}$ due to local constancy. While $df_{I_1 \ldots I_{n+1}} \ne 0$ on $\partial U_{I_1} \cap \ldots \cap U_{I_n}$ this surface is not in the support of $e_{I_1} \, de_{I_2} \wedge \ldots \wedge de_{I_{n+1}}$. Hence

$$d\alpha_{\{f\}} = f_{I_1 \ldots I_{n+1}} \, de_{I_1} \wedge \ldots \wedge e_{I_{n+1}} \qquad (23.1.15)$$

Next notice the relation $\chi_I e_I = \sum_I e_I = 1$ because $\mathrm{supp}(e_I) \subset U_I$. Hence $0 = de_I \chi_I + e_I d\chi_I = de_I \chi_I$ where we have used the fact that $d\chi_I$ is non-vanishing on ∂U_I only, which however is outside the support of e_I. Hence

$$\alpha_{\delta\{f\}} = (n+2)\chi_{[I}f_{I_1\ldots I_{n+1}]}\,e_I\,de_{I_1}\wedge\ldots\wedge de_{I_{n+1}}$$
$$= [\chi_I e_I]f_{I_1\ldots I_{n+1}}de_{I_1}\wedge\ldots\wedge de_{I_{n+1}}$$
$$+ \sum_{k=1}^{n+1}(-1)^k\,\chi_{I_k}\,f_{I_1\ldots\hat{i}_k I\ldots I_{n+1}}\,e_I\,de_{I_1}\wedge\ldots\wedge de_{I_{n+1}}$$
$$= d\alpha_{\{f\}} - \sum_{k=1}^{n+1}\chi_{I_k}\,f_{II_1\ldots\hat{i}_k\ldots I_{n+1}}\,e_I\,de_{I_1}\wedge\ldots\wedge de_{I_{n+1}}$$
$$= d\alpha_{\{f\}} + \sum_{k=1}^{n+1}(-1)^k\chi_{I_k}\,f_{II_1\ldots\hat{i}_k\ldots I_{n+1}}\,e_I\,de_{I_k}\wedge de_{I_1}\wedge\ldots\wedge d\hat{e}_{I_k}\wedge\ldots\wedge de_{I_{n+1}}$$
$$= d\alpha_{\{f\}} + \left[\sum_{k=1}^{n+1}(-1)^k\right][\chi_J de_J]\wedge\alpha_{\{f\}} = d\alpha_{\{f\}} \tag{23.1.16}$$

where in the last step we have relabelled $(I_k, I_{k+1},\ldots,I_{n+1})\leftrightarrow(J,I_k,\ldots,I_n)$.

Hence α maps coboundaries to closed forms. We now define $\alpha:\;H^n(\mathcal{U})\to H^n(M)$ by $\alpha_{[\{f\}]} := [\alpha_{\{f\}}]$ where the brackets denote the respective cohomology classes. This is well-defined, that is, independent of the representative, because for $\{f'\} = \{f\} + \{\delta\tilde{f}\}$ we obtain $[\alpha_{\{f\}}] = [\alpha_{\{f'\}}]$ due to (23.1.15). For the same reason the map is injective. To see that it is surjective assume that $\alpha\in Z^n(M)$ is given. We show this only for $n=2$, the general case is similar. Since U_I is contractible, by Poincaré's lemma we find $\beta_I\in C^1(U_I)$ such that $\alpha = d\beta_I$ on U_I. If $U_I\cap U_J\neq\emptyset$ we have $d(\beta_I-\beta_J)=0$ on $U_I\cap U_J$. Since $U_I\cap U_J$ is contractible we find $\gamma_{IJ} = -\gamma_{JI}\in C^0(U_I\cap U_J)$ such that $\beta_I-\beta_J = d\gamma_{IJ}$ on $U_I\cap U_J$. If $U_I\cap U_J\cap U_K\neq\emptyset$ then $d(\gamma_{IJ}+\gamma_{JK}+\gamma_{KI})=0$ on $U_I\cap U_J\cap U_K$, hence $f_{IJK} := \gamma_{IJ}+\gamma_{JK}+\gamma_{KI} = (\delta\gamma)_{IJK}$ is locally constant.

Notice that f_{IJK} is locally constant but not necessarily the γ_{IJ}, hence f is not exact in the sense of Čech cohomology. However, since purely algebraically $\delta^2 = 0$ we have that $\{f\}$ is closed and on $U_I\cap U_J\cap U_K\cap U_L$

$$(\delta f)_{IJKL} = f_{JKL} - f_{IKL} + f_{IJL} - f_{IJK} = 0 \tag{23.1.17}$$

Contracting (23.1.17) with $e_I\,de_J\wedge de_K$ we find the relation

$$\alpha_{\{f\}} = f_{IJK}de_J\wedge de_K = d(f_{IJK}e_J\wedge de_K) =: d\omega_I \tag{23.1.18}$$

on the interior of U_I. Contracting (23.1.17) with $e_J\,de_K$ we find

$$\omega_I - \omega_J = f_{IJK}de_K = d(f_{IJK}e_K) =: d\sigma_{IJ} \tag{23.1.19}$$

on the interior of $U_I\cap U_J$. Noticing that

$$\sigma_{IJ} = f_{IJ} + f_{JK}e_K + f_{KI}e_K \tag{23.1.20}$$

and defining on the interior of U_I

$$\lambda_I = \beta_I - \omega_I - d(f_{IJ}e_J) \tag{23.1.21}$$

we have on the interior of $U_I \cap U_J$

$$
\begin{aligned}
\lambda_I - \lambda_J &= \gamma_{IJ} - \omega_I + \omega_J - d(f_{IK}e_K - f_{JK}e_K) \\
&= -\omega_I + \omega_J + d(f_{IJ} + f_{JK}e_K + f_{KI}e_K) \\
&= -\omega_I + \omega_J + d\sigma_{IJ} = 0
\end{aligned}
\tag{23.1.22}
$$

hence $\lambda = \lambda_I$ is globally defined. Hence on U_I

$$
d\lambda = d\beta_I - d\omega_I = \alpha - \alpha_{\{f\}}
\tag{23.1.23}
$$

so that $[\alpha] = [\alpha_{\{f\}}]$ and α is a surjection. $\qquad\square$

After these preparations we can now state the main result of this section.

Theorem 23.1.4 (Weil's integrality criterion). *A prequantisation of (M, ω), that is, a principal $\mathbb{C} - \{0\}$ bundle B with global connection ∇ and ∇-compatible fibre metric ρ on an associated complex line bundle exists if and only if the Čech cohomology class of $\alpha^{-1}(\omega/(2\pi\hbar))$ is integral, that is, $[\alpha^{-1}(\omega/(2\pi\hbar))] \in \mathbb{Z}$ where $\alpha : H^2(\mathcal{U}) \to H^2(M)$ is the de Rham isomorphism.*

Moreover, the inequivalent choices of (P, ∇, ρ) are parametrised by $H^1(\mathcal{U})$ with values in $U(1)$.

Proof

$\Leftarrow$:

Suppose first that Weil's criterion is satisfied and let $[\omega] = [\alpha_{\{f\}}]$. From the proof of Theorem 23.1.3 we know that $f_{IJK} = \gamma_{IJ} + \gamma_{JK} + \gamma_{KI} = (\delta\gamma)_{IJK}$ on $U_I \cap U_J \cap U_K$ is locally constant with smooth functions $\gamma_{IJ} = -\gamma_{JI}$ on $U_I \cap U_J$. Moreover, by assumption $f_{IJK} = 2\pi\hbar n_{IJK}$ where n_{IJK} takes locally constant integer values on $U_I \cap U_J \cap U_K$. Define $g_{IJ} = \exp(i\gamma_{IJ}/\hbar)$, then $g_{IJ}g_{JI} = 1$ on $U_I \cap U_J$ and $g_{IJ}g_{JK}g_{KI} = 1$ on $U_I \cap U_J \cap U_K$ because n_{IJK} is integral, hence g_{IJ} is a cocycle with values in $\mathbb{C} - \{0\}$ and therefore qualifies as the transition function of a principal $(\mathbb{C} - \{0\})$ bundle. Moreover, if θ_I are the local potentials of ω then by definition $d\gamma_{IJ} = -i\hbar dg_{IJ}g_{IJ}^{-1} = \theta_I - \theta_J$ or with $A_I = i\theta_I/\hbar$ we find $A_J = A_I - dg_{IJ}g_{IJ}^{-1}$. Hence the A_I qualify as the pull-backs by local sections of a globally defined $\mathbb{C} - \{0\}$ connection ∇. Finally, since ω is real we may choose the θ_I to be real and hence $\rho = 1$ is a ∇-compatible fibre metric.

$\Rightarrow$: Suppose that (P, ∇, ρ) exist and let g_{IJ} be the transition functions of the bundle P with values in $\mathbb{C} - \{0\}$. We define

$$
\frac{f_{IJK}}{2\pi\hbar} := \frac{1}{2\pi i}[\ln(g_{IJ}) + \ln(g_{JK}) + \ln(g_{KL})]
\tag{23.1.24}
$$

where we choose the fundamental branch of the logarithm over each $U_I \cap U_J$ with cut at $\varphi = \pi$ so that $\ln(g_{IJ}) = -\ln(g_{JI})$. Hence (23.1.24) is completely skew and we have $g_{IJ} = \exp(\ln(g_{IJ} + 2\pi i n_{IJ}))$ for some $n_{IJ} \in \mathbb{Z}$. Since the g_{IJ} satisfy the cocycle condition, the right-hand side of (23.1.24) is integral $n_{IJK} \in \mathbb{Z}$

and obviously $\delta\{f\} = 0$. Since $A_J = A_I - dg_{IJ}g_{IJ}^{-1}$, $A_I = i\theta_I/\hbar$, $\omega = d\theta_I$ we conclude $[\omega/(2\pi\hbar)] = [\alpha_{\{f\}}/(2\pi\hbar)]$, hence Weil's criterion is satisfied.

Finally, if Weil's criterion is satisfied then $[\{f\}]$ is determined by γ_{IJ} only up to a coboundary $\delta\{x\}$ with real-valued, locally constant functions x_{IJ} over $U_I \cap U_J$ such that $(\delta x)_{IJK} = 2\pi\hbar m_{IJK}$ with m_{IJK} integral. That x_{IJ} is locally constant follows from $d(\gamma_{IJ} + x_{IJ}) = \theta_I - \theta_J = d\gamma_{IJ}$. Defining $h_{IJ} = \exp(ix_{IJ}/\hbar)$ we get new transition functions $g'_{IJ} = g_{IJ}h_{IJ}$ and the h_{IJ} satisfy the cocycle condition. Hence they define an element of $H^1(\mathcal{U})$ with values in U(1). $\qquad\square$

Recall that two bundles P, P' are equivalent if for their transition functions it holds that $g'_{IJ}(p)g_{IJ}^{-1}(p) = h_I(p)h_J(p)^{-1}$, hence $h_{IJ}(p) = h_I(p)h_J(p)^{-1}$ is a coboundary because then we have a bundle diffeomorphism (automorphism) $\phi'_I(p, g) = \phi_I(p, h_I(p)g)$ for their local trivialisations. Now if M is simply connected then $H^1(M) = \{0\}$, hence by the de Rham isomorphism also $H^1(\mathcal{U}) = \{0\}$. Thus in this case the prequantum bundle is unique once it exists.

Weil's criterion is not stated in the most practical form. The following criterion is equivalent.

Corollary 23.1.5. *Weil's criterion is equivalent with the requirement that for any closed two-surface S in phase space*

$$\int_S \frac{\omega}{2\pi\hbar} = \text{integer} \qquad\qquad (23.1.25)$$

Proof (Sketch). Suppose first that Weil's criterion holds. We will assume for simplicity that the contractible open cover $\mathcal{U}$ is such that the sets $D_I := S \cap U_I$ are open discs covering S such that no point of S lies in more than three different M_I (see Figure 23.1). In the more general case one has to introduce more notation, but the idea of the proof is the same. We partition S into sets S_n, $n = 1, 2, 3$ consisting of points which are contained in precisely n of the D_I. We obviously have

$$S_1 = \cup_I M_I, \quad M_I := D_I - \cup_{J \neq I}(D_I \cap D_J)$$
$$S_2 = \cup_{I<J} M_{IJ}, \quad M_{IJ} := D_I \cap D_J - \cup_{K \neq I,J}(D_I \cap D_J \cap D_K)$$
$$S_3 = \cup_{I<J<K} M_{IJK}, \quad M_{IJK} := D_I \cap D_J \cap D_K \qquad (23.1.26)$$

Of course we may restrict the sum to those I such that $S \cap D_I \neq \emptyset$. Here we have used the fact that the M_{IJK} are mutually disjoint by assumption and that $M_{IJ} = M_{JI}, M_{IJK} = M_{IKJ} = M_{JKI} = M_{JIK} = M_{KIJ} = M_{KJI}$. It is convenient to split M_I further into

$$M_I = D_I - \cup_{J \neq I} M_{IJ} - \cup_{J<K; \, J,K \neq I} M_{IJK} \qquad (23.1.27)$$

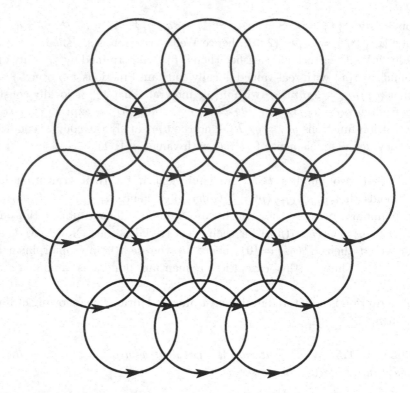

Figure 23.1 An open cover for a part of the two-surface in phase space for which no point is contained in more than three neighbourhoods.

Then with $\Omega := \omega/(2\pi\hbar)$

$$
\int_S \Omega = \sum_I \int_{M_I} \Omega + \sum_{I<J} \int_{M_{IJ}} \Omega + \sum_{I<J<K} \int_{M_{IJK}} \Omega
$$

$$
= \sum_I \int_{D_I} \Omega - \sum_{J\neq I} \int_{M_{IJ}} \Omega - \sum_{J<K,\,J,K\neq I} \int_{M_{IJK}} \Omega + \sum_{I<J} \int_{M_{IJ}} \Omega + \sum_{I<J<K} \int_{M_{IJK}} \Omega
$$

$$
= \sum_I \int_{D_I} \Omega - \sum_{I>J} \int_{M_{IJ}} \Omega - \sum_{J<K,\,J,K\neq I} \int_{M_{IJK}} \Omega + \sum_{I<J<K} \int_{M_{IJK}} \Omega
$$

$$
= \sum_I \int_{D_I} \Omega - \sum_{I>J} \int_{D_I\cap D_J} \Omega - \sum_{I>J,\,K\neq I,J} \int_{M_{IJK}} \Omega
$$
$$
\quad - \sum_{J<K,\,J,K\neq I} \int_{M_{IJK}} \Omega + \sum_{I<J<K} \int_{M_{IJK}} \Omega
$$

$$
= \sum_I \int_{D_I} \Omega - \sum_{I>J} \int_{D_I\cap D_J} \Omega + \sum_{I>J>K} \int_{D_I\cap D_J\cap D_K} \Omega
$$

$$
= \sum_I \int_{\partial D_I} \sigma_I - \sum_{I>J} \int_{\partial(D_I\cap D_J)} \sigma_J + \sum_{I>J>K} \int_{\partial(D_I\cap D_J\cap D_K)} \sigma_K \qquad (23.1.28)
$$

where in the third step we used $\sum_{J\neq I} = \sum_{I<J} + \sum_{I>J}$, in the fourth we used the definition of M_{IJ}, in the fifth we used $\sum_{J<K,\ J,K\neq I} = \sum_{I<J<K} + \sum_{J<I<K} + \sum_{J<K<I}$ and $\sum_{J<I,\ K\neq I,J} = \sum_{K<J<I} + \sum_{J<K<I} + \sum_{J<I<K}$ and finally in the sixth step we employed $\Omega = d\sigma_I$ on (any subset of) D_I.

We take all D_I with orientation such that the loop ∂D_I is counterclockwise for definiteness. Notice the disjoint partitions (up to boundary points)

$$\partial(D_I \cap D_J) = [(\partial D_I) \cap D_J] \cup [D_I \cap (\partial D_J)]$$
$$\partial(D_I \cap D_J \cap D_J) = [(\partial D_I) \cap D_J \cap D_K] \cup [D_I \cap (\partial D_J) \cap D_K] \cup [D_I \cap D_J \cap (\partial D_K)]$$
$$(23.1.29)$$

To convince oneself it is best to draw this on a sheet of paper. Notice also that

$$\partial^2(D_I \cap D_J) = \partial[(\partial D_I) \cap D_J] \cup \partial[D_I \cap (\partial D_J)] = \emptyset \qquad (23.1.30)$$

hence set-theoretically both contributions in (23.1.30) are equal to each other but as zero-dimensional manifolds we must pay attention to their orientation.

Since S is closed, the set ∂D_I is completely covered by the D_J, $J \neq I$ (otherwise we would have to take boundary points into account), so that we have the disjoint decomposition

$$\partial D_I = [\cup_{I\neq I}(\partial D_I) \cap M_{I,I}] \cup [\cup_{I<K,\ J,K\neq I}(\partial D_I) \cap M_{IJK}] \qquad (23.1.31)$$

Now use $\partial D_I \cap D_I = \partial D_I$, insert (23.1.31) into (23.1.28), utilise manipulations such as those applied in (23.1.28) and those displayed in (23.1.29) and perform some relabellings. A lengthy calculation shows

$$\int_S \Omega = \sum_{I>J} \int_{\partial[(\partial D_I)\cap D_J]} \gamma_{IJ}$$

$$+ \sum_{I>J>K} \left\{ \int_{\partial[(\partial D_I)\cap D_J\cap D_K]} \gamma_{KI} + \int_{\partial[D_I\cap(\partial D_J)\cap D_K]} \gamma_{KJ} \right\} \qquad (23.1.32)$$

where $\sigma_I - \sigma_J = d(\gamma_{IJ})$ on (any subset of) $D_I \cap D_J$ was used.

Now since S is closed, both endpoints of $\partial[(\partial D_I) \cap D_J]$ lie in precisely one of the D_K, $K \neq I, J$ so that

$$\partial[(\partial D_I) \cap D_J] = \cup_{K\neq I,J}(\partial[(\partial D_I) \cap D_J]) \cap D_K \qquad (23.1.33)$$

is a disjoint decomposition. Next one convinces oneself that

$$\partial[(\partial D_I) \cap D_J \cap D_K] = (\partial[(\partial D_I) \cap D_J]) \cap D_K \cup (\partial[(\partial D_I) \cap D_K]) \cap D_J$$
$$(23.1.34)$$

is a disjoint decomposition. Inserting (23.1.33) and (23.1.34) into (23.1.32) we arrive, after a lengthy calculation using the now familiar manipulations as well as $\gamma_{IJ} + \gamma_{JI} = 0$, at

$$\int_S \Omega = \sum_{I>J>K} \int_{(\partial[(\partial D_I)\cap D_J])\cap D_K} [\gamma_{IJ} + \gamma_{JK} + \gamma_{KI}]$$

$$= \sum_{I>J>K} n_{IJK} \in \mathbb{Z} \tag{23.1.35}$$

where we used the fact that $\gamma_{IJ} + \gamma_{JK} + \gamma_{KI} = n_{IJK}$ is integral and constant on (any subset of) $D_I \cap D_J \cap D_K$.

Conversely, fix some labels $I_0 > J_0 > K_0$ with $D_{I_0} \cap D_{J_0} \cap D_{K_0} \neq \emptyset$ and choose a closed S contained in $D_{I_0} \cap D_{J_0} \cap D_{K_0}$. Then the exact same calculation as above reveals $\int_S \Omega = n_{I_0 J_0 K_0}$ for some constant function with values in $\mathbb{R}$. Hence, if (23.1.25) holds then $n_{I_0 J_0 K_0}$ is integral. $\qquad\square$

Let us summarise our findings.

Definition 23.1.6. *A symplectic manifold (M, ω) is said to be prequantisable iff Weil's integrality criterion is satisfied. The associated structure (P, ∇, ρ) is called a prequantum bundle. The prequantum Hilbert space is $\mathcal{H}' = L_2(M, \Omega)$ with inner product between smooth sections of compact support of the associated line bundle E given by*

$$< \psi, \psi' > = \int_M \Omega \, \rho[\psi, \psi'], \quad \rho[\psi, \psi'] = \rho \overline{\psi} \psi' \tag{23.1.36}$$

where $d \ln(\rho) = \frac{2}{\hbar} \Im(\theta)$ and symmetric operators associated with real-valued functions $f \in C^\infty(M)$ are densely defined on E by

$$\hat{f} = i\hbar \nabla_{\chi_f} + f, \quad \nabla = d + \frac{1}{i\hbar}\theta, \quad \omega = d\theta \tag{23.1.37}$$

23.2 Polarisation

The prequantum Hilbert space is of course much too large because it is highly reducible. For instance, in ordinary quantum mechanics the famous Stone–von Neumann theorem [538] tells us that every irreducible, weakly continuous representation of the Weyl algebra generated by the operators $W(a, b) = \exp(i[a\hat{q} + b\hat{p}])$ is unitarily equivalent to the Schrödinger representation on $L_2(\mathbb{R}, dx)$ with $(W(a, 0)\psi)(x) = \exp(iax)\psi(x), (W(0, b)\psi)(x) = \psi(x - b)$ and weak continuity means that the one-parameter unitary groups $a \mapsto W(a, 0), b \mapsto W(0, b)$ are weakly continuous. In that case $M = T^*(\mathbb{R}) = T^1(\mathbb{R})$ and wave functions depend only on position. The wave functions of the associated prequantum Hilbert space would depend on momentum as well, which is why it must be highly reducible. This is the reason why one can define a homomorphism $f \mapsto \hat{f}$ for any function on M, thus circumventing the Groenwald–van Hove theorem [538] which states that no such homomorphism exists on $L_2(\mathbb{R}, dx)$ for all functions. It is possible to define a homomorphism for polynomials in p, q of at most second degree and for smooth functions at most linear in the momenta.

Reducible representations are undesirable because the complete physics takes place already on any of the irreducible subspaces (called superselection sectors since no operators can map between these spaces). In other words, a reducible representation indicates that the Hilbert space is too large or the algebra of operators is too small.

In geometric quantisation one deals with this problem by the process of polarisation. In broad terms, polarisation consists of the selection of a Lagrangian submanifold in M such as the configuration manifold of the q in the case of the ordinary phase space $T^*(\mathbb{R})$ coordinatised by the (q, p) with the symplectic structure $\omega = dp \wedge dq$. However, not every symplectic manifold has the topology of a cotangent bundle $M = T^1(C)$ over some configuration manifold and often it is of advantage to choose complex coordinates. Hence one is naturally led to the question of which symplectic manifolds possess which complex structures.

Definition 23.2.1

(i) *Let V be a vector space over $\mathbb{R}$. The complexification of $V_{\mathbb{C}}$ of V is a vector space over $\mathbb{C}$ consisting of vectors of the form $w = u + iv$, $u, v \in V$. The complex dimension of $V_{\mathbb{C}}$ then coincides with the real dimension of V because a basis for V continues to be a basis for $V_{\mathbb{C}}$, however, $V_{\mathbb{C}}$ is isomorphic to $V \oplus V$ which has twice the real dimension of V. The linear operations on $V_{\mathbb{C}}$ are defined as follows: $w_1 + w_2 := (u_1 + u_2) + i(v_1 + v_2)$, $zw := (xu - yv) + i(xv + yu)$ for $z = x + iy \in \mathbb{C}$ and we define $\bar{w} := u - iv$. Multilinear forms on tensor products of copies of V are extended in the obvious way, that is, $\lambda(\dots, w, \dots) := \lambda(\dots, u, \dots) + i\lambda(\dots, v, \dots)$. Since multilinear forms also form a vector space their complexification is defined analogously.*

(ii) *If M is a real manifold then we can define the complexification of the tensor fields $(T_b^a(M))_{\mathbb{C}}$ to be the tensor fields of the form $w(p) = t_1(p) + it_2(p)$, $t_1, t_2 \in T_b^a(M)$ defined by pointwise operations. In particular we can define multiplication with complex-valued functions. This can be done also if M does not have a complex structure.*

(iii) *Let (V, ω, J) be a symplectic vector space with ω-compatible complex structure J. The subspace $V^{\pm}$ of $V_{\mathbb{C}}$ consisting of vectors of the form $u^{\pm} := \frac{1}{2}(u \mp iJ[u])$ is called the subspace of holomorphic (antiholomorphic) vectors since $J[u^{\pm}] = \pm iu^{\pm}$. We set $P_J := V^+$.*

Lemma 23.2.2. *P_J is a Lagrangian subspace of $V_{\mathbb{C}}$ with the additional property that $P_J \cap \overline{P_J} = \{0\}$. Conversely, every Lagrangian subspace with this property determines a complex structure on V.*

Proof: Recall that a Lagrangian subspace F of a symplectic vector space V is defined by the property $F = F^{\perp} := \{v \in V; \ \omega(u, v) = 0 \ \forall \ u \in F\}$. We have for any $u, v \in V$

$$4\omega[u^+, v^\pm] = (\omega[u, v] \mp \omega[J[u], J[v]]) - i(\omega[J[u], v] \pm \omega[u, J[v]])$$
$$= [\omega[u, v] - i\omega[J[u], v]](1 \mp 1) \qquad (23.2.1)$$

by compatibility. Hence $P_J \subset (P_J)^\perp$ and $\overline{P_J} \cap (P_J)^\perp = \{0\}$ because ω is not degenerate. Since $V^+ = P_J, V^- = \overline{P_J}$ span $V_{\mathbb{C}}$ and satisfy $P_J \cap \overline{P_J} = \{0\}$ it follows that $P_J = (P_J)^\perp$. Obviously $\dim_{\mathbb{C}}(P_J) = m$ if $\dim_{\mathbb{R}}(V) = 2m$.

Conversely, given a Lagrangian subspace $P \subset V_{\mathbb{C}}$ with $P \cap \overline{P} = \{0\}$ we know that $V_{\mathbb{C}} = P \oplus \overline{P}$ and can decompose any $w \in V_{\mathbb{C}}$ uniquely as $w = w^+ + w^-$ with $w^+ \in P$, $w^- \in \overline{P}$. Notice that since ω is real and P is Lagrangian, also $\overline{P} = \{\overline{w}, w \in P\}$ is Lagrangian since $\omega[\overline{w}, \overline{w}'] = \overline{\omega[w, w']} = 0$. We now define $J[w] := i(w^+ - w^-)$ which determines J uniquely. Then

$$\omega[J[w_1], J[w_2]] = \omega[w_1^+, w_2^-] + \omega[w_1^-, w_2^+] = \omega[w_1, w_2] \qquad (23.2.2)$$

where we have used the Lagrangian subspace property. Hence J is ω-compatible. $\qquad \square$

Hence, given a complex structure it is easy to construct Lagrangian subspaces of $V_{\mathbb{C}}$ with $P_J \cap \overline{P_J} = \{0\}$. The other extreme are Lagrangian subspaces P with the property $P \cap \overline{P} = P$. These are complexifications of Lagrangian subspaces of V. We now study systematically all the cases in between these two extremes, utilising the following elementary results whose simple proof we leave to the reader.

Lemma 23.2.3. *Let $F \subset V$ be a subspace of the real vector space V and let $G \subset V_{\mathbb{C}}$ be a subspace of its complexification. We define the real subspace of G by $G_{\mathbb{R}} := G \cap V$. The annihilator subspaces $F^\perp$, $G^\perp$ of V, $V_{\mathbb{C}}$ respectively are defined by $F^\perp = \{v \in V; \omega[u, v] = 0 \,\forall\, u \in F\}$ and $G^\perp = \{v \in V_{\mathbb{C}}; \omega[u, v] = 0 \,\forall\, u \in G\}$ respectively. Then the following results hold:*

(i) $(F^\perp)_{\mathbb{C}} = (F_{\mathbb{C}})^\perp$.
(ii) If $G = \overline{G}$ then $(G^\perp)_{\mathbb{R}} = (G_{\mathbb{R}})^\perp$.
(iii) If $G = \overline{G}$ then $(G_{\mathbb{R}})_{\mathbb{C}} = G$.
(iv) Define $\tilde{F} = F/(F \cap F^\perp) = \{(u); u \in F\}$ where the rest classes are defined by $[u] = \{u + v; v \in F \cap F^\perp\}$. Then $\tilde{\omega}[(u), (u')] := \omega[u, u']$ is well-defined and $(\tilde{F}, \tilde{\omega})$ is a symplectic vector space.
(v) if F is co-isotropic $(F^\perp \subset F)$, $\pi : F \to \tilde{F} = F/F^\perp$ the canonical projection, P a Lagrangian subspace in V then $\tilde{P} := \pi(P \cap F)$ is a Lagrangian subspace in $(\tilde{F}, \tilde{\omega})$.

The following theorem is the key to the classification of polarisations.

Theorem 23.2.4. *Let $P \subset V^{\mathbb{C}}$ be a Lagrangian subspace and*

$$E := (P + \overline{P})_{\mathbb{R}}, \quad D := (P \cap \overline{P})_{\mathbb{R}} \qquad (23.2.3)$$

Then

(i) D is an isotropic $(D \subset D^{\perp})$ subspace of V.

(ii) E is a co-isotropic subspace of V and $E^{\perp} = D$.

(iii) $E_{\mathbb{C}} = P + \bar{P}$, $D_{\mathbb{C}} = P \cap \bar{P}$.

(iv) Let $\tilde{F} = E/D$. Then $\tilde{F}_{\mathbb{C}} = E_{\mathbb{C}}/D_{\mathbb{C}}$.

(v) If $\pi : E \to \tilde{F}$ is the canonical projection then $\tilde{P} := \pi(E_{\mathbb{C}} \cap P)$ is Lagrangian in $(\tilde{F}_{\mathbb{C}}, \omega)$.

(vi) $\tilde{P} \cap \overline{\tilde{P}} = \{0\}$.

We leave the simple proof (with the aid of Lemma 23.2.3, which is applicable since obviously $\bar{E} = E$, $\bar{D} = D$) to the reader.

The symplectic structure $\tilde{\omega}$ defined by P via Theorem 23.2.4 determines a $\tilde{\omega}$-symplectic structure $\tilde{J}$ via Lemma 23.2.2 which is denoted J_P. Let $\tilde{g}$ be the Hermitian structure associated with $(\tilde{F}, \tilde{\omega}, J_P)$, that is, $\tilde{g}[u, v] := \omega[u, J_P[v]]$ which defines a symmetric non-degenerate tensor since both $\tilde{\omega}, J_P$ are. Let $2\tilde{m}$ be the dimension of $\tilde{F}$ and let L be any Lagrangian subspace of $(\tilde{F}, \tilde{\omega})$ such that $\tilde{g}_{|L} = \tilde{g}_{|J_P[L]}$ is also non-degenerate. Then $L + J_P[L] = \tilde{F}$ because both $L, J_P[L]$ are Lagrangian subspaces and $L \cap J_P[L] = \{0\}$. In fact, $L, J_P[L]$ are orthogonal subspaces with respect to $\tilde{g}$. Since $\tilde{g}_{|L}$ defines a non-degenerate symmetric tensor it will have r positive and s negative eigenvalues with $\tilde{m} = r + s \leq m$. This signature does not depend on the choice of L because any two such Lagrangian subspaces give rise to different block diagonal matrices, which are related by a block diagonal, even canonical, transformation which does not change the signature of $\tilde{g}_{|L}$ by Sylvester's theorem. Since $E^{\perp} = D$ we have $2m = \dim(E) + \dim(D)$ and since $\tilde{F} = E/D$ with $D \subset E$ we have $2\tilde{m} = \dim(E) - \dim(D)$. Hence $\dim(E) = m + \tilde{m}$, $\dim(D) = m - \tilde{m}$.

Definition 23.2.5. *The type of the Lagrangian subspace $P \subset V_{\mathbb{C}}$ is the pair of integers (r, s) just defined. Special types are:*

(i) *Kähler:* $m = r + s$, *that is* $\dim(D) = 0$ *so* $P \cap \bar{P} = \{0\}$.

(ii) *Positive:* $m = r$, *that is* $s = 0$ *and Kähler, hence* $\omega = \tilde{\omega}$ *and the associated Kähler metric* $g[., .] = \omega[., .]$ *is positive definite.*

(iii) *Non-negative:* $s = 0$, *that is, the Kähler metric* $\tilde{g}$ *on* $\tilde{F}$ *is positive definite, however,* $\dim(D) > 0$ *is possible in which case* P *contains a real subspace.*

(iv) *Real:* $r = s = 0$, *that is,* $\dim(E) = \dim(D) = m$, *hence* $E = D$ *and* $P = \bar{P}$, *so* $P = L_{\mathbb{C}}$ *where* $L \subset V$ *is Lagrangian.*

Notice that canonical transformations of V which do not preserve E will change the type of P. In fact, an ω-compatible complex structure on a symplectic vector space is just an element of the symplectic group which squares to unity. Hence the symplectic group acts on the set of compatible complex structures by conjugation and every compatible complex structure can be obtained this way.

The reader is urged to check that on $(V = \mathbb{R}^{2m}, \omega = dp_a \wedge dq^a)$ an example for a Lagrangian subspace P of type (r, s) is given by

$$P = \text{span}_{\mathbb{C}} \left\{ Y_\mu := \left(\frac{\partial}{\partial p_\mu} \right)_{\mu=1}^{m-r-s}, \; Y_\mu^+ := \frac{1}{2} \left(\frac{\partial}{\partial p_\mu} + i \frac{\partial}{\partial q^\mu} \right)_{\mu=m-r-s+1}^{m-s}, \right.$$

$$\left. Y_\mu^- := \frac{1}{2} \left(\frac{\partial}{\partial p_\mu} - i \frac{\partial}{\partial q^\mu} \right)_{\mu=m-s+1}^{m} \right\} \tag{23.2.4}$$

D is the real span of the Y^μ, $\mu = 1, \ldots, m - \tilde{m}$, E is the real span of the Y^μ, $\mu = 1, \ldots, m - \tilde{m}$ and the $\partial/\partial p_\mu$, $\partial/\partial q^\mu$, $\mu = m - \tilde{m} + 1, \ldots, m$ and $\tilde{F}$ is spanned by the $\partial/\partial p_\mu$, $\partial/\partial q^\mu$, $\mu = m - \tilde{m} + 1, \ldots, m$. With respect to $L \subset \tilde{F}$ spanned by $\partial/\partial q^\mu$, $\mu = m - \tilde{m} + 1, \ldots, m$ we have $\tilde{g}_{|L} = \text{diag}(1_r, -1_s)$.

After this preparation we can now generalise from symplectic vector spaces to symplectic manifolds.

Definition 23.2.6

(i) *A complex distribution P on a real manifold M is an assignment of subspaces $p \mapsto P_p \subset (T_p(M))_{\mathbb{C}}$ whose complex dimension k is constant and which are spanned by k complex vector fields in a neighbourhood of each point p of M.*

(ii) *A complex polarisation of a symplectic manifold (M, ω) is a complex distribution P such that P_p is a Lagrangian subspace of $(T_p(M))_{\mathbb{C}}$ and such that the type of P_p is constant (equivalently, the real dimension of $D_p = (P_p \cap \overline{P}_p) \cap T_p(M)$ is constant).*

(iii) *A complex distribution is called integrable provided that in a neighbourhood U of each point p of M there are smooth complex-valued functions $f_{k+1}, \ldots, f_{2m}$ with linearly independent differentials df_j, $j = k + 1, \ldots, 2m$ such that $\overline{u}[f_j]$ for any vector field u tangential to P in U. It is said to be strongly integrable if in addition the real distribution $p \mapsto E_p = D_p^\perp$ is integrable.*

(iv) *A symplectic potential θ, $d\theta = \omega$ is said to be P-adapted provided that $i_{\overline{u}}\theta = 0$ for all u tangential to P. A polarisation is said to be admissible provided that local P-adapted symplectic potentials exist everywhere.*

(v) *A complex-valued function $\psi \in C^\infty(M)$ is said to be P-polarised provided that $\overline{u}[\psi] = 0$ for all u tangential to P. We use the notation $C_P^\infty(M)$ for such functions.*

The reason for using $\overline{u}$, $u \in P$ rather than u itself is because we want to get holomorphic functions as polarised functions rather than antiholomorphic ones. A very important example is the case that the symplectic manifold (M, ω) is also a Kähler manifold with ω-compatible complex structure. Then M carries two natural polarisations spanned by the holomorphic $\partial/\partial z^\mu$ and antiholomorphic $\partial/\partial \bar{z}^\mu$ vector fields selected by $J[u] = \pm iu$ respectively. We had seen that locally

$\omega = id \wedge \bar{d}K$ for a Kähler scalar K in Kähler manifolds. Hence $\theta = -idK$ is a local adapted symplectic potential and hence $P = \text{span}_{\mathbb{C}}\{\partial_\mu\}$ is admissible.

The importance of the notation of a strongly integrable polarisation is demonstrated by the following theorem:

Theorem 23.2.7. *Let P be a strongly integrable polarisation of a symplectic manifold (M, ω) of dimension $\dim(M) = 2(m' + \tilde{m}) = 2m$. Then in the neighbourhood of each point we find a system of coordinates $\{q^a, p_a, z^\alpha\}_{a=1 \, \alpha=1}^{m' \quad \tilde{m}}$ with q, p real and z complex such that $P = \text{span}_{\mathbb{C}}\{\partial/\partial p_a, \partial/\partial z^\alpha\}$ and there is a real-valued function $K(q, z, \bar{z})$ such that $\omega = d\theta$ where*

$$\theta = p_a dq^a - \frac{i}{2}(\partial_\alpha K)dz^\alpha + \frac{i}{2}(\partial_{\bar{\alpha}} K)dz^{\bar{\alpha}} \tag{23.2.5}$$

Proof: Since P is integrable we find m functions z^μ such that $\bar{u}[z^\mu] = dz^\mu[[\bar{u}]] = -i_{\bar{u}} i_{\chi_{z\mu}} \omega = 0$ for all u tangential to P. Hence $\chi_{z^\mu} \in \bar{P}^\perp = \bar{P}$. Hence $\bar{P} = \text{span}_{\mathbb{C}}\{\chi_{z^\mu}\}$. Since ω is non-degenerate and real we see that $P = \text{span}_{\mathbb{C}}\{\chi_{\bar{z}^\mu}\}$. By Frobenius' theorem 19.3.2 both systems of vector fields are involutive because the Pfaff system dz^μ; $\mu = 1, \ldots, m$ is closed.

Since ω is real, $D_p = (P_p \cap \bar{P}_p) \cap T_p(M)$ will be non-empty if and only if m' of the z^μ are actually real-valued. Let us therefore write $(z^\mu) = (q^a, z^\alpha)$, $a = 1, \ldots, m'$; $\alpha = 1, \ldots, \tilde{m}$. Then D is spanned by the χ_{q^a} and has constant dimension $m' = m - \tilde{m}$. By assumption $p \mapsto E_p = (\Gamma_p \cup \bar{\Gamma}_p) \cap T_p(M)$ is an integrable co-isotropic distribution of dimension $m + \tilde{m} = m' + 2\tilde{m}$, hence $D_p = (E_p)^\perp \subset E_p$. Since the statement of the theorem is local we may assume without loss of generality that E is reducible. Hence we find m real-valued functions f^a, $a = 1, \ldots, m$ on M such that the leaves I_f of the foliation are labelled by the constant values of the f^a. By definition $E_p = \text{span}_{\mathbb{R}}\{\chi_{q^a}, \frac{1}{2}(\chi_{z^\alpha} + \chi_{\bar{z}^\alpha}), \frac{-i}{2}(\chi_{z^\alpha} - \chi_{\bar{z}^\alpha})\}$. Since these vector fields are tangential to the leaves we must have $\chi_{q^a}[f^b] = \chi_{z^\alpha}[f^b] = \chi_{\bar{z}^\alpha}[f^b] = 0$. Since $P, \bar{P}$ are Lagrangian distributions we must have $i_{\chi_{z\mu}} i_{\chi_{z\nu}} \omega = i_{\chi_{\bar{z}\mu}} i_{\chi_{\bar{z}\nu}} \omega = 0$ for all $\mu, \nu = 1, \ldots, m$. Hence in particular $\chi_{q^a}[q^b] = \chi_{z^\alpha}[q^b] = \chi_{\bar{z}^\alpha}[q^b] = 0$ and we see that we may choose $f^b = q^b$.

Consider the vector spaces $\tilde{F}_p = E_p/D_p$ spanned by the real linear combinations of the $\chi_{z^\alpha}, \chi_{\bar{z}^\alpha}$ and denote by $\pi_p : E_p \to E_p/D_p$. Then it carries the polarisation defined by $\tilde{P}_p = \pi((E_p)_{\mathbb{C}} \cap P_p) = \pi(P_p)$ spanned by the χ_{z^α}. Hence within each leaf I_q the distribution $\tilde{F}_p$ integrates to an integral manifold $\tilde{F}_q$ which carries a Kähler polarisation P' because $P' \cap \bar{P}' = \{0\}$. As we have seen above, it carries a symplectic structure $\tilde{\omega}_q$ which, since $\tilde{F}_q$ is Kähler, takes the local form

$$\tilde{\omega}_q = i\partial_\alpha \partial_{\bar{\beta}} K_q(z, \bar{z})dz^\alpha \wedge dz^{\bar{\beta}} \tag{23.2.6}$$

for some local Kähler scalar K_q.

The leaves I_q have the structure of a fibre bundle over $\tilde{F}_q$ with typical fibre D_q which are the integral manifolds of the distribution D. Consider a $(2m - m')$-dimensional manifold S which intersects all the leaves I_q and within each I_q

it intersects every fibre D_q once. We define m' functions p_a, $a = 1, \ldots, m'$ by the requirement $\chi_{q^a}[p_b] = \delta_{ab}$, $(p_a)_{|S} = 0$. This defines p_a in a neighbourhood of S and thus in a region of m. It follows that the $(q^a, p_a, z^\alpha, \bar{z}^\alpha)$ are local coordinates for M. By writing $\chi_q, \chi_z, \chi_{\bar{z}}$ in terms of $\partial_q, \partial_p, \partial_z, \partial_{\bar{z}}$ and using that $\chi_{z^\mu}[z^\nu] = \chi_{\bar{z}^\mu}[\bar{z}^\nu] = 0$ from the fact that both $P, \bar{P}$ are Lagrangian we find that the χ_q and hence D is spanned by the ∂_p, that the χ_z and hence $\bar{P}$ is spanned by the $\partial_p, \partial_{\bar{z}}$ and that the $\chi_{\bar{z}}$ and hence P is spanned by the ∂_p, ∂_z. In fact $\chi_{q^a} = \partial_{p_a}$.

The general form of ω is now given by

$$\omega = dp_a \wedge dq^a + \sigma_{ab} dq^a \wedge dq^b + \sigma_{a\alpha} dq^a \wedge dz^\alpha + \overline{\sigma_{a\alpha}} dq^a \wedge d\bar{z}^\alpha$$
$$+ i(\partial_\alpha \partial_\beta K) dz^\alpha \wedge d\bar{z}^\beta \tag{23.2.7}$$

All functions displayed a priori depend on all coordinates except for K which is independent of p. The Ansatz obeys reality of ω and takes into account that $\omega[\partial_p, .]$ vanishes on $P \cup \bar{P}$. We now compute $d\omega = 0$ and learn by comparing coefficients that σ_{ab}, $\sigma_{a\alpha}$ are independent of p. Next $\partial_{[a} \sigma_{bc]} = 0$ implies locally that $\sigma_{ab} = \partial_{[a} f_{b]}$. Replacing p_a by $p_a + f_a$ we may set $f_a = 0$. With this choice we get the further relations $\partial_{[a} \sigma_{b]\alpha} = \sigma_{a[\alpha,\beta]} = 0$, which locally implies that $\sigma_{a\alpha} = -\frac{1}{2} \partial^2 g / \partial q^a \partial z^\alpha$ where g is a function unique up to adding $f(z, \bar{z})$. The final implication is

$$\frac{\partial^3}{\partial q^a \partial z^\alpha \partial \bar{z}^\beta} \left[iK - \frac{1}{2}(g - \bar{g}) \right] = 0 \tag{23.2.8}$$

which has the local solution

$$\Im(g) = K + h(q, z) + h(\bar{q}, z) + f(z, \bar{z}) \tag{23.2.9}$$

Since $K(q, z, \bar{z}) := K_q(z, \bar{z})$ is only unique up to some $h + \bar{h}$ where h is holomorphic and f can be absorbed into g we may choose $\Im(g) = K$. Reinserting into ω and replacing p_a by $p_a + \frac{1}{2} \partial \Re(g) / \partial q^a$ we arrive at the assertion. $\qquad \square$

We remark that every strongly integrable polarisation is admissible because upon replacing θ by $\theta - \frac{i}{2} dK$ there are no terms proportional $dp, d\bar{z}$ and hence θ is annihilated by the vector fields tangential to $\bar{P}$.

23.3 Quantisation

The idea of arriving at an irreducible representation is now by restricting wave functions to half of the degrees of freedom.

Definition 23.3.1. *Let P be a strongly integrable polarisation of a symplectic manifold (M, ω) and (B, ∇, ρ) a prequantum bundle over M. A smooth section $\psi : M \to E$ of the associated vector bundle E is said to be polarised provided that $\nabla_{\bar{u}} \psi = 0$ for all u tangential to P. The space of polarised sections is denoted by S_P.*

We remark that polarised sections always exist: by Theorem 23.2.7 P is spanned by $\partial_{p_a}, \partial_{z^\alpha}$ and θ depends only on dq^a, dz^α and is admissible. Hence $\nabla_{\bar{u}} = \bar{u}$ and so a polarised section is of the form $\psi(q, z)$. (In particular, in the Kähler case $P \cap \bar{P} = \{0\}$ wave functions are simply holomorphic.) Using that

$$\theta = p_a dq^a - i(\partial_\alpha K)dz^\alpha - \frac{i}{2}(\partial_a K)dq^a \qquad (23.3.1)$$

we find that $\Im(\theta) = -dK/2$ is exact and so the ∇-compatible fibre metric determined by $d\ln(\rho) = -\frac{2}{\hbar}\Im(\theta)$ is given by

$$\rho = \rho_0 \exp(K/\hbar), \quad K = K(q, z, \bar{z}) \qquad (23.3.2)$$

and is determined directly by the Kähler scalar.

In the quantisation step one would now simply like to restrict the prequantum Hilbert space $\mathcal{H}'$ to (the completion of) S_P, that is one would like to define the Hilbert space by $\mathcal{H}_P := \overline{\mathcal{H}' \cap S_P}$. This is problematic for the following reasons:

(A) *Normalisation*

For most polarisations it is not true that S_P contains any square integrable element. This is most evident in the case that P is not a Kähler polarisation because then neither the sections nor the Kähler scalar depend on the p_a, which when taking unbounded range will tend to make the p-integral diverge. Even in the case that P is Kähler but not positive there is a problem even if M is compact because the Kähler potential K is not necessarily bounded from above in this case. To see this, suppose, for example, that $\partial/\partial x^\alpha$ generate a real Lagrangian submanifold and that M is a complex manifold, that is, $J[\partial/\partial x^\alpha] = \partial/\partial y^\alpha$. It is then easy to see that

$$g_{\alpha\beta} = g[\partial_\alpha, \partial_\beta] = 2\frac{\partial^2 K}{\partial z^\alpha \partial \bar{z}^\beta} \qquad (23.3.3)$$

with $z^\alpha = x^\alpha + iy^\alpha$. Suppose that z_0 is a stationary point and make use of the ambiguity $K(z, \bar{z}) \mapsto K(z, \bar{z}) + h(z) + \overline{h(z)}$ in order to achieve $(\partial_z)^2 K = (\partial_{\bar{z}})^2 K = 0$ at z_0 as well. Then $(g_{\alpha\beta}(z_0))$ is the Hessian and its signature (r, s) shows that the integral converges at best when $s = 0$. Hence only for positive Kähler polarisations is $\mathcal{H}_P$ large enough.

(B) *Operators*

Of course the only admissible operators are those which preserve the polarisation. Hence we must have $\nabla_{\bar{u}} \hat{f}\psi = 0$ for every polarised section ψ and it is easy to check that this is the case if and only if $\nabla_{[\bar{u},\chi_f]}\psi = 0$, hence $[\bar{u}, \chi_f] \in \bar{P}$, so χ_f must preserve the polarisation. Hence not every function can be realised as a prequantum operator on the Hilbert space $\mathcal{H}_P$. One can sometimes deal with this difficulty by a procedure called 'pairing', the more sophisticated version of which is the so-called 'metaplectic correction'. For the case of a positive Kähler polarisation such that $\mathcal{H}_P \in \mathcal{H}'$ is a subspace it boils down to the following: denote by $\pi_P : \mathcal{H}' \to \mathcal{H}_P$ the orthogonal projection and define an operator $\hat{f}_P := \pi_P \hat{f} \pi_P$ on $\mathcal{H}_P$. While this is a symmetric operator, it is no longer necessarily true that $[\hat{f}_P, \hat{f}'_P] = \widehat{\{f, f'\}}'_P$ for any two functions f, f'. Another avenue is to write f in terms of functions whose operators do preserve the polarisation, but then of course ordering issues arise.

This ends our sketch of geometric quantisation. One of the many interesting issues which we did not discuss is how symplectic reduction and geometric quantisation work together (see, e.g., [890–894]. One can of course quantise after the reduction but of more practical use is quantisation before reduction. In that case it is not hard to believe that a critical condition for direct quantisation is that the Hamiltonian vector fields of the constraints should preserve the polarisation as otherwise the projections π_P could spoil the consistency of the constraint algebra. If that condition is not satisfied one then faces the just-mentioned ordering problems.

In summary, geometric quantisation provides a beautiful, geometric, general framework for the quantisation of a given symplectic manifold which is not necessarily of cotangent bundle type. However, while in the prequantisation step we can quantise all functions without picking up ordering corrections, once one chooses a polarisation and hence an irreducible representation, one is again faced with ordering corrections.

The Dirac algorithm for field theories with constraints

It is a crime that the subsequent analysis is not a standard ingredient of every course in theoretical mechanics. Every interaction that we know today underlies a gauge theory, that is, a field theory with constraints. However, constraints are generically at most mentioned in beginning theoretical mechanics courses. This is the more astonishing as this really important topic can be taught at a truly elementary level. Also quantum mechanics is not needed (at most for motivational purposes), the theory can be formulated in purely classical terms. We recommend the classic expositions by Dirac [219] and by Hanson *et al.* [895] as introductory texts. More advanced are the textbooks [659] and [263]. For geometrical quantisation with constraints see [890] and for a more mathematical formulation see [891–894].

24.1 The Dirac algorithm

We will consider only a finite number of degrees of freedom. The more general case can be treated straightforwardly, at least at a formal level. We will also not consider the most general actions but only those which lead to phase spaces with a cotangential bundle topology. For the more general cases see the cited literature.

Definition 24.1.1. *Consider a Lagrangian function $L : T_*(C) \to \mathbb{C}; (q^a, v^a) \mapsto L(q, v)$ on the tangential bundle over the configuration manifold C where $v := \dot{q}$ (velocity) defines the corresponding action principle.*

(i) The map

$$\rho_L : T_*(C) \to T^*(C); (q, v) \mapsto \left(q, p(q, v) := \frac{\partial L}{\partial v}(q, v) \right) \qquad (24.1.1)$$

is called a Legendre transformation.

(ii) A Lagrangian is called singular provided that ρ_L is not surjective, that is,

$$\det \left(\left(\frac{\partial^2 L}{\partial v^a \partial v^b} \right)^m_{a,b=1} \right) = 0 \qquad (24.1.2)$$

For singular Lagrangians it is not possible to solve the velocities in terms of the momenta, the underlying reason being that the Lagrangian is invariant under certain symmetries.

Let $m = \dim(\mathcal{C})$ and suppose that the rank of the matrix in (24.1.2) is $m - r$ with $0 < r \leq m$. By the inverse function theorem we can solve (at least locally) $m - r$ velocities for $m - r$ momenta and the remaining velocities, that is w.l.g.

$$p_A = \frac{\partial L}{\partial v^A}(q, v) \;\Rightarrow\; v^A = u^A(q^a, p_A, v^i) \tag{24.1.3}$$

where $a, b, \ldots = 1, \ldots m$; $A, B, \ldots = 1, \ldots, m - r$; $i, j, \ldots = m - r + 1, \ldots, m$. It follows that inserting (24.1.3) into the remaining equations $p_i = \partial L / \partial v^i$ cannot depend on the v^i any more, as otherwise the rank would exceed $m - r$. We therefore obtain r equations of the form

$$p_i = \left[\frac{\partial L}{\partial v^i}(q, v) \right]_{v^A = u^A(q^a, p_A, v^j)} =: \pi_i(q^a, p_A) \tag{24.1.4}$$

which show that the p_a are not independent of each other.

Definition 24.1.2

(i) *The functions*

$$\phi_i(q^a, p_a) := p_i - \pi_i(q^a, p_A) \tag{24.1.5}$$

are called primary constraints.

(ii) *The function*

$$H'(q^a, p_a, v^i) := [p_a v^a - L(q^a, p_a)]_{v^a = u^a(q^a, p_A, v^i)} \tag{24.1.6}$$

is called the primary Hamiltonian corresponding to L.

Lemma 24.1.3. *The primary Hamiltonian is linear in v^i with coefficients ϕ_i.*

Proof: Differentiating the expression

$$H'(q^a, p_a, v^i) = p_A u^A(q^a, p_B, v^j) + p_i v^i - L(q^a, u^A(q^a, p_B, v^j), v^i) \tag{24.1.7}$$

with respect to v^i we obtain

$$\frac{\partial H'(q^a, p_a, v^j)}{\partial v^i} = \left[p_A - \left(\frac{\partial L(q^a, v^a)}{\partial v^A} \right)_{v^A = u^A} \right] \frac{\partial u^A}{\partial v^i} + \left[p_i - \left(\frac{\partial L(q^a, v^a)}{\partial v^i} \right)_{v^A = u^A} \right]$$

$$= [p_i - \pi_i(q^a, p_A)] = \phi_i(q^a, p_a) \tag{24.1.8}$$

$\square$

We conclude that we may write

$$H'(q^a, p_a) = \tilde{H}(q^a, p_a) + v^i \phi_i(q^a, p_a) \tag{24.1.9}$$

where the new Hamiltonian $\tilde{H}$ is independent of the remaining velocities v^i.

Theorem 24.1.4. *The Hamiltonian equations*

$$\dot{q}^a = \frac{\partial H'}{\partial p_a}, \quad \dot{p}_a = -\frac{\partial H'}{\partial q^a}, \quad 0 = \frac{\partial H'}{\partial v^i} \tag{24.1.10}$$

are equivalent to the Euler–Lagrange equations

$$\dot{q}^a = v^a, \quad \frac{\partial L}{\partial q^a} = \left[\frac{d}{dt}\frac{\partial L}{\partial v^a}\right]_{v=\dot{q}} \tag{24.1.11}$$

We leave the simple proof (just use the definitions carefully) to the reader.

The phase space $\mathcal{M}$ of the constrained system is thus coordinatised by the q^a, p_a while the v^i are *Lagrange multipliers*, they do not follow any prescribed dynamical trajectory and are completely arbitrary. Our constrained phase space is equipped with the standard symplectic structure

$$0 = \{q^a, q^b\} = \{p_a, p_b\} = \{q^a, v^i\} = \{p_a, v^i\}, \ \{p_a, q^b\} = \delta_a^b \tag{24.1.12}$$

and the Hamiltonian H'.

The primary constraints force the system to the submanifold of the phase space defined by $\phi_i = 0$, $i = 1, \ldots, r$ for which we use the shorthand notation $\phi = 0$. This is consistent with the dynamics if and only if that submanifold is left-invariant, that is,

$$\dot{\phi}_i = \{H', \phi_i\} = \{\tilde{H}, \phi_i\} + v^j\{\phi_j, \phi_i\} \tag{24.1.13}$$

vanishes on the constraint surface $\bar{\mathcal{M}} := \mathcal{M}_{\phi=0}$ of the phase space. Now those primary constraints fall into the following three categories:

1. $[\dot{\phi}_i]_{\phi=0} \equiv 0$ for $i = 1, \ldots, \alpha$ is identically satisfied for any v^i.
2i. $[\dot{\phi}_i]_{\phi=0} \neq 0$ and $\{\phi_j, \phi_i\}_{\phi=0} = 0$ for all $j = 1, \ldots, r$ and $i = \alpha + 1, \ldots, \beta$.
2ii. $[\dot{\phi}_i]_{\phi=0} \neq 0$ for generic v^i but the matrix $\{\phi_j, \phi_i\}_{\phi=0}$ with $j = 1, \ldots, r$; $i = \beta + 1, \ldots, r$ has maximal rank $r - \beta$.

In case (2ii) we do not allow that the rank is smaller than $r - \beta$ since then we cannot find v^i in order to set $[\dot{\phi}_i]_{\phi=0} = 0$ and the theory would become inconsistent. Inconsistent theories have to be ruled out anyway.

Let us now extend the set of primary constraints by the $\phi_i := \dot{\phi}_{i-r+\alpha}$ with $i = r + 1, \ldots, r + \beta - \alpha$ and redefine r by $r \to r' := r + \beta - \alpha$. Now iterate the above case analysis (notice that H' always only contains the first r constraints while $\phi = 0$ means $\phi_i = 0$, $i = 1, \ldots, r'$) until case (2i) no longer appears ($\beta = \alpha$). The iteration stops after at most $2m - r$ steps because in each step the number of (automatically functionally independent) constraints increases by at least one and $2m$ constraints constrain the phase space to a discrete set of points.

Definition 24.1.5. *The constraints ϕ_i, $i = r' - r$ are called secondary constraints. Here r' is the value of the redefined r after the last iteration step.*

It follows that at the end of the procedure we have $[\dot{\phi}_i]_{\phi=0} \equiv 0$ identically for $i = 1, \ldots, \alpha$ for any choice of v^i and some $0 \leq \alpha \leq r'$ and the matrix $\{\phi_j, \phi_i\}_{\phi=0}$ with $j = 1, \ldots, r$; $i = \alpha + 1, \ldots, r'$ with $r' \geq r$ has maximal rank $r' - \alpha \leq r$. Let now $v^j = v_0^j + \lambda^\mu v_\mu^j$ where $v_0^j(q^a, p_a)$ is a special solution of the inhomogeneous linear equation

$$\{\tilde{H}, \phi_i\}_{\phi=0} + v^j \{\phi_j, \phi_i\}_{\phi=0} = 0 \tag{24.1.14}$$

and $v_\mu^j(q^a, p_a)$, $\mu = 1, \ldots, r - (r' - a)$ is a basis for the general solution of the homogeneous system. We define

$$H := \tilde{H} + v_0^j \phi_j, \quad \varphi_\mu := v_\mu^j \phi_j \tag{24.1.15}$$

24.2 First- and second-class constraints and the Dirac bracket

Definition 24.2.1. *A function $f \in C^\infty(\mathcal{M})$ is called of first class provided that $\{\phi_j, f\}_{\phi=0} = 0$ for all $j = 1, \ldots, r'$, otherwise of second class.*

Lemma 24.2.2

(i) *The functions φ_μ, H are of first class.*
(ii) *The first-class functions form a subalgebra of the Poisson algebra on $\mathcal{M}$.*

Proof

(i) is clear from the construction.
(ii) follows by realising that if f, f' are first class then there exist functions f_{ij}, f'_{ij} with $i, j = 1, \ldots, r'$ such that $\{\phi_i, f\} = f_{ij}\phi_j, \{\phi_i, f'\} = f'_{ij}\phi_j$. A short calculation then reveals that $\{\phi_i, \{f, f'\}\}_{\phi=0} = 0$.
$\square$

Let now $H_\lambda := H + \lambda^\mu \varphi_\mu$. Since at $\phi = 0$ the finite time evolution of a function f should be independent of the arbitrary parameters λ^μ, we require that $\{H_{\lambda_1}, \ldots, \{H_{\lambda_N}, f\}, \ldots\}_{\phi=0}$ is independent of the $\lambda_1, \ldots \lambda_N$ for any $N = 1, 2, \ldots$. It is easy to see from the above lemma that this is automatically the case if f is of first class. However, since the multiple Poisson brackets contain only the first-class constraints φ_μ it is actually sufficient that $\{f, \varphi_\mu\}_{\phi=0}$ for all μ.

This motivates us to extend the set of first-class constraints φ_μ already found to a maximal set C_μ, $\mu = 1, \ldots, k$ with $k \geq r - (r' - a)$ and to add them to the Hamiltonian with additional Lagrange multipliers. Denote the subset of the constraints ϕ_i functionally independent of the C_μ, that is, the second-class constraints, by ϕ_I, $I = 1, \ldots, r' - k$.

Definition 24.2.3

(i) *The set C_μ is called the set of generators of gauge transformations.*
(ii) *A function $f \in C^\infty(\mathcal{M})$ is called a weak Dirac observable provided that $\{f, C_\mu\}_{\phi=0}$ for all $\mu = 1, \ldots, k$. It is called a strong Dirac observable if this equation holds without the restriction $\phi = 0$, that is, everywhere on the phase space.[1]*

---

[1] One does not require $\{f, \phi_j\} = 0 \ \forall j = 1, \ldots, r'$ because this is inconsistent in general: by the Jacobi identity $0 = \{\phi_i, \{\phi_j, f\}\} - \{\phi_j, \{\phi_i, f\}\} = -\{f, \{\phi_i, \phi_j\}\}$. But $\{\phi_i, \phi_j\}$ is no constraint in general, so f would be overconstrained.

(iii) The extended Hamiltonian is defined by

$$H_\lambda = H + \lambda^\mu C_\mu \tag{24.2.1}$$

The nomenclature stems from the fact that $\{C_\mu, f\}$ can be interpreted as an infinitesimal motion generated by the flow of the Hamiltonian vector field associated with C_μ and an observable is invariant under this flow at least on $\bar{\mathcal{M}}$. That all first-class constraints C_μ should be considered as generators of gauge transformations (so-called Dirac conjecture) and not only the φ_μ which appear in H' is motivated by the fact that only the C_μ form a closed constraint algebra (see below), however, it does not follow strictly from the formalism. That it is physically correct to proceed that way has been confirmed in countless examples though and can even be proved under some restrictions [263].

Lemma 24.2.4. *We have that $r' - k = 2m'$ is even and that $(\{\phi_I, \phi_J\}_{\phi=0})$ is an invertible matrix.*

Proof: Suppose that $(\{\phi_I, \phi_J\}_{\phi=0})$ is singular, then there exist numbers $x^J \in \mathbb{C}$ such that $\{\phi_I, C_0\}_{\phi=0} = 0$ for all I where $C_0 = x^J \phi_J$. Since $\{C_\mu, C_0\}_{\phi=0} = 0$ anyway we find $\{\phi_i, C_0\}_{\phi=0} = 0$ for all i so that C_0 is a first-class constraint independent of the C_μ. This is a contradiction to the assumed maximality. It follows that $r' - k$ is even since $(\{\phi_I, \phi_J\}_{\phi=0})$ is an antisymmetric matrix. $\Box$

Definition 24.2.5. *Let $c^{IJ} := ((\{\phi_K, \phi_L\})^{-1})^{IJ}$. The Dirac bracket is defined by*

$$\{f, f'\}^* := \{f, f'\} + \{\phi_I, f\} c^{IJ} \{\phi_J, f'\} \tag{24.2.2}$$

Theorem 24.2.6. *The Dirac bracket defines a degenerate two-form on $\mathcal{M}$ with kernel spanned by χ_{ϕ_I} where χ_f denotes the Hamiltonian vector field of $f \in C^\infty(\mathcal{M})$ with respect to the symplectic structure determined by $\{.,.\}$. The Dirac bracket satisfies the Jacobi identity but the corresponding two-form is not necessarily gobally closed. It is possible to choose local canonical coordinates and equivalent sets of second-class constraints such that it is locally closed.*

Proof: Recall from Section 19.3 our conventions: $i_{\chi_f}\Omega + df = 0$ and $\{f, f'\} = -i_{\chi_f} i_{\chi_{f'}} \Omega = \chi_f(f') = i_{\chi_f}(df')$ defines the relation between a non-degenerate symplectic structure Ω, Hamiltonian vector field χ_f and Poisson bracket $\{.,.\}$. Also for a p-form $\omega = \omega_{\alpha_1 \dots \alpha_p} dx^{\alpha_1} \wedge \dots \wedge dx^{\alpha_p}$ we define exterior derivative, contraction with vector fields v and Lie derivative by

$$d\omega = [\partial_{\alpha_1} \omega_{\alpha_2 \dots \alpha_{p+1}}] dx^{\alpha_1} \wedge \dots \wedge dx^{\alpha_{p+1}}$$
$$i_v \omega = p \, v^\alpha \omega_{\alpha \alpha_1 \dots \alpha_{p-1}} dx^{\alpha_1} \wedge \dots \wedge dx^{\alpha_{p-1}}$$
$$\mathcal{L}_v \omega = [i_v \cdot d + d \cdot i_v]\omega \tag{24.2.3}$$

Let $\Omega = \frac{1}{2}\Omega_{\alpha\beta}dx^\alpha \wedge dx^\beta$ (here $\alpha, \beta, \ldots = 1, \ldots, 2m$). Define the inverse of $\Omega_{\alpha\beta}$ by $\Omega^{\alpha\gamma}\Omega_{\gamma\beta} = \delta^\alpha_\beta$. Then it is easy to verify that $\chi^\alpha_f = \Omega^{\alpha\beta}\partial_\beta f$ and therefore $\Omega^{\alpha\beta} = -\{x^\alpha, x^\beta\}$.

We first of all verify that a non-degenerate two-form is closed if and only if the associated Poisson bracket satisfies the Jacobi identity

$$\{f_{[1}, \{f_2, f_{3]}\}\} = 0 \tag{24.2.4}$$

To see this we just need to use the formula $\{f, f'\} = -\Omega^{\alpha\beta}(\partial_\alpha f)(\partial_\beta f')$ and the fact that $\delta\Omega^{-1} = -\Omega^{-1}(\delta\Omega)\Omega^{-1}$ to conclude that (24.2.4) is equivalent with $\partial_{[\alpha}\Omega_{\beta\gamma]} = 0$.

Next we verify directly from the definition for the Dirac bracket and by similar methods applied to c_{IJ} that on all of $\mathcal{M}$ the Jacobi identity

$$\{f_{[1}, \{f_2, f_{3]}\}^*\}^* = 0 \tag{24.2.5}$$

holds. Moreover

$$\{f, \phi_I\}^* = -\{\phi_I, f\}^* = 0 \tag{24.2.6}$$

for any $I = 1, \ldots, 2m'$ and $f \in C^\infty(\mathcal{M})$.

It is easy to see that the Dirac bracket corresponds to the two-form

$$\Omega^* = \Omega - c^{IJ} d\phi_I \wedge d\phi_J \tag{24.2.7}$$

via $\{f, f'\}^* =: -\Omega^*_{\alpha\beta}\chi^\alpha_f\chi^\beta_{f'}$. It follows from (24.2.6) that $i_{\chi_{\phi_I}}\Omega^* = 0$, hence Ω^* is degenerate and therefore not necessarily closed. Taking the exterior derivative we see that this two-form is closed if and only if $dc^{IJ} \wedge d\phi_I \wedge d\phi_J = 0$, which is not necessarily the case. However, by an appeal to Darboux' theorem [218] we can always replace the second-class constraints by equivalent ones such that $c^{IJ} = \text{const.}$ [659]. In such coordinates then Ω^* is locally closed. Moreover, we can find local Darboux coodinates $x^\alpha = (x^a, x^I := \phi_I)$ with respect to Ω with $a = 1, \ldots, 2(m - m')$, $I = 1, \ldots, 2m'$ such that $\Omega^{ab} = \Omega^{ab}_0$, $\Omega^{aI} = \Omega^{aI}_0 = 0$, $\Omega^{IJ} = \Omega^{IJ}_0$ where Ω_0 denotes the standard canonical form. It follows that then locally $\Omega^* = \pi^*\Omega$ where $\pi : \mathcal{M} \to \mathcal{M}' (x^a, x^I) \mapsto (x^a, 0)$ is the projection to the constraint manifold defined by second-class constraints. □

Notice that for the first-class constraints C_μ and the Hamiltonian H_λ we have for any $f \in C^\infty(\mathcal{M})$ that $\{C_\mu, f\}_{\phi=0} = \{C_\mu, f\}^*_{\phi=0}$ and $\{H_\lambda, f\}_{\phi=0} = \{H_\lambda, f\}^*_{\phi=0}$ (more generally this holds for any first-class function). Thus, on the constraint surface the Dirac bracket defines the same equations of motion as the original bracket. Notice, however, that in general $\{f, f'\}_{\phi=0} \neq \{f, f'\}^*_{\phi=0}$ even if f, f' are (weak) Dirac observables, unless at least one of f, f' is first class. Thus, the Dirac bracket changes the symplectic structure among observables, thus possibly complicating quantisation. However, the Dirac bracket is easily seen to have the important property

$$(\{f_{|\mathcal{M}'}, f'_{|\mathcal{M}'}\}^*)_{|\mathcal{M}'} = (\{f, f'\}^*)_{|\mathcal{M}'} \tag{24.2.8}$$

where $\mathcal{M}'$ is the constraint surface by setting the second-class constraints to zero. Thus, with respect to a Dirac bracket we may set the second-class constraints equal to zero before or after evaluating it. Thus, the virtue of the Dirac bracket is that one can either work on the full phase space or one can explicitly solve the second-class constraints without changing the symplectic structure.

Because of this and because the equations of motion and the gauge motions generated by the first-class constraints on the constraint surface are unaltered irrespective of whether we use the original Poisson bracket or the Dirac bracket, we may just forget about the second-class constraints for the rest of the analysis and work off the constraint surface defined by the second-class constraints while using the Dirac bracket. Notice, however, that the Dirac bracket generically alters the bracket between previously canonically conjugate quantities, which complicates the quantisation of the Dirac bracket tremendously in general. Hence one will try to avoid second-class constraints as much as possible.

The reason for treating the second-class constraints differently from the first-class constraints is as follows: the cleanest way to treat a constrained Hamiltonian system is to compute the full constraint surface $\bar{\mathcal{M}} = \{m \in \mathcal{M}; \ \phi_i(m) = 0 \ \forall \ i = 1, \ldots, r'\}$. Since the Hamiltonian is a first-class function, its Hamiltonian flow preserves the constraint surface. Since the Hamiltonian depends on arbitrary parameters, and physical observables must be independent of those, we have required that those observables must be independent of the Hamiltonian flow generated by the first-class constraints, at least on the constraint surface. This is, however, not possible to require for the second-class constraints because their Hamiltonian flow does not preserve the constraint surface.

Thus, what one should do is compute the gauge orbits $[m]$ with respect to the first-class constraints of points m on the full constraint surface (defined by both first- and second-class constraints). The manifold so obtained is called the reduced phase space $\widehat{\mathcal{M}}$ and observables (gauge-invariant functions) are naturally functions on $\widehat{\mathcal{M}}$. The reduced phase space is automatically equipped with a symplectic structure that one obtains locally by looking for a suitable set of first-class constraints and conjugate Darboux coordinates (together with a suitable choice of second-class constraints as Darboux coordinates). See [218] for details and Section 19.3.2 for basics of symplectic reduction. One would then quantise the reduced system.

The reason for why that is not always done is that for non-linear systems it is extremely difficult to compute $\bar{\mathcal{M}}, \widehat{\mathcal{M}}$ even classically and the reduced symplectic structure on the observables might be so complicated that it is very hard to find a representation of the associated canonical commutation relations in the quantum theory. Thus, in order to get started with the quantisation, Dirac has proposed solving the constraints not before but after the quantisation. Roughly speaking, we turn the constraints into operators and impose that physical states satisfy

$$\hat{C}_\mu \psi = 0 \qquad (24.2.9)$$

(this equation must actually be read in a generalised sense, see Chapter 30). Notice that we impose this only for the first-class constraints. To see why, notice that the first-class constraints must satisfy a subalgebra of the Poisson algebra (we know that $\{C_\mu, C_\nu\}_\phi = 0$ therefore $\{C_\mu, C_\nu\} = f_{\mu\nu}{}^\rho C_\rho + f_{\mu\nu}{}^I \phi_I$ for some structure functions $f_{\mu\ni}^\rho, f_{\mu\nu}^I$ and since the Poisson bracket is first class again we know that $f_{\mu\nu}^I = 0$). Therefore upon suitable operator ordering for a solution of (24.2.9) we have that

$$0 = [\hat{C}_\mu, \hat{C}_\mu]\psi = \hat{f}_{\mu\nu}{}^\rho \hat{C}_\rho \psi \qquad (24.2.10)$$

is a consistent equation. However, if we extend (24.2.9) to second-class constraints we get the contradiction

$$0 = [\hat{\phi}_I, \hat{\phi}_J]\psi \neq 0 \qquad (24.2.11)$$

since the commutator is proportional to a quantisation of c_{IJ}, which in the worst case is a constant (in general an operator which is not constrained to vanish). Thus, one solves the second-class constraints simply by restricting the argument of the wave function to the constraint surface $\mathcal{M}'$ defined by the second-class constraints (using the second-class constraints themselves as coordinates).

Two remarks are in order:

1. Notice that every second-class constraint classically removes one degree of freedom while every first-class constraint removes two since not only do we delete degrees of freedom, but we also compute gauge orbits. However, since the number of second-class constraints is always even, the reduced phase space has always again an even number of physical degrees of freedom (otherwise it would not have a non-degenerate symplectic structure). One may then wonder how it is possible that we just impose the constraint on the state and do not compute its gauge orbit in addition. The answer is that the wave function already depends only on half of the number of kinematical degrees of freedom (configuration space). The imposition of the constraint is actually the condition that the state be gauge-invariant and simultaneously the constraint operator is deleted.

2. One may also wonder why we do not simply remove the first-class constraints as well. The procedure to do this is called *gauge fixing*. Thus, we impose additional conditions $k_\mu = 0$ which ideally pick from each gauge orbit a unique representative and such that the matrix $(\{k_\mu, C_\nu\})$ is non-degenerate on the constraint surface. One may then remove the constraints C_μ by considering the system k_μ, C_μ as second-class constraints and by using the associated Dirac bracket. The reason for not doing this is that it is actually very problematic: usually functions with the required properties simply do not exist, for instance gauge orbits can be cut more than once, leading to the so-called Gribov copies [263, 659]. Also, the geometric structure of the system is very much veiled and different gauge conditions may lead to different physics.

Finally, let us display a simple example: consider the phase space $\mathcal{M} = T^*(\mathbb{R}^3)$ with constraints $\phi_1 = p_1$, $\phi_2 = q^2$, $\phi_3 = p_2$ where q^a, p_a, $a = 1, 2, 3$ are canonically conjugate configuration and momentum coordinates. It is easy to see that $C = \phi_1$ is the only first-class constraint and that ϕ_2, ϕ_3 is a pair of second-class constraints. For instance, functions which are independent of q^1, q^2, p_2 are first class but also the Hamiltonian $H = -(q^1)^2 + \sum_{a=1}^{3}[(q^a)^2 + (p_a)^2]$ and any function which is independent of q^1 is an observable but also the function $f = p_1 q^1$. The gauge motions generated by C are translations in the q^1 direction so that the value of q^1 is pure gauge. Obviously then the only second-class constraint reduced phase space is $\mathcal{M}' = T^*(\mathbb{R}^2)$, while the fully reduced phase space is $\widehat{\mathcal{M}} = T^*(\mathbb{R}^1)$.

25

Tools from measure theory

For an introduction to general measure theory see, for example, the beautiful textbook [552]. For more advanced topics concerning the extension theory of measures from self-consistent families of projections to σ-additive ones see, for example, [532].

25.1 Generalities and the Riesz–Markov theorem

Recall the notion of a topology and of continuous functions from Chapter 18.

Definition 25.1.1

(i) *Let X be a set. Then a collection of subsets $\mathcal{U}$ of X is called a σ-algebra provided that*

    1. $X \in \mathcal{U}$,

    2. $U \in \mathcal{U}$ implies $X - U \in \mathcal{U}$ and

    3. $\mathcal{U}$ is closed under countable unions, that is, if $U_n \in \mathcal{U}$, $n = 1, 2, \ldots$ then also $\cup_{n=1}^{\infty} U_n \in \mathcal{U}$.

    The sets $U \in \mathcal{U}$ are called measurable and a space X equipped with a σ-algebra a measurable space.

(ii) *Let X be a measurable space and let Y be a topological space. A function $f : X \to Y$ is said to be measurable provided that the pre-image $f^{-1}(V) \subset X$ of any open set $V \subset Y$ is a measurable subset in X.*

(iii) *Let X be a topological space. The smallest σ-algebra on X that contains all open (and due to (2) therefore all closed) sets of X is called the Borel σ-algebra of X. The elements of the Borel σ-algebra are called Borel sets.*

Given a collection $\mathcal{U}$ of subsets of X which is not yet a topology (σ-algebra) the weakest topology (smallest σ-algebra) containing $\mathcal{U}$ is obtained by adding to the collection the sets $X, \emptyset$ as well as arbitrary unions plus finite intersections (countable unions and intersections). Notice the similarity between a collection of sets $\mathcal{U}$ that qualify for a σ-algebra and a topology: in both cases the sets $X, \emptyset$ belong to $\mathcal{U}$ but while open sets are closed under arbitrary unions and finite intersections, measurable sets are closed under countable unions and intersections. Note also that if X, Y are topological spaces and $f : X \to Y$ is continuous then f is automatically measurable if X is equipped with the Borel σ-algebra.

Definition 25.1.2. *A complex measure μ on a measurable space $(X,\mathcal{U})$ is a function $\mu : \mathcal{U} \to \mathbb{C} - \{\infty\}$; $U \mapsto \mu(U)$ which is countably (or σ-)additive, that is,*

$$\mu\left(\bigcup_{n=1}^{\infty} U_n\right) = \sum_{n=1}^{\infty} \mu(U_n) \tag{25.1.1}$$

for any mutually disjoint measurable sets U_n. A positive measure is also a σ-additive map $\mu : \mathcal{U} \to \mathbb{R}^+ \cup \{0, \infty\}$ which however is positive semidefinite and may take the value ∞ with the convention $0 \cdot \infty = 0$ (which makes $[0, \infty]$ a set in which commutative, distributive and associative law hold). To avoid trivialities we assume that $\mu(U) < \infty$ for at least one measurable set U. A measure is called a probability measure if $\mu(X) = 1$. The triple $(X,\mathcal{U}, \mu)$ is called a measure space.

In what follows we will always assume that μ is a positive measure. Sets $B \in \mathcal{B}$ such that $\mu(B) = 0$ are called of μ-measure zero. We say that a property holds μ-a.e. (almost everywhere) on X if it holds strictly except on measure zero subsets.

A very powerful tool in measure theory are characteristic functions of subsets of X.

Definition 25.1.3. *A function $s : X \to \mathbb{C}$ on a measurable space $(X,\mathcal{U})$ is called simple provided its range consists of finitely many points only. If $z_k \in \mathbb{C}$, $k = 1, \ldots, N$ are these values and $S_k = s^{-1}(\{z_k\})$ then $s = \sum_{k=1}^{N} z_k \chi_{S_k}$, where χ_S with $\chi_S(x) = 1$ if $x \in S$ and $\chi_S(x) = 0$ otherwise, is called the characteristic function of the subset $S \subset X$. Obviously, a simple function is measurable if and only if the S_k are measurable.*

The justification for this definition lies in the following lemma.

Lemma 25.1.4. *Let $f : X \to [0, \infty]$ be measurable. Then there exists a sequence of measurable simple functions s_n such that*

(a) $0 \le s_1 \le s_2 \le \ldots \le f$
(b) $\lim_{n\to\infty} s_n(x) = f(x)$ pointwise in $x \in X$.

The proof can be found in [552], Theorem 1.17.

Definition 25.1.5

(i) *For a simple measurable function $s = \sum_{k=1}^{N} z_k \chi_{S_k}$ with $z_k > 0$ on a measure space $(X,\mathcal{U}, \mu)$ with positive measure μ we define*

$$\mu(s) := \int_X d\mu(x)s(x) := \sum_{k=1}^{N} z_k \mu(S_k) \tag{25.1.2}$$

For a general measurable function $f : X \to [0, \infty]$ we define

$$\mu(f) := \sup_{0 \le s \le f} \mu(s) \tag{25.1.3}$$

where the supremum is taken over the simple, positive measurable functions that are nowhere larger than f. The number $\mu(f)$ is called the Lebesgue integral of f. For a general complex-valued, measurable function f one can show that we have a unique split as $f = u + iv$, $u = u_+ - u_-$, $v = v_+ - v_-$ with non-negative measurable functions $u_\pm, v_\pm$ and the integral is defined as $\mu(f) = \mu(u_+) - \mu(u_-) + i[\mu(u_+) - \mu(u_-)]$. Also $|f|$ can be shown to be measurable.

(ii) A measure μ is called positive definite if for every non-negative measurable function f the condition $\mu(f) = 0$ implies $f = 0$ almost everywhere (a.e., i.e., up to measure zero sets).

Of fundamental importance are conditions under which one is allowed to exchange integration and taking limits or orders of integration.

Theorem 25.1.6. *Let $(X, \mathcal{U}, \mu)$ be a measure space with positive measure μ and let (f_n) be a sequence of measurable functions that converges pointwise on X to the function f.*

(i) Lebesgue monotone convergence theorem
Suppose that $0 \leq f_n(x) \leq f_{n+1}(x)$ for all $x \in X$. Then f is measurable and $\lim_{n\to\infty} \mu(f_n) = \mu(f)$.

(ii) Lebesgue dominated convergence theorem
A function F is said to be in $L_1(X, d\mu)$ if it is measurable and $\mu(|F|) < \infty$. Suppose now that there exists $F \in L_1(X, d\mu)$ such that $|f_n(x)| \leq |F(x)|$ for all $x \in X$. Then $f \in L_1(X, d\mu)$ and $\lim_{n\to\infty} \mu(|f - f_n|) = 0$.

(iii) Fubini's theorem
Let $(X, \mathcal{B}, \mu)$ and $(Y, \mathcal{C}, \nu)$ be σ-finite measure spaces (see 25.1.10). Consider the smallest σ-algebra $\mathcal{B} \times \mathcal{C}$ which contains all the 'rectangles' $B \times C, B \in \mathcal{B}, C \in \mathcal{C}$. Let $f : X \times Y \to \mathbb{C}$ be a measurable function on $\mathcal{B} \times \mathcal{C}$. Then $f_x : Y \to \mathbb{C}$; $f_x(y) := f(x, y)$ and $f^y : X \to \mathbb{C}$; $f^y(x) := f(x, y)$ are measurable functions on $\mathcal{C}$ and $\mathcal{B}$ respectively. For $D \in \mathcal{B} \times \mathcal{C}$ we set $D_x := \{y \in Y : (x, y) \in D\}$, $D^y := \{x \in X : (x, y) \in D\}$ and define the product measure $\mu \times \nu$ on $(X \times Y, \mathcal{B} \times \mathcal{C})$ by $\mu \times \nu(D) := \int_X d\mu(x)\, \nu(D_x) = \int_Y d\nu(y)\, \nu(D^y)$ where equality of these two expressions can be shown. The measure $\mu \times \nu$ is automatically σ-finite again.
1. If $F^* \in L_1(X, d\mu)$ where $F^*(x) := \int_Y d\nu|f_x|$ then $f \in L_1(X \times Y, d\mu \times d\nu)$.
2. If $f \in L_1(X \times Y, d\mu \times d\nu)$ then $f_x \in L_1(Y, d\nu)$ for a.a. $x \in X$ and $f^y \in L_1(X, d\mu)$ for a.a. $y \in Y$.
3. Let $F(x) := \int_Y d\nu(y) f_x$, $G(y) := \int_X d\mu(x) f^y$. Then $F \in L_1(X, d\mu)$, $G \in L_1(Y, d\nu)$ and $\int_X d\mu\, F = \int_Y d\nu\, G = \int_{X \times Y} d(\mu \times \nu)\, f$.

It is easy to see that $\lim_{n\to\infty} \mu(|f - f_n|) = 0$ implies $\lim_{n\to\infty} \mu(f_n) = \mu(f)$. Notice that (iii1), (iii2) and (iii3) imply that the order of iterated integrals is immaterial.

Another convenient observation is the following.

Theorem 25.1.7. *Let $(X, \mathcal{U}, \mu)$ be a measure space. Let $\mathcal{U}'$ be the collection of all $S \subset X$ such that there exist $U, V \in \mathcal{U}$ with $U \subset S \subset V$ and $\mu(V - U) = 0$ (in particular $\mathcal{U} \subset \mathcal{U}'$). Define $\mu'(S) = \mu(U)$ in that case. Then $(X, \mathcal{U}', \mu')$ is a measure space again, called the completion of $(X, \mathcal{U}, \mu)$.*

The theorem says that any measure can be completed. It means that if we have a set which is not measurable but can be sandwiched between measurable sets whose difference has zero measure, then we can add the set to the measurable sets and its measure is given by that of the sandwiching sets.

Definition 25.1.8

(i) *A set $Y \subset X$ in a measure space $(X, \mathcal{U}, \mu)$ is called thick or a support for μ provided that for any measurable set $U \in \mathcal{U}$ the condition $U \cap Y = \emptyset$ implies $\mu(U) = 0$. A support for μ will be denoted by $supp(\mu)$.*

(ii) *For two measures μ_1, μ_2 on the same measurable space we say that μ_1 is regular (or absolutely continuous) with respect to μ_2 iff $\mu_2(U) = 0$ for $U \in \mathcal{U}$ implies $\mu_1(U) = 0$. They are called mutually singular iff $supp(\mu_1) \cap supp(\mu_2) = \emptyset$.*

If Y is a measurable support then $X - Y$ is measurable and since $Y \cap (X - Y) = \emptyset$ we have $\mu(X - Y) = 0$, explaining the word support. If Y is a support not measurable with respect to μ one can define $\mathcal{U}' = [\mathcal{U} \cap Y] \cup Y$, $\mu'(U \cap Y) = \mu(U)$ and get a measure space $(Y, \mathcal{U}', \mu')$ for which Y is measurable, called the trace. A given support does not mean that there are not smaller sets which are still thick.

Theorem 25.1.9 (Radon–Nikodym). *Let μ_2 be a positive σ-finite (see below) measure and μ_1 a complex measure. Then there is a unique (so-called Lebesgue) decomposition into complex measures $\mu_1 = \mu_1^a + \mu_1^s$ such that μ_1^a, μ_1^s are respectively absolutely continuous and mutually singular with respect to μ_2. Moreover, there exists a unique function $f \in L_1(X, d\mu_2)$, called the Radon–Nikodym derivative, such that $d\mu_1^a = f \, d\mu_2$. If μ_2 is a positive, σ-finite measure, then f is non-negative.*

The following two definitions prepare to state the Riesz representation (or Riesz–Markov) theorem which will be of fundamental importance for our applications.

Definition 25.1.10

(i) *A topological space is said to be locally compact if every point $x \in X$ has an open neighbourhood whose closure is compact.*

(ii) *A subset $S \subset X$ of a topological space X is said to be σ-compact if it is a countable union of compact sets.*

(iii) *A subset $S \subset X$ in a measure space $(X, \mathcal{U}, \mu)$ with positive measure μ is said to be σ-finite if S is the countable union of measurable sets U_n with $\mu(U_n) < \infty$ for all $n \in \mathbb{N}$. The measure μ is said to be σ-finite if X itself is σ-finite.*

Definition 25.1.11. *Let X be a locally compact Hausdorff space and let $\mathcal{U}$ be its naturally defined Borel σ-algebra.*

(*i*) *A measure μ defined on the Borel σ-algebra is called a Borel measure.*

(*ii*) *A Borel set S is said to be outer regular with respect to a positive Borel measure μ provided that*

$$\mu(S) = \inf\{\mu(O); \ S \subset O; \ O \in \mathcal{U} \ open\} \tag{25.1.4}$$

(*iii*) *A Borel set S is said to be inner regular with respect to a positive Borel measure μ provided that*

$$\mu(S) = \sup\{\mu(K); \ S \supset K; \ K \in \mathcal{U} \ compact\} \tag{25.1.5}$$

(*iv*) *If μ is a positive Borel measure and every Borel set is both inner and outer regular then μ is called regular.*

Definition 25.1.12

(*i*) *Let X be a topological space. The support supp(f) of a function $f : X \to \mathbb{C}$ is the closure of the set $\{x \in X; \ f(x) \neq 0\}$. The vector space of continuous functions of compact support is denoted by $C_0(X)$.*

(*ii*) *A linear functional $\Lambda : \mathcal{F} \to \mathbb{C}$ on the vector space of functions $\mathcal{F}$ over a set X is called positive if $\Lambda(f) \in [0, \infty)$ for any $f \in \mathcal{F}$ such that $f(x) \in [0, \infty)$ for all $x \in X$.*

Theorem 25.1.13 (Riesz–Markov theorem)

(*i*) *Let X be a locally compact Hausdorff space and let $\Lambda : C_0(X) \to \mathbb{C}$ be a positive linear functional on the space of continuous, complex-valued functions of compact support in X. Then there exists a σ-algebra $\mathcal{U}$ on X which contains the Borel σ-algebra and a unique positive measure μ on $\mathcal{U}$ such that Λ is represented by μ, that is,*

$$\Lambda(f) = \int_X d\mu(x) f(x) \quad \forall f \in C_0(X) \tag{25.1.6}$$

Moreover, μ has the following properties:

1. *$\mu(K) < \infty$ if $K \subset X$ is compact.*
2. *For every $S \in \mathcal{U}$ property (25.1.4) holds.*
3. *For every open $S \in \mathcal{U}$ with $\mu(S) < \infty$ property (25.1.5) holds.*
4. *If $S' \subset S \in \mathcal{U}$ and $\mu(S) = 0$ then $S' \in \mathcal{U}$.*

(*ii*) *If, in addition to (i), X is σ-compact then μ has the following additional properties:*

5. *μ is regular.*
6. *For any $S \in \mathcal{U}$ and any $\epsilon > 0$ there exists a closed set C and an open set O such that $C \subset S \subset O$ and $\mu(O - C) < \epsilon$.*

7. For any $S \in \mathcal{U}$ there exist sets C' and O' which are respectively countable unions and intersections of closed and open sets respectively such that $C' \subset S \subset O'$ and $\mu(O' - C') = 0$.

A very instructive proof of this theorem can be found in [552]. It is also worth pointing out the following theorem (see, e.g., [552]) which underlines the prominent role that continuous functions play for Borel measures.

Theorem 25.1.14 (Lusin's theorem). *Let X be a locally compact Hausdorff space X with σ-algebra $\mathcal{U}$ and measure μ satisfying the properties (1), (2), (3) and (4) of Theorem 25.1.13 . Let f be a bounded measurable function with support in a measurable set of finite measure. Then there exists a sequence (f_n) of continuous functions of compact support, each of which is bounded by the same bound, such that $f(x) = \lim_{n \to \infty} f_n(x)$ almost everywhere with respect to μ (i.e., they coincide pointwise up to sets of measure zero).*

Let us also define the notion of faithfulness of measures:

Definition 25.1.15. *Let X be a locally compact Hausdorff space and let $\mathcal{U}, \mu$ have the properties of Theorem 25.1.13 . Then μ is called faithful if and only if the positive linear functional (25.1.6) determined by μ is positive definite, that is, if $f \in C_0(X)$ takes only values in $[0, \infty)$ and $\Lambda(f) = 0$ then $f = 0$.*

Notice that positive definiteness of a measure μ only allows us to conclude that $f = 0$ μ-a.e. from $\mu(f) = 0$ for positive measurable f. Faithfulness of the special kind of measures that come from positive definite linear functionals allows us to conclude $f = 0$ everywhere if f is continuous, non-negative and of compact support. This means that every open set must have positive measure for if a continuous function is positive at a point, it will be bounded away from zero in a whole open neighbourhood of that point.

The application that we have in mind is that X is not only locally compact but actually compact so that the set $C_0(X)$ coincides with $C(X)$. Hence, $C(X)$ contains the constant functions and we may w.l.g. assume that $\Lambda(1) = 1$, which is just a convenient choice of normalisation. (If X is compact, so is every closed subset, hence X is locally compact.) It is then trivially σ-compact being its own cover by compact sets. Therefore the stronger version (ii) of Theorem 25.1.13 applies and we see that by property (5) the measure μ is regular. Furthermore, property (7) tells us that every measurable set can be sandwiched between sets $C' \subset O'$ that belong to the Borel σ-subalgebra such that $C' - O'$ is of measure zero. In other words, every measurable set is a Borel set up to a set of measure zero: since $O' = S \cup (O' - S)$ we have from σ-additivity $\mu(S) = \mu(O')$ since $0 = \mu(O' - C') \geq \mu(O' - S)$ due to $O' - S \subset O' - C'$. Thus effectively the measure μ in (25.1.6) is a Borel measure and in that sense we have the following corollary.

Corollary 25.1.16. *Let X be a compact Hausdorff space and let $\Lambda : C(X) \to \mathbb{C}$ be a positive linear functional on the space of continuous functions on X with*

$\Lambda(1) = 1$. *Then there exists a unique, regular, Borel probability measure* μ *on the natural Borel* σ-*algebra* $\mathcal{U}$ *of* X *such that* μ *represents* Λ, *that is,*

$$\Lambda(f) = \int_X d\mu(x) f(x) \quad \forall f \in C(X) \tag{25.1.7}$$

Notice that regularity of μ on a compact Hausdorff space X reduces to the fact that the measure of every measurable set can be approximated arbitrarily well by open or compact (and hence closed, since in a Hausdorff space every compact subset is closed, see [533]) sets respectively. Also, Lusin's theorem simplifies to the statement that every bounded measurable function can be approximated arbitrarily well by continuous functions with the same bound up to sets of measure zero.

The notion of faithfulness actually comes from representation theory. Indeed, the origin of positive linear functionals in physics are usually states, that is, positive linear functionals ω on a unital C^*-algebra $\mathfrak{A}$ (see Chapter 27), which is not necessarily Abelian like the C^*-algebra $C(X)$ for a compact Hausdorff space X, such that $\omega(1) = 1$. Here a positive linear functional is a map $\omega : \mathfrak{A} \to \mathbb{C}; \ a \mapsto \omega(a)$ which satisfies $\omega(a^*a) \geq 0$ for any $a \in \mathfrak{A}$. Elements a of $\mathfrak{A}$ of the form b^*b are called positive, denoted $a \geq 0$ (equivalently, $a \geq 0$ iff for its spectrum $\sigma(a) \subset \mathbb{R}^+$ holds). One writes $a \geq a'$ if $a - a' \geq 0$, which equips $\mathfrak{A}$ with a partial order. We will see in Chapter 29 that positive linear functionals give rise to a representation π of the algebra on a Hilbert space via the GNS construction. If the unital C^*-algebra is Abelian then we can always think of it as an algebra of continuous functions on a compact Hausdorff space via the Gel'fand isomorphism and if the associated measure is faithful, that is, the state is positive definite then the representation is faithful (or non-degenerate), that is, $\pi(f) = 0$ if and only if $f = 0$.

Notice that every positive linear functional ω on a unital C^*-algebra $\mathfrak{A}$ is automatically bounded (continuous): if $||.||$ denotes the norm on $\mathfrak{A}$ and $*$ the involution then for any self-adjoint element $a = a^*$ we have $-||a|| \cdot 1 \leq a \leq ||a|| \cdot 1$ since $||a|| \geq r(a)$ (spectral radius). Hence $\omega(||a|| \cdot 1 \pm a) = ||a|| \omega(1) \pm \omega(a) \geq 0$. Since $\omega(1) \geq 0$ because $1 = 1^*1$ is positive, it follows that in particular $\omega(a) \in \mathbb{R}$ for self-adjoint a so that $|\omega(a)|/||a|| \leq \omega(1)$. If a is arbitrary we can decompose it uniquely into self-adjoint elements $a = a_+ + ia_-$ with $a_\pm = a_\pm^*$ and thus

$$4||a_\pm^2|| = ||(a^*)^2 + a^2 \pm (a^*a + aa^*)|| \leq ||(a^*)^2|| + ||a^2|| + ||a^*a|| + ||aa^*|| = 4||a||^2$$

where we have made use twice of the C^*-algebra property $||a^*a|| = ||a||^2$. It follows that

$$|\omega(a)|^2 = |\omega(a_+) + i\omega(a_-)|^2$$
$$= |\omega(a_+)|^2 + |\omega(a_-)|^2 \leq \omega(1)[||a_+||^2 + ||a_-||^2] \leq 2\omega(1)||a||^2$$

so a bound is given by $2\omega(1)$. One can actually show that a sharper bound is given by $\omega(1)$ even for unital Banach algebras with involution.

25.2 Measure theory and ergodicity

We now turn to another direction within measure theory.

Definition 25.2.1. *Let $(X, \mathcal{U}, \mu)$ be a measure space with a positive probability measure μ on X. Let $\lambda : G \times X \to X$; $(g, x) \mapsto \lambda_g(x)$ be a measure-preserving group action, that is, $(\lambda_g)_* \mu := \mu \circ \lambda_g^{-1} = \mu$ for all $g \in G$, in particular, λ_g preserves $\mathcal{U}$. The group action is called ergodic if the only invariant sets, that is, sets $S \in \mathcal{U}$ with $\lambda_g(S) = S$ for all $g \in G$, have measure zero or one.*

The definition captures exactly the intuitive idea of an ergodic group action, namely that it spreads any set all over X without changing its measure. It follows from the definition that a measure-preserving group action induces a unitary transformation on $L_2(X, d\mu)$ by the pull-back, that is,

$$(\hat{U}(g)f)(x) := (\lambda_g^* f)(x) = f(\lambda_g(x)) \tag{25.2.1}$$

Since the closed linear span of characteristic functions of measurable sets is all of $L_2(X, d\mu)$ as we have seen above, it follows that ergodicity is equivalent to the condition that $\hat{U}(g)f = f$ μ-a.e. for all $g \in G$ implies that $f = $ const. a.e. [*Proof:* If λ is ergodic let $f = \sum_k z_k \chi_{U_k}$. We may assume w.l.g. that the U_k are mutually disjoint since, e.g., $a\chi_U + \chi_V = a\chi_{U-V} + b\chi_{V-U} + (a+b)\chi_{U \cap V}$. Then $\hat{U}(g)f - \sum_k z_k \chi_{\lambda_{g^{-1}}(U_k)} = f$ a.e. for all $g \in G$ implies that all U_k must be invariant under λ, hence that all of them have measure zero or one. If U_k has measure zero then $\chi_{U_k} = 0$ a.e., if U_k has measure one then $X - U_k$ has measure zero so $\chi_{U_k} = \chi_X = 1$ a.e. The converse implication is similar.]

Theorem 25.2.2 (von Neumann mean ergodic theorem). *Let $\mathbb{R} \to G$; $l \mapsto y_t$ be a one-parameter group and $\hat{U} : G \to \mathcal{B}(L_2(X, d\mu))$ be a unitary representation of G. Let $\hat{P}$ be the projection on the closure of the set of a.e. invariant vectors under $\hat{U}(g_t)$, $t \in \mathbb{R}$. Then*

$$(\hat{P}f)(x) = \lim_{T \to \infty} \frac{1}{2T} \int_{-T}^{T} dt (\hat{U}(g_t)f)(x) \quad \mu\text{-a.e.} \tag{25.2.2}$$

For a proof see for instance [282]. We conclude that λ restricted to $t \mapsto g_t$ is ergodic if and only if

$$\lim_{T \to \infty} \frac{1}{2T} \int_{-T}^{T} dt f(\lambda_{g_t}(x)) = \left[\int_X d\mu(x') f(x') \right] \cdot 1 \quad \mu\text{-a.e.} \tag{25.2.3}$$

Namely, if $t \to \lambda_{g_t}$ is ergodic, then the set of a.e. invariant vectors is given by the constant functions whence $\hat{P}f \propto 1$, that is,

$$\hat{P}f = <1, \hat{P}f> \cdot 1 = <\hat{P}1, f> 1 = <1, f> \cdot 1 = \left[\int_X d\mu(x) f(x) \right] \cdot 1 \tag{25.2.4}$$

since $1(x) = 1$ and the definition of the inner product. Comparing with $\hat{P}f$ from (25.2.2) gives the claimed result (25.2.3). Conversely, if (25.2.3) holds then the right-hand side is constant almost everywhere and equals $\hat{P}f$, hence $t \to \lambda_{g_t}$ is ergodic by the above remark.

Criterion (25.2.3) is interesting for the following reason: suppose that $\mu_1 \neq \mu_2$ are different measures on the same measurable space $(X, \mathcal{U})$, and that $t \to \lambda_{g_t}$ is a measure-preserving, ergodic group action with respect to both of them. Then

$$\left[\int_X d\mu_1(x')f(x')\right] \cdot 1 =_{\mu_1-\text{a.e.}} \lim_{T \to \infty} \frac{1}{2T} \int_{-T}^{T} dt f(\lambda_{g_t}(x)) =_{\mu_2-\text{a.e.}} \left[\int_X d\mu_2(x')f(x')\right] \cdot 1$$

(25.2.5)

for any $f \in L_1(X, d\mu_1) \cap L_1(X, d\mu_2)$. Now the left- and right-hand sides in (25.2.5) do not depend at all on the point x on which the middle term depends. Thus, if we can find $f \in L_1(X, d\mu_1) \cap L_1(X, d\mu_2) \neq \emptyset$ such that the constants $[\int_X d\mu_1(x')f(x')] \neq [\int_X d\mu_2(x')f(x')]$ are different from each other then the middle term must equal the left-hand side whenever $x \in \text{supp}(\mu_1)$ and it must equal the right hand side whenever $x \in \text{supp}(\mu_2)$. This is no contradiction iff μ_1, μ_2 are mutually singular with respect to each other. Hence ergodicity gives a simple tool for investigating the singularity structure of measures with respect to each other and one easily shows that Gaußian measures with different covariances (e.g., scalar fields with different masses) are built on mutually singular measures.

Definition 25.2.3. *A one-parameter group of measure-preserving transformations $t \to \lambda_{g_t}$ is called mixing provided that*

$$\lim_{t \to \infty} <f, \hat{U}(g_t)f'> = <f, 1><1, f'>$$

(25.2.6)

It is easy to see that mixing implies ergodicity: suppose that f' is invariant a.e. under the one-parameter group. Then (25.2.6) reduces to $<f, f'> = <f, 1><1, f'>$ for any f. On the other hand, inserting the identity $1_{L_2} = \hat{Q} \oplus [1_{L_2} - \hat{Q}]$, where $\hat{Q} = |1><1|$ denotes the projection onto $\text{span}(\{1\})$, gives

$$<f, f'> = <f, 1><1, f'> + <f, [1_{L_2} - \hat{Q}]f'> = <f, 1><1, f'>$$
$$\Rightarrow <f, [1_{L_2} - \hat{Q}]f'> = 0 \ \forall \ f \in L_2(X, d\mu)$$

(25.2.7)

hence $[1_{L_2} - \hat{P}]f' = 0$ so that $f' = \text{const.}$ a.e., that is, ergodicity.

The notion of ergodicity or mixing is also fundamental for the Euclidean formulation of ordinary QFT where it replaces the Wightman axiom that requires the uniqueness of the vacuum. See, for example, [99].

26
Key results from functional analysis

Solid knowledge of functional analysis is mandatory in order to gain a proper understanding of the structure of modern quantum field theory. Here we can just give a tiny glimpse of this 'king's discipline' of mathematics by stating, without proof, the theorems more frequently used throughout the book. The reader is urged to work in detail through the standard reference [282] geared to mathematical physicists, especially volumes one and two.

26.1 Metric spaces and normed spaces

Depending on one's axiomatic starting point in set theory Zorn's lemma or, equivalently, the axiom of choice is an axiom or is derived from set theoretical axioms. This touches on the deep inconsistencies of mathematical logic and set theory which goes beyond the scope of this book.

Theorem 26.1.1 (axiom of choice). *Let $\mathcal{I}$ be an index set and $\{S_I\}_{I \in \mathcal{I}}$ be a collection of non-empty sets indexed by $\mathcal{I}$. Then there exists a choice function $f : \mathcal{I} \to \times_{I \in \mathcal{I}} S_I$, that is, an assignment of an element $f(I) \in S_I$ for all $I \in \mathcal{I}$ irrespective of the cardinality of $\mathcal{I}$.*

Theorem 26.1.2 (Zorn's lemma). *Let $X \neq \emptyset$ be a partially ordered set with the property that every linearly ordered subset $Y \subset X$ (i.e., $y \prec y'$ or $Y' \prec y$ for all $y, y' \in Y$) has an upper bound $x_Y \in X$ (i.e., $y \prec x_Y$ for all $y \in Y$). Then X has a maximal element $m \in X$ (i.e., $m \prec x$ for $x \in X$ implies $x = m$) which is a common upper bound for all linearly ordered subsets.*

Of particular importance in functional analysis are topological spaces whose topology derives from a metric or a norm. These are metric or normed spaces respectively.

Definition 26.1.3

(i) *A metric space (X, d) is a pair consisting of a set X and a function $d : X \times X \to \mathbb{R}^+$ called a metric which satisfies, for all $x, y, z \in X$: (1) $d(x, y) = 0$ iff $x = y$, (2) $d(x, y) = d(y, x)$ and (3) $d(x, z) \leq d(x, y) + d(y, z)$.*

(ii) *A sequence $(x_n)_{n \in \mathbb{N}}$ in a metric space is said to converge to an element x iff for all $\epsilon > 0$ there exists $n(\epsilon) \in \mathbb{N}$ such that $n > n(\epsilon)$ implies $d(x_n, x) < \epsilon$.*

(iii) *A sequence is called a Cauchy sequence iff for all $\epsilon > 0$ there exists $n(\epsilon) \in \mathbb{N}$ such that $m, n > n(\epsilon)$ implies $d(x_m, x_n) < \epsilon$.*

(iv) *A metric space in which every Cauchy sequence converges is called complete.*

Clearly every convergent sequence is Cauchy but not vice versa. Complete metric spaces contain all their Cauchy sequences, and this is how one completes an incomplete metric space.

Theorem 26.1.4. *For every incomplete metric space (X, d) there exists a complete metric space (X', d') with $X \subset X'$ such that $d'_{|X} = d$ and X is dense in X'.*

Proof: In general we call a bijection $b : X \rightarrow X'$ between metric spaces (X, d), (X', d') an isometry iff $d'(b(x), b(y)) = d(x, y)$. A subset Y' is said to be dense in X' if for all $\epsilon > 0$ and all $x' \in X'$ we find $y' \in Y'$ such that $d'(x', y') < \epsilon$.

The theorem says that X can be isometrically and densely embedded into its completion as a subset. To see this, define X' as the space of equivalence classes of all Cauchy sequences of (X, d) where $(x_n) \sim (y_n)$ iff $\lim_{n \to \infty} d(x_n, y_n) = 0$. Denoting the equivalence class of (x_n) by $[(x_n)]$ we define $d'([(x_n)], [(y_n)]) := \lim_{n \to \infty} d(x_n, y_n)$, which can be shown to converge and to be independent of the representative. Finally, define $b(x) = [(y_n)]$ with $y_n = x$ for all n. $\qquad\square$

The metric space (X', d') constructed in the preceding proof is called the Cauchy completion of (X, d).

The metric d on a metric space (X, d) defines a topology in the standard way familiar from $\mathbb{R}^n$.

Definition 26.1.5. *Let (X, d) be a metric space and consider for each $x \in X$, $\epsilon > 0$ the set $B_\epsilon(x) := \{y \in X; \, d(x, y) < \epsilon\}$ called the open ball of radius ϵ about x.*

(i) *A set $O \subset X$ is called open if for each $x \in O$ there exists some $B_\epsilon(x) \subset O$.*

(ii) *$x \in S \subset X$ is called a limit point of S if $[S - \{x\}] \cap B_\epsilon(x) \neq \emptyset$ for all $\epsilon > 0$. S is called closed if it includes all its limit points. The union of a set S with its limit points is called the closure $\bar{S}$ of S.*

By definition, a limit point x of S is the limit of a sequence (x_n) with $x_n \in S$ which converges in X to x but x may not lie in S. Since every convergent sequence is Cauchy, it follows that the Cauchy completion of S coincides with the closure of S.

An important special case of metric spaces are normed vector spaces.

Definition 26.1.6

(i) *A normed vector space is a pair $(X, ||.||)$ consisting of a vector space X and a function $||.|| : X \rightarrow \mathbb{R}^+$ subject to the following conditions for all*

$x, y \in X$: (1) $||x|| = 0$ iff $x = 0$, (2) $||\lambda x|| = |\lambda| \; ||x||$ for all $\lambda \in \mathbb{C}$ and (3) $||x + y|| \leq ||x|| + ||y||$.

(ii) A linear transformation $T : X \rightarrow Y$ between normed linear spaces $(X, ||.||)$, $(Y, ||.||')$ is called bounded iff $||Tx||' \leq K||x||$ for some $K > 0$ independent of $x \in X$.

(iii) The metric induced by the norm of a normed space $(X, ||.||)$ is defined by $d(x, y) := ||x - y||$. A normed space is complete when it is complete as a metric space in this induced metric and is then called a Banach space.

It is clear that linear transformations are continuous iff they are bounded: if it is bounded then $||T(y - x)||' \leq K||x - y||$ so it is continuous. If it is continuous, suppose that T is not bounded. Then for each $n \in \mathbb{N}$ we find $x_n \in X$ with $||Tx_n||' \geq n||x_n||$. Set $y_n = x_n/(n||x_n||)$. Then $||y_n|| = 1/n$ so $y_n \rightarrow 0$ while $||Ty_n||' \geq 1$ for all n. This contradicts continuity of T at $x = 0$, hence T is bounded.

Definition 26.1.7. *The topological dual of a Banach space X is the space X' of continuous linear functionals $l : X \rightarrow \mathbb{C}$. The dual space is also a normed linear space with norm $||l|| := \sup_{x \in X - \{0\}} |l(x)|/||x||$ and automatically complete, that is, a Banach space. One can show that the map $J : X \rightarrow X''$; $x \mapsto J_x$ where $J_x(l) = l(x)$ is an isometric injection. If it is a surjection then X is called reflexive.*

Theorem 26.1.8 (BLT theorem). *Suppose $T : X_1 \rightarrow X_2$ is a bounded linear transformation (BLT) from the normed linear space $(X_1, ||.||_1)$ to the complete normed linear space $(X_2, ||.||_2)$. Then T has a unique extension to the completion $\bar{X}_1$ of X_1 as a bounded linear transformation T' with the same bound.*

This theorem is convenient in that a bounded linear transformation between complete normed spaces is already completely determined by a dense subset in the domain of the map. It is clear that the completion of a dense subspace X of a complete space Y recovers the space Y.

26.2 Hilbert spaces

Definition 26.2.1

(i) A positive definite, sesquilinear form or inner product on a complex linear space X is a map $< ., . >: X \times X \rightarrow \mathbb{C}$ subject to: (1) $< x, x > \geq 0$, $< x, x > = 0$ iff $x = 0$, (2) $< x, y + z > = < x, y > + < x, z >$, (3) $< x, \lambda y > = \lambda < x, y >$ and (4) $< x, y > = \overline{< y, x >}$ for all $x, y, z \in X$, $\lambda \in \mathbb{C}$. The pair $(X, < ., . >)$ is called a pre-Hilbert space.

(ii) A collection of vectors (x_n) is said to be orthonormal iff $< x_m, x_n > = \delta_{mn}$.

Let us denote $||x|| := \sqrt{< x, x >}$ on a pre-Hilbert space $(X, < ., . >)$.

Lemma 26.2.2 (Bessel's inequality). *Let (x_n) be an orthonormal set. Then $||x||^2 \geq \sum_n | < x_n, x > |^2$ for all $x \in X$.*

Proof: The vectors x_n are orthogonal to the vector $z := x - \sum_n < x_n, x > x_n$. Set $y := x - z$ then z, y are orthogonal. Hence $||x||^2 = ||y||^2 + ||z||^2 = \sum_n | < x_n, x > |^2 + ||z||^2$. $\qquad\square$

Corollary 26.2.3 (Schwarz inequality). $| < x, y > | \leq ||x|| \, ||y||$ *for all x, $y \in X$ and equality is reached iff x, y are co-linear.*

Proof: For $y = 0$ there is nothing to show, so let us assume $y \neq 0$. Bessel's inequality applied to the orthonormal system consisting of the single vector $x_1 := y/||y||$ reveals the inequality. In order to reach equality we must have $z = x - < x_1, x > x_1 = 0$, that is, x, y are co-linear. $\qquad\square$

Definition 26.2.4. *It is not difficult to show that $||.||$ defines a norm on X. The completion of X with respect to this norm turns the pre-Hilbert space $(X, < ., . >)$ into a Hilbert space.*

It is possible to recover the inner product of a complex Hilbert space from the norm via the polarisation identity

$$< x, y > = \frac{1}{4} \sum_{\epsilon^4 = 1} \bar{\epsilon} \, ||x + \epsilon y||^2 \qquad (26.2.1)$$

Theorem 26.2.5 (Riesz lemma). *Let T be a continuous linear functional on a Hilbert space X, that is, a continuous linear map $T : X \to \mathbb{C}$. Then there exists a unique element $y_T \in X$ such that $T(.) = < y_T, . >$, moreover, $||T|| = \sup_{x \neq 0} \frac{|T(x)|}{||x||} = ||y_T||$.*

The space of continuous linear functionals on a linear topological space X is called the topological dual X' of X. The Riesz lemma says that Hilbert spaces are reflexive, that is, $X' = X$.

Definition 26.2.6. *An orthonormal system (x_n) in a Hilbert space $(X, < ., . >)$ is called an orthonormal basis (ONB) if the span of the x_n (i.e., finite complex linear combinations) is dense. Using the axiom of choice one can show that every Hilbert space admits an ONB. A Hilbert space is called separable if it admits a countable ONB.*

Definition 26.2.7. *A map U between Hilbert spaces which preserves inner products and is surjective is called unitary. The two Hilbert spaces are then called isomorphic.*

Due to the equivalence of norm and inner product via the polarisation identity, inner product-preserving means the same as norm-preserving, that is, isometric. It follows that unitary operators are injective, thus bijective and norm-preserving. It follows that all separable Hilbert spaces are isomorphic to the Hilbert space ℓ_2

of sequences of complex numbers λ_n with norm $||(\lambda_n)||^2 = \sum_n |\lambda_n|^2$ (just map the orthonormal bases bijectively).

Definition 26.2.8. *Given two Hilbert spaces $(X_I, < .,. >_I)$ we can construct the direct sum $X_1 \oplus X_2$ and direct (tensor) product $X_1 \otimes X_2$ respectively by setting*

$$< (x_1, x_2), (y_1, y_2) >_{X_1 \oplus X_2} := < x_1, y_1 >_1 + < x_2, y_2 >_2$$

$$< x_1 \otimes x_2, y_1 \otimes y_2 >_{X_1 \otimes X_2} := < x_1, y_1 >_1 < x_2, y_2 >_2 \qquad (26.2.2)$$

for $x_I, y_I \in X_I$ and completing the finite linear span of elements of the form (x_1, x_2) and $x_1 \otimes x_2$ respectively in those inner products. In particular, if $(b_n^{(I)})$ is a basis in X_I then $((b_n^{(1)}, 0), (0, b_n^{(2)}))$ is a basis for the direct sum and $(b_m^{(1)} \otimes b_n^{(2)})$ is a basis for the direct product.

26.3 Banach spaces

Recall that a Banach space is a complete, normed, linear space. Besides Hilbert spaces, an important example is given by the set $B(X, Y)$ of bounded linear transformations (operators) $T : X \to Y$ between normed linear spaces $(X, ||.||)$ and $(Y, ||.||')$. The operator norm is given by

$$||T||_{XY} := \sup_{x \neq 0} \frac{||Tx||'}{||x||} \qquad (26.3.1)$$

The normed linear space $B(X, Y), ||.||_{XY}$ is complete, that is, a Banach space, if Y is a Banach space. An element $T \in B(X, Y)$ is called an isomorphism if it is a bijection with bounded inverse. It is called an isometry if it is norm-preserving.

The most important theorem associated with Banach spaces is the following, which we state simultaneously in its real and complex version.

Theorem 26.3.1 (Hahn–Banach). *Let X be a real (complex) vector space and p a real-valued function such that $p(ax + by) \leq |a|p(x) + |b|p(y)$ for all $x, y \in X$ and positive real (arbitrary complex) numbers a, b with $|a| + |b| = 1$. Suppose that λ is a real (complex) linear functional on a subspace Y of X satisfying $\lambda(y) \leq p(y)$ $(|\lambda(y)| \leq p(y))$ for all $y \in Y$. Then there exists a real (complex) linear functional Λ on X such that $\Lambda(x) \leq p(x)$ $(|\Lambda(x)| \leq p(x))$ for all $x \in X$ and $\Lambda_{|Y} = \lambda$.*

We will mostly need the theorem in its complex version. Notice that in the complex version the function p must be positive, not only real-valued. An important application is the case of a subspace Y of a normed linear space X and $\lambda \in Y'$ a continuous linear functional on Y. Choose $p(x) := ||\lambda|| \, ||x||$. Then the extension $\Lambda \in X'$ guaranteed by the Hahn–Banach theorem satisfies $||\Lambda|| = ||\lambda||$.

Theorem 26.3.2 (Banach–Steinhaus; principle of uniform boundedness). *Let X be a Banach space and Y a normed linear space. Given a family*

$\mathcal{F}$ of elements of $B(X,Y)$ suppose that for each $x \in X$ the set $\{||Tx||_Y;\ T \in \mathcal{F}\}$ is a bounded set of positive real numbers. Then also the set of operator norms $\{||T||;\ T \in \mathcal{F}\}$ is bounded.

Theorem 26.3.3 (closed graph theorem). *Let X, Y be Banach spaces and $T : X \to Y$ linear. Then T is bounded if and only if the graph $\Gamma(T) := \{(x, Tx);\ x \in X\} \subset X \times Y$ of T is closed in $X \times Y$ (in the norm $||(x, y)|| = ||x||_X + ||y||_Y$).*

An application of this theorem is that an operator T on a Hilbert space X which together with its adjoint is everywhere defined is automatically bounded (Hellinger–Töplitz theorem). To see this suppose that $(x_n, Tx_n) \in \Gamma(T)$ converges to (x, y). We must show that the graph is closed, that is, $y = Tx$. But for any $z \in X$ $< z, Tx_n > = < T^\dagger z, x_n > \to < T^\dagger z, x > = < z, Tx >$ hence T is continuous and thus bounded.

26.4 Topological spaces

The results needed on topological spaces are already covered by Chapter 18.

26.5 Locally convex spaces

Locally convex spaces play an important role in the theory of distributions which typically arise as solutions of constraints.

Definition 26.5.1

(i) *A seminorm on a vector space X is a map $\rho : X \to [0, \infty)$ such that: (1) $\rho(x + y) \leq \rho(x) + \rho(y)$ and (2) $\rho(\lambda x) = |\lambda|\rho(x)$ for all $x, y \in X$, $\lambda \in \mathbb{C}$. Thus a seminorm becomes a norm if it is also positive definite.*

(ii) *A family of seminorms $(\rho_I)_{I \in \mathcal{I}}$ is said to separate the points of X if $\rho_I(x) = 0$ for all $I \in \mathcal{I}$ implies $x = 0$.*

(iii) *A locally convex space is a vector space X together with a family of seminorms $(\rho_I)_{I \in \mathcal{I}}$ separating the points. Its natural topology is the weakest topology in which all the ρ_I and the operation of addition is continuous. This topology is automatically Hausdorff, as one can show.*

(iv) *A locally convex space X whose underlying family of seminorms is countable can be equipped with the following metric*

$$d(x, y) := \sum_{n=1}^{\infty} 2^{-n} \frac{\rho_n(x - y)}{1 + \rho_n(x - y)} \tag{26.5.1}$$

which generates the same topology as the family of seminorms. If it is completed in this metric, it is called a Fréchet space.

An important application of these concepts is as follows: consider the space $\mathbb{R}^n$ with coordinates x_k and let $\alpha = (\alpha_1, \ldots, \alpha_n)$, $\alpha_k = 0, 1, 2, \ldots$ and $|\alpha| =$

$\sum_{k=1}^{n} \alpha_k$. Set $\partial_\alpha := \partial^{|\alpha|}/(\partial x_1^{\alpha_1} \ldots \partial x_n^{\alpha_n})$ and $x^\alpha = x_1^{\alpha_1} \ldots x_n^{\alpha_n}$. The space of smooth functions on $\mathbb{R}^n$ of rapid decrease $\mathcal{S}(\mathbb{R}^n)$ consists of those smooth functions f for which $\rho_{\alpha,\beta}(f) := \sup_x |x^\alpha \partial_\beta f(x)| < \infty$ for all α, β. They fall off together with their derivatives faster than any polynomial at infinity. One can show that this space with the countable family of seminorms $\rho_{\alpha,\beta}$ is a Fréchet space. Its topological dual $\mathcal{S}'(\mathbb{R}^n)$ is called the space of tempered distributions.

26.6 Bounded operators

Definition 26.6.1. *Let $T \in B(X,Y)$ be a bounded operator between Banach spaces X, Y (we will mostly be interested in the case that $X = Y$ is a Hilbert space). A net (T_α) in $B(X,Y)$ is said to converge to T in the uniform, strong or weak operator topology respectively iff*

$$\|T_\alpha - T\|_{B(X,Y)} := \sup_{x \in X - \{0\}} \frac{\|(T_\alpha - T)x\|_Y}{\|x\|_X} \to 0$$

$$\|(T_\alpha - T)x\|_Y \to 0 \ \forall \, x \in X$$

$$l[(T_\alpha - T)x] \to 0 \ \forall \, x \in X, \ l \in Y' \tag{26.6.1}$$

where Y' is the topological dual of Y.

The weak topology is weaker than the strong topology which is weaker than the uniform topology. For completeness we mention the weak $*$ topology with respect to a Hilbert space $Y = X$: this is similar to the weak topology, however, instead of $X' = X$ we now take a subspace $\mathcal{D}$ of X equipped with a finer topology and as $\mathcal{D}'$ the topological dual of that topological space. Physical applications are the topology in which the Hamiltonian constraint converges and the refined algebraic quantisation programme (RAQ) of Section 30.1.

Definition 26.6.2

(i) *Let $T \in B(X,Y)$. The adjoint $T' \in B(Y',X')$ of T is defined by $[T'l](x) := l(Tx)$ for all $x \in X$, $l \in Y'$.*

(ii) *In case that X is a Hilbert space we may identify $X = Y = X' = Y'$ via the Riesz lemma and write $T^\dagger$ or T^* instead of T'. We call a bounded operator self-adjoint if $T = T^*$.*

(iii) *A bounded operator T on a Hilbert space X is called positive if $< x, Tx > \geq 0$ for all $x \in X$. We denote this by $T \geq 0$ and given T_1, T_2 with $T_1 - T_2 \geq 0$ we write $T_1 \geq T_2$. Given any $T \in B(X,X)$ we define $|T| := \sqrt{T^*T}$ via the spectral theorem, see Section 29.2.*

(iv) *A bounded operator U is called a partial isometry if $\|Ux\| = \|x\|$ for all $x \in [\mathrm{Ker}(U)]^\perp$ where $\mathrm{Ker}(U)$ is the kernel of U, that is, the closure of the linear span of x with $Ux = 0$ and for any closed subspace Y of a Hilbert space X one defines the closed subspace $Y^\perp = \{x \in X; \ < x, y > = 0 \ \forall \, y \in Y\}$ called the orthogonal complement (it follows that $X = Y \oplus Y^\perp$). A partial* *isometry is called an isometry if* $\text{Ker}(U) = \{0\}$. *The space* $[\text{Ker}(U)]^{\perp}$ *is called the initial subspace of* U *and* $\text{Ran}(U)$, *the closure of the image of* U, *is called the final subspace.*

Theorem 26.6.3 (polar decomposition). *If* T *is a bounded operator on a Hilbert space then there exists a partial isometry* U *with* $\text{Ker}(U) = \text{Ker}(T)$ *and* $\text{Ran}(U) = \overline{\text{Ran}(T)}$ *such that* $T = U|T|$.

The theorem still holds for unbounded closed operators, see Definition 26.7.1 The only changes are that now the operator is only densely defined on a set $D(T)$ and one has $D(|T|) = D(T)$ where positive now means $< x, Tx > \geq 0$ for all $x \in D(T)$.

Definition 26.6.4. *An operator* $T \in B(X,Y)$ *is said to be compact if for every bounded sequence* (x_n) *in* X *the sequence* Tx_n *has a convergent subsequence.*

Things become especially nice if $X = Y$ is a separable Hilbert space. Notice that compact operators are always bounded by definition.

Theorem 26.6.5 (canonical form of compact operators)

(i) *Let* T *be a compact operator on a separable Hilbert space* X. *We denote the space of compact operators by* $K(X)$. *Then there exist orthonormal but not necessarily complete orthonormal systems* (x_n) *and* (y_n) *as well as a sequence of positive real numbers* λ_n *converging to zero, called the singular values of* T *such that* $T(.) = \sum_n \lambda_n y_n < x_n, . >$.

(ii) *The spectrum (see Section 26.7 for a precise definition) of a compact operator is a discrete set in the complex plane without accumulation points, except possibly at zero. Every non-zero eigenvalue has finite multiplicity (Riesz–Schauder theorem).*

(iii) *A compact self-adjoint operator has a complete orthonormal basis of eigenvectors* x_n *with real-valued eigenvalues* λ_n *which converge to zero (Hilbert–Schmidt theorem). It follows that the canonical form in this case simplifies to the* $|\lambda_n| \geq 0$ *as the singular values and* $y_n = \text{sgn}(\lambda_n)x_n$.

We see that compact operators are in a sense the closest analogue of finite-dimensional matrices and in fact the finite rank operators form a dense subset with respect to the uniform topology.

Theorem 26.6.6. *Let* X *be a separable Hilbert space and* T *a bounded positive operator. Let* (b_n) *be any ONB. Then* $\text{Tr}(T) := \sum_n < b_n, Tb_n >$ *is independent of the ONB and is called the trace of* T. *It satisfies the following properties: (1)* $\text{Tr}(T_1 + T_2) = \text{Tr}(T_1) + \text{Tr}(T_2)$, *(2)* $\text{Tr}(\lambda T) = \lambda \text{Tr}(T)$, *(3)* $\text{Tr}(UTU^{-1}) = \text{Tr}(U)$ *for all unitary operators* U *and (4)* $T \geq 0$ *implies* $\text{Tr}(T) \geq 0$ *for all positive* T_I *and* $\lambda \in \mathbb{C}$.

Two important subsets of compact operators based on the trace are the following.

Definition 26.6.7

(i) *A bounded operator T on a separable Hilbert space X is called trace class (Hilbert–Schmidt) iff $\mathrm{Tr}(|T|) < \infty$ ($\mathrm{Tr}(T^\dagger T) < \infty$). We denote the family of trace class (Hilbert–Schmidt) operators by $B_1(X)$ ($B_2(X)$).*

(ii) *For any $T \in B_1(X)$ we extend the trace by $\mathrm{Tr}(T) := \sum_n <b_n, Tb_n>$ which is independent of the ONB (b_n). The trace satisfies $\mathrm{Tr}(T^\dagger) = \overline{\mathrm{Tr}(T)}$ and $\mathrm{Tr}(AT) = \mathrm{Tr}(TA)$ for $T \in B_1(X)$ and $A \in B(X)$.*

These classes of operators play an important role in physics. For instance, mixed states or density matrices are nothing else than positive trace class operators of unit trace. Hilbert–Schmidt operators naturally appear in Bogol'ubov transformations (linear canonical transformations between different systems of annihilation and creation operators on Fock spaces) which are unitarily implementable if the corresponding linear transformation is Hilbert–Schmidt.

Theorem 26.6.8

(i) *Both $B_1(X)$, $B_2(X)$ are subsets of the set of compact operators. A compact operator T is in $B_1(X)$, $B_2(X)$ respectively iff $\sum_n \lambda_n$ or $\sum_n \lambda_n^2$ converges where λ_n are the singular values of T. It follows that every trace class operator is Hilbert–Schmidt, that is, $B_1(X) \subset B_2(X) \subset K(X) \subset B(X) := B(X, X)$.*

(ii) *Both $B_1(X)$, $B_2(X)$ are two-sided * ideals in $B(X) := B(X)$, that is, given $T \in B_I(X)$, $A \in B(X)$ then AT, TA, $T^\dagger \in B_I(X)$.*

(iii) *$B_2(X)$ is a Hilbert space with inner product $<T, T'> := \mathrm{Tr}(T^\dagger T')$. $B_1(X)$ is dense in $B_2(X)$ with respect to this inner product.*

26.7 Unbounded operators

Unbounded operators are not everywhere defined on the Hilbert space by the Hellinger–Töplitz theorem. The best one can achieve is that they can be defined on a dense domain D. That D is dense is sufficient because it means that we can define the operator on 'almost' every vector in the Hilbert space in the sense that every vector can be approximated arbitrarily closely by vectors in D. In order to distinguish from the bounded operators T on separable Hilbert spaces we allow here for unbounded operators a on not necessarily separable Hilbert spaces $\mathcal{H}$, unless otherwise stated.

We will define in detail the spectrum of (un)bounded operators. The spectrum allows for several different partitions which are useful in different contexts.

Definition 26.7.1

(i) *Let a be a densely defined operator with domain D. Let $D(a^\dagger) := \{\psi \in \mathcal{H} : \sup_{f \in D - \{0\}} |<\psi, af>|/\|f\| < \infty\}$. For $\psi \in D(a^\dagger)$ we define the bounded linear functional $f \mapsto <\psi, af>$ and by the Riesz lemma bounded linear* *functionals can be written in the form $f \mapsto\, <\psi', f>$ for some $\psi' \in \mathcal{H}$. We call $\psi' := a^\dagger \psi$ the adjoint of a.*

(ii) *Given a not necessarily densely defined operator a with domain D consider the set $\Gamma(a) := \{(\psi, a\psi) : \psi \in D\} \subset \mathcal{H} \times \mathcal{H}$, called the graph of a. The operator a is called closed if its graph is closed in the metric induced by the inner product $< (\psi_1, \psi_2), (\psi_1', \psi_2') > := <\psi_1, \psi_1'> + <\psi_2, \psi_2'>$. If a has a closed extension (i.e., an extension of its domain such that the associated graph is closed and such that the extended operator coincides with the original one on the original domain) it is called closable and the smallest such extension a is called the closure $\bar{a}$ of a. It is easy to see that if a is closable, then $\Gamma(\bar{a}) = \overline{\Gamma(a)}$.*

(iii) *An operator a is called symmetric if $D(a) \subset D(a^\dagger)$ and $a^\dagger_{|D(a)} = a$. It is called self-adjoint if it is symmetric and $D(a) = D(a^\dagger)$. A symmetric operator is called essentially self-adjoint if its closure (which can be shown to exist, see below) is self-adjoint.*

(iv) *The spectrum $\sigma(a)$ of a self-adjoint operator a is defined as the complement in $\mathbb{C}$ of the resolvent set $\rho(a) = \{z \in \mathbb{C} : \rho - z\mathbf{1}$ has bounded inverse$\}$. One can show that $\sigma(a)$ is a closed subset of $\mathbb{C}$.*

(v) *Let a be a self-adjoint operator.*

1. *The pure point spectrum $\sigma_{pp}(a)$ is the set of eigenvalues of a. It may not be closed.*

2. *By the Lebesgue decomposition theorem mentioned in Section 30.2, the Hilbert space decomposes as $\mathcal{H} = \mathcal{H}_{pp} \oplus \mathcal{H}_{ac} \oplus \mathcal{H}_{cs}$. The closed spaces $\mathcal{H}_{pp}, \mathcal{H}_{ac}, \mathcal{H}_{cs}$ respectively are characterised as follows: if $\psi \in \mathcal{H}_{pp}$ then the spectral measure μ_ψ has support on a countable set of points. If $\psi \in \mathcal{H}_{ac}, \mathcal{H}_{cs}$ then one-point sets are of measure zero with respect to μ_ψ. If $\psi \in \mathcal{H}_{ac}$ then μ_ψ is absolutely continuous with respect to Lebesgue measure and if $\psi \in \mathcal{H}_{cs}$ then μ_ψ is singular with respect to Lebesgue measure.*

   Consider the restricted operator $a_* := a_{|\mathcal{H}_*}, * \in \{pp, ac, cs\}$ and the spectrum of its restrictions $\sigma^* := \sigma(a_*)$. Then $\sigma_c(a) := \sigma^{ac}(a) \cup \sigma^{cs}(a)$ is called the continuous spectrum. Notice that the sets $\sigma^*(a)$ need not be disjoint, that $\sigma^{pp}(a) = \overline{\sigma_{pp}(a)}$ and $\sigma(a) = \sigma^{pp}(a) \cup \sigma^{ac}(a) \cup \sigma^{cs}(a)$.

3. *The discrete and essential spectrum of a respectively is defined as the subset $\sigma_d(a)$ or $\sigma_e(a)$ of $\sigma(a)$ respectively consisting of those points x such that $E((x - \epsilon, x + \epsilon))$ is a projection onto a finite- or infinite-dimensional subspace respectively as $\epsilon \to 0$. Here E is the projection-valued measure corresponding to a, see Theorem 29.2.3.*

Intuitively, in applications the spaces $\mathcal{H}_{pp}, \mathcal{H}_{ac}, \mathcal{H}_{cs}$ correspond to bound states, scattering states and states without physical interpretation. In physics, $\sigma^{cs}(a)$ is mostly absent. It is not difficult to show that $\sigma_d(a)$ consists of the isolated eigenvalues of finite multiplicity and that $\sigma_e(a)$ contains (1) $\sigma_c(a)$, (2) the limit points of $\sigma_{pp}(a)$ and the eigenvalues of infinite multiplicity. In particular $\sigma_e(a)$ contains the embedded eigenvalues, that is, those which have an open neighbourhood all of whose points belong to $\sigma(a)$.

The following theorem gives basic equivalent criteria for when the conditions of Definition 26.7.1 are met.

Theorem 26.7.2 (basic criterion for self-adjointness)

(i) A densely defined operator is closable iff its adjoint is densely defined. In particular, every symmetric operator is closable.

(ii) A symmetric operator is self-adjoint (s.a.) iff $[a \pm i 1_{\mathcal{H}}]D(a) = \mathcal{H}$. Equivalently, it is s.a. if a is closed and $\mathrm{Ker}(a^{\dagger} \pm i 1_{\mathcal{H}}) = \{0\}$.

(iii) A symmetric operator is essentially self-adjoint (e.s.a.) iff $\overline{[a \pm i 1_{\mathcal{H}}]D(a)} = \mathcal{H}$. Equivalently, it is e.s.a. if $\mathrm{Ker}(a^{\dagger} \pm i 1_{\mathcal{H}}) = \{0\}$.

In general, symmetric operators may or may not have self-adjoint extensions. One can show that this is possible if and only if the *deficiency indices* $n_{\pm} := \dim(\mathrm{Ker}([a^{\dagger} \pm i 1_{\mathcal{H}}]))$ are equal to each other and in this case the possible extensions are labelled by the points of the unitary group $U(n)$ (see, e.g., [649]). It follows from the theorem that for e.s.a. operators such a freedom does not exist: their self-adjoint extension is unique and given by their closure.

Theorem 26.7.3 (Stone). *Let $t \mapsto U(t)$ be a one-parameter, weakly continuous group of unitary operators. That is, (1) $U(t)$ is unitary, (2) $U(s)U(t) = U(s + t)$ and (3) $\lim_{t \to 0} < \psi, U(t)\psi' > = < \psi, \psi' >$ for all $s, t \in \mathbb{R}$ and $\psi, \psi' \in \mathcal{H}$. Then there exists a self-adjoint operator a such that $U(t) = \exp(ita)$, called the infinitesimal generator of the group. On its domain, a can be obtained as $\imath a \psi = (d/dt)_{t=0} U(t)\psi$.*

Notice that for unitary operators weak continuity and strong continuity are equivalent because $||[(U(t) - 1]\psi||^2 = 2[||\psi||^2 - \Re(< \psi, U(t)\psi >)]$.

Dealing with unbounded operators a, b is rather tricky because $D(a) \cap D(b)$ could fail to be dense so that $a + b$ is not obviously densely defined. Also, $bD(b)$ may not lie in $D(a)$ so that ab is ill-defined. This is especially bad in physical applications where we want to compute commutators. This is why for the self-adjoint operators that we are mostly interested in in physics it is convenient to pass to the unitary one-parameter groups or to its bounded spectral projections, to which we turn in Section 29.2.

26.8 Quadratic forms

In the construction of operators a in physics one often starts from its matrix elements $Q_a(\psi, \psi')$, which should equal $< \psi, a\psi' >$. However, this is not enough to define an operator in infinite dimensions because given an ONB (b_n) we must have $||a\psi|| = \sum_n |Q_a(b_n, \psi)|^2 < \infty$ in order that $\psi \in D(a)$. Hence it may happen that the so-called quadratic form $Q_a(\psi, \psi')$ exists for ψ, ψ' in a dense subset of $\mathcal{H}$ but $D(a) = \{0\}$. In what follows we give sufficient criteria for a quadratic form to give rise to an operator.

Definition 26.8.1. *A quadratic form Q on a Hilbert space $\mathcal{H}$ is a sesquilinear form on $D(Q) \times D(Q)$ where $D(Q)$ is a dense form domain. A quadratic form is called semibounded provided that $Q(l, l) \geq -c||l||^2$ for some $c \geq 0$ and positive if $c = 0$. A semibounded quadratic form Q is called closed provided that $D(Q)$ is complete in the norm $||l||_{+1} = \sqrt{Q(l, l) + c||l||^2}$. If Q is closed and $D'(Q) \subset D(Q)$ is dense then $D'(Q)$ is called a form core.*

Theorem 26.8.2 (Friedrich extension)

(i) Let a be a symmetric operator $(D(a) \subset D(a^*)$, $a^*_{|D(a)} = a)$. Then a is closable, however, its closure may not be self-adjoint $(D(\bar{a}) \neq D(\bar{a}^\dagger))$.

(ii) Let Q be a semibounded quadratic form. Then Q may not be closable, but if it is and the closure is semibounded, then Q is the quadratic form of a unique self-adjoint operator a according to $Q(l, l') = <l, al'> =: Q_a(l, l')$.

(iii) Let a be a positive, symmetric operator. Then the corresponding positive quadratic form Q_a has a positive closure $\overline{Q_a}$. The unique positive operator $\tilde{a}$ corresponding to that closure via $Q_{\tilde{a}} = \overline{Q_a}$ is called the Friedrich extension of a. It may extend the closure $\bar{a}$ of a and is the only self-adjoint extension which contains $D(\overline{Q_a})$.

27

Elementary introduction to Gel'fand theory for Abelian C*-algebras

There are many good mathematical textbooks on operator algebra and abstract C^*-algebra theory, see, for example, [167, 535–537]. The textbooks [649, 650] are more geared towards applications in mathematical physics. For a pedagogical introduction with elegant proofs, the beautiful review [651] is recommended.

27.1 Banach algebras and their spectra

Definition 27.1.1

(i) *An algebra $\mathfrak{A}$ is a vector space (taken over $\mathbb{C}$) together with a multiplication map $\mathfrak{A} \times \mathfrak{A} \to \mathfrak{A}$; $(a, a') \mapsto aa'$ which is associative, $(ab)c = a(bc)$ and distributive, $b(za + z'a') = zba + z'ba'$, $(za + z'a')b = zab + z'a'b$ for all $a, a', b \in \mathfrak{A}$, $z, z' \in \mathbb{C}$.*

(ii) *An algebra $\mathfrak{A}$ is called Abelian if all elements commute with each other and unital if it has a (necessarily unique) unit element 1 satisfying[1] $1a = a1 = a$ for all $a \in \mathfrak{A}$.*

(iii) *A vector subspace $\mathfrak{B}$ of $\mathfrak{A}$ is called a subalgebra if it is closed under multiplication. A subalgebra $\mathfrak{I}$ is called a left (right) ideal if $ab \in \mathfrak{I}$ ($ba \in \mathfrak{I}$) for all $a \in \mathfrak{A}$, $b \in \mathfrak{I}$ and a two-sided ideal (or simply ideal) if it is simultaneously a left and right ideal. An ideal of either kind is called maximal if there is no other ideal containing it except for $\mathfrak{A}$ itself.*

(iv) *An involution on an algebra $\mathfrak{A}$ is a map $* : \mathfrak{A} \to \mathfrak{A}; a \mapsto a^*$ satisfying*
   1. $(za + z'b)^* = \bar{z}a^* + \bar{z}'b^*$ *(conjugate linear),*
   2. $(ab)^* = b^*a^*$ *(reverses order) and*
   3. $(a^*)^* = a$ *(squares to the identity)*
      for all $a, b \in \mathcal{A}$, $z, z' \in \mathbb{C}$. An algebra with involution is called an $*$-algebra.

---

[1] For completeness we mention that a unital algebra is sometimes referred to as a ring $\mathfrak{R}$. A commutative ring such that $0 \neq 1$ and such that every element $a \neq 0$ is invertible is called a field $\mathfrak{F}$. A left $\mathfrak{R}$ module is an Abelian group (G, +) together with a left action $\mathfrak{R} \times G \to G$; $(r, x) \mapsto r \cdot x$ called scalar multiplication such that $(r + r') \cdot x = r \cdot x + r' \cdot x$, $r \cdot (x + x') = r \cdot x + r \cdot x'$, $(rr') \cdot x = r \cdot (r' \cdot x)$, $1 \cdot x = x$. A right $\mathfrak{R}$ module is defined analogously in terms of a right action. The most familiar example is the case that the ring is a field and the group a vector space. In general, a module is the generalisation of a representation of the ring.

(v) A homomorphism (*-homomorphism) is a linear map $\phi : \mathfrak{A} \to \mathfrak{B}$ between algebras (*-algebras) that preserves the multiplicative (and involutive) structure, that is, $\phi(ab) = \phi(a)\phi(b)$ (and $\phi(a^*) = (\phi(a))^*$).

(vi) A normed algebra $\mathfrak{A}$ is equipped with a norm $||.|| : \mathfrak{A} \to \mathbb{R}^+$ (that is $||a+b|| \leq ||a|| + ||b||$, $||za|| = |z|\,||a||$, $||a|| = 0 \Leftrightarrow a = 0$, if the last property is dropped, then $||.||$ is only a seminorm) whose compatibility with the multiplicative structure is contained in the submultiplicativity requirement $||ab|| \leq ||a||\,||b||$ for all $a, b \in \mathfrak{A}$. If $\mathfrak{A}$ has an involution we require $||a^*|| = ||a||$ and $\mathfrak{A}$ is called a normed *-algebra. If $\mathfrak{A}$ is unital we require $||1|| = 1$ (this is just a choice of normalisation).

(vii) A norm induces a metric $d(a, b) = ||a - b||$ and if the algebra $\mathfrak{A}$ is complete (every Cauchy sequence converges) then it is called a Banach algebra.

(viii) A C*-algebra $\mathfrak{A}$ is a Banach algebra with involution with the following compatibility condition between the involutive and metrical structure

$$||a^*a|| = ||a||^2 \tag{27.1.1}$$

The innocent-looking condition (27.1.1) determines much of the structure of C*-algebras. If a C*-algebra is not unital one can always embed it isometrically into a larger unital C*-algebra (see, e.g., [651]). While this does not remove all problems with C*-algebras without identity in our applications only unital C*-algebras will appear and this is what we will assume from now on. If $\mathfrak{J}$ is a two-sided ideal in an algebra $\mathfrak{A}$ we can form the quotient algebra $\mathfrak{A}/\mathfrak{J}$ which consists of the equivalence classes $[a] := \{a + b;\ b \in \mathfrak{J}\}$ for any $a \in \mathfrak{A}$ in which the rules for addition, multiplication and scalar multiplication are given by $[a] + [a'] = [a + a']$, $[a][a'] = [aa']$, $[za] = z[a]$ and it is easy to see that the condition that $\mathfrak{J}$ is an ideal is just sufficient for making these rules independent of the representative. Finally, if we think of $\mathfrak{A}$ as an algebra of operators on a Hilbert space and $||.||$ is the uniform operator norm then we see that we are dealing with algebras of bounded operators only, which trivialises domain questions.

Definition 27.1.2. *The spectrum $\Delta(\mathfrak{A})$ of a unital Banach algebra $\mathfrak{A}$ is the set of all non-zero *-homomorphisms $\chi : \mathfrak{A} \to \mathbb{C}$; $a \mapsto \chi(a)$, called the characters.*

Notice that $\mathbb{C}$ is itself a unital, Abelian C*-algebra in the usual metric topology of $\mathbb{R}^2$. Notice that $\chi(1) = 1$ since $\chi(a) = \chi(1a) = \chi(1)\chi(a)$ and if we choose $a \in \mathfrak{A}$ such that $\chi(a) \neq 0$ the claim follows. Similarly $\chi(a^{-1}) = \chi(a)^{-1}$ if a has an inverse in $\mathfrak{A}$, that is an element a^{-1} with $aa^{-1} = a^{-1}a = 1$. Finally $\chi(0) = 0$ since $1 = \chi(1) = \chi(1 + 0) = \chi(1) + \chi(0) = 1 + \chi(0)$.

Definition 27.1.3. *For a character in a unital Banach algebra $\mathfrak{A}$ define $\ker(\chi) := \{a \in \mathfrak{A};\ \chi(a) = 0\}$ to be its kernel.*

Clearly, $\ker(\chi)$ is a two-sided ideal in $\mathfrak{A}$ since $\chi(ab) = \chi(ba) = \chi(a)\chi(b) = 0$ for all $a \in \mathfrak{A}$, $b \in \ker(\chi)$. Since χ is in particular a linear functional on $\mathfrak{A}$ considered as a vector space, it follows that $\ker(\chi)$ is a vector subspace of $\mathfrak{A}$ of co-dimension one. After taking its closure in $\mathfrak{A}$ it is either still of co-dimension one or of co-dimension zero, the latter being impossible since then χ would be identically zero, which we excluded in the definition for a character. It follows that there exist elements $a \in \mathfrak{A} - \ker(\chi)$ and that $\mathfrak{A}$ is the closure of the span of $\{a, \ker(\chi)\}$. Thus, if there is an ideal $\mathfrak{I}$ of $\mathfrak{A}$ properly containing $\ker(\chi)$ then we can take such an $a \in \mathfrak{I} - \ker(\chi)$, from which it follows that $\mathfrak{I} = \mathfrak{A}$. We conclude that the kernel of a character determines a maximal ideal in $\mathfrak{A}$.

Definition 27.1.4. *Let $\mathfrak{A}$ be a normed, unital algebra. The spectrum $\sigma(a)$ of $a \in \mathfrak{A}$ is defined to be the complement $\mathbb{C} - \rho(a)$ where $\rho(a) := \{z \in \mathbb{C}; \ (a - z \cdot 1)^{-1} \in \mathfrak{A}\}$ is called the resolvent set of a. For $z \in \rho(a)$ one calls $r_z(a) := (a - z \cdot 1)^{-1}$ the resolvent of a at z. The number*

$$r(a) := \sup(\{|z|; \ z \in \sigma(a)\}) \tag{27.1.2}$$

is called the spectral radius of $a \in \mathfrak{A}$.

Notice that the condition $a^{-1} \in \mathfrak{A}$ implies that $||a^{-1}||$ exists, that is, the inverse has a norm ('is bounded'). If we are dealing with an algebra of possibly unbounded operators on a Hilbert space then Definition 27.1.4 must be made more precise: if a is a densely defined, closable (the adjoint $a^* \equiv a^\dagger$ is densely defined) linear operator on a Hilbert space $\mathcal{H}$ with dense domain $D(a)$ then $z \in \rho(a)$ iff $a - z \cdot 1$ is a bijection from $D(a)$ onto $\mathcal{H}$ with *bounded* inverse.

Later we will need the following technical result.

Lemma 27.1.5. *For the spectral radius the following identity holds*

$$r(a) = \lim_{n \to \infty} ||a^n||^{1/n} \tag{27.1.3}$$

Proof: First we show that the series of non-negative numbers $x_n = ||a^n||^{1/n}$ actually converges. For this purpose let $n \geq m \geq 1$ be any natural numbers and split n uniquely as $n = km + r$ for natural numbers k, r with $0 \leq r < m$. By submultiplicativity of the norm we have

$$||a^n||^{1/n} \leq ||a^{km}||^{1/n} \, ||a^r||^{1/n} \leq ||a^m||^{k/n} \, ||a^r||^{1/n} \tag{27.1.4}$$

Fix m and take $n \to \infty$ so that $k = (n - r)/m \to \infty$ while $r \in \{0, \ldots, m - 1\}$ stays bounded. Thus the right-hand side of (27.1.4) converges to $||a^m||^{1/m}$. It follows that the sequence (x_n), $x_n = ||a^n||^{1/n}$ is bounded and therefore must have an accumulation point, each of which must be smaller than x_m for any $m \geq 1$. Let $\lim_n \sup(x_n)$ be the largest accumulation point, then the inequality $\lim_n \sup(x_n) \leq x_m$ holds. Now take the infimum on the right-hand side which is also an accumulation point, then we get

$$\lim_n \sup(x_n) \leq \lim_m \inf(x_m) \tag{27.1.5}$$

which means that there is only one accumulation point, so the sequence converges. Denote $x := \lim_{n \to \infty} x_n$.

Now consider the geometrical (von Neumann) series for $z \neq 0$

$$r_z(a) = (a - z \cdot 1)^{-1} = -\frac{1}{z} \sum_{n=0}^{\infty} \left(\frac{a}{z}\right)^n \qquad (27.1.6)$$

which converges if there exists $0 \leq q < 1$ with $||(\frac{a}{z})^n||^{1/n} = ||a^n||^{1/n}/|z| < q$ for all $n > n(q)$. In other words, $z \in \rho(a)$ provided that $|z| > \lim_{n\to\infty} x_n$ or equivalently $z \in \sigma(a)$ provided that

$$|z| \leq x \qquad (27.1.7)$$

Taking the supremum in $\sigma(a)$ on the left-hand side of (27.1.7) we thus find

$$r(a) \leq x \qquad (27.1.8)$$

Suppose now that $r(a) < x$. Then there exists a real number R with $r(a) < R < x$ and since obviously $R \in \rho(a)$ it is clear that the resolvent $r_R(a)$ of a at R converges. Let ϕ be a continuous linear functional on $\mathfrak{A}$ then

$$\phi(r_R(a)) = -\frac{1}{R} \sum_{n=0}^{\infty} \phi\left(\left(\frac{a}{R}\right)^n\right) \qquad (27.1.9)$$

exists, which means that $\lim_{n\to\infty} \phi((\frac{a}{R})^n) = 0$. In other words, the function $n \mapsto \phi((\frac{a}{z})^n)$ is bounded for all continuous linear functionals ϕ.

Now the space $\mathfrak{A}'$ of continuous linear forms on $\mathfrak{A}$ is itself a Banach space with norm $||\phi|| := \sup_{0 \neq a \in \mathfrak{A}} |\phi(a)|/||a||$. Consider the family $\mathcal{F} := \{a^n/R^n; \, n \in \mathbb{N}\}$, then we have just shown that for each $b \in \mathcal{F}$ the set $\{|\phi(b)|; \, \phi \in \mathfrak{A}'\}$ is bounded. Let us consider each $b \in \mathcal{F}$ as a map $b: \mathfrak{A}' \to \mathbb{C}; \, \phi \mapsto \phi(b)$. We have $||b||' := \sup_{\phi \in \mathfrak{A}'} |\phi(b)|/||\phi|| = ||b||$ where the norm in the last equality is the one in $\mathfrak{A}$. By the principle of uniform boundedness [282] (or Banach–Steinhaus theorem) the set $\{||b||'; \, b \in \mathcal{F}\}$ is bounded. Therefore we know that the set of norms $||a^n/R^n||$ is bounded. But

$$||a^n/R^n|| = \left(\frac{||a^n||^{1/n}}{R}\right)^n = \left(\frac{x_n}{R}\right)^n \qquad (27.1.10)$$

Since x_n converges to x, for each $\epsilon > 0$ we find $n(\epsilon) \in \mathbb{N}$ such that $|x_n - x| < \epsilon$, that is, $-\epsilon < x_n - x < \epsilon$. Since $x_n, x > 0$ we may choose $x > \epsilon$ and thus $x_n > x - \epsilon$ for all $n > n(\epsilon)$. Hence (27.1.10) turns into

$$||a^n/R^n|| > \left(\frac{x - \epsilon}{R}\right)^n \qquad (27.1.11)$$

for all $n(\epsilon) < n$. Since by assumption $r(a) < R < x$ we can choose sufficiently small $\epsilon > 0$ such that $x - \epsilon > R$, say $x - \epsilon = Rq$ for some $q > 1$. Summarising, we find $x - Rq - \epsilon = 0$ for sufficiently small $q > 1$ such that q^n is bounded for all $n > n(\epsilon)$, which is a contradiction.

Thus in fact $r(a) = \lim_{n\to\infty} ||a^n||^{1/n}$. $\qquad \qquad \square$

We will now start establishing the relation between characters and maximal ideals.

Lemma 27.1.6. *If $\mathfrak{I}$ is an ideal in a unital Banach algebra $\mathfrak{A}$ then its closure $\overline{\mathfrak{I}}$ is still an ideal in $\mathfrak{A}$. Every maximal ideal is automatically closed.*

Proof: Recall that the closure of a subset Y in a topological space is Y together with the limit points of convergent nets in Y. Let now $\mathfrak{I}$ be an ideal in $\mathfrak{A}$ and let (a^α) be a net in $\mathfrak{I}$ converging to $a \in \overline{\mathfrak{I}}$. Then for any $b \in \mathfrak{A}$ we have $ba^\alpha \in \mathcal{I}$ since $\mathfrak{I}$ is an ideal and $\lim_\alpha ba^\alpha = ba$ since $||b(a^\alpha - a)|| \leq ||b|| \, ||a^\alpha - a|| \to 0$. Thus (ba^α) is a net in $\mathfrak{I}$ converging to $ba \in \mathfrak{A}$ and since all limit points of converging nets in $\mathfrak{I}$ by definition lie in $\overline{\mathfrak{I}}$ we actually have $ba \in \overline{\mathfrak{I}}$. Thus, $\overline{\mathfrak{I}}$ is an ideal.

Next we notice that every $a \in \mathfrak{A}$ such that $||a - 1|| < 1$ is invertible (use $a^{-1} = -(1 - (a-1))^{-1}$ and the geometric series representation for the latter with convergence radius 1). The set $\{a \in \mathfrak{A}; ||a - 1|| \geq 1\}$ is a closed subset of $\mathfrak{A}$ because if (a^α) is a convergent net in it then the net of real numbers $(||a^\alpha - 1||)$ belongs to the set $\{x \in \mathbb{R}; x \geq 1\}$ and since it converges to $||a - 1||$ it follows that $||a - 1|| \geq 1$ since $\{x \in \mathbb{R}; x \geq 1\}$ is closed (that $b^\alpha \to b$ implies $||b^\alpha|| \to ||b||$ follows from the triangle inequality $||a|| \leq ||a - b|| + ||b||, ||b|| \leq ||a - b|| + ||a||$). We conclude that every non-trivial (those not containing invertible elements) ideal $\mathfrak{I}$ must be contained in the closed set $\{a \in \mathfrak{A}; ||a - 1|| \geq 1\}$ and so must its closure $\overline{\mathfrak{I}}$. Obviously $1 \not\subset \{a \subset \mathfrak{A}; ||a - 1|| \geq 1\}$, hence, closures of non-trivial ideals are non-trivial.

Finally a maximal ideal must be closed as otherwise its closure would be a non-trivial ideal containing it. $\square$

Theorem 27.1.7 (Gel'fand). *If $\mathfrak{A}$ is an Abelian, unital Banach algebra and $\mathfrak{I}$ a two-sided, maximal ideal in $\mathfrak{A}$ then the quotient algebra $\mathfrak{A}/\mathfrak{I}$ is isomorphic with $\mathbb{C}$.*

Proof: By Lemma 27.1.6 $\mathfrak{I}$ is closed in $\mathfrak{A}$. We split the proof into three parts.

[i] If $\mathfrak{I}$ is a maximal ideal in a unital Banach algebra $\mathfrak{A}$ then $\mathfrak{A}/\mathfrak{I}$ is a Banach algebra.

The norm on $\mathfrak{A}/\mathfrak{I}$ is given by

$$||[a]|| := \inf_{b \in [a]} ||b|| \tag{27.1.12}$$

To see that this indeed defines a norm we check

$$||[za]|| = ||z[a]|| = \inf_{b \in [a]} ||zb|| = |z| \, ||[a]||$$

$$||[a + a']|| = ||[a] + [a']|| = \inf_{b \in [a]+[a']} ||b|| = \inf_{b \in [a], b' \in [a']} ||b + b'||$$

$$\leq \inf_{b \in [a], b' \in [a']} (||b|| + ||b'||) = ||[a]|| + ||[a']||$$

$$||[a]|| = \inf_{b \in [a]} ||b|| = 0 \Rightarrow [a] = [0] \tag{27.1.13}$$

In the second line we exploited the fact that every representative of $[a + a']$ can be written in the form $b + b'$ where b, b' are representatives of $[a], [a']$ and that the joint infimum is the same as the infimum. The conclusion in the last line means that $[a]$ contains elements of arbitrarily small norm. (Consider a net of elements $(a + b^\alpha)$ in $[a]$ whose norm converges to zero. The net (b^α) is a net in $\mathfrak{I}$ and since $\mathfrak{I}$ is closed it follows that the limit point $a + b$ lies in $[a]$. Since $||a + b|| = 0$ and $||.||$ is a norm it follows that $a + b = 0$, thus $0 \in [a]$ and so $[a] = [0]$.)

Suppose that $([a_n])$ is a Cauchy sequence in $\mathfrak{A}/\mathfrak{I}$. We may assume $||[a_{n+1}] - [a_n]|| = |||[a_{n+1} - a_n]||| < 2^{-n}$ (pass to a subsequence if necessary). Since

$$||[a_{n+1}] - [a_n]||| = \inf_{b_{n+1}\in[a_{n+1}], b_n\in[a_n]} ||b_{n+1} - b_n|| < 2^{-n} \qquad (27.1.14)$$

we certainly find representatives with $||c_{n+1} - c_n|| < 2^{-n+1}$. Then for $n > m$

$$||c_n - c_m|| = ||\sum_{k=m+1}^{n-1}(c_{k+1} - c_k)|| \leq \sum_{k=m+1}^{n-1} 2^{-k+1} = 2^{-m}\sum_{k=0}^{m-n-1} 2^k \leq 2^{-m+1}$$

$$(27.1.15)$$

which displays (c_n) as a Cauchy sequence in $\mathfrak{A}$. Since $\mathfrak{A}$ is complete this sequence converges to some $a \in \mathfrak{A}$. But then

$$||[a_n] - [a]||| = \inf_{b_n\in[a_n], b\in[a]} ||b_n - b|| \leq ||c_n - a|| \qquad (27.1.16)$$

so $([a_n])$ converges to $[a]$. It follows that $\mathfrak{A}/\mathfrak{I}$ is complete, that is, a Banach space with unit $[1]$.

[ii] *For an Abelian, unital algebra $\mathfrak{A}$ an ideal $\mathfrak{I}$ is maximal in $\mathfrak{A}$ iff $\mathfrak{A}/\mathfrak{I} - [0]$ consists of invertible elements only.*

$\Rightarrow$:
Suppose we find $[0] \neq [a] \in \mathfrak{A}/\mathfrak{I}$ but that $[a]^{-1}$ does not exist. This means that a^{-1} does not exist since $[a]^{-1} = [a^{-1}]$ as follows from $[a][a^{-1}] = [1]$. Consider now the ideal $\mathfrak{A} \cdot a = \{ba; b \in \mathfrak{A}\}$ (this is a two-sided ideal because $\mathfrak{A}$ is Abelian). Since $\mathfrak{I} \subset \mathfrak{A}$ we certainly have $\mathfrak{I} \cdot a \subset \mathfrak{A} \cdot a$ and since $\mathfrak{I} \cdot a = \mathfrak{I}$ because $\mathfrak{I}$ is in particular a right ideal we have $\mathfrak{I} \subset \mathfrak{A} \cdot a$. Now $a \in \mathfrak{A} \cdot a$ since $1 \in \mathfrak{A}$ and $a \notin \mathfrak{I}$ because otherwise $[a] = [0]$ which we excluded. It follows that $\mathfrak{I}$ is a proper subideal of $\mathfrak{A} \cdot a$. Finally, since $a^{-1} \notin \mathfrak{A}$, $\mathfrak{A} \cdot a$ cannot be all of $\mathfrak{A}$, for instance $1 \notin \mathfrak{A} \cdot a$ (an ideal that contains 1 or any invertible element is anyway the whole algebra). It follows that $\mathfrak{I}$ is not maximal.
$\Leftarrow$:
Suppose $\mathfrak{I}$ is not a maximal ideal. Then we find a proper subideal $\mathfrak{J}$ of $\mathfrak{A}$ of which $\mathfrak{I}$ is a proper subideal. Since every non-zero element of $\mathfrak{A}/\mathfrak{I}$ is invertible so is every element $[a]$ of $\mathfrak{J}/\mathfrak{I}$. But then $\mathfrak{J}$ contains the invertible element $a \in \mathfrak{A}$ and thus $\mathfrak{J}$ coincides with $\mathfrak{A}$, which is a contradiction.

[iii] A unital Banach algebra $\mathfrak{B}$ in which every non-zero element is invertible is isomorphic with $\mathbb{C}$.

Consider any $b \in \mathfrak{B}$, then we claim that $\sigma(b) \neq \emptyset$. Suppose that were not the case, then $\rho(b) = \mathbb{C}$. Let ϕ be a continuous linear functional on $\mathfrak{A}$ considered as a vector space with metric. Using linearity of ϕ and the expansion of $r_z(b)$ into an absolutely converging geometric series we see that $z \mapsto \phi(r_z(b))$ is an entire analytic function. Since ϕ is linear and continuous, it is bounded with bound $||\phi||$. Thus $|\phi(r_z(b))| \leq ||\phi|| \, ||r_z(b)||$. Since $\lim_{z \to \infty} ||r_z(b)|| = 0$ (use the geometric series) and $||r_z(a)||$ is everywhere defined in $\mathbb{C}$ we conclude that $z \mapsto \phi(r_z(b))$ is an entire bounded function which therefore, by Liouville's theorem, is a constant $c_a = \phi(r_z(b)) = \lim_{z \to \infty} \phi(r_z(b)) = 0$. Since ϕ was arbitrary it follows that $r_z(a) = 0$, implying that $b - z \cdot 1$ does not exist, which cannot be the case.

Thus we find $z_b \in \sigma(b)$, that is, $b - z_b \cdot 1$ is not invertible. By assumption, only zero elements are not invertible, hence $b = z_b \cdot 1$ for some $z_b \in \mathbb{C}$ for any $b \in \mathfrak{B}$. The map $b \mapsto z_b$ is then the searched for isomorphism $\mathfrak{B} \to \mathbb{C}$. Notice that $b = 0$ iff $z_b = 0$.

Let then $\mathfrak{I}$ be a maximal ideal in a unital, Abelian Banach algebra $\mathfrak{A}$. Then by [i] $\mathfrak{B} := \mathfrak{A}/\mathfrak{I}$ is a unital Banach algebra and by [ii] each of its non-zero elements is invertible. Thus by [iii] it is isomorphic with $\mathbb{C}$. $\qquad \square$

Corollary 27.1.8. *In an Abelian, unital Banach algebra $\mathfrak{A}$ there is a one-to-one correspondence between its spectrum $\Delta(\mathfrak{A})$ and the set $I(\mathfrak{A})$ of maximal ideals in $\mathfrak{A}$ via*

$$\Delta(\mathfrak{A}) \to I(\mathfrak{A}); \; \chi \mapsto \ker(\chi) \qquad (27.1.17)$$

Proof: That each character gives rise to a maximal ideal in $\mathfrak{A}$ through its kernel was already shown after Definition 27.1.3 . Conversely, let $\mathfrak{I}$ be a maximal ideal in a commutative unital Banach algebra then we can apply Theorem 27.1.7 and obtain a Banach algebra isomorphism $\chi : \mathfrak{A}/\mathfrak{I} \to \mathbb{C}; \; [a] \to \chi([a])$. We can extend this to a homomorphism $\chi : \mathfrak{A} \to \mathbb{C}$ by $\chi(a) := \chi([a])$. By construction $\chi(a) = 0$ iff $[a] = [0]$, that is, iff $a \in \mathfrak{I}$. In other words, the maximal ideal $\mathfrak{I}$ is the kernel of the character χ. $\qquad \square$

The subsequent lemma explains the word 'spectrum'.

Lemma 27.1.9. *Let $\mathfrak{A}$ be a unital, commutative Banach algebra and $a \in \mathfrak{A}$. Then $z \in \sigma(a)$ iff there exists $\chi \in \Delta(\mathfrak{A})$ such that $\chi(a) = z$.*

Proof: The requirement $\chi(a) = z$ is equivalent with $\chi(a - z \cdot 1) = 0$ so that $a - z \cdot 1 \in \ker(\chi)$. Since $\ker(\chi)$ is a maximal ideal in $\mathfrak{A}$ it cannot contain invertible elements, thus $(a - z \cdot 1)^{-1}$ does not exist, hence $z \in \sigma(a)$. $\qquad \square$

We now equip the spectrum with a topology. We begin by showing that the characters are in particular continuous linear functionals on the topological vector space $\mathfrak{A}$.

Definition 27.1.10. *For a character χ in an Abelian, unital Banach algebra we define its norm by*

$$||\chi|| := \sup_{0 \neq a \in \mathfrak{A}} \frac{|\chi(a)|}{||a||} \tag{27.1.18}$$

Lemma 27.1.11. *The characters of an Abelian, unital Banach algebra form a subset of the unit sphere in $\mathfrak{A}'$, the continuous linear functionals on $\mathfrak{A}$ considered as a topological vector space.*

Proof: By Lemma 27.1.9 we showed that $\sigma(a) = \{\chi(a); \chi \in \Delta(\mathfrak{A})\}$. It follows that

$$||\chi|| = \sup_{a \in \mathfrak{A}} \frac{|\chi(a)|}{||a||} \leq \sup_{a \in \mathfrak{A}} \frac{\sup\{|\chi'(a)|; \chi' \in \Delta(\mathfrak{A})\}}{||a||} = \sup_{a \in \mathfrak{A}} \frac{r(a)}{||a||} \leq 1 \tag{27.1.19}$$

since by Lemma 27.1.5 we have $r(a) = \lim_{n \to \infty} ||a^n||^{1/n} \leq ||a||$. On the other hand $\chi(1) = 1$, hence $||\chi|| = 1$ for every character χ. This shows that every character is a bounded linear functional on $\mathfrak{A}$, that is, $\Delta(\mathfrak{A}) \subset \mathfrak{A}'$. $\square$

Since we just showed that the characters are in particular bounded linear functionals it is natural to equip the spectrum with the weak *-topology of pointwise convergence induced from $\mathfrak{A}'$.

Definition 27.1.12

(i) *The weak *-topology on the topological dual X' of a topological vector space X (the set of continuous (bounded) linear functionals) is defined by pointwise convergence (that is, a net (ϕ^α) in X' converges to ϕ iff for any $x \in X$ the net of complex numbers $(\phi^\alpha(x))$ converges to $\phi(x)$). Equivalently, it is the weakest topology such that all the functions $x : X' \to \mathbb{C}; \phi \mapsto \phi(x)$ are continuous.*

(ii) *The Gel'fand topology on the spectrum of a unital, Abelian Banach algebra is the weak *-topology induced from $\mathfrak{A}'$ on its subset $\Delta(\mathfrak{A})$.*

We now show that in the Gel'fand topology the spectrum becomes a compact Hausdorff space. We need a preparational lemma.

Lemma 27.1.13. *Let X be a Banach space and X' its topological dual. Then the unit ball in X' is closed and compact in the weak *-topology.*

Proof: The unit ball B in X' is defined as the subset of elements ϕ with norm smaller than or equal to unity, that is, $||\phi|| := \sup_{x \in X} |\phi(x)|/||x|| \leq 1$. By Corollary 18.1.8 we must show that every universal net in B converges. Let ϕ^α be a universal net in B and consider for any given $x \in X$ the net of complex numbers $(\phi^\alpha(x))$ which are bounded by $||x||$. Our $x \in X$ defines a linear form $X' \to \mathbb{C}; \phi \to \phi(x)$ whence by Lemma 18.1.7(ii) the net $(\phi^\alpha(x))$ is universal. It is contained in the set $\{z \in \mathbb{C}; |z| \leq ||x||\}$ which is compact in $\mathbb{C}$ and therefore it converges. Define ϕ pointwise by the limit, that is, $\phi(x) := \lim_\alpha \phi^\alpha(x)$. Then

$$||\phi|| = \sup_{x \in X} \lim_\alpha |\phi^\alpha(x)|/||x|| \leq \lim_\alpha \sup_{x \in X} |\phi^\alpha(x)|/||x|| = \lim_\alpha ||\phi^\alpha|| \leq 1 \quad (27.1.20)$$

Thus ϕ^α converges pointwise to $\phi \in B$. In particular we have shown that B is closed. $\qquad\square$

Theorem 27.1.14. *In the Gel'fand topology, the spectrum $\Delta(\mathfrak{A})$ of a unital, Abelian Banach algebra is compact.*

Proof: Since we have shown (1) in Lemma 27.1.11 that $\Delta(\mathfrak{A})$ is a subset of the unit ball B in $\mathfrak{A}'$, (2) in Lemma 27.1.13 that B is compact in the weak $*$-topology and (3) in Lemma 18.1.10 that closed subspaces of compact spaces are compact in the subspace topology it will be sufficient to show that $\Delta(\mathfrak{A})$ is closed in B as the Gel'fand topology is the subspace topology induced from B.

Let then (χ^α) be a net in $\Delta(\mathfrak{A})$ converging to $\chi \in B$. We have, for example, $\chi(ab) = \lim_\alpha \chi^\alpha(ab) = \lim_\alpha \chi^\alpha(a)\chi^\alpha(b) = \chi(a)\chi(b)$ and similar for pointwise addition, scalar multiplication and involution in $\mathfrak{A}$. It follows that χ is a character, that is, $\chi \in \Delta(\mathfrak{A})$. $\qquad\square$

The Hausdorff property will be established in the next section.

27.2 The Gel'fand transform and the Gel'fand isomorphism

Definition 27.2.1. *The Gel'fand transform is defined by*

$$\bigvee : \mathfrak{A} \to \Delta(\mathfrak{A})'; \ a \mapsto \check{a} \ \text{where} \ \check{a}(\chi) := \chi(a) \qquad (27.2.1)$$

Here $\Delta(\mathfrak{A})'$ denotes the continuous linear functionals on $\Delta(\mathfrak{A})$ considered as a topological vector space.

It is clear that every $\check{a}$, $a \in \mathfrak{A}$ is a continuous linear functional on the spectrum since for any net (χ^α) in $\Delta(\mathfrak{A})$ which converges to χ we have $\lim_\alpha \check{a}(\chi^\alpha) = \lim_\alpha \chi^\alpha(a) = \chi(a) = \check{a}(\chi)$ because convergence of (χ^α) means pointwise convergence on $\mathfrak{A}$.

Theorem 27.2.2. *The Gel'fand transform extends to a homomorphism*

$$\bigvee : \mathfrak{A} \to C(\Delta(\mathfrak{A})); \ a \mapsto \check{a} \qquad (27.2.2)$$

with the following additional properties:

1. range($\check{a}$) $= \sigma(a)$.
2. $||\check{a}|| := \sup_{\chi \in \Delta(\mathfrak{A})} |\check{a}(\chi)| = r(a)$.
3. The image $\bigvee(\mathfrak{A})$ separates the points of $\Delta(\mathfrak{A})$.

Proof

0. Morphism and continuity

We have for example

$$(ab)^\vee(\chi) = \chi(ab) = \chi(a)\chi(b) = \breve{a}(\chi)\breve{b}(\chi) \tag{27.2.3}$$

for any $\chi \in \Delta(\mathfrak{A})$ and similarly for $(a+b)^\vee$. Thus multiplication and addition of functions are defined pointwise. That the functions $\breve{a}$ are continuous follows after Definition 27.2.1 from the fact that the weak *-topology on $\Delta(\mathfrak{A})$ *is defined by asking that all the Gel'fand transforms $\breve{a}$ be continuous and therefore is tautologous.

1. We have

$$\mathrm{range}(\breve{a}) = \{\breve{a}(\chi); \ \chi \in \Delta(\mathfrak{A})\} = \{\chi(a); \ \chi \in \Delta(\mathfrak{A})\} = \sigma(a) \tag{27.2.4}$$

as follows from Lemma 27.1.9 .

2. We have

$$\|\breve{a}\| = \sup_{\chi \in \Delta(\mathfrak{A})} |\breve{a}(\chi)| = \sup_{\chi \in \Delta(\mathfrak{A})} |\chi(a)| = \sup(\{|\chi(a)|; \ \chi \in \Delta(\mathfrak{A})\}) = r(a) \tag{27.2.5}$$

by definition of the spectral radius. Notice that the sup-norm is a natural norm on a space of continuous functions on a compact space.

3. Recall that a collection of functions $\mathcal{C}$ on a topological space X is said to separate its points iff for any $x_1 \neq x_2$ we find $f \in \mathcal{C}$ such that $f(x_1) \neq f(x_2)$. Consider then any $\chi_1, \chi_2 \in \Delta(\mathfrak{A})$ with $\chi_1 \neq \chi_2$. By definition of $\Delta(\mathfrak{A})$ there exists then $a \in \mathfrak{A}$ such that $\chi_1(a) = \breve{a}(\chi_1) \neq \chi_2(a) = \breve{a}(\chi_2)$.   □

To see that then $\Delta(\mathfrak{A})$ is a Hausdorff space recall the following lemma.

Lemma 27.2.3. *Let X be a topological space and $\mathcal{C} \subset C(X)$ a collection of continuous functions on X which separate the points of X. Then the topology on X is Hausdorff.*

Proof: Let $x_1, x_2 \in X$ with $x_1 \neq x_2$ be any two distinct points. Since $\mathcal{C}$ separates the points we find $f \in \mathcal{C}$ with $f(x_1) \neq f(x_2)$. Let $d := |f(x_2) - f(x_1)|$. Since f is continuous at x_I, for any $\epsilon > 0$ we find a neighbourhood $U_I(\epsilon)$ of x_I, $I = 1, 2$ such that $|f(x) - f(x_I)| < \epsilon$ for any $x \in U_I(\epsilon)$. Now $d = |f(x_2) - f(x_1)| \leq |f(x) - f(x_1)| + |f(x_2) - f(x)|$ for any $x \in X$. Thus $d - \epsilon < |f(x_2) - f(x)|$ for any $x \in U_1(\epsilon)$ and $d - \epsilon < |f(x_1) - f(x)|$ for any $x \in U_2(\epsilon)$. Choose $\epsilon < d/2$. Then $U_1(\epsilon) \cap U_2(\epsilon) = \emptyset$.   □

Corollary 27.2.4. *The Gel'fand topology on the spectrum of a unital, Abelian Banach algebra is Hausdorff.*

Proof: The proof follows trivially from the fact that by Theorem 27.2.2 $\mathcal{C} := \{\breve{a}; \ a \in \mathfrak{A}\}$ is a system of continuous functions separating the points of $\Delta(\mathfrak{A})$ together with Lemma 27.2.3 .   □

So far everything worked for an Abelian, unital Banach algebra $\mathfrak{A}$. We now invoke the further restriction that $\mathfrak{A}$ be an Abelian, unital C^*-algebra which makes the Gel'fand transform especially nice.

Theorem 27.2.5. *Let $\mathfrak{A}$ be a unital, commutative C^*-algebra (not only a Banach algebra). Then the Gel'fand transform is an isometric isomorphism between $\mathfrak{A}$ and the space of continuous functions on its spectrum.*

Proof: First of all, using the fact that in a commutative *-algebra every element is normal (meaning that $[a, a^*] = 0$) we have, making frequent use of the C^* property (27.1.1)

$$
\begin{aligned}
\|a^{2^n}\|^2 &= \|a^{2^n}(a^{2^n})^*\| = \|(aa^*)^{2^n}\| \\
&= \|(aa^*)^{2^{n-1}}((aa^*)^{2^{n-1}})^*\| = \|(aa^*)^{2^{n-1}}\|^2 \\
&= \|aa^*\|^{2^n} = \|a\|^{2^{n+1}}
\end{aligned}
\tag{27.2.6}
$$

where in the third equality we exploited that aa^* is self-adjoint and in the fifth equality we iterated the equality between the expressions at the end of the first and second line. We conclude that for any natural number n

$$
\|a\| = \|a^{2^n}\|^{1/2^n}
\tag{27.2.7}
$$

In Lemma 27.1.5 we proved the formula $r(a) = \lim_{n \to \infty} \|a^n\|^{1/n}$ meaning that every subsequence of the sequence $(\|a^n\|^{1/n})$ has the same limit $r(a)$ including the one displayed in (27.2.7). Thus we have shown that for Abelian C^*-algebras indeed

$$
r(a) = \|a\|
\tag{27.2.8}
$$

and not only $r(a) \leq \|a\|$. By item (2) of Theorem 27.2.2 we have therefore

$$
\|\breve{a}\| = \|a\|
\tag{27.2.9}
$$

that is, isometry.

Consider now the system of complex-valued functions on the spectrum given by $\mathcal{C} := \{\breve{a}; \ a \in \mathfrak{A}\}$. We claim that it has the following properties:

(i) $\mathcal{C} \subset C(\Delta(\mathfrak{A}))$.
(ii) $\mathcal{C}$ separates the points of $\Delta(\mathfrak{A})$.
(iii) $\mathcal{C}$ is a closed (in the sup-norm topology) *-subalgebra of $C(\Delta(\mathfrak{A}))$.
(iv) The constant functions belong to $\mathcal{C}$.

Properties (i), (ii) are the assertions (0) and (3) of Theorem 27.2.2 while (iv) follows from the fact that $\mathfrak{A}$ is unital, that is, $\breve{1}(\chi) = \chi(1) = 1$ so $\breve{1} = 1$. To show that (iii) $\mathcal{C}$ is a closed *-algebra in $C(\Delta(\mathfrak{A}))$ suppose that $(\breve{a}^\alpha)$ is a net in $\mathcal{C}$ converging to some $f \in C(\Delta(\mathfrak{A}))$. Thus, $(\breve{a}^\alpha)$ is in particular a Cauchy net, meaning that $\|\breve{a}^\alpha - \breve{a}^\beta\| = \|a^\alpha - a^\beta\|$ becomes arbitrarily small as α, β grow, where we have used isometry. It follows that (a^α) is a Cauchy net and therefore converges to some $a \in \mathfrak{A}$ since $\mathfrak{A}$ is in particular a Banach algebra and therefore complete. Therefore $f = \check{a} \in \mathcal{C}$, whence $\mathcal{C}$ is closed. Clearly $\mathcal{C}$ is also a *-subalgebra because $\mathfrak{A}$ is an algebra and $\bigvee$ a homomorphism.

Now recall from Theorem 27.1.14 and Corollary 27.2.4 that $\Delta(\mathfrak{A})$ is a compact Hausdorff space. Then properties (i)–(iii) of $\mathcal{C}$ enable us to apply the Stone–Weierstrass theorem, Theorem 18.1.11 (e.g., [282]) which tells us that either $\mathcal{C} = C(\Delta(\mathfrak{A}))$ or that there exists $\chi_0 \in \Delta(\mathfrak{A})$ such that $\check{a}(\chi_0) = 0$ for all $\check{a} \in \mathcal{C}$. By property (iv) the latter possibility is excluded, whence $\mathcal{C} = \bigvee(\mathfrak{A})$ is *all of* $C(\Delta(\mathfrak{A}))$. In other words, the Gel'fand transform is a surjection. Finally it is an injection since $\check{a} = \check{a}'$ implies $||\check{a} - \check{a}'|| = ||a - a'|| = 0$ by isometry, hence $a = a'$. $\qquad\qquad\qquad\qquad\qquad\qquad\qquad\qquad\qquad\qquad\qquad\qquad\qquad\qquad\square$

Corollary 27.2.6. *Every compact Hausdorff space X arises as the spectrum of an Abelian, unital C*-algebra $\mathfrak{A}$, specifically $\mathfrak{A} = C(X)$, $\Delta(\mathfrak{A}) = X$.*

Proof: Let X be a compact Hausdorff space and define $\mathfrak{A} := C(X)$ equipped with the sup-norm. Then $X \subset \Delta(C(X))$ by the definition $x(f) := f(x) =: \check{f}(x)$ for any $f \in \mathfrak{A}$, so the Gel'fand transform is the identity map on $C(X)$. Thus, if $\Delta(C(X)) - X \neq \emptyset$ then $\check{f}$ extends f continuously to $\Delta(C(X))$.

Next let (x^α) be a net in X which converges in $\Delta(C(X))$, then $\check{f}(x^\alpha)$ converges in $\mathbb{C}$ for any $\check{f} \in C(\Delta(C(X)))$, that is, $f(x^\alpha)$ converges in $\mathbb{C}$ for any $f \in C(X)$. It follows that (x^α) converges in X, that is, X is closed in $\Delta(C(X))$.

Suppose now that $\Delta(C(X)) - X \neq \emptyset$. Thus we find $\chi_0 \in \Delta(C(X)) - X$. Now in a Hausdorff space the one-point sets are closed [533]. Therefore the sets $X, \{\chi_0\}$ are disjoint closed sets in the compact Hausdorff space $\Delta(C(X))$. Since compact Hausdorff spaces are normal spaces [282] (i.e., one-point sets are closed and any two disjoint closed sets are contained in open disjoint sets) we may apply Urysohn's lemma [282] to conclude that there is a continuous function $F :$ $\Delta(C(X)) \to \mathbb{R}$ with range in $[0, 1]$ such that $F_{|X} = 0$ and $F|\{\chi_0\} = F(\chi_0) = 1$.

Consider then any $f \in C(X)$. Since $C(\Delta(C(X)))$ are *all* continuous functions on $\Delta(C(X))$, there exist different continuous extensions of f to $\Delta(C(X))$, for instance the functions $\check{f}, \check{f} + F$ where F is of the form just constructed. However, this contradicts the fact that $\bigvee$ is an isomorphism since it would not be surjective. $\qquad\qquad\qquad\qquad\qquad\qquad\qquad\qquad\qquad\qquad\qquad\qquad\qquad\qquad\square$

Corollary 27.2.6 tells us that a compact Hausdorff space can be reconstructed from its Abelian, unital C^*-algebra of continuous functions by constructing its spectrum. This is the starting point for generalisations to non-commutative topological spaces by using non-Abelian C^*-algebras [167].

28
Bohr compactification of the real line

In order to illustrate the notions of Abelian C^*-algebras, their spectra and corresponding measures thereon, we consider a simple example in which all these structures arise already at an elementary level. For more information about the Bohr compactification of topological groups and almost periodic functions, see [896].

28.1 Definition and properties

Definition 28.1.1

(i) *For any $k \in \mathbb{R}$ define the periodic functions of period $2\pi/k$ by*

$$T_k : \mathbb{R} \to \mathbb{C}; \; x \mapsto e^{ikx} \tag{28.1.1}$$

The algebra $\mathcal{C}$ of almost periodic functions is the finite complex linear span of the functions T_k, that is, functions of the form

$$f = \sum_{I=1}^{N} z_I \, T_{k_I} \; \text{where } N < \infty, \; k_I \in \mathbb{R}, \; z_I \in \mathbb{C} \tag{28.1.2}$$

These are obviously bounded functions on $\mathbb{R}$. They form a *-algebra because $T_k \, T_{k'} = T_{k+k'}$, $\overline{T_k} = T_{-k}$.

(ii) *Let $\overline{\mathcal{C}}$ be the closure of $\mathcal{C}$ in the sup-norm on $\mathbb{R}$. This is an Abelian C^*-algebra with respect to pointwise operations and complex conjugation as involution. The spectrum of this algebra $\overline{\mathbb{R}} := \Delta(\overline{\mathcal{C}})$ is called the Bohr compactification of $\mathbb{R}$.*

This definition of the Bohr compactification has a natural extension to any topological group. Notice that the Bohr compactification is in general different from the Stone–Čech compactification $\check{X}$ of a topological space X which is the spectrum of the C^*-algebra obtained as the norm closure of continuous bounded functions.

The notion of almost periodic functions results from the fact that $\mathbb{Q}$ is dense in $\mathbb{R}$, thus for any $\epsilon > 0$, $f = \sum_{I=1}^{N} z_I T_{k_I} \in \mathcal{C}$ we find $q_I = m_I/n_I$, $0 \neq n_I, m_I \in \mathbb{Z}$ relative prime such that $|k_I - q_I| < \epsilon$ and such that f behaves as if it was periodic with period $2\pi n_1 \ldots n_N$ for sufficiently small range of x. It is truly periodic only if the k_I are rationally dependent. In order to make the connection with the main text, consider the numbers $k \in \mathbb{R}$ as replacements for the spin network labels, the set $\mathbb{R}$ as $\mathcal{P}$, the periodic functions T_k as spin-network functions, the algebra $\mathcal{C}$ as the algebra of cylindrical functions on the space $\mathcal{A}$ of smooth connections, $\mathbb{R}$ as $\mathcal{A}$ and $\overline{\mathbb{R}}$ as $\overline{\mathcal{A}}$.

Let us describe $\overline{\mathbb{R}}$ in more detail. By definition its elements are arbitrary homomorphisms $\chi : \overline{\mathcal{C}} \to \mathbb{C}$ without any continuity assumptions. It is easy to see that any such character is determined once we know its values $X(k) := \chi(T_k)$, which are constrained by

$$X(k)\, X(k') = X(k + k'), \quad \overline{X(k)} = X(-k) \tag{28.1.3}$$

from which $|X(k)|^2 = 1$. Thus, $X : \mathbb{R} \to \mathrm{U}(1)$ is a group homomorphism which does not need to be continuous. This characterisation of the spectrum $\overline{\mathbb{R}}$ as the set of algebraic homomorphisms $\mathrm{Hom}(\mathbb{R}, \mathrm{U}(1))$ is precisely the analogue of the description of $\overline{\mathcal{A}}$ as $\mathrm{Hom}(\mathcal{P}, G)$ with G replaced by $\mathrm{U}(1)$.

If X is at least once differentiable then from (28.1.3) we get $X'(k) = X'(0)X(k)$ and the solution is of the form $X(k) = e^{ikx}$ for some $x \in \mathbb{R}$. Thus, $\mathbb{R} \subset \overline{\mathbb{R}}$ via $\chi_x(f) = f(x)$. However, $\overline{\mathbb{R}}$ is much larger than $\mathbb{R}$ as the following consideration reveals: our homomorphism $X(k)$ is $\mathrm{U}(1)$-valued and thus has the form $X(k) = \exp(if(k))$ where modulo 2π

$$f(k + k') = f(k) + f(k') \text{ and } f(-k) = -f(k) \tag{28.1.4}$$

We will consider the simpler case that f satisfies (28.1.4) exactly, not only modulo 2π. Then requirement (28.1.4) seems to imply that $f(k)$ is simply a linear map, but this is not the case since linearity also requires the scalar multiplication law that $f(\lambda k) = \lambda f(k)$ for all $\lambda \in \mathbb{R}$ so that actually $f(k) = k f(1)$ is already determined by the value $f(1)$. It is precisely this missing ingredient that enables us to construct maps $f(k)$ which are *everywhere discontinuous*. Here is a simple way of showing that.

Lemma 28.1.2. *A system of N real numbers k_I is called integrally (ILI) or rationally (RLI) linearly independent respectively provided that*

$$\sum_{I=1}^{N} q_I k_I = 0 \text{ implies } q_1 = \ldots = q_N = 0 \tag{28.1.5}$$

for $q_I \in \mathbb{Z}$ or $q_I \in \mathbb{Q}$ respectively (in particular, $k_I \neq 0$). Claim: rational and integral linear independence are equivalent.

Proof: That RLI implies ILI is trivial since $\mathbb{Z} \subset \mathbb{Q}$. Conversely, if k_I are ILI suppose that we find numbers $q_I = m_I/n_I \in \mathbb{Q}$, $0 \neq n_I, m_I \in \mathbb{Z}$ relative prime such that (28.1.5) holds. Multiplying the whole equation by $\prod_{I=1}^{N} n_I \neq 0$ implies

$$\sum_{I=1}^{N} \left[m_I \prod_{J \neq I} n_J \right] k_I = 0 \tag{28.1.6}$$

The numbers in the square brackets are now integral and due to ILI they must vanish for each I. Since the $n_J \neq 0$ we find $m_I = 0$, thus $q_I = 0$. $\qquad\square$

With this lemma we can now construct everywhere discontinuous characters. Step 1: Choose any $k_1 \neq 0$ and any value $f_1 \in \mathbb{R}$. Define the one-frequency *lattice* $S_1 := \{qk_1; \; q \in \mathbb{Q}\}$ and $f(qk_1) := qf_1$ for any $q \in \mathbb{Q}$. Requirement (28.1.4) is clearly satisfied on S_1 because $\mathbb{Q}$ is also an additive group

$$f(q_1 k_1 + q_2 k_1) = f([q_1 + q_2]k_1) = [q_1 + q_2]f_1 = f(q_1 k_1) + f(q_2 k_1) \text{ and}$$
$$f(-qk_1) = f([-q]k_1) = [-q]f_1 = -f(qk_1) \qquad (28.1.7)$$

Step 2: Next take any $k_2 \notin S_1$. Then k_1, k_2 are rationally independent. Choose any $f_2 \in \mathbb{R}$ and extend f to the two-frequency lattice $S_2 := \{q_1 k_1 + q_2 k_2; \; q_1, q_2 \in \mathbb{Q}\}$ by $f(q_1 k_1 + q_2 k_2) := q_1 f_1 + q_2 f_2$ which again satisfies (28.1.4) on S_2.

Step n: Given a set of rationally, linearly independent frequencies $k_1, \ldots, k_n$ and a set of real numbers $f_1, \ldots, f_n$ define the n-frequency lattice and the restriction of f to that lattice by

$$S_n := \left\{ \sum_{I=1}^{n} q_I k_I; \; q_I \in \mathbb{Q} \right\}, \quad f\left(\sum_{I=1}^{n} q_I k_I \right) := \sum_{I=1}^{n} q_I f_I \qquad (28.1.8)$$

The construction is completed by using the axiom of choice in order to iterate the procedure until all values of f have been defined.

Notice that all the sets S_n, $n = 1, 2, \ldots$ are dense in $\mathbb{R}$, they have the same cardinality as $\mathbb{Q}$. To see that for appropriate choice of k_I, f_I we obtain arbitrarily discontinuous maps, consider any $\epsilon > 0$ and any $q_2 \in \mathbb{Q}$. Since S_1 is dense in $\mathbb{R}$ we find $q_1 \in \mathbb{Q}$ such that $|q_2 k_2 - q_1 k_1| < \epsilon$, that is

$$q_2 \frac{k_2}{k_1} - \frac{\epsilon}{k_1} < q_1 < q_2 \frac{k_2}{k_1} + \frac{\epsilon}{k_1} \qquad (28.1.9)$$

It follows that

$$q_2 k_2 \left[\frac{f_2}{k_2} - \frac{f_1}{k_1} \right] - \epsilon \frac{f_1}{k_1} < f(q_2 k_2) - f(q_1 k_1)$$

$$= q_2 f_2 - q_1 f_1 < q_2 k_2 \left[\frac{f_2}{k_2} - \frac{f_1}{k_1} \right] + \epsilon \frac{f_1}{k_1} \qquad (28.1.10)$$

Taking $\epsilon \to 0$ for fixed q_2 (and of course fixed f_I, k_I) we get $|q_2 k_2 - q_1 k_1| \to 0$ while $|f(q_2 k_2) - f(q_1 k_1)| \to |q_2 k_2[\frac{f_2}{k_2} - \frac{f_1}{k_1}]|$. Thus, if $f_1/k_1 \neq f_2/k_2$, that is, if f is not linear, then f is discontinuous at all values $q_2 k_2$. Since the set of these values is dense in $\mathbb{R}$ we conclude that f is *discontinuous on a dense subset of* $\mathbb{R}$!

28.2 Analogy with loop quantum gravity

We conclude that $\overline{\mathbb{R}}$ is an incredibly much larger set than $\mathbb{R}$ itself, typical elements will consist of everywhere discontinuous homomorphisms $\mathbb{R} \to U(1)$ while the image of $\mathbb{R}$ in $\overline{\mathbb{R}}$ consists of the smooth homomorphisms. The explicit construction of these discontinuous homomorphisms hints at a projective description of the Bohr compactification: the label set $\mathcal{L}$ that we used for the projective description of $\overline{\mathcal{A}}$ consisted of subgroupoids of $\mathcal{P}$ generated by finite collections of holonomically independent *edges*. For $\overline{\mathbb{R}}$ we consider the label set $\mathcal{L}$ consisting of subgroups of $\mathbb{R}$ (considered as an additive group) generated by a *finite set of rationally independent frequencies*. Thus, any $l \in \mathcal{L}$ is determined by rationally independent numbers $k_1, \ldots, k_{n_l}$, $n_l < \infty$. The possible spin-network labels over the 'graph' l are then given by $\{\sum_I^{n_l} q_I k_I; \ q_I \in \mathbf{Z}\}$. We partially order $\mathcal{L}$ as follows: say that $l \prec l'$ if any k_I generating l can be integrally expressed by the k'_J generating l', that is, we find unique $q_{IJ} \in \mathbf{Z}$ such that $k_I = \sum_J^{n_{l'}} q_{IJ} k'_J$ for any $I = 1, \ldots, n_l$. The set $\mathcal{L}$ is then also directed: for l, l' simply consider the subgroup of $\mathbb{R}$ generated by the k_I and k'_J together. That is, consider the integral span of the combined set of frequencies and identify the smallest collection of rationally independent frequencies, denoted as $l \cup l'$, such that its integral span contains the integral span under consideration. (For instance, suppose that l is generated by k_1 and l' by k_2 but that they are not rationally independent, that is, $k_2 n_2 = k_1 n_1$ for some integers n_1, n_2 which are relative prime. Hence, define $k_3 := k_1/n_2 = k_2/n_1 \le k_1, k_2$ so the integral span of k_1, k_2 is contained in that of k_3. This procedure corresponds to subdividing edges of two original graphs by the edges of their union.) Consider the resulting l''. Then $l, l' \prec l''$.

Given $l \in \mathcal{L}$ we define $X_l := \mathrm{Hom}(l, \mathrm{U}(1)^{n_l})$. We can identify X_l with $\mathrm{U}(1)^{n_l}$ via the map $\rho_l : X_l \to \mathrm{U}(1)^{n_l}; \ x_l \mapsto \{x_l(k_I)\}_{I=1}^{n_l}$ because any homomorphism is already defined by the $x_l(k_I)$, whence $x_l(q_I k_I) = x_l(k_I)^{q_I}$. Using this identification, X_l becomes a compact Hausdorff space. Now for $l \prec l'$ we define $p_{ll'} : X_{l'} \to X_l; \ x_{l'} \mapsto (x_{l'})_{|l}$.(restriction map) which are certainly surjections and satisfy the consistency conditions $p_{l'l} \circ p_{l''l'} = p_{l''l}$ for $l \prec l' \prec l''$. We may then define the projective limit $\overline{X}$ as the closed subset of the direct product $X_\infty = \prod_{l \in \mathcal{L}} X_l$ defined in the usual way as

$$\overline{X} = \{x \in X_\infty; \ p_{l'l}(p_{l'}(x)) = p_l(x) \ \forall l \prec l'\} \tag{28.2.1}$$

where $x = (x_l)_{l \in \mathcal{L}}$ and $p_l(x) = x_l$ are the continuous projections (in the Tychonov topology on X_∞). Thus, $\overline{X}$ is a compact Hausdorff space.

To see that $\overline{\mathbb{R}}$ and $\overline{X}$ are homeomorphic we proceed similarly as in the main text: consider the map

$$\Phi : \overline{\mathbb{R}} := \mathrm{Hom}(\mathbb{R}, \mathrm{U}(1)) \to \overline{X}; \ \chi \mapsto x^\chi \text{ where } x_l^\chi := (\chi)_{|l} \tag{28.2.2}$$

Certainly $x^\chi \in \overline{X}$ because as a homomorphism it satisfies the consistency conditions encoded in (28.2.1). Φ is also an injection since $\Phi(\chi) = \Phi(\chi')$ implies $\chi(k) = \chi'(k)$ for all k, that is, $\chi = \chi'$. Conversely, given $x \in \overline{X}$ we define $\chi^x \in \overline{\mathbb{R}}$ as follows: for each $k \in \mathbb{R}$ use the axiom of choice to find some $l_k \in \mathcal{L}$ such that $k \in l_k$. Then define $\chi^x(k) := x_{l_k}(k)$. To see that this is well defined consider any other choice $k \to l'_k$. We must show that $x_{l_k}(k) = x_{l'_k}(k)$. We find $l_k, l'_k \prec \tilde{l}_k$. Then

$$x_{l_k}(k) = \left[p_{\tilde{l}_k l_k}(x_{\tilde{l}_k})\right](k)[x_{\tilde{l}_k}]_{|l_k}(k) = x_{\tilde{l}_k}(k) = \left[p_{\tilde{l}_k l'_k}(x_{\tilde{l}_k})\right](k) = x_{l'_k}(k) \tag{28.2.3}$$

so Φ is a surjection provided we can show that χ^x is a homomorphism. But this follows from x being a homomorphism on the various l. Concluding, Φ is a bijection and we must show that it together with its inverse is continuous. The proof of this fact follows the same reasoning as in the main text since that proof is completely categorial and can thus be omitted.

The Bohr compactification, considered as the spectrum of the closure of the algebra of almost periodic functions is equipped with the compact Hausdorff topology defined by saying that a net χ_α converges pointwise on $\overline{\mathcal{C}}$ (Gel'fand transforms of $f \in \overline{\mathcal{C}}$ are continuous). In particular, $\chi_\alpha(T_k) =: X_\alpha(k) \to X(k) =: \chi(T_k)$ for any $k \in \mathbb{R}$. Let us construct the analogue of the uniform measure on $\overline{\mathbb{R}}$:

Since the T_k play the role of spin-network functions, we may define μ_0 via the Riesz representation theorem from the positive linear functional on $C(\overline{\mathbb{R}})$ via

$$\Lambda(\check{T}_k) = \delta_{k,0} \tag{28.2.4}$$

where the right-hand side is a Kronecker symbol and not a δ-distribution. Notice that $\check{f}(\chi) = \chi(f)$. It follows that the $\check{T}_k$ form an orthonormal basis in the Hilbert space $L_2(\overline{\mathbb{R}}, d\mu_0)$ since

$$< \check{T}_k, \check{T}_{k'} > := \Lambda(\overline{\check{T}_k}\check{T}_{k'}) = \Lambda(\check{T}_{k'-k}) = \delta_{k'-k,0} = \delta_{k,k'} \tag{28.2.5}$$

so they form an orthonormal system and completeness follows from the fact that they form a subalgebra of $C(\overline{\mathbb{R}})$ which contains the constants and separates the points of $\overline{\mathbb{R}}$ (indeed $\chi(T_k) = \chi'(T_k)$ for all $k \in \mathbb{R}$ means $\chi(f) = \chi'(f)$ for all $f \in \overline{\mathcal{C}}$, i.e., $\chi = \chi'$) so that they are dense in $C(\overline{\mathbb{R}})$ by the Weierstrass theorem.

Cylindrical functions over the subgroups l are now defined as $f = p_l^* f_l$ for some $f_l : U(1)^{n_l} \to \mathbb{C}$, that is, $f(\chi) = f_l(\{\chi(k_I)\}_{I=1}^{n_l})$ with $\chi(k_I) := \chi(T_{k_I})$. The push-forward of the measure to the spaces $X_l = U(1)^{n_l}$ is easily checked to be

$$\mu_0(p_l^* f_l) = \mu_{0,l}(f_l) = \int_{U(1)^{n_l}} \prod_{I=1}^{n_l} d\mu_H(h_I) f_l(h_1, \ldots, h_{n_l}) \tag{28.2.6}$$

where μ_H is the Haar measure on $U(1)$. To see this, it is enough to check that (28.2.6) reproduces (28.2.5). Given subgroups l, l' generated by the rationally independent frequencies k_I, $I = 1, \ldots, N$ and k'_J, $J = 1, \ldots, N'$ respectively we find $l, l' \prec l''$ generated by rationally k''_L, $L = 1, \ldots, N''$ and integers n_{IL}, n'_{JL} such that $k_I = \sum_{L=1}^{N''} n_{IL} k''_L$ and $k'_J = \sum_{L=1}^{N''} n'_{JL} k''_L$. Now let $k = \sum_I n_I k_I$, $k' = \sum_J n_J k_J$ be given so that $\check{T}_k \in C(X_l)$, $\check{T}_{k'} \in C(X_{l'})$. Then the inner product according to (28.2.5) is given by $\delta_{k,k'}$, which due to the rational independence of the k''_j is equivalent to

$$\prod_{L=1}^{N''} \delta_{\Sigma_I n_I n_{IL}, \Sigma_J n'_J n'_{JL}} \tag{28.2.7}$$

But

$$[\overline{\tilde{T}_k \tilde{T}_{k'}}](\chi) = \prod_{L=1}^{N''} [\chi(k''_L)]^{\Sigma_J n'_J n'_{JL} - \Sigma_I n_I n_{IL}} \tag{28.2.8}$$

so that (28.2.6) gives precisely (28.2.7).

The measure μ_0 can also be considered as a Haar measure on $\mathbb{R}$ (that is, normalised and translation-invariant): define for $f \in \mathcal{C}$ its average or mean

$$\nu(f) := \lim_{R \to \infty} \frac{1}{2R} \int_{-R}^{R} dx f(x) \tag{28.2.9}$$

We claim that $\nu(f) = \mu_0(\check{f})$. To prove this it will be sufficient to check it for $f = T_k$. The function T_k has period $2\pi/k$ so we have with $Rk/(2\pi) - 1 < N_R \leq Rk/(2\pi)$ and any function with period $2\pi/k$

$$\int_{-R}^{R} dx f(x) = 2N_R \int_{0}^{2\pi/k} dx f(x) + \int_{-R}^{-N_R 2\pi/k} dx f(x) + \int_{N_R 2\pi/k}^{R} dx f(x)$$

$$=: 2N_R \int_{o}^{2\pi/k} dx f(x) + \delta_R \tag{28.2.10}$$

where the remainder δ_R is bounded uniformly in R. Since $N_R/R \to k/(2\pi)$ as $R \to \infty$ we find

$$\nu(f) := \frac{k}{2\pi} \int_{0}^{2\pi/k} dx f(x) = \frac{1}{2\pi} \int_{0}^{2\pi} dx f(x/k) = \int_{U(1)} d\mu_H(h) \check{f}(h) \tag{28.2.11}$$

where $f(x) = F(T_k(x))$ so $f(x/k) = F(T_1(x)) =: F(h)$ and $\check{f}(\chi) = F(\chi(T_k))$.

This construction is interesting because it provides a normalised and translation-invariant measure on a non-compact group (in this case $\mathbb{R}$). Unfortunately, non-Abelian (semisimple) non-compact groups are in general not menable, the averaging works only for so-called amenable groups [585].

29

Operator *-algebras and spectral theorem

As an application of Chapters 27 and 25 in addition to the general theory of the main text we present an elegant proof of the spectral theorem and sketch the GNS construction due to Gel'fand, Naimark and Segal. The GNS construction in turn is pivotal for the representation theory of the holonomy flux algebra of LQG.

29.1 Operator *-algebras, representations and GNS construction

We list the basic vocabulary of operator theory. See, for example, [535–537] for further information.

I. *Operator algebras*

An algebra $\mathfrak{A}$ is simply a vector space over $\mathbb{C}$ in which there is defined an associative and distributive multiplication. It is unital if there is a unit $\mathbf{1}$ which satisfies $a\mathbf{1} = \mathbf{1}a = a$ for all $a \in \mathfrak{A}$. It is a *-algebra if there is defined an involution satisfying $(ab)^* = b^*a^*$, $(a^*)^* = a$ which reduces to complex conjugation on the scalars $z \in \mathbb{C}$.

A Banach algebra is an algebra with norm $a \mapsto ||a|| \in \mathbb{R}^+$ which satisfies the usual axioms $||a + b|| \leq ||a|| + ||b||$, $||ab|| \leq ||a|| \, ||b||$, $||za|| = |z| \, ||a||$, $||a|| = 0 \Leftrightarrow a = 0$ and with respect to which it is complete.

A C^*-algebra is a Banach *-algebra whose norm satisfies the C^*-property $||a^*a|| = ||a||^2$ for all $a \in \mathfrak{A}$. Physicists are most familiar with the C^*-algebra $\mathcal{B}(\mathcal{H})$ of bounded operators on a Hilbert space $\mathcal{H}$.

A von Neumann algebra is a weakly closed subalgebra of the C^*-algebra of bounded operators on a Hilbert space.

II. *Representations*

A representation of a *-algebra $\mathfrak{A}$ is a pair $(\mathcal{H}, \pi)$ consisting of a Hilbert space $\mathcal{H}$ and a morphism $\pi : \mathfrak{A} \to \mathcal{L}(\mathcal{H})$ into the algebra of linear (not necessarily bounded) operators on $\mathcal{H}$ with common and invariant dense domain. This means that $\pi(za + z'a') = z\pi(a) + z'\pi(a')$, $\pi(ab) = \pi(a)\pi(b)$, $\pi(a^*) = [\pi(a)]^\dagger$ where † denotes the adjoint in $\mathcal{H}$.

The representation is said to be faithful if $\text{Ker}(\pi) = \{0\}$ and non-degenerate if $\pi(a)\psi = 0$ for all $a \in \mathfrak{A}$ implies $\psi = 0$.

A representation is said to be cyclic if there exists a normed vector $\Omega \in \mathcal{H}$ in the common domain of all the $a \in \mathfrak{A}$ such that $\pi(\mathfrak{A})\Omega$ is dense in $\mathcal{H}$. Notice that the existence of a cyclic vector implies that the states $\pi(b)\Omega$, $b \in \mathfrak{A}$

lie in the common dense and invariant domain for all $\pi(a)$, $a \in \mathfrak{A}$. A representation is said to be irreducible if every vector in a common dense and invariant (for $\mathfrak{A}$) domain is cyclic.

Two representations $\pi_I; \mathfrak{A} \to \mathcal{L}(\mathcal{H}_I)$; $I = 1, 2$ are called equivalent iff there exists a Hilbert space isomorphism $U : \mathcal{H}_1 \to \mathcal{H}_2$ such that $\pi_2(a) = U\pi_1(a)U^{-1}$ for all $a \in \mathfrak{A}$.

III. *States*

A state on a *-algebra is a linear functional $\omega : \mathfrak{A} \to \mathbb{C}$ which is positive, that is, $\omega(a^*a) \geq 0$ for all $a \in \mathfrak{A}$. If $\mathfrak{A}$ is unital we require that $\omega(1) = 1$. The states that physicists are most familiar with are vector states, that is, if we are given a representation $(\mathcal{H}, \pi)$ and an element ψ in the common domain of all the $a \in \mathfrak{A}$ then $a \mapsto< \psi, \pi(a)\psi >_{\mathcal{H}}$ evidently defines a state. These are examples of pure states, that is, those which cannot be written as convex linear combinations of other states. However, the concept of states is much more general and includes what physicists would call mixed (or temperature) states, see, for example, the notion of a folium below.

IV. *Automorphisms*

An automorphism of a *-algebra is an isomorphism of $\mathfrak{A}$ which is compatible with the algebraic structure. If G is a group then G is said to be represented on $\mathfrak{A}$ by a group of automorphisms $\alpha : G \to \mathrm{Aut}(\mathfrak{A})$; $g \mapsto \alpha_g$ provided that $\alpha_g \circ \alpha_{g'} = \alpha_{gg'}$ for all $g, g' \in G$. A state ω on $\mathfrak{A}$ is said to be invariant for an automorphism α provided that $\omega \circ \alpha = \omega$. It is said to be invariant for G if it is invariant for all α_g, $g \in G$.

The following two structural theorems combine the notions introduced above and are of fundamental importance for the construction and analysis of representations.

Theorem 29.1.1 (GNS construction). *Let ω be a state on a unital *-algebra $\mathfrak{A}$. Then there are GNS data $(\mathcal{H}_\omega, \pi_\omega, \Omega_\omega)$ consisting of a Hilbert space $\mathcal{H}_\omega$, a cyclic representation π_ω of $\mathfrak{A}$ on $\mathcal{H}_\omega$ and a normed, cyclic vector $\Omega_\omega \in \mathcal{H}_\omega$ such that*

$$\omega(a) =< \Omega_\omega, \pi_\omega(a)\Omega_\omega >_{\mathcal{H}_\omega} \tag{29.1.1}$$

Moreover, the GNS data are determined by (29.1.1) uniquely up to unitary equivalence.

The name GNS stands for Gel'fand–Naimark–Segal. The idea is very simple. The algebra $\mathfrak{A}$ is in particular a vector space and we can equip it with a sesquilinear form $< a, b >:= \omega(a^*b)$. (To see that it is sesquilinear, use the polarisation identity

$$a^*b = \frac{1}{4} \sum_{\epsilon^4=1} \epsilon[a + \epsilon b]^*[a + \epsilon b]$$

and positivity, which implies $\overline{\omega(c^*c)} = \omega(c^*c)$ as well as $a + \epsilon b = \epsilon(b + \bar{\epsilon}a)$, to conclude that $\overline{\omega(a^*b)} = \omega(b^*a)$.) This form is not necessarily positive definite. However, by exploiting the Cauchy–Schwarz inequality $|\omega(a^*b)|^2 \leq \omega(a^*a)\omega(b^*b)$ one convinces oneself that the set $\mathfrak{I}_\omega$ consisting of the elements of $\mathfrak{A}$ satisfying $\omega(a^*a) = 0$ defines a left ideal. We can thus pass to the equivalence classes $[a] = \{a + b; \ b \in \mathfrak{I}_\omega\}$ and define a positive definite scalar product by $< [a], [b] >:= \omega(a^*b)$ for which one checks independence of the representative. Since $\mathfrak{I}_\omega$ is a left ideal one checks that $[a] + [b] := [a + b]$, $z[a] := [za]$, $[a][b] := [ab]$ are well-defined operations. Then $\mathcal{H}_\omega$ is simply the Cauchy completion of the vectors $[a]$, the representation is simply $\pi_\omega(a)[b] := [ab]$ and the cyclic vector is just given by $\Omega_\omega := [1]$. Finally, if $(\mathcal{H}'_\omega, \pi'_\omega, \Omega'_\omega)$ are other GNS data then the operator $U : \mathcal{H}_\omega \to \mathcal{H}'_\omega$ defined densely by $U\pi_\omega(a)\Omega_\omega := \pi'_\omega(a)\Omega'_\omega$ is unitary.

Theorem 29.1.2. *Let ω be a state over a unital *-algebra $\mathfrak{A}$ which is invariant for an element $\alpha \in \mathrm{Aut}(\mathfrak{A})$. Then there exists a uniquely determined unitary operator U_ω on the GNS Hilbert space $\mathcal{H}_\omega$ such that*

$$U_\omega \, \pi_\omega(a) \, \Omega_\omega = \pi_\omega(\alpha(a))\Omega_\omega \tag{29.1.2}$$

The proof follows from the uniqueness part of Theorem 29.1.1 applied to the alternative data $(\mathcal{H}_\omega, \pi_\omega \circ \alpha, \Omega_\omega)$.

Corollary 29.1.3. *Let ω be a G-invariant state on a unital *-algebra. Then there is a unitary representation $g \mapsto U_\omega(g)$ of G on the GNS Hilbert space $\mathcal{H}_\omega$ defined by*

$$U_\omega(g) \, \pi_\omega(a) \, \Omega_\omega := \pi_\omega(\alpha_g(a)) \, \Omega_\omega \tag{29.1.3}$$

where $g \mapsto \alpha_g$ is the corresponding automorphism group.

Notice that this means that the group G is represented *without anomalies*, that is, there are, for example, no central extensions with non-vanishing obstruction cocycle.

An important concept in connection with a state ω is its *folium*. This is defined as the set of states ω_ρ on $\mathfrak{A}$ defined by

$$\omega_\rho(a) := \frac{\mathrm{Tr}_{\mathcal{H}_\omega}(\rho\pi_\omega(a))}{\mathrm{Tr}_{\mathcal{H}_\omega}(\rho)} \tag{29.1.4}$$

where ρ is a positive trace class operator (see, e.g., [282] and Definition 26.6.7) on the GNS Hilbert space $\mathcal{H}_\omega$.

If $\mathfrak{A}$ is not only a unital *-algebra but in fact a C^*-algebra then there are many more structural theorems available. For instance one can show, using the Hahn–Banach theorem, Theorem 26.3.1 (see [282]), that representations always exist, that every non-degenerate representation is a direct sum of cyclic representations, that every state is continuous so that the GNS representations are always by bounded operators and that for pure states the GNS representation is irreducible.

Of particular interest in the context of C^*-algebras is also the following universal result.

Theorem 29.1.4 (Fell's theorem). *The folium of a faithful state of a C^*-algebra is weakly dense in the set of all states.*

In other words, given $\epsilon > 0$, given any state ω, any faithful state ω_0 (i.e., its GNS representation is faithful) and any finite number of algebra elements $a_1, .., a_n$ we can find a state ω' in the folium of ω_0 such that $|(\omega - \omega')(a_k)| < \epsilon$ for all $k = 1, .., n$. This defines a weak neighbourhood $N(\epsilon; a_1, .., a_n)$ of ω in the space of continuous linear functionals of $\mathfrak{A}$. Physically this means that we cannot find out in which folium a state lies because in practice we can only perform a finite number of measurements.

While the C^*-norm implies this huge amount of extra structure, a reasonable C^*-norm on a $*$-algebra is usually very hard to guess unless one actually constructs a representation by bounded operators. An exception is given by the $*$-algebra generated by the Weyl elements of, say, a free scalar field theory which has a unique C^*-norm. This result is known as Slawny's theorem [548], an instructive proof of which can be found in [536], Theorem 5.2.8. For more general algebras, such as the one we are interested in here, uniqueness results are unknown. We have thus chosen to keep with the more general concept of $*$-algebras. To see that the algebra $\mathfrak{A} := \mathcal{B}(\mathcal{H})$ of bounded operators is a C^*-algebra with respect to the operator norm $||a|| := \sup_{\psi \neq 0} ||a\psi||/||\psi||$ we notice that by the Schwarz inequality

$$||a||^2 = \sup_\psi \frac{<\psi, a^\dagger a \psi>}{||\psi||^2} \leq ||a^\dagger a|| \leq ||a|| \, ||a^\dagger||,$$

$$||a^\dagger||^2 = \sup_\psi \frac{<\psi, a a^\dagger \psi>}{||\psi||^2} \leq ||a a^\dagger|| \leq ||a|| \, ||a^\dagger||$$

since $||ab|| \leq ||a|| \, ||b||$. It follows that $||a|| = ||a^\dagger||$ and thus from the first inequality $||a||^2 \leq ||a^\dagger a|| \leq ||a||^2$, hence $||a^\dagger a|| = ||a||^2$ which is the C^*-property.

In the rigorous algebraic approach to QFT [22] one uses the mathematical framework of operator algebras, the basics of which we just sketched and combines it with the physical concept of locality of nets of local algebras $\mathcal{O} \mapsto \mathfrak{A}(\mathcal{O})$. That is, given a background spacetime (M, η) consisting of a differentiable D-manifold and a background metric η, for each open region $\mathcal{O}$ one assigns a C^*-algebra $\mathfrak{A}(\mathcal{O})$. These are required to be mutually (anti)commuting for spacelike separated (with respect to η) regions. This is the statement of the most important one of the famous Haag–Kastler axioms. The framework is ideally suited to formulate and prove all of the structural theorems of QFT on Minkowski space and even to a large extent on curved spaces [27], at least perturbatively. In AQFT one cleanly separates the two steps of quantising a field theory, namely first to define a suitable algebra $\mathfrak{A}$ and then to study its representations in a second step. In LQG we follow the same logic.

29.2 Spectral theorem, spectral measures, projection valued measures, functional calculus

Let $\mathcal{H}$ be a Hilbert space and a a bounded, linear, normal operator on $\mathcal{H}$, that is $||a|| = \sup_{\psi \neq 0} ||a\psi||/||\psi|| < \infty$ where $||\psi||^2 =< \psi, \psi >$ denotes the Hilbert space norm and $[a, a^\dagger] = 0$ where the bounded operator $a^\dagger$ is defined by $< a^\dagger \psi, \psi' \cdot >:=< \psi, a\psi' >$. More precisely, consider the linear form on $\mathcal{H}$ defined by

$$l_\psi : \mathcal{H} \to \mathbb{C}; \psi' \to< \psi, a\psi' > \tag{29.2.1}$$

This linear form is continuous since $|l_\psi(\psi')| \leq ||\psi|| \, ||a|| \, ||\psi'||$ by the Schwarz inequality. Hence, by the Riesz lemma there exists $\xi_\psi \in \mathcal{H}$ such that $l_\psi =< \xi_\psi, . >$ and since l_ψ is conjugate linear in ψ it follows that $\psi \mapsto \xi_\psi := a^\dagger \psi$ actually defines a linear operator. Finally, $a^\dagger$ is bounded because

$$||a^\dagger \psi||^2 = | < \psi, aa^\dagger \psi > | \leq ||\psi|| \, ||aa^\dagger \psi|| \leq ||\psi|| \, ||a|| \, ||a^\dagger \psi|| \tag{29.2.2}$$

again by the Schwarz inequality.

Let $\mathfrak{A}$ be the unital, Abelian C^*-algebra generated by $1, a, a^\dagger$. It is Abelian since a is normal and the C^*-property follows from the following observation: let $b \in \mathfrak{A}$, then b is also normal and $||b\psi||^2 =< \psi, b^\dagger b\psi >= ||b^\dagger \psi||^2$ so that $||b|| = ||b^\dagger||$ for any $b \in \mathfrak{A}$. Now by the Schwarz inequality $||b\psi||^2 = | < \psi, b^\dagger b\psi > | \leq ||\psi|| \, ||b^\dagger b\psi||$ implying that $||b||^2 = ||b^\dagger||^2 \leq ||b^\dagger b||$. On the other hand, $||b^\dagger b|| \leq ||b|| \, ||b^\dagger||$ due to submultiplicativity.

Consider the spectrum $\Delta(\mathfrak{A}) = \mathrm{Hom}(\mathfrak{A}, \mathbb{C})$ and the map $z : \Delta(\mathfrak{A}) \to \mathbb{C}; \chi \mapsto \chi(a)$ which is continuous by the definition of the Gel'fand topology on the spectrum. We have seen already that the range of this map coincides with $\sigma(a)$. Moreover, z is injective because $\chi(a) - \chi'(a)$ implies that χ, χ' coincide on all polynomials of $a, a^\dagger$ since they are homomorphisms, and thus on all of $\mathfrak{A}$ by continuity whence $\chi = \chi'$. Thus, z is a continuous bijection between the spectra of $\mathfrak{A}$ and a respectively. Since a is bounded, both spectra are compact Hausdorff spaces. Now a continuous bijection between compact Hausdorff spaces is automatically a homeomorphism. (*Proof:* Let $f : X \to Y$ be a continuous bijection and let X be compact and Y Hausdorff. We must show that $f(U)$ is open in Y for every open subset $U \subset X$, or by taking complements, that images of closed sets are closed. Now since X is compact, it follows that every closed set U is also compact. Since f is continuous, it follows that $f(U)$ is compact. Since Y is Hausdorff it follows that $f(U)$ is closed. See Theorems 5.3 and 5.5 of [533].) We conclude that we can identify $\Delta(\mathfrak{A})$ topologically with $\sigma(a)$. We will denote points in $\sigma(a) \subset \mathbb{C}$ by λ in order to distinguish them from the points $\chi \in \Delta(\mathfrak{A})$.

By definition the polynomials p in $a, a^\dagger$ lie dense in $\mathfrak{A}$ and we have for $\chi \in \Delta(\mathfrak{A})$ that

$$\chi(p(a, a^\dagger)) = p(\chi(a), \overline{\chi(a)}) = p(z(\chi), \overline{z(\chi)}) = [p \circ (z, \bar{z})](\chi) = p(a, a^\dagger)^{\vee}(\chi) \tag{29.2.3}$$

It follows that by combining the Gel'fand isometric isomorphism $\vee : \mathfrak{A} \to C(\Delta(\mathfrak{A})); b \mapsto \check{b}$ with the map z^{-1} we obtain an isometric isomorphism $\sim : \mathfrak{A} \to C(\sigma(a)); b \mapsto \check{b} \circ z^{-1}$, that is $\tilde{b}(\lambda) := \chi(b)_{z(\chi)=\lambda}$.

Now consider any vector $\psi \in \mathcal{H}$ with $\|\psi\| = 1$. Then

$$\omega_\psi : \mathfrak{A} \to \mathbb{C}; \ b \mapsto \ <\psi, b\psi> \tag{29.2.4}$$

is obviously a state on $\mathfrak{A}$. Via the Gel'fand transform we obtain a positive linear functional on $C(\Delta(\mathfrak{A}))$ by

$$\Lambda_\psi : C(\Delta(\mathfrak{A})) \to \mathbb{C}; \ \check{b} \mapsto \omega_\psi(b) \tag{29.2.5}$$

and since $\Delta(\mathfrak{A})$ is a compact Hausdorff space we can apply the Riesz representation theorem in order to find a unique, regular Borel measure μ'_ψ on $\Delta(\mathfrak{A})$ such that

$$\omega_\psi(b) = \int_{\Delta(\mathfrak{A})} d\mu'_\psi(\chi)\check{b}(\chi) \tag{29.2.6}$$

Denoting $d\mu_\psi(\lambda) := d\mu'_\psi(z^{-1}(\lambda))$ we may change coordinates from χ to λ and replace μ'_ψ by μ_ψ as well as $\check{b}$ by $\tilde{b}$ in what follows so that (29.2.6) becomes

$$\omega_\psi(b) = \int_{\sigma(a)} d\mu_\psi(\lambda)\tilde{b}(\lambda) \tag{29.2.7}$$

The measure μ_ψ is called a *spectral measure*.

In the language of the previous chapter, the C^*-algebra $\mathfrak{A}$ generated by a normal, bounded operator $a \in \mathcal{B}(\mathcal{H})$ on a given Hilbert space is represented as $\pi(b) = b$ on $\mathcal{H}$. Notice that if ψ was cyclic for $\mathfrak{A}$, then (29.2.7) would show that $\mathcal{H}$ is unitarily equivalent to $L_2(\sigma(a), d\mu_\psi)$ and in that representation is realised as a multiplication operator which is already one of the versions of the spectral theorem. However, it may be impossible or difficult to find a cyclic vector. Therefore, in general more work is required, as follows.

Notice that the chosen Hilbert space representation is non-degenerate because $\mathfrak{A}$ contains the identity operator. Hence we may apply the result that such representations decompose directly into cyclic ones, see Lemma 8.1.1. We thus find an index set A, vectors ψ_α and closed, mutually orthogonal subspaces $\mathcal{H}_\alpha := \overline{\{b\psi_\alpha; \ b \in \mathfrak{A}\}}$ containing ψ_α such that $\mathcal{H} = \oplus_{\alpha \in A}\mathcal{H}_\alpha$. By construction, the subspaces $\mathcal{H}_\alpha$ are invariant for $\mathfrak{A}$. Then any vector $\psi \in \mathcal{H}$ is (in the closure of vectors) of the form $\psi = \sum_{\alpha \in A} b_\alpha \psi_\alpha$ with $b_\alpha \in \mathfrak{A}$ and we have

$$<\psi, \psi'> = \sum_{\alpha \in A} <\psi_\alpha, b_\alpha^\dagger b'_\alpha \psi_\alpha> \tag{29.2.8}$$

Using the result (29.2.6) we may write this as

$$<\psi, \psi'> = \sum_{\alpha \in A} \int_{\sigma(a)} d\mu_{\psi_\alpha}(\lambda)\overline{\tilde{b}_\alpha(\lambda)}\tilde{b}'_\alpha(\lambda) \tag{29.2.9}$$

where we have used the fact that $(b^\dagger b')^\vee = \bar{\tilde{b}}\tilde{b}'$. This formula suggests introducing the Hilbert spaces $L_2(\sigma(a), d\mu_{\psi_\alpha})$ as well as the space $\sigma := \bigcup_{\alpha \in A} \sigma(a)_\alpha$ (disjoint union of copies of $\sigma(a)$) and a measure μ on it defined by $\mu_{|\sigma(a)_\alpha} := \mu_{\psi_\alpha}$. Notice that measurable sets are of the form $\bigcup_{\alpha \in B \subset A} U_\alpha$ where U_α is measurable in $\sigma(a)_\alpha$, B can be any subindex set and unions, intersections and differences of measurable sets are performed componentwise. Let us now define the Hilbert space $L_2(\sigma, d\mu)$. An element $\tilde{\psi}$ of $L_2(\sigma, d\mu)$ is a square integrable function on σ with respect to the measure μ and may be defined in terms of an array of functions $\tilde{\psi}_\alpha \in L_2(\sigma(a)_\alpha, d\mu_{\psi_\alpha})$ which are its componentwise restriction, that is $\tilde{\psi}_\alpha := \tilde{\psi}_{|\sigma(a)_\alpha}$. Notice that indeed

$$<\tilde{\psi}, \tilde{\psi}'>_{L_2(\sigma, d\mu)} = \int_\sigma d\mu(\lambda)\overline{\tilde{\psi}(\lambda)}\tilde{\psi}'(\lambda)$$

$$= \sum_{\alpha \in A} \int_{\sigma(a)_\alpha} d\mu(\lambda)\overline{\tilde{\psi}(\lambda)}\tilde{\psi}'(\lambda)$$

$$= \sum_{\alpha \in A} \int_{\sigma(a)_\alpha} d\mu_{|\sigma(a)_\alpha}(\lambda)[\overline{\tilde{\psi}(\lambda)}\tilde{\psi}'(\lambda)]_{|\sigma(a)_\alpha}$$

$$= \sum_{\alpha \in A} \int_{\sigma(a)} d\mu_{\psi_\alpha}(\lambda)\overline{\tilde{\psi}_\alpha(\lambda)}\tilde{\psi}'_\alpha(\lambda) \tag{29.2.10}$$

explaining the requirement that $\tilde{\psi}_\alpha \in L_2(\sigma(a)_\alpha, d\mu_{\psi_\alpha})$. Here we have made use of σ-additivity, that is, $\mu(\bigcup_\alpha U_\alpha) = \sum_\alpha \mu(U_\alpha) = \sum_\alpha \mu_{\psi_\alpha}(U_\alpha)$ for the mutually disjoint sets $U_\alpha \subset U$. Comparing (29.2.9) and (29.2.10) we see that we can identify $L_2(\sigma, d\mu)$ with $\oplus_{\alpha \in A} L_2(\sigma(a)_\alpha, d\mu_{\psi_\alpha})$ and obtain a unitary transformation

$$U : \mathcal{H} \to L_2(\sigma, d\mu); \quad \psi = \sum_{\alpha \in A} h_\alpha \psi_\alpha \mapsto \tilde{\psi} \text{ where } \tilde{\psi}_{|\sigma(a)_\alpha} := \tilde{b}_\alpha \tag{29.2.11}$$

Moreover, we have

$$Ub\psi = U \sum_\alpha bb_\alpha \psi_\alpha = \tilde{\psi}' \text{ where } \tilde{\psi}'_{|\sigma(a)_\alpha} = \widetilde{bb_\alpha} = \tilde{b}\tilde{b}_\alpha \tag{29.2.12}$$

which means that on each subspace $L_2(\sigma(a)_\alpha, d\mu_{\psi_\alpha})$ the operator b is represented by multiplication by $\tilde{b}(\lambda)$. In particular, if $b = a$ or $b = a^\dagger$ it is represented by multiplication by λ or $\bar{\lambda}$ since

$$\tilde{a}(\lambda) = \check{a}(z^{-1}(\lambda)) = [z^{-1}(\lambda)](a) = z([z^{-1}(\lambda)]) = \lambda \tag{29.2.13}$$

This simple corollary from Gel'fand spectral theory and the Riesz representation theorem is the spectral theorem for bounded, normal operators which we just proved in a few lines above.

Theorem 29.2.1 spectral theorem; multiplication operator form. *Let a be a bounded, normal operator on a Hilbert space $\mathcal{H}$. Then there exists a unitary operator $U : \mathcal{H} \to L_2(\sigma, d\mu)$ where σ is a disjoint union of copies of the spectrum* *of a and μ is a direct sum of regular Borel measures on those copies such that for each measurable function f on the spectrum of a the operator $U f(a, a^\dagger) U^{-1}$ becomes the multiplication operator $f(\lambda, \bar{\lambda})$.*

The extension from polynomials to measurable functions is because polynomials on compact Hausdorff spaces are dense in the set of continuous functions by the Weierstrass theorem and the continuous functions are dense in the measurable functions by Lusin's theorem for Borel measures.

Notice that the spectral theorem is valid also if the Hilbert space is not separable. It obviously generalises to the case that we have a family (a_I) of bounded operators which together with their adjoints mutually commute with each other. The only difference is that we now get a homeomorphism between $\Delta(\mathfrak{A})$ and the joint spectrum $\prod_I \sigma(a_I)$ via $\chi \mapsto (\chi(a_I))_I$. We can also strip off the concrete Hilbert space context by considering an abstract unital C^*-algebra $\mathfrak{A}$ where instead of vector states ψ_α we use states ω_α on $\mathfrak{A}$ and apply the GNS construction. That for given $a \in \mathfrak{A}$ there is always a state ω with $\omega(a^*a) > 0$ follows from the Hahn–Banach theorem applied to the vector space $X := \mathfrak{A}$ and its one-dimensional subspace $Y := \text{span}(a^*a)$ with the bounding function required in the Hahn–Banach theorem given by the norm on X and by defining $\omega(a^*a) := ||a||^2$. The Hahn–Banach theorem guarantees that then ω can be extended as a positive linear functional to all of $\mathfrak{A}$.

Theorem 29.2.2. *Let (a_I) be a self-adjoint collection (closed under involution) of mutually commuting elements of a C^*-algebra $\mathcal{C}$. Then there exists a representation π of the sub-C^*-algebra $\mathfrak{A}$ generated by this collection on a Hilbert space $\mathcal{H}$ such that the $\pi(a_I)$ become multiplication operators.*

Let us mention the spectral resolution. Let a be a bounded self-adjoint operator then by the Riesz–Markov theorem we have $< \psi, f(a)\psi > = \int_{\sigma(a)} d\mu_\psi(\lambda) f(\lambda)$ for any measurable function f and μ_ψ is the spectral measure of ψ as above. Let $B \subset \mathbb{R}$ be measurable (i.e., a Borel set, $B \in \mathcal{B}$) and consider the operators $E_B := \chi_B(a)$ called the *spectral projections* where χ_B is the characteristic function of B. Then $< \psi, E_B \psi > = \int_{\sigma(a)} d\mu_\psi(\lambda) \chi_B(\lambda)$. This defines the so-called *projection-valued measures* (p.v.m.) $E: \mathcal{B} \to \mathcal{B}(\mathcal{H}); B \mapsto E_B$ which map Borel sets into projection operators, that is, they are 'measures' with values in the set of projection operators rather than the real numbers. Evidently $E_B E_{B'} = E_{B \cap B'}$ and $E_\mathbb{R} = 1_\mathcal{H}$.

The p.v.m. allow for an elegant formulation of the spectral theorem as follows: let $E(\lambda) := \chi_{(-\infty, \lambda]}(a)$ for $\lambda \in \mathbb{R}$ then we see that

$$< \psi, dE(\lambda)\psi > := d < \psi, E(\lambda)\psi > = d\mu_\psi(\lambda) \qquad (29.2.14)$$

whence

$$< \psi, f(a)\psi > = \int_\mathbb{R} < \psi, dE(\lambda)\psi > f(\lambda) \qquad (29.2.15)$$

for all $\psi \in \mathcal{H}$ or by the polarisation identity

$$f(a) = \int_{\mathbb{R}} dE(\lambda) \, f(\lambda) \qquad (29.2.16)$$

which is called the *spectral resolution* of $f(a)$. This works completely analogously for unitary (or any other normal) operators whose spectrum is a subset of $S^1 \equiv [0, 2\pi)$ where $0 \equiv 2\pi$ are identified. So one just has to replace $\mathbb{R}$ by S^1. To prove this, apply the inverse Caley transform, which we will discuss in a moment, to bounded self-adjoint operators.

The extension of the spectral theorem to unbounded self-adjoint operators on a Hilbert space can be traced back to the bounded case by using the following trick. (Recall that a densely defined operator a with domain $D(a)$ is called self-adjoint if $a^\dagger = a$ and $D(a^\dagger) = D(a)$ where

$$D(a^\dagger) := \{\psi \in \mathcal{H}; \quad \sup_{0 \neq \psi' \in D(a)} | < \psi, a\psi' > |/\|\psi'\| < \infty\}$$

and $a^\dagger$ is uniquely defined on $\psi \in D(a^\dagger)$ via $< a^\dagger \psi, \psi' > = < \psi, a\psi' >$ for all $\psi \in D(a)$ through the Riesz lemma): the spectrum of a will be an unbounded subset of the real line. Let f be a bijection $\mathbb{R}^* \to K$ where K is a compact one-dimensional subset of $\mathbb{C}$ and $\mathbb{R}^*$ the one-point compactification of $\mathbb{R}$. (The one-point compactification $X^* = X \cup \{\infty\}$ of a topological space X has as open sets (a) the open sets of X and (b) the sets $U \subset X^*$ containing $\{\infty\}$ such that $X - U$ is closed and compact in X. X^* is then compact and also Hausdorff iff X is locally compact and Hausdorff.) Suppose that $f(a)$ is a bounded operator. Then we can apply the spectral theorem for bounded normal operators to $f(a)$, which then becomes a multiplication operator and if f^{-1} is a measurable function then also a itself is a multiplication operator on a suitable domain. A popular tool is the Caley transform: consider the map $f : \mathbb{R} \to \mathbb{C}; \; x \mapsto \frac{x-i}{x+i}$. The image of f are all complex numbers z of modulus one except for $z = 1$ which would be the image of the point $x = \pm\infty$. Parametrising $z = e^{i\theta}$, $\theta \in [0, 2\pi)$ the full circle (with boundary points identified) we see that f is invertible with inverse $f^{-1}(z) = i\frac{1+z}{1-z} = -\cot(\theta/2)$, which is well-defined on the full circle except for the point $\theta = 0 \equiv 2\pi$ corresponding to $z = 1$ so that the image of f corresponds to the open interval $(0, 2\pi)$. Consider the unitary operator (the inverse of $a + i$ is well-defined because i is not in the spectrum of a, hence the inverse operator is bounded) $u := f(a) = (a - i)(a + i)^{-1}$ (with inverse $a = f^{-1}(u) = i(1 + u)(1 - u)^{-1}$) and let E be its projection-valued measure with spectral projections $E(\theta)$, $\theta \in [0, 2\pi)$. Then, since f^{-1} is measurable we have by the spectral theorem

$$a = \int_0^{2\pi} f^{-1}(e^{i\theta}) \, dE(\theta) = -\int_0^{2\pi} \cot(\theta/2) \, dE(\theta) = \int_{\mathbb{R}} x \, dE(2\arctan(-x))$$

where we changed coordinates in the last line. Hence the spectral projections for a are given by $E'(x) = E(2\arctan(-x))$.

Let us summarise our findings:

Theorem 29.2.3 (**spectral theorem; functional calculus form**). *Let   a be a self-adjoint, possibly unbounded, operator on a, possibly non-separable, Hilbert space $\mathcal{H}$. Then there is a system of bounded operators $\mathbb{R} \ni \lambda \mapsto E(\lambda)$ with the following properties:*

1. $E(-\infty) = 0$, $E(\infty) = 1_{\mathcal{H}}$,
2. $E(\lambda)E(\lambda') = E(\min(\lambda, \lambda'))$,
3. $s - \lim_{\lambda \to \lambda_0+} E(\lambda) = E(\lambda_0)$ (strong limit).

Moreover, for every measurable function f on the spectrum $\sigma(a)$ we have the strong equality

$$f(a) = \int_{\mathbb{R}} dE(\lambda)\, f(\lambda) \tag{29.2.17}$$

on the dense set of vectors ψ on which $\int d < \psi, E(\lambda)\psi > |f(\lambda)|^2$ converges.

Let us demonstrate how to do practical calculations with the functional calculus. Let f, f' be two measurable functions. We want to verify that $f(a)f'(a) = (ff')(a)$ in the notation (29.2.17) on the set of vectors ψ on which the product exists. We have

$$< \psi, f(a)\, f'(a)\psi > = \int_{\mathbb{R}} f(\lambda)\, d_\lambda \left[\int_{\mathbb{R}} f'(\lambda')d_{\lambda'} < \psi, E(\min(\lambda, \lambda'))\psi > \right]$$

$$= \int_{\mathbb{R}} f(\lambda)\, d_\lambda \left[\int_{-\infty}^{\lambda} f'(\lambda')d_{\lambda'} < \psi, E(\lambda')\psi > \right]$$

$$= \int_{\mathbb{R}} f(\lambda)f'(\lambda)d_\lambda < \psi, E(\lambda)\psi > \tag{29.2.18}$$

as claimed. Here d_λ denotes the differential with respect to λ and in the second step we used $d_{\lambda'}E(\min(\lambda, \lambda')) = 0$ for $\lambda' > \lambda$.

30

Refined algebraic quantisation (RAQ) and direct integral decomposition (DID)

In this chapter we describe two methods for solving the quantum constraints. They both make the original proposal by Dirac, to simply look for the common kernel of the quantum constraints, mathematically precise.

30.1 RAQ

RAQ provides strong guidelines for how to solve a given family of quantum constraints but unfortunately it is not an algorithm that one just has to apply in order to arrive at a satisfactory end result. In particular, as presently formulated it has its limitations since it does not cover the case that the constraints form an open algebra with structure functions rather than structure constants as would be the case for a Lie algebra. Unfortunately, quantum gravity belongs to the open algebra category of constrained systems and one has to resort to the second method, DID, presented in the next section. We mainly follow Giulini and Marolf in [277, 278].

Let $\mathcal{H}_{\text{kin}}$ be a Hilbert space, referred to as the kinematical Hilbert space because it is supposed to implement the adjointness and canonical commutation relations of the elementary kinematical degrees of freedom. However, these degrees of freedom are not observables (classically they do not have vanishing Poisson brackets with the constraints on the constraint surface) and the Hilbert space is not the physical one on which the constraint operators would equal the zero operators. The role of $\mathcal{H}_{\text{kin}}$ is to give the constraint operators $(\hat{C}_I)_{I \in \mathcal{I}}$ a home, that is, there is a common dense domain $\mathcal{D}_{\text{kin}} \subset \mathcal{H}_{\text{kin}}$ which is supposed to be invariant under all the $\hat{C}_I$ and we also require that the $\hat{C}_I$ be closable operators (i.e., their adjoint is densely defined as well). We do not require them to be bounded operators. The label set $\mathcal{I}$ is rather arbitrary and usually a combination of direct products of finite and infinite sets (e.g., tensor or gauge group indices times indices taking values in a separable space of smearing functions).

We will further require that the constraints form a first-class system and that they actually form a Lie algebra, that is, there exist complex-valued structure constants $f_{IJ}{}^K$ such that

$$[\hat{C}_I, \hat{C}_J] = f_{IJ}{}^K \hat{C}_K \tag{30.1.1}$$

where the summation over K performed here will involve an integral for generic $\mathcal{I}$. Notice that (30.1.1) makes sense due to our requirement on $\mathcal{D}_{\text{kin}}$. The case of an open algebra would correspond to the fact that the structure constants become operator-valued as well and then it becomes an issue how to choose the operator ordering in (30.1.1), in particular, if constraint operators and structure constant operators are chosen to be self-adjoint and anti-self-adjoint respectively (which would be natural if their classical counterparts are classically real- and imaginary-valued respectively) then one would have to order (30.1.1) symmetrically, which would be a disaster for solving the constraints, see below, which is why in the open case the constraints should not be chosen to be self-adjoint operators. Notice that there is no contradiction because self-adjointness is usually required to ensure that the spectrum (measurement values) of the operator lies in the real line, however, for constraint operators this requirement is void since we are only interested in their kernel and the only requirement is that the point zero belongs to the spectrum at all.

In order to allow for non-self-adjoint constraints, in what follows we will assume that the set $\mathcal{C} := \{\hat{C}_I;\ I \in \mathcal{I}\}$ is self-adjoint (i.e., contains with $\hat{C}_I$ also $\hat{C}_I^\dagger = \hat{C}_J$ for some J), which means that the dense domain $\mathcal{D}_{\text{kin}}$ is also a dense domain for the adjoints so that the constraints are explicitly closable operators. Let us now consider the self-adjoint set of kinematical observables $\mathcal{O}_{\text{kin}}$, that is, all operators on $\mathcal{H}_{\text{kin}}$ which have $\mathcal{D}_{\text{kin}}$ as common dense domain together with their adjoints. Obviously, $\mathcal{O}_{\text{kin}}$ contains $\mathcal{C}$. Consider the commutant of $\mathcal{C}$ within $\mathcal{O}_{\text{kin}}$, that is,

$$\mathcal{C}' := \{O \in \mathcal{O}_{\text{kin}};\ [C, O] = 0\ \forall\ C \in \mathcal{C}\} \tag{30.1.2}$$

It is clear that $\mathcal{C}'$ is a *-subalgebra of $\mathcal{O}_{\text{kin}}$ since $[O^\dagger, \mathcal{C}] = -([O, \mathcal{C}])^\dagger = 0$ and $[OO', \mathcal{C}] = O[O', \mathcal{C}] + [O, \mathcal{C}]O' = 0$ for any $O, O' \in \mathcal{C}'$ since $\mathcal{C}^\dagger = \mathcal{C}$ is a self-adjoint set. Moreover, $\mathcal{C}$ might have a non-trivial centre

$$\mathcal{Z} = \mathcal{C} \cap \mathcal{C}' \tag{30.1.3}$$

which generates a two-sided ideal $I_{\mathcal{Z}}$ in $\mathcal{C}'$ corresponding to classical functions that vanish on the constraint surface and is therefore physically uninteresting. Hence we will define the algebra of physical observables to be the quotient algebra

$$\mathcal{O}_{\text{phys}} := \mathcal{C}'/\mathcal{Z} \tag{30.1.4}$$

Usually the space $\mathcal{D}_{\text{kin}}$ comes with its own topology τ, different from the subspace topology inherited from the Hilbert space topology $||.||$ on $\mathcal{H}_{\text{kin}}$, generically a nuclear topology [280] so that $\mathcal{D}_{\text{kin}}$ becomes a Fréchet space.[1] The intrinsic topology τ is then finer than $||.||$ since $\mathcal{D}_{\text{kin}}$ is complete but also dense in $\mathcal{H}_{\text{kin}}$ (if it was coarser then a Cauchy sequence in $\mathcal{D}_{\text{kin}}$ with respect to the intrinsic topology would also be one in the Hilbert space topology and since $\mathcal{D}_{\text{kin}}$ is dense

---

[1] This is a space [282], see also Section 26.5, whose topology is generated by a countable family of seminorms that separates the points of $\mathcal{D}_{\text{kin}}$ and such that $\mathcal{D}_{\text{kin}}$ is complete in the associated norm; a general locally convex topological vector space is not necessarily complete and the family of seminorms need not to be countable (it is then not metrisable).

this completion would coincide with $\mathcal{H}_{\text{kin}}$). It follows that the space of continuous linear functionals $\mathcal{D}'_{\text{kin}}$ (with respect to the topology on $\mathcal{D}_{\text{kin}}$) or *topological dual* contains $\mathcal{H}_{\text{kin}}$ since a Hilbert space is reflexive, that is, $\mathcal{H}'_{\text{kin}} = \mathcal{H}_{\text{kin}}$ by the Riesz lemma so the elements of $\mathcal{H}_{\text{kin}}$ are in particular continuous linear functionals on $\mathcal{D}_{\text{kin}}$ with respect to $||.||$ so that they are also continuous with respect to τ (a function stays continuous if one strengthens the topology on the domain space). Let (l^α) be a net in $\mathcal{H}'_{\text{kin}}$ converging to l then

$$||l^\alpha - l||_{\mathcal{D}'_{\text{kin}}} = \sup_{f \in \mathcal{D}_{\text{kin}}} \frac{|<l_\alpha - l, f>|}{||f||_{\mathcal{D}_{\text{kin}}}} = \sup_{f \in \mathcal{D}_{\text{kin}}} \frac{||f||_{\mathcal{H}_{\text{kin}}}}{||f||_{\mathcal{D}_{\text{kin}}}} \frac{|<l_\alpha - l, f>|}{||f||_{\mathcal{H}_{\text{kin}}}}$$

$$\leq \sup_{f \in \mathcal{D}_{\text{kin}}} \frac{|<l_\alpha - l, f>|}{||f||_{\mathcal{H}_{\text{kin}}}} \leq \sup_{f \in \mathcal{H}_{\text{kin}}} \frac{|<l_\alpha - l, f>|}{||f||_{\mathcal{H}_{\text{kin}}}} = ||l^\alpha - l||_{\mathcal{H}'_{\text{kin}}}$$

$$(30.1.5)$$

where we used $||f||_{\mathcal{H}_{\text{kin}}}/||f||_{\mathcal{D}_{\text{kin}}} \geq 1$. Thus it converges in $\mathcal{D}'_{\text{kin}}$ as well, that is, the topology on $\mathcal{D}'_{\text{kin}}$ is weaker than that of $\mathcal{H}_{\text{kin}}$. We thus have topological inclusions

$$\mathcal{D}_{\text{kin}} \hookrightarrow \mathcal{H}_{\text{kin}} \hookrightarrow \mathcal{D}'_{\text{kin}} \qquad (30.1.6)$$

sometimes called a Gel'fand triple.

Unfortunately the definition of a Gel'fand triple requires a further input, the nuclear topology intrinsic to $\mathcal{D}_{\text{kin}}$ which we want to avoid since there seems no physical guiding principle (although then there are rather strong theorems available concerning the completeness of generalised eigenvectors [280]). We thus equip $\mathcal{D}_{\text{kin}}$ simply with the relative topology induced from $\mathcal{H}_{\text{kin}}$. The requirement that $\mathcal{D}_{\text{kin}}$ is dense is then no loss of generality since we may simply replace $\mathcal{H}_{\text{kin}}$ by the completion of $\mathcal{D}_{\text{kin}}$. Instead of the topological dual (which would coincide with $\mathcal{H}_{\text{kin}}$) we consider the *algebraic dual* $\mathcal{D}^*_{\text{kin}}$ of all linear functionals on $\mathcal{D}_{\text{kin}}$. This space is naturally equipped with the weak *-topology of pointwise convergence, that is, a net (l^α) converges to l iff the net of complex numbers $(l^\alpha(f))$ converges to $l(f)$ for any $f \in \mathcal{D}_{\text{kin}}$ (but not uniformly). Again we can consider $\mathcal{H}_{\text{kin}}$ as a subspace of $\mathcal{D}^*_{\text{kin}}$ and since a net converging in norm certainly converges pointwise we have again topological inclusions

$$\mathcal{D}_{\text{kin}} \hookrightarrow \mathcal{H}_{\text{kin}} \hookrightarrow \mathcal{D}^*_{\text{kin}} \qquad (30.1.7)$$

which in abuse of notation we will still refer to as a Gel'fand triple. Thus, the only input left is the choice of $\mathcal{D}_{\text{kin}}$ for which, however, there are no general selection principles available at the moment (see, however, [277, 278] for further discussion).

The reason for blowing up the structure beyond $\mathcal{H}_{\text{kin}}$ is that generically the point zero does not lie in the discrete part of the spectrum of $\mathcal{C}$, that is, if we look for solutions to the constraints in the form $\hat{C}_I \psi = 0$ for all $I \in \mathcal{I}$ for $\psi \in \mathcal{H}_{\text{kin}}$, then there are generically not enough solutions because ψ would be an eigenvector with eigenvalue zero but since zero does not lie in the discrete spectrum the eigenvectors do not form the entire solution space. This is precisely what happens with the diffeomorphism constraint for the case of quantum gravity where the only eigenvectors are the constant functions. We therefore look for generalised eigenvectors $l \in \mathcal{D}_{\mathrm{kin}}^*$ in the algebraic dual, for which we require

$$[(\hat{C}_I)'l](f) := l(\hat{C}_I^\dagger f) = 0 \ \forall \ I \in \mathcal{I}, \ f \in \mathcal{D}_{\mathrm{kin}} \tag{30.1.8}$$

where the dual action of an operator $\hat{O} \in \mathcal{O}_{\mathrm{kin}}$ on $l \in \mathcal{D}_{\mathrm{kin}}^*$ is defined by

$$[\hat{O}'l](f) := l(\hat{O}^\dagger f) \ \forall \ f \in \mathcal{D}_{\mathrm{kin}} \tag{30.1.9}$$

Notice that since we required $\mathcal{C}$ to be a self-adjoint set we can avoid taking the adjoint in (30.1.8) by passing to self-adjoint representatives $\hat{C}_I$. Due to the adjoint operation in (30.1.9) we have an antilinear representation of $\mathcal{O}_{\mathrm{kin}}$ on $\mathcal{D}_{\mathrm{kin}}^*$ which descends to an antilinear representation of $\mathcal{O}_{\mathrm{phys}}$ on the space of solutions $\mathcal{D}_{\mathrm{phys}}^* \subset \mathcal{D}_{\mathrm{kin}}^*$ to (30.1.8). The reason for taking the adjoint is that if a solution l was an element of the kinematical Hilbert space, $l \in \mathcal{H}_{\mathrm{kin}}$, then $\hat{C}_I l = 0$ would be equivalent to $[\hat{C}_I l](f) :=< \hat{C}_I l, f > = < l, \hat{C}_I^\dagger f >= l(\hat{C}_I^\dagger f) = 0$ for all $f \in \mathcal{D}_{\mathrm{kin}}$, hence (30.1.8) is the appropriate extension of this condition from $\mathcal{H}_{\mathrm{kin}}$ to $\mathcal{D}_{\mathrm{kin}}^*$.

At this point, the space $\mathcal{D}_{\mathrm{phys}}^*$ is just a subspace of $\mathcal{D}_{\mathrm{kin}}^*$. We would like to equip a subspace $\mathcal{H}_{\mathrm{phys}}$ of it with a Hilbert space topology. The reason for not turning all of $\mathcal{D}_{\mathrm{phys}}^*$ into $\mathcal{H}_{\mathrm{phys}}$ is that then $\mathcal{O}_{\mathrm{phys}}$ would be realised as an algebra of bounded operators on $\mathcal{H}_{\mathrm{phys}}$ since they would be defined everywhere on $\mathcal{D}_{\mathrm{phys}}^*$, which would be unnatural if the corresponding classical functions are unbounded. In particular, the topology on $\mathcal{H}_{\mathrm{phys}}$, as a complete norm topology, should be finer than the relative topology induced from $\mathcal{D}_{\mathrm{kin}}^*$. The idea is then to consider $\mathcal{D}_{\mathrm{phys}}^*$ as the algebraic dual of a dense subspace $\mathcal{D}_{\mathrm{phys}} \subset \mathcal{H}_{\mathrm{phys}}$ so that all of $\mathcal{O}_{\mathrm{phys}}$ is densely defined there. In other words we get a second Gel'fand triple

$$\mathcal{D}_{\mathrm{phys}} \hookrightarrow \mathcal{H}_{\mathrm{phys}} \hookrightarrow \mathcal{D}_{\mathrm{phys}}^* \tag{30.1.10}$$

with an antilinear representation of $\mathcal{O}_{\mathrm{phys}}$ on $\mathcal{H}_{\mathrm{phys}}$ defined by (30.1.9).

The choice of the inner product on $\mathcal{H}_{\mathrm{phys}}$ is guided by the requirement that the adjoint in the physical inner product, denoted by $\star$, represents the adjoint in the kinematical one, that is,

$$< \psi, \hat{O}'\psi' >_{\mathrm{phys}} = < (\hat{O}')^\star \psi, \psi' >_{\mathrm{phys}} = < (\hat{O}^\dagger)'\psi, \psi' >_{\mathrm{phys}} \tag{30.1.11}$$

for all $\psi, \psi' \in \mathcal{D}_{\mathrm{phys}}$. The canonical commutation relations among observables are automatically implemented because by construction $\mathcal{H}_{\mathrm{phys}}$ carries a representation of $\mathcal{O}_{\mathrm{phys}}$ on which the correct algebraic relations were already implemented as an abstract algebra.

A systematic construction of the physical inner product is available if we have an antilinear (so-called) rigging map

$$\eta : \ \mathcal{D}_{\mathrm{kin}} \to \mathcal{D}_{\mathrm{phys}}^*; \ f \mapsto \eta(f) \tag{30.1.12}$$

at our disposal which must be such that

1. The following is a positive semidefinite sesquilinear form (linear in f, antilinear in f')

$$< \eta(f), \eta(f') >_{\text{phys}} := [\eta(f')](f) \ \forall \ f, f' \in \mathcal{D}_{\text{kin}} \qquad (30.1.13)$$

2. For any $\hat{O} \in \mathcal{O}_{\text{phys}}$ we have

$$\hat{O}' \eta(f) = \eta(\hat{O}f) \ \forall \ f \in \mathcal{D}_{\text{kin}} \qquad (30.1.14)$$

which makes sure that the dual action preserves the space of solutions since $\hat{C}'\hat{O}'\eta(f) = 0$. Notice that both the left- and the right-hand side in (30.1.14) are antilinear in $\hat{O}$.

We could then define $\mathcal{D}_{\text{phys}} := \eta(\mathcal{D}_{\text{kin}})/\ker(\eta)$ (with the kernel being understood with respect to $||.||_{\text{phys}}$) and complete it with respect to (30.1.13) to obtain $\mathcal{H}_{\text{phys}}$. Notice that (30.1.11) is satisfied because for $\psi = \eta(f), \psi' = \eta(f')$ we have

$$< \psi, \hat{O}'\psi' >_{\text{phys}} = < \eta(f), \eta(\hat{O}f') >_{\text{phys}} = [\eta(\hat{O}f')](f)$$

$$= [\hat{O}'\eta(f')](f) = \eta(f')(\hat{O}^\dagger f)$$

$$= < \eta(\hat{O}^\dagger f), \eta(f') >_{\text{phys}} = < (\hat{O}^\dagger)'\psi, \psi' >_{\text{phys}} \quad (30.1.15)$$

To see that $\mathcal{H}_{\text{phys}}$ is a subspace of $\mathcal{D}^*_{\text{phys}}$ with a finer topology, notice that the map $J : \mathcal{H}_{\text{phys}} \to \mathcal{D}^*_{\text{phys}}$ defined by $[J(\psi)](f) := < \psi, \eta(f) >_{\text{phys}}$ is an injection because $J(\psi)$ vanishes iff ψ is orthogonal to all $\eta(f)$ with respect to $< ., . >_{\text{phys}}$, which means that $\psi = 0$ because the image of η is dense. Hence J is an embedding (injective inclusion) of linear spaces. Moreover, J is evidently continuous: if $||\psi^\alpha - \psi||_{\text{phys}} \to 0$ then $J(\psi^\alpha) \to J(\psi)$ in the weak *-topology iff $[J(\psi^\alpha)](f) \to [J(\psi)](f)$ for any $f \in \mathcal{D}_{\text{kin}}$, which is clearly the case. So convergence in $\mathcal{H}_{\text{phys}}$ implies convergence of $J(\mathcal{H}_{\text{phys}})$, hence the Hilbert space topology is stronger than the relative topology on $J(\mathcal{H}_{\text{phys}})$.

Thus, the existence of a rigging map solves the problem of defining a suitable inner product. A heuristic idea of how to construct η is through the group averaging proposal: since $\mathcal{C}$ is a self-adjoint set we may assume w.l.g. that the $\hat{C}_I$ are self-adjoint, and since they form a Lie algebra we may in principle exponentiate this Lie algebra (using the spectral theorem) and obtain a group of operators $t^I \to \exp(t^I \hat{C}_I)$ where $(t^I)_{I \in \mathcal{I}} \in T$ and T is some set depending on the constraints (the exponential map should be a bijection with the connected component of the associated group). Let then

$$\eta(f) := \overline{\int_T d\mu(t) \exp(t^I \hat{C}_I) f} \qquad (30.1.16)$$

with a bitranslation-invariant (Haar) measure μ on T. One easily sees that with

$$[\eta(f)](f') := \int_T d\mu(t) < \exp(t^I \hat{C}_I) f, f' >_{\text{kin}} \qquad (30.1.17)$$

formally $[\eta(f)](\hat{C}_I f') = 0$. Of course, one must check case by case whether T, μ exist and that η has the required properties. At present, the case of a finite-dimensional, locally compact Lie group is under complete control (existence and uniqueness) [278]. For the case of an infinite number of constraints, existence or uniqueness proofs of μ, T are not available yet. This case, however, can be treated by the method of DID, see the next section.

Let us make some short comments about the open algebra case: suppose that the classical constraint functions C_I and the structure functions $f_{IJ}{}^K$ are real- and imaginary-valued respectively. As mentioned already, it is now excluded to choose the corresponding operators to be (anti)-self-adjoint operators since this would require the ordering

$$[\hat{C}_I, \hat{C}_J] = \frac{i}{2}(\hat{f}_{IJ}{}^K \hat{C}_K + \hat{C}_K \hat{f}_{IJ}{}^K) \tag{30.1.18}$$

and could possibly lead to the following quantum anomaly: if we impose the condition (30.1.8) then we would find for an element $l \in \mathcal{D}^*_{\text{phys}}$ that

$$((\hat{f}_{IJ}{}^K)' \hat{C}'_K + \hat{C}'_K (\hat{f}_{IJ}{}^K)')l = [\hat{C}'_K, (\hat{f}_{IJ}{}^K)']l = 0 \tag{30.1.19}$$

which means that l is not only annihilated by the dual constraint operators but also by (30.1.19), which is not necessarily proportional to a dual constraint operator any longer, implying that the physical Hilbert space will be too small. If, on the other hand, we do not choose the $\hat{C}_I$ to be self-adjoint, the anomaly problem is potentially absent but now it is no longer true that $[\hat{C}'_I l](f) = l(\hat{C}_I f)$, in other words, the question arises whether it is $\hat{C}'_I l = 0$ or $(\hat{C}^\dagger_I)' l = 0$ that we should impose? The answer is that this just corresponds to a choice of operator ordering since the classical limit of both $\hat{C}_I$ and $\hat{C}^\dagger_I$ is given by the real-valued function C_I and thus the answer is that the correct ordering is the one in which the algebra is, besides being densely defined and closed, also free of anomalies. Thus, in the open algebra case we may proceed just as above with the additional requirement of anomaly freeness. Of course, group averaging does not work since we cannot exponentiate the algebra any longer.

We conclude this chapter with an example in order to illustrate the procedure: suppose $\mathcal{H}_{\text{kin}} = L_2(\mathbb{R}^2, d^2 x)$ and $\hat{C} = \hat{p}_1 = -i\partial/\partial x_1$. Obviously the kinematical Hilbert space implements the adjointness and canonical commutation relations among the basic variables x_1, x_2, p_1, p_2. A nuclear space choice would be $\mathcal{D}_{\text{kin}} = \mathcal{S}(\mathbb{R}^2)$ (test functions of rapid decrease). The functions l annihilated by $\hat{C}$ are those that do not depend on x^1 and are thus not normalisable. However, we can define them as elements of $\mathcal{D}^*_{\text{kin}}$ by $l(f) := <l, f>_{\text{kin}} = \int_{\mathbb{R}^2} d^2 x \overline{l(x)} f(x)$ which converges pointwise. Clearly $l(\hat{C}f) = 0$ if $l_{,x^1} = 0$. The physical observable algebra consists of operators not involving $\hat{x}_1$ and after taking the quotient with respect to the constraint ideal they involve only $\hat{p}_2, \hat{x}_2$. Obviously they leave the space $\mathcal{D}^*_{\text{phys}}$ invariant, consisting of those elements of $\mathcal{D}^*_{\text{kin}}$ that are x^1-independent. The physical Hilbert space that suggests itself (implementing the correct reality condition) is therefore $\mathcal{H}_{\text{phys}} = L_2(\mathbb{R}, dx_2)$, which is a proper subspace of $\mathcal{D}^*_{\text{phys}}$ and we have $D_{\text{phys}} = \mathcal{S}(\mathbb{R})$. Now an appropriate rigging map is obtained indeed by

$$\eta(f)(x_1, x_2) := \overline{\int_{\mathbb{R}} dt \exp(it\hat{p}_1) f(x_1, x_2)} = \overline{\int_{\mathbb{R}} dx_1 f(x_1, x_2)} = 2\pi \overline{\delta(\hat{C}) f(x_1, x_2)}$$

since $\hat{p}_1$ generates x_1 translations, produces functions independent of x_1 and dt is an invariant measure on $T = \mathbb{R}$. Notice that the integral converges because f is of rapid decrease. Notice also that we could define the delta distribution of the constraint, using the spectral theorem. We have

$$< \eta(f), \eta(f') >_{\text{phys}} := \eta(f')[\eta(f)] = \int_{\mathbb{R}} dt \int_{\mathbb{R}^2} d^2x \, \overline{f'(x_1 + t, x_2)} f(x_1, x_2)$$

$$= \int_{\mathbb{R}} dx_2 \left[\overline{\int dx'_1 f'(x'_1, x_2)} \right] \left[\int dx_1 f(x_1, x_2) \right] \quad (30.1.20)$$

which is the same inner product as chosen above.

In the case of an Abelian self-adjoint constraint algebra a reasonable Ansatz for a rigging map is always given by

$$\eta(f) = \overline{\prod_{I \in \mathcal{I}} \delta(\hat{C}_I) f} \quad (30.1.21)$$

30.2 Master Constraint Programme (MCP) and DID

The following construction is to date the only one that leads to the physical inner product when the constraints do not form a Lie algebra, that is, in the presence of structure functions rather than structure constants. In particular, RAQ is not possible in this case and therefore DID generalises RAQ.

Let there be given a phase space $\mathcal{M}$ with real-valued, first-class constraint functions $C_I(x) : \mathcal{M} \to \mathbb{R}; \; m \mapsto [C_I(x)](m)$ on $\mathcal{M}$. Here we let $I \in \mathcal{I}$ take discrete values while $x \in X$ belongs to some continuous index set. To be more specific, X is supposed to be a measurable space and we choose a measure ν on X. Then we consider the fiducial Hilbert space $\mathfrak{h} := L_2(X, d\nu)^{|\mathcal{I}|}$ with inner product

$$< u, v >_{\mathfrak{h}} = \int_X d\nu(x) \sum_{I \in \mathcal{I}} \overline{u_I(x)} v_I(x) \quad (30.2.1)$$

Finally we choose a positive operator-valued function $\mathcal{M} \to \mathcal{L}_+(\mathfrak{h}); \; m \mapsto K(m)$ where $\mathcal{L}_+(\mathfrak{h})$ denotes the cone of positive linear operators on $\mathfrak{h}$.

Definition 30.2.1. *The*   *for the system of constraints* $m \mapsto [C_I(x)](m)$ *corresponding to the choice* μ *of a measure on* X *and the operator-valued function* $m \mapsto K(m)$ *is defined by*

$$(m) = \frac{1}{2} < C(m), K(m) \cdot C(m) >_{\mathfrak{h}} \quad (30.2.2)$$

Of course ν, K must be chosen in such a way that (30.2.2) converges and defines a differentiable function on $\mathcal{M}$, but apart from that the definition of a   allows a great deal of flexibility, which we will exploit in the examples to be discussed. It is clear that the infinite number of constraint equations $C_I(x) = 0$ for a.a. $x \in X$ and all $I \in \mathcal{I}$ is equivalent with the single equation $= 0$ so that classically all admissible choices of μ, K are equivalent.

Notice that we have explicitly allowed $\mathcal{M}$ to be infinite-dimensional. In case that we have only a finite-dimensional phase space, simply drop the structures x, X, ν from the construction. We compute for any function $O \in C^2(\mathcal{M})$ that

$$\{\{O, \quad\}, O\}_{=0} = [< \{O, C\}, K \cdot \{O, C\} >_{\mathfrak{h}}]_{=0} \tag{30.2.3}$$

hence the single equation $\{\{O, \quad\}, O\}_{=0} = 0$ is equivalent to $\{O, C_I(x)\}_{=0} = 0$ for a.a. $x \in X$ and $I \in \mathcal{I}$.

Among the set of all weak Dirac observables satisfying (30.2.3) the strong Dirac observables form a subset. These are the twice differentiable functions on $\mathcal{M}$ satisfying $\{O, \quad\} \equiv 0$ identically on all of $\mathcal{M}$. They can be found as follows: let $t \mapsto \alpha_t$ be the one-parameter group of automorphisms of $\mathcal{M}$ defined by time evolution with respect to   . Then the **ergodic mean** of $O \in C^\infty(\mathcal{M})$

$$[O] := \lim_{T \to \infty} \frac{1}{2T} \int_{-T}^{T} dt\, \alpha_t\,(O) \tag{30.2.4}$$

has a good chance of being a strong Dirac observable if twice differentiable, as one can see by formally commuting the integral with the Poisson bracket with respect to   . In order that the limit in (30.2.4) is non-trivial, the integral must actually diverge. Using l'Hospital's theorem we therefore find that if (30.2.4) converges and the integral diverges (the limit being an expression of the form ∞/∞) then it equals

$$[O] := \lim_{T \to \infty} \frac{1}{2}[\alpha_T(O) + \alpha_{-T}(O)] \tag{30.2.5}$$

which is a great simplification because, while one can often compute the time evolution α_t for a bounded function O (for bounded functions the integral will typically diverge linearly in T so that the limit exists), doing the integral is impossible in most cases. Hence we see that the MCP even provides some insight into the structure of the classical Dirac observables for the system under consideration.

Now we come to the quantum theory. We assume that a judicious choice of ν, K has resulted in a positive, self-adjoint operator $\hat{\quad}$ on some kinematical Hilbert space $\mathcal{H}_{\text{kin}}$ which we assume to be separable. Following (a slight modification of) a proposal due to Klauder [87], if zero is not in the spectrum of $\hat{\quad}$ then compute the finite, positive number $\lambda_0 := \inf(\sigma(\hat{\quad}))$ and redefine $\hat{\quad}$ by $\hat{\quad} - \lambda_0 \mathrm{id}_{\mathcal{H}_{\text{kin}}}$. Here we assume that λ_0 vanishes in the $\hbar \to 0$ limit so that the modified operator still qualifies as a quantisation of   . This is justified in all examples encountered so far where λ_0 is usually related to some reordering of the operator. Hence in what follows we assume w.l.g. that $0 \in \sigma(\widehat{})$.

Under these circumstances we can completely solve the $\widehat{} = 0$ and explicitly provide the physical Hilbert space and its physical inner product. Namely, as we recall below, the Hilbert space $\mathcal{H}_{\text{kin}}$ is unitarily equivalent to a direct integral

$$\mathcal{H}_{\text{kin}} \cong \int_{\mathbb{R}^+}^{\oplus} d\mu(\lambda) \, \mathcal{H}_{\text{kin}}^{\oplus}(\lambda) \tag{30.2.6}$$

where μ is a so-called spectral measure and $\mathcal{H}_{\text{kin}}^{\oplus}(\lambda)$ is a separable Hilbert space with inner product induced from $\mathcal{H}_{\text{kin}}$. This simply follows from spectral theory. The operator $\widehat{}$ acts on $\mathcal{H}_{\text{kin}}^{\oplus}(\lambda)$ by multiplication by λ, hence the physical Hilbert space is simply given by

$$\mathcal{H}_{\text{phys}} = \mathcal{H}_{\text{kin}}^{\oplus}(0) \tag{30.2.7}$$

The Hilbert spaces $\mathcal{H}_{\text{kin}}^{\oplus}(\lambda)$ are unique up to sets of μ-measure zero and do not depend on the choice of μ.

Strong quantum Dirac observables can be constructed in analogy to (30.2.4), (30.2.5), namely for a given bounded operator on $\mathcal{H}_{\text{kin}}$ we define, if the uniform limit exists

$$\widehat{[O]} := \lim_{T \to \infty} \frac{1}{2}[U(T)\hat{O}U(T)^{-1} + U(T)^{-1}\hat{O}U(T)] \tag{30.2.8}$$

where

$$U(t) = e^{it\,\widehat{}} \tag{30.2.9}$$

is the unitary evolution operator corresponding to the self-adjoint $\widehat{}$ via Stone's theorem. One must check whether the spectral projections of the bounded operator (30.2.8) commute with those of $\widehat{}$ but if they do then $\widehat{[O]}$ defines a strong quantum Dirac observable.

This concludes our sketch of the general theory. We will now provide the corresponding mathematical theory which also gives a concrete algorithm for how to construct the physical Hilbert space from a given Master Constraint Operator $\widehat{}$.

Before we do this, let us check that the MCP gives sensible results in the simplest case, namely a countable collection of, not necessarily self-adjoint but closable (so that all $\hat{C}_I, \hat{C}_I^{\dagger}$ are densely defined on a common dense domain $\mathcal{D}$) operators $\hat{C}_I$ which do not necessarily form a (infinite-dimensional) Lie algebra but which are such that $\{0\}$ lies only in their common point spectrum. We now show that the infinite number of equations (A) $\hat{C}_I \psi = 0 \; \forall \; I$ (meaning that ψ is a common, normalisable zero eigenvector) is equivalent to the single equation (B) $\widehat{} \psi = 0$ where $\widehat{} = \sum_I k_I \, \hat{C}_I^{\dagger} \hat{C}_I$ and $k_I > 0$ are certain constants which converge sufficiently fast in order that $\widehat{}$ is densely defined on $\mathcal{D}$. The implication (A) $\Rightarrow$ (B) is obvious. Conversely, $\hat{}\,\psi = 0$ implies $< \psi, \hat{}\,\psi >= \sum_I k_I \,||\hat{C}_I \,\psi||^2 = 0$ hence (A).

Moving on to the general case, we need a few preparations. See, for example, [536] for more details on direct integrals and [282] for the multiplicity theory for operators. The following theory is also summarised more precisely in the first reference of [252], where we follow closely [897].

Definition 30.2.2. *Let $\mathfrak{X}$ be a locally compact space, μ a measure on $\mathfrak{X}$ and $x \mapsto \mathcal{H}_x$ an assignment of Hilbert spaces such that the function $x \mapsto m_x$, where m_x is the countable dimension of $\mathcal{H}_x$, is measurable. It follows that the sets $\mathfrak{X}_m = \{x \in \mathfrak{X}; \; m_x = m\}$, where m denotes any cardinality, are measurable. Since Hilbert spaces whose dimensions have the same cardinality are unitarily equivalent we may identify all the $\mathcal{H}_x$, $m_x = m$ with a single $\mathcal{H}_m = \mathbb{C}^m$. We now consider maps*

$$\psi : \mathfrak{X} \to \prod_{x \in \mathfrak{X}} \mathcal{H}_x; \quad x \mapsto (\psi(x))_{x \in \mathfrak{X}} \tag{30.2.10}$$

subject to the following two constraints:

1. *The maps $x \mapsto < \psi, \psi(x) >_{\mathcal{H}_m}$ are measurable for all $x \in \mathfrak{X}_m$ and all $\psi \in \mathcal{H}_m$.*
2. *If*

$$< \psi_1, \psi_2 > := \sum_m \int_{\mathfrak{X}_m} d\mu(x) < \psi_1(x), \psi_2(x) >_{\mathcal{H}_m} \tag{30.2.11}$$

then $< \psi, \psi > < \infty$.

The completion of the space of maps (30.2.10) in the inner product (30.2.11) is called the direct integral of the $\mathcal{H}_x$ with respect to μ and one writes

$$\mathcal{H}^\oplus = \int_{\mathfrak{X}}^{\oplus} d\mu(x) \, \mathcal{H}_x, \quad < \psi_1, \psi_2 > = \int_{\mathfrak{X}} d\mu(x) \; < \psi_1(x), \psi_2(x) >_{\mathcal{H}_x} \tag{30.2.12}$$

Definition 30.2.3

(i) *Two measures μ, ν are said to be equivalent if they have the same measure zero sets. The corresponding measure classes are denoted by $< \mu >$.*

(ii) *Two measure classes $< \mu >, < \nu >$ are said to be disjoint if any two representatives $\mu_1 \in < \mu >$, $\nu_1 \in < \nu >$ are mutually singular.*

In the terminology of Chapter 25 we see that equivalent measures are mutually absolutely continuous. For disjoint measure classes any two representatives have disjoint support.

Definition 30.2.4. *A self-adjoint operator a is said to be of uniform multiplicity m provided that there is a unitary operator U such that UaU^{-1} is represented on some $\oplus_{k=1}^{m} L_2(\mathbb{R}, d\mu)$ as a multiplication operator by λ (every term has the same measure).*

The following result shows that the multiplicity and the measure class is a unitary invariant of representation theory.

Lemma 30.2.5. *If a self-adjoint operator is unitarily equivalent to a multiplication operator on $\oplus_{k=1}^m L_2(\mathbb{R}, d\mu)$ and $\oplus_{k=1}^n L_2(\mathbb{R}, d\nu)$ respectively then $m = n$, $< \mu >=< \nu >$.*

Theorem 30.2.6 (commutative multiplicity theorem). *Let a be a self-adjoint operator on a Hilbert space $\mathcal{H}$. Then there exists a unitary operator U such that $U\mathcal{H} = \oplus_{m=1}^\infty \mathcal{H}_m$ and*

1. *$\mathcal{H}_m$ is an invariant subspace for UaU^{-1},*
2. *$a_{|\mathcal{H}_m}$ has uniform multiplicity m,*
3. *The measure classes $< \mu_m >$ underlying $\mathcal{H}_m = \oplus_{k=1}^m L_2(\mathbb{R}, d\mu_m)$ are mutually singular.*

Moreover, the unitary equivalence classes of the subspaces $\mathcal{H}_m$ (some of which might be empty) and the measure classes are uniquely determined by these three properties.

We come now to the main result of this chapter.

Theorem 30.2.7 (Direct Integral Decomposition (DID)). *Let a be a self-adjoint operator on a separable Hilbert space $\mathcal{H}$. Then there is a unitary operator U such that $U\mathcal{H} = \mathcal{H}^\oplus = \int_{\mathbb{R}}^\oplus d\mu(\lambda)\, \mathcal{H}^\oplus(\lambda)$ where μ is a probability measure and UaU^{-1} is represented on $\mathcal{H}^\oplus(\lambda)$ by multiplication by λ. Moreover, the measure class $< \mu >$ and the Hilbert spaces $\mathcal{H}^\oplus(\lambda)$ are uniquely determined.*

Proof: We use the spectral theorem to arrive at a constructive proof.

Let the projection-valued measure of a be denoted by $E(\lambda)$. Consider the bounded, weakly continuous unitary groups $W_t = \exp(ita) = \int_{\mathbb{R}} e^{it\lambda} dE(\lambda)$. Given $\psi \in \mathcal{H}$ and a smooth function of compact support $f \in C_0^\infty(\mathbb{R})$, let

$$\psi_f := \int_{\mathbb{R}} dt\, f(t) W_t \psi \tag{30.2.13}$$

It follows from Stone's theorem that ψ_f is a C^∞-vector for a, that is, it is in the invariant domain of a. Specifically $ia\psi_f = -\psi_{\dot{f}}$. Moreover, the span of the ψ_f as ψ, f vary is dense in $\mathcal{H}$.

Choose any vector ψ_1 and function $f_1 \in C_0^\infty(\mathbb{R})$ and set $\Omega_1 = \psi_{1,f_1}$. Denote by $\mathcal{H}_1$ the closure of the finite linear span of the W_t, that is, elements of the form $p(W) := \sum_{k=1}^N z_k W_{t_k}$. If $\mathcal{H}_1 \neq \mathcal{H}$ choose $\psi_2 \in \mathcal{H}_1^\perp$ and $f_2 \in C_0^\infty(\mathbb{R})$. Then also $\Omega_2 = \psi_{2,f_2} \in \mathcal{H}_1^\perp$ because $W_t \Omega_1 \in \mathcal{H}_1$ by construction, hence $< \Omega_2, \Omega_1 >= \int dt\, \overline{f_2(t)} < \psi_2, W_{-t}\Omega_1 >= 0$. Iterating, since $\mathcal{H}$ is separable we arrive at the direct sum

$$\mathcal{H} = \oplus_{n=1}^\infty \mathcal{H}_n \tag{30.2.14}$$

A dense set of vectors can be presented in the form $(p_n(W)\,\Omega_n)_{n=1}^{\infty}$. We claim that for each m, n the vectors $a^m\Omega_n$ are elements of $\mathcal{H}_n$. To see this we just need to show that they can be approximated arbitrarily well by a dense set of vectors of $\mathcal{H}_n$. However, this is clear from the construction of the Ω_n. Hence the $a^n\Omega_m$ belong to the closure $\mathcal{H}_m$ and by the same argument the $a^m\Omega_n$ are dense in $\mathcal{H}_n$. Thus, a dense set of vectors can be given in the form $(p_n(a)\Omega_n)_{n\in\mathbb{N}}$ where p_n is a polynomial in a and the Ω_n are C^{∞}-vectors for a.[2]

By the functional calculus of Chapter 29

$$< \psi, \psi' >_{\mathcal{H}} = \sum_{n=1}^{\infty} < \Omega_n, p_n(a)^{\dagger} p'_n(a)\Omega_n > = \int_{\mathbb{R}} d\mu_n(\lambda)\,\overline{p_n(\lambda)}\,p'_n(\lambda) \quad (30.2.15)$$

where $\mu_n(\lambda) = < \Omega_n, E(\lambda)\Omega_n >$ are the corresponding probability spectral measures.

Consider the probability Borel measure (all the μ_n are Borel measures)

$$\mu(\lambda) := \sum_{n=1}^{\infty} c_n\,\mu_n(\lambda) \quad (30.2.16)$$

where $c_n > 0$ are any non-negative numbers such that $\sum_{n=1}^{\infty} c_n = 1$. A popular choice is $c_n = 2^{-n}$. It is clear that for any measurable set S the condition $\mu(S) = 0$ implies $\mu_n(S) = 0$ for all n since $c_n > 0$. Therefore μ_n is absolutely continuous with respect to μ for all n. It is for this reason that we have restricted ourselves to a separable Hilbert space: if $\mathcal{H}$ was not separable then it might happen that the labels n take an uncountable range. But then $\sum_n c_n = 1$ would imply that $c_n = 0$ except for countably many and the absolute continuity of μ_n with respect to μ could not be concluded.

Since μ_n, μ are *finite*, positive measures, the unique Radon–Nikodym derivative $\rho_n = d\mu_n/d\mu$ (a non-negative $L_1(\mathbb{R}, d\mu)$ function) exists, see Chapter 25. The function values $\rho_n(\lambda)$ are, of course, only defined up to sets of μ measure zero. To reduce this ambiguity, one demands that the set of Ω_n be minimal (this is always possible and measure theoretically unique, see [253]) and for those resulting ρ_n we demand that the *algebra* of (weak or strong) Dirac observables, which as one can show preserve the fibres $\mathcal{H}^{\oplus}(\lambda)$, be represented irreducibly on the physical Hilbert space $\mathcal{H}^{\oplus}(0)$. This tends to fix the values $\rho_n(0)$ numerically, not only up to measure zero, because one cannot arbitrarily change the $\rho_n(\lambda)$ on measure zero sets without destroying the algebra structure. See [253] where an explicit action of the Dirac observables on the physical Hilbert space is derived. Fortunately, in practice the following simpler rule leads to the correct result: choose any representative which is everywhere non-negative and continuous at $\lambda = 0$ from the right (provided it exists).[3]

---

[2] One can avoid the C^{∞}-vectors by working with the bounded spectral projections, see [253].
[3] The choice of the representative is completely irrelevant for the direct integral decomposition of $\mathcal{H}$, however, it affects the size of the subspaces $\mathcal{H}^{\oplus}(\lambda)$ and hence of $\mathcal{H}_{\text{phys}} = \mathcal{H}^{\oplus}(0)$. This is why we specify a suitable class of representative here.

Let S_n be the support of the so-determined ρ_n. We define the function m : $\mathbb{R} \to \mathbb{N}$ by $m(\lambda) = m$ provided that λ lies in precisely m of the S_n. The function m is measurable because the natural numbers carry the discrete topology (every subset is open) and the pre-images $X_m := \{\lambda \in \mathbb{R};\ m(\lambda) = m\}$ of the one-point sets $\{m\}$ are given by $\cup_{n_1 < \ldots n_m} \cap_{k=1}^m S_{n_k}$ which is measurable because the S_n are and this is a countable union of measurable sets.

We now compute

$$\|\psi\|^2 = \sum_{n=1}^\infty \int_\mathbb{R} d\mu_n(\lambda)\ \overline{p_n(\lambda)}\ p_n'(\lambda)$$

$$= \sum_{n=1}^\infty \int_\mathbb{R} d\mu(\lambda)\ \overline{\left[\sqrt{\rho_n(\lambda)}p_n(\lambda)\right]}\ \left[\sqrt{\rho_n(\lambda)}p_n'(\lambda)\right]$$

$$= \int_\mathbb{R} d\mu(\lambda) \sum_{n=1}^\infty \overline{\left[\sqrt{\rho_n(\lambda)}p_n(\lambda)\right]}\ \left[\sqrt{\rho_n(\lambda)}p_n'(\lambda)\right]$$

$$= \sum_{m=1}^\infty \int_{X_m} d\mu(\lambda) \sum_{n=1}^\infty \overline{\left[\sqrt{\rho_n(\lambda)}p_n(\lambda)\right]}\ \left[\sqrt{\rho_n(\lambda)}p_n'(\lambda)\right] \quad (30.2.17)$$

In the third step we have made use of the fact that the Hilbert spaces $L_2(X, d\mu; \ell_2^M)$ of square integrable vector-valued functions with values in the Hilbert space $\ell_2^M = \mathbb{C}^M$, $M = 1, 2, \ldots, \infty$ of square summable sequences is unitarily equivalent to the Hilbert spaces $\ell_2^M(L_2(X, d\mu))$ of square summable sequences of square integrable functions (see, e.g., [282], theorem II.10). This follows for instance by showing that both spaces are isomorphic to the tensor product Hilbert space $L_2(X, d\mu) \otimes \ell_2^M$, see Definition 26.2.8, or directly from the Lebesgue dominated convergence theorem. In any case, this is why we were allowed to interchange the integral with the sum. See also [253] for more details.

For $\lambda \in X_m$ we may order the m indices $n = n_k(\lambda)$ for which $\rho_n(\lambda) \neq 0$ according to $n_1(\lambda) < \ldots < n_m(\lambda)$. For $\lambda \in X_m$ let

$$\psi_k^{(m)}(\lambda) := \sqrt{\rho_{n_k(\lambda)}(\lambda)}p_{n_k(\lambda)} \quad (30.2.18)$$

introduce an orthonormal basis $e_k^{(m)}$, $k = 1, \ldots, m$ of the Hilbert space $\mathcal{H}_\mu^\oplus(\lambda) := \mathbb{C}^m$ with standard inner product and set

$$\psi_\mu(\lambda) := \sum_{k=1}^m \psi_k^{(m)}(\lambda)\ e_k^{(m)} \quad (30.2.19)$$

Therefore

$$\|\psi_\mu(\lambda)\|^2_{\mathcal{H}_\mu^\oplus(\lambda)} = \sum_{k=1}^m \left|\psi_k^{(m)}(\lambda)\right|^2 = \sum_{n=1}^\infty \rho_n(\lambda)|p_n(\lambda)|^2 \quad (30.2.20)$$

and thus

$$\|\psi\|^2 = \sum_{m=1}^\infty \int_{X_m} d\mu(\lambda)\|\psi_\mu(\lambda)\|^2_{\mathcal{H}_\mu^\oplus(\lambda)} \quad (30.2.21)$$

Comparison with Definition 30.2.2 reveals that the unitary map $U : \psi \mapsto (\psi(\lambda))_{\lambda \in \mathbb{R}}$ displays $\mathcal{H}^{\oplus} := U\mathcal{H}$ precisely as a direct integral of Hilbert spaces

$$U\mathcal{H} = \mathcal{H}_{\mu}^{\oplus} = \int_{\mathbb{R}}^{\oplus} d\mu(\lambda) \, \mathcal{H}_{\mu}^{\oplus}(\lambda) \qquad (30.2.22)$$

and that, by the spectral theorem, the operator UaU^{-1} acts in $\mathcal{H}_{\mu}^{\oplus}(\lambda)$ by multiplication by λ.

Now let $d\mu^{(m)} := \chi_{X_m} d\mu$ where χ_{X_m} denotes the characteristic function of X_m. Consider the unitary map

$$V : \mathcal{H} \to \oplus_{m=1}^{\infty} \mathcal{H}_{\mu,m}^{\oplus}, \quad \mathcal{H}_{\mu,m}^{\oplus} := \oplus_{k=1}^{m} L_2(\mathbb{R}, \, d\mu^{(m)}); \quad \psi \mapsto (\psi_k^{(m)})_{k=1}^{m} \quad (30.2.23)$$

Then $\mathcal{H}_{\mu,m}^{\oplus}$ is a subspace of $V\mathcal{H}$ which is (1) invariant under VaV^{-1}, (2) on which it acts by multiplication by λ and on which it has uniform multiplicity m and (3) the measure classes of the $\mu^{(m)}$ are mutually disjoint because the χ_{X_m} have disjoint support. We may therefore apply Theorem 30.2.6 and conclude that the measure classes $< \mu^{(m)} >$ and the Hilbert spaces $\mathcal{H}_m^{\oplus}$ are unique. In other words, if we have a second direct integral $\mathcal{H} \cong \int^{\oplus} d\nu \, \mathcal{H}_{\nu}^{\oplus}(\lambda)$ then we know that $< \mu^{(m)} > = < \nu^{(m)} >$ and that $\mathcal{H}_{\mu,m}^{\oplus}$, $\mathcal{H}_{\mu,m}^{\oplus}$ are unitarily equivalent for all m. All of this is reviewed in detail including proofs in [253].

It follows that if $d\mu^{(m)} = f_m d\nu^{(m)}$ is the corresponding Radon–Nikodym derivative then both of these measures must be supported on X_m and

$$d\mu = \sum_m d\mu^{(m)} = \left[\sum_m f_m \chi_{X_m} \right] d\nu =: f d\nu \qquad (30.2.24)$$

so that f is strictly positive. Hence we obtain a corresponding unitary map

$$\mathcal{H}_{\mu}^{\oplus}(\lambda) \to \mathcal{H}_{\nu}^{\oplus}(\lambda); \quad \psi_{\nu}(\lambda) = f(\lambda)\psi_{\nu}(\lambda) \qquad (30.2.25)$$

so the inner product between the two Hilbert spaces differs by a positive constant and both are isomorphic to $\mathbb{C}^m$ when $\lambda \in X_m$. $\qquad \square$

One usually drops the dependence of the direct integral decomposition on the choice of μ since different choices just give rise to unitarily equivalent presentations. This implies, in particular, that measure theoretically nothing depends on the choice of c_n, Ω_n. What is important is that for each point $\lambda \in \sigma(a)$ the dimensionality m of $\mathcal{H}_{\mu}^{\oplus}(\lambda)$ and its inner product, given by the standard inner product on $\mathbb{C}^m$ for $\lambda \in X_m$, are uniquely determined once the $\rho_n(0)$ have been fixed.

Whatever choice, the operator a acts by multiplication by λ and therefore $\mathcal{H}_{\text{phys}} = \mathcal{H}_{\text{kin}}^{\oplus}(0)$ for any choice. It is interesting to see that the support of the measure μ is the union of the supports S_n of the $\mu_n = < \Omega_n, E(\lambda)\Omega_n >$. Hence, if zero does not belong to the spectrum of $\hat{\ }$, that is, if there exists $\epsilon > 0$ such that $E(\epsilon) = 0$ then $0 \notin S_n$ for all n and thus $\mathcal{H}_{\text{phys}} = \emptyset$. In this case one will shift $\hat{\ }$ by a quantum correction such that this does not happen.

Let us again illustrate this procedure for a simple case: two commuting constraints $C_1 = p_1$, $C_2 = p_2$ for a particle moving in the plane. We work in the momentum representation, thus $\mathcal{H}_{\mathrm{kin}} = L_2(\mathbb{R}^2, d^2 k)$. Consider $\Omega_0 = \exp(-r^2/2)/\sqrt{\pi}$ where $r^2 = k_1^2 + k_2^2$ and notice that the set of vectors $k_1^{n_1} k_2^{n_2} \Omega_0$ is dense in $\mathcal{H}_{\mathrm{kin}}$ since they can be decomposed into Hermite polynomials. Using polar coordinates we have $k_\pm = k_1 \pm i k_2 = r e^{\pm i\phi}$ and see that the set of vectors $k_+^{n_+} k_-^{n_-} \Omega_0 = r^{n_+ + n_-} e^{i(n_+ - n_-)\phi} \Omega_0$ is dense. Since $\mathcal{H}_{\mathrm{kin}} \cong L_2(\mathbb{R}^+, r dr) \otimes L_2(S^1, d\phi)$ we see that the vectors proportional to $e^{in\phi}$, $n \in \mathbb{Z}$ are mutually orthogonal. In order to generate such a vector with given n we must choose $n_+ - n_- = n$, hence for $\pm n > 0$ we may choose $n_\mp = 0$ and so get mutually orthogonal $\Omega_n = c_n r^{|n|} e^{in\phi}$ where c_n is a normalisation. Moreover, the vectors of the form $r^{2m} \Omega_n \propto \hat{\ }^m \Omega_n$ are dense. The spectral projections are $E(\lambda) = \theta(\lambda - r^2/2)$ and the spectral measures are

$$\mu_n(\lambda) = \frac{|c_n|^2}{\pi} \int_{\mathbb{R}^+} r\, dr r^{2|n|} e^{-r^2} \theta(\lambda - r^2/2)\, (2\pi) = \frac{|c_n|^2}{2^{|n|}} \int_0^{2\lambda} dx x^{|n|} e^{-x}$$

$$(30.2.26)$$

which shows that $|c_n|^2 = 2^{|n|}/(|n|)!$. Thus, since $\mu_n = \mu_{-n}$ we choose coefficients

$$\mu(\lambda) = \frac{1}{2} \sum_{n=0}^{\infty} 2^{-n} \mu_n(\lambda) = \frac{1}{2} \sum_{n=0}^{\infty} \int_0^{2\lambda} dx \frac{(x/2)^n}{n!} e^{-x} = \frac{1}{2} \int_0^{2\lambda} dx e^{-x/2} \quad (30.2.27)$$

The Radon–Nikodym derivatives are

$$\rho_n(\lambda) = d\mu_n(\lambda)/d\mu(\lambda) = \frac{2^{|n|+1}}{|n|!} e^{-\lambda} \lambda^{|n|} \qquad (30.2.28)$$

so that the supports are $S_0 = \mathbb{R}^+$ and $S_n = \mathbb{R}^+ - \{0\}$ for $n \neq 0$ respectively. We have $m(\lambda) = \infty$ for $\lambda > 0$ and $m(0) = 1$. Hence $X_1 = \{0\}$ and $X_\infty = \mathbb{R}^+ - \{0\}$. In particular, if $\psi = (\rho_n(\hat{\ })\Omega_n)$ we have $\psi(0) = \sqrt{2}\rho_0(0)e_0^{(1)}$ and $\psi(\lambda) = \sum_{n \in \mathbb{Z}} \sqrt{\rho_n(\lambda)} \rho_n(\lambda) e_n^{(\infty)}$. The physical Hilbert space therefore is one-dimensional as it should be. In terms of the Hilbert space $\mathcal{H}_{\mathrm{kin}} = L_2(\mathbb{R}^2, d^2 k)$ we see, by formally using the functional calculus for distributions, that the physical Hilbert space corresponds to the distributions $\delta(\hat{\ })\psi$ for suitable ψ and the inner product is $< \delta(\hat{\ })\psi, \delta(\hat{\ })\psi' >_{\mathrm{phys}} = < \psi, \delta(\hat{\ })\psi' >_{\mathrm{kin}}$.

Many more, less trivial examples have been worked out in [254–257]. In all cases one gets exact agreement with independent approaches. There one also sees that one has to refine the procedure in the following sense: recall [253] that any Hilbert space can be uniquely split into the direct sum

$$\mathcal{H} = \mathcal{H}_{\mathrm{pp}} \oplus \mathcal{H}_{\mathrm{ac}} \oplus \mathcal{H}_{\mathrm{cs}} \qquad (30.2.29)$$

This decomposition relies on the Lebesgue decomposition theorem which says that any Borel measure μ on the real line can be decomposed uniquely into the following pieces: $\mu(B) = \mu_{\mathrm{pp}}(B) + \mu_{\mathrm{ac}}(B) + \mu_{\mathrm{cs}}(B)$ for any Borel set B. The measure μ_{pp} is characterised by the fact that it has support on a countable set of points, it is called the pure point measure. The discrete sets are measure zero sets for the measures μ_{ac}, μ_{cs} and one also calls $\mu_c = \mu_{ac} + \mu_{cs}$ the continuous part of μ. Finally, μ_{ac} is absolutely continuous with respect to the Lebesgue measure dx while μ_{cs} is continuous singular with respect to dx. Hence all measures are mutually singular. In practice, μ_{cs} rarely appears. The decomposition (30.2.29) is now characterised by the fact that all vectors $\psi \in \mathcal{H}_*$ have spectral measure $\mu_\psi(B) := \; < \psi, E(B)\psi >$ of type $* \in \{\mathrm{pp}, \mathrm{ac}, \mathrm{cs}\}$. The spaces $\mathcal{H}_*$ are invariant subspaces under the $E * \lambda$.

The refinement of the DID programme now consists in the fact that one *first* performs the split (30.2.29) and *then* performs the direct integral decomposition. The reason for this is as follows: suppose that $\lambda = 0$ is an embedded eigenvalue, that is, $\lim_{\epsilon \to 0} E([-\epsilon, \epsilon]) \neq 0$ and $\lim_{\epsilon \to 0}(E([-\delta, \delta]) - E([-\epsilon, \epsilon])) \neq 0$ for any $\delta > 0$. This means that the spectral measures with respect to the Ω_n will have non-trivial pure point and continuous part. In the decomposition $\mu = \sum_n c_n \mu_n$ it is easy to see that the corresponding Lebesgue decomposition leads to $\mu^* = \sum_n c_n \mu_n^*$ for $* = \mathrm{pp}$, ac (assuming for simplicity that there is no continuous singular part). Now by definition $\rho_n(0) = \lim_{B \to \{0\}} \mu_n(B)/\mu(B) = \mu_n^{\mathrm{pp}}(\{0\})/\mu^{\mathrm{pp}}(\{0\})$ while $\rho_n^{\mathrm{pp}}(0) = \lim_{B \to \{0\}} \mu_n^{\mathrm{pp}}(B)/\mu^{\mathrm{pp}}(B) = \mu_n^{\mathrm{pp}}(\{0\})/\mu^{\mathrm{pp}}(\{0\})$ and $\rho_n^{\mathrm{ac}}(0) = \lim_{B \to \{0\}} \mu_n^{\mathrm{ac}}(B)/\mu^{\mathrm{ac}}(B) = \sigma_n(0)/\sigma(0)$ where $d\mu_n^{\mathrm{ac}}(x) = \sigma_n(x)dx$, $d\mu^{\mathrm{ac}}(x) = \sigma(x)dx$. Hence we see that ρ_n obtained from DID without performing the split (30.2.29) will only capture $\rho_n^{\mathrm{pp}}(0)$ while doing it after the split captures also $\rho_n^{\mathrm{ac}}(0)$. In general physical systems one needs both contributions, an example being a constraint of the form $C = p_1 p_2$ for a particle moving on a cylinder with momenta p_1, p_2. The classical constraint forces the particle to move either along the axis or along the circle. The spectrum of p_1, p_2 is continuous and discrete respectively, hence zero is an embedded eigenvalue of C. If Ω_j is the (generalised) zero eigenvector of p_j on $\mathcal{H}_1 := L_2(\mathbb{R}, dx_1)$ and $\mathcal{H}_2 := L_2(S^1, dx_2)$ respectively then the (generalised) zero eigenvectors on the tensor product $\mathcal{H}_1 \otimes \mathcal{H}_2$ are respectively $\Omega_1 \otimes \psi_2$, $\psi_1 \otimes \Omega_2$ with $\psi_j \in \mathcal{H}_j$. The physical Hilbert space is then the closed linear span of these vectors and thus isomorphic to the direct sum of $\mathcal{H}_2$ and $\mathcal{H}_1$ respectively. Keeping only the pure point part would mean that quantum mechanically the particle is only allowed to move along the axis, which is physically wrong. See [253] for more details.

Finally, let us compare the RAQ and DID programmes:

1. *Uniqueness*

    The RAQ programme depends on a choice $\mathcal{D}_{\mathrm{kin}}$ of dense subspace of $\mathcal{H}_{\mathrm{kin}}$ and different choices can lead to different $\mathcal{H}_{\mathrm{phys}}$ [277,278]. One can show [253] that this introduces more ambiguities than the DID programme does, even when employing the machinery of rigged Hilbert spaces.

2. *Structure functions*

    Only DID is applicable in the presence of structure functions. Also, even if there are only structure constants, DID seems to be more easily applicable in the case that the gauge group is non-compact or infinite-dimensional. While there exist proposals to do RAQ in such situations, it is no longer clear that the proposed physical inner product is positive, see [253] and references therein.

3. *Non-separable Hilbert spaces*

Only RAQ is applicable if $\mathcal{H}_{\text{kin}}$ is non-separable unless $\mathcal{H}_{\text{kin}}$ is a direct sum of $\hat{\,}$-invariant separable subspaces or the function $m(\lambda)$ is still measurable even if it takes values in any Cantor aleph.

4. *Mixed spectrum*

Of course, one can apply RAQ to a given Master Constraint $\hat{\,}$ itself and compare the results. This is the case of a single, Abelian constraint to which RAQ certainly applies. Here the MCP supplies a prescription for how to choose $\mathcal{D}_{\text{kin}}$. Namely, we define the rigging map as

$$\eta_{\text{DID}} : \mathcal{D}_{\text{kin}} \rightarrow \mathcal{H}_{\text{phys}}; \; \psi \mapsto \psi(0) \tag{30.2.30}$$

and $\mathcal{D}_{\text{kin}}$ is defined to be the domain of this map, that is, all those elements of $\mathcal{H}_{\text{kin}}$ for which a square integrable representative $\psi(0)$ of $\mathcal{H}_{\text{phys}}$ exists. Notice that this cannot be all of $\mathcal{H}_{\text{kin}}$ because the $\psi(\lambda)$ and hence their norms are only defined μ-a.e. while we require that $\|\psi(\lambda)\|^2_{\mathcal{H}^{\oplus}(\lambda)}$ is a finite number.

In order that this rigging map coincides with the heuristic RAQ programme for which

$$\eta_{\text{RAQ}}(\psi) := \int_{\mathbb{R}} d\rho(t) < e^{it\hat{\,}} \psi, . >_{\text{kin}} \tag{30.2.31}$$

we must choose a measure $d\rho(t)$ such that

$$\int_{\mathbb{R}} d\rho(t) < \psi, e^{-it\hat{\,}} \psi' >_{\text{kin}}$$

$$= \int_{\mathbb{R}^+} d\mu(\lambda) < \psi(\lambda), \psi'(\lambda) >_{\mathcal{H}^{\oplus}(\lambda)} \left[\int_{\mathbb{R}} d\rho(t) \, e^{-it\lambda} \right]$$

$$=< \psi(0), \psi'(0) >_{\text{phys}} \tag{30.2.32}$$

hence

$$\int_{\mathbb{R}} d\rho(t) \, e^{-it\lambda} = \delta_{\mu}(\lambda, 0) \tag{30.2.33}$$

must be the δ-distribution with respect to μ. In the above example we have $d\mu(\lambda) = e^{-\lambda} d\lambda$ hence it is appropriate to choose $d\rho(t) = dt/\pi$ (twice the δ-distribution measure because the range of λ is positive).

However, one can show that the prescription (30.2.31) does not always lead to the correct result compared with (30.2.30), especially not in the case of embedded eigenvalues [253]. Even in more fortunate cases, detailed knowledge of the spectrum of $\hat{\,}$ is necessary in order to guess the correct measure ρ.

31

Basics of harmonic analysis on compact Lie groups

Due to the importance of spin-network functions for the general theory developed in the main text, we recall here for the convenience of the reader some essential ingredients of the representation theory of compact Lie groups. We follow the exposition in [551].

31.1 Representations and Haar measures

Let us begin with some general notation.

Definition 31.1.1

   (i) *A representation of a group G is a map $\pi : G \to \mathcal{B}(V); g \mapsto \pi(g)$ where $\mathcal{B}(V)$ denotes the bounded linear operators on some Hilbert space V, called the representation space, satisfying*

$$\pi(g_1 g_2) = \pi(g_1)\pi(g_2) \ \forall \ g_1, g_2 \in G \tag{31.1.1}$$

It follows that the operators $\pi(g)$ have an inverse given by $\pi(g^{-1})$ and that $\pi(1_G) = 1_V$. In particular, π is a homomorphism between G and a group of non-singular operators on V.

   (ii) *A representation π is called faithful if it is injective (equivalently: $\pi(g) = 1_V \Rightarrow g = 1_G$) and it is called trivial if $\pi(g) = 1_V$ for all $g \in G$. If $\dim(V) < \infty$ then π is called finite-dimensional.*

   (iii) *Two representations $\pi_I : G \to \mathcal{B}(V_I)$, $I = 1, 2$ are called (unitarily) equivalent iff there exists an invertible (unitary) operator $U : V_1 \to V_2$ such that $\pi_2(.) = U\pi_1(.)U^{-1}$. By $[\pi]$ we denote the equivalence class of the representation π.*

   (iv) *Let V' be the space dual to V, that is, the space of continuous linear functionals on V (since V is a Hilbert space, $V' = V$ by the Riesz lemma). Then the representation π' dual (or contragredient) to π is defined by*

$$[\pi'(g)f](v) := f(\pi(g^{-1})v) \tag{31.1.2}$$

In a Hilbert space V we have $f(.) = <f, .>$ so that $\pi'(g) = [\pi(g^{-1})]^\dagger$ where $\dagger$ denotes the adjoint with respect to $<., .>$.

   (v) *A representation π is called unitary if $\pi(g)$ is a unitary operator on V for all $g \in G$, that is, $[\pi(g)]^\dagger = [\pi(g)]^{-1}$. It follows that $\pi'(g) = [\pi(g^{-1})]^{-1} = \pi(g)$, that is, unitary representations equal their dual representations.*

(vi) A closed subspace $V_1 \subset V$ is called invariant for a representation π iff $\pi(g)V_1 \subset V_1$ for all $g \in G$. A representation is called irreducible if it has no invariant subspaces except for the trivial invariant subspaces $V, \{0\}$, otherwise reducible. Then $\pi_1 := \pi_{|V_1} : G \to \mathcal{B}(V_1)$ is called the restriction of π. The orthogonal complement $V_2 := V_1^\perp$ does not need to be invariant as well but if it is then π can be written as the direct sum of the restricted representations $\pi = \pi_1 \oplus \pi_2$ with $V = V_1 \oplus V_2$.

(vii) A representation π is said to be completely reducible if it decomposes into a direct sum of irreducible representations π_I on the spaces V_I, that is,

$$\pi = \oplus_I \pi_I \tag{31.1.3}$$

where $V = \oplus_I V_I$ and the set of indices I is countable (more generally one has to consider direct integrals but we will not need that for compact groups).

(viii) Let $\pi_I : G \to \mathcal{B}(V_I)$, $I = 1, 2$ be representations. The tensor product $\pi_1 \otimes \pi_2 : G \to \mathcal{B}(V_1 \otimes V_2)$ is defined by

$$[\pi_1 \otimes \pi_2](g) \cdot v_1 \otimes v_2 := (\pi_1(g) \cdot v_1) \otimes (\pi_2(g) \cdot v_2) \tag{31.1.4}$$

(ix) Suppose that for all $g \in G$ the operator $\pi(g)$ is trace class.[1] Then

$$\chi_\pi(g) := \mathrm{Tr}(\pi(g)) \tag{31.1.5}$$

is called the character of π. It actually depends only on $[\pi]$ and the conjugacy class of $g \in G$.

The following lemma is needed for the proof of the Peter and Weyl theorem.

Lemma 31.1.2. *Every finite-dimensional unitary representation is completely reducible.*

Proof: Let π be a unitary representation of G on the finite-dimensional Hilbert space V. If π is not irreducible, choose a non-trivial invariant subspace V_1 and any $v_2 \in V_2 := V_1^\perp$. Since π is unitary we have $\pi^\dagger(g) = \pi^{-1}(g) = \pi(g^{-1})$. Consider any $v_1 \in V_1$. Since $\pi(g^{-1})v_1 \in V_1$ we have

$$0 = <\pi(g^{-1})v_1, v_2> = <v_1, \pi(g)v_2> \quad \forall \; v_1 \in V_1, \; g \in G \tag{31.1.6}$$

---

[1] A bounded operator A on a separable Hilbert space $\mathcal{H}$ is said to be trace class if $|A| = \sqrt{A^\dagger A}$ has finite trace, that is $\mathrm{Tr}(|A|) := \sum_I <b_I, |A|b_I> < \infty$ where e_I is any orthonormal basis of $\mathcal{H}$. Then $\mathrm{Tr}(A)$ is independent of the basis chosen. An operator A is called Hilbert–Schmidt if $A^\dagger A$ is trace class. Trace class operators are dense in a Hilbert space with inner product $<A, B> = \mathrm{Tr}(A^\dagger B)$ and its completion are the Hilbert–Schmidt operators. Hence, every trace class operator is Hilbert–Schmidt but not vice versa. Both are compact operators, that is, they have a pure point spectrum with only eigenvalues of finite multiplicity except, possibly, for the eigenvalue zero and they have finite trace. See Definition 26.6.7 for details.

hence $\pi(g)v_2 \in V_2$ for all $g \in G$. Since v_2 was arbitrary we conclude that V_2 is an invariant subspace for π. Now iterate with V_1 replaced by V_2. The process must come to an end since $\dim(V) < \infty$. $\qquad\square$

Definition 31.1.3. *Let G be a locally compact group. A left (right) Haar measure μ_H^l (μ_H^r) is a positive measure on G invariant under left (right) translations $h \mapsto L_g(h) = gh$ ($R_g(h) = hg$) for all $g \in G$, that is, $\mu_H^l \circ L_g^{-1} = \mu_H^l$ ($\mu_H^r \circ R_g^{-1} = \mu_H^r$) for all $g \in G$.*

Theorem 31.1.4. *Let G be a finite-dimensional Lie group. Then left and right Haar measures exist which are unique up to positive constants. If G is compact then both measures coincide and are unique if fixed to be probability measures. In this case the resulting Haar measure μ_H is also invariant under inversions $h \mapsto I(h) := h^{-1}$.*

Proof: Let G be any Lie group to begin with. If G is not connected, then it has connected components G_n where n can be from any index set. The component of the identity G_0 is the image of the exponential map $\exp : \mathrm{Lie}(G) \to G_0; A \mapsto \exp(A)$ which is a group by itself. By definition of a connected component, if G_0 is the component of the identity in G then any element of G_n is of the form $g_n h$ for some $h \in G_0$ and for an arbitrary choice $g_n \in G_n$. It follows that G_n is invariant under right translations from G_0.

Suppose that $\mu_{H,0}^l$ is a left-invariant Borel measure on G_0. Then any $f \in C(G)$ is measurable and we can define the integral

$$\mu_H^l(f) := \int_G d\mu_H^l(g)f(g) := \sum_n \int_{G_n} \mu_{H,n}^l(g)f(g) := \sum_n \int_{G_0} d\mu_{H,0}^l(h)f(g_n h)$$

(31.1.7)

To see that (31.1.7) is independent of the choice of g_n, consider any other choice $g_n' \in G_n$. Then $f(g_n' h) = f(g_n[g_n^{-1}g_n']h)$. But since $L_{g_n}, L_{g_n'}$ map G_0 onto G_n, it follows that $g_n^{-1}g_n' \in G_0$. Choice independence of (31.1.7) thus follows from left invariance of $\mu_{H,0}^l$.

We claim that the measure μ_H^l defined in (31.1.7) is left-invariant. Let $g_0 = g_m h_0 \in G_m$ where $h_0 \in G_0$. Then by definition

$$[(L_{g_0})_*\mu_H^l](f) = \mu_H^l(L_{g_0}^* f) = \int_G d\mu_H^l(g)f(g_0 g) = \sum_n \int_{G_0} d\mu_{H,0}^l(h)f(g_m h_0 g_n h)$$

$$= \sum_n \int_{G_0} d\mu_{H,0}^l(h)f(g_m[h_0 g_n h_0^{-1}]h)$$

(31.1.8)

where in the last step we have used left invariance of $\mu_{H,0}^l$. Let $g_n' = h_0 g_n h_0^{-1}$. Now the g_n' belong to mutually different $G_{n'}$ for suppose that was not the case then we find $n_1 \neq n_2$ and some $h_1 \in G_0$ such that $g_{n_2}' = g_{n_1}' h_1$, that is, $g_{n_2} h_0 = g_{n_1} h_0^{-1} h_1$ which is a contradiction because the left-hand side belongs to G_{n_2} while the right-hand side belongs to G_{n_1}. By the same argument, also the

$\tilde{g}_{n'(m,n)} := g_m g'_n$ belong to mutually different components. In other words, the map $n \mapsto n'(m, n)$ is a bijection. Hence

$$[(L_{g_0})_* \mu_H^l](f) = \sum_n \int_{G_0} d\mu_{H,0}^l(h) f(\tilde{g}_{n'(m,n)} h) = \sum_n \int_{G_0} d\mu_{H,0}^l(h) f(\tilde{g}_n h)$$

$$= \sum_n \int_{G_0} d\mu_{H,0}^l(h) f(g_n h) = \mu_H^l(f) \tag{31.1.9}$$

by choice independence. Conversely, suppose that μ_H^l is a left-invariant measure on G then using the disjoint union $G = \cup_n G_n$ we have by left invariance

$$\mu_H^l(f) = \sum_n \int_{G_n} d\mu_H^l(g) f(g) = \sum_n \int_{G_0} d\mu_H^l(h) f(g_n h) \tag{31.1.10}$$

so that a left-invariance measure on G is necessarily of the form (31.1.7).

These considerations reveal that a left-invariant measure is known once we know its restriction to G_0. Let us construct the latter. The component G_0 is in bijection, via the exponential map, with some subset D of $\mathbb{R}^N$ where $N = \dim(G)$. Thus there exists an explicit parametrisation $h(x) = \exp(x^j \tau_j)$ where $x \in D$ and $x = 0$ corresponds to the identity 1_G. Since G_0 is a closed subgroup of G it follows that there is a composition map $c : D \times D \to D$; $(x, y) \mapsto c(x, y)$ uniquely defined by $h(x)h(y) = h(c(x, y))$. Changing coordinates from $G_0 = \exp(D)$ to D we have

$$\int_{G_0} d\mu_{H,0}^l(h) f(h) = \int_D d^N y J(y) f(h(y)) \tag{31.1.11}$$

where the Jacobean $J(y) = d\mu_H^l(g(y))/d^N y$ is just the Radon–Nikodym derivative. Let now $h_0 - h(x)$. The left invariance condition reads

$$\int_{G_0} d\mu_{H,0}^l(h) f(h) = \int_D d^N y J(y) f(h(y)) = \int_{G_0} d\mu_{H,0}^l(h) f(h_0 h)$$

$$= \int_D d^N y J(y) f(h(c(x, y))) \tag{31.1.12}$$

Relabelling $y \to y'$ and introducing the new integration variable y defined by (notice that $c(x, .)$ is a bijection of D) $y'(y) = c(x, y)$ for the left-hand side of (31.1.12) we find

$$\int_D d^N y J(y'(y)) |\det((\partial y'(y)/\partial y))| f(h(y'(y))) = \int_D d^N y J(y) f(h(y'(y))) \tag{31.1.13}$$

Since f is arbitrary this implies

$$J(y) = J(c(x, y)) |\det((\partial c(x, y)/\partial y))| \tag{31.1.14}$$

for any $x \in D$. Evaluating (31.1.14) at $y = 0$ we find

$$J(x) = \frac{K}{|\det((\partial c(x, y)/\partial y))|_{y=0}} \tag{31.1.15}$$

where $K = J(0)$ is some positive constant. To see that (31.1.15) indeed solves (31.1.14) for all $y \in D$ we insert (31.1.15) into (31.1.14) and find the condition

$$|\det((\partial c(c(x,y),z)/\partial z))|_{z=0} = |\det((\partial c(y,z)/\partial z))|_{z=0}|\det((\partial c(x,y)/\partial y))|$$
(31.1.16)

Noticing that

$$|\det((\partial c(x,y)/\partial y))| = [|\det((\partial c(x,u)/\partial))|_{u=c(y,z)}]_{z=0}$$
(31.1.17)

(31.1.16) is an identity following from associativity of group multiplication $[h(x)h(y)]h(z) = h(x)[h(y)h(z)]$ which translates into $c(c(x,y),z) = c(x,c(y,z))$ and when differentiated at $z = 0$ gives precisely (31.1.16).

Concluding, we see that a left-invariant measure exists on every finite-dimensional Lie group and that measure is unique up to a positive constant factor. By the same reasoning we also see that every finite-dimensional Lie group admits a right-invariant measure which is unique up to a positive constant.

We now show that the left- and right-invariant measures so constructed coincide if G is compact and if we fix the constant K in (31.1.15) in such a way that $\int_G \mu_H^l = \int_G \mu_H^r = 1$, which amounts to $\int_{G_0} d\mu_{H,0}^l = \int_{G_0} d\mu_{H,0}^r = 1/Z$ where $Z < \infty$ is the number of connected components (here we have used compactness which in this case means closed, bounded and without boundary [left or right translations are transitive, so there cannot be any boundary]). Consider for any $h_0 \in G_0$ the measure μ on G_0 defined by

$$\int_{G_0} d\mu(h)f(h) := \int_{G_0} d\mu_H^r(h)f(h_0 h h_0^{-1})$$
(31.1.18)

By right invariance of μ_H^r we have for any $h_1 \in G_0$

$$\int_{G_0} d\mu(h)f(hh_1) = \int_{G_0} d\mu(h)[R_{h_1}f](h) = \int_{G_0} d\mu_H^r(h)[R_{h_1}f](h_0 h h_0^{-1})$$

$$= \int_{G_0} d\mu_H^r(h)f(h_0 h h_0^{-1}h_1) = \int_{G_0} d\mu_H^r(h)f(h_0 h)$$

$$= \int_{G_0} d\mu_H^r(h)f(h_0 h h_0^{-1}) = \int_{G_0} d\mu(h)f(h)$$
(31.1.19)

for any h_1. Thus, μ is a right-invariant whence $\mu = K_r(h_0)\mu_{H,0}^r$ for some $K_r(h_0) > 0$ by the uniqueness statement shown above. We can in fact compute the constant by using the fact that for compact groups the constants are integrable: setting $f = 1$ we immediately find from (31.1.18) that $K_r(h_0)/Z = 1/Z$ so that $K_r(h_0) = 1$.

Inserting this result back into (31.1.18) we get

$$\int_{G_0} d\mu_{H,0}^r(h)f(h) = \int_{G_0} d\mu_H^r(h)f(h_0 h h_0^{-1}) = \int_{G_0} d\mu_H^r(h)f(h_0 h)$$
(31.1.20)

for any $h_0 \in G_0$ where we have used right invariance again. It follows that $\mu_{H,0}^r$ is left-invariant whence $\mu_{H,0}^r = K\mu_{H,0}^l$ due to uniqueness again and since both measures are normalised to the same constant we get $\mu^r_{H,0} = \mu^l_{H,0} = \mu_{H,0}$. The fact that constants are integrable (compactness) was essential in this result and equality of left- and right-invariant measures does not hold in general for non-compact groups.

Finally, the measure ν on G_0 defined by

$$\int_{G_0} d\nu(h) f(h) := \int_{G_0} d\mu_{H,0}(h) f(h^{-1}) \qquad (31.1.21)$$

is also bi-invariant and normalised, thus by uniqueness $\nu = \mu_{H,0}$ and $\mu_{H,0}$ is also inversion-invariant. $\qquad \square$

We remark that given local coordinates x^i on G (angles) and a parametrisation $D \to G_0$; $x \mapsto g(x)$ we have up to normalisation the following explicit formula for the Haar measure

$$d\mu_H(g(x)) = \sqrt{\det(q)}(x) \, d^N x, \quad q_{ij}(x) := \mathrm{Tr}\left(\frac{\partial g(x)}{\partial x^i}(g(x))^{-1} \frac{\partial g(x)}{\partial x^j}(g(x))^{-1} \right)$$

$$(31.1.22)$$

This can be verified by calculating $g_0 g = g(x_0)g(x) = g(c(x_0, x))$ which defines $y(x) := c(x_0, x)$. Then by changing coordinates

$$\int_{G_0} d\mu_H(g) \, f(g_0 g) = \int_D d^N y \, f(g(y)) \, |\det(\partial x(y)/\partial y)| \, \sqrt{\det(q)}(x(y))$$

$$= \int_D d^N y \, f(g(y)) \, \sqrt{\det(x^* q)}(y) \qquad (31.1.23)$$

But $g(y(x)) = g_0^{-1} g(y)$ and

$$(x^* q)_{ij}(y) = \mathrm{Tr}\left(\frac{\partial g(x(y))}{\partial y^i}(g(x(y)))^{-1} \frac{\partial g(x(y))}{\partial y^j}(g(y))^{-1} \right)$$

$$= \mathrm{Tr}\left(\left[g_0^{-1} \frac{\partial g(y)}{\partial y^i} \right] [g_0^{-1} g(y)]^{-1} \left[g_0^{-1} \frac{\partial g(y)}{\partial y^j} \right] [g_0^{-1} g(y)]^{-1} \right)$$

$$= q_{ij}(y) \qquad (31.1.24)$$

Corollary 31.1.5. *If G is a compact, finite-dimensional Lie group and $\pi : G \to \mathcal{B}(V)$ a continuous representation on some linear space V then V can always be equipped with an inner product such that π is a unitary representation.*

Proof: If $< ., . >$ denotes the inner product on V and μ_H the unique Haar measure on G, consider the new inner product

$$< u, v >' := \int_G d\mu_H(g) < \pi(g)u, \pi(g)v > \qquad (31.1.25)$$

which is well-defined because G is compact (hence the integrand is uniformly bounded due to its continuity in the topology of G). The statement now easily follows from the left invariance of the measure. $\qquad \square$

By combining Lemma 31.1.2 and Corollary 31.1.5 we infer that the finite-dimensional representations of compact Lie groups are completely reducible and that one can consider them as unitary without loss of generality. In other words, every irreducible representation of a compact Lie group can be considered as an irreducible subgroup of some $U(N)$ with N sufficiently large.

31.2 The Peter and Weyl theorem

Definition 31.2.1. *Let $H \subset G$ be a subgroup and $\pi_H : H \to \mathcal{B}(V_H)$ a representation of H. Consider the Hilbert space V_G of functions $f : G \to V_H$ satisfying $(\pi_H(h)f)(g) = f(gh)$ for all $h \in H$ which are normalisable with respect to the scalar product*

$$< f, f' >_G := \int_G d\mu_H^l(g) < f(g), f'(g) >_H \tag{31.2.1}$$

where μ_H^l is a left-invariant measure on G. Consider the representation $\pi_G : G \to \mathcal{B}(V_G)$ defined by

$$(\pi_G(g)f)(g') := f(g^{-1}g') \tag{31.2.2}$$

Operation (31.2.2) maps normalisable functions to normalisable functions due to left invariance of the measure and

$$[\pi_G(g)f](g'h) = f(g^{-1}g'h) = [\pi_H(h)f](g^{-1}g') = [\pi_G(g)\pi_H(h)f](g')$$
$$= [\pi_H(h)(\pi_G(g)f)](g') \tag{31.2.3}$$

since left and right translations commute. Therefore π_G is a representation of G called the representation induced by the representation π_H of H. In particular, if $H = \{e\}$ and π_H is the trivial representation, then π_G is called the left regular representation.

One can show that the representation space of infinite-dimensional unitary representations of compact Lie groups decomposes into a direct sum of finite-dimensional invariant subspaces in which irreducible representations of G are induced.

Lemma 31.2.2 (Schur). *Suppose that $\pi_I : G \to \mathcal{B}(V_I)$ are finite-dimensional irreducible representations of G and that there exists an intertwiner $A : V_1 \to V_2$ such that $\pi_2(.)A = A\pi_1(.)$. Then either (1) $A = 0$ or (2) A is invertible and π_1, π_2 are equivalent. In case (2) the operator A is determined up to a multiplicative factor.*

Proof: Consider the subspaces $\mathrm{Ker}(A) \subset V_1$, $\mathrm{Im}(A) \subset V_2$. We have for $x \in \mathrm{Ker}(A)$, $y = Ax \in \mathrm{Im}(A)$

$$A\pi_1(g)x = \pi_2(g)Ax = 0 \text{ and } \pi_2(g)y = \pi_2(g)Ax = A\pi_1(g)x \in \mathrm{Im}(A) \tag{31.2.4}$$

whence $\mathrm{Ker}(A), \mathrm{Im}(A)$ are invariant for π_1, π_2 respectively. Since π_1, π_2 are irreducible, we must have $\mathrm{Ker}(A) = \{0\}, V_1$ and $\mathrm{Im}(A) = \{0\}, V_2$.

1. If $\mathrm{Ker}(A) = V_1$ then $A = 0$ which gives possibility (1).
2. If $\mathrm{Ker}(A) = \{0\}$ then $\mathrm{Im}(A) \neq \{0\}$, therefore $\mathrm{Im}(A) = V_2$ so that A is invertible and π_1, π_2 are equivalent. This is possibility (2).

For case (2) consider any other intertwiner B. Then for $z \in \mathbb{C}$ also $A - zB$ is an intertwiner and we may choose z in such a way that $A - zB = B[B^{-1}A - z1_{V_1}]$ is singular (simply choose z to be an eigenvalue of the operator $B^{-1}A$). But then $A - zB = 0$ by (1). In particular, if $\pi_1 = \pi_2$ we see that $A = 0$ or $A = \lambda 1_{V_1}$. $\square$

We now come to the most important theorem in this subject.

Theorem 31.2.3 (Peter and Weyl). *Fix once and for all a representative π_j from each equivalence class $j \in J$ of finite-dimensional, unitary, irreducible representations of a compact Lie group G on representation spaces V_j. Let d_j be the dimension of π_j and*

$$g \mapsto b_{jmn}(g) := \sqrt{d_j}[\pi_j(g)]_{mn}; \quad m, n = 1, \ldots, d_j \qquad (31.2.5)$$

multiples of the matrix element functions. Consider the Hilbert space $\mathcal{H} := L_2(\mathrm{G}, d\mu_H)$ where μ_H is the unique Haar measure on G. Then the system of functions b_{jmn} is a complete orthonormal basis for $\mathcal{H}$.

Proof

1. *Orthonormal system*

Let $E^{jj'}_{n_0 n'_0}$ be a $d_j \times d_{j'}$ matrix with entries $(E^{jj'}_{n_0 n'_0})_{nn'} = \delta_{nn_0}\delta_{n'n'_0}$. Consider the matrix

$$A^{jj'}_{n_0 n'_0} := \int_{\mathrm{G}} d\mu_H(g) b_j(g) \cdot E^{jj'}_{n_0 n'_0} \cdot b_{j'}(g^{-1}) \qquad (31.2.6)$$

Using the bi-invariance of the Haar measure it is easy to see that $A_{jj'}$ is an intertwiner between representations $\pi_j, \pi_{j'}$. If $j \neq j'$ then the irreducible representations are inequivalent and therefore $A^{jj'}_{n_0 m'_0} = 0$ by Schur's lemma. Since the representations are unitary, the representation matrices are unitary matrices, that is, $\pi_j(g^{-1}) = [\pi(g)]^{-1} = [\pi(g)]^{\dagger} = \overline{[\pi(g)]^T}$ where $(.)^T$ denotes the transpose. Hence for $j \neq j'$

$$0 = \left(A^{jj'}_{n_0 n'_0}\right)_{m_0 m'_0} := \int_{\mathrm{G}} d\mu_H(g) b_{jm_0 n_0}(g)\overline{b_{j'm'_0 n'_0}(g)} = <b_{j'm'_0 n'_0}, b_{jmn}> = 0$$

$$(31.2.7)$$

If $j = j'$ then again by Schur's lemma, $A^{jj}_{n_0 n'_0} = \lambda^j_{n_0 n'_0} 1_{V_j}$ for some $\lambda^j_{n_0 n'_0} \in \mathbb{C}$. To compute $\lambda^j_{n_0 n'_0}$ we take the trace of (31.2.6) which reveals $\lambda^j_{n_0 n'_0} = \delta_{n_0 n'_0}$. Summarising

$$<b_{jmn}, b_{jm'n'}> = \delta_{jj'}\delta_{mm'}\delta_{nn'} \qquad (31.2.8)$$

2. Completeness

Consider the subalgebra $\mathcal{B} \subset C(\mathrm{G})$ of the Abelian C^*-algebra of continuous functions on G (with sup-norm) defined by finite linear combinations of the b_{jmn}. To see that this is indeed an algebra we notice that the product of such functions can be regarded as matrix elements of a tensor product of two representations which is again finite-dimensional and therefore completely reducible by Corollary 31.1.5. This function algebra contains the identity (through the trivial representation) and it separates the points of G (already the fundamental representation does). By the Stone–Weierstrass theorem therefore $\mathcal{B}$ is dense in $C(\mathrm{G})$ because G is a compact Hausdorff group. Now $C(\mathrm{G})$ is dense in $L_2(\mathrm{G}, d\mu_H)$ because $L_2(\mathrm{G}, d\mu_H)$ is the GNS Hilbert space generated from the positive linear functional (state) on the C^*-algebra $C(\mathrm{G})$ defined by $\omega(f) := \int_{\mathrm{G}} d\mu_H(g) f(g)$. To see this just choose $\pi_\omega(f) = f$, $\Omega_\omega = 1$, $\mathcal{H}_\omega = \mathcal{H}$. It follows that $\mathcal{B}$ is dense in $L_2(\mathrm{G}, d\mu_H)$.

(*Alternatively*: It is a general result that every Borel-measurable function on locally compact Hausdorff spaces can be approximated a.e. by continuous functions, see Lusin's theorem in Chapter 25. But for a probability measure we have $L_2 \subset L_1$ by the Schwarz inequality $\|f\|_{L_1}^2 = |<f>|^2 = |<1, f>|^2 \leq\leq 1, 1><f, f> = \|f\|_{L_2}^2$. Since $C(\mathrm{G})$ is dense in $L_1(\mathrm{G})$ it follows that it is also dense in $L_2(\mathrm{G})$.) $\qquad\qquad\square$

32

Spin-network functions for SU(2)

The name spin-network function actually derives from the case $G = SU(2)$. For this case we can construct the intertwiners rather explicitly using simple angular momentum quantum mechanics (see [288, 289] for more details and derivations).

32.1 Basics of the representation theory of SU(2)

We denote the edges of a graph by e as usual. The irreducible representations of $SU(2)$ are labelled by half-integral spin quantum numbers $j_e = \frac{1}{2}, 1, \frac{3}{2}, \ldots$. We introduce magnetic quantum numbers $m_e, n_e \in \{-j_e, -j_e + 1, \ldots, j_e\}$ and label the matrix elements of the $(2j_e + 1)$-dimensional representation by $[\pi_{j_e}(h)]_{m_e n_e}$. These matrix elements can be expressed explicitly in terms of the fundamental two-dimensional representation by

$$[\pi_j(h)]_{mn} = c_{jm} c_{jn} h_{(A_1 B_1} \cdots h_{A_{2j}) B_{2j}}, \quad c_{jm} := \sqrt{\binom{2j}{j+m}} \quad (32.1.1)$$

where the round bracket denotes total symmetrisation corresponding to the fact that the representation space of the spin j representation are the totally symmetric spinors of rank $2j$ and the labels $A_k, B_k = \pm\frac{1}{2}$ are arbitrary subject only to the constraints $A_1 + \cdots + A_{2j} = m$, $B_1 + \cdots + B_{2j} = n$. To see this, notice that a spinor $\psi_{A_1 \ldots A_{2j}}$ of rank $2j$ transforms in the $(2j)$-fold tensor product of the fundamental representation. Now we can write

$$\psi_{A_1 \ldots A_{2j}} = \psi_{(A_1 A_2) A_3 \ldots A_{2j}} + \psi_{[A_1 A_2] A_3 \ldots A_{2j}}$$
$$= \psi_{(A_1 A_2) A_3 \ldots A_{2j}} + \frac{1}{2} \epsilon_{A_1 A_2} \epsilon^{B_1 B_2} \psi_{B_1 B_2 A_3 \ldots A_{2j}} \quad (32.1.2)$$

The totally antisymmetric 2-spinor ϵ_{AB} is $SU(2)$-invariant. Iterating (32.1.2) we see that we can decompose any spinor of rank $2j$ into totally symmetric spinors of rank equal to and lower than $2j$ times powers of invariant 2-spinors. The 2-spinors transform in the trivial representation as we just saw, hence we just proved by elementary means that any tensor product of fundamental representations is completely reducible, the irreducible subspaces corresponding to totally symmetric spinors (further hypothetical invariant subspaces would require additional antisymmetrisations which would vanish on totally symmetric spinors. See, e.g., [898] for the required theory of Young tableaux). Now the components of a totally symmetric spinor $\psi_{A_1...A_{2j}} = \psi_{(A_1...A_{2j})}$ of rank $2j$ are obviously completely characterised by how many, say l, of the A_k take the value $+1/2$. There are $2j+1$ possibilities and we have for the magnetic quantum number $m := A_1 + \cdots + A_{2j} = l/2 - (2j-l)/2 = l - j$, $l = 0,\ldots,2j$ or $l = j+m$. Hence a symmetric spinor is also completely characterised by its magnetic quantum number. The combinatorial factor c_{jm} in (32.1.2) is to ensure the representation property

$$\sum_k [\pi_j(h)]_{mk}[\pi_j(g)]_{kn}$$

$$= c_{jm}c_{jn} \sum_{k=-j}^{j} c_{jk}^2 \, [h_{(A_1C_1} \cdots h_{A_{2j})C_{2j}} \, g_{C_1(B_1} \cdots g_{C_{2j}B_{2j})}]|_{C_1+...+C_{2j}=k}$$

$$= c_{jm}c_{jn} \sum_{C_1,...,C_{2j}=\pm 1/2} h_{(A_1C_1} \cdots h_{A_{2j})C_{2j}} \, g_{C_1(B_1} \cdots g_{C_{2j}B_{2j})}$$

$$= c_{jm}c_{jn}(hg)_{(A_1B_1} \cdots (hg)_{A_{2j})B_{2j}} = [\pi_j(hg)]_{mn} \qquad (32.1.3)$$

where $A_1 + \cdots + A_{2j} = m$, $B_1 + \cdots + B_{2j} = n$.

In order to work out the expression for $[\pi_j(h)]_{mn}$ in terms of the complex numbers a,b,c,d in $h = \begin{pmatrix} a & b \\ c & d \end{pmatrix}$ with $h_{1/2,1/2} = a$ we consider the special symmetric spinor of rank $2j$ given by $\psi_m := c_{jm}u_{A_1} \cdots u_{A_{2j}}$, $A_1 + \cdots + A_{2j} = m$ where u is a spinor in the $j = 1/2$ representation. Then

$$\sum_n [\pi_j(h)]_{mn}\psi_n$$

$$= c_{jm} \sum_{B_1,...,B_{2j}=\pm 1/2} h_{A_1B_1} \cdots h_{A_{2j}B_{2j}} u_{B_1} \cdots u_{B_{2j}}$$

$$= c_{jm}(hu)_{A_1} \cdots (hu)_{A_{2j}} = c_{jm}[(hu)_{1/2}]^{j+m}[(hu)_{-1/2}]^{j-m}$$

$$= c_{jm}[au_{1/2} + bu_{-1/2}]^{j+m}[cu_{1/2} + du_{-1/2}]^{j-m}$$

$$= c_{jm} \sum_{k=0}^{j+m}\sum_{l=0}^{j-m} \binom{j+m}{k}\binom{j-m}{l} a^k b^{j+m-k} c^l d^{j-m-l} u_{1/2}^{k+l} u_{-1/2}^{2j-(k+l)}$$

$$= \sum_{n=-j}^{j} \psi_n \sum_l \binom{j+m}{n+j-l}\binom{j-m}{l} \frac{c_{jm}}{c_{jn}} a^{n+j-l} b^{m-n+l} c^l d^{j-m-l}$$

$$(32.1.4)$$

from which we read off

$$[\pi_j(h)]_{mn} = \sum_l \frac{\sqrt{(j+m)!\,(j-m)!\,(j+n)!\,(j-n)!}}{(j+n-l)!\,(m-n+l)!\,(j-m-l)!\,l!} a^{j+n-l}\, b^{m-n+l}\, c^l\, d^{j-m-l}$$

$$(32.1.5)$$

where the sum over l is over all non-negative integers such that all factorials have non-negative arguments. Formula (32.1.5) is also valid for $G = SL(2, \mathbb{C})$ (recall that all the spinor fields transform in tensor products of the defining representation and its complex conjugate).

32.2 Spin-network functions and recoupling theory

Let us now turn to spin-network functions. In order to make the analogy with angular momentum quantum mechanics clearer, let us introduce the states

$$< h|jm >_{m'} := \sqrt{2j+1}[\pi_j(h)]_{mm'} \tag{32.2.1}$$

which provide an orthonormal basis in the Hilbert space $L_2(SU(2), d\mu_H)$ as shown explicitly in Chapter 31. On the other hand, consider the usual angular momentum eigenstates $|jm>$. In terms of the usual angular momentum operators J_k subject to the algebra $[J_k, J_l] = i\epsilon_{klm}J_m$ we have the eigenstate relations $\sum_{k=1,2,3} J_k^2|jm> = j(j+1)|jm>$, $J_3|jm> = m|jm>$ and the ladder operator equations $J_3 J_\pm|jm> = (m \pm 1)J_\pm|jm>$ where $J_\pm = J_1 \pm iJ_2$. From these relations one finds as usually the normalisation $J_\pm|jm> = \sqrt{j(j+1) - m(m \pm 1)}|j, m \pm 1>$. Now since we think of h as the holonomy of a connection along some path e we find that under gauge transformations at $v = b(e)$ we have

$$< h|U(g)|jm >_{m'} = \sqrt{2j+1}[\pi_j(g(v)h)]_{mm'} = [\pi_j(g(v))]_{mn} < h|jn >_{m'} \tag{32.2.2}$$

The right-invariant vector fields are defined by

$$< h|R_k|jm >_{m'} := \left[\frac{d}{dt}\right]_{t=0} < \exp(t\tau_k)h|jm >_{m'}$$

$$\Rightarrow R_k|jm >_{m'} = \left[\frac{d}{dt}\right]_{t=0} U(\exp(t\tau_k)) \tag{32.2.3}$$

where $i\tau_k = \sigma_k$ are the Pauli matrices. It is easy to check that in terms of $Y_k := -iR_k/2$ we have $[Y_k, Y_l] = i\epsilon_{klm}Y_m$, hence these vector fields satisfy the same algebra as the J_k so that for fixed m' there must be a unitary transformation between the Hilbert space $\mathcal{H}_{m'}^j$ spanned by the $|jm>_{m'}$ and the Hilbert space $\mathcal{H}^j$ spanned by the $|jm>$. In order to determine this transformation we notice that since the states $|jm>$ are basis states while the ψ_m are components of spinor states, the basis states transform in the transpose of (32.1.4), that is, with $\psi = \sum_m \psi_m|jm>$ we have $\psi' = \sum_{m,n}[\pi_j(h)]_{mn}\psi_n|jm> = \sum_{m,n}[\pi_j(h)]_{nm}\psi_m|jn>$ so that we get

$$V(h)|jm> = [\pi_j(h)]_{nm}|jn> \tag{32.2.4}$$

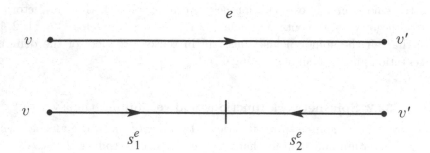

Figure 32.1 Breaking an edge into two pieces for the purpose of constructing the intertwiner.

which should be compared with (32.2.2). One can now explicitly verify that

$$J_k|jm> = \frac{i}{2}\left[\frac{d}{dt}\right]_{t=0} V(\exp(t\tau_k))|jm> \qquad (32.2.5)$$

The unitary transformation $W^j_{m'} : \mathcal{H}^j \to \mathcal{H}^j_{m'}$ which accomplishes $WJ_kW^{-1} = Y_k$ can be derived to be given by

$$W^j_{m'}|jm> := \sum_n [\pi_j(\epsilon)]_{mn}\,|jn>_{m'} \qquad (32.2.6)$$

where $\epsilon = -\tau_2$ (use $\epsilon^T = \epsilon^{-1} = -\epsilon$, $\epsilon g^T \epsilon^{-1} = g^{-1}$, valid for all $g \in \mathrm{SL}(2,\mathbb{C})$ to see this as well as $\tau_K^{-1} = -\tau_k = \overline{\tau_k}^T$; notice that $\tau_k \in \mathrm{SL}(2,\mathbb{C})$).

The purpose of these derivations is that now we may use standard recoupling theory for abstract spin systems with N degrees of freedom where N is the valence of the vertex v in question. Recall that a vertex of a spin-network state is an intersection of at least two edges. If the valence of the vertex is two, then it is either a non-differentiable intersection or, if it is an at least $C^{(1)}$ intersection, then the intertwiner must be non-trivial. Given a graph γ, let us denote by $V(\gamma)$ only two-valent non-differentiable vertices and higher-valent vertices. Let $E(\gamma)$ be the set of edges of γ. We split each edge $e \in E(\gamma)$ into two halves $e = s^e_1 \circ (s^e_2)^{-1}$ where s^e_1, s^e_2 are outgoing from $b(e)$, $f(e)$ respectively. The arbitrary intersection point $s^e_1 \circ s^e_2$, which is an interior point of e, is at least $C^{(1)}$ and the segments s^e_1, s^e_2 are ingoing here (see Figure 32.1). Denote the graph defined by the collection of segments s^e_1, s^e_2, $e \in E(\gamma)$ by γ'. By $E(\gamma')$ we mean the collection of these segments and by $V(\gamma')$ we mean the union of $V(\gamma)$ with the set of the interior points just introduced. We start from a gauge-variant spin-network state

$$T_{\gamma',\vec{j},\vec{m},\vec{n}}(A) := \prod_{s\in E(\gamma')} <A(s)|j_s m_s>_{n_s} = \prod_{v\in V(\gamma)} \prod_{s\in E(\gamma');b(s)=v} <A(s)|j_s m_s>_{n_s}$$

$$(32.2.7)$$

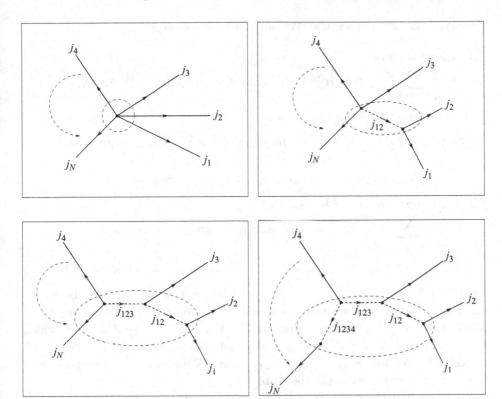

Figure 32.2 Graphical visualisation of a recoupling scheme: by virtually blowing up a vertex to a neighbourhood one can think of a recoupling quantum number as a virtual edge. The collection of these quantum numbers defines the intertwiner.

where it is understood that $j_{s_1^e} - j_{s_2^e} = j_e$. Fix a vertex v and choose some labelling $s_1^v, \ldots, s_{N_v}^v$ of the edges adjacent to the N_v-valent vertex v. We introduce the following recoupling scheme (see also Figure 32.2)

$$(j_1, j_2) \to j_{12}, (j_{12}, j_3) \to j_{123}, \ldots (j_{12\ldots k}, j_{k+1}) \to j_{12\ldots k+1}, \ldots (j_{12\ldots N-1}, j_N) \to J$$

$$(32.2.8)$$

where $j_k = j_{s_k^v}$, $N = N_v$. Here $j_{12} \in \{|j_1 - j_2|, |j_1 - j_2| + 1, \ldots, j_1 + j_2\}$, etc. take values in the possible irreducible representations into which the tensor product $j_1 \otimes j_2$ can be decomposed.[1] The $j_{12\ldots k}, J$ are called recoupling momenta and

---

[1] This is known as the Clebsch–Gordan theorem. The proof is simple: take two completely symmetric spinors of rank $2j_1$ and $2j_2$ respectively and apply the symmetrising and antisymmetrising process to $\psi^1_{A_1 \ldots A_{2j_1}} \psi^2_{B_1 \ldots B_{2j_2}}$. It is easy to see that we get at least one completely symmetric spinor of rank $2k$ where $k = |j_1 - j_2|, \ldots, j_1 + j_2$. To see that we get precisely one contribution for each such k, verify that the dimension $(2j_1 + 1)(2j_2 + 1)$ of the tensor product coincides with the dimension $\sum_{k=|j_1-j_2|}^{j_1+j_2}(2k + 1)$ of the decomposition.

they label the intertwiner at v. We now define successively

$$|j_1 j_2; j_{12} m_{12} >_{n_1, n_2}$$

$$:= \sum_{m_1 + m_2 = m_{12}} < j_1 j_2; j_{12} 2 m_{12} | j_1 m_1, j_2 m_2 > |j_1 m_1 >_{n_1} \otimes |j_2 m_2 >_{n_2}$$

$$|j_1 \cdots j_k; j_{12} \cdots j_{1 \ldots k} m_{1 \ldots k} >_{n_1, \ldots, n_k}$$

$$:= \sum_{m_{1 \ldots k-1} + m_k = m_{1 \ldots k}} < j_{1 \ldots k-1} j_k; j_{1 \ldots k} m_{1 \ldots k} | j_{1 \ldots k-1} m_{1 \ldots k-1}, j_k m_k >$$

$$\times |j_1 \cdots j_{k-1}; j_{12} \cdots j_{1 \ldots k-1} m_{1 \ldots k-1} >_{n_1, \ldots, n_{k-1}} \otimes |j_k m_k >_{n_k}$$

$$|j_1 \cdots j_N; j_{12} \cdots j_{1 \ldots N-1} J M >_{n_1, \ldots, n_N}$$

$$:= \sum_{m_{1 \ldots N-1} + m_N = M} < j_{1 \ldots N-1} j_N; J M | j_{1 \ldots N-1} m_{1 \ldots N-1}, j_N m_N >$$

$$\times |j_1 \cdots j_{N-1}; j_{12} \cdots j_{1 \ldots N-1} m_{1 \ldots N-1} >_{n_1, \ldots, n_{N-1}} \otimes |j_N m_N >_{n_N} \quad (32.2.9)$$

Here $< j_1 j_2; J M | j_1 m_1, j_2 m_2 >$ are the Clebsch–Gordan coefficients familiar from quantum mechanics that relate the basis $|j_1 m_1 > \otimes |j_2 m_2 >$ in which $(J_1^k)^2, (J_2^k)^2, J_1^3, J_2^3$ are diagonal to the basis $|j_1, j_2; J M >$ in which $(J_1^k)^2, (J_2^k)^2, (J_1^k + J_2^k)^2, J_1^3 + J_2^3$ are diagonal. We see that the intertwiner I^v is explicitly labelled by $j_{12}^v, \ldots, j_{1 \ldots N-1}^v, J^v, M^v$. We apply the above procedure to every $v \in V(\gamma)$ resulting in the state $T_{\gamma', \vec{j}, \vec{I}, \vec{n}}$.

We claim that the state $|j_1 \cdots j_N; j_{12} \cdots j_{1 \ldots N-1} J M >_{n_1 \ldots n_N} >$ transforms in the representation corresponding to total angular momentum J at v. To see this it is sufficient to calculate

$$U(g) |j_1 j_2; j_{12} m_{12} >_{n_1 n_2}$$

$$= \sum_{m_1, m_2, m'1, m'2} < j_1 j_2; j_{12} m_{12} | j_1 m_1, j_2 m_2 >$$

$$\times [\pi_{j_1}(g)]_{m_1 m_1'} [\pi_{j_2}(g)]_{m_2 m_2'} |j_1 m_1' >_{n_1} \otimes |j_2 m_2' >_{n_2}$$

$$= \sum_{m'1, m'2} < j_1 j_2; j_{12} m_{12} | V(g) | j_1 m_1', j_2 m_2' > |j_1 m_1' >_{n_1} \otimes |j_2 m_2' >_{n_2}$$

$$= \sum_{m'1, m'2} (V(g^{-1}) |j_1 j_2; j_{12} m_{12} >, |j_1 m_1', j_2 m_2' >) |j_1 m_1' >_{n_1} \otimes |j_2 m_2' >_{n_2}$$

$$= \sum_{m'1, m'2, m'_{12}} \overline{[\pi_{j_{12}}(g^{-1})]_{m'_{12} m_{12}}} < j_1 j_2; j_{12} m'_{12} >,$$

$$|j_1 m_1', j_2 m_2' > | |j_1 m_1' >_{n_1} \otimes |j_2 m_2' >_{n_2}$$

$$= \sum_{m'_{12}} [\pi_{j_{12}}(g)]_{m_{12} m'_{12}} |j_1 j_2; j_{12} m'_{12} >_{n_1 n_2} \quad (32.2.10)$$

and to iterate. Here we have used unitarity twice.

In order to complete the definition of the spin-network state we must contract the magnetic quantum numbers n_s. We do this, for every $e \in E(\gamma)$, by multiplying by $< j_e j_e; 0 0 | j_1 n_{s_1^e}, j_e n_{s_2^e} >$ and summing over $n_{s_1^e}, n_{s_2^e} \in -j_e, \ldots, j_e$.

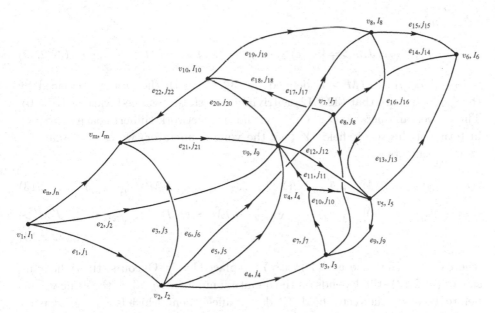

Figure 32.3 The complete set of labels of a spin-network defining a spin-network function and consisting of an oriented graph, a choice of spins for its edges and a choice of intertwiners for its vertices. There is no upper bound on the number of edges or the valence of a vertex.

Equivalently we construct

$$T_{\gamma,\vec{j},\vec{I}} := \sum_{\{n_s\}} \left[\prod_{e \in E(\gamma)} < j_e j_e; 00 | j_1 n_{s_1^e}, j_e n_{s_2^e} > \right] T_{\gamma',\vec{j},\vec{I},\vec{n}} \qquad (32.2.11)$$

Of course, $< j_e j_e; 00 | j_1 n_{s_1^e}, j_e n_{s_2^e} > = [\pi_{j_e}(\epsilon)]_{n_{s_1^e} n_{s_2^e}}$ and gauge invariance at the interior points is now obvious due to the SL(2, $\mathbb{C}$) identity $g^T \epsilon g = \epsilon$. Notice that the whole construction is independent of the breaking point of the edges e because implicitly any other interior point of every edge is contracted with precisely the same intertwiner. In fact, that intertwiner enables us to make it explicit that the state (32.2.11) just depends on the $A(e)$ and not on the $A(s)$. The orthonormality of the states $T_{\gamma,\vec{j},\vec{I}}$ follows from the fact that switching between the tensor product basis and the recoupling basis is a unitary transformation. Finally, if we want to introduce an at least $C^{(1)}$ bivalent vertex on one of the edges, then instead of contracting with $< jj; 00 | j_1 n_1, j_2 n_2 >$ we contract with $< jj; JM | j_1 n_1, j_2 n_2 >$ with $J > 0$ for the breaking point, which is chosen to be that bivalent and at least $C^{(1)}$ vertex. See Figure 32.3 for an example of a spin-network.

In order to carry out concrete calculations it is convenient to make use of the unitary transformation W and to translate everything from the spin-network Hilbert space to the angular momentum Hilbert space of an abstract multispin system. By exactly the same calculation as in (32.2.10) one verifies that for fixed

$m'_1, \ldots m'_N$

$$W|j_1 j_N; j_{12} \ldots JM >= [\pi_J(\epsilon)]_{MM'}|j_1 \ldots j_N; j_{12} \ldots JM >_{m'_1 \ldots m'_N} \quad (32.2.12)$$

where $|j_1 j_N; j_{12} \ldots JM >$ is defined by (32.2.9) with the $|jm >_n$ replaced by the $|jm >$. Notice that (32.2.12) is trivial if the state is gauge-invariant ($J = 0$). Thus, given an operator $O = O(Y)$ on the spin-network Hilbert space expressed in terms of the vector fields Y (say the volume operator) we can calculate its matrix elements as

$$(|j_1 \ldots j_N; j_{12} \ldots JM >_{n_1 \ldots n_N}, O(Y)|j'_1 \ldots j'_{N'}; j'_{12} \ldots J'M' >_{n'_1 \ldots n'_{N'}}) \quad (32.2.13)$$

$$= [\pi_J(\epsilon)]_{M\tilde{M}} [\pi_{J'}(\epsilon)]_{M'\tilde{M}'} (|j_1 \ldots j_N; j_{12} \ldots J\tilde{M} >, O(J)|j'_1 \ldots j'_{N'}; j'_{12} \ldots J'\tilde{M}' >)$$

$$\quad (32.2.14)$$

where $O(J)$ is the same as $O(Y)$ with Y^k_e replaced by J^k_e. Of course the right-hand side of (32.2.13) still depends on the quantum numbers n_k, n'_k in the same way as before, however, the vector fields Y^k_e do not affect them, which is why we have not displayed them. Notice that if O is gauge-invariant, then we have the selection rule $J = J', M = M'$ and moreover (32.2.13) is independent of M because O commutes with $V(g)$. Then (32.3.1) is $\delta_{MM'}\delta_{JJ'}$ times the average $\frac{1}{2J+1} \sum_M < \ldots M|O(J)| \ldots M >$ and thus by elementary properties of the ϵ matrix, the $\pi_J(\epsilon)$ matrices disappear from (32.2.13). In this way, all matrix element calculations are reduced to multispin system calculations.

32.3 Action of holonomy operators on spin-network functions

We need an additional formula in order to carry out calculations with spin-network functions: if we apply to a spin-network function a holonomy operator along an edge which does not overlap with the underlying graph, then we simply get a (gauge-variant, up to normalisation) spin-network function on the bigger graph consisting of the old graph and the additional edge where the spin on that edge coincides with the one of the holonomy operator. If, however, the edge does overlap with the graph, then in order to express the result in terms of spin-network functions we must perform a Clebsch–Gordan decomposition. By suitably subdividing the edges of the graph in question we may reduce the analysis to the case that the edge of the holonomy operator coincides with an edge of the graph. We consider first the case that the holonomy operator carries spin 1/2, which is the most important one in the applications. The more general case can be reduced to iterations of this case because higher-spin holonomies are polynomials of spin 1/2 holonomies as we showed above. We also give a more elegant derivation at the end of this section.

Our task is then to decompose the function $h \mapsto h_{AB}[\pi_j(h)]_{mn}$ into irreducible representations. It is clear from Clebsch–Gordan theory that the occurring spins are $j \pm 1/2$, however, we are interested in the precise coefficients. We have with $A_1 + \cdots + A_{2j} = m$, $B_1 + \cdots + B_{2j} = n$ that

$$
h_{A_0 B_0} \frac{[\pi_j(h)]_{mn}}{c_{jm} c_{jn}} - h_{A_0(B_0} \cdots h_{A_{2j} B_{2j})}
$$

$$
= h_{A_0 B_0} h_{A_1(B_1} \cdots h_{A_{2j} B_{2j})} - h_{A_0(B_0} \cdots h_{A_{2j} B_{2j})}
$$

$$
= \frac{1}{(2j+1)!} \left[(2j+1) h_{A_0 B_0} \sum_{\pi \in S_{2j}} h_{A_1 B_{\pi(1)}} \cdots h_{A_{2j} B_{\pi(2j)}} \right.
$$

$$
\left. - \sum_{\pi \in S_{2j+1}} h_{A_0 B_{\pi(0)}} \cdots h_{A_{2j} B_{\pi(2j)}} \right]
$$

$$
= \frac{1}{(2j+1)!} \left[(2j+1) h_{A_0 B_0} \sum_{\pi \in S_{2j}} h_{A_1 B_{\pi(1)}} \cdots h_{A_{2j} B_{\pi(2j)}} \right.
$$

$$
\left. - \sum_{\pi \in S_{2j+1}} h_{A_0 B_{\pi(0)}} \cdots h_{A_{2j} B_{\pi(2j)}} \right]
$$

$$
= \frac{1}{(2j+1)!} \sum_{k=1}^{2j} \left[\sum_{\pi \in S_{2j}} h_{A_0 B_0} h_{A_1 B_{\pi(1)}} \cdots h_{A_{2j} B_{\pi(2j)}} \right.
$$

$$
\left. - \sum_{\pi \in S_{2j+1} | \pi(0) = k} h_{A_0 B_k} h_{A_1 B_{\pi(1)}} \cdots h_{A_{2j} B_{\pi(2j)}} \right]
$$

$$
= \frac{1}{(2j+1)!} \sum_{k,l=1}^{2j} \left[\sum_{\pi \in S_{2j} | \pi(l) = k} h_{A_0 B_0} h_{A_l B_k} h_{A_1 B_{\pi(1)}} \cdots \widehat{h_{A_l B_{\pi(l)}}} \cdots h_{A_{2j} B_{\pi(2j)}} \right.
$$

$$
\left. - \sum_{\pi \in S_{2j+1} | \pi(0) = k, \pi(l) = 0} h_{A_0 B_{\pi(k)}} h_{A_l B_0} h_{A_1 B_{\pi(1)}} \cdots \widehat{h_{A_l B_{\pi(l)}}} \cdots\cdots h_{A_{2j} B_{\pi(2j)}} \right]
$$

$$
= \frac{1}{(2j+1)!} \sum_{k,l=1}^{2j} \epsilon_{B_0 B_k} \epsilon_{A_0 A_l}
$$

$$
\sum_{\pi \in S_{2j} | \pi(l) = k} h_{A_0 B_0} h_{A_l B_k} h_{A_1 B_{\pi(1)}} \cdots \widehat{h_{A_l B_{\pi(l)}}} \cdots h_{A_{2j} B_{\pi(2j)}}
$$

$$
= \frac{1}{(2j+1)(2j)} \sum_{k,l=1}^{2j} \epsilon_{B_0 B_k} \epsilon_{A_0 A_l} \frac{1}{c_{j-1/2, m-A_l} c_{j-1/2, n-B_k}} [\pi_{j-1/2}(h)]_{m-A_l, n-B_l}
$$

$$
= \frac{1}{(2j+1)(2j)} \sum_{k,l=1}^{2j} \epsilon_{B_0 B_k} \epsilon_{A_0 A_l} \frac{1}{c_{j-1/2, m-A_l} c_{j-1/2, n-B_k}} [\pi_{j-1/2}(h)]_{m-A_l, n-B_l}
$$

$$= \frac{(-1)^{1+A_0+B_0}}{(2j+1)(2j)} \frac{1}{c_{j-1/2m+A_0}c_{j-1/2n+B_0}} [\pi_{j-1/2}(h)]_{m+A_0,n+B_0}$$

$$\times \left[\sum_{k=1}^{2j} \delta_{B_0+B_k,0}\right] \left[\sum_{l=1}^{2j} \delta_{A_0+A_l,0}\right]$$

$$= \frac{(-1)^{1+A_0+B_0}}{(2j+1)(2j)} \frac{(j-2A_0m)(j-2B_0n)}{c_{j-1/2,m+A_0}c_{j-1/2,n+B_0}} [\pi_{j-1/2}(h)]_{m+A_0,n+B_0} \qquad (32.3.1)$$

where in the fourth step we have used the fact that $S_{2j+1} = \cup_{k=0}^{2j}[S_{2j+1}]_{\pi(0)=k}$, in the fifth we have factored out the identical $(2j-1)$ monomials in the sums over the two symmetric groups (the hat over h means omission of that factor), in the seventh we used the unimodularity of SU(2) and realised that $[S_{2j}]_{|\pi(l)=k} = [S_{2j+1}]_{\pi(0)=k,\pi(l)=0}$ so that in both $(2j-1)$ monomials of the h's no A_l, B_k appears, in the eighth we have used the definition of π_j in terms of symmetrised monomials again as well as $\sum_{k=1}^{2j} A_k = m$, $\sum_{k=1}^{2j} B_k = n$, in the ninth step we employed the identity $\epsilon_{AB} = (-1)^{A-1/2}\delta_{A+B,0}$ and finally in the last step we used the fact that $j - 2A_0m$ is the number of A_l subject to $\sum_l A_l = m$ which satisfy $A_l = -A_0$.

Correspondingly

$$h_{A_0B_0}[\pi_j(h)]_{mn} = \frac{c_{jm}c_{jn}}{c_{j+1/2,m+A_0}c_{j+1/2,n+B_0}} [\pi_{j+1/2}(h)]_{m+A_0,n+B_0}$$

$$+ \frac{(-1)^{A_0+B_0+1}(j-2A_0m)(j-2B_0n)}{2j(2j+1)} \frac{c_{jm}c_{jn}}{c_{j-1/2,m+A_0}c_{j-1/2,n+B_0}}$$

$$\times [\pi_{j-1/2}(h)]_{m+A_0,n+B_0}$$

$$= \frac{\sqrt{(j+2A_0m+1)(j+2B_0n+1)}}{2j+1} [\pi_{j+1/2}(h)]_{m+A_0,n+B_0}$$

$$+ 4A_0B_0 \frac{\sqrt{(j-2A_0m)(j-2B_0n)}}{2j+1} [\pi_{j-1/2}(h)]_{m+A_0,n+B_0}$$

$$\qquad (32.3.2)$$

This is the end result which now can be cast into the language of spin-networks again.

We now treat the general case. We consider an abstract two-spin system on which we have a unitary representation of SU(2) defined by

$$U(h)|j_1m_1> \otimes |j_2m_2> = [\pi_{j_1}(h)]_{m_1n_1} [\pi_{j_2}(h)]_{m_2n_2} |j_1n_1> \otimes |j_2n_2> \qquad (32.3.3)$$

where $h = \exp(\theta^j \tau_j)$ and $U(h) = \exp(i\theta[J_1^j + J_2^j]) = U_1(h) \otimes U_2(h)$ since the individual angular momenta are mutually commuting. We now expand the tensor basis into the recoupling basis and find

$$[\pi_{j_1}(h)]_{m_1 n_1}[\pi_{j_2}(h)]_{m_2 n_2}$$
$$= (<j_1 n_1| \otimes <j_2 n_2|, U(h)|j_1 m_1 > \otimes |j_2 m_2 >)$$

$$= \sum_{j_{12}=|j_1-j_2|}^{j_1+j_2} <j_{12}m_1 + m_2|j_1 m_1; j_2 m_2 > (<j_1 n_1| \otimes <j_2 n_2|, U(h)|j_{12}m_1 + m_2 >)$$

$$= \sum_{j_{12}=|j_1-j_2|}^{j_1+j_2} <j_{12}m_1 + m_2|j_1 m_1; j_2 m_2 > [\pi_{j_{12}}(h)]_{m_1+m_2,n} (<j_1 n_1| \otimes <j_2 n_2|j_{12}n >)$$

$$= \sum_{j_{12}=|j_1-j_2|}^{j_1+j_2} <j_{12}m_1 + m_2|j_1 m_1; j_2 m_2 >< j_1 n_1; j_2 n_2|j_{12}n_1 + n_2 > [\pi_{j_{12}}(h)]_{m_1+m_2,n}$$

$$(32.3.4)$$

One can indeed verify that the coefficients displayed in (32.3.2) are the products of CGCs given in (32.3.4).

32.4 Examples of coherent state calculations

As we have seen in Chapter 11, kinematical coherent states for the currently most studied classes of complexifiers take the form $\psi_{\gamma,m} = \otimes_{e \in E(\gamma)} \psi_{e,m}$ where γ is a graph and m is a point in the classical phase space. The states on the edges take the form

$$\psi_{m,e} := \psi^{t_e}_{g_e(m)}, \quad \psi^t_g(h) = \sum_{\pi} d_\pi \, e^{-t\lambda_\pi/2} \, \chi_\pi(gh^{-1}) \qquad (32.4.1)$$

where the sum is over equivalence classes of irreducible representations π of the compact gauge group G, $-\lambda_\pi \leq 0$ is the eigenvalue of the Laplacian $(R^j/2)^2$ (right-invariant vector fields) in that representation, d_π, χ_π are respectively its dimension and character, t is the classicality parameter, $g \in G^{\mathbb{C}}$ is an element of the complexification of G and $h \in G$.

The physics is in the maps $m \mapsto g_e(m)$, $e \mapsto t_e$ which depend on the complexifier used, as explained in Chapter 11. Here we will focus on how to do computations with (32.4.1). Our presentation will be brief, many more details can be found in [488, 489]. We will write (32.4.1) explicitly for U(1) and SU(2)

$$\psi_g(h) = \begin{cases} \sum_{n \in \mathbb{Z}} e^{-tn^2/2} (gh^{-1})^n & \text{U(1)} \\ \sum_{2j=0}^{\infty} (2j+1) e^{-tj(j+1)/2} \, \text{Tr}(\pi_j(gh^{-1})) & \text{SU(2)} \end{cases} \qquad (32.4.2)$$

In order to write the SU(2) character $\chi_j = \text{Tr}(\pi_j)$ more explicitly we can make use of (32.1.5) as follows: every element $g \in \text{SU}(2)^{\mathbb{C}} = \text{SL}(2, \mathbb{C})$ can be diagonalised by an element $S \in \text{GL}(2, \mathbb{C})$, that is, $g = SDS^{-1}$ where D is diagonal. The entries of $D = \text{diag}(\lambda_1, \lambda_2)$ can be determined by $\lambda_1 \lambda_2 = \det(g) =$

1, $\lambda_1 + \lambda_2 = \mathrm{Tr}(g)$. The solution is

$$\lambda := \lambda_1 = \frac{1}{2}\mathrm{Tr}(g) + \sqrt{\left[\frac{1}{2}\mathrm{Tr}(g)\right]^2 - 1}, \ \lambda_2 = \frac{1}{\lambda_1} = \frac{1}{2}\mathrm{Tr}(g) - \sqrt{\left[\frac{1}{2}\mathrm{Tr}(g)\right]^2 - 1}$$

(32.4.3)

or $\lambda_1 \leftrightarrow \lambda_2$. This sign ambiguity is a reflection of the Weyl group of SU(2). Since the character is invariant under similarity transformations (conjugations), we may take the sum over $m, n \in \{-j, \ldots, j\}$ in (32.1.5) with $b = c = 0$ and $d = 1/a$, $a = \lambda$. This requires $m = n$, $l = 0$ and we find

$$\chi_j(g) = \sum_{m=-j}^{j} \lambda^{2m} = \frac{\mathrm{sh}([2j+1]z)}{\mathrm{sh}(z)}$$

(32.4.4)

Notice that (32.4.4) is manifestly invariant under $\lambda \leftrightarrow \lambda^{-1}$. Here we have introduced the complex number z through $\mathrm{ch}(z) := \mathrm{Tr}(g)/2$ so that $\lambda = e^z$, which helps to compute the geometric sum in (32.4.4).

Indispensable for practical calculations is the following.

Theorem 32.4.1 (Poisson resummation formula). *Let f be an $L_1(\mathbb{R}, dx)$ function such that the series $g_s(x) := \sum_{n \in \mathbb{Z}} f(x + ns)$ converges absolutely and uniformly for all $x \in [0, s]$ for some $s > 0$. Then*

$$\sum_{n \in \mathbb{Z}} f(ns) = \frac{1}{s} \sum_{n \in \mathbb{Z}} \int_{\mathbb{R}} dx \, e^{2\pi i n x/s} f(x)$$

(32.4.5)

See [488, 646] for the proof.

The significance of this theorem becomes evident when we ask, for instance, for the maximum of probability amplitude

$$p_g^t(h) := \frac{\left|\psi_g^t(h)\right|^2}{\left\|\psi_g^t\right\|^2}$$

(32.4.6)

or for the maximum of the overlap function

$$j_{g_1, g_2}^t := \frac{\left| <\psi_{g_1}^t, \psi_{g_2}^t> \right|^2}{\left\|\psi_{g_1}^t\right\|^2 \left\|\psi_{g_2}^t\right\|^2}$$

(32.4.7)

Here the norms occur because the ψ_g^t are not normalised and scalar products are of course evaluated in the Hilbert space $L_2(\mathrm{G}, d\mu_H)$ using the Peter and Weyl theorem. We will compute only (32.4.7) for our illustrative purposes, (32.4.6) works similarly. Also we will restrict ourselves to the mathematically more difficult and physically more interesting case SU(2). We find with

$$< \pi_{jmn}, \pi_{j'm'n'} > = \frac{\delta_{jj'}\delta_{mm'}\delta_{nn'}}{2j+1} \tag{32.4.8}$$

that

$$j^t_{g_1,g_2} = \frac{\left| \sum_{2j=0}^{\infty} (2j+1) e^{-tj(j+1)} \chi_j(g_1^{\dagger}g_2) \right|^2}{||\psi^t_{g_1}||^2 \, ||\psi^t_{g_2}||^2} \tag{32.4.9}$$

where the norms squared are given by the square root of the numerator of (32.4.9) for $g_1 = g_2$.

Now recall that t is a tiny number, therefore (32.4.9) is a fraction of two numbers involving slowly converging series which are therefore difficult to deal with in analytical investigations. The Poisson resummation formula now enables us to transform those series into rapidly converging ones because it basically interchanges $s := \sqrt{t}$ with $1/s$ as follows: in order to apply Theorem 32.4.1 we must bring (32.4.9) into the form (32.4.5). For U(1) this would already be the case. The dimension of π_j is given by $n = 2j + 1 \in \mathbb{N}_0$ and this combination also appears in $\chi_j(g)$. Now observe that $j(j+1) = [n^2 - 1]/4$. It follows that

$$< \psi^t_{g_1}, \psi^t_{g_2} > = e^{t/4} \sum_{n=0}^{\infty} n \, e^{-tn^2/4} \frac{\text{sh}(nz_{12})}{\text{sh}(z_{12})}$$

$$= \frac{e^{t/4}}{2\text{sh}(z_{12})} \sum_{n\in\mathbb{Z}} n \, e^{tn^2/4} \, e^{nz_{12}} \tag{32.4.10}$$

where in the second step we decomposed $\text{sh}(nz) = [e^{nz} - e^{-nz}]/2$, observed that the second term at n is obtained from the first one by changing n into $-n$ and that the term at $n = 0$ vanishes anyway. Here $\text{ch}(z_{12}) = \text{Tr}(g_1^{\dagger}g_2)/2$ and for the norms of $\psi^t_{g_j}$, $j = 1, 2$ exactly the same formula holds just with z_{12} replaced by $\text{ch}(z_j) = \text{Tr}(g_j^{\dagger}g_j)/2$.

Consider now for complex z the function $f_z(x) = x \, e^{-x^2/4} \, e^{xz}$. Then we see that Theorem 32.4.1 can be applied because of the Gaußian damping factor and we obtain with $s = \sqrt{t}$

$$j^t_{g_1,g_2} = \frac{\text{sh}(z_1)\text{sh}(z_2)}{|\text{sh}(z_{12})|^2}$$

$$\frac{\left| \sum_{n\in\mathbb{Z}} \int_{\mathbb{R}} dx \, e^{2\pi i n x/s} f_{z_{12}/s}(x) \right|^2}{\left[\sum_{n\in\mathbb{Z}} \int_{\mathbb{R}} dx \, e^{2\pi i n x/s} f_{z_1/s}(x) \right] \left[\sum_{n\in\mathbb{Z}} \int_{\mathbb{R}} dx \, e^{2\pi i n x/s} f_{z_2/s}(x) \right]} \tag{32.4.11}$$

The integrals that appear in (32.4.11) can be computed by standard Gaußian integral techniques

$$\sum_{n\in\mathbb{Z}}\int_{\mathbb{R}} dx\, e^{2\pi inx/s}\, f_{z/s}(x)$$

$$=4\sum_{n\in\mathbb{Z}}\int_{\mathbb{R}} dx\, x\, e^{2(2\pi in+z)x/s}e^{-x^2}$$

$$=4\sum_{n\in\mathbb{Z}} e^{(2\pi in+z)^2/t}\int_{\mathbb{R}} dx\, e^{-x^2}[x+(2\pi in+z)/s]$$

$$=4\sqrt{\pi/t}\sum_{n\in\mathbb{Z}} e^{(2\pi in+z)^2/t}\,[2\pi in+z]$$

$$=4\sqrt{\pi/t}\,e^{z^2/t}\sum_{n\in\mathbb{Z}} e^{-4\pi^2 n^2/t}\, e^{4\pi inz/t}\,[2\pi in+z]$$

$$=8\sqrt{\pi/t}\,e^{z^2/t}\left\{z\left[1+2\sum_{n=1}^{\infty} e^{-4\pi^2 n^2/t}\,\cos(4\pi nz/t)\right]\right.$$

$$\left.-\left[4\pi\sum_{n=1}^{\infty} e^{-4\pi^2 n^2/t}\,\sin(4\pi nz/t)\right]\right\} \qquad (32.4.12)$$

where in the second step we changed variables from x to $2x$ and in the third we applied a contour argument. The point of performing this calculation is that the two terms in the square brackets of the last line of (32.4.12) are rapidly converging for $t\to 0$, in fact it is not difficult to show that they approach zero faster than any power of t. Thus, for purposes of $\hbar$ expansions these $O(t^\infty)$ terms can be neglected and (32.4.11) becomes

$$j^t_{g_1,g_2}=\frac{|z_{12}|^2\mathrm{sh}(z_1)\mathrm{sh}(z_2)}{|\mathrm{sh}(z_{12})|^2 z_1 z_2}e^{[z_{12}^2+\bar{z}_{12}^2-z_1^2-z_2^2]/t} \qquad (32.4.13)$$

up to $O(t^\infty)$ corrections. Expression (32.4.13) is very easy to analyse and it is not difficult to show, using convex function techniques and Riemannian geometry on S^3, that (32.4.13) is strongly peaked at $g_1=g_2$ in a Gaußian fashion where the peak of the Gaussian has width $\sqrt{t}$, see [488, 489] for all details.

Most coherent state calculations can be reduced to the above manipulations. Consider, for instance, the expectation value of an electric flux operator $\hat{E}_j(S)$ whose surface is transversal to one edge e only. Then the expectation value $<\psi_{\gamma,m},\hat{E}_j(S)\psi_{\gamma,m}>/||\psi_{\gamma,m}||^2$ becomes essentially (up to multiplicative constants) $<\psi_g^t,R_j\psi_g^t>/||\psi_g^t||^2$. Now it is not difficult to see that $R_j\psi_g^t=[d/dr]_{r=0}\,\psi_{\exp(-r\tau_j)g}^t$ and that $||\psi_{\exp(-r\tau_j g)}^t||=||\psi_g^t||$. Thus, if we set $g_1=g$, $g_2=\exp(-r\tau_j)g$ then we find

$$\frac{<\psi_g^t,R_j\psi_g^t>}{||\psi_g^t||^2}=[d/dr]_{r=0}\,j^t_{g_1,g_2} \qquad (32.4.14)$$

The computation of the expectation value of (matrix entries of) a holonomy operator $[\hat{A}(e)]_{mn}$ reduces in a similar way to the computation of $<\psi_g^t,\hat{h}_{mn}\psi_g^t>$ $/||\psi_g^t||^2$. In order to evaluate this one can use, for instance, formula (32.3.4) which makes it possible to work out the appearing product of spin 1/2 and spin j matrices as linear combination of spin $j \pm 1/2$ matrices and then to use the Poisson resummation formula again.

Similar manipulations can be applied to polynomials of those elementary operators which appear in the semiclassical analysis of the Master Constraint operator where it is understood that the fractional powers of the volume operator that are involved are replaced by polynomials by the method described in Section 13.4.4. See [488, 489, 591] for all details and [590, 637, 638, 835] for concrete physical applications of these techniques.

$^+$ Functional analytic description of classical connection dynamics

This chapter is for the benefit of the reader who wants to get a glimpse of the functional analytic questions that arise when properly defining the function spaces of (gauge) fields on infinite-dimensional symplectic manifolds. It turns out to be rather difficult to consistently restrict the space of classical fields on a given differential manifold in such a way that the classical action remains functionally differentiable, usually critically depending on the boundary conditions that one imposes, while keeping 'enough' solutions of the field equations. Usually the simplest solutions, those with a high degree of symmetry, are at the verge of lying outside the space of fields that the variational principle was based on. Fortunately, these issues will not be too important for us as the space of quantum fields tends to be even much larger and generically is of a distributional kind without leading to any problems. Those issues will, however, be of some interest again when we discuss the classical limit. We can therefore be brief here and will just sketch some of the main ideas. The interested reader is referred to the exhaustive treatment in [886].

33.1 Infinite-dimensional (symplectic) manifolds

Let G be a compact gauge group, σ a D-dimensional manifold which admits a principal G-bundle with connection over σ. Let us denote the pull-back to σ of the connection by local sections by A_a^i where $a, b, c, \ldots = 1, \ldots, D$ denote tensorial indices and $i, j, k, \ldots = 1, \ldots, \dim(G)$ denote indices for the Lie algebra of G. We will denote the set of all smooth connections by $\mathcal{A}$ and endow it with a globally defined metric topology of the Sobolev kind

$$d_\rho[A, A'] := \sqrt{-\frac{1}{N} \int_\sigma d^D x \sqrt{\det(\rho)(x)} \mathrm{tr}([A_a - A'_a](x)[A_b - A'_b](x)) \rho^{ab}(x)}$$

(33.1.1)

where $\mathrm{tr}(\tau_i \tau_j) = -N \delta_{ij}$ is our choice of normalisation for the generators of a Lie algebra Lie(G) of rank N and our conventions are such that $[\tau_i, \tau_j] = 2 f_{ij}{}^k \tau_k$ define the structure constants of Lie(G). Here ρ_{ab} is a fiducial metric on σ of everywhere Euclidean signature. In what follows we assume that either $D \neq 2$ (for $D = 2$, (33.1.1) depends only on the conformal structure of ρ and cannot guarantee convergence for arbitrary fall-off conditions on the connections) or that $D = 2$ and the fields A are Lebesgue integrable.

Let F_i^a be a Lie algebra-valued vector density test field of weight one and let f_a^j be a Lie algebra-valued co-vector test field. Let, as before, A_a^j be the pull-back of a connection to σ and consider a vector bundle of electric fields, that is, of Lie algebra-valued vector densities of weight one whose bundle projection to σ we denote by E_i^a. We consider the smeared quantities

$$F(A) := \int_\sigma d^D x F_i^a A_a^i \text{ and } E(f) := \int_\sigma d^D x E_i^a f_a^i \qquad (33.1.2)$$

While both are diffeomorphism-covariant it is only the latter which is gauge-covariant, one reason to consider the singular smearing through holonomies discussed below. The choice of the space of pairs of test fields $(F, f) \in S$ depends on the boundary conditions on the space of connections and electric fields, which in turn depends on the topology of σ and will not be specified in what follows.

We now want to select a subset $\mathcal{M}$ of the set of all pairs of smooth functions (A, E) on σ such that (33.1.2) is well-defined (finite) for any $(F, f) \in S$ and endow it with a manifold structure and a symplectic structure, that is, we wish to turn it into an infinite-dimensional symplectic manifold.

We define a topology on $\mathcal{M}$ through the metric:

$$d_{\rho,\sigma}[(A, E), (A', E')]$$

$$:= \sqrt{-\frac{1}{N} \int_\sigma d^D x \left[\sqrt{\det(\rho)} \rho^{ub} \text{tr}([A_a - A'_a][A_b - A'_b]) + \frac{\sigma_{ab} \text{tr}([E^a - E^{a\prime}][E^b - E^{b\prime}])}{\sqrt{\det(\sigma)}} \right]}$$

$$(33.1.3)$$

where ρ_{ab}, σ_{ab} are again fiducial metrics on σ of everywhere Euclidean signature. Their fall-off behaviour has to be suited to the boundary conditions of the fields A, E at spatial infinity. Notice that the metric (33.1.3) is gauge-invariant (and thus globally defined, i.e., is independent of the choice of local section) and diffeomorphism-covariant and that $d_{\rho,\sigma}[(A, E), (A', E')] = d_\rho[A, A'] + d_\sigma[E, E']$ (recall (1.1.1)).

Now, while the space of electric fields in Yang–Mills theory is a vector space, the space of connections is only an affine space. However, as we also have applications in General Relativity with asymptotically Minkowskian boundary conditions in mind, the space of electric fields will in general not be a vector space. Thus, in order to induce a norm from (33.1.3) we proceed as follows: consider an atlas of $\mathcal{M}$ consisting only of $\mathcal{M}$ itself and choose a fiducial background connection and electric field $A^{(0)}, E^{(0)}$ (for instance $A^{(0)} = 0$). We define the global chart

$$\varphi : \mathcal{M} \mapsto \mathcal{E}; \ (A, E) \mapsto \left(A - A^{(0)}, E - E^{(0)} \right) \qquad (33.1.4)$$

of $\mathcal{M}$ onto the vector space of pairs $(A - A^{(0)}, E - E^{(0)})$. Obviously, φ is a bijection. We topologise $\mathcal{E}$ in the norm

$$\left\| (A - A^{(0)}, E - E^{(0)}) \right\|_{\rho\sigma} := \sqrt{d_{\rho\sigma}[(A, E), (A^{(0)}, E^{(0)})]} \qquad (33.1.5)$$

The norm (33.1.5) is of course no longer gauge- and diffeomorphism-covariant since the fields $A^{(0)}, E^{(0)}$ do not transform, they are background fields. We need it, however, only in order to encode the fall-off behaviour of the fields which are independent of gauge and diffeomorphism covariance.

Notice that the metric induced by this norm coincides with (33.1.3). In the terminology of weighted Sobolev spaces the completion of $\mathcal{E}$ in the norm (33.1.5) is called the Sobolev space $H^2_{0,\rho} \times H^2_{0,\sigma-1}$ (see, e.g., [899]). We will call the completed space $\mathcal{E}$ again and its image under φ^{-1}, $\mathcal{M}$ again (the dependence of φ on $(A^{(0)}, E^{(0)})$ will be suppressed). Thus, $\mathcal{E}$ is a normed, complete vector space, that is, a Banach space, in fact it is even a Hilbert space. Moreover, we have modelled $\mathcal{M}$ on the Banach space $\mathcal{E}$, that is, $\mathcal{M}$ acquires the structure of a (so far only topological) Banach manifold. However, since $\mathcal{M}$ can be covered by a single chart and the identity map on $\mathcal{E}$ is certainly C^∞, $\mathcal{M}$ is actually a smooth manifold. The advantage of modelling $\mathcal{M}$ on a Banach manifold is that one can take over almost all the pleasant properties from the finite-dimensional case to the infinite-dimensional one (in particular, the inverse function theorem).

Next we study differential geometry on $\mathcal{M}$ with the standard techniques of calculus on infinite-dimensional manifolds (see, e.g., [900]). We will not repeat all the technicalities of the definitions involved, the interested reader is referred to the literature quoted.

(i) A function $f : \mathcal{M} \mapsto \mathbb{C}$ on $\mathcal{M}$ is said to be differentiable at m if $g := f \circ \varphi^{-1} : \mathcal{E} \mapsto \mathbb{C}$ is differentiable at $u = \varphi(m)$, that is, there exist *bounded* linear operators $Dg_u, Rg_u : \mathcal{E} \mapsto \mathbb{C}$ such that

$$g(u+v) - g(u) = (Dg_u) \cdot v + (Rg_u) \cdot v \text{ where } \lim_{||v|| \to 0} \frac{|(Rg_u) \cdot v|}{||v||} = 0 \quad (33.1.6)$$

$Df_m := Dg_u$ is called the functional derivative of f at m (notice that we identify, as usual, the tangent space of $\mathcal{M}$ at m with $\mathcal{E}$). The definition extends in an obvious way to the case where $\mathbb{C}$ is replaced by another Banach manifold. The equivalence class of functions differentiable at m is called the germ $G(m)$ at m. Here two functions are said to be equivalent provided they coincide in a neighbourhood containing m.

(ii) In general, a tangent vector v_m at $m \in \mathcal{M}$ is an equivalence class of triples (U, φ, v_m) where (U, φ) is a chart of the atlas of $\mathcal{M}$ containing m and $v_m \in \mathcal{E}$. Two triples are said to be equivalent provided that $v'_m = D(\varphi' \circ \varphi^{-1})_{\varphi(m)} \cdot v_m$. In our case we have only one chart and equivalence becomes trivial. Tangent vectors at m can be considered as derivatives on the germ $G(m)$ by defining

$$v_m(f) := (Df_m) \cdot v_m = \left(D(f \circ \varphi^{-1})_{\varphi(m)} \right) \cdot v_m \quad (33.1.7)$$

Notice that the definition depends only on the equivalence class and not on the representative. The set of vectors tangent at m defines the tangent space $T_m(\mathcal{M})$ of $\mathcal{M}$ at m.

(iii) The cotangent space $T'_m(\mathcal{M})$ is the topological dual of $T_m(\mathcal{M})$, that is, the set of *continuous* linear functionals on $T_m(\mathcal{M})$. It is obviously isomorphic with $\mathcal{E}'$, the topological dual of $\mathcal{E}$. Since our model space $\mathcal{E}$ is reflexive (it is a Hilbert space) we can naturally identify tangent and cotangent spaces (by the Riesz lemma), which also makes the definition of contravariant tensors less ambiguous. We will, however, not need them for what follows. Similarly, one defines the space of p-covariant tensors at $m \in \mathcal{M}$ as the space of *continuous* p-linear forms on the p-fold tensor product of $T_m(\mathcal{M})$.

(iv) So far the fact that $\mathcal{E}$ is a Banach manifold was not very crucial. But while the tangent bundle $T(\mathcal{M}) = \cup_{m \in \mathcal{M}} T_m(\mathcal{M})$ carries a natural manifold structure modelled on $\mathcal{E} \times \mathcal{E}$ for a general Fréchet space (or even locally convex space) $\mathcal{E}$, the cotangent bundle $T'(M) = \cup_{m \in \mathcal{M}} T'_m(\mathcal{M})$ carries a manifold structure only when $\mathcal{E}$ is a Banach space as one needs the inverse function theorem to show that each chart is not only a differentiable bijection but that also its inverse is differentiable. In our case again there is no problem. We define differentiable vector fields and p-covariant tensor fields as cross sections of the corresponding fibre bundles.

(v) A differential form of degree p on $\mathcal{M}$ or p-form is a cross-section of the fibre bundle of completely skew continuous p-linear forms. Exterior product, pull-back, exterior differential, interior product with vector fields and Lie derivatives are defined as in the finite-dimensional case.

Definition 33.1.1. *Let $\mathcal{M}$ be a differentiable manifold modelled on a Banach space $\mathcal{E}$. A weak, respectively strong, symplectic structure Ω on $\mathcal{M}$ is a closed 2-form such that for all $m \in \mathcal{M}$ the map*

$$\Omega_m : T_m(\mathcal{M}) \to T'_m(\mathcal{M}); \ v_m \mapsto \Omega(v_m, .) \tag{33.1.8}$$

is an injection, respectively a bijection.

Strong symplectic structures are more useful because weak symplectic structures do not allow us to define Hamiltonian vector fields through the definition $DL + i_{\chi_L}\Omega = 0$ for differentiable L on $\mathcal{M}$ and Poisson brackets through $\{f, g\} := \Omega(\chi_f, \chi_g)$ (see, e.g., [220] for details).

Thus we define finally a strong symplectic structure for our case by

$$\Omega((f, F), (f', F')) := \int_\sigma d^D x \left[F_i^a f_a'^i - F_i'^a f_a^i \right](x) \tag{33.1.9}$$

for any $(f, F), (f', F') \in \mathcal{E}$. To see that Ω is a strong symplectic structure we observe first that the integral kernel of Ω is constant so that Ω is clearly exact, so, in particular, closed. Next, let $\theta \in \mathcal{E}' \equiv \mathcal{E}$. To show that Ω is a bijection it suffices to show that it is a surjection (injectivity follows trivially from linearity). We must find $(f, F) \in \mathcal{E}$ so that $\theta(.) = \Omega((f, F), .)$ for any one-form θ. Now by the Riesz lemma there exists $(f_\theta, F_\theta) \in \mathcal{E}$ such that $\theta(.) = < (f_\theta, F_\theta), . >$ where

$< \cdot, \cdot >$ is the inner product induced by (33.1.5). Comparing (33.1.3) and (33.1.9) we see that we have achieved our goal provided that the functions

$$F_i^a := \rho^{ab}\sqrt{\det(\rho)}f_{b\theta}^i, \quad f_a^i := -\frac{\sigma_{ab}}{\sqrt{\det(\rho)}}F_{i\theta}^b \qquad (33.1.10)$$

are elements of $\mathcal{E}$. Inserting the definitions we see that this will be the case provided that the functions $\rho^{cd}\sigma_{ca}\sigma_{db}/\sqrt{\det(\rho)}$ and $\det(\rho)\sigma_{cd}\rho^{ca}\rho^{db}/\sqrt{\det(\sigma)}$ respectively fall off at least as $\sigma_{ab}/\sqrt{\det(\sigma)}$ and $\rho^{ab}\sqrt{\det(\rho)}$ respectively. In physical applications these metrics are usually chosen to be of the form $1 + O(1/r)$ where r is an asymptotic radius function so that these conditions are certainly satisfied. Therefore, $(f, F) \in \mathcal{E}$.

Let us compute the Hamiltonian vector field of a function L on our $\mathcal{M}$. By definition, for all $(f, F) \in \mathcal{E}$ we have at $m = (A, E)$

$$DL_m \cdot (f, F) = \int_\sigma d^D x \left[(DL_m)_i^a f_a^i + (DL_m)_a^i F_i^a \right]$$

$$= -\int_\sigma d^D x \left[(\chi_{Lm})_i^a f_a^i - (\chi_{Lm})_a^i F_i^a \right] \qquad (33.1.11)$$

thus $(\chi_L)_i^a = -(DL)_i^a$ and $(\chi_L)_a^i = (DL)_a^i$. Obviously, this defines a bounded operator on $\mathcal{E}$ in our case if the components of DL themselves define an element of $\mathcal{E}'$ (by the Schwarz inequality). Finally, the Poisson bracket is given by

$$\{L, L'\}_m = \Omega(\chi_L, \chi_{L'}) = \int_\sigma d^D x \left[(DL_m)_a^i (DL'_m)_i^a - (DL_m)_i^a (DL'_m)_a^i \right]$$

$$(33.1.12)$$

It is easy to see that Ω has the symplectic potential Θ, a one-form on $\mathcal{M}$, defined by

$$\Theta_m((f, F)) = \int_\sigma d^D x F_i^a f_a^i \qquad (33.1.13)$$

since

$$D\Theta_m((f, F), (f', F')) := (D(\Theta_m) \cdot (f, F)) \cdot (f', F') - (D(\Theta_m) \cdot (f', F')) \cdot (f, F)$$

and $DE_i^a(x)_m \cdot (f, F) = F_i^a(x)$ as follows from the definition.

Coming back to the choice of $\mathcal{S}$, it will in general be a subspace of $\mathcal{E}$ so that (33.1.9) still converges. We can now compute the Poisson brackets between the functions $F(A), E(f)$ on $\mathcal{M}$ and find

$$\{E(f), E(f')\} = \{F(A), F'(A)\} = 0, \quad \{E(f), A(F)\} = F(f) \quad (33.1.14)$$

Remark: In physicists' notation one often writes $(DL_m)_a^i(x) := \frac{\delta L}{\delta A_i^a(x)}$, called a functional derivative, etc. and, abusing the notation, one writes the symplectic structure as $\Omega = \int d^D x \, DE_i^a(x) \wedge DA_a^i(x)$.

References

[1] C. Beetle and A. Corichi. Bibliography of publications related to classical and quantum gravity in terms of connection and loop variables. [gr-qc/9703044]

[2] A. Hauser and A. Corichi. Bibliography of publications related to classical self-dual variables and loop quantum gravity. [gr-qc/0509039]

[3] C. Rovelli. *Quantum Gravity* (Cambridge University Press, Cambridge, 2004).

[4] C. Rovelli. Loop quantum gravity. *Living Rev. Rel.* **1** (1998) 1. [gr-qc/9710008]

[5] C. Rovelli. Strings, loops and others: a critical survey of the present approaches to quantum gravity. Plenary lecture given at 15th Intl. Conf. on Gen. Rel. and Gravitation (GR15), Pune, India, Dec 16–21, 1997. [gr-qc/9803024]

[6] M. Gaul and C. Rovelli. Loop quantum gravity and the meaning of diffeomorphism invariance. *Lect. Notes Phys.* **541** (2000) 277–324. [gr-qc/9910079]

[7] G. Horowitz. Quantum gravity at the turn of the millennium. [gr-qc/0011089]

[8] C. Rovelli. Notes for a brief history of quantum gravity. [gr-qc/0006061]

[9] S. Carlip. Quantum gravity: a progress report. *Rept. Prog. Phys.* **64** (2001) 885. [gr-qc/0108040]

[10] A. Ashtekar. Quantum mechanics of geometry. [gr-qc/9901023]

[11] T. Thiemann. Introduction to modern canonical quantum general relativity. [gr-qc/0110034]

[12] T. Thiemann. Lectures on loop quantum gravity. *Lect. Notes Phys.* **631** (2003) 41–135. [gr-qc/0210094]

[13] A. Ashtekar and J. Lewandowski. Background independent quantum gravity: a status report. *Class. Quant. Grav.* **21** (2004) R53. [gr-qc/0404018]

[14] L. Smolin. An invitation to loop quantum gravity. [hep-th/0408048]

[15] A. Ashtekar with invited contributions. *New Perspectives in Canonical Gravity* (Bibliopolis, Napoli, 1988).

[16] A. Ashtekar. *Non-Perturbative Canonical Gravity*. Lectures notes prepared in collaboration with R. S. Tate (World Scientific, Singapore, 1991).

[17] A. Ashtekar. In *Gravitation and Quantisations*, B. Julia (ed.) (Elsevier, Amsterdam, 1995).

[18] A. Liddle. *An Introduction to Modern Cosmology* (John Wiley & Sons, Chichester, 1999).

[19] S. Dodelson. *Modern Cosmology* (Academic Press, Amsterdam, 2003).

[20] V. Mukhanov. *Physical Foundations of Cosmology* (Cambridge University Press, Cambridge, 2006).

[21] R. F. Streater and A. S. Wightman. *PCT, Spin and Statistics, and all that* (Benjamin, New York, 1964).

[22] R. Haag. *Local Quantum Physics*, 2nd edn. (Springer-Verlag, Berlin, 1996).

[23] S. A. Fulling. *Aspects of Quantum Field Theory in Curved Space-Time* (Cambridge University Press, Cambridge, 1989).

[24] R. M. Wald. *Quantum Field Theory in Curved Space-Time and Black Hole Thermodynamics* (Chicago University Press, Chicago, 1995).

[25] M. J. Radzikowski. *The Hadamard Condition and Kay's Conjecture in (Axiomatic) Quantum Field Theory on Curved Space-Time*, PhD Thesis (Princeton University, October, 1992).

[26] W. Junker. Hadamard states, adiabatic vacua and the construction of physical states for scalar quantum fields on curved spacetime. *Rev. Math. Phys.* **8** (1996) 1091–159.

[27] R. Brunetti, K. Fredenhagen and M. Kohler. The microlocal spectrum condition and Wick polynomials of free fields on curved spacetimes. *Commun. Math. Phys.* **180** (1996) 633–52. [gr-qc/9510056]

[28] R. Brunetti and K. Fredenhagen. Microlocal analysis and interacting quantum field theories: renormalisation on physical backgrounds. *Commun. Math. Phys.* **208** (2000) 623–61. [math-ph/9903028]

[29] S. Hollands and R. M. Wald. Local Wick polynomials and time ordered products of quantum fields in curved spacetime. *Commun. Math. Phys.* **223** (2001) 289–326. [gr-qc/0103074]

[30] S. Hollands and R. M. Wald. Existence of local covariant time ordered products of quantum fields in curved spacetime. *Commun. Math. Phys.* **231** (2002) 309–45. [gr-qc/0111108]

[31] S. Hollands and R. M. Wald. On the renormalisation group in curved spacetime. *Commun. Math. Phys.* **237** (2003) 123–60. [gr-qc/0209029]

[32] S. Hollands. A general PCT theorem for the operator product expansion in curved spacetime. *Commun. Math. Phys.* **244** (2004) 209–44. [gr-qc/0212028]

[33] W. Junker and E. Schrohe. Adiabatic vacuum states on general spacetime manifolds: definition, construction and physical properties. *Annales Poincare Phys. Theor.* **3** (2002) 1113–82. [math-ph/0109010]

[34] G. Scharf. *Finite Quantum Electrodynamics: The Causal Approach* (Springer-Verlag, Berlin, 1995).

[35] R. Brunetti, K. Fredenhagen and R. Verch. The generally covariant locality principle: a new paradigm for local quantum field theory. *Commun. Math. Phys.* **237** (2003) 31–68. [math-ph/0112041]

[36] R. Verch. A spin statistics theorem for quantum fields on curved spacetime manifolds in a generally covariant framework. *Commun. Math. Phys.* **223** (2001) 261–88. [math-ph/0102035]

[37] E. E. Flanagan and R. M. Wald. Does backreaction enforce the averaged null energy condition in semiclassical gravity? *Phys. Rev.* **D54** (1996) 6233–83. [gr-qc/9602052]

[38] N. Marcus and A. Sagnotti. The ultraviolet behaviour of $N = 4$ Yang–Mills and the power counting of extended superspace. *Nucl. Phys.* **B256** (1985) 77.

[39] M. H. Goroff and A. Sagnotti. Quantum gravity at two loops. *Phys. Lett.* **B160** (1985) 81.

[40] P. Van Nieuwenhuizen. Supergravity. *Phys. Rep.* **68** (1981) 189.

[41] M. H. Goroff and A. Sagnotti. The ultraviolet behaviour of Einstein gravity. *Nucl. Phys.* **B266** (1986) 709.

[42] S. Deser. Two outcomes of two old (super)problems. [hep-th/9906178]

[43] S. Deser. Infinities in quantum gravities. *Ann. Phys.* **9** (2000) 299–307. [gr-qc/9911073]

[44] S. Deser. Non-realisability of (last hope) $D = 11$ supergravity with a terse survey of divergences in quantum gravities. [hep-th/9905017]

[45] M. B. Green, J. Schwarz and E. Witten. *Superstring Theory*, Vols 1, 2, (Cambridge University Press, Cambridge, 1987).

[46] J. Polchinski. *String Theory*, Vol. 1: An introduction to the bosonic string, Vol. 2: Superstring theory and beyond (Cambridge University Press, Cambridge, 1998).

[47] E. D'Hoker and D. H. Phong. Lectures on two loop superstrings. [hep-th/0211111]

[48] L. Smolin. How far are we from a quantum theory of gravity. [hep-th/0303185]

[49] W. Lerche. Recent developments in string theory. [hep-th/9710246]

[50] A. Bilal. (M)atrix theory: a pedagogical introduction. [hep-th/9710136]

[51] Washington Taylor IV. Lectures on D-branes, gauge theory and (m)atrices. [hep-th/9801182]

[52] M. Haack, B. Kors and D. Lüst. Recent developments in string theory: from perturbative dualities to M theory. [hep-th/9904033]

[53] I. Antoniadis, K. Benakli and M. Quiros. Direct collider signatures of large extra dimensions. *Phys. Lett.* **B460** (1999) 176–83. [hep-ph/9905311]

[54] B. Zwiebach. Closed string field theory: an introduction. In *Gravitation and Quantisations*, B. Julia (ed.) (Elsevier, Amsterdam, 1995). [hep-th/9305026]

[55] R. Helling and H. Nicolai. Supermembranes and matrix theory. [hep-th/9809103]

[56] A. W. Peet. TASI lectures on black holes in string theory. [hep-th/0008241]

[57] M. Dine, D. O'Neil and Z. Sun. Branches of the landscape. *JHEP* **0507** (2005) 014. [hep-th/0501214]

[58] F. Denef and M. Douglas. Distributions of flux vacua. *JHEP* **0405** (2004) 072. [hep-th/0404116]

[59] B. S. Acharya and M. R. Douglas. A finite landscape? [hep-th/0606212]

[60] J. Shelton, W. Taylor and B. Wecht. Generalised flux vacua. [hep-th/0607015]

[61] L. Susskind. The anthropic landscape of string theory. [hep-th/0302219]

[62] L. Smolin. Scientific alternatives to the anthropic principle. [hep-th/0407213]

[63] L. Smolin. The case for background independence. [hep-th/0507235]

[64] J. Maldacena. The large N limit of superconformal field theories and supergravity. *Adv. Theor. Math. Phys.* **2** (1998) 231–52. [hep-th/9711200]

[65] E. Witten. Anti-de Sitter space and holography. *Adv. Theor. Math. Phys.* **2** (1998) 253–91. [hep-th/9802150]

[66] P. D. Francesco, P. Mathieu and D. Sénéchal. *Conformal Field Theory* (Springer-Verlag, New York, 1997).

[67] G. 't Hooft. The holographic principle: opening lecture. In *Erice 1999: Basics and Highlights in Fundamental Physics*, pp. 397–413. [hep-th/0003004]

[68] N. Beisert and M. Staudacher. The $N = 4$ SYM integrable super spin chain. *Nucl. Phys.* **B670** (2003) 439–63. [hep-th/0307042]

[69] K. H. Rehren. Algebraic holography. *Annales Henri Poincare* **1** (2004) 607–23. [hep-th/9905179]

[70] K. H. Rehren. Proof of the AdS/CFT correspondence. In *Quantum Theory and Symmetries*, H.-D. Doebner *et al.* (eds) (World Scientific, Singapore, 2000) pp. 278–84. [hep-th/9910074]

[71] K.-H. Rehren. Local quantum observables in the AdS/CFT correspondence. *Phys. Lett.* **B493** (2000) 383–8. [hep-th/0003120]

[72] B. Schroer. Facts and fictions about anti-De Sitter space-times with local quantum matter. *Commun. Math. Phys.* **219** (2001) 57–76. [hep-th/9911100]

[73] P. A. M. Dirac. Quantum theory of localisable dynamical systems. *Phys. Rev.* **73** (1948) 1092–103.

[74] P. A. M. Dirac. Forms of relativistic dynamics. *Rev. Mod. Phys.* **21** (1949) 392–9.

[75] P. A. M. Dirac. Generalised Hamiltonian dynamics. *Proc. Roy. Soc.* **A246** (1958) 326.

[76] P. A. M. Dirac. The theory of gravitation in Hamiltonian form. *Proc. Roy. Soc.* **A246** (1958) 333.

[77] J. L. Anderson and P. G. Bergmann. Constraints in covariant field theories. *Phys. Rev.* **83** (1951) 1018–25.

[78] P. G. Bergmann and I. Goldberg. Dirac bracket transformations in phase space. *Phys. Rev.* **98** (1955) 531–8.

[79] A. Komar. Covariant conservation laws in general relativity. *Phys. Rev.* **113** (1959) 934–6.

[80] P. G. Bergmann and A. Komar. Poisson brackets between locally defined observables in general relativity. *Phys. Rev. Lett.* **4** (1960) 432–3.

[81] R. Arnowitt, S. Deser and C. W. Misner. Canonical variables for general relativity. *Phys. Rev.* **117** (1960) 1595–602.

[82] B. S. DeWitt. Quantum theory of gravity. I. The canonical theory. *Phys. Rev.* **160** (1967) 1113–48.

[83] B. S. DeWitt. Quantum theory of gravity. II. The manifestly covariant theory. *Phys. Rev.* **162** (1967) 1195–238.

[84] B. S. DeWitt. Quantum theory of gravity. III. Applications of the covariant theory. *Phys. Rev.* **162** (1967) 1239–56.

[85] J. A. Wheeler. *Geometrodynamics* (Academic Press, New York, 1962).

[86] K. Kuchař. In *Quantum Gravity II: A second Oxford Symposium*, C. J. Isham, R. Penrose, D. W. Sciama (eds) (Clarendon Press, Oxford, 1981).

[87] J. Klauder. The affine quantum gravity program. *Class. Quant. Grav.* **19** (2002) 817–26. [gr-qc/0110098]

[88] J. Klauder. Noncanonical quantisation of gravity I. Foundations of affine quantum gravity. *J. Math. Phys.* **40** (1999) 5860–82. [gr-qc/9906013]

[89] J. Klauder. Noncanonical quantisation of gravity II. Constraints and the physical Hilbert space. *J. Math. Phys.* **42** (2001) 4440–65. [gr-qc/0102041]

[90] J. Klauder. Universal procedure for enforcing quantum constraints. *Nucl. Phys.* **B547** (1999) 397–412. [hep-th/9901010]

[91] A. Kempf and J. R. Klauder. On the implementation of constraints through projection operators. *J. Phys.* **A34** (2001) 1019–36. [quant-ph/0009072]

[92] A. Ashtekar. New variables for classical and quantum gravity. *Phys. Rev. Lett.* **57** (1986) 2244–7.

[93] A. Ashtekar. New Hamiltonian formulation of general relativity. *Phys. Rev.* **D36** (1987) 1587–602.

[94] J. B. Hartle and S. W. Hawking. Wave function of the universe. *Phys. Rev.* **D28** (1983) 2960–75.

[95] L. J. Garay, J. J. Halliwell and G. A. Mena Marugan. Path integral quantum cosmology: a class of exactly soluble scalar field minisuperspace models with exponential potentials. *Phys. Rev.* **D43** (1991) 2572–89.

[96] J. J. Halliwell and J. Louko. Steepest descent contours in the path integral approach to quantum cosmology. 1. The De Sitter minisuperspace model. *Phys. Rev.* **D39** (1989) 2206.

[97] J. J. Halliwell and J. Louko. Steepest descent contours in the path integral approach to quantum cosmology. 2. Microsuperspace. *Phys. Rev.* **D40** (1989) 1868.

[98] J. J. Halliwell and J. Louko. Steepest descent contours in the path integral approach to quantum cosmology. 3. A general method with applications to anisotropic minisuperspace models. *Phys. Rev.* **D42** (1990) 3997–4031.

[99] J. Glimm and A. Jaffe. *Quantum Physics* (Springer-Verlag, New York, 1987).

[100] S. Weinberg. *The Quantum Theory of Fields*, Vols 1–3 (Cambridge University Press, Cambridge, 1996).

[101] O. Lauscher and M. Reuter. Towards nonperturbative renormalisability of quantum Einstein gravity. *Int. J. Mod. Phys.* **A17** (2002) 993–1002. [hep-th/0112089]

[102] O. Lauscher and M. Reuter. Is quantum gravity nonperturbatively normalisable? *Class. Quant. Grav.* **19** (2002) 483–92. [hep-th/0110021]

[103] M. Reuter and H. Weyer. Running Newton constant, improved gravitational actions and galaxy rotation curves. *Phys. Rev.* **D70** (2004) 124028. [hep-th/0410117]

[104] M. Reuter and H. Weyer. Quantum gravity at astrophysical distances. *JCAP* **0412** (2004) 001. [hep-th/0410119]

[105] A. Bonanno and M. Reuter. Proper time flow equation for gravity. *JHEP* **0502** (2005) 035. [hep-th/0410191]

[106] M. Reuter and F. Saueressig. From big bang to asymptotic de Sitter: complete cosmologies in a quantum gravity framework. *JCAP* **0509** (2005) 012. [hep-th/0507167]

[107] O. Lauscher and M. Reuter. Fractal spacetime structure in asymptotically safe gravity. *JHEP* **0510** (2005) 090. [hep-th/0508202]

[108] M. Reuter and H. Weyer. Do we observe quantum gravity effects at galactic scales? [astro-ph/0509163]

[109] P. Forgacs and M. Niedermaier. A fixed point for truncated quantum Einstein gravity. [hep-th/0207028]

[110] M. Niedermaier. On the renormalisation of truncated Einstein gravity. *JHEP* **0212** (2002) 066. [hep-th/0207143]

[111] M. Niedermaier. Dimensionally reduced gravity theories are asymptotically safe. *Nucl. Phys.* **B673** (2003) 131–69. [hep-th/0304117]

[112] R. Loll. Discrete approaches to quantum gravity in four dimensions. *Living Rev. Rel.* **1** (1998) 13. [gr-qc/9805049]

[113] T. Regge. General relativity without coordinates. *Nuovo Cimento* **19** (1961) 558–71.

[114] R. M. Williams and P. Tuckey. Regge calculus: a bibliography and brief review. *Class. Quant. Grav.* **9** (1992) 1409–22.

[115] R. M. Williams. Recent progress in Regge calculus. *Nucl. Phys. Procs. Suppl.* **57** (1997) 73–81. [gr-qc/9702006]

[116] J. Ambjorn, M. Carfora and A. Marzuoli. *The Geometry of Dynamical Triangulations* (Springer-Verlag, Berlin, 1998).

[117] J. Ambjorn and R. Loll. Nonperturbative Lorentzian quantum gravity and topology change. *Nucl. Phys.* **B536** (1998) 407–34. [hep-th/9805108]

[118] J. Ambjorn, R. Loll and J. L. Nielsen. Euclidean and Lorentzian quantum gravity: lessons from two dimensions. *Chaos Solitons Fractals* **10** (1999) 177–95. [hep-th/9806241]

[119] J. Ambjorn, K. N. Anagnostopoulos and R. Loll. On the phase diagram of 2-D Lorentzian quantum gravity. *Nucl. Phys. Proc. Suppl.* **83** (2000) 733–5. [hep-lat/9908054]

[120] J. Ambjorn, K. N. Anagnostopoulos and R. Loll. Crossing the $C = 1$ barrier in 2-D Lorentzian quantum gravity. *Phys. Rev.* **D61** (2000) 044010. [hep-lat/9909129]

[121] J. Ambjorn, J. Correia, C. Kristjansen and R. Loll. On the relation between Euclidean and Lorentzian 2-D quantum gravity. *Phys. Lett.* **B475** (2000) 24–32. [hep-th/9912267]

[122] J. Ambjorn, J. Jurkiewicz and R. Loll. Lorentzian and Euclidean quantum gravity: analytical and numerical results. [hep-th/0001124]

[123] J. Ambjorn, J. Jurkiewicz and R. Loll. A nonperturbative Lorentzian path integral for gravity. *Phys. Rev. Lett.* **85** (2000) 924–7. [hep-th/0002050]

[124] J. Ambjorn, J. Jurkiewicz and R. Loll. Nonperturbative 3-D Lorentzian quantum gravity. *Phys. Rev.* **D64** (2001) 044011. [hep-th/0011276]

[125] A. Dasgupta and R. Loll. A proper time cure for the conformal sickness in quantum gravity. *Nucl. Phys.* **B606** (2001) 357–79. [hep-th/0103186]

[126] J. Ambjorn, J. Jurkiewicz and R. Loll. Emergence of a 4D world from causal quantum gravity. *Phys. Rev. Lett.* **93** (2004) 131301. [hep-th/0404156]

[127] J. Ambjorn, J. Jurkiewicz and R. Loll. Spectral dimension of the universe. *Phys. Rev. Lett.* **95** (2005) 171301. [hep-th/0505113]

[128] J. Ambjorn, J. Jurkiewicz and R. Loll. Reconstructing the universe. *Phys. Rev.* **D72** (2005) 064014. [hep-th/0505154]

[129] D. Marolf and C. Rovelli. Relativistic quantum measurement. *Phys. Rev.* **D66** (2002) 023510. [gr-qc/0203056]

[130] A. Ashtekar, L. Bombelli and O. Reula. The covariant phase space of asymptotically flat gravitational fields. In *Analysis, Geometry and Mechanics: 200 Years after Lagrange*, M. Francaviglia and D. Holm (eds) (North-Holland, Amsterdam, 1991).

[131] R. E. Peierls. The commutation laws of relativistic field theory. *Proc. R. Soc. Lond.* **A214** (1952) 143.

[132] B. S. DeWitt. *Dynamical Theory of Groups and Fields* (Gordon & Breach, New York, 1965).

[133] M. J. Gotay, J. Isenberg and J. E. Marsden. Momentum maps and classical relativistic fields. Part 1: Covariant field theory (with the collaboration of Richard Montgomery, Jedrzej Sniatycki and Philip B. Yasskin). [physics/9801019]

[134] H. A. Kastrup, Canonical theories of dynamical systems in physics. *Phys. Rept.* **101** (1983) 1.

[135] I. V. Kanatchikov. Canonical structure of classical field theory in the polymomentum phase space. *Rept. Math. Phys.* **41** (1998) 49–90. [hep-th/9709229]

[136] I. V. Kanatchikov. On the field theoretic generalisations of a Poisson algebra. *Rept. Math. Phys.* **40** (1997) 225. [hep-th/9710069]

[137] C. Rovelli. A note on the foundation of relativistic mechanics. [gr-qc/0111037]

[138] C. Rovelli. A note on the foundation of relativistic mechanics 2. Covariant Hamiltonian general relativity. [gr-qc/0202079]

[139] C. Rovelli. Covariant Hamiltonian formalism for field theory: symplectic structure and Hamilton–Jacobi equation on the space G. [gr-qc/0207043]

[140] I. V. Kanatchikov. Precanonical perspective in quantum gravity. *Nucl. Phys. Proc. Suppl.* **88** (2000) 326–30. [gr-qc/0004066]

[141] I. V. Kanatchikov. Precanonical quantisation and the Schrödinger wave functional. *Phys. Lett.* **A283** (2001) 25–36. [hep-th/0012084]

[142] I. V. Kanatchikov. Precanonical quantum gravity: quantisation without the spacetime decomposition. *Int. J. Theor. Phys.* **40** (2001) 1121–49. [gr-qc/0012074]

[143] R. B. Griffiths. Consistent histories and the interpretation of quantum mechanics. *J. Stat. Phys.* **36** (1984) 219–72.

[144] R. B. Griffiths. The consistency of consistent histories: a reply to d'Espagnat. *Found. Phys.* **23** (1993) 1601–10.

[145] R. Omnés. Logical reformulation of quantum mechanics. 1. Foundations. *J. Stat. Phys.* **53** (1988) 893–932.

[146] R. Omnés. Logical reformulation of quantum mechanics. 2. Interferences and the Einstein–Podolsky–Rosen experiment. *J. Stat. Phys.* **53** (1988) 933–55.

[147] R. Omnés. Logical reformulation of quantum mechanics. 3. Classical limit and irreversibility. *J. Stat. Phys.* **53** (1988) 957–75.

[148] R. Omnés. Logical reformulation of quantum mechanics. 4. Projectors in semiclassical physics. *J. Stat. Phys.* **57** (1989) 356–82.

[149] R. Omnés. Consistent interpretations of quantum mechanics. *Rev. Mod. Phys.* **64** (1992) 339–82.

[150] M. Gell-Mann and J. B. Hartle. Classical equations for quantum systems. *Phys. Rev.* **D47** (1993) 3345–82. [gr-qc/9210010]

[151] M. Gell-Mann and J. B. Hartle. Equivalent sets of histories and multiple quasiclassical domains. [gr-qc/9404013]

[152] M. Gell-Mann and J. B. Hartle. Strong decoherence. [gr-qc/9509054]

[153] J. B. Hartle. Space-time quantum mechanics and the quantum mechanics of space-time. In *Gravitation and Quantisations*, B. Julia (ed.) (Elsevier, Amsterdam, 1995).

[154] F. Dowker and A. Kent. Properties of consistent histories. *Phys. Rev. Lett.* **75** (1995) 3038–41. [gr-qc/9409037]

[155] F. Dowker and A. Kent. On the consistent histories approach to quantum mechanics. *J. Statist. Phys.* **82** (1996) 1575–646. [gr-qc/9412067]

[156] C. J. Isham and N. Linden. Quantum temporal logic and decoherence functionals in the histories approach to generalised quantum theory. *J. Math. Phys.* **35** (1994) 5452–76. [gr-qc/9405029]

[157] C. J. Isham and N. Linden. Continuous histories and the history group in generalised quantum theory. *J. Math. Phys.* **36** (1995) 5392–408. [gr-qc/9503063]

[158] C. J. Isham, N. Linden, K. Savvidou and S. Schreckenberg. Continuous time and consistent histories. *J. Math. Phys.* **39** (1998) 1818–34. [quant-ph/9711031]

[159] K. N. Savvidou. General relativity histories theory: space-time diffeomorphisms and the Dirac algebra of constraints. *Class. Quant. Grav.* **18** (2001) 3611–28. [gr-qc/0104081]

[160] K. N. Savvidou. Poincaré invariance for continuous time histories. *J. Math. Phys.* **43** (2002) 3053–73. [gr-qc/0104053]

[161] K. N. Savvidou. The action operator in continuous time histories. *J. Math. Phys.* **40** (1999) 5657. [gr-qc/9811078]

[162] K. N. Savvidou. Continuous time in consistent histories. PhD thesis. [gr-qc/9912076]

[163] K. N. Savvidou and C. Anastopoulos. Histories quantisation of parametrised systems. 1. Development of a general algorithm. *Class. Quant. Grav.* **17** (2000) 2463–90. [gr-qc/9912077]

[164] I. Kouletsis and K. Kuchař. Diffeomorphisms as symplectomorphisms in history phase space: bosonic string model. *Phys. Rev.* **D65** (2002) 125026. [gr-qc/0108022]

[165] C. J. Isham and K. N. Savvidou. Quantising the foliation in history quantum field theory. [quant-ph/0110161]

[166] C. J. Isham and K. N. Savvidou. The foliation operator in history quantum field theory. *J. Math. Phys.* **43** (2002) 5493–513.

[167] A. Connes. *Noncommutative Geometry* (Academic Press, London, 1994).

[168] G. Landi. *An Introduction to Non-Commutative Spaces and their Geometries* (Springer-Verlag, Berlin, 1997).

[169] N. Seiberg and E. Witten. String theory and noncommutative geometry. *JHEP* **9909** (1999) 032. [hep-th/9908142]

[170] C. Isham. Topos theory and consistent histories: the internal logic of the set of all consistent sets. *Int. J. Theor. Phys.* **36** (1997) 785–814. [gr-qc/9607069]

[171] C. J. Isham and J. Butterfield. A topos perspective on the Kochen–Specker theorem. 1. Mathematical development. [quant-ph/9803055]

[172] C. J. Isham and J. Butterfield. A topos perspective on the Kochen–Specker theorem. 2. Conceptual aspects. *Int. J. Theor. Phys.* **38** (1999) 827–59. [quant-ph/9808067]

[173] C. J. Isham and J. Butterfield, A topos perspective on the Kochen–Specker theorem. 4. Interval valuations. *Int. J. Theor. Phys.* **41** (2002) 613–39. [quant-ph/0107123]

[174] C. J. Isham and J. Butterfield. Some possible roles of topos theory in quantum theory and quantum gravity. *Found. Phys.* **30** (2000) 1707–35. [gr-qc/9910005]

[175] R. Penrose. Twistor theory and the Einstein vacuum. *Class. Quant. Grav.* **16** (1999) A113–30.

[176] R. Penrose. A brief introduction into twistors. *Surveys High Energ. Phys.* **1** (1980) 267–88.

[177] R. Penrose and M. A. H. MacCallum. Twistor theory: an approach to the quantisation of fields and space-time. *Phys. Rept.* **6** (1972) 241–316.

[178] R. Penrose. The twistor programme. *Rept. Math. Phys.* **12** (1977) 65–76.

[179] R. Penrose. Nonlinear gravitons and curved twistor theory. *Gen. Rel. Grav.* **7** (1976) 31–52.

[180] R. Penrose. Twistor algebra. *J. Math. Phys.* **8** (1967) 345.

[181] D. Rideout and R. Sorkin. Evidence for a continuum limit in causal set dynamics. *Phys. Rev.* **D63** (2001) 104011. [gr-qc/0003117]

[182] G. Brightwell, H. Fay Dowker, R. S. Garcia, J. Henson and R. D. Sorkin. Observables in causal set cosmology. *Phys. Rev.* **D67** (2003) 084031. [gr-qc/0210061]

[183] R. D. Sorkin. Indications of causal set cosmology. *Int. J. Theor. Phys.* **39** (2000) 1731–6. [gr-qc/0003043]

[184] D. P. Rideout and R. D. Sorkin. A classical sequential growth dynamics for causal sets. *Phys. Rev.* **D61** (2000) 024002. [gr-qc/9904062]

[185] L. Bombelli, J.-H. Lee, D. Meyer and R. Sorkin. Space-time as a causal set. *Phys. Rev. Lett.* **59** (1987) 521.

[186] L. Bombelli, R. K. Koul, J.-H. Lee and R. D. Sorkin. A quantum source of entropy for black holes. *Phys. Rev.* **D34** (1986) 373.

[187] G. 't Hooft. Quantum gravity as a dissipative deterministic system. *Class. Quant. Grav.* **16** (1999) 3263–79. [gr-qc/9903084]

[188] G. 't Hooft. Determinism and dissipation in quantum gravity. [hep-th/0003005]

[189] K. Pohlmeyer. A group theoretical approach to the quantisation of the free relativistic closed string. *Phys. Lett.* **B119** (1982) 100.

[190] K. Pohlmeyer and K. H. Rehren. Algebraic properties of the invariant charges of the Nambu–Goto theory. *Commun. Math. Phys.* **105** (1986) 593.

[191] K. Pohlmeyer and K. H. Rehren. The algebra formed by the charges of the Nambu–Goto theory: identification of a maximal Abelian subalgebra. *Commun. Math. Phys.* **114** (118) 55.

[192] K. Pohlmeyer and K. H. Rehren. The algebra formed by the charges of the Nambu–Goto theory: their geometric origin and their completeness. *Commun. Math. Phys.* **114** (1988) 177.

[193] K. Pohlmeyer. The invariant charges of the Nambu–Goto theory in WKB approximation. *Commun. Math. Phys.* **105** (1986) 629.

[194] K. Pohlmeyer. The algebra formed by the charges of the Nambu–Goto theory: Casimir elements. *Commun. Math. Phys.* **114** (1988) 351.

[195] K. Pohlmeyer. Uncovering the detailed structure of the algebra formed by the invariant charges of closed bosonic strings moving in $(1 + 2)$-dimensional Minkowski space. *Commun. Math. Phys.* **163** (1994) 629–44.

[196] K. Pohlmeyer. The invariant charges of the Nambu–Goto theory: non-additive composition laws. *Mod. Phys. Lett.* **A10** (1995) 295–308.

[197] K. Pohlmeyer. The Nambu–Goto theory of closed bosonic strings moving in $(1 + 3)$-dimensional Minkowski space: the quantum algebra of observables. *Annal. Phys.* **8** (1999) 19–50. [hep-th/9805057]

[198] K. Pohlmeyer and M. Trunk. The invariant charges of the Nambu–Goto theory: quantisation of non-additive composition laws. *Int. J. Mod. Phys.* **A19** (2004) 115–48. [hep-th/0206061]

[199] G. Handrich and C. Nowak. The Nambu–Goto theory of closed bosonic strings moving in $(1 + 3)$-dimensional Minkowski space: the construction of the quantum algebra of observables up to degree five. *Annal. Phys.* **8** (1999) 51–4. [hep-th/9807231]

[200] G. Handrich. Lorentz covariance of the quantum algebra of observables: Nambu–Goto strings in $3 + 1$ dimensions. *Int. J. Mod. Phys.* **A17** (2002) 2331–49.

[201] G. Handrich, C. Paufler, J. B. Tausk and M. Walter. The representation of the algebra of observables of the closed bosonic string in $1 + 3$ dimensions: calculation to order $\hbar^7$. [math-ph/0210024]

[202] G. Handrich, C. Paufler, J. B. Tausk and M. Walter. The presentation of the quantum algebra of observables of the closed bosonic string in $1 + 3$ dimensions: the exact quantised generating relations of orders $\hbar^6$ and $\hbar^7$. *Mod. Phys. Lett.* **A17** (2002) 2611–15.

[203] C. Meusburger and K. H. Rehren. Algebraic quantisation of the closed bosonic string. *Commun. Math. Phys.* **237** (2003) 69–85. [math-ph/0202041]

[204] D. Bahns. The invariant charges of the Nambu–Goto string and canonical quantisation. *J. Math. Phys.* **45** (2004) 4640–60. [hep-th/0403108]

[205] T. Thiemann. The LQG string: loop quantum gravity quantisation of string theory I: Flat target space. *Class. Quant. Grav.* **23** (2006) 1923–70. [hep-th/0401172]

[206] R. Arnowitt, S. Deser and C. W. Misner. In *Gravitation: An Introduction to Current Research*, L. Witten (ed.) (Wiley, New York, 1962).

[207] R. M. Wald. *General Relativity* (The University of Chicago Press, Chicago, 1989).

[208] S. Hawking and G. Ellis. *The Large Scale Structure of Spacetime* (Cambridge University Press, Cambridge, 1989).

[209] R. Geroch. The domain of dependence. *Math. Phys.* **11** (1970) 437–509.

[210] A. Anderson and B. Dewitt. Does the topology of space fluctuate? *Found. Phys.* **16** (1986) 91–105.

[211] F. Dowker. Topology change in quantum gravity. In *Cambridge 2002: The Future of Theoretical Physics and Cosmology*, pp. 436–52 (Cambridge University Press, Cambridge,2002). [gr-qc/ 0206020]

[212] F. Dowker and S. Surya. Topology change and causal continuity. *Phys. Rev.* **D58** (1998) 124019. [gr-qc/9711070]

[213] H. F. Dowker and R. S. Garcia. A handlebody calculus for topology change. *Class. Quant. Grav.* **15** (1998) 1859–79. [gr-qc/9711042]

[214] R. Loll and W. Westra. Sum over topologies and double scaling limit in 2-D Lorentzian quantum gravity. *Class. Quant. Grav.* **23** (2006) 465–72. [hep-th/0306183]

[215] R. Loll and W. Westra. Spacetime foam in 2-D and the sum over topologies. *Acta Phys. Polon.* **B34** (2003) 4997. [hep-th/0309012]

[216] C. J. Isham and K. Kuchař. Representations of space-time diffeomorphisms. 1. Canonical parametrised field theories. *Ann. Phys.* **164** (1985) 288.

[217] C. J. Isham and K. Kuchař. Representations of space-time diffeomorphisms. Canonical geometrodynamics. *Ann. Phys.* **164** (1985) 316.

[218] N. M. J. Woodhouse. *Geometric Quantisation*, 2nd edn. (Clarendon Press, Oxford, 1991).

[219] P. A. M. Dirac. *Lectures on Quantum Mechanics* (Belfer Graduate School of Science, Yeshiva University Press, New York, 1964).

[220] J. E. Marsden and P. R. Chernoff. *Properties of Infinite Dimensional Hamiltonian Systems* (Lecture Notes in Mathematics, Springer-Verlag, Berlin, 1974).

[221] P. G. Bergmann and A. Komar. The phase space formulation of general relativity and approaches towards its canonical quantisation. *Gen. Rel. Grav.* **1** (1981) 227–54.

[222] A. Komar. General relativistic observables via Hamilton Jacobi functionals. *Phys. Rev.* **D4** (1971) 923–7.

[223] A. Komar. Commutator algebra of general relativistic observables. *Phys. Rev.* **D9** (1974) 885–8.

[224] A. Komar. Generalised constraint structure for gravitation theory. *Phys. Rev.* **D27** (1983) 2277–81.

[225] A. Komar. Consistent factor ordering of general relativistic constraints. *Phys. Rev.* **D20** (1979) 830–33.

[226] P. G. Bergmann and A. Komar. The coordinate group symmetries of general relativity. *Int. J. Theor. Phys.* **5** (1972) 15.

[227] B. Dittrich and T. Thiemann. Facts and fiction about Dirac observables. In preparation.

[228] J. Pons. Generally covariant theories: the Noether obstruction for realising certain spacetime diffeomorphisms in phase space. *Class. Quant. Grav.* **20** (2003) 3279. [gr-qc/0306035]

[229] J. A. Gracia and J. Pons. Lagrangian Noether symmetries as canonical transformations. *Int. J. Mod. Phys.* **A16** (2001) 3897. [hep-th/0012094]

[230] J. A. Gracia and J. Pons. Rigid and gauge Noether symmetries for constrained systems. *Int. J. Mod. Phys.* **A15** (2000) 4681. [hep-th/9908151]

[231] J. A. Gracia and J. Pons. Canonical Noether symmetries and commutativity properties for gauge systems. *J. Math. Phys.* **41** (2000) 7333. [math-ph/0007037]

[232] J. M. Pons, D. C. Salisbury and L. C. Shepley. Gauge transformations in the Lagrangian and Hamiltonian formalisms of generally covariant theories. *Phys. Rev.* **D55** (1997) 658. [gr-qc/9612037]

[233] S. A. Hojman, K. Kuchař and C. Teitelboim. Geometrodynamics regained. *Annal. Phys.* **96** (1976) 88–135.

[234] Y. Choquet-Bruhat and C. DeWitt-Morette. *Analysis, Manifolds and Physics*, Part I (North-Holland, Amsterdam, 1989).

[235] Y. Choquet-Bruhat and C. DeWitt-Morette. *Analysis, Manifolds and Physics*, Part II (North-Holland, Amsterdam, 1989).

[236] Y. Choquet-Bruhat. Positive-energy theorems. In *Relativity, Groups and Topology II*, B. S. DeWitt and R. Stora (eds), p. 739 (Elsevier Science Publishers B. V., Amsterdam, 1984).

[237] R. Schoen and S. T. Yau. On the proof of the positive mass conjecture in general relativity. *Commun. Math. Phys.* **65** (1979) 45–76.

[238] R. Schoen and S. T. Yau. The energy and the linear momentum of space-times in general relativity. *Commun. Math. Phys.* **79** (1981) 47–51.

[239] R. Schoen and S. T. Yau. Proof of the positive mass theorem. II. *Commun. Math. Phys.* **79** (1981) 231–60.

[240] E. Witten. A new proof of the positive energy theorem. *Commun. Math. Phys.* **80** (1981) 381–402.

[241] A. Ashtekar and R. O. Hansen. A unified treatment of null and spatial infinity in general relativity. I. Universal structure, asymptotic symmetries and conserved quantities at spatial infinity. *J. Math. Phys.* **19** (1978) 1542.

[242] R. Beig and B. G. Schmidt. Einstein's equations near spatial infinity. *Commun. Math. Phys.* **87** (1982) 65.

[243] R. Penrose. *The Emperor's New Mind* (Oxford University Press, Oxford, 1990).

[244] R. Beig and O. Murchadha. The Poincaré group as the symmetry group of canonical general relativity. *Ann. Phys.* **174** (1987) 463.

[245] T. Regge and C. Teitelboim. Role of surface integrals in the Hamiltonian formulation of general relativity. *Annal. Phys.* **88** (1974) 286.

[246] C. G. Torre and I. M. Anderson. Symmetries of the Einstein equations. *Phys. Rev. Lett.* **70** (1993) 3525–9. [gr-qc/9302033]

[247] C. G. Torre and I. M. Anderson. Classification of generalised symmetries for the vacuum Einstein equations. *Commun. Math. Phys.* **176** (1996) 479–539. [gr-qc/9404030]

[248] C. Rovelli. What is observable in classical and quantum gravity? *Class. Quant. Grav.* **8** (1991) 297–316.

[249] C. Rovelli. Quantum reference systems. *Class. Quant. Grav.* **8** (1991) 317–332.

[250] C. Rovelli. Time in quantum gravity: physics beyond the Schrödinger regime. *Phys. Rev.* **D43** (1991) 442–56.

[251] C. Rovelli. Quantum mechanics without time: a model. *Phys. Rev.* **D42** (1990) 2638–46.

[252] T. Thiemann. The phoenix project: master constraint programme for loop quantum gravity. *Class. Quant. Grav.* **23** (2006) 2211–48. [gr-qc/0305080]

[253] B. Dittrich and T. Thiemann. Testing the master constraint programme for loop quantum gravity: I. General framework. *Class. Quant. Grav.* **23** (2006) 1025–66. [gr-qc/0411138]

[254] B. Dittrich and T. Thiemann. Testing the master constraint programme for loop quantum gravity: II. Finite-dimensional systems. *Class. Quant. Grav.* **23** (2006) 1067–88. [gr-qc/0411139]

[255] B. Dittrich and T. Thiemann. Testing the master constraint programme for loop quantum gravity: III. SL(2R) models. *Class. Quant. Grav.* **23** (2006) 1089–1120. [gr-qc/0411140]

[256] B. Dittrich and T. Thiemann. Testing the master constraint programme for loop quantum gravity: IV. Free field theories. *Class. Quant. Grav.* **23** (2006) 1121–42. [gr-qc/0411141]

[257] B. Dittrich and T. Thiemann. Testing the master constraint programme for loop quantum gravity: V. Interacting field theories. *Class. Quant. Grav.* **23** (2006) 1143–62. [gr-qc/0411142]

[258] B. Dittrich. Partial and complete observables for Hamiltonian constrained systems. [gr-qc/0411013]

[259] B. Dittrich. Partial and complete observables for canonical general relativity. [gr-qc/0507106]

[260] T. Thiemann. Reduced phase space quantisation and Dirac observables. *Class. Quant. Grav.* **23** (2006) 1163–80. [gr-qc/0411031]

[261] J. Pullin and R. Gambini. Making classical and quantum canonical general relativity computable through a power series expansion in the inverse cosmological constant. *Phys. Rev. Lett.* **85** (2000) 5272–5. [gr-qc/0008031]

[262] R. Gambini and J. Pullin. The large cosmological constant approximation to classical and quantum gravity: model examples. *Class. Quant. Grav.* **17** (2000) 4515–40. [gr-qc/0008032]

[263] M. Henneaux and C. Teitelboim. *Quantisation of Gauge Systems* (Princeton University Press, Princeton, 1992).

[264] T. Thiemann. Solving the problem of time in general relativity and cosmology with phantoms and k-essence. [astro-ph/0607380]

[265] D. Giulini, C. Kiefer, E. Joos, J. Kupsch, I. O. Stamatescu and H. D. Zeh. *Decoherence and the Appearance of a Classical World in Quantum Theory* (Springer-Verlag, Berlin, 1996).

[266] A. Ashtekar, J. Lewandowski, D. Marolf, J. Mourão and T. Thiemann. Quantisation of diffeomorphism invariant theories of connections with local degrees of freedom. *J. Math. Phys.* **36** (1995) 6456–93. [gr-qc/9504018]

[267] A. Higuchi. Quantum linearisation instabilities of de Sitter space-time. 1 *Class. Quant. Grav.* **8** (1991) 1961–81.

[268] A. Higuchi. Quantum linearisation instabilities of de Sitter space-time. 2. *Class. Quant. Grav.* **8** (1991) 1983–2004.

[269] A. Higuchi. Linearised gravity in de Sitter space-time as a representation of SO(4,1). *Class. Quant. Grav.* **8** (1991) 2005–21.

[270] A. Higuchi. Linearised quantum gravity in flat space with toroidal topology. *Class. Quant. Grav.* **8** (1991) 2023–34.

[271] D. Marolf. Path integrals and instantons in quantum gravity: minisuperspace models. *Phys. Rev.* **D53** (1996) 6979.

[272] D. Marolf. Almost ideal clocks in quantum cosmology: a brief derivation of time. *Class. Quant. Grav.* **12** (1995) 2469–86.

[273] D. Marolf. The spectral analysis inner product for quantum gravity. In *Proceedings of the VIIth Marcel Grossman Conference*, R. Ruffini and M. Keiser (eds) (World Scientific, Singapore, 1995). [gr-qc/9409036]

[274] D. Marolf. Quantum observable and recollapsing dynamics. *Class. Quant. Grav.* **12** (1995) 1199–220. [gr-qc/9404053]

[275] D. Marolf. Group averaging and refined algebraic quantisation: where are we now? In *Proceedings of 9th Marcel Grossman Meeting on Recent Developments in Theoretical and Experimental General Relativity, Gravitation and Relativistic Field Theories (MG 9), Rome, Italy, 2–9 Jul 2000.* [gr-qc/0011112]

[276] A. Gomberoff and D. Marolf. On group averaging for SO(N,1). *Int. J. Mod. Phys.* **D8** (1999) 519–35. [gr-qc/9902069]

[277] D. Giulini and D. Marolf. A uniqueness theorem for constraint quantisation. *Class. Quant. Grav.* **16** (1999) 2489–505. [gr-qc/9902045]

[278] D. Giulini and D. Marolf. On the generality of refined algebraic quantisation. *Class. Quant. Grav.* **16** (1999) 2479–88. [gr-qc/9812024]

[279] H. Araki. Hamiltonian formalism and the canonical commutation relations in quantum field theory. *J. Math. Phys.* **1** (1960) 492.

[280] I. M. Gel'fand and N. Ya. Vilenkin. *Generalised Functions, Vol. 4: Applications of Harmonic Analysis* (Academic Press, New York, 1964).

[281] C. Isham. In *Relativity Groups and Topology II*, B. DeWitt and R. Stora (eds) (North-Holland, Amsterdam, 1984).

[282] M. Reed and B. Simon. *Methods of Modern Mathematical Physics*, Vols 1–4 (Academic Press, Boston, 1980).

[283] P. Hajíček and K. Kuchař. Constraint quantisation of parametrised relativistic gauge systems in curved space-times. *Phys. Rev.* **D41** (1990) 1091–104.

[284] P. Hajíček and K. Kuchař. Transversal affine connection and quantisation of constrained systems. *J. Math. Phys.* **31** (1990) 1723–32.

[285] A. Sen. On the existence of neutrino 'zero modes' in vacuum spacetimes. *J. Math. Phys.* **22** (1981) 1781–6.

[286] A. Sen. Gravity as a spin system. *Phys. Lett* **B119** (1982) 89–91.

[287] A. Sen. Quantum theory of spin 3/2 field in Einstein spaces. *Int. J. Theor. Phys.* **21** (1982) 1–35.

[288] R. Sexl and K. Urbantke. *Relativität, Gruppen, Teilchen* (Springer-Verlag, Berlin, 1982).

[289] R. Penrose and W. Rindler. *Spinors and Spacetime*, Vols 1, 2 (Cambridge University Press, Cambridge, 1990).

[290] A. Ashtekar. In *Mathematics and General Relativity* (American Mathematical Society, Providence, Rhode Island, 1987).

[291] A. Ashtekar. Old problems in the light of new variables. *Contemporary Math.* **71** (1988) 39.

[292] J. Samuel. A Lagrangian basis for Ashtekar's formulation of canonical gravity. *Pramana J. Phys.* **28** (1987) L429–32.

[293] T. Jacobson and L. Smolin. The left-handed spin connection as a variable for canonical gravity. *Phys. Lett.* **B196** (1987) 39–42.

[294] T. Jacobson and L. Smolin. Covariant action for Ashtekar's form of canonical gravity. *Class. Quant. Grav.* **5** (1987) 583.

[295] T. Jacobson. Fermions in canonical gravity. *Class. Quantum Grav.* **5** (1987) L143.

[296] T. Jacobson. New variables for canonical supergravity. *Class. Quant. Grav.* **5** (1988) 923.

[297] A. Ashtekar, J. D. Romano and R. S. Tate. New variables for gravity: inclusion of matter. *Phys. Rev.* **D40** (1989) 2572.

[298] J. N. Goldberg. Triad approach to the Hamiltonian of general relativity. *Phys. Rev.* **D37** (1987) 2116–20.

[299] M. Henneaux, J. E. Nelson and C. Schomblond. Derivation of Ashtekar variables from tetrad gravity. *Phys. Rev.* **D39** (1989) 434–7.

[300] R. Capovilla, T. Jacobson and J. Dell. Gravitational instantons as su(2) gauge fields. *Class. Quant. Grav.* **7** (1990) L1–3.

[301] R. Capovilla, T. Jacobson and J. Dell. General relativity without the metric. *Phys. Rev. Lett.* **63** (1989) 2325.

[302] R. Capovilla, T. Jacobson and J. Dell. A pure spin connection formulation of gravity. *Class. Quant. Grav.* **8** (1991) 59–73.

[303] R. Capovilla, T. Jacobson. J. Dell and L. Maison. Selfdual two forms and gravity. *Class. Quant. Grav.* **8** (1991) 41–57.

[304] P. Peldan and I. Bengtsson. Ashtekar's variables, the theta term, and the cosmological constant. *Phys. Lett.* **B244** (1990) 261–4.

[305] P. Peldan and I. Bengtsson. Another 'cosmological' constant. *Int. J. Mod. Phys.* **A7** (1992) 1287–308.

[306] P. Peldan. Legendre transforms in Ashtekar's theory of gravity. *Class. Quant. Grav.* **8** (1991) 1765–84.

[307] P. Peldan. Actions for gravity, with generalisations: a review. *Class. Quant. Grav.* **11** (1994) 1087–132. [gr-qc/9305011]

[308] P. Peldan. Ashtekar's variables for arbitrary gauge group. *Phys. Rev.* **D46** (1992) 2279–82. [hep-th/9204069]

[309] P. Peldan. Unification of gravity and Yang–Mills theory in $(2 + 1)$-dimensions. *Nucl. Phys.* **B395** (1993) 239–62. [gr-qc/9211014]

[310] F. Barbero. Real Ashtekar variables for Lorentzian signature space times. *Phys. Rev.* **D51** (1995) 5507–10. [gr-qc/9410014]

[311] F. Barbero. Reality conditions and Ashtekar variables: a different perspective. *Phys. Rev.* **D51** (1995) 5498–506. [gr-qc/9410013]

[312] C. Rovelli and T. Thiemann. The Immirzi parameter in quantum general relativity. *Phys. Rev.* **D57** (1998) 1009–14. [gr-qc/9705059]

[313] G. Immirzi. Quantum gravity and Regge calculus. *Nucl. Phys. Proc. Suppl.* **57** (1997) 65. [gr-qc/9701052]

[314] A. Corichi and K. Krasnov. Ambiguities in loop quantisation: area versus electric charge. *Mod. Phys. Lett.* **A13** (1998) 1339–46.

[315] T. Thiemann. Reality conditions inducing transforms for quantum gauge field theories and quantum gravity. *Class. Quant. Grav.* **13** (1996) 1383–403. [gr-qc/9511057]

[316] T. Thiemann. An account of transforms on $\overline{\mathcal{A}/\mathcal{G}}$. *Acta Cosmol.* **21** (1995) 145–67. [gr-qc/9511049]

[317] A. Ashtekar. A generalised Wick transform for gravity. *Phys. Rev.* **D53** (Rapid Communications) (1996) R2865–9. [gr-qc/9511083]

[318] B. C. Hall. The Segal–Bargmann coherent state transform for compact Lie groups. *J. Funct. Anal.* **122** (1994) 103–151.

[319] B. C. Hall and J. J. Mitchell. Coherent states on spheres. *J. Math. Phys.* **43** (2002) 1211–36. [quant-ph/0109086]

[320] B. C. Hall and J. J. Mitchell. The large radius limit for coherent states on spheres. [quant-ph/0203142]

[321] A. Ashtekar, J. Lewandowski, D. Marolf, J. Mourão and T. Thiemann. Coherent state transforms for spaces of connections. *J. Funct. Anal.* **135** (1996) 519–51. [gr-qc/9412014]

[322] G. A. Mena Marugán. Geometric interpretation of Thiemann's generalised Wick transform. [gr-qc/9705031]

[323] G. A. Mena Marugán. Thiemann transform for gravity with matter fields. *Class. Quant. Grav.* **15** (1998) 3763–75. [gr-qc/9805010]

[324] B. Hartmann and J. Wisniewski. Generalised Wick transform in dimensionally reduced gravity. *Class. Quant. Grav.* **21** (2004) 697–728. [gr-qc/0309081]

[325] T. Thiemann. Anomaly-free formulation of non-perturbative, four-dimensional Lorentzian quantum gravity. *Phys. Lett.* **B380** (1996) 257–64. [gr-qc/9606088]

[326] S. Holst. Barbero's Hamiltonian derived from a generalised Hilbert–Palatini action. *Phys. Rev.* **D53** (1996) 5966. [gr-qc/9511026]

[327] N. Barros e Sá. Hamiltonian analysis of general relativity with the Immirzi parameter. *Int. J. Mod. Phys.* **D10** (2001) 261–72. [gr-qc/0006013]

[328] R. Capovilla, M. Montesinos, V. A. Prieto and E. Rojas. BF gravity and the Immirzi parameter. *Class. Quant. Grav.* **18** (2001) L49. Erratum: **18** (2001) 1157. [gr-qc/0102073]

[329] J. Samuel. Canonical gravity, diffeomorphisms and objective histories. *Class. Quant. Grav.* **17** (2000) 4645–54. [gr-qc/0005094]

[330] J. Samuel. Is Barbero's Hamiltonian formulation a gauge theory of Lorentzian gravity? *Class. Quant. Grav.* **17** (2000) L141. [gr-qc/00050095]

[331] S. Alexandrov, I. Grigentch and D. Vassilevich. SU(2) invariant reduction of the (3 + 1)-dimensional Ashtekar's gravity. *Class. Quant. Grav.* **15** (1998) 573–80. [gr-qc/9705080]

[332] S. Alexandrov and D. V. Vassilevich. Path integral for the Hilbert–Palatini and Ashtekar gravity. *Phys. Rev.* **D58** (1998) 124029. [gr-qc/9806001]

[333] S. Alexandrov. SO(4,C) covariant Ashtekar–Barbero gravity and the Immirzi parameter. *Class. Quant. Grav.* **17** (2000) 4255–68. [gr-qc/0005085]

[334] S. Alexandrov and D. Vassilevich. Area spectrum in Lorentz covariant loop gravity. *Phys. Rev.* **D64** (2001) 044023. [gr-qc/0103105]

[335] S. Melosch and H. Nicolai. New canonical variables for D = 11 supergravity. *Phys. Lett.* **B416** (1998) 91–100. [hep-th/9709227]

[336] T. Thiemann. Generalised boundary conditions for general relativity for the asymptotically flat case in terms of Ashtekar's variables. *Class. Quant. Grav.* **12** (1995) 181–98. [gr-qc/9910008]

[337] M. Nakahara. *Geometry, Topology and Physics* (Institute of Physics Publishing Ltd, Bristol, 1990).

[338] J. W. Milnor and J. D. Stasheff. *Characteristic Classes* (Princeton University Press, Princeton, 1974).

[339] R. Gambini and A. Trias. Second quantisation of the free electromagnetic field as quantum mechanics in the loop space. *Phys. Rev.* **D22** (1980) 1380.

[340] C. Di Bartolo, F. Nori, R. Gambini and A. Trias. Loop space quantum formulation of free electromagnetism. *Lett. Nuov. Cim.* **38** (1983) 497.

[341] R. Gambini and A. Trias. Gauge dynamics in the C representation. *Nucl. Phys.* **B278** (1986) 436.

[342] R. Giles. The reconstruction of gauge potentials from Wilson loops. *Phys. Rev.* **D8** (1981) 2160.

[343] A. Ashtekar and J. Lewandowski. Completeness of Wilson loop functionals on the moduli space of SL(2,C) and SU(1,1) connections. *Class. Quant. Grav.* **10** (1993) L69. [gr-qc/9304044]

[344] T. Jacobson and L. Smolin. Nonperturbative quantum geometries. *Nucl. Phys.* **B299** (1988) 295.

[345] C. Rovelli and L. Smolin. Loop space representation of quantum general relativity. *Nucl. Phys.* **B331** (1990) 80.

[346] R. Gambini and J. Pullin. *Loops, Knots, Gauge Theories and Quantum Gravity* (Cambridge University Press, Cambridge, 1996).

[347] B. Brügmann and J. Pullin. Intersecting N loop solutions of the Hamiltonian constraint of quantum gravity. *Nucl. Phys.* **B363** (1991) 221–46.

[348] B. Brügmann, J. Pullin and R. Gambini. Knot invariants as nondegenerate quantum geometries. *Phys. Rev. Lett.* **68** (1992) 431–4.

[349] B. Brügmann, J. Pullin and R. Gambini. Jones polynomials for intersecting knots as physical states of quantum gravity. *Nucl. Phys.* **B385** (1992) 587–603.

[350] L. H. Kauffman. *Knots and Physics* (World Scientific, Singapore, 1991).

[351] H. Kodama. Holomorphic wave function of the universe. *Phys. Rev.* **D42** (1990) 2548–65.

[352] T. Thiemann and H. A. Kastrup. Canonical quantisation of spherically symmetric gravity in Ashtekar's self-dual representation. *Nucl. Phys.* **B399** (1993) 211–58. [gr-qc/9310012]

[353] T. Thiemann. The reduced phase space of spherically symmetric Einstein–Maxwell theory including a cosmological constant. *Nucl. Phys.* **B436** (1995) 681–720. [gr-qc/9910007]

[354] T. Thiemann. Reduced models for quantum gravity. *Lecture Notes in Physics* **434** (1994) 289–318. [gr-qc/9910010]

[355] A. Ashtekar, V. Husain, C. Rovelli, J. Samuel and L. Smolin. $(2+1)$-quantum gravity as a toy model for the $(3+1)$ theory. *Class. Quant. Grav.* **6** (1989) L185.

[356] V. Husain and L. Smolin. Exactly solvable quantum cosmologies from two Killing field reductions of general relativity. *Nucl. Phys.* **327** (1989) 205–38.

[357] H. Kodama. Specialisation of Ashtekar's formalism to Bianchi cosmology. *Progr. Theor. Phys.* **80** (1988) 1024–40.

[358] V. Husain and J. Pullin. Quantum theory of spacetimes with one Killing field. *Mod. Phys. Lett.* **A5** (1990) 733–41.

[359] A. Ashtekar and J. Pullin. Bianchi cosmologies: a new description. *Annal. Israel Phys. Soc.* **9** (1990) 65–76.

[360] A. Ashtekar and V. Husain. Symmetry reduced Einstein gravity and generalised sigma and chiral models. *Int. J. Mod. Phys.* **D7** (1998) 549–66. [gr-qc/9712053]

[361] V. Husain and K. Kuchař. General covariance, new variables and dynamics without dynamics. *Phys. Rev.* **D42** (1990) 4070–77.

[362] A. Ashtekar and C. J. Isham. Representations of the holonomy algebras of gravity and non-Abelian gauge theories. *Class. Quant. Grav.* **9** (1992) 1433. [hep-th/9202053]

[363] A. Rendall. Comment on a paper of Ashtekar and Isham. *Class. Quant. Grav.* **10** (1993) 605–8.

[364] A. Ashtekar and J. Lewandowski. Representation theory of analytic holonomy $C^\star$ algebras. In *Knots and Quantum Gravity*, J. Baez (ed.) (Oxford University Press, Oxford, 1994). [gr-qc/9311010]

[365] D. Marolf and J. M. Mourão. On the support of the Ashtekar–Lewandowski measure. *Commun. Math. Phys.* **170** (1995) 583–606. [hep-th/9403112]

[366] A. Ashtekar, D. Marolf and J. Mourão. Integration on the space of connections modulo gauge transformations. In *Proceedings of the Lanczos International Centenary Conference*, J. D. Brown *et al.* (eds) (SIAM, Philadelphia, 1994). [gr-qc/9403042]

[367] A. Ashtekar and J. Lewandowski. Projective techniques and functional integration for gauge theories. *J. Math. Phys.* **36** (1995) 2170–91. [gr-qc/9411046]

[368] J. Lewandowski. Topological measure and graph differential geometry on the quotient space of connections. *Int. J. Mod. Phys.* **D3** (1994) 207–10. [gr-qc/9406025]

[369] A. Ashtekar and J. Lewandowski. Differential geometry on the space of connections via graphs and projective limits. *J. Geo. Physics* **17** (1995) 191–230. [hep-th/9412073]

[370] J. Baez. Generalised measures in gauge theory. *Lett. Math. Phys.* **31** (1994) 213–24. [hep-th/9310201]

[371] J. Baez. Diffeomorphism invariant generalised measures on the space of connections modulo gauge transformations. In *Proceedings of the Conference 'Quantum Topology'*, D. Yetter (ed.) (World Scientific, Singapore, 1994). [hep-th/9305045]

[372] Millennium Price Problems. *Clay Mathematics Institute.* [http://www.claymath.org/prizeproblems/ index.htm]

[373] T. Balaban, J. Imbrie and A. Jaffe. Exact renormalisation group for gauge theories. In *Proceedings of the 1983 Cargèse Summer School.*

[374] T. Balaban and A. Jaffe. Constructive gauge theory. In *Proceedings of the 1986 Erichi Summer School.*

[375] T. Balaban. In *Constructive Gauge Theory II*, G. Velo and A. S. Wightman (eds) (Plenum Press, New York, 1990).

[376] T. Balaban. Regularity and decay of lattice Green's functions. *Commun. Math. Phys.* **89** (1983) 571.

[377] T. Balaban. Propagators and renormalisation transformations for lattice gauge theories 1. *Commun. Math. Phys.* **95** (1984) 17.

[378] T. Balaban. Propagators and renormalisation transformations for lattice gauge theories 2. *Commun. Math. Phys.* **96** (1984) 223.

[379] T. Balaban. Averaging operations for lattice gauge theories. *Commun. Math. Phys.* **98** (1985) 17.

[380] T. Balaban. Spaces of regular gauge field configurations on a lattice and gauge fixing conditions. *Commun. Math. Phys.* **99** (1985) 75.

[381] T. Balaban. Propagators for lattice gauge theories in a background field. *Commun. Math. Phys.* **99** (1985) 398.

[382] T. Balaban. The variational problem and background fields in renormalisation group methods for lattice gauge theories. *Commun. Math. Phys.* **102** (1985) 277.

[383] T. Balaban. Renormalisation group approach to lattice gauge fields theories. 1. Generation of effective actions in a small fields approximation and a coupling constant renormalisation in four-dimensions. *Commun. Math. Phys.* **109** (1987) 249.

[384] T. Balaban. Effective action and cluster properties of the Abelian Higgs model. *Commun. Math. Phys.* **114** (1988) 257.

[385] T. Balaban. Convergent renormalisation expansions for lattice gauge theories. *Commun. Math. Phys.* **119** (1988) 243–85.

[386] T. Balaban. Large field renormalisation. 1: The basic step of the R operation. *Commun. Math. Phys.* **122** (1989) 175–202.

[387] T. Balaban. Large field renormalisation. 2: Localisation, exponentiation, and bounds for the R operation. *Commun. Math. Phys.* **122** (1989) 355–92.

[388] P. Federbush. On the quantum Yang–Mills field theory. *Can. Math. Soc., Conf. Proc.* **9** (1987) 29–36.

[389] M. Göpfert and G. Mack. Proof of confinement of static quarks in three-dimensional U(1) lattice gauge theory for all values of the coupling constant. *Commun. Math. Phys.* **82** (1981) 545.

[390] E. Seiler. Gauge theory as a problem in constructive quantum field theory and statistical mechanics. *Lect. Notes Phys.* **159** (1982) 1–192.

[391] M. Asorey and P. K. Mitter. Regularised, continuum Yang–Mills process and Feynman–Kac functional integral. *Commun. Math. Phys.* **80** (1981) 43.

[392] M. Asorey and F. Falceto. Geometric regularisation of gauge theories. *Nucl Phys.* **B327** (1989) 427.

[393] P. K. Mitter and P. Viallet. On the bundle of connections and the gauge orbit manifold in Yang–Mills theory. *Commun. Math. Phys.* **79** (1981) 457.

[394] V. Rivasseau. *From Perturbative to Constructive Renormalisation* (Princeton University Press, Princeton, 1991).

[395] A. Ashtekar, J. Lewandowski, D. Marolf, J. Mourão and T. Thiemann. A manifestly gauge invariant approach to quantum theories of gauge fields. In *Geometry of Constrained Dynamical Systems*, J. Charap (ed.), pp. 60–86 (Cambridge University Press, Cambridge, 1995). [hep-th/9408108]

[396] T. Thiemann. An axiomatic approach to constructive quantum gauge field theory. *Banach Centre Publ.* **39** (1996) 389–403. [hep-th/9511122]

[397] K. Osterwalder and R. Schrader. Axioms for Euclidean Green's functions. *Commun. Math. Phys.* **31** (1973) 83–112.

[398] K. Osterwalder and R. Schrader. Axioms for Euclidean Green's functions. 2. *Commun. Math. Phys.* **42** (1975) 281.

[399] A. Ashtekar, J. Lewandowski, D. Marolf, J. Mourão and T. Thiemann. SU(N) quantum Yang–Mills theory in two dimensions: a complete solution. *J. Math. Phys.* **38** (1997) 5453–82. [hep-th/9605128]

[400] Y. M. Makeenko and A. A. Migdal. Exact equation for the loop average in multicolour QCD. *Phys. Lett.* **B88** (1979) 135, Erratum. **B89** (1980) 437.

[401] Y. M. Makeenko and A. A. Migdal. Selfconsistent areas law in QCD. *Phys. Lett.* **B97** (1980) 253.

[402] Y. M. Makeenko and A. A. Migdal. Quantum chromodynamics as dynamics of loops. *Nucl. Phys.* **B188** (1981) 269.

[403] Y. M. Makeenko and A. A. Migdal. Quantum chromodynamics as dynamics of loops. *Yad. Phys.* **32** (1980) 838–54.

[404] Y. M. Makeenko and A. A. Migdal. Dynamics of loops: asymptotic freedom and quark confinement. *Yad. Phys.* **33** (1981) 1639.

[405] Y. M. Makeenko. Conformal operators in quantum chromodynamics. *Sov. J. Nucl. Phys.* **33** (1981) 440. *Yad. Fiz.* **33** (1981) 842–7.

[406] A. A. Migdal. Momentum loop dynamics and random surfaces in QCD. *Nucl. Phys.* **B265** (1986) 594–614.

[407] L. Gross, C. King and A. Sengupta. Two-dimensional Yang–Mills theory via stochastic differential equations. *Ann. Phys.* **194** (1989) 65–112.

[408] B. K. Driver. YM(2): Continuum expectations, lattice convergence, and lassos. *Commun. Math. Phys.* **123** (1989) 575–616.

[409] N. Bralic. Exact computation of loop averages in two-dimensional Yang–Mills theory. *Phys. Rev.* **D22** (1980) 3090.

[410] S. Klimek and W. Kondracki. A construction of two-dimensional quantum chromodynamics. *Commun. Math. Phys.* **113** (1987) 389–402.

[411] E. Witten. On quantum gauge theories in two-dimensions. *Commun. Math. Phys.* **141** (1991) 153–209.

[412] E. Witten. Two-dimensional gauge theories revisited. *J. Geom. Phys.* **9** (1992) 303–68.

[413] D. Gross and W. Taylor. Twists and Wilson loops in the string theory of two-dimensional QCD. *Nucl. Phys.* **B403** (1993) 395–452.

[414] M. Douglas and V. A. Kazakov. Large *N* phase transition in continuum QCD in two-dimensions. *Phys. Lett.* **B319** (1993) 219–30.

[415] C. Fleischhack. A new type of loop independence and SU(N) quantum Yang–Mills theory in two dimensions. *J. Math. Phys.* **40** (1999) 2584–610.

[416] C. Fleischhack. On the structure of physical measures in gauge theories. *J. Math. Phys.* **41** (2000) 76–102. [math-ph/0107022]

[417] A. Ashtekar, D. Marolf, J. Mourão and T. Thiemann. Constructing Hamiltonian quantum theories from path integrals in a diffeomorphism invariant context. *Class. Quant. Grav.* **17** (2000) 4919–40. [quant-ph/9904094]

[418] J. M. Mourão, T. Thiemann and J. M. Velhinho. Physical properties of quantum field theory measures. *J. Math. Phys.* **40** (1999) 2337–53. [hep-th/9711139]

[419] R. Penrose. In *Quantum Gravity and Beyond*, T. Bastin (ed.) (Cambridge University Press, Cambridge, 1971).

[420] C. Rovelli and L. Smolin. Spin networks and quantum gravity. *Phys. Rev.* **D53** (1995) 5743–59. [gr-qc/9505006]

[421] J. Baez. Spin networks in non-perturbative quantum gravity. In *The Interface of Knots and Physics*, L. Kauffman (ed.) (American Mathematical Society, Providence, Rhode Island, 1996). [gr-qc/9504036]

[422] T. Thiemann. The inverse loop transform. *J. Math. Phys.* **39** (1998) 1236–48.
[hep-th/9601105]

[423] R. De Pietri. On the relation between the connection and the loop representation of
quantum gravity. *Class. Quant. Grav.* **14** (1997) 53–70. [gr-qc/9605064]

[424] T. Thiemann. A length operator for canonical quantum gravity. *J. Math. Phys.* **39**
(1998) 3372–92. [gr-qc/9606092]

[425] C. Rovelli and L. Smolin. Discreteness of volume and area in quantum gravity. *Nucl.
Phys.* **B442** (1995) 593–622. Erratum: **B456** (1995) 753. [gr-qc/9411005]

[426] A. Ashtekar and J. Lewandowski. Quantum theory of geometry I: Area operators.
Class. Quant. Grav. **14** (1997) A55–82. [gr-qc/9602046]

[427] A. Ashtekar and J. Lewandowski. Quantum theory of geometry II: Volume operators.
Adv. Theor. Math. Phys. **1** (1997) 388–429. [gr-qc/9711031]

[428] J. Lewandowski. Volume and quantisations. *Class. Quant. Grav.* **14** (1997) 71–6.
[gr-qc/9602035]

[429] S. Major and M. Seifert. Modeling space with an atom of geometry. *Class. Quant.
Grav.* **19** (2002) 2211–28. [gr-qc/0109056]

[430] S. Major. Operators for quantised directions. *Class. Quant. Grav.* **16** (1999) 3859–77.
[gr-qc/9905019]

[431] J. Baez and S. Sawin. Functional integration on spaces of connections. [q-alg/9507023]

[432] J. Baez and S. Sawin. Diffeomorphism invariant spin-network states. [q-alg/9708005]

[433] J. Lewandowski and T. Thiemann. Diffeomorphism invariant quantum field theories of
connections in terms of webs. *Class. Quant. Grav.* **16** (1999) 2299–322. [gr-qc/9901015]

[434] C. Fleischhack. Proof of a conjecture by Lewandowski and Thiemann. *Commun. Math.
Phys.* **249** (2004) 331–52. [math-ph/0304002]

[435] J.-A. Zapata. A combinatorial approach to diffeomorphism invariant quantum gauge
theories. *J. Math. Phys.* **38** (1007) 5663–81. [gr-qc/9703037]

[436] J.-A. Zapata. A combinatorial space from loop quantum gravity. *Gen. Rel. Grav.* **30**
(1998) 1229–45. [gr-qc/9703038]

[437] T. Thiemann. Quantum spin dynamics (QSD): III. Quantum constraint algebra and
physical scalar product in quantum general relativity. *Class. Quant. Grav.* **15** (1998)
1207–47. [gr-qc/9705017]

[438] T. Thiemann. Quantum spin dynamics (QSD). *Class. Quant. Grav.* **15** (1998) 839–73.
[gr-qc/9606089]

[439] T. Thiemann. Quantum spin dynamics (QSD): II. The kernel of the Wheeler–DeWitt
constraint operator. *Class. Quant. Grav.* **15** (1998) 875–905. [gr-qc/9606090]

[440] T. Thiemann. Quantum spin dynamics (QSD): IV. 2 + 1 Euclidean quantum gravity as
a model to test 3 + 1 Lorentzian quantum gravity. *Class. Quant. Grav.* **15** (1998)
1249–80. [gr-qc/9705018]

[441] T. Thiemann. Quantum spin dynamics (QSD): V. Quantum gravity as the natural
regulator of the Hamiltonian constraint of matter quantum field theories. *Class. Quant.
Grav.* **15** (1998) 1281–314. [gr-qc/9705019]

[442] T. Thiemann. Quantum spin dynamics (QSD): VI. Quantum Poincaré algebra and a
quantum positivity of energy theorem for canonical quantum gravity. *Class. Quant.
Grav.* **15** (1998) 1463–85. [gr-qc/9705020]

[443] T. Thiemann. Kinematical Hilbert spaces for fermionic and Higgs quantum field
theories. *Class. Quant. Grav.* **15** (1998) 1487–512. [gr-qc/9705021]

[444] C. Rovelli and H. Morales-Tecótl. Fermions in quantum gravity. *Phys. Rev. Lett.* **72**
(1995) 3642–5. [gr-qc/9401011]

[445] C. Rovelli and H. Morales-Tecótl. Loop space representation of quantum fermions and
gravity. *Nucl. Phys.* **B331** (1995) 325–61.

[446] J. Baez and K. Krasnov. Quantisation of diffeomorphism invariant theories with
fermions. *J. Math. Phys.* **39** (1998) 1251–71. [hep-th/9703112]

[447] A. Ashtekar, J. Lewandowski and H. Sahlmann. Polymer and Fock representations for
a scalar field. *Class. Quant. Grav.* **20** (2003) L11. [gr-qc/0211012]

[448] W. Kaminski, J. Lewandowski and A. Okolow. Background independent quantisations: the scalar field II. *Class. Quant. Grav.* **23** (2006) 2761–70. [gr-qc/0508091]

[449] W. Kaminski, J. Lewandowski and A. Okolow. Background independent quantisations: the scalar field II. [gr-qc/0604112]

[450] M. Han and Y. Ma. Dynamics of scalar field in polymer-like representation. *Class. Quant. Grav.* **23** (2006) 2741–60. [gr-qc/0602101]

[451] R. Borissov, R. De Pietri and C. Rovelli. Matrix elements of Thiemann's Hamiltonian constraint in loop quantum gravity. *Class. Quant. Grav.* **14** (1997) 2793–823. [gr-qc/9703090]

[452] M. Gaul and C. Rovelli. A generalised Hamiltonian constraint operator in loop quantum gravity and its simplest Euclidean matrix elements. *Class. Quant. Grav.* **18** (2001) 1593–624. [gr-qc/0011106]

[453] M. Reisenberger and C. Rovelli. Sum over surfaces form of loop quantum gravity. *Phys. Rev.* **D56** (1997) 3490–508. [gr-qc/9612035]

[454] J. W. Barrett and L. Crane. Relativistic spin networks and quantum gravity. *J. Math. Phys.* **39** (1998) 3296–302. [gr-qc/9709028]

[455] J. W. Barrett and L. Crane. A Lorentzian signature model for quantum general relativity. *Class. Quant. Grav.* **17** (2000) 3101–18. [gr-qc/9904025]

[456] L. Freidel and E. R. Livine. Spin networks for noncompact groups. *J. Math. Phys.* **44** (2003) 1322–56. [hep-th/0205268]

[457] E. R. Livine. Projected spin networks for Lorentz connection: linking spin foams and loop gravity. *Class. Quant. Grav.* **19** (2002) 5525–42. [gr-qc/0207084]

[458] S. Alexandrov and E. R. Livine. SU(2) loop quantum gravity seen from covariant theory. *Phys. Rev.* **D67** (2003) 044009. [gr-qc/0209105]

[459] K. Noui and A. Perez. Dynamics of loop quantum gravity and spin foam models in three dimensions. [gr-qc/0402112]

[460] K. Noui and A. Perez. Three dimensional loop quantum gravity: physical scalar product and spin foam models. *Class. Quant. Grav.* **22** (2005) 1739–62. [gr-qc/0402110]

[461] A. Perez and C. Rovelli. A spin foam model without bubble divergences. *Nucl. Phys.* **B599** (2001) 255–82. [gr-qc/0006107]

[462] A. Perez. Finiteness of a spin foam model for Euclidean quantum general relativity. *Nucl. Phys.* **B599** (2001) 427–34. [gr-qc/0011058]

[463] D. V. Boulatov. A model of three-dimensional lattice gravity. *Mod. Phys. Lett.* **A7** (1992) 1629–46. [hep-th/9202074]

[464] H. Ooguri. Topological lattice models in four-dimensions. *Mod. Phys. Lett.* **A7** (1992) 2799–810. [hep-th/9205090]

[465] R. De Pietri, L. Freidel, K. Krasnov and C. Rovelli. Barrett–Crane model from a Boulatov–Ooguri field theory over a homogeneous space. *Nucl. Phys.* **B574** (2000) 785–806. [hep-th/9907154]

[466] R. De Pietri and C. Petronio. Feynman diagrams of generalised matrix models and the associated manifolds in dimension 4. *J. Math. Phys.* **41** (2000) 6671–88. [gr-qc/0004045]

[467] K. Krasnov. On statistical mechanics of gravitational systems. *Gen. Rel. Grav.* **30** (1998) 53–68. [gr-qc/9605047]

[468] C. Rovelli. Black hole entropy from loop quantum gravity. *Phys. Rev. Lett.* **77** (1996) 3288–91. [gr-qc/9603063]

[469] A. Ashtekar, J. C. Baez and K. Krasnov. Quantum geometry of isolated horizons and black hole entropy. *Adv. Theor. Math. Phys.* **4** (2001) 1–94. [gr-qc/0005126]

[470] A. Ashtekar and A. Corichi. Laws governing isolated horizons: inclusion of dilaton couplings. *Class. Quant. Grav.* **17** (2000) 1317–32. [gr-qc/9910068]

[471] A. Ashtekar, A. Corichi and D. Sudarsky. Hairy black holes, horizon mass and solitons. *Class. Quant. Grav.* **18** (2001) 919–40. [gr-qc/0011081]

[472] S. Fairhurst and B. Krishnan. Distorted black holes with charge. *Int. J. Mod. Phys.* **D10** (2001) 691–710. [gr-qc/0010088]

[473] J. Baez. Spin network states in gauge theory. *Adv. Math.* **117** (1996) 253–72. [gr-qc/9411007]

[474] J. Velhinho. A groupoid approach to spaces of generalised connections. *J. Geom. Phys.* **41** (2002) 166–80. [hep-th/0011200]

[475] C. Fleischhack. Gauge orbit types for generalised connections. *Commun. Math. Phys.* **214** (2000) 607–49. [math-ph/0001006]

[476] C. Fleischhack. Hyphs and the Ashtekar–Lewandowski measure. [math-ph/0001007]

[477] C. Fleischhack. Stratification of the generalised gauge orbit pace. *Commun. Math. Phys.* **214** (2000) 607–49. [math-ph/0001008]

[478] C. Fleischhack. On the Gribov problem for generalised connections. *Commun. Math. Phys.* **234** (2003) 423–54. [math-ph/0007001]

[479] T. Thiemann and O. Winkler. Gauge field theory coherent states (GCS): IV. Infinite tensor product and thermodynamic limit. *Class. Quant. Grav.* **18** (2001) 4997–5033. [hep-th/0005235]

[480] M. Arnsdorf. Loop quantum gravity on noncompact spaces. *Nucl. Phys.* **B577** (2000) 529–46. [gr-qc/9909053]

[481] A. Döring and H. F. de Groote. The kinematical frame of loop quantum gravity I. [gr-qc/0112072]

[482] M. C. Abbati, A. Mania and E. Provenzi. Inductive construction of the loop transform for Abelian gauge theories. *Lett. Math. Phys.* **57** (2001) 69–81. [math-ph/0105041]

[483] M. C. Abbati and A. Mania. On the spectrum of holonomy algebras. *J. Geom. Phys.* **44** (2002) 96–114. [math-ph/0202004]

[484] A. Ashtekar, C. Rovelli and L. Smolin. Weaving a classical geometry with quantum threads. *Phys. Rev. Lett.* **69** (1992) 237–40. [hep-th/9203079]

[485] T. Thiemann. Quantum spin dynamics (QSD): VII. Symplectic structures and continuum lattice formulations of gauge field theories. *Class. Quant. Grav.* **18** (2001) 3293–338. [hep-th/0005232]

[486] T. Thiemann. Gauge field theory coherent states (GCS): I. General properties. *Class. Quant. Grav.* **18** (2001) 2025–64. [hep-th/0005233]

[487] T. Thiemann. Complexifier coherent states for canonical quantum general relativity. *Class. Quant. Grav.* **23** (2006) 2063–118. [gr-qc/0206037]

[488] T. Thiemann and O. Winkler. Gauge field theory coherent states (GCS): II. Peakedness properties. *Class. Quant. Grav.* **18** (2001) 2561–636. [hep-th/0005237]

[489] T. Thiemann and O. Winkler. Gauge field theory coherent states (GCS): III. Ehrenfest theorems. *Class. Quant. Grav.* **18** (2001) 4629–81. [hep-th/0005234]

[490] H. Sahlmann, T. Thiemann and O. Winkler. Coherent states for canonical quantum general relativity and the infinite tensor product extension. *Nucl. Phys.* **B606** (2001) 401–40. [gr-qc/0102038]

[491] M. Varadarajan. Fock representations from U(1) holonomy algebras. *Phys. Rev.* **D61** (2000) 104001. [gr-qc/0001050]

[492] M. Varadarajan. Photons from quantised electric flux representations. *Phys. Rev.* **D64** (2001) 104003. [gr-qc/0104051]

[493] M. Varadarajan. Gravitons from a loop representation of linearised gravity. *Phys. Rev.* **D66** (2002) 024017. [gr-qc/0204067]

[494] M. Varadarajan. The graviton vacuum as a distributional state in kinematic loop quantum gravity. *Class. Quant. Grav.* **22** (2005) 1207–38. [gr-qc/0410120]

[495] A. Ashtekar and J. Lewandowski. Relation between polymer and Fock excitations. *Class. Quant. Grav.* **18** (2001) L117–28. [gr-qc/0107043]

[496] J. M. Velhinho. Denseness of Ashtekar–Lewandowski states and a generalised cut-off in loop quantum gravity. *Class. Quant. Grav.* **22** (2005) 3061–72. [gr-qc/0502038]

[497] M. Bojowald and H. Kastrup. Quantum symmetry reduction for diffeomorphism invariant theories of connections. *Class. Quant. Grav.* **17** (2000) 3009. [hep-th/9907042]

[498] M. Bojowald and H. Kastrup. The area operator in the spherically symmetric sector of loop quantum gravity. *Class. Quant. Grav.* **17** (2000) 3044. [hep-th/9907043]

[499] M. Bojowald. Abelian BF theory and spherically symmetric electromagnetism. *J. Math. Phys.* **41** (2000) 4313. [hep-th/9908170]

[500] S. Tsujikawa, P. Singh and R. Maartens. Loop quantum gravity effects on inflation and the CMB. *Class. Quant. Grav.* **21** (2004) 5767–75. [astro-ph/0311015]

[501] V. Husain and O. Winkler. On singularity resolution in quantum gravity. *Phys. Rev.* **D69** (2004) 084016. [gr-qc/0312094]

[502] S. Hofmann and O. Winkler. The spectrum of fluctuations in inflationary cosmology. [astro-ph/0411124]

[503] G. Amelino-Camelia. Are we at dawn with quantum gravity phenomenology? *Lect. Notes Phys.* **541** (2000) 1–49. [gr-qc/9910089]

[504] G. Amelino-Camelia. Gravity wave interferometers as probes of a low energy effective quantum gravity. *Phys. Rev.* **D62** (2000) 024015. [gr-qc/9903080]

[505] G. Amelino-Camelia. An interferometric gravitational wave detector as a quantum gravity apparatus. *Nature* **398** (1999) 216–18. [gr-qc/9808029]

[506] G. Amelino-Camelia, J. R. Ellis, N. E. Mavromatos, D. V. Nanopoulos and S. Sarkar. Potential sensitivity of gamma ray burster observations to wave dispersion in vacuo. *Nature* **393** (1998) 763–5. [astro-ph/9712103]

[507] C. Laemmerzahl and H. J. Dittus. Fundamental physics in space: a guide to present projects. *Ann. Phys. (Leipzig)* **11** (2002) 95–150.

[508] R. Gambini and J. Pullin. Nonstandard optics from quantum spacetime. *Phys. Rev.* **D59** (1999) 124021. [gr-qc/9809038]

[509] R. Gambini and J. Pullin. Quantum gravity experimental physics? *Gen. Rel. Grav.* **31** (1999) 1631–7.

[510] J. Alfaro, H. A. Morales-Tecotl and L. F. Urrutia. Quantum gravity corrections to neutrino propagation. *Phys. Rev. Lett.* **84** (2000) 2318–21. [gr-qc/9909079]

[511] J. Alfaro, H. A. Morales-Tecotl and L. F. Urrutia. Loop quantum gravity and light propagation. *Phys. Rev.* **D65** (2002) 103509. [hep-th/0108061]

[512] T. Jacobson, S. Liberati and D. Mattingly. Lorentz violation at high energy: concepts, phenomena and astrophysical constraints. *Annal. Phys.* **321** (2006) 150–96. [astro-ph/0505267]

[513] H. S. Snyder. Quantised spacetime. *Phys. Rev.* **71** (1947) 38.

[514] H. Sahlmann. When do measures on the space of connections support the triad operators of loop quantum gravity. [gr-qc/0207112]

[515] H. Sahlmann. Some comments on the representation theory of the algebra underlying loop quantum gravity. [gr-qc/0207111]

[516] A. Okolow and J. Lewandowski. Diffeomorphism covariant representations of the holonomy flux algebra. *Class. Quant. Grav.* **20** (2003) 3543–68. [gr-qc/03027112]

[517] A. Okolow and J. Lewandowski. Automorphism covariant representations of the holonomy flux *-algebra. *Class. Quant. Grav.* **22** (2005) 657–80. [gr-qc/0405119]

[518] H. Sahlmann, T. Thiemann. On the superselection theory of the Weyl algebra for diffeomorphism invariant quantum gauge theories. [gr-qc/0302090]

[519] H. Sahlmann and T. Thiemann. Irreducibility of the Ashtekar–Isham–Lewandowski representation. *Class. Quant. Grav.* **23** (2006) 4453–72. [gr-qc/0303074]

[520] C. Fleischhack. Representations of the Weyl algebra in quantum geometry. [math-ph/0407006]

[521] J. Lewandowski, A. Okolow, H. Sahlmann and T. Thiemann. Uniqueness of diffeomorphism invariant states on holonomy–flux algebras. *Commun. Math. Phys.* **267** (2006) 703–33. [gr-qc/0504147]

[522] T. Thiemann. Quantum spin dynamics (QSD): VIII. The master constraint. *Class. Quant. Grav.* **23** (2006) 2249–66. [gr-qc/0510011]

[523] M. Han and Y. Ma. Master constraint operator in loop quantum gravity. *Phys. Lett.* **B635** (2006) 225–31. [gr-qc/0510014]

[524] A. Ashtekar, A. Corichi and J. A. Zapata. Quantum theory of geometry III: Non-commutativity of Riemannian structures. *Class. Quant. Grav.* **15** (1998) 2955–72. [gr-qc/9806041]

[525] H. Whitney. Differentiable manifolds. *Ann. Math.* **37** (1936) 648–80.

[526] J. Velhinho. On the structure of the space of generalised connections. *Int. J. Geom. Meth. Mod. Phys.* **1** (2004) 311–34. [math-ph/0402060]

[527] N. Biggs. *Algebraic Graph Theory*, 2nd edn. (Cambridge University Press, Cambridge, 1993).

[528] D. Stauffer and A. Aharony. *Introduction to Percolation Theory*, 2nd edn. (Taylor and Francis, London, 1994).

[529] D. M. Cvetovic, M. Doob and H. Sachs. *Spectra of Graphs* (Academic Press, New York, 1979).

[530] T. Nowotny and M. Requardt. Dimension theory of graphs and networks. *J. Phys.* **A31** (1998) 2447–63. [hep-th/9707082]

[531] T. Filk. Random graph gauge theories as toy models for non-perturbative string theories. *Class. Quant. Grav.* **17** (2000) 4841–54. [hep-th/0010126]

[532] Y. Yamasaki. *Measures on Infinite Dimensional Spaces* (World Scientific, Singapore, 1985).

[533] J. R. Munkres. *Toplogy: A First Course* (Prentice Hall, Englewood Cliffs (NJ), 1980).

[534] J. M. Velhinho. Functorial aspects of the space of generalised connections. *Mod. Phys. Lett.* **A20** (2005) 1299. [math-ph/0411073]

[535] R. V. Kadison and J. R. Ringrose. *Fundamentals of the Theory of Operator Algebras*, Vols 1, 2 (Academic Press, London, 1983).

[536] O. Bratteli and D. W. Robinson. *Operator Algebras and Quantum Statistical Mechanics*, Vols 1, 2 (Springer-Verlag, Berlin, 1997).

[537] J. Dixmier. *Les Algebres d' Operateurs dans l' Espace Hilbertien (Algebres de von Neumann)* (Gauthiers-Villars, Paris, 1957).

[538] G. B. Folland. *Harmonic Analysis in Phase Space* (Ann. Math. Studies, No. 122, Princeton University Press, Princeton (NJ), 1989).

[539] H. Narnhofer and W. Thirring. Covariant QED without indefinite metric. *Rev. Math. Phys.* **SI1** (1992) 197–211.

[540] K. Kuchař. Parametrised scalar field on $R \times S^1$: dynamical pictures, spacetime diffeomorphisms and conformal isometries. *Phys. Rev.* **D39** (1989) 1579–93.

[541] K. Kuchař. Dirac quantisation of a parametrised field theory by anomaly-free operator representations of spacetime diffeomorphisms. *Phys. Rev.* **D39** (1989) 2263–80.

[542] C. G. Torre and M. Varadarajan. Quantum fields at any time. *Phys. Rev.* **D58** (1998) 064007. [hep-th/9707221]

[543] C. G. Torre and M. Varadarajan. Functional evolution of free quantum fields. *Class. Quant. Grav.* **16** (1999) 2651–68. [hep-th/9811222]

[544] D. H. Cho and M. Varadarajan. Functional evolution of quantum cylindrical waves. [gr-qc/0605065]

[545] M. Varadarajan. Dirac quantisation of parametrised field theory. [gr-qc/0607068]

[546] G. Mack. Introduction to conformally invariant quantum field theory in two and more dimensions. In *Nonperturbative Quantum Field Theory* (Cargese, 1987); Preprint DESY 88-120.

[547] R. C. Helling and G. Policastro. String quantisation: Fock vs. LQG representations. [hep-th/0409182]

[548] J. Slawny. On factor representations and the C*-algebra of canonical commutation relations. *Commun. Math. Phys.* **24** (1972) 151–70.

[549] A. Ashtekar, S. Fairhurst and J. L. Willis. Quantum gravity, shadow states and quantum mechanics. *Class. Quant. Grav.* **20** (2003) 1031. [gr-qc/0207106]

[550] K. Fredenhagen and F. Reszewski. Polymer state approximations of Schrödinger wave functions. [gr-qc/0606090]

[551] N. J. Vilenkin. *Special Functions and the Theory of Group Representations* (American Mathematical Society, Providence, Rhode Island, 1968).

[552] W. Rudin. *Real and Complex Analysis* (McGraw-Hill, New York, 1987).

[553] D. Giulini. The group of large diffeomorphisms in classical and quantum gravity. *Helv. Phys. Acta* **69** (1996) 333–4.

[554] D. Giulini. Determination and reduction of large diffeomorphisms. *Nucl. Phys. Proc. Suppl.* **57** (1997) 342–5. [gr-qc/9702021]

[555] D. Giulini. The group of large diffeomorphisms in general relativity. *Banach Centre Publ.* **39** (1997) 303–15. [gr-qc/9510022]

[556] D. Giulini. Properties of three manifolds for relativists. *Int. J. Theor. Phys.* **33** (1994) 913–30. [gr-qc/9308008]

[557] N. Grot and C. Rovelli. Moduli space structure of knots with intersections. *J. Math. Phys.* **37** (1996) 3014–21. [gr-qc/9604010]

[558] W. Fairbairn and C. Rovelli. Separable Hilbert space in loop quantum gravity. *J. Math. Phys.* **45** (2004) 2802–14. [gr-qc/0403047]

[559] T. Thiemann. Closed formula for the matrix elements of the volume operator in canonical quantum gravity. *J. Math. Phys.* **39** (1998) 3347–71. [gr-qc/9606091]

[560] D. Marolf, J. Mourão and T. Thiemann. The status of diffeomorphism superselection in Euclidean 2+1 gravity. *J. Math. Phys.* **38** (1997) 4370–40. [gr-qc/9701068]

[561] H. Sahlmann. Exploring the diffeomorphism invariant Hilbert space of a scalar field. [gr-qc/0609032]

[562] A. Okolow. Hilbert space built over connections with a non-compact structure group. *Class. Quant. Grav.* **22** (2005) 1329–60. [gr-qc/0406028]

[563] M. Blencowe. The Hamiltonian constraint in quantum gravity. *Nucl. Phys.* **B341** (1990) 213–51.

[564] B. Brügmann and J. Pullin. On the constraints of quantum gravity in the loop representation. *Nucl. Phys.* **B390** (1993) 399–438.

[565] R. Gambini. Loop space representation of quantum general relativity and the group of loops. *Phys. Lett.* **B255** (1991) 180–88.

[566] R. Gambini, A. Garat and J. Pullin. The constraint algebra of quantum gravity in the loop representation. *Int. J. Mod. Phys.* **D4** (1995) 589. [gr-qc/9404059]

[567] C. Rovelli and L. Smolin. The physical Hamiltonian in nonperturbative quantum gravity. *Phys. Rev. Lett.* **72** (1994) 446. [gr-qc/9308002]

[568] C. Rovelli. A generally covariant quantum field theory and a prediction on quantum measurements of geometry. *Nucl. Phys.* **B405** (1993) 797–816.

[569] L. Smolin. Finite diffeomorphism invariant observables in quantum gravity. *Phys. Rev.* **D49** (1994) 4028–40. [gr-qc/9302011]

[570] R. Loll. Making quantum gravity calculable. *Acta Cosmol.* **21** (1995) 131–44. [gr-qc/9511080]

[571] R. Loll. A real alternative to quantum gravity in loop space. *Phys. Rev.* **D54** (1996) 5381–4. [gr-qc/9602041]

[572] J. Baez. Matters of gravity 1996. URL html://www.phys.psu.edu/PULLIN/. [gr-qc/9609008]

[573] K. Giesel and T. Thiemann. Consistency check on volume and triad operator quantisation in loop quantum gravity. I. *Class. Quant. Grav.* **23** (2006) 5667–91. [gr-qc/0507036]

[574] K. Giesel and T. Thiemann. Consistency check on volume and triad operator quantisation in loop quantum gravity. II. *Class. Quant. Grav.* **23** (2006) 5693–771. [gr-qc/0507037]

[575] R. Loll. Spectrum of the volume operator in quantum gravity. *Nucl. Phys.* **B460** (1996) 143–54. [gr-qc/9511030]

[576] M. Creutz. *Quarks, Gluons and Lattices* (Cambridge University Press, Cambridge, 1983).

[577] H. Nicolai, K. Peeters and M. Zamaklar. Loop quantum gravity: an outside view. *Class. Quant. Grav.* **22** (2005) R193. [hep-th/0501114]

[578] T. Thiemann. Loop quantum gravity: an inside view. [hep-th/0608210]

[579] D. Marolf and J. Lewandowski. Loop constraints: a habitat and their algebra. *Int. J. Mod. Phys.* **D7** (1998) 299–330. [gr-qc/9710016]

[580] R. Gambini, J. Lewandowski, D. Marolf and J. Pullin. On the consistency of the constraint algebra in spin network gravity. *Int. J. Mod. Phys.* **D7** (1998) 97–109. [gr-qc/9710018]

[581] Y. Ma and Y. Ling. The Q operator for canonical quantum gravity. *Phys. Rev.* **D62** (2000) 104021. [gr-qc/0005117]

[582] L. Smolin. The classical limit and the form of the Hamiltonian constraint in non-perturbative quantum general relativity. [gr-qc/9609034]

[583] A. Perez. On the regularisation ambiguities in loop quantum gravity. *Phys. Rev.* **D73** (2006) 044007. [gr-qc/0509118]

[584] L. Freidel and L. Smolin. Linearisation of the Kodama state. *Class. Quant. Grav.* **21** (2004) 3831–44. [hep-th/0310224]

[585] J.-P. Pier. *Amenable Locally Compact Groups* (John Wiley & Sons, New York, 1984).

[586] K. V. Kuchař and J. D. Romano. Gravitational constraints which generate an algebra. *Phys. Rev.* **D51** (1995) 5579–82. [gr-qc/9501005]

[587] F. Markopoulou. Gravitational constraint combinations generate a Lie algebra. *Class. Quant. Grav.* **13** (1996) 2577–84. [gr-qc/9601038]

[588] F. Antonsen and F. Markopoulou. 4-Diffeomorphisms in canonical gravity and Abelian deformations. [gr-qc/9702046]

[589] K. Giesel and T. Thiemann. Algebraic quantum gravity (AQG) I. Conceptual setup. *Class. Quant. Grav.* **24** (2007) 2465–97. [gr-qc/0607099]

[590] K. Giesel and T. Thiemann. Algebraic quantum gravity (AQG) II. Semiclassical analysis. *Class. Quant. Grav.* **24** (2007) 2499–564. [gr-qc/0607100]

[591] K. Giesel and T. Thiemann. Algebraic quantum gravity (AQG) III. Semiclassical perturbation theory. *Class. Quant. Grav.* **24** (2007) 2565–88. [gr-qc/0607101]

[592] A. Corichi and J. A. Zapata. On diffeomorphism invariance for lattice theories. *Nucl. Phys.* **B493** (1997) 475–90. [gr-qc/9611034]

[593] P. Hasenfratz. The theoretical background and properties of perfect actions. [hep-lat/9803027]

[594] S. Hauswith. Perfect discretisations of differential operators [hep-lat/0003007]

[595] S. Hauswith. The perfect Laplace operator for non-trivial boundaries. [hep-lat/0010033]

[596] M. Bobienski, J. Lewandowski and M. Mroczek. A two surface quantisation of Lorentzian gravity. [gr-qc/0101069]

[597] C. Kiefer. Conceptual issues in quantum cosmology. *Lect. Notes Phys.* **541** (2000) 158–87. [gr-qc/9906100]

[598] A. O. Barvinsky. Quantum cosmology at the turn of the millennium. [gr-qc/0101046]

[599] J. B. Hartle. Quantum cosmology: problems for the 21st century. [gr-qc/9701022]

[600] S. Carlip. Lectures in $(2+1)$-dimensional gravity. *J. Korean Phys. Soc.* **28** (1995) S447–67. [gr-qc/9503024]

[601] S. Carlip. The statistical mechanics of the three-dimensional Euclidean black hole. *Phys. Rev.* **D55** (1997) 878–82. [gr-qc/9606043]

[602] E. Witten. $(2+1)$-dimensional gravity as an exactly solvable system. *Nucl. Phys.* **B311** (1988) 46.

[603] A. Mikovic and N. Manojlovic. Ashtekar formulation of $(2+1)$ gravity on a torus. *Nucl. Phys.* **B385** (1992) 571–86. [hep-th/9204022]

[604] A. Mikovic and N. Manojlovic. Remarks on the reduced phase space of $(2+1)$ gravity on a torus in the Ashtekar formulation. *Class. Quant. Grav.* **15** (1998) 3031–9. [gr-qc/9712011]

[605] F. Barbero and M. Varadarajan. The phase space of $(2+1)$-dimensional gravity in the Ashtekar formulation. *Nucl. Phys.* **B415** (1994) 515–32. [gr-qc/9307006]

[606] D. Marolf. Loop representations for $(2+1)$ gravity on a torus. *Class. Quant. Grav.* **10** (1993) 2625–48. [gr-qc/9303019]

[607] D. Marolf. An illustration of $(2+1)$ gravity loop transform troubles. *Can. Gen. Rel.* **14** (1993) 256. [gr-qc/9305015]

[608] J. Louko and D. Marolf. Solution space of $(2+1)$ gravity on $R \times T^2$ in Witten's connection formulation. *Class. Quant. Grav.* **11** (1994) 311–30. [gr-qc/9308018]

[609] A. Ashtekar and R. Loll. New loop representations for $(2+1)$ gravity. *Class. Quant. Grav.* **11** (1994) 2417–34. [gr-qc/9405031]

[610] R. Loll. Independent loop invariants for $(2+1)$ gravity. *Class. Quant. Grav.* **12** (1995) 1655–62. [gr-qc/9408007]

[611] T. Jacobson. $(1 + 1)$ sector of $(3 + 1)$ gravity. *Class. Quant. Grav.* **13** (1996) L111–16.
       Erratum. **13** (1996) 3269. [gr-qc/9604003]

[612] J. Lewandowski and J. Wisniewski. Degenerate sectors of Ashtekar gravity. *Class.
       Quant. Grav.* **16** (1999) 3057–69. [gr-qc/9902037]

[613] C. Di Bartolo, R. Gambini, J. Griego and J. Pullin. Canonical quantum gravity in the
       Vasiliev invariants arena: I. Kinematical structure. *Class. Quant. Grav.* **17** (2000)
       3211–37.

[614] C. Di Bartolo, R. Gambini, J. Griego and J. Pullin. Canonical quantum gravity in the
       Vasiliev invariants arena: II. Constraints, habitats and consistency of the constraint
       algebra. *Class. Quant. Grav.* **17** (2000) 3239–64.

[615] R. Gambini and J. Pullin. Consistent discretisations for classical and quantum gravity.
       [gr-qc/0108062]

[616] R. Gambini and J. Pullin. Canonical quantisation of constrained theories on discrete
       spacetime lattices. *Class. Quant. Grav.* **19** (2002) 5275–69. [gr-qc/0205123]

[617] R. Gambini and J. Pullin. Canonical quantisation of general relativity in discrete
       spacetimes. *Phys. Rev. Lett.* **90** (2003) 021301. [gr-qc/0206055]

[618] R. Gambini and J. Pullin. Discrete gravity: applications to cosmology. *Class. Quant.
       Grav.* **20** (2003) 3341. [gr-qc/0212033]

[619] R. Gambini, R. A. Porto and J. Pullin. Decoherence from discrete quantum gravity.
       Class. Quant. Grav. **21** (2004) L51–7. [gr-qc/0305098]

[620] R. Gambini, R. A. Porto and J. Pullin. A relational solution of the problem of time in
       quantum mechanics and quantum gravity induces a fundamental mechanism for
       quantum decoherence. *New J. Phys.* **6** (2004) 45. [gr-qc/0402118]

[621] R. Gambini, R. A. Porto and J. Pullin. No black hole information puzzle in a relational
       universe. *Int. J. Mod. Phys.* **D13** (2004) 2315–20. [hep-th/0405183]

[622] R. Gambini, R. A. Porto and J. Pullin. Realistic clocks, universal decoherence and the
       black hole information paradox. *Phys. Rev. Lett.* **93** (2004) 240401. [hep-th/0406260]

[623] R. Gambini, R. A. Porto and J. Pullin. Fundamental decoherence from relational time
       in discrete quantum gravity: Galilean covariance. *Phys. Rev.* **D70** (2004) 124001.
       [gr-qc/0408050]

[624] R. Gambini, R. A. Porto and J. Pullin. Consistent discretisation and loop quantum
       geometry. *Phys. Rev. Lett.* **94** (2005) 101302. [gr-qc/0409057]

[625] R. Gambini, R. A. Porto and J. Pullin. Fundamental gravitational limitations to
       quantum computing. [quant-ph/0507262]

[626] C. Di Bartolo, R. Gambini, R. A. Porto and J. Pullin. Dirac-like approach for
       consistent discretisations of classical constrained theories. *J. Math. Phys.* **46** (2005)
       012901. [gr-qc/0405131]

[627] R. Gambini, M. Ponce and J. Pullin. Consistent discretisations: the Gowdy spacetimes.
       Phys. Rev. **D72** (2005) 024031. [gr-qc/0505043]

[628] J. Iwasaki and C. Rovelli. Gravitons as embroidery of the weave. *Int. J. Mod. Phys.*
       D1 (1993) 533–57.

[629] J. Iwasaki and C. Rovelli. Gravitons from loops: nonperturbative loop space quantum
       gravity contains the graviton physics approximation. *Class. Quant. Grav.* **11** (1994)
       1653–76.

[630] Y. Ma and Y. Ling. The classical geometry from a physical state in canonical quantum
       gravity. *Phys. Rev.* **D62** (2000) 064030. [gr-qc/0004070]

[631] M. Arnsdorf and S. Gupta. Loop quantum gravity on noncompact spaces. *Nucl. Phys.*
       B577 (2000) 529–46. [gr-qc/9909053]

[632] M. Arnsdorf. Approximating connections in loop quantum gravity. [gr-qc/9910084]

[633] M. Varadarajan and J. A. Zapata. A proposal for analysing the classical limit of
       kinematic loop gravity. *Class. Quant. Grav.* **17** (2000) 4085–110. [gr-qc/0001040]

[634] L. Bombelli. Statistical Lorentzian geometry and the closeness of Lorentzian manifolds.
       J. Math. Phys. **41** (2000) 6944–58. [gr-qc/0002053]

[635] A. Ashtekar and L. Bombelli. Statistical weaves and semiclassical quantum gravity. In
       preparation.

[636] L. Bombelli, A. Corichi and O. Winkler. Semiclassical quantum gravity: statistics of combinatorial Riemannian geometries. *Annal. Phys.* **14** (2005) 499–519. [gr-qc/0409006]

[637] H. Sahlmann and T. Thiemann. Towards the QFT on curved spacetime limit of QGR. 1. A general scheme. *Class. Quant. Grav.* **23** (2006) 867–908. [gr-qc/0207030]

[638] H. Sahlmann and T. Thiemann. Towards the QFT on curved spacetime limit of QGR. 2. A concrete implementation. *Class. Quant. Grav.* **23** (2006) 909–54. [gr-qc/0207031]

[639] J. Velhinho. Invariance properties of induced Fock measures for U(1) holonomies. *Commun. Math. Phys.* **227** (2002) 541–50. [math-ph/0107002]

[640] J. Zegwaard. Weaving of curved geometries. *Phys. Lett.* **B300** (1993) 217–22. [hep-th/9210033]

[641] J. Zegwaard. Gravitons in loop quantum gravity. *Nucl. Phys.* **B378** (1992) 288–308.

[642] C. Itzykson and J.-M. Drouffe. *Statistical Field Theory*, Vol. 2 (Cambridge University Press, Cambridge, 1989).

[643] J. Klauder and B.-S. Skagerstam. *Coherent States* (World Scientific, Singapore, 1985).

[644] A. Perelomov. *Generalised Coherent States and their Applications* (Springer-Verlag, Berlin, 1986).

[645] F. Bayen, M. Flato, C. Fronsdal, A. Liechnerowicz and D. Sternheimer. Deformation theory and quantisation. *Annal. Phys.* **111** (1978) 61–110, 111–51.

[646] S. Bochner. *Vorlesungen über Fouriersche Integrale* (Chelsea Publishing Company, New York, 1948).

[647] J. von Neumann. On infinite direct products. *Comp. Math.* **6** (1938) 1–77.

[648] W. Thirring. *Lehrbuch der Mathematischen Physik*, Vol. 4 (Springer-Verlag, Wien, 1994).

[649] W. Thirring. *Lehrbuch der Mathematischen Physik*, Vol. 3 (Springer-Verlag, Berlin, 1978).

[650] H. Baumgärtel and M. Wollenberg. *Causal Nets of Operator Algebras* (Akademie Verlag, Berlin, 1992).

[651] O. Landford III. Selected topics in functional analysis. In *Proceedings of Les Houches Summer School 'Statistical Mechanics and Quantum Field Theory'*, C. DeWitt and R. Stora (eds) (Gordon and Broach Science Publishers, London, 1971).

[652] M. Varadarajan. The graviton vacuum as a distributional state in kinematic loop quantum gravity. *Class. Quant. Grav.* **22** (2005) 1207–38. [gr-qc/0410120]

[653] F. Conrady. Free vacuum for loop quantum gravity. *Class. Quant. Grav.* **22** (2005) 3261–93. [gr-qc/0409036]

[654] A. Ashtekar and C. Isham. Inequivalent observable algebras: another ambiguity in field quantisation. *Phys. Lett.* **B274** (1992) 393–8.

[655] A. Ashtekar, C. Rovelli and L. Smolin. Gravitons and loops. *Phys. Rev.* **D44** (1991) 1740–55. [hep-th/9202054]

[656] B. DeWitt. *Supermanifolds* (Cambridge University Press, Cambridge, 1992).

[657] J. Velhinho. Comments on the kinematical structure of loop quantum cosmology. *Class. Quant. Grav.* **21** (2004) L109. [gr-qc/0406008]

[658] I. L. Buchbinder and S. L. Lyahovich. *Class. Quant. Grav.* **4** (1987) 1487.

[659] D. M. Gitman and I. V. Tyutin. *Quantisation of Fields with Constraints* (Springer-Verlag, Berlin, 1990).

[660] L. Smolin. Recent developments in non-perturbative quantum gravity. [hep-th/9202022]

[661] R. Loll. Further results on geometric operators in quantum gravity. *Class. Quant. Grav.* **14** (1997) 1725–41. [gr-qc/9612068]

[662] R. Loll. Simplifying the spectral analysis of the volume operator. *Nucl. Phys.* **B500** (1997) 405–20. [gr-qc/9706038]

[663] R. Loll. The volume operator in discretised quantum gravity. *Phys. Rev. Lett.* **75** (1995) 3048–51. [gr-qc/9506014]

[664] R. De Pietri and C. Rovelli. Geometry eigenvalues and scalar product from recoupling theory in loop quantum gravity. *Phys. Rev.* **D54** (1996) 2664–90. [gr-qc/9602023]

[665] J. Brunnemann and T. Thiemann. Simplification of the spectral analysis of the volume operator in loop quantum gravity. *Class. Quant. Grav.* **23** (2006) 1289–346. [gr-qc/0405060]

[666] A. R. Edmonds. *Angular Momentum in Quantum Mechanics* (Princeton University Press, Princeton, 1974).

[667] A. Alekseev, A. P. Polychronakos and M. Smedback. On area and entropy of a black hole. *Phys. Lett.* **B574** (2003) 296–300. [hep-th/0004036]

[668] A. Corichi. Comments on area spectra in loop quantum gravity. *Rev. Mex. Fis.* **50** (2005) 549–52. [gr-qc/0402064]

[669] R. De Pietri. Spin networks and recoupling in loop quantum gravity. *Nucl. Phys. Proc. Suppl.* **57** (1997) 251–4. [gr-qc/9701041]

[670] R. Loll. Imposing $\det(E) > 0$ in discrete quantum gravity. *Phys. Lett.* **B399** (1997) 227–32. [gr-qc/9703033]

[671] J. C. Baez. An introduction to spin foam models of quantum gravity and BF theory. *Lect. Notes Phys.* **543** (2000) 25–94. [gr-qc/9905087]

[672] J. C. Baez. Spin foam models. *Class. Quant. Grav.* **15** (1998) 1827–58. [gr-qc/9709052]

[673] J. W. Barrett. State sum models for quantum gravity. [gr-qc/0010050]

[674] J. W. Barrett. Quantum gravity as topological quantum field theory. *J. Math. Phys.* **36** (1995) 6161–79. [gr-qc/9506070]

[675] A. Perez. Spin foam models for quantum gravity. *Class. Quant. Grav.* **20** (2003) R43. [gr-qc/0301113]

[676] D. Oriti. *Spin foam models of quantum space-time*, PhD thesis. [gr-qc/0311066]

[677] G. Ponzano and T. Regge. Semiclassical limit of Racah coefficients. In *Spectroscopy and Group Theoretical Methods in Physics*, F. Bloch (ed.) (North-Holland, New York, 1968).

[678] V. Turarev and O. Viro. State sum invariants of 3-manifolds and quantum 6j symbols. *Topology* **31** (1992) 865–902.

[679] H. Ooguri. Topological lattice models in four dimensions. *Mod. Phys. Lett.* **A7** (1992) 2799–810.

[680] L. Crane and D. Yetter. A categorical construction of 4D TQFTs. In *Quantum Topology*, pp. 120–30, L. Kauffman and R. Baadhio (eds) (World Scientific, Singapore, 1993).

[681] L. Crane, L. Kauffman and D. Yetter. State-sum invariants of 4-manifolds. *J. Knot Theory & Ram.* **6** (1997) 177–234.

[682] M. F. Atiyah. Topological quantum field theories. *Publ. Math. IHES.* **68** (1989) 175–86.

[683] M. F. Atiyah. *The Geometry of Physics and Knots* (Cambridge University Press, Cambridge, 1990).

[684] C. Kassel. *Quantum Groups* (Springer-Verlag, Berlin, 1995).

[685] L. Kauffman. *Knots and Physics* (World Scientific Press, Singapore, 1993).

[686] L. Kauffman and S. Lins. *Temperley–Lieb Recoupling Theory and Invariants of 3-Manifolds* (Princeton University Press, Princeton, 1994).

[687] V. Tuarev. *Quantum Invariants of Knots and 3-Manifolds* (de Gruyter, New York, 1994).

[688] E. Witten. Quantum field theory and the Jones polynomial. *Commun. Math. Phys.* **121** (1989) 351–99.

[689] N. Reshetikhin. Invariants of 3-manifolds via link polynomials and quantum groups. *Invent. Math.* **103** (1991) 547–97.

[690] D. Birmingham, M. Blau, M. Rakowski and G. Thompson. Topological field theories. *Phys. Rep.* **209** (1991) 129–40.

[691] R. Friedman and J. Morgan. *Gauge Theory and the Topology of Four-Manifolds* (AMS, Providence, 1998).

[692] M. Reisenberger. World sheet formulations of gauge theories and gravity. [gr-qc/9412035]

[693] M. P. Reisenberger. A lefthanded simplicial action for Euclidean general relativity. *Class. Quant. Grav.* **14** (1997) 1753–70. [gr-qc/9609002]

[694] L. Freidel, K. Krasnov and R. Puzio. BF description of higher dimensional gravity theories. *Adv. Theor. Math. Phys.* **3** (1999) 1289–324. [hep-th/9901069]

[695] L. Freidel and K. Krasnov. Spin foam models and the classical action principle. *Adv. Theor. Math. Phys.* **2** (1999) 1183–247. [hep-th/9807092]

[696] L. Freidel and D. Louapre. Nonperturbative summation over 3-d discrete topologies. *Phys. Rev.* **D68** (2003) 104004. [hep-th/0211026]

[697] L. Freidel and D. Louapre. Diffeomorphisms and spin foam models. *Nucl. Phys.* **B662** (2003) 279. [gr-qc/0212001]

[698] A. Barbieri. Space of vertices of relativistic spin networks. [gr-qc/9709076]

[699] A. Barbieri. Quantum tetrahedra and simplicial spin networks. *Nucl. Phys.* **B518** (1998) 714–28. [gr-qc/9707010]

[700] J. C. Baez and J. W. Barrett. The quantum tetrahedron in three dimensions and four dimensions. *Adv. Theor. Math. Phys.* **3** (1999) 815–50. [gr-qc/9903060]

[701] M. P. Reisenberger. On relativistic spin network vertices. *J. Math. Phys.* **40** (1999) 2046–54. [gr-qc/9809067]

[702] D. Yetter. Generalised Barrett–Crane vertices and invariants of embedded graphs. [qa/9801131]

[703] J. W. Barrett. The classical evaluation of relativistic spin networks. *Adv. Theor. Math. Phys.* **2** (1998) 593–600. [math.qa/9803063]

[704] J. W. Barrett and R. M. Williams. The asymptotics of an amplitude for the four simplex. *Adv. Theor. Math. Phys.* **3** (1999) 209–15. [gr-qc/9809032]

[705] S. Sen, J. C. Sexton and D. H. Adams. A geometric discretisation scheme applied to the Abelian Chern–Simons theory. [hep-th/0001030]

[706] H. Whitney. *Geometric Integration Theory* (Princeton University Press, Princeton, 1957).

[707] D. H. Adams. R-torsion and linking numbers from simplicial Abelian gauge theories. [hep-th/9612009]

[708] J. Ambjorn, B. Durhuus and T. Jonnson. *Quantum Geometry: A Statistical Field Theory Approach* (Cambridge University Press, Cambridge, 1997).

[709] L. Freidel. Group field theory: an overview. *Int. J. Theor. Phys.* **44** (2005) 1769–83. [hep-th/0505016]

[710] M. Reisenberger and C. Rovelli. Spin foams as Feynman diagrams. [gr-qc/0002083]

[711] M. P. Reisenberger and C. Rovelli. Space time as a Feynman diagram: the connection formulation. *Class. Quant. Grav.* **18** (2001) 121–40. [gr-qc/0002095]

[712] I. M. Gel'fand and M. A. Naimark. Unitary representations of the proper Lorentz group. *Izv. Akad. Nauk. SSSR.* **11** (1947) 411.

[713] J. C. Baez and J. W. Barrett. Integrability of relativistic spin networks. *Class. Quant. Grav.* **18** (2001) 4683–700. [gr-qc/0101107]

[714] A. Perez and C. Rovelli. Spin foam model for Lorentzian general relativity. *Phys. Rev.* **D63** (2001) 041501. [gr-qc/0009021]

[715] L. Crane, A. Perez and C. Rovelli. A finiteness proof for the Lorentzian state sum spin foam model for quantum general relativity. [gr-qc/0104057]

[716] L. Crane, A. Perez and C. Rovelli. Perturbative finiteness in spin-foam quantum gravity. *Phys. Rev. Lett.* **87** (2001) 181301.

[717] E. Buffenoir, M. Henneaux, K. Noui and Ph. Roche. Hamiltonian analysis of Plebanski theory. *Class. Quant. Grav.* **21** (2004) 5203–20. [gr-qc/0404041]

[718] A. Perez and C. Rovelli. Observables in quantum gravity. [gr-qc/0104034]

[719] D. Oriti and H. Pfeiffer. A spin foam model for pure gauge theory coupled to quantum gravity. *Phys. Rev.* **D66** (2002) 124010. [gr-qc/0207041]

[720] D. Oriti. Boundary terms in the Barrett–Crane spin foam model and consistent gluing. *Phys. Lett.* **B532** (2002) 363–72. [gr-qc/0201077]

[721] H. Pfeiffer. Dual variables and a connection picture for the Euclidean Barrett–Crane model. *Class. Quant. Grav.* **19** (2002) 1109–38. [gr-qc/0112002]

[722] L. Freidel and D. Louapre. Ponzano–Regge model revisited I: Gauge fixing, observables and interacting spinning particles. *Class. Quant. Grav.* **21** (2004) 5685–726. [hep-th/0401076]

[723] L. Freidel and D. Louapre. Ponzano–Regge model revisited II: Equivalence with Chern–Simons. [gr-qc/0410141]

[724] L. Freidel and D. Louapre. Ponzano–Regge model revisited III: Feynman diagrams and effective field theory. *Class. Quant. Grav.* **23** (2006) 2021–62. [hep-th/0502106]

[725] E. Livine and R. Oeckl. Three-dimensional quantum supergravity and supersymmetric spin foam models. *Adv. Theor. Math. Phys.* **7** (2004) 951–1001. [hep-th/0307251]

[726] K. Noui and A. Perez. Observability and geometry in three-dimensional quantum gravity. [gr-qc/0402113]

[727] K. Noui and A. Perez. Three dimensional loop quantum gravity: coupling to point particles. *Class. Quant. Grav.* **22** (2005) 4489–514. [gr-qc/0402111]

[728] K. Noui and A. Perez. Dynamics of loop quantum gravity and spin foam models in three dimensions. [gr-qc/0402112]

[729] K. Noui and A. Perez. Three dimensional loop quantum gravity: physical scalar product and spin foam models. *Class. Quant. Grav.* **22** (2005) 1739–62. [gr-qc/0402110]

[730] J. C. Baez, J. D. Christensen, T. R. Halford and D. C. Tsang. Spin foam models of Riemannian quantum gravity. *Class. Quant. Grav.* **19** (2002) 4627–48. [gr-qc/0202017]

[731] J. C. Baez, J. D. Christensen and G. Egan. Asymptotics of 10j symbols. *Class. Quant. Grav.* **19** (2002) 6489. [gr-qc/0208010]

[732] J. C. Baez and J. D. Christensen. Positivity of spin foam amplitudes. *Class. Quant. Grav.* **19** (2002) 2291–306. [gr-qc/0110044]

[733] A. Perez. Spin foam quantisation of Plebanski's action. *Adv. Theor. Math. Phys.* **5** (2002) 947–68. [gr-qc/0203058]

[734] L. Freidel and D. Louapre. Asymptotics of 6j and 10j symbols. *Class. Quant. Grav.* **20** (2003) 1267–94. [hep-th/0209134]

[735] H. Pfeiffer. Positivity of relativistic spin network evaluations. *Adv. Theor. Math. Phys.* **6** (2003) 827. [gr-qc/0211106]

[736] M. Bojowald and A. Perez. Spin foam quantisation and anomalies. [gr-qc/0303026]

[737] L. Smolin and A. Starodubtsev. General relativity with a topological phase: an action principle. [hep-th/0311163]

[738] L. Freidel and A. Starodubtsev. Quantum gravity in terms of topological observables. [hep-th/0501191]

[739] L. Freidel, J. Kowalski-Glikman and A. Starodubtsev. Particles as Wilson lines of gravitational field. [gr-qc/0607014]

[740] F. Markopoulou and L. Smolin. Causal evolution of spin networks. *Nucl. Phys.* **B508** (1997) 409–30. [gr-qc/9702025]

[741] F. Markopoulou. Dual formulation of spin network evolution. [gr-qc/9704013]

[742] F. Markopoulou and L. Smolin. Quantum geometry with intrinsic local causality. *Phys. Rev.* **D58** (1998) 084032. [gr-qc/9712067]

[743] F. Markopoulou. The internal description of a causal set: what the universe is like from inside. *Commun. Math. Phys.* **211** (2000) 559–83. [gr-qc/9811053]

[744] F. Markopoulou. Quantum causal histories. *Class. Quant. Grav.* **17** (2000) 2059–72. [hep-th/9904009]

[745] F. Markopoulou. An insider's guide to quantum causal histories. *Nucl. Phys. Proc. Suppl.* **88** (2000) 308–13. [hep-th/9912137]

[746] E. R. Livine and D. Oriti. Implementing causality in the spin foam quantum geometry. *Nucl. Phys.* **B663** (2003) 231–79. [gr-qc/0210064]

[747] E. R. Livine and D. Oriti. Causality in spin foam models for quantum gravity. [gr-qc/0302018]

[748] H. Pfeiffer. On the causal Barrett–Crane model: measure, coupling constant, Wick rotation, symmetries and observables. *Phys. Rev.* **D67** (2003) 064022. [gr-qc/0212049]

[749] F. Markopoulou. An algebraic approach to coarse graining. [hep-th/0006199]

[750] F. Markopoulou. Coarse graining in spin foam models. *Class. Quant. Grav.* **20** (2003) 777–800. [gr-qc/0203036]

[751] A. Connes and D. Kreimer. Renormalisation in quantum field theory and the Riemann–Hilbert problem. *JHEP* **9909** (1999) 024. [hep-th/9909126]

[752] A. Connes and D. Kreimer. Renormalisation in quantum field theory and the Riemann–Hilbert problem. 1. The Hopf algebra structure of graphs and the main theorem. *Commun. Math. Phys.* **210** (2000) 249–73. [hep-th/9912092]

[753] A. Connes and D. Kreimer. Renormalisation in quantum field theory and the Riemann–Hilbert problem. 2. The beta function, diffeomorphisms and the renormalisation group. *Commun. Math. Phys.* **216** (2001) 215–41. [hep-th/0003188]

[754] R. Oeckl. Renormalisation of discrete models without background. *Nucl. Phys.* **B657** (2003) 107–38. [gr-qc/0212047]

[755] H. Pfeiffer. Four-dimensional lattice gauge theory with ribbon categories and the Crane–Yetter state sum. *J. Math. Phys.* **42** (2001) 5272–305. [hep-th/0106029]

[756] H. Pfeiffer and R. Oeckl. The dual of non Abelian lattice gauge theory. *Nucl. Phys. Proc. Suppl.* **106** (2002) 1010–12. [hep-lat/0110034]

[757] H. Pfeiffer and R. Oeckl. The dual of pure non Abelian lattice gauge theory as a spin foam model. *Nucl. Phys.* **B598** (2001) 400–26. [hep-th/0008095]

[758] R. Oeckl. Generalised lattice gauge theory, spin foams and state sum invariants. *J. Geom. Phys.* **46** (2003) 308. [hep-th/0110259]

[759] H. Pfeiffer. Quantum general relativity and the classification of smooth manifolds. [gr-qc/0404088]

[760] H. Pfeiffer. Diffeomorphisms from finite triangulations and absence of 'local' degrees of freedom. *Phys. Lett.* **B591** (2004) 197–201. [gr-qc/0312060]

[761] F. Girelli and H. Pfeiffer. Higher gauge theory: differential versus integral formulation. *J. Math. Phys.* **45** (2004) 3949–71. [hep-th/0309173]

[762] H. Pfeiffer. Higher gauge theory and a non-Abelian generalisation of 2-form electrodynamics. *Annal. Phys.* **308** (2003) 447. [hep-th/0304074]

[763] A. Mikovic. Spin foam models of matter coupled to gravity. *Class. Quant. Grav.* **10** (2202) 2335–54. [hep-th/0108099]

[764] D. Oriti and J. Ryan. Group field theory formulation of 3-D quantum gravity coupled to matter fields. [gr-qc/0602010]

[765] D. Oriti and T. Tlas. Causality and matter propagation in 3-D spin foam quantum gravity. [gr-qc/0608116]

[766] K. Noui and P. Roche. Cosmological deformation of Lorentzian spin foam models. *Class. Quant. Grav.* **20** (2003) 3175–214. [gr-qc/0211109]

[767] E. Buffenoir, K. Noui and P. Roche. Hamiltonian quantisation of Chern–Simons theory with SL(2,C) group. *Class. Quant. Grav.* **19** (2002) 4953. [hep-th/0202121]

[768] D. Oriti, C. Rovelli and S. Speziale. Spinfoam 2D quantum gravity and discrete bundles. *Class. Quant. Grav.* **22** (2005) 85–108. [gr-qc/0406063]

[769] E. Livine and D. Oriti. About Lorentz invariance in a discrete quantum setting. *JHEP* **0406** (2004) 050. [gr-qc/0405085]

[770] F. Girelli, R. Oeckl and A. Perez. Spin foam diagrammatics and topological invariance. *Class. Quant. Grav.* **19** (2002) 1093–108. [gr-qc/0111022]

[771] R. Oeckl. A 'general boundary' formulation for quantum mechanics and quantum gravity. *Phys. Lett.* **B575** (2003) 318. [hep-th/0306025]

[772] F. Conrady, L. Doplicher, R. Oeckl and C. Rovelli. Minkowski vacuum in background independent quantum gravity. *Phys. Rev.* **D69** (2004) 064019. [gr-qc/0307118]

[773] C. Rovelli. Graviton propagator from background-independent quantum gravity. [gr-qc/0508124]

[774] E. Bianchi, L. Modesto, C. Rovelli and S. Speziale. Graviton propagator in loop quantum gravity. [gr-qc/0604044]

[775] J. D. Bekenstein. Black holes and entropy. *Phys. Rev.* **D7** (1973) 2333–46.

[776] J. D. Bekenstein. Generalised second law for thermodynamics in black hole physics. *Phys. Rev.* **D9** (1974) 3292–300.

[777] S. W. Hawking. Particle creation by black holes. *Commun. Math. Phys.* **43** (1975) 199–220.

[778] S. Hayward. Marginal surfaces and apparent horizons. [gr-qc/9303006]

[779] S. Hayward. On the definition of averagely trapped surfaces. *Class. Quant. Grav.* **10** (1993) L137–40. [gr-qc/9304042]

[780] S. Hayward. General laws of black hole dynamics. *Phys. Rev.* **D49** (1994) 6467–74.

[781] S. Hayward, S. Mukohyama and M. C. Ashworth. Dynamic black hole entropy. *Phys. Lett.* **A256** (1999) 347–50. [gr-qc/9810006]

[782] A. Ashtekar, C. Beetle, O. Dreyer, S. Fairhurst, B. Krishnan, J. Lewandowski and J. Wisniewski. Isolated horizons and their applications. *Phys. Rev. Lett.* **85** (2000) 3564–7. [gr-qc/0006006]

[783] A. Ashtekar. Classical and quantum physics of isolated horizons: a brief overview. *Lect. Notes Phys.* **541** (2000) 50–70.

[784] A. Ashtekar. Interface of general relativity, quantum physics and statistical mechanics: some recent developments. *Annal. Phys.* **9** (2000) 178–98. [gr-qc/9910101]

[785] A. Ashtekar, C. Beetle and S. Fairhurst. Isolated horizons: a generalisation of black hole mechanics. *Class. Quant. Grav.* **16** (1999) L1–7. [gr-qc/9812065]

[786] A. Ashtekar and B. Krishnan. Dynamical horizons and their properties. *Phys. Rev.* **D68** (2003) 104030. [gr-qc/0308033]

[787] A. Ashtekar and B. Krishnan. Isolated and dynamical horizons and their applications. *Living Rev. Rel.* **7** (2004) 10. [gr-qc/0407042]

[788] A. Ashtekar and K. Krasnov. Quantum geometry and black holes. [gr-qc/9804039]

[789] A. Ashtekar, A. Corichi and K. Krasnov. Isolated horizons: the classical phase space. *Adv. Theor. Math. Phys.* **3** (2000) 419–78. [gr-qc/9905089]

[790] A. Ashtekar, C. Beetle and S. Fairhurst. Mechanics of isolated horizons. *Class. Quant. Grav.* **17** (2000) 253–98. [gr-qc/9907068]

[791] A. Ashtekar, S. Fairhurst and B. Krishnan. Isolated horizons: Hamiltonian evolution and the first law. *Phys. Rev.* **D62** (2000) 104025. [gr-qc/0005083]

[792] L. Smolin. Linking topological quantum field theory and non-perturbative quantum gravity. *J. Math. Phys.* **36** (1995) 6417. [gr-qc/9505028]

[793] S. Axelrod, S. D. Pietra and E. Witten. Geometric quantisation of Chern–Simons gauge theory. *J. Diff. Geo.* **33** (1991) 787–902.

[794] D. Mumford. *Tata Lectures on Theta I* (Birkäuser, Boston, 1983).

[795] M. Domagala and J. Lewandowski. Black hole entropy from quantum geometry. *Class. Quant. Grav.* **21** (2004) 5233–44. [gr-qc/0407051]

[796] K. Meissner. Black hole entropy in loop quantum gravity. *Class. Quant. Grav.* **21** (2004) 5245–52. [gr-qc/0407052]

[797] A. Ghosh and P. Mitra. A bound on the log correction to the black hole area law. *Phys. Rev.* **D71** (2005) 027502. [gr-qc/0401070]

[798] A. Ghosh and P. Mitra. An improved lower bound on black hole entropy in the quantum geometry approach. *Phys. Lett.* **B616** (2005) 114–17. [gr-qc/0411035]

[799] A. Ghosh and P. Mitra. Counting of isolated horizon states. [hep-th/0605125]

[800] A. Ghosh and P. Mitra. Counting of black hole microstates. [gr-qc/0603029]

[801] A. Corichi, J. Diaz-Polo and E. Fernandez-Borja. Entropy counting for microscopic black holes in LQG. [gr-qc/0605014]

[802] I. B. Khriplovich and R. V. Korkin. How is the maximum entropy of a quantised surface related to its area? *J. Exp. Theor. Phys.* **95** (2002) 1. [gr-qc/0112074]

[803] S. Fairhurst. Table of lowest hundred eigenvalues for the area operator. Unpublished.

[804] J. Bekenstein and V. Mukhanov. Spectroscopy of the quantum black hole. *Phys. Lett.* **B360** (1995) 7–12. [gr-qc/9505012]

[805] A. Corichi, J. Diaz-Polo and E. Fernandez-Borja. Black hole entropy quantisation. [gr-qc/0609122]

[806] K. D. Kokkotas and B. G. Schmidt. Quasinormal modes of stars and black holes. *Liv. Rev. Rel.* **2** (1999) 2. [gr-qc/9909058]

[807] H. P. Nollert. About the significance of quasinormal modes of black holes. *Phys. Rev.* **D53** (1996) 4397–402. [gr-qc/9602032]

[808] S. Hod. Bohr's correspondence principle and the area spectrum of quantum black holes. *Phys. Rev. Lett.* **81** (1998) 4293. [gr-qc12002]

[809] L. Motl. An analytical computation of asymptotic Schwarzschild quasinormal frequencies. *Adv. Theor. Math. Phys.* **6** (2003) 1135. [gr-qc/0212096]

[810] L. Motl and A. Neitzke. Asymptotic black hole quasinormal frequencies. *Adv. Theor. Math. Phys.* **7** (2003) 307–30. [hep-th/0301173]

[811] O. Dreyer. Quasinormal modes, the area spectrum and black hole entropy. *Phys. Rev. Lett.* **90** (2003) 081301. [gr-qc/0211076]

[812] A. Ashtekar, C. Beetle and J. Lewandowski. Mechanics of rotating isolated horizons. *Phys. Rev.* **D64** (2001) 044016. [gr-qc/0103026]

[813] A. Ashtekar, J. Engle, T. Pawlowski and C. Van Den Broeck. Multipole moments of isolated horizons. *Class. Quant. Grav.* **21** (2004) 2549–70. [gr-qc/0401114]

[814] A. Ashtekar, J. Engle and C. Van Den Broeck. Quantum horizons and black hole entropy: inclusion of distortion and rotation. *Class. Quant. Grav.* **22** (2005) L27. [gr-qc/0412003]

[815] K. Krasnov. Quantum geometry and thermal radiation from black holes. *Class. Quant. Grav.* **16** (1999) 563–78. [gr-qc/9710006]

[816] M. Barreira, M. Carfora and C. Rovelli. Physics with nonperturbative quantum gravity: radiation from a black hole. *Gen. Rel. Grav.* **28** (1996) 1293–9. [gr-qc/9603064]

[817] S. Carlip. Liouville lost, Liouville regained: central charge in a dynamical background. *Phys. Lett.* **B508** (2001) 168–72. [gr-qc/0103100]

[818] S. Carlip. Entropy from conformal field theory at Killing horizons. *Class. Quant. Grav.* **16** (1999) 3327–48. [gr-qc/9906126]

[819] S. Carlip. Black hole entropy from horizon conformal field theory. *Nucl. Phys. Proc. Suppl* **88** (2000) 10–16. [gr-qc/9912118]

[820] O. Dreyer, A. Ghosh and J. Wisniewski. Black hole entropy calculations based on symmetries. *Class. Quant. Grav.* **18** (2001) 1929–38. [hep-th/0101117]

[821] S. Carlip. Near horizon conformal symmetry and black hole entropy. *Phys. Rev. Lett.* **88** (2002) 241301. [gr-qc/0203001]

[822] L. Modesto. Disappearance of black hole singularity in quantum gravity. *Phys. Rev.* **D70** (2004) 124009. [gr-qc/0407097]

[823] L. Modesto. Loop quantum black hole. [gr-qc/0509078]

[824] L. Modesto. Quantum gravitational collapse. [gr-qc/0504043]

[825] V. Husain and O. Winkler. Quantum black holes. *Class. Quant. Grav.* **22** (2005) L135–42. [gr-qc/0412039]

[826] V. Husain and O. Winkler. Quantum resolution of black hole singularities. *Class. Quant. Grav.* **22** (2005) L127–34. [gr-qc/0410125]

[827] V. Husain and O. Winkler. Quantum black holes from null expansion operators. *Class. Quant. Grav.* **22** (2005) L135–41.

[828] V. Husain and O. Winkler. Quantum Hamiltonian for gravitational collapse. *Phys. Rev.* **D73** (2006) 124007. [gr-qc/0601082]

[829] A. Dasgupta. Semiclassical quantisation of spacetimes with apparent horizons. *Class. Quant. Grav.* **23** (2006) 635–72. [gr-qc/0505017]

[830] A. Dasgupta. Counting the apparent horizon. [hep-th/0310069]

[831] A. Dasgupta. Coherent states for black holes. *JCAP* **0308** (2003) 004. [hep-th/0305131]

[832] A. Ashtekar and M. Bojowald. Black hole evaporation: a paradigm. *Class. Quant. Grav.* **22** (2005) 3349–62. [gr-qc/0504029]

[833] A. Ashtekar and M. Bojowald. Quantum geometry and the Schwarzschild singularity. *Class. Quant. Grav.* **23** (2006) 391–411. [gr-qc/0509075]

[834] J. Brunnemann and T. Thiemann. On (cosmological) singularity avoidance in loop quantum gravity. *Class. Quant. Grav.* **23** (2006) 1395–428. [gr-qc/0505032]

[835] J. Brunnemann and T. Thiemann. Unboundedness of triad-like operators in loop quantum gravity. *Class. Quant. Grav.* **23** (2006) 1429–84. [gr-qc/0505033]

[836] L. Modesto and C. Rovelli. Particle scattering in loop quantum gravity. *Phys. Rev. Lett.* **95** (2005) 191301. [gr-qc/0502036]

[837] C. Rovelli. F. Mattei, C. Rovelli, S. Speziale and M. Testa. From 3-geometry transition amplitudes to graviton states. *Nucl. Phys.* **B739** (2006) 234–53. [gr-qc/0508007]

[838] C. Rovelli and S. Speziale. On the perturbative expansion of a quantum field theory around a topological sector. [gr-qc/0508106]

[839] S. Speziale. Towards the graviton from spinfoams: the 3-D toy model. *JHEP* **0605** (2006) 039. [gr-qc/0512102]

[840] A. Baratin and L. Freidel. Hidden quantum gravity in 3-D Feynman diagrams. [gr-qc/0604016]

[841] E. R. Livine, S. Speziale and J. L. Willis. Towards the graviton from spinfoams: higher order corrections in the 3-D toy model. [gr-qc/0605123]

[842] B. Bolen, L. Bombelli and A. Corichi. Semiclassical states in quantum cosmology: Bianchi I coherent states. *Class. Quant. Grav.* **21** (2004) 4087–106. [gr-qc/0404004]

[843] A. Ashtekar, L. Bombelli and A. Corichi. Semiclassical states for constrained systems. *Phys. Rev.* **D72** (2005) 025008. [gr-qc/0504052]

[844] M. Bojowald. Loop quantum cosmology. I. Kinematics. *Class. Quant. Grav.* **17** (2000) 1489–508. [gr-qc/9910103]

[845] M. Bojowald. Loop quantum cosmology. II. Volume operators. *Class. Quant. Grav.* **17** (2000) 1509–26. [gr-qc/9910104]

[846] M. Bojowald. Loop quantum cosmology. III. Wheeler–DeWitt operators. *Class. Quant. Grav.* **18** (2001) 1055–70. [gr-qc/0008052]

[847] M. Bojowald. Loop quantum cosmology. IV. Discrete time evolution. *Class. Quant. Grav.* **18** (2001) 1071–88. [gr-qc/0008053]

[848] M. Bojowald. Absence of singularity in loop quantum cosmology. *Phys. Rev. Lett.* **86** (2001) 5227–30. [gr-qc/0102069]

[849] M. Bojowald. Dynamical initial conditions in quantum cosmology. *Phys. Rev. Lett.* **87** (2001) 121301. [gr-qc/0104072]

[850] M. Bojowald and G. Date. Consistency conditions for fundamentally discrete theories. *Class. Quant. Grav.* **21** (2004) 121–43. [gr-qc/0307083]

[851] M. Bojowald, D. Cartin and G. Khanna. Generating function techniques for loop quantum cosmology. *Class. Quant. Grav.* **21** (2004) 4495. [gr-qc/0405126]

[852] M. Bojowald. Isotropic loop quantum cosmology. *Class. Quant. Grav.* **19** (2002) 2717–42. [gr-qc/0202077]

[853] M. Bojowald. The inverse scale factor in isotropic quantum geometry. *Phys. Rev.* **D64** (2001) 084018. [gr-qc/0105067]

[854] M. Bojowald and F. Hinterleitner. Isotropic loop quantum cosmology with matter. *Phys. Rev.* **D66** (2002) 104003. [gr-qc/0207038]

[855] M. Bojowald, G. Date and K. Vandersloot. Homogeneous loop quantum cosmology: the role of the spin connection. *Class. Quant. Grav.* **21** (2004) 1253–78. [gr-qc/0311004]

[856] M. Bojowald. Quantisation ambiguities in isotropic quantum geometry. *Class. Quant. Grav.* **19** (2002) 5113–20. [gr-qc/0206053]

[857] M. Bojowald. Inflation from quantum geometry. *Phys. Rev. Lett.* **89** (2002) 261301. [gr-qc/0206054]

[858] M. Bojowald. The semiclassical limit of loop quantum cosmology. *Class. Quant. Grav.* **18** (2001) L109–16. [gr-qc/0105113]

[859] M. Bojowald and K. Vandersloot. Loop quantum cosmology, boundary proposals and inflation. *Phys. Rev.* **D67** (2003) 124023. [gr-qc/0303072]

[860] M. Bojowald, J. Lidsey, D. Mulryne, P. Singh and R. Tavakol. Inflationary cosmology and quantisation ambiguities in semiclassical loop quantum gravity. *Phys. Rev.* **D70** (2004) 043530. [gr-qc/0403106]

[861] M. Bojowald, G. Date and G. M. Hossain. The Bianchi IX model in loop quantum cosmology. *Class. Quant. Grav.* **21** (2004) 3541–70. [gr-qc/0404039]

[862] M. Bojowald and G. Date. Quantum suppression of the generic chaotic behaviour close to cosmological singularities. *Phys. Rev. Lett.* **92** (2004) 071302. [gr-qc/0311003]

[863] M. Bojowald and H. A. Morales-Tecotl. Cosmological applications of loop quantum gravity. *Lect. Notes Phys.* **646** (2004) 421–62. [gr-qc/0306008]

[864] A. Ashtekar, M. Bojowald and J. Lewandowski. Mathematical structure of loop quantum cosmology. *Adv. Theor. Math. Phys.* **7** (2003) 233. [gr-qc/0304074]

[865] B. Bahr and T. Thiemann. Approximating the physical inner product of Loop Quantum Cosmology. [gr-qc/0607075]

[866] T. Damour, M. Henneaux, A. Rendall and M. Weaver. Kasner like behaviour for subcritical Einstein matter systems. *Annales Henri Poincare* **3** (2002) 1049–111. [gr-qc/0202069]

[867] T. Damour, M. Henneaux and H. Nicolai. Cosmological billiards. *Class. Quant. Grav.* **20** (2003) R145–200. [hep-th/0212256]

[868] A. Ashtekar, T. Pawlowski and P. Singh. Quantum nature of the big bang. *Phys. Rev. Lett.* **96** (2006) 141301. [gr-qc/0602086]

[869] A. Ashtekar, T. Pawlowski and P. Singh. Quantum nature of the big bang: an analytical and numerical investigation. I. *Phys. Rev.* **D73** (2006) 124038. [gr-qc/0604013]

[870] A. Ashtekar, T. Pawlowski and P. Singh. Quantum nature of the big bang: improved dynamics. [gr-qc/0607039]

[871] V. F. Mukhanov, H. A. Feldman and R. H. Brandenberger. Theory of cosmological perturbations; Part 1. Classical perturbations; Part 2. Quantum Theory of Perturbations; Part 3. Extensions. *Phys. Rept.* **215** (1992) 203–333.

[872] F. Markopoulou. Planck scale models of the universe. [gr-qc/0210086]

[873] S. D. Biller *et al.* Limits to quantum gravity effects from observations of TeV flares in active galaxies. *Phys. Rev. Lett.* **83** (1999) 2108–11. [gr-qc/9810044]

[874] J. Kowalski-Glikman. Introduction to doubly special relativity. *Lect. Notes Phys.* **669** (2005) 131–59. [hep-th/0405273]

[875] J. Kowalski-Glikman and S. Nowak. Doubly special relativity theories as different bases of kappa Poincaré algebra. *Phys. Lett.* **B539** (2002) 126–32. [hep-th/0203040]

[876] J. Lukierski and A. Nowicki. Doubly special relativity versus kappa deformation of relativistic kinematics. *Int. J. Mod. Phys.* **A18** (2003) 7–18. [hep-th/0203065]

[877] J. Kowalski-Glikman and S. Nowak. Noncommutative spacetime of doubly special relativity theories. *Int. J. Mod. Phys.* **D12** (2003) 299–316. [hep-th/0204245]

[878] L. Freidel, J. Kowalski-Glikman and L. Smolin. 2 + 1 gravity and doubly special relativity. *Phys. Rev.* **D69** (2004) 044001. [hep-th/0307085]

[879] J. Collins, A. Perez, D. Sudarsky L. Urrutia and H. Vucetich. Lorentz invariance: an additional fine tuning problem. *Phys. Rev. Lett.* **93** (2004) 191301. [gr-qc/0403053]

[880] S. Hossenfelder. Interpretation of quantum field theories with a minimal length scale. *Phys. Rev.* **D73** (2006) 105013. [hep-th/0603032]

[881] S. Majid. Noncommutative model with spontaneous time generation and Planckian bound. *J. Math. Phys.* **46** (2005) 103520. [hep-th/0507271]

[882] L. Freidel and S. Majid. Noncommutative harmonic analysis, sampling theory and the Duflo map in 2 + 1 quantum gravity. [hep-th/0601004]

[883] S. Majid. Algebraic approach to quantum gravity. II. Noncommutative spacetime. [hep-th/0604130]

[884] S. Majid. Algebraic approach to quantum gravity. III. Noncommutative Riemannian geometry. [hep-th/0604132]

[885] J. L. Kelley. *General Topology* (Springer-Verlag, Berlin, 1975).

[886] E. Binz, J. Sniatycki and H. Fischer. *Geometry of Classical Fields* (North-Holland, Amsterdam, 1988).

[887] S. Kobayashi and K. Nomizu. *Foundations of Differential Geometry*, Vols 1, 2 (Interscience, New York, 1963).

[888] S. Lojasiewicz. Triangulation of semianalytic sets. *Ann. Scuola. Norm. Sup. Pisa.* **18** (1964) 449–74.

[889] E. Bierstone and P. D. Milman. Semianalytic and subanalytic sets. *Publ. Maths. IHES* **67** (1988) 5–42.

[890] A. Ashtekar and M. Stillerman. Geometric quantisation and constrained systems. *J. Math. Phys.* **27** (1986) 1319–30.

[891] M. Gotay. Constraints, reduction and quantisation. *J. Math. Phys.* **27** (1986) 2051–66.

[892] V. Guillemin and S. Sternberg. Geometric quantisation and the multiples of group
       representations. *Invent. Math.* **67** (1982) 515–38.
[893] J. Sniatycki. Constraints and quantisation. In *Non-linear Partial Differential Operators
       and Quantisation Procedures*, S. Anderson and H.-D. Doebner (eds) (Lecture Notes in
       Mathematics **1037**, Springer-Verlag, Berlin, 1983).
[894] M. Blau. On the geometric quantisation of constrained systems. *Class. Quant. Grav.* **5**
       (1988) 1033–44.
[895] A. Hanson, T. Regge and C. Teitelboim. *Constrained Hamiltonian Systems*
       (Accademia Nazionale dei Lincei, Roma, 1978).
[896] E. Hewitt and K. A. Ross. *Abstract Harmonic Analysis I* (Springer-Verlag, Berlin,
       1987).
[897] M. S. Birman and M. Z. Solomjak. *Spectral Theory of Self-adjoint Operators in Hilbert
       Space* (D. Reidel, Dordrecht, 1987).
[898] H. Boerner. *Representations of Groups* (North-Holland, Amsterdam, 1970).
[899] E. Hebey. *Sobolev Spaces on Riemannian Manifolds* (Lecture Notes in Mathematics
       1635, Springer-Verlag, Berlin, 1996).
[900] S. Lang. *Differential Manifolds* (Addison-Wesley, Reading (MA), 1972).

Index

Printed in the United States
By Bookmasters